FROM DEPOSITIONAL SYSTEMS TO SEDIMENTARY SUCCESSIONS ON THE NORWEGIAN CONTINENTAL MARGIN

Other publications of the International Association of Sedimentologists

SPECIAL PUBLICATIONS

45 **Linking Diagenesis to Sequence Stratigraphy**
Edited by Sadoon Morad, J. Marcelo Ketzer and Luiz F. De Ros
2012, 522 pages, 246 illustrations

44 **Sediments, Morphology and Sedimentary Processes on Continental Shelves**
Advances in Technologies, Research and Applications
Edited by Michael Z. Li, Christopher R. Sherwood and Philip R. Hill
2012, 440 pages, 231 illustrations

43 **Quaternary Carbonate and Evaporite Sedimentary Facies and Their Ancient Analogues**
A Tribute to Douglas James Shearman
Edited by Christopher G. St. C. Kendall and Abdulrahman S. Alsharhan
2010, 494 pages, 263 illustrations

42 **Carbonate Systems During the Olicocene-Miocene Climatic Transition**
Edited by Maria Mutti, Werner E. Piller and Christian Betzler
2010, 304 pages, 154 illustrations

41 **Perspectives in Carbonate Geology**
A Tribute to the Career of Robert Nathan Ginsburg
Edited by Peter K. Swart, Gregor P. Eberli and Judith A. McKenzie
2009, 387 pages, 230 illustrations

40 **Analogue and Numerical Modelling of Sedimentary Systems**
From Understanding to Prediction
Edited by P. de Boer, G. Postma, K. van der Zwan, P. Burgess and P. Kukla
2008, 336 pages, 172 illustrations

39 **Glacial Sedimentary Processes and Products**
Edited by M.J. Hambrey, P. Christoffersen, N.F. Glasser and B. Hubbard
2007, 416 pages, 181 illustrations

38 **Sedimentary Processes, Environments and Basins**
A Tribute to Peter Friend
Edited by G. Nichols, E. Williams and C. Paola
2007, 648 pages, 329 illustrations

37 **Continental Margin Sedimentation**
From Sediment Transport to Sequence Stratigraphy
Edited by C.A. Nittrouer, J.A. Austin, M.E. Field, J.H. Kravitz, J.P.M. Syvitski and P.L. Wiberg
2007, 549 pages, 178 illustrations

36 **Braided Rivers**
Process, Deposits, Ecology and Management
Edited by G.H. Sambrook Smith, J.L. Best, C.S. Bristow and G.E. Petts
2006, 390 pages, 197 illustrations

35 **Fluvial Sedimentology VII**
Edited by M.D. Blum, S.B. Marriott and S.F. Leclair
2005, 589 pages, 319 illustrations

34 **Clay Mineral Cements in Sandstones**
Edited by R.H. Worden and S. Morad
2003, 512 pages, 246 illustrations

33 **Precambrian Sedimentary Environments**
A Modern Approach to Ancient Depositional Systems
Edited by W. Altermann and P.L. Corcoran
2002, 464 pages, 194 illustrations

32 **Flood and Megaflood Processes and Deposits**
Recent and Ancient Examples
Edited by I.P. Martini, V.R. Baker and G. Garzón
2002, 320 pages, 281 illustrations

31 **Particulate Gravity Currents**
Edited by W.D. McCaffrey, B.C. Kneller and J. Peakall
2001, 320 pages, 222 illustrations

30 **Volcaniclastic Sedimentation in Lacustrine Settings**
Edited by J.D.L. White and N.R. Riggs
2001, 312 pages, 155 illustrations

29 **Quartz Cementation in Sandstones**
Edited by R.H. Worden and S. Morad
2000, 352 pages, 231 illustrations

28 **Fluvial Sedimentology VI**
Edited by N.D. Smith and J. Rogers
1999, 328 pages, 280 illustrations

27 **Palaeoweathering, Palaeosurfaces and Related Continental Deposits**
Edited by M. Thiry and R. Simon Coinçon
1999, 408 pages, 238 illustrations

26 **Carbonate Cementation in Sandstones**
Edited by S. Morad
1998, 576 pages, 297 illustrations

25 **Reefs and Carbonate Platforms in the Pacific and Indian Oceans**
Edited by G.F. Camoin and P.J. Davies
1998, 336 pages, 170 illustrations

24 **Tidal Signatures in Modern and Ancient Sediments**
Edited by B.W. Flemming and A. Bartholomä
1995, 368 pages, 259 illustrations

REPRINT SERIES

4 **Sandstone Diagenesis: Recent and Ancient**
Edited by S.D. Burley and R.H. Worden
2003, 648 pages, 223 illustrations

3 **Deep-water Turbidite Systems**
Edited by D.A.V. Stow
1992, 479 pages, 278 illustrations

2 **Calcretes**
Edited by V.P. Wright and M.E. Tucker
1991, 360 pages, 190 illustrations

Special Publication Number 46 of the International Association of Sedimentologists

From Depositional Systems to Sedimentary Successions on the Norwegian Continental Margin

Edited by

A. W. Martinius

Statoil Research Centre, Rotvoll, 7005 Trondheim, Norway

R. Ravnås

AS Norske Shell, PO Box 40, 4098 Tananger, Norway
Department of Petroleum Engineering, University of Stavanger, Stavanger, Norway

J. A. Howell

University of Aberdeen, Department of Geology & Petroleum Geology
King's College, Aberdeen, AB24 3UE, UK

R. J. Steel

Department of Geological Sciences, University of Texas at Austin, 1 University Station, C-1100, Austin, Texas 78712, USA

J. P. Wonham

Total Norge A.S., Dusavika, 4029 Stavanger, Norway

SERIES EDITOR

Thomas Stevens

Department of Geography
Royal Holloway, University of London
Egham, Surrey
TW20 0EX
UK

IAS

WILEY Blackwell

Registered Office
John Wiley & Sons, Ltd, The Atrium, Southern Gate, Chichester, West Sussex, PO19 8SQ, UK

Editorial Offices
9600 Garsington Road, Oxford, OX4 2DQ, UK
The Atrium, Southern Gate, Chichester, West Sussex, PO19 8SQ, UK
111 River Street, Hoboken, NJ 07030-5774, USA

For details of our global editorial offices, for customer services and for information about how to apply for permission to reuse the copyright material in this book please see our website at www.wiley.com/wiley-blackwell.

Library of Congress Cataloging-in-Publication Data

From depositional systems to sedimentary successions on the Norwegian continental margin /
edited by A. W. Martinius [and four others].
pages cm
Includes bibliographical references and index.
ISBN 978-1-118-92046-6 (cloth)
1. Sedimentary structures–Norway. 2. Sedimentology–Norway. 3. Continental shelf–Norway.
4. Geology, Stratigraphic–Eocene. 5. Geology, Stratigraphic–Paleocene. 6. Petroleum–Geology–Norway.
I. Martinius, A. W.
QE471.F737 2014
552′.509481–dc23

2014017296

A catalogue record for this book is available from the British Library.

Wiley also publishes its books in a variety of electronic formats. Some content that appears in print may not be available in electronic books.

Cover image: Fan delta, Jameson Land, East Greenland

Set in 10/12pt Melior by SPi Publisher Services, Pondicherry, India
Printed and bound in Singapore by Markono Print Media Pte Ltd

1 2014

Contents

Preface

The Norwegian Continental Shelf (NCS; Fig. 1) is a prolific hydrocarbon region and both exploration and production activity remains high to this day. Due to the wealth of data that have been collected and analyzed over the past half-a-century, with exploration starting in 1963, the region is relatively well understood. Exploration activity peaked during the first 20 years or so during which the bulk of the resources on the NCS were proven. However, the exploration revival of the last ten years has resulted in exciting results with significant new hydrocarbon discoveries. In addition, dedicated operators are pushing the recovery rates of producing fields to exceptionally high values, in some cases extending the originally estimated life time of a field by 10 to 20 years. Production outlook as presented by the Norwegian Petroleum Directorate (Petroleum Resources on the Norwegian Continental Shelf, NPD, 2013) is positive but critically depends, among other factors, on continued technological developments. Importantly, this includes improving the geological understanding of the genesis and nature of the sedimentary units in which the hydrocarbons were generated and accumulated.

A key element today and in the future is to couple technological developments to improving our understanding of specific geological situations. The theme of this volume, 'From Depositional Systems to Sedimentary Successions on the Norwegian Continental Margin' reflects the immense efforts made by all industry operators and their academic partners on the NCS to understand in detail the structural setting, sedimentology and stratigraphy of the hydrocarbon bearing units and their source and seal. It is a tribute to the fact that industry and academy are mutually dependent on each other to push both fields of human endeavour to a high level of performance.

The NCS stretches for c. 2500 km along the coastline of Norway from the North Sea in the south to the Barents Sea in the north (Fig. 1) and includes hydrocarbon plays ranging from Devonian to Pleistocene in age. Reservoir types vary from alluvial fans to deepwater fans, in almost every climate type from arid through humid to glacial, in tectonic settings ranging from intra-montane through compressional/transpressional to extensional/transtensional basins and passive margins. This range of depositional systems and tectonic settings provides an exciting challenge for petroleum geologists trying to understand and predict stratigraphy and lithological properties. The initial focus of sequence stratigraphy on eustasy as the key control on depositional architecture has evolved to encompass tectonic, climatic and geomorphologic controls on sediment delivery and preserved stratal architecture.

The papers in this volume were presented at a three-day conference on the sequence stratigraphy of the NCS which was initiated, organised and hosted by the Norwegian Petroleum Society (NPF) in Stavanger in May 2010. The thematics outlined above and their impact on our understanding of the NCS was the focus of the conference. Special attention was given to the integration of both analogue studies and process-based models with the insights gained from the interpretation and visualisation of high quality subsurface well and seismic datasets and associated case studies. The conference followed on from previous successful NPF events on the subject, building further on advances in our understanding of the role of climate and tectonics in controlling sediment delivery from source to sink and its impact on stratigraphic architecture and petroleum exploration offshore Norway. The aim of the conference was to bring together sedimentologists and stratigraphers working on all aspects of sedimentary deposits on the NCS and present an up-to-date overview of current issues and a future outlook. More than seventy oral and poster presentations were accommodated and together they covered topics ranging from Triassic dryland river systems and reservoirs to Late Cretaceous deep marine systems. Moreover, core workshops as well as virtual geology workshops were organised with

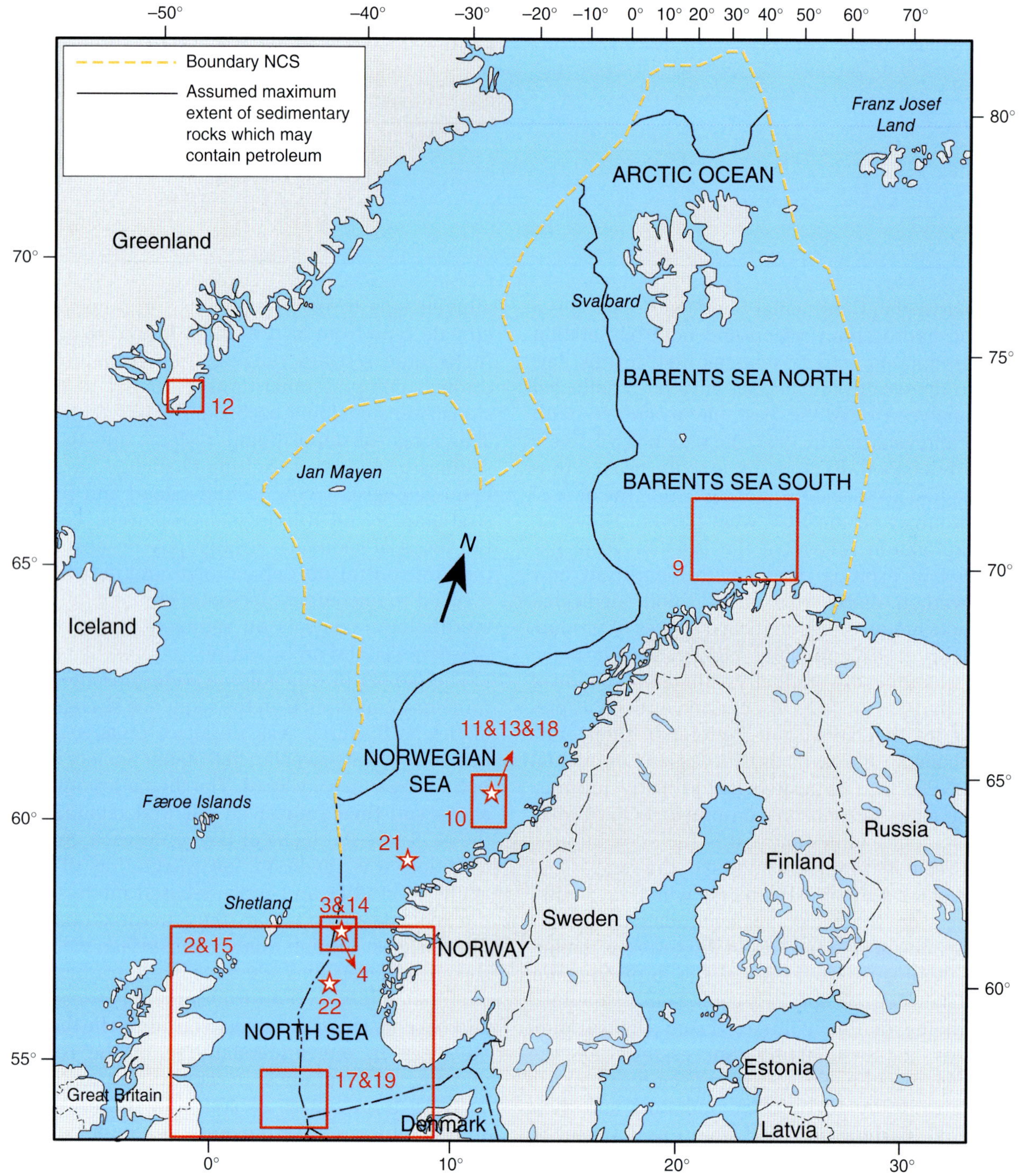

Fig. 1. The Norwegian Continental Shelf (NCS); the focus of this special publication (after Norwegian Petroleum Directorate report, 2013). The numbered boxes and stars refer to the location of the study area of the respective papers in this volume. Papers with a more generic topic are not included in this map.

special sessions on computer and flume tank modelling of stratigraphy and sedimentation.

The contributions to this volume are organised according to their approximate location on the hypothetical depositional profile from up-dip fluvial environments through paralic to deep marine environments down dip. The contribution by Postma presents a perspective on generic autogenic behaviour in fluvial systems as resulting from landscape-scale experimental research.

Steep-gradient, moderate-gradient and low-gradient systems display different autogenic behaviour and it is concluded that it is more logical to predict autogenic behaviour directly from aggradation rate and changes thereof than indirectly from sea-level, climate or tectonic forcing. This statement is supported by Martinius *et al.* in their sequence stratigraphic analysis of the fluvial Eiriksson Fm in the Statfjord Field in the Norwegian Sea (Fig. 1). They propose a practical definition of fluvial base level which is applied to identify changes in aggradation rate in the well data set leading to a reservoir zonation. McKie analyses the Triassic succession in the central North Sea (Fig. 1; Skagerrak Fm) deposited in structurally active basins identifying signals for alternating arid and pluvial depositional conditions related to warming and cooling periods in the Tethys region. Fluvial systems of the successive Lunde, Raude and Eiriksson fms at the Jurassic-Triassic boundary in the Norwegian Sea (Fig. 1) are examined by Nystuen *et al.* documenting systematic changes in fluvial style influenced by an overall low-frequency climatic change from more arid to more humid conditions. Allostratigraphic units are inferred to have formed in response to high-frequency climatic cycles (or base level changes).

Autogenic processes under controlled conditions at the delta top in flume tank experiments are investigated by Kim *et al.* to examine quantitatively sediment storage and release associated with river pattern changes as well as changes in fluvial autocyclic behaviour that are driven by external forces. The main concepts of autostratigraphy in river deltas are reviewed by Muto & Steel, who suggest that steady dynamic external forcing generally results in unsteady stratigraphic response; and that steady stratigraphic configuration can result from unsteady dynamic external forcing. Autogenic processes in deltas are also the topic of the paper by Olariu, who concludes that pervasive and rapid process changes were common in ancient deltas and particularly well-expressed on the delta front. It is suggested that the main intervals bounded by spatially extensive flooding surfaces are successive deltaic complexes, reflecting the allogenic response. The great internal variability reflects autogenic spreading of the lobes during the cross-shelf transits.

The contributions of wave forcing on the delta front to delta morphodynamics and time-varying stratigraphy are investigated by Hillen *et al.* using a high-resolution physics-based numerical model. The simulated stratigraphy is subsequently applied to a standard reservoir modelling workflow to calculate rock properties. Fluvial-dominated delta plain deposition at the Jurassic-Triassic boundary in the Barents Sea (Fig. 1) is shown by Ryseth to be controlled dominantly by climate change, analogous to the conclusions of Nystuen *et al.* for the same time interval in the Norwegian Sea (Fig. 1).

The Lower to Middle Jurassic succession of the Halten Terrace in the Norwegian Sea (Fig. 1) is the topic of three studies. Ravnås *et al.* recognise three megasequences, the Tilje, the Tofte-Ile and the Garn megasequences, with a common architectural motif and stratigraphic structure. These are interpreted to have formed in response to uplift of rift margins and hinterland and determine the distribution of high-quality sheet-like reservoir bodies. Thrana *et al.* present a revised depositional model for the fluviodeltaic overall transgressive Åre Fm (underlying the Tilje megasequence of Ravnås *et al.*) which forms the basis for a new stratigraphic framework and revised reservoir zonation in the Heidrun Field. The nature of the stratigraphic surfaces confirms the findings of Olariu in that these spatially extensive flooding surfaces bound successive deltaic complexes, each of which displays significant internal variability. A similar conclusion is reached by Ichaso & Dalrymple for the Tilje Fm in the Smørbukk Field based upon a detailed facies and stratigraphic analysis, identifying two major sequences (both within the Tilje megasequence of Ravnås *et al.*) resulting from two main tectonic pulses. Propagation of blind normal faults and the generation of isolated syn-depositional shallow hanging-wall depocentres exerted a subtle control on facies distribution patterns within both Tilje sequences. Contemporaneously with the Lower to Middle Jurassic Halten Terrace succession, along the western side of the 400 km wide proto-Norwegian-Greenland Sea, the Neill Klinter Group succession was deposited. Ahokas *et al.* suggest that it is an outcrop analogue for the Tilje and the Tofte-Ile megasequences, in particular for depositional environments, external and internal factors controlling depositional architecture and resulting three-dimensional heterogeneity of sandstone body architecture and rock properties.

The influence of developing structures on shallow marine sedimentation patterns and resulting stratal architectures was studied by a number of contributors to the volume. The Middle Jurassic succession in the northern North Sea (Fig. 1) was studied in the Gullfaks-Kvitebjørn area by Folkestad *et al.*, who showed that funnel-shaped hangingwall areas of rotated fault-blocks formed local depocentres typified by enhancement of tidal currents. A wider perspective was taken by Jarsve *et al.* who reviewed the Mesozoic and Cenozoic basin configurations in the North Sea in order to understand changes in style of basin extension, geometry and infill dynamics in the context of source-to-sink relationships. Four main basin configurations were recognised associated with asymmetrical uplift and erosion of basin flanks. An alternative way to better understand and predict the effects of controlling factors on regional scale stratigraphic architecture is to employ three-dimensional stratigraphic forward modelling as shown by Granjeon. It is suggested that the numerical modelling technique can be applied as an industrial guide to reduce subsurface uncertainties. Wonham *et al.*, in a case study from the Upper Jurassic of the North Sea Central Graben, developed three tectono-sedimentary models to explain and predict sedimentation, erosion and redeposition in the basin. The nature of each of these models depends on the evolving geometry of the rift, the location in the rift and the timing of underlying salt displacement.

Repetitive cycles of tidal sand ridge deposition in a structurally confined and subsiding basin were analyzed by Messina *et al.* and related to quantified reservoir architecture and rock property distributions and anisotropy.

Patterns of chalk deposition in the Maastrichtian Tor Fm were studied in the Ekofisk Field and adjacent areas (North Sea; Gennaro & Wonham). An extensive channel, oriented along the bathymetric contours of major structural features, is identified and interpreted to have formed by bottom currents that originated from the deepest part of a nearby graben.

At the down-dip end of the depositional profile, Basani *et al.* constructed a three-dimensional model using deterministic process modelling software for the simulation of turbidity currents with the aim to fill the gap between small-scale and large-scale modelling of turbidite architecture. The simulations show a close fit to experimental sandy turbidity current dynamics. Large-scale architecture in the Upper Maastrichtian to Danian Ormen Lange turbidite system were analysed by Ravnås *et al.* Multiple phases of repeated hinterland uplift, basin-margin tilting and sandy fan development occurred; and the turbidite systems evolved through fill-spill processes on the slope terraces or in minibasin(s) to backfilling and upslope onlapping. Similarly, the architecture of a mid to outer turbidite fan setting of the Late Palaeocene to Early Eocene Hermod Fan in the North Sea was analysed by Bryn & Ackers to understand its three-dimensional architecture. A network of fairways was mapped using advanced geophysical techniques. The combination of primary depositional patterns and extensive post-depositional sand remobilisation resulted in a complex three-dimensional architecture in which numerous thick units of clean sandstone occur.

We are sincerely indebted to the authors who have contributed to the volume and to the reviewers whose comments and insight have invariably and significantly improved the quality and presentation format of the papers. As always, without the expertise, dedication and efforts of the reviewers (acknowledged in each paper except where they prefer to remain anonymous) this volume would not have been published. Finally, we wholeheartedly thank Thomas Stevens, IAS Special Publications Editor, for his high standards of editing, quality assurance as well as his perseverance and patience. We are also grateful for all the prompt technical assistance from Adam Corres, Editorial Assistant for the IAS Special Publications.

The reviewers are (in addition to three anonymous reviewers):

Jennifer Aschoff
Bryan Bracken
Massimo Dall'Asta
Rory Dalman
Josh Dixon
Trevor Elliott (deceased)
Michael Engkilde
Anjali Fernandes
Atle Folkestad
Lars-Magnus Fält
Rob Gawthorpe
Liviu Giosan
Gary Hampson
William Helland-Hansen
David Hodgson
Chris Jackson
Howard Johnson
Ian Kane
Wonsuck Kim
Kjell Sigve Lervik
Ian Lunt
James MacEachern
Tom McKie
Donatella Mellere
Andrea Moscariello
Cornel Olariu
Snorre Olaussen
Henrik Olsen
Chris Paola
Per Pedersen
Piret Plink-Björklund
George Postma
David Pyles
Erik Rasmussen
Rodmar Ravnås
Manasij Santra
Ron Steel
Esther Stouthamer
Kyle Straub
Finn Surlyk
David Uličný
Allard van der Molen
David Waltham
Brian Willis
Mike Young

Dedication to John Gjelberg, Michael Talbot and Trevor Elliott

It is with considerable sadness that we dedicate this volume on Norwegian sedimentary geology to three of our colleagues who have recently passed away.

John Gjelberg received his doctorate from the University of Bergen in 1982. He began immediately to work for Norsk Hydro and worked there for the rest of his career, apart from a short period in Statoil/Hydro and a final few years in North Energy. John spent his entire working career researching and publishing papers on the sedimentology and petroleum geology of all parts of the Norwegian Continental Shelf, as well as on the tectono-stratigraphic development of the Congo Basin, offshore Angola, and on outcrop geology in France and Spain. Some of the highlights of John's work include study of the Carboniferous alluvial and marine strata of Bear Island and Svalbard, early exploration reservoir models on the Barents shelf, the tectono-stratigraphic development of the mid-Norwegian margin and seismic modelling of the Ainsa turbidite system. On the academic front, John is known for his 1994 paper (with W. Helland-Hansen) on the early sequence stratigraphy debate concerning the variability of systems tracts, with ideas that have gained new traction in the stratigraphic community in recent years. John also taught courses and supervised graduate students at the University of Bergen. John Gjelberg is greatly missed but remembered for his huge contribution to Norwegian Geology.

Mike Talbot received his Ph.D. in geology from the University of Bristol in 1968. He then worked on the coral reefs and associated sediments of the Seychelles as a NERC Post-Doctoral Research Assistant, University of Dundee, U.K. From 1972 to 1978 he was employed as a Lecturer in Geology at the University of Ghana's Department of Earth Sciences. He subsequently worked as a Lecturer at the Department of Earth Sciences, University of Leeds from 1978 to 1981 before accepting a post as Senior Lecturer at the Geological Institute in Bergen, Norway. In 1984 he was appointed professor at the same institute (now Department of Earth Science). Mike had various research interests that were focused on limnogeology, modern and ancient rift sedimentation, carbonate sedimentology and diagenesis. His mid-career research effort was concentrated on the large lakes of East Africa. During the last 10 years, Mike returned to carbonates, focusing on sedimentological and diagenetic studies of collapse breccias associated with the Late Palaeozoic carbonates and evaporites in central Spitsbergen. Mike was an efficient and well-liked Editor-in-Chief of the journal Sedimentology from 1990 to 1994 and was European representative on the Lake Drilling Panel, Science Advisory Group of the International Continental Scientific Drilling Program. In 2007, Mike received the W.H. Bradley medal from The International Association of Limnogeology for "his stellar research career as well as his dedicated services to the Limnogeology community". Mike also supervised some 74 Masters and doctoral students and will be long remembered for the outstanding quality of his work and for his dedication to the University of Bergen.

Trevor Elliott, one of the most influential geoscientists of his generation, received his D.Phil. at the University of Oxford under the supervision of Prof. Harold Reading. After postdoctoral studies at Leiden and Reading, he became Lecturer in Sedimentology at the University College of Swansea (1976) and later the George Herdman Professor of Geology at the University of Liverpool, a position he held for more than 20 years. Trevor's long and productive teaching and advising career resulted in numerous undergraduate, postgraduate and postdoc theses and research projects. One of his substantial legacies is the impact that his students have made and are still making in their companies and geology departments around the world. Trevor authored the chapters on Deltas and Shorelines in the seminal book: *Sedimentary*

Environments and Facies that was first published in 1978 and subsequently updated and revised. Trevor's research interests were broad and ranged from fluvial to deepwater sedimentology, stratigraphy and basin analysis. He was awarded the Bigsby Medal from the Geological Society of London and was a Distinguished Lecturer of the AAPG. In the 1980s, Trevor researched shallow marine sandstone outcrops for application to the wave-influenced and tidally-influenced reservoirs of the Norwegian margin. This Norwegian industry collaboration led to further support for Trevor's PhD students on the sedimentological study of Norwegian sector cores. Trevor enjoyed much collaboration with the petroleum industry. His long term study of the Clare Basin in Ireland and his dream of drilling boreholes behind the Clare cliff line to develop a three-dimensional model were eventually realised recently through Statoil and University College Dublin collaboration. From 2006, Trevor developed and taught industry classes in sedimentology and stratigraphy, engaging a large new community of geoscientists and engineers. Trevor's greatest strengths were his scientific method and teaching style. His keen insight, rigorous analytical mind and ability to engage at all levels were inspiring to students and earth science professionals alike.

Int. Assoc. Sedimentol. Spec. Publ. (2014) **46**, 1–18

Generic autogenic behaviour in fluvial systems: lessons from experimental studies

GEORGE POSTMA

Utrecht University, Faculty of Geosciences, Department of Geosciences, PO Box 80.021, 3508 TA Utrecht, The Netherlands (E-mail: g.postma@uu.nl)

ABSTRACT

Substantial advancement in the recognition of generic autogenic behaviour in fluvial systems stems from recent landscape-scale experimental research, where features of stratigraphic architecture can be verified along known input and boundary conditions. Combining experimental work, numerical modelling and field data indicate different autogenic behaviour for 1) steep-gradient, stream-flow dominated alluvial fan-type systems, 2) moderate-gradient braided-river type systems and 3) low-gradient single-river meandering and anastomosing systems. Autogenic behaviour is by cyclic alternating sheet and channelised flow on alluvial fan surfaces, by avulsions of small bifurcating streams around migrating mid channel bars in the braided channel belt and by splitting of the flow from one into two channel belts in case of single thread rivers. Experimental studies indicate that aggradation rate is directly related to the frequency of autogenic behaviour and that absolute frequency values depend on the backfilling rate in the channel and the available accumulation space. Since aggradation rate is the direct result of the interplay of all allogenic controls, it is more logical to predict autogenic behaviour directly from aggradation rate than indirectly from sea-level, climate or tectonic forcing. It is shown that if the change in allogenic forcing is fast relative to the time that a river needs to come to grade (equilibrium time), then the change in aggradation rate and the frequency of autogenic behaviour is highest. If the change in allogenic forcing is slow relative to the river's equilibrium time, then there is also little change in the frequency of autogenic behaviour.

Keywords: Autogenic, avulsion, aggradation, alluvial fan, braided river, single thread river, backfilling.

INTRODUCTION

Generic autogenic behaviour of alluvial systems refers to the common intrinsic property of alluvial systems to shift their streams and rivers laterally by avulsion and bifurcation (Beerbower, 1964; Allen, 1965; Slingerland & Smith, 2004). In this way the river spreads its sediment load evenly over the available surface area to reach the lowest potential energy level and highest stability (Paola *et al.*, 2009). This intrinsic property can be observed everywhere in nature. The sediment spread is commonly more even for high-gradient and moderate-gradient river systems that carry little fine-grained suspension load and experience high discharges (alluvial fans, outwash plains, braided river systems), than for low-gradient rivers that carry abundant suspension load. In the first case, bank stability is low and in the latter it is high, which keeps the river better in position. If all alluvial accumulation space is filled and the river has reached its grade (base level), no net erosion or deposition along the channel belt occurs and the spread of sediment and thus avulsion of streams is halted, while all supplied sediment bypasses the region.

In nature, boundary conditions that define base level fluctuate continuously by changes in discharge, sediment load, grain size etc., affecting the river's slope and its buttress (the point

From Depositional Systems to Sedimentary Successions on the Norwegian Continental Margin, First Edition. Edited by A. W. Martinius, R. Ravnås, J. A. Howell, R. J. Steel, and J. P. Wonham.

where the river profile grades to, e.g. shoreline, see Holbrook *et al.* (2006) and further discussion below). Hence, fluvial systems are rarely at grade and usually alternate between spreading their sediment laterally, if still aggradational, and funnelling their sediment by incision down slope, if erosional. The degree of spreading is thus determined by upslope and downslope boundary conditions, which change continuously with changes in sea-level, climate and tectonics.

The aim of this contribution is to review the work that has been done over the last decades towards understanding autogenic behaviour in stream-flow dominated alluvial fan and fluvial systems. In particular, the results of new experimental, theoretical and numerical studies are used to arrive at a new working hypothesis to predict frequency of autogenic change from basic principles (thus by abductive inference, see Kleinhans *et al.*, 2010). The usefulness of the hypothesis is discussed along a few well-researched natural examples.

AUTOGENIC PROCESSES

A river shifts its course by lateral migration, avulsion and bifurcation (see review by Jones & Schumm, 1999). Kleinhans (2010) distinguishes avulsion from bifurcation simply by the shift of the course of a river over several meander bend wavelengths. At bifurcations, water and sediment are divided over two downstream branches. Avulsions can be instantaneous or gradual, whilst an avulsion site is at least temporarily a bifurcation, because the new channel develops while the old one is still active. For practical reasons in stratigraphic reconstructions and as a consequence of the limited resolving power of ^{14}C dating, Stouthamer & Berendsen (2000) defined instantaneous avulsion as an avulsion where two adjacent channel belts coexisted for less than 200 years. If two adjacent channel belts coexisted for more than 200 years, avulsion was defined as gradual (cf. Törnqvist, 1994).

Process-oriented studies discussed below reveal different autogenic processes for 1) steep-gradient alluvial fan (with slopes in the order of more than 2 degrees), 2) moderate-gradient braided river (slopes ~ 0.4 degrees) and 3) low-gradient, meandering or anastomosing river; so these will be dealt with separately.

Steep-gradient alluvial fan

Autogenic processes in a steep-gradient, stream-flow dominated, alluvial fan were studied from analogue experimental studies of small alluvial fan models. The slope of such a steep-gradient fluvial system builds up by alternations of sheet and channelised flow (Schumm *et al.*, 1987; Bryant *et al.*, 1995; Whipple *et al.*, 1998). In Eurotank, at Utrecht University, the autogenic behaviour was analysed in detail by Van Dijk *et al.* (2009, 2011). Alluvial fans and fan deltas (the latter defined by Nemec & Steel, 1988, as alluvial fans prograding into a standing body of water) were formed by feeding water and sediment through a narrow (4.5 cm wide) duct. The water jet that issued from the duct could expand freely on a large sediment table 2.5 m wide and 2.7 m downslope. The observed morphodynamics are governed by cyclic alternating sheet flow and channelised flow. The sheet flow builds a convex-shaped fan apex cascading down onto the mid-fan region (Fig. 1). When the slope of the apex reaches its critical threshold value, the water flow incises creating progressively expanding channelised flow through fan incision and headward erosion. The trench so created in the fan apex funnels the sediment down the fan and forms telescoping fan lobes and bifurcating channels at the slope break in case of the alluvial fan and at the shoreline in case of the fan delta, while deposition is forced by reduction of the stream gradient. The backfilling of the trench starts with mid-channel bar formation, ultimately bringing the system back to the sheet flow stage and its critical slope; then the process of fan incision and channel formation will start again. In the experiments by Van Dijk *et al.* (2009, 2012) each channel incision was stacked on top of the previous one, a phenomenon which was ascribed to the upper boundary condition: a fixed 4.5 cm wide duct, through which the stream is debouching onto the fan apex (Van Dijk *et al.* 2009). When using wider ducts, bars can be seen developing both in and at the outlet avulsing the flow towards the right and left of the fan body producing compensation cycles and 'fanning' (e.g. Bryant *et al.* 1995; Whipple *et al.* 1998).

Moderate-gradient braided stream systems

The autogenic process in braided streams was studied in detail by Ashworth *et al.* (2004, 2007) through distorted Froude-scaled models (see

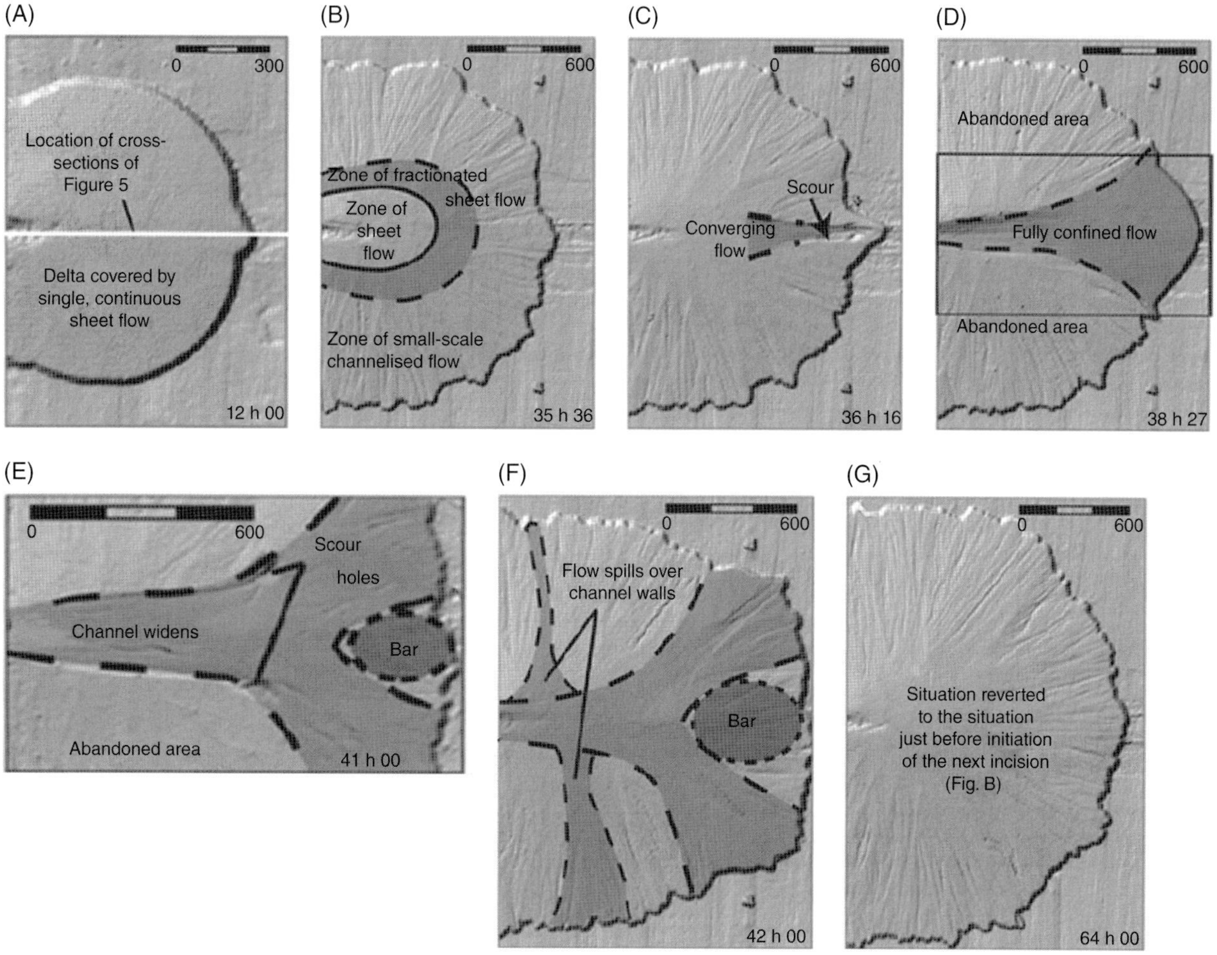

Fig. 1. Shaded relief maps of an autogenic cycle. Run time is shown in the lower right corner. The scale of the plots is indicated in the upper right corner; the position of panel (E) is given by the inset in panel (D). A. The initial jet transformed into an expanding sheet flow. Single sheet flow producing smooth delta plain topography; B. With growth of the delta plain sheet flow fractionates causing small-scale channelised flow near the shoreline; C. Progressive aggradation at the apex increased the gradient of the delta plain up to the point when a scour hole was initiated along the centre line of the fan delta; D. The scour hole developed quickly into a knickpoint that moved upstream connecting the scour with the feeder channel; E. Backfilling starts with the deposition of a mouth-channel or mid-channel bar; F. Progressive backfilling while the flow gradually started to exceed the confining channel walls and increasingly spilled over the margin in the course of the backfilling process; G. When the entire channel had been filled, fractionated sheet flow and aggradation of the apex were restored (from Van Dijk *et al.*, 2009).

Peakall *et al.* 1996) of a braid plain with characteristic channel and mid-channel bar configurations. The set up differed from those of Bryant *et al.* (1995) and Van Dijk *et al.* (2009, 2011) in that the feeder channel was uplifted, herewith creating accumulation space (terminology *sensu* Blum & Törnqvist, 2000) in the braid plain basin. Ashworth and co-workers arbitrarily defined stream avulsion around the numerous mid channel bars by the sudden, lateral shift of minimal 30 cm of a channel, while the new channel position must be maintained for at least a 15 min period. Each avulsion starts off from a bifurcation, where the mid-channel bar splits the active channel flow over two branches following similar processes as pictured in Fig. 1D and 1E. The experiments of Sheets *et al.* (2002) and Hickson *et al.* (2005) conducted at Saint Anthony Falls Laboratory had multiple entries, which produced a kind of braid plain showing autogenic processes probably more akin to those occurring on stream-dominated coalescing alluvial fans, each with characteristic alternating channelised and sheet flow processes. Sheet flow does occur in braided river systems, in particular during floods, yet their possible contribution to the avulsion processes on the braid plain itself is not tackled by the experimental studies mentioned in this section.

Low-gradient rivers

The autogenic behaviour in the low-gradient river category includes the behaviour of all single thread leveed anastomosing and meandering channel systems. Experimental studies for this category focusing on effective aggradation rates and flow occupancy are almost non-existent, with the exception of the revolutionary cohesive-delta experiments by Hoyal & Sheets (2009). They find, on the basis of their experiments for cohesive delta plains, that avulsion of channels (and their lobes) happens in three steps: The first step involves bar aggradation above the point where the incipient topography affects the flow which leads to flow widening and flow bifurcation leaving a V-shaped, subaerial region on the bar surface and ending the bar cycle. The second step, of negative feedback, involves a morphodynamically mediated backwater effect that is created by the mid-channel bar. As the bar grows, a hydraulic backwater effect propagates slowly upstream in the delta distributaries and is followed immediately by a wave of channel bed aggradation. As the lobe continues to grow and channel bed aggradation increases, overbank flow drives accelerated subaerial levee growth. This drives the system to step 3, where the combined effect of bed aggradation and progressively upstream levee growth leads to super-elevation of the channel and ultimately to the 'discovery' of a more favourable path to the shoreline, i.e. avulsion.

Additional insight into the autogenic behaviour of low gradient rivers is mainly based on historical and sedimentological reconstructions and on numerical modelling. These reconstructions have led to the common belief that avulsion of single thread rivers is driven by 1) local super elevation of some part of a channel or channel complex above its surroundings by the ratio between cross-valley and down-valley gradient (gradient advantage) and 2) the occurrence of a trigger event, commonly a flood (see review by Jones & Schumm, 1999; Stouthamer & Berendsen, 2007) or storm surge, the latter being important in delta distributaries. The river flood may cause avulsion by blockage of the flow by local reduction in channel capacity or by local obstruction. Flow blockage may also be caused by storm surge migrating up river (backwater effect).

For low-gradient, subcritical (Froude < 1), flowing rivers, the backwater effect is defined by the distance L at which the water level has adapted to 67% of its upstream normal flow depth and is estimated by

$$L = \frac{h}{3s} \quad [\mathrm{m}] \tag{1}$$

with h = flow depth [m] and s = channel slope [-] (e.g. Van Rijn, 1994). Hoyal & Sheets (2009) found in their experiments that the real morphodynamic backwater effect may easily be twice as much of the calculated effect, which could bring the avulsion node that much farther upstream, theoretically. Several important examples of avulsions triggered by various means of channel blockages have been documented by King & Martini (1984), Schumann (1989), McCarthy *et al.* (1992) and Harwood & Brown (1993). The interaction of both drivers for autogenic change (i.e. gradient advantage and triggering events) was tested by numerical modelling of river behaviour (e.g. Mackey & Bridge, 1995; Törnqvist & Bridge, 2002; Karssenberg & Bridge, 2008). Recent numerical modelling by Kleinhans *et al.* (2008) demonstrated that during the initial bifurcation of the river, when water and sediment are split over two branches, the choice of which bifurcate channel becomes more important than the other is determined by a number of factors of which local gradient advantage is just one. The other factors are the position of the avulsion node relative to the upstream meander bend (Kleinhans *et al.*, 2008), the channel width-depth ratio of the bifurcate channels or the breach (e.g. Slingerland & Smith, 1998), the grain size sorting and the presence of local obstructions (bars and bank irregularities, see Kleinhans *et al.*, 2008). The factors together offer an explanation of why some bifurcations were destabilised in decades and others in centuries in the Rhine Meuse system (Kleinhans, 2010).

Although avulsion drives the single thread rivers to distribute their sediment evenly over the coastal lowlands, bank stability and differential compaction rates between the fine grained and peaty floodplains and silty to sandy channel belts makes the surface area of such systems highly irregular, even at high avulsion rates (e.g.

Stouthamer & Berendsen, 2001). Peat formation potentially influences avulsion by inhibiting lateral migration and increasing aggradation in the channel belt. Peat compaction and oxidation in flood basins also leads to relief amplification and to super-elevation of channel belts (Van Asselen *et al.* 2009).

Cyclic avulsion processes in the Yellow River delta were forced by rapid delta progradation causing the river to adjust its channel belt profile by aggradation (Kriele *et al.*, 1998). At some point the aggradation led to an increase of transverse slopes causing the channel to avulse in another direction. It is noted here that this process is in contrast with the initiation of avulsion on stream-dominated alluvial fans as determined from our experimental studies, where sheet flow increases the gradient of the entire apex to levels of instability and new channel incision. In braid plains, gradient advantage in bifurcations does play a similar role, yet occurs on much smaller temporal and spatial scales.

In summary, autogenic behaviour in alluvial fans and braided river systems is different from that in moderate-gradient and low-gradient river systems and is most strikingly different by the effect that backwater has on the channel belt aggradation. The effect of backwater is virtually lacking in the moderate and steeply graded systems (see equation 1), while the water flow in the channels is often close to supercritical (Sheets *et al.* 2002; CGER, 1996; Hoyal & Sheets, 2009); so caution is needed when applying experimental studies of those systems to low gradient rivers. However, in all cases backfilling of the channels is a prerequisite for avulsion, since backfilling elevates the channel above its surroundings. In the case of alluvial fan systems, the avulsion trigger is clearly related to steepening of the apex by the sheet flow, which can only commence if backfilling is completed. In the case of braided river systems, the trigger is by a growing advantage of one bifurcate over the other. The avulsion process of the braided river is thus, in this respect, similar to that of the single river, where avulsion also starts with a bifurcation but where the change from bifurcation to avulsion is up to three orders of magnitude slower. Avulsion frequencies vary greatly among modern river systems, with a lowest rate of 28 years for the Kosi River in India and up to 1400 years for the Mississippi River (Slingerland & Smith, 2004).

FREQUENCY OF AUTOGENIC PROCESSES

The rate of backfilling (aggradation in the channel belt) defines the frequency of autogenic processes (cf. Van Dijk *et al.* 2009). Backfilling commences where there is sufficient reduction in channel slope to force deposition. If the lower boundary is the shoreline, then reduction of slope is obtained by progradation of the shoreline and the creation of a mouth bar (e.g. Kriele *et al.* 1998). If the fluvial system is prograding over a (flood) plain, it must also be the reduction in slope by progradation of the system that forces deposition and formation of a mid-channel bar, heralding the backfilling. The accumulation space and the rate at which the required sediment volume aggrades the channel both determine the avulsion frequency. Bryant *et al.* (1995), on the basis of their experimental results, were the first to link avulsion rate with aggradation rate, although they did not measure the aggradation but simply took sediment yield at the apex of the fan as a proxy for aggradation rate. Here, it is important to realise that it is not the total of the supplied sediment is important, but how much of the supply is used for aggradation of the bed. Supplied sediment that is not used for aggradation bypasses the channel and is used for progradation of the system at the channel mouth (telescoping fans). If the rate of backfilling is slow, observations from experiments (Van Dijk *et al.* 2009) indicate that the river valley has time to deepen and widen, which increases its accumulation space causing a negative feedback to avulsion frequency, thus reducing it. Incipient relief, channel length and fan-produced local surface irregularities (lobes, scours, channels and bars) are most probably causes for observed deviations in cycle duration and the timing of re-incisions.

Van Dijk *et al.* (2012) showed how frequency of autogenic processes in their steep gradient systems is related to channel backfill rate. The frequency in their experimental alluvial fans, which prograded over a near horizontal plain, appeared to be much higher than for fan deltas. Since the upper boundary conditions for the alluvial fans were the same in both experimental set ups, the change in frequency must have been caused by the only difference between the experimental fans, i.e. the presence of a shoreline. They demonstrated that the presence of the shoreline caused different aggradation rates on

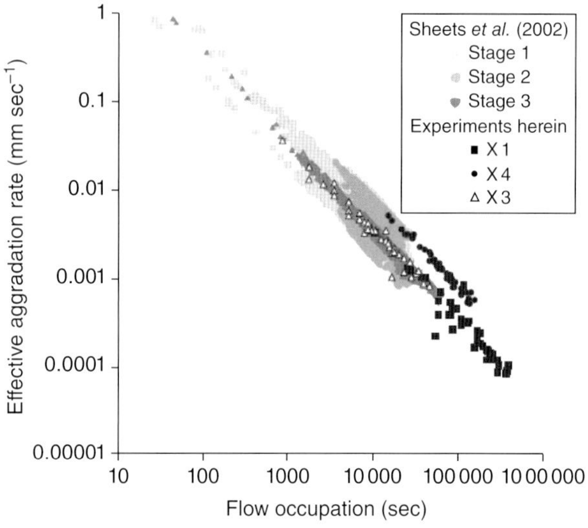

Fig. 2. Change in effective aggradation rate as a function of duration of flow occupancy for all sections in the three experiments. From Ashworth *et al.* 2007 with data from stages 1 to 3 from Sheets *et al.* (2002).

the alluvial fan surface. While in the first case all supplied sediment aggraded on the fan surface, in the second case part of the supplied sediment bypassed the fan surface and aggraded in the subaqueous delta.

Ashworth *et al.* (2007) plotted flow occupancy against effective aggradation rates including the data from Sheets *et al.* (2002). The plot shows a near-inverse relationship and demonstrates that with the highest aggradation rates the channels still have time to distribute sediment sufficiently over the braid plain. With the lowest aggradation rates, the channel can be seen to migrate predominantly laterally with, rare, sudden shifts. Ultimately, a perfect inverse relationship between effective aggradation rates and flow occupancy must exist if the braidplain aggrades evenly over its entire surface. Ashworth *et al.* (2007) compiled their data and that of Sheets *et al.* (2002) in a diagram (Fig. 2) to demonstrate that aggradation rate is an important driver of flow occupancy and thus frequency of autogenic processes. In spite of the fact that the avulsion process is of a different kind than in the alluvial fan studies of Sheets *et al.* (2002), the correlation is excellent.

For low-gradient rivers, Karssenberg & Bridge (2008) modelled bifurcation and avulsion frequency three dimensionally by simulating sediment transport by the diffusion equation. The timing and location of channel bifurcation is controlled stochastically as a function of the cross-valley slope of the floodplain adjacent to the channel belt relative to the down-valley slope and of annual flood discharge. To examine how the model responds to extrinsic controls, the model was run under conditions of changing base level and increasing sediment supply. Rises and falls in base level and increases in sediment supply occurred over 10,000 years. Rising base level caused a wave of aggradation to move up-valley, until aggradation occurred over the entire valley. Frequency of bifurcations and avulsions increased with rate of base-level rise and aggradation rate. Kleinhans *et al.* (2008) modelled numerically duration of the avulsion time, which is assumed to be determined by the width to depth ratio of the channel, the upstream bend radius determining the gradient advantage for one bifurcate over the other and the length of the bifurcates. According to Kleinhans *et al.* (2008) the model explains how combination of variables may result in the observed large variation of avulsion duration in historical and geological data.

In summary, experimental and numerical studies show that aggradation rate is an important driver for the frequency at which autogenic processes occur. Furthermore, these process studies show that absolute values of frequencies depend on the backfilling rate of the channel, which depends strongly on the accumulation space that needs to be filled, but also depends on local factors that determine the hydrology. It is important to note at this point that all experimental modelling discussed here showed that avulsion frequency is certainly not constant with aggradation rate or varies around a mean value, as assumed in the modelling of Leeder (1978).

ALLOGENIC CONTROLS ON GENERIC AUTOGENIC BEHAVIOUR

The experiments on three-dimensional alluvial stratigraphy by Hickson *et al.* (2005) were designed to investigate the influence of allogenic controls on architecture in alluvial successions. They concluded that the alluvial architecture of their models is controlled, very strongly, by externally forced facies migrations, hence by changes in sediment supply, base level or subsidence. Sea-level, climate and tectonics change the accumulation

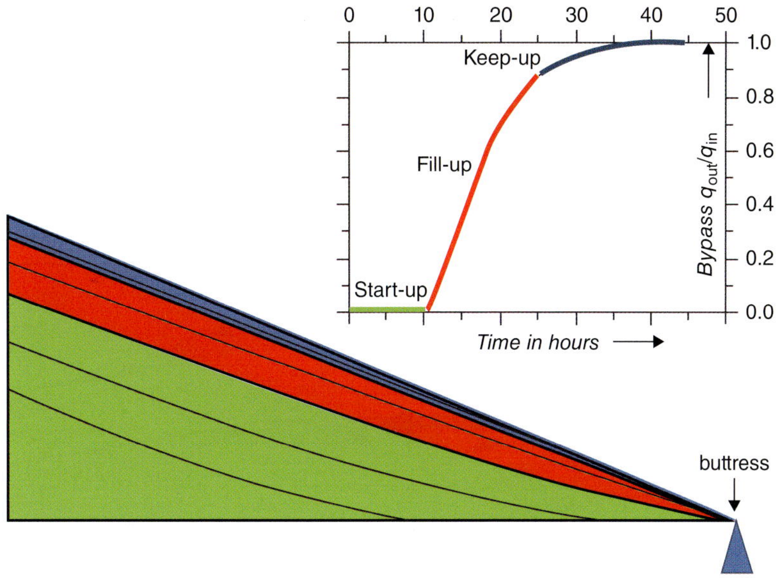

Fig. 3. Schematic illustration of the stratigraphic development of a two-dimensional, fluvial sediment wedge in a duct of 0.11 m width and 4.5 m length. The changes from start-up, to fill-up and from fill-up to keep-up stages have been marked by colours. The black lines are time lines at about 5 hour intervals. During the start-up stage the system progrades to base level. Once its toe has reached base level the system will come in its fill-up stage and finally in its keep-up stage. The graph in the inset shows the percentage of sediment bypass relative to what enters the system (based on Postma *et al.*, 2008).

space in the fluvial realm continuously and together control the ultimate gradient of the fluvial system. Detailed studies of fluvial architecture in the Rhine Meuse delta system in combination with good age control have led Stouthamer & Berendsen (2000, 2001, 2004 and 2007) and Van Asselen *et al.* (2009) to relate avulsion frequency to these allocyclic controls. However, since the interplay of sea-level, climate, local tectonics and regional tectonics together defines the aggradation rate, it will always remain challenging to unravel the relative contributions of each from that which drives autogenic behaviour directly: aggradation rate. For the geologist who wishes to predict fluvial architecture and sandstone body connectivity, the direct relationship between aggradation rate and frequency of autogenic behaviour is thus an interesting one (cf. Leeder, 1978), because it simplifies questions about cause and effect. Aggradation rate can reasonably be measured and bounding surfaces can be dated, so prediction of autogenic behaviour can be done on the basis of quantitative criteria.

Aggradation rate

Aggradation or deposition rate is not to be confused with sediment supply rate, since aggradation rates do not vary linearly with sediment supply, as was demonstrated by simple 2-dimensional experiments performed in a duct of 0.11 m width and 6 m length (Postma *et al.*, 2008). The purpose of these experiments was to produce fluvial stratigraphy by adding water and sediment to the duct. It was found that channel aggradation is predicted best by non-linear diffusion (Fig. 3). For the two dimensional channel belt case, there is increasingly more bypass with steepening of the channel gradient, when the channel system is building up to grade. Depending on the amount of bypass, each channel system can be seen to pass through three development stages: 1) a start-up stage, in which the system aggrades towards base level and during which no sediment can bypass base level, 2) a fill-up stage, where the system both aggrades and progrades beyond base level, hence with sediment bypass up to the arbitrarily chosen 90% level and 3) a keep-up stage, in which less than 10% of the sediment input is used for aggradation, whilst the rest bypasses the system. Allogenic controls will force the system back and forth between the start-up and keep-up stages resulting in variation in aggradation rate and related avulsion frequency. In the section below, an estimate is made of this variation.

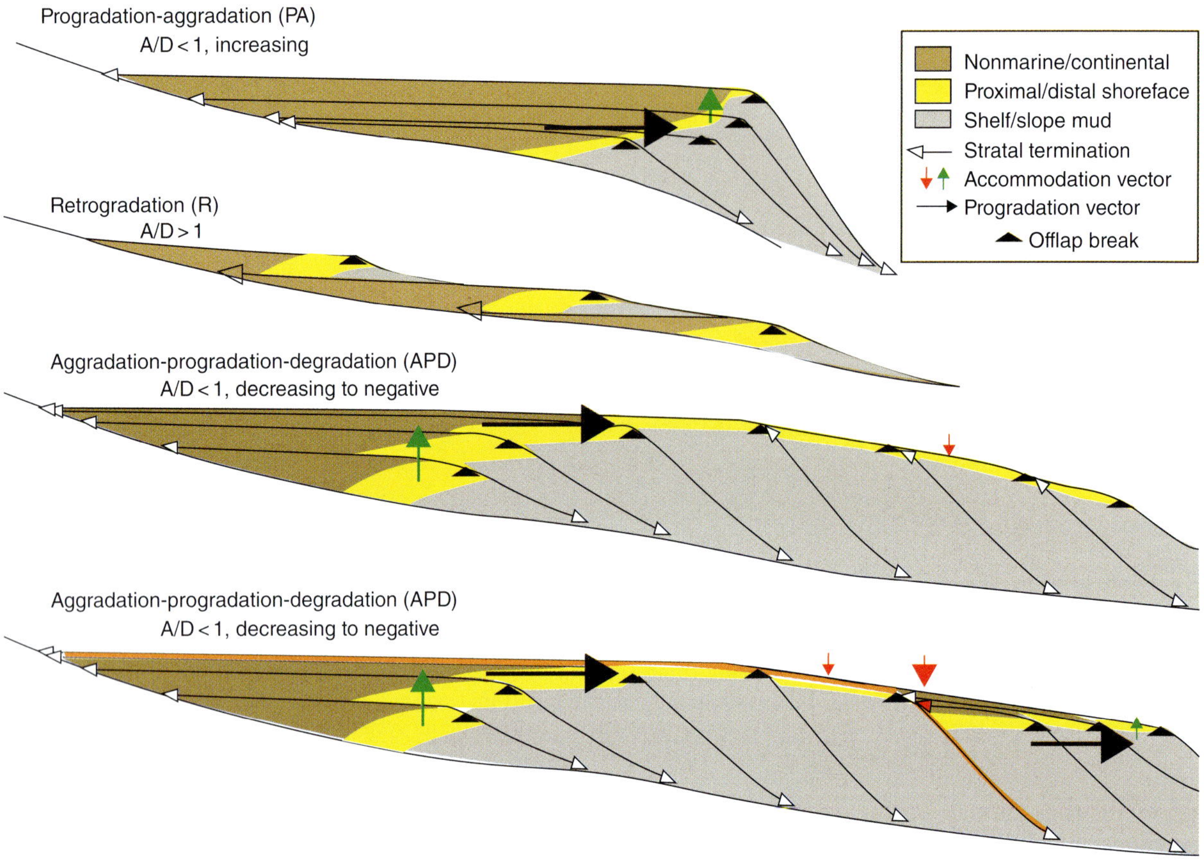

Fig. 4. Stratal stacking patterns associated with changing rates of coastal accommodation creation (A) and sediment fill (S), referred to as accommodation succession. Following a sequence boundary, the stratigraphic motif observed in the geologic record is progradation to aggradation (PA), retrogradation (R) and aggradation to progradation to degradation (APD), representing stratal geometries of lowstand, transgressive and highstand systems tracts, respectively. APD systems tract indicates a decrease in accommodation on the shelf through time. Toward the end of the APD systems tract, accommodation on the shelf may become negative, generating degradational stacking, not showing downward shift in coastal onlap. When downward shift in coastal onlap is observed, onlapping of proximal facies over distal ones would indicate formation of sequence boundary and initiation of another PA, R and APD succession (from Neal & Abreu, 2010).

Sea-level

Large-scale, sea-level forced architectural styles of river-delta systems comprise progradation-aggradation, progradation-degradation and retrogradation-aggradation stacking styles (Curray, 1964). These styles are recognised widely in seismic sections (e.g. Neal & Abreu, 2010) and related to normal regression, forced regression and transgression, respectively (Catuneanu *et al.*, 2009) (Fig. 4). Curray (1964) and Jervey (1988) related the three stacking styles to the rate of change of accumulation space (at the shoreline) [A] and the rate of sediment supply [S], thus A/S. The filling of accommodation, however, is not directly controlled by S, but by the aggradation (deposition) rate (D) (as was also argued by Muto & Steel (1997; 2001), so that the ratio A/D defines the fluvial deltaic architecture.

During normal regression there is accumulation space to fill in the alluvial system (PA, see Fig. 4). During the fill, there is significant bypass of sediment causing the progradation of the clinoform. Hence, the alluvial system will start somewhere in the fill-up stage and, depending on the progradation rate, reaches the keep-up stage. Under these conditions slow back filling in the channels occurs (significant bypass of sediment), so avulsion rate can be expected to be low.

During forced regression there is deposition, erosion and down stepping of the alluvial system (APD, see Fig. 4). Above the knickpoint, fluvial systems can still aggrade, as has been observed in landscape evolution experimental studies

(Van Heijst & Postma, 2001). Muto & Swenson (2005) quantified the maintenance of the fluvial grade by the specific square-root-of-time dependent rate of relative sea-level fall. The specific coefficient depends on sediment-water supply and system geometry. Hence, upslope of the knickpoint the alluvial river system can remain aggradational for a wide range of relative sea-level fall rates and channels can still back fill and avulse. Downslope of the knickpoint the river profile steepens, which hinders the back-fill process.

During transgression the shoreline steps back (PA, see Fig. 4). During the retrogradation brief stages of progradation occur (delta lobe building), during which the fluvial system can aggrade (see the experimental results of Muto & Steel (2001) and Hoyal & Sheets (2009). The development of a coastal barrier system forces the shoreline seaward bringing out the base-level point towards which the fluvial system is going to adjust itself. In this period of time the system is brought back close to the start-up stage leading to maximal aggradation in the alluvial realm and hardly any sediment bypass. During these periods, backfilling in channels is maximal and avulsion rates must be at their highest. The regular avulsions cause regular delta lobe progradation and shifting that is recognised as parasequences: shallowing upward sequences developed on top of flooding surfaces. The precise development of these coastal sequences can depend strongly on rate of sea-level rise (e.g. Cattaneo & Steel, 2003).

Climate

Holbrook *et al.* (2006) described the river profile as being highly variable due to changes in discharge and supply, i.e. by climate change. All potential river profiles are bounded above by a profile of highest possible aggradation and below by the profile of maximum possible incision. These upper and lower profiles are called 'buffers' and they envelop the available fluvial preservation space (Fig. 5). Thickness of the buffer zone is determined by variability in upstream controls and should increase up dip to the limit of downstream profile dominance.

The buffer model considers fluvial preservation to be limited to some space between upper and lower maximum possible profiles: 'buffers' that move and/or alter shape with downstream base-level shifts. Downstream, base level is considered to be controlled by movement of some physical 'buttress' (e.g. sea-level) below which streams cannot incise and above which streams cannot aggrade substantially. Upper and lower buffers are both anchored to this buttress and may diverge for some distance up-dip as profile variability is introduced by increasing influence of upstream base level controls. Upstream controls like climate and tectonics primarily determine spacing trends between these upper and lower buffers.

The change in river profile as a consequence of climate change is relatively fast in the case of a change in average discharge and much slower in the case of averaged change in sediment yield, as shown by experimental studies by Van den Berg van Saparoea & Postma (2008). These experiments demonstrate a fundamental difference between the response of the sediment flux at the river mouth due to changes in discharge and due to changes in sediment flux and differences between the total mass accumulation history in response to changes in discharge and sediment flux. The first fundamental difference between a response to either discharge or sediment input change is the total sediment budget at the valley outlet, which is much larger in case of a discharge change. The second fundamental difference is that the gradient of the valley floor is correlated positively with sediment influx and negatively with discharge (cf. also Mackin, 1948). The third difference is that the response to changes of discharge is very rapid, whilst the response to sediment flux changes is much slower (Van den Berg van Saparoea & Postma, 2008).

Hence, aggradation rates of the channel belt (and thus avulsion frequencies) would decrease at high discharges, as a consequence of reduction of accumulation space by lowering of the river profile; yet, backwater effects and channel blocking may temporarily increase the avulsion rate, silting up adjacent floodplains. If the river system would be near its grade, deviations in accumulation space forced by climate change are not likely to be very large, so the system will remain in the fill-up stage.

Tectonics

As also hypothesised by Holbrook *et al.* (2006), regional tectonics result in tilting of the river profile, while more local tectonics cause sagging, all with direct consequences for aggradation rate (Fig. 5). Channel belts appear not to be attracted to the subsidence maximum, unless subsidence

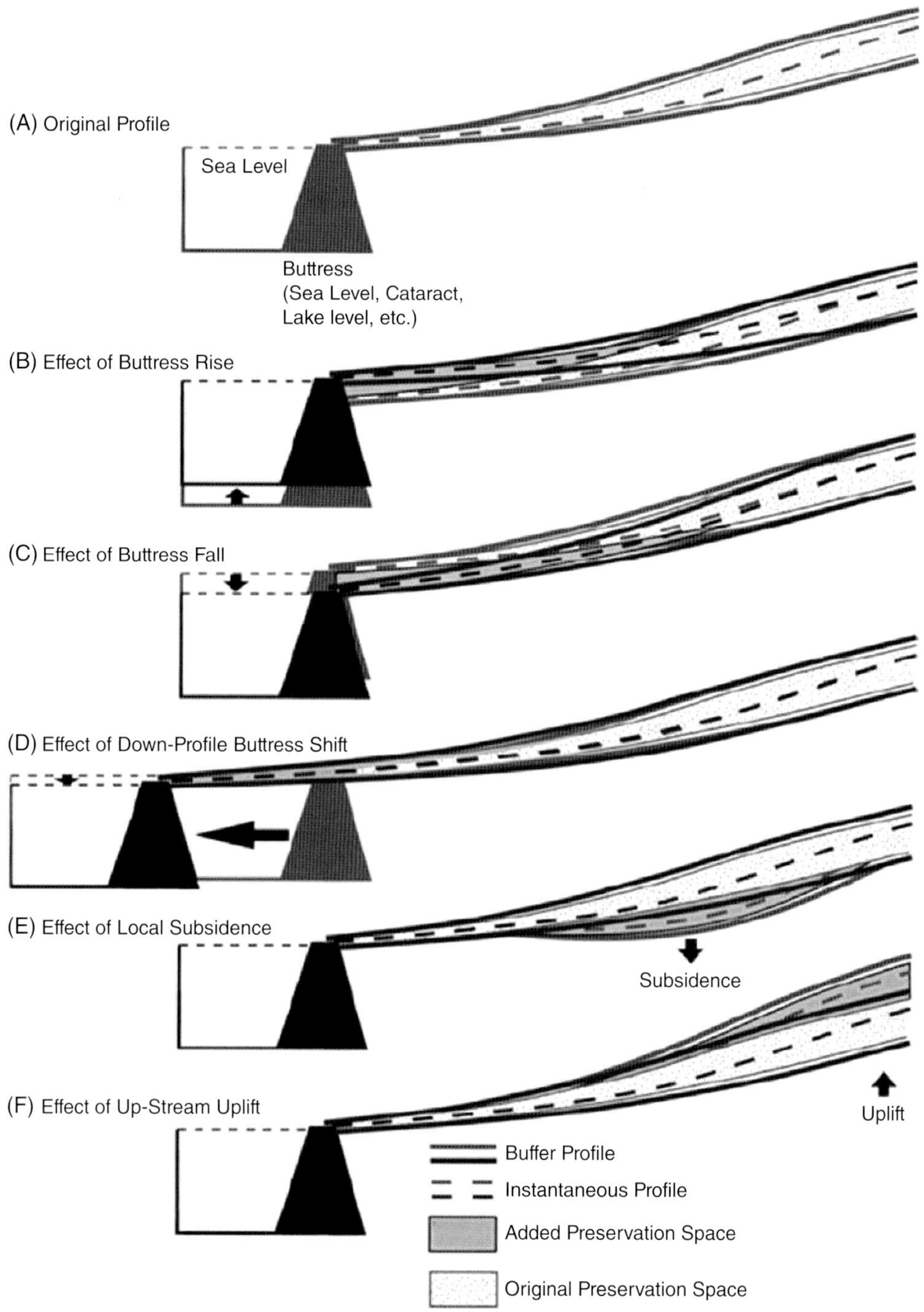

Fig. 5. Preservation space added as a result of shifts in initial buffer profiles (A) because of either buttress movement or tectonic adjustment. Fluvial preservation space may be added as a result of a simple buttress rise (B) or fall (C). Sediments deposited in added preservation space resulting from a buttress fall (C) are generally sequestered as easily eroded terraces hanging from the valley wall. They thus tend to have less long-term preservation potential than deposits buried by aggradation during a buttress rise (B). Movement of the buttress along the trajectory of the original longitudinal profile (D) tends to lengthen preservation space but otherwise adds minimal room for sediment accumulation. Subsidence beneath reaches of the lower buffer profile (E) tends to lower sediments deposited within the prior preservation space beneath active erosion. Long-term preservation potential of these sediments is high. Uplift beneath buffer profiles (F) tends to leave deposits from previous preservation spaces stranded as terraces where they could potentially be preserved long term but have high probability of erosion before eventual burial. In each of the above cases B through to F, the total space for potential accumulation of a fluvial unit is the integral of all preservation spaces produced over the period through which the depositing fluvial system was actively preserving sediment (from Holbrook *et al.*, 2006).

proceeds faster than aggradation of the river so that the latter can adjust to the formation of a topographic low by the deposition of overbank material in the form of splays and sheet sands (see experiments of Hickson *et al.* 2005).

Kim & Paola's (2007) experimental studies of sedimentation in an experimental relay ramp showed that autogenic cycles developed stratal packages of subaerial prograding lacustrine delta deposits bounded by fluvial aggradation units under constant discharge and sediment yield. These cycles were formed by strong variations in sediment delivery associated with tectonically-driven routing of river flow across and around the footwall uplift. Flow patterns of sheet flow and channelised flow ('avulsion cycles') became five times longer during the active subsidence (delayed the backfilling process). The period of the tectonic-driven autogenic processes was inferred to be of the order of 10 kyr to 100 kyr, which would be much lower than the normal autogenic behaviour.

Hence, the response in aggradation rate to tectonic change varies strongly with the kind of kinematics. Active fault scarps could make a fluvial stretch to subside instantly, bringing the system from fill-up to start-up stage, herewith increasing aggradation rates instantly. Basinward tilting of the fluvial profile as occurs, for instance, in passive margin settings would decrease aggradation rates, because the profile is tilted towards its grade.

DISCUSSION

In a discussion about how well fluvial architecture can be predicted in surface and subsurface analyses, Miall (2006) concluded that little can be expected beyond the provision of a general starting point. He argued that the variety of fluvial forms in modern rivers and the ancient record is vast, making the choice of an appropriate analogue very difficult. Fluvial style varies laterally or vertically through most real stratigraphic units because of the constant interplay of several allogenic controls acting on different time scales. Given the complex-response character of fluvial systems to allogenic forcing and including the tendency for systems to lag behind changes in forcing functions at varying rates, the predictability of fluvial architecture aerially and stratigraphically must be considered quite limited.

However, the experimental research mentioned here give reasons toward a more positive attitude. In spite of the fact that the experiments are not scaled hydraulically, the experimentalist has the great advantage of looking at a natural 'forward' model with similarity of process where the product can be studied in relation to input conditions (Paola, 2000; Paola *et al.* 2009).

The existence of scale-invariable morphological features like channels, bars and lobes hints to the similarity of process that is obtained in laboratory models. Sediment transport averaged over sufficiently long time periods can be predicted by diffusion (Paola *et al.* 1992). The crude fluvial architecture stemming from aggradation as well as from variations in depositional slope characteristic for the various river types can be simulated easily by using different exponents in a non-linear diffusion equation (Postma *et al.* 2008). Fig. 6 shows a dimensionless plot of aggradation rate by normalised sediment yield q_{in}/q_{out} against time (T) relative to the timescale that the fluvial system requires to reach grade (T_{eq}). The equilibrium timescale is the ratio of L^2/k, with L being a length scale which is given by the river's active depositional trajectory and k the diffusivity coefficient, which is related to the discharge (Paola *et al.*, 1992). The active depositional trajectory relevant for autogenic behaviour (avulsion) would be the backfill trajectory. With mean diffusivities of the order of 0.01 km^2/yr (Paola *et al.*, 1992), channel depth of 7 m to 10 m and slopes of the order of 0.0001 (from Kleinhans *et al.*, 2008), most low gradient rivers in the delta plain have a backwater length of approximately 25 km, so that T_{eq} for the reach is about 60 kyr. For low gradient rivers a linear diffusion equation for simulation of sediment transport over long time intervals is justified (e.g. Paola *et al.*, 1992), so that the start-up stage is almost non-existent (Fig. 6). However, it should be noted that both the length scale and the diffusivity coefficient vary dynamically and with that the calculated equilibrium time. Hence, its value should be treated with caution and only in a first order of approach.

Allogenic forcing brings the system continuously out of balance and changes its accumulation space and herewith the aggradation rate, as was discussed above. In asking 'is it possible to predict the change in aggradation rate?' the author believes it is possible to predict the change in a first order of approach. If the time period for the change in accumulation space is much faster than T_{eq},

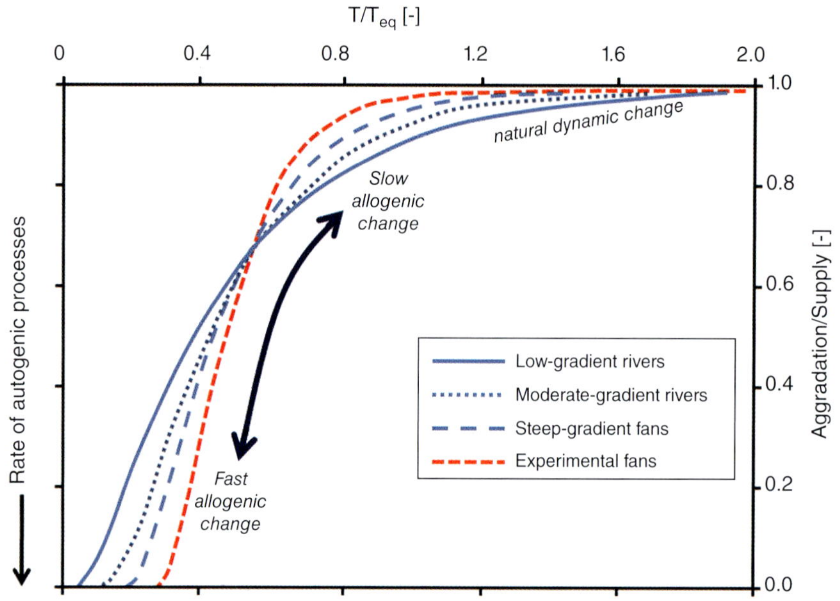

Fig. 6. Dimensionless plot of aggradation rate by normalised sediment yield q_{in}/q_{out} against time (T) relative to the time that the fluvial system requires to reach grade (T_{eq}). Sediment transport is calculated with a non-linear diffusion equation using different exponents. The low-gradient rivers are simulated with the linear diffusion equation (exponent m = 1, based on Paola *et al.* 1992). The steepest curve is calibrated against experimental results, for which the equilibrium slope is much steeper than for the deeper natural streams. The dotted curves intermediate of experimental fans and low-gradient rivers are estimations based on numerical interpolation between the steep experimental slopes and those of low gradient rivers (see Postma *et al.*, 2008). If the time period of allogenic forcing is much faster than equilibrium time of the river system then the frequency of avulsion will change with it. If it is slow then there will be no significant change (see text for further explanation).

aggradation rates are enhanced and if the period of change is much slower than T_{eq}, there will be little change in aggradation rate (see also Paola *et al.* 1992; Van Heijst & Postma, 2001). Hence, slow changes, as imposed for instance by regional tectonics, will hardly affect the aggradation rate so that the system remains in, or close to, the keep-up stage. Yet, rapid progradation of a delta lobe and subsidence near a fault scarp can have a significant effect on the accumulation space of the fluvial system and may bring it back into the start-up stage (Fig. 6). The experiments by Hickson *et al.* (2005) illustrate this point beautifully: fast subsidence is counterbalanced by high aggradation rates and slow subsidence rates by low aggradation rates. This causes fluvial systems not to migrate towards places with highest subsidence rates unless aggradation rates cannot keep up with the subsidence.

The analysis above leads us to a new working hypothesis that predicts the change in autogenic frequency: the rate of change (i.e. fast or slow change) in allogenic forcing relative to the equilibrium time related to the morpho-dynamically active part of the river system is the dominant driver of the rate of change in aggradation and herewith the change in frequency of autogenic behaviour. Slow changes in aggradation rate do not change autogenic behaviour significantly, whereas fast change does. However, it is not to say that other parameters, like peat growth in adjacent floodplains, flood frequency, storm surge frequencies and others cannot be important in causes for a change in avulsion rate. Yet, it would be interesting to test the launched hypothesis and measure aggradation rates in delta plains and plot them against reconstructions of avulsion frequency.

IMPLICATIONS

At the scale of the channel belt, Leeder (1978) attempted to establish fundamental connections amongst subsidence, avulsions and channel belt sandstone bodies stacking density. He suggested that channel-belt stacking density and hence connectivity is *inversely* correlated to temporal (vertical) changes in sedimentation rate and that

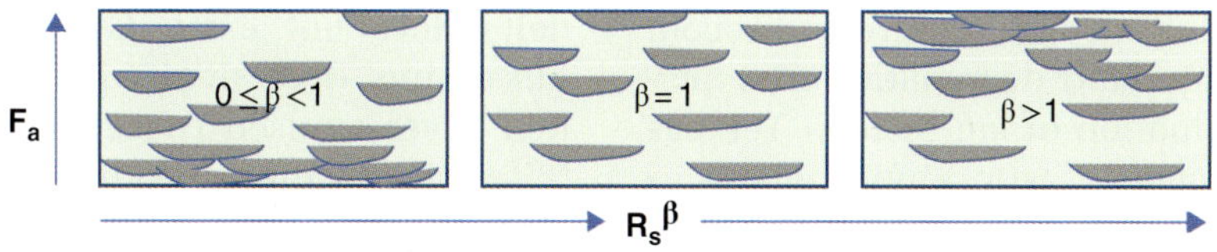

Fig. 7. Relation between channel-belt stacking density and sedimentation rate for three possible regimes defined by the exponent β in a power law relation between avulsion frequency and sedimentation rate. Dark grey indicates channel belt sand bodies and yellow indicates flood plain fines (redrawn from Bryant *et al.* 1995). Experiments suggest that the relationship pictured in the right hand side panel is most likely to occur in nature, which is an increase of avulsion rate with aggradation rate causing channels to stack more densely resulting in higher connectivity. F_a is the frequency of avulsions and R_s is the aggradation rate.

channel-belt stacking density and hence connectedness is directly correlated to lateral (horizontal) changes in sedimentation rate. Leeder (1978) suggested that reduction in subsidence rate with time increases the stacking density by allowing channel belts more time to remove floodplain fines.

Bryant *et al.* (1995) examined various forms of coupling between avulsion frequency and aggradation rate by examining their exponential relationship. If F_a is the frequency of avulsions and R_s is the aggradation rate then $F_a \approx R_s^{\beta}$, where β is a positive, real valued exponent. This leads to three qualitatively different regimes (Fig. 7) with $\beta=0$ resulting in a constant avulsion frequency, as assumed in Leeder's (1978) model. For $\beta=1$ the stacking pattern is independent of aggradation rate and for $\beta>1$ the autogenic behaviour would increase with aggradation rate; this case is evident for all laboratory models presently known. This means that maximal removal of floodplain fines and greatest connectivity of channel bodies would occur if aggradation rates are highest.

Hickson *et al.* (2005) conclude, on the basis of their findings, that the two-dimensional variation in alluvial architecture is controlled very strongly by externally forced sedimentary facies migrations such as changes in sediment supply, base level or subsidence. However, the three variables together control the aggradation rate, the basic control on facies change. If the imposed variations are slow then facies migrations are kept at a minimum but if they are relatively fast (as in some of Hickson *et al.*, 2005 runs), then they become a dominant control on alluvial architecture. Leeder's (1978) point about the effect of the lateral changes in sedimentation rate (stating that avulsion rates must be highest at subsidence maxima and lowest at subsidence minima, while the overall lateral stacking density of channel belts may remain unchanged) agrees well with the experimental findings of Hickson *et al.* (2005), who state that only if subsidence is faster than aggradation rate, will the river adjust and migrate to the topographic low that is formed.

Reconstructions of generic avulsion behaviour

For reconstructions of generic avulsion behaviour, detailed surface and subsurface mapping in combination with good age control is needed. Much of the hypothesis launched here still needs to be tested by thorough fieldwork, which at present gives ambiguous results. The cases dealt with below are nothing more than examples that help to demonstrate the frequency of avulsion and its relation to aggradation rates and are not meant as an exhaustive review.

Steep-gradient and moderately-gradient systems

Scott and Erskine (1994) studied twelve similarly sized Australian alluvial fans all subjected to the same catastrophic, rain-triggered, floods. The fans and catchment areas involved have similar sizes and gradients and were all located in a zone which received very similar rainfall intensities. Hence, the fans were subject to similar but significant flood discharges. Of the 12 fans, seven were entrenched and five were not before the storm event. The fans reacted in a different way to the storm event. Effects ranged from no change at all to trench incision or backfilling. Scott & Erskine (1994) propose that each fan showed a different stage of a similar autogenic cycle. The cycle consists of: (i) aggradation of the fan; (ii) the initiation of a fan-head trench due to exceeding the threshold slope; (iii) coalescence of scour pools to a continuous trench; and (iv) backfilling of the trench due to its widening and slope reduction.

Careful mapping of fan architecture and reconstruction of the hinterland degradation may permit a complete evaluation of fan history. Yet, age control in these coarse clastic environments is often not sufficient to determine details about the periods of autogenic processes in these coarse clastic systems (e.g. Nemec & Postma, 1993; Mack *et al.* 2008) and much more field work in combination with improved age control of the rock record is needed to confirm the findings of experimental studies.

Low-gradient river systems

No doubt the most detailed avulsion reconstructions come from the Rhine-Meuse delta complex. Although a complete review of these works is beyond the scope of this paper, it is worthwhile to point out how much fast and slow changes relative to equilibrium time affect the avulsion frequency.

From a detailed reconstruction of the Holocene avulsion history of the whole Rhine-Meuse delta Stouthamer & Berendsen (2000, 2001, 2007) determined quantitative values for the avulsion parameters, avulsion frequency, avulsion duration and inter-avulsion period. In the Rhine-Meuse delta the number of coeval channels is related to avulsion frequency supported by high resolution age control that helps to analyse beginning and ending of river activity. The data show that instantaneous and gradual avulsions were almost equally important in the Rhine-Meuse delta, with two dominant avulsion styles: (1) regional avulsion: the new channel followed an entirely new course and (2) avulsion leading to reoccupation of existing channels. In the case of reoccupation, they found two possibilities: the new channel reoccupied (a) its previous channel (local avulsion) or (b) a pre-existing channel. Stouthamer *et al.* (2010) found that, over the Holocene time scale, avulsion style was related to aggradation rate and coastal evolution. Initial high avulsion frequencies decreased with dropping rates of sea-level rise (dropping rates of accumulation space) from 9000 years ago until ~ 3000 years ago. Thereafter avulsion frequency increased again (1.89 avulsions/100 yrs) during an inferred period of increased delivery of fine sediment and slightly altered discharge regimes (Stouthamer *et al.*, 2011), which increased back barrier aggradation rates significantly.

For the Mississippi delta, a database on avulsion frequency like that available for the Rhine–Meuse delta does not exist. Over the past 5 kyr, the Mississippi River avulsed only four times and the Red River avulsed twice in the southern Lower Mississippi Valley, as found by Aslan *et al.* (2006). Relocation of the Mississippi River eastward shifted local base level and led to the Red River avulsion. Mississippi and Red River avulsion occurred primarily through channel reoccupation. Aslan *et al.* (2006) argue that gradient advantages did not necessarily lead to the avulsions, although gradient advantages along the Mississippi River floodplain are widespread.

Fielding *et al.* (2006) found climate change to be a primary control on the Holocene Burdekin delta architecture, inhibiting a shorter avulsion period by facilitating extreme variability of discharge. They suggested further that more frequent avulsion may also have been facilitated by the lengthening of the delta-plain channels as the system progrades seaward, which would increase the accumulation space rapidly. The Mfolozi River Floodplain (South Africa) is characterised by avulsions in the floodplain head that occur primarily during extremely infrequent large flood events, where the stream flow capacity of the channel is insufficient (Grenfell *et al.* 2009).

Other factors that control avulsion more indirectly have been forwarded from the study of natural examples, including neotectonics, subsidence, substrate composition, sinuosity changes and human activities (Fisk, 1952; Schumann, 1989; Schumm *et al.*, 1996; Jones & Harper, 1998; Smith *et al.*, 1998; Stouthamer & Berendsen, 2000) and meander bend properties (Kleinhans, 2010). Schumm *et al.* (1996) describe how reductions in channel gradient caused by increased channel sinuosity lead to decreased sediment transport capacity and avulsion (e.g. Makaske, 2001). It is challenging to relate the control of all these factors back to the control they have on aggradation rate and to relate that to frequency in autogenic behaviour. Yet, future stratigraphic studies in ancient fluvial and delta settings should be conducted to evaluate the significance of the change in aggradation rate on channel stacking density and connectedness further. That evaluation of change in aggradation rates is more practical than applying sequence stratigraphical principles in surface and subsurface stratigraphic analysis is shown by Moscariello (2003) and Martinius *et al.* (this volume) in correlating stratigraphic sequences in the Triassic-Jurassic Statfjord Formation (Norwegian Sea).

CONCLUSIONS

Autogenic behaviour is discussed for steep, moderate and low gradient fluvial systems and is found to be different for each fluvial type. In the first two types sediment is spread evenly over wide areas, in contrast with the low-gradient rivers. Other differences are that:

1. Steep-gradient systems show alternations of sheet flow and relief steepening, incision and progradation followed by backfilling. Cycle duration depends on the total volume of the fan incision and the backfilling rate, ranging from years to decades, depending on fan activity;
2. Moderate-gradient systems show bifurcations of short duration during high aggradation rates and bifurcations with slowly lateral migrating channels during low aggradation rates. Cycle duration is relative short ranging from days to years;
3. Low-gradient rivers build their channel belts up to grade while their floodplains lag significantly behind. It is inferred on the basis of numerical modelling that the origin of avulsion lies in gradient advantage, channel plugging and local hydrology in river bends. Details of the avulsion process are least understood in this type because experiments that allow study of their autogenic behaviour are absent. Cycle durations are relatively long and of the order of decades to centuries and strongly dependant on the backwater adaptation length.
4. Aggradation rate in the channel belts is the most important driver for the frequency of autogenic behaviour. Aggradation rate decreases non-linearly when the fluvial system builds up to grade, so that a full spectrum from high to low frequency autogenic behaviour is to be expected in natural systems that build up to grade.
5. Fast and slow changes in allogenic forcing relative to the equilibrium time (here related to the backwater length) of delta plain river systems is suggested here as the dominant driver of changes in the rate of aggradation and herewith the frequency of autogenic behaviour (Fig. 6). Fast change will increase the frequency of autogenic processes and will force rapid vertical changes in fluvial architecture, whilst a slow change will result in little or very gradual change of architecture.

Significant advancement in recognition and quantification of generic autogenic behaviour stems from experimental research on a landscape scale, where drivers for stratigraphic architecture can be verified along known input and boundary conditions. Scaling is by similarity of process, which is justified by scale invariant architecture and morphology as channels, sheet flow deposits, bars and lobes. Such landscape experiments are, as yet, non-existent for low gradient river systems, although stretches of low gradient river systems are presently studied by groups experimenting in more detail. In addition, field studies with high resolution age control are required to learn about the forcing mechanisms of avulsion in fluvial systems.

ACKNOWLEDGEMENTS

The ideas presented in this paper evolved during the PhD study of Maurits van Dijk and through discussions with Kim Cohen, Esther Stouthamer and Maarten Kleinhans. In January, 2011 first draft of the MS was read critically by Kick Kleverlaan, Maarten Kleinhans and Allard Martinius, who provided useful comments for improvement. I would also like to thank journal reviewers Andrea Moscariello and an anonymous reviewer for their critical reading, which helped to clarify the text in places.

REFERENCES

Allen, J.R.L. (1965) A review of the origin and characteristics of recent alluvial sediments. *Sedimentology*, **5**, 89–101.

Allen, J.R.L. (1979) Studies in fluviatile sedimentation: An elementary geometrical model for the connectedness of avulsion-related channel sand bodies. *Sed. Geol.*, **24**, 253–267.

Ashworth, P.J., Best, J.L. and **Jones, M.A.** (2004) Relationship between sediment supply and avulsion frequency in braided rivers. *Geology*, **32**, 21–24.

Ashworth, P.J., Best, J.L. and **Jones, M.A.** (2007) The relationship between channel avulsion, flow occupancy and aggradation in braided rivers: insights from an experimental model. *Sedimentology*, **54**, 497–513.

Aslan, A. Whitney, J.A. and **Blum, M.D.** (2006). Causes of river avulsion: insights from the late Holocene avulsion history of the Mississippi River, U.S.A. *J. Sed. Res.*, **75**, 650–664.

Beerbower, J.R. (1964) Cyclothems and cyclic depositional mechanisms in alluvial plain sedimentation. In: *Symposium on Cyclic Sedimentation* (Ed. Merriam, D.F.), *Bull. Kansas Geol. Survey*, **169.2.1.6**, 31–42.

Blum, M.D. and **Törnqvist, T.E.** (2000) Fluvial responses to climate and sea-level change: a review and look forward. *Sedimentology*, **47**, 2–48.

Bridge, J.S. and **Leeder, M.R.** (1979) A simulation model of alluvial stratigraphy. *Sedimentology*, **26**, 617–644.

Bryant, M., Falk, P. and **Paola, C.** (1995) Experimental-study of avulsion frequency and rate of deposition. *Geology*, **23**, 365–368.

Cattaneo, A. and **Steel, R.J.** (2003) Transgressive deposits: a review of their variability. *Earth-Science Reviews*, **62**, 187–228.

Catuneanu, O., Abreu, V., Bhattacharya, J.P., Blum, M.D., Dalrymple, R.W., Eriksson, P.G., Fielding, C.R., Fisher, W.L., Galloway, W.E., Gibling, M.R., Giles, K.A., Holbrook, J.M., Jordan, R., Kendall, C.G. St.C., Macurda, B., Martinsen, O.J., Miall, A.D., Neal, J.E., Nummedal, D., Pomar, L., Posamentier, H.W., Pratt, B.R., Sarg, J.F., Shanley, K.W., Steel, R.J., Strasser, A., Tucker, M.E. and **Winker, C.** (2009) Towards the standardization of sequence stratigraphy. *Earth-Sci. Rev.*, **92**, 1–33.

Commission on Geosciences, Environment and Resources (CGER) (1996) *Alluvial fan flooding*. The National Academies Press.

Curray, J.R. (1964) Transgressions and regressions. In: *Papers in Marine Geology* (Ed. R. L. Miller), pp 175–203. Shepard commemorative volume, MacMilland, New York.

Ethridge, F.G., Skelly, R.L. and **Bristow, C.S.** (1999) Avulsion and crevassing in the sandy, braided Niobrara River: complex response to base-level rise and aggradation. In: *Fluvial Sedimentology VI* (Eds N.D. Smith, N.D. and J. Rogers), *Int. Assoc. Sedimentol. Spec. Publ.*, **28**, 179–191.

Fielding, C.R., Trueman, J.D. and **Alexander, J.** (2006) Holocene depositional history of the Burdekin river delta of north eastern Australia: A model for a low-accommodation, highstand delta. *J. Sed. Res.*, **76**, 411–428.

Fisk, H.N. (1952) *Geological Investigation of the Atchafalaya Basin and the Problem of the Mississippi River Diversion: Vicksburg, Mississippi.* U.S. Army Corps of Engineers, Waterways Experiment Station, 145 pp.

Grenfell, S.E., Ellery, W.N. and **Grenfell, M.C.** (2009) Geomorphology and dynamics of the Mfolozi River floodplain, KwaZulu-Natal, South Africa. *Geomorphology*, **107**, 226–240.

Harwood, K. and **Brown, A.G.** (1993) Fluvial processes in a forested anastomosing river; flood partitioning and changing flow patterns. *Earth Surf. Proc. Land.*, **18**, 741–748.

Hickson, T.A., Sheets, B.A., Paola, C. and **Kelberer, M.** (2005) Experimental test of tectonic controls on three dimensional alluvial facies architecture. *J. Sed. Res.*, **75**, 710–722.

Holbrook, J., Scott, R.W. and **Oboh-Ikuenobe, F.E.** (2006) Base-level buffers and buttresses: A model for upstream versus downstream control on fluvial geometry and architecture within sequences. *J. Sed. Res.*, **76**, 162–174.

Hoyal, D. and **Sheets, B.** (2009) Morphodynamic evolution of experimental cohesive deltas. *J. Geophys. Res.*, **114**, F02009, DOI: 10.1029/2007JF000882.

Jervey, M.T. (1988) Quantitative geological modeling of siliciclastics rock sequences and their seismic expression. In: *Sea-Level Changes: An Integrated Approach* (Eds C.K. Wilgus, B.S. Hastings, C.G.St.C. Kendall, H.W. Posamentier, C.A. Ross and J.C. Van Wagoner), SEPM Spec. *Publ.*, **42**, 47–70.

Jones, L.S. and **Harper, J.T.** (1998) Channel avulsions and related processes and large-scale sedimentation patterns since 1875, Rio Grande, San Luis Valley, Colorado. *Geol. Soc. Am. Bull.*, **110**, 411–421.

Jones, L.S. and **Schumm, S.A.** (1999) Causes of avulsion: an overview. In: *Fluvial Sedimentology VI* (Eds N.D. Smith, N.D. and J. Rogers), *Int. Assoc. Sedimentol. Spec. Publ.*, **28**, 171–178.

Karssenberg, D. and **Bridge, J.S.** (2008) A three-dimensional model of sediment transport, erosion and deposition within a network of channel belts, floodplain and hill slope: extrinsic and intrinsic controls on floodplain dynamics and alluvial architecture. *Sedimentology*, **55**, 1717–1745.

Kim, W. and **Paola, C.** (2007) Long-period cyclic sedimentation with constant tectonic forcing in an experimental relay ramp. *Geology*, **35**, 331–334.

King, W.A. and **Martini, I.P.** (1984) Morphology and Recent sediments of the lower anastomosing reaches of the Attawapiskat River, James Bay, Ontario, Canada. *Sed. Geol.*, **37**, 295–320.

Kleinhans, M.G. (2010) Sorting out river channel patterns. *Prog. Phys. Geogr.*, **34**, 287–326.

Kleinhans, M.G., Jagers, H., Mosselman, E. and **Sloff, C.** (2008) Bifurcation dynamics and avulsion duration in meandering rivers by one-dimensional and three-dimensional models. *Water Resour. Res.*, **44**, W08454. doi:10.1029/2007WR005912.

Kleinhans, M.G., Weerts, H.J.T. and **Cohen, K.M.** (2010) Avulsion in action: reconstruction and modelling sedimentation pace and upstream flood water levels following a Medieval tidal-river diversion catastrophe (Biesbosch, The Netherlands, 1421-1750 AD), *Geomorphology*, doi:10.1016/j.geomorph.2009.12.009.

Kriele, H., Wang, Z. and **De Vries, M.** (1998) Morphological interaction between the Yellow River and its estuary. In: *Physics of Estuaries and Coastal Seas* (Eds J. Dronkers and M. Scheffers), 8th International Biennial Conference on Physics of Estuaries and Coastal Seas, The Hague, Netherlands, 287–295, Taylor and Francis/Balkema, London, U.K.

Leeder, M.R. (1978) A quantitative stratigraphic model for alluvium, with special reference to channel deposit density and interconnectedness. In: *Fluvial sedimentology* (Ed. Miall, A.D.), *Can. Soc. Petrol. Geol. Mem.*, **5**, 587–596.

Mack, G.H., Leeder, M.R. and **Carothers-Durr, M.** (2008) Modern flood deposition, erosion and fan-channel avulsion on the semi-arid Red Canyon and Palomares Canyon alluvial fans in the southern Rio Grande rift New Mexico, U.S.A., *J. Sed. Res.*, **78**, 432–442.

Mackey, S. and **Bridge, J.** (1995) Three-dimensional model of alluvial stratigraphy: theory and application. *J. Sed. Res.*, **B65**, 7–31.

Mackin, J.H. (1948) Concept of the graded river. *Bull. Geol. Soc. America*, **59**, 463–512.

Makaske, B. (2001) Anastomosing rivers: a review of their classification, origin and sedimentary products: *Earth-Sci. Rev.*, **53**, 149–196.

Martinius, A.W., Elfenbein, C. and **Keogh, K.J.** (2014) Applying A/S concepts to stratigraphic analysis and zonation of fluvial reservoirs. In: *From Depositional Systems to Sedimentary Successions on the Norwegian Continental Shelf Sedimentation* (Eds A.W. Martinius, R. Ravnas, J. Howell, R.J. Steel and J.P. Wonham), *Int. Assoc. Sedimentol. Spec. Publ.*, **46**, (this volume).

McCarthy, T.S., Ellery, W.N. and **Stanistreet, I.G.** (1992) Avulsion mechanisms on the Okavango fan, Botswana: the control of a fluvial system by vegetation. *Sedimentology*, **39**, 779–795.

McCarthy, T.S., Ellery, W.N. and **Stanistreet, I.G.** (2002) Avulsion mechanisms on the Okavango fan, Botswana: the control of a fluvial system by vegetation. *Sedimentology*, **39**, 779–796.

Miall, A.D. (2006) Reconstructing the architecture and sequence stratigraphy of the preserved fluvial record as a tool for reservoir development: A reality check. *AAPG Bull*, **90**, 989–1002.

Moscariello, A. (2003) The Schooner Field, Blocks 44/26a and 43/30a, UUK Southern North Sea. In:, *United Kingdom Oil & Gas Fields* (Eds J. Gluyas, H. Hichens, J. Evans), *Commemorative Millenium Volume, Geol. Soc. Mem.*, **20**, 811–824.

Muto, T. and **Steel R.J.** (1997) Principles of regression and transgression: The nature of the interplay between accommodation and sediment supply. *J. Sed. Res.*, **67**, 994–1000.

Muto, T. and **Steel R.J.** (2001) Autostepping during the transgressive growth of deltas: Results from flume experiments. *Geology*, **29**, 771–774.

Muto, T. and **Swenson, J.B.** (2005) Large-scale fluvial grade as a nonequilibrium state in linked depositional systems: Theory and experiment. *J. Geophys. Res.*, **110**, F03002, oi:10.1029/2005JF000284.

Neal, J. and **Abreu, V.** (2010) Sequence stratigraphy hierarchy and the accommodation method. *Geology*, **37**, no 9, 779–782; doi: 10.1130/G25722A.

Nemec, W. and **Postma, G.** (1993) Quaternary alluvial fans in southwestern Crete: sedimentation processes and geomorphic evolution. In: *Alluvial Sedimentation* (Eds Marzo, M. and Puigdefabregas, C.), *Int. Assoc. Sedimentol. Spec. Publ.*, **17**, 235–276.

Nemec, W. and **Steel, R.J.** (1988) What is a fan-delta and how do we recognize it? In: *Fan Deltas: Sedimentology and Tectonic Settings* (Eds W. Nemec & R.J. Steel), pp 3–13. Blackie, Glasgow.

Paola, C. (2000) Quantitative models of sedimentary basin filling. *Sedimentology*, **47**, 121–178.

Paola, C., Heller, P. and **Angevine, C.** (1992) The large scale dynamics of grain-size variation in alluvial basins. I: Theory. *Basin Res.*, **4**, 73–90.

Paola, C., Straub, K., Mohrig, D. and **Reinhardt, L.** (2009) The "unreasonable effectiveness" of stratigraphic and geomorphic experiments. *Earth-Sci. Rev.*, doi:10.1016/2009.05.003.

Peakall, J., Ashworth, P.J. and **Best, J.L.** (1996) Physical modelling in fluvial geomorphology: principles, applications and unresolved issues. In: *The scientific nature of geomorphology* (Eds B.L. Rhoads and C.E. Thorn) *Proceedings of the 27th Binghamton symposium in geomorphology*, 221–254. John Wiley & Sons, Chichester.

Postma, G., Kleinhans, M.G., Meijer, P.T. and **Eggenhuisen, J.** (2008) Sediment transport in analogue flume models compared with real world sedimentary systems: a new look at scaling sedimentary systems evolution in a flume. *Sedimentology*, **55**, 1–17.

Schumann, R.R. (1989) Morphology of Red Creek, Wyoming, an arid-region anastomosing channel system. *Earth Surf. Proc. Land.*, **14**, 277–288.

Schumm, S.A., Mosley, P.M. and **Weaver, P.H.** (1987) *Experimental Fluvial Geomorphology*. John Wiley & Sons, New York.

Schumm, S.A., Erskin, W.D. and **Tilleard, J.W.** (1996) Morphology, hydrology and evolution of the anastomosing Ovens and King Rivers, Victoria, Australia. *Geol. Soc. Am. Bull.*, **108**, 1212–1224.

Scott, P.F. and **Erskine, W.D.** (1994) Geomorphic Effects of a Large Flood on Fluvial Fans. *Earth Surf. Proc. Land.*, **19**, 95–108.

Sheets, B., Hickson, T.A. and **Paola, C.** (2002) Assembling the stratigraphic record: depositional patterns and time-scales in an experimental alluvial basin. *Basin Res.*, **14**, 287–301.

Slingerland, R. and **Smith, N.** (1998) Necessary conditions for a meandering-river avulsion. *Geology*, **26**, 435–438.

Slingerland, R. and **Smith, N.D.** (2004) River avulsions and their deposits. *Annu. Rev. Earth Planet. Sci.*, **32**, 254–285.

Smith, N., Slingerland, R., Pérez-Arlucea, M. and **Morozova, G.** (1998) The 1870's avulsion of the Saskatchewan River, *Canadian Journal of Earth Sciences*, **35**, 453–66.

Stouthamer, E. and **Berendsen, H.J.A.** (2000) Factors Controlling the Holocene Avulsion History of the Rhine-Meuse Delta (The Netherlands). *J. Sed. Res.*, **70**, 1051–1064.

Stouthamer, E. and **Berendsen, H.J.A.** (2001) Avulsion frequency, avulsion duration and interavulsion period of Holocene channel belts in the Rhine-Meuse delta, the Netherlands. *J. Sed. Res.*, **71**, 589–598.

Stouthamer, E. and **Berendsen, H.J.A.** (2007) Avulsion: the relative roles of autogenic and allogenic processes. *Sed. Geol.*, **198**, 309–325.

Stouthamer, E., Cohen, K.M. and **Gouw, M.J.P.** (2010) Avulsion and its implications for fluvial-deltaic architecture: insights from the Holocene Rhine-Meuse delta. In: *From River to Rock Record: The Preservation of Fluvial Sediments and Their Subsequent Interpretation* (Eds S.K. Davidson, S. Leleu, and C.P. North), *SEPM Spec. Publ.*, **97**, 215–231.

Stouthamer, E., Cohen, K.M., and **Gouw, M.J.P.** (2011) Avulsion and its implications for fluvial-deltaic architecture: insights from the Holocene Rhine–Meuse delta. In: *From River To Rock Record: The Preservation Of Fluvial Sediments And Their Subsequent Interpretation, SEPM Special Publication*, **97**, 215–231.

Törnqvist, T.E. (1994), Middle and late Holocene avulsion history of the River Rhine (Rhine-Meuse delta, Netherlands), *Geology*, **22** (8), 711–714.

Törnqvist, T.E. and **Bridge, J.S.** (2002) Spatial variation of overbank aggradation rate and its influence on avulsion frequency. *Sedimentology*, **49**, 891–905.

Van Asselen, S., **Stouthamer, E.** and **van Asch, Th.W.J.** (2009) Effects of peat compaction on delta evolution: a review on processes, responses, measuring and modeling. *Earth-Sci. Rev.*, **92**, 35–51.

Van den Berg van Saparoea, A.-P. and **Postma, G.** (2008) Control of climate change on the yield of river systems. In: *Recent Advances in Models of Siliciclastic Shallow-marine Stratigraphy* (Eds G. Hampson, R.J. Steel, P.M. Burgess and R.W. Dalrymple), *Int. Assoc. Sedimentol. Spec. Publ.*, **40**, 191–205.

Van Dijk, M., **Postma, G.** and **Kleinhans, M.G.** (2009) Autocyclic behaviour of fan deltas: an analogue experimental study. *Sedimentology*, **56**, 1569–1589.

Van Dijk, M., **Kleinhans, M.G.**, **Postma, G.** and **Kraal, E.** (2012) Contrasting morphodynamics in alluvial fans and fan deltas: effect of the downstream boundary. *Sedimentology* **59**, 2125–2145.

Van Heijst, M.I.W.M. and **Postma, G.** (2001) Fluvial response to sea-level changes: a quantitative analogue, experimental approach. *Basin Res.*, **13**, 269–292.

Van Rijn, L.C. (1994) *Principles of Fluid Flow and Surface Waves in Rivers, Estuaries, Seas and Oceans.* Aqua Publications, Oldemarkt, the Netherlands, 334 pp.

Whipple, K.X., **Parker, G.**, **Paola, C.** and **Mohrig, D.** (1998) Channel dynamics, sediment transport and the slope of alluvial fans: experimental study. *J. Geol.*, **106**, 677–693.

Int. Assoc. Sedimentol. Spec. Publ. (2014) **46**, 19–58

Climatic and tectonic controls on Triassic dryland terminal fluvial system architecture, central North Sea

TOM McKIE

Shell UK Exploration and Production, 1 Altens Farm Road, Nigg, Aberdeen, AB12 3FY, United Kingdom (E-mail: Tom.Mckie@Shell.com)

ABSTRACT

The Triassic succession in the central North Sea was deposited in a dryland alluvial setting during the early phases of rifting which marked the break-up of Pangaea. Extension was episodic and triggered widespread halokinesis in the underlying Zechstein Group halite, particularly during the Early and Late Triassic Hardegsen and Cimmerian I tectonic events. These movements created a tortuous basin topography which temporarily inhibited fluvial sand dispersal and resulted in a dominance of floodbasin fines. Halokinesis also exerted a particularly strong control on stratigraphic preservation on the basin margins by conveying shallow-buried successions, which were vulnerable to erosion, into long-term preservation space below the truncation level of subsequent Jurassic unconformities. During periods of relative fault quiescence and reduced aridity Skagerrak Formation terminal fluvial systems were able to expand across the basin as large transverse Fennoscandian fluvial fans which interacted with both axial and transverse systems draining off the Scottish Highlands. These systems, driven by the Tethyan monsoon, ultimately flowed southward into the northern margin of the Southern Permian Basin to terminate in widespread playa and coastal sabkha. Expansion of these terminal fluvial systems probably took place during pluvial climate episodes and the prevalence of vegetation and burrowing activity on the Skagerrak Formation floodplain suggests an availability of year-round soil moisture, indicating that monsoonal flooding was augmented by intermittent to perennial, dry season, flow. The pluvial episodes potentially to correspond to episodes of warming in the Tethys region, which may have resulted in enhanced evaporation, increased thermal convection and monsoon intensification. The consequent reduction in temperature gradients may also have weakened north-easterly trade winds, allowing more frequent impingement of moist Tethyan air masses into the region during the dry season. In addition to such regional climatic changes, flooding of the Southern Permian Basin by the Muschelkalk Sea during a period of widespread Anisian aridity briefly altered the local climate, allowing the maintenance of marshes and floodplain lakes adjacent to this interior seaway. Contraction of the Skagerrak Formation terminal fluvial systems apparently occurred during cooling, as the climate reverted to the prevailing Pangaean aridity and resulted in the deposition of basin-wide playa. By the Middle Triassic, such playa facies had blanketed much of north-west Europe as the Keuper Formation and Mercia Mudstone Group and the contemporaneous Skagerrak Formation fluvial systems stand in marked contrast to these arid, mud-dominated, basin-fills due to their access to runoff from large Fennoscandian catchments located outwith the arid rift interior.

Keywords: dryland fluvial, Triassic, Skagerrak Formation, Smith Bank Formation, climate, tectonics, halokinesis.

From Depositional Systems to Sedimentary Successions on the Norwegian Continental Margin, First Edition. Edited by A. W. Martinius, R. Ravnås, J. A. Howell, R. J. Steel, and J. P. Wonham.

INTRODUCTION

The architecture and preservation of terminal, dryland fluvial systems such as the Triassic in the central North Sea are relatively under-researched compared to that of paralic fluvial systems where relative sea-level fluctuations exert an over-arching control (Shanley & McCabe, 1994; Blum & Törnqvist, 2000). Early models of the sequence stratigraphy of such hinterland successions (e.g. Steel & Ryseth, 1990; McKie & Garden, 1996; Legarreta & Uliana, 1998) emphasised the predictable expansion and contraction of generally terminal fluvial systems in response to changes in the accommodation: supply ratio of the depositional system and fluctuations in stratigraphic base level (Martinsen *et al.*, 1999), but lacked the regional palaeoclimatic information to disentangle the complex fluvial response to basinal and catchment controls on base level and sediment yield (Blum & Törnqvist, 2000). The aim of this paper is to explore the multivariate controls on the alluvial architecture of the central North Sea Triassic by placing the evolution of these terminal fluvial depositional systems within a wider framework of the tectonics and climate change which were occurring on the northern Tethyan margin of Pangaea, and to examine their relative contribution on the source-to-sink fluvial routing into the central North Sea region. Although the Triassic in the central North Sea has limited well penetrations, a patchy preservation (Fig. 1) and relatively low-resolution chronostratigraphic framework, the Southern Permian Basin immediately to the south, with its classical 'Germanic' succession of Bunter, Muschelkalk and Keuper formations, is relatively well understood (Aigner & Bachmann, 1992; Geluk, 2005; Bachmann *et al.*, 2010) and provides a robust, down-palaeoflow framework of tectonics, climate change and marine influence which can be extended both southward into the marine Tethyan margin and northward into the fully terrestrial central North Sea area (Michelsen & Clausen, 2002). The overall stratigraphic evolution between these regions can be compared and similarities and departures from the regional climate and tectonic setting used to make inferences on the controls on deposition and preservation in the generally endorheic central North Sea region.

Central North Sea stratigraphic framework

The stratigraphic subdivision of the Triassic across the North Sea region varies both by basin and across international boundaries (Lervik, 2006). In the UK central North Sea (Fig. 2) the Triassic succession is generally divided into the mud-prone Smith Bank and overlying, sand-prone, Skagerrak formations (Cameron *et al.*, 1992). These subdivisions can be correlated locally into the Norwegian (Lervik, 2006) and Danish (Michelsen & Clausen, 2002) sectors in the area to the north of the mid-North Sea and Ringkøbing-Fyn highs (Fig. 1B). South of these highs, in the Southern Permian Basin, the more marine influenced Germanic succession prevails (Bertelsen, 1980; Michelsen & Clausen, 2002; Geluk, 2005; Geluk, 2007). In the UK sector of the central North Sea the Skagerrak Formation has been further subdivided into biostratigraphically constrained, sand-prone and mud-prone members (Fig. 2; Goldsmith *et al.*, 1995, 2003), but in the more sparsely penetrated Norwegian sector these subdivisions are locally more difficult to identify. Biostratigraphic recovery is generally poor, particularly in the Early Triassic to Anisian (Smith Bank and Bunter Sandstone formations and lower Judy Sandstone Member), but the Late Anisian-Early Ladinian and Late Ladinian intervals are locally highly productive (Goldsmith *et al.*, 1995; Lindström *et al.*, 2009), allowing the regional correlation of the upper Judy Sandstone, Julius Mudstone and lower Joanne Sandstone members. The Joanne Sandstone Member forms a widespread alluvial package, but this has been variably eroded by an intra-Skagerrak Formation unconformity (Archer *et al.*, 2010). The uppermost parts of the Triassic succession, corresponding to the Jonathan Mudstone, Josephine Sandstone and Joshua Mudstone members (Fig. 2), are also very poorly preserved due to multiple Jurassic erosional events and are only well developed in UK blocks 30/7, 30/12 and 30/13 (Goldsmith *et al.*, 1995). Equivalent successions may be present in the Norwegian and Danish sectors, but these can lack the well-defined shale packages which differentiate individual members.

Whilst a robust stratigraphic succession has been defined in key areas where palynologically productive horizons are preserved, correlation of the Skagerrak Formation members is hampered by the generally poor biostratigraphic recovery,

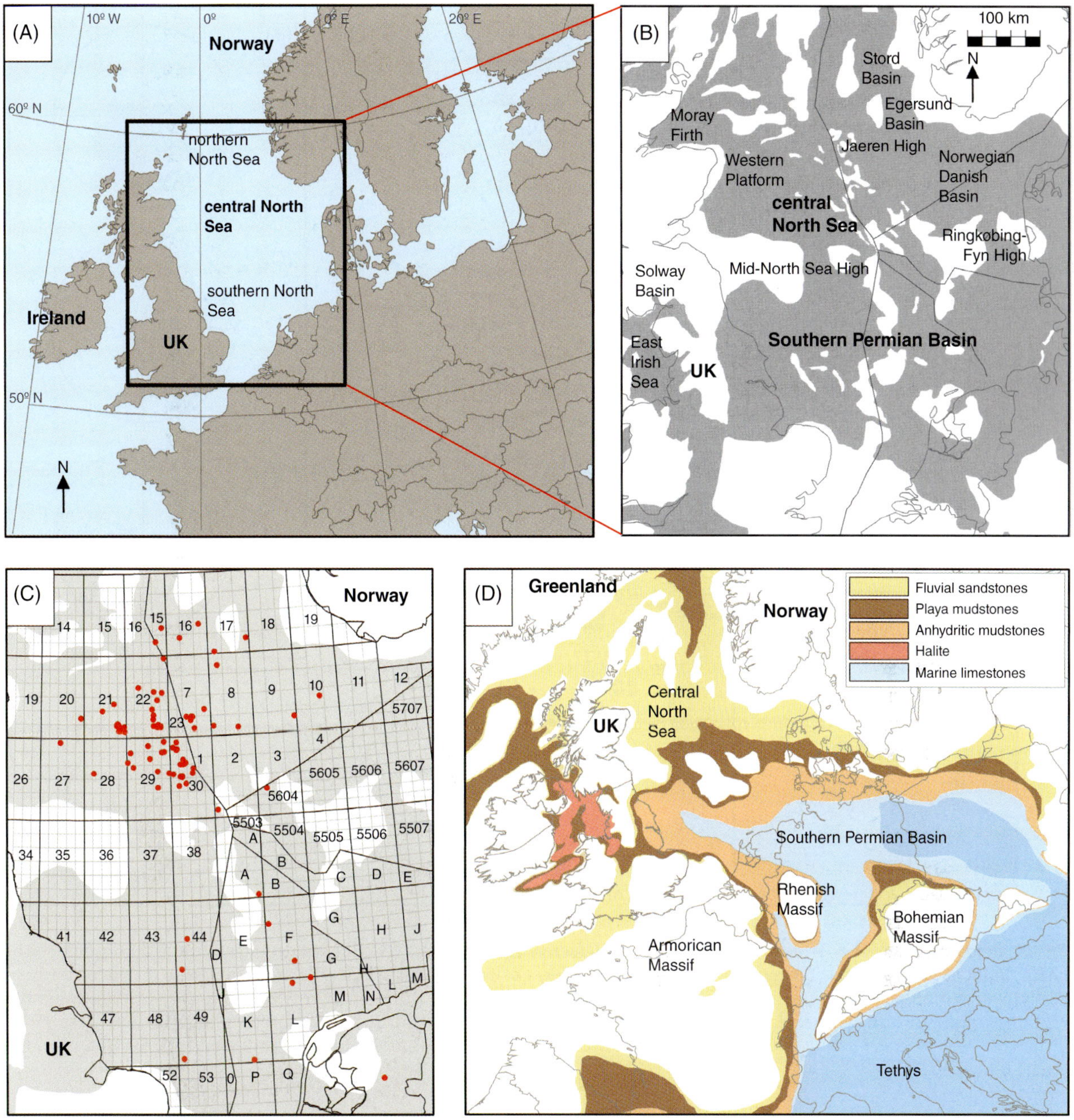

Fig. 1. Central North Sea location and distribution of Triassic basins. (A) General North Sea location map. (B) Distribution of Triassic basin-fills across the North Sea region (based on Goldsmith *et al.*, 2003 and Geluk, 2005). Erosional gaps within the Triassic basin distribution mark the positions of footwall unconformities which trace the overprint of the cross-cutting Jurassic rift system. (C) Location of wells referred to in the text. (D) Gross palaeogeographic setting of the central North Sea (after McKie & Shannon, 2011) in relation to the Southern Permian Basin and the margin of the Tethys Sea. The map broadly corresponds to the Late Anisian when Tethyan flooding of the Southern Permian Basin created the Muschelkalk seaway adjacent to the central North Sea. Marine access to the Southern Permian Basin was via restricted straits which dissected the chain of massifs marking the remnants of the Variscan mountains.

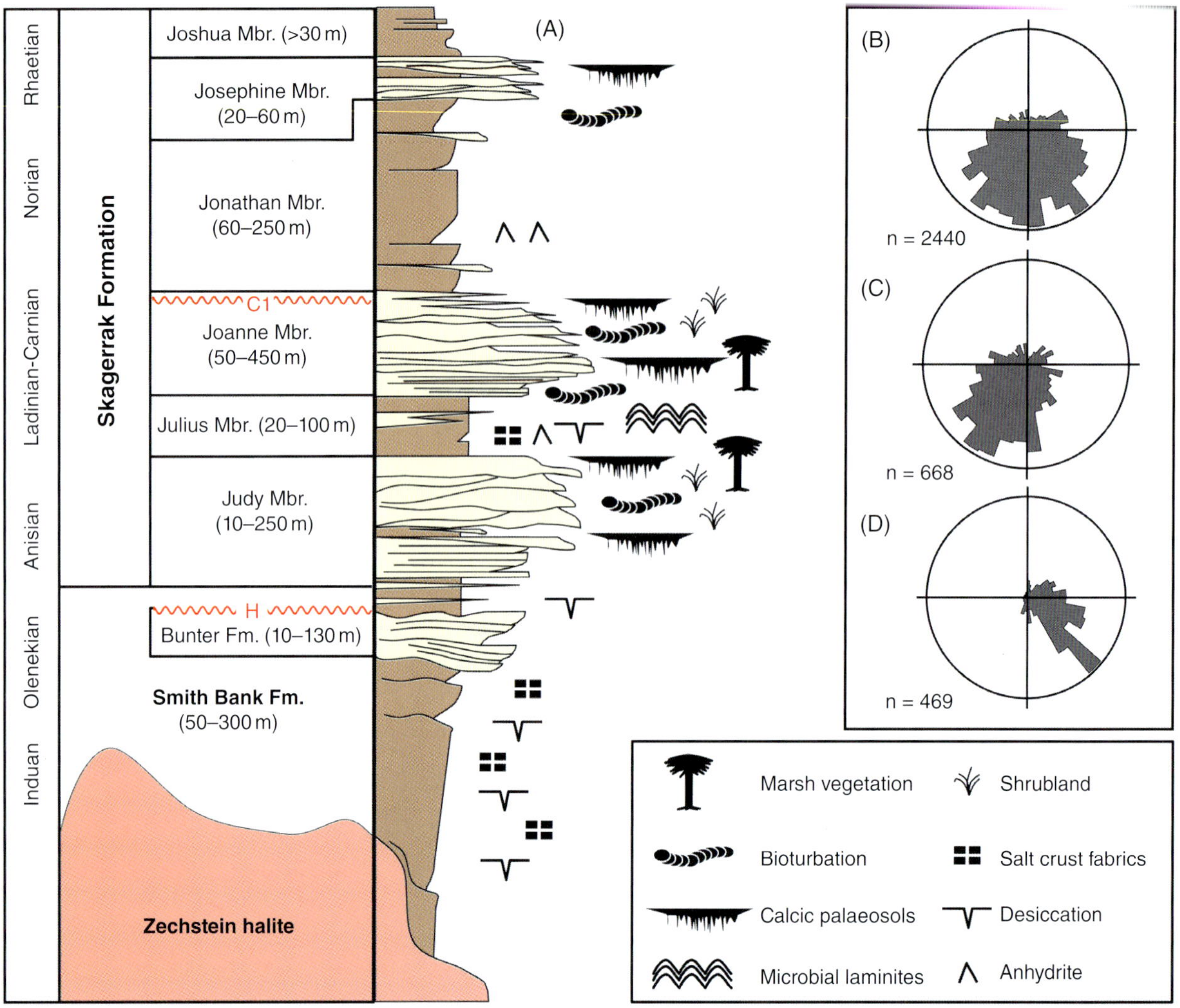

Fig. 2. Central North Sea Triassic stratigraphy and facies architecture. (A) Schematic stratigraphic column of the Smith Bank and Skagerrak formation section. H- Hardegsen Unconformity, C1- Early Cimmerian 1 Unconformity. (B) Palaeocurrent rose for Skagerrak Formation fluvial cross-bedding, illustrating a wide scatter in orientations, but with an overall southward trend. This trend can locally be differentiated into two discrete modes: a transverse mode to the south-west (e.g. C, data from the Heron Field, UKCS) and an axial mode to the south-east (e.g. D, data from the Skua Field, UKCS).

particularly in the Early Triassic and in deep, high temperature parts of the basin, and the broadly similar facies composition of many of the sandstone and mudstone members. Regional correlation therefore relies on support from non-biostratigraphical techniques including use of heavy minerals (Mange-Rajetzky, 1995), chemostratigraphy, cyclostratigraphy (De Jong *et al.*, 2006) and magnetostratigraphy (Turner *et al.*, 1996). Correlation is also facilitated by distinct stratigraphic markers such as well dated marine limestone and evaporite intervals in the Southern Permian Basin (e.g. Muschelkalk limestones and evaporites, Röt and Keuper evaporites) which can be traced into their northern facies equivalents (Michelsen & Clausen, 2002).

Regional tectonic setting

Deposition of Triassic sediments in the north-west European region occurred during the punctuated rifting episodes that initiated the break-up of Pangaea. Multiple extension directions were involved in this process (Coward, 1995; Coward *et al.*, 2003) as rifting propagated from the Tethys Sea to the south (Fig. 1) and the Boreal Ocean to the north. Rifting between Greenland and Fennoscandia, beginning in the latest Permian to Early Triassic (Wignall & Twitchett, 2002; Seidler *et al.*, 2004; Müller *et al.*, 2005), propagated southward along the Atlantic seaboard of the UK and western Ireland, onshore UK and East Irish Sea (Stolfová & Shannon, 2009) and into the northern and central North Sea (Steel & Ryseth, 1990). Northward directed rifting propagated into the Southern Permian Basin and central North Sea from the Tethyan margin via the Polish Trough (Geluk, 2005). The location and trend of these rift basins was influenced strongly by the structural grain of the Precambrian basement, position of former Variscan thrusts and the location of pre-existing Carboniferous and Early Permian basins (Peacock, 2004). Two dominant basin orientations developed, controlled by north-west to south-east oriented Arctic-Atlantic extension and north-east to south-west extension related to the opening of Neotethys. This resulted in a broadly orthogonal fault pattern across the region (Coward, 1995). Earliest Triassic deposition commonly infilled remnant Permian rift topography (Stolfová & Shannon, 2009), but as the break-up of Pangaea progressed two main extensional pulses occurred in the late Early Triassic and in the early Late Triassic which can be identified from the Southern Permian Basin to mid-Norway (Geluk, 2005; Müller *et al.*, 2005). These generated the widespread Hardegsen and Early Cimmerian unconformities, although it is likely that these episodes were not precisely synchronous across the region (e.g. Hounslow & McKintosh, 2003). As a result of these events, the Permo-Triassic basin-fills commonly show a broadly half-graben form (although this is commonly masked by subsequent Jurassic rifting which may not have the same fault polarity) and contain a fill showing multiple extensional episodes across migrating fault arrays (Tomasso *et al.*, 2002; Stolfová & Shannon, 2009). In the central North Sea there has also been widespread denudation of the Triassic during Middle Jurassic thermal doming (Underhill & Partington, 1993) and footwall/rift flank erosion during Late Jurassic rifting (Erratt *et al.*, 1999). However, regional sections locally give a general sense of a deeply truncated half graben wedge (Fig. 3A and B), with the active basin-bounding faults probably located on the eastern, Norwegian, margin (Faerseth, 1996; Erratt *et al.*, 1999).

In the central North Sea, mobilisation of the underlying Zechstein Group halite due to sediment loading, faulting and dissolution resulted in the formation of salt withdrawal minibasins (Fig. 3C, D). The most prominent of these appear to be Early Triassic in age and filled by Smith Bank Formation mudstones. Stewart & Clark (1999) and Penge *et al.* (1999) note the localised presence of an intra-Triassic unconformity separating the main episode of differential minibasin subsidence from the younger Skagerrak Formation. Whilst these authors suggest that this unconformity may be Anisian to Norian in age, it is likely that this boundary, which occurs close to the Smith Bank-Skagerrak formation boundary, represents the late Early Triassic Hardegsen Unconformity recognised across the Southern Permian Basin (Geluk, 2005), Wessex Basin (Hounslow & McKintosh, 2003), East Irish Sea (Colter & Barr, 1975) and Cheshire Basin (Evans *et al.*, 1993). Higher in the succession, Archer *et al.* (2010) note the presence of an unconformity within the Carnian Joanne Sandstone Member. This unconformity, which shows onlap by the Jonathan Mudstone member, probably represents the Early Cimmerian I unconformity which, although widely recognised in the Southern Permian Basin (Geluk, 2005) and towards the Ringkøbing-Fyn High (Clausen & Pedersen, 1999), is not well represented in the deeply truncated central North Sea succession where there is limited preservation of Carnian and younger sections.

DEPOSITIONAL SETTING

On a regional scale, the facies architecture of the Skagerrak and Smith Bank formations in the central North Sea is dominated by dryland fluvial deposits. To the south, crossing the mid-North Sea and Ringkøbing-Fyn highs and entering the Southern Permian Basin, the Triassic succession becomes more mud-prone, with episodic marine incursions recorded by carbonate and evaporite facies (Michelsen & Clausen, 2002; Geluk, 2005). This facies transition records the downstream (Fig. 2B) terminal shale-out of the Skagerrak Formation fluvial systems into playa and coastal sabkha facies. In addition to this regional facies transition, there is a fundamental stratigraphic change in depositional character between the Early and Middle to Late Triassic successions in

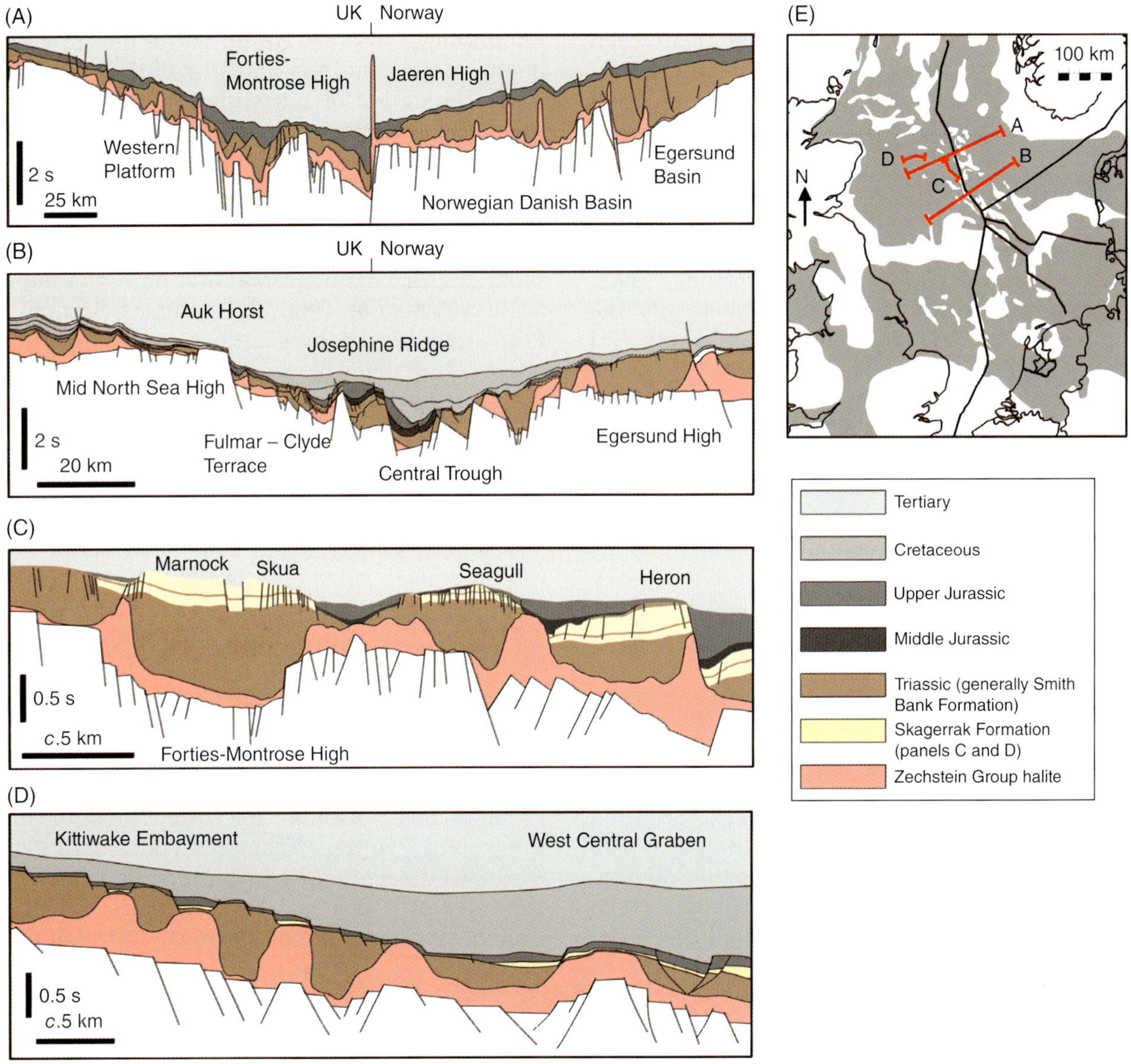

Fig. 3. Geoseismic sections illustrating the tectono-stratigraphic setting of the Triassic in the central North Sea region. (A) Regional geoseismic section from the Egersund Basin in the east, cutting across the Forties-Montrose High and onto the UK Western Platform (after Zanella & Coward, 2003). (B) Geoseismic section from the Egersund High in the east, through the Josephine Ridge and onto the mid-North Sea High (after Erratt, 1999). (C) Section across the eastern flank of the Forties-Montrose High structure running through the Marnock, Skua, Seagull and Heron Fields. (D) Section across the UK Western Platform (after Wakefield *et al.*, 1993). (E) Location of sections A-D. In (C) pod grounding did not occur until the Jurassic and interpod accommodation was created during Late Jurassic extension and infilled by syn-rift sediments. Continuous sections of Smith Bank and Skagerrak formations are preserved, although truncated by later Jurassic unconformities. In (D) pod grounding probably occurred early in the Triassic and although deposition of the Skagerrak Formation was formerly widespread, only interpod areas tend to have preserved the basal remnants of these successions due to on-going halokinetic subsidence and dissolution conveying these remnants below the level of deep Jurassic truncation.

terms of the presence and abundance of biogenic structures, palaeosol development, degree of aeolian influence and fluvial depositional style, which requires the Early to early Middle Triassic succession to be treated as distinct from the later Skagerrak Formation successions (Fig. 4).

Early to early Middle Triassic facies associations

The Early Triassic succession in the central North Sea can be subdivided into four end-member facies associations (Figs 5 and 6). These associations

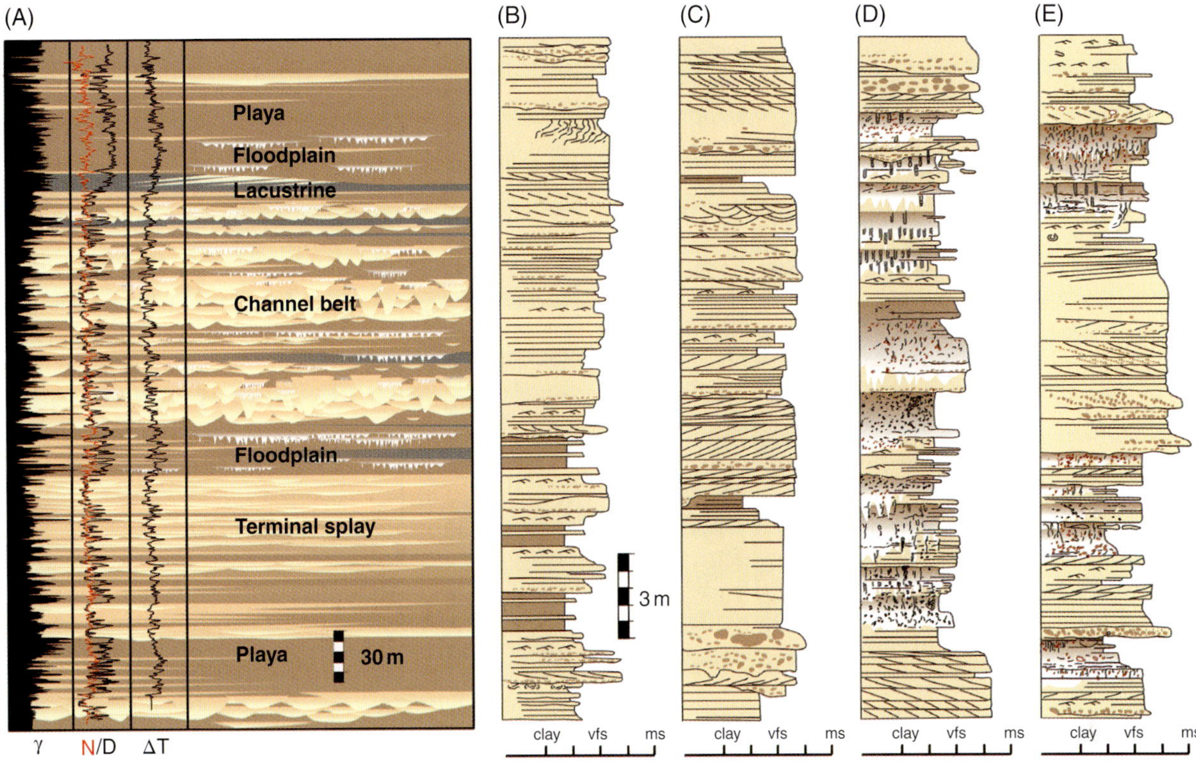

Fig. 4. Typical facies architecture of the Skagerrak Formation. (A) Architecture in UK well 22/24d-10. The lower part of the section comprises playa mudrocks passing upwards into terminal splay deposits and records deposition in a largely vegetation-free dryland setting. The upper part of the section is dominated by vegetated floodplain deposits containing calcic palaeosols, interbedded with channel belt sandstones. The uppermost part of the section is playa-dominated, but with minor lacustrine mudrocks. This uppermost mud-prone interval corresponds to the Julius Mudstone Member, overlying a near-complete Judy Sandstone member succession. (B) Sedimentary log illustrating a coarsening-upward cycle ranging from basal playa mudrocks to terminal splay sandstones (UK 22/24b-9). Such facies are typical of the Early Triassic to Early Anisian). (C) Sedimentary log of fluvial channel sandstones associated with terminal splay sandstones (UK 22/24a-3), again characteristic of the Early Triassic to early Anisian). (D) Sedimentary log of floodplain deposits with extensive palaeosol disruption of mud-prone heterolithic units (UK 22/24b-5S1). Palaeosol and root trace disruption is most common in the Anisian to Norian section in association with fluvial channel sandstones. (E) Sedimentary log of fluvial sandstones punctuated by minor palaeosol disruption in the upper parts of discrete and amalgamated, multi-storey channel bodies (UK 22/24a-3).

are commonly interbedded on a scale of metres to tens of metres, but show a gradual transition in overall dominance over lateral distances of the order of tens of kilometres.

Dry playa deposits (Snyder, 1962) typically form heterolithic intervals 1 m to 20 m thick and are most common in the Smith Bank Formation, in the muddy intervals immediately overlying the Bunter Sandstone and into the lowermost parts of the Judy Sandstone Member. The heteroliths are composed of millimetre-scale to centimetre-scale alternations of very fine-grained to fine-grained sandstone and mudstone (Fig. 5A), organised into metre-scale cleaning-upward and coarsening-upward cycles. Wave-ripple and current-ripple lamination, dewatering and desiccation fabrics are common and minor adhesion and salt crust fabrics (cf. Goodall *et al.*, 2000) are locally developed. Burrows and root traces are present but extremely rare in these deposits, in marked contrast to Middle and Late Triassic playa deposits. Locally, limited core data suggest the presence of well sorted, homogeneous siltstone to very fine-grained sandstone (Fig. 5B) which occurs in intervals up to 5 m thick. These intervals lack bedding or internal stratification, although a very indistinct, irregular, dish-like, wispy lamination is rarely discernible. Individual packages of massive siltstone are separated by thin (<0.1 m), current ripple laminated, fine-grained sandstones with mud chips.

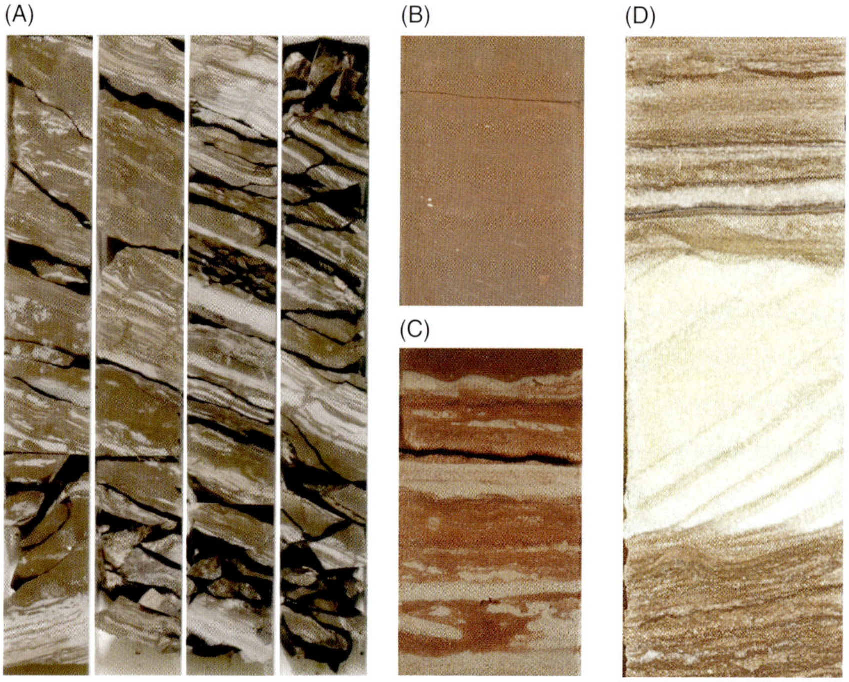

Fig. 5. Lower Triassic facies. (A) Smith Bank Formation dry playa heterolithic deposits, UK 31/26a-8. Cores 1 m long. (B) Structureless, well sorted siltstone within the Smith Bank Formation, UK 20/25-1 (image 10 cm wide). (C) Smith Bank Formation wet playa deposits, UK 22/14b-4, with irregular salt crust fabric and isolated deflation granules (image 10 cm wide). (D) mud-prone, Bunter Formation, wet playa deposits showing an irregularly laminated salt crust fabric enclosing an isolated aeolian dune set defined by grain flow cross-stratification and pinstripe lamination, northern Dutch sector well NL F18-11 (image 10 cm wide).

These mud-prone deposits represent low-energy deposition in a setting characterised by episodic sand deposition and erratic water-table conditions that fluctuated between subaqueous and desiccated in response to ephemeral flood events. No significant thicknesses of fully subaqueous, lacustrine mudstones or halite beds are present and it is likely that these deposits represent extensive dry playa rather than the damp, fluctuating margins to perennial lakes or salina. The origin of the structureless siltstones is enigmatic but, given their sorting and lack of structure, they could represent the accumulation of wind-blown dust or record the rapid deposition of fluvially reworked aeolian material, with the locally discernible wispy fabric the product of clay infiltration or dewatering fabrics.

Wet playa deposits (Snyder, 1962) occur in intervals tens of metres thick within the Smith Bank and Bunter Sandstone formations and become particularly prevalent in the broadly time-equivalent sections in the Southern Permian Basin. The dominant facies within these sections consists of irregularly laminated and deformed, bimodally-sorted, mud-prone sandstones with common granule lags and scattered grains (Fig. 5C and D). The distinctive, irregular fabric is composed of adhesion lamination, together with salt ridge and crust structures (Fryberger *et al.*, 1983; Goodall *et al.*, 2000). These wavy bedded deposits are punctuated by thin (<0.5 m) intervals of laminated mudstone. Fine-grained to medium-grained cross-stratified sandstones (Fig. 5D) showing pinstripe lamination (Fryberger & Schenk, 1988) and grain flow-toe pinch-outs form isolated cross-sets (<1 m thick). Fine-grained to medium-grained, mud clast bearing, climbing current ripple laminated and plane bedded sandstones up to 0.5 m thick occur sporadically throughout these sections, together with isolated sandstones of a similar composition to the cross-stratified sandstones but lacking any internal structure, or with indistinct deformation structures.

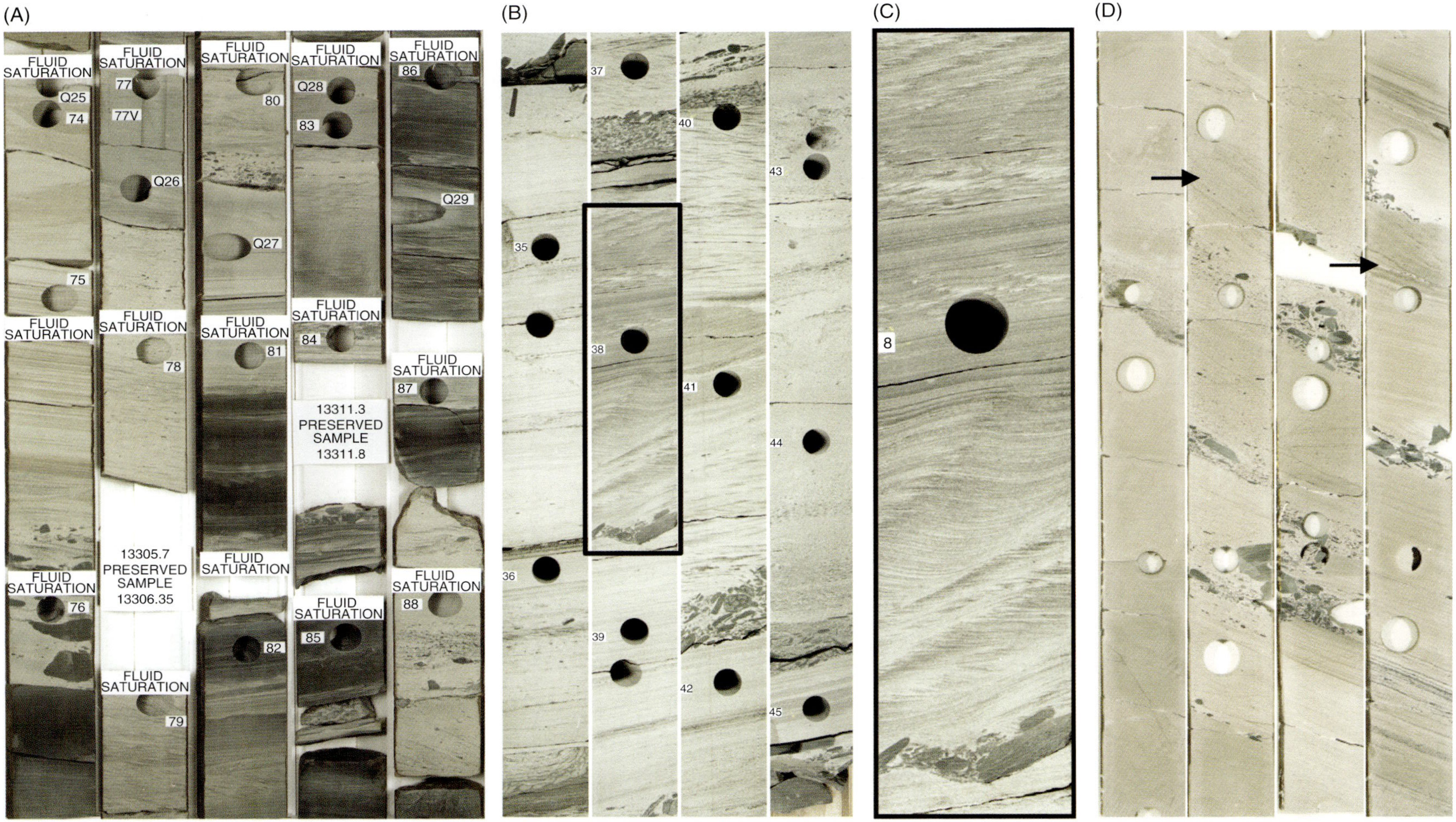

Fig. 6. Lower Triassic facies. (A) Terminal splay deposits comprising interbedded shales and ripple laminated to plane bedded sandstones, UK 29/8-3. (B) Weakly confined to unconfined fluvial deposits composed of current ripple laminated, very fine-grained to fine-grained sandstones with abundant mud clast lags, UK 21/18-5. (C) Close-up image of highlighted area in (B) illustrating mud clast-floored scour and steeply climbing current ripple lamination. (D) Plane bedded and cross-bedded (arrowed), fine- to medium-grained fluvial sandstones with mud clast lags, UK 21/19-6. Cores 1 m long.

These interbedded facies record a variety of depositional processes. The wavy bedded deposits record extensive disruption by haloturbation under evaporitic conditions close to the water table, with episodic adhesion as sand was blown across damp substrates. The laminated mudstones record brief episodes of raised water tables and the development of localised subaqueous ponds. The pinstripe cross-stratified sandstones represent the migration of isolated aeolian dunes across this wet playa surface, with the current ripple laminated sandstones representing episodic, low energy fluvial deposition. The structureless and dewatered sandstones are texturally identical to the dune sands and may record fluvial reworking of aeolian dunes or mass wasting by rainfall during convective desert storms.

Terminal splay deposits consist of micaceous, very fine-grained to fine-grained sandstones and mudstones organised into coarsening-upward cycles up to 10 m thick (and locally grouped into composite cycles up to 40 m thick; Fig. 4). At the bases of these cycles are thin shales and heterolithic units (typically <0.5 m thick), which are wave and current ripple laminated or structureless, commonly desiccated and very rarely burrowed. The contact with the overlying sandstones is either sharp (with mud clasts or desiccation cracks) or gradational and passes upwards into successions dominated by planar-lamination and climbing-current-ripple lamination, with scattered mud-clast concentrations and locally developed dewatering fabrics (Fig. 6A).

The basal shales to these cycles record subaqueous deposition and the paucity of biogenic activity and common desiccation suggest that these were ephemeral water bodies. The overlying plane-bedded and climbing-ripple-laminated sandstones represent the product of deposition from shallow, sediment-charged, unconfined flows and their arrangement into coarsening-upward cycles is suggestive of progradation. The variably sharp (with evidence of desiccation) or gradational contacts suggest that these progradational sand systems alternated between entering standing water bodies or covering subareally exposed surfaces. The overall setting appears to have been of flood-dominated systems characterised by rapid deposition into ephemeral water bodies or across dry lake floors.

Fluvial deposits occur in close association with the terminal splay deposits described above (Fig. 4) and comprise up to 3 m thick, typically vertically isolated, sharply or erosionally based, fining-upward or trendless packages of fine-grained to medium-grained sandstones with abundant mud-flake lags (Fig. 6). Mudclast-filled gutter casts and scours are common throughout. The sandbodies have an internal fabric varying from structureless to indistinctly dewatered, or dominated by horizontal to gently inclined planar lamination (with abundant mica concentrations defining laminae) and climbing current ripple lamination. Cross-bedding is generally uncommon but locally dominant in discrete stratigraphic intervals, typically at the top of large-scale, terminal splay coarsening-upward successions (Fig. 4C).

These deposits record deposition under a range of variably channel-confined conditions. The abundance of mud-flake lags may either be indicative of channel reworking of overbank muds, or local reworking of desiccated, intra-channel, mud drapes. The structureless to dewatered sandstones suggest catastrophic deposition, either through bank collapse or rapid deposition from waning, sediment-laden flows, whilst the plane-bedded and current ripple-laminated intervals, in association with common scours and gutter casts, indicate shallow, high-velocity, waning flows with common scouring and erosion. These features indicate a dominance of ephemeral flood processes (cf. Fielding, 2006). The vertically isolated nature of these sandbodies, encased within splay and playa deposits, together with evidence of rapid cut and fill, suggest that these fluvial systems did not form established channel belts and instead comprised discrete ribbons prone to sediment plugging and avulsion and that floodwaters expanded and shallowed into lateral and terminal splays on the margins of playa.

Overall, the Early to early Middle Triassic succession appears to have been deposited under relatively arid conditions as terminal fluvial systems which discharged into desiccated and evaporitic playas (Fig. 7). Aeolian processes are recorded, but generally high water tables and low sand supply in distal reaches appears to have acted to suppress dune field development. The overall setting may have been broadly analogous to the present day Lake Eyre basin (cf. Lang *et al.*, 2004; Fisher *et al.*, 2008), although with considerably less vegetation cover. The provenance of these Early Triassic sediments appears to have been from Caledonian metasediments from the UK margin (Mange-Rajetzky, 1995) and reworked Carboniferous

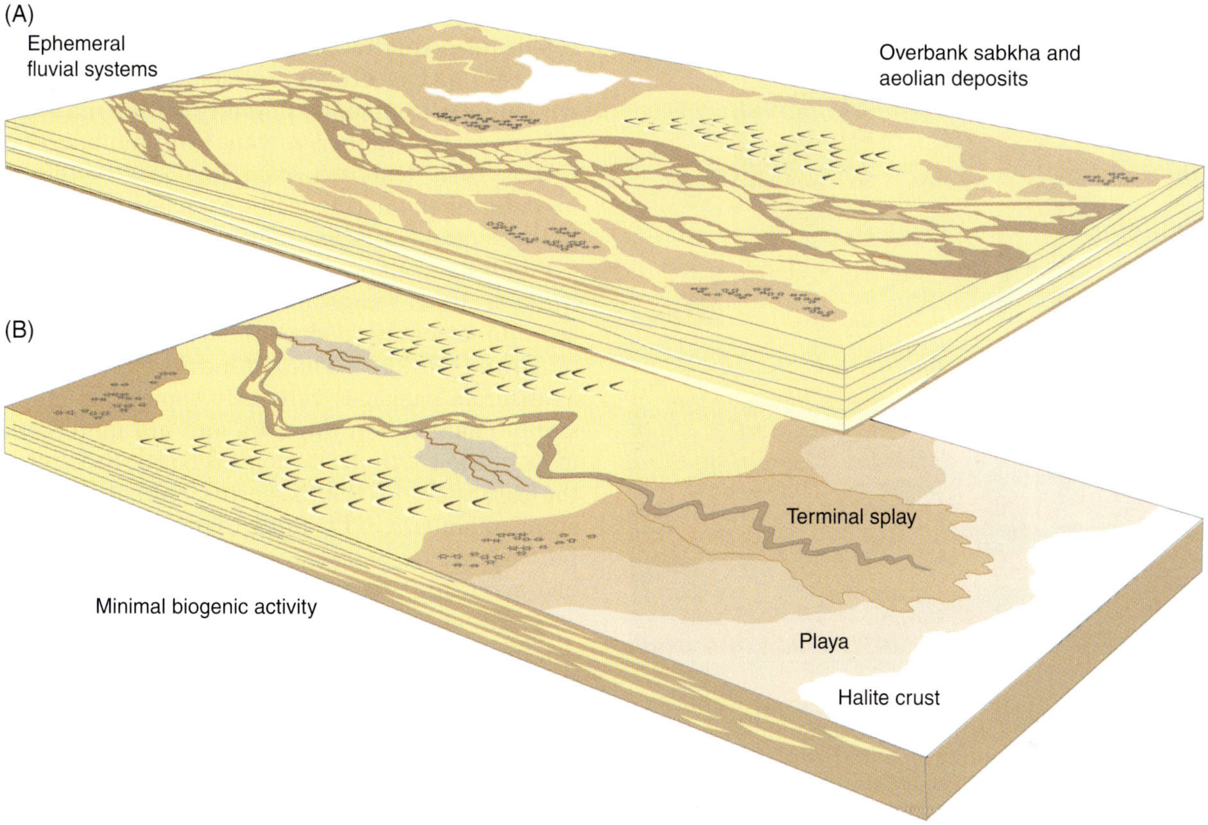

Fig. 7. Conceptual depositional model for Early Triassic deposition in the central North Sea. The overall setting was arid and likely devoid of biogenic activity as a result of both a lack of soil moisture and an impoverished terrestrial population in the wake of the end Permian extinction event. (A) Proximal regions were dominated by channel-confined facies and associated overbank playa. (B) Distal regions were dominated by wet and dry playa deposits, with isolated aeolian sand patches and terminal splay sandstones.

sediments (based on palynology). The presence of rounded and frosted grains has been taken as indicative of reworking of Permian aeolian sandstones (Mange-Rajetzky, 1995), but these could equally be a product of fluvial reworking of Triassic aeolianites.

Middle to Late Triassic facies associations

Within a poorly dated interval within the Anisian Judy Sandstone Member there is a marked change in depositional character from the dryland deposits as described above towards widespread evidence of biogenic activity (Fig. 8) and loss of aeolian facies. Calcrete also becomes widespread at this stratigraphic level and is present within palaeosols as rhizoliths, pedogenic nodules and as groundwater calcrete. These calcretes were commonly reworked as allochthonous clasts into fluvial deposits. Despite this intra-Anisian change in character towards less evidence of aridity, salt crust fabrics are preserved within some shale members and aeolian facies are present locally, together with episodic halite deposition in the broadly equivalent sections in the Southern Permian Basin (Aigner & Bachmann, 1992; Geluk, 2005) and as far north as east Greenland until the Carnian (Surlyk, 1990). Four main facies associations can be recognised within this upper section.

Interbedded playa and marsh deposits form intervals c. 1 m to 30 m thick, composed of metre-scale alternations between black mudrocks; rooted, burrowed and stratified heterolithic beds; localised micrite horizons showing laminar carbonate cementation and haloturbated, muddy sandstone intervals (Fig. 9A). This facies association interfingers with marginal Muschelkalk lagoonal facies in the Anisian part of the section (Fig. 9B). The black mudrocks are laminated and generally lacking biogenic traces or sandstone partings. The heterolithic sections show the widespread development

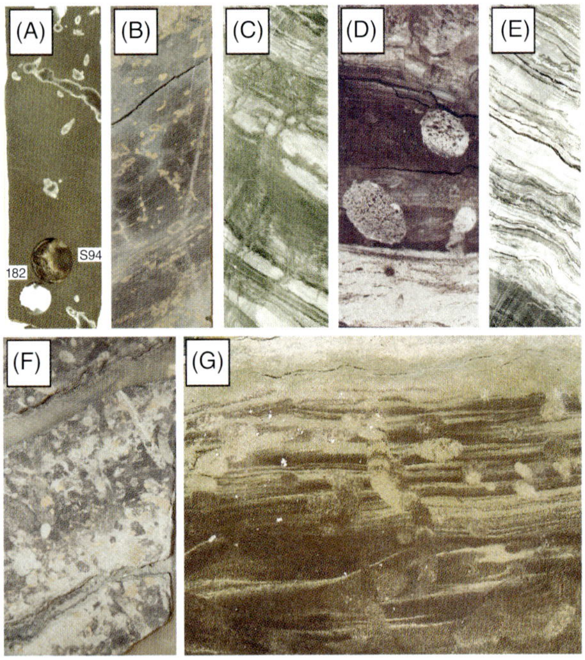

Fig. 8. Biogenic indicators appear relatively abruptly in the Middle Triassic (possibly in the Late Anisian) section and remain common throughout the overlying Middle to Late Triassic interval. Diversity however is relatively low and dominated by common calcareous rhizoliths (A and B, UK 22/29-5RE), root traces (C, UK 22/29-5RE), glossifungites burrows in playa heterolithic deposits (D, NO 7/8-4), microbial laminites in playa and marsh deposits (E, UK 22/29-5RE) and pervasive adhesive meniscate burrows (F and G, UK 29/10-3) which occur in floodplain and fluvial upper bar deposits. Cores 10 cm wide.

of current and small-scale wave ripple lamination, indeterminate burrow mottling, glossifungites and adhesive meniscate burrows, microbial laminites, root-traces and flat-pebble conglomerates (Figs 8 and 9). Deeply penetrating (up to 0.3 m) desiccation cracks are common. The disrupted muddy sandstones locally show fabrics typical of evaporitic salt crusts (Goodall *et al.*, 2000) and rare chicken wire anhydrite. Palynological recovery is locally common in Middle Triassic examples of this facies association, assemblages of which show a dominance of pteridophytes, bryophytes and arborescent lycopods (Goldsmith *et al.*, 1995).

The black mudrocks record subaqueous conditions in relatively persistent water bodies, either as floodplain lakes or small ponds. In the heterolithic sections more marginal, distal sheetflood conditions predominated, with the plant rooting, desiccation, microbial structures and early carbonate cementation within the mudstones indicative of episodically stressed, palustrine and marsh conditions. These apparently alternated with desiccated and evaporitic playa. The palynological data suggest a mix of drought-resistant shrubland and wetland environments, possibly with a maritime influence (cf. Kelber & van Konijnenburg-van Cittert, 1998; Hubbard & Boulter, 2000) although no definitively marine facies have been identified in the central North Sea region.

Floodplain deposits form mud-rich, heterolithic sections 1 m to 30 m thick (Fig. 4D) which are ubiquitously pedogenically modified and burrowed (Fig. 9D). These deposits tend to occur as discrete packages vertically separating multistorey channel-belt deposits and are more common in the south of the central North Sea and in the basal parts of the Skagerrak Formation sandstone members. The mudstones are disrupted pervasively by palaeosols which are variable in character, ranging from weakly developed entisols, through aridisols with relatively immature calcrete profiles and rooted alfisols with abundant illuviated clay. Ped fabrics with crystallaria and soil slickensides are locally common in the aridisol intervals. Interbedded sand-prone intervals are metre-scale in thickness, composed of poorly sorted, very fine-grained to fine-grained sandstone containing a high proportion of intraformational clay and carbonate material and commonly have a structureless to disrupted appearance with abundant adhesive meniscate burrows and root traces.

The alfisol deposits record vegetated, shrubland, floodplain conditions, with the interbedded aridisol intervals showing calcrete, ped fabrics and slickensides suggesting an alternation with dry, well-drained, but seasonally wet conditions. The overprinting by burrows suggests episodically higher soil-moisture conditions favourable for insect larvae (cf. Smith & Hasiotis, 2008; Smith *et al.*, 2008) and indicates the former presence of abundant organic detritus derived from plant material.

The interbedded sandstones represent weakly confined to unconfined splay deposition which was subject to pervasive, post-depositional palaeosol modification. The sharp bases and localised lags suggest some scouring and preservation of small-scale bedforms, but in general these deposits were probably deposited as unconfined lateral and terminal splays to the channel facies (cf. Tooth, 2005; Fisher *et al.*, 2007).

Terminal splay-lacustrine delta deposits comprise 1 m to 5 m of normally graded, very fine-grained to fine-grained sandstones and mudrocks interbedded on a scale of 0.01 m to 0.1 m (Fig. 10).

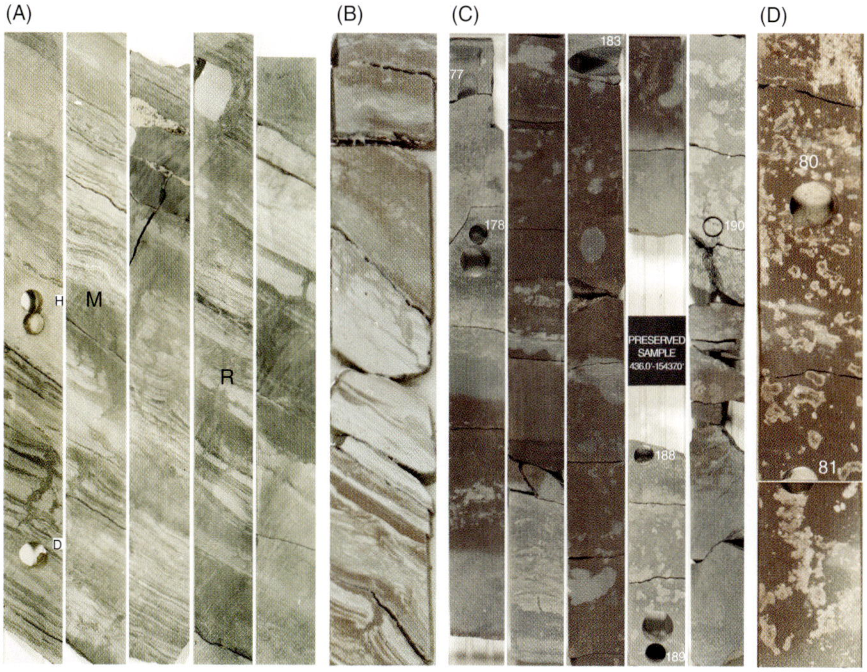

Fig. 9. Middle and Late Triassic facies. (A) Heterolithic, calcareous marsh deposits with microbial laminites (M) and root traces (R), UK 22/29b-5RE. (B) Marginal Muschelkalk interbedding of rooted, pedogenic dolomarlstones and desiccated algal dolomarlstones, NL WYK-5B (cf Borkhataria *et al.*, 2006). (C) Mottled, calcareous, well-drained floodplain palaeosols, UK 22/29-1S1. (D) Calcic palaeosol profile with prominent rhizoliths, UK 22/30a-2. Cores 1 m long.

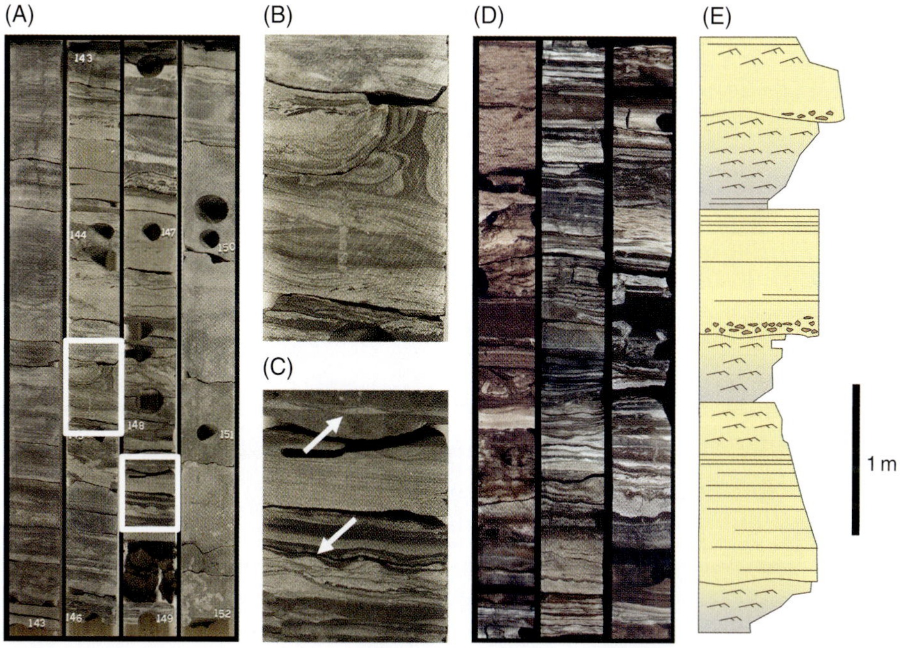

Fig. 10. Middle and Late Triassic facies. (A) Dewatered and loaded, graded beds interbedded with laminated shales, representing episodic hyperpycnite deposition on the margins of floodplain ponds and lakes, UK 22/30a-6. (B) Detail of deformation feature in (A), cross-cut by burrows. (C) Detail of small-scale wave ripple forms in (A). (D) Climbing current and wave-ripple laminated heterolithics, NO 7/8-4. Cores 1 m long. (E) Graphic log of lacustrine delta deposits in the Josephine Member, comprising climbing current-ripple laminated, coarsening-upward cycles erosively truncated by mud-clast mantled scour-fills and channel-fills, UK 30/1c-3.

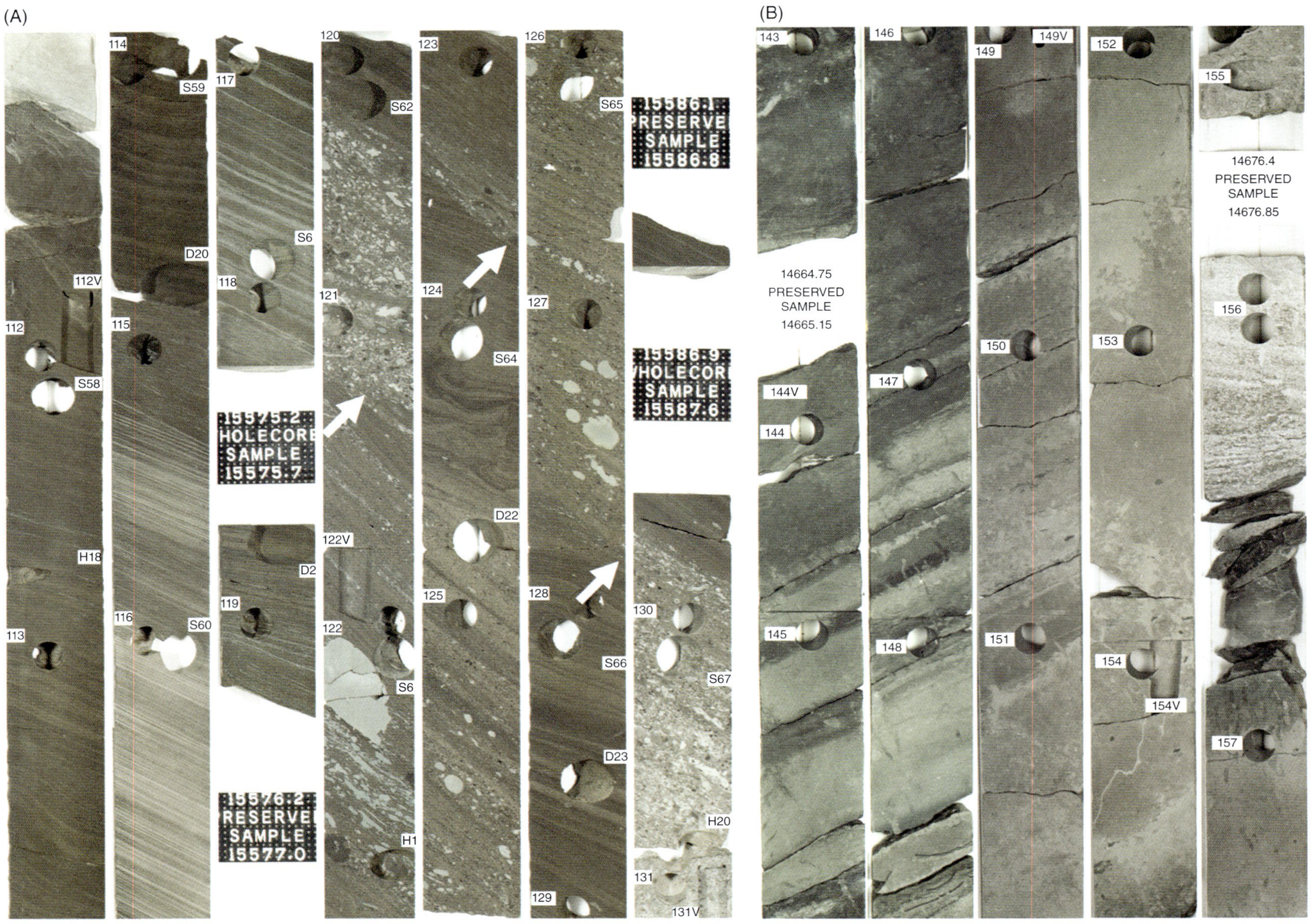

Fig. 11. Middle and Late Triassic facies. (A) Proximal cross-stratified and low angle stratified fluvial sandstones with common mud-clast lags. The composition of the mud-clast lags commonly differs from floodplain fines in being finer grained and well laminated and may represent erosion of in-channel slack water mud drapes, UK 22/29-5RE. Truncation surfaces defining low angle cross bedding are highlighted by arrows. (B) Distal, mud-prone and heterolithic fluvial deposits representing muddy accretionary bar deposits with vegetated and burrowed upper bar surfaces, UK 29/10-3. Such facies are more common in the south of the basin and in the stratigraphic transition into the regional Skagerrak Formation shale members. Both examples have carbonate cemented, calcrete clast basal lags. Cores 1 m long.

The sandstone beds are erosionally based and variable in internal character, ranging from structureless to plane bedded, dominated by wave-ripple and climbing current-ripple lamination, or more rarely exhibiting humpback antidune lamination (Fielding, 2006). Soft-sediment deformation and dewatering fabrics are common (Fig. 10B). The interbedded mudrocks are typically black coloured and laminated. Generally there is little organisation to these deposits beyond a weakly developed cleaning-upward motif on a scale of 0.5 m to 1 m, but intervals with well-developed coarsening-upward motifs tend to be characterised by a predominance of climbing current and wave-ripple lamination (Fig. 10E). However, interbedding of organised and disorganised end-member types is also common.

These heterolithic deposits record a continuum between two end-member depositional settings. The poorly organised intervals characterised by interbedded laminated mudstones and dewatered, massive and normally graded sandstones record episodic density-current deposition into standing water and their association with playa margin facies suggest that they represent hyperpycnal flow of flood events directly into floodplain lakes and ponds (cf. Sáez *et al.*, 2007). In contrast, the well-developed coarsening-upward cycles with climbing current ripple lamination probably record lacustrine delta progradation into standing water. Together, the interbedding of these end-members suggest deposition within subareal terminal splay complexes which episodically formed deltas during lake-filling episodes, in a manner similar to that observed on the margins of modern Lake Eyre (Lang *et al.*, 2004; North & Warwick, 2007).

Fluvial deposits are generally sand-prone and expressed by sharp-based, fining-upward (or locally trendless) cycles 1 m to 6 m thick composed of very coarse-grained to very fine-grained sandstone (Fig. 4). The cycles are stacked into multi-storey intervals up to 20 m thick. The basal surfaces of these cycles are erosional and mantled by intraformational mud clasts, reworked calcic palaeosol and rhizolith material (Fig. 11A). The upper parts of the cycles commonly grade upwards into pedogenically modified mudstones and very fine-grained heterolithic facies, with common adhesive meniscate bioturbation (cf. Smith *et al.*, 2008), root traces and calcic rhizoliths which penetrate into the underlying sands. Low-angle accretionary deposits dominate this association and cross-stratified fills are subordinate (McKie, 2011). Less common mud-prone fills (Fig. 11B) are characterised by up to 4 m of heterolithic, locally rooted, burrowed and slumped siltstones and very fine-grained sandstones which grade upwards from sharp based lags of intraclast-bearing, fine-grained to medium-grained sandstone. These mud-prone deposits tend to be more common in the south of the central North Sea area and in stratigraphic intervals transitional into the regional shale members. Palaeocurrent data derived from oriented core and borehole images show a wide scatter, but with an overall dominance of flow towards the south (Fig. 2B). More locally, discrete stratigraphic intervals show orientations to the south-west and south-east (Fig. 2C, D), suggesting that the overall data are a composite of axial flow down the central North Sea and transverse flow off the Norwegian margin (McKie *et al.*, 2010; McKie, 2011), in keeping with the heavy mineral provenance data (Mange-Rajetzky, 1995).

These coarse-grained, erosionally based and upward-fining intervals show widespread evidence of floodplain soil erosion and are interpreted as the products of channel migration across the Skagerrak Formation alluvial plain. Such facies are widely distributed, with the lack of evidence for areas of sustained floodplain deposition possibly indicating a dominance of mobile, potentially multiple channel systems and frequent avulsion leading to a uniformly channel-dominated architecture over wide areas (up to 10 km). The predominance of low-angle planar lamination and low-angle cross-bedding suggests that bar accretion took place under shallow, high-velocity flows transitional between lower and upper flow regimes, possibly analogous to the examples described by Allen (1982). The muddy upper sections of the fining-upward cycles represent accretion of the upper parts of bars and these were commonly subaerially exposed, sparsely vegetated and supported a fauna of burrowing insects. This may reflect a variable discharge regime, with dry-season colonisation or desiccation of upper bar surfaces and wet-season, bank-full conditions reworking desiccated mud drapes and soil material. The more heterolithic and mud-prone deposits are indicative of channels which experienced infrequent bank-full conditions and sustained periods of reduced or limited flow (albeit with sufficient year-round water availability to maintain biogenic activity and avoid complete channel desiccation). The increased frequency of such deposits in a down-palaeoflow direction

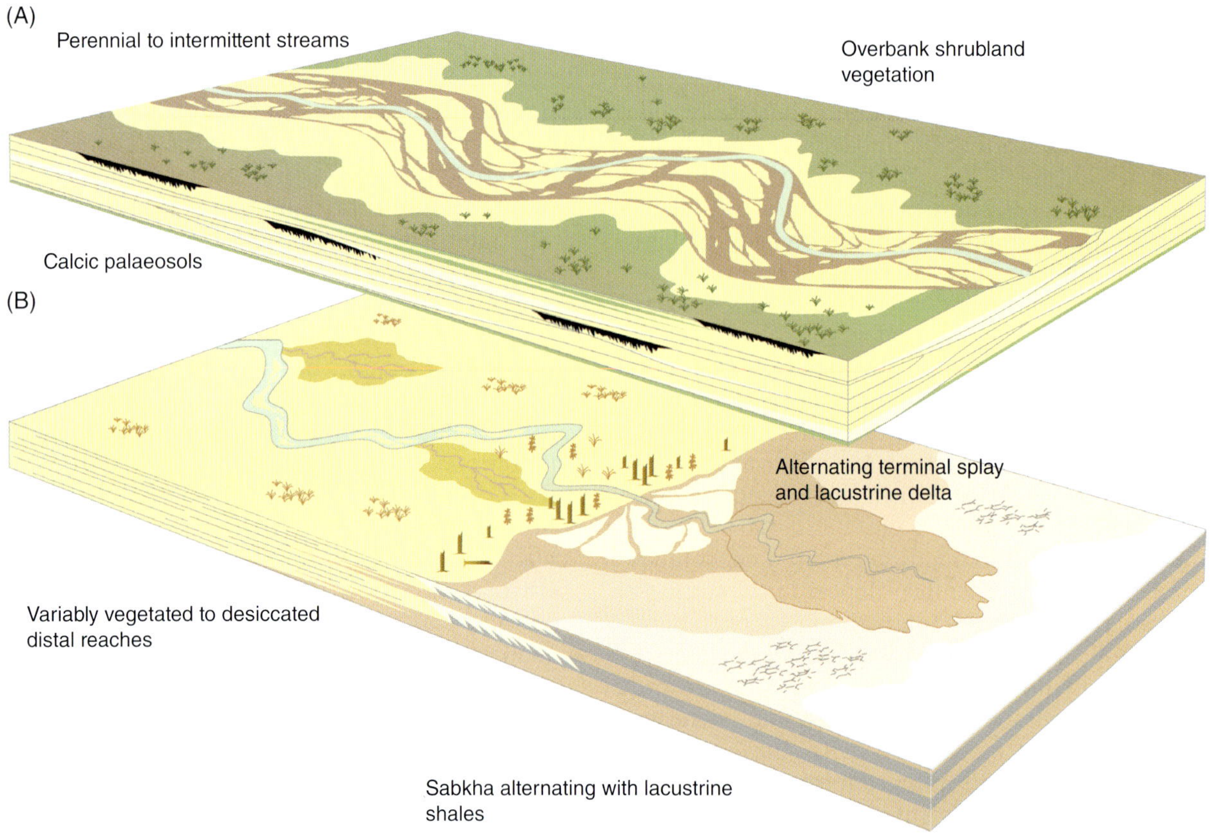

Fig. 12. Conceptual depositional model for the Middle and Late Triassic in the central North Sea. (A) Proximal facies were typically dominated by sand-prone accretionary bar deposits from large, intermittent river systems flanked by well drained shrubland floodplain. (B) Distal regions were characterised by interbedded floodplain, lacustrine and playa facies representing conditions of fluctuating desiccation.

may be indicative of the reduced frequency of sand transporting discharges in the lower reaches of the Skagerrak Formation terminal systems.

In summary, the general facies composition of the middle and upper Triassic succession in the central North Sea is suggestive of a dryland depositional system which was typically wetter than that of the lower Triassic (Fig. 12), alternated between episodic playa desiccation and periods of deposition along active river courses which sourced perennial lakes and marshes and supported a moderately diverse plant and animal population (McKie & Audretsch, 2005; McKie, 2011).

Terminal character of the Triassic fluvial systems

The provenance and palaeocurrent data for the central North Sea fluvial systems are indicative of lateral drainage from the UK and Fennoscandia, which converged into an axial drainage towards the south (Mange-Rajetzky, 1995; McKie *et al.*, 2010; McKie, 2011). There is very little evidence of northward directed palaeoflow. The extent to which the Triassic fluvial systems exited the central North Sea is debatable due to the paucity of data adjacent to and within the mid-North Sea High region. However, well data from the northern part of the Dutch sector of the North Sea, which would be immediately downstream of the central North Sea fluvial systems, generally shows limited evidence of major fluvial activity and typically comprises mud-prone playa (Bunter Shale and Keuper formations) or marginal marine, mixed carbonate-clastic sabkha or lagoonal facies (Muschelkalk Formation). These facies either represent the distal termination of the central North Sea systems or, if the mid-North Sea High region had acted as a physical barrier, then the systems must have been fully contained within the central North Sea and terminated within this region. There is no evidence that the river systems

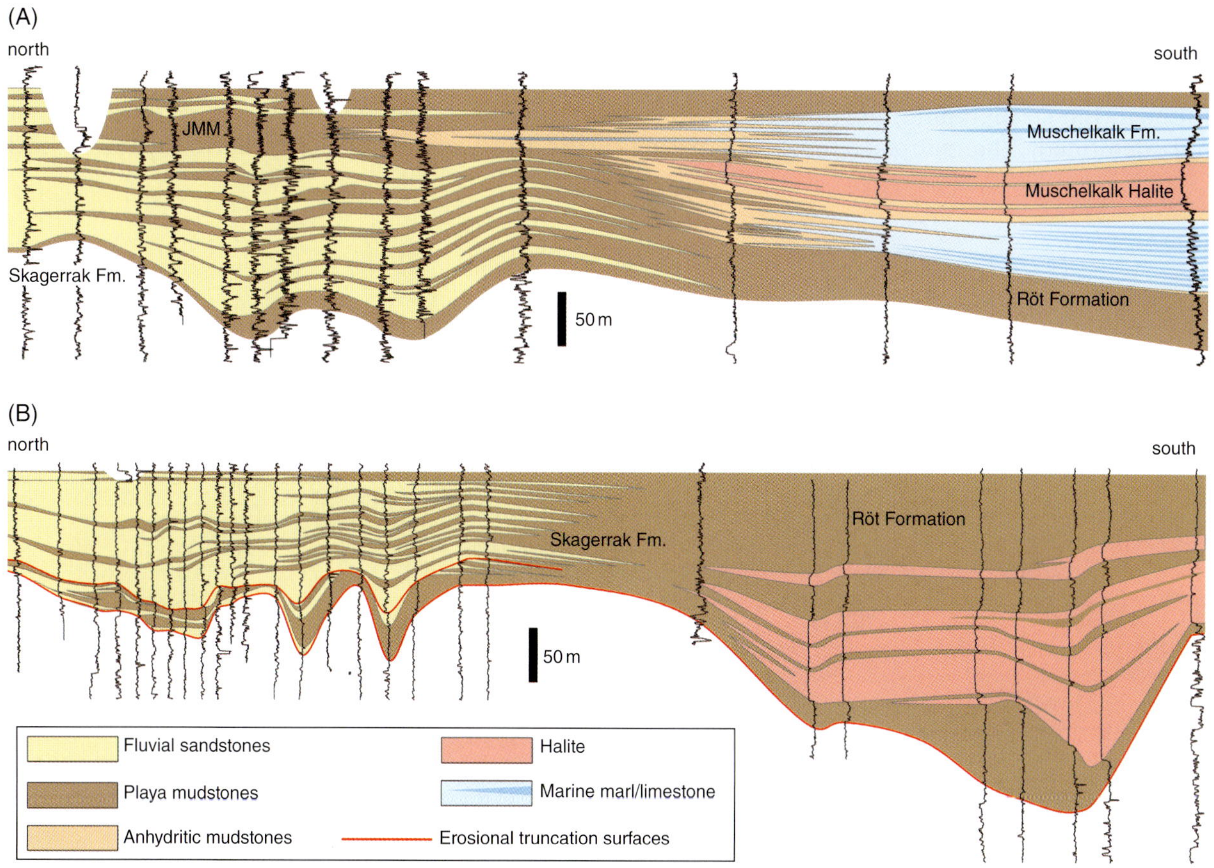

Fig. 13. Down-palaeoflow trends in Skagerrak Formation facies architecture. (A) North-south section through the upper Judy Sandstone Member illustrating the progressive shale-out of the Skagerrak Formation towards the south and the northward penetration of marginal Muschelkalk Formation dolomitic facies into the Julius Mudstone Member (JMM). Wells, from north to south comprise: NO 17/10-1, NO 15/12-1, NO 7/8-3, NO 7/11-6, UK 22/24d-10, UK 22/30A-6, UK 29/5b-F3Z, UK 29/14b-2, UK 30/7a-8, UK 30/13-3, DK West Lulu-1, UK 44/12-1, UK 44/21-2, UK 49/26-1, NL K17-1. (B) North-south section through the lower Judy Sandstone Member showing the southward shale-out into the Röt Formation and equivalent halites. Wells, from north to south, comprise: NO 16/10-3, UK 22/12a-2, UK 22/15-1, UK 22/19-2, UK 22/19-1, UK 22/24a-1, UK 22/24b-7, UK 22/24b-4S1, UK 22/29-2S1, UK 22/29-1S1, UK 22/24d-10, UK 22/30a-6, UK 29/3b-4, UK 29/8-3, UK 29/5a-7, UK 29/14b-2, UK 29/15-1, UK 29/19-1A, UK 30/7a-7, UK 30/12b-2, UK 49/26-1, UK 44/12-1, UK 44/21-2, NL A18-1, NL F04-3, NL F14-7, NL L02-1, NL K17-1.

debouched into a perennial interior lake and evidence of subaqueous conditions is limited to rare examples of sub-metre-scale laminated mudrocks, indicating that standing water bodies were not a common or dominant feature of the Triassic landscape in this region. Within the central North Sea the down-palaeoflow termination of the fluvial systems can be traced towards the mid-North Sea High and this is expressed by a general decrease in sand:shale ratio, decrease in sandbody thickness and increased proportion of splay facies relative to channel-confined deposits (Fig. 13). Evidence for this is best expressed in the Judy and Joanne Sandstone members (McKie & Shannon, 2011); the Josephine Sandstone is very poorly preserved due to Jurassic erosion and the deeper parts of the succession (Bunter Formation) are poorly penetrated. However, although data for these intervals are more limited, the lack of evidence of fluvial systems in the adjacent Dutch offshore sector again suggests that these sand-prone fluvial systems did not extend into this region.

TECTONIC CONTROLS ON FACIES ARCHITECTURE

On a regional scale the preserved architecture of these extensional basin-fills is dominated by the effects of halokinesis and minibasin formation,

Middle Jurassic thermal doming and erosion, and the Late Jurassic fragmentation of these minibasin remnants by extensional faulting. However, although the stratigraphic record is fragmentary and deformed, the succession of alternating sand-prone and mud-prone members (Fig. 2) is recognisable and broadly correlateable across large parts of the central North Sea with limited evidence of major downstream shale-out of the sand-prone members until the mid-North

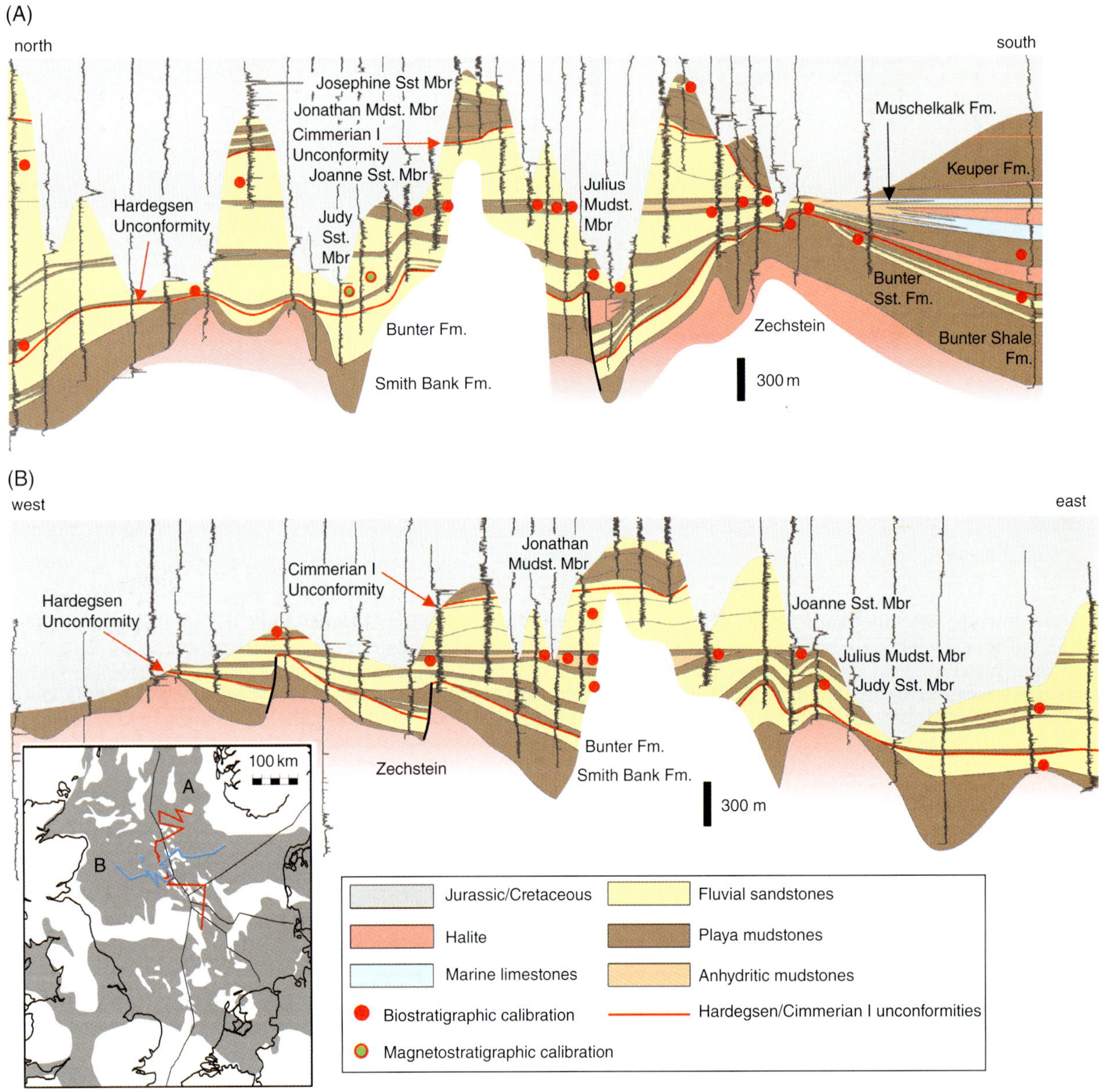

Fig. 14. Regional well log correlations of the Triassic succession in the central North Sea. (A) North-south section through Norwegian sector wells NO 17/12-1, NO 16/9-1, NO 17/10-1, NO 15/9-9, NO 16/10-3, NO 8/1-1, NO 6/3-1, UK sector wells UK 22/15-1, UK 22/14b-4, UK 22/19-1, UK 22/24b-5, UK 22/30a-2, UK 22/29-1S1, UK 29/5b-F3Z, UK 30/1c-3, UK 30/1c-5A, UK 30/7a-3, UK 30/7a-8, UK 30/7a-7, UK 30/7a-6, UK 30/11b-1, UK 29/15-1, UK 30/13-3, UK 30/13-1, UK 30/12b-4, UK 30/12b-2, UK 30/12b-3, UK 30/17b-5, UK 30/17a-4, DK West Lulu 1 (Danish sector) and NL A18-1 (Dutch sector). (B) East-west section through UK sector wells UK 27/3-1, UK 28/12-1, UK 29/6b-5, UK 29/11a-1, UK 29/6a-3, UK 29/19-1A, UK 29/14b-2, UK 29/8a-3, UK 29/15-1, UK 30/12b-4, UK 30/13-3, UK 30/7a-4A, UK 30/7a-8, UK 30/7a-7, UK 30/7a-9, UK 30/1c-3, UK 30/1c-4, UK 30/1c-2A, UK 29/5b-F3Z, Norwegian sector wells NO 7/11-8, NO 7/8-4, NO 7/8-3, NO 7/9-1, NO 8/10-1, NO 8/11-1, NO 9/12-1 and NO 10/8-1. Line of sections shown in the inset map.

Sea High to the south and limited evidence of basin margin pinch-out of shale members until the more marginal areas, e.g. the Egersund and Stord basins. However, deep erosion leaves limited evidence of these trends across the major Jurassic structural highs such as the Forties-Montrose and Jaeren highs.

Extensional basin architecture

Seismic resolution of the Triassic is hampered by the poor impedance contrast between the Smith Bank Formation and the underlying Zechstein Group halite. However, a broad eastward thickening pattern is suggestive of the former presence of a large-scale, half graben wedge geometry across the central North Sea, which has been subsequently dissected by a narrower Jurassic rift (Fig. 3), with rift flank erosion on the Norwegian margin locally creating the impression of thinning and wedging (Fig. 3A). However, stratigraphic thickening patterns on east-west well sections (Fig. 14) locally confirm the gross half graben geometry seen on seismic, indicating that the thicker successions seen on the Norwegian margin were probably the sites of active faulting and the depositionally thinner (albeit heavily eroded) UK margin was less active. Internal unconformities can be recognised within the Skagerrak Formation succession, but are expressed subtly by thinning and truncation beneath stratigraphic markers correlated using well logs, and are rarely visible in seismic sections (cf. Tomasso *et al.*, 2002; Archer *et al.*, 2010). The Hardegsen Unconformity can be traced as the variable truncation of Bunter and Smith Bank formation successions beneath a widespread shale package at the base of the Judy Sandstone Member (Fig. 14), in a remarkably similar style to the onlap of the Solling Formation across the erosionally truncated Bunter, Detfurth and Hardegsen formation successions in the adjacent Southern Permian Basin (e.g. Geluk, 2005; Radies *et al.*, 2005). This feature is locally visible on seismic, but resolution of the deeper Triassic succession is insufficient to enable regional mapping. The lower, pre-Hardegsen Unconformity, succession is typically highly variable in total thickness and relative proportion and thickness of Smith Bank and Bunter formation sandstones, reflecting both unconformity erosion and the strong effects of halokinesis and minibasin formation (which is largely smoothed-out in the widely spaced well log correlations in Fig. 14, but cf. Fig. 3). The Cimmerian I Unconformity at the top of the Joanne Sandstone Member (Archer *et al.*, 2010) is constrained by fewer well data, but does appear to have involved some erosion of the upper parts of the succession.

The Judy and Joanne Sandstone members show moderate thickness variations, reflecting on-going halokinesis during deposition. However, thickness changes are gradational, the sandstones remain sheet-like across the basin and they retain their internal wireline log and facies character across wide areas with little evidence of abrupt facies changes or changes in sand : shale ratio. The smooth and gradational changes in character suggest that there was minimal surface expression of any underlying salt movement during deposition of these fluvial sheets and that sediment supply was sufficient to infill any emerging floodplain irregularities. Onlap of Middle to Late Triassic deposits onto Zechstein salt walls does however occur locally, indicating that the alluvial plain, although largely unconfined, locally experienced salt deformation at surface.

Pods and interpods – deposition versus preservation

Although salt controlled facies changes are not common in the post-Hardegsen Unconformity succession, salt tectonics clearly exerted a significant control on the stratigraphic accommodation and preservation of Early Triassic sediments. Regional seismic data show the widespread development of minibasins infilled by Smith Bank Formation (Fig. 3), particularly on the basin flanks where Zechstein halite was thinner and more deformed. Minibasin formation was likely to have been driven by a combination of salt dissolution, sediment loading and regional evacuation (Clark *et al.*, 1999; Stewart & Clark, 1999; Cartwright *et al.*, 2001). It has been proposed that the geometry of these minibasins exerted a fundamental control on the deposition of the Skagerrak Formation fluvial systems, defining accommodation, local palaeoslope and sediment routing (e.g. Hodgson *et al.*, 1992; Smith *et al.*, 1993). In areas of reduced salt thickness, such as the UK Western Platform and Norwegian Jaeren High, it has also been suggested that the Smith Bank Formation minibasins grounded at a relatively early stage in the Triassic basin evolution and that subsequent deposition of the Skagerrak Formation occurred within interpod palaeovalleys which provided the conduits to feed Late Triassic fluvial sediment into the deeper basin (cf. Stewart & Clark, 1999). Critical to these interpretations is the assumption that when pod grounding occurred there was no further accommodation available in the pod locations and that deposition was permanently re-located to interpod areas characterised by salt dissolution, fluvial incision and sediment by-pass. Whilst plausible, this model does not take into account the accommodation available due to regional post-rift basin subsidence following Late Permian and Early Triassic extension (Steel & Ryseth, 1990; Müller *et al.*, 2005) and Early Permian thermal events (van Wees *et al.*, 2000) which would have been available for fluvial aggradation above the crests of grounded pods. As a result, whilst the interpod palaeovalley model is realistic, it can only represent an end-member scenario where stratigraphic base level (Muto & Steel, 2000) was permanently located at or below the level of the grounded pod crests (Fig. 15A). In addition to this situation it is also possible to envisage additional scenarios in which on-going regional subsidence sustained accommodation creation for continuous aggradation above the level of grounded pods (Fig. 15B) or a transitional situation in which fluctuations in stratigraphic base level above and below the pod crests forced alternations between regional aggradation and focussed incision and back-filling within interpod palaeovalleys (Fig. 15C). Subsequent erosion of the thinner, more elevated, successions above the grounded pods during Middle Jurassic thermal

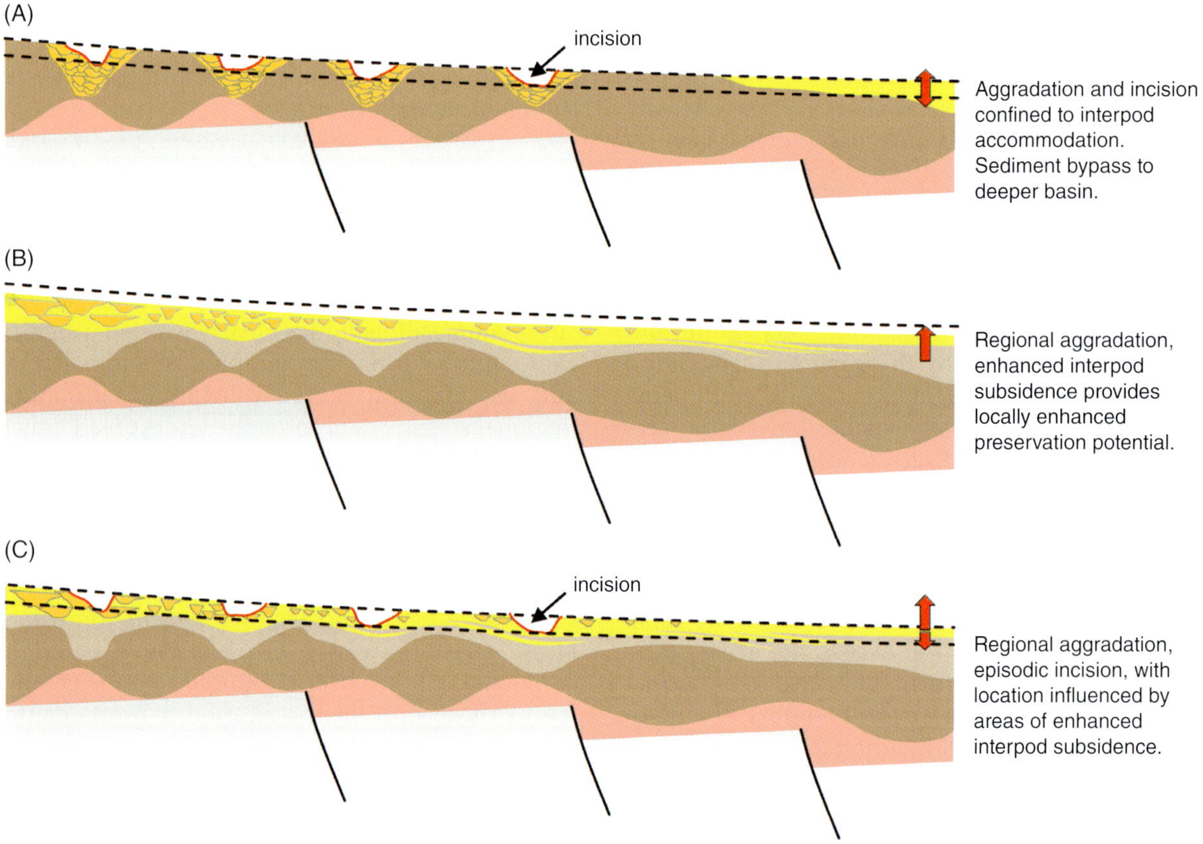

Fig. 15. Model for the stratigraphic fill of interpod regions under varying relationships between the fluvial profile relative to the crest of the grounded Smith Bank Formation pods and interpod salt highs. (A) Following Smith Bank Formation pod grounding deposition was permanently relocated to interpod areas where on-going salt evacuation and/or dissolution created the only available accommodation for accumulation of Skagerrak Formation sands. Interpod areas became conduits for sediment supply to the deeper basin. (B) Continuous aggradation above the level of the pods and interpods due to regional basin subsidence, with no falls in geomorphic base level. (C) On-going basin subsidence allowed continuous aggradation above pod level, with interpod areas supplying localised zones of higher subsidence. Episodic fluvial incision was focussed into these interpod areas during falls in geomorphic base level. Model B appears to be the most common situation on the UK Western Platform.

doming would leave an apparently similar record to that of the palaeovalley model (Fig. 16), with Bunter and Skagerrak formation sandstones restricted to zones between grounded Smith Bank Formation minibasins. However, this configuration would not be a reflection of the original geomorphology, but would be entirely the product of the distribution of localised, long-term preservation space (Kocurek & Havholm, 1993). The UK Western Platform may provide some clues to differentiate these models and provide support for a biased preservation model. In this area core from interpod areas is entirely composed of fine-grained, distal sheetflood and playa margin deposits of Bunter Formation-lower Judy Sandstone Member character, which show no evidence of fluvial confinement. Facies changes are gradational across successive interpod regions, suggesting that although now preserved in areally restricted fairways, these Skagerrak Formation sandstones were originally deposited on an unconfined alluvial plain and in a relatively distal position on the fluvial drainage rather than channel confined, proximal feeder zones. Their current distribution within a complex network of interpod lows is more likely to be a reflection of their localised preservation due to continued subsidence of the interpod areas as salt continued to be evacuated (McKie & Audretsch, 2005), both during the Triassic and subsequently into the Upper Jurassic (following Middle Jurassic erosion) when interpod areas again became sites of preferential preservation of open marine Fulmar

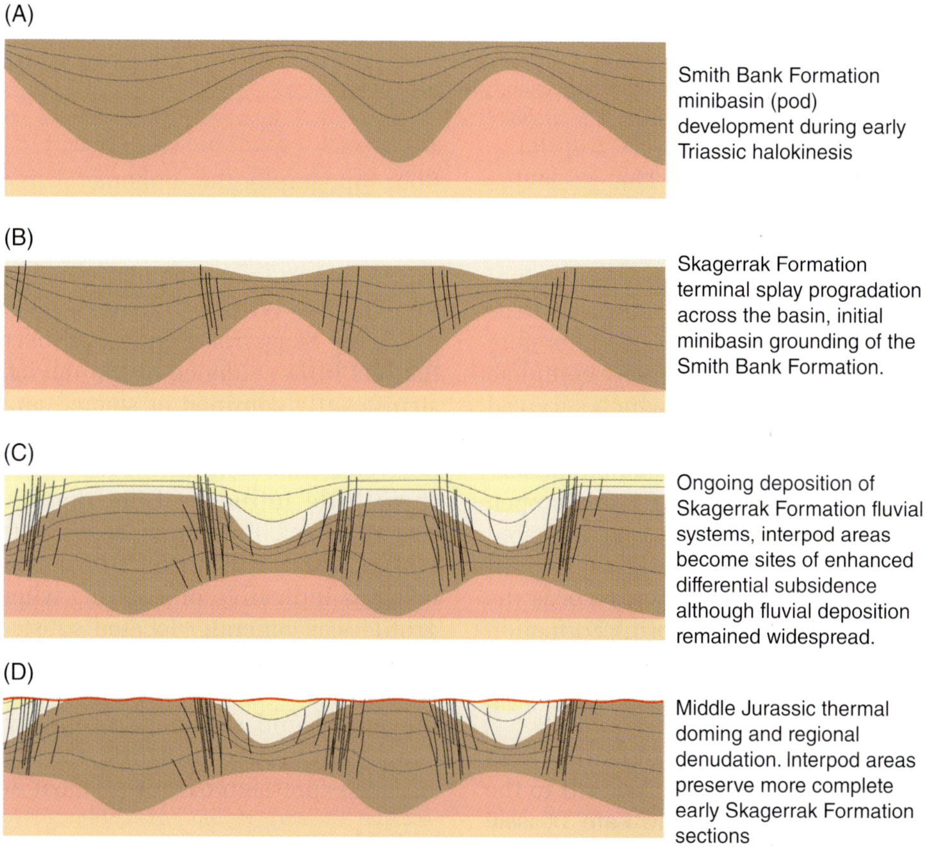

Fig. 16. Conceptual model for pod-interpod accumulation and preservation in areas of early pod grounding such as the UK Western Platform. Initial minibasin formation during deposition of the Smith Bank Formation (A) evacuated the underlying halite, resulting in pod grounding on the Rotliegend Group (B). Subsequent progradation of Bunter Formation and lower Judy Sandstone Member terminal splay complexes occurred across the region, with localised thickening in interpod areas where salt continued to be evacuated and/or dissolved (B). This process continued through the Middle and Late Triassic as the region was blanketed by sediment as the basin continued to subside (C), but subsequent erosion during Middle Jurassic thermal doming (D) eroded much of the Triassic stratigraphy, leaving erosional remnants of Skagerrak Formation terminal splays in interpod troughs. The geometries of the troughs do not reflect the depositional geomorphology and are sites of preferential preservation of formerly unconfined distal sheets.

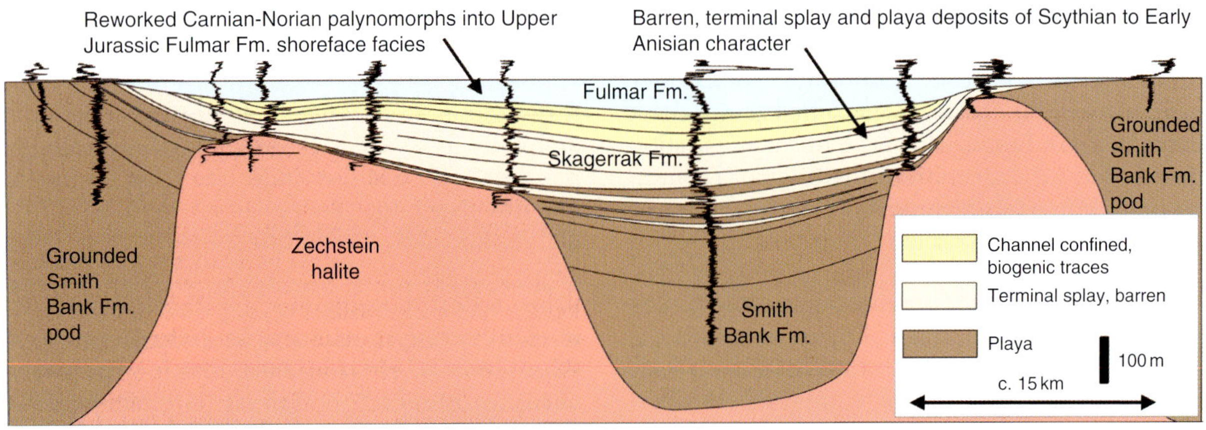

Fig. 17. Smith Bank and Skagerrak formation well log correlation across a single interpod on the UK Western Platform in the area of the Gannet and Guillemot Fields. The section runs through wells UK 21/24-2, UK 21/29b-7, UK 21/30-8, UK 21/25-2, UK 21/25-3, UK 21/30-3, UK 21/30-6A, UK 21/30-19, UK 21/30-4 and UK 21/30-13S1. Interpod Triassic facies are composed of distal terminal splay facies. The overlying Fulmar Formation records renewed salt withdrawal from interpod regions during Late Jurassic extension. These open marine deposits commonly contain reworked Late Triassic palynomorphs and erosively overlie in-situ Lower and Middle Triassic deposits. The restricted distribution of these open marine sandstones is likewise interpreted to be a product of the same preservational bias as the underlying Skagerrak Formation, rather than a reflection of syn-depositional bathymetry or geomorphology.

Formation sandstones (Fig. 17). The overlying Fulmar Formation locally contains reworked Triassic palynomorphs of Ladinian to Rhaetic age, indicating erosion of a former cover of upper Triassic and reworking into the interpod areas during Jurassic halokinesis. It is likely therefore that there had formerly been a relatively complete Triassic section, the lower parts of which survived erosion due to on-going salt withdrawal and subsidence below stratigraphic base level. Both unconfined Triassic sheetflood and upper Jurassic open marine deposits are now preserved in a shared geometry inconsistent with their original depositional setting and the Triassic section records the oldest and deepest parts of the Skagerrak Formation succession which subsided into long-term preservation space below the erosion level of regional Middle Jurassic denudation.

Across the central North Sea the transition from the mudstone-dominated Early Triassic fill to the widespread dispersal of Middle and Late Triassic fluvial sediments in basin-wide sheets is probably indicative of a fundamental change in the extent to which salt controlled the routing of sediment into and across the central North Sea. The mud-prone Smith Bank Formation, with its highly variable thickness and minibasin geometry, probably records sediment loading and dissolution during initial fluvial sediment supply into the desiccated Zechstein basin, which created a tortuous basin floor topography that ephemeral river systems in an arid climate were unable to breach or traverse. Basin filling occurred by distal muddy floodwaters infiltrating and ponding in the evolving minibasin topography. Whilst thickness variations in the overlying Skagerrak Formation succession occur, there is little evidence of the fluvial facies being significantly confined or steered on a basin scale by the underlying salt and palaeocurrent data derived from borehole images and oriented cores show little influence from the underlying salt structures (cf. Prochnow *et al.*, 2006; Matthews *et al.*, 2007). This pattern of deposition can be taken as indicative of a setting where the fluvial profile was generally located substantially above the level of salt deformation, which was responding passively to sediment loading and subdued fault movement and sediment supply was generally capable of quickly infilling and smoothing any surface expression of movement at depth.

CLIMATIC CONTROLS

Through the Permian and Triassic Western Europe was located within the equatorial arid and semi-arid climatic belts (Hounslow & Ruffell, 2006; Feist-Burkhardt *et al.*, 2008). The vast continental interior of Pangaea created extreme continentality, with a pronounced monsoonal seasonality and dry winters and summer wet conditions along the northern Tethyan margin (Manzpeizer, 1982; Kutzbach & Gallimore, 1989; Parrish, 1993; Sellwood & Valdes, 2006, 2007). Episodic marine flooding of the Southern Permian Basin in the Middle Triassic may have buffered the local climate and elevated humidity levels in the central North Sea (McKie & Williams, 2009). In order to assess the relative impact of climatic changes on the Skagerrak Formation fluvial systems it is important to identify whether any apparent change in fluvial depositional character within the central North Sea was a product of a regional climatic change within the basin, changes originating in the catchments, or a response to more local factors, e.g. fluvial avulsions creating locally wet conditions within an otherwise arid basin (c.f. van der Zwan & Spaak, 1992; Visscher *et al.*, 1994). Regional facies associations provide a general sense of basinal climate at a broad scale across the region (Bourquin *et al.*, 2009), which can be combined with palynology and a range of climate proxies to develop a regional framework of the evolving climatic situation across north-west Europe.

An arid rift interior

To the west of the North Sea region, in basins that were remote from Tethyan and Muschelkalk marine influence and located within the heart of the rift system, the Triassic succession is characterised by widespread evidence of sustained aridity. Brookfield (2004, 2008) documents a Triassic succession within the Solway Basin (Fig. 1), partially isolated from the main rift axes, which was deposited under consistently arid to hyper-arid conditions with little evidence of dramatic climate oscillations. Supply of sand into the basin did not occur via large river systems and was likely to have been dominated by aeolian transportation and subsequent reworking by small desert streams generated by local rainstorms. Outside of this isolated basin large Early Triassic fluvial systems flowed northwards through the UK and offshore Ireland, but these were largely sourced from remote, wetter catchments and terminated within these arid interior basins (cf Péron *et al.*, 2005). The terminal regions of these systems (e.g. East Irish Sea) lack evidence of significant vegetation and were subject to widespread aeolian reworking in evaporite-prone, terminal basins (McKie & Williams, 2009). The Middle and Late Triassic fills of these interior basins show very little evidence of fluvial activity and are largely composed of mud-prone playa and evaporites and show limited evidence of significant standing water (Ruffell & Shelton, 1999; McKie & Williams, 2009). Exceptions to this pattern of general aridity occurred in the Middle Carnian, Late Ladinian and possibly Late Norian, with the brief development of more perennial lacustrine conditions (Milroy & Wright, 2000; Porter & Gallois, 2008) during wetter climatic episodes. However, evidence of aeolian activity, evaporite precipitation and desiccated playa can be traced as far north as eastern Greenland until the Carnian (Surlyk, 1990). The general impression from these regions is one of an arid continental interior, with any significant fluvial input derived from distant, wetter catchments or during discrete pluvial events of regional to global significance.

The evidence of aridity, based on the dryland facies within the continental interior is supported by general circulation models which show the region as one of net evaporation (Kutzbach & Gallimore, 1989; Parrish, 1993; Péron *et al.*, 2005). These models also suggest that the central North Sea region itself was probably relatively arid. This aridity is broadly indicated by the downstream shale-out of the Skagerrak Formation fluvial systems, indicating that flood waters were entirely dissipated through evaporation and transmission losses, resulting in terminal, endorheic fluvial systems. However, despite these arid conditions there is little evidence in the central North Sea of significant evaporite precipitation beyond isolated instances of salt crust fabrics and no evidence of significant aeolian reworking after the Early Triassic. Anhydrite is present within the Middle Triassic Julius Mudstone in the southern part of the central North Sea (Goldsmith *et al.*, 1995; Michelsen & Clausen, 2002), which may reflect the development of coastal sabkha facies on the northern margin of the Muschelkalk Sea, but the overall prevalence of plant rooting and burrows is more indicative of an availability of year-round soil moisture, suggesting that runoff from the catchments was perennial (albeit seasonal) and that groundwater was delivered into the central North Sea, even during dry seasons, to suppress evaporitic and aeolian processes and maintain biogenic activity.

Skagerrak Formation fluvial expansion during pluvial phases

Whilst the latitudinal position of north-west Europe within Pangaea dictated a generally arid climatic regime, fluctuations in aridity are recorded

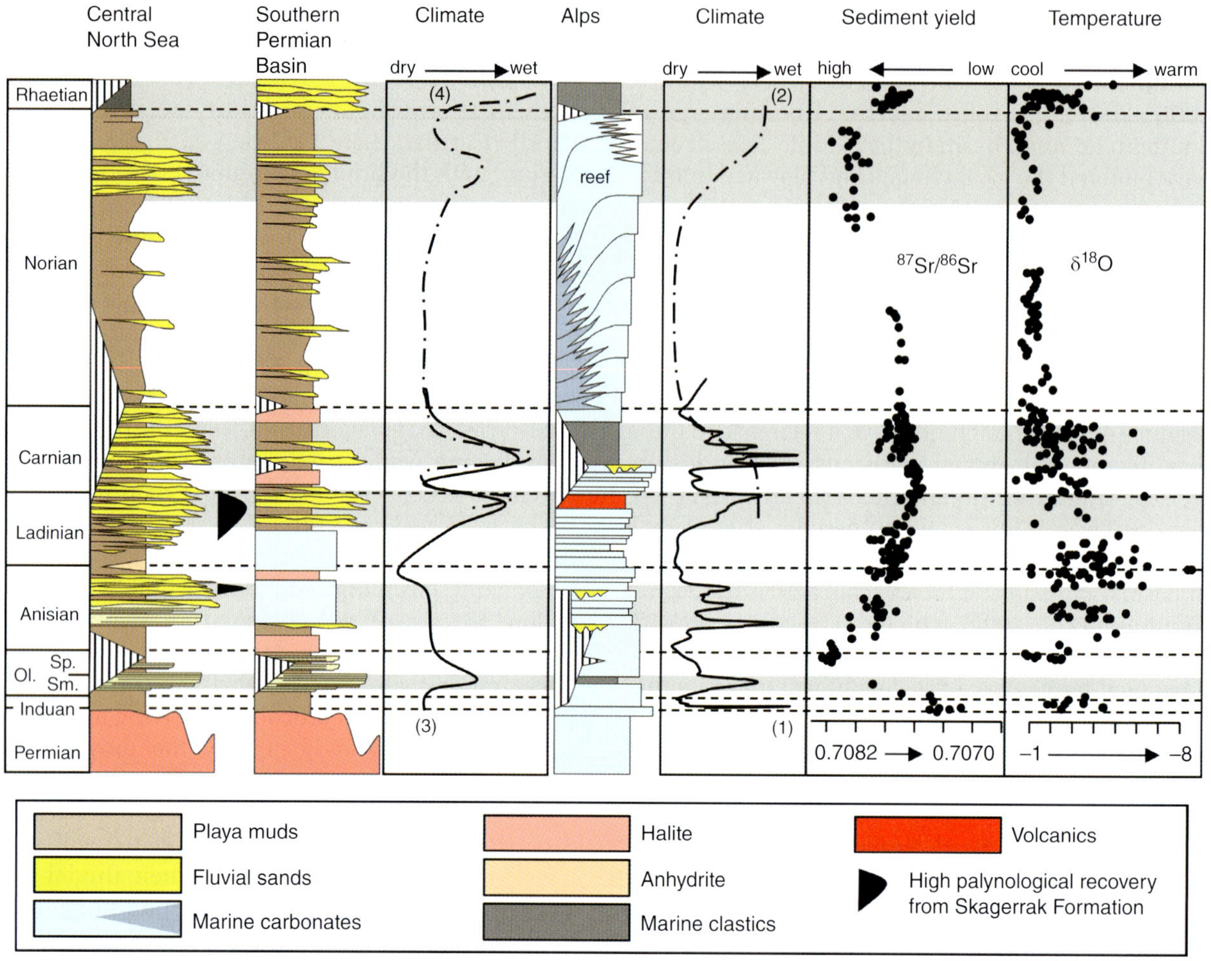

Fig. 18. Comparative climate and sediment yield proxies for a transect from the central North Sea to the northern margin of Tethys (Southern Permian Basin stratigraphy after Aigner & Bachmann, 1992, Geluk, 2005; Alpine stratigraphy after Gianolla *et al.*, 1998). Climate curves from (1) Stefani *et al.* (2010); (2) Mutti & Weissert (1995); (3) Kürschner & Herngreen (2010); (4) Hornung & Aigner (2002). Time scale follows Preto *et al.* (2010). Ol. – Olenekian, Sm. – Smithian, Sp. – Spathian. Dark spikes in the central North Sea profile highlight intervals of enhanced palynological recovery (Goldsmith *et al.*, 1995). Strontium isotope data from Korte *et al.* (2003), oxygen isotope data from Korte *et al.* (2005). Sand-prone fluvial deposition in the central North Sea broadly corresponds to pluvial episodes which can be traced across the northern Tethyan region and into the Southern Permian Basin. These episodes appear to have occurred during climatic warming, which may have weakened trade winds, allowing moist Tethyan air masses to penetrate farther into the European basins and increased monsoon intensity. Major increases in sediment yield in the Early Triassic may have been a product of devegetation in the aftermath of the end Permian extinction and in the Carnian as a result of the widespread development of exorheic fluvial systems. However, for much of the Triassic clastic sediment yield to Tethys was limited by the prevailing aridity and prevalence of endorheic basins.

across the region. The climate record from marine and marginal marine successions deposited along the northern Tethyan margin (Kustatscher *et al.*, 2010; Preto *et al.*, 2010; Stefani *et al.*, 2010) shows a generally dry background climate punctuated by wetter periods, with pluvial events occurring in the Smithian, Anisian, Ladinian and Carnian (Fig. 18). In particular, the Early-Middle Anisian, Late Ladinian and Middle Carnian appear to have been intervals of relatively frequent or sustained humidification, whilst the Late Anisian-Early Ladinian and Late Carnian-Norian show long term aridity (Mutti & Weissert, 1995).

Between the northern Tethyan margin and the central North Sea, the Southern Permian Basin region provides an intermediate palaeogeographic link between the climatic record seen on the northern Tethyan margin (the source of monsoonal rainfall and general maritime precipitation) and the central North Sea. The climatic trends observed in this basin (Kürschner & Herngreen, 2010) record an Early Triassic period dominated by opportunist vegetation populating dry, well drained floodplain environments, with a possible increase in humidity during the latter part of the Early Triassic that may correspond to the Smithian humid phase recorded on the Tethyan margin (Stefani, 2010). Sporomorph assemblages indicate generally dry to arid conditions through the Anisian (van der Zwan & Spaak, 1992; Kürschner & Herngreen, 2010), with the Early Ladinian (upper Muschelkalk) characterised by more semi-arid conditions, again broadly following a similar pattern to that seen to the south (Fig. 18). More humid conditions returned during the Late Ladinian, with a peak in palynomorph diversity recording the widespread expansion of wetland floras. This Late Ladinian event is also well expressed in the Norwegian-Danish Basin (Lindström *et al.*, 2009) and central North Sea (Goldsmith *et al.*, 1995). The Early Carnian was marked by a return to aridity, although a second peak in species diversity occurred during the Middle Carnian pluvial event when conditions became wetter across the entire north-west European region (Simms & Ruffell, 1989, 1990). During the Late Carnian and Norian palynofloral abundance declined as conditions returned to aridity. Local recovery within the Stubensandstein Formation (Hornung & Aigner, 2002) is indicative of generally arid conditions, with minor humidification in the Late Norian (Fig. 18). This event may also equate to enhanced clastic transport and wetter conditions across the margins of the Tethyan carbonate shelf (Berra *et al.*, 2010) and, although poorly dated, to deposition of the Josephine Sandstone Member in the central North Sea. The transition into the Rhaetian was marked by widespread humidification across the region (Ahlberg *et al.*, 2002) in response to the marine inundation of the interior of Pangaea.

The climate records from the Tethyan margin (Stefani, 2010) and Southern Permian Basin (van der Zwan & Spaak, 1992; Kürschner & Herngreen, 2010) appear broadly similar (Fig. 18), although there are differences in detail and resolution due to the variety of approaches used. The entire region to the south of the central North Sea appears to have been subject to widespread increases in rainfall and fluvial runoff in the Smithian, Late Ladinian and Middle Carnian and sustained drought during the Spathian, Anisian and Norian. These fluctuations were not necessarily global in extent – palynological data from the Barents Sea region (Hochuli & Vigran, 2010) record a different climatic pattern of warm temperatures and minimal seasonality, with only the Spathian and Late Anisian drying events, together with the mid-Anisian tendency towards wetter conditions being expressed this far north. The Late Ladinian pluvial phase is absent and the Carnian increase in humidity was sustained through the Late Triassic rather than reverting back to drier conditions. This partially similar record may indicate a combination of global temperature changes impacting all latitudes (Galfetti *et al.*, 2007a) and more regional (i.e. Tethyan) climate changes (Hornung *et al.*, 2007); or possibly latitudinal differences in atmospheric response to global change as the migration range of the inter-tropical convergence zone varied and the scale of Hadley cells expanded and contracted (cf. Perlmutter & Matthews, 1990). It is also possible that Triassic atmospheric circulation patterns were dissimilar to the present, with a breakdown of Hadley circulation into numerous small cells, or collapse of the Polar cell and development of a two cell system leading to humidity at the poles (Parrish, 1993; Kidder & Worsley, 2004). These would give a less predictable response than migration of a simple latitudinal zonation. However, Olsen & Kent (2000) note that Late Triassic facies distributions do appear to follow latitudinal climate zones suggesting, at least for this period, that Hadley circulation was in operation.

Comparison of the Tethyan and Southern Permian Basin climatic records with the Triassic succession in the central North Sea indicate a broad correspondence between wetter climatic phases and the basinward expansion of the Skagerrak Formation fluvial systems (Fig. 18). In this framework the Judy and Joanne Sandstone members (and possibly the Josephine Sandstone Member) appear to have extended basinwards during episodes of more frequent humidification across the region, with the Skagerrak Formation shale members representing terminal fluvial contraction during aridification. This would be in keeping with the evidence of salt crust fabrics and microbial mats within the shale-prone members and well developed soil profiles and bioturbation within the sand-prone members. This scenario contradicts the climatic model suggested by Goldsmith *et al.* (2003), who used palynological indications of marsh and lacustrine conditions in the Julius Mudstone Member, together with variations in shale colour, as a guide to water table conditions and to infer a wetter climate during mudstone member deposition. However, shale colour in the Skagerrak Formation is unlikely to be a reliable indicator of palaeogroundwater conditions and invariably shows reduction within hydrocarbon columns and reddening in dry holes and aquifers. Archer *et al.* (2010) also propose that fluvial expansion was more likely to have taken place during wetter conditions and cite the absence of lacustrine deltas and presence of evaporites within the shale members as evidence that fluvial runoff was reduced and the climate was drier during shale deposition.

Fluvial expansion driven by catchment wettening

Hinterland pollen and spores from the upper part of the Julius Mudstone and lower Joanne Sandstone members are dominated by gymnosperms (Goldsmith *et al.*, 1995, 2003), indicating semi-arid conditions in the UK and Fennoscandian headwaters in the Early Ladinian. However, outside of these intervals there is little palynological evidence of catchment conditions, particularly during peak fluvial expansion when conditions may have been wetter. The barren nature of Early Triassic fluvial sections may have been the result of the prevailing aridity at this time (Péron *et al.*, 2005), combined with deposition during recovery from the end Permian extinction (Payne *et al.*, 2004). Hochuli & Vigran (2010) document diverse spore and pollen assemblages in the Early Triassic of the Barents Sea region, indicating that low latitude aridity may well have been the key controlling factor in limiting recovery from the central North Sea. The barren Norian section in the central North Sea may likewise be indicative of general aridity in both the basin and catchment regions.

General circulation models for Pangaea predict a strong seasonality to the precipitation across the north-west European region, with arid, evaporative conditions across the central North Sea during the dry season (Kutzbach & Gallimore, 1989; Sellwood & Valdes, 2006, 2007; Goddéris *et al.*, 2008). This is consistent with the regional evidence of aeolian activity, evaporite precipitation, paucity of perennial lacustrine facies and the terminal nature of the fluvial drainage, but does not predict the vegetated, burrowed character of the Skagerrak Formation (indicating year-round soil moisture), laminated mudstone channel abandonment-fills (indicating standing water; McKie, 2011), common rooted alfisols (pointing to widespread shrubland vegetation) and lack of evaporite and aeolian facies after the Early Triassic. One possible explanation for the year-round availability of water within the Skagerrak Formation fluvial drainage network could be that the headwaters extended into higher latitude, generally wetter catchments, allowing some runoff or groundwater flow into the basin during the dry season to maintain riparian vegetation. This hypothesis can be tested by estimating the possible size of the Skagerrak Formation catchments. Davidson & North (2009) outline a simple, albeit imprecise, method for estimating ancient catchment size based on bankfull depth and using regional relationships between depth and catchment area based on modern streams from a range of climate settings. For the Skagerrak Formation streams two estimates of bankfull depth of 3 m and 6 m were applied based on the typical thickness of fining-upward cycles seen in core and validated by bankfull depth estimates calculated from the mean thickness of stacked cross-sets using the method outlined by Bridge & Tye (2000). Two regional curves from modern dryland regions were used to relate bankfull depth and catchment area (U.S. Western Cordillera and Basin and Range Province; cf. Davidson & North, 2009). The two depth estimates and regional curves combine to give drainage areas ranging from c. 14,800 to 69,500 km^2 to c. 129,400 to 1,083,000 km^2. As Davidson & North (2009) note, there is a paucity of regional curves for modern dryland climatic zones and the results can only be used to give a general sense of the possible scale of fluvial catchment size in these settings. However, the latter range is excessively large for the scale of the Skagerrak Formation fluvial systems within the central North Sea and, more critically, fails to fit within the available area of uplifted Caledonian basement on the UK margin of the basin, for which there is established provenance data (Mange-Rajetzky, 1995). The smaller range is more realistic, with the advantage of scaling with the UK Caledonian catchments and is more likely to be representative of Skagerrak Formation catchment size (Fig. 19). In neither case, however, does the calculated area (using plausible catchment aspect ratios) extend the drainage sufficiently into a higher latitude region that would have been consistently subject to more perennial precipitation (Kutzbach & Gallimore, 1989; Sellwood & Valdes, 2006, 2007). This would appear to suggest that the year-round maintenance

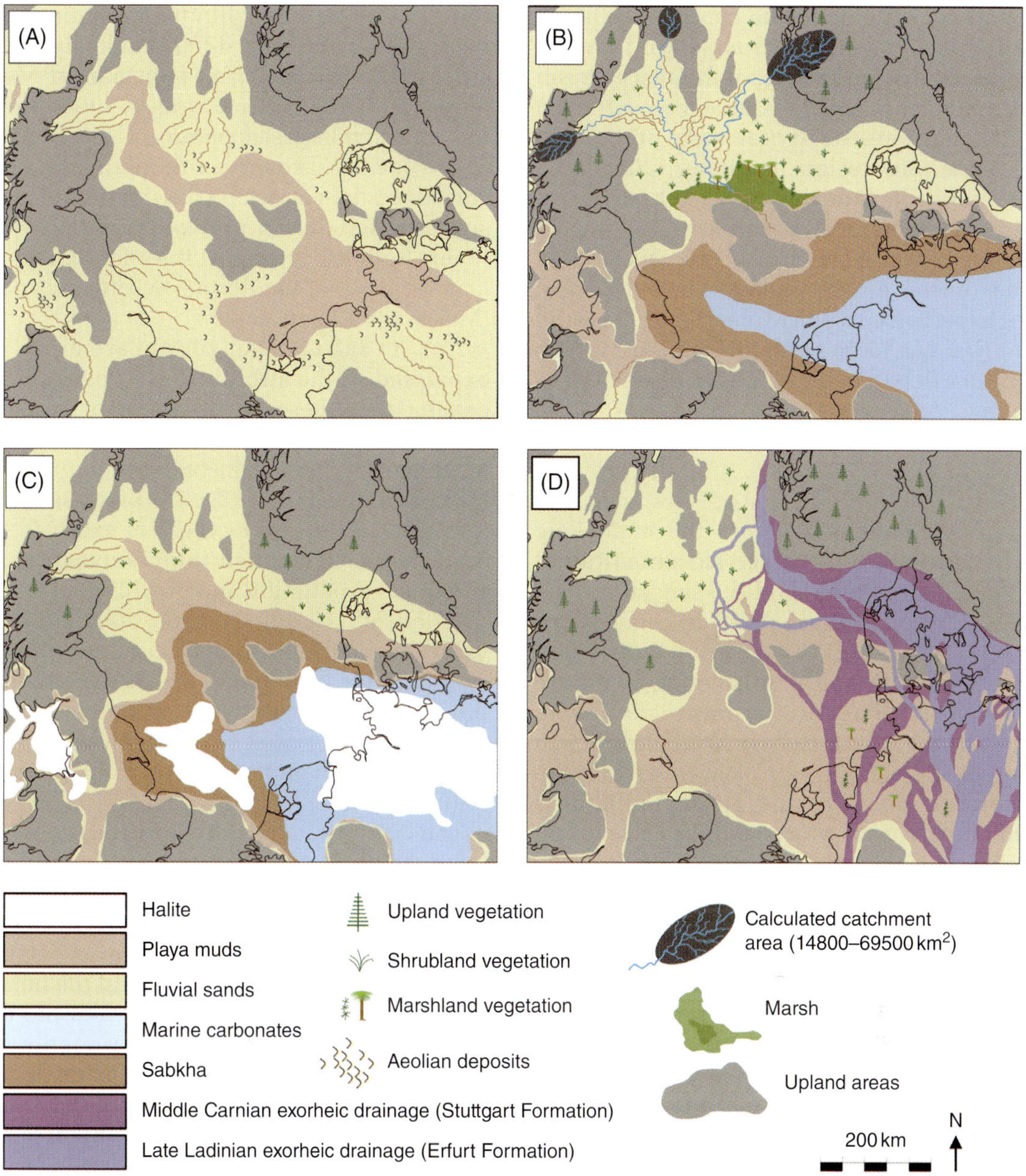

Fig. 19. Palaeogeographic response to regional climate wettening. (A) Early Triassic basins were arid and characterised by endorheic drainage which discharged into playa-dominated basin centres during pluvial phases. Aeolian reworking was common on the distal fringes of the fluvial systems. (B) During Middle Triassic pluvial phases large terminal fluvial drainage systems developed across the central North Sea, with widespread riparian vegetation cover. These may have locally exited the central North Sea, but were still terminal in character. (C) During Middle Triassic arid phases fluvial activity was reduced. Terminal fluvial systems contracted and playa expanded. However, the presence of the Muschelkalk seaway may have elevated local humidity, allowing the maintenance of vegetation cover. (D) During major pluvial episodes in the Late Ladinian and Middle Carnian fluvial systems formed an exorheic drainage towards Tethys, with rift flank uplift potentially diverting Fennoscandian drainage away from the central North Sea and across the Southern Permian Basin towards Tethys. The Southern Permian Basin became a site of fluvial incision as Erfurt Formation and Stuttgart Formation rivers initially flowed towards a lower marine base level. The central North Sea record is fragmentary and may have been entirely denudational during these episodes, but evidence from adjacent areas suggests that perennial lakes formed in isolated, sand-starved, depocentres.

of dryland ecosystems in the central North Sea was not driven by runoff from distant, wetter regions but by discharge sourced from local, basin margin catchments which experienced sufficient precipitation during the dry season to prevent the central North Sea region becoming fully desiccated, evaporitic and unable to support plants and animals. This perennial runoff within a region broadly characterised as one of net evaporation occurred during pluvial pulses when the basin margins experienced a departure from widespread aridity towards a more semi-arid dry season with groundwater flow and a wet season characterised by flooding and overland flow.

Sediment yield

The Triassic mudstones across the region are generally lacking kaolinite and smectite (Ruffell *et al.*, 2002; Jeans, 2006) and the sandstones are typically arkosic to lithic arkoses in composition and highly feldspathic (10% to 50%), indicating a dominance of mechanical weathering in source areas subject to long term aridity. The pluvial episodes do not appear to have been of sufficient duration or intensity to significantly modify the weathering products, except during the Middle Carnian pluvial phase when kaolinite became episodically more abundant (Simms & Ruffell, 1990). Although sediment supply to these basins rarely reached Tethys or the Muschelkalk seaway, $^{87}Sr/^{86}Sr$ data from these marine basins (Korte *et al.*, 2003) may provide an indication as to the broader conditions of sediment yield during pluvialisation by exorheic river systems across the region, which can be compared to the nearby endorheic basins (Fig. 18). These data indicate three major episodes characterised by marked increases in sediment yield. An Early Triassic increase is likely to reflect increased weathering and erosion resulting from widespread devegetation during the recovery from the end Permian extinction, with the subsequent long-term reduction in yield broadly corresponding to the Anisian recovery and resumption of sediment binding by plants (Newell *et al.*, 1999; Korte *et al.*, 2003). Evidence for this high sediment yield appears to be lacking in the central North Sea, which may be due to several factors, including the general aridity of the region and infrequent sediment supply. Although it would be tempting to ascribe Early Triassic halokinesis to unusually rapid sediment loading, the long-term sedimentation rates for the interval comprising the Smith Bank and Bunter formations (up to 600 m thick) do not differ markedly over the c. 5 Ma of Early Triassic time from 'normal' Mesozoic terrestrial rates (Sadler & Strauss, 1990). However, the more widespread nature of the Bunter and Sherwood sandstones across north-west Europe compared to younger Triassic fluvial systems suggests that during pluvial episodes, with sufficient water to transport sediment, dispersal of large volumes of sand-prone detritus was possible (McKie & Williams, 2009).

Anisian and Ladinian pluvial events are poorly expressed in the strontium isotope record, suggesting that these events were of insufficient duration or magnitude to disturb Tethyan seawater chemistry, but the major Carnian pluvial event is marked by a minor $^{87}Sr/^{86}Sr$ increase reflecting the widespread development of exorheic fluvial systems which drained into Tethys, terminated carbonate production and precipitated a Tethyan 'reef crisis' (Hornung *et al.*, 2007). Increasing values through the Late Triassic are interpreted to reflect increasing erosion of the Cimmeride-Indosinian orogens, marking the closure of the Palaeotethys (Korte *et al.*, 2003), but could plausibly have some enhancement from climatically induced runoff.

The $^{87}Sr/^{86}Sr$ data provide a record of major trends in sediment yield into Tethys which highlight continental-scale sediment supply pulses in the Early Triassic, Carnian and Late Norian-Rhaetic by exorheic rivers. However, this record provides only limited insights into the behaviour of endorheic systems in basins like the central North Sea or the higher frequency supply pulses, which might be expected given the evidence of a highly variable climatic record. These events may be more subtly expressed. For example, the upper part of the Anisian Judy Sandstone Member shows an abrupt increase in vegetation cover and proportion of channel-fills, marking a major increase in sediment supply into the basin, availability of year-round soil moisture and expansion of the fluvial catchments into Fennoscandia (McKie, 2011). This event is composite in nature, being marked initially by a facies and provenance change from terminal splay to cross-stratified fluvial deposits indicative of well organised, Fennoscandian, channel systems, followed by a change in floodplain style from barren playa mudstones to rooted floodplain mudstone with calcrete (Fig. 4A). The onset of palynological recovery from the Skagerrak Formation (Goldsmith *et al.*, 1995) broadly occurs at this level and it is likely that this mid-Judy Sandstone Member event was the combined product of one or more Anisian pluvial episodes, combined with local climatic humidification associated with the marine flooding of the Muschelkalk seaway (Fig. 18) and marks a major change in the Fennoscandian sediment supply to the basin. This contribution was sustained through the Ladinian and Carnian as the subsequent progradation of the Joanne Sandstone Member and marks a major departure from the more regional dominance of Keuper Formation and Mercia Mudstone Group playa environments.

Base level

A definition of geomorphic base level within ephemeral, terminal systems is problematic because in many cases the fluvial discharge fails to reach the terminus of the fluvial system due to evaporation and infiltration. However, over geological time scales the ultimate base level for endorheic systems must be the playa or lake floor, depending on the balance of runoff versus evaporation/infiltration and of sediment supply versus subsidence (cf. Nichols, 2007). Low runoff will result in distal fringes dominated by aeolian processes (and preservation dictated by water table fluctuations; Kocurek & Havholm, 1993), with increasing runoff leading to an increasing dominance of fluvial facies, until runoff exceeds evaporation and transmission losses and perennial lakes form. At this point lake level fluctuations dominate the fluvial architecture within the basin in a similar manner to coastal plain systems.

In general the Triassic fluvial systems appear to have drained towards a desiccated playa floor in which geomorphic base level changes would have largely been the product of a changing balance between sediment supply and subsidence, with the former dictated by climatic changes in the headwaters of the fluvial system and the latter by regional thermal subsidence, episodic extension and more localised salt movement. There is little evidence of fluvial incision in the form of readily identifiable incised valleys or mature interfluve palaeosols (e.g. Fig. 15C), suggesting that the systems were dominantly aggradational in character (Fig. 15B) and did not experience systematic, high frequency base level fluctuations in the same manner as fluvial systems discharging into perennial lakes or marine water bodies. Mature calcic palaeosols and major lags of reworked calcrete occur locally, indicating that terracing and incision did occur, but not in a cyclical or repetitive manner that can be correlated over large distances. Although evidence of base level fall is limited, possible exceptions may be during the Late Ladinian and Carnian pluvial events, when Fennoscandian drainage reached Tethys and incised Erfurt and Stuttgart formation fluvial systems drained across the Southern Permian Basin towards an initially lowered marine base level (Aigner & Bachmann, 1992; Kozur & Bachmann, 2010). The expression of these events in the central North Sea, however, is not clear-cut.

Erosional by-pass of exorheic fluvial systems?

The Late Ladinian and Middle Carnian are characterised regionally by evidence of a widespread increase in humidity and fluvial run-off (Simms & Ruffell, 1989, 1990; Hornung *et al.*, 2007; Bachmann *et al.*, 2010; Kozur & Bachmann, 2010), with the main expression in the Southern Permian Basin being the regional development of the fluvio-deltaic Erfurt (Lettenkeuper) and Stuttgart (Schilfsandstein) formations (Geluk, 2005; Kozur & Bachmann, 2010). These deposits record the drainage off Fennoscandia of large, southward flowing, incised river systems (e.g. Shukla *et al.*, 2010; Fig. 19D) which supported widespread riparian vegetation (Visscher *et al.*, 1994). The local evidence of brackish facies and tidal influence (Kozur & Bachmann, 2010; Bachmann *et al.*, 2010; Shukla *et al.*, 2010) indicates that the rivers reached Tethys as exorheic fluvial systems and in the Carnian supplied sufficient clastic detritus to precipitate a 'reef crisis' (Hornung *et al.*, 2007). Provenance data (Köppen & Carter, 2000; Paul *et al.*, 2008) indicate derivation from the Fennoscandian and Caledonian basement region of western Norway – the same region from which the westward draining Skagerrak Formation streams flowed. Regional maps of the incised channels of these fluvial systems in Germany typically project the southward drainage pattern northwards directly onto the Fennoscandian Shield in a departure from the typical westward drainage seen towards the central North Sea. If this change in drainage is genuine it is conceivable that in the Early Carnian the Cimmerian 1 extensional episode induced sufficient rift flank uplift of the eastern margin of the central North Sea to divert the Norwegian drainage away from the central North Sea and southward to directly enter the Southern Permian Basin. Widespread basin margin uplift has been noted in the Early Carnian of the mid-Norway region (Müller *et al.*, 2005), extending to southern Norway (Paul *et al.*, 2009), which would lend some support to this model. This mechanism of basin margin uplift has been appealed to as an on-going process by Kozur & Bachmann (2010), in which the brief episode of Schilfsandstein drainage is attributed to a period of unusually high precipitation on a persistently uplifted rift flank. However, the persistence of westward fluvial drainage into the central North Sea from Fennoscandia would indicate that if the Schilfsandstein rivers did indeed drain southward then this brief departure from the typical drainage pattern would potentially require an equally brief tectonic deflection. In the absence of conclusive evidence that drainage flowed directly southwards off the Fennoscandian region it is also plausible that this Fennoscandian drainage, which traversed the central North Sea from Anisian to Early Carnian times, was extended in the Late Ladinian and Middle Carnian via the Norwegian-Danish Basin into the Southern Permian Basin, at which point it began to flow southwards towards Tethys. However, although both models are plausible, there is little evidence of an abrupt change to more humid facies in the central North Sea to support a correlation of Erfurt and Schilfsandstein formation fluvial facies into the time equivalent Ladinian-Carnian Joanne Sandstone Member (except possibly in the Norwegian-Danish Basin; Lindström *et al.*, 2009). This paucity of humid facies would suggest either that there was no expression of these climatic changes in the central North Sea or that the time equivalent section is absent due to erosion or non-deposition and is instead represented by an unconformity or hiatus. This might be explicable for the Middle Carnian pluvial event if it followed Cimmerian 1 rifting and large areas were eroded by uplifting fault blocks (although hangingwall areas might be expected to have preserved some record). However, there is no comparable record of tectonism in the Late Ladinian which could explain the absence of humid facies indicators for the broadly similar Erfurt Formation drainage. The absence of diagnostic facies may therefore be a result of non-deposition and erosional bypassing during these episodes. In both cases the Erfurt and Stuttgart formation fluvial systems temporarily drained towards a marine base level and were commonly widely incised across the Southern Permian Basin, suggesting that Tethys at this time represented a lowered base level compared to the preceding endorheic drainage contained within the Southern Permian Basin. Further upstream, in the central North Sea, it is also likely that the fluvial profiles were similarly incised, with conditions of even less accommodation than the basins to the south. The record of the Late Ladinian and Middle Carnian pluvial events in the central North Sea may therefore be entirely degradational as a result of fluvial incision across the region towards a lowered marine base level. Whilst subsequent backfilling in the Southern Permian basin took place under on-going humid conditions, backfilling further north in the central North Sea may not have occurred until later when the regional climate had reverted to dryer conditions and the central North Sea converted to an endorheic drainage with its own internal geomorphic base level.

DISCUSSION

Deposition of the Skagerrak Formation occurred as terminal fluvial systems which debouched into an arid central North Sea region (Fig. 19). Discharge was seasonal, with flood waters dissipating through evaporation and transmission losses towards distal playa and sabkha (when the Southern Permian Basin was marine influenced). During pluvial episodes these terminal systems expanded as a result of increased flood frequency and/or magnitude and dry season runoff was sufficient to maintain year-round vegetation and burrowing organisms. Only during two major pluvial episodes, in the Late Ladinian and Middle Carnian, did the fluvial systems expand outside the central North Sea region to reach Tethys as exorheic systems. A variety of mechanisms have been proposed for these pluvial events, including rift flank uplift to form an orographic barrier which captured moisture-laden monsoon circulation (Kozur & Bachmann, 2010), orogenic events altering atmospheric circulation patterns (Hornung & Brandner, 2005) and increased injection of CO_2 during volcanic episodes (Korte *et al.*, 2005; Galfetti *et al.*, 2007b; Feist-Burkhardt *et al.*, 2008). Of these mechanisms volcanism is probably the most likely process to have induced the relatively abrupt climatic swings seen in these Triassic successions, although whether these resulted in cooling (Feist-Burkhardt *et al.*, 2008) or warming (Hornung *et al.*, 2007; Kozur & Bachmann, 2010) is debated. The Tethyan and Muschelkalk oxygen isotopic record (Fig. 18; Korte *et al.*, 2005) shows a clear correspondence between lighter values, the general timing of pluvial phases and the expansion of the Skagerrak Formation fluvial systems. These lighter values potentially reflect warmer sea surface temperatures, but will also have been influenced by a variety of additional mechanisms including enhanced freshwater input and pH (Korte *et al.*, 2005). However, there remains the general impression of warmer, albeit highly variable (Nützel *et al.*, 2010), temperatures during the pluvial dominated Anisian to Carnian episodes, but with substantial cooling to long term stability through the Norian (Fig. 18). This stability is reflected in the comparatively monotonous terrestrial and marine successions of the Mercia Mudstone Group, Keuper and Hauptdolomit formations and cooling in the latitudinal restriction of Late Triassic reefs (Kiessling, 2010). However, whilst volcanism may explain some Ladinian and Carnian pluvial episodes when Tethyan magmatism was active, it remains to be proven that there is in all cases a direct correspondence between volcanic episodes and pluvialisation. In particular, the more global Spathian drying and Carnian wettening episodes are likely to have been driven by more widespread changes than localised volcanic episodes on the margins of Tethys (e.g. eruption of the large igneous province of Wrangellia; Furin *et al.*, 2006). Also, the influence of the Muschelkalk Sea on the local climate is also evidenced by the onset of wetland environments in the upper Julius Mudstone-lower Joanne Sandstone member interval during a time of general aridity (Fig. 18). In addition to volcanism, tectonic events in the Tethys Sea may have altered marine circulation patterns, with the Carnian closure of Palaeotethys possibly allowing greater influx of cooler oceanic waters into a more open Neotethys, thereby altering temperature gradients across the region and contributing to the apparent stability and arid nature of the Norian interval.

Studies of Holocene climate change offer some clues as to the likely contributing variables which may have influenced Triassic climate, although this post-Pleistocene period was still marked by icehouse conditions and a markedly different continental configuration to the Triassic Pangaean greenhouse situation described here. The Holocene record is one of relatively complex climate change as a result of strongly non-linear feedback processes (cf. Su & Neelin, 2005) between solar insolation (as the Earth reached perihelion during the northern hemisphere summer), changing oceanic and atmospheric circulation (albeit partly driven by fluctuating polar ice volumes) and surface albedo (resulting from the changing distribution of perennial water bodies and vegetation cover). Arid and pluvial episodes in the equatorial regions appear to have had a strong component of orbital forcing, which resulted in fluctuating solar insolation and latitudinal migration of the extent of the summer monsoon. Arid episodes correspond to cooler periods when monsoon systems weakened (Prell & Kutzbach, 1987), evaporation was lower from cooler oceans and thermal convection over tropical landmasses reduced (Mayewski *et al.*, 2004). In addition, increased sea surface temperature gradients resulted in strengthened trade winds, which inhibited the inflow of moist ocean air into some continental regions (Schefuβ *et al.*, 2005; Brook *et al.*, 2007).

In the Triassic a reduced temperature gradient during warming episodes may also have resulted in a weakening of the north-easterly trade winds, allowing inflow of moist Tethyan air masses into the central North Sea to maintain water availability in the dry season (Fig. 20), which combined with enhanced flood magnitude in the wet season as the monsoon intensified to maintain year-round moisture. However, the Tethyan monsoon might also be expected to have generated substantial fluvial transport off the Variscan massifs into the Southern Permian Basin, which does not always appear to have been the case. Reinhardt & Ricken (2000) and Vollmer *et al.* (2008) report a detectable monsoon influence in the Keuper Formation, with evidence of intensification in the Milankovitch frequency band, indicating that a monsoon was maintained and influenced deposition within Carnian-Norian playa in the Southern Permian Basin. The absence of significant sand transport off these massifs when a monsoon was present could be attributed to a loss of relief across this formerly mountainous region (cf. Lelue & Hartley, 2010), coupled with a lack of significant thermal convection so close to the Tethyan margin, or could be an indication that monsoon precipitation from Tethys was not sufficient to generate large sand-transporting fluvial systems (which would call into question whether the Skagerrak Formation was also the product of enhanced monsoon precipitation). The regional distribution

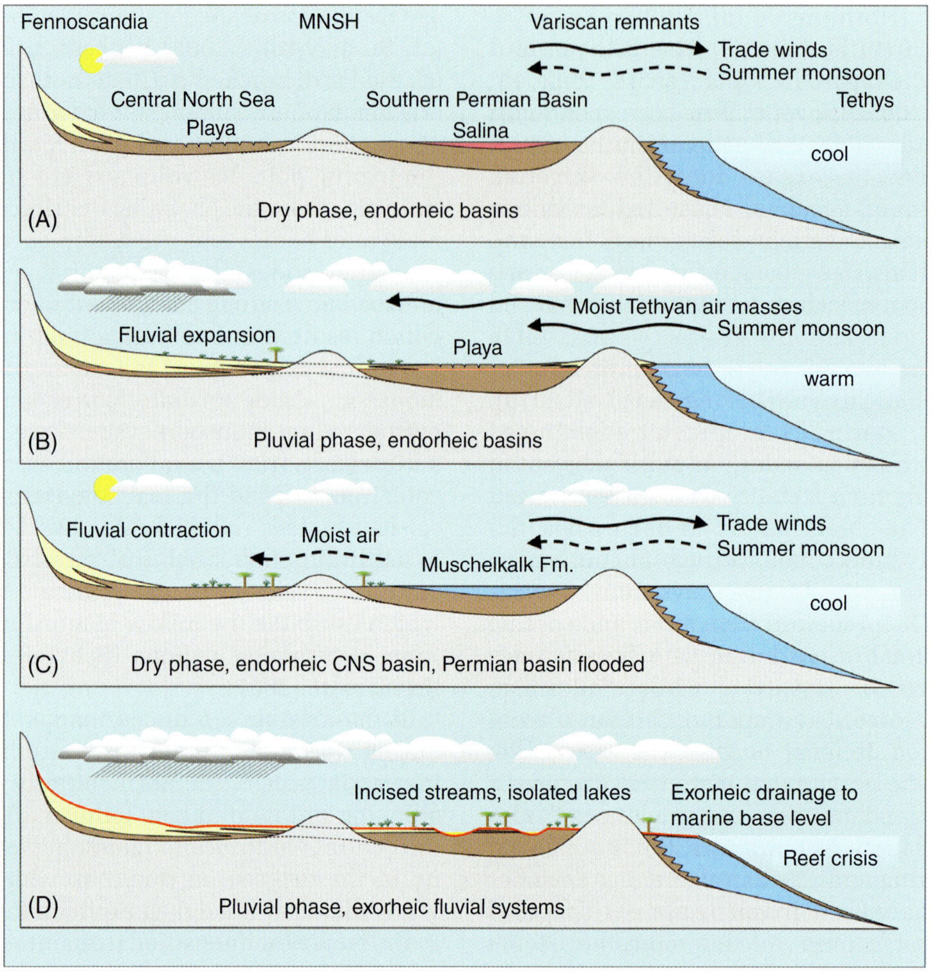

Fig. 20. End-member climatic scenarios. During conditions of general background aridity the central North Sea and Southern Permian Basin were characterised by widespread playa deposition and seasonal monsoon precipitation (A). During pluvial episodes (B) the summer monsoon may have intensified, resulting in the basinward expansion of Skagerrak Formation terminal fluvial systems. Weakened trade winds may have allowed the northward transport of moist Tethyan air masses, which helped maintain year-round soil moisture and intermittent fluvial transport. During the Middle Triassic vegetation cover and standing water were maintained through arid episodes (C) as a result of the flooding of the Southern Permian Basin, which elevated humidity and ground water tables in the adjacent central North Sea. During the major Ladinian and Carnian pluvial phases (D) large exorheic systems drained off Fennoscandia into the Tethys Sea. This drainage to an initially lower base level resulted in widespread fluvial incision and sediment by-pass in the Southern Permian Basin. In the central North Sea this pluvial event is apparently absent, although this is likely to be due to a combination of fluvial incision, drainage diversion away from the basin centre and a limited dataset biased towards structural highs which may have become areas of by-pass during this period of lowered base level. MNSH – mid-North Sea High, CNS – central North Sea.

of sand-prone fluvial facies in the North Sea and UK Atlantic margin is indicative of a persistent sediment supply from Greenland, Fennoscandia and the Scottish Highlands (McKie & Williams, 2009), which contrasts with the marked reduction in fluvial sand transport seen across southern Britain, offshore Ireland and the Southern Permian Basin (Fig. 21). This reduction occurred relatively rapidly through the Anisian, resulting in the deposition of widespread Keuper and Mercia Mudstone playa and limited clastic input to the Muschelkalk seaway. Such a reduction is unlikely to have been the product of slow denudation processes. Instead it appears that there was a general reduction in sediment yield in southern latitudes after the Early Triassic, whilst fluvial transport was maintained further north through much of the Triassic. Given the continuity of sandy, vegetated fluvial

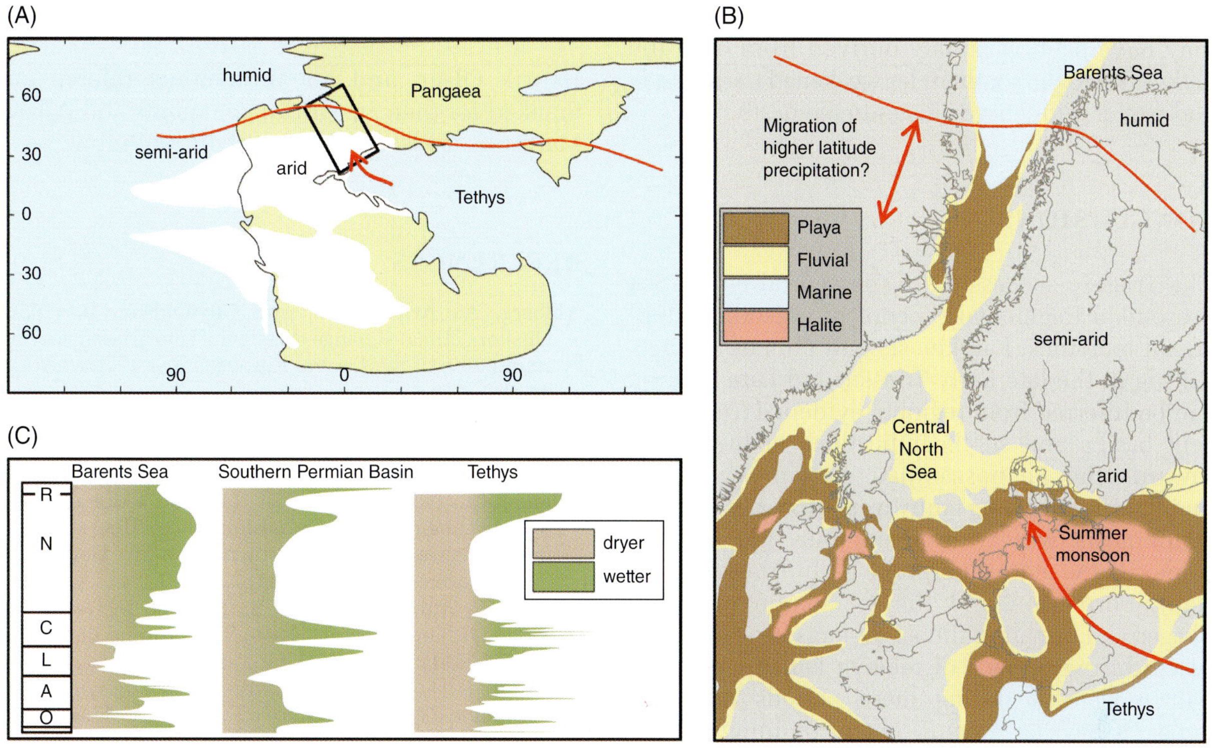

Fig. 21. Regional climatic setting of the central North Sea. (A) Late Triassic distribution of climatic belts across the northern hemisphere of Pangaea (with the region of little or no winter precipitation highlighted in white; after Sellwood & Valdes, 2006). The region from the Barents Sea to Tethys traverses a zone of southward increasing aridity. (B) Regional distribution of fluvial and playa facies during the Early Carnian (after McKie & Williams, 2009; Müller *et al.*, 2005). This facies distribution is also broadly representative of the Middle Triassic (excluding the period of Muschelkalk flooding) and illustrates that the central North Sea area remained part of a latitudinal region which sustained fluvial activity across the North Sea and Atlantic basins, whilst to the south the basins were increasingly playa dominated and evaporitic. During arid periods the Skagerrak Formation shale members record expansion of these playa facies into the central North Sea, but during pluvial episodes vegetated alluvial systems expanded off the basin-flanking catchments. This may have been driven by intensified monsoon activity, but could also reflect a contribution from the southward migration of higher latitude precipitation. However, southward migration is difficult to reconcile with increased runoff during warming (Fig. 18) when the reverse situation might be expected. In addition, the Barents Sea climate record at higher latitude (C) shows limited coherence with the records to the south to support either synchronous movement or expansion/contraction of climate zones. Curve sources as per Fig. 18, Barents Sea curve based on hygrophyte vs. xerophyte palynological data documented by Hochuli & Vigran (2010).

systems extending northwards from the central North Sea, into the northern North Sea and westward into the north Atlantic region (McKie & Williams, 2009) and the presence of generally wetter conditions through the boreal and polar regions (Golonka *et al.*, 1994; Ziegler *et al.*, 2003; Hochuli & Vigran, 2010) it could be argued that the central North Sea area was more influenced by the southward migration of mid-latitude precipitation (given that the fluvial catchments were not large enough to traverse climatic zones). However, fluvial expansion appears to have occurred independently of Boreal trends of wettening and drying (Fig. 21C), with the Ladinian expansion of the Joanne Sandstone Member occurring during pronounced Boreal drying. Such differences in climate trends, together with the similarity to the climate record to the south, would tend to favour a predominantly Tethyan origin for the precipitation which drove fluvial expansion. The development of pluvial conditions during warming episodes might also be expected to be characterised by a northward migration of mid-latitude precipitation rather than southward expansion. However, dry season runoff may have been episodically augmented by northern sources in the Anisian and Carnian when all regions appear to have been wetter. The fluvial systems remained terminal and flood-dominated throughout, suggesting dominance of seasonal monsoon precipitation, but with moisture supply during the dry winter season which may have been variously derived from both the Tethys Sea to the south under weakened trade winds and wetter mid-latitude zones to the north.

CONCLUSIONS

The Triassic succession in the central North Sea presents a fragmentary record of episodic extensional tectonics, halokinesis and climate change. Rifting in the late Early Triassic and Late Triassic can be inferred from regional evidence from adjacent basins and subtle indications of thickening towards the Fennoscandian margin where the main basin bounding faults were probably located. Halokinesis was widespread, but this affected primarily the more mud-prone Early Triassic Smith Bank Formation and the expansion of major Skagerrak Formation fluvial systems probably took place during periods of relative fault quiescence. Salt movement during these periods does not appear to have been significant in confining or guiding the fluvial systems, which probably had profiles which were elevated above the level of subdued salt movement and which supplied sufficient sediment to infill any slowly evolving topography. However, halokinesis was a significant contributor to the local preservation of stratigraphy through the conveying of successions below the erosion level of successive Jurassic unconformities. The apparently simple Skagerrak Formation succession of alternating sand-prone and mud-prone members was a product of fluctuating climatic conditions within the headwaters of the fluvial drainage. Fluvial expansion occurred during pluvial events, which increased runoff from the catchments and extended intermittent discharge streams into the dry basin. During periods of sustained drought the fluvial systems contracted as the region reverted to the background aridity that characterised much of the north-west European region throughout the Triassic. These pluvial events appear to have been driven by climatic warming, which intensified the Tethyan monsoon, enhanced evaporation and weakened the north-easterly trade winds which normally directed moisture laden air masses away from the region. In addition, precipitation from wetter, higher latitude climate zones may have episodically migrated southward, leading to a latitudinal belt of sandy, vegetated fluvial systems to the north of widespread Keuper and Mercia Mudstone playa.

ACKNOWLEDGEMENTS

Henrik Olsen and an anonymous referee are thanked for their insightful comments and Allard Martinius is also thanked for editorial input.

REFERENCES

Ahlberg, A., Arndorff, L. and **Guy-Ohlson, D.** (2002) Onshore climate change during the Late Triassic marine inundation of the Central European Basin. *Terra Nova*, **14**, 241–248.

Aigner, T. and **Bachmann, G.H.** (1992) Sequence stratigraphic framework of the German Triassic. *Sed. Geol.*, **80**, 115–135.

Allen, J.R.L. (1982) Studies in fluviatile sedimentation: bars, bar-complexes and sandstone sheets (low-sinuosity braided steams) in the Brownstones (L. Devonian), Welsh Borders. *Sed. Geol.*, **33**, 237–293.

Archer, S., Ward, S., Menad, S., Shahim, I., Grant, N., Sloan, H. and **Cole, A.** (2010) The Jasmine discovery, central North Sea, UKCS. In: *Petroleum Geology: From Mature Basins to New Frontiers – Proceedings of the 7th Petroleum Geology Conference* (Eds B.A. Vining and S.C. Pickering), pp 225–243. Geol. Soc. London.

Bachmann, G.H., Geluk, M.C., Warrington, G., Becker-Roman, A., Beutler, G., Hagdorn, H., Hounslow, M.W., Nitsch, E., Röhling, H.-G., Simon, T. and **Szulc, A.** (2010) Triassic. In: *Petroleum Geological Atlas of the Southern Permian Basin Area* (Eds J.C. Doornenbal and A.G. Stevenson), pp 149–173. EAGE Publications b.v., Houten.

Berra, F., Jadoul, F. and **Anelli, A.** (2010) Environmental control on the end of the Dolomia Principale/ Hauptdolomit depositional system in the central Alps: Coupling sea-level and climate changes. *Palaeogeogr. Palaeoclimatol. Palaeoecol.*, **290**, 138–150.

Bertelsen, F. (1980) Lithostratigraphy and depositional history of the Danish Triassic. *Dan. Geol. Unders.*, Serie B, 4, 59 pp.

Blum, M.D. and **Törnqvist, T.E.** (2000) Fluvial responses to climate and sea-level change: a review and look forward. *Sedimentology*, **47**, 2–48.

Borkhataria, R., Aigner, T. and **Pipping, K.J.C.P.** (2006) An unusual, muddy, epeiric carbonate reservoir: The Lower Muschelkalk (Middle Triassic) of the Netherlands. *AAPG Bull.*, **90**, 61–89.

Bourquin, S., Guillocheau, F. and **Péron, S.** (2009) Braided rivers within an arid alluvial plain (example from Lower Triassic, western German Basin): recognition criteria and expression of stratigraphic cycles. *Sedimentology*, **56**, 2235–2264.

Bridge, J.S. and **Tye, R.S.** (2000) Interpreting the dimensions of ancient fluvial channel bars, channels and channel belts from wireline-logs and cores. *AAPG Bull.*, **84**, 1205–1228.

Brook, G.A., Marais, E., Srivastava, P. and **Jordan, T.** (2007) Timing of lake-level changes in Etosha Pan, Namibia, since the Middle Holocene from OSL ages of relict shorelines in the Okondeka region. *Quatern. Int.*, **175**, 29–40.

Brookfield, M.E. (2004) The enigma of fine-grained alluvial basin fills: the Permo-Triassic (Cumbrian Coastal and Sherwood Sandstone Groups) of the Solway Basin, NW England and SW Scotland). *Int. J. Earth Sci.*, **93**, 282–296.

Brookfield, M.E. (2008) Palaeoenvironments and palaeotectonics of the arid to hyperarid intracontinental latest Permian- Late Triassic Solway basin (U.K.). *Sed. Geol.*, **210**, 27–47.

Cameron, T.D.J., Crosby, A., Balson, P.S., Jeffery, D.H., Lott, G.K., Bulat, J. and **Harrison, D.H.** (1992) *United Kingdom Offshore Regional Reports: The Geology of the Southern North Sea.* HMSO for British Geological Survey, 152 pp.

Cartwright, J., Stewart, S. and **Clark, J.** (2001) Salt dissolution and salt-related deformation of the Forth Approaches Basin, UK North Sea. *Mar. Petrol. Geol.*, **18**, 757–778.

Clark, J., Cartwright, J. and **Stewart, S.** (1999) Mesozoic dissolution tectonics on the West Central Shelf, UK central North Sea. *Mar. Petrol. Geol.*, **16**, 283–300.

Clausen, O.R. and **Pedersen, P.K.** (1999) Late Triassic structural evolution of the southern margin of the Ringkøbing-Fyn High, Denmark. *Mar. Petrol. Geol.*, **16**, 653–665.

Colter, V.S. and **Barr, K.W.** (1975) Recent developments in the geology of the Irish Sea and Cheshire Basins. In: *Petroleum and the Continental Shelf of North-West Europe, 1, Geology* (Ed. A.W. Woodland), pp 61–75. Applied Science Publishing, London.

Coward, M.P. (1995) Structural and tectonic setting of the Permo-Triassic basins of north-west Europe. In: *Permian and Triassic Rifting in North-west Europe* (Ed. S.A.R. Boldy), *Geol. Soc. London Spec. Publ.*, **91**, 7–39.

Coward, M.P., Dewey, J.F., Hempton, M. and **Holroyd, J.** (2003) Tectonic evolution. In: *The Millennium Atlas: Petroleum Geology of the Central and Northern North Sea* (Eds D. Evans, C. Graham, A. Armour and P. Bathurst), pp 17–33. Geological Society, London.

Davidson, S.K. and **North, C.P.** (2009) Geomorphological regional curves for prediction of drainage area and screening modern analogues for rivers in the rock record. *J. Sed. Res.*, **79**, 773–792.

De Jong, M., Smith, D., Nio, S.D. and **Hardy, N.** (2006) Subsurface correlation of the Triassic of the UK southern Central Graben: new look at an old problem. *First Break*, **24**, 103–109.

Erratt, D., Thomas, G.M. and **Wall, G.R.T.** (1999) The evolution of the central North Sea Rift, In: *Petroleum Geology of North-west Europe: Proceedings of the 5th Conference* (Eds A.J. Fleet and S.A.R. Boldy), pp 63–82. Geological Society, London.

Evans, D.J., Rees, J.G. and **Holloway, S.** (1993) The Permian to Jurassic stratigraphy and structural evolution of the central Cheshire Basin. *J. Geol. Soc. London*, **150**, 857–870.

Faerseth, R.B. (1996) Interaction of Permo-Triassic and Jurassic extensional fault-blocks during the development of the northern North Sea. *J. Geol. Soc. London*, **153**, 931–944.

Feist-Burkhardt, S., Götz, A.E., Szulc, J., Borkhataria, R., Geluk, M., Haas, J., Hornung, J., Jordan, P., Kempf, O., Michalík, J., Nawrocki, J., Reinhardt, L., Ricken, W., Röhling, H.-G., Rüffer, T., Török, Á. and **Zühlke, R.** (2008) Triassic. In: *The Geology of Central Europe, Volume 2 – Mesozoic and Cenozoic* (Ed. T. McCann), pp 749–822. Geological Society, London.

Fielding, C.R. (2006) Upper flow regime sheets, lenses and scour fills: extending the range of architectural elements for fluvial sediment bodies. *Sed. Geol.*, **190**, 227–240.

Fisher, J.A., Nichols, G.J. and **Waltham, D.A.** (2007) Unconfined flow deposits in the distal sectors of fluvial distributary systems: examples from the Luna and Huesca Systems, northern Spain. *Sed. Geol.*, **195**, 55–73.

Fisher, J.A., Krapf, C.B.E., Lang, S.C., Nichols, G.J. and **Payenberg, T.D.** (2008) Sedimentology and architecture of the Douglas Creek terminal splay, Lake Eyre, central Australia. *Sedimentology* **55**, 1915–1930.

Fryberger, S.G., Al-Sari, A.M. and **Clisham, T.J.** (1983) Eolian dune, interdune, sand sheet and siliciclastic sabkha sediments of an offshore prograding sand sea, Dhahran area, Saudi Arabia. *AAPG Bull.*, **67**, 80–312.

Fryberger, S.G. and **Schenk, C.J.** (1988) Pin stripe lamination: a distinctive feature of modern and ancient eolian sediments. *Sed. Geol.*, **55**, 1–15.

Furin, S., Preto, N., Rigo, M., Roghi, G., Gianolla, P., Crowley, J.L. and **Bowring, S.A.** (2006) High-precision U-Pb zircon age from the Triassic of Italy: Implications for the Triassic time scale and the Carnian origin of calcareous nannoplankton and dinosaurs. *Geology*, **34**, 1009–1012.

Galfetti, T., Hochuli, P.A., Brayard, A., Bucher, H., Weissert, H. and **Vigran, J.** (2007a) Smithian-Spathian boundary event: Evidence for global climatic change in the wake of the end-Permian biotic crisis. *Geology*, **35**, 291–294.

Galfetti, T., Bucher, H., Brayard, A., Hochuli, P.A., Weissert, H., Guodun, K., Atudorei, V. and **Guex, J.** (2007b) Late Early Triassic climate change: insights from carbonate carbon isotopes, sedimentary evolution and ammonoid paleobiogeography. *Palaeogeogr. Palaeoclimatol. Palaeoecol.*, **243**, 394–411.

Geluk, M.C. (2005) *Stratigraphy and tectonics of Permo-Triassic basins in the Netherlands and surrounding areas.* Unpublished Ph.D. thesis, University of Utrecht. 171 pp.

Geluk, M.C. (2007) Triassic. In: *Geology of the Netherlands* (Eds Th.E. Wong, D.A.J. Batjes and J. de Jager), pp. 85–106. Royal Netherlands Academy of Arts and Sciences.

Gianolla, P., de Zanche, V. and **Mietto, P.** (1998) Triassic sequence stratigraphy in the southern Alps (northern Italy): definition of sequences and basin evolution. In: *Mesozoic–Cenozoic Sequence Stratigraphy of European Basins* (Eds P.C. de Gracianscky, J. Hardenbol, T. Jacquin and P.R. Vail), *SEPM Spec. Publ.*, **60**, 723–751.

Goddéris, Y., Donnadieu, Y, de Vargas, C., Pierrehumbert, R.T., Dromart, G. and **van de Schootbrugge, B.** (2008) Causal or casual link between the rise of nannoplankton calcification and a tectonically-driven massive decrease in Late Triassic atmospheric CO_2? *Earth Planet. Sci. Lett.*, **267**, 247–255.

Goldsmith, P.J., **Rich, B.** and **Standring, J.** (1995) Triassic correlation and stratigraphy in the South Central Graben, UK North Sea. In: *Permian and Triassic Rifting in North-west Europe* (Ed. S.A.R. Boldy), *Geol. Soc. London Spec. Publ.*, **91**, 123–143.

Goldsmith, P.J., **Hudson, G.** and **Van Veen, P.** (2003) Triassic. In: *The Millennium Atlas: Petroleum Geology of the Central and Northern North Sea* (Eds D. Evans, C. Graham, A. Armour and P. Bathurst), pp 105–127. *The Geological Society of London*, London, UK.

Golonka, J., **Ross, M.I.** and **Scotese, C.R.** (1994) Phanerozoic paleogeographic and paleoclimatic modeling maps. In: *Pangea: Global Environments and Resources* (Eds A.F. Embry, B. Beauchamp and D.J. Glass), *Can. Soc. Petrol. Geol. Mem.*, **17**, 1–47.

Goodall, T.M., **North, C.P.** and **Glennie, K.W.** (2000) Surface and sub-surface sedimentary structures produced by salt crusts. *Sedimentology*, **47**, 99–118.

Hochuli, P.A. and **Vigran, J.O.** (2010) Climate variations in the Boreal Triassic inferred from palynological records from the Barents Sea. *Palaeogeogr. Palaeoclimatol. Palaeoecol.*, **290**, 20–42.

Hodgson, N.A., **Farnsworth, J.** and **Fraser, A.J.** (1992) Salt-related tectonics, sedimentation and hydrocarbon plays in the Central Graben, North Sea, UKCS. In: *Exploration Britain: Geological Insights for the Next Decade* (Ed R. F. P. Hardman), *Geol. Soc. London Spec. Publ.*, **67**, 31–63.

Hornung, J. and **Aigner, T.** (2002) Reservoir architecture in a terminal alluvial plain: an outcrop analogue study (upper Triassic, southern Germany): Part II, Cyclicity, controls and models. *J. Petrol. Geol.*, **25**, 151–178.

Hornung, T. and **Brandner, R.** (2005) Biostratigraphy of the Reingraben Turnover (Hallstatt Facies Belt): local black shale events controlled by the regional tectonics, climatic change and plate tectonics. *Facies*, **51**, 460–479.

Hornung, T., **Brandner, R.**, **Krystyn, L.**, **Joachimski, M.M.** and **Keim, L.** (2007) Multistratigraphic constraints on the NW Tethyan "Carnian Crisis". In: *The Global Triassic* (Eds S.G. Lucas and J.A. Spielmann), *New Mexico Museum of Natural History and Science Bulletin*, **41**, 59–67.

Hounslow, M.W. and **McIntosh, G.** (2003) Magnetostratigraphy of the Sherwood Sandstone Group (lower and middle Triassic), south Devon, U.K.: detailed correlation of the marine and non-marine Anisian. *Palaeogeogr. Palaeoclimatol. Palaeoecol.*, **193**, 325–348.

Hounslow, M.W. and **Ruffell, A.H.** (2006) Triassic: seasonal rivers, dusty deserts and saline lakes. In: *Geology of England and Wales* (Eds P.J. Brenchley and P.F. Rawson), pp 295–324. *Geol. Soc. London*.

Hubbard, N.L.B. and **Boulter, M.C.** (2000) Phytogeography and paleoecology in Western Europe and Eastern Greenland near the Triassic-Jurassic boundary. *Palaios*, **15**, 120–31.

Jeans, C.V. (2006) Clay mineralogy of the Permo-Triassic strata of the British Isles: onshore and offshore. *Clay Mineral.*, **41**, 309–354.

Kelber, K.-P. and **van Konijnenburg-van Cittert, J.H.A.** (1998) *Equisetites arenaceus* from the upper Triassic of Germany with evidence for reproductive strategies. *Rev. Palaeobot. Palynol.*, **100**, 1–26.

Kidder, D. and **Worsley, T.R.** (2004) Causes and consequences of extreme Permo-Triassic warming to globally equable climate and relation to the Permo-Triassic extinction and recovery. *Palaeogeogr. Palaeoclimatol. Palaeoecol.*, **203**, 207–237.

Kiessling, W. (2010) Reef expansion during the Triassic: spread of photosymbiosis balancing climatic cooling. *Palaeogeogr. Palaeoclimatol. Palaeoecol.*, **290**, 11–19.

Kocurek, G. and **Havholm, K.G.** (1993) Eolian sequence stratigraphy – a conceptual framework. In: Siciliclastic Sequence Stratigraphy (Eds P. Weimer and H.W. Posamentier). *AAPG Mem.*, **58**, 393–409.

Köppen, A. and **Carter, A.** (2000) Constraints on provenance of the central European Triassic using detrital zircon fission track data. *Palaeogeogr. Palaeoclimatol. Palaeoecol.*, **161**, 193–204.

Korte, C., **Kozur, H.W.**, **Bruckschen, P.** and **Veizer, J.** (2003) Strontium isotope evolution of Late Permian and Triassic seawater. *Geochim. Cosmochim. Acta*, **67**, 47–62.

Korte, C., **Kozur, H.W.** and **Veizer, J.** (2005) $\delta^{13}C$ and $\delta^{18}O$ values of Triassic brachiopods and carbonate rocks as proxies for coeval seawater and palaeo-temperature. *Palaeogeogr. Palaeoclimatol. Palaeoecol.*, **226**, 287–306.

Kozur, H.W. and **Bachmann, G.H.** (2010) The Middle Carnian wet intermezzo of the Stuttgart Formation (Schilfsandstein), Germanic basin. *Palaeogeogr. Palaeoclimatol. Palaeoecol.*, **290**, 107–119.

Kürschner, W.M. and **Herngreen, G.F.W.** (2010) Triassic palynology of central and north-western Europe: a review of palynofloral diversity patterns and biostratigraphic subdivisions. In: *The Triassic Timescale* (Ed. S.G. Lucas), *Geol. Soc. London Spec. Publ.*, **334**, 263–283.

Kustatscher, E., **van Konijnenburg-van Cittert, J.H.A.** and **Roghi, G.** (2010) Macrofloras and palynomorphs as possible proxies for palaeoclimatic and palaeoecological studies: a case study from the Pelsonian (Middle Triassic) of Kühwiesenkopf/Monte Prà della Vacca (Olang Dolomites, N-Italy). *Palaeogeogr. Palaeoclimatol. Palaeoecol.*, **290**, 71–80.

Kutzbach, J.E. and **Gallimore, R.G.** (1989) Pangean climates: megamonsoons of the megacontinent. *J. Geophys. Res.*, **94**, 3341–3357.

Lang, S.C., **Payenberg, T.H.D.**, **Reilly, M.R.W.**, **Hicks, T.**, **Benson, J.** and **Kassan, J.** (2004) Modern analogues for dryland sandy fluvial-lacustrine deltas and terminal splay reservoirs. *Aust. Petrol. Prod. Explor. Assoc. J.*, **44**, 329–356.

Legarreta, L. and **Uliana, M.A.** (1998) Anatomy of hinterland depositional sequences: upper Cretaceous fluvial strata, Neuquen Basin, west-central Argentina. In: *Relative Role of Eustasy, Climate and Tectonism in Continental Rocks* (Eds K.W. Shanley and P.J. McCabe), *SEPM Spec. Publ*, **59**, 83–92.

Lelue, S. and **Hartley, A.J.** (2010) Controls on the stratigraphic development of the Triassic Fundy Basin, Nova Scotia: implications for the tectonostratigraphic evolution of Triassic Atlantic rift basins. *J. Geol. Soc. London*, **167**, 437–454.

Lervik, K.-S. (2006) Triassic lithostratigraphy of the northern North Sea basin. *Nor. J. Geol.*, **86**, 93–116.

Lindström, S., Vosgerau, H., Piasecki, S., Nielsen, L.H., Dybkjær, K. and **Erlström, M.** (2009) Ladinian palynofloras in the Norwegian–Danish Basin: a regional marker reflecting a climate change. *Bull. Geol. Surv. Denmark and Greenland.*, **17**, 21–24.

Mange-Rajetzky, M. (1995) Subdivision and correlation of monotonous sandstone sequences using high-resolution heavy mineral analysis, a case study: the Triassic of the Central Graben. In: *Non-biostratigraphical Methods of Dating and Correlation* (Eds R.E. Dunay and E.A. Hailwood). *Geol. Soc. London Spec. Publ.*, **89**, 23–30.

Manspeizer, W. (1982) Triassic-Liassic Basins and climate of the Atlantic passive margins. *Geol. Rundsch.*, **71**, 895–917.

Martinsen, O.J., Ryseth, A., Helland-Hansen, W., Flesche, H., Torkildsen, G. and **Idil, S.** (1999) Stratigraphic base level and fluvial architecture: Ericson Sandstone (Campanian), Rock Springs Uplift, SW Wyoming, USA. *Sedimentology*, **46**, 235–259.

Matthews, W.J., Hampson, G.J., Trudgill, B.D. and **Underhill, J.R.** (2007) Controls on fluvio lacustrine reservoir distribution and architecture in passive salt-diapir provinces: insights from outcrop analogs. *AAPG Bull.*, **91**, 1367–1403.

Mayewski, P.A., Rohling, E.E., Stager, J.C., Karlén, W., Maasch, K.A., Meeker, L.D., Meyerson, E.A., Gasse, F., van Kreveld, S., Holmgren, K., Lee-Thorp, J., Rosqvist, G., Rack, F., Staubwasser, M., Schneider, R.R. and **Steig, E.J.** (2004) Holocene climate variability. *Quatern. Res.*, **62**, 243–255.

McKie, T. (2011) Architecture and behavior of dryland fluvial reservoirs, Triassic Skagerrak Formation, central North Sea. In: *From River to Rock Record: The Preservation of Fluvial Sediments and their Subsequent Interpretation* (Eds S.K. Davidson, C.P. North and S. Leleu), *SEPM Spec. Publ.* **97**, 189–214.

McKie, T. and **Audretsch, P.** (2005) Depositional and structural controls on Triassic reservoir performance in the Heron Cluster, ETAP, central North Sea. In: *Petroleum Geology of North-west Europe and Global Perspectives – Proceedings of the Sixth Petroleum Geology Conference* (Eds A.G. Dore and B.A. Vining), Geol. Soc. London, pp 285–297.

McKie, T. and **Garden, R.** (1996) Hierarchical stratigraphic cycles in the non-marine Clair Group (Devonian) UKCS. In: *High Resolution Sequence Stratigraphy: Innovations and Applications* (Eds J.A. Howell and J.F. Aitken). *Geol. Soc. London Spec. Publ.* **104**, 139–157.

McKie, T. and **Shannon, P.M.** (2011) Comment on "The Permian–Triassic transition and the onset of Mesozoic sedimentation at the north-western peri-Tethyan domain scale: Palaeogeographic maps and geodynamic implications" by S. Bourquin, A. Bercovici, J. López-Gómez, J. B. Diez, J. Broutin, A. Ronchi, M. Durand, A. Arché, B. Linol and F. Amour. *Palaeogeogr. Palaeoclimatol. Palaeoecol.*, **299**, 265–280.

McKie, T. and **Williams, B.** (2009) Triassic palaeogeography and fluvial dispersal across the north-west European Basins. *Geol. J.*, **44**, 711–741.

McKie, T., Jolley, S. and **Kristensen, M.B.** (2010) Stratigraphic compartmentalisation in dryland fluvial reservoirs- examples from the Triassic Skagerrak Formation, UKCS. In: *Reservoir Compartmentalization* (Eds S.J. Jolley, Q.J. Fisher, A.B. Ainsworth, P. Vrolijk and S. Delisle), *Geol. Soc. London Spec. Publ.*, **347**, 165–198.

Michelsen, O. and **Clausen, O.R.** (2002) Detailed stratigraphic subdivision and regional correlation of the southern Danish Triassic succession. *Mar. Pet. Geol.*, **19**, 563–587.

Milroy, P.G. and **Wright, V.P.A.** (2000) Highstand oolitic sequence and associated facies from a Late Triassic lake basin, south-west England. *Sedimentology*, **47**, 187–209.

Müller, R., Nystuen, J.P., Eide, F. and **Lie, H.** (2005) Late Permian to Triassic basin infill history and palaeogeography of the Mid-Norwegian shelf – East Greenland region. In: *Onshore-Offshore Relationships on the North Atlantic Margin* (Eds B.T.G Wandås, J.P. Nystuen, E. Eide and F. Gradstein), *Nor. Pet. Soc. Spec. Publ.*, **12**, 165–189.

Muto, T. and **Steel, R.J.** (1997) Principles of regression and transgression: the nature of the interplay between accommodation and sediment supply. *J. Sed. Res.*, **67**, 994–1000.

Mutti, M. and **Weissert, H.** (1995) Triassic monsoonal climate and its signature in Ladinian-Carnian carbonate platforms (southern Alps, Italy). *J. Sed. Res.*, **B65**, 357–367.

Newell, A.J., Tverdokhlebov, V.P. and **Benton, M.J.** (1999) Interplay of tectonics and climate on a transverse fluvial system, upper Permian, Southern Uralian Foreland Basin, Russia. *Sed. Geol.*, **127**, 11–29.

Nichols, G.J. (2007) Features of fluvial systems in desiccating endorheic basins. In: *Sedimentary processes, environments and basins – a tribute to Peter Friend* (Eds G.J. Nichols, E.A. Williams and C. Paola). *Int. Assoc. Sedimentol. Spec. Publ.*, **38**, 567–587.

North, C.P. and **Warwick, G.L.** (2007) Fluvial fans: myths, misconceptions and the end of the terminal-fan model. *J. Sed. Res.*, **77**, 693–701.

Nützel, A., Joachimski, M. and **López Correa, M.** (2010) Seasonal climatic fluctuations in the Late Triassic tropics- high-resolution oxygen isotope records from aragonitic bivalve shells (Cassian Formation, Northern Italy). *Palaeogeogr. Palaeoclimatol. Palaeoecol.*, **285**, 194–204.

Olsen, P.E. and **Kent, D.V.** (2000) High-resolution Early Mesozoic Pangean climatic transect in lacustrine environments. *Zbl. Geol. Pälontol. Teil*, I (H11/12), 1475–1495.

Parrish, J.T. (1993) Climate of the supercontinent Pangea. *J. Geol.*, **101**, 215–33.

Paul, J., Wemmer, K. and **Ahrendt, H.** (2008) Provenance of siliciclastic sediments (Permian to Jurassic) in the Central European Basin. *Z. Deut. Ges. Geowiss.*, **159**, 641–650.

Paul, J., Wemmer, K. and **Wetzel, F.** (2009) Keuper (Late Triassic) sediments in Germany- indicators of rapid uplift of Caledonian rocks in southern Norway. *Nor. J. Geol.*, **86**, 193–202.

Payne, J.L., Lehrmann, D.J., Wei, J., Orchard, M.J., Schrag, D.P. and **Knoll, A.H.** (2004) Large perturbations of the carbon cycle during recovery from the end-Permian extinction. *Science*, **305**, 506–509.

Peacock, D.C.P. (2004) The post-Variscan development of the British Isles within a regional transfer zone influenced by orogenesis. *J. Struct. Geol.*, **26**, 2225–2231.

Penge, J., Munns, J.W., Taylor, B. and Windle, T.M.F. (1999) Rift-raft tectonics: examples of gravitational tectonics from the Zechstein basins of north-west Europe. In: *Petroleum Geology of North-west Europe: Proceedings of the 5th Conference* (Eds A. Fleet and S.A.R. Boldy), pp 201–213. Geological Society, London.

Perlmutter, M.A. and **Matthews, M.D.** (1990) Global cyclostratigraphy – a model. In: *Quantitative Dynamic Stratigraphy* (Ed. T. Cross), pp 233–60. Prentice Hall, New Jersey.

Péron, S., Bourquin, S., Fluteau, F. and **Guillocheau, F.** (2005) Paleoenvironment reconstructions and climate simulations of the Early Triassic: impact of the water and sediment supply on the preservation of fluvial system. *Geodin. Acta.*, **18**, 431–446.

Porter, R.J. and **Gallois, R.W.** (2008) Identifying fluvio–lacustrine intervals in thick playa-lake successions: an integrated sedimentology and ichnology of arenaceous members in the Mid–Late Triassic Mercia Mudstone Group of south-west England, UK. *Palaeogeogr. Palaeoclimatol. Palaeoecol.*, **270**, 381–398.

Prell, W.L. and **Kutzbach, J.E.** (1987) Monsoon variability over the past 150,000 years. *J. Geophys. Res.*, **92**, 8411–8425.

Preto, N., Kustatscher, E. and **Wignall, P.B.** (2010) Triassic climates – state of the art and perspectives. *Palaeogeogr. Palaeoclimatol. Palaeoecol.*, **290**, 1–10.

Prochnow, S.J., Atchley, S.C., Boucher, T.E., Nordt, L.C. and **Hudec, M.R.** (2006) The influence of salt withdrawal subsidence on palaeosol maturity and cyclic fluvial deposition in the upper Triassic Chinle Formation: Castle Valley, Utah. *Sedimentology*, **53**, 1319–1345.

Radies, D., Stollhofen, H., Hollmann, G. and **Kukla, P.** (2005) Synsedimentary faults and amalgamated unconformities: insights from 3D-seismic and core analysis of the Lower Triassic middle Buntsandstein, Ems Trough, north-western Germany. *Int. J. Earth Sci.*, **94**, 863–875.

Reinhardt, L. and **Ricken, W.** (2000) The stratigraphic and geochemical record of playa cycles: monitoring a Pangaean monsoon-like system (Triassic, middle Keuper, S. Germany). *Palaeogeogr. Palaeoclimatol. Palaeoecol.*, **161**, 205–227.

Ruffell, A., McKinley, J.M. and **Worden, R.H.** (2002) Comparison of clay mineral stratigraphy to other proxy palaeoclimate indicators in the Mesozoic of NW Europe. *Phil. Trans. R. Soc. London. A: Math. Phys. Eng. Sci.* **360**, 675–693.

Ruffell, A. and **Shelton, R.** (1999) The control of sedimentary facies by climate during phases of crustal extension: examples from the Triassic of onshore and offshore England and Northern Ireland. *J. Geol. Soc. London*, **156**, 779–789.

Sadler, P.M. and **Strauss, D.J.** (1990) Estimation of completeness of stratigraphical sections using empirical data and theoretical models. *J. Geol. Soc. London*, **147**, 471–485.

Schefuβ, E., Schouten, S. and **Schneider, R.R.** (2005) Climatic controls on central African hydrology during the past 20,000 years. *Nature*, **437**, 1003–1006.

Seidler, L., Steel, R.J., Stemmerik, L. and **Surlyk, F.** (2004) North Atlantic marine rifting in the Early Triassic: new evidence from east Greenland. *J. Geol. Soc. London*, **161**, 583–592.

Sellwood, B.W. and **Valdes, P.J.** (2006) Mesozoic climates: general circulation models and the rock record. *Sed. Geol.*, **190**, 269–287.

Sellwood, B.W. and **Valdes, P.J.** (2007) Mesozoic climate. In: *Deep-time perspectives on climate change: marrying the signal from computer models and biological proxies* (Eds M. Williams, A.M. Haywood, F.J. Gregory and D.N. Schmidt), *Micropal. Soc. Spec. Publ.*, TMS002, pp 201–224. Geological Society, London.

Shanley, K.W. and **McCabe, P.J.** (1994) Perspectives on the sequence stratigraphy of continental strata: report of a working group at the 1991 NUNA Conference on high resolution sequence stratigraphy. *AAPG Bull.*, **78**, 544–568.

Shukla, U.K., Bachmann, G.H. and **Singh, I.B.** (2010) Facies architecture of the Stuttgart Formation (Schilfsandstein, upper Triassic), central Germany and its comparison with modern Ganga system, India. *Palaeogeogr. Palaeoclimatol. Palaeoecol.*, **297**, 110–128.

Simms, M.J. and **Ruffell, A.H.** (1989) Synchroneity of climatic change in the Late Triassic. *Geology*, **17**, 265–268.

Simms, M.J. and **Ruffell, A.H.** (1990) Climatic and biotic change in the Late Triassic. *J. Geol. Soc. London*, **147**, 321–327.

Smith, J.J. and **Hasiotis, S.T.** (2008) Traces and burrowing behaviors of the cicada nymph *Cicadetta calliope*: neoichnology and paleoecological significance of extant soil-dwelling insects. *Palaios*, **23**, 503–513.

Smith, J.J., Hasiotis, S.T., Kraus, M.J. and **Woddy, D.T.** (2008) *Naktodemasis bowni*: new ichnogenus and ichnospecies for adhesive meniscate burrows (amb) and paleoenvironmental implications, Paleogene Willwood Formation, Bighorn Basin, Wyoming. *J. Paleont.*, **82**, 267–278.

Smith, R.I., Hodgson, N. and **Fulton, M.** (1993) Salt control on Triassic reservoir distribution, UKCS central North Sea. In: *Petroleum Geology of North-west Europe: Proceedings of the 4th Conference* (Ed J.R. Parker), pp 547–557. Geological Society, London.

Snyder, C.T. (1962) A hydrologic classification of valleys in the Great Basin, western United States. *Int. Assoc. Sci. Hydrol. Bull.*, **7**, 53–59.

Steel, R. and **Ryseth, A.** (1990) The Triassic-Early Jurassic succession in the northern North Sea: megasequence stratigraphy and intra-Triassic tectonics. In: *Tectonic Events Responsible for Britain's Oil and Gas Reserves* (Eds R.F.P Hardman and J. Brookes), *Geol. Soc. London Spec. Publ.*, **55**, 139–168.

Stefani, M., Furin, S. and **Gianolla, P.** (2010) The changing climate framework and depositional dynamics of Triassic carbonate platforms from the Dolomites. *Palaeogeogr. Palaeoclimatol. Palaeoecol.*, **290**, 43–57.

Stewart, S.A. and **Clark, J.A.** (1999) Impact of salt on the structure of the central North Sea hydrocarbon fairways, In: *Petroleum Geology of North-west Europe: Proceedings of the 5th Conference* (Eds A.J. Fleet and S.A.R. Boldy), pp 179–200. Geological Society, London.

Stolfová, K. and **Shannon, P.M.** (2009) Permo-Triassic development from Ireland to Norway: basin architecture and regional controls. *Geol. J.*, **44**, 652–676.

Su, H. and **Neelin, J.D.** (2005) Dynamical mechanisms for African monsoon changes during the mid-Holocene. *J. Geophys. Res.*, **110**, D19105.

Surlyk, F. (1990) Timing, style and sedimentary evolution of Late Palaeozoic-Mesozoic extensional basins of East Greenland. In: *Tectonic Events Responsible for Britain's Oil and Gas Reserves* (Eds R.F.P. Hardman and J. Brooks), *Geol. Soc. London Spec. Publ.*, **55**, 107–125.

Szurlies, M. (2007) Latest Permian to Middle Triassic cyclo-magnetostratigraphy from the Central European Basin, Germany: implications for the geomagnetic polarity timescale. *Earth Planet. Sci. Lett.*, **261**, 602–619.

Tomasso, M., Underhill, J.R., Hodgkinson, R.A. and **Young, M.J.** (2002) Structural styles and depositional architecture in the Triassic of the Ninian and Alwyn North fields: implications for basin development and prospectivity in the northern North Sea. *Mar. Pet. Geol.*, **25**, 588–605.

Tooth, S. (2005) Splay formation along the lower reaches of ephemeral rivers on the Northern Plains of arid Central Australia. *J. Sed. Res.*, **75**, 636–649.

Turner, P., Heywood, M.L. and **Chandler, P.** (1996) Magnetostratigraphy and depositional sequence analysis of Triassic fluvial sediments: UK central North Sea. *Cuad. Geol. Ibérica*, **21**, 117–147.

Underhill, J.R. and **Partington, M.A.** (1993) Jurassic thermal doming and deflation in the North Sea: implications of the sequence stratigraphical evidence. In: *Petroleum Geology of North-west Europe, Proceedings of the 4th Conference* (Ed. J.R. Parker), pp 337–346. Geological Society, London.

Van der Zwan, C.J. and **Spaak, P.** (1992) Lower to middle Triassic sequence stratigraphy and climatology of the Netherlands, a model. *Palaeogeogr. Palaeoclimatol. Palaeoecol.*, **91**, 277–290.

Van Wees, J.D., Stephenson, R.A., Ziegler, P.A., Bayer, U., McCann, T., Dadlez, R., Gaupp, R., Narkiewicz, M., Bitzer, F. and **Scheck, M.** (2000) On the origin of the Southern Permian Basin, Central Europe. *Mar. Pet. Geol.*, **17**, 43–59.

Visscher, H., van Houte, M., Brugman, W.A. and **Poort, R.J.** (1994) Rejection of a Carnian (Late Triassic) "pluvial event" in Europe. *Rev. Palaeobot. Palynol.*, **83**, 217–226.

Vollmer, T., Werner, R., Weber, M., Tougiannidis, N., Röhling, H.-G. and **Hambach, U.** (2008) Orbital control on Upper Triassic playa cycles of the Steinmergel-Keuper (Norian): a new concept for ancient playa cycles. *Palaeogeogr. Palaeoclimatol. Palaeoecol.*, **267**, 1–16.

Wakefield, L.L., Droste, H., Giles, M.R. and **Janssen. R.** (1993) Late Jurassic plays along the western margin of the Central Graben. In: *Petroleum Geology of North-west Europe: Proceedings of the 4th Conference* (Ed. J.R. Parker), pp 459–468. Geological Society, London.

Wignall, P.B. and **Twitchett, R.J.** (2002) Permian–Triassic sedimentology of Jameson Land, East Greenland: incised submarine channels in an anoxic basin. *J. Geol. Soc. London*, **159**, 691–703.

Zanella, E. and **Coward, M.P.** (2003) Structural Framework. In: *The Millenium Atlas: Petroleum Geology of the Central and Northern North Sea* (Eds D. Evans, C. Graham, A. Armour and P. Bathurst), pp 45–59. Geological Society, London.

Ziegler, A.M., Eshel, G., McAllister Rees, P., Rothfus, T.A., Rowley, D.B. and **Sunderlin, D.** (2003) Tracing the tropics across land and sea: Permian to present. *Lethaia*, **36**, 227–254.

Int. Assoc. Sedimentol. Spec. Publ. (2014) **46**, 59–100

Late Triassic to Early Jurassic climatic change, northern North Sea region: impact on alluvial architecture, palaeosols and clay mineralogy

JOHAN P. NYSTUEN*, AUDUN V. KJEMPERUD*†, REIDAR MÜLLER*‡, VICTORIA ADESTÅL*§ and EDWIN R. SCHOMACKER*‖

**Department of Geosciences, University of Oslo, 0316 Oslo, Norway (E-mail: j.p.nystuen@geo.uio.no)*
†Idemitsu Petroleum Norge, Fillipstad Brygge 1, 0250 Oslo, Norway
‡Tullow Norway, Tordenskiold gate 6b, 0160 Oslo, Norway
§BayernGas Norge, Lilleakerveien 8, 0283 Oslo, Norway
‖Statoil ASA, Martin Linges vei 33, 1330 Fornebu, Norway

ABSTRACT

The European continental plate drifted northwards in Late Triassic to Early Jurassic time. The change in plate position gave rise to a long-term shift from arid/semi-arid to humid climate in the North Sea region, in addition to a global warming effect of the igneous activity in the Central Atlantic Magmatic Province. In the Tampen Spur area in the northern North Sea, this climatic change profoundly influenced the development of alluvial successions represented by the Lunde Formation (Norian-early Rhaetian) and the overlying Raude and Eiriksson formations (Rhaetian-Hettangian) in the Statfjord Group. Arid to semi-arid climate during the Norian and early Rhaetian is expressed in the lower part of the Lunde Formation by ephemeral braided stream sandstones, reddish-brown entisols/inceptisols and floodplain mudstones characterised by diverse clay mineral suites and high feldspar contents. A change towards semi-humid climate in the late Rhaetian is reflected in the upper part of the Lunde Formation by fluvial sandstones representing meandering rivers, moderately developed cumulative vertisols and higher contents of smectite and kaolinite resulting from increased chemical weathering. Increased production of clays in the hinterland, lowered relief and decreased rate of sediment supply gave rise to a fining-upward trend with an increasing degree of palaeosol development in the upper part of the Lunde Formation and the lowermost part of the Statfjord Group. Increasing humidity during the latest Rhaetian and Hettangian is witnessed in the lower part of the Statfjord Group by a change from reddish-brown to greenish-grey colour in mudstones, disappearance of calcrete nodules and smectite as the dominating clay mineral formed on wet floodplains. Quartz-arenitic braided stream sandstones and kaolinite-dominated floodplain mudstones in the upper part of the Statfjord Group reveal that humid and warm climate conditions prevailed, with intense chemical weathering in the hinterland. Monsoonal rainfall gave high surface runoff, producing sediment-laden discharge within well-drained alluvial fans or plains. Coupled with low accommodation space, this resulted in a coarsening-upwards trend in the Statfjord Group, braided-stream sandstones with wide lateral extent and floodplain mudstones with palaeosols reflecting highly varying degrees of pedogenic alteration. Allostratigraphic units in the Lunde Formation and the Statfjord Group may have formed in response to more high-frequency climatic cycles or base-level changes.

Keywords: Palaeoclimate, alluvial architecture, palaeosols, clay mineralogy, Lunde Formation, Statfjord Group, Late Triassic, Early Jurassic, North Sea.

From Depositional Systems to Sedimentary Successions on the Norwegian Continental Margin, First Edition.
Edited by A. W. Martinius, R. Ravnås, J. A. Howell, R. J. Steel, and J. P. Wonham.

INTRODUCTION

Climate is a major factor controlling the sedimentary infill of alluvial basins. River dimension, channel morphology and fluvial deposits are related to overall discharge, the run-off pattern and variation in hydraulic energy, rate of sediment supply, sediment composition and vegetation cover; all these variables are controlled and influenced by precipitation and temperature. Tectonics and eustasy influence base level and thereby influence accommodation space and sediment influx. Thus, the depositional style of an alluvial succession may be the result of complex interactions between climate and other external (extrinsic, allogenic) factors as well as internal (intrinsic, autogenic) factors (Blum & Törnqvist, 2000; Schumm, 2005; Bridge, 2005).

Alluvial architecture and facies models for ephemeral dryland streams and perennial streams of humid climatic regions have been widely applied to infer palaeoclimate (Miall, 1996; Tooth, 2000; Patterson *et al.*, 2006; Roberts, 2007; Howard, 2009); recently, facies models have also been established for rivers in seasonal tropical regions (Fielding *et al.*, 2011).

Palaeosols are particularly important features for recording palaeoclimate, both in terms of soil-forming processes controlled by humidity, temperature, drainage and vegetation, floodplain dynamics, rate of sedimentation and resulting alluvial architecture (Retallack, 1986; Kraus, 1987, 1999, 2002; Marriott & Wright, 1993; Tandon & Gibling, 1994; Bestland, 1997; Hamer *et al.*, 2007; Terry, 2007; Moore *et al.* 2008).

Clay mineralogy represents a third category of applied palaeoclimate indicators (Blanc-Valleron & Thiry, 1997; Gingele & Deckker, 2004; Robert, 2004; Suresh *et al.*, 2004; Wan *et al.*, 2011).

The architectural style of alluvial basins is the combined result of several extrinsic and intrinsic factors operating on various time scales ranging from some hundreds to several millions of years. Most studies concerning the impact of climate on fluvial architecture have dealt with climatic variations of Milankovitch-scale, or even shorter duration (Blum, 1993; Yang & Baumfalk, 1994; Olsen, 1994; Gibling & Bird, 1994; Sinha & Sarkar, 2009). Long-term changes in alluvial architecture have been related mainly to tectonism and eustasy (Wright & Marriott, 1993; Leeder *et al.*, 1996; Miall 1996). Long-term climate shifts as the main controlling factor on changes in depositional style have only been discussed in a few studies (Perlmutter & Matthews, 1989; Olsen & Larsen, 1993; Van der Zwan, 2002; Feldman *et al.* 2005). However, the impact of long-term climatic changes on palaeosol development has been pointed out by several authors (Hubert, 1977; Wright, 1990; Vanstone, 1991; Mack, 1992; Joeckel, 1999; Ahlberg *et al.*, 2002). The restricted use of climate in alluvial models of ancient fluvial deposits was emphasised by Blum & Törnqvist (2000), who proposed that much more discussion was needed on the effects of climate on the stratigraphic record in general.

The continental Permo-Triassic to Lower Jurassic succession in the northern North Sea region is usually thought to record a change in climate from warm and arid to warm and humid (Evans *et al.* 2003). This climate change has been attributed to the northward drift of the Eurasian plate, with the central northern North Sea position migrating from about 40°N in the Late Triassic (Carnian-Norian) to about 50°N in the Early Jurassic (Rhaetian-Sinemurian) (Goldsmith *et al.*, 2003). Global increase in temperature at the Triassic/Jurassic transition (Hesselbo *et al.*, 2002; Bonis & Kürschner, 2012) may also have contributed to climatic changes in the northern North Sea area. Recent studies of the alluvial Triassic of the northern North Sea, particularly dryland depositional systems, are presented by Müller *et al.* (2002), McKie & Audretsch (2005), McKie & Williams (2009) and McKie *et al.* (2010).

The present study deals with a rock record comprising the Upper Triassic Lunde Formation and the Lower Jurassic Statfjord Group (elevated to group status by Lervik, 2006) in the Tampen Spur area, northern North Sea, particularly in the Snorre oilfield (Fig. 1). These lithostratigraphic units are characterised by variously stacked fluvial sandstone bodies alternating with units of floodplain mudstones with palaeosols and clay mineralogy that change up through the succession (Nystuen *et al.*, 1989; 2008; Steel & Ryseth, 1990; Nystuen & Fält, 1995; Adestål, 2002; Müller, 2003; Müller *et al.*, 2004; Kjemperud, 2008; Schomacker, 2008). The object of the present paper is to integrate data on alluvial facies, palaeosols and clay mineralogy in a synthesis of climatic impact on basin infill dynamics and alluvial architecture from Late Triassic to Early Jurassic time in the northern North Sea (Fig. 1).

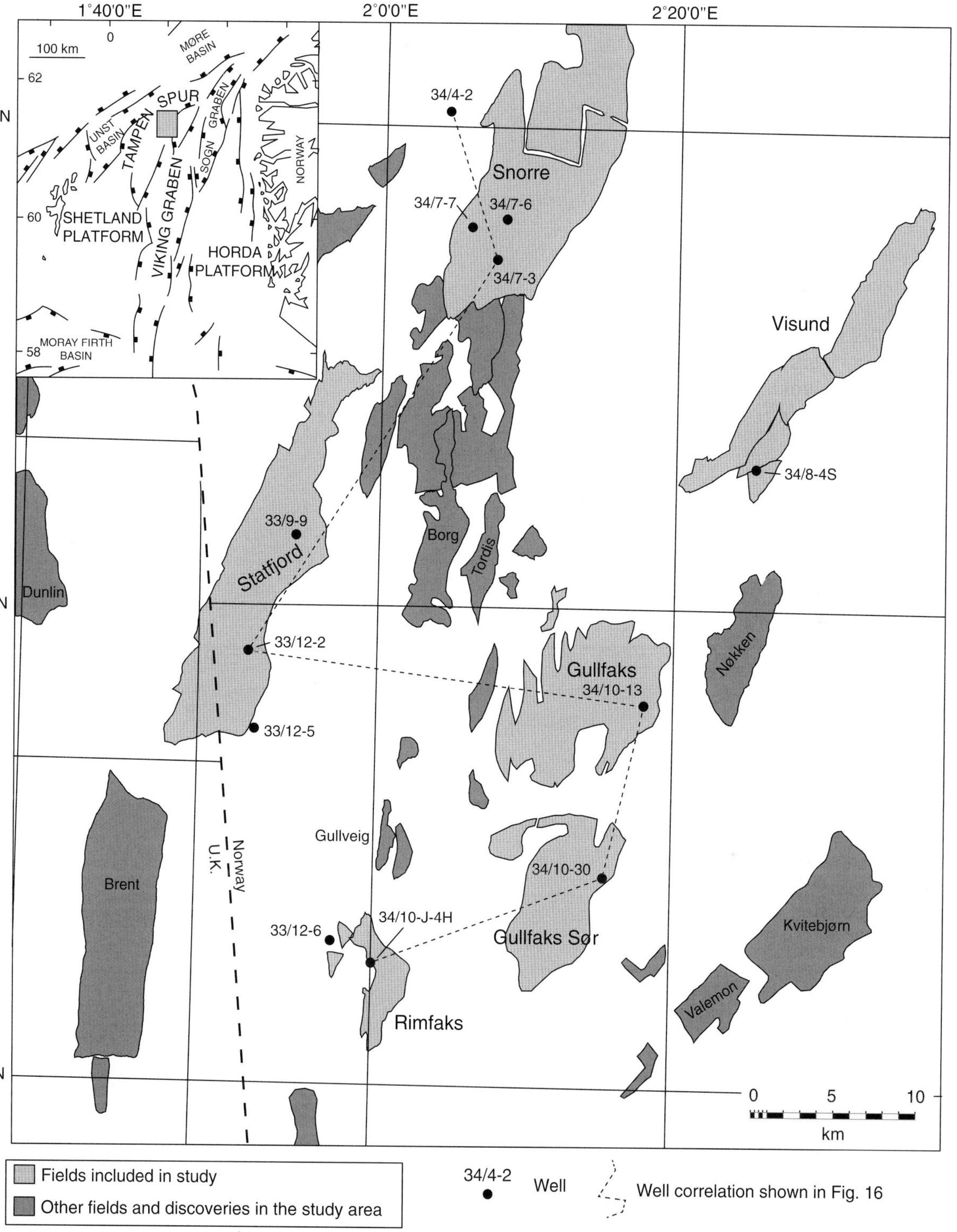

Fig. 1. Structural features and hydrocarbon fields in the Tampen Spur area, northern North Sea (insert map.) Positions of wells used in regional correlation (Fig. 16) are shown.

GEOLOGICAL FRAMEWORK

Stratigraphy and lithology

The Norian to early Rhaetian Triassic Lunde Formation and the late Rhaetian to earliest Pliensbachian Statfjord Group (Figs 2 and 3) were deposited in the northern part of a continental basin that covered most of the present North Sea area, until the region was transgressed in the Late Sinemurian to Early Pliensbachian (Røe & Steel, 1985; Nystuen *et al.*, 1989; Steel & Ryseth, 1990; Steel, 1993; Nystuen & Fält, 1995; Charnock *et al.*, 2001; Ryseth, 2001; Goldsmith *et al.*, 2003; Mckie & Williams, 2009). The Lunde Formation was defined originally as a part of the continental Cormorant Formation by Deegan & Scull (1977). Vollset & Doré (1984) redefined the Cormorant Formation in the Norwegian sector of the northern North Sea as the Hegre Group and subdivided it into the Teist, Lomvi and Lunde formations. Nystuen *et al.* (1989) subsequently subdivided the Lunde Formation into three informal members: the lower, middle and upper Lunde members. Lervik (2006) redefined the Lunde Formation to comprise only the upper Lunde member of Nystuen *et al.* (1989) and established the Alke Formation, consisting of the previous lower and middle Lunde members, located above the Lomvi Formation (Fig. 2).

The Statfjord lithostratigraphic unit was originally given formation status and subdivided into three members; in ascending order Raude, Eiriksson and Nansen (Deegan & Scull, 1977). Vollset & Doré (1984) redefined the lower boundary of the Statfjord Formation (see below), but maintained its formation status and the three formal members. Lervik (2006) raised the rank of the Statfjord

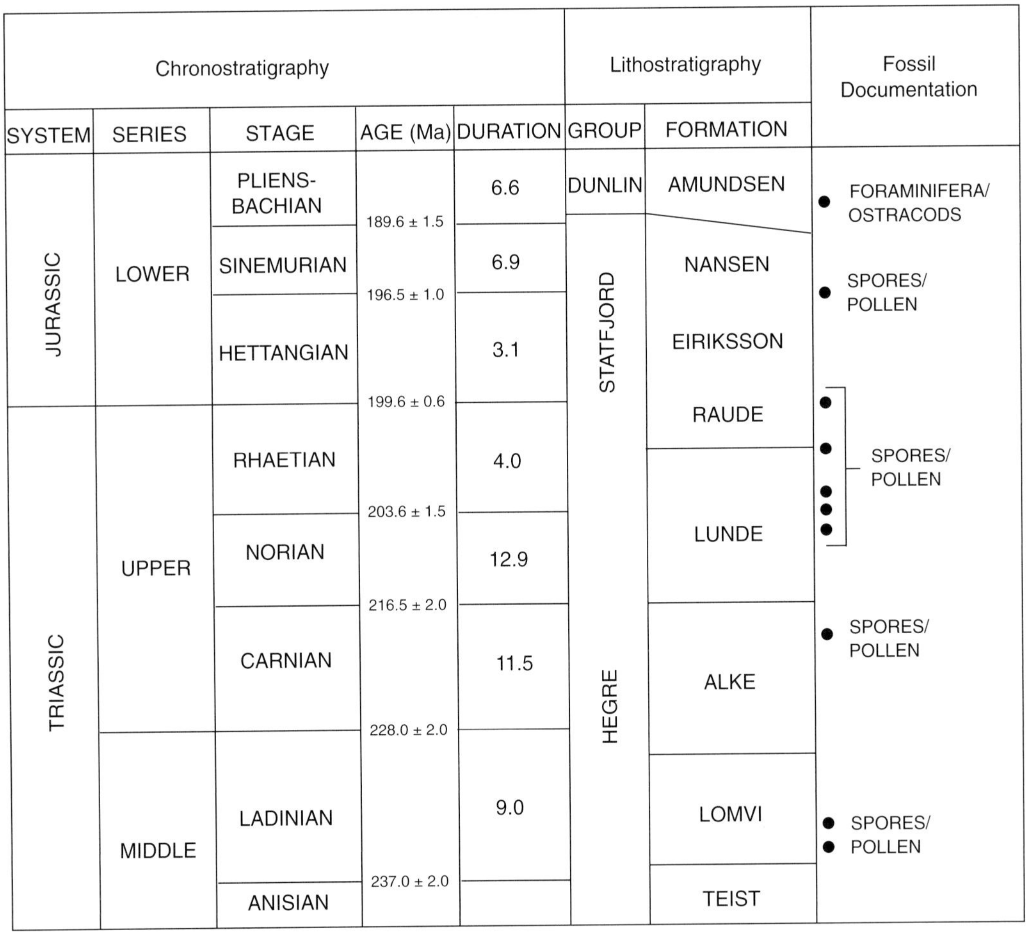

Fig. 2. Lithostratigraphy and chronostratigraphy of the Upper Triassic Hegre Group and the Lower Jurassic Statfjord and Dunlin groups in the Norwegian sector of the northern North Sea region. Modified from Deegan & Scull (1977) and Lervik (2006), with ages from Gradstein *et al.* (2004) and fossil documentation from Eide (1989) and Lervik *et al.* (1989).

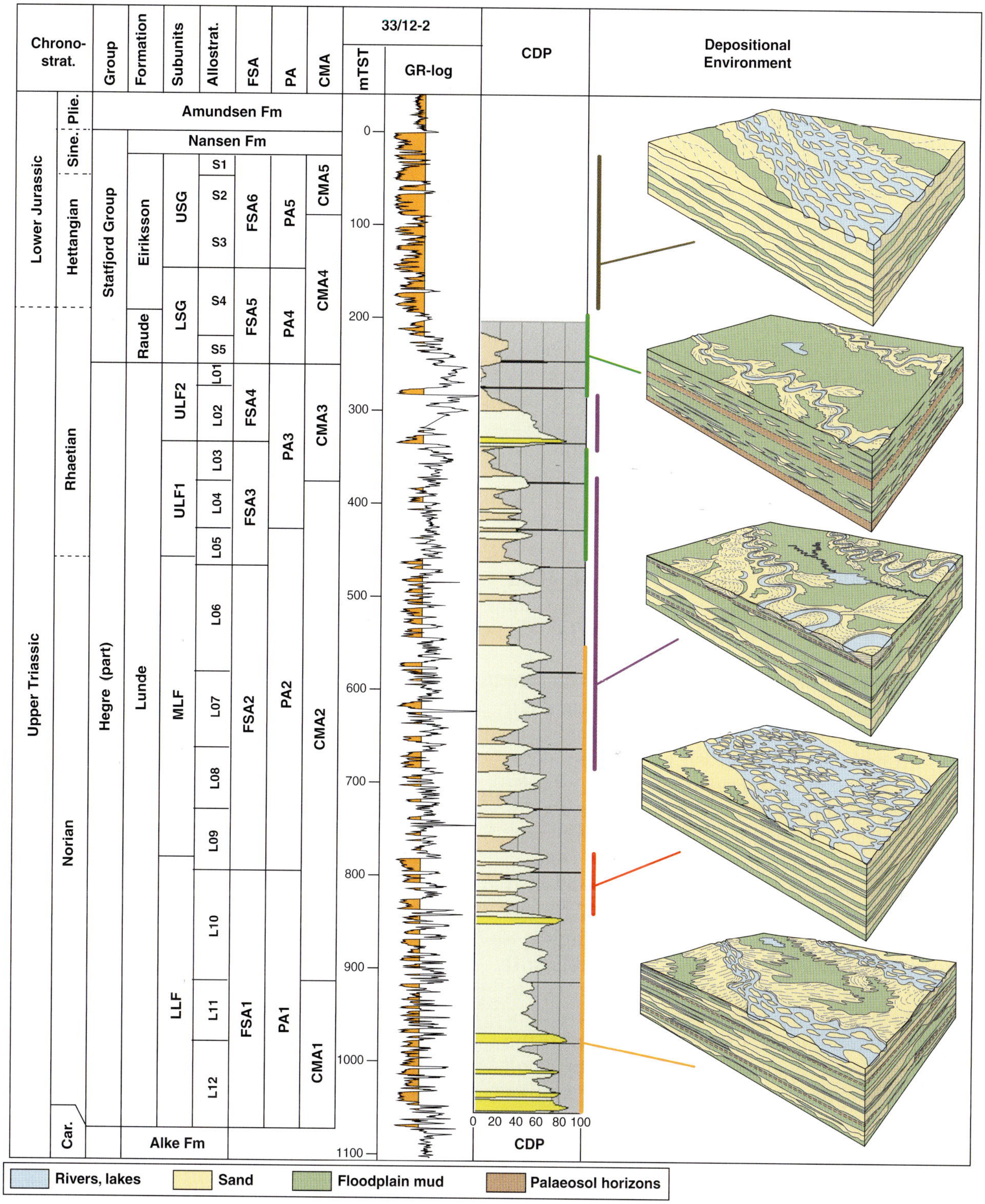

Fig. 3. Stratigraphy of the Lunde Formation and the Statfjord Group in the Tampen Spur area, northern North Sea, illustrated by a gamma-log from well 33/12-2, Statfjord Field, with true stratigraphic thickness in metres (m TST) and channel deposit proportion (CDP). Allostratigraphic units (L12 to L01, S5 to S1) and informal subunits of the Lunde Formation and the Statfjord Group as applied in this paper are shown together with stratigraphic intervals of fluvial sandstone assemblages (FSA1 to 6), palaeosol assemblages (PA1 to 5), clay mineral assemblages (CMA1 to 5) and an overall interpretation of the depositional environment (modified from Nystuen *et al.* 2008). The Nansen Formation is shallow-marine and is overlain by the open-marine Amundsen Formation of the Dunlin Group. LLF = lower Lunde Formation, MLF = middle Lunde Formation, ULF = upper Lunde Formation, LSG = lower Statfjord Group, USG = Upper Statfjord Group. Depositional models: blue = rivers and lakes, yellow = sand, green = floodplain mud, red = palaeosol horizons. The lower boundary of the Statfjord Group is according to Deegan & Scull (1977) and Vollset & Doré (1984) (2951 m below KB).

Formation to group and the previous members to formations. In our paper we apply the lithostratigraphic definitions and terminology of Lervik (2006) for the Hegre and Statfjord groups (Fig. 2).

The chronostratigraphy of the Hegre Group and the continental part of the Statfjord Group is based on terrestrial palynomorphs; in addition, ostracods and foraminifera have been employed for the upper marine part of the Statfjord Group and the overlying Amundsen Formation of the Dunlin Group (Eide, 1989; Lervik *et al.*, 1989; Charnock *et al.*, 2001; Goldsmith *et al.*, 2003) (Fig. 2). Due to poor preservation conditions in the continental redbed facies, the biostratigraphic data of the Hegre and Statfjord groups are generally scarce and widely dispersed, both stratigraphically and geographically. The chronostratigraphic subdivision and regional correlation is consequently uncertain (Eide, 1989; Lervik *et al.*, 1989; Goldsmith *et al.* 2003).

Eide (1989) showed that in the Snorre Field (Fig. 1) the Alke Formation and the Lunde Formation (previous upper Lunde member) reveal various palynomorph assemblages, with Assemblage I comprising approximately the present Lunde Formation and Assemblage II the present Alke Formation. The palynomorph assemblages are characterised by lacustrine/brackish water green algae, with *Plaesiodictyon mosellanum* always being present in Assemblage II and *Botryococcus* sp. sporadically in Assemblage I. Eide (1989) suggested that the difference in microfloral composition between Assemblage II and Assemblage I might be the result of different depositional environments (discussed below) and assigned a Norian to Early Rhaetian age to Assemblage II (Alke Formation) and an Early Rhaetian age to Assemblage I (Lunde Formation).

The palynological data of the Triassic in the North Sea were evaluated by Goldsmith *et al.* (2003). Their chronostratigraphic reinterpretation is adopted in this paper with slight modifications made by Lervik (2006). According to Goldsmith *et al.* (2003) a mid Norian to late Norian age for the Lunde Formation (the previous upper Lunde member) is indicated by a palynomorph assemblage that includes *Kraeuselisporites reissingerii*, *Enzonalasporites vigens* and *Riccisporites tuberculata*. In addition to this Norian age, the Lunde Formation probably also includes the lower Rhaetian (Eide, 1989; Lervik *et al.* 1989) (Fig. 2).

The location of the Triassic–Jurassic boundary in the Tampen Spur area is very uncertain due to lack of diagnostic fossil assemblages within the actual interval (Deegan & Scull, 1977; Vollset & Doré, 1984; Nystuen & Fält, 1995). The Rhaetian–Hettangian boundary is generally considered to be located in the upper part of the Raude Formation (Goldsmith *et al.*, 2003) (Fig. 2). The basal Jurassic palynomorph assemblages lack the typical Triassic suite of species and are dominated by *Chasmatosporites* sp. and *Cerebropollenites thiergartii* (Goldsmith *et al.*, 2003). The Hettangian to Sinemurian and Sinemurian to Pliensbachian boundaries are not well constrained by palynological data; the fluvial Eiriksson Formation is likely to be dominantly Hettangian and the marine Nansen Formation Sinemurian to Early Pliensbachian (Charnock *et al.* 2001). Large numbers of the palynomorph *Deltoidospora spp.* are recorded in the Statfjord Formation in the Snorre Field well 34/7-7 and also in wells 31/4-4A and 34/10-30 in the Tampen Spur area (Müller, 2003). Recent outlines of Triassic to Early Jurassic palynology in north-western Europe and its significance for palaeoclimate are given by Hubbard & Boulter (2000), Kürschner & Herngreen (2010) and Bonis & Kürschner (2012).

The Hegre Group consists of interchanging mudstone and sandstone. The Induan to Carnian Teist Formation consists of interbedded sandstone, claystone, mudstone and marl. The depositional environment is interpreted as ephemeral fluvial channels migrating into distal mud basins (Nystuen *et al.*, 1989; Lervik, 2006). The Ladinian Lomvi Formation is a massive sandstone unit inferred to be aeolian and partly fluvial (Nystuen *et al.*, 1989). The Alke Formation is dominated by reddish-brown, grey and grey-green mudstones and marls with thin, fine-grained, sheet-like sandstone beds. The depositional environment of the Alke Formation in the northern North Sea area is considered to have been typified by lacustrine and terminal mud basins with proximal alluvial fans (Nystuen *et al.*, 1989, 2008; Lervik *et al.* 1989; Lervik, 2006; McKie & Williams, 2009), possibly with some intermittent marine incursions (Eide, 1989; Goldsmith *et al.*, 2003) (see below).

The lower boundary of the Lunde Formation in the Tampen Spur area is marked by fluvial channel-sandstone bodies resting on lacustrine and terminal mud-basin deposits of the Alke Formation. Pebbly conglomerate beds at the base of the Lunde Formation contain extrabasinal granitic and quartzitic clasts. The boundary, correlated through several wells in the Tampen Spur area (Nystuen *et al.*, 1989), represents a marked change in depositional environment, from lacustrine or terminal basins to alluvial fans and wide alluvial plains (Nystuen *et al.*, 1989; Steel & Ryseth, 1990; Steel, 1993; Nystuen & Fält, 1995).

The Lunde Formation is composed of interchanging fluvial channel sandstone bodies and reddish-brown floodplain mudstones, altogether about 800 m to 850 m thick. The Lunde Formation is an overall fining-upwards (FU) succession and the Statfjord Group a coarsening-upwards (CU) succession, about 150 m to 330 m thick in the Tampen Spur area. The lower boundary of the Statfjord Group was defined by Deegan & Scull (1977) as the base of the coarsening-upwards succession. Later, Vollset & Doré (1984) redefined the lower boundary as the base of the first thick channel sandstone body occurring above the mudstone-dominated interval that was defined as the upper part of the Lunde Formation. The original definition is applied in the present study, as the trend change from FU to CU can be identified in most wells penetrating this interval in the northern North Sea area (Nystuen *et al.*, 1989). The turnaround from fining-upwards to coarsening-upwards in grain size-trend is likely to reflect a major regional change in the depositional system (Nystuen *et al.*, 1989; Steel & Ryseth, 1990; Steel, 1993; Nystuen & Fält, 1995) and is discussed further below.

The Raude Formation is dominated by reddish-brown mudstone and single-storey fluvial channel sandstone bodies and associated thin sheet sandstones of crevasse splay origin. The Eiriksson Formation has a higher frequency of fluvial channel sandstone bodies compared with the Raude Formation. Individual channelised sandstone bodies commonly form composite multistorey sandstone bodies. Mudstones are dominantly greenish-grey. The Nansen Formation is predominantly sandstone of shallow-marine origin, overlain by marine mudstone and sandstone of the Amundsen Formation in the Dunlin Group (Charnock *et al.*, 2001; Goldsmith *et al.*, 2003).

Basin configuration and structural framework

Most of the Hegre Group and the whole of the Statfjord Group were deposited during the post-rift phase of the Late Permian-Early Triassic rifting (Badley *et al.*, 1988; Gabrielsen *et al.*, 1990; Odinsen *et al.*, 2000). Throughout most of the Triassic, alluvial systems fed into a large terminal lake system located in the central and northern North Sea area (Mckie & Williams, 2009). The Lunde Formation has been interpreted as having formed within a wide alluvial plain and the continental part of the Statfjord Group as deposited as alluvial fans, alluvial plains or braidplains (Røe & Steel, 1985; Nystuen *et al.*, 1989; Steel & Ryseth, 1990; Nystuen & Fält, 1995; Ryseth, 2001) (Fig. 3). The ultimate base level during deposition of the Lunde Formation is believed to have been the Borealic Sea to the north (Müller *et al.*, 2005) and the Tethys to the south-east (McKie & Williams, 2009), both seaways several hundred kilometres from the 'Lunde basin'. The transgression during deposition of the upper part of the Statfjord Group took place both from the south and the north (Nystuen & Fält, 1995; Ryseth, 2001; Goldsmith *et al.*, 2003).

The total thickness of the Lunde Formation (about 800 m to 850 m) is relatively uniform in the Tampen Spur area. The thickness of the Statfjord Group, about 150 m to 330 m in the Tampen Spur area, increases further to the south; the variation may be due to syndepositional tectonics or differential subsidence (Vollset & Doré, 1984; Fält, 1987; Steel & Ryseth, 1990; Ryseth & Ramm, 1996). Estimated ages of the Lunde Formation and the Statfjord Group (Fig. 2) indicate an overall marked decline in sedimentation rates from the Norian-Rhaetian to the late Sinemurian. By applying the time scale of Gradstein *et al.* (2004) (Fig. 2), the Lunde Formation tentatively spans about 15 million years, giving an average rate of vertical deposition in the order of 0.055 mm/year. The Statfjord Group may represent about 12 million years, giving an average depositional rate in the Tampen Spur area of about 0.012 mm/year to 0.027 mm/year. These numbers are only approximate but they do illustrate a marked decline in average sedimentation rate and/or accommodation rate from late Triassic to earliest Jurassic. The source areas of clastic debris supplied to the Lunde Formation and the Statfjord Group, i.e. the Shetland Platform to the west and north-west and south-western Norway to the east, were composed of Precambrian gneisses, Caledonian metamorphic rocks and Devonian sandstones (Mearns *et al.*, 1989, Nystuen & Fält, 1995; Knudsen, 2001). The Triassic to Middle Jurassic succession in the Snorre Field was uplifted and partly eroded during a major rifting episode in the North Sea region during Late Jurassic to Early Cretaceous times. The uplift was followed by subsidence and burial during crustal cooling from the Early Cretaceous to Recent (Dahl & Solli, 1993; Nøttvedt *et al.*, 1995). The rocks of the Lunde Formation and the Statfjord Group were then buried to their present depths of about 2500 m to 3000 m.

Subdivision of the Lunde Formation and the Statfjord Group

The Lunde Formation and the Raude and Eiriksson formations in the Statfjord Group are subdivided into allostratigraphic units (cf. North American Commission on Stratigraphic Nomenclature, 1983, p. 865). These are numbered stratigraphically from below: L12 to L01 in the Lunde Formation and S5 to S1 in the Statfjord Group (Diesen *et al.*, 1995; Nystuen & Fält, 1995) (Fig. 3).

The allostratigraphic units in the Lunde, Raude and Eiriksson formations are generally fining-upwards (FU) successions with a lower part dominated by multi-storey channelised sandstone bodies and an upper part with a decreasing frequency of channel sandstone bodies and increasing abundance of overbank sandstones and floodplain mudstones. Individual allostratigraphic units are some tens of metres thick. The internal organisation of individual units commonly changes laterally due to facies variation across the depositional basin. Two or more allostratigraphic units form higher-order fining-upwards and coarsening-upwards (CU) sets (Fig. 3). We refer in the description and discussion below to an informal subdivision of the Lunde Formation in a lower (LLF), middle (MLF) and upper part (ULF) and the fluvial part of the Statfjord Group into a lower (LSG) and upper part (USG) (Fig. 3). ULF is subdivided into ULF 1 and 2. This subdivision of the Lunde Formation should not be confused with the former informal member subdivision of Nystuen *et al.* (1989) (see above).

Material and methods

This study presents data from 36 wells from the Snorre Field and adjacent oil fields in the Tampen Spur area (Table 1, Fig. 1). Approximately 3500 m of core from 22 selected wells have been studied in detail with respect to stratigraphic variation in fluvial style, using traditional sedimentological logging techniques. Additional wells have been used for correlation within the Tampen Spur area (Fig. 1, Table 1).

Detailed information from mudrock units comes from continuously cored intervals with high recovery (Table 1). Standard polished thin sections were prepared from selected samples and employed to quantify fine material and non-clay minerals. Palaeosol features encompass colour, pedogenic aggregates (peds) and cutans, roots and other bioturbation structures, striated pedogenic fault surfaces, desiccation cracks and various types of calcrete nodules and coalesced nodules (Müller *et al.*, 2004).

The fluvial channel sandstones of the Lunde Formation and the Statfjord Group have an average porosity of about 25% according to core plug analyses. The mineralogical composition of the sandstones was made by modal analyses in thin sections (Sørlie 1996) (Table 2).

The mineralogy of 124 mudrock samples was studied from seven wells (Tables 1, 3 and 4). Clay mineralogy was determined by XRD. Powder samples were prepared by crushing in a hand mortar to a particle diameter of approximately 1 mm or less, then transferred to a beaker, covered with distilled water and 10 ml of the deflocculant Calgon and placed in an ultrasonic bath for 2 minutes. The resulting suspension was poured into a settling column and allowed to stand until the required size fraction of ≤ 2 µm had separated and could be removed. Samples were made by sedimentation under vacuum onto a porous ceramic tile. Samples were treated with $MgCl_2$ (aq) so that all the smectites would have the same *d* spacing during the analyses. Air-dried samples were analysed with a Philips PW 1710 X-ray diffractometer in the interval 3° to 50° 2θ, with Cu (Kα) radiation. Five major clay mineral groups were identified by their basal spacing (Table 4). Smectite was distinguished from chlorite by subsequently treating the sample with ethylene glycol, which expands the basal spacing to about 17 Å. Heating to 550° destroyed kaolinite and certain chlorites (Hardy & Tucker 1995). The iron oxides hematite and goethite were also determined by XRD. The computer program Mac Diff 4.2.5 and correlation factors for qualitative and quantitative analysis of the different clay minerals (Ferrell & Dypvik, 2009) were applied.

Sedimentary facies and alluvial architecture

The Lunde Formation and the Statfjord Group consist of repeated successions of (1) channelised sandstone bodies interbedded with (2) proximal

Table 1. Wells in the Tampen Spur area from which mineralogy, sedimentary logs and palaeosol analysis have been obtained and used in the present study.

Well database with core	Core in Lunde Fm	Core in Statfjord Gp	Channel sandstone mineralogy	Mudstone mineralogy	Sedimentary logs	Palaeosol analysis
34/4-4	x		x		x	x
34/4-6	x		x		x	x
34/4-7					x	
34/4-9 S	x			x	x	x
34/4-C-6 H	x			x	x	x
34/7-1	x		x	x	x	x
34/7-3	x	x	x	x	x	x
34/7-6			x			
34/7-7	x	x	x		x	x
34/7-9	x		x	x	x	x
34/7-10		x	x			
34/7-A-2 H	x				x	x
34/7-A-3 H	x			x	x	x
34/7-A-4 H	x			x	x	x
34/7-A-5 H	x				x	x
34/7-A-9 H	x			x	x	x
34/7-P-8	x			x	x	x
34/7-P-12	x				x	x
34/7-P-13				x		
34/7-P-14	x	x		x	x	
34/7-P-18	x			x	x	
34/7-P-28				x		
34/7-P-29	x			x	x	x
34/7-P-31	x				x	x
34/7-P-33				x		
34/7-P-37		x	x			
34/10-15	x				x	
34/10-C-2	x	x			x	
34/10-C-4	x	x			x	
34/10-C-6	x				x	
34/10-C-8	x	x			x	
34/10-C-14	x	x			x	
34/10-C-29	x				x	
34/10-30	x	x			x	
34/10-J-4 H	x				x	
34/8-4 S	x				x	

Table 2. Average content of major minerals in fluvial channel sandstones in the Lunde Formation (L12-L01) and the Statfjord Group (S5-S1). Data from wells 34/4-4, 34/4-6, 34/7-1, 34/7-3, 34/7-7, 34/7-6, 34/7P-37, 34/7-9, 34/7-10 (Sørlie, 1996).

Unit	Number of samples	Quartz	K-feldspar	Plagioclase	Kaolinite	Mica
S2 - S1	13	69.0	6.3	7.1	12.6	5.0
S3 - S4	5	70.4	8.3	8.9	10.4	2.0
L01 - S5	2	51.3	9.2	17.1	14.5	7.9
L05 - L02	12	54.0	10.1	19.4	12.2	4.3
L08 - L06	11	50.4	11.0	21.3	11.8	5.5
L09	11	52.3	12.3	25.4	4.6	5.4
L12 - L10	11	50.4	13.0	14.4	6.9	5.3

Table 3. Average content of non-clay minerals in percentage of bulk composition in mudrocks with palaeosols in the Lunde Formation (L12-L01) and the Statfjord Group (S5-S1), recorded from Snorre Field wells 34/7-P-13, 34/7-P-18, 34/7-P-28, 34/7-P-29, 34/7-P-33 and 34/7-A-4H (Saigal & Nystuen, in Adestål, 2002).

Unit	Number of samples	Quartz	K-feldspar	Plagioclase	Calcite	Dolomite	Siderite
S1	5	46.6	3.8	2.0	0.1	0.0	3.7
S2	4	26.6	5.9	10.4	0.3	0.2	0.2
S3	13	45.6	8.9	8.8	0.4	1.4	1.1
S4	6	33.4	8.5	25.3	1.1	9.2	0.1
S5	3	31.6	2.4	0.4	0.1	0.2	0.3
L01	2	13.3	2.2	18.5	0.8	0.2	0.6
L02	23	28.4	6.6	19.4	1.5	5.5	0.1
L03	12	20.8	5.4	22.2	21.4	1.9	0.3
L04	14	18.5	6.6	34.2	8.6	6.2	0.6
L05	6	23.1	7.8	19.5	0.7	5.8	0.1
L06	7	23.6	7.2	19.8	5.1	5.3	0.1
L07	31	21.8	10.1	20.5	1.2	2.4	0.0
L08	17	21.1	5.9	18.8	22.8	2.4	1.1
L09	12	14.6	12.1	25.0	17.4	1.0	1.8
L10	15	14.5	6.3	18.2	28.9	11.0	1.4
L11	7	20.2	7.1	22.4	1.8	0.7	0.0
L12	5	18.9	8.3	19.5	3.0	0.0	0.0

Table 4. Average content of clay minerals in mudrocks within allostratigraphic intervals L12-L11, L10-L05, L03-L01 in the Lunde Formation (L12-L01) and S5-S3 and S1 in the Statfjord Group. From wells 34/4-C-6H, 34/4-9S, 34/7-1, 34/7-3, 34/7-9, 34/7-P-8, 34/7A-3H, 34/7A-9H, 34/7-P-14.

CMA	Allostrati-graphic units]	Smectite 17 Å	Swelling mixed-layer (ML) 17 Å	Chlorite 14 Å	Illite 10 Å	Kaolinite 7 Å	Goethite	Hematite
5	S1 (n=8)	8	4	0	2	75	11	0
4	S5–S3 (n=38)	82	1	0	3	11	2	2
3	L03–L01 (n=23)	29	13	8	15	18	3	15
2	L10–L05 (n=22)	33	2	9	23	11	1	22
1	L12–L11 (n=24)	1	13	15	36	22	1	13

overbank facies (levee and crevasse-splay sandstones) and (3) distal overbank and floodplain mudrocks with beds of sandy to silty sheet-flood deposits. The facies associations have been interpreted to represent different types of fluvial channels based on sedimentological well core data and lateral extension of sandstone bodies inferred from well log correlation and seismic mapping (Table 5). The great variability in vertical and lateral facies distribution of fluvial channel systems in general, coupled with the difficulties in interpreting well cores and well logs in terms of channel types and alluvial environments (Bridge, 1993; Miall, 1995, 2006, 2010; Weissmann *et al.*, 2011; Fielding *et al.* 2012; Horn *et al.*, 2012), leads to considerable uncertainty in the fluvial style and sandstone body geometries inferred from the present dataset. However, because of a lack of outcrops of the late Triassic to Early Jurassic reservoir rocks in the Snorre Field (and other oil fields in the northern North Sea), sedimentological data obtained from well cores and electric logs were used to construct reservoir models (Diesen *et al.*, 1995; Nystuen & Fält, 1995). The reservoir model made from interpretation of fluvial style and sandstone geometry has been confirmed generally by the production history of the Snorre Field since 1992.

Table 5. Characteristic features and criteria used in the interpretation of three major fluvial channel types in the Lunde Formation and the Statfjord Group: single, straight or low-sinuous (steer head), high sinuous (meandering), low sinuous (braided) (see Fig. 3).

Channel type	Common thickness range	Characteristic features and facies	Vertical trend and correlation	Example
Low sinuous (braided)	6 to 7 m	Erosive lower boundary, commonly with channel base conglomerate with mudstone clasts. Sharp upper boundary commonly draped by floodplain mudstone Extrabasinal pebbles (quartz, quartzite, granite) in some stratigraphic intervals Dominantly plane-parallel and low-angle cross-stratification, interchanging with less frequent and thinner beds of trough cross-stratification Frequent internal erosive surfaces and cut-and fill structures Dominant medium sand grain size varies considerably in sorting within and between adjacent beds No to minor internal bioturbation	Bell-shaped to box-shaped grain-size and gamma ray trends with sharp base and top Individual channel units commonly form multistorey sandstone bodies Multistorey channel sandstone bodies are commonly correlated between wells at distances up to some kilometres	34/4-C-6 H 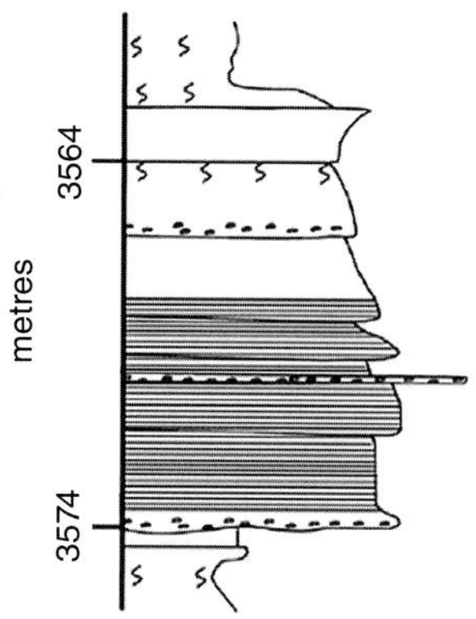
High sinuous (meandering)	8 to 13 m	Erosive lower boundary, frequently with mudstone and calcrete clasts at the base. The upper boundary is transitional into overlying floodplain mudstone Lower part dominated by thick bedsets with trough cross-stratification, separated by slightly inclined accretion surfaces frequently lined with small mudstone and calcrete clasts Upper part dominated by decreasing dimension of trough cross-stratification, passing into ripple cross lamination, accretion surfaces draped with some cm-thick silt and mudstone beds Dominant sand grain size from medium to fine, slight variation in sorting within and between adjacent beds Bioturbation common in mudstone beds in upper part	Overall upward fining grain-size trend and gamma ray values increasing into background values of overlying floodplain mudstone Clear separation between RHOB/NPHI curves in uppermost part of upward dirtying well log trend Individual channel units form multistorey sandstone bodies in upper part of the Lunde Formation Channel belts above zonation boundaries are correlated some kilometres between wells	34/7-P-29 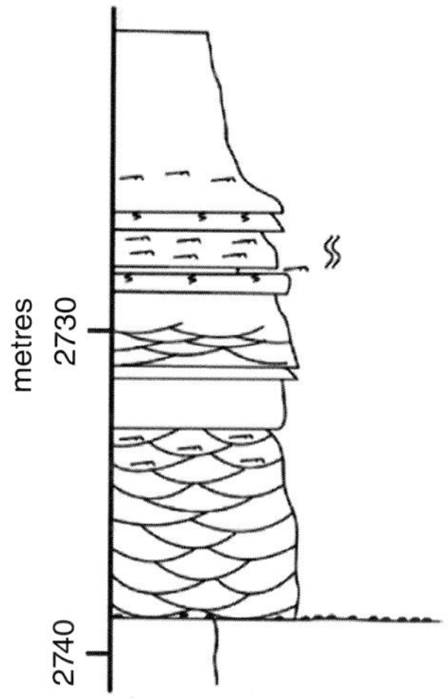
Single, straight or low-sinuous (steer head)	3 to 7 m	Erosive lower boundary, commonly covered with intra-basinal mudstone clasts Lower part dominated by trough cross-stratified or plane parallel-stratified sandstone beds Upper part is commonly ripple cross-laminated Dominant sand grain size fine to very fine Bioturbation is abundant, particularly in upper part	Distinct single upward fining and gamma ray trends Occur as single channel infill units within mudstone-dominated stratigraphic intervals Channel sandstone units commonly not correlated between wells	34/7-P-29 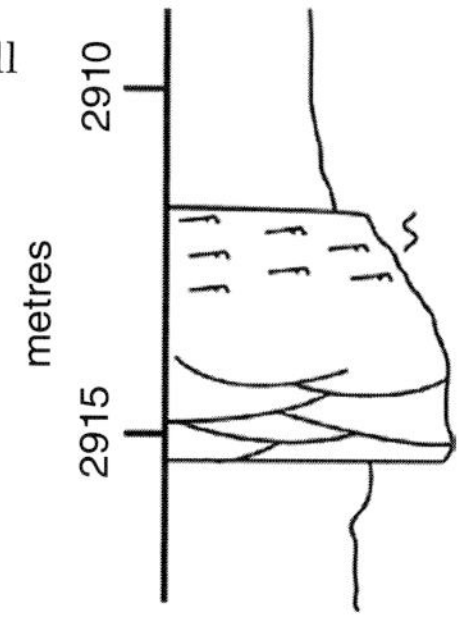

The vertical organisation of fluvial facies associations within the Lunde Formation and the Statfjord Group can be characterised in three major motifs (Table 5). These major types of vertical facies trends have been interpreted to represent (a) braided stream deposits dominating in the lower part of the Lunde Formation (LLF) and in the upper part of the Statfjord Group (USG), (b) sinuous to meandering major streams in the upper part of the Lunde Formation (ULF) and (c) single, isolated sinuous to straight stream deposits primarily in the interval comprising the ULF and the lower part of the Statfjord Group (LSG). Soils formed on floodplains and proximal slopes of alluvial ridges are represented in well cores by various types of palaeosols (Müller *et al.*, 2004). No significant changes are recorded in the grain size of channel sandstones through the Lunde Formation and the LSG, but the sand grain-size increases to coarse and very coarse in the USG.

In the Lunde Formation, conglomerates with extrabasinal clasts are extremely scarce, except in the lower part (L11-L09) where pebble-sized clasts of granite, gneiss, quartz and quartzite occur in thin conglomerate beds at, or close to, the base of the allostratigraphic units. Similar extrabasinal clasts also occur in the coarse-grained sandstones in the upper part of the Statfjord Group. Channel lag conglomerates are otherwise composed of intrabasinal clasts of pedogenic carbonates and mudrock.

The channel sandstones throughout the Lunde Formation and the lowermost part of the Statfjord Group are feldspathic with an average feldspar/quartz ratio of 0.56. In the USG, the sandstones are quartz arenites with an average feldspar/quartz ratio of 0.18 (Table 2). Clay minerals, namely illite, smectite, kaolinite and chlorite, occur as clastic grains or diagenetic products from the alteration of feldspar and mica, while calcite fills pore spaces (Sørlie, 1996; Khanna *et al.*, 1997).

The 'channel deposit proportion' (CDP), defined as the proportion of channel deposits relative to bulk volume (Bridge & Mackey, 1993), generally decreases upwards through the Lunde Formation, reaching a minimum within the transitional boundary zone to the overlying Statfjord Group (Fig. 4); from here the CDP increases once more. The fluvial deposits of the Lunde and Statfjord formations are subdivided into six 'fluvial sandstone assemblages' (FSA1-6), each consisting of one or several allostratigraphic units characterised by particular fluvial facies and facies associations, CDPs and inferred channel types. Core photographs of representative facies are presented in Fig. 5.

Fluvial sandstone assemblage 1 (FSA1) – mobile braided stream channel belts

FSA1 comprises the allostratigraphic units L12-L10 in LLF (Figs 3 and 6). The lowermost channel sandstone body of unit L12 rests with a marked erosional boundary on the Alke Formation (Fig. 6). This is a regional unconformity in the Tampen Spur area (see above). The lower boundary of L11 is characterised in several wells by an extrabasinal conglomerate bed with pebbles of quartz, quartzite, granite and gneiss. This type of conglomerate is also present in the upper part of unit L11. In the L12-L10 units, the CDP exceeds 40%, the average varying from about 66% in L12 to 53% in L10 (Fig. 4). Floodplain fines occupy about 35% on average and fine-grained overbank sandstones about 10%.

Channel sandstone bodies, inferred to represent individual channel infills, range in thickness between 6 m and 9.5 m and are composed dominantly of plane-parallel to low-angle cross-stratified sandstone, with mainly box-shaped vertical grain-size curve, or slight upward fining in the uppermost part towards the top. Internal erosional surfaces, occasionally draped with mudstone fragments, are frequent in some sandstone bodies. Calcrete and mudstone clasts are common at the base of many of the channel deposits in L12; a small coal fragment has been recorded at the base of one channel sandstone. Individual channel sandstone bodies are frequently stacked, forming composite 20 m to 40 m thick sandstone units, or multi-storey sandstone bodies (Fig. 6). A few channel infill successions reveal distinct fining-upward grain-size trends. Floodplain mudstones between channel sandstone bodies are commonly reworked by bioturbation and have root structures along palaeosol horizons (see below). The bioturbation is simple burrows of various orientation, most commonly with diameters of about 1 cm; some burrows are irregularly shaped as chamber-like structures. No detailed studies of the bioturbation have been performed; its origin may be related to the burrowing of worms and/or insects in moist or dry mud.

The fluvial sandstone bodies in L12-L10 are characterised by the association of plane-parallel and low-angle cross-stratified beds, no marked change in vertical grain size except at the very top

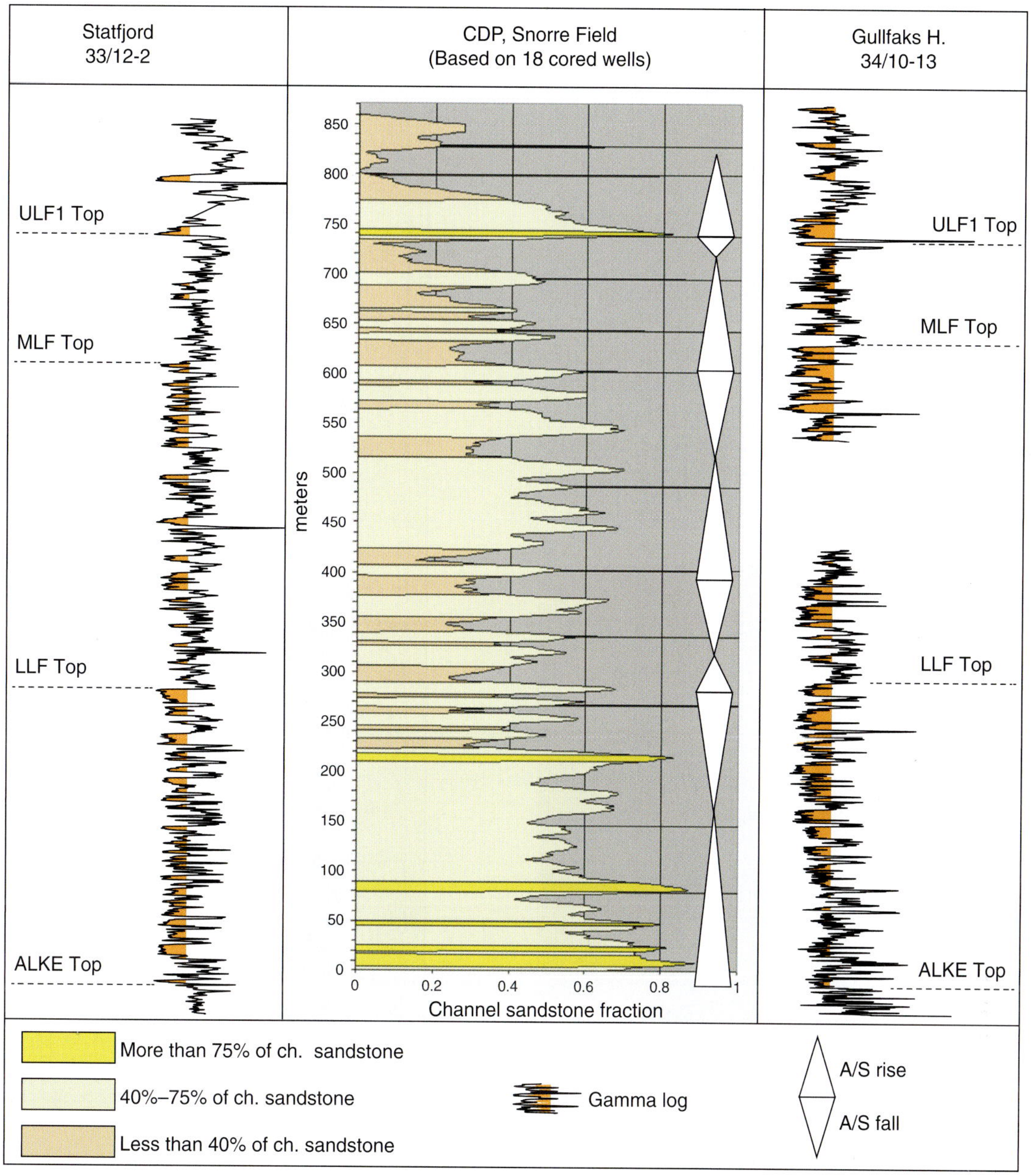

Fig. 4. Channel deposit proportion (CDP) in the Lunde Formation from the Snorre Field, compared with well logs from the Statfjord and Gullfaks H fields. The CDP trend line illustrates the general fining upward in the Lunde Formation and how the rate of accommodation (A) *versus* rate of sedimentation (S) (A/S) can be inferred to have changed during deposition of all the allostratigraphic units in the Lunde Formation.

of individual channel infill successions, frequent internal erosional surfaces and pebble beds. They are interpreted to have been formed in highly mobile braided stream channels and channel belts, forming sheet-like multi-storey and multi-lateral composite sandstone units with lateral extents of several kilometres (Table 5, Figs 3 and 6). Channel infill successions floored by extrabasinal pebble beds are inferred to represent high-energy flood events and internal erosional surfaces as cut-and-fill structures caused by fluctuating flow energy. Occasional distinct fining-upward channel infill successions may indicate periodical dissection of the floodplain by single, narrow stream channels. Coal fragments probably represent plant debris that has been coalified during periods with high groundwater level in parts of the alluvial basin.

The braided stream channel deposits of FSA1 probably formed during violent ephemeral flash-flood events, since their facies range and association with reddish-brown mudstones indicates an arid to semi-arid climate.

Fluvial sandstone assemblage 2 (FSA2) – braided to meandering streams

The fluvial sandstone assemblage 2 (FSA2), seen only in the middle part of the Lunde Formation (L09 to L06), shows approximately equal amounts (40%) of channel deposits and floodplain fines, with approximately 20% to 25% crevasse splay sandstones (Figs 3, 4 and 7). Unlike the underlying allostratigraphic units with FSA1, the channel

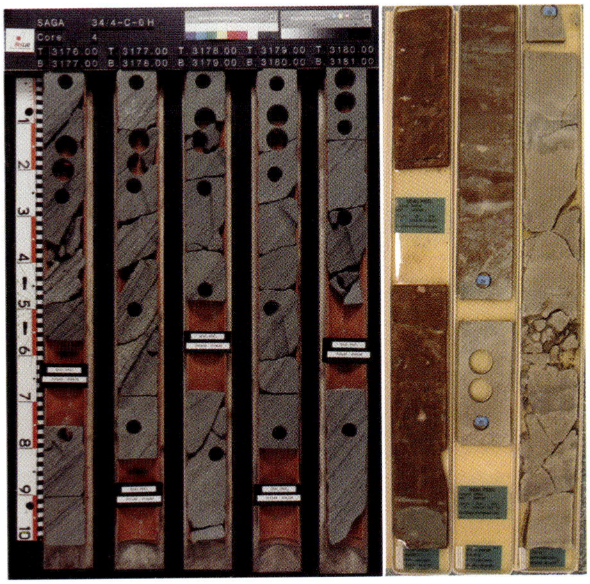

Fig. 5. Core photographs of representative alluvial facies of the Lunde Formation. To the left, braided stream sandstone bodies with typical bedsets of trough cross-stratified, low-angle-stratified and parallel-stratified sandstone, from the lower part of the Lunde Formation, well 34/4-C-6H. To the right, the upper part of a point bar sandstone with intercalations of mudstone, passing upwards into overlying floodplain mudstone with palaeosol, upper part of the Lunde Formation, well 34/4-9S.

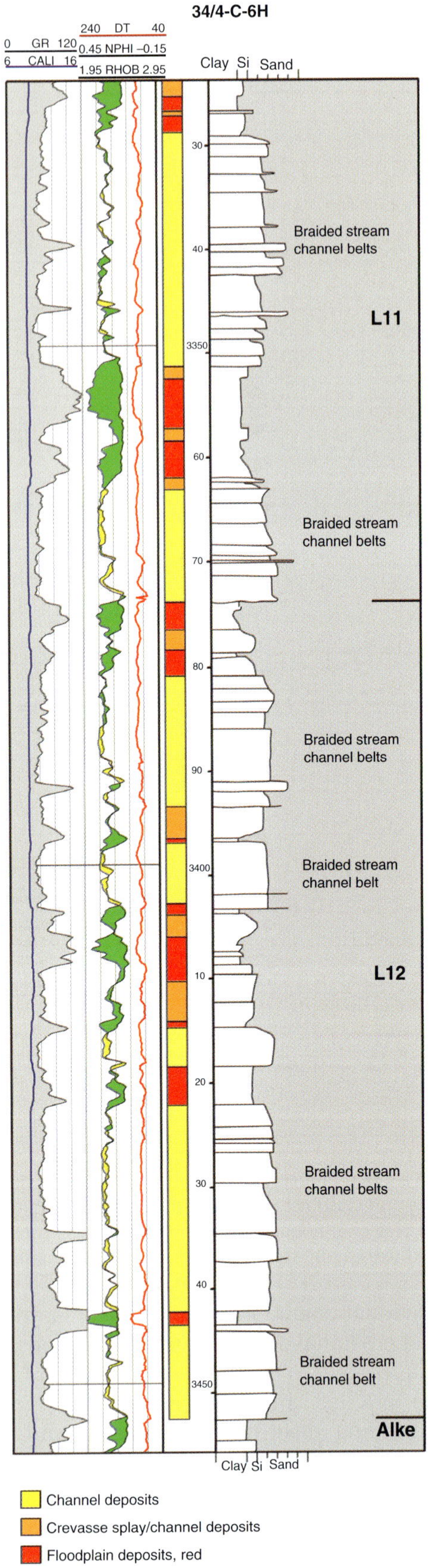

Fig. 6. Sedimentary log from well 34/4-C-6H, Snorre Field, showing characteristic stratigraphic organisation of fluvial sandstone assemblage FSA1 in the lower part of the Lunde Formation (LLF). Channel sandstone bodies are interpreted to have been deposited by ephemeral braided streams. Gr = Gamma ray, CALIP = Caliper, DT = Velocity, NPHI = Neutron porosity, RHOB = Density.

infill deposits of FSA2 are evenly distributed, separated by floodplain mudrock units. Thin conglomerate beds with extrabasinal pebbles are present in L09, L08 and L07. Unit L07 is the highest stratigraphic level in the Lunde Formation where extrabasinal clasts have been recorded.

The channel sandstones in the lower part of L09 and L08 are mainly structureless or plane-parallel stratified. Rip-up mudstone clasts are common and small calcrete clasts occur. Trough cross-stratification with internal thin mudstone beds becomes increasingly frequent in the channel sandstones upwards through L09 to L06. In well core sections, stratification shows that cross-stratified beds commonly occur in several metres thick bedsets that may be separated by thin mudstone beds and laminae or by erosional surfaces. Upward-fining of channel infill successions is generally more common than in FSA1. Individual upward-fining channel infill units, 8 m to 9 m thick (Table 5), are separated by bioturbated floodplain mudstone. The bioturbation is of same style as in FSA1 with burrows up to about 1 cm in diameter and some with chamber structure. Structureless and plane-parallel sandstone units, up to 5 m thick, also occur, predominantly in the upper part of L06 (Fig. 7).

The fluvial style in the lower part of FSA2 is interpreted to be dominated by braided streams, formed during ephemeral floods competent to carry extrabasinal clasts far out into the alluvial basin. The several metres thick trough cross-stratified sandstone bedsets, separated by thin mudstone beds or laminae, or erosional surfaces, are interpreted as large bars composed of 3D sand dunes, formed by lateral accretion and upward fining in sinuous to meandering channels, probably as point bars, as in L06. Structureless or plane-parallel sandstone units alternating with floodplain mudstone are interpreted as crevasse channel and crevasse splay deposits. The interval L07 is interpreted to contain channel types of mixed braided and sinuous character. Ephemeral flooding is inferred to have decreased with increasing sinuosity (discussed below).

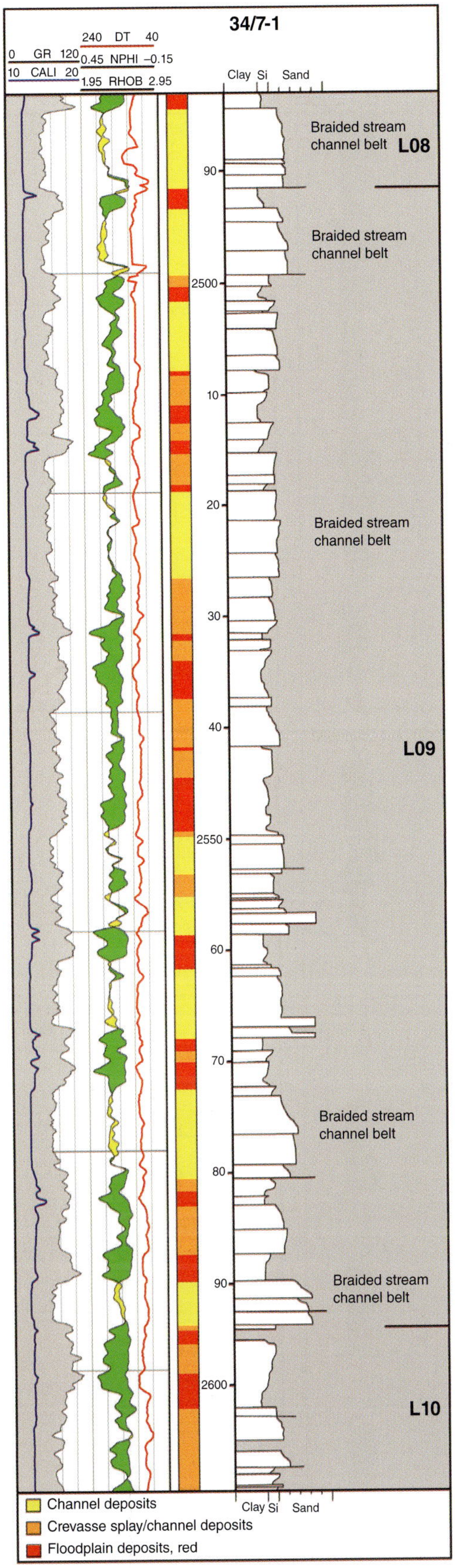

Fig. 7. Sedimentary log from well 34/7-1, Snorre Field, showing characteristic stratigraphic organisation of fluvial sandstone assemblage FSA2 in the middle part of the Lunde Formation (MLF). Sandstone bodies are interpreted to represent ephemeral mobile braided streams, meandering streams and crevasse sheets and crevasse channels. Gr = Gamma ray, CALIP = Caliper, DT = Velocity, NPHI = Neutron porosity, RHOB = Density.

Fluvial sandstone assemblage 3 (FSA3) – isolated meandering or straight streams

The fluvial sandstone assemblage FSA3 within the allostratigraphic units L05 to L03 in the upper part of the Lunde Formation (ULF1) is characterised by a marked increase in floodplain facies, while the average CDP ranges from about 36% in L05 to 24% in L03 (Fig. 4). The uppermost part of FSA3 is represented by L03 in Fig. 8. Channel sandstone bodies generally occur as up to 7 m thick, upward-fining units, with distinct erosional lower boundaries towards underlying floodplain fines and upper boundaries transitional into proximal and distal overbank facies and overlying floodplain mudstone. The lower part of FSA3 sandstone bodies commonly consists of a coarse-grained sandstone or a granule basal bed, passing upwards into medium-grained to fine-grained trough cross-stratified beds and fine-grained to very fine-grained sandstone that is parallel stratified and laminated, or structureless and strongly bioturbated. The bioturbation is represented by up to c. 1 cm thick burrows of various orientation.

The internal organisation of facies, upward-fining vertical grain-size distribution, abundance of bioturbation and lack of stacked multi-storey composite sandstone bodies have been interpreted in terms of isolated single channels formed by fluvial incision in floodplain mud, succeeded by vertical aggradation of sediments during waning stream energy and coalescence of the uppermost bioturbated channel fines with the adjacent floodplain mud. The channels may have been straight, or meandering with restricted lateral migration (Fig. 3, Table 5).

Fluvial sandstone assemblage 4 (FSA4) – meandering streams

Fluvial sandstone assemblage FSA4 occurs in the uppermost part of the Lunde Formation, ULF 2 and is comprised of allostratigraphic units L02 and L01. However, these two units differ markedly in their content of channel sandstone units, with CDPs of

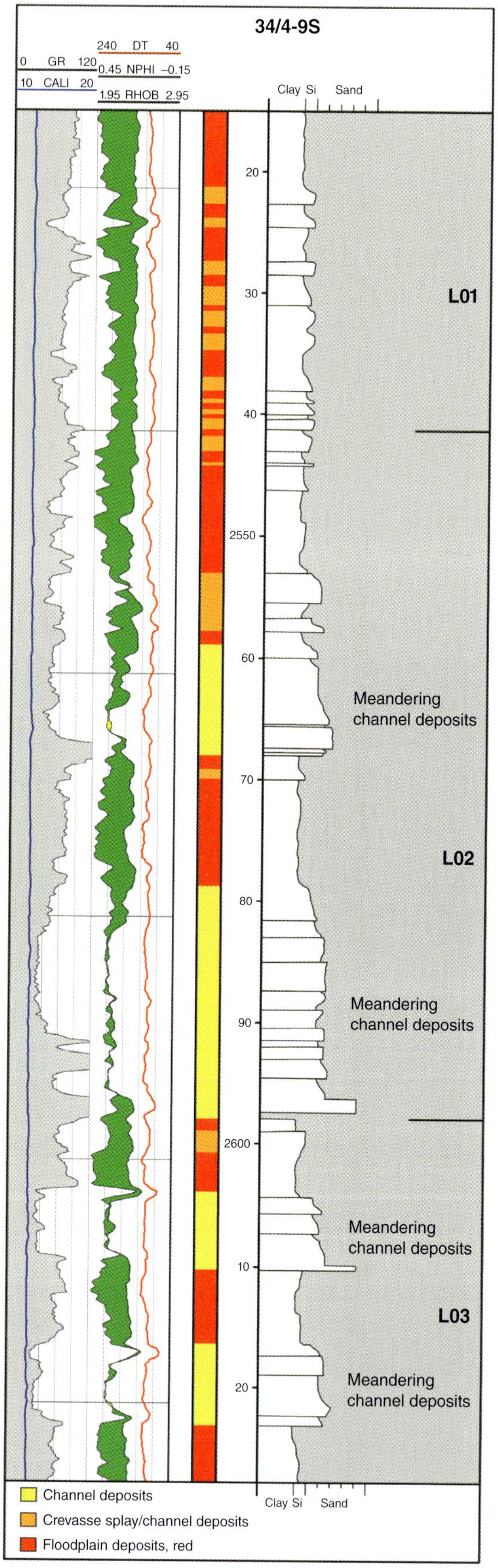

Fig. 8. Sedimentary log from well 34/4-9S, Snorre Field, showing characteristic stratigraphic organisation of fluvial sandstone assemblage FSA4 in the upper part of theLunde Formation (ULF). The sandstone bodies are interpreted as having formed in meandering rivers on mud-dominated floodplains. Gr = Gamma ray, CALIP = Caliper, DT = Velocity, NPHI = Neutron porosity, RHOB = Density.

48.2% and 4.1%, respectively (Fig. 4). In unit L02 the thickness range of individual channel sandstone deposits is about 5 m to 15 m; amalgamated channel successions have given rise to multi-storey bodies up to about 40 m in thickness. The lower boundary of L02 is thus defined as the base of a 40 m thick to 60 m thick interval dominated by several stacked composite sandstone bodies, underlain by a 25 m to 35 m interval dominated by floodplain mudstone in the upper part of L03 (Figs 4 and 8). L02 and L01 together form a marked upward-fining trend in the uppermost part of the Lunde Formation (Figs 3, 4 and 8). The sandstone bodies of L01 occur as single-storey channel infill successions.

Individual channel infill successions in both L02 and L01 are generally upward fining, dominated by trough cross-stratification. The cross-stratified beds are organised in several metres thick bedsets with practically parallel bounding surfaces that dip at a steeper angle than the structural dip in the cored sections. Cross-stratified beds usually pass into current-ripple laminated fine-grained and very fine-grained sandstone, alternating with thin beds of bioturbated mudstone in the uppermost part of individual upward-fining successions. Coalified plant debris and rip-up clasts of dark, organic-rich mudstone are commonly present within the lower part of the sandstone bodies in L02 (Fig. 8).

The individual upward-fining channel successions are thought to represent large compound bedforms of superimposed 3D sand dunes formed as lateral accretion or point bars, overlain by levée and overbank to floodplain fines. The individual FSA4 channel sandstone bodies are thus interpreted as meandering river deposits (Fig. 3, Table 5). In L01, individual fining-upwards channel sandstone bodies within thick mudstone units are interpreted as isolated, sinuous or straight channels within mud-dominated floodplains (Table 5). Coalified plant debris and thin organic-rich mudstone beds point to at least periodically high groundwater level in depressions and abandoned channel segments.

Fluvial sandstone assemblage 5 (FSA5) – sand sheets and single floodplain channels

Sandstone bodies of FSA5 occur in the allostratigraphic units S5 to S4 in the lower Statfjord Group (LSG). They are typically fine-grained and generally thin, up to some decimetres in thickness displaying internal subparallel lamination or current-ripple lamination, unless bioturbated and structureless (Fig. 9). A few fining-upward sandstone units are 3 m thick to 4 m thick, underlain by erosional channel-floor surfaces. In the upper part of the interval, in the S4 unit, this type of channel sandstone body increases in frequency. Mudstone units are dominantly reddish-brown with some calcrete nodules, but grey and greenish-grey mudstone units without calcrete occur too (see below).

S5 and the lower part of S4, together with all of L01, represent the stratigraphic interval in which a turnaround takes place from fining- to coarsening-upward grain-size trends and thus the transition from the Lunde Formation to the Statfjord Group (Figs 3, 8 and 9).

The dominating type of sandstone bodies of FSA5 have been interpreted as having formed as sheets of sand, spread onto the muddy floodplain during floods. The less abundant, FU sandstone bodies are considered to represent single and narrow, straight, low-sinuosity or meandering streams on the floodplain. Overall, the CDP index of this interval is very low, generally below c. 0.2. Variation in mudstone colour is thought to reflect changes in soil wetness from undersaturated and aerated (reddish-brown) to saturated and wet (grey and greenish-grey).

Fluvial sandstone assemblage 6 (FSA6) – mobile braided stream channels

Fluvial sandstone assemblage 6 (FSA6) is present in the USG in the allostratigraphic units S3 to S1 (Figs 3 and 9). The fluvial sandstone units here are dominated by multi-storey bodies of medium-grained to coarse-grained quartz-rich sandstone (Table 5). Individual channel infill successions are characterised by an erosional channel base, frequently covered by pebbles of quartz, quartzite, metamorphic crystalline rocks, intrabasinal mudstone clasts and coalified plant fragments, succeeded by cross-stratified, massive and parallel-stratified beds, forming blocky or slightly fining-upward grain-size trends.

Grey and greyish-green mudstone are preserved as thin beds or laminae within individual sandstone bodies or up to some metres thick units between channel infill successions and with some thin sandstone bed intercalations. Carbonaceous root and rootlet structures penetrate a few centimetres down into some of the mudstone beds; a few beds have high organic content. This organic-rich mudstone facies is associated with thin coal

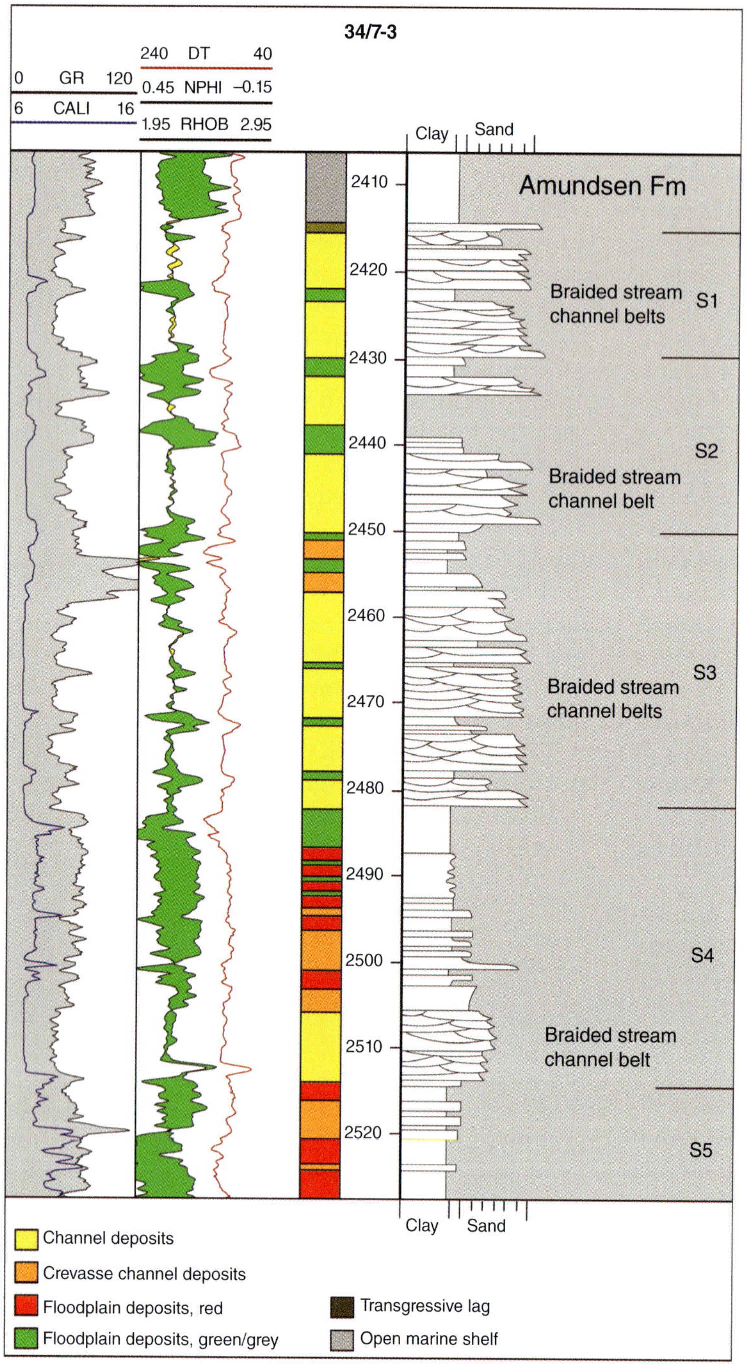

Fig. 9. Sedimentary log from well 34/7-3, Snorre Field, showing characteristic stratigraphic organisation of fluvial sandstone assemblages FSA5 in S5 and S4 and FSA6 in S3 to S1, in the Statfjord Group. Allostratigraphic unit S1 is overlain by a marine transgressive ravinement conglomerate at the base of the Amundsen Formation in the Dunlin Group. Gr=Gamma ray, CALIP=Caliper, DT=Velocity, NPHI=Neutron porosity, RHOB=Density.

beds in the Eiriksson Formation in other parts of the northern North Sea area (Ryseth, 2001).

The fluvial sandstone bodies of USG are attributed to high-energy braided streams. The braided channel style was most likely triggered by the influx of coarse-clastic debris (cf. Schumm, 1981); an increase in precipitation and river discharge combined with overall low accommodation space could have resulted in high transport capacity and enhanced stream competency during flood events. Cross-stratified sandstone beds are inferred to represent 3D sand dunes and massive and parallel-stratified beds formed during high discharge floods (Nystuen & Fält, 1995; Ryseth, 2001). These sandstone bodies provide high quality oil reservoirs and their wide lateral extent, proposed in the sedimentological model of the Snorre Field (Diesen *et al.*, 1995), has been confirmed by the production history since 1992. The mudstone beds that occur between the fluvial channel sandstone bodies are attributed to mud settling on floodplains and in shallow lakes (Nystuen & Fält, 1995; Ryseth, 2001). Many mudstone beds have been affected by pedogenic alteration to a greater or lesser extent (see below).

Palaeosols and mudrock facies

Five palaeosol assemblages (PA1 to 5) have been defined (Fig. 3). Most of the palaeosols are developed in floodplain mudstones and only to a minor extent in mudrocks and associated fine-grained sandstones formed as channel levées and crevasse splays. The palaeosol assemblages are defined on their content of mudrock facies, types of pedogenic structures, content and types of calcrete and colour and thickness of palaeosol profile. Five mudrock facies (MF1 to 5), including palaeosol types, are defined on the basis of relative proportion of clay, silt and sand, colour and mottling, degree of preserved primary lamination, occurrence of reworked mudrocks, pedogenic mud aggregates (peds), pedogenic slickensided surfaces and desiccation cracks (Table 6). Further information on palaeosol types is given by Müller *et al.* (2004). Most attention is given to the 'redbed facies' in the Lunde Formation and lower part of the Statfjord Group (allostratigraphic units S5 to 4).

Table 6. Mudrock facies and palaeosols recorded in the Lunde Formation and lower part of the Statfjord Group (Raude Formation).

Mudrock facies	Mudrock characteristics
MF1 Laminated mudrock	Variable content of clay, silt, sand and mud aggregates. Well preserved plane-parallel lamination. In LF and LSG: Red-brown colour, small isolated rhizocretions, up to 10 cm-deep desiccation sand-filled cracks. In USG: Greenish-grey, some few coalified root structures and some gleyed palaeosols. Interpretation: Deposited in small temporal ponds or shallow lakes on floodplain. Inceptisols.
MF2 Unstratified and structureless	Clay/sand ratio relative low and lower than in MF3 and MF4. In LF and LSG: brownish red colour, discrete and isolated carbonate nodules, isolated rhizocretions, burrows, 10 to 60 cm-deep sand-filled desiccation cracks. No mud aggregates, peds or pedogenic slickensided surfaces. In USG: Greenish-grey, some coalified root structures. Interpretation: Mud altered by pedogenic processes to entisols/inceptisols.
MF3 Mudrock with slickensided fault surfaces	Clay/sand ratio higher than in MF2. In LF and LSG: Brownish-red colour, frequent arcuate and smooth randomly orientated slickensided surfaces, spheroidal and angular peds, ptygmatic folds, rhizocretions, isolated carbonate nodules. In USG: Greenish-grey colour, some few slickensided surfaces, coalified root structures. Interpretation: Mud altered by pedogenic processes to high chroma vertisols in LF and LSG and low chroma vertisols in USG.
MF4 Greyish-green and mottled mudrocks	Generally high content of clays, some mudrocks with high contents of fine-grained to medium-grained sand. Greenish-grey and mottling red-greenish grey colour. Slickensided horizons, spheroidal peds, mud aggregates, 0.1 m to 1.5 m-deep, sand-filled desiccation cracks, complex mottling patterns. Interpretation: Mud with colour formed by reduction processes of iron oxides due to fluctuating groundwater level and periods of waterlogging. Altered high and low chroma vertisols.
MF5 Reworked mudrock	Texturally varied reworked pedogenic mud aggregates (0.1 mm to 5 mm) and carbonate rip-up clasts (1 cm to 6 cm). Brownish-red colour, some places mottled green-red. Interpretation: Reworked mud and calcrete nodules deposited as infill of shallow channels and in crevasse splay and sheet floods. Reworked palaeosol horizons.

Definition of palaeosols: Gleyed soil and palaeosol: Soil or palaeosol with bluish grey and greenish grey colour, accumulation of organic matter common, root structures may be preserved in top of gleyed beds. Vertisol: Clay-rich soils with wide desiccation cracks and peds formed by repeated seasonal dry-wet cycles, calcrete nodules common, no remains of primary sedimentary features. Inceptisol: Relicts of primary sedimentary features may be present. May have calcrete nodules. Entisol: Slight degree of soil formation. Primary sedimentary features little altered. (Modified from Retallack, 2001.)

The palaeosols were formed as soils beneath distinct land surfaces and reflect breaks in sedimentation and erosion and the intensity and duration of pedogenic processes. Representative core photographs of mudrocks and palaeosols are presented in Fig. 10. The relative frequency of mudrock facies in the LF and LSG, PA1 to 4, is shown in Fig. 11.

Pedocomplexes are vertically stacked individual palaeosol horizons and are compound, composite, or cumulative (Fig. 11). Compound and in part composite, pedocomplexes commonly represent weakly developed palaeosols formed during rather frequent, episodic aggradation of the floodplain. Cumulative pedocomplexes represent better developed palaeosols associated with lower rates of floodplain aggradation, longer periods of non-deposition and continuous pedogenesis (Marriott & Wright, 1993; Wright & Marriott, 1996; Kraus & Aslan, 1999).

Fig. 10. Core photographs of mudrocks and palaeosol types in the Lunde Formation. (A) Sub-vertical stacked carbonate nodules which represent rhizocretions (2548.7 m RkB, well 34/4-4). (B) Several hierarchies of peds defined by slickensides. Pedogenic mud aggregates postdate the slickensides (2524.5 m RkB, well 34/7A-9H). (C) Sand-filled desiccation cracks filled in with sand from overlying sand bed have been oxidised and later mottled by reduction of iron-oxide in the uppermost part. Both wells from the Snorre Field. Scale bars are 5 cm. From Müller *et al.* (2004).

Palaeosol assemblage 1 (PA1)

Palaeosol assemblage 1(PA1) correlates entirely with FSA1 and occurs in the allostratigraphic units L12 to L10 in the lower part of the Lunde Formation (Fig. 3). It comprises red or reddish-brown mudrocks with up to 50% to 60% sand and silt. The content of clays (<2 μm) is relatively low. The floodplain units are composed dominantly of laminated mudrocks (MF1) and unstratified, structureless mudrocks (MF2), both facies with entisols/inceptisols (Table 6, Fig. 12).

In PA1, intervals with thin (<2 m) entisols/inceptisols are commonly mixed with laminated beds and capped by laminated mudrocks or sheet-flood sandstones. Small-scale sand-filled desiccation cracks occur infrequently. Scattered calcrete nodules, nodular carbonate horizons, vertically stacked calcrete nodules, rhizocretions, root structures and burrows are common, but not abundant. The percentage of carbonate nodules reaches 20% to 30% of the total mudrock volume within a few thin horizons (0.2 m to 0.4 m). Vertisols with slickensided surfaces are rare (<5%) within PA1. Within the unit L10 a transition takes place from PA1 to PA2, with vertisols comprising up to 15% of the total floodplain succession (Fig. 11). Reworked carbonate nodules indicate that some alluvial mud with calcrete has been eroded. PA1 is dominated by compound pedocomplexes (Fig. 11).

Palaeosol assemblage 2 (PA2)

In PA2, within L09 to L05, floodplain successions constitute approximately 45% to 50% of the total rock volume (Figs 6, 8 and 13). Red and reddish-brown laminated mudrocks (MF1) and entisols/inceptisols (MF2) dominate, but high chroma vertisols (in MF3) are also common (Table 6). Individual palaeosol horizons can be up to 4 m thick before they are succeeded by stratified mudrocks or sheet-flood sandstones. Reworked mudrocks (MF5) are also present in L09 to L05, as reworked mud aggregates and reworked carbonate

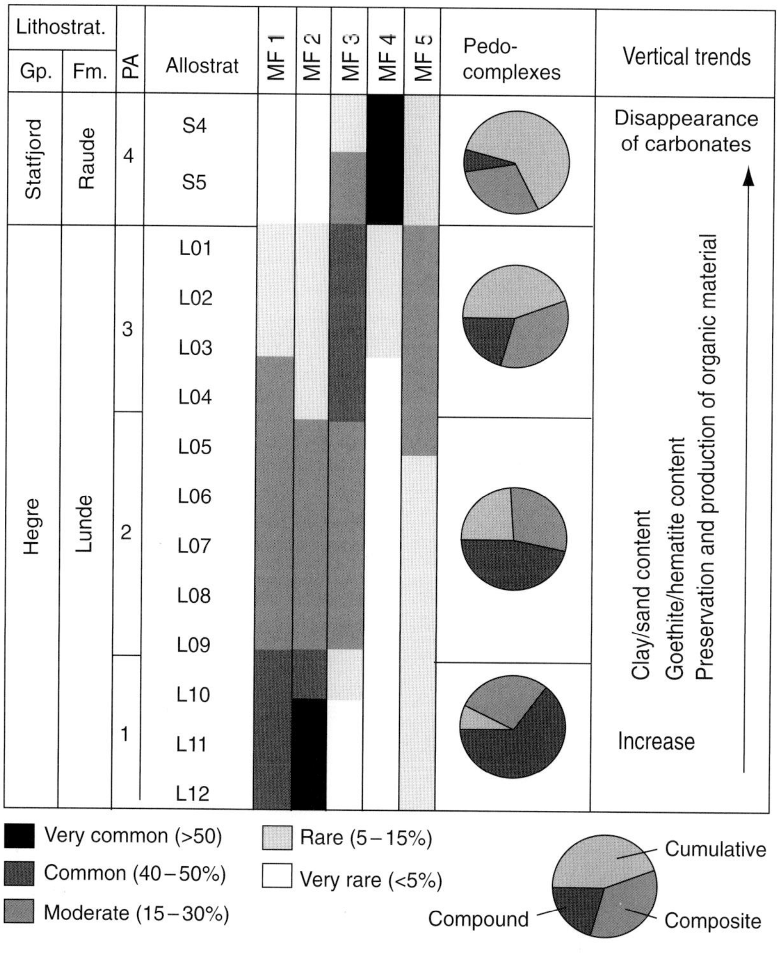

Fig. 11. Stratigraphic distribution of mudrock facies MF1-5 and pedocomplexes in the Lunde and Raude formations. PA = palaeosol assemblage.

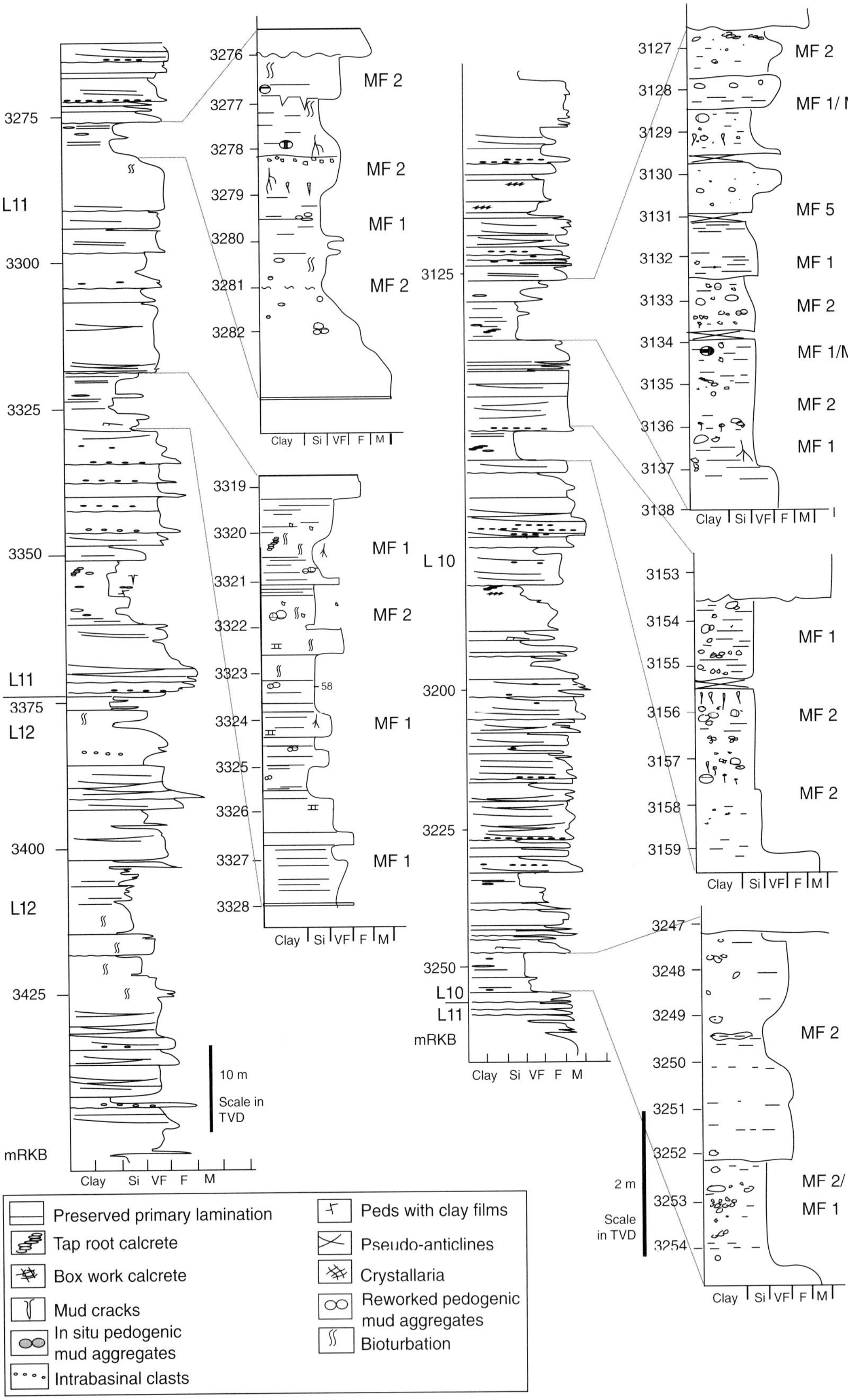

Fig. 12. Sediment and palaeosol logs from allostratigraphic units L12, L11 and L10 in the lower part of the Lunde Formation (LLF) showing distribution of mudstone facies MF1, 2 and 5 and various pedogenic features. Most palaeosols are organised as compound pedocomplexes. Colour is indicated by numbers referring to the scale of the Munsell Soil Color Chart (Munsell Color 1975) (example 10 R 3/4, brownish-red). Well 34/4-C-6H. VF, F and M refer to very fine, fine and medium sand respectively. RKB = Rotary Kelly Bushing, TVD = True Vertical Depth.

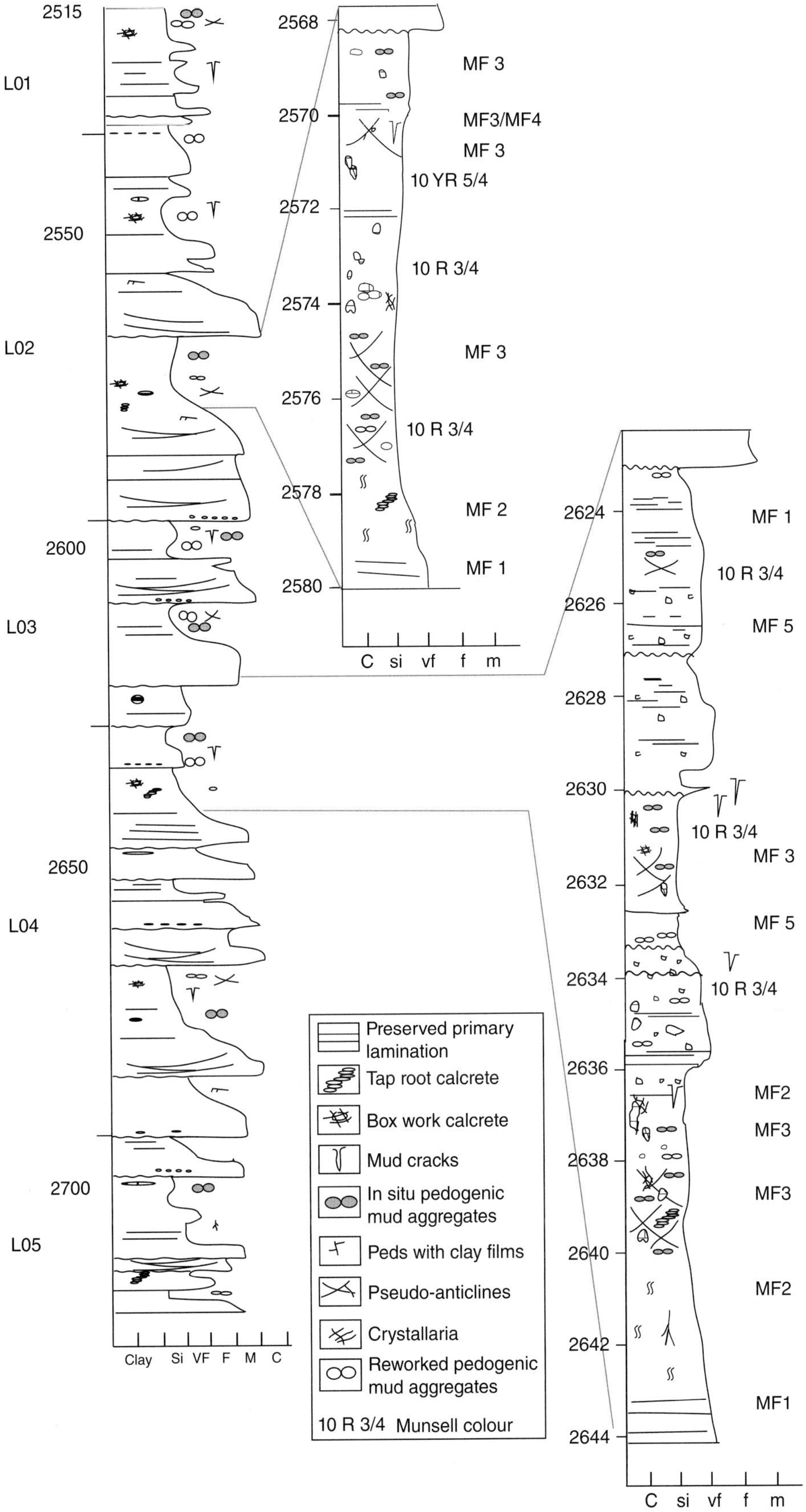

Fig. 13. Sediment and palaeosol logs from allostratigraphic units L05 to L01 in upper part of the Lunde Formation (LLF) showing distribution of mudstone facies MF1, 2, 3, 4 and 5 and various pedogenic features. Palaeosols form mainly cumulative pedocomplexes with high chroma vertisols (with desiccation cracks, slickensided surfaces, pseudo-anticlines, crystallaria). Colour is indicated by numbers referring to the scale of the Munsell Soil Color Chart (Munsell Color 1975) (example 10 R 3/4, brownish-red). Well 34/4-9 S. Scale is in m RKB (Rotary Kelly Bushing).

nodules. PA2, like PA1 contains scattered small calcrete nodules, whereas nodular carbonate horizons, vertically stacked carbonate nodules, rhizocretions, root structures and burrows are common, although not abundant. Compound pedocomplexes are slightly dominating, whereas composite and cumulative pedocomplexes are nearly equally represented (Fig. 11).

Palaeosol assemblage 3 (PA3)

Palaeosol assemblage PA3 characterises the thick floodplain successions in ULF (L04 to L01). The mudstones are mostly red and reddish-brown and dominated by mudrock facies MF3 with high chroma vertisols (Table 6, Figs 8, 10 and 13). Entisols/Inceptisols (MF2) and reworked mudrocks (MF5) are also common. Individual Vertisol profiles can be up to 6 m to 7 m thick and are commonly truncated and overlain by laminated mudrocks, reworked mudrocks or sheet-flood sandstones (Fig. 13). In PA3 some horizons (0.4 m to 0.7 m thick) with nearly coalesced carbonate nodules (c.50% to 60% of total mudrock volume) occur, often with complex overprinting of crystallaria (Müller *et al.* 2004). Some of the vertisols contain intensively slickensided horizons. Vertically stacked calcrete nodules, carbonate box-work structures, rhizocretions, root structures, burrows, soils peds (Müller *et al.*, 2004) and deep and wide sand-filled desiccation cracks, are all abundant features of PA3.

The vertical arrangement of mudrocks above channel sandstones is, in general, laminated mudrocks (MF1) overlain by entisols/inceptisols (MF2) and vertisols (MF3) (Fig. 12). Up to 2 m to 3 m thick zones of reworked mud aggregates are associated with vertisols, particularly in the transition zone from ULF to LSG. In well 34/4-9S in the Snorre Field, a bone of the plant-eating dinosaur *Plateosarus* has been found in deep-red floodplain mudstone in the allostratigraphic unit L03 (Hurum *et al.*, 2006).

Mudrocks with complex mottling patterns (MF4) also occur in the uppermost part of the Lunde Formation and in the lower part of the Statfjord Group (LSG). The mottling is mainly related to the presence of pseudo-gleying (reduction) along desiccation cracks and peds. A few thin (<0.2 m), olive-green, restricted gleyed horizons are identified (Table 6). These olive-green horizons have a generally higher goethite content than the adjacent mudrocks and are also present in the upper part of the Statfjord Group (USG) (see below). The mudrock facies of L04 to L01 are characterised by relatively high clay/sand ratios and percentages of very fine-grained material and are dominated by cumulative pedocomplexes (Fig. 11).

Palaeosol assemblage 4 (PA4)

Palaeosol assemblage PA4 is characteristic of floodplain successions in the allostratigraphic units S5 to S4 in the lower Statfjord Group (LSG). The mudrocks generally change from dominantly reddish-brown in S5 to increasingly mottled and greenish to grey upwards in S4. PA4 mudstones are mottled with low-chroma vertisols (MF4), replacing the red-coloured entisols/inceptisols or high chroma vertisols of PA3 in the underlying ULF.

Individual palaeosol horizons are up to 5m to 7 m thick. Laminated mudrocks (MF1) and reworked mudrocks (MF5) are very rare. The mottling pattern is mainly related to gleying and oxidation along cracks and ped margins. Carbonate nodules and rhizocretions become generally absent in unit S4. Slickensided horizons and soil peds are common, though not abundant, whereas desiccation cracks and root structures are rare. Coalified plant debris is common in mudrocks and channel sandstones. The palaeosols of LSG are typified by a high content (up to 9%) of goethite in some mudrock units (see below).

PA4 has a highly bimodal grain-size distribution, dominated by medium to coarse-grained sand and fine clays (<2 μm). The assemblage is characterised by some intensively slickensided horizons and complex mottling patterns and a dominance of cumulative pedocomplexes (Fig. 11).

Palaeosol assemblage 5 (PA5)

PA5 refers to the allostratigraphic units S3 to S1 in the upper part of the Statfjord Group (USG). The palaeosols in these mudrocks are dominantly grey and greenish-grey; a reddish-brown colour is very rare and observed only within S3. Slickensided surfaces are present, though not abundant and some angular soil peds have been recorded. As described above from FSA1, coalified root and rootlet structures occur, but are not abundant, whereas coalified plant debris (wood) is rather common at the base of fluvial channels. Dark grey organic-rich thin mudstone beds are present. Primary sedimentary lamination can be preserved to varying degrees in the mudstone beds. Sand-filled desiccation cracks are rare and only recorded in the lowermost part of the USG. Carbonate nodules, rhizocretions and soil peds are absent.

The palaeosols of PA5 can be characterised as inceptisols, entisols and low-chroma vertisols all of which form pedocomplexes. Most pedocomplexes are compound and composite. The dark greenish-grey palaeosols with high organic content are inferred to contain iron in the reduced state (Fe^{2+}), formed by gleying processes (Table 6). This palaeosol type occurs only in S2 and S1.

Clay mineralogy and iron oxides

The mineralogy of mudrocks changes through the Lunde Formation and the Statfjord Group. This is reflected in both the non-clay and clay mineral contents (Tables 2 and 3, Fig. 14). The quartz and feldspar content of the mudrocks (Table 3) reflects a similar stratigraphic trend to that in the fluvial channel sandstones (Table 2), with a moderate content of quartz in the Lunde Formation and a high content in the Statfjord Group, whereas the opposite trend is shown by the feldspars. As in the sandstones (Sørlie, 1996; Khanna *et al.*, 1997), albitic plagioclase feldspar dominates relative to potassium feldspar. No significant trend with burial depth has been shown for the mudrock minerals, whereas the stratigraphic trends are distinct (see below); for this reason it is concluded that the recorded variation in the mineral content of mudstones is primarily due to factors related to sediment source and depositional environment and not to burial diagenesis. The Lunde-Statfjord succession can be subdivided into five 'clay mineral assemblages' (CMA1-5, Fig. 3), according to the total amounts and relative occurrence of the minerals smectite, swelling mixed-layer clay minerals (ML), chlorite, illite and kaolinite (Table 4). Units L04 and S2 were not sampled as the mudstones of these units were anticipated to be very similar to those in L05 and S1, respectively. In addition, the relative amounts of the iron oxides goethite and hematite contribute to the characteristics of the clay mineral assemblages (Table 4, Fig. 14).

Samples from the upper part of the Alke Formation contain a clay mineral assemblage comprised of illite and kaolinite in practically equal amounts and a slight lesser amount of chlorite. Smectite is present only in negligible amounts and the swelling ML mineral content is also low. Goethite and hematite have not been detected in these samples.

Clay mineral assemblage 1 (CMA1)

The general distribution pattern of clay minerals in the Alke Formation continues through the mudrock intervals in the allostratigraphic units L12 and L11 in the lowermost part of the Lunde Formation (LLF). This is clay mineral assemblage CMA1 (Fig. 14). There appears to be a peak in the content of swelling mixed-layer (ML) minerals within L12. The combined content of illite and chlorite, though fluctuating, is higher than the kaolinite content. Illite is prominent in CMA1 with an average of 41% of the total clay mineral content, whereas kaolinite occupies on average 26% of all clay minerals. Hematite is the dominant iron oxide in this clay mineral assemblage (Table 4, Fig. 14).

Clay mineral assemblage 2 (CMA2)

CMA2 occurs within the allostratigraphic units L10 to L04, from the lower to the upper part of the Lunde Formation. The boundary between L11 and L10 shows a marked change in the relative proportion of the clay minerals. Most noticeable is the abrupt increase in smectite content, from practically zero in L12 and L11 to about 90% of all clay minerals in the lower part of L10 (Table 4, Fig. 14). Upwards from this maximum the smectite fluctuates, but reveals a distinct decreasing trend towards zero at the top of the assemblage, uppermost in unit L05. The relative content of swelling ML minerals is low in units L10 to L05 and does not reveal any trend through CMA2.

The relative amount of chlorite varies greatly through CMA2; there may be a slight upward-decreasing trend. The change in the smectite content at the lower boundary of CMA2 is accompanied by a drop in the illite content in the lower part of L10, succeeded by a general increase upwards to a maximum of about 80% of the clays within unit L05. The kaolinite content is generally low through the whole L10 to L05 interval, showing no specific trend through CMA2. Hematite is the totally dominant iron oxide in L05, showing a vertical increase to approximately 40% of clays and iron oxides in the uppermost part of L05, though followed by a distinct decrease in the uppermost part of CMA2, at the transition to CMA3 (Fig. 14).

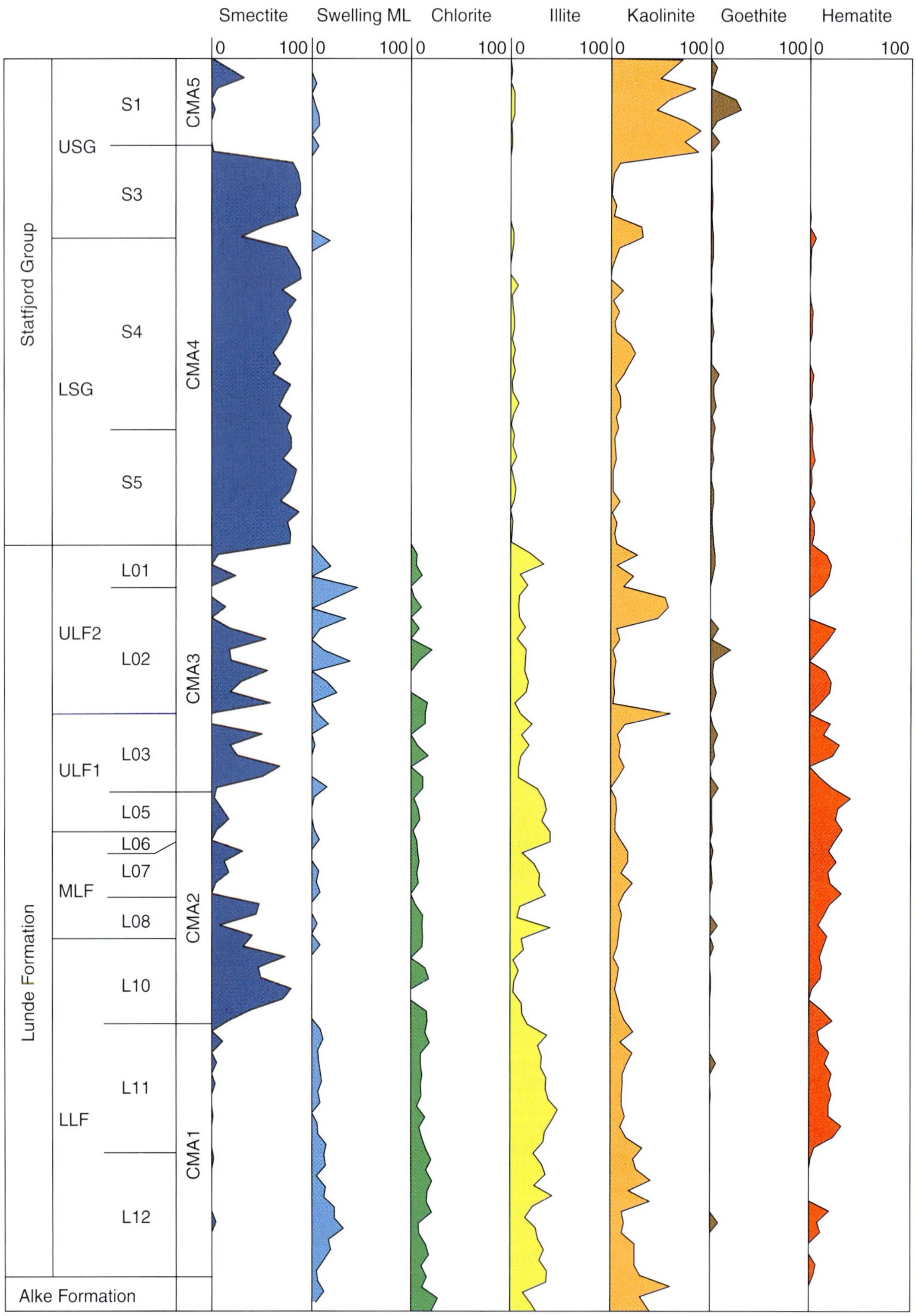

Fig. 14. Stratigraphic distribution of clay minerals (%) smectite, swelling mixed-layer (ML), chlorite, illite and kaolinite and iron oxides hematite and goethite through the Lunde Formation and the Statfjord Group.

Clay mineral assemblage 3 (CMA3)

Clay mineral assemblage CMA3, comprising the interval L03 to L01, is again marked by some characteristic trend patterns. The content of smectite varies a lot up through CMA3, from zero in some mudstone units up to about 80% of the total clay mineral content in some other floodplain mudstones. The overall smectite trend, though, is upward-decreasing, like that in the underlying CMA2 (Fig. 14). The content of swelling ML minerals shows generally an increasing trend from close to zero in the lower part of unit L03 to a maximum in the lower part of L01.

The chlorite content varies from zero to about 25% of clays + iron oxides (Fig. 14). The relative content of illite decreases from about 80% of clays + iron oxides in the upper part of L05 down to about 15% in L03 and L02. A slight increase may be present in L01 at the top of CMA3. The relative content of kaolinite varies considerably, with peaks in the upper parts of L03 and L02, separated by a low in the lower part of L02. There is a slight increase in goethite in part of CMA3, compared to the underlying and overlying strata. A goethite peak is recorded in a floodplain mudstone in the middle part of unit L02. The hematite content also varies within clay mineral assemblage 3 (Fig. 14).

Clay mineral assemblage 4 (CMA4)

Clay mineral assemblage CMA4 occurs in the allostratigraphic units S5 to S3, corresponding to LSG and the lower part of USG (Figs 3 and 14). CMA4 reveals the most dramatic change in clay mineralogy within the whole Lunde-Statfjord succession (Table 4, Fig. 14). The change takes place very abruptly from L01 to S5 and is reflected by all the clay minerals. Smectite dominates totally with a content that varies between 80% and 100% of all clays + iron oxides; however, the smectite content drops to zero at the top of CMA4. The swelling ML minerals disappear (except in one sample), chlorite is absent and the relative content of illite drops well below 10%. The content of kaolinite is generally low but increases abruptly in the upper part of S3, here accompanying an increasing frequency in channel sandstone bodies that coincides with the dramatic drop in the smectite content. Goethite and hematite are both practically absent within clay mineral assemblage 4.

Clay mineral assemblage 5 (CMA5)

The uppermost clay mineral assemblage CMA5 comprises the upper part of the Statfjord Group (USG), S2 and S1, also including FSA6, see above. The grey and greenish-grey mudstone and palaeosols of S1 are characterised by a low smectite content. The kaolinite content is generally high, varying between about 60% and 100% of all clay minerals. Swelling ML minerals, chlorite and illite are scarce or absent and goethite is the only iron oxide recorded (Table 4, Fig. 14).

Vertical trends and regional correlation

The Lunde-Statfjord succession can be subdivided into two major parts, each defined by characteristic vertical grain-size trends. The lower part, comprising the allostratigraphic units L10 to L01/S5, is fining-upwards, whereas the upper part, consisting of the allostratigraphic units S4 to S1, is coarsening-upwards. As mentioned above, the turnaround from fining-upward to coarsening-upward, located in the L01-S5 interval, was applied as the lithostratigraphic boundary between the Lunde Formation and the Statfjord Group by Deegan and Scull (1977).

The Lunde-Statfjord succession reveals several vertical trends in alluvial features, pedogenic features and mudrock mineralogy (Fig. 15). The fluvial sandstones in the Lunde Formation are interpreted as ephemeral streams in LLF and MLF, increasingly large meandering streams from the middle part of MLF leading to dominance of this stream type in ULF and then small, single storey meandering or/and straight streams in the floodplain-dominated uppermost part of the ULF. In the lower part of the Statfjord Group (Raude Formation) the fluvial style of small floodplain streams continues to dominate, whereas the fluvial style of the upper part of the Statfjord Group (Eiriksson Formation) is characterised by laterally extensive braided stream sandstone sheets, deduced to have been more perennial than ephemeral.

The main change in palaeosol features is from entisols/inceptisols and compound pedocomplexes in LLF and MLF to high chroma vertisols and cumulative pedocomplexes in the upper part of the Lunde Formation and lower part of the Statfjord Group and finally to inceptisols, entisols and low chroma vertisols, but decreased frequency of cumulative pedocomplexes in the Eiriksson Formation in the upper part of the Statfjord Group. Horizons with coalesced carbonate nodules

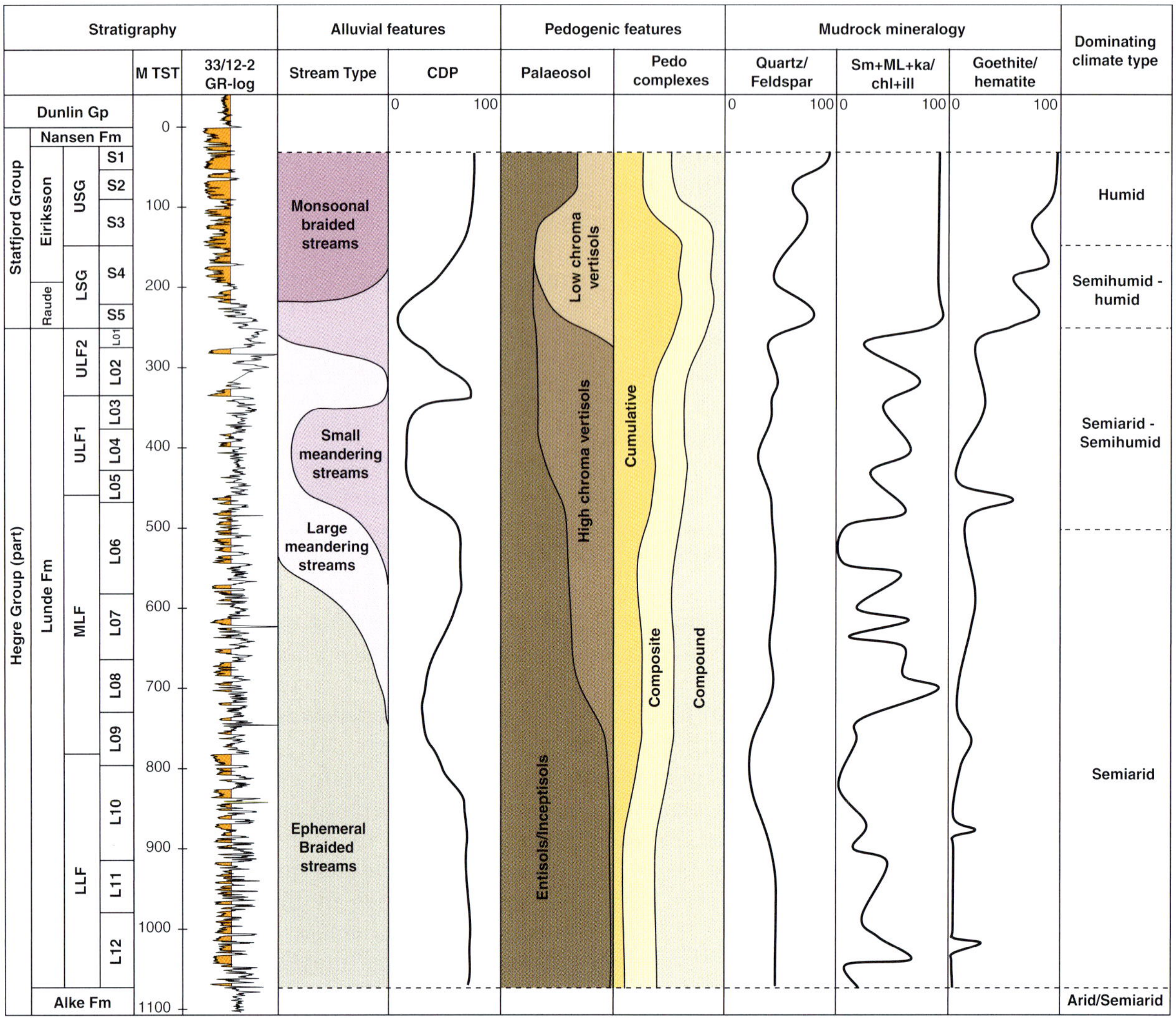

Fig. 15. Summary of trend variations through the Lunde Formation and the Statfjord Group in the Snorre Field. Stratigraphy and reference well as in Fig. 3. CDP = channel deposit proportion, Sm = smectite, ML = mixed layer clay minerals, ka = kaolinite, chl = chlorite, ill = illite.

increase upwards through the Lunde Formation, then decrease and disappear up through the Statfjord Group (Fig. 15).

In addition to distinct vertical trends in the mudrock mineralogy (Fig. 15), there is also an increasing content in the clay fraction of mudstones upward through the Lunde Formation and of organic-rich mudstone in the upper part of the Lunde Formation and the Statfjord Group. The Lunde Formation is characterised by a polymineralogical suite of clay minerals, varying in an apparent rhythmic pattern. At the boundary between the Lunde and Raude formations the ratio of smectite + mixed layer and smectite + kaolinite to chlorite + illite increases abruptly, with smectite dominating in units S5 to S3 and kaolinite in S1 (Figs 14 and 15). The increasing goethite/hematite ratio upwards from ULF corresponds to a change in mudstone colour from reddish-brown to mottled greenish-grey. There is a very strong correspondence in the stratigraphic positions of trend changes in alluvial features, pedogenic features and mudrock mineralogy (Figs 3 and 15).

The trend variation in architectural style and sandstone : mudstone ratio established within the Lunde Formation and the Statfjord Group in the Snorre Field area has been used for correlation within the Tampen Spur area (Fig. 16). The recorded trends reflect the response of various processes and mechanisms that controlled the alluvial

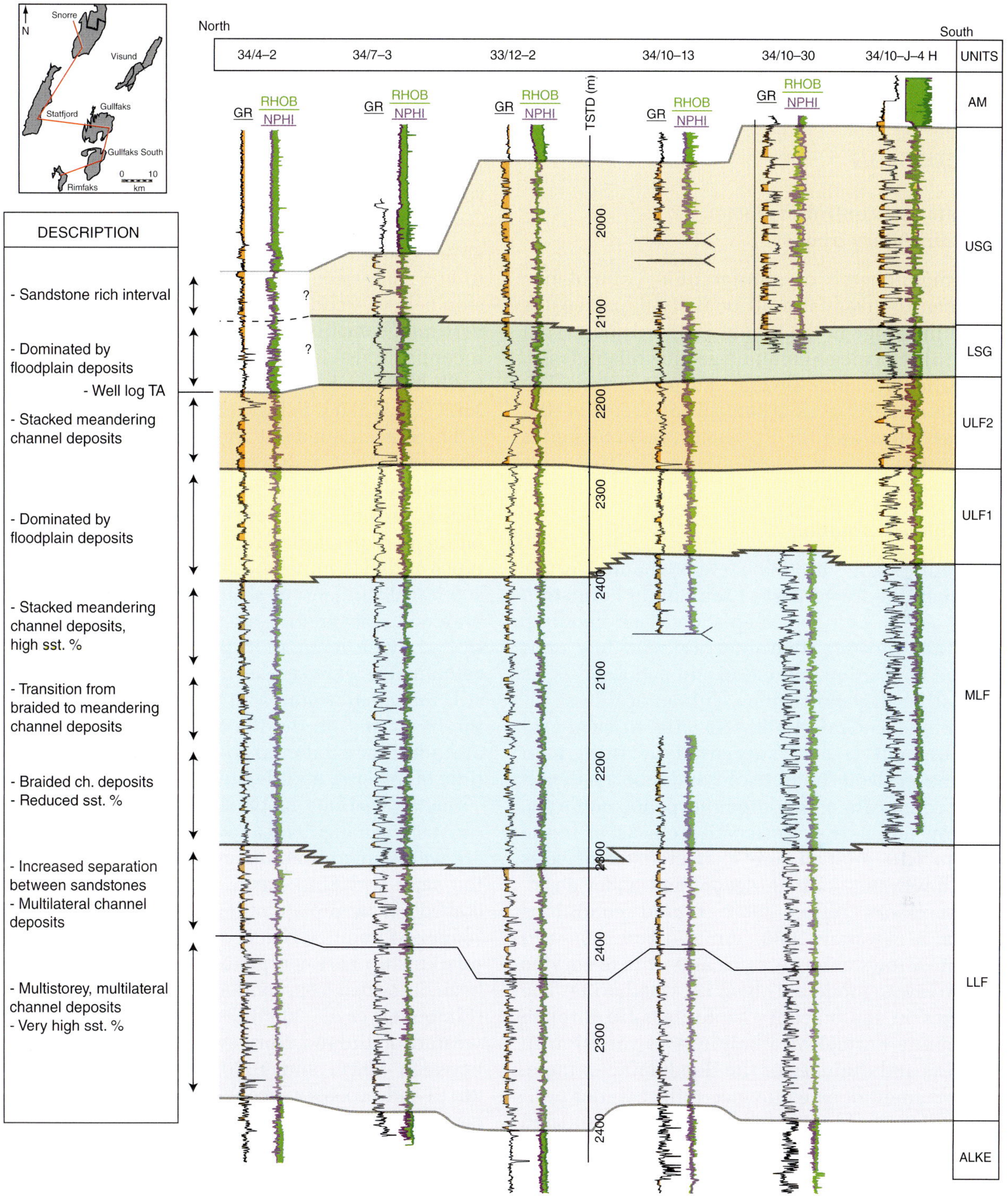

Fig. 16. Allostratigraphic correlation within the Lunde Formation and the Stafjord Group in the Tampen Spur area, northern North Sea, from the Snorre Field in the north to the Rimfaks Field in the south. LLF = lower Lunde Formation, MLF = middle Lunde Formation, ULF = upper Lunde Formation, LSF = lower Statfjord Group (Raude Formation), USG = upper Statfjord Group (Eiriksson Formation) and AM = Amundsen Formation.

depositional systems during latest Triassic and Early Jurassic in this part of the northern North Sea region.

DISCUSSION

Climate as a controlling factor on sedimentary trends

The high degree of stratigraphic coincidence between the trends in alluvial style, palaeosols and mineralogy in the Lunde-Statfjord succession (Fig. 15) indicates a common genetic relationship, controlled by one or several extrinsic factors. The impact of climate as a controlling factor is discussed below and then climate *versus* tectonics and eustasy.

The Lunde Formation

The ephemeral braided stream channel sandstones in the lower part of the Lunde Formation (LLF) can be ascribed to episodic flash flooding, separated by long time intervals with low rates of mud deposition, oxidation and pedogenesis, as typical for dryland regions with an arid to semi-arid climate (Graf, 1988). An arid to semi-arid environment is also supported by thin, high chroma entisols/inceptisols and poorly developed composite and compound pedocomplexes. Shallow and narrow desiccation cracks, as in the L12 to L10 interval, are characteristic of soils with low water content; deeper and wider desiccation cracks reflect more humid conditions (Dudal & Eswaran, 1988). An arid and semi-arid climate allows chlorite, mica and illite to survive due to low to moderate chemical weathering. The changes in the smectite/kaolinite ratio through the Lunde Formation likely reflect variation in wetness and drainage of the floodplain; an increasing content of smectite may reflect wetter periods and/or higher groundwater levels compared to periods when kaolinite dominated (Table 4, Fig. 14). A general arid to semi-arid climate would also explain the rather high content of feldspar in both mudrocks and channel sandstones found through most of the Lunde Formation (Tables 2 and 3, Fig. 15).

Mobile braided streams continued to dominate during deposition of the LLF and MLF; some violent floods carried extrabasinal clasts far out into the basin. The transition into meandering stream channel sandstones towards the top of the MLF may indicate a change from ephemeral to more perennial streams. Increased perennial flow would have transported more clay material in suspension to be deposited on levees and floodplains; together with denser river bank vegetation, this would have stabilised river channels and formed alluvial ridges. A high clay production by chemical weathering in the hinterland is reflected with the total of 45% to 50% floodplain mudrocks within the L09 to L05 succession (Figs 3 and 4) (see also discussion below). Reddish-brown laminated mudstone, very high hematite/goethite ratios and entisols and inceptisols also indicate semi-arid conditions and well-drained plains during the deposition of the MLF (Fig. 15) (cf. Schwertmann, 1988).

There is an apparent contradiction in the middle part of the Lunde Formation between various climatic signals: an increasing frequency of meandering streams indicates rising humidity, precipitation and perennial runoff, whereas palaeosol types and clay minerals still reflect a dominantly semi-arid climate. A shift from ephemeral and braided to meandering streams was probably controlled by an increase in precipitation in upland catchments, whereas clay minerals and palaeosols still reflect an overall arid to semi-arid climate in the basin and on the floodplain. The hinterland altitude is not inferred to have risen during the time of lower to middle Lunde Formation (Fig. 3).

Such variations in precipitation and temperature between highland and adjacent lowland areas are well known from the modern world; rivers in the semi-arid Ebro Basin in northern Spain are controlled by precipitation in the high Pyrenees (López-Moreno & Garcia-Ruiz, 2004). A similar relationship between precipitation and relief has been described from semi-arid regions in Morocco (Elmouden *et al.*, 2005). During the late Triassic, western mainland Norway was at an altitude of between 500 m and 1000 m (Gabrielsen *et al.*, 2010) and probably received more rain than adjacent lowland basins. A general shift from an overall semi-arid to a semi-humid climate continued through the late Rhaetian, as revealed in the upper part of the MLF and the ULF by the increasing frequency of major meandering stream belts, small single storey floodplain rivers (FSA3 and 4) and high chroma vertisols (PA3) (Figs 3 and 15). The overall increase in the clay/sand ratio upward through the ULF may indicate an increase in production of clay relative to sand in the catchment area, or a general increase in the accommodation space. The degree of soil development increased from the lower part of the middle Lunde Formation (MLF) towards the lower part of the Statfjord Group (LSG) (Fig. 15). This may be explained by a decrease in rate of accommodation and sediment accumulation, giving more time for pedogenic processes to operate. This is substantiated by the decline in average rate of sedimentation calculated for the Lunde Formation and the Statfjord Group (see above). The overall fining-upward trend of the Lunde Formation, particularly in the ULF and lowermost part of the Statfjord Group is thus interpreted to have been primarily controlled by increasing chemical weathering and clay production (Fig. 11). Subordinate fluctuations in accommodation space may have controlled formation of allostratigraphic units, as expressed in the vertical variation in the CDP value (Fig. 4). Alternatively, these cycles may also have been controlled by climatic variations, for example on an astronomic scale (see below).

The thick, multi-storey, sandstone body at the base of L02 most probably represents a time interval with increased sand influx. The great vertical thickness of the sandstone body could also imply an increase in accommodation space giving rise to aggradational stacking of several channel successions. However, this unit has been correlated over a wide area (Figs 1 and 16), showing that the rate of creation of accommodation space was not higher than it allowed the fluvial channel system to develop a wide meandering stream belt rather than isolated meander channel belts as would be the case with a high rate of subsidence (cf. Ethridge *et al.*, 1998).

Modern vertisols occur commonly in semi-arid to semi-humid regions with a pronounced dry season (Wilding & Tessier, 1988; Retallack, 2001) (Table 6). However, vertisols are highly complex and other factors than duration of dry and wet seasons are important; parent material (clay content, clay mineralogy, grain-size distribution) and time available for pedogenic processes may impact greatly on vertisol development (McGarry, 1996). Thick cumulative pedocomplexes of high chroma vertisols and with thin horizons with coalesced carbonate nodules, as in the uppermost part of the Lunde Formation (Figs 3 and 15), are favoured by low rates in creation of accommodation and sediment aggradation (see above). Vertisols of high chroma (reddish colour) are presumed to correspond to well-drained floodplains and increased duration of the dry season, whereas deep and wide desiccation cracks associated with intensively slickensided high chroma vertisols, as in ULF2, can be explained by a deep wet zone in soils (Dudal & Eswaran, 1988). The abrupt increase in smectite at the base of ULF (Fig. 14), followed by a fluctuating decline in the smectite content, may indicate an increase in water content of the soil, followed by a time interval with drier climate or improved floodplain drainage.

Coalesced carbonate nodules, comprising c. 50% to 60% of the total mudrock volume at certain horizons in PA3 in the ULF, represent the most advanced calcareous palaeosols. Pedogenic carbonate is useful for understanding palaeoclimatic conditions (Mack, 1992, Driese & Mora, 1993; Retallack, 2001). Calcic horizons are mainly restricted to areas with less than 1000 mm annual precipitation and are most common where the yearly precipitation is below 750 mm (Cerling, 1984). Royer (1999) documented that soils with more than 95% pedogenic carbonate horizons correlate with an annual precipitation below 760 mm. Some carbonate pedofeatures though, like micritic root structures, can form in environments with up to 1500 mm annual precipitation (Reeves, 1970).

The dominance of hematite relative to goethite in mudrocks in the upper part of the Lunde Formation (Figs 14 and 15) favours that the floodplain was dry and the soil aerated most of the year. However, the presence of complex mottling with pseudo-gleying and the occurrence of goethite along desiccation cracks and peds show that there was a high water content in the soil and reduction of some Fe^{3+} to Fe^{2+}. The mottling pattern is probably caused by fluctuating groundwater and/or periodic stagnant surface water (cf. Piipujol & Buurman, 1994). In modern vertisols decreasing soil colour value and chroma generally reflect an increase in soil moisture (Dudal & Eswaran, 1988). The presence of lower chroma (greyish-green) vertisols in the ULF may thus correlate to periods of more prolonged flooding associated with wetter soil conditions. The wet season of plant growth on the floodplain must have been long enough to supply food for herds of plant-eating dinosaurs on migration through the lowland, like *Plateosarus* (Hurum *et al.* 2006). Altogether, palaeosols and clay minerals within the ULF indicate that as well as seasonal variation in precipitation there was periodic, longer term, change at the basin floor from an overall semi-arid to semi-humid climate.

The Statfjord Group

The fluvial style of the uppermost part of the Lunde Formation is continued into S5 and S4 of the Statfjord Group by predominantly isolated fluvial channel sandstones, inferred to represent single and isolated rivers within muddy floodplains. The high mudrock: sandstone proportion in S5 to S4 and clay mineral composition reflect a continuing high rate of chemical weathering (Table 6, Fig. 14).

The transition from the mostly reddish-brown mudrocks in S5 to the increasingly mottled and greenish to grey mudrocks upwards in S4 is accompanied by dominantly mottled and low-chroma vertisols (MF 4), replacing red-coloured entisols/inceptisols or high chroma vertisols. All these features indicate increasing water content in the soil. The rare occurrence of aggregated horizons in LSG may also imply wetter conditions, as the depth and degree of self-mulching and soil aggregation are closely related to soil moisture (Probert *et al.*, 1987; Dudal & Eswaran, 1988, McGarry, 1996).

The presence of organic material and coalified plant debris in mudrocks and channel sandstones in the S4 unit indicates a high groundwater table on the floodplain and thus an increase in vegetation cover and preservation of organic material in swamps and depressions. The rich occurrence of *Deltoidospora spp.* in the SG is particularly indicative of humid conditions (Hubbard & Boulter, 2000). The influx of these palynomorphs might also reflect changes in vegetation type. Increasing humidity is also reflected by the decrease in carbonate nodules and rhizocretions through S5 to their absence in S4. The LSG is also typified by a high content (up to 9%) of goethite in some mudrock units (Table 6).

Palaeosol development up through the ULF and LSG generally shows better developed palaeosol types and thicker cumulative pedocomplexes. This general trend is also reflected in the clay mineralogy by the increase in abundance of kaolinite and smectite at the expense of mixed-layer clay minerals, chlorite and illite. The marked increase in the quartz: feldspar ratio in fluvial channel sandstones and mudrocks from the ULF to the LSG is probably the result of increased chemical weathering, triggered by increased humidity in a warm climate (Tables 2 and 3, Figs 14 and 15).

The characteristic CU pattern of the upper part of the Statfjord Group and the increase in braided stream channel sandstone bodies with sheet-like geometry in FSA6 demonstrate a decrease in the A/S ratio towards the upper termination of the unit S1. The quartz-arenitic composition of sandstones in the upper part of the Statfjord Group (Sinemurian) is interpreted to reflect advanced chemical weathering in a humid and warm climate in the hinterland. The palaeosol types with overall greenish-grey colour, absence of carbonate nodules, rhizocretions and mottling reflect a humid climate with overall water-saturated soils on the alluvial plains. This is also documented by coal fragments in fluvial channel conglomerates and mudstones with high contents of plant debris.

The clay mineralogy, being characterised by a sudden drop in smectite followed by an abrupt increase in kaolinite (Figs 14 and 15), can be explained by a warm humid climate promoting production of kaolinite through advanced chemical weathering of hinterland bedrocks and/or of rather well-drained floodplain sediments (cf. Girty, 1991). Improved drainage in the USG, relative to LSG, would result from an increasing influx of sand, forming steeper-sloping and very wide alluvial fans or braidplains. Ryseth (2001) also emphasised that the highest depositional gradients in the Statfjord Formation were in the Tampen Spur area. The relatively high occurrence of greenish-grey entisols/inceptisols *versus* low chroma vertisols in the allostratigraphic units S2 and S1 (Fig. 15) may be explained by low sediment stability on alluvial plains/fans due to reworking from highly mobile braided streams during annual monsoonal seasons.

Tectonics and eustasy *versus* climate as cause of change in depositional trends

In most alluvial basins, tectonics, eustasy and climate interact. Unless the basin is terminal and beyond the reach of sea-level fluctuations, eustasy would also to some degree influence the sedimentation patterns in such basins. Interpreting the relative contribution of the various factors can be difficult. This problem is well demonstrated by the Lunde-Statfjord succession in the northern North Sea.

Steel & Ryseth (1990) and Steel (1993) explained the vertical grain-size trends in the Lunde-Statfjord succession primarily as controlled by variation in tectonic subsidence. The increase in coarse sand in the Statfjord Group has mostly been explained in terms of tectonic uplift and consequent increased run-off, forced by increased precipitation in the hinterlands (Nystuen *et al.*, 1989; Steel & Ryseth, 1990; Steel, 1993; Nystuen & Fält, 1995; Ryseth & Ramm, 1996; Ryseth, 2001; Goldsmith *et al.*, 2003). Differential subsidence, driven by tectonics and/or by compaction, may account for thickness variations in the Lunde Formation and the Statfjord Group in the Tampen Spur area (Fig. 16). Regional thickness variation is particularly pronounced in the USG (Fig. 16) and this has been explained by syn-depositional growth faulting (Ryseth & Ramm, 1996).

Weltje *et al.* (1998) pointed out that in most basin studies changes in depositional stacking pattern were ascribed to changes in accommodation driven by tectonics and eustasy, whereas much less attention was given to the role of sediment supply controlled by climate. Leeder *et al.* (1998) presented examples of how climate and climatic changes controlled basin infill architecture. Tectonics as a major factor in the formation of fining-upwards and coarsening-upwards trends within the Lunde-Statfjord succession may have been less influential than previously considered. The high quartz content of sandstones and mudstones in the Statfjord Group (Table 2, Fig. 15) favours that sand was formed by advanced chemical weathering of granitic bedrock and not by erosion of juvenile granite or gneiss exposed due to rapid tectonic uplift. This is supported by the association of the quartz-arenitic sandstone with smectite-rich or kaolinite-rich mudstones, products of increased chemical weathering (Fig. 15). The weathered quartz-rich sand may have been stored in the hinterlands over long geologic times, trapped by depositional and/or geomorphic thresholds (cf. Schumm, 1977) and then delivered to the basin by breakthrough of thresholds during events of very high precipitation and flooding in Sinemurian times. It cannot be excluded that tectonic movements in the hinterland may have affected the run-off pattern.

As pointed out by McKie & Williams (2009), it is likely that the fluvial systems of the Triassic basins in the northern North Sea region were terminal until the onset of the Rhaetian transgression. However, the Norian-Rhaetian Lunde Formation has not revealed any signatures of a hydrologically closed system, in contrast to the underlying Alke Formation (Lervik, 2006). The difference in microfloral composition between the Alke and Lunde formations, recognised by Eide (1989) and ascribed by her as the result of different depositional environments (see above), might be the signature of a change from a closed to an open fluvial system. Eide (1989, p. 296) further suggested that "local and temporary marginal marine influence may be reflected in the sporadic presence of single and questionable specimens of the acritarch *Schizocystia* sp" (her Assemblage I, Lunde Formation). Goldsmith *et al.* (2003) recognised that both Assemblage II (Alke Formation) and Assemblage I of Eide (1989) include rare marine and brackish microplankton and concluded from a regional correlation that the transgression in the North Sea area started at the sequence stratigraphic surface Tr42, corresponding to an interval in the upper part of the Alke Formation/lower part of the Lunde Formation.

The regional unconformity at the base of the Lunde Formation, documenting the establishment of large fluvial channel systems (see above), may represent a change in the Tampen Spur area from a closed to an open regional fluvial drainage system. There is no other candidate surface indicating such a fundamental shift in type of base level control in the whole Lunde-Statfjord succession. The ultimate base level of the Lunde-Statfjord fluvial system is thus inferred to have been embayments of the sea encroaching onto the northern North Sea region; Tethys in the south-east and/or the Borealic Sea in the north. However, shorelines must have been several hundreds of kilometres away from the alluvial basin during deposition in the Rheatian and early Hettangian (Ryseth, 2001; Goldsmith *et al.* 2003; Müller *et al.* 2005; McKie & Williams, 2009).

Effects of eustatic sea-level fluctuations can nevertheless be felt up to some hundred kilometres landward from the shoreline (Schumm, 1993; Shanley & McCabe, 1994; Ethridge *et al.*, 1998; Posamentier & Allen, 1999; Catuneanu, 2006). Repeated rise and fall in relative sea-level during the Triassic has been documented as transgressive-regressive (T/R) cycles in shallow marine basins in Arctic Canada, Spitsbergen and the western Barents Sea (Johannessen & Embry, 1989; Mørk *et al.*, 1989; Glørstad-Clark *et al.*, 2010, 2011). Embry (1997) suggested that some of these cycles were global. The Rhaetian-Hettangian continental basins in the North Sea region may thus have been affected by several stratigraphic base level fluctuations that controlled the rhythmitic allostratigraphy in the Tampen Spur area. It is beyond the stratigraphic resolution of the present biostratigraphic data to correlate high-frequency cyclic sedimentation in shallow-marine and alluvial basins in northern Pangaea during the Late Triassic-Early Jurassic. The rhythmites can equally well be explained by cyclic changes in climate, like the allostratigraphic valley fill successions described from the Pleistocene (Blum, 1993), without invoking any change in base level.

Cause and wider regional aspects of the Triassic–Jurassic climate change

The Hettangian-Sinemurian climatic change from arid/semi-arid to humid conditions has been recorded further south in the North Sea (Frostick *et al.*, 1992; Goldsmith *et al.*, 2003), in the Mid-Norwegian shelf (Müller *et al.*, 2005), in East Greenland (Clemmensen *et al.*, 1998) and in southern Scandinavia (Arndorff, 1993; Ahlberg *et al.*, 2002). McKie & Williams (2009) suggested that arid basins in the northern North Sea region during Late Triassic were supplied by sediments from Fennoscandia and Greenland by winter runoff and from the Variscan Mountains in the south by summer monsoon flooding. The Shetland Platform area was also a hinterland furnishing the Tampen Spur area with clastic material, in addition to a hinterland to the north-west of the northern North sea area, as indicated by samarium-neodymium isotope studies (Mearns *et al.*, 1989). Bedrock geology, relief, drainage morphology and local climatic conditions in the hinterland areas were of crucial significance for sediment production and sediment transport to the alluvial basin.

In Late Triassic times the central northern North Sea region was located about 33°N to 35°N; and in Early Jurassic about 42°N to 43°N (Torsvik *et al.*, 2002; Goldsmith *et al.* 2003). During this northward drift the North Sea region moved out of the high pressure belts and into temperate regions with humid climate. The change in climate might also have been influenced by the Rhaetian marine transgression from the Tethys (Ahlberg *et al.*, 2002). Superimposed on this regional climatic change, induced by plate tectonic movements, there appears to have been a global temperature increase at the Triassic-Jurassic boundary, due to the release of large volumes of CO_2 to the atmosphere brought about by the igneous activity in the Central Atlantic Magmatic Province (Hesselbo *et al.*, 2002; Ruhl & Kürschner; 2011 and references therein). Cycles of temperature change and associated monsoonal precipitation, triggered by variation in the CO_2 content of the atmosphere (Hesselbo *et al.*, 2002; Bonis *et al.*, 2012) and Milankovitch-scale changes in temperature (Ruhl *et al.*, 2010; Bonis *et al.*, 2010), may thus also have influenced runoff during the Rhaetian-Hettangian in the northern North Sea region, modified by local catchment morphology. However, studies of modern runoff in Europe have shown that low-frequency variance (long-term variations) in runoff on a continental scale appears to be positively correlated with mean climatic conditions (temperature and precipitation), more than with catchment properties (size, elevation and geomorphology) (Gudmundsson *et al.*, 2011). This may also have been the case in North-west-Europe during the Late Triassic-Early Jurassic climatic change.

A climatic depositional model for the Lunde Formation and lower part of the Statfjord Group in the northern North Sea region is summarised in Fig. 17. A depositional model of the upper part of the Statfjord Group (S3 to S1) is comparable with that of L12 to L08, but in a humid climate, with monsoonal precipitation and high-energy flooding in mobile braided streams instead of in an arid or semi-arid climate characterised by ephemeral floods.

Impact on alluvial architecture and consequences for hydrocarbon exploration

The change in fluvial style in response to climatic changes impacted the sand:mud ratio and the alluvial architecture and palaeosol development in the Lunde-Statfjord succession. Variation in channel belt width/thickness and avulsion frequency on the geometric properties of fluvial sandstone bodies has been modelled in several studies (Leeder, 1978; Bridge & Leeder, 1979; Bridge & Mackey, 1993; Mackey & Bridge, 1995) and also tested in outcrops (Zaleha, 1997; Bridge *et al.*, 2000; Kjemperud *et al.* 2008). In these studies, variation in channel belt proportion (CDP) has been primarily related to different avulsion frequencies and channel belt dimensions (width and thickness). The decrease in CDP upward through the Lunde Formation and the Statfjord Group (Fig. 5) may be ascribed to reduced avulsion frequencies caused by an overall change from ephemeral, braided rivers to more perennial, meandering rivers. The return to braided streams during deposition of the upper part of the Statfjord Group is here considered to have been forced by high periodic influx of coarse-clastic debris and high runoff caused by increased

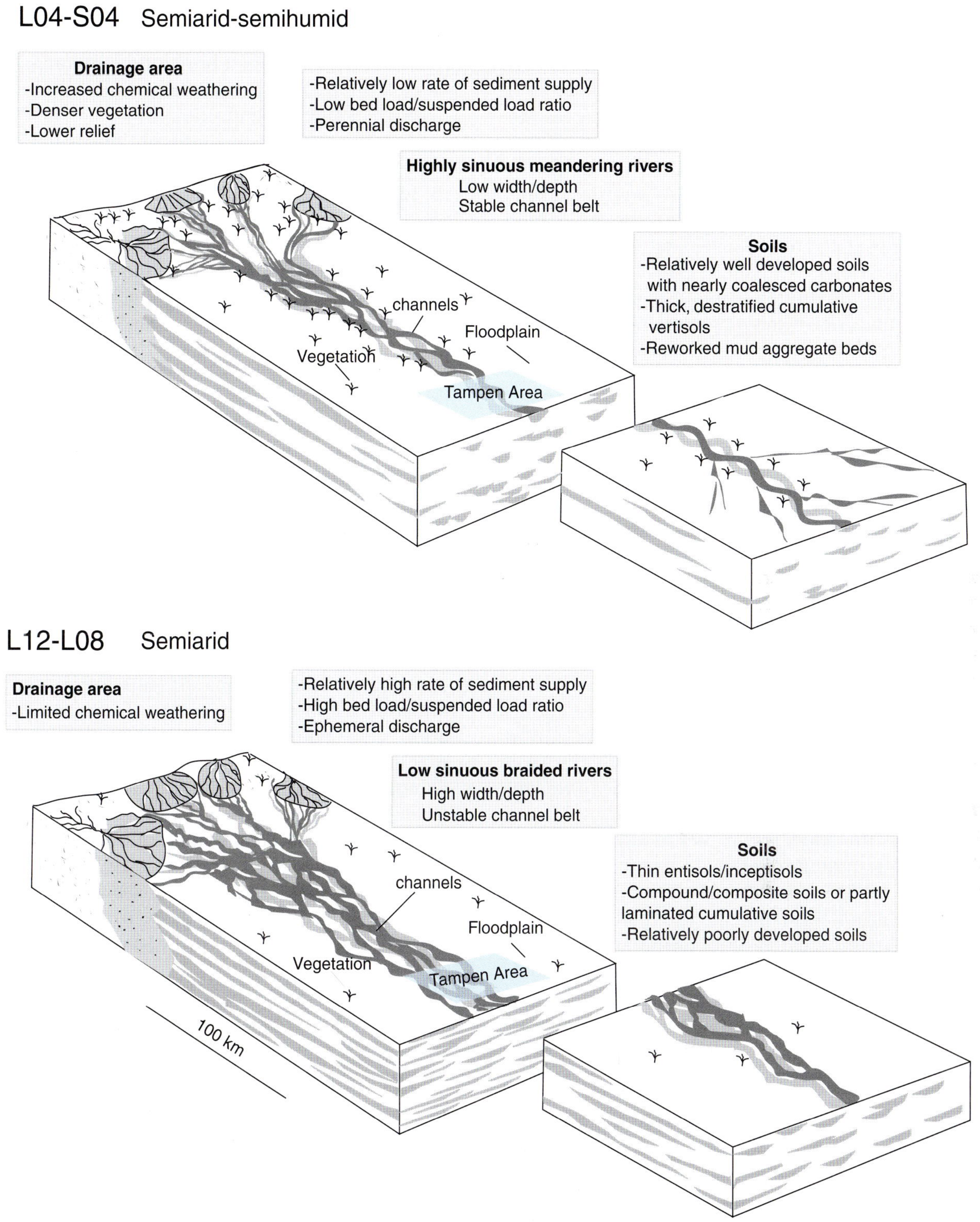

Fig. 17. Climatic depositional models for the fluvial Lunde Formation and the fluvial part of the Stafjord Group (Raude and Eiriksson formations), integrating variation in fluvial style and palaeosol types with characteristics of alluvial architecture and soil types.

precipitation in hinterland areas during monsoon seasons. The decrease in CDP upward through the Lunde Formation and in the lower part of the Statfjord Group is obviously a function of increased clay/sand ratio of the sediment load, whereas the opposite relation explains the increase in CDP in the upper part of the Statfjord Group. Changes in sediment supply changed the A/S ratio; a shift from $A/S < 1$ to $A/S > 1$ would have caused a change from amalgamated to isolated channel belts deposits and *vice versa*.

Properties of fluvial Upper Triassic to Lower Jurassic sandstone reservoirs in the northern North Sea region have been reviewed by Diesen *et al.* (1995), Nystuen & Fält (1995) and Lundschien & Mørch (1998). The climatic control on depositional architecture, sandstone facies and textural and mineralogical composition is of great importance for further exploration for Triassic reservoirs in the northern North Sea region. The climate, as an extrinsic control factor on alluvial sedimentation, has given rise to deposition of allostratigraphic units of high correlative potential.

CONCLUSIONS

1 A long term shift from arid/semi-arid to humid climate in the northern North Sea region during the Late Triassic (Rhaetian) to Early Jurassic (Hettangian-Sinemurian) was primarily related to drifting of the European continental plate towards the north and is recorded in the continental Lunde Formation (Rhaetian) and the Raude (Rhaetian-Hettangian) and Eiriksson (Sinemurian) formations in the Statfjord Group.

2 The climatic change had profound influence on variation in weathering, sediment production, runoff and sediment discharge and is reflected by changes in fluvial style, palaeosol development and of sandstone and mudrock mineralogy.

3 Climatic signals and changes in these given by alluvial style, palaeosols and mineralogy generally coincide. Responses of the sedimentary system to the climate change gave rise to a long-term upward-fining (Lunde Formation) followed by an upward-coarsening (Statfjord Group) succession with corresponding changes in channel belt proportion and alluvial architecture.

4 During deposition of the lower part of the Lunde Formation (Rhaetian), arid to semi-arid climate gave rise to ephemeral braided streams, entisols/inceptisols with scattered calcrete nodules and hematite in reddish-brown floodplain mudstones, high amounts of detrital feldspar and a diversified clay mineral suite.

5 In late Rhaetian times, increased precipitation gave rise to dominantly meander channel belts in the upper part of the Lunde Formation, reflecting cyclic changes in hinterland precipitation, whilst paleosols and clay minerals in floodplain mudstones were controlled by climate and drainage on the alluvial plain.

6 The Raude Formation represents a marked climatic change from semi-arid to semi-humid climatic conditions, reflected by thick mudstone beds and small single storey channel sandstone bodies and wetplains with mudrocks dominated by smectite as clay mineral and generally mottled and low-chroma vertisols.

7 The Eiriksson Formation reflects a change to humid climate with quartz-arenitic braided stream sandstones formed during monsoonal high runoff and flushing of hinterland storages of chemically weathered sand and greenish-grey mudstones dominated by kaolinite formed in well-drained alluvial fans or plains.

8 Rhythmitic successions of sets of alluvial allostratigraphic units in the Tampen Spur area were controlled by cycles of varying precipitation in the hinterland area, possibly in combination with changes in base level or differential subsidence. Regional correlation of allostratigraphic successions and their sandstone reservoir properties are of fundamental importance for further hydrocarbon exploration in the continental Upper Triassic-Lower Jurassic in the northern North Sea.

ACKNOWLEDGEMENTS

The authors would thank the former Saga Petroleum and the present Statoil for allowing us to perform this study and giving permission for publication of the results. We owe gratitude to a number of people that have been working with the Upper Triassic and the Lower Jurassic on the Tampen Spur in the northern North Sea area through many years for cooperation and a series of discussions. Two anonymous reviewers have given very valuable input to the final version of this paper. We thank Adrian Read for having corrected the English of the manuscript and also for scientific discussions of the topic. We also thank Editor Ron Steel for his work with the manuscript.

REFERENCES

Adestål, V. (2002) *Paleoclimatic conditions during late Triassic to early Jurassic, northern North Sea: evidence from clay minerals.* Master of Science Thesis. University of Oslo and University of Uppsala, 37 pp.

Ahlberg, A., Arndorff, L. and **Guy-Ohlson, D.** (2002) Onshore climate change during the Late Triassic marine inundation of the Central European Basin. *Terra Nova*, **14**, 241–248.

Arndorff, L. (1993) Lateral relations of deltaic palaeosols from the Lower Jurassic Rønne Formation on the Island of Bornholm, Denmark. *Palaeogeogr., Palaeoclimatol., Palaeoecol.*, **100**, 235–250.

Badley, M.E., Price, J.D., Rambech Dahl, C. and **Agdestein, T.** (1988) The structural evolution of northern Viking Graben and its bearing upon extensional models of basin formation. *J. Geol. Soc. London*, **145**, 455–472.

Bestland, E. (1997) Alluvial terraces and palaeosols as indicators of early Oligocene climate change (John Day Formation, Oregon). *J. Sed. Res.*, **67**, 840–855.

Blanc-Valleron, M.M. and **Thiry M.** (1997) Clay minerals, paleoweatherings, paleolandscapes and climatic sequences: the Paleogene continental deposits in France. In: *Soils and Sediments: Mineralogy and Geochemistry* (Eds H. Paquet and N. Clauer), Springer Verlag, Heidelberg, 223–247.

Blum, M.D. (1993) Genesis and architecture of incised valley fill sequences: a late Quaternary example from the Colorado River, Guld Coastal Plain of Texas. In: *Siliciclastic Sequence Stratigraphy: Recent Developments and Applications* (Eds P. Weimer and H.W. Posamentier), *AAPG Mem.*, **58**, 259–283.

Blum, M.D. and **Törnqvist, T.** (2000) Fluvial response to climate and sea-level change: a review and look forward. *Sedimentology*, **47**, 2–48.

Bonis, N.R. and **Kürschner, W.M.** (2012) Vegetation history, diversity patterns and climate change across the Triassic/Jurassic boundary. *Paleobiology*, **38**, 240–264.

Bonis, N.R., Ruhl, M. and **Kürschner, W.M.** (2010) Milankovitch-scale palynological turnover across the Triassic-Jurassic transition at St. Audrie's Bay, SW UK. *J. Geol. Soc. London*, **167**, 877–888.

Bridge, J.S. (1993) Description and interpretation of fluvial deposits: a critical perspective. *Sedimentology*, **40**, 801–810.

Bridge, J.S. (2005) *Rivers and Floodplains – Forms, Processes and Sedimentary Record.* Blackwell Publishing, 491 pp.

Bridge, J.S. and **Leeder, M.R.** (1979) A simulation model of alluvial stratigraphy. *Sedimentology*, **26**, 617–644.

Bridge, J.S. and **Mackey, S.D.** (1993) A revised alluvial stratigraphy model. In: *Alluvial Sedimentation* (Ed. by M. Marzo & C. Puigdefabregas), *Int. Assoc. Sedimentol. Spec. Publ.*, **17**, 319–336.

Bridge, J.S. Jalfin, G.A. and **Georgieff, S.M.** (2000) Geometry, lithofacies and spatial distribution of Cretaceous fluvial sandstone bodies, San Jorge Basin, Argentina: Outcrop analog for the hydrocarbon-bearing Chubut Group. *J. Sed. Res.*, **70**, 341–359.

Catuneanu, O. (2006) *Principles of Sequence Stratigraphy.* Elsevier, 375 pp.

Cerling, T.E. (1984) The stable isotopic composition of modern soil carbonate and its relation to climate. *Earth Planet. Sci. Lett.*, **71**, 229–240.

Charnock, M.A., Kristiansen, I.L., Ryseth, A. and **Fenton, J.P.G.** (2001) Sequence stratigraphy of the Lower Jurassic Dunlin Group, northern North Sea. In: *Sedimentary Environments Offshore Norway – Paleozoic to Recent* (Eds O.J. Martinsen and T. Dreyer), *Nor. Petrol. Soc. (NPF), Spec. Publ.*, **10**, Elsevier, Amsterdam, 145–174.

Clemmensen, L., Kent, D. V. and **Jenkins, F. A., Jr.** (1998) A late Triassic lake system in East-Greenland: facies, depositional cycles and palaeoclimate. *Paleogeogr., Paleoclimatol., Paleoecol.*, **140**, 135–159.

Dahl, N. and **Solli, T.** (1993) The structural evolution of the Snorre Field and surrounding areas. In: *Petroleum Geology of NW Europe: Proceedings of the 4th Conference* (Ed. J.R. Parker), Geological Society, London, 1159–1166.

Deegan, C.E. and **Scull, B.J.** (compilers) (1977) A standard lithostratigraphic nomenclature for the Central and Northern North Sea. Report, Institute of Geological Sciences London, **77/25**; and *Norw. Petrol. Direct. Bull.*, **1**, 36 pp.

Diesen, G.W., Edvardsen, A., Nystuen, J.P., Sverdrup, E. and **Tollefsrud, J.I.** (1995) Geophysical and geological tools and methods used for reservoir characterization and modelling of the Snorre Field - North Sea, Norway. In: *Reservoir Characterization: Integration of Geology, Geophysics and Reservoir Engineering* (Ed. H. Miyazaki), The third JNOC-TRC International Symposium February 20–23, 1995, TRC, Chiba, Japan, *Technology Research Center Special Publication*, **5**, Japan National Oil Corporation, 69–90.

Driese, S.G. and **Mora, C.I.** (1993) Physico-chemical environment of pedogenic carbonate formation in Devonian vertic palaeosols, central Appalachians, U.S.A. *Sedimentology*, **40**, 199–216.

Dudal, R. and **Eswaran, H.** (1988) Distribution, properties and classification of Vertisols. In: *Vertisols: Their Distribution, Properties, Classification and Management* (Eds L.P. Wilding & R. Puentes). College Station, Texas, Texas A & m University Printing Center, 1–22.

Eide, F. (1989) Biostratigraphic correlation within the Triassic Lunde Formation in the Snorre area. In: *Correlation in Hydrocarbon Exploration* (Ed. J.D. Collinson), Norw. Petrol. Soc. (NPF), Graham & Trotman, London, 291–297.

Elmouden, A., Bouchaou, L. and **Snoussi, M.** (2005) Constraints on alluvial clay mineral assemblages in semiarid regions; the Souss Wadi basin (Marocco, north-western Africa). *Geol. Acta*, **3**, 3–13.

Embry, A.F. (1997) Global sequence boundaries of the Triassic and their identification in the Western Canada sedimentary basin. *Can. Petrol. Geol. Bull.*, **45**, 415–433.

Ethridge, F.G., Wood, L.J. and **Schumm, S.A.** (1998) Cyclic variables controlling fluvial sequence development: problems and perspectives. In: *Relative Role of Eustasy, Climate and Tectonism in Continental Rocks* (Eds K.W. Shanley and P.J. McCabe), *SEPM Spec. Publ.*, **59**, 17–29.

Evans, D., **Graham, C.** and **Bathurst, P.** (Eds) (2003). *The Millenium Atlas: Petroleum Geology of the Central and Northern North Sea.* Geol. Soc. London, 389 pp.

Fält, L.M. (1987) Tectonic control on sedimentation in the Statfjord Formation, Statfjord Field, N. North Sea. *Nor. Geol. Tidsskr.*, **67**, 433–434.

Feldman, H.R., **Franseen, E.K.**, **Joeckel, R.M.** and **Heckel, P.H.** (2005) Impact of longer-term modest climate shifts on architecture of high-frequency sequences (cyclothems), Pennsylvanian of Midcontinent U.S.A. *J. Sed. Res.*, **75**, 350–368.

Ferrell, R.E. Jr. and **Dypvik, H.** (2009) The mineralogy of the Exmore beds-Chickahominy Formation boundary section of the Chesapeake Bay impact structure revealed in the Eyreville core. *Geol. Soc. Am. Spec. Pap.*, **458**, 723–746.

Fielding, C.R., **Allen, J.P.**, **Alexander, J.** and **Gibling, M.R.** (2011) Facies model for fluvial systems in the seasonal tropics and subtropics. *Geology*, **37**, 623–626.

Fielding, C.R., **Ashworth, P.J.**, **Best, J.L.**, **Prokocki, E.W.** and **Sambrook Smith, G.H.** (2012) Tributary, distributary and other fluvial patterns: What *really* represents the norm in the continental rock record? *Sed. Geol.*, 261–262, 15–32.

Frostick, L.E., **Linsey, T.K.** and **Reid, I.** (1992) Tectonic and Climatic control of Triassic sedimentation in the Beryl basin, Northern North Sea. *J. Geol. Soc. London*, **149**, 13–26.

Gabrielsen, R.H., **Færseth, R.B.**, **Steel, R.J.**, **Idil, S.** and **Kløvjan, O.S.** (1990) Architectural styles of basin fill in the Northern Viking Graben. In: *Evolution of the North Sea Rifts* (Eds D.J. Blundell and A.D. Gibbs), Clarendon, Oxford , 158–179.

Gabrielsen, R.H., **Faleide, J.I.**, **Pascal, C.**, **Braathen, A.**, **Nystuen, J.P.**, **Etzelmuller, B.** and **O'Donnell, S.** (2010) Latest Caledonian to Present tectonomorphological development of southern Norway. *Mar. Petrol. Geol.*, **27**, 709–723.

Gibling, M.R. and **Bird, D.J.** (1994) Late Carboniferous cyclothems and alluvial paleovalleys in the Sydney Basin, Nova Scotia. *Geol. Soc. Am. Bull.*, **106**, 105–117.

Gingele, F.X. and **Deckker, P.** (2004) Fingerprinting Australia's rivers with clay minerals and the application for the marine record of climate change. *Aust. J. Earth Sci.*, **51**, 339–348.

Girty, G.H. (1991) A note on the composition of plutoniclastic sand produced in different climatic belts. *J. Sed. Petrol.*, **61**, 428–433.

Glørstad-Clark, E., **Faleide, J.I.**, **Lundschien, B.A.** and **Nystuen, J.P.** (2010) Triassic sequence stratigraphy and paleogeography of the western Barents Sea area. *Mar. Petrol. Geol.*, **27**, 1448–1475.

Glørstad-Clark, E., **Birkeland, E.P.**, **Nystuen, J.P.**, **Faleide, J.I.** and **Midtkandal, I.** (2011) Triassic platform-margin deltas in the western Barents Sea. *Mar. Petrol. Geol.*, **28**, 1294–1414.

Goldsmith, P.J., **Hudson, G.** and **Van Veen, P.** (2003) Triassic. In: *The Millenium Atlas: petroleum geology of the central and northern North Sea* (Eds D. Evans, C. Graham, A. Armour and P. Bathurst), Geol. Soc. London, 105–127.

Gradstein, F.M., **Ogg, J.G.** and **Smith, A.G.** (2004) *A Geologic Time Scale 2004.* Cambridge University Press, 589 pp.

Graf, W.L. (1988) *Fluvial Processes in Dryland Rivers.* Springer Verlag, Berlin, 346 pp.

Gudmundsson, L., **Tallaksen, L.M.**, **Stahl, K.** and **Fleig, A.K.** (2011) Low-frequency variability of European runoff. *Hydrol. Earth Syst. Sci. Discuss*, **8**, 1705–1727.

Hamer, J.M.M., **Sheldon, N.D.**, **Nichols, G.J.** and **Collinson, M.E.** (2007) Late Oligocene-Early Miocene palaeosols of distal fluvial systems, Ebro Basin, Spain. *Palaeogeogr., Palaeoclimatol., Palaeoecol.*, **247**, 220–235.

Hardy, R. and **Tucker, M.** (1995) X-ray powder diffraction of sediments. In: *Techniques in Sedimentology* (Ed. M. Tucker), Blackwell Science, Oxford, 191–228.

Hesselbo, S.P., **Robinson, S.A.**, **Surlyk, F.** and **Piasecki, S.** (2002) Terrestrial and marine extinction at the Triassic-Jurassic boundary synchronized with major carbon-cycle perturbation: A link to initiation of massive volcanism. *Geology*, **30**, 251–254.

Horn, J.H., **Fielding, C.R.** and **Joeckel, R.M.** (2012) Revision of Platte River alluvial facies model through observations of extant channels and barform and subsurface alluvial valley fills. *J. Sed. Res.*, **82**, 72–91.

Howard, C.S. (2009) A fluvial depositional system as an indicator of paleoclimate. The Lower Cretaceous Kootenai Formation in south-western Montana. *Geol. Soc. Am. Abstr. Progr.*, **41**, 274

Hubbard, R. and **Boulter, M.C.** (2000) Phytogeography and paleoecology in Western Europe and Eastern Greenland near the Triassic-Jurassic boundary. *Palaios*, **15**, 120–131.

Hubert, J.F. (1977) Palaeosol caliche in the New Haven Arkose, Newark Group, Connecticut. *Palaeogeogr., Palaeoclimatol., Palaeoecol.*, **24**, 151–168.

Hurum, J.H., **Bergan, M.**, **Müller, R.**, **Nystuen, J.P.** and **Klein, N.** (2006) Late Triassic dinosaur bone, offshore Norway. *Nor. J. Geol.*, **86**, 117–123.

Joeckel, R.M. (1999) Palaeosols in Galesburg Formation (Kansas City Group, Upper Pennsylvanian), northern Midcontinent, U.S.A.: Evidence for climate change and mechanisms of marine transgression. *J. Sed. Res.*, **69**, 720–737.

Johannesen, E.P. and **Embry, A.F.** (1989) Sequence correlation: Upper Triassic to Lower Jurassic succession, Canadian and Norwegian Arctic. In: *Correlation in Hydrocarbon Exploration* (Ed. J.D. Collinson), Norw. Petrol. Soc., Graham & Trotman, 155–170.

Kjemperud, A.V. (2008) *Sequence Stratigraphy of Alluvial Successions: Depositional Style and Controlling Factors.* Ph.D. Thesis, Department of Geosciences, University of Oslo, Norway, 270 pp.

Kjemperud, A.V., **Schomacker, E.** and **Cross, T.A.** (2008) Architecture and stratigraphy of alluvial deposits, Morrison Formation (Upper Jurassic), Utah. *AAPG Bull.*, **92**, 1055–1076

Khanna, M., **Saigal, G.C.** and **Bjørlykke, K.** (1997) Kaolinitization of Upper Triassic-Lower Jurassic sandstones of the Tampen Spur Area, North Sea. In: *Implications for Early Diagenesis and Fluid Flow, in Basin-Wide Diagenetic Patterns – Integrated Petrologic, Geochemical and Hydrologic Considerations* (Eds I.P. Montaneez, J.M. Gregg and K.L. Shelton), *SEPM Spec. Publ.*, **57**, 253–269.

Knudsen, T.L. (2001) Contrasting provenance of Triassic/Jurassic sediments in North Sea Rift; a single zircon

(SIMS), Sm-Nd and trace element study. *Chem. Geol.*, **171**, 273–293.

Kraus, M.J. (1987) Integration of channel and floodplain suites, II. Vertical relations of alluvial palaeosols. *J. Sed. Petrol.*, **57**, 602–612.

Kraus, M. (1999) Palaeosols in clastic sedimentary rocks: their geologic applications. *Earth-Sci. Rev.*, **47**, 41–70.

Kraus, M. (2002) Basin-scale changes in floodplain palaeosols: implications for interpreting alluvial architecture. *J. Sed. Res.*, **72**, 500–509.

Kraus, M. and Aslan, A. (1999) Palaeosols sequences in floodplain environments: a hierarchical approach. In: *Paleoweathering, Paleosurfaces and Related Continental Deposits* (Ed. M. Thiry), *Int. Assoc. Sedimentol. Spec. Publ.*, **27**, 303–321.

Kürschner, W.M. and **Herngreen, G.F.W.** (2010) Triassic palynology of Central and Northwest-Europe: a review of palynofloral patterns and biostratigraphic subdivisions. In: *The Triassic Timescale* (Ed. S. Lucas), *Geol. Soc. London Spec. Publ.*, **334**, 263–283.

Leeder, M.R. (1978) A quantitative stratigraphic model for alluvium, with special reference to channel deposits density and interconnectedness. In: *Fluvial Sedimentology* (Ed. A.D. Miall), *Can. Soc. Petrol. Geol. Mem.*, **5**, 587–596.

Leeder, M.R., Mack, G.H. and **Salyards, S.L.** (1996) First quantification test of alluvial stratigraphic models: southern Rio Grande rift, New Mexico. *Geology*, **24**, 87–90.

Leeder M.R., Harris, T. and **Kirkby, M.J.** (1998) Sediment supply and climate change: implication for basin stratigraphy. *Basin Res.*, **10**, 7–18.

Lervik, K.-S. (2006) Triassic lithostratigraphy of the Northern North Sea Basin. *Nor. J. Geol.*, **86**, 93–116.

Lervik, K.-S., Spencer, A.M. and **Warrington, G.** (1989) Outline of Triassic stratigraphy and structure in the central and northern North Sea. In: *Correlation in Hydrocarbon Exploration* (Ed. J.D. Collinson), Nor. Petrol. Soc. (NPF), Graham & Trotman, London, 291–297.

López-Moreno, J.I. and **García-Ruiz, J.M.** (2004) Influence of snow accumulation and snowmelt on streamflow in the central Spanish Pyrenees. *Hydrolog. Sci. Journal.*, **49**, 787–802.

Lundschien, B.A. and **Mørch, T.** (1998) Use of predictive high resolution sequence stratigraphy in reservoir modelling: an example from the Upper Triassic-Lower Jurassic Statfjord Formation, Snorre Field, northern North Sea. In: *Sequence Stratigraphy – Concepts and Applications* (Eds F.M. Gradstein, K.O. Sandvik and N.J. Milton). Norw. Petrol. Soc. (NPF) Spec. Publ. **8**, 229–249, Elsevier.

Mack, G.H. (1992) Palaeosols as an indicator of climatic change at the early-late Cretaceous boundary, southwestern New Mexico. *J. Sed. Petrol.*, **62**, 483–494.

Mackey, S.D. and **Bridge, J.S.** (1995) Three-dimensional model of alluvial stratigraphy; theory and applications. *J. Sed. Res. (B)*, **65**, 7–31.

Marriott, S.B. and **Wright, V.P.** (1993) Palaeosols as indicator of geomorphic stability in two Old Red Sandstone alluvial suites, South Wales. *J. Geol. Soc. London*, **150**, 1109–1120.

McGarry, D. (1996) Structure and grain size distribution of Vertisols. In: *Vertisols and Technologies for Their Management*, (Eds N. Ahmad and A. Mermut), Elsevier, Amsterdam, 231–302.

McKie, T. and **Audretsch, P.** (2005) Depositional and structural controls on Triassic reservoir performance in the Heron Cluster, ETAP, Central North Sea. In: *Petroleum Geology: North-West Europe and Global Perspectives* (Eds A.G. Doré and B. Vinings). Proceedings of the 6th Petroleum Geology Conference, Geol. Soc. London, 285–298.

McKie, T. and **Williams, B.** (2009) Triassic palaeogeography and fluvial dispersal across the northwest European Basin. *Geol. J.*, **44**, 711–741.

McKie, T., Jolley, S.J. and **Kristensen, M.B.** (2010) Stratigraphic and structural compartmentalization of dryland fluvial reservoirs: Triassic Heron Cluster, Central North Sea. *Geol. Soc. London Spec. Publ.*, **347**, 165–198.

Mearns, E.W., Knarud, R., Ræstad, N., Stanley, K.O. and **Stockbridge, C.P.** (1989) Samarium-neodymium isotope stratigraphy of the Lunde and Statfjord Formations of Snorre Oil Field, northern North Sea. *J. Geol. Soc. London*, **146**, 217–228.

Miall, A.D. (1995) Discussion. Description and interpretation of fluvial deposits: a critical perspective. *Sedimentology*, **42**, 379–389.

Miall, A.D. (1996) *The Geology of Fluvial Deposits.* Springer Verlag, 582 p.

Miall, A.D. (2006) Reconstructing the architecture and sequence stratigraphy of the preserved fluvial record as a tool for reservoir development: a reality check. *AAPG Bull.*, **90**, 989–1002.

Miall, A.D. (2010) Alluvial deposits. In: *Facies Models 4* (Ed. N.P. James & R.W. Dalrymple), *GEOtext 6*, Geol. Assoc. of Canada, 105–137.

Moore, T., Scotese, C. and **Goswami, A.** (2008) Predicting palaeosol distributions using paleoclimate simulations: Results for the Late Devonian, Early Permian, Late Triassic, Late Jurassic, Late Cretaceous (Cenomanian) and Early Eocene. *Geol. Soc. Am. Abstr. Progr.*, **40**, 259.

Mørk, A., Embry, A.F. and **Weitschat, W.** (1989) Triassic transgressive-regressive cycles in the Sverdrup Basin, Svalbard and the Barents Shelf. In: *Correlation in Hydrocarbon Exploration* (Ed. J.D. Collinson), Nor. Petrol. Soc., Graham & Trotman, 113–130.

Müller, R. (2003) *Basin Infill Dynamics of the Triassic of the Northern North Sea and Mid-Norwegian Shelf: Control of Autogenic and Allogenic Factors.* Dr. Scientarium Thesis. University of Oslo. 262 pp.

Müller, R., Nystuen, J.P. and **Adestål, V.** (2002) Dryland mudrocks and palaeosols of the Upper Triassic – Lower Jurassic Lunde and Statfjord Formations, northern North Sea: implications for palaeoclimate and basin infill dynamics. Programme and Abstracts, Dryland Rivers: Process and Product. University of Aberdeen. 8-9. August 2002, 19.

Müller, R., Nystuen, J.P. and **Wright, V.P.** (2004) Pedogenic mud aggregates and palaeosol development in ancient dryland river systems: criteria for interpreting alluvial mudrocks and floodplain dynamics. *J. Sed. Res.*, **74**, 537–551.

Müller, R., Nystuen, J.P., Eide, F. and **Lie, H.** (2005) Late Permian to Triassic basin infill history and palaegeography of the Mid-Norwegian Shelf - East Greenland region. In: *Onshore-Offshore Relationships on the*

North Atlantic Margin (Eds B.T.G. Wandås, J.P. Nystuen, E.A. Eide and F.M. Gradstein), *Nor. Petrol. Soc., Spec. Publ.*, **12**, Elsevier, Amsterdam, 165–189.

Munsell Color (1975) Munsell Color Charts. Munsell Color, Baltimore, MD.

North American Commission on Stratigraphic Nomenclature (NACSN) (1983) North American stratigraphic code, *AAPG Bull.*, **67**, 841–875.

Nøttvedt, A., Gabrielsen, R. H. and **Steel, R.J.** (1995) Tectonostratigraphy and sedimentary architecture of rift basins, with reference to the northern North Sea. *Mar. Petrol. Geol.*, **12**, 881–901.

Nystuen, J.P. and **Fält, L.M.** (1995) Upper Triassic-Lower Jurassic reservoir rocks in the Tampen Spur area, Norwegian North Sea. In: *Petroleum Exploration and Exploitation in Norway* (Ed. S. Hanslien), *Nor. Petrol. Soc. (NPF) Spec. Publ.*, **4**, Elsevier, Amsterdam, 141–185.

Nystuen, J.P., Knarud, R., Jorde, K. and **Stanley, K.O.** (1989) Correlation of Triassic to Lower Jurassic sequences, Snorre Field and adjacent areas, northern North Sea. In: *Correlation in Hydrocarbon Exploration* (Ed. J.D. Collinson), Nor. Petrol. Soc. (NPF), Graham & Trotman, London, 273–289.

Nystuen, J.P., Mørk, A., Müller, R. and **Nøttvedt, A.** (2008) From desert to alluvial plain – from land to sea. In: *The Making of a Land – Geology of Norway* (Eds I.B. Ramberg, I. Bryhni, A. Nøttvedt and K. Rangnes), Norsk geologisk forening (The Norwegian Geological Association), Trondheim, Norway, 330–355.

Odinsen, T., Reemst, P., Van Der Beek, P., Faleide, J.I. and **Gabrielsen, R.** (2000) Permo-Triassic and Jurassic extension in the northern North Sea: results from tectonostratigraphic forward modelling. In: *Dynamics of the Norwegian Margin* (Eds A. Nøttvedt, B.T. Larsen, S. Olaussen, B. Tørudbakken, J. Skogseid, R.H. Gabrielsen, H. Brekke and Ø. Birkeland), *Geol. Soc. London, Spec. Publ.*, **167**, 83–103.

Olsen, H. (1994) Orbital forcing on continental depositional systems – lacustrine and fluvial cyclicity in the Devonian of East Greenland. In: *Orbital Forcing and cyclic sequences* (Eds P.L. de Boer and D.G. Smith), *Int. Assoc. Sediment. Spec. Publ.*, **19**, 429–438.

Olsen, H. and **Larsen, P.H.** (1993) Structural and climatic controls on fluvial depositional systems: Devonian of East Greenland. In: *Alluvial Sedimentation* (Eds M. Marzo and C. Puigdefabregas), *Int. Assoc. Sedimentol. Spec. Publ.*, **17**, 401–423.

Patterson, P.E., Jones, C. and **Skelly, R.** (2006) Climatic controls on alluvial architecture, Doba Basin, Chad. *AAPG Ann. Met. Abstr.*, **15**, 82.

Perlmutter, M.A. and **Matthews, M.D.** (1989) Global cyclostratigraphy: a model. In: *Quantitative Dynamic Stratigraphy* (Ed. T.A. Cross), Prentice Hall, Englewood Cliffs, 233–260.

Piipujol, M.D. and **Buurman, P.** (1994) The distinction between ground-water gley and surface water gley phenomena in Tertiary palaeosols of the Ebro basin, NE Spain. *Palaeogeogr., Palaeoclimatol., Palaeoecol.*, **110**, 103–113.

Posamentier, H.W. and **Allen, G.P.** (1999) *Siliciclastic Sequence Stratigraphy – Concepts and Applications.* SEPM Concepts in Sedimentology and Paleontology, **7**, 210 pp.

Probert, M.E., Fergus, I.F., Bridge, B.J., McGarry, D., Thompson, C.H. and **Russel, J.S.** (1987) *The Properties and Management of Vertisols.* CAB International/IBSRAM, Wallington, Oxon., UK, 20 pp.

Reeves, C.C. (1970) Origin, classification and geologic history of caliche in Southern High Plains, Texas and eastern New Mexico. *J. Geol.*, **78**, 352–362.

Retallack, G.J. (1986) Fossil soils as grounds for interpretation long-term controls on ancient rivers. *J. Sed. Petrol.*, **56**, 1–18.

Retallack, G.J. (2001) *Soils of the Past – An Introduction to Palaeopedology.* Blackwell Science, Oxford, 404 pp.

Robert, C. (2004) Late Quaternary variability of precipitation in Southern California and climatic implications; clay mineral evidence from the Santa Barbara Basin, ODP Site 893. *Quatern. Sci. Rev.*, **23**, 1029–1040.

Roberts, E.M. (2007) Facies architecture and depositional environments of the Upper Cretaceous Kaiporowits Formation, southern Utah. *Sed. Geol.*, **197**, 207–233.

Røe, S.-L. and **Steel, R.** (1985) Sedimentation, sea-level rise and tectonics at the Triassic-Jurassic boundary (Statfjord Formation), Tampen Spur, northern North Sea. *J. Petrol. Geol.*, **8**, 103–186.

Royer, D.L. (1999) Depth to pedogenic carbonate horizon as a paleoprecipitation indicator? *Geology*, **27**, 1123–1126.

Ruhl, M. and **Kürschner, W.M.** (2011) Multiple phases of carbon cycle disurbance from large igneous province formation at the Triassic-Jurassic transition. *Geology*, **39**, 431–434.

Ruhl, M., Deenen, M.H.L., Abels, H.A., Bonis, N.R., Krijgsman, W. and **Kürschner, W.M.** (2010) Astronomical constraints on the duration of the early Jurassic Hettangian stage and recovery rates following the end-Triassic mass extinction (St. Audrie´s Bay/East Quantoxhead, U.K.). *Earth Planet. Sci. Lett.*, **259**, 262–276.

Ryseth, A. (2001) Sedimentology and palaeogeography of the Statfjord Formation (Rhaetian-Sinemurian), North Sea. In: *Sedimentary Environments Offshore Norway – Paleozoic to Recent* (Eds O.J. Martinsen and T. Dreyer), *Norw. Petrol. Soc. (NPF), Spec. Publ.*, **10**, Elsevier, Amsterdam, 67–85.

Ryseth, A. and **Ramm, M.** (1996) Alluvial architecture and differential subsidence in the Statfjord Formation, North Sea: prediction of reservoir potential. *Petrol. Geosci.*, **2**, 271–287.

Schomacker, E.R. (2008) *Geological Reservoir Characterisation of Alluvial Deposits: Architectural Style and Variability.* Ph.D. Thesis, Department of Geosciences, University of Olso, Norway, 262 pp.

Schumm, S.A. (1981) Evolution and response of the fluvial system, sedimentologic implications. In: *Recent and Ancient Nonmarine Depositional Environments: Models for Exploration* (Eds F.G. Ethridge and R.M. Flores), *SEPM Spec. Public.*, **31**, 19–29.

Schumm, S.A. (1977) *The Fluvial System.* The Blackburn Press, Caldwell, New Jersey, USA, 338 pp.

Schumm, S.A. (1993) River Response to Baselevel Change: Implications for Sequence Stratigraphy. *J. Geol.*, **101**, 279–294.

Schumm, S.A. (2005) *River Variability and Complexity.* Cambridge University Press, Cambridge, UK, 220 pp.

Schwertmann, U. (1988) Occurrence and formation of iron-oxides in various pedoenvironments. In: *Iron in*

Soils and Clay Minerals (Eds J.W. Stucki, B.A. Goodman and U. Schwertmann), Dordrecht, The Netherlands, Reidel, 267–308.

Shanley, K.W. and **McCabe, P.J.** (1994) Perspectives on the sequence stratigraphy of continental strata. *AAPG Bull.*, **78**, 544–568.

Sinha, R. and **Sarkar, S.** (2009) Climate-induced variability in the late Pleistocene-Holocene fluvial and fluvio-deltaic successions in the Ganga plains, India: a synthesis. *Geomorphology*, **113**, 173–188.

Sørlie, R. (1996) *Diagenetisk utvikling i Lunde- og Statfjord-formasjonens slam- og sandsteiner på Snorrefeltet.* Cand.Scient, oppgave i geologi. Institutt for Geologi, Universitetet i Oslo (In Norwegian), 297 pp.

Steel, R. (1993) Triassic-Jurassic megasequence stratigraphy in the Northern North Sea: rif to post-rift evolution. In: *Petroleum Geology of NW Europe: Proceedings of the 4th Conference*, (Ed. J.R. Parker), **1**, Geol. Soc. London, 299–315.

Steel, R.J. and **Ryseth, A.** (1990) The Triassic-early Jurassic succession in the northern North Sea: megasequence stratigraphy and intra-Triassic tectonics. In: *Tectonic Events Responsible for Britain's Oil and Gas Reserves* (Eds R.F.P. Hardman and J. Brooks). *Geol. Soc. London Spec. Publ.*, **55**, 139–168.

Suresh, N., **Ghosh, S.K.**, **Kumar, R.** and **Sangode, S.J.** (2004) Clay-mineral distribution patterns in late Neogene fluvial sediments of the Subathu sub-basin, central sector of Himalayan foreland basin: implications for provenance and climate. *Sed. Geol.*, **163**, 265–278.

Tandon, S.K. and **Gibling, M.R.** (1994) Calcrete and coal in late Carboniferous cyclothems of Nova Scotia, Canada: Climate and sea-level changes linked. *Geology*, **22**, 755–758.

Terry, D.O. (2007) Palaeosol-based interpretations of basin history and paleoclimatic change: the Eocene-Oligocene White River Group of Nebraska and South Dakota. *Geol. Soc. Am. Abstr. Progr.*, **39**, 193.

Tooth, S. (2000) Process, form and change in dryland rivers: a review of recent research. *Earth-Sci. Rev.*, **51**, 67–107.

Torsvik, T.H., **Carlos, D.**, **Mosar, J.**, **Cocks, R.M.** and **Malme, T.N.** (2002) Global reconstructions and North Atlantic palaeogeography 440 Ma to Recent. In: *Atlas-Mid Norway Plate Reconstructions Atlas with Global and Atlantic Perspectives* (Ed. E. Eide) Geol. Surv. Nor., Trondheim, 18–39.

Van der Zwan, C.J. (2002) The impact of Milankovitch-scale climatic forcing on sediment supply. *Sed. Geol.*, **147**, 271–294.

Vanstone, S.D. (1991) Early Carboniferous (Mississippian) palaeosols from southwest Britain: influence of climatic change on soil development. *J. Sed. Petrol.*, **61**, 445–457.

Vollset, J. and **Doré, A.G.** (Eds) (1984) *A revised Triassic and Jurassic lithostratigraphic nomenclature for the Norwegian North Sea.* Norw. Petrol. Dir., Bull., **3**, 53 pp.

Wan, S., **Tian, J.**, **Steinke, S. Li, A.** and **Li, T.** (2011) Evolution and variability of the east Asian summer monsoon during Pliocene; evidence from clay mineral records of the South China Sea. *Palaeogeogr., Palaeoclimatol., Palaeoecol.*, **293**, 237–247.

Weissman, G.S., **Hartley, A.J.**, **Nichols, G.J.**, **Scuderi, L.A.**, **Olson, M.E.**, **Buehler, H.A.** and **Massengil, L.C.** (2011) Alluvial facies distributions in continental sedimentary basins – distributive fluvial systems. In: *From River to Rock Record: the Preservation of Fluvial Sediments and their Subsequent Interpretation* (Eds S.K. Davidson, S. Leleu and C.P. North). *SEPM Spec. Publ.*, **97**, 327–355.

Weltje, G.J., **Meijer, X.D.** and **de Boer, P.L.** (1998) Stratigraphic inversions of siliciclastic basin fills: a note on the distinction between supply signals resulting from tectonic and climatic forcing. *Basin Res.*, **10**, 129–153.

Wilding, L.P. and **Tessier, D.** (1988) Genesis of vertisols: shrink-swell phenomena. In: *Vertisols: Their Distribution, Properties, Classification and Management: College Station, Texas* (Eds L.P. Wilding and R. Puentes). Soil Management Support Services Technical Monograph, **18**, 55–81.

Wright, V.P. (1990) Equatorial aridity and climatic oscillations during the early Carboniferous, southern Britain. *J. Geol. Soc. London*, **147**, 359–363.

Wright, V.P. and **Marriot, S.B.** (1993) The sequence stratigraphy of fluvial depositional systems: the role of the floodplain storage. *Sed. Geol.*, **86**, 203–210.

Wright, V.P. and **Marriott, S.** (1996) A quantitative approach to soil occurrence in alluvial deposits and its application to the Old Red Sandstone in Britain. *J. Geol. Soc. London*, **153**, 907–913.

Yang, C.S. and **Baumfalk, Y.A.** (1994) Milankovitch cyclicity in the Upper Rotliegend Group of The Netherlands offshore. In: *Orbiral Forcing and Cyclic Sequences* (Eds P.L. de Boer and D.G. Smith). *Int. Assoc. Sediment. Spec. Publ.*, **19**, 47–61.

Zaleha, M.J. (1997) Intra- and extrabasinal controls on fluvial deposition in the Miocene Indo-Gangetic foreland basin, northern Pakistan. *Sedimentology*, **44**, 369–390.

Int. Assoc. Sedimentol. Spec. Publ. (2014) **46**, 101–126

Applying accommodation versus sediment supply ratio concepts to stratigraphic analysis and zonation of a fluvial reservoir

ALLARD W. MARTINIUS[†], CARSTEN ELFENBEIN[†*] and KEVIN J. KEOGH[†‡]

[†] *Statoil Research Centre, Rotvoll, 7005 Trondheim, Norway (E-mail: awma@statoil.com)*
[*] *Present address: Det Norske Oljeselskap ASA, Nedre Bakklandet 58c, 7014 Trondheim, Norway*
[‡] *Present address: Statoil ASA, 4035 Stavanger, Norway*

ABSTRACT

The use of sequence stratigraphic principles as a correlation tool within fluvial depositional systems has been a topic of debate for a long time. Changes in facies stacking patterns caused by allocyclic events are often masked or completely overprinted by autocyclic events and therefore the recognition of correlative time-significant markers can be difficult. These problems are more difficult to solve when trying to apply sequence stratigraphic principles to subsurface well datasets. A method is presented here based on the concepts of accommodation versus sediment supply interactions (A/S) and their variations at a hierarchy of temporal scales. These changes lead to the development of a pattern of overall increasing or decreasing A/S values reflected in the sedimentary record. The key addition to existing ideas about continental fluvial sequence stratigraphy is the notion that the rate of fluvial aggradation through time, as controlled by the accommodation versus sediment supply rate, is dynamic and not static. The use of the concepts of A/S interactions allows for the fluvial stratigraphic system to be analysed not only (or solely) in terms of regional to local base level changes that affect accommodation but as an integrated parameter with regional to local sediment supply changes. A practical definition of base level (or stratigraphic reference level) in fluvial systems is outlined and formulated as the lower boundary of real-time accommodation or upper boundary of net accommodation (lower limit to erosion). In areas not directly affected by laterally migrating fluvial channels or channel belts, the upper boundary of net accommodation at a specific time scale can in some cases be genetically coupled to the most mature palaeosols. In other cases, the upper boundary of net accommodation may be picked at a laterally extensive caliche layer formed at a depth relatively unaffected by short-term erosion and deposition. The method presented here is not meant as a substitute for previous stratigraphic methods or models for fluvial systems but as an alternative to be used as seen appropriate given the type and amount of data available. The application and limitations of this method are highlighted in a case study from the Statfjord Group, Statfjord Field, in the Norwegian North Sea.

Keywords: Fluvial stratigraphy, A/S hierarchy, reservoir zonation, Statfjord Group, base level.

INTRODUCTION

One of the most important recent advances in sequence stratigraphy is the increased awareness of the influence of temporal and spatial variations in primary sedimentary processes and geomorphological variability on the stratigraphic development of depositional systems and resulting three-dimensional architecture. Temporal and spatial variations in multiple interdependent controlling factors such as sediment supply, hydraulic characteristics and autogenic responses of the depositional system (Puigdefàbregas, 1993; Hovius, 1998; Leeder *et al.*, 1998; Ethridge *et al.*, 1998; Muto & Steel, 2002, 2004; Yoshida *et al.*, 2007; Muto *et al.*, 2007) are now also considered

From Depositional Systems to Sedimentary Successions on the Norwegian Continental Margin, First Edition.
Edited by A. W. Martinius, R. Ravnås, J. A. Howell, R. J. Steel, and J. P. Wonham.

important controls. Sediment supply, hydraulic characteristics and autogenic responses and their importance for sequence stratigraphic concepts and models, are studied using experimental models (Koss *et al.*, 1994; Wood *et al.*, 1992; Van Heijst, 2001; Van Heijst *et al.*, 2002; Muto & Steel, 2004; Swenson & Muto, 2005) and numerical models (De Vriend, 2005; Eggenhuisen *et al.*, 2005; Postma, 2014) and through numerous dedicated studies of outcrop analogues and modern systems (see references in Schumm, 1993; Bridge, 2003; Catuneanu, 2006; Rygel & Gibling, 2006). This combined approach is particularly important for continental depositional systems and the resulting three-dimensional architectural patterns because of their sensitivity to controlling factors in both source and sink areas, in particular the role of sediment supply and water discharge.

Geomorphic models for continental fluvial deposition, erosion and preservation at short to intermediate time scales (1 yr to 10^5 yr), outside the direct control of relative sea-level changes (see overviews in Schumm, 1977; Knighton, 1998; Bridge, 2003), are only to a limited extent used in sequence stratigraphy to better understand the mechanisms that control long-term sediment preservation and the formation of erosional surfaces in the non-marine stratigraphic record. A number of non-marine (sequence) stratigraphic models exist (Legarreta *et al.*, 1993; Wright & Marriott, 1993; Shanley & McCabe, 1994; Olsen *et al.*, 1995; Allen *et al.*, 1996; Currie, 1997; Dahle *et al.*, 1997; Dong *et al.*, 1997; Fu, 1997; Wang *et al.*, 1997; Cai & Zhang, 1999; Diessel *et al.*, 1999; Martinsen *et al.*, 1999; Boyd *et al.*, 2000; Yu *et al.*, 2000; Plint *et al.*, 2001; Weissmann *et al.*, 2002; Holbrook *et al.*, 2006) in which either an attempt is made to apply existing marine sequence stratigraphic terminology (correlative surfaces, systems tracts) to non-marine settings or to define new concept or correlative surfaces and non-marine systems tracts. No unity in non-marine sequence stratigraphy has yet been achieved and it is argued that the application of systems tracts as defined for marine environments to non-marine successions is difficult to justify (Ethridge *et al.*, 1998).

The practical application of continental fluvial sequence stratigraphic concepts and methods to subsurface datasets as used in hydrocarbon exploration and particularly in production-related work tasks is often complex due to a lack of data. In general, correlation surfaces are commonly not well developed and difficult to recognise and correlate across the succession. For example, in combined wireline and core fluvial datasets it is often difficult to differentiate between regional unconformities formed during stratigraphic degradation and local channel base scours formed during stratigraphic aggradation. In these instances, correlation of wells based on erosion surfaces surpasses an acceptable level of uncertainty and is better done based on trends. In most published cases, low and high accommodation zones are recognised. These have to be of sufficient resolution and an acceptable uncertainty level to be able to build geologically realistic 3D reservoir models and to justify the investments associated with the drilling of infill production wells particularly in the later phases of hydrocarbon field development.

AIMS

The aim of this study is to construct an improved stratigraphic correlation framework for reservoir property modelling of the fluvial Statfjord Group (Rhaetian to early Sinemurian; Norwegian North Sea; Figs 1 to 3; Lervik, 2006) in the Statfjord Field. This is achieved by integrating two key additions to existing ideas on continental fluvial sequence stratigraphy: 1) the notion that the rate of fluvial aggradation through time, as controlled by the accommodation versus sediment supply rate, is dynamic and not static and ii) a modified definition of base level in fluvial successions to be able to estimate changes in aggradation rate. The A/S (in which A is the rate of creation of net accommodation and S is the rate of sediment input as a function of sediment liberation and discharge) and autostratigraphy approaches are applied to a spatially restricted and limited subsurface well dataset at reservoir scale (0.1 km to 20 km; Statfjord Field; Fig. 2) that covers a time scale of several million years. The approach is used as a tool for organising and mapping changes in depositional trends as well as reservoir property trends from the viewpoint of an integrated subsurface team while preserving scientific objectivity.

We do not aim for a generalised approach or hypothesis to describe fluvial successions by presenting an overarching argument for organisation of all fluvial systems. However, the study does aim at contributing to further developing a methodology and technique for continental fluvial sequence stratigraphy that has practical use in subsurface

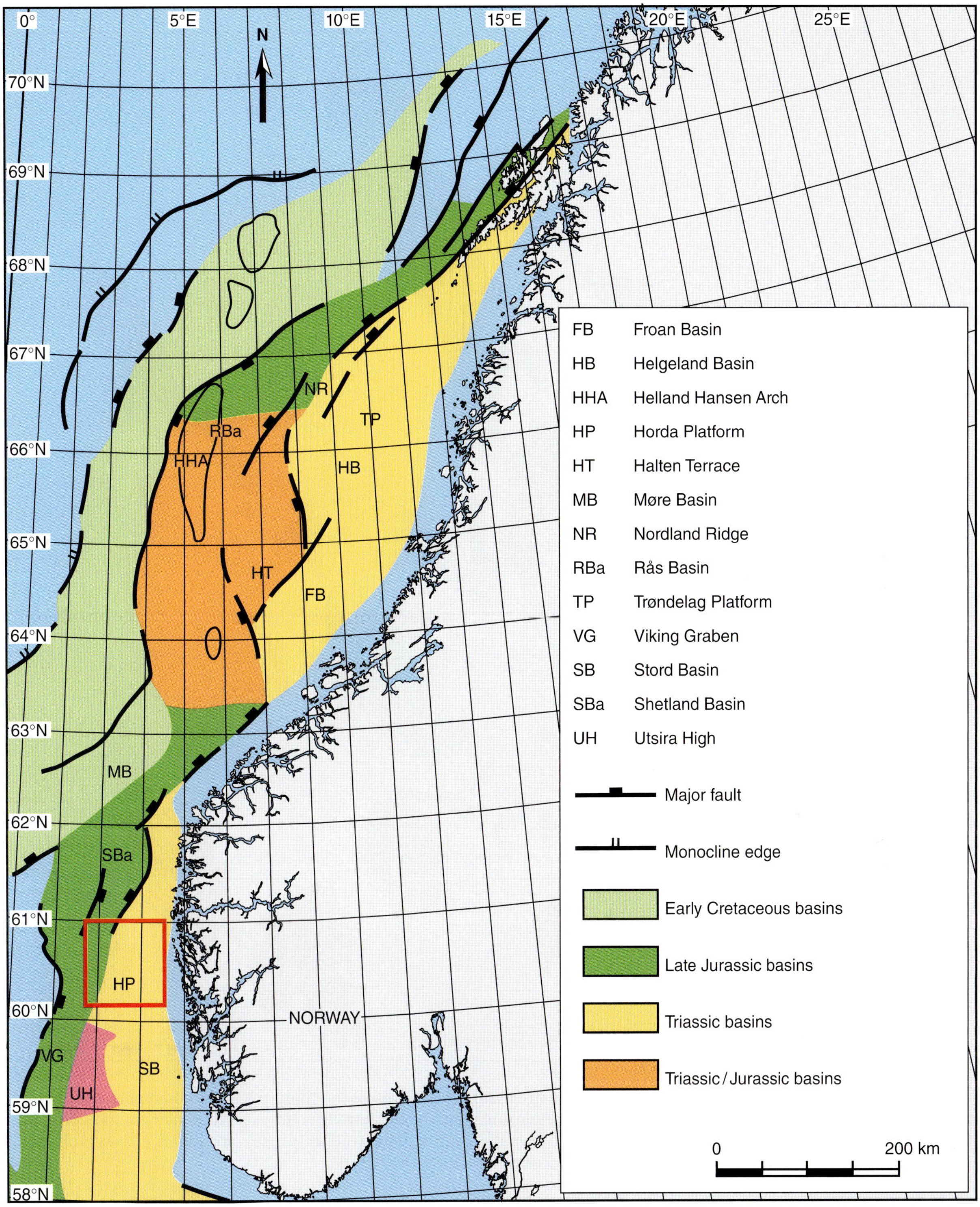

Fig. 1. Schematic location map of the basins and associated tectonic main structures on the Norwegian Continental Shelf. The red box indicates the location of the study area within the Statfjord licence. Modified after Coward *et al.* (2003).

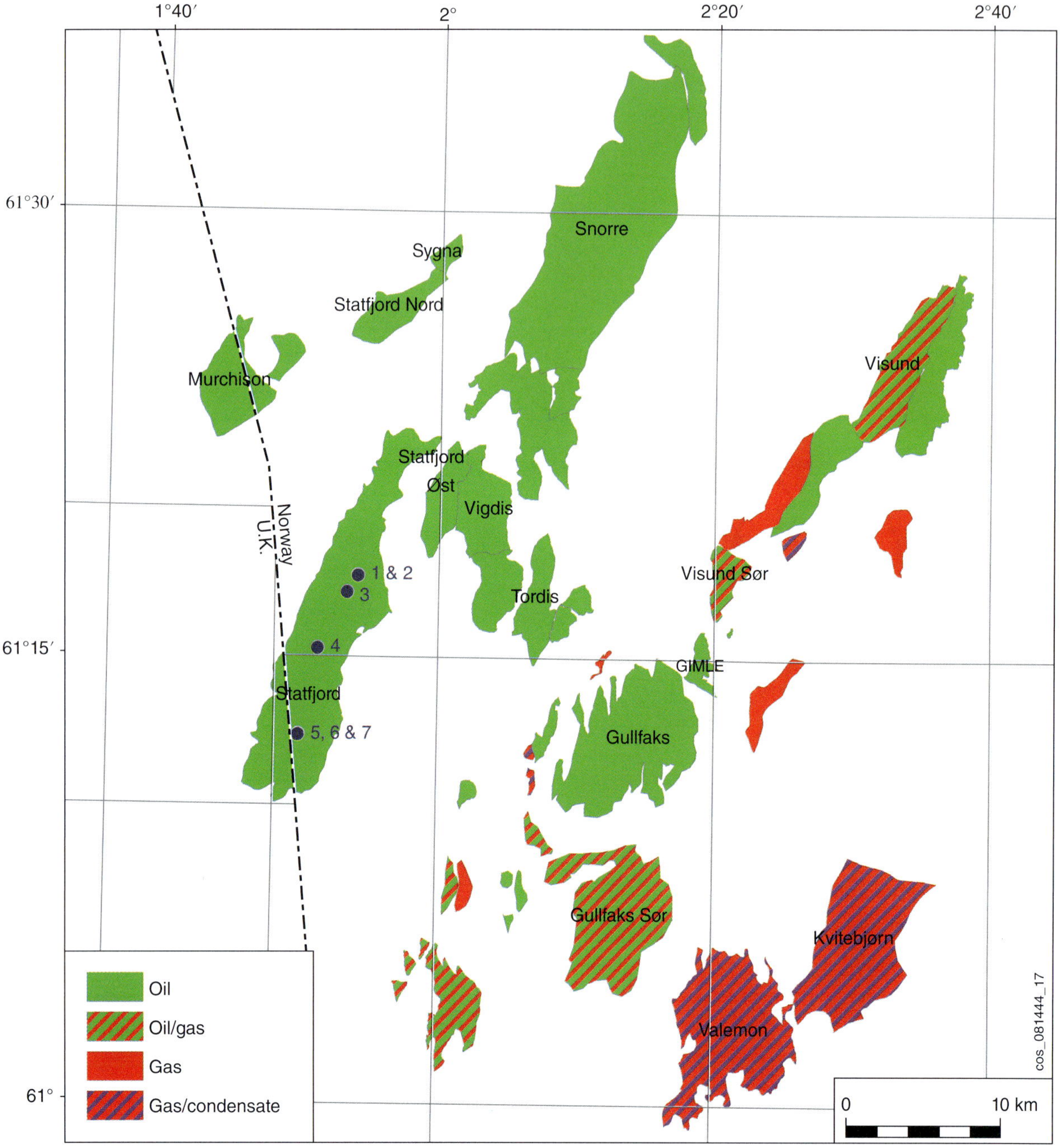

Fig. 2. Map of the hydrocarbon fields on the Tampen area with the outline of the Statfjord licence and surroundings; the study area is part of the Statfjord Field. The sea bottom locations of the cored wells used to construct the correlation of Fig. 12 are indicated (numbers refer to well numbers). Approximate well distances are given at reservoir depth (top Statfjord Group). Modified after Norwegian Petroleum Directorate fact maps.

well data analysis. Consequently, an alternative way of dividing continental fluvial stratigraphy is proposed that is in contrast to a more static subdivision in low and high accommodation zones. Accommodation is defined following Muto & Steel (2000) as ‘the thickness, measured at a specific site and time, of a space which becomes filled with sediments during a specified time interval’.

It is suggested that the theoretical foundation that justifies this approach lies in the allogenic forcing mechanisms on a fluvial system. For example, climate control on long-term water discharge

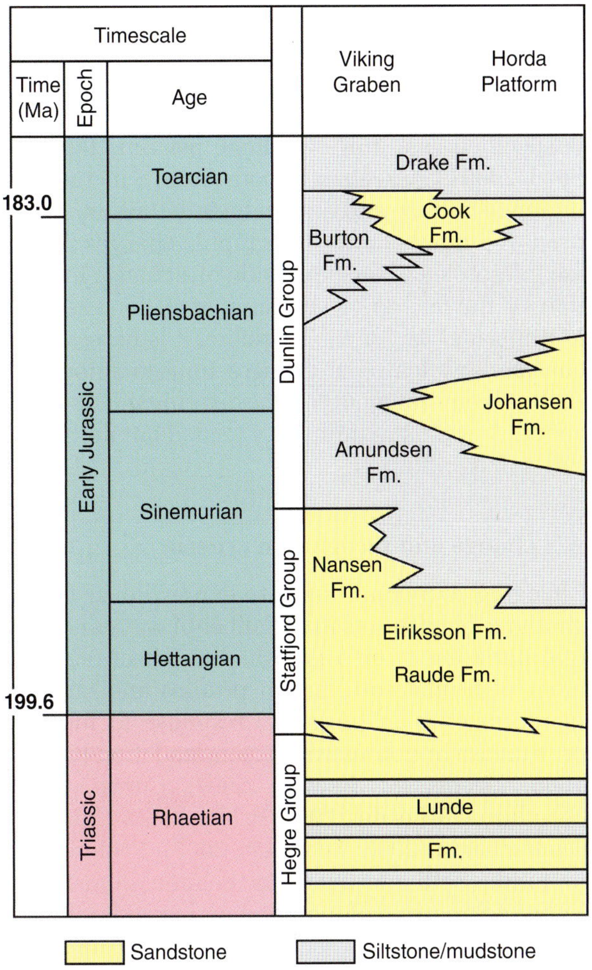

Fig. 3. Early to Middle Jurassic lithostratigraphy of the Viking Graben and Horda Platform structures (see Fig. 1) in the Tampen area with the Statfjord Group indicated. Modified after Husmo *et al.* (2003) and Lervik (2006).

and sediment flux variations causes an effect on physical processes of fluvial sedimentation. Long-term climatic variations (>10^4 yr) as expressed in fluvial stratigraphy are ultimately controlled by Milankovitch scale variations in insolation patterns and their effect on, for example, vegetation (Perlmutter & Matthews, 1990; Matthews & Perlmutter, 1994; Fraticelli *et al.*, 2004; Weissmann *et al.*, 2002, 2005; Gibling *et al.*, 2005). In addition, allogenic forcing keeps a fluvial system continuously out of balance and changes its accumulation space and consequently the aggradation rate (Postma, 2014). Particularly, if the time period for the change in accumulation space is much shorter than the time period that the fluvial system requires to reach grade, the change in aggradation rate is relatively large (Paola *et al.*, 1992; Van Heijst & Postma, 2001; Postma, 2014).

ACCOMMODATION TO SEDIMENT SUPPLY RATIO: SIGNIFICANCE, RECOGNITION AND TRENDS

Allocyclic and autocyclic control

Although the method advocated here is data driven, it is required to embrace a conceptual element to be able to construct a stratigraphic zonation. It is therefore necessary to apply 1) the concept of accommodation (see Appendix I) to sediment supply ratio (A/S) based on the regime theory of Swift & Thorne (1991) and Thorne & Swift (1991) and 2) principles of autostratigraphy as defined by Muto *et al.* (2007) for the definition, recognition and correlation of time equivalent stratigraphic surfaces or zones that potentially improve the ability to correlate (cf. Catuneanu, 2006).

At time scales of 10 ka to 1 Ma, scale-dependant erosional-depositional thresholds can be defined in fluvial systems that separate the modes of fluvial aggradation and degradation (Bull, 1991). These thresholds are controlled, at the largest scale, by two groups of allocyclic factors: climatic and tectonic processes. These two groups interact with each other in a complex manner, making it often difficult to ascribe changes in continental fluvial architecture to a particular tectonic or climatic cause. A significant change in one or both of the external controls will lead to a change in fluvial behaviour (Schumm, 1977, 1993; Bull, 1991). Bull (1991) argued that aggradation or degradation occurs during climate changes with the fluvial system attaining equilibrium during much of the remaining time. During climate changes, vegetation shifts on hill slopes and the fluvial system responds by releasing sediment which leads to aggradation. Examples of this fluvial response to climate change are given by Weissmann *et al.* (2002, 2005) for a fluvial system developed during glacial and interglacial periods and by Gibling *et al.* (2005) for a monsoonal system. These studies show that degradation and exposure of the fluvial system covers most of the time recorded by the fluvial sediments, with relatively short bursts of aggradation due to climate change. For purposes of (sequence) stratigraphic analysis, allocyclic factors, operating at long spatial and temporal scales, need to be separated from autocyclic processes (the natural cyclic progression of events in a fluvial system, or the evolution of landforms to a condition of relative instability; Stouthamer & Berendsen, 2000, 2007; Postma, 2014) that operate over small spatial and temporal scales.

Sequence stratigraphic analysis is included in the concept of autostratigraphy developed for fluvio-deltaic systems. Autostratigraphy is defined as the stratigraphic non-equilibrium response generated by large-scale autogenic processes (Muto *et al.*, 2007). These authors suggest that sequence stratigraphy represents a limiting case of the more general model of autostratigraphy and will be applicable only if the periodicity of the unsteady allocyclic factor is much smaller than the steady dynamic allocyclic forcing. Some abrupt breaks in the stratigraphic record are not necessarily associated with changes in allogenic conditions but can result from purely autogenic processes of the system. Therefore, any fluvial sequence stratigraphic model needs to incorporate the combined effects of A/S changes and large-scale autogenic processes as well as short-term autogenic processes such as avulsion.

Fluvial base level

The upper boundary of net accommodation (lower limit to erosion or fluvial base level; for definition see Appendix 1) is used as a measure of stratigraphic change and as a stratigraphic correlation level (base level). Its recognition is complicated by the fact that the graded stream profile is never in equilibrium at any spatial and temporal scale (Schumm, 1977; Bull, 1991; Holbrook *et al.*, 2006; Postma, 2014) and consequently local autogenic factors have the potential to overprint the stratigraphic signature (called 'autogenic shadowing' by Kjemperud, 2008). It is therefore unlikely that the upper boundary of net accommodation will have a tangible expression in the fluvial rock record. Additionally, through time several upper boundaries of net accommodation will be preserved at different hierarchical scales and therefore stratigraphic intervals have to be identified that express similar stratigraphic patterns of aggradation and/or degradation using the concept of A/S.

The A/S concept is a useful way of analysing spatial and temporal variations in depositional systems (Jervey, 1988; Thorne & Swift, 1991; Schlager, 1993; Shanley & McCabe, 1994; Allen *et al.*, 1996; Martinsen *et al.*, 1999; Muto & Steel, 2002) and it is used commonly in fluvial sequence stratigraphic models. Accommodation (A) has a unit of length per time and Sediment Flux (S) reflects the total volume of sediments supplied for a given time interval and is specified as a unit of volume per time (Muto & Steel, 2000). Consequently, both A and S include the notion of volume and time in their definition. A/S is used as a number and stratigraphic changes in A/S are interpreted as an expression of a change in position of fluvial base level (similar to, for example, Martinsen *et al.*, 1999). Therefore, it is necessary to analyse how temporal changes in accommodation in fluvial environments can be determined. As noted above, A and S are intimately related but not necessarily in equilibrium.

A/S change and recognition criteria

Changing A/S conditions are expressed in fluvial stratigraphic datasets in a number of ways and a set of criteria has to be established to detect readjustment of fluvial equilibrium profiles and channel planform style in (possible) response to increase or decrease in non-marine accommodation (Ouchi, 1985; Leeder & Stewart, 1997; Currie, 1997; Gawthorpe & Leeder, 2000; Bridge, 2003; Adams & Bhattacharya, 2005; Schumm, 2005). Facies interpretation (both interfluves and channel sandstones) linked to analysis of temporal and spatial changes in three-dimensional facies architecture are a key issue as the most important tool for non-marine correlation is to detect organised changes (trends) in preserved facies. Several related parameters can be used to detect these changes. However, care must be taken because many of these parameters are controlled by upstream as well as downstream factors of source area, climate, differential subsidence and eustacy (Schumm, 2005). It is therefore important that they are not used in isolation. The parameters include (see also Table 1):

1 Changes in the degree of palaeosol maturation. These are related to variations in climatic factors such as total precipitation, seasonality, temperature and the activities of organisms. Cumulative palaeosols with only modest degrees of palaeosol development are typical of alluvial basins with rapid continuous sedimentation.
2 Variations in the thickness of preserved palaeosol profiles. These are also related to variations in sedimentation rate and/or changes in accommodation generation.

Table 1. Parameters indicating A/S change (increasing or decreasing) used to analyse the Statfjord Group in the Statfjord Field; their relative importance and degree of applicability to the Statfjord succession.

	Parameter indicating A/S change	Importance	Degree of applicability
1	Changes in the degree of palaeosol maturation	High	High
2	Variations in the thickness of preserved palaeosol profiles	High	High
3a	Variations in the type and relative proportions of preserved fluvial sandstone lithofacies	High	High
3b	Variations in the type and relative proportions of preserved fluvial sandstone facies associations	Moderate	Moderate
4	Variations in the degree of channel belt thicknesses (1) and width (2) compared to single channel fills	High	1: High to 2: Low
5	Variations in the amount of fine-grained deposits preserved in the succession	High under moderate to low subsidence rate	High under moderate to low subsidence rate
6	Frequency of occurrence and characteristics of (erosion) surfaces between facies associations on a depositional profile	Moderate	Moderate
7	Variations in long-term avulsion frequency coupled to variations in channel stability and location and variations in the width of the floodplain	Moderate	Low

3 Variations in the type and relative proportions of preserved fluvial sandstone lithofacies and facies associations indicating changes in channel style in response to upstream changes in discharge and/or sediment load.

4 Variations in channel belt thicknesses and width as compared to the average thickness and width of single channel fills. This gives an indication of changes in vertical and lateral sandstone connectivity related to changes in accommodation generation with respect to the rate of sediment flux and water discharge. In cases where the rate of deposition was (very) high (with a sufficient rate of accumulation space generation) two types of (combined) responses are expected to occur: 1) much higher avulsion frequencies can be expected (Bryant *et al.*, 1995; Postma, 2014) which will lead to the (partial) removal of floodplain fines by erosion and a denser stacking of channels resulting in a higher connectivity and 2) higher preservation completeness (as expressed by the preservation of barform tops) of individual channel facies associations. In such cases, systematic channel planform style changes may indicate longer-term (>20 ka) extrinsically driven variations in hydrodynamic flow conditions during deposition.

5 Variations in the amount of fine-grained deposits preserved in the succession indicating changes in accommodation generation with respect to the rate of sediment delivery.

6 Frequency of occurrence and characteristics of (erosion) surfaces between facies associations within channel belts and on the overall depositional profile, such as the basal surfaces of local individual channel fills, both in the coarse (sand-filled channels) and well as the fine (mud pellet-filled channels) sediment fraction. Key issues include recognising the differences between, for example, erosion related to i) down cutting of a single channel or a distributary, ii) deep thalweg or a confluence scour, or iii) a channel belt creating a fluvial valley. Alternatively, the erosion may be the product of floodplain-wide degradation.

7 Variations in long-term avulsion frequency (Stouthamer & Berendsen, 2000, 2007; Postma, 2014) coupled to variations in channel stability (higher aggradation rates are expected to result in a higher avulsion frequency; lower aggradation rates are expected to result in increased channel stability and incision) and location and variations in the width of the floodplain.

Trends in A/S change

It is emphasised that it is not so much the absolute value of A/S but the A/S rate of change at a particular time scale that is important. Steady state A/S can occur when either A or S or both (but independently) are unsteady and fluctuate with a short periodicity randomly around an average (steady) value at short time scales so that the average over a longer time scale is constant. It can also be the result of steady dynamic allocyclic forcing (constant A/S) as a result of the autostratigraphic response (Muto *et al.*, 2007). No indication has been found in the Statfjord case study that justifies the systematic use of steady state A/S and therefore only positive and increasing or decreasing (non-steady state) A/S responses were applied in addition to degradation periods.

Consequently, two cases of fluvial A/S trends are defined associated with three key fluvial stratigraphic levels of which one can be identified as a surface (comparable to concepts applied by Allen *et al.*, 1996 and Diessel *et al.*, 1999; methods for their identification are outlined in subsequent sections):

1 Describing the situation in which degradation does not occur and no subaerial unconformity is formed. For $0 < A/S \leq 1$, an increasing A/S trend and a decreasing A/S trend is recognised (Fig. 4A; see also Sweet *et al.*, 2005). Either the turn-around from increasing to decreasing A/S (relative maximum aggradation surface) or the turn-around from decreasing to increasing A/S (relative minimum aggradation surface) can be chosen as the correlation level.
2 Describing the situation in which both aggradation ($A/S > 0$) and degradation occur ($A/S < 0$) within an A/S cycle. In this case, the subaerial unconformity (formed during the time span from the turn-around from aggradation to degradation to aggradation associated with no deposition during a time period at the studied locality) serves as the correlation surface.

Case 1 can occur as a higher frequency sequence formed within the time represented by the hiatus in case 2. A relative minimum aggradation correlation level or a subaerial unconformity can become eroded by a successive subaerial unconformity if the decreasing A/S trend is of such magnitude that the foregoing sequence is completely eroded away (Fig. 4B).

The relative minimum and maximum aggradation levels may be difficult or impossible to recognise as a single surface in high-accommodation settings associated with a rapid sedimentation rate (*cf.* Rygel & Gibling, 2006) or due to the way the turn-arounds are expressed, the scale of resolution (vertically and/or laterally) as well as autogenic shadowing (*cf.* Kjemperud, 2008). More commonly, a zone can be recognised in which the turn-around occurs. For example, decreasing palaeosol maturation values stratigraphically upward from a mature caliche horizon suggest increasing A/S values whereas increasing palaeosol maturation values suggest decreasing A/S values (Fig. 5). The subaerial unconformity may form at the base of extensive sandstone/conglomeratic units with or without combination with fine-grained clastics or entirely within fine-grained clastics (Sweet *et al.*, 2005).

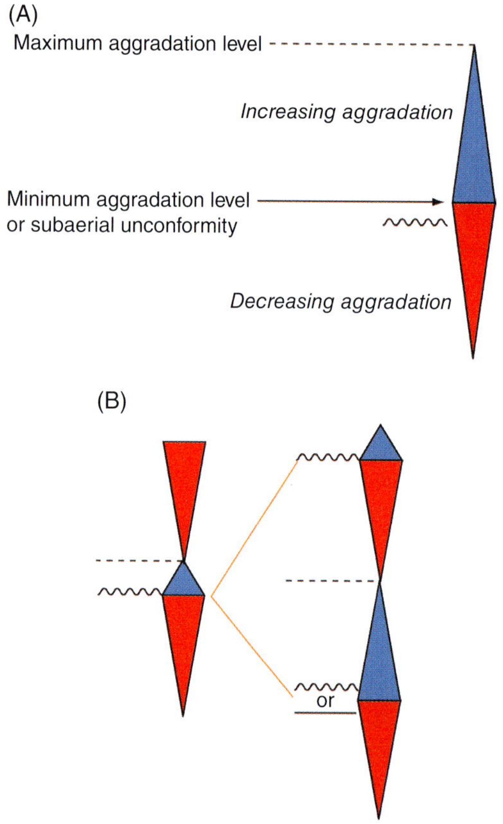

Fig. 4. Accommodation to sediment supply (A/S) relationships. (A) The direction of change of the accommodation to sediment supply (A/S) value can result in either an increasing value (blue triangle) or a decreasing value (red triangle). Together, they form an A/S cycle. The relative maximum aggradation surface is picked at the turn-around from increasing to decreasing A/S. The relative minimum aggradation surface is picked at the turn-around from decreasing to increasing A/S. If the value of A/S is not equal to zero or does not fall below zero a subaerial unconformity is not formed. (B) Considering a fluvial stratigraphic succession in three dimensions, a relative minimum aggradation surface related to a specific time scale in one position may represent a full A/S cycle developed at a shorter time scale in another location.

The definition of the low accommodation and high accommodation systems tracts of Dahle *et al.* (1997) and Martinsen *et al.* (1999), defined based on changing fluvial accommodation conditions through time (see also the discussion in Catuneanu, 2006, p. 230 and 231), is similar to the usage of the

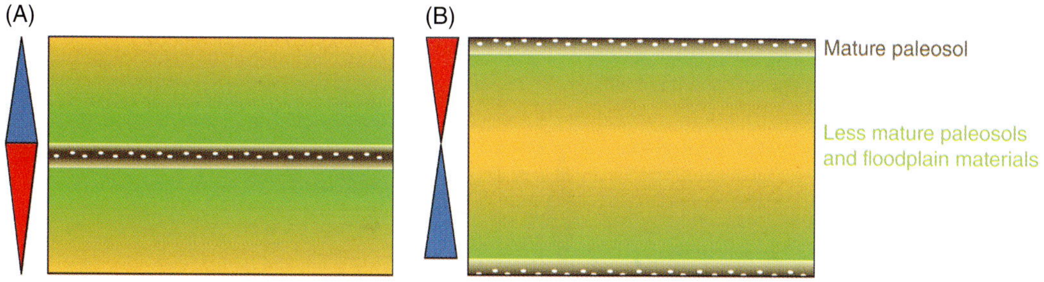

Fig. 5. Palaeosol maturation and accommodation/sediment supply (A/S) relationships. In the stratigraphic analysis of fluvial successions, the degree of maturation of palaeosols developed in floodplain mudstone deposits and their relation with underlying and overlying palaeosols needs to be taken into account. Decreasing palaeosol maturation values (indicated by the colour transition from green to yellow) stratigraphically upward from a mature caliche horizon suggests increasing A/S values (A). Increasing palaeosol maturation values (indicated by the colour transition from yellow to green) suggest decreasing A/S values (B). White dots in the brown-coloured interval indicate caliche in a mature palaeosol.

two trends defined herein. However, here, trends are defined based on a number of criteria that define changing A/S conditions through time (although stratigraphic intervals typified by increasing and decreasing A/S values were also mapped by Martinsen *et al.*, 1999). Deposition and preservation of sediment operate at different rates and time scales. Different time scales of interacting controlling mechanisms (tectonic and climatic) result in a hierarchy of cyclicity and related sequence stratigraphic surfaces formed by fluvial processes.

Long-term A/S change versus short-term autogenic processes

The stratigraphic analysis of changes in A/S value can only be applied to sediment stored in the net accommodation 'window' (see definition in Appendix I). It cannot be applied to sediments preserved in the real-time accommodation 'window' (see definition in Appendix I) because of the temporary nature of sediment storage. Sediment deposited in the real-time accommodation 'window' as a result of avulsions can only be permanently preserved when the channel(belt) stays away long enough from the avulsion site for the sediment to subside into the net accommodation 'window' (Fig. 6). For example, avulsions on the Holocene Rhine-Meuse lower delta plain are both autogenically as well as allogenically controlled and are associated with periods of aggradation (Stouthamer & Berendsen, 2000, 2007). Avulsion frequencies occur at various time scales. Shorter periodicities are likely to be autogenically controlled whereas longer periodicities appear to be controlled by allogenic factors. Stouthamer & Berendsen (2000, 2007) suggest that avulsion frequencies may decrease or increase during periods of decreased aggradation depending on the time scale and that avulsion locations are primarily related to allogenic factors interacting over time and space. Flume studies on the other hand suggest that avulsion frequencies increase during increased aggradation rate and vice versa (Bryant *et al.*, 1995; Postma, 2014). This is here interpreted to be possibly related to the fact that these flume studies primarily address the effects of changes in aggradation rate over longer time scales. Individual avulsions are commonly locally occurring events and, given the above mentioned factors, do not necessarily have sequence stratigraphic correlation value. Stouthamer & Berendsen (2007) reported a short-term autogenically controlled avulsion frequency for the Rhine-Meuse delta of 500 yr and an allogenic controlled long-term trend at a time scale of approximately 10^4 yr. Given this behaviour and the uncertain nature of factors controlling avulsions in continental fluvial systems, it is therefore important to recognise single avulsion products in core and exclude these from usage in stratigraphic correlation. Although problematic, single avulsion products may be recognised by higher preservation completeness of an individual sandstone body as expressed by the preservation of barform tops of individual channel facies associations. The occurrence of avulsion products in a stratigraphic unit bounded by correlatable surfaces can be used for analysis of environmental and stratigraphic change. An example of these principles of correlation in fluvial successions is shown in Fig. 7.

An expansion surface (Dahle *et al.*, 1997; Martinsen *et al.*, 1999) may develop during

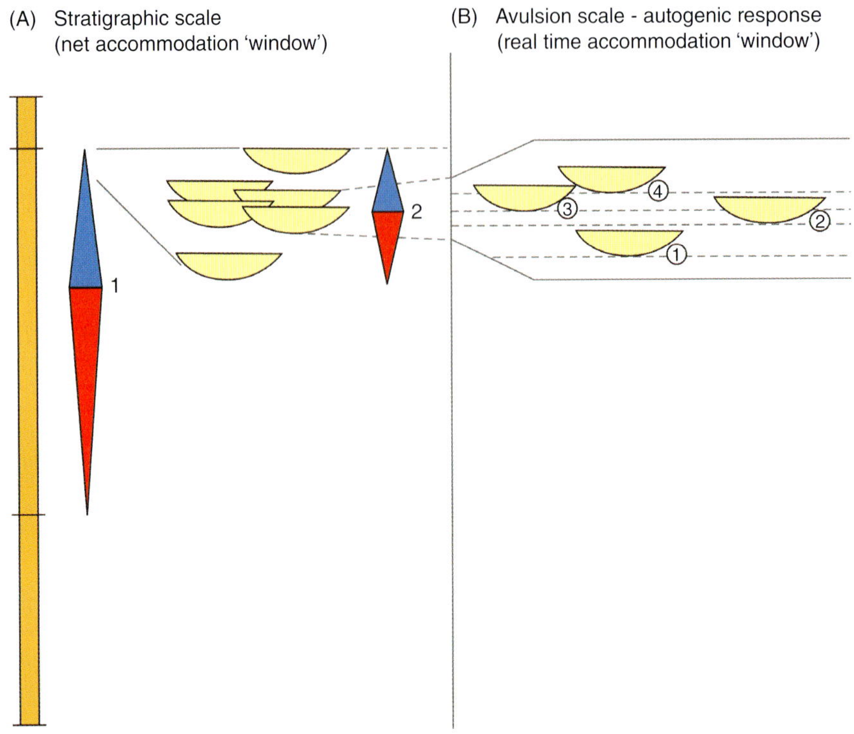

Fig. 6. Illustration of the principle of superposition of scales of A/S sequences. (A) Based on 1) the definition of real-time accommodation and net accommodation and their use in stratigraphic analysis (see text) and 2) avulsion data presented by Stouthamer & Berendsen (2007) it is suggested that the highest resolution related to the shortest time scale that can be used in the A/S analysis of fluvial deposits (level 2 in the figure) is approximately 10^4 yr. These authors suggested that individual avulsion deposits may or may not be the result of autocyclic processes and are initially deposited in the real-time accommodation window. Consequently, these deposits cannot be used for rock record. The orange bar indicates the stratigraphic column. (B) Avulsion scale. This cycle scale can be autocyclic controlled and can therefore not be used for reservoir zonation purposes. Numbers schematically refer to individual avulsion events.

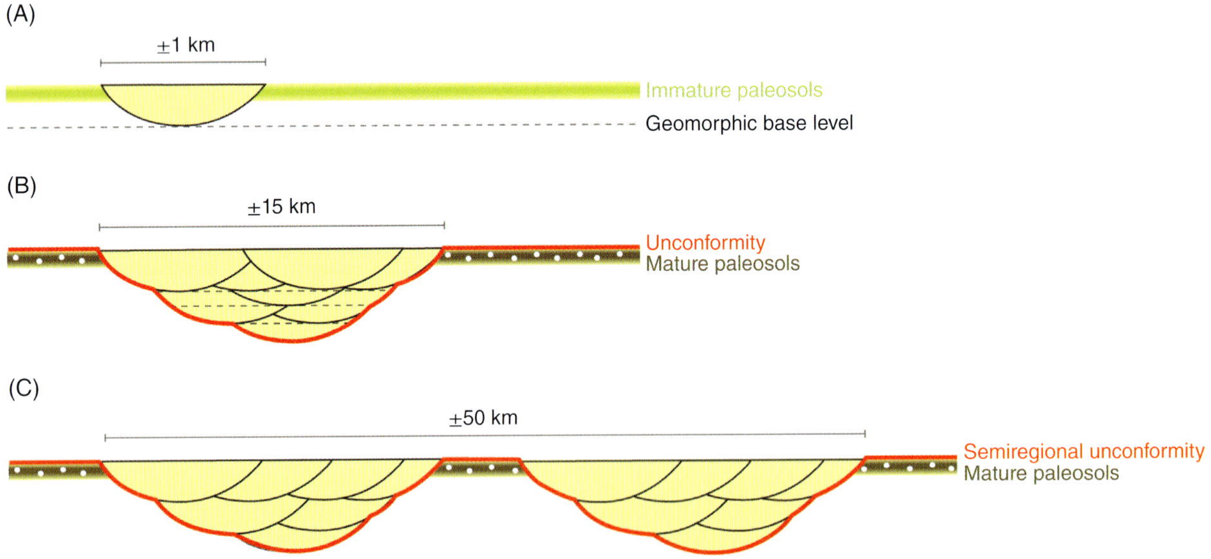

Fig. 7. Principles of correlation in fluvial successions. (A) Individual channel deposits cannot be used for stratigraphic correlation purposes (see text for explanation). (B), (C) The most mature palaeosols on the floodplain correlate with the unconformable basal surfaces of alluvial valleys to form stratigraphic turn-around locations from decreasing to increasing accumulation space and can be used for subsurface correlation purposes.

increasing A/S when sedimentation, initially confined to previously formed incised or fluvial valley (when A/S dropped below zero), overflows the valley margins and spreads across a much larger area. This is a local geomorphological effect – no sudden change in the rate of A/S increase is necessarily associated with this event (the rate of A/S increase can remain stable) but sedimentation is not confined anymore and resulting sandstone body stacking patterns suddenly change to less accommodation.

APPLICATION TO THE STATFJORD GROUP

General setting and characteristics

The largely fluvial Statfjord Group on the Tampen area of the North Sea (Deegan & Scull, 1977; Røe & Steel, 1985; Nystuen & Fält, 1995; Ryseth & Ramm, 1996; Ryseth, 2001; Lervik, 2006; Figs 1–3 and Fig. 8) was deposited in a rift basin that experienced relatively high subsidence rates during deposition. It was deposited on large subsiding 'mega-blocks' in a structurally confined area set up by thermally driven post-rift subsidence (Badley *et al.*, 1988; Steel & Ryseth, 1990; Steel, 1993; the North Sea Viking Graben, Fig. 8). In a SW to NE direction from the Utsira High into the Viking Graben (a distance of approximately 140 km) the formation changes in thickness from about 200 m to 600 m. This change in formation thickness is not reflected in significant overall changes in depositional environment (Ryseth, 2001) indicating that large-scale subsidence patterns and differential compaction generally did not affect reservoir architecture.

Main sediment source areas as identified from Sm/Nd studies (Mearns *et al.*, 1989) were located north and north-west of the Tampen Spur. Possible source terrains include the Rona Ridge to the west of the Shetland Islands which was part of the larger Lewisian basement of western Scotland. This source terrain might be reactivated several times due to early Cimmerian tectonics. Other workers (Dalland *et al.*, 1995; Morton *et al.*, 1996) included Caledonide foldbelt areas to the north of Tampen Spur as possible sources for at least parts of the Statfjord Group, which also makes eastern sources probable.

There is a clear proximal-distal trend from North (Tampen Spur) to South (Utsira High) for contemporaneous deposits of the Statfjord Group, visible in grain-size distributions, sand contents, organic matter contents and 'wetness' of overbank deposits. The Statfjord Group has been described as dominated by braided (Tampen area) to meandering (Utsira High area) channel systems with coal-bearing overbank (floodplain) deposits. South of the Tampen area, on the Utsira High, fluvial sandstones are finer grained and the associated floodplain deposits contain higher proportions of non-pedogenic deposits (Nystuen & Fält, 1995; Ryseth & Ramm, 1996; Ryseth, 2001).

In parts of the Tampen area, the Statfjord Group is divided into three lithostratigraphic formations (Deegan & Scull, 1977; Lervik, 2006; Fig. 3): The calcrete bearing red beds of the Raude Formation (uppermost Triassic to lowermost Jurassic age), the coal-bearing grey-beds with coarse-grained sandstones of the Eiriksson Formation (lowermost Jurassic age) and the finer-grained and partially heterolithic sandstones and mudstones of the Nansen Formation (lower Jurassic age; Deegan & Scull, 1977). This transition is interpreted as a progressive change towards more humid climatic conditions and possibly hinterland rejuvenation (Røe & Steel, 1985; Ryseth, 2001).

The Raude Fm is typified by an upward trend from fewer, relatively stationary and narrower channel belts to broader and possibly more mobile braid-belts. Overall N/G is around 45%. In the lower part, isolated sandstone bodies are encapsulated in extensive mudstones with palaeosol horizons. Channel sandstone thicknesses are quite variable and both multi-storey and possible single-storey channels occur. The latter are interpreted as possibly anastomosing trunk channels on the alluvial plain, while the multi-storey sandstone bodies represent braid belts (Nystuen & Fält, 1995; Ryseth & Ramm, 1996; Ryseth, 2001).

The lower part of the Eiriksson Fm is composed of highly amalgamated sandstone bodies, formed in sinuous channels, with good lateral and vertical connectivity and a sandstone N/G fraction of more than 80%. Most of the upper part of the Eiriksson Fm is interpreted as being deposited in a lower coastal plain environment with fluvial-dominated distributary mouth bars and bay fills in the Tampen area.

The uppermost part of the Eiriksson Fm and the Nansen Fm are considered to be influenced by marine processes and witness the onset of transgression that culminates in the deposition of the

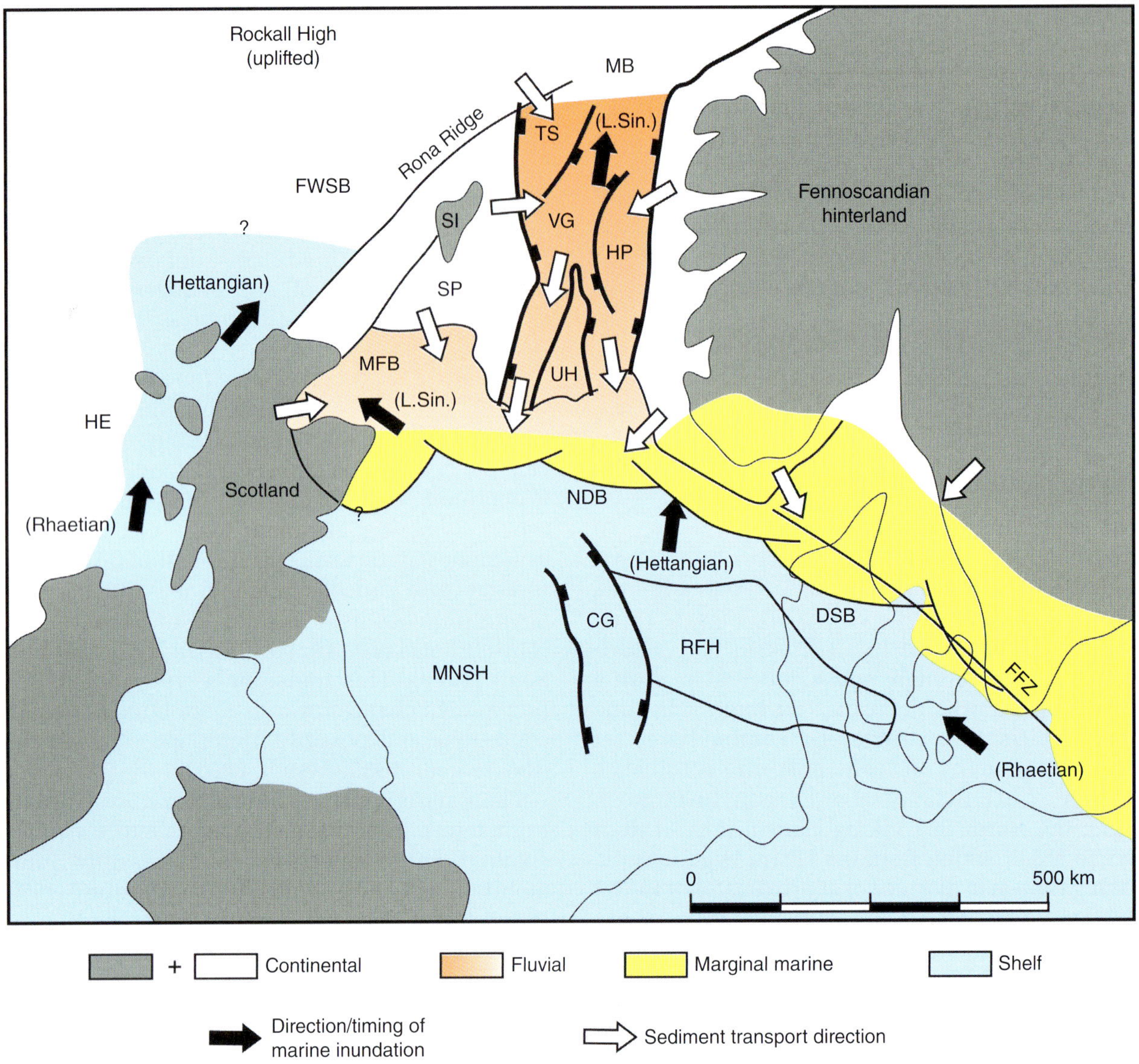

CG	Central Graben	NDB	Norwegian-Danish Basin
DSB	Danish Subbasin	MNSH	Mid North Sea High
FFZ	Fjerritslev Fault Zone	RFH	Ringkøping-Fyn-High
FWSB	Faroe-West Shetland Basin	SI	Shetland Islands
HE	Hebrides Basin	SP	Shetland Platform
HP	Horda Platform	TS	Tempen Spur
MB	Møre Basin	UH	Utsira High
MFB	Moray Firth Basin	VG	Viking Graben

Fig. 8. Schematic palaeogeographic map of the Statfjord Group (Raude and Eiriksson fms) in the Viking Graben (see Fig. 1) and the area south of it during the Hettangian showing the inferred sediment transport patterns and general depositional environments. Modified after Ryseth (2001) and Michelsen *et al.* (2003).

fully marine Dunlin Shale. Final drowning of the system is considered to have been rapid over a large area causing a relatively sudden facies transition at the base of the Amundsen Fm (base Dunlin Group) to fully offshore marine deposits. Continental Statfjord Group deposits are linked to age equivalent marine sediments in the southern North Sea implying that the palaeo-coastline for most of 'Statfjord times' was located south of the Utsira High; i.e. approximately 300 km to 350 km south of the Statfjord Field area (Ryseth, 2001; Fig. 8). The top of the Statfjord Group on the Statfjord Field is essentially flat and a backstepping of the uppermost Statfjord facies with time-equivalent Dunlin deposits interfingering can only be seen to the north and west of the Statfjord Field area.

For this study, facies observations from 2990 m of core and associated wireline information from 20 wells from the Statfjord, Snorre, Gullfaks Sør and Gullfaks Fields was used. A lithofacies description scheme was established which was used to define nine facies associations. Subsequently, the Statfjord Group reservoir correlation and zonation work discussed here was limited to 11 cored wells in the Statfjord Field.

Correlation and zonation challenges

Analysis of depositional environments suggests strongly that no marked proximal-distal trends can be identified on Statfjord Field-scale (approximately 5 x 24 km; Fig. 2). Stratigraphic trends in sandstone body connectivity are currently difficult to detect and only apparent at low resolution, large-scale correlations. In addition, reliable field-wide correlation zones, formed by fine-grained floodplain sediment, are nearly absent.

The succession also suffers from a severe lack of biostratigraphic data. Miospores occur abundantly in some fine-grained intervals but abundance events are local due to variations in source, dispersal and deposition. In addition, very few, good, age-diagnostic marker species for correlation occur. Megaspores are the only microfossil group with potential for Upper Triassic-Jurassic stratigraphic analysis of continental strata (Morris *et al.*, 2009). Additionally, chemostratigraphy and magnetostratigraphy do not have the required resolution in the Statfjord Group. The inability to recognise time lines though the fluvial succession, in addition to the difficulty to physically trace key stratigraphic surfaces, imposes a significant uncertainty related to the correlation of surfaces with stratigraphic significance as well as establishing the relative importance of these surfaces in a hierarchical framework.

Seismic data for the Statfjord Group on Statfjord Field are of generally poor quality and of limited use for correlation. Furthermore, after many years of injection and production, the Eiriksson Fm is more or less homogenised pressure-wise, while the more isolated Raude sandstone bodies show more variable pressure regimes. There are limited areas of pressure communication across the Raude to Eiriksson boundary but, in general, a pressure jump exists at this stratigraphic location which was used to support the Raude to Eiriksson boundary identification based on core and log data.

Stratigraphic analysis

Identification of A/S change in the Statfjord Group relies on core observations (Figs 9 and 10). Following the lines of reasoning presented above, for a given interval, minimum A/S periods are identified by either a combination of most vertisol-like features and highest maturity in the palaeosols (for example, most intense red colouration, layered calcretes, intense rooting) or, where present, the base of the multi-storey channel sandstone unit with the coarsest and/or most mineralogically immature lag deposits (Figs 6 and 9 to 11). The development and preservation of a palaeosol horizon implies that the net sediment accumulation rate in the location where the soil developed became neutral or weakly aggradational for a long time causing burial of the soil without significant erosion. Consequently, the site of pedogenic alteration on the floodplain was either distal and/or topographically raised above the fluvial channel (Leeder, 1975). Mature palaeosols, especially those featuring mature calcretes, require considerable periods of stability for growth (Todd, 1996) in the order of 10^5 yr for the most mature soils. During this time, the nearest channel belt has to remain at a distance sufficiently large to allow the development of a stable interfluve environment and to not affect palaeosol formation processes. A situation like this is considered a local A/S relative minimum (Fig. 12, Well 1). Observing these sedimentary features at the same stratigraphic level in multiple neighbouring wells suggests that the occurrence of these mature palaeosols is a semi-regional, stratigraphic phenomenon (Fig. 11A) controlled by external forcing factors, rather than local, autogenic

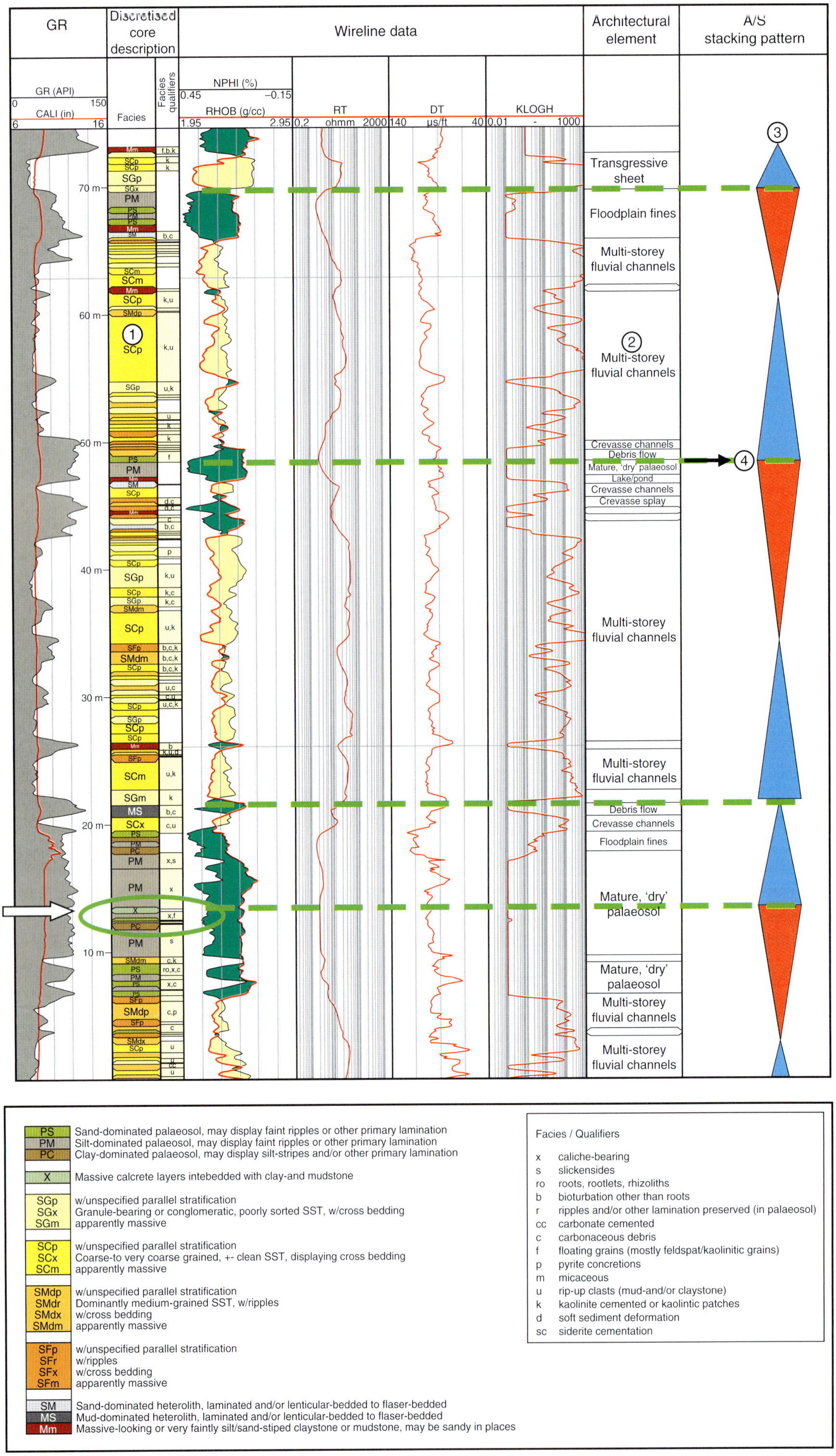

GR
Discretised core description
Wireline data
Architectural element
A/S stacking pattern
GR (API)
0
150
CALI (in)
6
16
Facies
Facies qualifiers
NPHI (%)
0.45
−0.15
RHOB (g/cc)
1.95
2.95
RT
0.2
ohmm
2000
DT
140
μs/ft
40
KLOGH
0.01
-
1000
70 m
60 m
50 m
40 m
30 m
20 m
10 m
Transgressive sheet
Floodplain fines
Multi-storey fluvial channels
Multi-storey fluvial channels
Crevasse channels
Debris flow
Mature, 'dry' palaeosol
Lake/pond
Crevasse channels
Crevasse splay
Multi-storey fluvial channels
Multi-storey fluvial channels
Debris flow
Crevasse channels
Floodplain fines
Mature, 'dry' palaeosol
Mature, 'dry' palaeosol
Multi-storey fluvial channels
Multi-storey fluvial channels
1
2
3
4
PS Sand-dominated palaeosol, may display faint ripples or other primary lamination
PM Silt-dominated palaeosol, may display faint ripples or other primary lamination
PC Clay-dominated palaeosol, may display silt-stripes and/or other primary lamination
X Massive calcrete layers intebedded with clay-and mudstone
SGp w/unspecified parallel stratification
SGx Granule-bearing or conglomeratic, poorly sorted SST, w/cross bedding
SGm apparently massive
SCp w/unspecified parallel stratification
SCx Coarse-to very coarse grained, +- clean SST, displaying cross bedding
SCm apparently massive
SMdp w/unspecified parallel stratification
SMdr Dominantly medium-grained SST, w/ripples
SMdx w/cross bedding
SMdm apparently massive
SFp w/unspecified parallel stratification
SFr w/ripples
SFx w/cross bedding
SFm apparently massive
SM Sand-dominated heterolith, laminated and/or lenticular-bedded to flaser-bedded
MS Mud-dominated heterolith, laminated and/or lenticular-bedded to flaser-bedded
Mm Massive-looking or very faintly silt/sand-stiped claystone or mudstone, may be sandy in places
Facies / Qualifiers
x caliche-bearing
s slickensides
ro roots, rootlets, rhizoliths
b bioturbation other than roots
r ripples and/or other lamination preserved (in palaeosol)
cc carbonate cemented
c carbonaceous debris
f floating grains (mostly feldspat/kaolinitic grains)
p pyrite concretions
m micaceous
u rip-up clasts (mud-and/or claystone)
k kaolinite cemented or kaolintic patches
d soft sediment deformation
sc siderite cementation

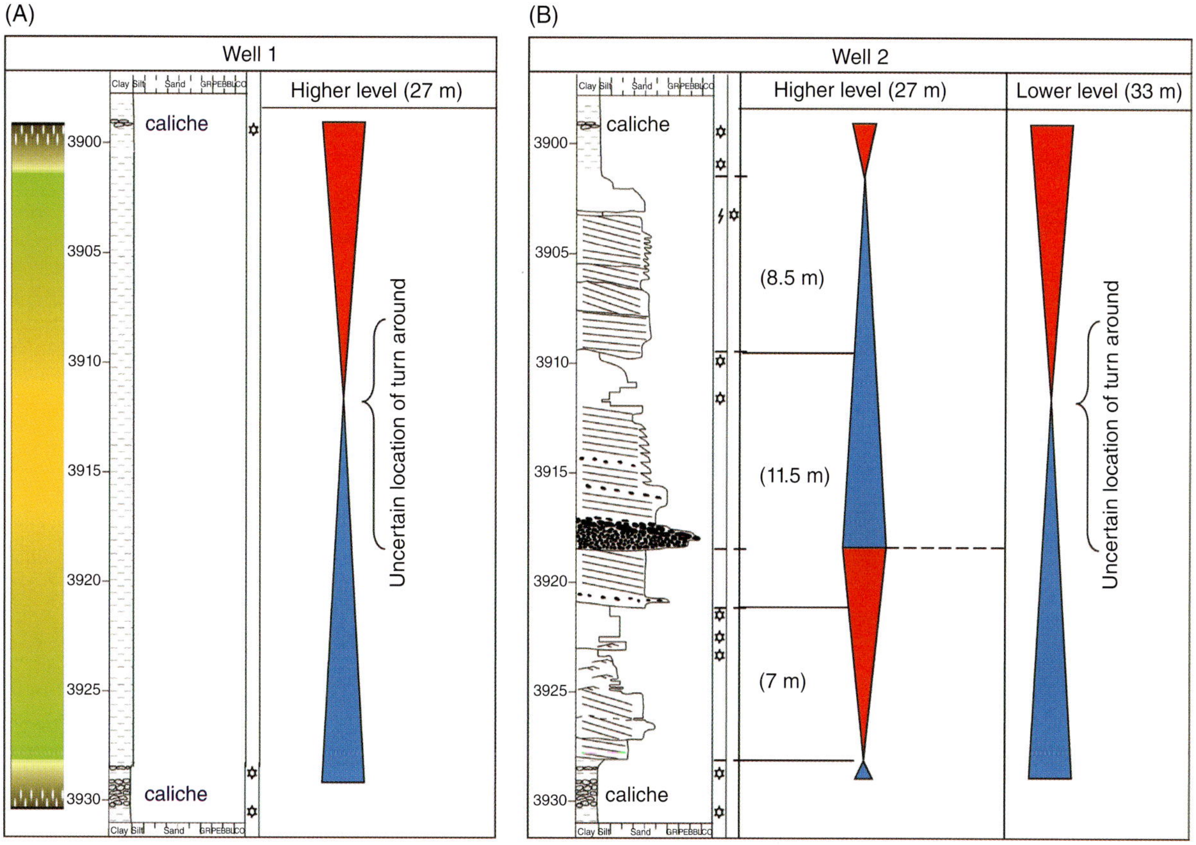

Fig. 10. Principle of subsurface correlation using cores. (A) A mudstone-dominated A/S cycle in Well 1 formed by decreasing palaeosol maturation values stratigraphically upward from a mature caliche horizon (indicated by the colour transition from green to yellow) suggesting increasing A/S values. Subsequent increasing palaeosol maturation values (indicated by the colour transition from yellow to green) suggest decreasing A/S values towards the next caliche horizon. White dots in the brown-coloured interval indicate caliche in a mature palaeosol. (B) A higher-order sandstone-dominated A/S cycle in Well 2 is superposed on the lower order mudstone-dominated cycle. Decreasing A/S values are interpreted from the combination of i) increasing grain size, ii) decreasing sorting and roundness and iii) decreasing preserved set thicknesses. Increasing A/S values are indicated by the reversed trend of the aforementioned parameters.

phenomenon. From the A/S relative minimum upwards, an increasing A/S value is indicated by either less mature palaeosol development (Fig. 12, Well 1 overall trend) or by overall upward fining of channel deposits (Fig. 12, Well 2 overall trend).

Correlation of stratigraphic sequences defined by changes in A/S may occur across sedimentary units with laterally marked lithological changes. This can result in significant lateral variations in A/S cycle thicknesses for higher-order cycles because the base of sand-rich channel intervals, interpreted to indicate the onset of a minimum aggradation trend, are correlated to relatively thin, time equivalent, palaeosols in other locations (Fig. 11A). The

Fig. 9. An example of an interpreted cored well interval (well 3 of Fig. 12) covering reservoir zones 5 to 9 (uppermost part of the Raude Fm and the Eiriksson Fm.; for location see Fig. 2). The stratigraphic understanding of the Statfjord Group is built up stepwise. Architectural elements (2) are defined based on a descriptive subdivision of all core samples into lithofacies (1) comprising the smallest definable building blocks of the reservoir and facies associations. Cyclic stacking patterns of architectural elements (3) are interpreted in terms of changes in accumulation space and sediment supply as well as depositional breaks and erosion during sedimentation. This makes identification of marker horizons possible, which later define reservoir zone boundaries (4). Reservoir zones defined in this way will consequently honour geological developments through time instead of similar-looking lithologies.

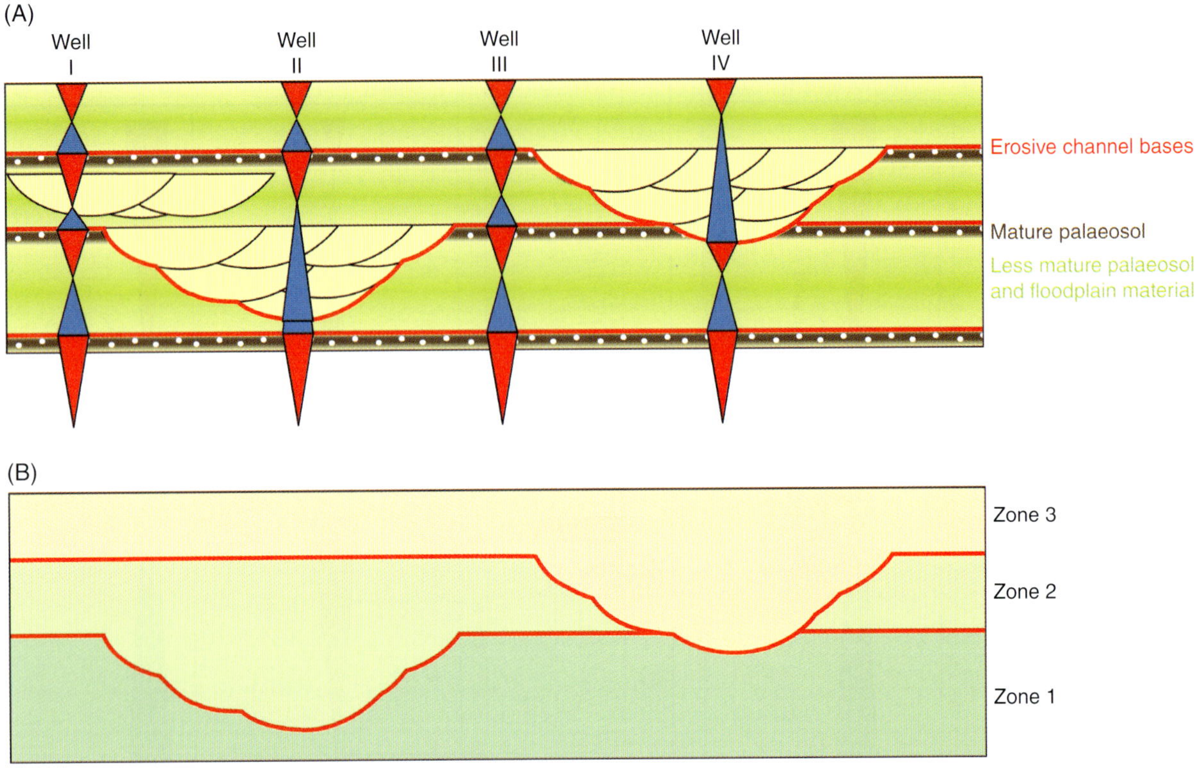

Fig. 11. (A) Simplified sketch of multi-well correlation principles and; (B) resulting reservoir zonation. Note that in (A) i) three full A/S cycles and 1 half A/S cycle are recognised (well 3) and ii) lateral correlation of the 2nd mature palaeosols is discontinuous in the area of wells 2 and 4 due to channel belt incision (with an unconformity at the base) removing the first complete A/S increasing-to-decreasing cycle in well 2 and the second complete A/S increasing-to-decreasing cycle in well 4.

resulting architecture within reservoir zones is reflecting the A/S related stratigraphic development of the reservoir more realistically than when using lithostratigraphic correlation techniques because the concept of time-equivalence is being taken into account. Correlation lines are assumed to be quasi-time lines that create envelopes containing genetically related depositional elements of approximately equal ages (Fig. 11B). Connectivity between reservoir zones in this case depends on local sand-to-sand contacts across boundaries, not along boundaries as in the case of lithostratigraphic zonation.

In this Statfjord Group example three different hierarchical orders of A/S cyclicity are identified (Fig. 12). The lowest order of cyclicity is marked by the overall increase in A/S from the lower part of the Statfjord Group into the lower Jurassic Dunlin Group. This A/S increase is attributed to an overall base-level rise and is represented by the transition from terrestrial alluvial plain deposits (the Raude Fm) through coastal plain deposits (Eiriksson Fm) and transgressive reworking (Nansen Fm) into the marine sediments of the Amundsen Fm. The Statfjord Group thus comprises only the increasing A/S trend of this lowest order cyclical development.

An intermediate, intra Statfjord Group-scale A/S order of cyclicity is interpreted to be expressed by the vertical development of facies and channel belt stacking patterns. An upward increase in channel belt connectivity for the sandstone bodies of the Raude Fm, both vertically and laterally, is interpreted to reflect a decrease in A/S up to the base of the Eiriksson Fm (Fig. 12, A/S cycle 1). Whilst the lower parts of the Raude Fm comprise fewer, relatively stationary, channel belts, resulting in more isolated sandstone bodies and limited connectivity, the upper Raude Fm shows a higher degree of amalgamation and increased connectivity (but still relatively discontinuous sandstone bodies compared to the Eiriksson Fm). An overall upward increase in sand content associated with decreasing channel-type facies variability is seen also in this intermediate decreasing A/S half-cycle. Palaeosol horizons within the Raude Fm

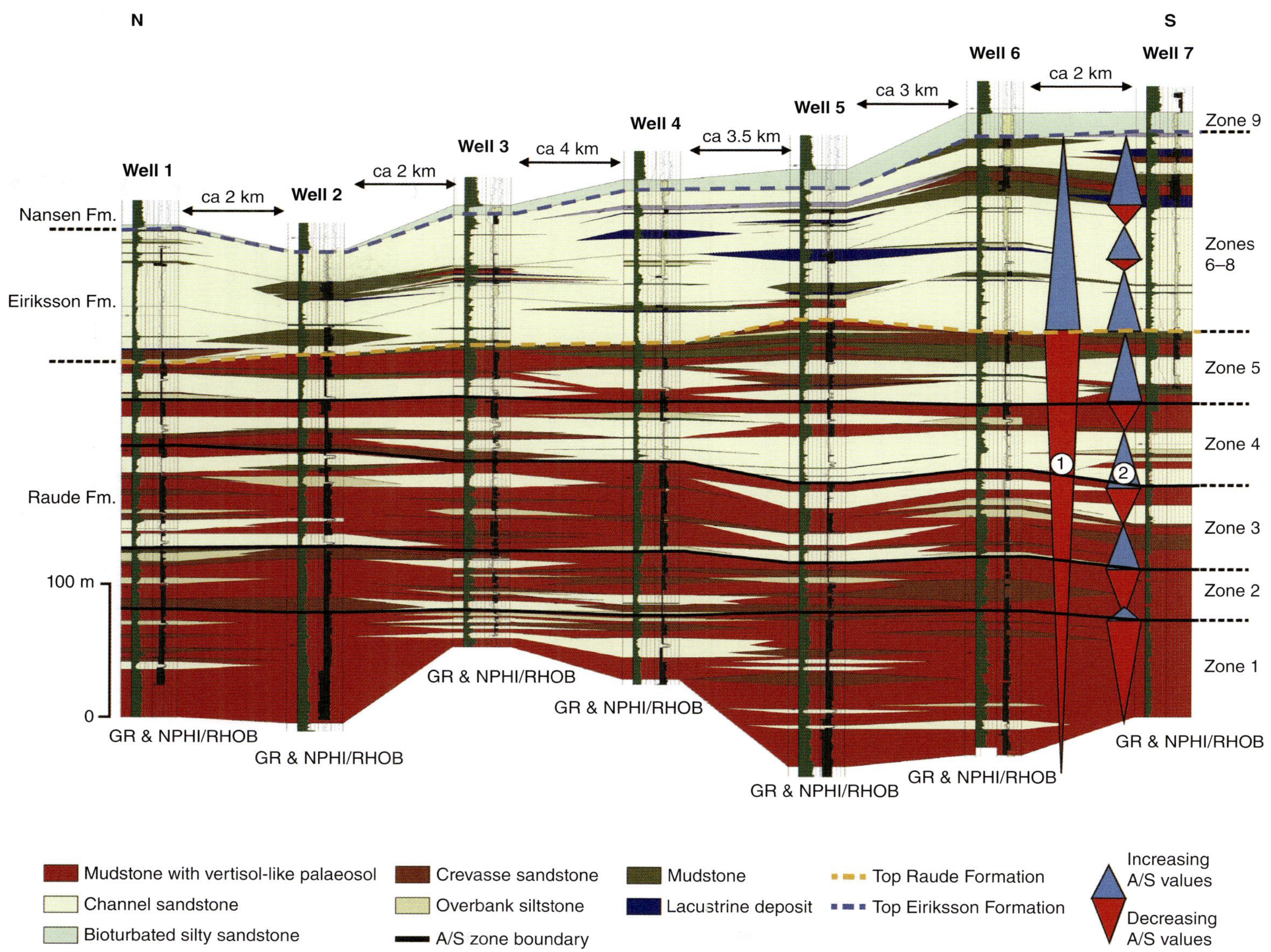

Fig. 12. N-S correlation panel of cored wells of the Statfjord Group within the Statfjord Field. Horizontal A/S zone boundaries mark reservoir subdivision (zone 1 to 9). A/S sequence symbols are equivalent to Fig. 5. 1 = Intermediate (intra Statfjord-) scale A/S cycle: upward more and more amalgamated and laterally extensive channel belt deposits in the Raude Fm mark a decrease in A/S (although the long-term development displays the opposite trend) and the minimum A/S is set equal to base Eiriksson Fm. 2 = Correlative A/S sequences on reservoir zone-scale (see text).

are generally much thicker and more correlative than in the Eiriksson Fm. For the Eiriksson Fm, the character of the fine-grained floodplain sediments change markedly and the generally vertisol-like palaeosols of the Raude Fm are replaced by immature or almost non-altered, mostly hydromorphic soils. In addition, petrographic analysis show that the content of kaolinite clay increases at the expense of smectite clays and other mixed layer minerals, which is interpreted as a sign of changing climatic conditions. The level of relative minimum aggradation for this intermediate A/S cycle is interpreted at base Eiriksson Fm. As shown in Fig. 12, the character of this boundary in terms of underlying higher-order cycles is non-uniform, i.e. there is not always a complete higher-order cycle beneath and the development of these is in places interrupted inside the respective A/S rise trend. This observation confirms that the magnitude and significance of the A/S minimum (and negative if $A/S < 0$) at the boundary between the Raude-Eiriksson formations is different from the intra-Raude A/S minimum situations (see discussion below). A certain erosive relief (at the scale of approximately 30 metres) of the surface between the Raude-Eiriksson formations and resulting variable thickness of the Eiriksson Fm (and thereby also the Raude Fm), together with the arguments noted above, seems to support an interpretation of this surface as a sequence boundary. Correlation of this boundary away from the erosive event reveals the existence of an interfluve (Fig. 12, Well 5). Here, mature palaeosols are interpreted to be time-equivalent to erosive channel bases in neighbouring wells, implying that the palaeosols matured while multiple, long-lasting erosional events and subsequent infilling occurred in other locations.

The third order of A/S cyclicity corresponds to the smallest correlative units that can be identified using channel-belt stacking patterns and facies variations as described above (Fig. 12, A/S cycle 2). Channel belt stacking patterns potentially comprise higher frequency cyclicity than the A/S cyclicity derived from palaeosol maturities, because the erosive base of the channel belt reflects a more local A/S minimum and subsequent increase in A/S allows for preservation of the channel-belt deposits. However, for a given interpretation of channel-belt interval cyclicity, the A/S maximum is a rather hypothetical situation and the exact position is highly speculative. One possibility is to pick the turn-around from increasing to decreasing A/S where the variability in sandstone facies and the proportion of fine-grained material inside the channels is highest. Because of the uncertainty related to the correct recognition of its stratigraphic location, especially from core data, the relative maximum aggradation level is not considered a good correlative marker and the relative minimum aggradation position is generally picked as the candidate correlation level (Fig. 12, zone boundaries). The A/S maximum in intervals consisting of mainly palaeosols is picked where the least mature palaeosols occur (Fig. 12, well 7, zone 2, based on core and gamma-ray log response).

In the Statfjord Group biostratigraphy cannot provide any intra-formational time control due to the lack of good quality recoverable biostratigraphic data in the deposits that can be calibrated to absolute time needed to assess the diachronity of subaerial unconformities and to build a reliable reservoir zonation. Correlation of thick, mature palaeosol intervals is assumed to give the most robust framework for more detailed definition of correlative marker horizons (Fig. 11A). It is these correlative markers that are then used further in the zonation of the reservoir (Fig. 11B).

Reservoir zonation improvements

The notion that the rate of fluvial aggradation through time is dynamic and not static, in combination with the use of A/S concepts and a more practical definition of fluvial base level to measure A/S change has led to the recognition of a suite of parameters that was used in combination to recognise variations in the rate of A/S change and turn around positions in core, to analyse stratigraphic architecture and to define reservoir zones. Together, these improvements in stratigraphic concept and the zonation workflow have led to a more robust treatment of the natural depositional variability and mapping of connectivity patterns. Moreover, it has also significantly decreased the level of uncertainty associated with the reservoir zonation framework and the 3D distribution of static reservoir properties. Consequently, static reservoir property trends for each zone are better characterised which is expected to lead to improved predictions of flow properties and production profiles.

CONCLUSIONS

A revised definition of base level, or stratigraphic reference level, for non-marine successions is introduced as the lower boundary of real-time accommodation or upper boundary of net accommodation (lower limit to erosion). Net accommodation lies below the maximum depth of channel incision (upper boundary) and is related to the lower limit to erosion and time scale of observation that need to be sufficiently large. Preservation in the stratigraphic record occurs when the deposits become buried below possible depths of channel incision and removal, that is, below the lower boundary of real-time accommodation, i.e. base level. A/S value can only be applied to sediment stored in the net accommodation 'window'. It cannot be applied to sediments preserved in the real-time accommodation 'window' (defined as the accumulation space at the scale of present-day fluvial process regimes) because of the temporary nature of sediment storage.

Two types of effective accommodation to sediment supply value (A/S) response exist: 1) positive and increasing A/S equivalent to increasing rate of aggradation (fluvial base level can be picked at subsequently higher stratigraphic positions) and 2) positive but decreasing A/S equivalent to decreasing rate of aggradation (fluvial base level can be picked at subsequently lower stratigraphic positions).

Three key fluvial stratigraphic surfaces are defined: 1) the relative maximum aggradation level which marks the turn-around from increasing aggradation to decreasing aggradation, 2) the relative minimum aggradation level which marks the turn-around from decreasing aggradation to increasing aggradation (no degradation is reached) and 3) the subaerial unconformity surface which marks the turn-around from aggradation to degradation to aggradation (no deposition during a time period at the studied locality). Accordingly, an increasing A/S trend followed by a decreasing A/S trend is defined together forming a cycle. A/S trend bounding levels are chosen as the stratigraphic zone boundaries. If degradation occurs ($A/S < 0$), the subaerial unconformity is the zone boundary.

It is proposed that the concept of increasing and decreasing rates of A/S during the aggradational part of a fluvial succession may work to divide largely aggradational successions into zones. The notion that the rate of fluvial aggradation through time, as controlled by the accommodation versus sediment supply rate, is dynamic and not static (classified as high or low) can be applied to spatially and temporarily restricted (limited) subsurface datasets at reservoir scale (0.1 km to 20 km). Facies interpretation (both fines and sandstones) linked to analysis of temporal and spatial changes in three-dimensional facies architecture are a key issue to detect organised changes (trends) in preserved facies that can be used to establish a reservoir zonation. The alternative reservoir zonation approach for dividing fluvial stratigraphy that is proposed herein enables the organisation of fluvial successions in a hierarchical framework incorporating zone property trends for, for example, sandstone body connectivity and rock properties such as average permeability. It is successfully applied to the uppermost Triassic to lower Jurassic Statfjord Group (Norwegian continental shelf).

ACKNOWLEDGEMENTS

We acknowledge critical but constructive and helpful comments on earlier versions of the manuscript by Frank Ethridge, Lars-Magnus Fält, Martin Gibling, Colin North, Tobi Payenberg, Ron Steel and Gary Weissmann. The manuscript was formally reviewed by Cornel Olariu and George Postma. Responsibility for remaining errors remains with the authors. Statoil ASA is thanked for permission to publish.

APPENDIX

Definitions for measuring change in A/S

In order to recognise changes in the rate of A/S change in core and consequently changes in the position of fluvial base level, a working concept coupled with a number of definitions is required. The argument is constructed by specifying real time versus net accommodation. Subsequently, fluvial base level, as used in this study, is defined followed by the definition of change in position of fluvial base level, which enables the recognition of successive accumulated sediment packages. This is considered as a measure for the change in A/S at specific time scales.

Real time versus net accommodation

The original definition of accommodation of Jervey (1988) included present-day and time-averaged time scales. Following ideas of Blum & Törnqvist (2000; and also Kocurek & Havholm, 1993; Kocurek, 1998), accumulation space at the scale of present-day alluvial process regimes is

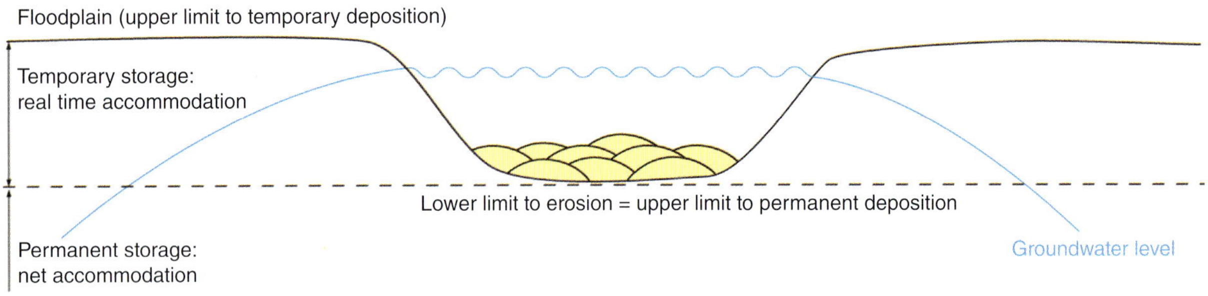

Fig. 13. Definition of real-time accommodation (sensu Blum & Törnqvist, 2000) and net accumulation space. Real-time accommodation is the accumulation space in which sediment is initially stored but can be removed again at time scales less that 10^3 yr. The upper boundary of real-time accommodation is defined by the upper limit to accumulation (equilibrium floodplain heights) and the lower boundary of real-time accommodation is defined by the lower limit to erosion (maximum depth of channel incision). Net accommodation (*sensu* Blum & Törnqvist, 2000) is defined as the accumulation space that lies below the maximum depth of channel incision (upper boundary) and is related to the lower limit to erosion. The lower boundary of real-time accommodation, equivalent to the upper boundary of net accommodation (lower limit to erosion), is chosen as the stratigraphic reference level or base level.

the volume of space that can be filled. Filling of the available accumulation space is governed by the relationship between stream power and sediment load and how this changes in response to the lower limit to erosion. From this it follows that sediment initially stored in the accumulation space can be removed again and it is therefore defined as real-time accommodation (cf. Blum & Törnqvist, 2000; Fig. 13) estimated here to operate at time scales of 10^3 yr to 10^4 yr. Increases and decreases in real-time accumulation space at process scale occur in response to (1) changes in the lower limit to erosion and (2) changes in discharge regimes and sediment supply which control equilibrium floodplain heights (upper limits to accumulation). Consequently, the upper boundary of real-time accommodation is defined by the upper limit to accumulation (equilibrium floodplain heights) and the lower boundary of real-time accommodation is defined by the lower limit to erosion (maximum depth of channel incision). Note that the upper limit to accumulation varies in elevation due to local build-up of levees of up to several metres. Consequently, the upper boundary of real-time accommodation varies transiently in height between levees and floodplain surfaces.

It is important to note that deposition and preservation operate at different rates and time scales. Preservation in the stratigraphic record occurs when the deposits become buried below possible depths of channel incision and removal, that is, below the lower boundary of real-time accommodation. Preservation space, or net accommodation (a term proposed by Blum & Törnqvist, 2000), therefore, lies below the maximum depth of channel incision (upper boundary) and is related to the lower limit of erosion (Fig. 13). It is also related to the time scale of observation and to aerial space because the maximum depth of incision will vary through time as a result of base level changes suggesting different fluvial response times following a perturbation (such as, for example, a single channel incision and relocation or the relocation of an entire channel belt).

Fluvial base level

The use of sea-level (upper boundary of potential accommodation) as stratigraphic reference level (or base level) has no practical meaning in subaerial environments far away from the contemporaneous shoreline such as the Statfjord Group in the Statfjord Field. Fluvial systems are only influenced by marine base level at their distal terminations and therefore issues associated with changes in relative sea-level are not discussed.

The equilibrium profile of graded rivers (Mackin, 1948) is used commonly as base level in sequence stratigraphic studies of fluvial and alluvial plain successions. Mackin (1948) defined the graded profile as a conceptual surface which marks equilibrium with prevailing discharge and channel characteristics. At time scales of 1 yr to 10^5 yr, studies of Quaternary deposits have indicated that upstream controls (for example, water discharge and sediment influx) cause 1 m to 10 m of profile adjustment within the same drainage (Goodbred & Kuehl, 2000; Blum & Törnqvist, 2000; Schumm *et al.*, 2000). However, it is problematic to recognise

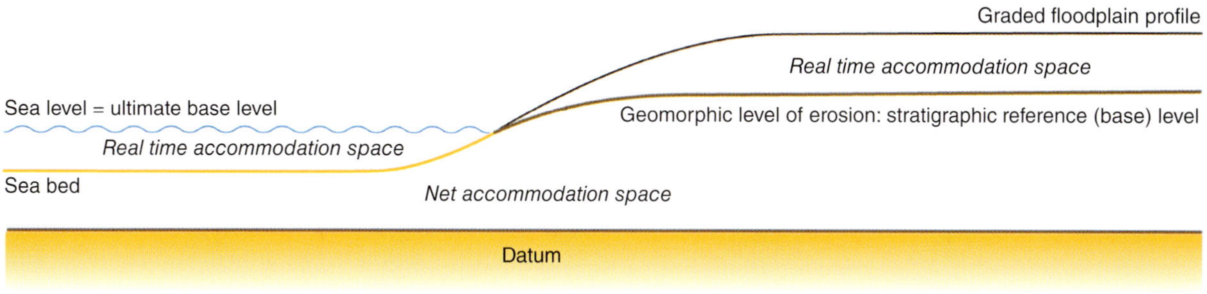

Fig. 14. Position of the stratigraphic reference, or base level, in alluvial and fluvial depositional systems and its relation with sea-level and the sea bed.

the physical location of the equilibrium profile in the alluvial subsurface rock record (i.e. core) and how it changes with time and location. Factors that influence sedimentation and erosion in alluvial plain and fluvial environments include variations in (i) dispersive sediment transport in response to the stream power, (ii) rate of sediment supply, (iii) grain size of sediment delivered and (iv) compaction (Thorne & Swift, 1991; Schumm, 1993). The boundary between the aggradational and degradational river states is the graded state (Davis, 1902; Mackin, 1948) and Swenson & Muto (2005) showed that the graded state is linked to sediment supply and the parameters discussed above.

From studies on modern river systems it is clear that no universally accepted set of criteria exists for determining whether all or part of a river system is in equilibrium and the 'average' character of a stream is uncertain from observation at a specific geographic location (Knighton, 1998). It is therefore almost impossible to determine whether an ancient stream was in equilibrium, in fact it is most likely that the system was not in equilibrium at the point in time of preservation as each channel is subject to short-term fluctuations and long-term evolutionary tendencies (Knighton, 1998). The co-occurrence of different channel types in specific stratigraphic intervals, for example, could point to a state of disequilibrium or non-equilibrium of streams at medium (10^3 yr to 10^4 yr) to long ($>10^5$ yr) time scales.

It is proposed here to use the lower boundary of real-time accommodation or upper boundary of net accommodation (lower limit to erosion) as the stratigraphic reference level or fluvial base level (Figs 13 and 14). In areas not directly affected by laterally migrating fluvial channels or channel belts, the upper boundary of net accommodation (lower limit to erosion) at a specific time scale can in some cases be genetically coupled to the most mature palaeosols (Weissmann *et al.*, 2002; Bennett *et al.*, 2006). However, in other cases large interfluve areas can have a complex nature with co-existing lakes, small rivers and aeolian dunes in addition to areas experiencing, for example, gully erosion (Gibling *et al.*, 2005; Tandon *et al.*, 2006). In such cases, the active upper zone of the interfluve area lies within the envelope of short-term accumulation. The stratigraphic reference level (the upper boundary of net accommodation equivalent to the lower limit to erosion) may in such cases be picked at a laterally extensive caliche layer formed at a depth relatively unaffected by short-term erosion and deposition (the base of the envelope of short-term accumulation). The upper limit of real-time accumulation space (or temporary deposition) is not considered a valid choice for the stratigraphic reference level because the upper limit to deposition will not be preserved systematically in the stratigraphic record. Note that the concept of stratigraphic base level was originally defined as the equilibrium surface that 'describes the direction in which a stratigraphic system is likely to move, toward sedimentation and stratigraphic preservation or sediment bypass and erosion' (Barrell, 1917; Sloss, 1962; Wheeler, 1964; Cross, 1988). Blum & Törnqvist (2000) noted that the concept of stratigraphic base level is really the same as the upper limit of accommodation. They considered that it has outlived its usefulness and should be abandoned.

The nature of the relatively most mature palaeosols depends on the location of the alluvial system with respect to the global (and zonal) system of convective atmospheric circulation ('Hadley' system controlling both temperature and humidity) and the local climatic regime (controlled by local geographical factors). For example, in a wet-dry regime thick well-developed caliche horizons represent the upper boundary of net accommodation. Note that over time periods longer than 20 ka the position and size of 'Hadley' cells will change systematically due to orbitally controlled changes in insolation (Perlmutter & Matthews, 1990; Matthews & Perlmutter, 1994).

Change in the rate of A/S change

To estimate changes in the rate of A/S change, the variations in thickness of preserved fluvial sediment packages have to be measured. Subaerial accommodation for a given time period is defined as the space developed (and filled) between two successive recognisable upper boundaries of net accommodation (lower limit to erosion) of different ages but with a similar hierarchical age allocation that define that time period. Consequently, the change in rate of accumulation space creation versus sediment supply (A/S) at a specific hierarchical scale can be appraised by mapping a series of successive upper boundaries of net accommodation in the stratigraphic successions and measuring the stratigraphic thicknesses between these successive surfaces.

REFERENCES

Adams, M.M. and **Bhattacharya, J.P.** (2005) No change in fluvial style across a sequence boundary, Cretaceous Blackhawk and Castlegate Formations of central Utah, U.S.A. *J. Sed. Res.*, **75**, 1038–1051.

Allen, G., Lang, S., Musakti, O. and **Chirinos, A.** (1996) Application of sequence stratigraphy to continental successions. In: *Implications for Mesozoic cratonic interior basins of Eastern Australia.* Mesozoic Geology of the Eastern Australian Plate Conference, Brisbane, Australia, Extended Abstracts, 22–26.

Badley, M.E., Price, J.D., Rambech Dahl, C. and **Agdestein, T.** (1988) The structural evolution of the northern Viking Graben and its bearing upon extensional modes of basin formation. *J. Geol. Soc. London*, **145**, 455–472.

Barrell, J. (1917) Rhythms and the measurement of geologic time. *Geol. Soc. Am. Bull.*, **28**, 745–924.

Bennett, V.G.L., Weissmann, G.S., Baker, G.S. and **Hyndman, D.W.** (2006) Regional-scale assessment of a sequence-bounding palaeosol on fluvial fans using ground-penetrating radar, eastern San Joaquin Valley, California. *Geol. Soc. Am. Bull.*, **118**, 724–732.

Blum, M.D. and **Törnqvist, T.E.** (2000) Fluvial responses to climate and sea-level change; a review and look forward. *Sedimentology*, **47**, Supplement **1**, 2–48.

Boyd, R., Diessel, C.F.K., Wadsworth, J., Little, M., Leckie, D. and **Zaitlin, B.** (2000) Developing a model for non-marine sequence stratigraphy. In: *15th Australian Geological Convention*, Sydney, Australia, Abstracts, p. 48.

Bridge, J.S. (2003) *Rivers and Floodplains: Forms, Processes and Sedimentary Record.* Blackwell Publishing, Oxford, 491 pp.

Bryant, M., Falk, P. and **Paola, C.** (1995) Experimental study of avulsion frequency and rate of deposition. *Geology*, **23**, 365–368.

Bull, W.B. (1991) *Geomorphic Responses to Climatic Change.* Oxford University Press, London, 312 p.

Cai, X. and **Zhang, M.** (1999) Several problems in sequence stratigraphy in continental facies basin; case of Songliao Basin, China. *Xiandai Dizhi/Geoscience*, **13**, 287–290.

Catuneanu, O. (2006) *Principles of Sequence Stratigraphy.* Elsevier, Amsterdam, 375 p.

Coward, M.P., Dewey, J.F., Hempton, M. and **Holroyd, J.** (2003) Tectonic evolution. In: *The Millennium Atlas: Petroleum Geology of the Central and Northern North Sea* (Eds D. Evans, C. Graham, A. Armour and P. Bathurst), Geol. Soc. London, 17–33.

Cross, T.A. (1988) Controls on coal distribution in transgressive-regressive cycles, Upper Cretaceous, Western Interior, U.S.A.. In: *Sea-level changes: An integrated approach* (Eds C.K. Wilgus, B.S. Hastings, C.G.St.C. Kendall, H.W. Posamentier, C.A. Ross and J.C. Van Wagoner), *SEPM Spec. Publ.*, **42**, 371–380.

Currie, B.S. (1997) Sequence stratigraphy of nonmarine Jurassic-Cretaceous rocks, central Cordilleran foreland-basin system. *Geol. Soc. Am. Bull.*, **109**, 1206–1222.

Dahle, K., Flesja, K., Talbot, M.R. and **Dreyer, T.** (1997) Correlation of fluvial deposits by the use of Sm-Nd isotope analysis and mapping of sedimentary architecture in the Escanilla Formation (Ainsa Basin, Spain) and the Statfjord Formation (Norwegian North Sea). In: *6th Fluvial Sedimentology International Conference*, Cape Town, South Africa, Abstracts, p. 15–16.

Dalland, A., Mearns, E.W. and **McBride, J.J.** (1995) The application of Samarium-Neodymium provenance ages to correlation of biostratigraphically barren data: a case study of the Statfjord Formation on the Gullfaks oilfield, Norwegian North Sea. In: *Non-Biostratigraphical Methods of Dating and Correlation* (Eds R.D. Dunnay and E.A. Hailwood), *Geol. Soc. London Spec. Publ.*, **89**, 201–222.

Davis, W.M. (1902) Base level, grade and peneplain. *J. Geol.*, **10**, 77–111.

De Vriend, H.J. (2005) Present-day river modeling. In: *8th International Conference on Fluvial Sedimentology* (Eds G.J. Weltje, P.L. de Boer, J. VandenBerghe, K. van der Zwan, E. Stouthamer and H. Wolfert), 15–16. Delft, The Netherlands.

Deegan, C.E. and **Scull, B.J.** (1977) A proposed standard lithostratigraphic nomenclature for the central and northern North Sea. Report Institute Geological Science, London, 77/25, *Norwegian Petroleum Directorate Bull.*, **1**, 36 pp.

Diessel, C.F.K., Boyd, R., Chalmers, G. and **Wadsworth, J.** (1999) New significant surfaces in onshore sequence stratigraphy. In: *Proceedings of the Thirty Third Newcastle Symposium on Advances in the Study of the Sydney Basin* (Eds C.F.K. Diessel, E. Swift and S. Francis), 169–181.

Dong, Q., Cui, B., Li, X. and **Guo, W.** (1997) The division of continental sequence stratigraphy and the identification of its interface with high resolution core analyses and well-logging, *Shiyou Shiyan Dizhi/Experimental Petrol. Geol.*, **19**, 121–126.

Eggenhuisen, J.T., Postma, G., Meijer, P.T. and **Van den Berg van Saparoea, A.P.H.** (2005) Stratigraphic response to climatic pulses and sea-level fluctuations in the fluvial part of the sedimentary system. In: *8th International Conference on Fluvial Sedimentology* (Eds G.J. Weltje, P.L. de Boer, J. VandenBerghe, K. van der Zwan, E. Stouthamer and H. Wolfert), 96–97. Delft, The Netherlands.

Ethridge, F.G., Wood, L.J. and **Schumm, S.A.** (1998) Cyclic variables controlling fluvial sequence development: Problems and perspectives. In: *Relative Role of Eustacy, Climate and Tectonism in Continental Rocks* (Eds K.W. Shanley and P.J. McCabe), *SEPM Spec. Publ.*, **59**, 17–29.

Fraticelli, C.M., West, B.P., Bohacs, K.M., Patterson, P.E. and **Heins, W.A.** (2004) Vegetation-precipitation interactions drive paleoenvironmental evolution. *Eos Transactions AGU*, **85**(47), Fall Meeting Supplement, Abstract H54A-06.

Fu, Q. (1997) Approaches to sequence stratigraphy of continental foreland basins; an example from the Early Cretaceous northern Tarim Basin. Yanxiang Gudili/Sed. *Facies and Palaeogeography*, **17**, 1–10.

Gawthorpe, R.L. and **Leeder, M.** (2000) Tectono-sedimentary evolution of active extensional basins. *Basin Res.*, **12**, 195–218.

Gibling, M.R., Tandon, S.K., Sinha, R. and **Jain, M.** (2005) Discontinuity-bounded alluvial sequences of the southern Gangetic Plains, India: Aggradation and degradation in response to monsoonal strength. *J. Sed. Res.*, **75**, 369–385.

Goodbred, S.L. and **Kuehl, S.A.** (2000) Enormous Ganges–Brahmaputra sediment discharge during strengthened early Holocene monsoon. *Geology*, **28**, 1083–1086.

Heller, P.L. and **Paola, C.** (1992) The large-scale dynamics of grain-size variation in alluvial basins, 2: Application to syntectonic conglomerate. *Basin Res.*, **4**, 91–102.

Holbrook J., Scott, R.W. and **Oboh-Ikuenobe, F.E.** (2006) Base-level buffers and buttresses: A model for upstream versus downstream control on fluvial geometry and architecture within sequences. *J Sed. Res.*, **76**, 162–174.

Hovius, N. (1998) Controls on sediment supply by large rivers. In: *Relative Role of Eustacy, Climate and Tectonism in Continental Rocks* (Eds K.W. Shanley, K.W. and P.J. McCabe), *SEPM Spec. Publ.*, **59**, 3–16.

Husmo, T., Hamar, G.P., Høiland, O. Johannessen, E.P., Rømuld, A., Spencer, A.M. and **Titterton, R.** (2003) Lower and Middle Jurassic. In: *The Millennium Atlas: Petroleum geology of the central and northern North Sea* (Eds D. Evans, C. Graham, A. Armour and P. Bathurst), Geol. Soc. London, 129–155.

Jervey, M.T. (1988) Quantitative geological modeling of siliciclastic rock sequences and their seismic expression. In: *Sea-level changes: An integrated approach* (Eds C.K. Wilgus, B.S. Hastings, C.G.St.C. Kendall, H.W. Posamentier, C.A. Ross and J.C. Van Wagoner), *SEPM Spec. Publ.*, **42**, 47–69.

Kjemperud, A.V. (2008) *Sequence stratigraphy of alluvial successions: Depositional style and controlling factors.* PhD dissertation, University of Oslo, 289 p.

Knighton, D. (1998) *Fluvial Forms and Processes.* Arnold Publishers, London, 383 p.

Kocurek, G. (1998) Aeolian system response to external forcing factors – a sequence stratigraphic view of the Saharan region. In: *Quaternary Deserts and Climate Change* (Eds A.S. Alsharhan, K. Glennie, G.L. Whittle and C.G.St.C. Kendall), Balkema Press, Dordrecht, 327–337.

Kocurek, G. and **Havholm, K.G.** (1993) Eolian sequence stratigraphy: A conceptual framework. In: *Siliciclastic Sequence Stratigraphy: Recent Developments and Applications* (Eds P. Weimer and H.W. Posamentier), *AAPG Mem.*, **58**, 393–410.

Koss, J.E., Ethridge, F.G. and **Schumm, S.A.** (1994) An experimental study of the effects of base-level change on fluvial, coastal plain and shelf sediments. *J. Sed. Petrol.*, **B64**, 90–98.

Leeder, M.R. (1975) Pedogenic carbonates and flood sediment accretion rates: a quantitative model for alluvial arid-zone lithofacies. *Geol. Mag.*, **112**, 257–270.

Leeder, M.R. (1993) Tectonic controls upon drainage basin development, river channel migration and alluvial architecture: Implications for hydrocarbon reservoir development and characterization. In: *Characterization of Fluvial and Aeolian Reservoirs* (Eds C.P. North and D.J. Prosser), *Geol. Soc. London Spec. Publ.*, **73**, 7–22.

Leeder, M.R. and **Stewart, M.D.** (1997) Fluvial incision and sequence stratigraphy: alluvial responses to relative sea-level fall and their detection in the geological record. In: *Sequence Stratigraphy in British Geology* (Eds S.P. Hesselbo and D.N. Parkinson), *Geol. Soc. London Spec. Publ.*, **103**, 25–39.

Leeder, M.R., Harris, T. and **Kirkby, M.J.** (1998) Sediment supply and climate change: implications for basin stratigraphy. *Basin Res.*, **10**, 7–18.

Legarreta, L., Uliana, M.A., Larotonda, C.A. and **Meconi, G.R.** (1993) Approaches to nonmarine sequence stratigraphic-theoretical models and examples from Argentine basins. In: *Subsurface Reservoir Characterization from Outcrop Observations* (Eds R. Eschard and B. Doligez), Éditions Technip, Paris, 125–143.

Lervik, K.-S. (2006) Triassic lithostratigraphy of the Northern North Sea Basin. *Norwegian J. Geol.*, **86**, 93–116.

Mackin, J.H. (1948) Concept of the graded river. *Geol. Soc. Am. Bull.*, **59**, 463–512.

Martinsen, O.J., Ryseth, A., Helland-Hansen, W., Flesche, H., Torkildsen, G. and **Idil, S.** (1999) Stratigraphic base level and fluvial architecture: Ericson Sandstone (Campanian), Rock Springs Uplift, SW Wyoming, USA. *Sedimentology*, **46**, 235–259.

Matthews, M.D. and **Perlmutter, M.A.** (1994) Global cyclostratigraphy: an application to the Eocene Green River Basin. In: *Orbital-forcing and Cyclic Sequences* (Eds P.L. de Boer and D.G. Smith), *Int. Assoc. Sedimentol. Spec. Publ.*, **19**, 459–481.

Mearns, E.W., Knarud, R., Raestad, N., Stanley, K.O. and **Stockbridge, C.P.** (1989) Samarium-Neodymium isotope stratigraphy of the Lunde and Statfjord Formations of Snorre oilfield, northern North Sea. *J. Geol. Soc. London*, **146**, 217–228.

Michelsen, O., **Nielsen, L.H.**, **Johannessen, P.N. Andsbjerg, J.** and **Surlyk, F.** (2003) Jurassic lithostratigraphy and stratigraphic development onshore and offshore Denmark. In: *The Jurassic of Denmark and Greenland* (Eds J.R. Ineson and F. Surlyk), Geol. Survey of Denmark and Greenland Bull., **1**, 147–216.

Morris, P.H., **Cullum, A.**, **Pearce, M.A.** and **Batten, D.J.** (2009) Megaspore assemblages from the Åre Formation (Rheatian-Pliensbachian) offshore mid-Norway and their value as field and regional stratigraphic markers. *J. Micropalaeontology*, **28**, 161–181.

Morton, A.C., **Claoue-Long, J.** and **Berge, C.** (1996) SHRIMP constraints on sediment provenance and transport history on the Mesozoic Statfjord Formation, *North Sea. J. Geol. Soc. London*, **153**, 915–929.

Muto, T. and **Steel, R.J.** (2000) The accommodation concept in sequence stratigraphy: some dimensional problems and possible redefinition. *Sed. Geol.*, **130**, 1–10.

Muto, T. and **Steel, R.J.** (2002) Role of autoretreat and A/S changes in the understanding of deltaic shoreline trajectory: a semi-quantitative approach. *Basin Res.*, **14**, 303–318.

Muto, T. and **Steel, R.J.** (2004) Autogenic response of fluvial deltas to steady sea-level fall: Implications from flume-tank experiments. *Geology*, **32**, 401–404.

Muto, T., **Steel, R.J.** and **Swenson, J.B.** (2007) Autostratigraphy: a framework norm for genetic stratigraphy. *J. Sed. Res.*, **77**, 2–12.

Nystuen, J.P. and **Fält, L.M.** (1995) Upper Triassic-Lower Jurassic reservoir rocks in the Tampen Spur area, Norwegian North Sea. In: *Petroleum Exploration and Exploitation in Norway* (Ed. S. Hanslien), *Norwegian Petrol. Soc. Spec. Publ.*, **4**, Elsevier, Amsterdam, 135–179.

Olsen, T., **Steel, R.J.**, **Høgseth, K.**, **Skar, T.** and **Røe, S.** (1995) Sequential architecture in a fluvial succession: Sequence stratigraphy in the Upper Cretaceous Mesaverde Group, Price Canyon, *Utah. J. Sed. Res.*, **B62**, 265–278.

Ouchi, S. (1985) Response of alluvial rivers to slow active tectonic movement. *Geol. Soc. Am. Bull.*, **96**, 504–515.

Paola, C., **Heller, P.** and **Angevine, C.** (1992) The large scale dynamics of grain-size variation in alluvial basins. *I: Theory. Basin Res.*, **4**, 73–90.

Plint, A.G., **McCarthy, P.J.** and **Faccini, U.F.** (2001) Nonmarine sequence stratigraphy: Updip expression of sequence boundaries and systems tracts in a high-resolution framework, Cenomanian Dunvegan Formation, Alberta foreland basin, *Canada. AAPG Mem.*, **85**, 1967–2001.

Postma, G. (2014) Generic autogenic behaviour in fluvial systems: lessons from experimental studies. In: *From Depositional Systems to Sedimentary Successions on the Norwegian Continental Shelf* (Eds A.W. Martinius, J. Howell, R. Ravnås, R.J. Steel and J.P. Wonham). *Int. Assoc. Sedimentol. Spec. Publ.*, **46**, 1–18.

Perlmutter, M.A. and **Matthews, M.D.** (1990) Global cyclostratigraphy – a model. In: *Quantitative Dynamic Stratigraphy* (Ed. T.A. Cross), Prentice Hall, 233–260.

Puigdefàbregas, C. (1993) Controls on fluvial sequence architecture. In: *Proceedings 5th International Conference on Fluvial Sedimentology, Brisbane, Australia* (Eds B. Yu and C.R. Fielding), Extended Abstracts, K42–K48.

Rygel, M.C. and **Gibling, M.R.** (2006) Natural geomorphic variability recorded in a high-accommodation setting: fluvial architecture of the Pennsylvanian Joggins Formation of atlantic Canada. *J. Sed. Res.*, **76**, 1230–1251.

Ryseth, A. (2001) Sedimentology and palaeogeography of the Statfjord Formation (Rhaetian – Sinemuran), North Sea. In: *Sedimentary Environments Offshore Norway – Palaeozoic to Recent* (Eds O.J. Martinsen and T. Dreyer), *Norwegian Petrol. Soc. Spec. Publ.*, **10**, 67–85.

Ryseth, A. and **Ramm, M.** (1996) Alluvial architecture and differential subsidence in the Statfjord Formation, North Sea: prediction of reservoir potential. *Petrol. Geosc.*, **2**, 271–287.

Røe, S.-L. and **Steel, R.** (1985) Sedimentation, sea-level rise and tectonics at the Triassic-Jurassic boundary (Statfjord Formation), Tampen Spur, northern North Sea. *J. Petrol. Geol.*, **8**, 163–186.

Schlager, W. (1993) Accommodation and supply – a dual control on stratigraphic sequences. *Sed. Geol.*, **86**, 111–136.

Schumm, S.A. (1977) *The Fluvial System*. Wiley, Chichester, 338 pp.

Schumm, S.A. (1993) River response to baselevel change: Implications for sequence stratigraphy. *J. Geol.*, **101**, 279–294.

Schumm, S.A. (2005) *River Variability and Complexity*. Cambridge University Press, New York, 220 pp.

Schumm, S.A., **Dumont, J.F.** and **Holbrook, J.M.** (2000) *Active Tectonics and Alluvial Rivers*. Cambridge, U.K., Cambridge University Press, 276 p.

Shanley, K.W. and **McCabe, P.J.** (1994) Perspectives on the sequence stratigraphy of continental strata. *AAPG Bull.*, **78**, 544–568.

Sloss, L.L. (1962) Stratigraphic models in exploration. *J. Sed. Petrol.*, **32**, 415–422.

Steel, R.J. (1993) Triassic – Jurassic megasequence stratigraphy in the Northern North Sea: rift to post-rift evolution. In: *Petroleum Geology of Northwest Europe: Proceedings of the 4th Conference* (Ed. J.R. Parker), Geol. Soc. London, 299–315.

Steel, R.J. and **Ryseth, A.** (1990) The Triassic-Early Jurassic succession in the northern North Sea: megasequence stratigraphy and intra-Triassic tectonics. In: *Tectonic Events Responsible for Britain's Oil and Gas Reserves* (Eds R.F.P. Hardman and J. Brooks), *Geol. Soc. London Spec. Publ.*, **55**, 139–168.

Stouthamer, E. and **Berendsen, H.J.A.** (2000) Factors controlling the Holocene avulsion history of the Rhine-Meuse Delta (The Netherlands). *J. Sed. Res.*, **70**, 1051–1064.

Stouthamer, E. and **Berendsen, H.J.A.** (2007) Avulsion: The relative roles of autogenic and allogenic processes. *Sed. Geol.*, **198**, 309–325.

Sweet, A.R., **Catuneanu, O.** and **Lerbekmo, J.F.** (2005) *Uncoupling the position of sequence bounding unconformities from lithological criteria in fluvial systems. In: American Association of Petroleum Geoscientists Annual Convention*, Calgary, Canada, Abstracts, p. A136.

Swenson, J.B. and **Muto, T.** (2005) Fluvial grade as a non-equilibrium state: theory and flume experiments. In: *8th International Conference on Fluvial Sedimentology*

(Eds G.J. Weltje, P.L. de Boer, J. VandenBerghe, K. van der Zwan, E. Stouthamer and H. Wolfert), p. 272. Delft, The Netherlands.

Swift, D.J.P. and **Thorne, J.A.** (1991) Sedimentation on continental margins, I: a general model for shelf sedimentation. In: *Shelf Sand and Sandstone Bodies* (Eds D.J.P. Swift, G.F. Oertel, R.W. Tillman and J.A. Thorne), *Int. Assoc. Sedimentol. Spec. Publ.*, **14**, 3–31.

Tandon, S.K., Gibling, M.R., Sinha, R., Singh, V., Ghazanfari, P., Dasgupta, A., Jain, M. and **Jain, V.** (2006) Alluvial valleys of the Gangetic Plains, India: causes and timing of incision. *In: Dalrymple, R.W., Leckie, D.A., Tillman, R.W. (Eds.), Incised Valleys, vol. 85., SEPM Special*, Tulsa, pp. 15–35.

Thorne, J.A. and **Swift, D.J.P.** (1991) Sedimentation on continental margins, II: application of the regime concept. In: *Shelf Sand and Sandstone Bodies* (Eds D.J.P. Swift, G.F. Oertel, R.W. Tillman and J.A. Thorne), *Int. Assoc. Sedimentol. Spec. Publ.*, **14**, 33–58.

Todd, S.P. (1996) Process Deduction from Fluvial Sedimentary Structures. In: *Advances in Fluvial Dynamics and Stratigraphy* (Eds P.A. Carling and M.R. Dawson), Wiley and Sons Ltd., New York, 299–350.

Van Heijst, M.I.W.M. and **Postma, G.** (2001) Fluvial response to sea-level changes: a quantitative analogue, experimental approach. *Basin Res.*, **13**, 269–292.

Van Heijst, M.I.W.M., Postma, G., Van Kesteren, W.P. and **De Jongh, R.G.** (2002) Control of syndepositional faulting on systems tract evolution across growth-faulted shelf margins: An analog experimental model of the Miocene Imo River field, Nigeria. *AAPG Bull*, **86**, 1335–1366.

Wang, D., Liu, L. and **Li, J.** (1997) Research of sequence stratigraphy in a continental rift basin (as exemplified from the Songliao Basin, China). *Geol. Pacific Ocean*, **13**, 1065–1079.

Weissmann, G.S., Mount, J.F. and **Fogg, G.E.** (2002) Glacially driven cycles in accumulation space and sequence stratigraphy of a stream-dominated alluvial fan, San Joaquin Valley, California, U.S.A.. *J. Sed. Res.*, **72**, 240–251.

Weissmann, G.S., Bennett, G.L. and **Lansdale, A.L.** (2005) Factors controlling sequence development on Quaternary fluvial fans, San Joaquin Basin, California, U.S.A. In: *Alluvial Fans: Geomorphology, Sedimentology, Dynamics* (Eds A. Harvey, A. Mather and M. Stokes), *Geol. Soc. London Spec. Publ.*, **251**, 169–186.

Wheeler, H.E. (1964) Baselevel, lithostratigraphic surface and time stratigraphy. *Geol. Soc. Am. Bull.*, **75**, 599–610.

Wood, L.J., Ethridge, F.G. and **Schumm, S.A.** (1992) The effects of rate of base-level fluctuation on coastal-plain, shelf and slope depositional systems: an experimental approach. In: *Sequence Stratigraphy and Facies Associations* (Eds H.W. Posamentier, C.P. Summerhayes, B.U. Haq and G.P. Allen), *Int. Assoc. Sedimentol. Spec. Publ.*, **18**, 43–53.

Wright, V.P. and **Marriott, S.B.** (1993) The sequence stratigraphy of fluvial depositional systems: the role of floodplain sediment storage. *Sed. Geol.*, **86**, 203–210.

Yoshida, S., Steel, R.J. and **Dalrymple, R.W.** (2007) Changes in depositional process – An ingredient in a new generation of sequence stratigraphic models. *J. Sed. Res.*, **77**, 447–460.

Yu, W., Ding, B. and **Wie, N.** (2000) Thoughts on the study of continental sequence stratigraphy. *Dizhi Lunping / Geol. Rev.*, **46**, 347–354.

Int. Assoc. Sedimentol. Spec. Publ. (2014) **46**, 127–138

Investigating the autogenic process response to allogenic forcing: experimental geomorphology and stratigraphy

WONSUCK KIM*, ANDREW PETTER†, KYLE STRAUB‡ and DAVID MOHRIG*

**Department of Geological Sciences, Jackson School of Geosciences, University of Texas at Austin, Austin, 78712 TX, USA (E-mail: delta@jsg.utexas.edu)*
†St. Anthony Falls Laboratory, University of Minnesota, Minneapolis, MN 55414, USA
‡Department of Earth and Environmental Sciences, Tulane University, New Orleans, LA 70118, USA

ABSTRACT

Decoupling the preserved signal of environmental (allogenic) forcing from those of internally generated (autogenic) processes is at the centre of understanding the evolution of the Earth's surface preserved in the sedimentary record. A major stumbling block for distinguishing between allogenic versus autogenic signatures in the stratigraphic record is the lack of quantitative understanding of autogenic processes and their interactions with allogenic forcing. Physical experiments with moving sediment and water construct topography through dynamic self-organised fluvial systems, thus providing an opportunity to investigate autogenic processes under controlled boundary conditions (e.g. sediment supply and tectonics). This paper presents a set of tank experiments that are used to examine quantitatively 1) the autocyclic storage and release of sediment in the delta top surface associated with river-pattern changes between channel and sheet flow and 2) changes in fluvial autocyclic behaviour driven by external forcing (e.g. sea-level change and tectonics). The time and event scales of the autogenic processes observed in the experiments conducted without external forcing have provided the first-order quantitative understanding of the autogenic processes. Changes in the frequency of autogenic processes through base-level changes and lateral ground tilting have provided a new view into the coupled allogenic and autogenic controls on stratigraphic development. Coupled experiments that test the effects of allogenic forcing on autogenic process are presented here: one experiment was conducted with constant external forcing and the other was conducted with cyclic changes in external controls. This review provides 1) quantitative measurements of fluvial autogenic processes and thorough comparisons of cyclic strata attributed to allogenic versus autogenic controls and 2) suggestions for future experimental studies of fluvial autogenic processes that will enhance our ability to interpret the mixed signals of environmental variation and internal dynamics in the sedimentary record.

Keywords: Autogenic process, stratigraphy, experiment, geomorphology, shoreline.

INTRODUCTION

Sedimentary deposits in basins are sensitive indicators of environmental variations, e.g. global sea-level change, river floods, coastal storms and earthquakes. These environmental conditions exert a crucial influence on the shape of Earth's surface which, in turn, influences the development of subsurface architecture (stratigraphy). Decoupling the products of environmental (allogenic) forcing from the internally generated (autogenic) 'noise' written in the sedimentary record remains as a fundamental goal of sedimentary geosciences. Classically allogenic deposits have been recognised by their cyclic nature. Their cyclicity is interpreted as the signature of periodic changes in climate, base-level, sediment supply, and/or tectonics. On the other hand, it has also been generally accepted that non-linearity in sediment transport (e.g. delta lobe switching, channel avulsion) results in random or scale-free (fractal) deposits, i.e. autogenic-driven deposits (Jerolmack & Paola, 2007; Jerolmack &

From Depositional Systems to Sedimentary Successions on the Norwegian Continental Margin, First Edition.
Edited by A. W. Martinius, R. Ravnås, J. A. Howell, R. J. Steel, and J. P. Wonham.

Paola 2010; Jerolmack & Swenson, 2007; Kim & Jerolmack, 2008).

Even with current advanced technology for imaging the subsurface in high-resolution and more than a half-century of research aimed at developing techniques to disentangle autogenic noise from allogenic signals in stratigraphic data, accurate reconstruction of allogenic variations preserved in the stratigraphic record remains challenging. An important foundation for many steps to overcome this challenge is actually quantitatively understanding the 'noise' of autogenic processes (Jerolmack, 2011). This includes the understanding of 1) the autogenic processes and their stratigraphic products and 2) complicated reactions of the autogenic processes to basinal forcing and the effects of these interactions on stratigraphic products. A fundamental understanding of autogenic processes and their stratigraphic products that are isolated from external controls should be achieved first and then changes in autogenic processes due to external controls imposed on the system should be investigated systematically. These steps will facilitate the decoupling of stratigraphic signatures into allogenic and autogenic components in basin interpretations. Field studies always deal with geological data that record multiple environmental controls and must deduce palaeo-environmental changes from a complicated mixture. This traditional method of interpreting basin history will be significantly improved by the quantitative understanding of individual causes (external basin controls) and effects (stratigraphic products composed of autogenic and allogenic signatures) under controlled conditions through physical tank experiments.

This paper presents a review of the previous experimental studies. Firstly, fluvial autogenic processes isolated from the effect of basin forcing (i.e. no sea-level change or tectonic activities) are introduced. Then, a set of tank experiments that were conducted with changes in basin forcing is reviewed. This paper also presents two experiments that both show similar cyclic sedimentation; although one is caused by autogenic fluvial cycles under a steady tectonic forcing and the other is caused by cyclic tectonic variations. Finally, suggestions for potential studies using physical and mathematical experiments, to enhance the current results and understanding of the autogenic processes, are discussed.

QUANTIFYING FLUVIAL AUTOGENIC PROCESSES

Recent experimental studies with sediment and water (Fig. 1) clearly demonstrate self-organisation due to internally driven sediment transport processes (e.g. Ashworth *et al.*, 2004; Bryant *et al.*, 1995; Cazanacli *et al.*, 2002; Heller *et al.*, 2001; Hickson *et al.*, 2005; Jerolmack & Mohrig, 2005; Kim & Jerolmack, 2008; Kim & Muto, 2007; Kim & Paola, 2007; Kim *et al.*, 2006a; Muto & Steel, 2001; Paola, 2000; Paola *et al.*, 2001; Paola *et al.*, 2009) and suggest an opportunity to investigate quantitatively the morphodynamic and stratigraphic representations of autogenic processes. The current advancement in experimental technology allows for better control in boundary conditions; and therefore a simpler mixture of autogenic processes with environmental variations for investigation. This section details previous experimental studies that focus on autogenic processes under either no or minimal external controls (i.e. base-level change and tectonic variation).

Fluvial autogenic sediment storage and release

Recent publications in tank experiments have observed strong landward-to-seaward fluctuations in shoreline position from an overall progradational experimental delta under both constant

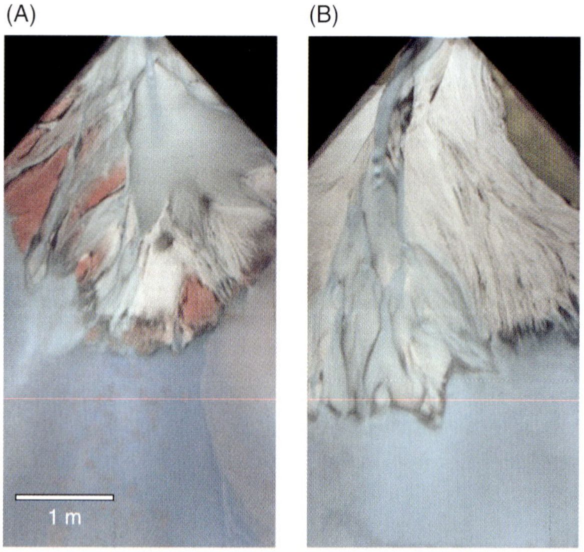

Fig. 1. Experimental fluvial surface taken during XES 02 showing (A) a sheet-flow dominated sediment storage event and (B) strong channelisation with a release event.

sediment supply (Q_s) and water discharge (Q_w) with either no or slight relative sea-level changes (e.g. Kim & Jerolmack, 2008; Van Dijk *et al.*, 2009). These shoreline fluctuations were caused by autocyclic fluvial sediment storage and release associated with changes in the fluvial planform pattern between channelised flow and sheet flow (Fig. 1). Sheet flow covers most of the delta top surface and forces strong aggradation of the fluvial system as hardly any sediment gets delivered to the shoreline. Channelisation is initiated by focusing flow in a narrow corridor on the delta top. As a result, transport capacity increases due to increase in both channel depth and flow velocity. Supplied sediment, as well as eroded sediment from the channel bed, is transported to the shoreline, thus producing a strong pulse of shoreline regression. Quantitative measurements of 1) the time frequency of the autogenic process and 2) magnitude of the autogenic shoreline fluctuation (i.e. shoreline signature of the fluvial autogenic process) have been reported (e.g. Kim & Jerolmack, 2008; Van Dijk *et al.*, 2009).

Autogenic shoreline fluctuation

Two sets of experimental data are presented in the current analysis. One set is from the eXperimental EarthScape (henceforth XES) facility at St. Anthony Falls Laboratory, University of Minnesota and the other is from the Eurotank Flume facility at Utrecht University. The individual results have been published (Kim & Jerolmack, 2008; Van Dijk *et al.*, 2009), so only a brief summary is given below.

Kim & Jerolmack (2008) used data from two experiments in XES. The XES basin is 6 m long, 3 m wide and 1.5 m deep. A detailed description of the XES facility can be found in Paola *et al.* (2001). The experiment conducted in 2002 (XES 02) had roughly five times greater sediment discharge than the second experiment conducted in 2005 (XES 05). However, both experiments had roughly the same sediment to water supply ratio (Q_s/Q_w) at 0.01 in volume influx. Both experiments used a sediment mixture with roughly 70% quartz sand (D = 110 µm) and 30% coal sand (bimodal D = 460 and 190 µm). The coal sediment is much less dense than the quartz sediment and thus works as a proxy for fine material. Sediment supply and water discharge were kept constant during the experiments. Data used for the analysis were taken from the initial stages of the two experiments (10 to 18 hours of experimental run time in XES 02 and 80 to 100 hours in XES 05), during which the shorelines in both experiments prograded 20 to 40 cm basinward (Fig. 2). The overall lengths of both deltas were around 3 m but XES 02 had a ~ 10 cm water depth and XES 05 had ~ 0.5 cm water depth in front of the deltas.

Autogenic sediment storage and release events in the fluvial surface were observed twice during the period that data were collected. During the sediment storage events on the fluvial surface associated with sheet/widespread flows, the shoreline migration was reduced to a rate below the long-term averaged progradation rate. However, the shoreline rapidly advanced during release events by strong channelisation. When the surface is degraded and the shoreline progrades basinward enough to decrease the topographic slope, a new storage process is reinitiated. Autogenic processes in XES 02 showed the shoreline migration with high frequency and high magnitude fluctuations (even with a deeper water depth at the delta front) in comparison to the XES 05, which showed low frequency and low magnitude shoreline fluctuations (Fig. 2).

Three experiments conducted in the Eurotank Flume facility at Utrecht University also exhibited autocyclic incisions by channelised flow that were then progressively backfilled by sediments in the incised channels (Van Dijk *et al.*, 2009). Two of the three experiments were conducted side by side with the same Q_s but different Q_w values. As a result, the sediment to water discharge ratio in these two experiments varied between 0.002 (A004-1) and 0.003 (A004-2). For these experiments the sediment mixture was composed of grains with D = ~200 to 250 µm and the deltas were built in a basin with a total size of 2.7 m × 2.7 m. The decrease of Q_w in A004-2 resulted in higher topographic slopes than in the A004-1 experiment, which reduced overall progradation of the shoreline in A004-2. Roughly, A004-1 had a range of delta-top slopes between 0.02 and 0.06 but A004-2 had a range between 0.04 and 0.07. The high Q_w experiment shows more distinct shoreline regression periods (total shoreline distance travelled during each release event is longer) than those periods shown in the low Q_w experiment (See Fig. 7 in van Dijk *et al.* (2009)), which would be caused by more channelised incisions.

Reitz *et al.* (2010) reported a smaller-scale tank experiment (3 m length × 1 m width × 1 m depth)

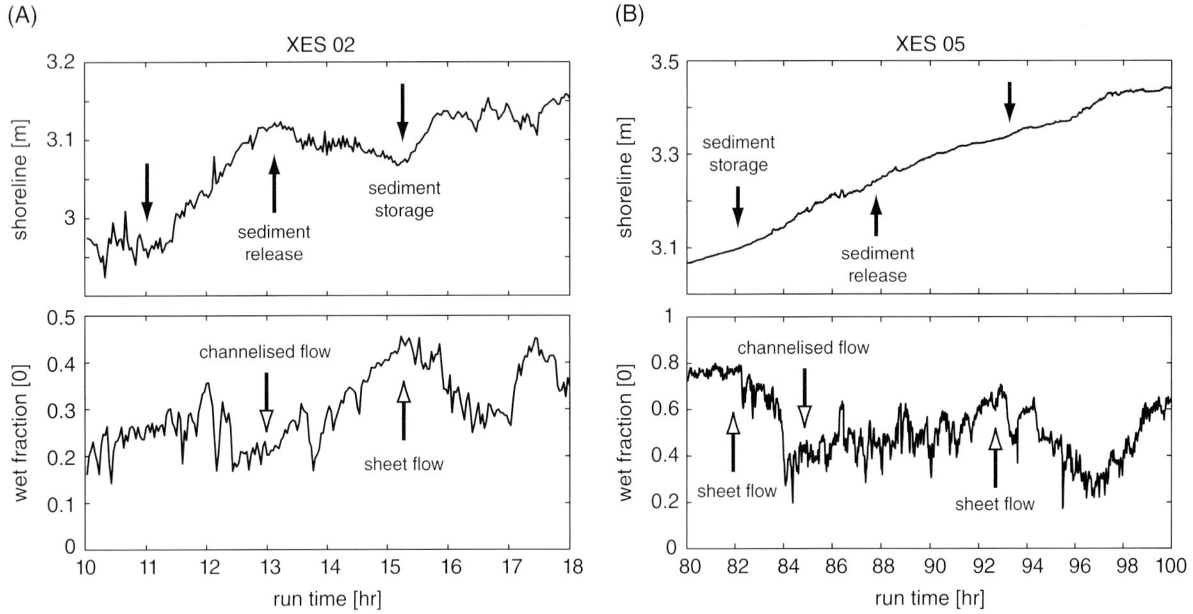

Fig. 2. Time series data of shoreline migration (top panels) and wetted fraction of the deltaic surface (bottom panels) in (A) XES 02 and (B) XES 05.

using a bimodal sediment mixture at the University of Pennsylvania. The sediment mixture is composed of 80% acrylic sand (D = 300 μm) and 20% granite chips (D = 2 mm). This bimodal mixture was designed to maximise the contrast between grain size and density, thus enhancing the threshold effect for sediment transport. Fluctuation in the wetted fraction associated with strong channelisation and backfilling in the fluvial surface is consistent with the other experiments presented above. The time necessary for the surface to reorganise through refilling incised channels and searching for new paths was shown to scale with the following form (Reitz *et al.*, 2010):

$$T_{ap} \sim \frac{HBs}{Q_s} \tag{1}$$

where T_{ap} denotes the characteristic avulsion (autogenic process) time scale, H is the vertical length of alluvial cutting by the release events, which can be roughly scaled with flow depth, B is the horizontal length of basin, scaled with total channel width, s denotes the downstream shoreline position as a function of time equivalent with the basin size in the downbasin direction and Q_s is sediment supply. This time scale captured the sediment storage and release cycles in their experiment as well as correctly predicting autogenic frequencies in the XES 02 and 05 experiments. The XES 02 experiment had five times greater sediment supply, which would cause a shorter time duration between the autogenic cycles, than that shown in XES 05. XES 02 showed an increase in the wetted fraction over 2 to 3 hour backfilling/storage events whilst the XES 05 shows 8 to 10 hour events (Fig. 2).

The sediment to water discharge ratios for the XES 02 and 05 experiments were kept constant at ~0.01, which generated similar delta top surface slopes of 0.036 and 0.048, respectively. However, this small increase in delta topset slope induces a fair decrease in the size of the fluvial sediment buffer (i.e. potential volume capacity to hold sediment during a storage event) and forces less sediment on the topset potion of the delta to be accommodated and released to offshore (Kim & Jerolmack, 2008). Direct and indirect measurements for the slope fluctuations in XES 02 and 05, due to the fluvial autogenic processes, indicate averaged 0.004 and 0.0027 slope changes across the basin, respectively (11% and 6% changes compared to the averaged topset slopes), to account for the observed shoreline fluctuations. The amount of sediment that is either stored or released in the delta top surface thus defines the magnitude of the shoreline fluctuation over the autogenic process (Kim & Jerolmack, 2008). XES 02 shows a larger amount of sediment reworked during the storage and release events, in comparison to XES 05. The high water discharge in XES 02 also took a relatively longer period for channelisation and led to larger degradation in the delta top surface. More symmetrical cycles of the wetted fraction were shown in XES 02 while asymmetrical (positively skewed) fluctuations were dominant in XES 05, which indicates a relatively different period for the release event. XES 05 shows quicker release of sediment and a longer period for sediment storage, whilst XES 02 shows a very symmetric distribution of times for storage and release events (Fig. 2).

Suggestions for further work

The experimental studies discussed above provide a suite of quantitative measurements of autogenic processes; however, more thorough investigations are required to understand the physical links between the sediment and water discharge conditions and autogenic processes. The experimental studies discussed above dealt with 1) the same Q_s/Q_w ratio but different absolute amounts of sediment and water (Kim & Jerolmack, 2008) and 2) changes in Q_w whilst keeping Q_s the same (Van Dijk *et al.*, 2009). However, experiments dealing with changes in Q_s while keeping Q_w constant are missing. Additionally, 1) the current suite of experiments was limited to only a few cycles of the autogenic events; and 2) the effects of different grain sizes, sediment mixture with multiple grain sizes and cohesiveness have not been thoroughly tested yet. More experiments should be done with experimental duration sufficient enough to capture events with ranges of Q_s and Q_w, over which the experiment can produce enough numbers of the autogenic events for a meaningful statistical analysis, multiple grain-size mixtures and cohesiveness. These experiments will provide insight into quantitative understanding of the fluvial autogenic processes and implications to field-scale stratigraphy.

CONTROLS OF ALLOGENIC FORCING ON AUTOGENIC PROCESSES

Basin response time scale and high-frequency stratigraphic signals

Sedimentary systems are sensitive to environmental changes. Changing architectural-stacking patterns in stratigraphy have generally been accepted as signatures of palaeo-environmental variations. The causes of these changes are frequently inferred from episodic active tectonics separated by relatively quiescent periods (e.g. Blair & Bilodeau, 1988; Dorsey *et al.*, 1997), climatically controlled variation in sediment yield (e.g. Smith, 1994), sea-level change (e.g. Posamentier *et al.*, 1988; Posamentier & Vail, 1988) and global climate changes by the Earth's orbital cyclicity (e.g. House, 1985) and solar and oceanic variations (e.g. Roth & Reijmer, 2005).

However, stratigraphic interpretations of high-frequency signals have been questioned in the rock record. It has been argued that depositional systems might not fully record high-frequency external changes in the stratigraphy if the time that sedimentary systems need to respond to the external changes is longer than the period over which the external forcing cycles. It has also been argued that depositional systems might exaggerate high-frequency autogenic signals in the stratigraphy, depending on the nature of the external forcing. A characteristic response time for depositional basins was first theoretically derived by Paola *et al.* (1992) and Paola (2000) and implications of this scaling to natural systems and flume analogues were further discussed in Postma *et al.* (2008). The basin equilibrium time serves to define what 'high frequency' is here; high-frequency external controls are environmental basin forcing that cycles over a smaller time period compared to the basin equilibrium timescale. This time scale takes the following form:

$$T_{eq} = \frac{L^2}{\upsilon} \tag{2}$$

where L is the length of the depositional system and υ is the diffusivity coefficient. Castelltort & Van Den Driessche (2003) applied this basin diffusional relaxation time to modern worldwide rivers and reported that the basin response time scale ranges from 10^4 to 10^6 years. Allen (2008) conducted a similar analysis on large Asian rivers and suggested that this time scale is in the range of 10^5 to 10^6 years. Large river systems with extensive floodplains therefore tend to buffer any external variations when the external forcing period is less than the response time scale. The basin response time scale overlaps with the time scales for many geological processes, e.g. Earth orbital cycles, which calls into question the origin and interpretations of high-frequency patterns in stratigraphic architecture.

The following sections describe the complex responses of autogenic processes to external forcing (i.e. base-level change, tectonic tilting) in physical experiments. Both time scale and event size of the fluvial autogenic processes are different when base-level and tectonic controls are manipulated in the experiments compared to those conducted without the external forcing. Some elongated and magnified internal processes observed from the previous studies suggest stratigraphic signals exhibiting periodicity close to or longer than the equilibrium time scale could also be attributed to autogenic processes that have comparable stratigraphic results to allogenic controls.

Base-level change

Changes in the magnitude of autogenic fluctuations in shoreline migration rates during base-level rise and fall were first quantified in Kim *et al.* (2006a) using the XES 02 experiment. The deltaic shoreline responded to multiple sinusoidal base-level cycles applied in the experiment and exhibited high-frequency shoreline fluctuations superimposed over long-term responses (see studies about stratigraphic (allogenic) responses to the base-level cycles in Kim *et al.* (2006b) and Martin *et al.* (2009a)). The first base-level cycle started at 26 hours in run time and lasted for 108 hours which is a longer duration than a rapid cycle that was applied for 18 hours after the initial slow cycle. The amplitude of the first slow base-level cycle was 0.11m, which is ten times larger than the averaged channel depth in the experiment. In the study, shoreline migration rates were calculated using 10 min laterally averaged downstream positions of the shoreline and roughly show a 3-fold increase in autogenic variability during the base-level rise than that during the base-level fall (Fig. 3).

Topographic scans reveal that the slope of the delta top surface varies cyclically during the autogenic processes: The slope decreases during release events and increases during storage events. This slope fluctuation is casted in a geometric model in Kim *et al.* (2006a), the results of which

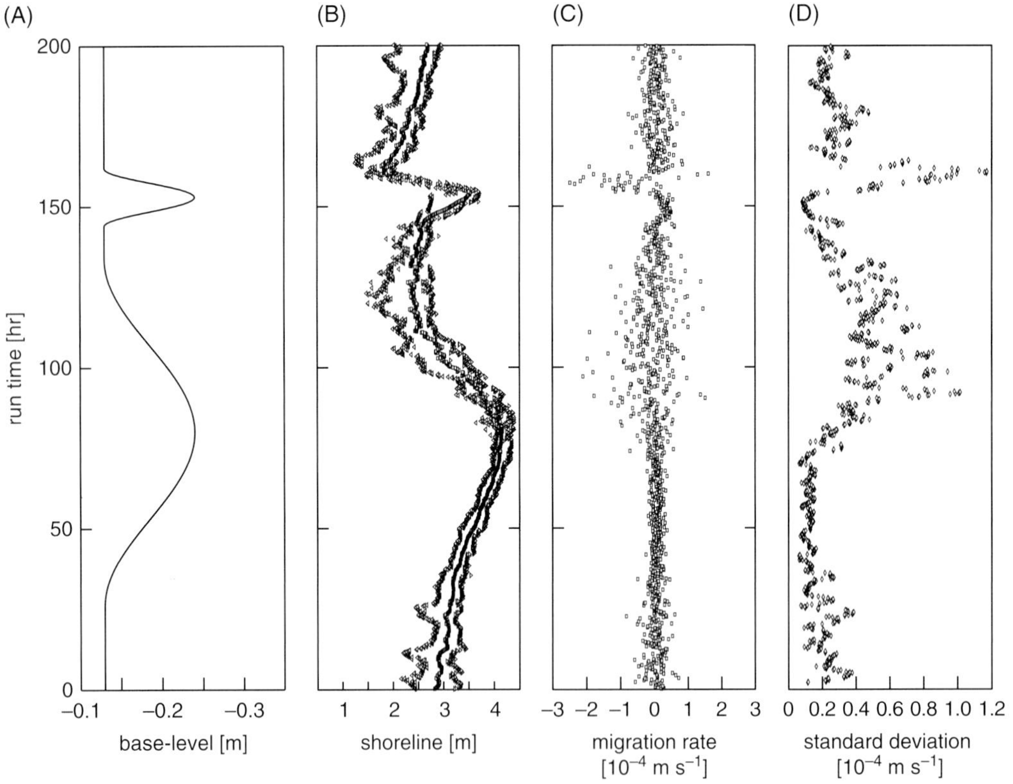

Fig. 3. Shoreline data in XES 02 (note that the time series includes only the first 200 hours of the total 310 run time). (A) Imposed base-level cycles; (B) maximum, minimum and laterally averaged shoreline downstream positions; (C) migration rate calculated using the averaged shoreline positions; and (D) standard deviations of the shoreline migration rates.

accounted well for the difference in variability in the shoreline migration rate observed during the falling and rising rims of the sinusoidal base-level cycle (see Fig. 11 in Kim *et al.* (2006a)). In the model, the magnitude of the slope increase within a single modelling time step during the storage events are constrained by the supplied sediment discharge; and thus the slope increment is smaller for a larger delta. A fraction of the topset slope averaged over the total experiment (i.e. 1% to 4% in XES 02) was given to limit the total range of the slope fluctuation in the model. The modelling results show that autogenic event signatures (i.e. variability in shoreline migration rate) differ according to the direction of base-level change even though the autogenic process size (i.e. the range of topset slope fluctuation between maximum and minimum threshold slopes) is assumed to be constant. Base-level fall enhances the sediment release process, forcing the shoreline seaward migration, but regression is inevitable in this setting, thus diminishing the effect on varying the shoreline migration. However, a sediment release event during overall transgression easily reverses the direction of shoreline migration because the delta front develops over the shallow submerged topset surface. In the Kim *et al.* (2006) study, it is clear that the delta geometry plays a role in the footprint of the shoreline autogenic fluctuation during base-level rise and fall but it is not clear if the event size (angle between threshold maximum and minimum topset slopes) varies by the direction of the base-level change due to the lack of high-res (in time) topographic scans. The following experiments address in part how autogenic event size varies during base-level rise and fall.

A research team in the ExxonMobil Upstream Research Company conducted a series of experiments using a sediment mixture with a polymer that improves deposit cohesiveness (Hoyal & Sheets, 2009; Martin *et al.*, 2009b). This sediment mixture restricts both channel sidewall erosion and channel widening, thus allowing for relatively stable, distributary channel networks to form, in contrast to the typical braided system in other experiments discussed in this paper. Martin *et al.* (2009b) produced a cohesive delta in an experiment with two stages: stage 1 without base-level rise; stage 2 with base-level rise. In the second stage, the rate of base-level rise was kept constant at a rate designed to nearly maintain the same size of the delta top surface (i.e. strongly aggrading delta with minor progradation). The base-level rise in the second stage forced a two-fold increase in fluvial deposition and caused a two-fold increase in frequency of the fluvial autogenic process compared to the first stage. Roughness of the shoreline along the experiment was also measured and characterised, thus showing statistical saturation at an opening angle $\theta = 16.5^{\circ}$ (distributary lobes are scaled with 2θ). These shoreline roughness and lobe scales are consistent across both the delta progradation (with no base-level rise) and aggradation stages (with base-level rise), supporting the two-fold increase in the autogenic channel time scale due to enhanced fluvial deposition of the supplied sediment. High-resolution topographic data are still missing in this experiment, which might allow for detecting changes in the threshold slopes due to base-level forcing in more detail. However, the consistent roughness in the shoreline across the two stages of constant base-level and linear rise of base-level hints that no major changes in the autogenic event size occurred due to the base-level control.

The migration of the upstream end of the experimental deposit in these experiments (XES 02 and Martine *et al.* experiment) is restricted due to the vertical tank wall. As a result, the upstream boundary at the transition between alluvium and bedrock exposure (henceforth alluvial-bedrock boundary) could not migrate freely. Kim & Muto (2007) presented a series of experiments that allowed for free migration of the upstream end of the deposit. An isolated delta free from the tank walls developed over a sloped non-erodible basement in each of their experiments. The basement slope was set at a higher slope than the steepest delta topset slopes so that sediments bypassed the exposed upstream bedrock surface. Either constant base-level rise or fall was applied over the last half of an experiment after the first half of stationary base-level. The overall trend of the variability change in the shoreline migration rate during base-level rise and fall is in good agreement with the previous results: that is maximised when the mean shoreline migration direction is against the base-level change, but minimised if they are aligned. Thus there is no difference in the process for sediment internal buffering by allowing a free moving alluvial-bedrock boundary. However, the amplitude of the autogenic signal in migration rate of the alluvial-bedrock transition increases during base-level fall and diminishes during base-level rise, which is the opposite trend to that at the shoreline. This change in amplitude of the autogenic variability is mainly controlled by changes in the axial length of the alluvial system, i.e. shortening the length results in a decrease in the alluvial-bedrock transition autogenic variability; and vice versa. A simple geometric, mass-balance model similar to one in Kim *et al.* (2006a) was also employed in this study. The model again used varying fluvial slopes to express sediment transport efficiency in the fluvial system and captured the patterns of the autogenic signature in the both moving boundaries. The dimensionless variability of the shoreline and alluvial-bedrock boundaries, enhanced by the base-level forcing, indicated that the autogenic response has the same order of magnitude as the allogenic response. This suggests the possibility of overlap in time and event scales across the autogenic and allogenic stratigraphic products, even though the allogenic signal is mostly expected to be coherent over much longer length and time scales.

Tectonics: Lateral ground tilting

The response of rivers to tectonic activity is generally accepted as a key control for spatial distribution of the subsurface channel sandbodies. Stacking density of channel deposits therefore has been used to infer changes in 1) tectonic activity and 2) sediment supply (as a proxy for climate changes) from catchments (e.g. Alexander *et al.*, 1994; Alexander & Leeder, 1987; Allen, 2008; Bridge, 1993; Bridge & Leeder, 1979; Heller & Paola, 1996; Kim *et al.*, 2010; Leeder, 1978; Leeder *et al.*, 1996). However, stratigraphic evolution in active tectonic basins cannot be properly understood without consideration of the dynamic interactions of autogenic process with tectonic forcing. The following experimental example shows a complex autogenic response to tectonic activity.

The XES 05 experiment further considered the influence of subsidence on autogenically-driven stratigraphy. XES 05 had an initial stage with no tectonic component, in which the results for quantifying the autogenic process were presented in the previous sections. The experiment was composed of two separated normal fault segments that set up a relay-ramp. Constant fault slip rates were applied to both faults in the experiment (Kim *et al.*, 2010). Topographic displacement occurred across the upstream footwall and downstream hangingwall basins, which substantially lengthened the time for the fluvial channel system to redistribute sediment and to reach a quasi-steady-state landscape. Analysis using time integrated maps of wetted surface by channel flow in the tectonic stage (Fig. 4) indicates an increase in the characteristic time scale of the fluvial autogenic variation from 13 hours (non-tectonic stage) to 65 hours (tectonic stage) (Kim & Paola, 2007).

The five-fold increase in the autogenic time scale (i.e. slow channel lateral migration) caused temporal variation in sediment supply to the position of the maximum subsidence in the hangingwall basin. The slow redistribution of

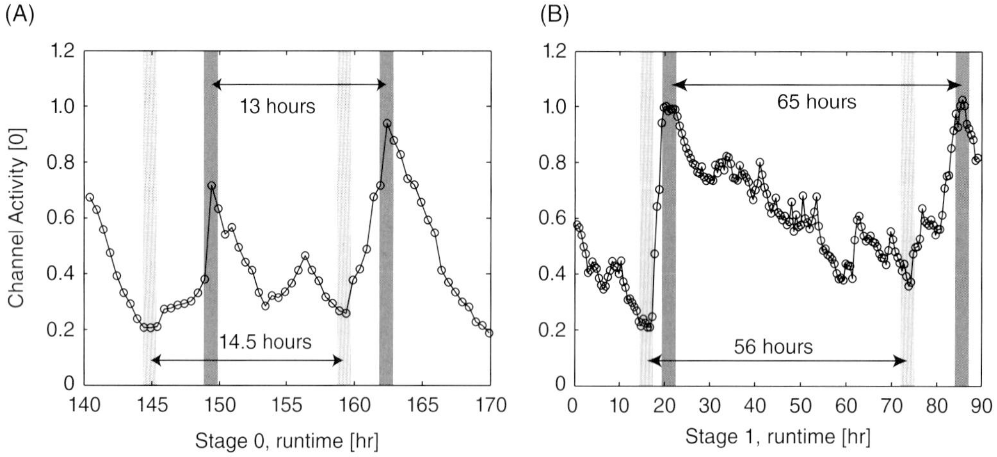

Fig. 4. Characteristic measurement of channel activity in the delta top surface (low value represents rapid migration (and/or avulsion) of channels). (A) Channel activity cycles every ~ 13 hours in the first non-tectonic stage, but (B) the cycle period increases to 65 hours due to lateral ground tilting imposed during the second tectonic stage.

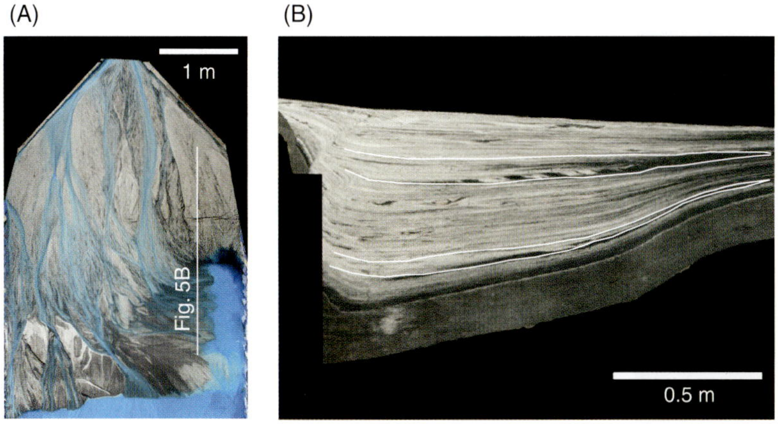

Fig. 5. (A) Surface Image taken from XES 05 when the hangingwall basin developed a lake autogenically. (B) A sliced dip section (location is indicated in (A)) across the footwall and hangingwall showing cyclic sedimentation between fluvial and lake facies (bounded by white lines).

sediment over an area with variable subsidence rates allows enough time to develop a tectonic depression in the surface topography, causing periodic opening of a lake (Fig. 5A). Opening and closing of the autogenic lake cycled twice in XES 05. During the opening of the lake, the channels flowed away from the hangingwall basin, whereas the channels flowed into the hangingwall basin around and across the fault during the closing period. Cyclic alternation between foreset and fluvial stratification was observed in sliced deposit sections, which developed even without time variations in tectonic controls (i.e. no changes in fault slip rate or sediment supply from the upstream source) (Fig. 5B).

These experimental results can be extrapolated to field conditions using scour depths as a reference scale. During the 60 hour exchange period between foreset and fluvial deposits, about 20 scour-depths (1 scour depth ≅ 2 cm) developed in the subsidence maximum. Assuming a 5 m scour depth in natural depositional basins and a sedimentation rate of 0.001 m/yr, the record of the full 60 hour autogenic event would be equivalent to 50 to 100 m thick strata and the entire autogenic cycle time would be approximately 10^5 years (Kim & Paola, 2007). This unexpected modification of scale of the internal process suggests the possibility of very long-period, high-magnitude autogenic variability associated with the coupling of channel reorganisation and active tectonic deformation.

Suggestions for future work

The Earth's surface is active over all time scales due to external perturbations and internally organised dynamics. In nature, depositional systems respond actively to multiple external controls with a wide range of time and space scales. The complex mixture of allogenic and autogenic signatures recorded in sedimentary records presents a puzzle; whose pieces consist of the causes and effects responsible for basin-fill history. Physical experiments in sediment transport and sedimentary basin evolution for understanding the relationship between external and internal changes along with associated stratigraphic architecture have advantages over numerical modelling and field study because 1) the depositional system in experiments naturally self-evolves and 2) precise managing of key external controls and monitoring of basin evolution can be achieved.

Initial efforts should focus on isolating individual causes and examining the complex effects associated with autogenic dynamics. Studying the modification of time and event scales by autogenic processes from linear external controls (e.g. Kim & Muto, 2007; Martin *et al.*, 2009b) should be continued in order to define their fundamental relationship under simple conditions. High-resolution topographic measurements in time and space should be taken for robust investigation of event size of autogenic processes. Coupled allo-autogenic study should also be expanded to investigating 1) the effect of base-level change over a wide range of rates, 2) various tectonic styles such as passive-margin and foreland basin styles and 3) long-term increase and/or decrease in sediment supply.

One of the hurdles to overcome for improving our ability to decouple allogenic signals from the rock record is the suppression of signals that come from external controls in stratigraphy. An experiment in the XES basin was performed in 2008 with an identical tectonic geometry as

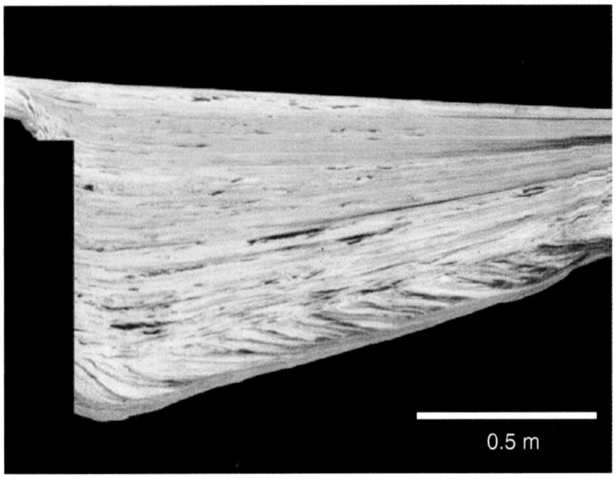

Fig. 6. A sliced dip section across the footwall and hangingwall in XES 08. The blue and top yellow layers were developed during the 4th stage with two tectonically quiescent periods.

the XES 05 experiment but with cyclic variations of the fault slip rate in time. Unlike the XES 05 experiment, the XES 08 experiment did not develop clear cyclic patterns for tectonic variation in the strata (Fig. 6). The 4th stage of the XES 08 experiment was composed of three substages with two tectonically quiescent periods in the first (16 hours) and last (24 hours) stages and a middle tectonic stage. The middle tectonic stage was applied for 6 hours with a slip rate almost double that of the other stages, both in the XES 05 and 08 experiments (Straub *et al.*, 2009). However, the 6 hour duration (cf. 65 hours modified autogenic channel time scale in XES 05) was too short to develop a topographic low significant enough to be recorded in the final deposit and the two 16 and 24 hour non-tectonic stages had long enough duration (compared to the 13 hour autogenic frequency in XES 05 non-tectonic stage) to reorganise the tectonically deformed surface.

A numerical rice-pile model in Jerolmack & Paola (2010) demonstrated a damping of cyclic sediment (rice grains) input signal in the output flux at the downstream end. Cyclic input applied at the upstream end, with a time period less than a characteristic relaxation time, is shredded through the rice-pile transport system, unless the magnitude of the perturbation is very large (Fig. 4 in Jerolmack & Paola (2010)).

Stratigraphic products attributed to periodic changes in base-level, tectonic movement and sediment supply should be better examined. Advances in experimental techniques allow for more complete tests of hypotheses related to sensitivity of transport systems to imposed frequency of external forcing. Experiments with a wide range of frequencies in the basinal forcing will provide a fundamental understanding of origin of cyclic sedimentation in natural systems and thus a way to better decouple environmental signatures from rock records.

CONCLUSIONS

1 The fluvial autogenic sediment storage and release process occurs naturally and periodically changes transport rates in fluvial systems. Channel patterns alternate between strong channelisation and sheet flow to modulate fluvial sediment transport processes. The characteristic time for this internal alternation is scaled with the sediment supply rate, basin length, flow depth and width, as shown in equation (1).
2 Moving boundaries, such as the shoreline and alluvial-bedrock transition, records internal variability. Experiments with base-level change produce increase in the variability in the shoreline migration rate for base-level rise and decrease in the variability of the shoreline migration rate for base-level fall. The variability in the alluvial-bedrock transition shows an opposite trend to the base-level forcing.
3 Lateral tectonic tilting lengthens the autogenic channel time scale by decreasing the channel lateral migration rate. An experiment with a lateral subsidence variation slows down the fluvial reorganisation process and induces long-term cyclic sedimentation without time-varying allogenic controls.
4 Cyclic changes in basinal forcing at a frequency less than the characteristic relaxation (equilibrium) time in equation (2) were significantly blurred during the stratal development in sedimentation.
5 Future projects should be specifically designed to measure quantitatively the fluvial autogenic processes and investigate associated stratigraphic products under precisely controlled boundary conditions (tectonic subsidence rate, sediment and water supplies etc.). The level of contribution of autogenic processes to stratigraphic development almost certainly varies

non-linearly with sediment transport conditions and mixings with basinal forcing; and thus the control of fluvial autogenic processes on the resulting strata can be more significant than what has been normally accepted. Results of the previous experiments and proposed next steps will fundamentally improve the understanding of autogenic processes and their stratigraphic products. Further investigations of the coupled allogenic-autogenic signatures in experiments, but starting with simplified external controls in the experiments, will aid in disentangling allogenic and autogenic effects quantitatively.

REFERENCES

Alexander, J., **Bridge, J.S.**, **Leeder, M.R.**, **Collier, R.E.L.** and **Gawthorpe, R.L.** (1994) Holocene meander-belt evolution in an active extensional basin, southwestern Montana. *J. Sed. Res.*, **64**, 542–559.

Alexander, J. and **Leeder, M.R.** (1987) Active tectonic control on alluvial architecture; Recent developments in fluvial sedimentology. In: *Third International Fluvial Sedimentology Conference, Fort Collins, CO, United States, Aug. 1985* (Eds F.G. Ethridge, R.M. Flores, M.D. Harvey and J.N. Weaver), **39**, 243–252, United States (USA).

Allen, P.A. (2008) Time scales of tectonic landscapes and their sediment routing systems. *Geol. Soc. London Spec. Publ.*, **296**, 7–28. doi: 10.1144/sp296.2.

Ashworth, P.J., **Best, J.L.** and **Jones, M.** (2004) Relationship between sediment supply and avulsion frequency in braided rivers. *Geology* (Boulder), **32**, 21–24.

Blair, T.C. and **Bilodeau, W.L.** (1988) Development of tectonic cyclothems in rift, pull-apart and foreland basins: Sedimentary response to episodic tectonism. *Geology*, **16**, 517–520.

Bridge, J.S. (1993) Description and interpretation of fluvial deposits – a critical perspective. *Sedimentology*, **40**, 801–810.

Bridge, J.S. and **Leeder, M.R.** (1979) A simulation model of alluvial stratigraphy. *Sedimentology*, **26**, 617–644.

Bryant, M., **Falk, P.** and **Paola, C.** (1995) Experimental study of avulsion frequency and rate of deposition. *Geology* (Boulder), **23**, 365–368.

Castelltort, S. and **Van Den Driessche, J.** (2003) How plausible are high-frequency sediment supply-driven cycles in the stratigraphic record? *Sed. Geol.*, **157**, 3–13.

Cazanacli, D., **Paola, C.** and **Parker, G.** (2002) Experimental steep, braided flow: Application to flooding risk on fans. *J. Hydraul. Eng.*, **128**, 322–330.

Dorsey, R.J., **Umhoefer, P.J.** and **Falk, P.D.** (1997) Earthquake clustering inferred from Pliocene Gilbert-type fan deltas in the Loreto Basin, Baja California Sur, Mexico. *Geology* (Boulder), **25**, 679–682.

Heller, P.L. and **Paola, C.** (1996) Downstream changes in alluvial architecture; an exploration of controls on channel-stacking patterns. *J. Sed. Res.*, **66**, 297–306.

Heller, P.L., **Paola, C.**, **Hwang, I.-G.**, **John, B.** and **Steel, R.** (2001) Geomorphology and sequence stratigraphy due to slow and rapid base-level changes in an experimental subsiding basin (XES 96–1). *AAPG Bull.*, **85**, 817–838.

Hickson, T.A., **Sheets, B.A.**, **Paola, C.** and **Kelberer, M.** (2005) Experimental test of tectonic controls on three-dimensional alluvial facies architecture. *J. Sed. Res.*, **75**, 710–722.

House, M.R. (1985) A new approach to an absolute timescale from measurements of orbital cycles and sedimentary microrhythms. *Nature*, **315**, 721–725.

Hoyal, D. and **Sheets, B.A.** (2009) Morphodynamic evolution of experimental cohesive deltas. *J. Geophys. Res.*, **114**, 18. doi: F0200910.1029/2007jf000882.

Jerolmack, D.J. (2011) Causes and effects of noise in landscape dynamics. *Eos*, **92**, 385–386.

Jerolmack, D.J. and **Mohrig, D.** (2005) Frozen dynamics of migrating bedforms. *Geology*, **33**, 57–60.

Jerolmack, D.J. and **Paola, C.** (2007) Complexity in a cellular model of river avulsion. *Geomorphology*, **91**, 259–270.

Jerolmack, D.J. and **Paola, C.** (2010) Shredding of environmental signals by sediment transport. *Geophys. Res. Lett.*, **37**, L19401. 10.1029/2010gl044638.

Jerolmack, D.J. and **Swenson, J.B.** (2007) Scaling relationships and evolution of distributary networks on wave-influenced deltas. *Geophys. Res. Lett.*, **34**, 23402.

Kim, W. and **Jerolmack, D.J.** (2008) The Pulse of Calm Fan Deltas. *J. Geol.*, **116**, 315–330. doi: 10.1086/588830.

Kim, W. and **Muto, T.** (2007) Autogenic response of alluvial-bedrock transition to base-level variation: Experiment and theory. *J. Geophys. Res.*, **112**. doi: 10.1029/2006JF000561.

Kim, W. and **Paola, C.** (2007) Long-period cyclic sedimentation with constant tectonic forcing in an experimental relay ramp. *Geology*, **35**, 331–334.

Kim, W., **Paola, C.**, **Swenson, J.B.** and **Voller, V.R.** (2006a) Shoreline response to autogenic processes of sediment storage and release in the fluvial system. *J. Geophys. Res.*, **111**. doi: 10.1029/2006JF000470.

Kim, W., **Paola, C.**, **Voller, V.R.** and **Swenson, J.B.** (2006b) Experimental measurement of the relative importance of controls on shoreline migration. *J. Sed. Res.*, **76**, 270–283. doi: 10.2110/jsr.2006.019.

Kim, W., **Sheets, B.A.** and **Paola, C.** (2010) Steering of experimental channels by lateral basin tilting. *Basin Res.*, **22**, 286–301.

Leeder, M.R. (1978) A quantitative stratigraphic model for alluvium, with special reference to channel deposit density and interconnectedness; Fluvial sedimentology. First international symposium on fluvial sedimentology, Calgary, Alberta, Canada, Oct. 20–22, 1977, 587–596.

Leeder, M.R., **Mack, G.H.**, **Peakall, J.** and **Salyards, S.L.** (1996) First quantitative test of alluvial stratigraphic models: southern Rio Grande Rift, New Mexico. *Geology*, **24**, 87–90.

Martin, J., **Paola, C.**, **Abreu, V.**, **Neal, J.** and **Sheets, B.** (2009a) Sequence stratigraphy of experimental strata under known conditions of differential subsidence and variable base-level. *AAPG Bull.*, **93**, 503–533.

Martin, J., **Sheets, B.**, **Paola, C.** and **Hoyal, D.** (2009b) Influence of steady base-level rise on channel mobility, shoreline migration and scaling properties of a cohesive experimental delta. *J. Geophys. Res.*, **114**, F03017. 10.1029/2008jf001142.

Muto, T. and **Steel, R.J.** (2001) Autostepping during the transgressive growth of deltas; results from flume experiments. *Geology* (Boulder), **29**, 771–774.

Paola, C. (2000) Quantitative models of sedimentary basin filling; Millennium reviews. *Sedimentology*, **47**, Suppl. 1, 121–178.

Paola, C., Heller, P.L. and **Angevine, C.L.** (1992) The large-scale dynamics of grain-size variation in alluvial basins; 1, *Theory. Basin Res.*, **4**, 73–90.

Paola, C., Mullin, J., Ellis, C., Mohrig, D.C., Swenson, J.B., Parker, G.S., Hickson, T., Heller, P.L., Pratson, L., Syvitski, J., Sheets, B. and **Strong, N.** (2001) Experimental stratigraphy. *GSA Today*, **11**, 4–9.

Paola, C., Straub, K., Mohrig, D. and **Reinhardt, L.** (2009) The "unreasonable effectiveness" of stratigraphic and geomorphic experiments. *Earth-Sci. Rev.*, **97**, 1–43.

Posamentier, H.W. and **Vail, P.R.** (1988) Eustatic controls on clastic deposition; II, Sequence and systems tract models. In: *Sea-Level Changes: An Integrated Approach* (Eds C.K. Wilgus, B.S. Hastings, C.A. Ross, H.W. Posamentier, J. Van Wagoner and C.G.S.C. Kendall), **42**, pp. 125–154, Houston, TX, USA.

Posamentier, H.W., Jervey, M.T. and **Vail, P.R.** (1988) Eustatic controls on clastic deposition; I, Conceptual framework. In: *Sea-Level Changes: An Integrated Approach* (Eds C.K. Wilgus, B.S. Hastings, C.A. Ross, H.W. Posamentier, J. Van Wagoner and C.G.S.C. Kendall), **42**, pp. 109–124, Houston, TX, USA.

Postma, G., Kleinhans, M.G., Meijer, P.T. and **Eggenhuisen, J.T.** (2008) Sediment transport in analogue flume models compared with real-world sedimentary systems: a new look at scaling evolution of sedimentary systems in a flume. *Sedimentology*, **55**, 1541–1557.

Reitz, M.D., Jerolmack, D.J. and **Swenson, J.B.** (2010) Flooding and flow path selection on alluvial fans and deltas. *Geophys. Res. Lett.*, **37**. doi: 10.1029/2009GL041985.

Roth, S. and **Reijmer, J.J.G.** (2005) Holocene millennial to centennial carbonate cyclicity recorded in slope sediments of the Great Bahama Bank and its climatic implications. *Sedimentology*, **52**, 161–181.

Smith, G.A. (1994) Climatic influences on continental deposition during late-stage filling of an extensional basin, southeastern Arizona. *Geol. Soc. Am. Bull.*, **106**, 1212–1228.

Straub, K.M., Paola, C., Kim, W. and **Sheets, B.** (2009) Controls on steering of channels in laterally tilting basins: an experimental study: Eos Trans. *AGU*, **90**, no. 52, Fall meet. suppl., Abstract EP53A-0612.

Van Dijk, M., Postma, G. and **Kleinhans, M.G.** (2009) Autocyclic behaviour of fan deltas: an analogue experimental study. *Sedimentology*, **56**, 1569–1589.

Int. Assoc. Sedimentol. Spec. Publ. (2014) **46**, 139–148

The autostratigraphic view of responses of river deltas to external forcing: a review of the concepts

TETSUJI MUTO* and RONALD J. STEEL†

* *Graduate School of Fisheries Science and Environmental Studies, Nagasaki University, 1-14 Bunkyomachi, Nagasaki 852-8521, Japan (E-mail: tmuto@nagasaki-u.ac.jp)*

† *Department of Geological Sciences, University of Texas at Austin, 1 University Station, C-1100, Austin, Texas, 78712, USA*

ABSTRACT

The geology of river deltas has long been based on the notion, here referred to as the 'hypothesis of equilibrium response', that steady external dynamic forcing gives rise to steady stratigraphic configuration. Though this equilibrium response is physically possible, the hypothesis does not hold true in general and can cause serious misinterpretation of the stratigraphic record. Autostratigraphy takes a totally different view which takes full account of non-equilibrium responses and accompanying deterministic autogenesis, suggesting that (1) steady dynamic external forcing generally results in unsteady stratigraphic response and (2) steady stratigraphic configuration can result from unsteady dynamic external forcing. Further exploration of non-equilibrium responses of river deltas to external forcing will bring more clarity to their stratigraphy.

Keywords: Autogenesis, autostratigraphy, deterministic, equilibrium, forcing, non-equilibrium, river deltas, sea level, stratigraphic response.

INTRODUCTION

A critical advance in our understanding of the geology of river deltas during the last 10 years comes partly from the clarification of the intrinsic stratigraphic responses resulting from external forcing (sea-level forcing, particularly) (Muto *et al.*, 2007) and partly from rapid developments of experimental stratigraphy and geomorphology (Paola *et al.*, 2009). In addition, the finding of deterministic autogenesis (e.g. Muto & Steel, 2002a) has been crucial because it clearly indicates the general lack of a quantitatively balanced state between the effect of relative sea-level rise and the effect of sediment supply to the depositional system (Muto, 2001; Swenson & Muto, 2007). The opposite has been incorporated as a core hypothesis in most genetic stratigraphy models including sequence stratigraphy.

Genetic stratigraphy that takes full account of deterministic autogenesis is referred to as autostratigraphy (Muto *et al.*, 2007). The autostratigraphic model of river deltas implies that (1) abrupt stratigraphic breaks are not necessarily associated with sudden changes in rate of base-level movement but can result from purely autogenic responses of the system and (2) stratigraphic responses that fall on a break-free curve can result from an unsteady change in external forcing. These are expressions of non-equilibrium response. We review here some key aspects of the response of river deltas to external forcing (particularly dynamic sea-level forcing).

AUTOGENESIS AND EXTERNAL FORCING

Autogenesis has conventionally been associated with responses that are local (a small part of the system), stochastic and cyclic, such as typically illustrated with river avulsion or delta-lobe switching (Beerbower, 1965; Edmonds *et al.*, 2009; Straub *et al.*, 2009). However, there is also another type of autogenesis that is global (i.e. the entire system), deterministic and non-cyclic, as has been noticed recently (Muto & Steel, 2002a; Muto *et al.*, 2007; Paola *et al.*, 2009). It is this latter type of

From Depositional Systems to Sedimentary Successions on the Norwegian Continental Margin, First Edition.
Edited by A. W. Martinius, R. Ravnås, J. A. Howell, R. J. Steel, and J. P. Wonham.

autogenesis that is considered primarily in autostratigraphy.

Whenever a depositional system is considered, it must have a distinct spatial extent with a boundary to its outside. Any action by the outside upon the depositional system is regarded as external forcing. Two kinds of external forcing can be distinguished: static and dynamic. In the case of river deltas, static external forcing includes stationary base level and unchanged basement topography; dynamic external forcing includes base level change and basin tectonics. Sediment and water supply is also part of dynamic external forcing since it involves input of material from the outside of the system. In the present context, it hardly makes sense to presume a situation under which no sediment (and no water) is supplied to the depositional system and thus no strata accumulate there. In this sense, any active depositional system cannot be free from either dynamic external forcing or static external forcing. What matters with an active depositional system is whether the dynamic external forcing is steady (i.e. rate constant) or is unsteady (i.e. rate variable). From this viewpoint, the term autogenic should refer to the origin of stratigraphic or geomorphic features that arise as responses from both steady dynamic external forcing and static external forcing (Fig. 1). Conversely, the term allogenic refers to stratigraphic or geomorphic responses generated as a result of the dynamic external forcing being unsteady. It should be noted, contrary to some popular usage, that the terms autogenic and allogenic are therefore best used to refer to stratigraphic responses and not as adjectives for the main controls themselves.

One might claim that a perfect sediment-bypass (and water-bypass too) system, where neither aggradation nor degradation occurs in spite of continuing sediment supply, could be regarded as being effectively free from dynamic external forcing. The state of such a system, in case of river deltas, would be grade. However, the graded state of an alluvial-deltaic river having moving boundaries is possible only when relative sea-level drops in a particular pattern depending on basin topography (Muto & Swenson, 2005, 2006; Petter & Muto, 2008). The attainment of grade rather reflects significant exertion of dynamic external forcing.

THE HYPOTHESIS OF EQUILIBRIUM RESPONSE

Two primary factors controlling regression and transgression in deltaic settings are sediment supply to the basin and relative sea-level rise or fall (basin subsidence+base level fluctuation). An open question to this proposition is how these two factors function in building the stratigraphic architectures of river deltas. With respect to this question, there is a long-standing notion, here referred to as ‘the hypothesis of equilibrium response’. Equilibrium response is a type of response by which steady external forcing gives rise to steady stratigraphic configuration or steady sedimentary behaviour of the depositional system (Fig. 2). In fact, equilibrium response is physically possible but limited only to a special circumstance (Muto & Swenson, 2006). Nonetheless, the hypothesis of equilibrium response is a popular

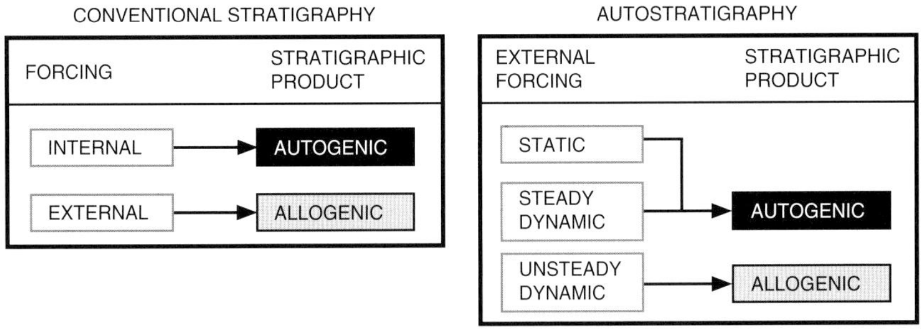

Fig. 1. Usage of the terms autogenic and allogenic in terms of the relationship between forcing and resulting stratigraphic products. In conventional stratigraphy, autogenic and allogenic products correspond to internal and external forcing of the system, respectively. However, in the autostratigraphic view, autogenic refers to the origin of stratigraphic or geomorphic features that arise as responses from both steady dynamic external forcing and static external forcing, whereas allogenic refers to stratigraphic or geomorphic responses to unsteady dynamic external forcing.

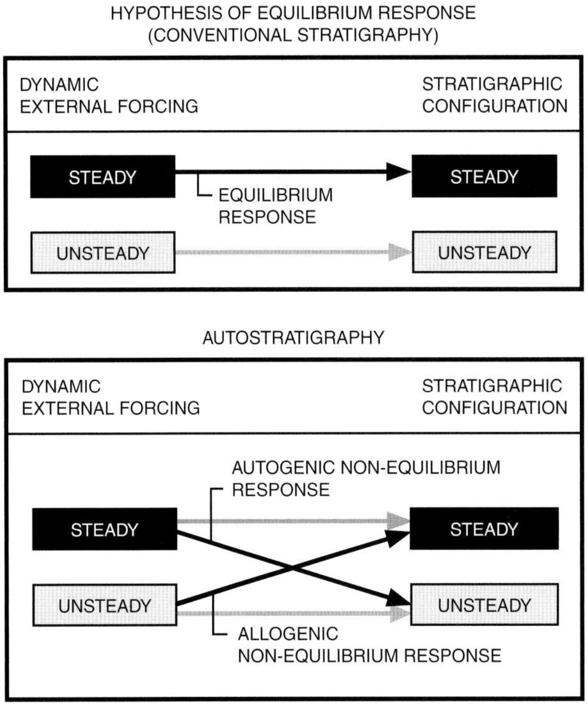

Fig. 2. Different views of the stratigraphic response of a depositional system to dynamic external forcing. Conventional stratigraphy has been developed based upon the hypothesis of equilibrium response that to steady dynamic external forcing, equilibrium response holds true in general. Here any unsteady stratigraphic features are commonly attributed to unsteady dynamic external forcing. Autostratigraphy suggests two more types of responses: autogenic non-equilibrium response (unsteady stratigraphic configuration by steady forcing) and allogenic non-equilibrium response (steady stratigraphic configuration by unsteady forcing).

and common assumption, as though equilibrium response holds true in general. The corollary of this popular assumption is that any significant unsteady stratigraphic trends would be attributed to unsteady dynamic external forcing (Fig. 2).

The hypothesis of equilibrium response suggests that regression and transgression reflect imbalances between the effect of relative sea-level rise and the effect of sediment flux; and that coastal aggradation without significant shoreline migration reflects a balanced state between these two controls. Alongside this, it has also been generally assumed that, given steady external forcing by constant sediment supply (rate Q_s) and constant relative sea-level rise (rate R_{slr}), a river delta grows to achieve an equilibrium configuration, produces a particular sediment-stacking pattern and maintains a constant rate of shoreline migration in a particular direction. Furthermore, if Q_s and R_{slr} do not change with time, the deltaic system maintains its stratigraphic response; for example, if sediment supply is large enough to overcome relative sea-level rise, regression will occur and will then be maintained as long as there is no change in Q_s and R_{slr}. Most existing models of genetic stratigraphy including sequence stratigraphy have been developed on the basis of this hypothesis, explicitly or implicitly (Weller, 1960; Sloss, 1962; Curray, 1964; Curtis, 1970; Vail *et al.*, 1977; Posamentier *et al.*, 1988; Galloway, 1989; Shanley & McCabe, 1994; Neal & Abreu, 2009). It is not that these examples are necessarily wrong, but we believe that the assumption of equilibrium response leads to a lack of rigor in the analysis and can cause a misleading result, as discussed further below.

AUTOGENIC NON-EQUILIBRIUM RESPONSES

There are two other main types of stratigraphic response that need to be considered along with the equilibrium response option. The first is autogenic non-equilibrium response, by which steady external forcing produces an unsteady stratigraphic configuration (Fig. 2). A clear example of autogenic non-equilibrium response is the autoretreat of deltaic shorelines (Muto & Steel, 1992, 1997; Swenson *et al.*, 2000; see Fig. 3). Even when all basin dynamic conditions including R_{slr}, Q_s and upstream water discharge are kept constant, all deltaic systems experiencing regression eventually become transgressed. During autoretreat, there inevitably occurs a critical moment, the 'autobreak' of Muto (2001) (see also Parker *et al.*, 2008), after which the depositional system loses its original geometry and deltaic character. This sequential transition from early deltaic regression to later non-deltaic rapid transgression is inevitable, as long as relative sea-level continues to rise at a constant rate.

The fundamental cause for this non-equilibrium response in river deltas is their progressive growth during rising sea-level. As the delta grows, an increasing fraction of sediment (with constant Q_s) supplied by the feeder river system is consumed for alluvial aggradation, whereby sediment available for progradation of the delta tends to

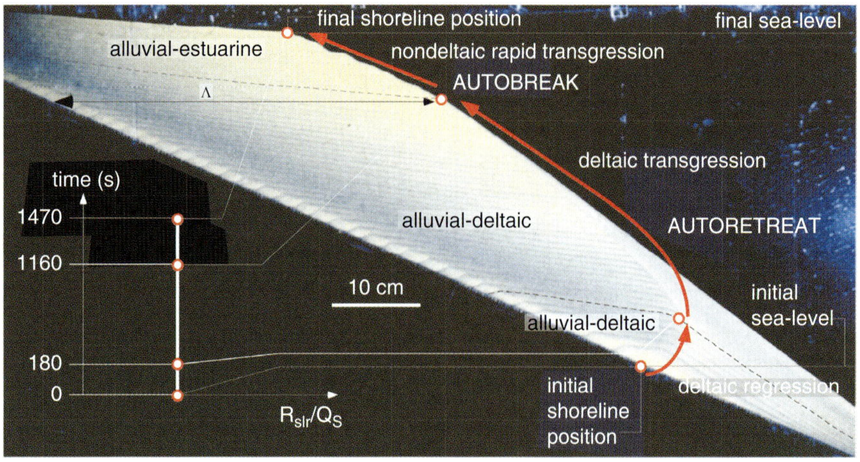

Fig. 3. Shoreline autoretreat and subsequent autobreak, illustrating an autogenic non-equilibrium response to steady sea-level rise. This river delta was built during an experimental run conducted with constant sediment supply and constant relative sea-level rise. Λ is autostratigraphic length scale defined with Q_s/R_{slr} (Q_s: rate of sediment supply, R_{slr}: rate of relative sea-level rise). See Muto (2001) for details of the experimental runs.

decrease with time (Muto & Steel, 1992, 1997; Milton & Bertram, 1995). As a result of this, including continuing sediment supply, the regressing delta becomes unable to maintain its areally expanding forward growth, despite the steady dynamic external forcing including continuing sediment supply. Eventually, the delta inevitably meets a critical moment at which it can no longer retain its original sedimentation style (i.e. autobreak). It was emphasised by Muto & Steel (1997) that autoretreat would be delayed in settings where there were very low rates of sea-level rise and hastened where these rates were higher. A good stratigraphic example where shoreline progradational length (before transgressive turnaround) can be seen to relate to degree of aggradation in the coastal system is shown in Fig. 4 (from Kirschbaum & Hettinger, 2004) where regressive-shoreline transit lengths in the 'higher accommodation' setting (upper half of diagram with Rollins Sandstone shorelines) rarely exceed 20 km to 40 km, in contrast to the significantly longer (60 km to 80 km) shoreline lengths of the 'low-accommodation' Sego-Corcoran-Cozette sandstones (lower half of Fig. 4). This of course does not prove that the termination of individual shoreline regressions in this part of the Western Interior Seaway was caused by autoretreat but it does provide a simple alternative to more complex allogenic explanations. Aschoff & Steel (2011) suggest a general tectonic explanation (initiating Laramide tectonic movements) for the anomalous, 'low-accommodation' Sego-Corcoran-Cozette shorelines as a group, though not for the individual shoreline turnarounds.

Other examples arising from autogenic non-equilibrium response include autodrowning of a deltaic system with rising sea-level under a special geomorphic condition (Tomer & Muto, 2010; Tomer *et al.*, 2011), autoincision of an alluvial river with falling sea-level (Muto & Steel, 2002b; Swenson & Muto, 2007) and autodetachment of a deltaic system with falling sea-level (Petter & Muto, 2008). These are all global, deterministic and non-cyclic autogenesis that inevitably occur given a particular set of conditions of steady dynamic external forcing.

It should be noted that the term 'accommodation' mentioned above is only in the context of illustrating autogenic non-equilibrium responses with the existing papers. The framework of autostratigraphy itself is passive to this conventional concept (Muto *et al.*, 2007), partly because accommodation, as space for potential sediment accumulation (Jervey, 1988), can hardly be treated quantitatively, specifically and objectively (Muto & Steel, 2000; for special treatment see Kim *et al.*, 2006). Another major reason is that this concept is based on a misunderstanding of alluvial grade, which, in sequence stratigraphy, has been regarded as the final stable state of the alluvial downstream reaches of a river system that is attained by equilibrium response to stationary base level (e.g. Posamentier *et al.*, 1988). In natural

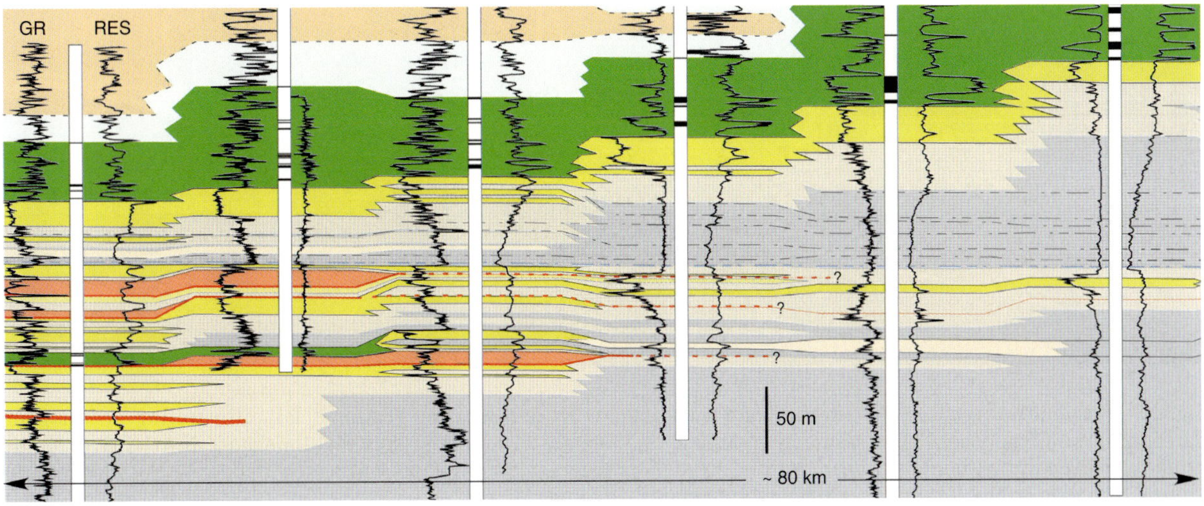

Fig. 4. Shoreline progradational patterns along eastern Book Cliffs, near Pallisades, Colorado (from Kirschbaum & Hettinger, 2004). The thin yellow-coloured shorelines of Sego/Corcoran (lower quarter of diagram) and Cozette (group of shorelines just below the middle of the diagram) members prograded rapidly and extensively, in contrast to the aggradational and short shoreline lengths of the overlying Rollins sandstone (see Aschoff & Steel, 2011 for details). Associated thicknesses of coastal plain deposits (green) reflect differing aggradational styles and suggest that individual shoreline turnarounds may well be non-equilibrium autogenic responses to steady external forcing. Red colours reflect transgressive estuarine valleys. The difference in external forcing between the two groups of shorelines was suggested to be tectonic by Aschoff & Steel (2011). GR = gamma ray and RES = resistivity.

fluvio-deltaic systems having moving boundaries, the alluvial grade can be attained and sustained only during base level fall and never functions to limit alluvial aggradation (Muto & Swenson, 2005, 2006). Even with falling sea-level, alluvial rivers can aggrade steadily without valley incision if a particular set of geomorphic conditions is available (Petter & Muto, 2008). In terms of this recent notion of alluvial grade, which plays a key role in the autostratigraphy of river deltas, the accommodation concept meets a serious self-contradiction that sediment accumulates even if there is no accommodation (i.e. even if the space is part of 'anti-accommodation'; see Muto & Steel, 2000). In autostratigraphy, the accommodation concept is not useful to explore the understanding of responses of river deltas to sea-level changes.

ALLOGENIC NON-EQUILIBRIUM RESPONSES

The second type of non-equilibrium response is where steady stratigraphic configuration can result from unsteady external forcing. This type of stratigraphic response is referred to as allogenic non-equilibrium response (Fig. 2). The autoretreat-autobreak process is unavoidable if sea-level continues to rise at a constant rate. For a deltaic shoreline to migrate in a particular constant direction and not to experience autoretreat-autobreak, R_{slr} is required to decrease inversely proportional to the square-root to cubic-root of time, otherwise Q_s is required to increase in proportion to the square to cubic of time (Muto & Steel, 2002a). Such unsteady dynamic external forcing would illustrate allogenic non-equilibrium response.

A clear example of allogenic non-equilibrium response is allogenic grade (Muto & Swenson, 2005), the graded state of any alluvial river attained and sustained under unsteady dynamic external forcing. Fig. 5 shows a longitudinal profile of a one-dimensional deltaic clinoform built with sea-level fall decelerating at a rate inversely proportional to the square-root of time. Although an alluvial river on the topset of the prograding clinoform has always existed, it has simply extended basinward without significant deposition and erosion to sustain the grade. If sea-level falls at a constant rate in the same geomorphic condition, the river cannot attain grade but inevitably undergoes an irreversible change from aggradation to degradation (Swenson & Muto, 2007).

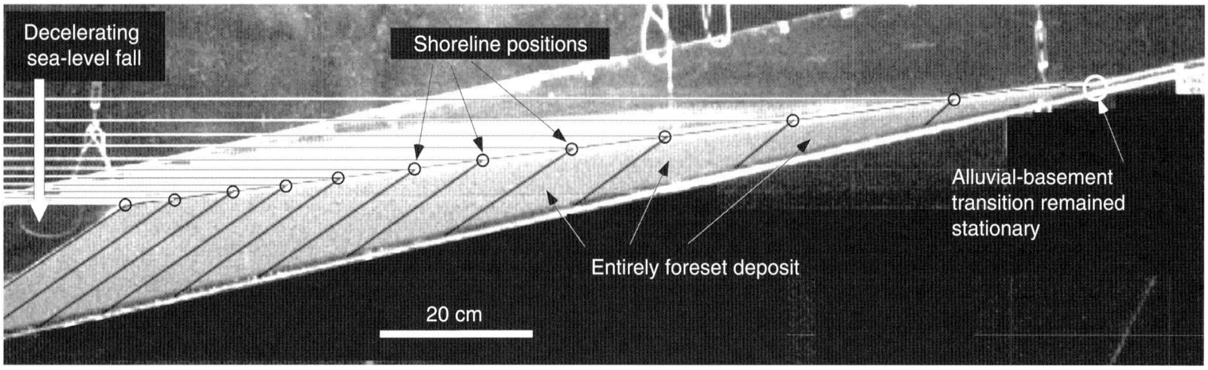

Fig. 5. Allogenic alluvial grade attained and sustained with decelerating sea-level fall, illustrating an allogenic non-equilibrium response. The timeline interval is approximately 11 minutes. Note that the feeder alluvial river simply extended basinward without significant topset aggradation and degradation, representing the graded state of the river. See Muto & Swenson (2005) for details of the experimental runs.

CONTROL OF MANIFESTATION OF NONEQUILBIRUM RESPONSE

If one is unaware of non-equilibrium response (and thus deterministic autogenesis), the conventional explanation of regression and transgression simply follows the hypothesis of equilibrium response, i.e. in terms of balance or imbalance between the effect of sea-level rise and the effect of sediment supply. Full recognition of non-equilibrium response, on the other hand, provides a totally different explanation. Regression, aggradation (with stationary shoreline) and transgression are simply the non-equilibrium response of the deltaic system to a constant rise of relative sea-level. As long as relative sea-level continues to rise, none of these represent an equilibrium configuration but rather transient states of the depositional system evolving toward a non-deltaic transgressive system, the final state.

Magnitudes of R_{slr} and Q_s themselves do not determine which one of regression, aggradation and transgression occurs. The primary function of the two factors is to control length and time scales of the depositional system. Any river delta growing during sea-level change has a particular time and length scale that is specified with R_{slr} and Q_s (Muto *et al.*, 2007). Autostratigraphic length scale Λ and time scale τ are usually defined as:

$$\Lambda = \frac{Q_S}{|R_{slr}|} \quad (1)$$

$$\tau = \frac{\Lambda^2}{\upsilon} = \alpha \frac{Q_S}{R_{slr}{}^2} \quad (2)$$

where υ is linear diffusivity in alluvial sedimentation and α is averaged alluvial slope (Muto & Swenson, 2005). Λ reflects, for example, the horizontal distance between shelf basement and shoreline position at the attainment of autobreak (Muto, 2001; Fig. 3).

The stratigraphic meaning of τ is as follows. Let T be the time interval for external dynamic forcing of relative sea-level rise. If $\tau << T$, prominent signals of the non-equilibrium response are expected. The river delta cannot avoid experiencing autoretreat and autobreak within the period of the sea-level rise. The non-equilibrium response of such a case is referred to as 'large-scale' in autostratigraphy. However, if $\tau >> T$, only a very small window of time is available for the non-equilibrium response and the river delta can behave steadily, as if making an equilibrium response. In the case of the experimental run shown in Fig. 3 for example, T (= run time) was 1470s, whereas the calculated magnitude of τ is a much shorter 179s. This is the reason why autoretreat and autobreak were attained during the run of the experiment. If the experiment had been stopped one minute after the beginning of the run, only the initial regressive part would have been seen. In that case, it would be hard to recognise any non-equilibrium response.

APPLICATION TO STRATIGRAPHIC RECORDS

A primary aim of autostratigraphy is to explore deterministic autogenic processes and their stratigraphic responses, thereafter to identify allogenic stratigraphic products and responsible unsteady dynamic external forcing (Fig. 6). Although stratigraphic records are generally composed of both autogenic and allogenic products, conventional stratigraphy has been apt to ignore the importance of autogenesis and thus to overrate allogenic processes. Any large-scale stratigraphic analysis without paying attention to the autostratigraphic view can lead to a serious misinterpretation of stratigraphic successions (Muto *et al.*, 2007).

The deterministic nature of autostratigraphic analysis is advantageous for detecting unsteady dynamic external forcing because any significant unsteadiness should come out as a discrepancy from the theoretical prediction. An example of the quantitative application of autostratigraphic analysis has been given from an Eocene stratigraphic succession of regressive shorelines in the Central Tertiary Basin on Spitsbergen (Muto & Steel, 2002a). This regressive wedge developed as a delta-driven, shelf-margin system under a regime of overall (i.e. long-term) rise of relative sea-level, with a corresponding shelf-edge trajectory available for analysis (Fig. 7). The reciprocal relationship between R_{slr} and Q_s (Eq. 1) indicates that an increase in Λ is possible both by increasing Q_s and decreasing R_{slr}; and a decrease in Λ can result from both decreasing Q_s and increasing R_{slr}. Thus, unless a sufficient number of time lines are provided in the stratigraphic section, any change in R_{slr} cannot be calculated without specifying or assuming Q_s beforehand; and vice versa. Fig. 8 shows two different scenarios obtained from autoretreat-based, numerical simulations; one for variable R_{slr} with constant Q_s and another for variable Q_s with constant R_{slr}. On the assumption that Q_s was constant, the analysis of field data obtained from three sites (P: proximal, M: middle, D: distal) within the basin suggests a spectrum of temporal patterns of R_{slr} that can account for the observed shoreline trajectory (Fig. 8). A notion to be remembered here is that even if R_{slr} is constant, the resulting shoreline trajectory will be a concave-upward and concave-landward curve (Fig. 3). The deterministic prediction by the theory of shoreline autoretreat suggests that the initial water depth in the basin was around 0.45 km as a most likely scenario and that the overall relative sea-level rise (c. 0.80 km) happened largely during an early time period and was followed by a longer period of much lower rate of rise. This pattern of relative sea-level rise is consistent with the Palaeogene

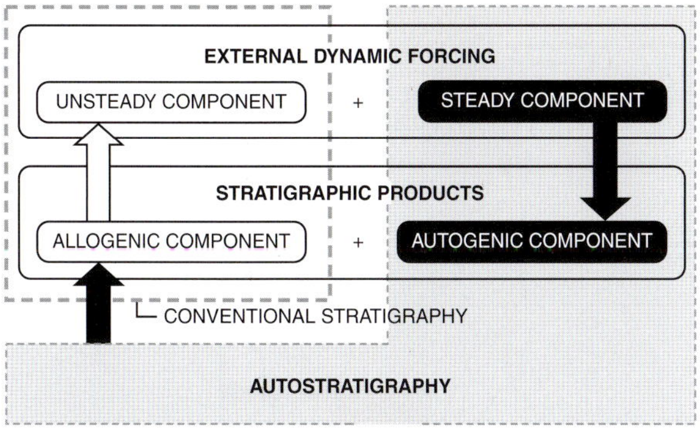

Fig. 6. The strategy of autostratigraphy. Autostratigraphy firstly tries to explain any stratigraphic features only in terms of autogenic processes or steady dynamic external forcing. Allogenic interpretation is not incorporated as long as autogenic interpretation is valid. Without this procedure, the effect of unsteady external dynamic forcing can be overrated. Conventional stratigraphy, usually assuming a sinusoidal pattern for sea-level changes, tends to overestimate the allogenic effect of external dynamic forcing.

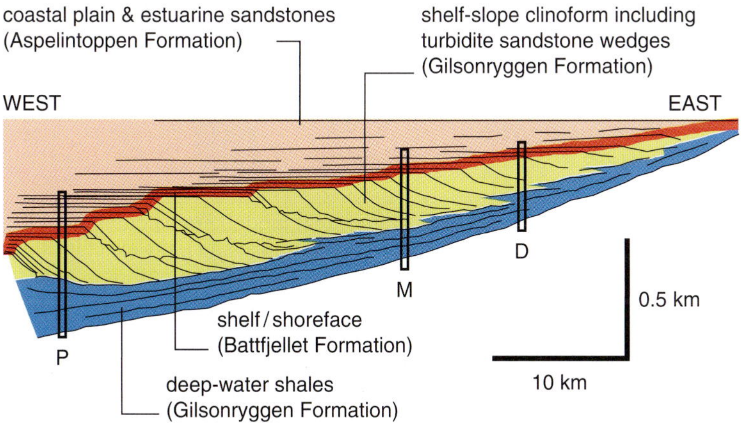

Fig. 7. Shelf-edge trajectory within the upper part (Eocene) of the van Mijenfjorden Group, the Central Tertiary Basin, Spitsbergen. P (proximal), M (middle) and D (distal) are the observation sites in which the parameters needed for numerical simulations were measured. Adopted from Muto & Steel (2002a).

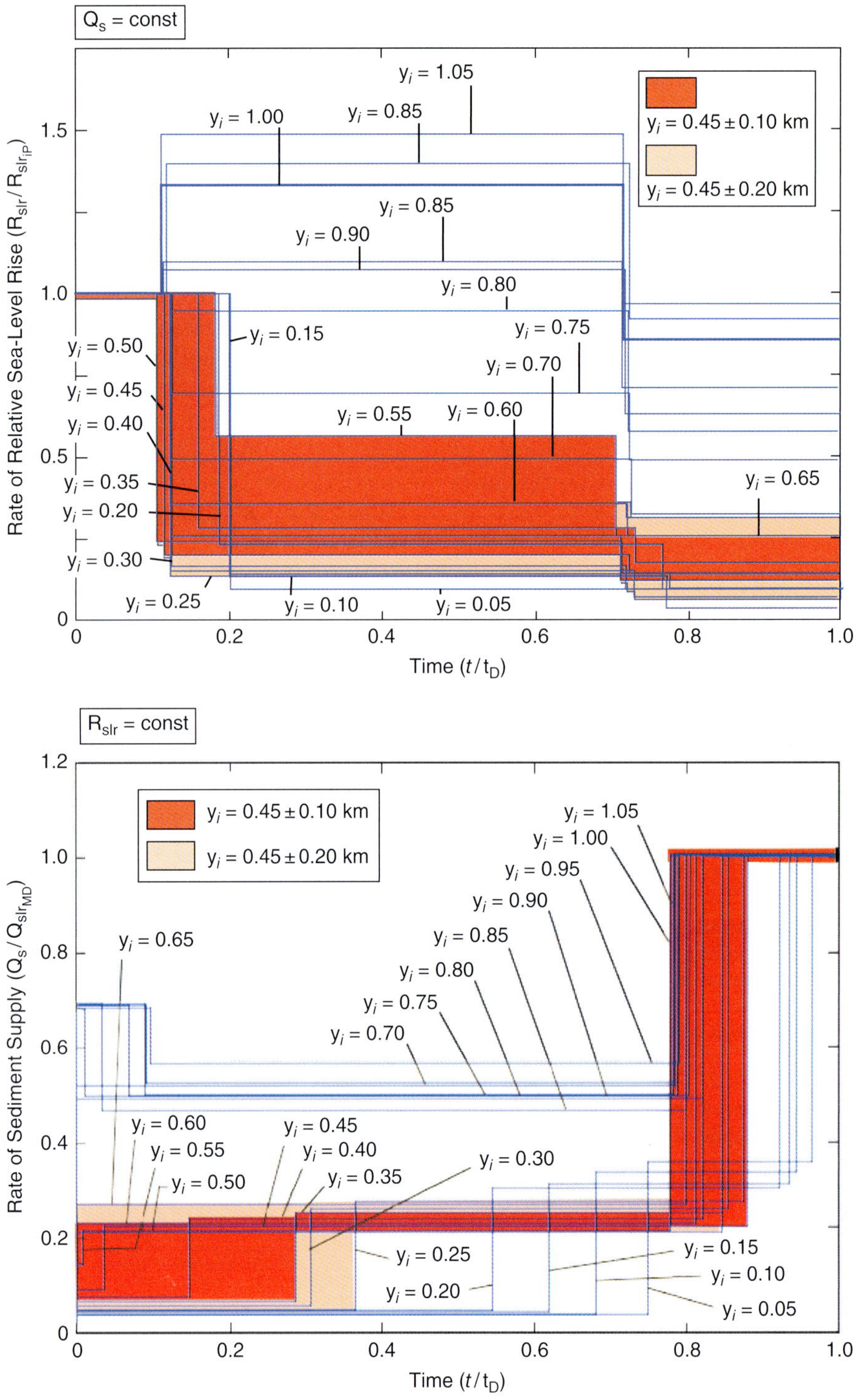

Fig. 8. Time patterns of R_{slr} (top) and Q_s (bottom) reconstructed on the basis of the autoretreat model. For calculation of R_{slr}, Q_s is assumed to have been constant. y_i (km) is an assumed height of the initial shoreline position at site i, the exact location of which is not given a priori. Magnitudes of R_{slr} and time t are here non-dimensionalised by dividing with $R_{slr_{0P}}$ and t_D, respectively, where $R_{slr_{iP}}$ is R_{slr} averaged for time interval during which shelf edge migrated from site i to site P and t_D is time elapse for shelf-edge migration from the initial point i to site D. On the other hand, for calculation of Q_s, R_{slr} is assumed to have been constant. Magnitudes of Q_s are non-dimensionalised by dividing with Q_{sMD}, where Q_{sMD} is Q_s averaged for time interval during which shelf edge migrated from site M to site D. Adopted from Muto & Steel (2002a).

tectonic subsidence trend of the basin which was determined independently through geohistory analysis (Schellpeper, 2002). On the other hand, if relative sea-level had been rising steadily (R_{slr} = const), Q_s would have had to increase very rapidly, as exponential with time (Fig. 8). These possibilities of unsteady changes in R_{slr} and Q_s would not have been recognised quantitatively until the deterministic autogenesis was fully explored in the analysis of the stratigraphic succession.

REMAINING PROBLEMS

Autostratigraphy is essential to the understanding of any depositional system, not simply because it deals with autogenic processes that have been unknown or insufficiently understood but primarily because it explores the inherent responses of depositional systems to external forcing. Autostratigraphy encompasses both equilibrium and non-equilibrium responses and takes full account of both steady and unsteady dynamic external forcing (Muto *et al.*, 2007). Further exploration of non-equilibrium responses in river deltas will further clarify the stratigraphic responses to both dynamic and static external forcing.

The present framework of autostratigraphy treats only river deltas unaffected by waves and tides. There is no reason that non-equilibrium responses are limited in this way. The challenge is thus to extend the autostratigraphic view to other sedimentary systems/environments including wave-dominated systems, tide-dominated systems, inland fluvial basins which have no influence of sea-level forcing, deep water systems and even non-clastic depositional systems. Different environments would have a wide variety of stratigraphic responses to external forcing. The key to further exploration of autostratigraphy will be the recognition of non-equilibrium response in rock successions.

ACKNOWLEDGMENTS

The present work was financially supported in part by JSPS (20340140) and the RioMAR Industry Consortium at University of Texas in Austin.

REFERENCES

Aschoff, J.L. and **Steel, R.J.** (2011) Anomalous clastic progradation in the Cordilleran Foreland Basin: implications for the Sevier-Laramide Transition. *Geol. Soc. Am. Bull.*, **123**, 1822–1835.

Beerbower, J.R. (1965) Cyclothems and cyclic depositional mechanisms in alluvial plain sediments. *Kansas State Geological Survey Bulletin*, **169**, 31–42.

Curray, J.R. (1964) Transgressions and regressions. In: *Papers in Marine Geology* (Ed. R.L. Miller), pp. 175–203. Macmillan, New York.

Curtis, D.M. (1970) Miocene deltaic sedimentation, Louisiana Gulf Coast. In: *Deltaic Sedimentation, Modern and Ancient* (Ed. J.P. Morgan), *SEPM Spec. Publ.*, **15**, 293–308.

Edmonds, D.A., **Hoyal, D.C.J.D.**, **Sheets, B.A.** and **Slingerland, R.L.** (2009) Predicting delta avulsions: Implications for coastal wetland restoration. *Geology*, **37**, 759–762.

Galloway, W.E. (1989) Genetic stratigraphic sequences in basin analysis – part I: architectures and genesis of flooding-surface bounded depositional units. *Amer. Assoc. Petrol. Geol. Bull.*, **73**, 125–142.

Jervey, M.T. (1988) Quantitative geological modeling of siliciclastic rock sequences and their seismic expression: In: *Sea-Level Changes: An Integrated Approach* (Eds C.K. Wilgus, B.S. Hastings, C.G.St.C. Kendall, H.W. Posamentier, C.A. Ross and J.C. Van Wagoner), *SEPM Spec. Publ.*, **42**, 47–69.

Kim, W., **Paola, C.**, **Voller, V.R.** and **Swenson, J.B.** (2006) Experimental measurement of the relative importance of controls on shoreline migration. *J. Sed. Res.*, **76**, 270–283.

Kirschbaum, M.A. and **Hettinger, R.D.** (2004) *Facies Analysis and Sequence Stratigraphic Framework of Upper Campanian Strata (Neslen and Mount Garfield Formations, Bluecastle Tongue of the Castlegate Sandstone and Mancos Shale)*, Eastern Book Cliffs, Colorado and Utah: USGS Digital Data Report DDS-69-G.

Milton, N.J. and **Bertram, G.T.** (1995) Topset play types and their controls: In: Sequence Stratigraphy of Foreland Basin Deposits, Outcrop and Subsurface Examples from the Cretaceous of North America (Eds J.C. Van Wagoner and G.T. Bertram), *Amer. Assoc. Petrol. Geol. Mem.*, **64**, 1–9.

Muto, T. and **Steel, R.J.** (1992) Retreat of the front in a prograding delta. *Geology*, **20**, 967–970.

Muto, T. and **Steel, R.J.** (1997) Principles of regression and transgression: the nature of the interplay between accommodation and sediment supply. *J. Sed. Res.*, **67**, 994–1000.

Muto, T. and **Steel, R.J.** (2000) The accommodation concept in sequence stratigraphy: some dimensional problems and possible redefinition. *Sed. Geol.*, **130**, 1–10.

Muto, T. (2001) Shoreline autoretreat substantiated in flume experiments. *J. Sed. Res.*, **71**, 246–254.

Muto, T. and **Steel, R.J.** (2002a) Role of autoretreat and A/S changes in the understanding of deltaic shoreline trajectory: a semi-quantitative approach. *Basin Res.*, **14**, 303–318.

Muto, T. and **Steel, R.J.** (2002b) In defense of shelf-edge delta development during falling and lowstand of relative sea-level. *J. Geol.*, **110**, 421–436.

Muto, T. and **Swenson, J.B.** (2005) Large-scale fluvial grade as a non-equilibrium state in linked depositional systems: Theory and experiment. *J. Geophys. Res.*, **110** (F), F03002, DOI: 10.1029/2005JF000284.

Muto, T. and **Swenson, J.B.** (2006) Autogenic attainment of large-scale alluvial grade with steady sea-level fall: An analog tank/flume experiment. *Geology*, **34**, 161–164.

Muto, T., **Steel, R.J.** and **Swenson, J.B.** (2007) Autostratigraphy: a framework norm for genetic stratigraphy. *J. Sed. Res.*, **77**, 2–12.

Neal, J. and **Abreu, V.** (2009) Sequence stratigraphy hierarchy and the accommodation succession method. *Geology*, **37**, 779–782.

Paola, C., **Straub, K.**, **Mohrig, D.** and **Reinhardt, L.** (2009) The "unreasonable effectiveness" of stratigraphic and geomorphic experiments. *Earth-Sci. Rev.*, **97**, 1–43.

Parker, G., **Muto, T.**, **Akamatsu, Y.**, **Dietrich, W.** and **Lauer, J.W.** (2008) Unravelling the conundrum of river response to rising sea-level from laboratory to field. Part I. Laboratory experiments. *Sedimentology*, **55**, 1643–1655.

Petter, A.L. and **Muto, T.** (2008) Sustained alluvial aggradation and autogenic detachment from shoreline in response to steady fall of relative sea-level. *J. Sed. Res.*, **78**, 98–111.

Posamentier, H.W., **Jervey, M.T.** and **Vail, P.R.** (1988) Eustatic controls on clastic deposition I – Conceptual framework. In: *Sea-level Changes: An Integrated Approach* (Eds C.K. Wilgus, B.S. Hastings, C.G.St.C. Kendall, H.W. Posamentier, C.A. Ross, and J.C. Van Wagoner), *SEPM Spec. Publ.*, **42**, 109–124.

Schellpeper, M.E. (2000) *Basin Analysis of the Central Tertiary Basin and Sequence Stratigraphy of an Eocene Clinoform in the Battfjellet Formation, Spitsbergen.* MSc Thesis, Department of Geology and Geophysics, University of Wyoming, 63 p.

Shanley, K.W. and **McCabe, P.J.** (1994) Perspectives on the sequence stratigraphy of continental strata. *Amer. Assoc. Petrol. Geol. Bull.*, **78**, 544–568.

Sloss, L.L. (1962) Stratigraphic models in exploration. *J. Sed. Petrol.*, **32**, 415–422.

Straub, K.M., **Paola, C.**, **Mohrig, D.**, **Wolinsky, M.A.** and **George, T.** (2009) Compensational stacking of channelized sedimentary deposits. *J. Sed. Res.*, **79**, 673–688.

Swenson, J.B. and **Muto, T.** (2007) Response of coastal-plain rivers to falling relative sea-level: allogenic controls on the aggradational phase. *Sedimentology*, **54**, 207–221.

Swenson, J.B., **Voller, V.R.**, **Paola, C.**, **Parker, G.** and **Marr, J.G.** (2000) Fluvio-deltaic sedimentation: A generalized Stefan problem. *Eur. J. Appl. Math.*, **11**, 433–452.

Tomer, A. and **Muto, T.** (2010) Emergence and drowning of fluviodeltaic systems during steady rise of sea-level: implication from geometrical modeling and tank experiments. *J. Sed. Soc. Japan*, **69**, 63–72.

Tomer, A., **Muto, T.** and **Kim, W.** (2011) Autogenic hiatus in fluviodeltaic successions: geometrical modeling and physical experiments. *J. Sed. Res.*, **81**, 207–217.

Vail, P.R., **Mitchum, R.M., Jr.** and **Thompson, S.** III (1977) Seismic Stratigraphy and global changes of sea-level, Part 3: relative changes of sea-level from coastal onlap. In: *Seismic Stratigraphy: Applications to Hydrocarbon Exploration* (Ed. C.E. Payton), *Amer. Assoc. Petrol. Geol. Mem.*, **26**, 63–81.

Weller, J.M. (1960) *Stratigraphic Principles and Practice.* Harper & Row, New York, 725 p.

Int. Assoc. Sedimentol. Spec. Publ. (2014) **46**, 149–166

Autogenic process change in modern deltas: lessons for the ancient

CORNEL OLARIU

The University of Texas at Austin, Department of Geological Sciences, 1 University Station, C1100, Austin, TX 78712 (E-mail: cornelo@mail.utexas.edu); National Institute for Marine Geology and Geoecology – GeoEcoMar, 23–25 Dimitrie Onciul Street, Bucharest 024053, Romania

ABSTRACT

River deltas are usually interpreted in terms of fluvial-dominated, wave-dominated or tidal-dominated process regimes and are thought to undergo change which is mainly by unsteady external forcing conditions. However, modern deltas have shown significant autogenic process changes over relatively short temporal and spatial scales (hundreds to thousands of years; km to 10s of km) during their Holocene progradation. These deltaic stratigraphic responses were autogenic because late Holocene external forcing was relatively steady (constant rates of relative sea-level rise and sediment supply). Responses over longer time intervals would probably be allogenic due to the increased probability of unsteady external forcing over longer time scales. Nevertheless, the short temporal and spatial scales by themselves are not a condition for the autogenic response of deltaic environments, but the steady external forcing is. The dominant depositional process can change down depositional dip as the delta progrades across the shelf and into slightly deeper water, or can change laterally because of shifting fluvial discharge or because of oceanographic differences in front of distributaries. The three most common types of process change seen in Holocene deltas are from fluvial-dominance to wave-dominance (Mississippi and Danube examples) from tidal-dominance to wave-dominance (Mekong example) or from fluvial-dominance to tidal-dominance (Mahakam example). Particular segments of large delta complexes can also show these changes. The pervasive and rapid process changes seen in Holocene deltas suggest that such changes were also common in ancient deltas, expressed by changes in the character of the deltaic succession, especially on the delta front; which is the regime-defining segment of the delta. Such changes should be considered more as a rule than an exception. Campanian and Maastrichtian delta deposits from the Western Interior Seaway and Laramide Washakie Basin show clear evidence of such changing process regimes. A reasonably good time framework in the rock record is needed to aid discussion of whether observed stratigraphic responses are autogenic or allogenic and it is important to make this distinction because of local or regional implications for stratigraphy. A common theme seen in shelf-delta stratigraphy is spatially extensive flooding surfaces bounding a deltaic complex of 1 ka to 300 ka duration but with great lateral, between-lobe variability within such units. The main flooding-surface bounded interval, the allogenic response, is the overall cross-shelf regression and transgression of the delta complex, whereas the great internal variability reflects autogenic spreading of the lobes during the cross-shelf transits.

Keywords: Autogenic, river deltas, stratigraphy, process change, fluvial-dominated delta, tidal-dominated delta, wave-dominated delta.

INTRODUCTION

Ancient delta deposits are usually interpreted as fluvial-dominated (Bhattacharya & Walker, 1991; Barton, 1994; Mellere *et al.*, 2002; van den Berg & Garrison 2004; Petter & Steel, 2006; Uroza & Steel 2008; Olariu *et al.*, 2010), wave-dominated (Plink-Bjorklund, 2008; Charvin *et al.*, 2010) or tide-dominated deltas (Mellere & Steel 2000; Willis *et al.*, 1999; Bhattacharya & Willis, 2001; Willis & Gabel, 2001;

From Depositional Systems to Sedimentary Successions on the Norwegian Continental Margin, First Edition. Edited by A. W. Martinius, R. Ravnås, J. A. Howell, R. J. Steel, and J. P. Wonham.

Pontén & Plink-Björklund, 2009). Ancient delta deposits are only rarely interpreted in terms of mixed energy (Gani & Bhattacharya, 2007; Carvajal & Steel, 2009) and the criteria used to recognise transitions between process regimes are not always clear. Using observations and published studies on modern and Holocene deltas, models of autogenic process change from fluvial to wave and to tide process dominance are suggested in this study. Although the models are based on modern deltas they should also be generally valid for ancient delta deposits. Two main points that are emphasised here are (1) that process change in deltas are commonly autogenic responses (changes in stratigraphy or morphology that happen despite constant forcing of the external variables, see Muto *et al.* 2007) and (2) that process change most often happens (though not always) at short time scales (tens to thousands of years) and over short distances (kilometres to 10s of km).

Delta process changes

The tripartite classification of deltas (Galloway, 1975) based on morphology (Fig. 1A) and dominant processes (fluvial, wave, tidal) is widely used because it is simple to understand and easy to apply. The same classification is used here to separate delta types or segments of deltas. Although not explicit within the fluvial-wave-tide process classification, there are other controlling factors inherent to the classification. For example, river discharge, yearly hydrograph variability and grain size distribution will contribute to the fluvial regime (e.g. Orton & Reading, 1993) whereas coastal morphology, shelf bathymetry, basin area and shape have some control on wave and tidal regimes (Orton & Reading, 1993; Ainsworth *et al.*, 2008). In his delta process classification article, Galloway (1975) suggested the tendency for the regressive part of a sequence to be fluvial-dominated whereas the back-stepping or transgressive parts tend to be wave-dominated.

The implications of sea-level change on delta type was also discussed by Porebski & Steel (2006) and by Yoshida *et al.* (2007) who argued that tidal processes tend to be more common in highstand shorelines and inner-shelf deltas, whereas lowstand deltas, especially those at the shelf edge, tend to be wave-dominated. Although some of these indirect generalities are likely to hold, the most direct controls on process dominance are nevertheless the more local ones such as shoreline morphology, position of the main sediment routing and accommodation (Ainsworth *et al.*, 2008). The process changes in Holocene deltas have been described in terms of transitions from one type to another (McManus, 2002) and were presented as a 'dynamic' application of Galloway's classification but no link with the autogenic responses was made. Holocene process changes of the deltas described below are also shown in Fig. 1B but here we emphasise that these are autogenic process changes.

The fluvial, wave or tide processes dominate the delta shoreline at a given time or at a given location but the process dominance can change in time, as the delta evolves, or in space along the same shoreline. The process changes in deltas have been averaged in the delta classification (Galloway, 1975) and this is useful when the entire delta complex is described but does not address the architectural complexities at lobe level and smaller scale. Realisation of the multiple possibilities of the process dominance, or influence, led to a classification scheme with 22 delta types (Aisworth *et al.*, 2011). This detailed classification divides the ternary classification into multiple fields within which deltas are ascribed by their dominant process and two levels of subordinate process. For example, the Mitchell River Delta in Australia is described as a tide-dominated, fluvial-influenced, wave-affected delta. Despite improved understanding that comes with such a classification, it still does not address the dynamics of the delta system in space and time.

Autogenic responses

An understanding of autogenic response accompanies the realisation that not all changes seen in rivers and deltas are equilibrium responses (i.e. to changes in A/S ratio). Accommodation (A) and supply (S) ratio was defined as a key descriptor for the behaviour of the deltaic shoreline but the two terms are linked in a non-linear manner and generate complex shoreline behaviour (Muto & Steel, 1997; Muto *et al.*, 2007). Some changes happen without changes in the external variables A and S and these are non-equilibrium or autogenic responses (Muto & Steel, 2014). The reference to external variables A and S is for an entire river-delta system, for example the sediment supply (S) delivered by a river is constant but its delivery might vary laterally along the delta

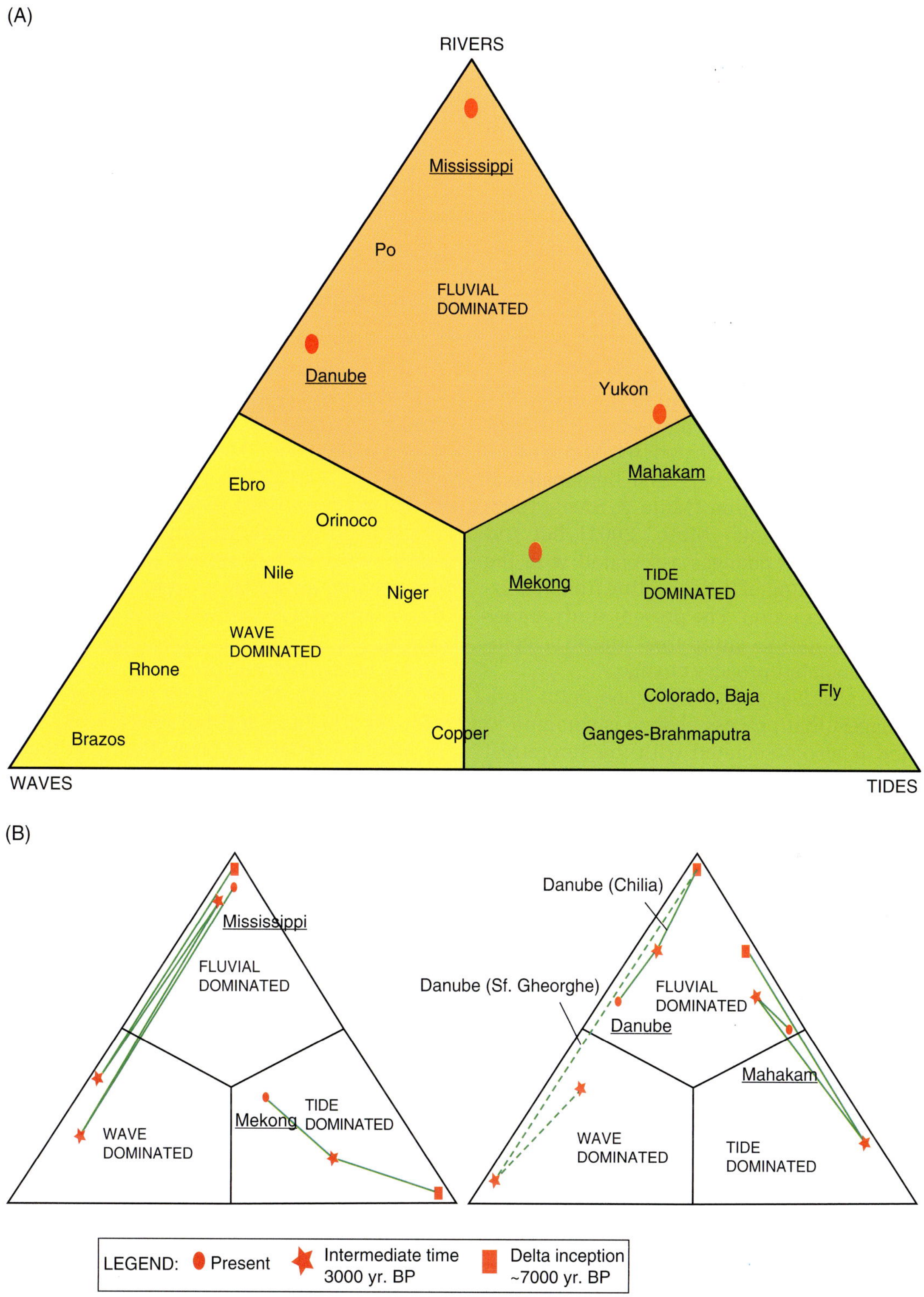

Fig. 1. Delta morphology classification. (A) Delta ternary diagram based on dominant processes (Galloway, 1975). Red dots indicate modern deltas that are analysed further below. (B) Variation of the delta processes during the last 7 ka for selected Holocene deltas. Note that process changes happen under steady external forcing conditions.

shoreline due to channel avulsions. The resulting variability of deposits under such conditions is considered as autogenic responses because there is no unsteadiness in the external forcing (to the given river-delta system). Furthermore, these autogenic responses in system morphology or in stratigraphy can be stochastic (e.g. channel avulsions or lobe-shifting and their stratigraphic responses) or deterministic. The latter response, exemplified by delta autoretreat (Muto & Steel, 1992) is of particular importance because it can be extracted from stratigraphic data (method shown by Muto & Steel, 2002), thereby allowing the residual stratigraphic response to be interpreted as 'allogenic', a result of change in the rate of the external forcing variables. This is the essence of 'autostratigraphy', as advocated by Muto *et al.* (2007). Delta autoretreat was first demonstrated from numerical modelling (Muto & Steel, 1992) and flume experiments (Muto, 2001) but did not include process changes associated with the autogenic turnaround of the shoreline from regression to net transgression. The present study argues that during shoreline autoretreat there is commonly an associated process change.

Based on observations of Holocene delta studies, we suggest that process change is common in deltas and is also associated with the autoretreat process. The data collected suggest that deltas commonly become wave-dominated during the turnaround from regression to transgression, during shoreline autoretreat.

Why are Holocene deltaic changes attributed to autogenic responses?

External forcing during the last 7.5 ka of Holocene time is considered conducive to autogenic responses because sea-level rose at a relatively constant rate of about 2.5 mm per year (Fairbanks, 1989, fig. 2) and also climate (a proxy for sediment supply) did not commonly undergo dramatic changes. There are variable rates of relative sea-level rise because deltas occur in tectonic settings with different subsidence rates. Sediment supply also probably varied at decadal to centennial scales linked to global climate patterns because the deltas are in different climatic zones (Syvitski *et al.*, 2003). Goodbred & Keuhle (2000) demonstrated a high supply peak in the Ganges-Brahmaputra delta because of monsoon intensification during the early Holocene (11 ka to 7 ka BP) before the inception of most Holocene deltas (Stanley & Warrne, 1994). However, despite some local Holocene variability, sea-level rise (Fig. 2) and sediment supply can be considered constant over this period for a given delta system. The Holocene can therefore be considered a period of 'steady forcing' with deltas showing primarily autogenic responses.

Autogenic progradation of modern deltas

Holocene delta formation

Most modern deltas started to prograde about 7.5 ka ago (Stanley & Warne, 1994) when eustatic sea-level rise rate decreased from 11 mm/yr to about 2.5 mm/yr (Fairbanks, 1989, fig. 2). Many deltas (Danube, Mekong, Godavari, Rhone, Mississippi) started to prograde when they were far inland from their present shoreline positions, at the head of embayments. They built multiple lobes as the deltaic system prograded across the pre-existing shelf. These deltas were fed by relatively large rivers and were able to prograde even during slight sea-level rise. The examples presented below focus on process changes as reflected in delta morphology and on the controls on these changes. The temporal and spatial scale of changes in modern deltas has been described herein and the observations compared with outcrop examples of delta process changes.

Fluvial-dominated to wave-dominated process change: modern examples

For fluvial-dominated to wave-dominated process change, the examples of the Mississippi and Danube deltas are instructive. The Mississippi Delta has had an extensively studied evolution (Fisk *et al.*, 1954; Frazier, 1967; Coleman, 1988; Tornqvist *et al.*, 1996; Roberts, 1997, 1998). The Mississippi Delta started to form, like most large deltas, around 7.5 ka when the shoreline was about 300 km landward of the present shoreline. The delta is composed of multiple lobes (Frazier, 1967) and is considered the type example for fluvial-dominated cases, with multiple coeval distributary channels (Galloway, 1975; Coleman & Wright, 1975). Despite the fluvial dominance of the last lobe (Plaquemines-Modern), it is probable that the same lobe after abandonment will turn into a wave-dominated delta and then into a shoreface and a shoal (Penland *et al.*, 1985). The

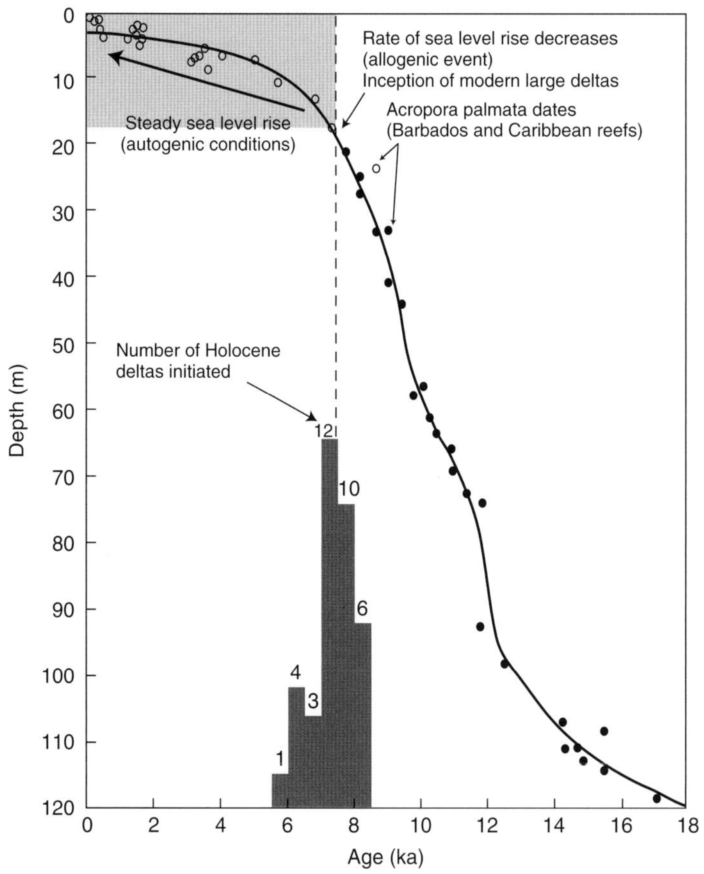

Fig. 2. Eustatic sea-level variation during last 17 ka (Fairbanks, 1989) and the timing of Holocene delta formation (Stanley & Warne, 1994). Note that most modern deltas started to form around 7.5 ka to 8 ka and sea-level has risen steadily since then.

change of one lobe from fluvial-dominated to wave-dominated is caused by the decrease in discharge to the active delta during avulsion (Fig. 3) followed by compactional subsidence. Older Mississippi delta lobes (Sale-Cypremort, Teche, St. Bernard, Lafourche) that were fluvial-dominated at the time of progradation also turned into wave-dominated deltas and eventually to barrier islands. When the present Balize delta lobe is abandoned and becomes wave-dominated, the most recent delta lobe will be the fluvial-dominated Atchafalaya Delta. Thus, each past lobe was initially a fluvial-dominated delta lobe of the Mississippi delta complex that was abandoned and turned into a wave-dominated delta lobe and later into a wave-dominated estuary, as autogenic responses. The transition from fluvial-dominated to complete abandonment and transgression is a gradual process that is likely to take hundreds of years (Roberts, 1997, his fig. 2). The abandonment time (gradual decrease in fluvial discharge to one lobe) is the interval when the delta will become wave-dominated before being transgressed. Fig. 3 shows that the oldest lobe (Sale-Cypremort) was abandoned about 4 ka BP and was then transgressed for a greater distance than the younger lobes. Furthermore, it has a larger area of sandy transgressive deposits than the more recently abandoned St. Bernard lobe (abandoned about 1.5 ka) and Lafourche lobe (newly abandoned).

The Danube Delta also initiated about 7 ka ago (Giosan *et al.*, 2006) at the head of an embayment in the north-western Black Sea. It formed multiple lobes during overall progradation (Panin, 1989) but at present has three main distributaries supplying lobes (Fig. 4) of varying character because of differing fluvial discharge (Panin & Jipa, 2002). The modern northern lobe (Chilia) is fluvial-dominated and a historical map from 1856 (Fig. 4) shows multiple distributary channels, each bifurcated by mouth bars as is typical for fluvial-dominated delta morphology (Olariu & Bhattacharya, 2006). On satellite images acquired in 2003 (Fig. 4) the number of distributary channels was reduced

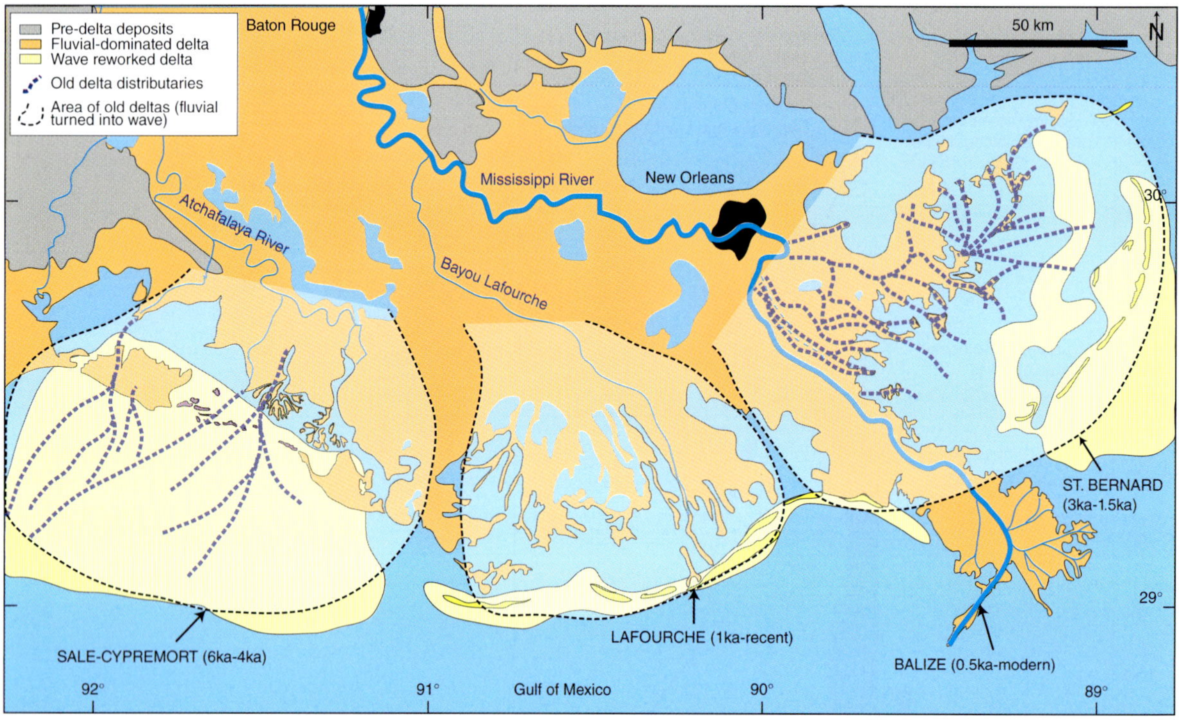

Fig. 3. Fluvial-dominated to wave-dominated autogenic process change, Mississippi Delta example (data from Coleman & Gagliano, 1964; Frazier, 1967; Coleman *et al.*, 1998). Note that the oldest abandoned lobe (Sale-Cypremort) underwent the most significant transgression and contains the most widespread wave reworked sediments when compared with younger abandoned lobes.

and shore-parallel sandy spits were forming between the distributaries, suggesting the transition to a wave-influenced morphology.

The southern Danube lobe (Sf. Gheorghe), which is coeval with the northern lobe (Fig. 4) has always been wave-dominated (Panin, 1989; Giosan *et al.*, 2006). The southern lobe is wave-dominated because it was shown that the southern distributary has less than half the water discharge (and presumably sediment flux) compared to the northern distributary (Panin & Jipa, 2002). The wave regime is also different, as the dominant wave front approaches the shoreline of the southern lobe from the north-east at a high angle with a resulting strong effect on sediment transport (Giosan *et al.*, 1999; Dan *et al.*, 2009).

Tide-dominated to wave-dominated process change: modern example

The Mekong Delta in the South China Sea (Fig. 5) has prograded more than 250 km during the last 6 ka (Ta *et al.*, 2002, 2005). The main distributary channels have maintained the same position and their locations are controlled by deeper faults (Nguyen *et al.*, 2000). The modern delta is mesotidal with a maximum tidal range of 3 m to 4 m (Ta *et al.*, 2002). During the interval 6 Ka to 3 Ka, the delta had a tide-dominated morphology characterised by radial distributary channels without shore-parallel sandy ridges (Ta *et al.*, 2005). From 3 ka the development of sandy ridges that are spaced closer and closer together (Fig. 5) indicate the increased influence of the waves on the shoreline. Presently the delta is transitional between tide-dominated and wave-dominated (Tamura *et al.*, 2012). The transition between the entirely tidal and the tidal-wave part of the delta is delineated by its plan-view shoreline trajectory (Helland-Hansen & Hampson, 2009, fig. 12). The location of the boundary between their tide-dominated and wave-dominated segments suggests that the autogenic process change was caused by progradation of the delta from a protected setting into a higher energy, wave regime (Fig. 5). Another tidal to wave process change has been described during the last 5.4 ka evolution of Baram Delta in South China Sea (Caline & Huong, 1992; Sandal, 1996).

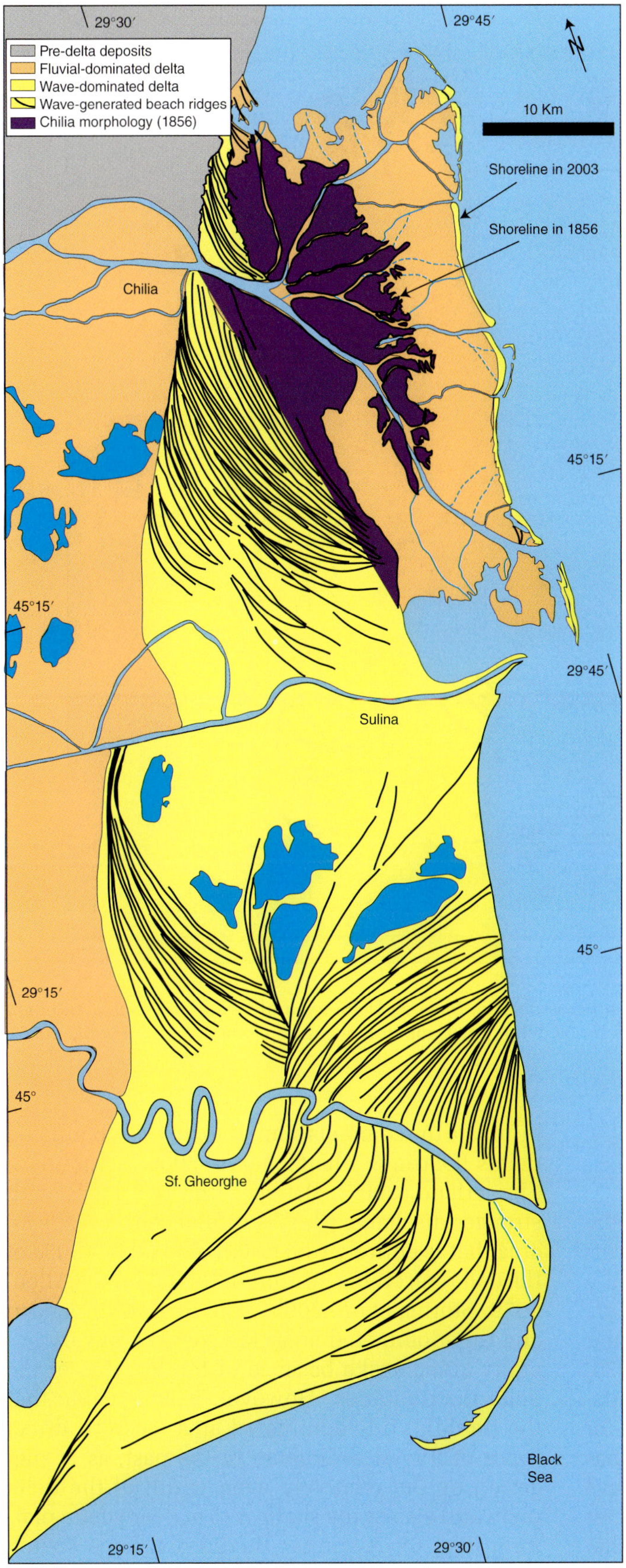

Fig. 4. Fluvial-dominated to wave-dominated autogenic process change, Danube Delta example. The 1856 morphology of northern lobe (Chilia Delta) is drawn after Rosetti and Rey (1931). The actual morphology of the Danube was drawn after an ASTER image collected on June 2003.

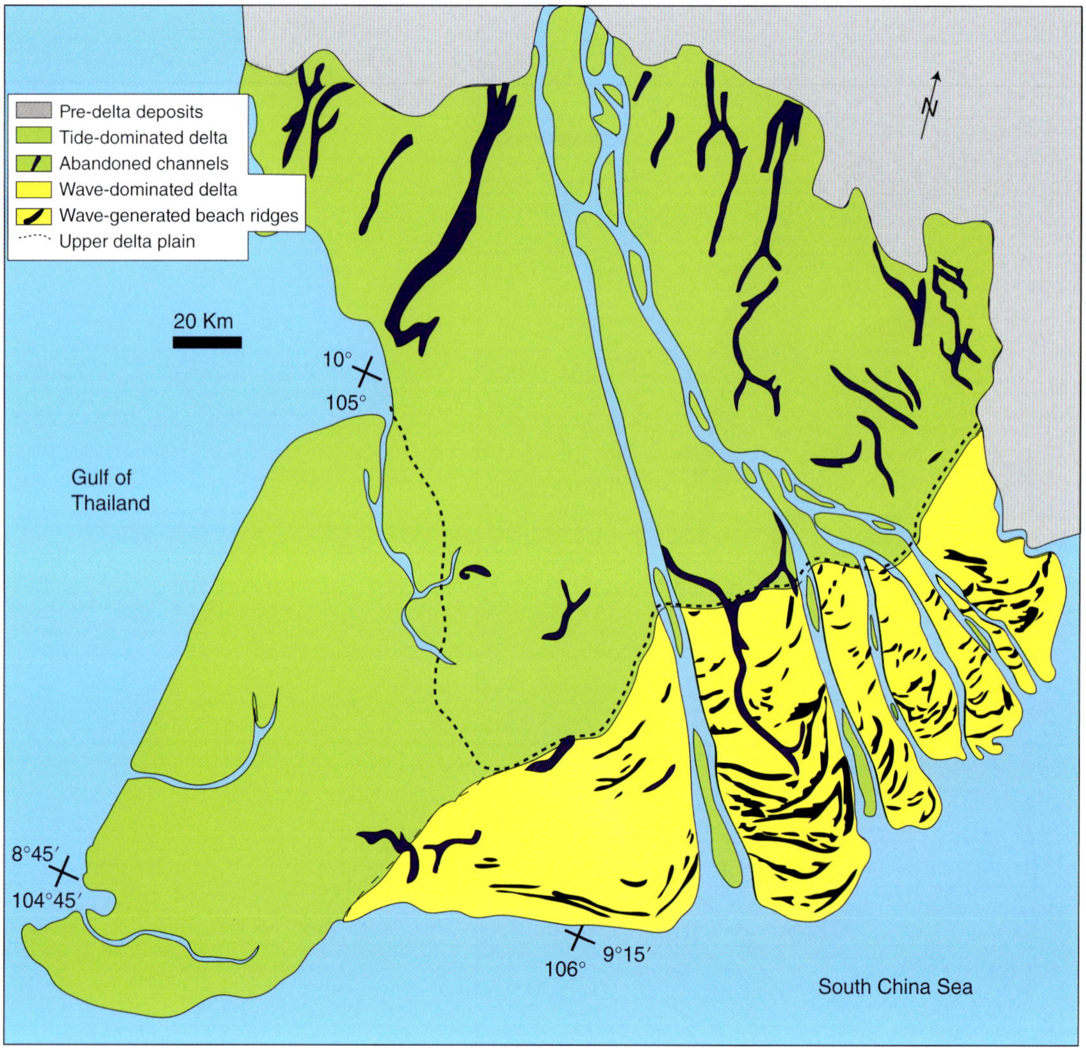

Fig. 5. Tide-dominated to wave-dominated autogenic process change, Mekong Delta example. Note the sharp boundary between the tide-influenced and wave-influenced parts of the delta. The transition occurred at about 3 ka (Ta *et al.*, 2005).

Fluvial-dominated to tide-dominated process change: modern example

The Mahakam Delta prograded 60 km into the Makassar Strait (Fig. 6) during the last 5 ka (Storms *et al.*, 2005). The tidal range is mesotidal with a maximum spring tidal range of 3 m (Storms *et al.*, 2005). The Mahakam Delta has a semi-circular distribution of delta plain facies (Allen *et al.*, 1979) suggesting a relatively smooth shoreline progradation without distinct delta lobes, as is typical for tide-dominated deltas. Nowadays the delta has northern and southern distributaries that have a fluvial-influenced to tide-dominated character (Fig. 6) with sediments forming mouth bars that bifurcate the channels similar to fluvial-dominated deltas (Olariu & Bhattacharya, 2006; Edmonds & Slingerland, 2007). Between the two areas with stronger discharge (fluvial influenced) there is an area that is entirely tide dominated (about 30% of the delta) receiving minimum fluvial input (Allen, 1979; Storms *et al.*, 2005). The fluvial-influenced and tide-dominated areas are coeval and vary laterally along strike because of the location of the delta distributaries. However, it is possible that the (middle) tide-dominated area received direct input of fluvial sediments in the past, as is suggested by the relatively even width of the delta plain and also of the shallow (5 m deep) delta platform (Fig. 6). A similar fluvial discharge switch

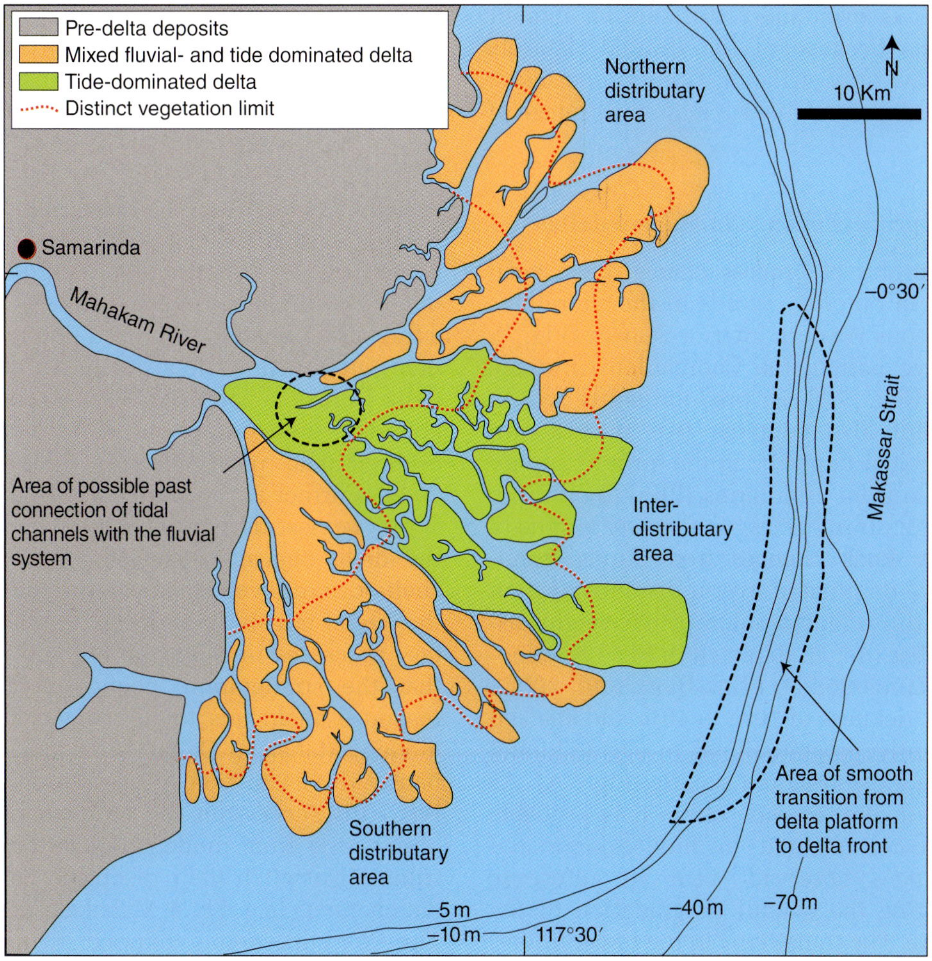

Fig. 6. Tide-dominated to fluvial-influenced autogenic process change, Mahakam Delta example. The northern and southern distributaries are fluvial-dominated whereas the central (inter-distributary) area is tide-dominated (Storms *et al.*, 2005). Note that the platform in front of the delta is relatively even, suggesting that the tide-dominated central area also received fluvial sediment in the past.

between long-lived distributaries in a tide-dominated delta was observed in the Fly Delta (Slingerland *et al.*, 2008).

To summarise observations on the studied Holocene deltas, the mapped process changes are interpreted as autogenic responses that occur down depositional dip or laterally, along depositional strike. The spatial and temporal changes are intimately linked because the fluvial sediment supply and the wave and the tide energy change at a given location as the shoreline progrades. In the examples presented, both spatial and temporal autogenic changes occur but in some deltas only one of the spatial or temporal process changes dominates. The spatial autogenic process changes dominate coeval Danube Delta lobes while the spatial (lobe avulsion) and temporal (older lobes gradually lost the fluvial input) process changes are present in the case of the Mississippi Delta. During the Mekong Delta progradation, the temporal process changes explain delta variation from tide-dominated to wave-dominated. In the Mahakam Delta process changes between distinct tide-dominated and fluvial-influenced lobes are explained by spatial process changes generated by the location of the fluvial distributaries.

It is not clear how much of the variability in process regime is explained by spatial changes (i.e. due to lateral switching of distributaries or the lateral change of the wave regime because of irregularities in the shoreline) and how much by temporal changes (i.e. the process regime of the overall delta system changes with time). The understanding of the process changes, e.g. spatial vs. temporal, is even more difficult in ancient delta deposits because of the limited datasets (outcrops show only strike or dip-oriented faces, lack detailed dating and lack a map view image of the delta).

Examples of process change in ancient deltas

Ancient examples of process change have been documented in outcrop records of the shorelines of the Cretaceous Western Interior Seaway (Fig. 7). The regressive part of the Campanian Chimney Rock delta (Fig. 7A) is predominantly wave-dominated but also contains two intervals of fluvial-dominated deposits, in the form of gravity flows on delta-front clinoforms (Plink-Bjorklund, 2008). The delta-front facies of the lower tongue of the Chimney Rock changes from fluvial-dominated to wave-dominated over a distance of 2 km to 3 km. The dip-oriented outcrops have eight distinct delta lobes (not all depicted in Fig. 7A, clinoform sets documented by Plink-Bjorklund, 2008), each with a thickness of 10 m to 40 m and stacked in a dominantly progradational to aggradational pattern along a dip-oriented section (Fig. 7A). As the lobes are lateral to each other, it is probable that the wave-dominated to fluvial-dominated process changes observed were an autogenic response during the overall progradation of the delta complex. The transitions in the lower (older) lobes from fluvial-dominated to wave-dominated character are probably analogous to the northern distributary of the Danube Delta (Chilia lobe) where process change from fluvial-dominated to wave-dominated occurred because of progradation into a more open, stronger wave regime. However, the Chimney Rock lobes were mainly wave-dominated, with only locally occurring fluvial dominance, as illustrated by the presence of fluvial mouth bar and gravity flow facies. Another ancient delta shoreline which shows an interplay between fluvial-dominated and wave-dominated shoreline that is probably similar to Danube (Sf. Gheorghe lobe) or Mississippi deltas is the Cretaceous Aberdeen Member of the Blackhawk Formation in Utah (Charvin *et al.*, 2010). As the wave-dominated to fluvial-dominated changes within the Aberdeen-1 Member were interpreted to be in the same parasequence (Charvin *et al.*, 2010) it is possible that the deposits record autogenic responses of the sedimentary system. Shoreline morphological changes without process changes (wave-dominated) have been interpreted in Cretaceous Blackhawk Fm. shallow marine deposits of Cretaceous Interior Seaway (Howell & Flint, 2003). The Maastrichtian Fox Hills Formation in the Washakie Basin, S. Wyoming (Fig. 7B) also shows significant process changes. The outcrops of the Fox Hills delta in Fig. 7B correspond to clinothem 10 of the fill of the Maastrichtian Washakie Basin (Carvajal & Steel, 2009; Olariu *et al.*, 2012). Clinothem 10 topset contains Fox Hills Delta deposits that change over a distance of < 1 km from a fluvial-dominated delta front dominated by sediment gravity flow deposits to a tide dominated delta front characterised by stacked cross-strata, abundant mud drapes and fluid mud layers and bi-directional palaeocurrents (Carvajal & Steel, 2009). Further to the south the same outcrop belt shows large-scale syn-depositional deformation of the delta deposits suggesting proximity to the shelf-edge rollover. However, the short distance of the lateral transition between dominant sedimentary processes in the same outcrop belt suggests that the stratigraphic change is an autogenic response analogous to that noted above for the modern Mahakam Delta. The change from dominantly fluvial to tidal processes observed in the Fox Hills succession can be explained using the observation from modern deltas that fluvial distributaries switch their discharge between different channels (e.g. Fly & Mahakam deltas) (Fig. 6). The Fox Hills deltas mapped in the subsurface, using several hundred wells, show the change of the sand geometries from dip elongated to strike elongated sand bodies as deltas migrated from the inner to outer shelf (Olariu *et al.*, 2012).

In the Upper Cretaceous Wall Creek Member of the Frontier Formation (Fig. 7C) delta-front deposits have been documented that contain tidal-dominated and fluvial-dominated deposits with changes to minor wave-dominated deposits over a distance less than 500 m (Gani & Bhattacharya, 2007). As in the previous examples, the transition between different facies regimes on the delta front occurs over a short distance, suggesting that this was also an autogenic response of the delta. The Wall Creek Delta had wave influence on the proximal part of its delta front and this turned to fluvial influence and eventually into intervals with tidal domination toward south-east. The combination of proximal wave and tidal process alternation resembles that noted above for the modern Mekong delta (Fig. 5) but the presence of fluvial and tidal processes is similar to that in the Mahakam delta (Fig. 6). However, the Wall Creek outcrops are of

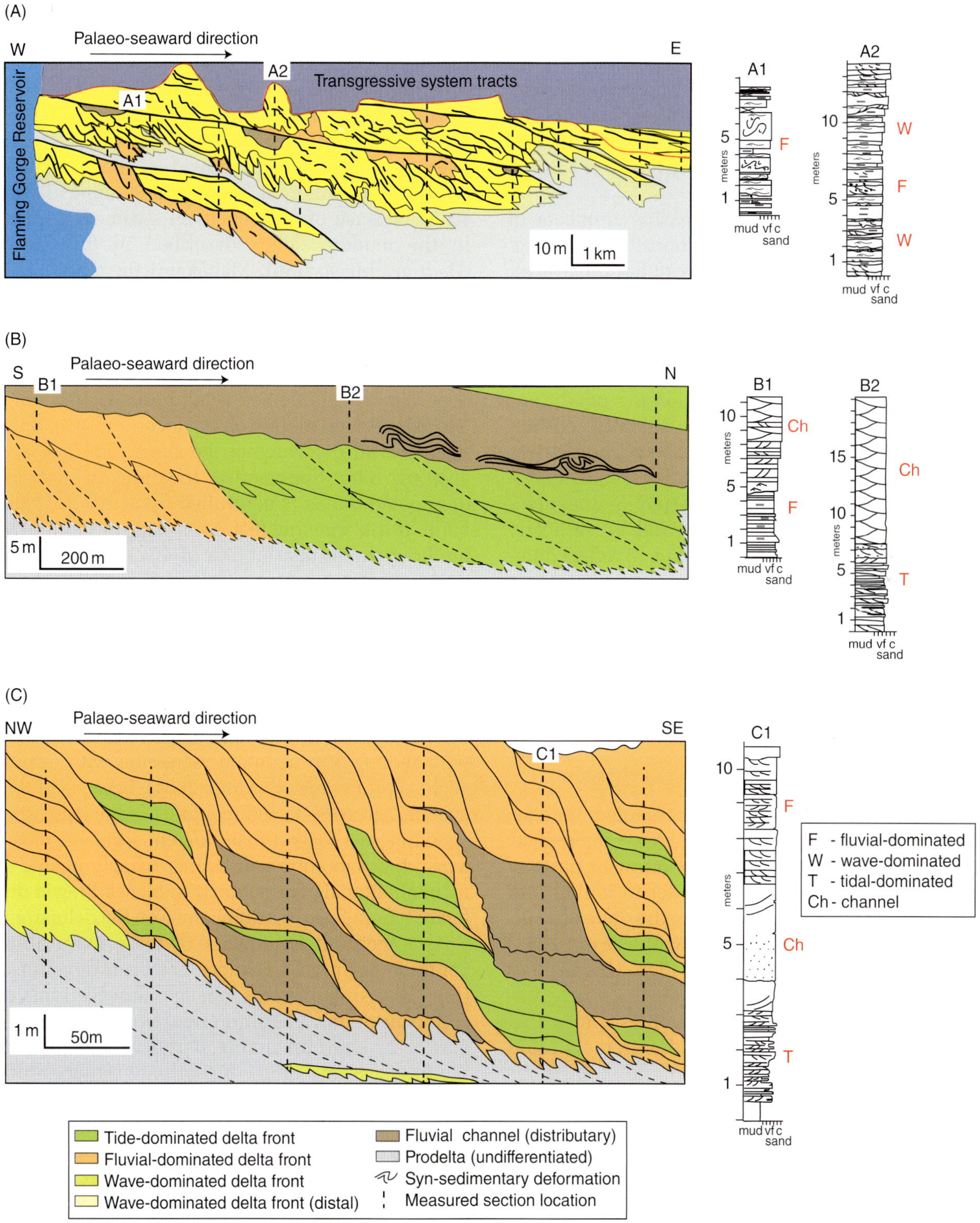

Fig. 7. Ancient outcrop examples of autogenic delta process change. (A) Chimney Rock Member (Plink-Bjorlund, 2008). (B) Fox Hills sandstone (Carvajal & Steel, 2009). (C) Wall Creek Member (Gani & Bhattacharya, 2007). Note that changes between delta front deposits with different dominant processes occur over short distances (hundreds of metres).

limited extent and it is difficult to interpret facies variation in comparison with modern deltas that extend tens of kilometres. Turronian Ferron Delta in Utah is another delta where some parasequences have been interpreted to change the depositional processes, between fluvial and wave, within the same parasequence (Dewey & Morris, 2004; Li *et al.*, 2011).

In summary, there are increasing numbers of descriptions of ancient delta deposits that reflect process changes along the same outcrop belt over distances from a few hundred metres (Wall Creek Member) to a few kilometres (Chimney Rock Tongue) (Fig. 7). The delta process changes can represent either allogenic or autogenic stratigraphic responses to external forcing conditions but where the delta deposits are laterally continuous along the same outcrop belt and occur over short distances within the same parasequence, an autogenic interpretation is preferred. Sedimentary facies that change laterally as well as vertically within the same parasequence (Fig. 7) reflect dominant process changes within a single allogenic unit. It is difficult to distinguish whether any given process change is spatial or temporal but based on comparison with modern systems both may occur.

DISCUSSION

Models for autogenic delta process changes

The Holocene delta examples presented are classified as 'mixed-influence' (Fig. 1A) which reflects the mixture of processes that dominate parts of the same delta. However, in order to understand delta evolution and architecture, the delta classifications (Galloway, 1975; Ainsworth *et al.*, 2011) should also consider temporal (Figs. 1B and C) as well as spatial changes.

Various Holocene delta examples show that process changes have happened extremely rapid (hundreds to thousands years), the transitions occur over relatively short distances (100s of metres to kilometres) and are laterally extensive at the scale of the delta system or of its smaller components (Figs. 3 to 6). These changes are autogenic because the allocyclic controlling factors are in a steady state (i.e. developed during uniform rates of sea-level change and constant sediment supply rate). For example, tidal conditions in the case of inner Mekong Delta and fluvial conditions in the case of the inner Danube Delta persisted during progradation of the delta for 200 km and 30 km respectively, with no significant change in process. For the Danube and Mekong deltas, waves have a stronger influence as the delta prograded into more open and deeper water (Dan et al., 2009; Tamura et al., 2012). The transition from one type of delta to another happened relatively fast and over a short distance. However, the transition occurred at different times along depositional strike because of differences in sediment supply. In the modern deltas described in this study (Danube, Mississippi, Mahakam) the differences between coeval lobes happens because the strength of fluvial discharge, and/or wave and tidal regime, varies along depositional strike. These lateral, along depositional strike, regime changes occur over distance of kilometres to hundreds of kilometres and depend mainly on the locus and size of the main fluvial discharge at the shoreline (Figs 3, 4 and 6). The depocentre location in fluvial-dominated deltas is controlled by avulsion of the main 'trunk' channel (e.g. Mississippi example) or by the unequal distribution of discharge between the main distributaries (e.g. Danube example).

The autoretreat process in deltas is an inbuilt autogenic response, an inevitable shoreline turnaround from regression to net transgression during steady external forcing (Muto & Steel, 1992). The shoreline turnaround is explained by the ever-extending area on the delta that needs to be filled with sediment to maintain regression with even a slight rise of sea-level (Muto, 2001). Because of this eventual decrease in sediment flux to the regressing shoreline, we suggest that different segments of the expanding shoreline will be increasingly unable to maintain fluvial domination (Fig. 8), resulting in different types of process dominance on the delta front.

A conceptual framework of three main autogenic delta process changes has been developed (Fig. 8) using process change observations of Holocene deltas. The autoretreat and transgressive part of the shoreline in Fig. 8 used Mississippi for the first case and then conceptional estuary models for the two other cases. The first is a change from an initial fluvial-dominated delta to one that becomes wave-dominated and then to a wave-dominated estuary (such as the Atchafalaya Bay or west of modern Mississippi Delta on Fig. 3) during the subsequent transgression (Fig. 8A). This case is well exemplified by different growth stages of the Mississippi and Danube deltas (Fig. 8A). The second case is where an initial fluvial-dominated delta has a reduced sediment supply and turns into a tide-dominated

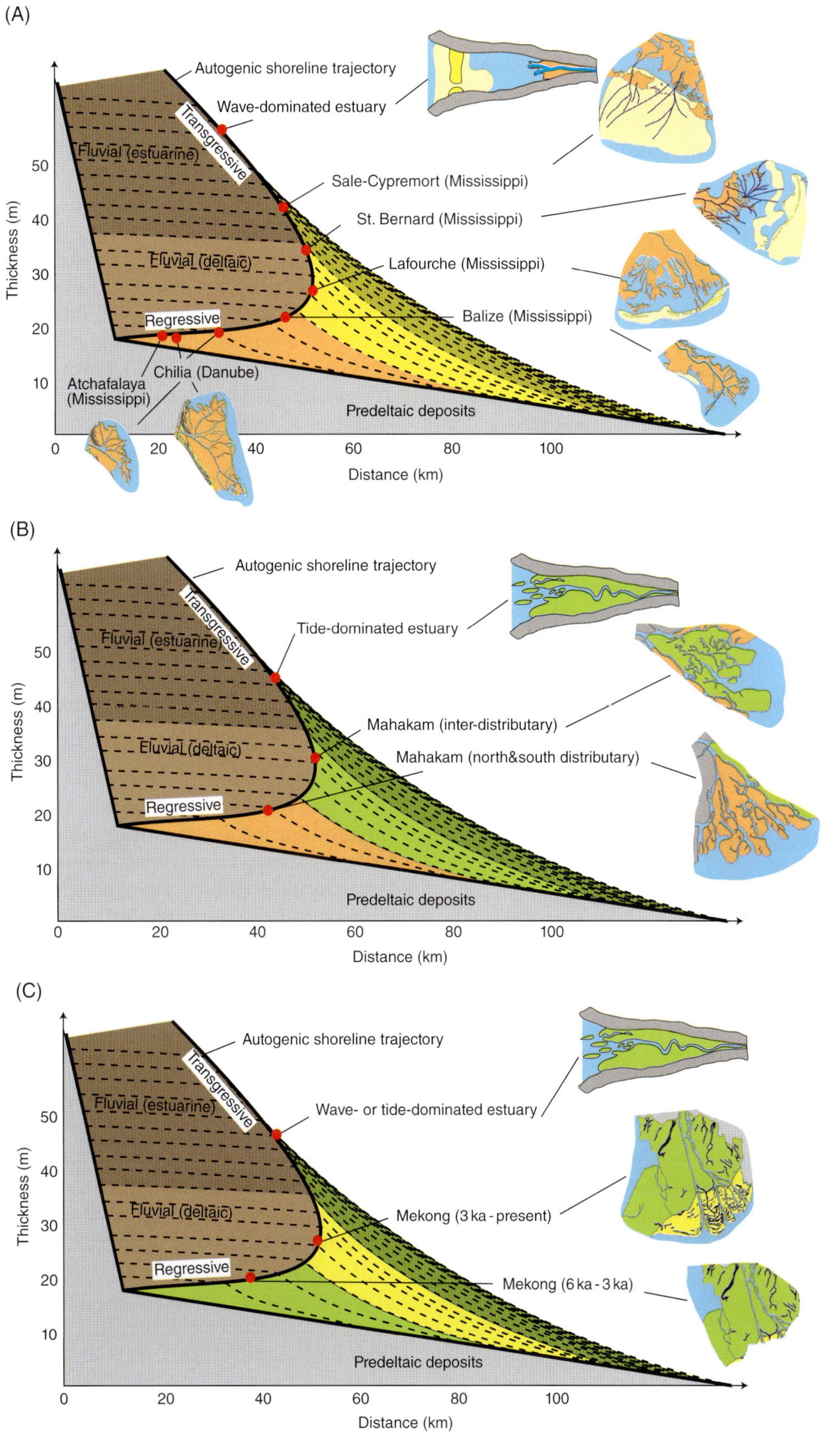

Fig. 8. Delta process change models. (A) Autogenic process changes from fluvial-dominated to wave-dominated during shoreline autoretreat. (B) Autogenic process changes from fluvial-dominated to tide-dominated during shoreline autoretreat. (C) Autogenic process changes from tide-dominated to wave-dominated during shoreline autoretreat.

delta and eventually into a tide-dominated estuary during transgression (Fig. 8B). The third case is when an initial tide-dominated delta changes into a wave-dominated delta because of a decrease in sediment supply or an increase in wave regime and later into a wave-dominated (or tide-dominated) estuary (Fig. 8C). There are two other cases without regime change during autoretreat. One case is when the delta is wave-dominated initially and continues to be wave-dominated until it is transgressed and turns into a wave-dominated estuary. Another case is when a tide-dominated delta maintains its character during progradation until it later transforms into a tide-dominated estuary.

The autogenic process-change model depicted in Fig. 8 is conceptual but the regressive part of the model is based on the observations of the Holocene deltas studied (Figs 3 to 6). The transgressive part of the model presented in Fig. 8 is less well documented but is based on the late Pleistocene histories of the studied deltas which show that they initiated as the rate of sea-level rise decrease (Fig. 2; during late Pleistocene to early Holocene) within lowstand incised estuaries or embayments that formed during late Pleistocene to early Holocene transgression which subsequently were flooded (Coleman, 1988; Panin, 1989; Storms *et al.*, 2005; Ta *et al.*, 2005).

Implications for ancient delta studies

It is difficult to reconstruct the plan view morphology of ancient deltas using outcrops data but delta process changes are well recorded in the facies stratigraphy, through distinct delta-front facies successions (Gani & Bhattacharya, 2007; Plink-Bjorklund, 2008; Carvajal & Steel, 2009). In the outcrop examples presented herein (Fig. 7), the facies changes occur over short distances (hundreds of metres to kilometres), even shorter than inferred from modern deltas. Observations of modern deltas strongly suggest that such process changes in the rock record are autogenic responses. It has been shown that delta process changes can be allogenic responses to the unsteadiness of external drivers (Porebski & Steel, 2006, Yoshida *et al.*, 2007). However, there is not always a need to invoke an increase (or decrease) of the rate of sea-level change to produce the observed stratigraphy in different delta types, as shown by modern deltas (Figs 4 to 6). The autogenic process change model developed herein (Fig. 8) suggests that facies changes within the same regressive-transgressive cycle are likely to be an autogenic responses.

Ancient deltas are commonly mapped and classified in the subsurface based on the morphology of sand isopach maps (Galloway, 1975; Fisher & McGowen, 1967; Busch, 1971; Xue & Galloway, 1995). With the increased seismic resolution of dense 2-D datasets (e.g. Sydow & Roberts, 1994) and acquisition of larger 3-D datasets, it is possible to do a seismic morphological analysis of deltas (e.g. Jackson *et al.*, 2010, Perov & Bhattacharya, 2011) to distinguish delta process changes. Autogenic process changes (at short time scales), as observed during the evolution of the Holocene deltas or in ancient deltaic deposits at outcrop, are difficult to infer from subsurface maps. However, the model of autogenic process changes described above (Fig. 8) might serve as a rule for delta evolution under steady forcing conditions for the subsurface datasets that have low resolution (e.g. seismic).

Why is it important to differentiate autogenic and allogenic delta process changes?

The interpretation of autogenic process changes in deltas has implications for regional stratigraphy and for local correlations and reservoir architecture; there is no need to infer changes in the external variables in order to have significant stratigraphic changes within deltaic units. However, allogenic process changes in deltas are also known to be common and certainly at time scales of eustatic change (10s of ka). This was pointed out by Porebski & Steel (2006) and Yoshida *et al.* (2007), who noted that straight wave-dominated coastlines had some tendency to be associated with rising and highstands of sea-level in the sedimentary geology literature and for tide-influenced shorelines to be associated with limited and protected (within bays and estuaries) accommodation, especially during falling and lowstands of sea-level. Allogenic process changes of this type are probably fairly widespread laterally and commonly associated with the development of large-scale flooding surfaces and sequence boundaries, developed, say, at the time scale of cross-shelf regressive-transgressive eustatic cycles (1 ka to 300 ka, as shown by Burgess & Hovius, 1998 and by Muto & Steel, 2002). Autogenic behaviour of delta system, on the other hand, would apply at the scale of segments of a delta

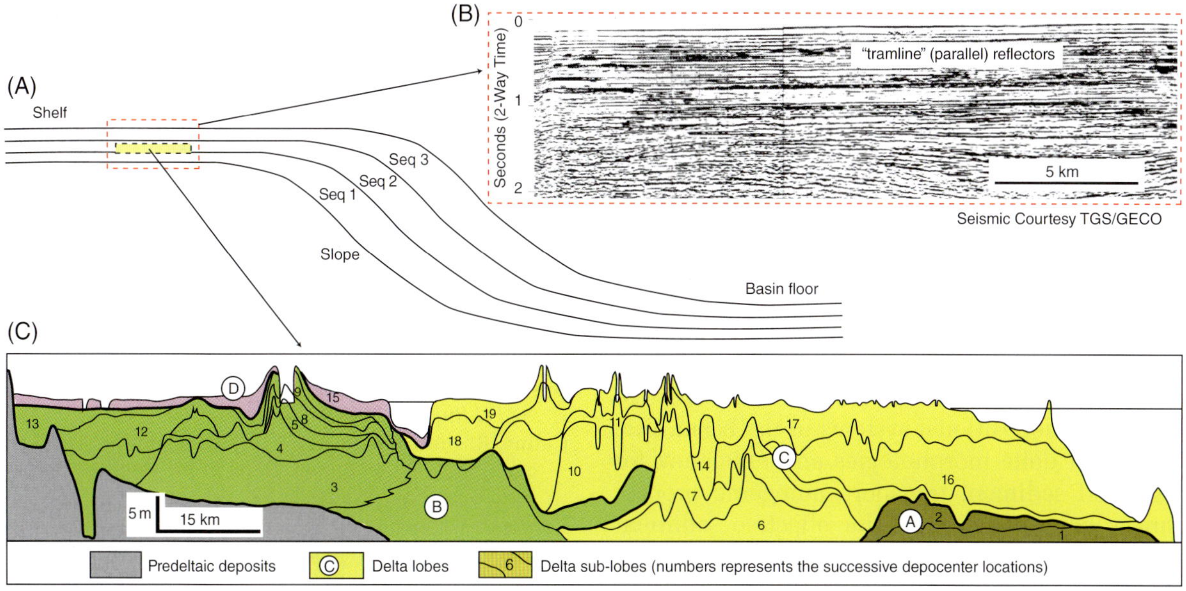

Fig. 9. Distinction between autogenic and allogenic generated deposits. (A) Sketch with the shelf to basin floor clinoform. (B) Typical 'tramlines': seismic reflectors over the shelf of the Gulf of Mexico (modified after Mitchum *et al.*, 1990). These reflect the transgression-regression cycles at the large (shelf) scale. (C) Internal architecture of the Holocene Mississippi Delta (modified after Frazier, 1967). Note the complex architecture generated by autogenic response with multiple locations of the delta depocentres at large scale (letters) or smaller scale (numbers). These changes happen over 8 ka and within 15 m thick deposits.

front, individual delta lobes or of smaller components, and coeval changes will most probably not be recognised in other delta systems in the same basin. The lateral extent of the process changes observed in Holocene deltas tends to scale with the system; 5 km to 10 km (Mahakam), 20 km to 30 km (Danube), 40 km to 50 km (Mekong) or hundreds of kilometres (Mississippi). The timing of the process change is also system specific, as for the specific time scale for shoreline autoretreat (Muto, 2001). Nevertheless, we emphasise that these apparently chaotic (Fig. 9C) autogenic responses (causing great difficult for correlation at this spatial and time scale) are usually contained by more widespread flooding surfaces and sequence boundaries (and more easily correlated) caused by the longer time scale (few 100 ka) regressive movement to transgressive movement of the entire delta system across the shelf, in response to allogenic unsteadiness of sea-level, tectonics or climate (Figs. 9A, B). The landward-seaward extent of these latter 'sequences' is dependent mainly on the landward extent of marine and coeval brackish-water flooding back across the shelf, which is especially probable to vary with Icehouse and Greenhouse climates (Steel *et al.*, 2008; Sømme *et al.*, 2009). These allogenic units create the 'tramline' character of many shelf or topset segments of shelf-margin sedimentary prisms (Fig. 9B). Fig. 9 suggests the spatial and temporal contrast between the allogenic and autogenic components of shelf stratigraphy, at least until it is demonstrated that autogenic responses can occur at a time scale of several 100 ka.

CONCLUSIONS

Autogenic process changes are a common behaviour in delta systems during large-scale cross-shelf progradation and retrogradation, as shown by Holocene river deltas (Mississippi, Danube, Mahakam, Mekong) that displayed autogenic responses in the form of changes in dominant process and attendant changes in morphology despite steady Holocene external forcing conditions. The process changes interpreted in these deltas happened with relative rapidity (hundreds to thousands of years) and are demonstrably autogenic in nature because, the external forcing factors were constant for the last 7 ka to 8 ka. Process change from fluvial-dominated or tidal-dominated to wave-dominated deltas have been documented across depositional dip or between fluvial-dominated to tidal-dominated or wave-dominated deltas laterally along depositional strike.

Autogenic process change can be inferred in ancient deltas when distinct changes in delta-front facies character are observed in the same regressive-transgressive cycle parasequence. The facies change from one dominant process to another is considered to be an autogenic response of the system if the changes are recorded over a relatively small area but not at basin scale. It is important to distinguish between autogenic and allogenic delta process responses because the former are confined to parts of a single system, whereas the latter are likely to apply to all depositional systems in the basin. Due to different delta morphologies and efficiency for cross-shelf sediment transport, autogenic process changes have implications for effective sediment dispersal, not only across the shelf but also across the shelf break and into deepwater areas.

ACKNOWLEDGEMENTS

I acknowledge the RioMAR Consortium sponsors (BG, BHPBilliton, BP, Devon, ENI, Shell, Statoil, PDVSA, Petrobras, Woodside) for their continuous support and lively discussions during consortium meetings. I thank associate editor John Howell and reviewers Gary Hampson and Brian Bracken who improved an initial version of the manuscript through constructive comments. Ron Steel greatly improved the clarity of the ideas and of the manuscript through multiple discussions with the author.

REFERENCES

Ainsworth, R.B., **Flint, S.S.** and **Howell, J.A.** (2008) Predicting coastal depositional style: influence of basin morphology and accommodation to sediment supply ratio within a sequence stratigraphic framework. In: *Recent Advances in Models of Siliciclastic Shallow-Marine Stratigraphy* (Eds G.J. Hampson, R.J. Steel, P.M. Burgess and R.W. Dalrymple), *SEPM Spec. Publ.*, **90**, 237–263.

Ainsworth, R.B., **Vakarelov, B.K.** and **Nanson, R.A.** (2011) Dynamic temporal and spatial prediction of changes in depositional processes on clastic shorelines: toward improved subsurface uncertainty reduction and management: *AAPG Bull.*, **95**, 267–297.

Allen, G.P., **Laurier, D.** and **Thouvenin, J.** (1979) *Étude Sédimentologique du delta de la Mahakam: Notes et Mémoirs*. Compagnie Fraçaise des Pétroles, **15**, 156 p.

Barton, M.D. (1994) Outcrop characterization of architecture and permeability structure in fluvial–deltaic sandstones, Cretaceous Ferron Sandstone, Utah [Ph.D. thesis], University of Texas, Austin, 259 p.

Bhattacharya, J.P. and **Walker, R.G.** (1991) River- and wave-dominated depositional systems of the Upper Cretaceous Dunvegan Formation, northwestern Alberta. *Bull. Can. Petrol. Geol.*, **39**, 165–191.

Bhattacharya, J.P. and **Willis, B.J.** (2001) Lowstand deltas in the Frontier Formation, Powder River basin, Wyoming: implications for sequence stratigraphic models. *AAPG Bull.*, **85**, 261–294.

Burgess, P.M. and **Hovius, N.** (1998) Rates of delta progradation during highstands: Consequences for timing of deposition in deep-marine systems. *Journal of the Geological Society*, **155**, p. 217–222.

Busch, D.A. (1971) Genetic units in delta prospecting. *AAPG Bull.*, **55**, 1137–1154.

Caline, B. and **Huong J.** (1992) New insight into the recent evolution of the Baram Delta from satellite imagery. *Bull. Geol. Soc. Malaysia*, **32**, 1–13.

Carvajal, C. and **Steel, R.** (2009) Shelf-edge architecture and bypass of sand to deep water: influence of sediment supply, sea-level and shelf-edge processes. *J. Sed. Res.*, **79**, 652–672.

Charvin, K., **Hampson, G.J.**, **Gallagher, K.L.** and **Labourdette, R.** (2010) Intra-parasequence architecture of an interpreted asymmetrical wave-dominated delta. *Sedimentology*, **57**, 760–785.

Coleman, J.M. (1988) Dynamic changes and processes in the Mississippi River delta. *Geol. Soc. Am. Bull.*, **100**, 999–1015.

Coleman, J.M. and **Gagliano, S.M.** (1964) Cyclic sedimentation in the Mississippi River deltaic plain. *Trans. Gulf Coast Assoc. Geol. Soc.*, **14**, 67–80.

Coleman, J.M. and **Wright, L.D.** (1975) Modern river deltas; variability of processes and sand bodies. In: *Deltas, Models for Exploration* (Ed. M.L. Broussard), 99–149.

Coleman, J.M., **Roberts, H.H.** and **Stone, G.W.** (1998) Mississippi river delta: an overview. *J. Coastal Res.*, **14**, 698–716.

Dan, S., **Stive, M.J.F.**, **Walstra, D-J.** and **Panin, N.** (2009) Wave climate, coastal sediment budget and shoreline changes for the Danube Delta. *Mar. Geol.*, **262**, 39–49.

Dewey, J. A., Jr. and **Morris T.H.** (2004) Geologic framework of the lower portion of the ferron sandstone in the willow springs wash area, Utah: facies, reservoir continuity and the importance of recognizing allocyclic and autocyclic processes. In: *Regional to Wellbore Analog for Fluvial-Deltaic Reservoir Modeling: The Ferron Sandstone of Utah* (Eds T.C. Chidsey, Jr., R.D. Adams and T.H. Morris), *AAPG Studies in Geology*, **50**, 305–329.

Edmonds, D.A. and **Slingerland, R.L.** (2007) Mechanics of river mouth bar formation: Implications for the morphodynamics of delta distributary networks. *J. Geophys. Res.*, **112**, F02034, doi: 10.1029/2006JF000574.

Fairbanks, R.G. (1989) A 17,000-year glacio-eustatic sea-level record: influence of glacial melting rates on the Younger Dryas event and deep-ocean circulation. *Nature*, **342**, 637–641.

Fisher, W.L. and **McGowen, J.H.** (1967) Depositional systems in the Wilcox group of Texas and their relationship to occurrence of oil and gas. *Trans. Gulf Coast Assoc. Geol. Soc.*, **XVII**, 105–125.

Fisk, H.N., **Kolb, C.R.**, **McFarlan, E.**, Jr. and **Wilbert, L.J.**, **Jr.** (1954) Sedimentary framework of the modern Mississippi delta [Louisiana]. *J. Sed. Petrol.*, **24**, 76–99.

Frazier, D.E. (1967) Recent deltaic deposits of the Mississippi river: their development and chronology. *Trans. Gulf Coast Assoc. Geol. Soc.*, **XVII**, 287–315.

Galloway, W.E. (1975) Process framework for describing the morphologic and stratigraphic evolution of deltaic depositional systems. In: *Deltas, Models for Exploration* (Ed. M.L. Broussard), 87–98.

Gani, M.R. and **Bhattacharya, J.P.** (2007) Basic building blocks and process variability of a cretaceous delta: internal facies architecture reveals a more dynamic interaction of river, wave and tidal processes than is indicated by external shape. *J. Sed. Res.*, **77**, 284–302.

Giosan L., **Bokuniewicz, H.**, **Panin, N.** and **Postolache, I.** (1999) Longshore sediment transport pattern along the Romanian Danube Delta coast. *J. Coastal Res.*, **15**, 859–871.

Giosan, L., **Donnelly, J.P.**, **Vespremeanu, E.**, **Constantinescu, S.**, **Filip, F.**, **Ovejanu, I.**, **Vespremeanu-Stroe, A.** and **Duller, G.A.T.** (2006) Young Danube delta documents stable Black Sea-level since the middle Holocene: Morphodynamic, paleogeographic and archaeological implications. *Geology*, **34**, 757–760.

Goodbread, S.L. and **Kuehl, S.A.** (2000) Enormous Ganges-Brahmaputra sediment discharge during strengthened early Holocene monsoon. *Geology*, **28**, 12, 1083–1086.

Helland-Hansen, W. and **Hampson, G.J.** (2009) Trajectory analysis: concepts and applications. *Basin Res.*, **21**, 454–483.

Howell, J.A. and **Flint S.S.** (2003) The parasequences of the Book Cliffs succession. In: *The Sedimentary Record of Sea-Level Change* (Ed. A.E. Coe), pp 158–178. Cambridge University Press.

Jackson, C.A.L., **Grunhagen, H.**, **Howell, J.A.**, **Larsen, A.L. Andersson, A.**, **Boen, F.** and **Groth, A.** (2010) 3D seismic imaging of lower delta-plain beach ridges: lower Brent Group, northern North Sea. *J. Geol. Soc.*, **167**; issue.6, 1225–1236.

Li, W., **Bhattacharya, J.P.**, **Zhu, Y.**, **Garza, D.**, and **Blankenship, E.** (2011) Evaluating delta asymmetry using three-dimensional facies architecture and ichnological analysis, Ferron 'Notom Delta', Capitan Reef, Utah, USA. *Sedimentology*, **58**, 478–507.

McManus, J. (2002) Deltaic responses to changes in river regimes. *Mar. Chem.*, **79**, 155–170.

Mellere, D. and **Steel, R.J.** (2000) Style contrast between forced regressive and lowstand/transgressive wedges in the Campanian of south-central Wyoming (Hatfield Member of the Haystack Mountains Formation). In: *Sedimentary Responses to Forced Regressions* (Eds D. Hunt and R.L. Gawthorpe), *Geol. Soc. Spec. Publ.*, **172**, 141–162.

Mellere, D., **Plink-Bjorklund, P.** and **Steel, R.** (2002) Anatomy of shelf deltas at the edge of a prograding Eocene shelf margin, Spitsbergen. *Sedimentology*, **49** (6), 1181–1206.

Mitchum, R.M. Jr., **Sangree, J.B.**, **Vail, P.R.** and **Wornardt, W.W.** (1990) Sequence stratigraphy in late Cenozoic expanded sections, Gulf of Mexico: Sequence stratigraphy as exploration tool – Concepts and practices in the Gulf Coast: GCS-SEPM Foundation 11th Annual Research Conference, 237–256.

Muto, T. and **Steel, R.J.** (1992) Retreat of the front in a prograding delta. *Geology*, **20**, 967–970.

Muto, T. and **Steel, R.J.** (1997) Principles of regression and transgression: the nature of the interplay between accommodation and sediment supply. *J. Sed. Res.*, **67**, 994–1000.

Muto, T. (2001) Shoreline autoretreat: substantiated in flume experiments. *J. Sed. Res.*, **71**, 246–254.

Muto, T. and **Steel, R.J.** (2002) In Defense of Shelf-Edge Delta Development during Falling and Lowstand of Relative Sea-level. *J. Geol.*, **110**, 421–436.

Muto, T., **Steel, R.J.** and **Swenson, J.B.** (2007) Autostratigraphy: a framework norm for genetic stratigraphy. *J. Sed. Res.*, **77**, 2–12.

Muto, T. and **Steel, R.J.** (2014) The autostratigraphic view of responses of river deltas to external forcing: Development of the concept. *Int. Assoc. Sedimentol. Spec. Publ.*, **47**, 139–148.

Nguyen, V.L., **Ta, T.K.O.** and **Tateishi, M.** (2000) Holocene delta evolution and sediment discharge of the Mekong River, Southern Vietnam. *J. Asian Sci.*, **18**, 427–439.

Olariu, C. and **Bhattacharya, J.P.** (2006) Terminal distributary channels and delta front architecture of river-dominated delta systems. *J. Sed. Res.*, **76**, 212–233.

Olariu, C., **Steel, R.J.** and **Petter, A.L.** (2010) Delta-front hyperpycnal bed geometry and implications for reservoir modeling: Cretaceous Panther Tongue Delta Utah., *AAPG Bull.*, **94**, 819–845.

Olariu, M.I., **Carvajal, C.**, **Olariu, C.** and **Steel, R.J.** (2012) Deltaic process and architectural evolution during cross-shelf transits – Fox Hills Deltas, Washakie Basin, Wyoming. *AAPG Bull*, **96**, 10, 1931–1956.

Orton, G.J. and **Reading, H.G.** (1993) Variability of deltaic processes in terms of sediment supply, with particular reference on grain size. *Sedimentology*, **40**, 475–512.

Panin, N. (1989) Danube Delta genesis, evolution and sedimentology. *Rev. Roum. Géol.*, **33**, 25–36.

Panin N. and **Jipa, D.** (2002) Danube River sediment input and its interaction with the north-western Black Sea. *Estuar. Coast. Shelf Sci.*, **54**, 551–562.

Penland, S., **Suter, J.R.** and **Boyd, R.** (1985) Barrier island arcs along abandoned Mississippi River deltas. *Mar. Geol.*, **63**, 197–233.

Perov, G. and **Bhattacharya, J.P.** (2011) Pleistocene shelf-margin delta: intradeltaic deformation and sediment bypass, northern Gulf of Mexico. *AAPG Bull.*, **9**, 1617–1641.

Petter, A.L. and **Steel, R.J.** (2006) Hyperpycnal flow variability and slope organization on an Eocene shelf margin, Central Basin, Spitsbergen. *AAPG Bull.*, **90**, 1451–1472.

Plink-Björklund, P. (2008) Wave-to-tide facies change in a Campanian shoreline complex, Chimney Rock Tongue, Wyoming-Utah, USA: In: *Recent Advances in Models of Siliciclastic Shallow-Marine Stratigraphy* (Eds G.J. Hampson, R.J. Steel, P.M. Burgess and R.W. Dalrymple), *SEPM Spec. Publ.*, **90**, 265–291.

Pontén, A. and **Plink-Björklund, P.** (2009) Process regime changes across a regressive to transgressive turnaround in a shelf–slope basin, Eocene Central Basin of Spitsbergen. *J. Sed. Res.*, **79**, 2–23.

Porebski, S. and **Steel, R.J.** (2006) Deltas and sea-level change. *J. Sed. Res.*, **76**, 390–403.

Roberts, H.H. (1997) Dynamic Changes of the Holocene Mississippi River Delta Plain: The Delta Cycle. *J. Coastal Res.*, **13**, 605–637.

Roberts, H.H. (1998) Delta switching, early responses to the Atchafalaya River diversion. *J. Coastal Res.*, **14**, 882–899.

Rossetti, C. and **Rey, F.** (Eds), (1931) *La Commission Européenne du Danube et son oeuvre de 1856 a 1931*, Imprimerie Nationale, Paris.

Sandal, S.T. (1996) *Geology and Hydrocarbon Resources of Negara Brunei Darussalam: Bandar Seri Bagawen*, Brunei Shell Petroleum Company, 243 p.

Slingerland, R., Driscoll, N.W., Milliman, J.D., Miller, S.R. and **Johnstone, E.A.** (2008) Anatomy and growth of a Holocene clinothem in the Gulf of Papua. *J. Geophys. Res.*, **113**, F01S13, doi: 10.1029/2006JF000628.

Somme, T.O., Helland-Hansen, W. and **Granjeon, D.** (2009) Impact of eustatic amplitude variations on shelf morphology, sediment dispersal and sequence stratigraphic interpretation: Icehouse vs Greenhouse Systems. *Geology*, **37**, 587–590.

Stanley, D.J. and **Warne, A.G.** (1994) Worldwide initiation of Holocene marine deltas by deceleration of sea-level rise. *Science*, **265**, 228–231.

Steel, R.J., Carvajal, C., Peter, A. and **Uroza, C.** (2008) The growth of shelves and shelf margins. In: *Recent Advances in Models of Siliciclastic Shallow-Marine Stratigraphy* (Eds G.L. Hampson, R.J. Steel, P.M. Burgess and R.W. Dalrymple), *SEPM Spec. Publ.*, **90**, 47–71.

Storms, J.E.A., Hoogendoorn, R.M., Dam R.A.C., Hoitink, A.J.F. and **Kroonenberg, S.B.** (2005) Late-Holocene evolution of the Mahakam delta, East Kalimantan, Indonesia. *Sed. Geol.*, **180**, 149–166.

Sydow, J. and **Roberts, H.H.** (1994) Stratigraphic framework of a late Pleistocene shelf-edge delta, northeast Gulf of Mexico. *AAPG Bull.*, **78**, 1276–1312.

Syvitski, J.P.M., Peckham, S.D., Hilberman, R. and **Mulder T.** (2003) Predicting the terrestrial flux of sediment to the global ocean: a planetary perspective. *Sed. Geol.*, **162**, 5–24.

Ta, T.K.O., Nguyen, V.L., Tateishi, M., Konayashi, I. and **Saito, Y.** (2005) Holocene delta evolution and depositional models of the Mekong River delta, southern Vietnam. In: *River Deltas: Concepts, Models and Examples* (Eds L. Giosan and J.P. Bhattacharya), *SEPM Spec. Publ.*, **83**, 453–466.

Ta, T.K.O., Nguyen, V.L., Tateishi, M., Konayashi, I., Saito, Y. and **Nakamura T.** (2002) Sediment facies and Late Holocene progradation of the Mekong River Delta in Bentre Province, southern Vietnam: an example of evolution from a tide-dominated to a tide- and wave-dominated delta. *Sed. Geol.*, **152**, 313–325.

Tamura, T., Saito, Y., Nguyen, V.L., Ta, T.K.O., Le, M.D., Bateman, M.D., Matsumoto, D. and **Yamashita, S.** (2012) Origin and evolution of interdistributary delta plains; insights from Mekong River delta. *Geology* **40**, 303–306.

Tornqvist, T.E., Kidder, T.R., Autin, W.J., van der **Borg, K.,** de **Jong, A.F.M. Cornelis, Klerks, J.W., Snijders, E.M.A., Storms, J.E.A., van Dam, R.L.** and **Wiemann, M.C.** (1996) A Revised Chronology for Mississippi River Subdeltas. *Science*, **273**, 1693–1696.

Uroza, C.A. and **Steel, R.J.** (2008) A highstand shelf-margin delta system from the Eocene of West Spitsbergen, Norway. *Sed. Geol.*, **203**, 229–245.

Van den Berg, T.C.V. and **Garrison, J.R.** (2004) Effects of changes in rate of sedimentation and relative change in sea-level on the geometry, architecture and sedimentology of fluvial and deltaic sandstones within the Upper Ferron Sandstone Last Chance Delta, East-Central Utah: implications for reservoir modeling, In: *Analog for Fluvial– Deltaic Reservoir Modeling: Ferron Sandstone of Utah* (Eds T.C. Chidsey, Jr., R.D. Adams and T.H. Morrison), *AAPG Studies in Geology*, **50**, 451–498.

Willis, B.J. and **Gabel, S.** (2001) Sharp-based, tide-dominated deltas of the Sego Sandstone, Book Cliffs, Utah, USA. *Sedimentology*, **48**, 479–506.

Willis, B.J., Bhattacharya, J.P., Gabel, S.L. and **White, C.D.** (1999) Architecture of a tide influenced river delta in the Frontier Formation of central Wyoming, USA. *Sedimentology*, **46**, 667–688.

Xue, L. and **Galloway, W.E.** (1995) High-Resolution Depositional Framework of the Paleocene Middle Wilcox Strata, Texas Coastal Plain. *AAPG Bull.*, **79**, 205–230.

Yoshida, S., Steel, R. J. and **Dalrymple, R.W.** (2007) Changes in Depositional Processes – an ingredient of a new generation of stratigraphic models. *J. Sed. Res.*, **77**, 447–460.

Int. Assoc. Sedimentol. Spec. Publ. (2014) **46**, 167–186

Morphodynamic modelling of wave reworking of an alluvial delta and application of results in the standard reservoir modelling workflow

MARTEN M. HILLEN*, NATHANAËL GELEYNSE‡, JOEP E.A. STORMS‡, DIRK JAN R. WALSTRA*§ and REMCO M. GROENENBERG‡

* *Section of Hydraulic Engineering, Department of Hydraulic Engineering, Delft University of Technology, PO Box 5048, 2600 GA, Delft, The Netherlands*
‡ *Geoscience & Engineering Department, Delft University of Technology, PO Box 5048, 2600 GA, Delft, The Netherlands (E-mail: j.e.a.storms@tudelft.nl)*
§ *Hydraulic Engineering Department, Deltares, PO Box 177, 2600 MH, Delft, The Netherlands*

This contribution relates to conference theme: III - Autogenic processes and products - learning from numerical modelling.

ABSTRACT

Wave forcing has been recognised as an important mechanism of transformation in modern and ancient delta systems. However, relatively few attempts have quantitatively assessed its contribution to deltaic morphodynamics and time-varying stratigraphy. Here, a high-resolution physics-based numeric model (Delft3D) is employed to study the influence of wave forcing on a hypothetical clastic, subaerial delta. Three cases are presented that differ from each other in the degree of fluvial input relative to a schematised time-invariant quiet-weather offshore wave field. In all cases, infill of higher-order distributaries on the delta plain is observed as well as delta front erosion in the absence of fluvial input. Furthermore, all cases show sediment sorting: river-derived fines, stored at the delta front, are entrained and transported in an offshore direction, to finally settle in deeper waters. However, at some stage coarser deposits prevent underlying fines from erosional wave stirring (armouring), thus increasing local preservation potential. In the case of appreciable fluvial input, single-channel dominance gives rise to the formation of bars that are laterally (coast-parallel) deformed due to wave action. Following the physics-based modelling, the simulated stratigraphy is applied to a standard reservoir modelling workflow to calculate properties including sandbody connectivity.

Keywords: Delta, wave, numerical model, Delft3D, morphodynamics, stratigraphy, reworking.

INTRODUCTION

Social and scientific relevance

Deltaic deposits have been studied intensively over the past decades for the purpose of hydrocarbon exploration. At the same time, our understanding of processes that govern present-day delta development has increased significantly. Recent studies have focussed on the effects of human activity on deltas (e.g. Ericson *et al.*, 2006), such as river-modification and catchment modification, which generally reduces sediment flux to deltas (e.g. Syvitski & Saito, 2007). This knowledge also contributes to reconstruct palaeodeltas in oil-reservoirs and gas reservoirs and their outcrop analogues. However, modern-day reservoir modelling software relies on either stochastic or geometric methods to populate the grid cells of a deltaic reservoir model that are not intersected by wells. Recent developments in forward (process-imitating) numerical models provide reservoir geologists with a third option to populate

From Depositional Systems to Sedimentary Successions on the Norwegian Continental Margin, First Edition.
Edited by A. W. Martinius, R. Ravnås, J. A. Howell, R. J. Steel, and J. P. Wonham.

the grid cells in a reservoir model: physics-based numerical modelling. Physics-based numerical models apply basic hydrodynamic and sediment transport algorithms to simulate delta formation and the associated stratigraphic record.

Recent work on physics-based models clearly shows their potential in reproducing essential fluvio-deltaic processes as described from modern-day systems (e.g. Geleynse *et al.*, 2010; Edmonds & Slingerland, 2010). Whilst most work focuses on morphodynamics by sole river forcing, Geleynse *et al.* (2011) have also shown that it is possible to produce a detailed stratigraphic record and include both wave and tide forcing during delta progradation. However, so far, little attention has been paid to reworking processes following the formation of deltaic strata by wind-generated surface gravity waves (hereafter abbreviated as 'waves'). Although numerical models are able to produce realistic deltaic sediment bodies, their preservation potential is determined by many other factors such as wave reworking, subsidence, sea-level and sediment supply. Here, we focus on the effects of wave reworking following the formation of a fluvial-dominated delta to the preservation potential of deltaic stratigraphy.

Wave reworking is a key factor in the development of present-day systems and ancient deltaic reservoirs. Syvitski (2008) found that many deltas become increasingly affected by waves due to shoreline adjustment, water-extraction and hydrocarbon extraction and Nicholls *et al.* (2008) state that the impact of wave reworking on deltas will increase due to the expected increase in frequency and amplitude of waves and storm surges. As such, a better understanding of the effects of wave reworking on deltas is required. Surprisingly, few attempts have been made to simultaneously model the morphodynamic and stratigraphic development of wave-influenced deltas (Geleynse *et al.*, 2011), in part due to the limited accessibility of sufficient computational power. However, the recent increasing computational power of desktop computers does allow for advanced physics-based numerical models such as Delft3D (e.g. Lesser *et al.*, 2004) to be run at time scales relevant to stratigraphic modelling. Hence, qualitative concepts regarding behaviour of retrograding deltas can be tested using quantitative physics-based models.

The process-response shoreface model of Storms (2003) and Storms & Hampson (2005) has already shown that wave height regime and sediment characteristics can be of equal significance to stratal geometry of shallow-marine environments as sea-level fluctuations and sediment supply. In addition, it is widely recognised that significant morphologic changes, particularly in the shallow-marine regime, can be established during low-magnitude, high-frequency events as well as by high-magnitude, low-frequency events, thus casting doubt on whether effective representation of an 'event' is achieved in quantitative models (e.g. Storms, 2003; Swenson *et al.*, 2005).

Goal and present numerical model scope

The work presented here continues along these lines but shifts focus towards reworking of river-derived sands and silts in deltas under wave action. Schematised water and sediment flow patterns in a river-dominated delta with characteristic stratigraphy (Geleynse *et al.*, 2010) change when subjected to quiet-weather short-crested waves, in turn deforming delta morphology and internal stratigraphy. Hence, we focus on the transgressive part of the delta cycle model (Scruton, 1960; Roberts, 1997) by applying a physics-based numerical model (Delft3D) to understand transgressive morphodynamics and preservation potential under conditions of constant high-frequency, low-amplitude wave forcing due to wind action, without the complicating effect of long-term sea-level fluctuations.

PREVIOUS WORK ON QUALITATIVE AND QUANTITATIVE DELTA MODELS

Qualitative delta models

Several qualitative delta classification schemes have been proposed throughout the past decades (Nemec, 1990). Their inherent limitations and practical implications were addressed in review papers by various authors (Nemec, 1990; Reading & Collison, 1996). The principles underlying these schemes vary considerably; from the gross character of the feeder system (Holmes, 1965) to the pattern of delta thickness distribution (Coleman & Wright, 1975), tectono-physiographic setting (Ethridge & Wescott, 1984), delta-front hydrologic regime (Galloway, 1975), sediment characteristics (Orton, 1988), delta-face slope (Corner *et al.*, 1990) and main physiographic attributes (Postma, 1990). The most commonly employed classification scheme (Galloway, 1975) is based on the morphologic expression of the delta interpreted as originating from the balance between fluvial, tidal and wave forcing. The accompanying diagram thereby includes the degree of delta front reworking by waves and tides.

Quantitative delta models

Quantitative delta models based on physics considerations vary in terms of their governing spans of space and time and associated aggregated level of landform representation. For an elaborate overview of present-day delta and shelfal models, the reader is referred to Paola (2000) and Overeem *et al.* (2005). Here, we constrain ourselves to a short overview of quantitative models on wave-influenced deltas to contextualise the present model.

2D-planform models

Several quantitative models have focused on the dynamics of deltas in planform for a varying degree of dominance of fluvial versus marine forcing. Komar (1973) already addressed deltaic shoreline reorientation for different ratios of wave energy flux and river sand supply, different offshore wave field angles and different shoreface slopes with a relatively simple computer model. Bhattacharya & Giosan (2003) suggested using the ratio of net longshore sediment transport rate at a river mouth to average river water discharge to determine the degree of planform symmetry of wave-influenced deltas and deltalobes. Following up on a numerical model on generation of large-scale shoreline instabilities (Ashton *et al.*, 2001), Ashton & Giosan (2007, 2011) pointed at characteristics of the wave climate, particularly spatial distribution of wave-angles and riverine bed-load sediment influx to govern wave-influenced deltaic shoreline configuration. Different combinations of these parameters enabled them to identify five 'prototype' planform geometries of deltas that differ in small-scale shoreline roughness (spits) and large-scale shoreline roughness (across-shoreline to along-shoreline topset length ratio and topset symmetry with respect to river-basin axis). While these models provide insights into the dynamics of the delta enclosure (shoreline), the third spatial dimension is only represented implicitly.

2D-profile models

Many quantitative models have been proposed that focus on the dynamics of deltas in cross-section. For example, Swenson *et al.* (2005) studied subaerial-subaqueous delta couplets (compound clinoforms) and underlined the importance of autogenic controls. Based on different combinations of three dimensionless numbers, a phase diagram for fluvio-deltaic clinoforms was suggested, entailing two end-members that correspond to river flood dominance and coastal storm dominance, respectively, and supported by natural field examples (Swenson *et al.*, 2005). Inherent to their coupled fluvio-marine model, surf zone geometry is collapsed to a vertical shoreface with a certain breaker depth, rather than equated as a moving boundary. Consequently, three-dimensional surf-zone dynamics are represented only implicitly, though in principle this is equivalent to assigning the shoreface any (equilibrium) geometry. Other quantitative profile models focus on the stratigraphic response to (time-dependent) forcing conditions but these generally do not rigorously couple fluvial and marine morphodynamics and often neglect wave action (e.g. Hoogendoorn *et al.*, 2008 and references therein).

3D models

Recent work of Storms *et al.* (2007) was devoted to morphodynamic-stratigraphic modelling of hypothetical deltas under riverine forcing, including transport of both sand and silt. Also, Edmonds & Slingerland (2007) provided detailed insights into the formation of sand mouth bars, the basic fabric of many fluvial networks. Concurrently, research has been undertaken with respect to the stability of bifurcations, another vital component of fluvial distributary networks (e.g. Bertoldi & Tubino, 2007; Kleinhans *et al.*, 2008; Edmonds & Slingerland, 2008). Geleynse *et al.* (2011) addressed the role of the feeder system to distributary network formation in a semi-unconfined basin under wave and tide influenced conditions. While these studies address general delta formation and behaviour in a genetic way, van Maren (2005) used a morphodynamic model to specifically investigate plausible explanations for barrier island formation on the prograding Ba Lat delta platform (Vietnam), in line with studies of Stutz & Pilkey (2002) and Bhattacharya & Giosan (2003), that downplayed the necessity that barrier formation be associated with a transgressive shoreline setting. Van Maren (2005) concluded that onshore transport of sediment, induced by wave asymmetry, is of prime importance to the initiation of subaqueous barriers. Subsequent (subaerial) stages of barrier(–spit) transformation were hypothesised to be characterised by alongshore currents and overwashes, while river (flood) flow was suggested as a destabilising mechanism to barrier settlement. The role of typhoons in barrier destruction was considered less significant (van Maren, 2005). Although van Maren (2005) investigated the morphodynamics of a real-world delta under riverine and wave forcing, enabling the formulation of a concept for cyclic constitution of barriers on delta plains, the stratal response of the sediment surface was not addressed.

MODEL DESCRIPTION AND SET-UP

In this study, a high-resolution depth-averaged (2DH) hydrodynamic model is combined with a three-dimensional stratigraphic module (Fig. 1), which are both part of the Delft3D numerical modelling suite. Delft3D is a physics-based numerical model for applications in coastal, river and estuarine areas (Lesser *et al.*, 2004). For the hydrodynamics of this study, non-steady flow and transport phenomena (flow-module) and wind-generated short-crested waves (wave-module) are simulated. The stratigraphy is simulated on the same computational grid as the hydrodynamics as the product of deposition and erosion over time. Flow, sediment transport and bottom updating are all executed at each time step (Roelvink, 2006).

Hydrodynamic processes are simulated by solving the unsteady depth-averaged shallow-water equations. Fluvial discharge and wave climate, imposed at the upstream and downstream boundaries of the computational domain, drive the system. The shallow water equations are discretised on a staggered grid (cell size 50 m × 50 m) with the water level points defined at the cell centres and the velocity components at the midpoints of the cell faces. An alternating direction implicit time integration (ADI) method is used to solve the shallow water equations (Stelling & Leendertse, 1991). A fixed time step of 30 s is applied, which meets the requirements of the Courant number for turbulent flow.

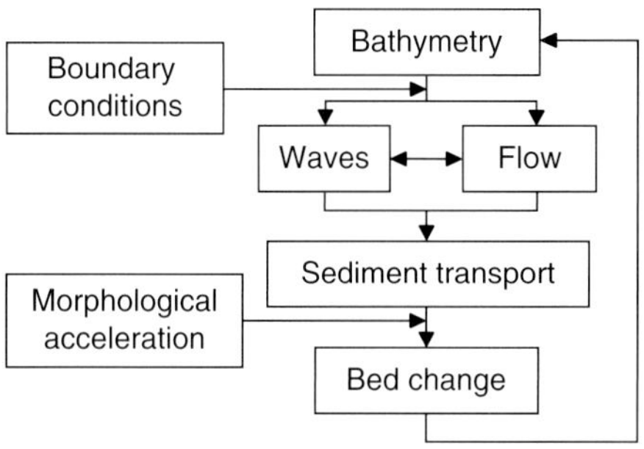

Fig. 1. Schematic overview of Delft3D –online (adjusted after Roelvink, 2006).

A wave climate is imposed as an offshore boundary condition at the downstream boundary of the computational domain. Simulation of wind-generated short-crested waves is performed by the third-generation SWAN model (Booij *et al.*, 1999). SWAN (Simulating WAves Nearshore) computes the propagation of short waves in coastal regions based on the discrete spectral balance of action density and driven by boundary conditions and winds. The flow and wave modules of Delft3D communicate via the 'online' coupling method, in which the flow and wave modules communicate with each other at each time step, establishing a two-way wave-current interaction. Wave conditions are updated each (hydrodynamic) hour. The lateral boundaries of the basin are defined as Neumann boundaries, thus imposing an alongshore water level gradient as an open boundary condition to allow water levels and currents to develop freely (Roelvink & Walstra, 2004). The ocean side of the basin (alongshore direction) is an open water level boundary, which defines the water level (Fig. 2).

For the simulations including fluvial input (low and high fluvial input cases are Q_{low} and Q_{high} respectively) the river discharge and sediment load are defined as an upstream boundary condition (Fig. 2). The river discharge is taken constant (500 m^3 s^{-1} and 2000 m^3 s^{-1} of respectively the Q_{low} and Q_{high} cases) during a simulation. The sediment load is defined as an equilibrium concentration depending on the hydrodynamic conditions in the cells immediately downstream of the inflow boundary.

The sediment dynamics are calculated for two sediment fractions; a fine sediment fraction with weakly cohesive characteristics, which represents silt, and a coarser sediment fraction that represents non-cohesive fine sand. The sediment fractions are defined by specific and dry bed density and the median grain diameter (d_{50}). The fine silt-like sediment fraction has a specific density of

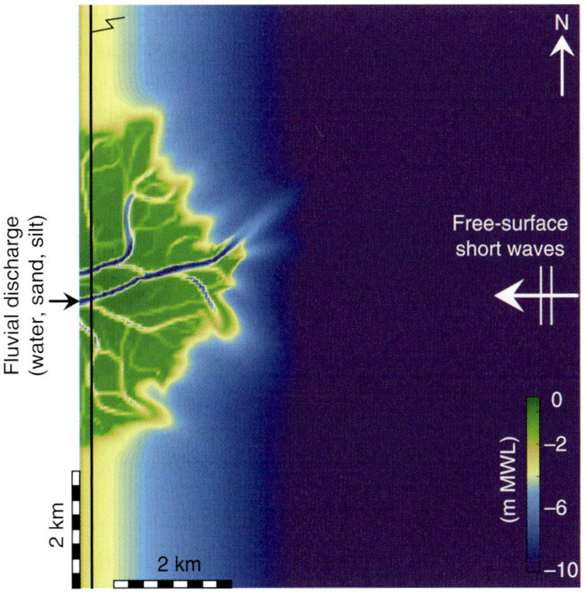

Fig. 2. Model domain and initial morphology as input to subsequent wave reworking. MWL is beneath Mean Water Level.

2650 kg m^{-3}, a dry bed density of 500 kg m^{-3} and a d_{50} of 50 μm. The sand sediment fraction has a specific density of 2650 kg m^{-3}, a dry bed density of 1600 kg m^{-3} and a d_{50} of 125 μm.

Sediment transport for both suspended and bed-load is calculated with the sediment transport formulation TRANSPOR2004 (van Rijn, 2007a; van Rijn, 2007b). This transport formulation provides one unified framework for both types of sediment transport and is applicable for a broad range of sediment sizes (fine silt to coarse sand) subject to varying conditions. It accounts for cohesive effects of fine sediments (<62 μm), as well as for the combined effects of waves and currents, particle-particle interaction, bed slope effects, flocculation and hindered settling. Current-induced and wave-induced sediment transport can be scaled with (calibration) transport factors. Since wave-related transport is often over-predicted in Delft3D (Lesser *et al.*, 2004), the wave-related transport factors in our simulations are scaled accordingly (0.3). Bed shear stresses are calculated with van Rijn's roughness predictor (van Rijn, 2007a), which is based on four types of roughness; grain size roughness, wave-related bed form roughness, current-related bed form roughness and the apparent roughness. The apparent roughness is the bed roughness resulting from wave-current interaction. Bed level changes are determined based on bed load sediment flux and the net result of deposition and entrainment. At every time step the bathymetry is updated with a morphologic acceleration factor of 60 (Lesser *et al.*, 2004; Roelvink, 2006) which implies that constant hydrodynamic conditions (including sediment transport) are assumed for 60 subsequent time steps. This allows for increased computational efficiency.

The stratigraphy is defined according to the multi-layer concept of Ribberink (1987). Individual beds are specified by thickness and sediment composition (sand/silt by volume). On top of the beds a transport layer of fixed height (here, 0.2 m) allows exchange of sediment with the water flow. The lowest bed is the base layer, which contains the sediment from beds with numbers that exceed the given bed tracking capacity. The substrate below the base layer is assumed to be unerodible. Together, the individual beds (0.1 m) and the substrate represent the stratigraphy. It is a vertical array of cells which records the sedimentary composition of morphodynamic elements such as bars, ridge and distributaries.

An extensive sensitivity analysis has been performed to test the robustness of the simulations and optimise the values of several parameters with respect to formulations of sediment transport and sediment characteristics, bed roughness and morphological acceleration (Hillen *et al.*, 2009). The optimised settings are summarised in Table 1. The geometry and stratigraphy of a delta modelled with a comparable Delft3D setup (Geleynse *et al.*, 2010) is used as an initial condition at the start of the simulations. Wave reworking is investigated by subjecting this predefined delta, which formed under mere fluvial forcing, to a time-invariant quiet-weather wave climate.

SCENARIOS

The setup of the numerical experiments represents an abandoned delta lobe following an upstream avulsion. Sea-level is presumed to be stable. For the sake of simplicity the simulated initial delta is a fully fluvial-dominated delta, implying no wave influence during its formation. However, with the abrupt decrease of river discharge, the delta enters the retrogradational phase of the delta cycle (Roberts, 1997) and waves become the dominant process. The simulations start at that exact moment. Such a model setting is comparable to delta switching examples known from the Mississippi delta (Coleman & Gagliano, 1964; Scruton, 1960) but human engineering can also cause this reduction of fluvial input (McManus, 2002; Syvitski & Saito, 2007). Both subsidence and reduced upstream sediment supply (and associated deterioration of wetlands, e.g. due to distorted balance of nutrient supply and salt water intrusion) induce topographic changes that, in turn, will change local characteristics of the incoming wave field, which itself may be constant in deep water. Both processes are the main causes for increased wave power in the Mississippi delta (Day *et al.*, 2007) and presumably also in other deltas. The increased water depth and lack of natural buffer allow the waves to propagate and cause further retrogradation. No compaction of the deposits and no other sediment reworking processes are included in the simulation, hence the focus is on deltaic wave reworking only.

Several scenarios are set up to analyse the effect of wave reworking under fluvial discharge conditions with different magnitudes (Table 2). All scenarios have equal parameter settings and initial conditions but the imposed boundary conditions vary. The Base case scenario is defined by sediment reworking under perpendicular incident quiet-weather short-crested waves (Hs = 1 m) only. No fluvial discharge is taken into account. The Base case simulation is repeated with a range of different fluvial discharge boundary conditions to mimic a gradual or partially successful upstream avulsion (cases Q_{low} and Q_{high}). For the scenario with low fluvial input (case Q_{low}), a simulation with a discharge of 500 m³/s is conducted and for the scenario with high fluvial input (case Q_{high}), a discharge of 2000 m³/s is applied. The latter discharge is identical to the discharge during the formation of the initial delta. Sediment transport in river channels is mainly governed by flow velocity and therefore increases with river discharge. For these scenarios, the grid of the Base

Table 1. Model parameters.

Parameter	Value
Timespan simulated	44 months
Timestep	30 s
Morphologic acceleration factor	60
Grid resolution	50 m × 50 m
Flow grid size	167 × 200
Wave grid size	170 × 382
Morphological characteristics	
Sediment transport formula	TRANSPOR2004 for both sediment fractions
Specific density (both sediment fractions)	2650 kg m^{-3}
Median grain size (sediment fraction 1)	125 μm
Dry bed density (sediment fraction 1)	1600 kg m^{-3}
Median grain size (sediment fraction 2)	50 μm
Dry bed density (sediment fraction 2)	500 kg m^{-3}
Number of underlayers	75
Maximum thickness underlayer	0.1 m
Thickness of transport layer	0.2 m
Suspended transport factor	1.0
Bed-load transport factor	1.0
Wave-related suspended transport factor	0.3
Wave-related bed-load transport factor	0.3
Spin-up interval before morphological changes	60 min

Table 2. Model scenarios.

Overview case settings				
		Boundary conditions		
Scenario	Description	Wave climate (offshore)	River discharge (upstream)	River sediment load (upstream)
Base case	Wave reworking with no river discharge	Perpendicular, $H_s = 1$ m, $T_p = 5$ s	–	–
Case Q_{low}	Wave reworking with low fluvial input	Perpendicular, $H_s = 1$ m, $T_p = 5$ s	500 m^3 s^{-1}	Equilibrium sediment concentration for silt – and sand fraction
Case Q_{org}	Wave reworking with high fluvial input	Perpendicular, $H_s = 1$ m, $T_p = 5$ s	2000 m^3 s^{-1}	Equilibrium sediment concentration for silt – and sand fraction

case scenario is extended in the landward direction to include part of the river system (Fig. 2). The discharge is imposed directly at an already present (main) channel.

RESULTS

Base case scenario

The initial fluvio-deltaic morphology of the delta, which contains multiple distributaries dissecting the delta plain (Fig. 2), rapidly adjusts to the imposed wave conditions by infilling of distributaries with silt caused by onshore transport that is driven by wave asymmetry (Fig. 3). Large volumes of silt are entrained by waves at the delta front. Overall, wave reworking and associated wave-induced currents cause erosion and smoothing of the delta front (Fig. 3). As expected, the highest transport rates are observed along the delta front due to longshore currents as a result of longshore gradients in wave height (Fig. 4). Consequently, the shoreline perturbation gradually diminishes as the delta front diffuses alongshore with respect to the incoming waves.

Fig. 4 displays the significant wave height, depth-averaged flow velocity field and the depth-averaged suspended load for both silt and sand fractions, for two specific time steps (t=2 and t=45 months). After 2 months, all distributaries, except for the main one, are disconnected from the retrograding shoreline (Fig. 4A), which is characterised by relatively strong, front-aligned currents (Fig. 4B). The remnants of the decoupled distributaries can be identified from the depth-averaged flow velocity field, by weakly-channelised flow zones on the delta plain and complex small-scale gyres at their termini. The depth-averaged suspended-load transport fields for the sand and silt fractions (Fig. 4C) show an overall dominance of the silt fraction; a large amount of silt is transported in an offshore direction by two counter-rotating gyres on either side of the longitudinal delta axis (cf. longitudinal drift divides commonly present at major headlands). Maximum suspended sand transport is concentrated at the tip of the longitudinal delta axis, while maximum suspended silt transport is distributed along the delta front, at either side of the longitudinal delta axis. After 45 months, overall flow field and suspended transport fields at the delta front have become more axially symmetric, though the signature of the initial delta front can still be observed from the wave height distribution (Fig. 4D). Furthermore, the major distributary is now fully disconnected from deeper water; however, its former configuration can be traced from the local wave height gradient, which is lower compared to the overall steeper decline. In addition, note that the roughness of the wave height distribution decreases with simulated time.

Wave reworking and transport leads to typical sediment sorting patterns. Silt is transported as suspended load over a broad area in both the longshore direction and offshore direction and settles in deeper waters, mainly off-axis at the delta front

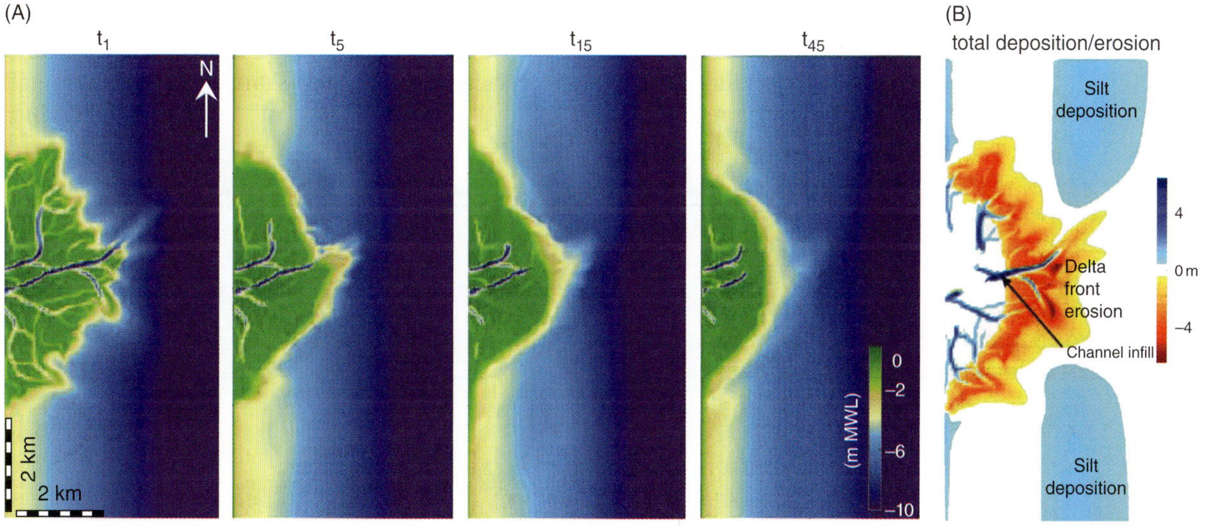

Fig. 3. Morphologic development of the Base case scenario at four distinct time intervals. Timesteps (t) are given in months. To the right: cumulative sedimentation and erosion at t=45 months.

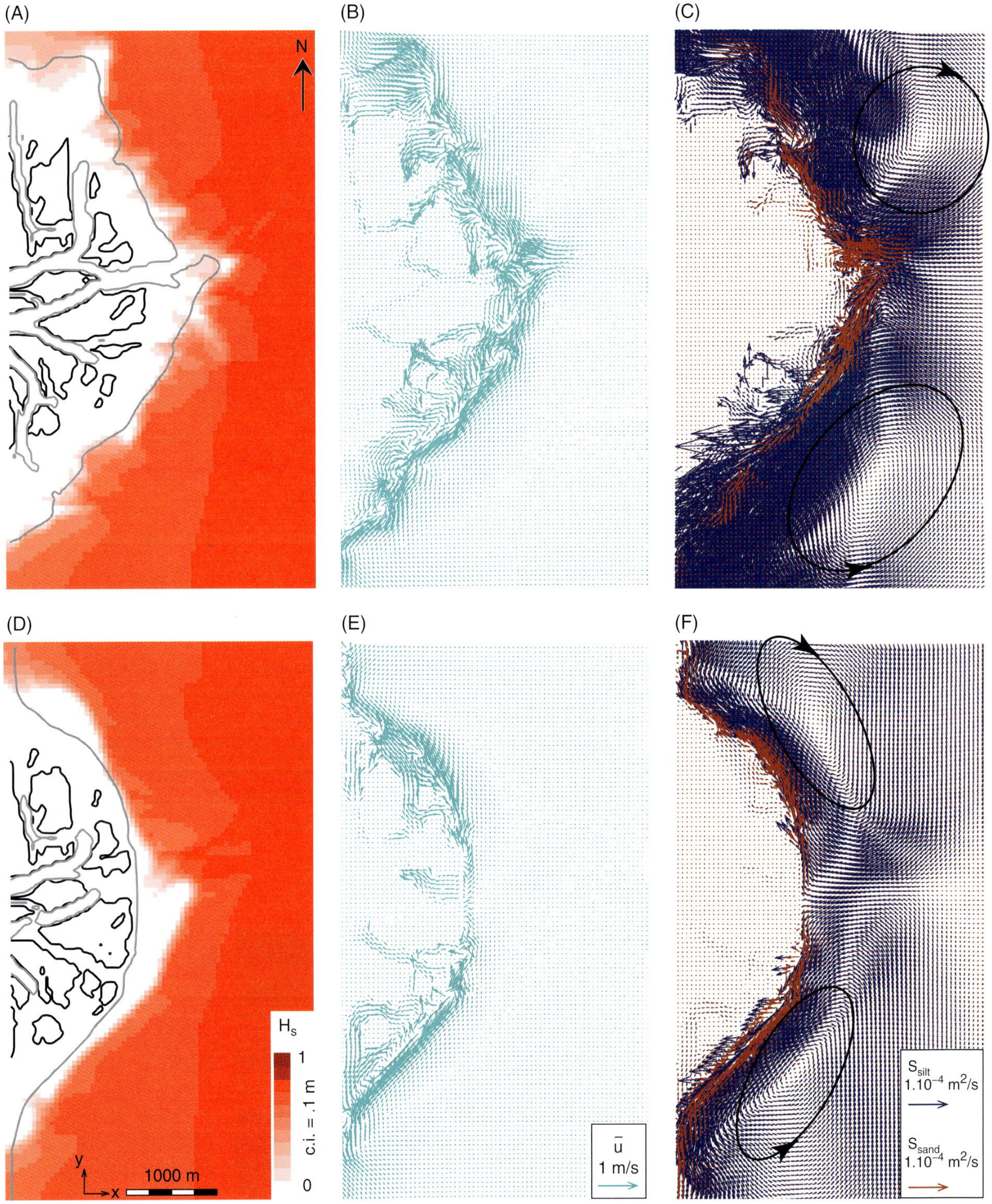

Fig. 4. Simulation of the Base case scenario showing after 2 simulated months (top): (A) Significant wave height (H_s) distribution. Here, simulated bed levels are represented by the 0 m (black) and −1 m (grey) MWL contours. (B) depth-averaged flow velocity field (ū), and (C) depth-averaged suspended load transport field for the sand (S_{sand}) and silt (S_{silt}) fractions. (D to F) show similar plots after 45 simulated months.

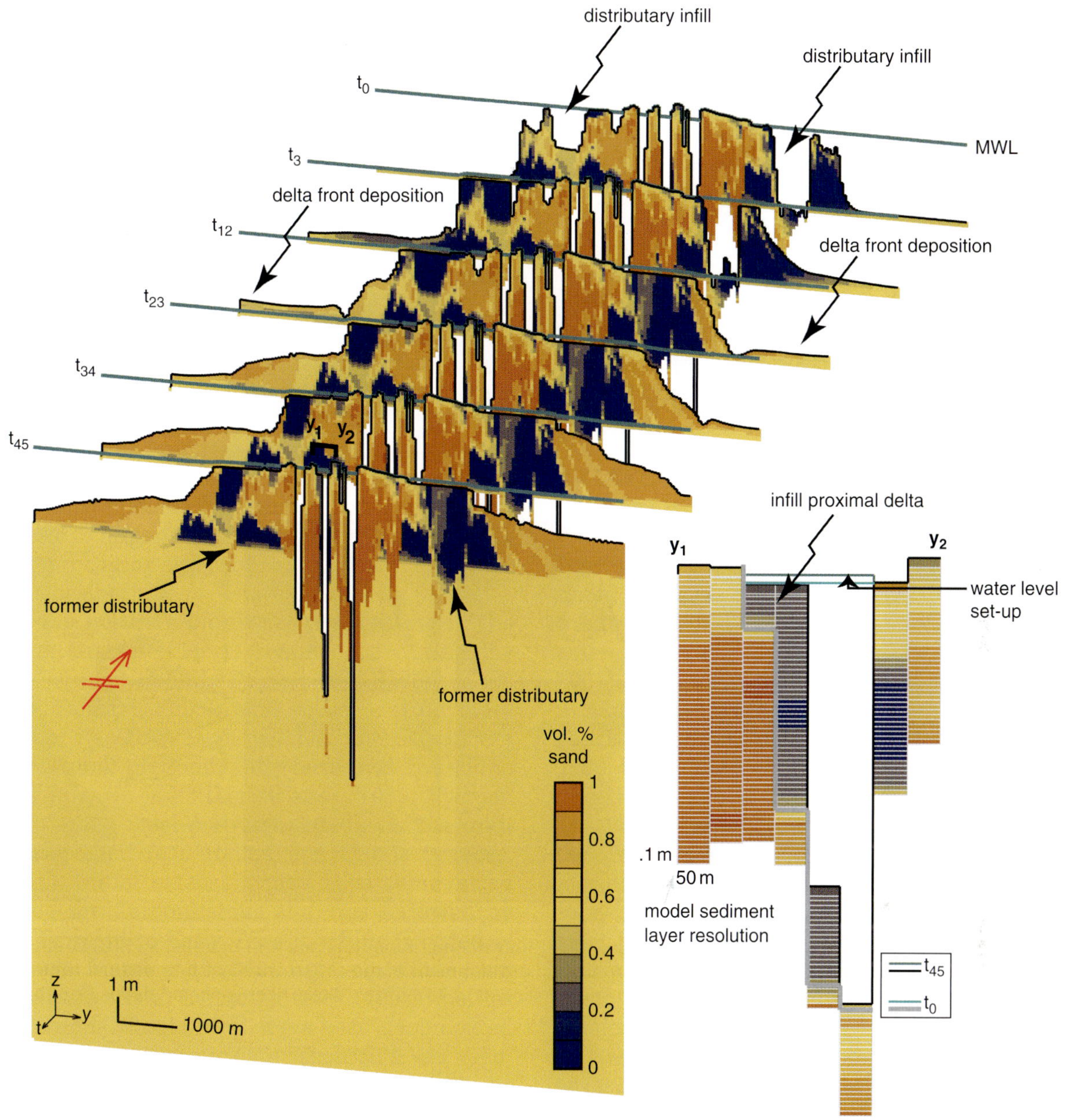

Fig. 5. Morphodynamics and associated stratal response of a cross-section in the proximal delta (see Fig. 2 for location) under sole wave forcing (Base case scenario). Subscripts of t refer to simulated number of months. The red arrow indicates the direction of the waves. The cross section inset shows in detail the deposition at a main distributary channel (location indicated at t_{45} cross-section to the left).

(Fig. 3). Suspended load transport is much higher than bed load transport for both sediment fractions. The bed load transport consists mainly of sand but sand transport along the delta front is only marginal. Fig. 5 shows the morphologic and stratigraphic adjustment in time (t_0 to t_{45}) of a cross section in detail illustrating both the significant sand deposition along the flank of the delta as well as the infill of distributary channels by silt. Note the large-scale lateral coarsening associated with the retrograding delta front and the infill of delta plain distributaries by fines. After the final time step, even the most proximal distributary is filled with fines by waning wave-induced currents. Consequently, some distributaries that were formerly incised into the substrate are preserved.

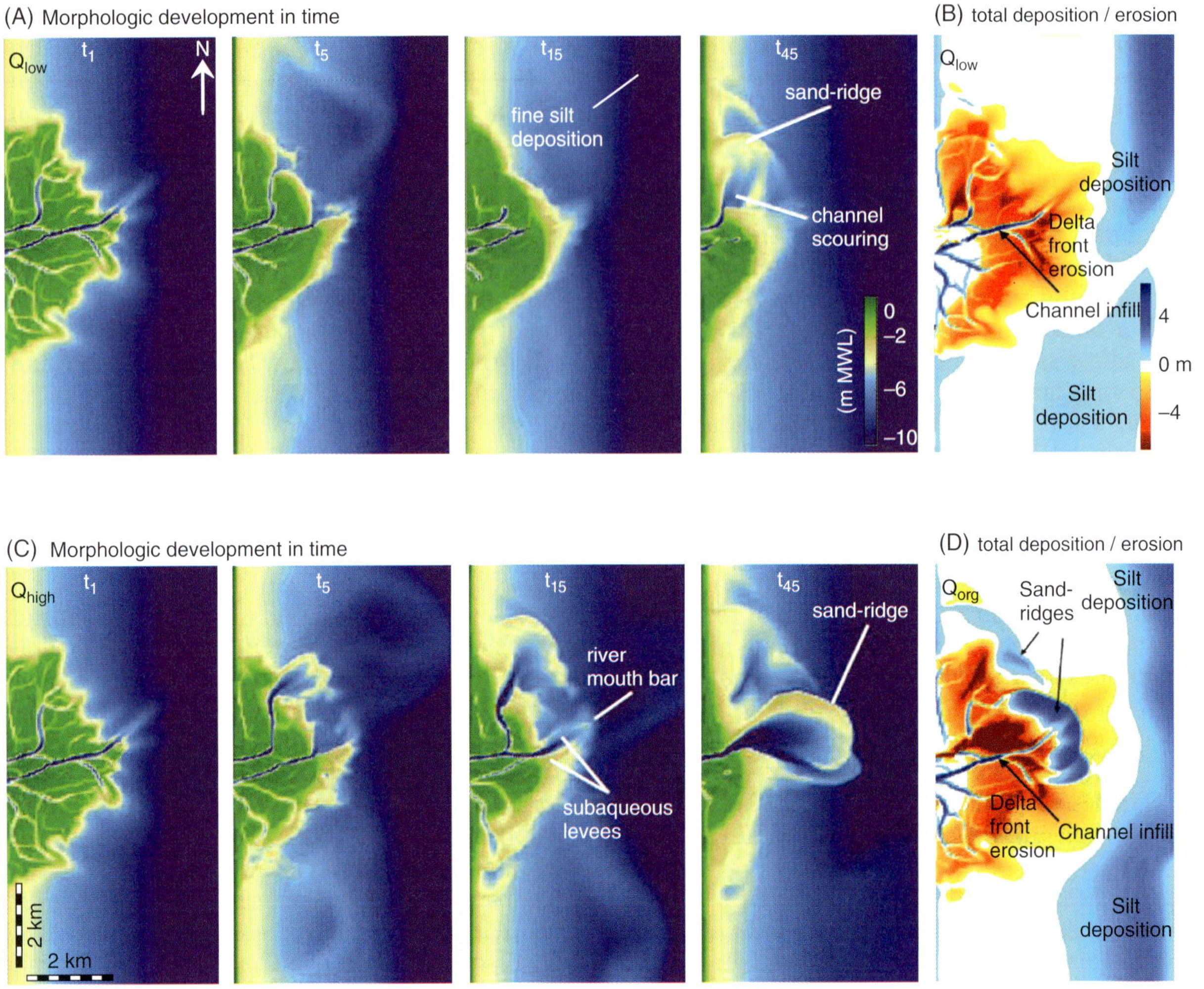

Fig. 6. (A) Morphologic development of the Q_{low} scenario at four distinct time intervals (t in months) and (B) cumulative sedimentation and erosion at t = 45 months. (C) Morphologic development of the Q_{high} scenario at four distinct time intervals (t in months) and (D) cumulative sedimentation and erosion at t = 45 months. Areas of erosion and deposition that are notated on the plots are described in further detail in the text.

Wave reworking with low fluvial input (Q_{low})

For the Q_{low} scenario, the small distributaries are abandoned and rapidly filled in by silt (Fig. 6A and B), similar to the Base case scenario. However, the ongoing discharge in the main distributary channel prevents it from being filled in by sediments derived from onshore directed wave-induced transport. Due to the low discharge, depth-averaged flow velocities remain under 1 m s^{-1} and little scouring occurs.

The initial morphology of the delta is such that the main distributary is oriented in the northern direction and it therefore debouches into the basin under an angle. Asymmetric sand-ridges consisting predominantly of reworked sand from the delta are formed at the river mouth; the main distributary channel carries only little bedload. The offshore transport of silt by suspended load occurs more distinctively in the northern area due to the distributary outflow. Compared to the Base case scenario, a larger part of the delta plain area is affected by reworking (compare the red area in Fig. 6B to that in Fig. 3B). Accordingly, the results suggest that reworking is more efficient when a wave-influenced delta is still fed by riverine discharge.

Wave reworking with original fluvial input (case Q_{high})

For a discharge of 2000 m^3/s, which is equal to the original discharge applied during delta progradation, some of the smaller distributary channels initially remain open but over time only the main distributary channel stays active (Fig. 6C and D). All other distributary channels are filled-in with silt. The relatively high discharge combined with a steady decrease in the number of distributary channels results in increasing flow velocities (1.3 m/s to 1.5 m/s) in the main distributary channel. This causes severe scouring. Consequently, the pathway of the main distributary channel orientation switches to the hydraulically most efficient route, i.e. through the basin centre.

The main distributary channel also transports sand, as bed load, and silt, as suspended load, offshore. Additional sand transported by the river is deposited immediately downstream of the river mouth, thereby forming a subaqueous levee and incipient mouth bar (Fig. 6C; centre-right panel). As the sand-ridge aggrades it gradually blocks the outgoing current causing enhanced deposition, thereby deflecting the main distributary channel. The southern delta front is not influenced by fluvial input and mainly silt is transported in a southern direction. The degree of wave-influence thus not only changes temporally, due to changes in riverine supply, but also spatially, in agreement with field observations (Bhattacharya & Giosan, 2003).

Sediment sorting occurs more distinctively than in the Base case and the Q_{low} scenarios. The removal of large volumes of silt from the delta front by the scouring distributary channel leads to a more sandy character of the retrograding delta. When channel switching occurs, abandoned sand ridges are gradually reworked in landward direction and lateral direction and are expected to reconnect to the shoreline (Fig. 6C and D). These ridges help to prevent further erosion by reducing wave impact on landward deposits. Furthermore, the formation of sandy shoreface-like deposits prevents the silty substrate from further erosion as observed in the Base case scenario (Fig. 3).

Sediment transport directions

Our model results indicate that cross-shore sediment transport is an order of magnitude lower than long shore sediment transport. Long shore transport gradients are large because of the curved deltaic shoreline. Cross-shore transport is most relevant in areas where the cross-shore profile (shoreface) is out of equilibrium, either becoming too gentle or too steep in relation to the induced waves. The constant wave forcing ensures that the cross-shore profiles are generally in equilibrium, except for areas with local features such as channel infills or areas where coastline realignment takes place by long shore transport. Therefore, the large-scale deltaic development of the scenario presented above is driven by long shore transportation of sand.

LINKED PROCESS AND RESERVOIR MODELLING

The simulated initial and wave-reworked deltaic stratigraphy can be imported into reservoir modelling software for post processing to obtain reservoir characteristics. Here we used the Petrel 2008 seismic and simulation software by Schlumberger. The simulated stratigraphy, which has a resolution of 50 m by 50 m by 0.1 m, is up-scaled to 50 m by 50 m by 0.25 m. We have isolated the deltaic sediments by assuming that the final stratigraphy, including the channels, is overlain by silt thus creating a heterogeneous synthetic deltaic reservoir of 10,000 m by 5000 m by 15 m.

Synthetic wells

Fig. 7 shows the net-to-gross ratio (N/G) at the palaeo-surface of the reworked delta deposit for the Base case scenario as well as the position of five synthetic wells (D3D01 to D3D05). A series of fence diagrams for all scenarios shows the stratigraphic heterogeneity of the deposits (Fig. 8). The substrate is a homogeneous mixture of sand and silt. Fig. 9 shows synthetic stratigraphy encountered in the five wells for the four scenarios. The synthetic wells all show both fining and coarsening upward cycles which are typical for fluvio-deltaic stratigraphy. Well D3D01 is characterised by substrate-overlying fine-grained prodelta sediments that gradually change into delta front sediments. This coarsening-upward succession is overlain by delta plain sediments which typically fine upward. Well D3D01 is unaffected by wave reworking in the Base case scenario and hardly affected by wave reworking in the Q_{low} scenario. However, it is fully reworked during the Q_{high} scenario where the progradational delta sequence is replaced by a typical coarsening upward shoreface succession.

Well D3D02 shows a similar stratigraphic pattern as that in well D3D01 but in this case it is more proximal and therefore has a higher N/G ratio. As mentioned before, significant reworking occurred only in the Q_{high} scenario; however, in this well the

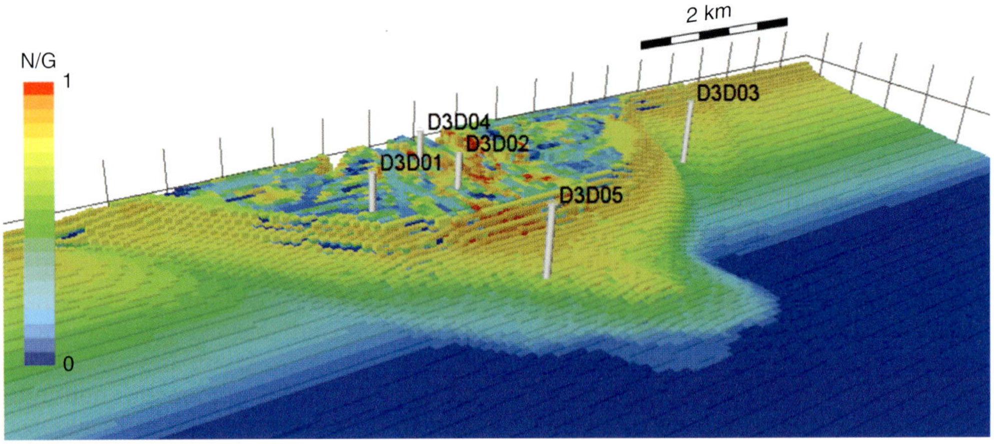

Fig. 7. Static reservoir model of the simulated Base case scenario indicating the net-to-gross ratio (N/G) at the palaeo delta top as well as the location of five synthetic wells (D3D01 – D3D05).

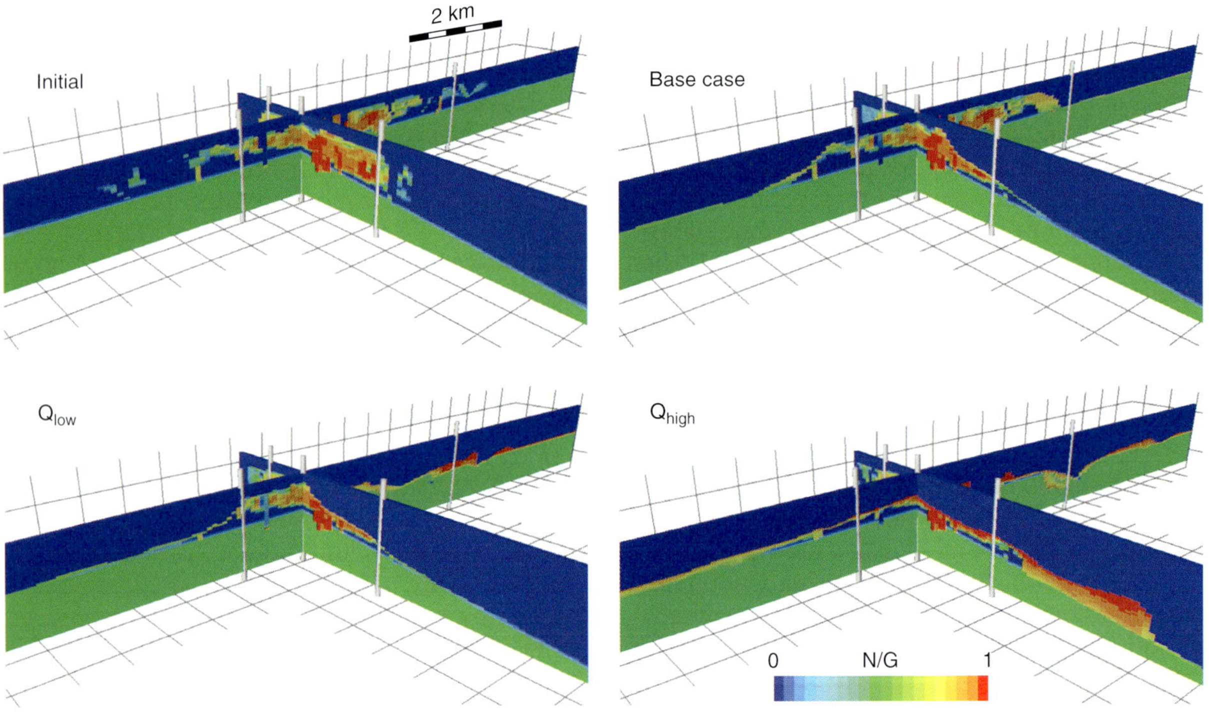

Fig. 8. Fence diagrams of all four scenarios (Initial, Base case, Q_{low} and Q_{high}) showing stratal heterogeneity in net-to-gross ratio (N/G) and well locations (see Fig. 7 for well labels).

lower part of the delta succession is preserved and capped with a thin shoreface lag deposit. Well D3D03 for the initial scenario shows a distal delta progradational succession, similar to well D3D01. This succession is eroded by waves in the Base case scenario. The increase in sand input by the distributary channel for the Q_{low} and Q_{high} scenarios leads to well-sorted deposits of sand that has been transported by wave-induced longshore currents (Fig. 4C and F).

In contrast, well D3D04 displays a vertical succession near the delta apex which is dominated by channelised flow. Channel migration during delta progradation has resulted in a thick succession of well-sorted sand. Due to its inland location, it is unaffected by wave reworking. Furthermore, the

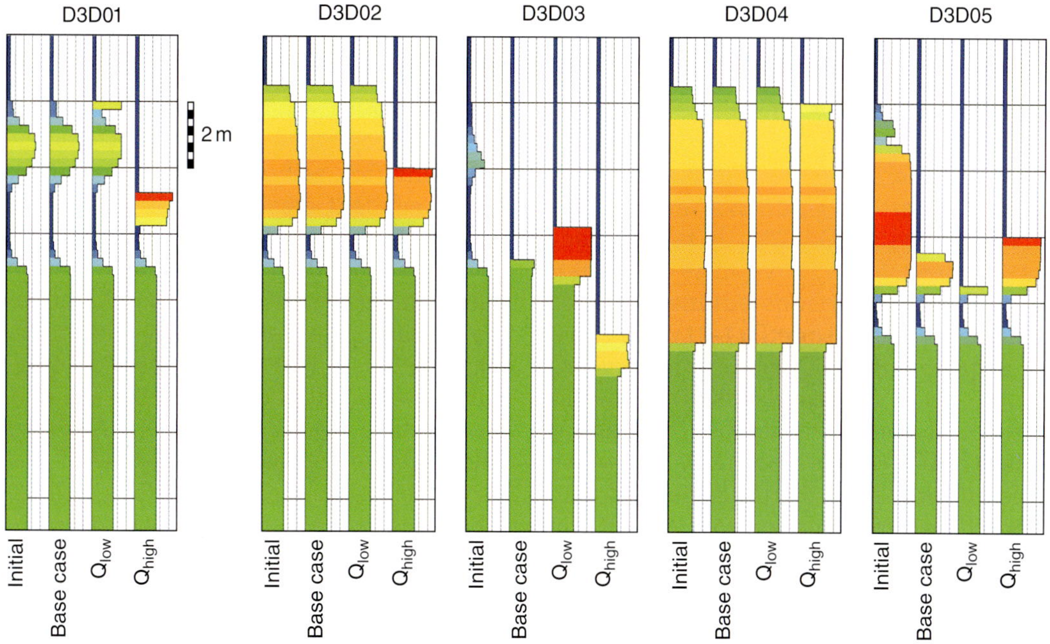

Fig. 9. Synthetic wells (D3D01 – D3D05) of all four scenarios (Initial, Base case, Q_{low} and Q_{high}) showing net-to-gross ratio (N/G; see Fig. 8 for colour scale). The synthetic wells show typical deltaic features such as coarsening and fining upward as well as lag deposits. See text for further explanation. For locations see Figs 7 and 8.

Table 3. Overview reservoir modelling. Individual sand bodies are represented by representative colours in Fig. 10.

Overview reservoir modelling							
		Connected volumes of NRV (%) for individual sand bodies (SB)					
Scenario	Net Rock Volume ($*10^6 m^3$)	SB1	SB2	SB3	SB4	SB5	sum
Initial case	10.46	55.6	24.8	15.7	1.2	1.0	98.3
Base case	9.41	74.9	17.5	3.6	1.3	1.0	98.3
Case Q_{low}	12.75	96.8	1.0	0.7	0.5	0.4	99.4
Case Q_{high}	28.30	100	–	–	–	–	100

succession is also unaffected by ongoing low discharge (Q_{low} scenario) but some minor fluvial reworking occurs at the top of the succession during the Q_{high} scenario. Well D3D05 is located in front of the main distributary channel and therefore is characterised by a sand-rich coarsening-upward succession representing delta progradation, capped by a fining-upward delta plain deposit. Wave reworking in this deltaic headland is effective, leading to significant erosion of the delta succession for all three wave scenarios. In case of the Q_{high} scenario, a sand bed associated with alongshore transport has been deposited on top of the original prodelta fines.

Overall it is evident that wave reworking changes the original delta succession quite dramatically. The changes differ in character and intensity depending on location, which is valuable information when trying to infer process-control from subsurface wells or outcrop logs.

Net Rock Volume and connected reservoirs

Not all deltaic deposits have good reservoir properties. We have applied a net-to-gross ratio criterion of >0.8 to identify sands with good reservoir properties, which results in Net Rock Volumes (NRV) with different values for the four scenarios. A higher NRV was obtained for the Q_{low} and Q_{high} scenarios (Table 3). This may be attributed to the effective removal of fines during the Q_{low} and Q_{high}

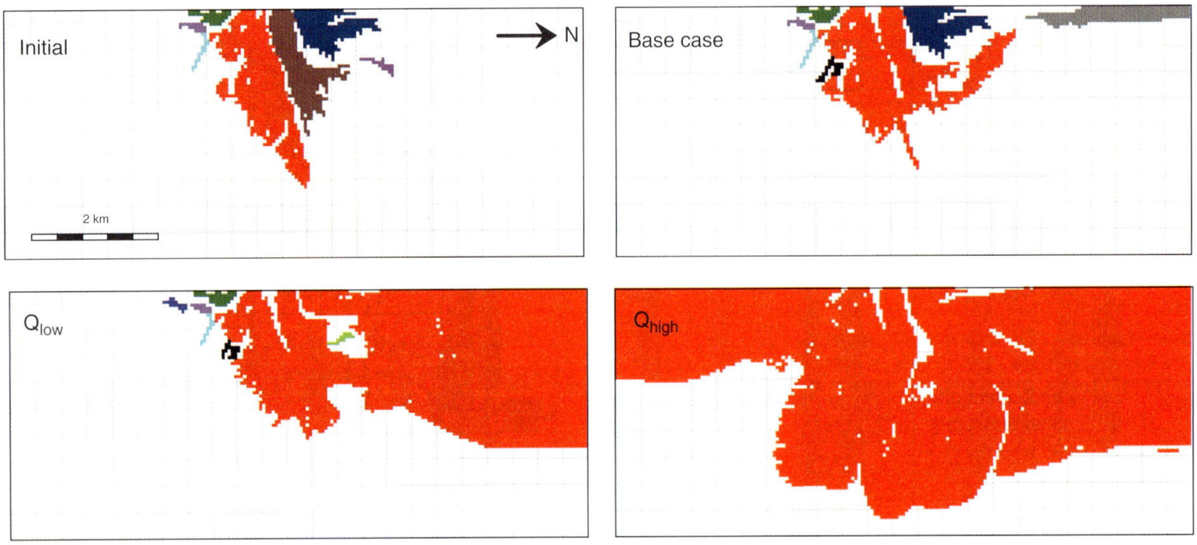

Fig. 10. Maps indicating the connected reservoir compartments. Each colour represents an individual reservoir compartment (See Table 3 for details) for all four scenarios (Initial, Base case, Q_{low} and Q_{high}). See text for further explanation.

scenarios. The initial scenario results in a compartmentalised reservoir consisting of three main interfluve areas directed in an on-offshore direction (Fig. 10). The silt-filled channels form potential flow barriers between the high-quality reservoir areas. The connectivity between reservoir volumes is increased by the formation of a wave-induced sand-rich shoreface that improves the connectivity of the isolated sands.

DISCUSSION

This study highlights the effects of wave reworking on a fluvial-dominated delta based on numerical simulations. Relevant physical processes underlying flow transport and sediment transport are included. When comparing the simulation results to real-world data, several similarities are observed. For example, in data from the Mississippi delta (e.g. Atchafalaya Bay, Louisiana; Neill & Allison, 2005; Roberts *et al.*, 2005) similar silt and sand transport conditions are observed. Furthermore, the partial preservation of silts by reworked sand deposits is observed along the 'Isles Dernieres' barrier islands in the Mississippi delta, where a protective sand cap prevented erosion of the underlying fines (Dingler & Reiss, 1990). A comparable sand cap also plays a key role in a conceptual morpho-sedimentary model for Louisiana coastal restoration and barrier retreat developed by Campbell (2005).

The decrease in the rate of morphological change over time in the three cases suggests that they move towards an equilibrium situation (Fig. 11A, C and E). The hypsometric curves, defined as the cumulative distribution of bed level in the modelled area, tend towards stabilisation when the rate of change decreases per time step. Erosion of the delta front in the Base case occurs to a depth of 7 metres (Fig. 11B). This erosion is solely due to wave reworking. Beyond 7 m water depth aggrading conditions prevail as the eroded sediments are re-deposited in deeper water. In the fluvial input cases (Q_{low} and Q_{high}) the erosion extends to 10 m in depth (Fig. 11C and E). The increased retrogradation in shallow water compared to the base case scenario is due to the higher efficiency of the transport mechanism for silt induced by the offshore-directed fluvial discharge. The net offshore transport of silt results in an increased sediment loss at the shallow shoreface. Channel scouring, migration and extension related to fluvial processes (Q_{low} and Q_{high}) are responsible for the hypsometric changes in areas where the water depth exceeds 7 m.

The initial rapid adaptation of the shoreface profile to the imposed wave climate suggests that deltas may be very vulnerable, especially if their deposits are composed of fine grained sediments. However, our simulations show that, in time, the sorting mechanism induced by wave reworking results in a gradual increase of shoreface sands that act as a protective layer to wave-induced erosion,

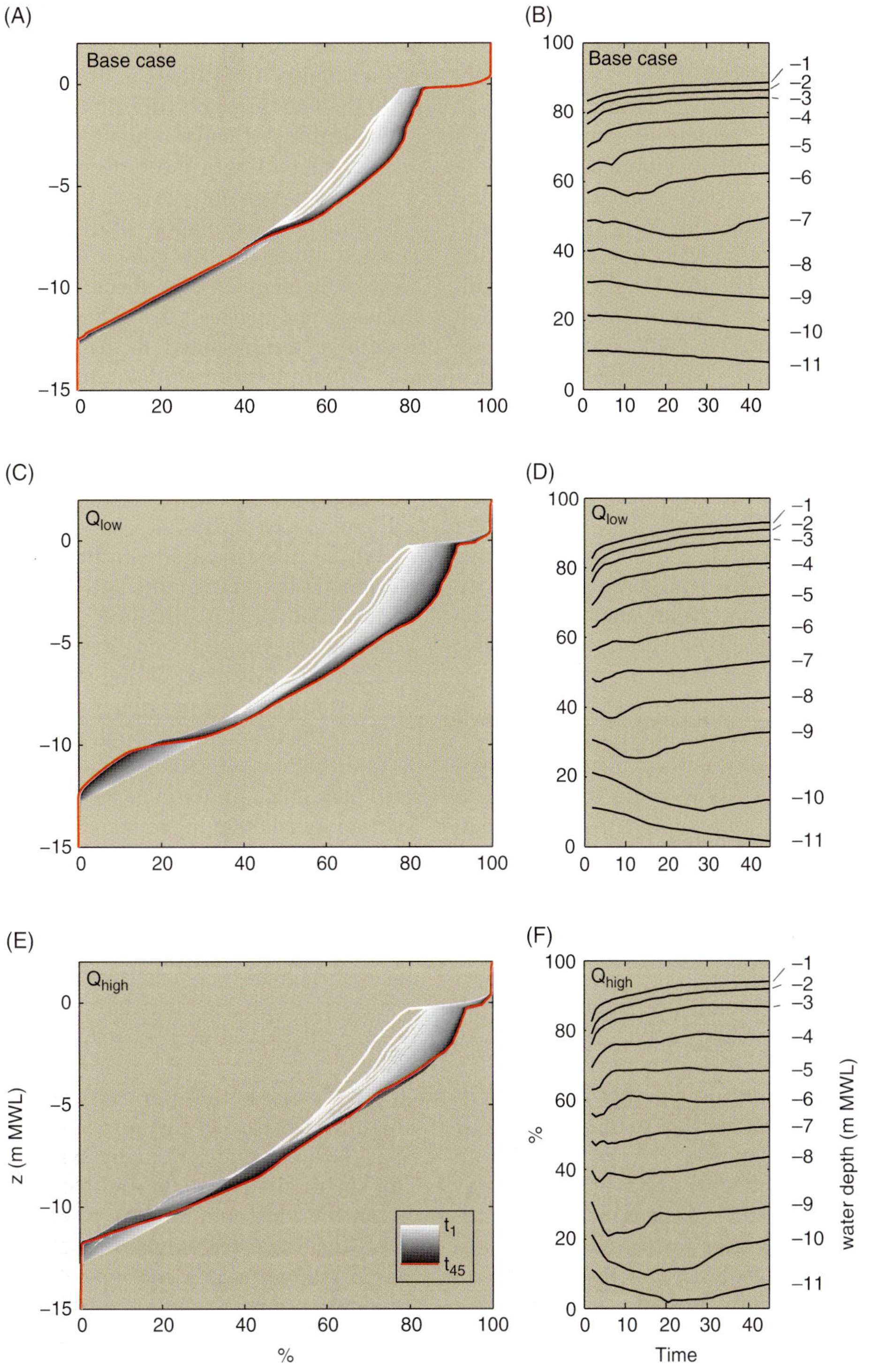

Fig. 11. Hypsometric curves for the model domain showing cumulative distribution (%) bed-level changes in time (lines graded from white to grey) with the final hypsometric curve after 44 simulated months (in red) of wave-induced reworking superimposed for (A) no fluvial discharge (Base case), (C) low fluvial discharge (Q_{low}), (E) high fluvial discharge (Q_{high}). Figures B, D and F show the temporal development of sediment volume (in %) present beneath specific bed level intervals (−1 m to −11 m MWL).

thus stabilising the shoreface profile. One important condition for this stabilising mechanism to work is that sand is present in the system in order to reduce or halt the retrogradation rate due to wave-induced erosion. As mentioned by van Maren (2005) for the Ba Lat delta in Vietnam, cross-shore sediment transport may play a major role in the formation of a new stage in delta development. However, most delta formation models that include wave influence mainly focus on longshore sediment transport to reproduce characteristic deposition patterns (Ashton & Giosan, 2007; Bhattacharya & Giosan, 2003). This study shows that both transport modes account for the morphodynamics of a retrograding delta.

Fig. 6A and C shows both the subaereal (in green colour) and subaqueous (in yellow to blue colours) morphology. In time, the subaereal area of the delta gradually reduces by shoreline retreat. Wave induced sediment re-suspension combined with wave and river induced currents generate a strong morphological response with developing sand ridges oriented perpendicular to the coastline. Deflection of river plumes by wave action occurs in nature, as reported via observations and our computational modelling. At this stage, results from our model experience may show some limitations in morphological development which can be linked to three factors: 1) Simulated sediment transport only applies to subaqueous areas; no mechanism for subareal sediment transport is included in the model, although there is bank erosion that can erode higher subareal areas, 2) the forcing (boundary conditions) in these model experiments are constant, which may exaggerate local morphological features or may even produce unrealistic results and 3) there is a high gradient in the forcing conditions from fully fluvial to fully marine at the start of the simulations. We also know that deposits of river plumes are deflected into barriers and spits by (obliquely-incident) windwaves and associated longshore currents. At this stage, barrier formation is probably not well-simulated in Delft3D.

While the numerical experiments shown in this paper are based on a fluvial delta that is subsequently reworked by waves after abandonment, additional simulations were run to evaluate the effects of waves during delta formation (Geleynse *et al.*, 2011). These experiments showed that wave-induced currents limit the settling of fines in shallow water, which leads to a significant reduction in delta plain area when compared to the fluvial (no wave) case. In addition, there is a reduction in the number of active channels, as described in literature (e.g. Bhattacharya & Giosan, 2003). The present findings, based on explicit descriptions of wave-related sediment transports, are consistent with these results.

Partial preservation is something rarely addressed in numerical simulation studies. From the experiments shown in this paper it is clear that wave reworking significantly affects reservoir properties. Wave reworking leads to a change in sediment distribution and associated facies when compared to the initial undisturbed delta. In addition, the wave-induced sorting mechanism results in cleaner sands and improved reservoir properties. Also, reservoir connectivity is improved by wave reworking because the sandy shoreface that results from it connects the different lobes of the delta.

Whilst physics-based simulation of sedimentary systems is becoming more and more useful in reservoir modelling, it is still limited by calculation time. Simulations like the ones described above take a few days to finish. Applying such models to better understand stacked parasequence sets is still unfeasible at the moment although numerical efficiency is rapidly improving due to the use of adaptive grids and parallel computations. Nevertheless, it remains extremely challenging to take the next step: to condition a physics-based numerical model to field data. A common method for conditioning a numerical forward model to real-world data is inversion, where each forward model realisation is compared against the real-world data and evaluated based on a number of criteria (e.g. Charvin *et al.*, 2011, Miller *et al.*, 2008). This approach, though successful, generally takes many forward runs, ranging from 10^2 in the case of the approach of Miller *et al.* (2008) to 10^4 in the case of the approach of Charvin *et al.* (2011). Such conditioning by means of inversion is not yet possible for delta simulations using Delft3D, given the long computation times associated with such inversion techniques. Nevertheless, physics-based modelling is a very useful tool for reservoir modelling when using the so-called 'Integrated Process-Aided Reservoir Modelling' approach (Miller *et al.*, 2008), where geometric and statistical information of sedimentary bodies are retrieved from the simulations. Such synthetic information complements real-world outcrop analogue data and can be easily used when a reservoir model is built based on statistical methods (Michael *et al.*, 2010). The advantage of physics-based modelling results over outcrop analogues is that the former are custom made, i.e. they are based on understanding of the processes that drive the system and their interaction. Whilst real-world outcrops are limited in number and exposure, never exactly fitting the reservoir data, synthetic (numerical) outcrops are fully three dimensional and represent the known process and boundary conditions of the specific reservoir. In addition, by varying the boundary conditions (e.g. fluvial discharge, basin geometry etc.), uncertainty can be quantified.

CONCLUSION

With increased computational power and a larger need for quantitative and case-specific information, physics-based numerical modelling shows great potential. The conceptual setting of this study illustrates the possibilities of the application of the Delft3D model to investigate the changes in both the morphology and stratigraphy of a delta subject to wave reworking. Here, we evaluate the preservation potential of an abandoned delta lobe by an upstream avulsion under different fluvial regimes. Wave-induced currents lead to specific sediment redistribution patterns. Furthermore, combined fluvial and wave-induced currents lead to an increase in offshore transport of fine sediments. Overall, the preserved delta, subsequent to being subjected to wave reworking, shows an increased amount of sand-rich deposits and an increased connectivity.

ACKNOWLEDGMENTS

We thank Statoil for funding this work. This paper greatly benefitted from discussions with Marcel Stive (Delft University of Technology), Bert Jagers (Deltares) and Allard Martinius (Statoil).

REFERENCES

Ashton, A., Murray, A.B. and **Arnoult, O.** (2001) Formation of coastline features by large-scale instabilities induced by high-angle waves. *Nature*, **414**, 296–300.

Ashton, A.D. and **Giosan, L.** (2007) Investigating plan-view asymmetry in wave-influenced deltas. In: *5th IAHR Symposium on River, Coastal, and Estuarine Morphodynamics*, pp. 681–688. International Association of Hydraulic Researchers, Enschede.

Ashton, A.D. and **Giosan, L.** (2011) Wave-angle control of delta evolution, *Geophys. Res. Lett.*, **38**, L13405, doi:10.1029/2011GL047630.

Bertoldi, W. and **Tubino, M.** (2007) River bifurcations: Experimental observations on equilibrium configurations. *Water Resour. Res.*, **43**, 1–10.

Bhattacharya, J.P. and **Giosan, L.** (2003) Wave-influenced deltas: geomorphological implications for facies reconstruction. *Sedimentology*, **50**, 187–210.

Booij, N., Ris, R.C. and **Holthuijsen, L.H.** (1999) A third-generation wave model for coastal regions. Part 1: Model description and validation. *J. Geophys. Res.*, **104**, 7649–7666.

Campbell, T. (2005) Development of morphosedimentary model for design of coastal restoration projects along the Louisiana coast. *J. Coast. Res.*, **44**, 234–244.

Charvin, K., Hampson, G.J., Gallagher, K., Storms, J.E.A. and **Labourdette, R.** (2011) Characterization of controls on high-resolution stratigraphic architecture in wave-dominated shoreface-shelf parasequences using numerical modelling. *J. Sed. Res.*, **81**, 562–578. DOI: 10.2110/jsr.2011.48

Coleman, J.M. and **Gagliano, S.M.** (1964) Cyclic sedimentation in the Mississippi River deltaic plain. *Trans. Gulf Coast Assoc. Geol. Soc.*, **14**, 67–80.

Coleman, J.M. and **Wright, L.D.** (1975) Modern River Deltas: Variability of Processes and Sandbodies. In: *Deltas: Models for Exploration* (Ed. M.L. Broussard), pp. 99–149. Houston Geological Society.

Corner, G.D., Nordahl, E., Munch-Ellingsen, K. and **Robertsen, K.R.** (1990) Morphology and Sedimentology of an Emergent Fjord-Head Gilbert-Type Delta: Alta Delta, Norway. In: *Coarse Grained Deltas* (Eds A. Colella and D.B. Prior), *Int. Assoc. Sedimentol. Spec. Publ.*, **10**, 3–12.

Day, J.W., Boesch, D.F., Clairain, E.J., Kemp, G.P., Laska, S.B., Mitsch, W.J., Orth, K., Mashriqui, H., Reed, D.J., Shabman, L., Simenstad, C.A., Streever, B.J., Twilley, R.R., Watson, C.C., Wells, J.T. and **Whigham, D.F.** (2007) Restoration of the Mississippi Delta: lessons from hurricanes Katrina and Rita. *Science*, **315**, 1679–1684.

Dingler, J.R. and **Reiss, T.E.** (1990) Cold-front driven storm erosion and overwash in the central part of the Isles Dernieres, a Louisiana barrier-island arc. *Mar. Geol.*, **91**, 195–206.

Edmonds, D.A. and **Slingerland, R.L.** (2007) Mechanics of river mouth bar formation: Implications for the morphodynamics of delta distributary networks. *J. Geophys. Res.*, **112**, F02034, doi:10.1029/2006JF000574.

Edmonds, D.A. and **Slingerland, R.L.** (2008) Stability of delta distributary networks and their bifurcations. *Water Resour. Res.*, **44**, doi:10.1029/2008WR006992.

Edmonds, D.A. and **Slingerland, R.L.** (2010) Significant effect of sediment cohesion on delta morphology. *Nature Geosci.*, **3**, 105–109.

Ericson, J.P., Vorosmarty, C.J., Dingman, S.L., Ward, L.G. and **Meybeck, M.** (2006) Effective sea-level rise and deltas: causes of change and human dimension implications. *Global Planet. Change*, **50**, 63–82.

Ethridge, F.G. and **Wescott, W.A.** (1984) Tectonic setting, recognition and hydrocarbon reservoir potential of fan-delta deposits. In: *Sedimentology of Gravels and Conglomerates* (Eds E.H. Koster and R.J. Steel). *Can. Soc. Petrol. Geol. Mem.*, **10**, 217–235.

Fagherazzi, S. and **Overeem, I.** (2007) Models of deltaic and inner continental shelf evolution. *Annu. Rev. Earth Planet. Sci.*, **35**, 685–715.

Galloway, W.E. (1975) Process framework for describing the morphologic and stratigraphic evolution of deltaic depositional systems. In: *Deltas: Models for Exploration* (Ed. M.L. Broussard), pp. 99–149. Houston Geol. Soc.

Geleynse, N., Storms, J.E.A., Stive, M.J.F., Jagers, H.R.A. and **Walstra, D.J.R.** (2010) Modelling of a mixed-load fluvio-deltaic system. *Geophys. Res. Lett.*, **37**, L05402, doi:10.1029/2009GL042000.

Geleynse, N., Storms, J.E.A., Walstra, D.J.R., Jagers, H.R.A., Wang, Z.B. and **Stive, M.J.F.** (2011) Controls on river delta formation; insights from numerical modelling. *Earth Planet. Sci. Lett.*, **302**, 217–226.

Hillen, M.M., Geleynse, N., Storms, J.E.A., Walstra, D.J.R. and **Stive, M.J.F.** (2009) Morphology and stratigraphy of a degrading delta. *Proc. Coastal Dynamics*, pp. 1–2.

Holmes, A. (1965) *Principles of Physical Geology*. 3rd edn. Thomas Nelson, London, 730 pp.

Hoogendoorn, R.M., Overeem, I. and **Storms, J.E.A.** (2008) Process response modelling of fluvial deltaic systems, simulating evolution and stratigraphy on geological timescales. *Comp. and Geosci.*, **34**, 10, 1394–1416.

Kleinhans, M.G., Jagers, H.R.A., Mosselman, E. and **Sloff, C.J.** (2008) Bifurcation dynamics and avulsion duration in meandering rivers by one-dimensional and three-dimensional models. *Water Resour. Res.*, **44**, doi:10.1029/2007WR005912.

Komar, P.D. (1973) Computer models of delta growth due to sediment input from rivers and longshore transport. *Geol. Soc. Am. Bull.*, **84**, 2217–2226.

Lesser, G.R., Roelvink, J.A., van **Kester, J.A.T.M.** and **Stelling, G.S.** (2004) Development and validation of a three-dimensional morphological model. *Coast. Eng.*, **51**, 883–915.

McManus, J. (2002) Deltaic responses to changes in river regimes. *Mar. Chem.*, **79**, 155–170.

Michael, H.A., Li, H., Boucher, A., Sun, T., Caers, J. and **Gorelick, S.M.** (2010) Combining geologic process models and geostatistics for conditional simulation of 3D subsurface heterogeneity, *Water Resour. Res.* **46**, W05527, doi:10.1029/2009WR008414.

Miller, J.K., Sun, T., Li, H., Stewart, J., Genty, C., Li, D. and **Lyttle, C.** (2008) Direct modelling of reservoirs through forward process based models: Can we get there?, Paper 12729 presented at International Petroleum Technology Conference, Kuala Lumpur.

Neill, C.F. and **Allison, M.A.** (2005) Subaqueous deltaic formation on the Atchafalaya Shelf, Louisiana. *Mar. Geol.*, **214**, 411–430.

Nemec, W. (1990) Deltas - remarks on terminology and classification. In: *Coarse Grained Deltas* (Eds A. Colella and D.B. Prior), *Int. Assoc. Sedimentol. Spec. Publ.*, **10**, 3–12.

Nicholls, R.J., Wong, P.P., Burkett, V., Woodroffe, C.D. and **Hay, J.** (2008) Climate change and coastal vulnerability assessment: scenarios for integrated assessment. *Sustainability Science*, **3**, 89–102.

Orton, G.J. (1988) A spectrum of Middle Ordovician fan deltas and braidplain deltas, North Wales: a consequence of varying fluvial clastic input. In: *Fan Deltas: Sedimentology and Tectonic Settings* (Eds W. Nemec and R.J. Steel), pp. 23–49. Blackie and Son, London.

Overeem, I., Syvitski, J.P.M. and **Hutton, E.W.H.** (2005) Three-dimensional numerical modelling of deltas. In: *River Deltas: Concepts, Models and Examples* (Eds L. Giosan and J.P. Bhattacharya), *Soc. Sed. Geol. Spec. Issue*, **83**, 13–30.

Paola, C. (2000) Quantitative models of sedimentary basin filling. *Sedimentology*, **47**, 121–178.

Postma, G. (1990) Depositional architecture and facies of river and fan deltas: a synthesis. In: *Coarse Grained Deltas* (Eds A. Colella and D.B. Prior), *Int. Assoc. Sedimentol. Spec. Publ.*, **10**, 3–12.

Reading, H.G. and **Collinson, J.D.** (1996) Clastic coasts. In: *Sedimentary Environments: Processes, Facies and Stratigraphy* (Ed H.G. Reading) Third edn, 154–231. Blackwell Science, Oxford.

Ribberink, J.S. (1987) *Mathematical Modelling of One-Dimensional Morphological Changes in Rivers with Non-Uniform Sediment.* PhD dissertation, Delft University of Technology, Delft.

Roberts, H.H. (1997) Dynamic changes of the Holocene Mississippi River delta plain: the delta cycle. *J. Coast. Res.*, **13**, 605–627.

Roberts, H.H., Walker, N.D., Sheremet, A. and **Stone, G.W.** (2005) Effects of cold fronts on bayhead delta development: Atchafalaya Bay, Louisiana, USA. In: *High resolution morphodynamics and sedimentary evolution of estuaries* (Eds D.M. FitzGerald and J. Knight), *Coastal Systems and Continental Margins*, pp. 269–298. Springer, Dordrecht.

Roelvink, J.A. (2006) Coastal morphodynamic evolution techniques. *Coast. Eng.*, **53**, 277–287.

Roelvink, J.A. and **Walstra, D.J.R.** (2004) Keeping it simple by using complex models. In: *Advances in Hydroscience and Engineering* (Eds M.S. Altinakar, S.S.Y. Wang, K.P. Holz and M. Kawahara), pp. 335–346. National Center for Computational Hydroscience and Engineering, Oxford.

Scruton, P.C. (1960) Delta building in the deltaic sequence. In: *Recent sediments, Northwest Gulf of Mexico* (Eds F.P. Shepard, F.B. Phleger and T.H. van Andel), pp. 82–102, AAPG, Tulsa.

Stelling, G.S. and Leendertse, J.J. (1991) Approximation of convective processes by cyclic AOI methods. In: *The 2nd ASCE Conference on Estuarine and Coastal Modelling*, pp. 771–782. Am. Soc. Civ. Eng., Tampa.

Storms, J.E.A. (2003) Event-based stratigraphic simulation of wave-dominated shallow-marine environments. *Mar. Geol.*, **199**, 83–100.

Storms, J.E.A. and Hampson, G.J. (2005) Mechanisms for forming discontinuity surfaces within shoreface-shelf parasequences: sea-level, sediment supply, or wave regime? *J. Sed. Res.*, **53**, 67–81.

Storms, J.E.A., Stive, M.J.F., Roelvink, J.A. and **Walstra, D.J.R.** (2007) Initial morphologic and stratigraphic delta evolution related to buoyant river plumes. In: *Coastal Sediments '07* (Eds N.C. Kraus and J.D. Rosati), **1**, pp. 736–748. Am. Soc. Civ. Eng., New Orleans.

Stutz, M.L. and **Pilkey, O.H.** (2002) Global distribution of deltaic barrier island systems. *J. Coast. Res.*, **S136**, 694–707.

Swenson, J.B., Paola, C., Pratson, L., Voller, V.R. and **Murray, A.B.** (2005) Fluvial and marine controls on combined subaerial and subaqueous delta progradation: Morphodynamic modelling of compound-clinoform development. *J. Geophys. Res.*, **110**, doi:10.1029/2004JF000265.

Syvitski, J.P.M. (2008) Deltas at risk. *Sustainability Sci.*, **3**, 23–32.

Syvitski, J.P.M. and **Saito, Y.** (2007) Morphodynamics of deltas under the influence of humans. *Glob. Planet. Change*, **57**, 261–282.

van Maren, D.S. (2005) Barrier formation on an actively prograding delta system: The Red River Delta, Vietnam. *Mar. Geol.*, **224**, 123–143.

van Rijn, L.C. (2007a) Unified view of sediment transport by currents and waves. I: Initiation of motion, bed roughness, and bed-load transport. *J. Hydrol. Eng.*, **133**, 649–667.

van Rijn, L.C. (2007b) Unified view of sediment transport by currents and waves. II: Suspended transport. *J. Hydrol. Eng.*, **133**, 668–689.

Int. Assoc. Sedimentol. Spec. Publ. (2014) **46**, 187–214

Sedimentation at the Jurassic–Triassic boundary, south-west Barents Sea: indication of climate change

ALF RYSETH

Statoil Exploration and Production, Norway, PO Box 10, N-9414 Harstad
(E-mail: alery@statoil.com)

ABSTRACT

Facies analysis of Upper Triassic to Lower Jurassic (Carnian to Sinemurian) strata in the Barents Sea area shows that deposition took place in two widespread, fluvial-dominated, delta plain settings, separated by a regional pro-deltaic mudrock of Early Norian age. Comparison of alluvial architectural trends, combined with mineralogical data and other published data, indicates that subsidence rates decreased across the Triassic to Jurassic boundary in the south-west Barents Sea area, accompanied by palaeogeographic re-configurations and rejuvenation of hinterlands to the south. Comparison with age-equivalent strata in the North Sea and on the Halten Terrace to the south allows for a discussion of possible regional climatic effects on deposition. It is inferred that the Late Triassic climate comprised two zones; a dry and hot zone to the south and a much more humid zone to the north. The Early Jurassic climate appears to be equable and generally humid. It is suggested that hinterland rejuvenation, larger annual precipitation and reduced subsidence rates combined to increase the grain size and sandstone content of Early Jurassic delta plain deposits relative to the Late Triassic units in the Barents Sea area.

Keywords: Triassic, Jurassic, sedimentology, climate, Barents Sea.

INTRODUCTION

Late Triassic to Early Jurassic strata on the Norwegian shelf (Fig. 1) are documented in the North Sea to the south (e.g. Vollset & Doré, 1984), from basins offshore mid-Norway (Jacobsen & van Veen, 1984; Dalland *et al.* 1988), and from the Barents Sea to the north (Dalland *et al.* 1988; Dallmann *et al.* 1999). Currently, the northern North Sea (59° N to 62° N) and the south-west Barents Sea (approximately 71° N to 72° N) are separated by more than 1500 km and by approximately 11° to 13° in latitudinal distance (Fig.1). However, these basins are linked in time and space; particularly, the Early Palaeozoic Caledonian collision between Laurentia and Baltica (Gee *et al.*, 2008) led to the development of a fold belt between Norway and Greenland, with an arm extending towards Svalbard (Worsley, 2008). Subsequently, Late Palaeozoic collapse of the Caledonian Orogen, followed by rift-related subsidence, led to the evolution of the North Atlantic rift system, of which the basins listed above are integral parts (e.g. Doré, 1992). However, the Barents Sea area (Fig. 2A) differs from the basins in the North Sea and on the mid-Norway shelf by being influenced by the younger (Late Palaeozoic to Mesozoic) folding in the Uralides to the east (Smelror *et al.* 2009).

Subsidence and sedimentation in the Barents Sea area (Fig. 2A) commenced after the Caledonian phase, with a general flow of sediment from westerly hinterlands to basins located to the east (Worsley, 2008). By Carboniferous times, extensional basins had started to form (Gudlaugsson *et al.*, 1998; Worsley, 2008; Smelror *et al.*, 2009), with possible initiation of several basins, including the Tromsø and Bjørnøya basins to the west and the Hammerfest and Nordkapp basins to the east (Nilsen *et al.*, 1995; Breivik *et al.*, 1998; Gudlaugsson *et al.* 1998). However, to the east, continent collision between Baltica and the West Sibirian Craton commenced during the Middle

From Depositional Systems to Sedimentary Successions on the Norwegian Continental Margin, First Edition.
Edited by A. W. Martinius, R. Ravnås, J. A. Howell, R. J. Steel, and J. P. Wonham.

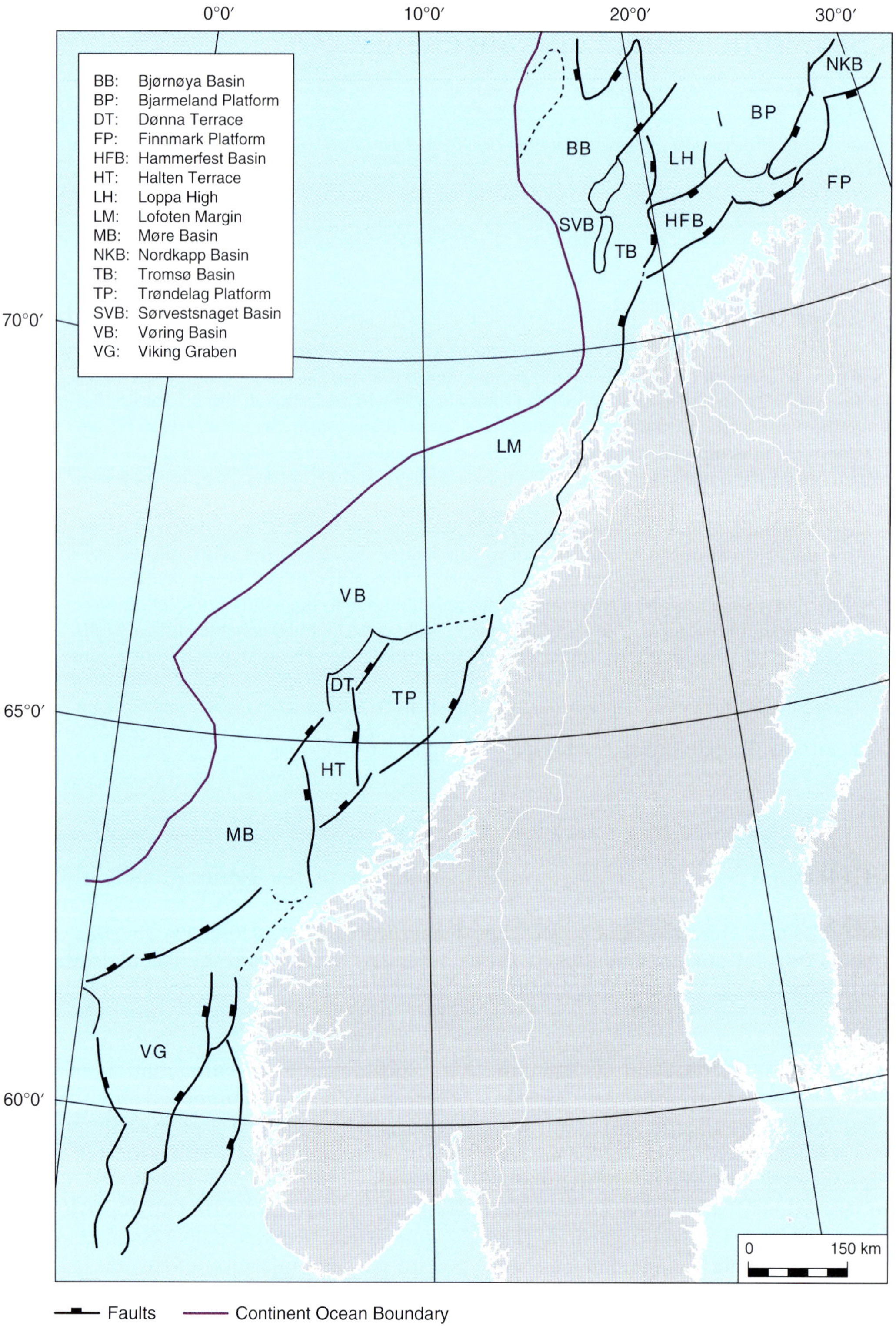

Fig. 1. Main structural features and sedimentary basins of the Norwegian Continental Shelf (based on Gabrielsen *et al.* 1990; Blystad *et al.* 1995 and Færseth, 1996).

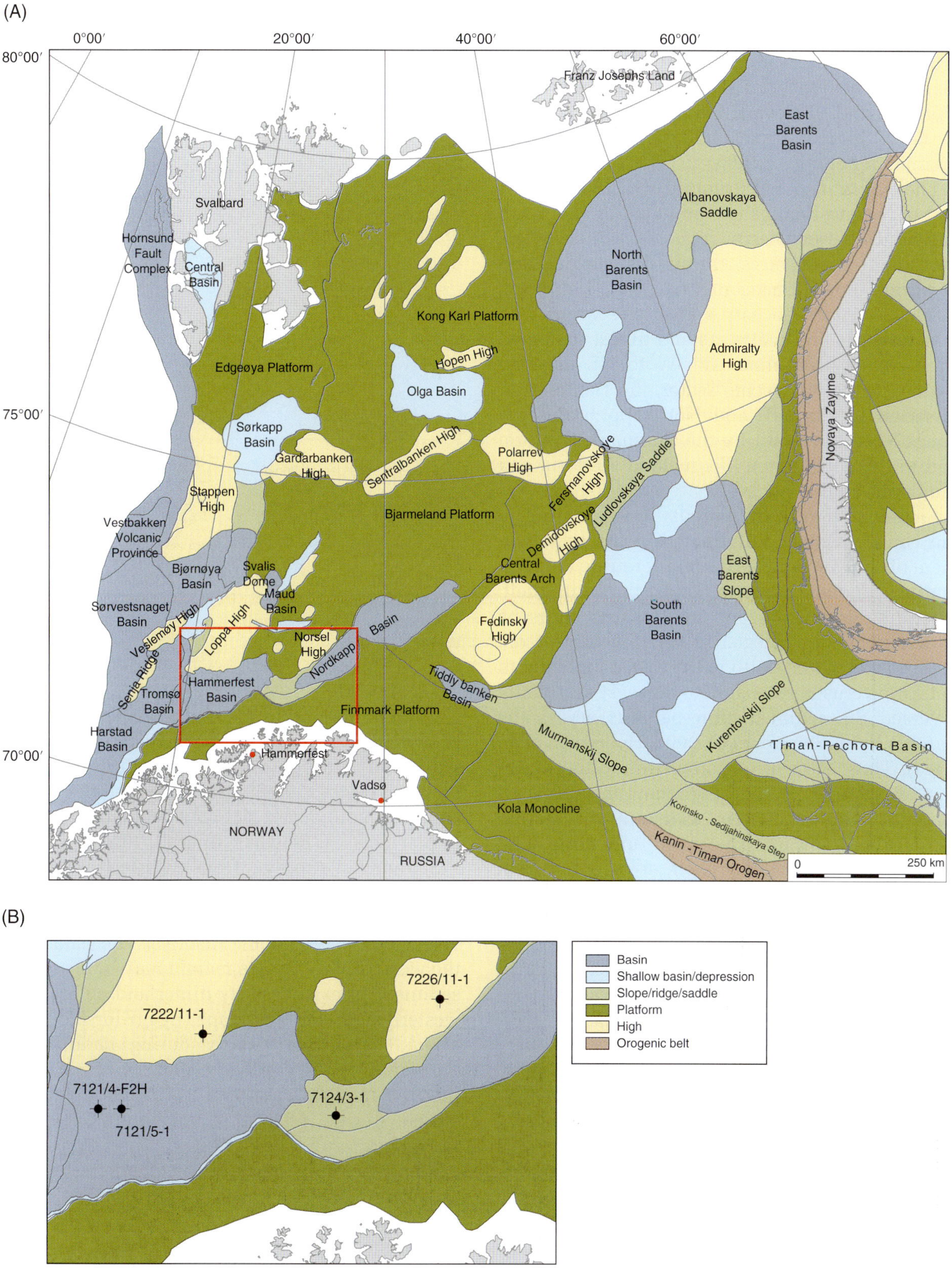

Fig. 2. (A) Regional structure of the Barents Sea (adopted from Henriksen *et al.* 2011). (B) Well locations, present study.

Carboniferous, eventually initiating the Uralide Orogeny (Smelror *et al.* 2009), which successively propagated northwards. The final folding took place to the east of the Timan Pechora Basin (Fig. 2A) and in Novaya Zemlya during Permian to Early Jurassic times (O'Leary *et al.*, 2004; Ritzman & Faleide, 2009).

The current regional structure of the Barents Sea east of Novaya Zemlya is dominated by the North and South Barents basins (Fig. 2A), which where the main depocentres during Palaeozoic and Mesozoic times (Johansen *et al.*, 1993; Henriksen *et al.*, 2011). These basins pass westwards into broader structural platforms and highs, separated by a set of predominantly SW to NE oriented basins. In the Norwegian sector, the Hammerfest (detailed in Fig. 2B) and Nordkapp basins are bounded to the south by the Finnmark Platform and to the north by the Bjarmeland Platform and the Loppa High.

Sedimentation in the Barents Sea area prior to the Uralide Orogen culminated in the establishment of widespread Late Carboniferous to Permian marine carbonate platforms across the entire shelf, extending eastwards and southwards into the Tethyan domain (Ziegler, 1988; Smelror *et al.*, 2009). The Uralide uplift to the east, coupled with persisting crustal extension to the west eventually reversed the sedimentation patterns across the Barents shelf (Smelror *et al.* 2009; Henriksen *et al.* 2011). Thus, the marine connection to the south-east was closed by end-Permian time and vast amounts of clastic material were transported from the Uralides towards the north and north-west across the entire Barents shelf, as recorded from seismic data revealing the progradation pattern of Early to Late Triassic shelf edges (Riis *et al.* 2008; Glørstad-Clark *et al.* 2010). Basins and platforms in the south-west Barents Sea area experienced significant subsidence during the Triassic, with accumulation of more than 3 km of clastic sediment (e.g. Riis *et al.* 2008).

Fig. 3 shows a schematic palaeogeographic model for the Late Triassic in the Barents Sea area. A widespread coastal plain/delta plain environment stretched westwards from Novaya Zemlya and Timan Pechora towards a marine seaway between the western Barents shelf and the Greenland landmass. The most significant clastic hinterlands were located to the east, within the Uralides, with a more limited flux from the Fennoscandian hinterland to the south (Riis *et al.* 2008). Triassic sandstones, derived from the Uralide hinterland to the south-east, are typically characterised by abundant plagioclase feldspars, which can be recognised, e.g. across the Bjarmeland Platform (Mørk, 1999). Jurassic sandstones are, in contrast, more quartz-rich and generally depleted in plagioclases (Bergan & Knarud, 1993).

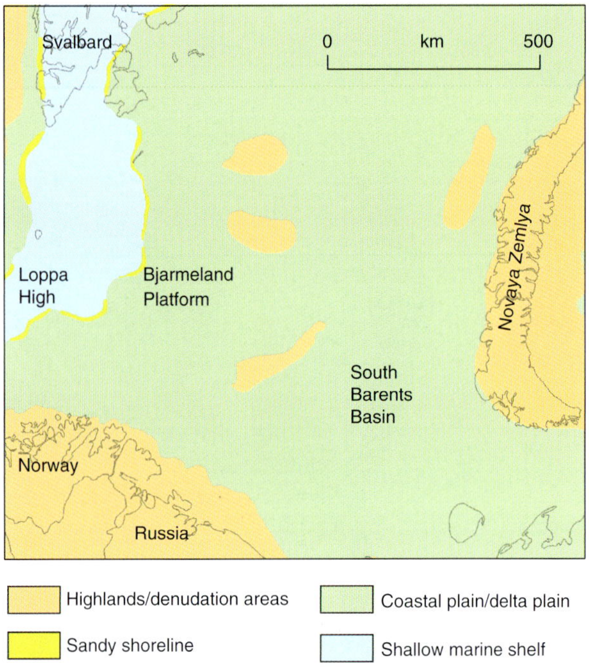

Fig. 3. Late Triassic palaeogeographic setting in the Barents Sea (modified from Henriksen *et al.* 2011). The south-west Barents Sea may have been located at a palaeo-latitude of about 50° to 55°N during the Late Triassic (Golonka, 2007; Worsley, 2008).

Sediment supply from westerly sources (Greenland) has been documented from Triassic strata in western Svalbard (e.g. Steel & Worsley, 1984; Riis *et al.*, 2008). Also, seismic data from the western Barents Sea margin may indicate that sediment from a westerly hinterland entered the western margin of the Loppa High (Glørstad-Clark *et al.*, 2010), particularly during Ladinian to Carnian times. Accordingly, some evidence exists that the marine seaway in the Atlantic rift was closed during phases of maximum shoreline regression.

The Early Jurassic palaeogeographic setting in the Barents Sea area seems to follow similar distribution patterns as inferred for the Late Triassic, with a marine seaway to the west and sediment supply from the south and east across the Barents shelf (Smelror *et al.*, 2009; Henriksen *et al.*, 2011). However, marine incursions into the Russian

Barents basins to the west of Novaya Zemlya may have restricted the sediment flux from the south-east since the latest Triassic (Norian) (see maps in Smelror *et al.* 2009). Such Early Jurassic reconfiguration of sediment supply can also be inferred from western Svalbard, where marginal marine sediment are sourced from the east, as opposed to the westerly source recorded from Triassic strata (Steel & Worsley, 1984; Smelror *et al.* 2009).

Analyses of uplift and denudation from apatite fission tracks in onshore areas to the south of the Barents shelf (eastern Finnmark and the Kola Peninsula) are indicative of significant and rapid Late Triassic to Early Jurassic uplift (Hendriks & Andriessen, 2002; Hendriks, 2003). These observations are of importance for the understanding of offshore sedimentation patterns, as they imply some rejuvenation of the hinterlands at or near the Triassic to Jurassic boundary, possibly influencing the composition and calibre of sediment delivered to the basins.

In the northern North Sea area, rifting in the Viking Graben (Fig. 1) commenced during the Late Permian to Early Triassic (e.g. Badley *et al.*, 1988; Færseth, 1996), with subsequent development of a post-rift basin which accumulated more than 2 km of predominantly terrestrial Triassic to Early Jurassic strata (Lervik *et al.*, 1989; Nystuen *et al.*, 1989; Steel & Ryseth, 1990). On the mid-Norway shelf (e.g. Halten and Dønna terraces; Fig. 1), initial extension may have commenced during the Carboniferous and was followed by Late Permian to Early Triassic rifting (Blystad *et al.*, 1995; Marsh *et al.* 2009). Here, more than 3 km of Triassic strata, including thick (400 m) halites at the mid-Triassic level, have been documented from wells and seismic data (Jacobsen & van Veen, 1984; Müller *et al.* 2005).

A Late Jurassic to Early Cretaceous rift climax seems to have affected the entire Norwegian shelf area from the North Sea (e.g. Badley *et al.*, 1988; Færseth, 1996) and the mid-Norway shelf (Blystad *et al.*, 1995; Marsh *et al.*, 2009) to the western parts of the Barents Sea area (Gabrielsen *et al.*, 1990). This extensional phase was eventually followed by regional post-rift subsidence, associated with a general westwards shift of Cretaceous to Cenozoic crustal extension to the Møre and Vøring basins (Fig. 1) off the mid-Norway shelf (Doré, 1992; Doré *et al.* 1997) and to the deep Tromsø, Bjørnøya and Sørvestsnaget basins in the western margin of the Barents Sea (Figs 1 and 2A; Breivik *et al.*, 1998; Ryseth *et al.*, 2003).

The present study is primarily related to the Late Triassic to Early Jurassic (Late Carnian to Sinemurian) succession in the Barents Sea, which will be discussed in terms of depositional environments and variations in sedimentary architecture, with some considerations of possible palaeogeographic and climatic changes. A comparison of time-equivalent sedimentary facies from the North Sea (Viking Graben) and the mid-Norway shelf (Halten Terrace) is also included, to provide some more detail to the possible regional variation in the Late Triassic to Early Jurassic climates and on the possible climatic effects on deposition.

TRIASSIC TO EARLY JURASSIC CLIMATIC SETTINGS

During Triassic time, landmasses were concentrated into a single continent (Pangaea) and distributed around the equator, with continental areas extending towards the poles (Golonka, 2007). During the Late Triassic, the North Sea area was located at around 40° N and the Barents Sea basins at approximately 50° N to 55° N (e.g. Ziegler 1988; Golonka, 2007; Worsley; 2008), with the mid-Norway shelf in an intermediate position. The Barents shelf bordered the Boral Sea during the Triassic, whereas the North Sea area was located well within the continental realm and away from the Boral and Tethyan seas (Ziegler, 1988; McKie & Williams, 2009). Consequently, terrestrial strata are dominant in the Triassic of the North Sea area (e.g. McKie & Williams, 2009) whereas age-equivalent strata in the Barents Sea area are characterised by intercalated marine and terrestrial facies (e.g. Henriksen *et al.*, 2011).

Modelling of global Triassic climates (Sellwood & Valdes, 2006) indicates that the continental positions between 40° N and 40° S (e.g. including the current North Sea area and the mid-Norway shelf) were generally warm, with sustained hot periods during which temperatures may have exceeded 30 °C for most of the year. The Early Triassic may have been one of the hottest phases in geological time, with development of extensive calcretes (soil carbonate), e.g. throughout Europe. Also, during Middle and Late Triassic times, the equatorial zone was generally hot and dry, although with a tendency towards more semi-arid conditions during the Late Triassic (Sellwood & Valdes, 2006; McKie & Williams, 2009).

Considerations of Triassic precipitation patterns (Sellwood & Valdes, 2006) suggests that large areas of Pangaea, extending almost to 50° N and 50° S, were characterised by a large excess of evaporation, with only short monsoonal periods. Wetter climates supporting coal formation existed in higher latitudes to the north and south, particularly during Late Triassic time (Sellwood & Valdes, 2006). Reconstruction of Triassic climates in the south-west Barents Sea (Hochuli & Vigran, 2010) also indicates that generally warm and humid conditions persisted throughout Late Carnian to Norian times, in contrast to the drier conditions to the south.

The Triassic to Jurassic boundary seems to associate with the establishment of a more humid climate in the present NW European basins (Hallam, 1985). This climatic change has been related to north-directed plate-tectonic drift combined with marine incursions (e.g. Frostick *et al.* 1992; Ahlberg *et al.* 2002). Other studies also suggests that widespread volcanism and massive release of isotopic light CO_2 to oceans and the atmosphere during the initial fragmentation of Pangaea at the Triassic to Jurassic boundary triggered significant climatic change and both marine and terrestrial biological mass extinction on a global scale (e.g. Hesselbo *et al.*, 2002, 2007; Schoene *et al.*, 2010). The sedimentary successions along the Norwegian shelf should perhaps reflect such a climatic change? This point will be addressed below in relation to the evaluation of possible climatic controls on deposition.

STUDY AREA AND WELL DATABASE

This study is based on core and wire-line log data from five wells (Fig. 2B), located in and around the Hammerfest Basin. Environmental interpretations are based on macroscopic descriptions of approximately 310 m of core, which are presented on 1:200 or 1:500 scales. Biostratigraphic data are available for all wells, but stratigraphic age assignments are primarily referred to the Hammerfest Basin, about which the most detailed records exist.

The stratigraphically oldest described core interval is obtained from well 7222/11-1 in the south-east corner of the Loppa High (Fig. 2B). Here a section of Late Carnian age, combined with 3D seismic data, illustrates the nature of the Late Triassic depositional system. Well 7124/3-1 is located in the structural transition zone between the Hammerfest and Nordkapp basins. Biostratigraphy data indicate that the cored section is of general Norian–Rhaetian age. Accordingly, it is a key well for the understanding of end-Triassic deposition in the area.

Further to the east, well 7226/11-1 is located on the Norsel High, in the structural transition zone between the Bjarmeland Platform and the Nordkapp Basin. The cored section is dated as Early Norian to Sinemurian and is a relevant reference for the Early Jurassic succession in a more easterly position relative to the other wells. Well 7121/4-F-2H is located in the central part of the Hammerfest Basin (Snøhvit field) and is a key reference well for the Early Jurassic (Hettangian to Early Pliensbachian) depositional systems in the central part of the Hammerfest Basin. Wire-line log data from well 7121/5-1 in the central part of the Hammerfest Basin are included for correlation purposes. In addition to the core and wire-line data from the listed wells, some petrographic data from a separate well on the Bjarmeland Platform will be discussed.

LATE TRIASSIC TO EARLY JURASSIC STRATIGRAPHY AND FACIES

Fig. 4 outlines the stratigraphy of Carnian to Early Pliensbachian strata in the Barents Sea, with correlations to the Halten Terrace offshore mid-Norway and to the Viking Graben in the North Sea. The Late Triassic to Early Jurassic succession in the Hammerfest Basin and adjacent areas in the south-west Barents Sea comprises strata deposited in fluvial, marginal marine (paralic) and offshore, pro-deltaic, environments and are contained within the lithostratigraphic Storfjorden and Realgrunnen subgroups and corresponding formations (Fig. 4; see also Dalland *et al.*, 1988 and Mørk *et al.*, 1999 for definitions). The present study is entirely related to the upper part of the Snadd Formation and to the overlying Fruholmen and Tubåen formations, for which observed formation tops and thicknesses in the studied wells are listed in Table 1.

The Snadd Formation (Ladinian to Early Norian; lower part not included in Fig. 4) is widely distributed across the Barents shelf, with a thickness commonly exceeding 1000 m to 1200 m (Mørk *et al.* 1999). The basal boundary with the underlying unit (Anisian to Ladinian Kobbe Formation; Mørk *et al.*, 1999) is defined by an organic

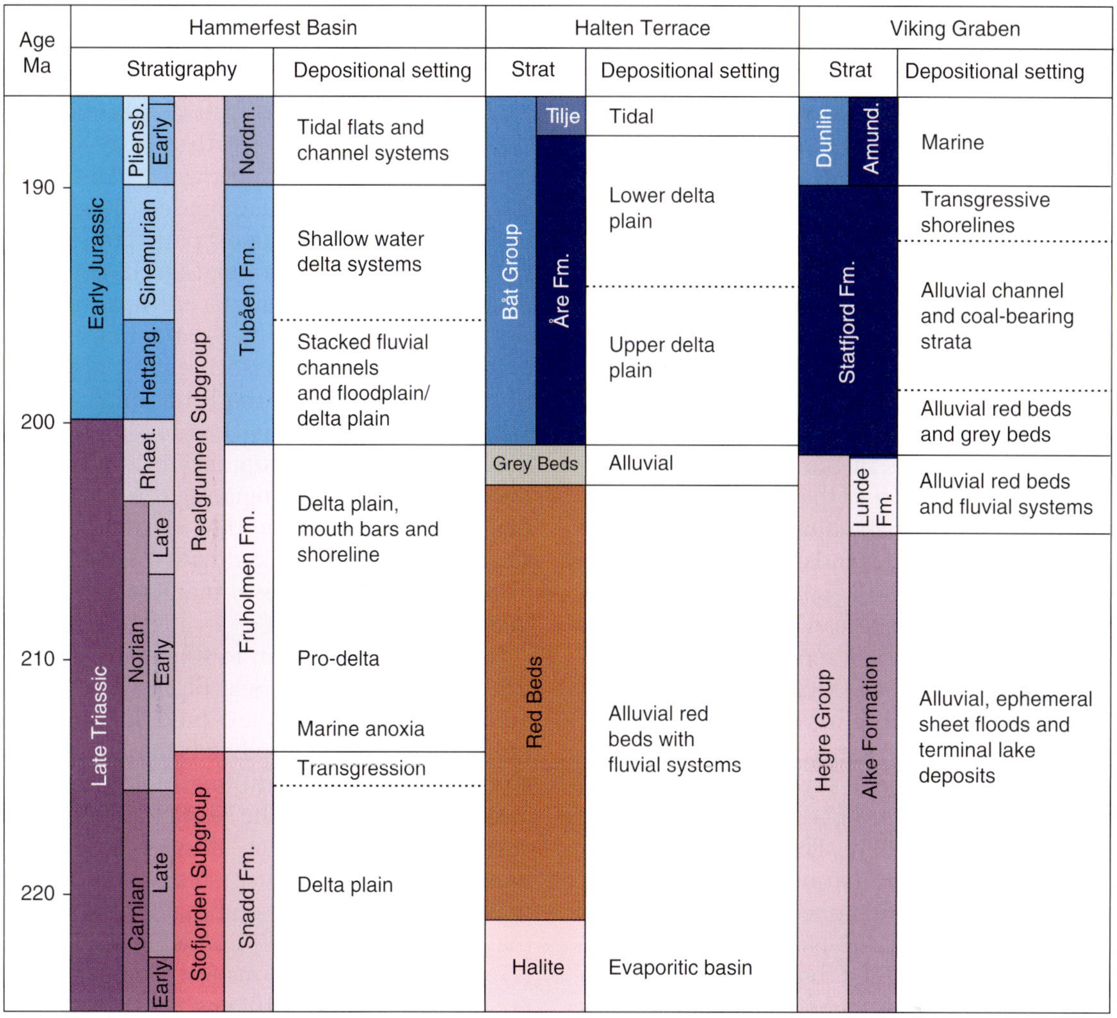

Fig. 4. Late Triassic to Middle Jurassic stratigraphy of the Hammerfest Basin (after Mørk *et al.* 1999), with correlations to the Halten Terrace and North Viking Graben. The stratigraphy from the Halten Terrace follows Dalland *et al.* (1988) and the recent updates of Müller *et al.* (2005) for the Triassic succession and Morris *et al.* (2009) for the Åre Formation. The lithostratigraphy in the Viking Graben follows the nomenclature of Vollset and Doré (1984) and the updated definitions of Lervik (2006) and Nystuen *et al.* (this volume). The absolute time scale (Ma) follows Ogg *et al.* (2008), with the Triassic to Jurassic boundary at 199.6 Ma.

Table 1. Well database with stratigraphic picks and formation thicknesses.

	Formation tops (m)			Formation thickness (m)		
Well	Snadd	Fruholmen	Tubåen	Snadd	Fruholmen	Tubåen
7121/4-F-2H		2800	2678			109*
7121/5-1	2793	2572	2507	407**	221	65
7124/3-1	1438	1304	1285	455	75	20
7222/11-1	627			1413		
7226/11-1	1296	1233.5	1199.5	582	62.5	34

*Corrected for well deviation.
**Formation not penetrated by well.

mudrock representing a regional transgression (Henriksen *et al.*, 2011). Above this boundary, the formation comprises a thick coarsening upwards unit of mudrocks and sandstones, followed by heterolithic and commonly coal-bearing strata. The whole interval is related to large-scale delta progradation, with the establishment of a widespread delta plain during Carnian to Early Norian times. The contact with the overlying Fruholmen Formation is marked by another organic mudrock, recording a major Early Norian transgression in the area (Henriksen *et al.* 2011).

The Fruholmen Formation (Early Norian to Rhaetian) shows variable thickness, but can exceed 250 m thickness in the Hammerfest Basin (Mørk *et al.*, 1999). Akin to the Snadd Formation, it comprises a lower coarsening upwards interval of dark mudrocks and sandstones, overlain by coal-bearing, heterolithic deposits. The formation has been related to widespread delta progradation, with the coal-bearing strata representing the delta plain (Gjelberg *et al.*, 1987; Mørk *et al.*, 1999).

The Tubåen Formation (Rhaetian to Sinemurian) is dominated by sandstones with subordinate mudrocks and thin coal layers and is up to 150 m thick in the western part of the Hammerfest Basin (Mørk *et al.*, 1999). The lower boundary with the Fruholmen Formation is marked by an abrupt increase in sandstone content, whereas the upper contact with the Nordmela Formation (Pliensbachian) is defined by a transition into finer grained sandstones and siltstones. Sedimentological studies indicate that the Tubåen Formation is of fluvial, deltaic, origin (Berglund *et al.*, 1986; Gjelberg *et al.*, 1987), with the main sandstone bodies representing a variety of fluvial and distributary channel types. The formation seems to mark a significant change in fluvial style across the Barents Sea and comprises more densely stacked channel sandstones with less inter-bedded fine-grained deposits than the underlying units. Gjelberg *et al.* (1987) noted that coarsening upwards sandstone bodies in the uppermost part of the Tubåen Formation represent deltaic mouth bars marking initial Early Jurassic marine incursion from the west. Sandstones in the Tubåen Formation are used for CO_2 storage in the Snøhvit petroleum development in the axial part of the Hammerfest Basin (Spencer *et al.*, 2008).

Together, the Fruholmen and Tubåen formations record significant Late Triassic to Early Jurassic delta progradation and re-establishment of a widespread fluvial dominated delta plain. It is assumed that the maximum regression was reached during the Hettangian to Sinemurian stages (Tubåen Fm.; Henriksen *et al.*, 2011). Nøttvedt *et al.* (1993) indicated that the boundary between Fruholmen and Tubåen formations is partly erosive. Available biostratigraphic data from well 7121/4-F-2H and other wells in the Hammerfest Basin yield a Rhaetian age for the topmost part of the Fruholmen Formation in the basin position. In contrast, Late Norian and younger strata are apparently truncated or missing in wells 7124/3-1 and 7226/11-1 to the east, where Early Norian strata are preserved in the topmost part of the Fruholmen Formation. Furthermore, the basal strata in the Tubåen Formation are of Rhaetian age in the Hammerfest Basin, but yield a Hettangian age on the Bjarmeland Platform and towards the Finnmark Platform. These observations accord with a Rhaetian unconformity towards the basin margins and across the Bjarmeland Platform.

Late Triassic (Carnian to Rhaetian) strata in the Viking Graben (North Sea; Fig. 1) and the Halten Terrace (mid-Norway) are predominantly terrestrial (Fig. 4). Alluvial red beds deposited in generally dry and hot conditions, dominate the Late Triassic succession in the North Sea Viking Graben and seem to reflect fluvial run-off from Fennoscandia and other source areas, draining into terminal basins dominated by ephemeral playa lakes (Alke and Lunde fms.; Fig. 4; e.g. Nystuen *et al.*, 1989; Frostick *et al.*, 1992; Morad *et al.*, 1998; McKie & Williams, 2009). On the Halten Terrace, evaporitic salt layers (halite) of inferred Ladinian to Carnian age are interbedded with reddish mudrocks and are eventually succeeded by alluvial red beds which seem to dominate the Late Triassic succession (Müller *et al.*, 2005).

Early Jurassic strata from the Viking Graben comprise the Statfjord Formation (Rhaetian to Sinemurian) and the Dunlin Group (Pliensbachian to Toarcian). Of these, the Statfjord Formation is generally related to widespread alluvial deposition (Nystuen *et al.*, 1989; Steel & Ryseth; 1990; Ryseth, 2001). Furthermore, it shows a gradual transition from alluvial red beds in its lower part, to coal-bearing grey beds in the upper part. However, the uppermost part of the Statfjord Formation is marginal marine and records Middle to Late Sinemurian marine incursion in the Viking Graben (Steel & Ryseth, 1990; Ryseth, 2001). Overlying strata in the Dunlin Group are predominantly of offshore and shallow marine origin (Charnock *et al.*, 2001).

On the Halten Terrace, the Triassic red beds pass vertically into alluvial grey beds of Rhaetian age (Morris *et al.*, 2009) and eventually coal-bearing grey beds in the Åre Formation (Rhaetian to Pliensbachian; Dalland *et al.*, 1988; Morris *et al.*, 2009). Sedimentological studies of the Åre Formation (Svela, 2001) indicate that the lower part is of fluvial/alluvial origin, whereas the upper part comprises paralic facies of interbedded wave-dominated mouth bars, distributary channels and wave-influenced bay fill deposits. Based on biostratigraphic data, a general Hettangian to Sinemurian age can be inferred for the lower, fluvial part of the Åre Formation (Morris *et al.*, 2009).

The Late Triassic red bed associations from the Viking Graben and Halten Terrace are approximately time equivalent with the upper Snadd Formation and the Fruholmen Formation in the Barents Sea (Fig. 4). It also follows from the presented age assignments that the Tubåen Formation is more or less time-equivalent with the Statfjord and lower Åre formations to the south. Accordingly, terrestrial and dominantly fluvial conditions seem to have persisted over most of the Norwegian shelf during the Carnian to Sinemurian stages, although with some more marine influence to the north.

Regional climatic scenarios were indicated above and will be addressed at the end of this study. However, the dominance of red beds in the Late Triassic successions to the south can clearly be related to generally dry and hot conditions (e.g. McKie & Williams, 2009). In contrast, concurrent coal-bearing successions in the Barents Sea testify to a much more humid climate to the north, e.g. during deposition of the Snadd and Fruholmen formations (Fig. 4). Also, the common occurrence of coal-bearing strata in the Statfjord and lower Åre formations to the south and in the Tubåen Formation to the north, indicate a more or less equable and generally humid climate across the entire area from the North Sea to the Barents Sea during the Early Jurassic (see also Hallam, 1985).

In the following paragraphs, the Late Triassic to Early Jurassic depositional setting in the Barents Sea area will be outlined, with a discussion of possible mechanisms controlling deposition. Subsequently, some effects of different Late Triassic climatic settings will be addressed from a regional perspective, with additional comments on the inferred increase in humidity at the Triassic to Jurassic boundary.

Late Carnian to Early Norian delta plain (Snadd Formation)

The Late Carnian to Early Norian succession in the Barents Sea area represents a possible regressive maximum, with a widespread delta plain environment extending across large parts of the shelf (Riis *et al.*, 2008; Glørstad-Clark *et al.*, 2010; Henriksen *et al.*, 2011). Fig. 5 shows a graphic core log from the uppermost Snadd Formation in well 7222/11-1 (Loppa High) together with a Gamma Ray curve. Core photographs from the same interval are shown in Fig. 6.

The cored section comprises a lower unit (Fig. 5; below app. 798 m core depth) of thin coal beds and inter-bedded mudrocks and sandstones (Fig. 6A to C). Thin root horizons are usually seen below the coal beds, occasionally with siderite concretions. The associated fine-grained strata comprise laminated siltstones and very fine-grained sandstones with various types of ripple lamination and bioturbation (Fig. 6B). Coarser sandstone beds range in thickness from approximately 1 m to 2.5 m. They are characterised by rather flat upper and lower boundaries (Fig. 5) and are dominated by planar lamination and ripple lamination.

The overlying thicker sandstone unit in Fig. 5 (app. 778 m to 798 m core depth) is sharply based and comprises a basal intraclast conglomerate (Fig. 6D), followed by fine-grained to medium-grained sandstones. Primary sedimentary features include dm-scale cross-stratification with possible paired drapes (Fig. 6E and F) and associated massive sandstones. The log data (Gamma Ray curve; Fig. 5) show that the sandy unit has a total thickness of approximately 25 m. Furthermore, available wire-line log data from well 7222/11-1 (not included in Fig. 5) show that it is the only significant sandstone body present within a more than 500 m thick, mudrock-dominated succession. Seismic amplitudes calibrated to well data (Fig. 7) indicate that the cored sandstone in Fig. 5 (labelled A0 in Fig. 7) forms a low-sinuosity ribbon with a general SW to NE orientation and a width of approximately 1000 m. The amplitude map also show that other ribbon-like features of variable width, sinuosity and orientation are present away from the well.

Data from the uppermost part of the Snadd Formation in well 7124/3-1 are shown in Fig. 8. Here, the cored section is divided into two main facies associations. The lower association comprises thin-bedded sandstone layers and grey muddy strata with a thin coal bed, abundant roots

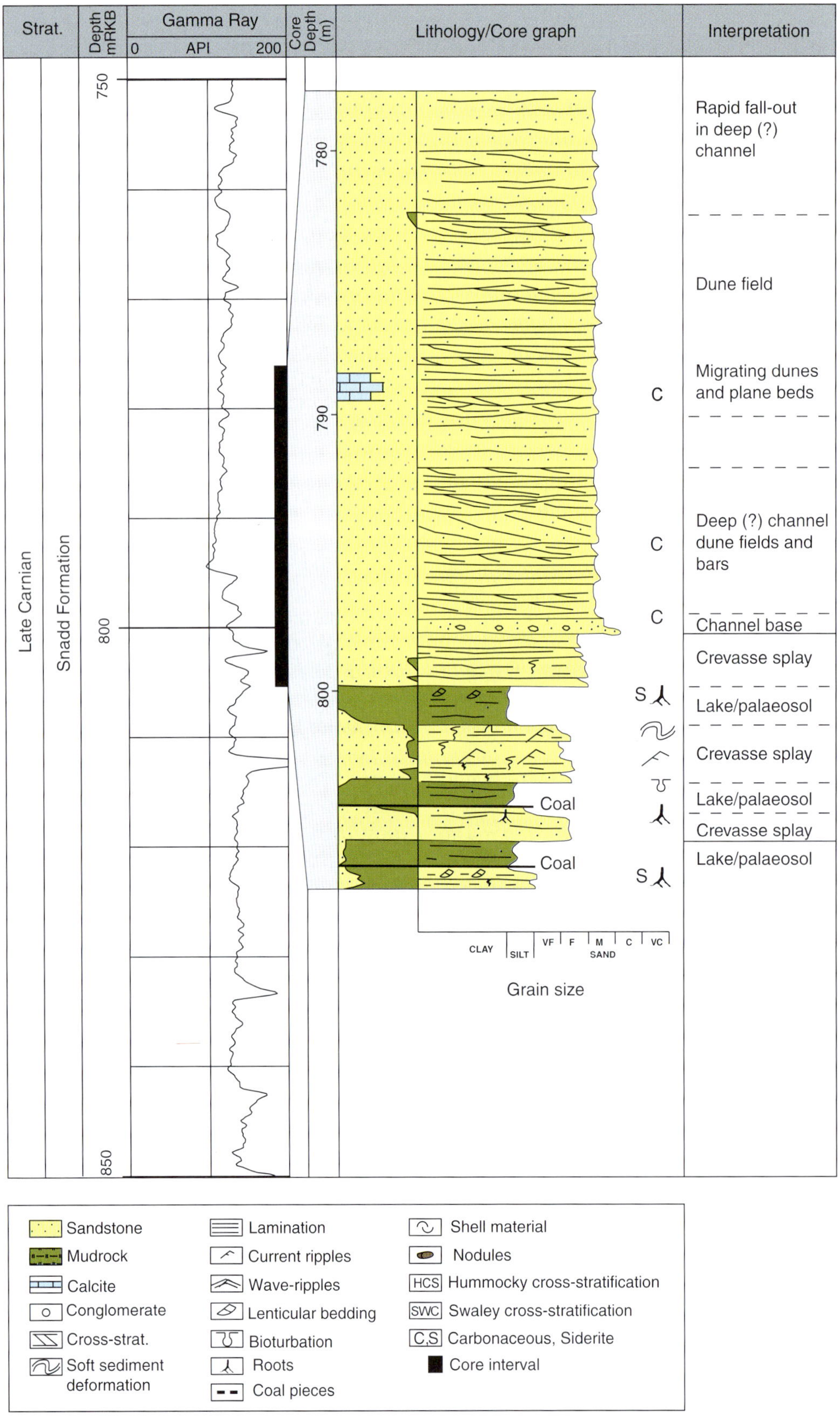

Fig. 5. Cored section and Gamma Ray curve from the Late Carnian upper Snadd Formation, well 7222/11-1, south-east Loppa High. Note that the cored section is plotted on a different vertical scale than the wire-line curve.

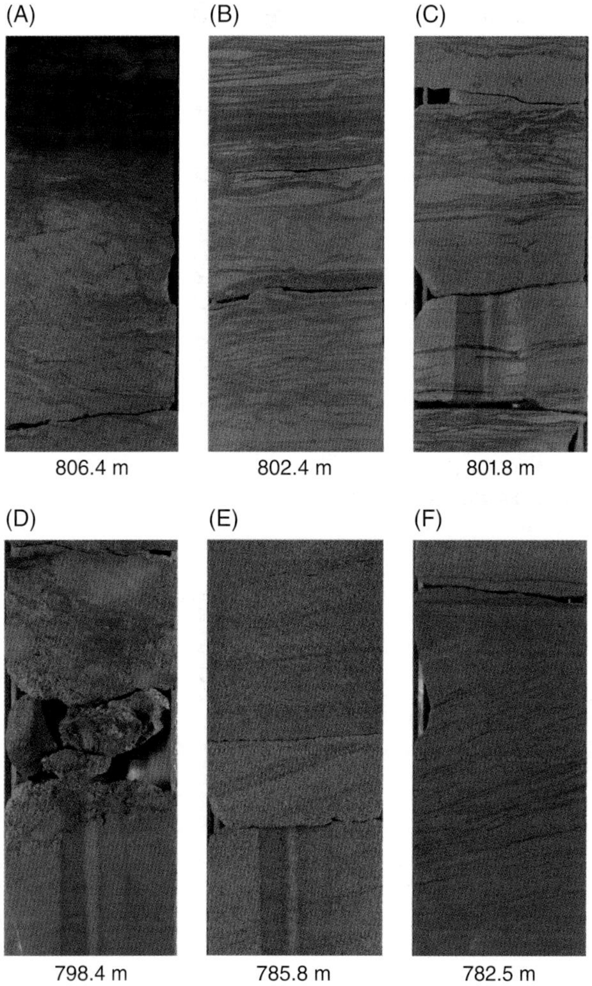

Fig. 6. Core photographs, Snadd Formation, well 7222/11-1. (A) Bioturbated/rooted siltstone capped by thin coal bed; (B) Moderately bioturbated sandy mudrock with wave-ripple lamination; (C) Thin-bedded sandstone beds in weakly bioturbated background sediment; (D) Sharply based intraclast conglomerate/channel base lag deposit; (E) Medium-grained sandstone with dm-scale cross-stratification; (F) Cross-stratified sandstone with paired (?), muddy/micaceous drapes on fore-sets. Core width is ~12 cm.

and other pedogenic features, as was also seen in well 7222/11-1 (Fig. 5). The overlying association is approximately 12 m thick and composed of very fine-grained to fine-grained, micaceous sandstone, dominated by wave-generated primary structures and some intensively bioturbated intervals.

As shown in Fig. 5 (well 7222/11-1), the heterolithic deposits below the main sandstone body are related to deposition in a lower delta plain environment. The coal beds and root horizons reflect soil formation and accumulation of organic material. Some marine influence can be inferred from the observed bioturbation and deposition probably took place in lagoons and bays. The sandstone beds seem to record rather abrupt events of deposition and are thought to represent crevasse splays spilled from adjacent channels. A similar interpretation is also favoured for the lower part of the cored section in well 7124/3-1 (Fig. 8), which is related to deposition in the overbank and smaller distributary channels of the delta plain during the final phase of terrestrial deposition in the Snadd Formation.

Based on its large thickness (app. 25 m), primary sedimentary features and the ribbon-like geometry (Fig. 7), the main sandstone body in Fig. 5 is thought to represent a major distributary channel on the lower delta plain, with the basal conglomerate forming an eroded lag deposit. Some tidal influence may be inferred from the observed paired drapes on the cross-stratified sets and it is assumed that the channel filled due to dune migration under the influence of both fluvial and tidal currents. The massive sandstone beds may record rapid fall-out of suspended sandy sediment during waning flows (e.g. Johnson, 1984).

All the ribbon-like features seen in the seismic data (Fig. 7) may represent channelised sandstone bodies isolated within the fine-grained delta plain deposits. The large variation in orientation and sinuosity between individual channels may relate to a very low depositional gradient typical for lower delta plain settings (e.g. Reading & Collinson, 1996). Similar seismic features are observed in several 3D seismic surveys throughout the Snadd Formation. Sandstone thicknesses are highly variable and occasionally exceeding 50 m, with widths reaching maximum values of up to 20 km (Henriksen *et al.*, 2011). Such data clearly show that some significant channel systems were present, possibly funnelling large amounts of sediment through the Late Triassic delta plain of the Barents shelf.

Wave-rippled and bioturbated strata in the uppermost part of the Snadd Formation (Fig. 8) are related to marine reworking of the underlying delta plain during the final transgression/marine incursion of the delta plain. These strata are in turn succeeded by mudrocks in the Fruholmen Formation.

Norian to Rhaetian delta progradation (Fruholmen Formation)

Above the Snadd Formation, the basal mudrocks of the Fruholmen Formation in well 7124/3-1 (Fig. 8) are dark grey and laminated, with abundant

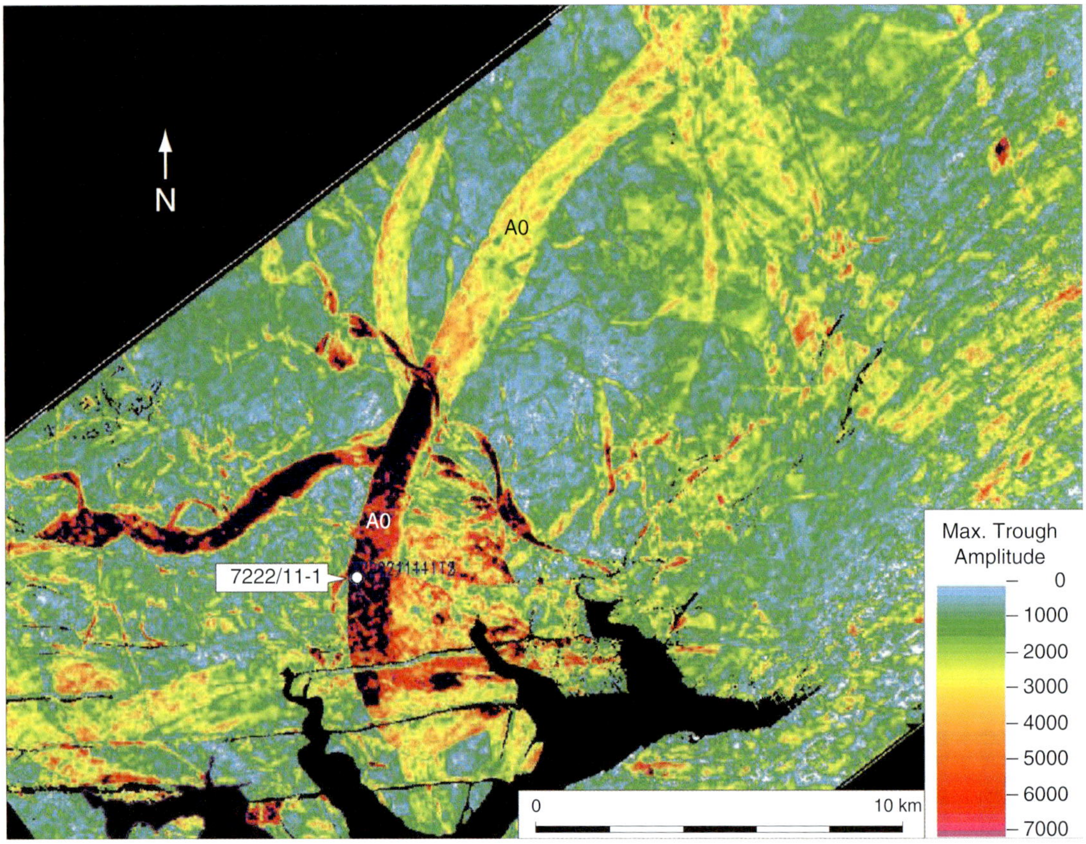

Fig. 7. 3D seismic representation of channelised sandstone bodies in the upper part of the Snadd Formation, SE Loppa High. A0 corresponds to the cored section (channel) shown in Fig. 5. Seismic image (SG9803) by Inger Laursen (Statoil).

siderite concretions. However, these fine-grained strata are replaced vertically by an 8 m thick, sharply based, unit of bioturbated and wave-rippled sandstones with interbedded muddy layers, thus giving the lower part of the Fruholmen Formation a general coarsening upwards grain size profile.

The bioturbated sandstone is subsequently overlain by a 25 m thick coarse-grained to fine-grained sandstone unit (app. 1335 m to 1360 m in Fig. 8), comprising two, stacked, fining upwards units, each with very sharp and apparently erosive bases. This major sandstone body is otherwise characterised by common intra-clast lag conglomerates (mainly coal pieces and mudrock fragments), dm-scale cross-stratification and massive sandstone sections, capped by fine-grained, current-rippled sandstone.

Intercalated thin sandstone units, laminated and weakly bioturbated mudrock and thin coal beds with associated root horizons characterise the upper 30 m of the Fruholmen Formation in well 7124/3-1 (app. 1305 m to 1335 m; Fig. 8). A single, sharp-based and fining upwards, 3 m thick sandstone unit with some dm-scale cross-stratification and current ripple lamination is present in this interval (app. 1325 m; Fig. 8). Other sandstone layers are generally thin and inter-bedded with muddy deposits. A very similar succession is also seen in Fruholmen Formation in well 7226/11-1 (Fig. 9) on the Norsel High (Fig. 2). Here, the uppermost part of the formation is cored (app. 1233 m to 1247 m core depth in Fig. 9) and comprises a mudrock-dominated succession with associated thin-bedded sandstones. Rooted horizons are generally grey with associated carbonaceous rootlets, however with a notable thin red/grey mottled zone (app. 1240 m core depth in Fig. 9).

Regionally, in the Barents Sea, deposition in the Fruholmen Formation was initiated by a major marine incursion during Early Norian times (Henriksen *et al.* 2011). The basal mudrock in the Fruholmen Formation is related to deposition in a marine, pro-deltaic environment with the dark

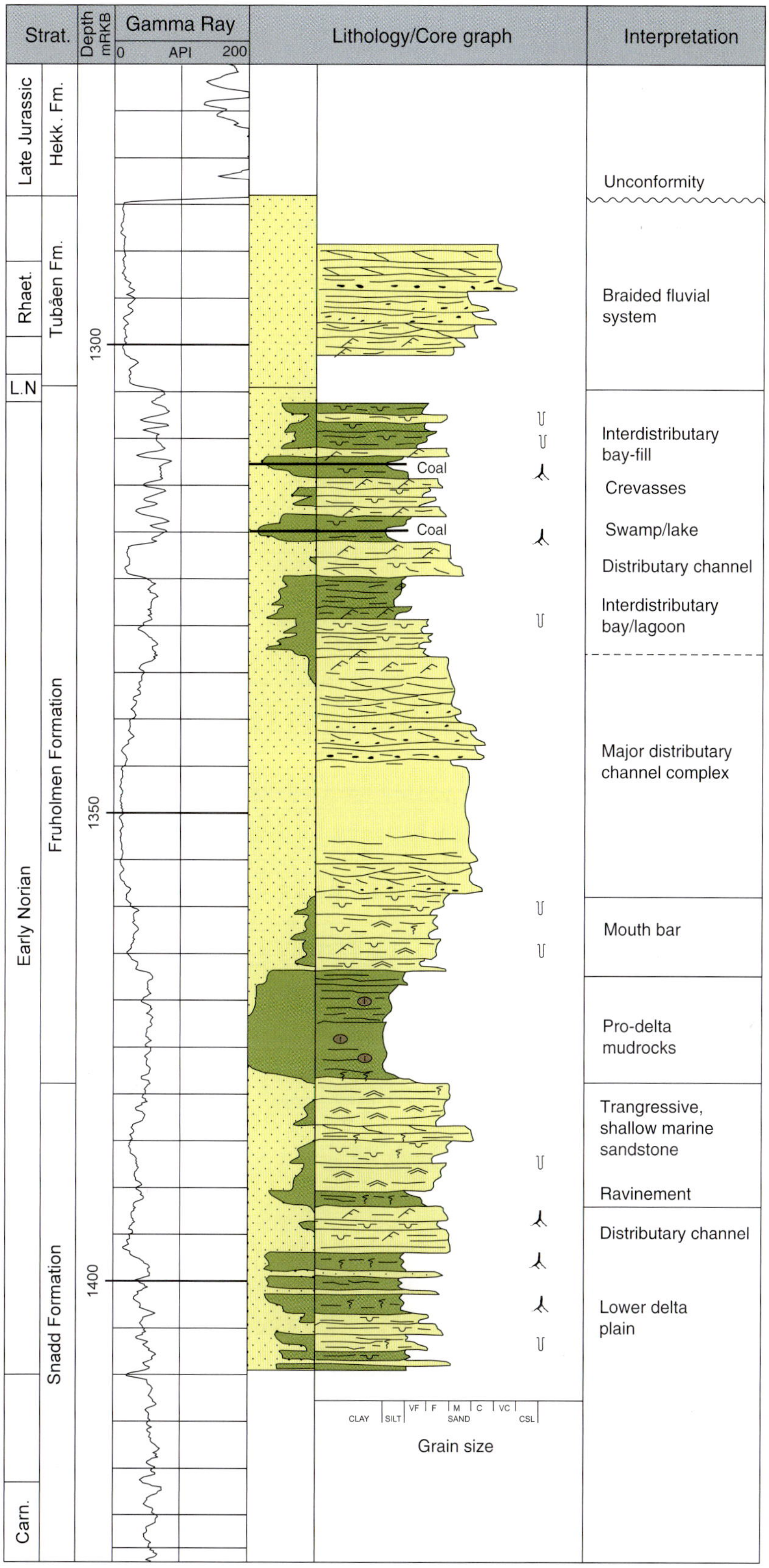

Fig. 8. Cored section and Gamma Ray curve from Early Norian to Rhaetian strata (Snadd, Fruholmen and lowermost Tubåen formations) in well 7124/3-1, located in the transition zone between the Hammerfest Basin and the Nordkapp Basin. For legend, see Fig. 5.

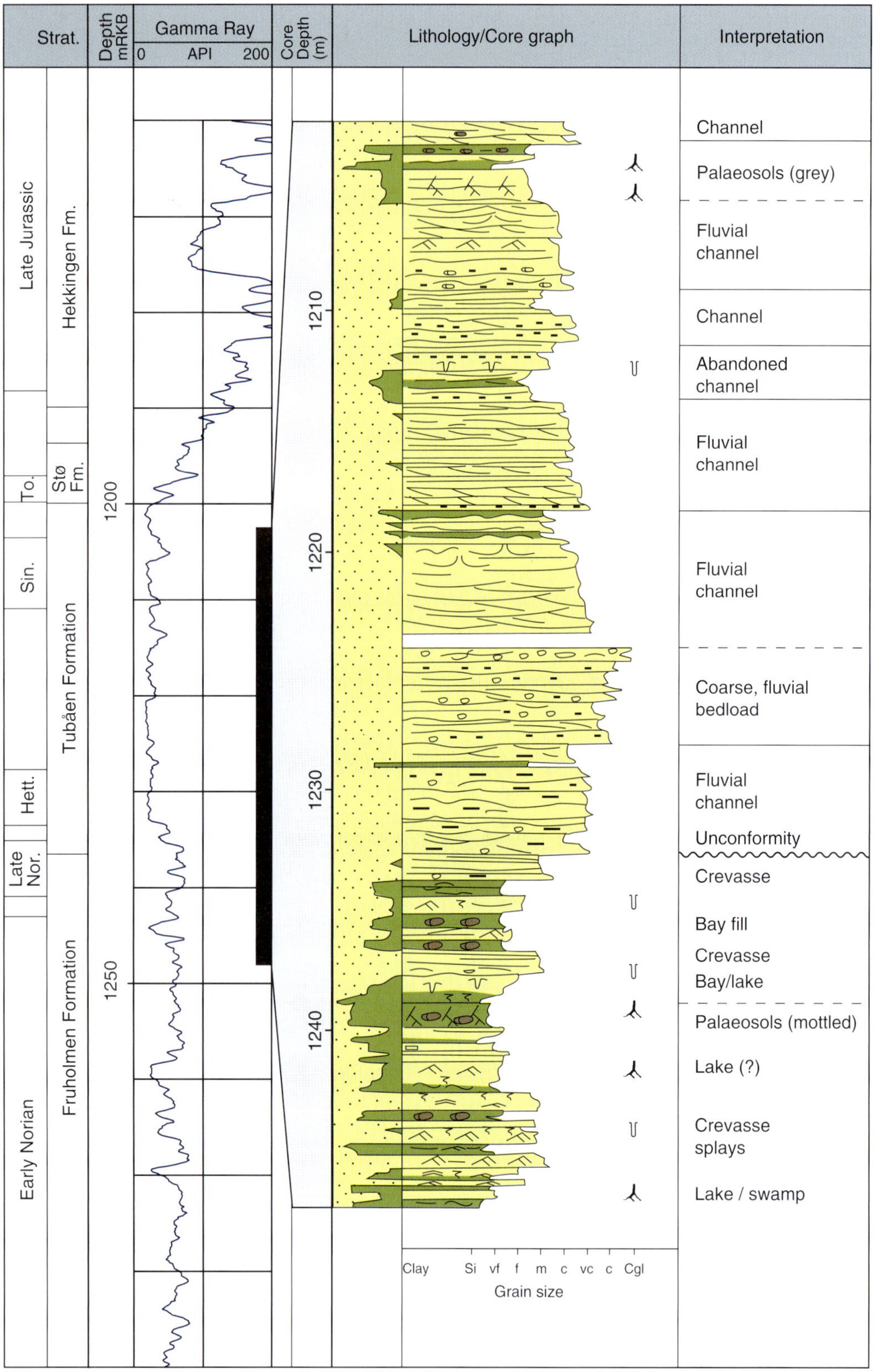

Fig. 9. Cored section of the upper Fruholmen and Tubåen formations in well 7226/11-1, Norsel High. Note that the cored section is plotted on a different vertical scale than the wire-line curve. For legend, see Fig. 5.

grey colour indicating a slightly depleted oxygen level in the bottom sediment. The recorded wave-generated structures and bioturbation in the overlying sandstone unit, coupled with its relatively sharp base and the inter-bedding of sandy and muddy layers, seem most compatible with deposition in a delta mouth bar environment. The relatively sharp base of the interpreted mouth bar unit accords with a sudden incursion of sand and may reflect a lateral delta lobe shift driven, for example, by river avulsion. The coarsening upwards pattern in the lower part of the Fruholmen Formation is widespread (Mørk *et al.*, 1999; Henriksen *et al.*, 2011), hence the interval is related to significant delta progradation following the Early Norian transgression.

Strata in the middle and upper parts of the Fruholmen Formation can be related to deposition in a variety of sub-environments of a delta plain (as summarised in Figs 8 and 9). All the primary sedimentary features (e.g. cross-stratification and current ripple lamination) in the 25 m-thick sandstone body capping the mouth bar deposits (Fig. 8) can be related to deposition from sustained traction currents in a channelised setting (e.g. Collinson, 1996), with the massive sandstone beds recording rapid sediment fall-out of suspended material during waning flows (as was also inferred for the Snadd Formation channel; Fig. 5). Accordingly, this sandstone body is related to deposition from a major fluvial distributary channel complex, possibly representing a proximal feeder system for the underlying mouth bar system. The organisation of the main sandstone body (Fig. 8) into two stacked fining upwards units may also indicate that it formed in response to lateral migration and aggradation within a channel belt.

Common rootlet horizons and thin coal beds in the upper part of the Fruholmen Formation record soil formation and peat accretion in a terrestrial setting and thus confirm the delta plain environment. In this context, the 3 m-thick sandstone unit with observed dm-scale cross-stratification (Fig. 8; 1325 m) may represent a smaller distributary channel. Associated muddy intervals with thin-bedded sandstone layers are related to deposition within lakes and small bays, with the sandy interbeds representing crevasse systems and possible levees derived from nearby channels. Observations of bioturbation in these beds probably reflect some marine impact, thus implying a lower delta plain setting for the investigated strata, as was also inferred for the underlying Snadd Formation. The grey colour of the fine-grained sediment testifies to deposition in generally water-saturated and stagnant conditions. However, scattered occurrences of red/grey coloured mottling point to infrequent (?) events of drying and partial oxidation of the delta plain. Such changes can be driven by vertical shifts in the ground water level (e.g. Besly & Fielding, 1989) and may reflect seasonal variations in precipitation.

Late Rhaetian to Sinemurian fluvial system (Tubåen Formation)

The Tubåen Formation is observed in three well sections, which will be described and interpreted separately, followed by a short discussion of the depositional system.

Well 7124/3-1

The Tubåen Formation is approximately 20 m-thick in well 7124/3-1 and is truncated beneath Late Jurassic mudrocks (Fig. 8). The basal boundary is apparently sharp and characterised by an abrupt vertical transition from heterolithic deposits (Fruholmen Formation) to thick-bedded sandstones in the Tubåen Formation. The cored section comprises fine-grained to very coarse-grained sandstones with dm-scale cross-stratification, planar lamination and current ripple lamination. Quartzite pebbles and other rock fragments (<0.5 cm), angular mud-clasts (up to 10 cm), cm-size coal fragments and fine-ground organic material are also present throughout the sandstone unit. The logged section (Fig. 8) comprises at least two stacked and sharply based fining upwards units.

The relatively coarse nature of the sandstones and the association of primary sedimentary structures, with abundant cross-stratification, planar lamination and current ripple lamination, all testify to deposition from strong and sustained traction currents in a channelised setting (Collinson, 1996). Also, the abundance of eroded organic material is indicative of a more or less fully terrestrial, alluvial setting. The organisation of strata into stacked fining upwards units is related to juxtaposition of several channel units, possibly within mobile channel belts. As shown in Fig. 8, a braided fluvial setting is inferred for the Tubåen Formation in well 7124/3-1; a point that will be further discussed below.

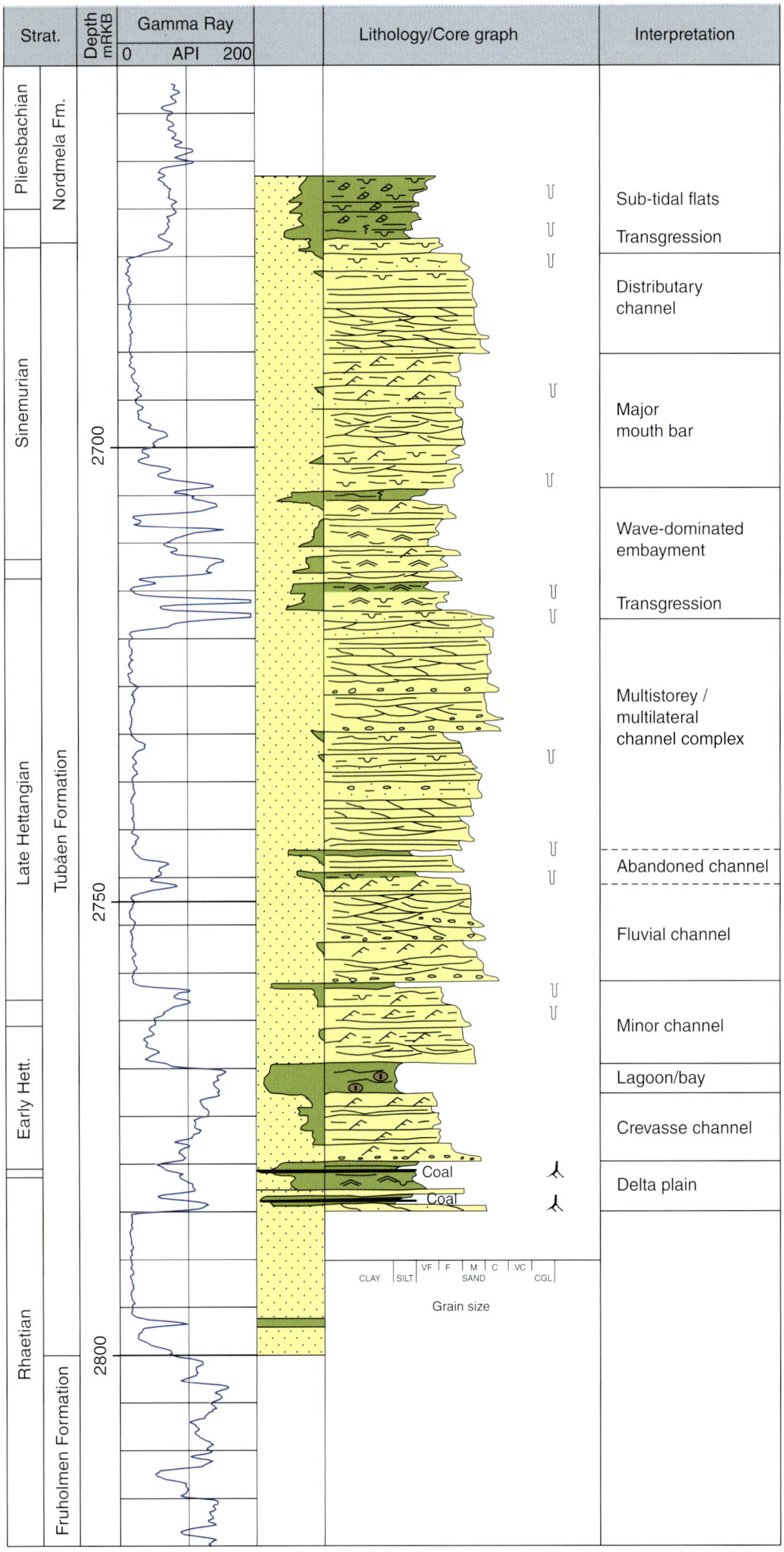

Fig. 10. Cored section of the Tubåen and lowermost Nordmela formations in well 7121/4-F-2H, Hammerfest Basin (Snøhvit field). For legend, see Fig. 5.

Well 7226/11-1 (Norsel High)

The Tubåen Formation is 34 m thick in well 7226/11-1 (Fig. 9, cored section above app. 1233 m) and is more or less completely sandstone-dominated, akin to the section in well 7124/3-1. As illustrated (Fig. 9), the formation can be divided into a series of sharply based fining upwards units, ranging in thickness from approximately 3 m and up to 10 m. The coarsest material consists of rounded quartzite clasts with long axes up to 10 cm and the sandstones also contain abundant eroded coal fragments and muddy intra-clasts. Furthermore, the primary structures include stratified, gravelly, sandstone and dm-scale cross-stratified sandstone, with current ripples and horizontal laminations in the upper parts of each unit. Finer-grained intervals occur in the upper part of the cored section, comprising laminated very fine-grained sandstones and siltstones, occasionally with preserved carbonaceous roots.

A high-energy fluvial setting is inferred for the Tubåen Formation in well 7226/11-1. Particularly, the coarse-grained nature of the sandstones and the dominance of traction-generated sedimentary features, such as cross-stratification and horizontal lamination, may indicate a braided fluvial style (as in well 7124/3-1). The preserved finer material is related to events of channel abandonment and/or the establishment of plant cover (palaeosol in Fig. 9) in poorly preserved, associated floodplain deposits.

Well 7121/4-F-2H (Hammerfest Basin)

Fig. 10 summarises the stratigraphy and sedimentology of the Tubåen and lowermost Nordmela formations in the central part of the Hammerfest Basin. Here, the Tubåen Formation shows a thickness of 121.5 m, of which approximately 110 m are covered by the core. As in other wells, the basal contact with the Fruholmen Formation is marked by an abrupt increase in sandstone content, although a stratigraphically more complete transition is observed from biostratigraphy (e.g. compared to well 7226/11-1; Fig. 9), with inclusion of Rhaetian strata in the lowermost part of the formation. The upper contact with the Nordmela Formation is marked by a rapid transition into finer-grained sandstones and siltstones.

Below approximately 2720 m (Fig. 10), stacked sandstone units with predominant fining upwards grain size motifs dominate, and are associated with fine-grained units and thin coal beds. Sandstone units are typically sharply based (Fig. 11A) and are associated with eroded organic debris, mudstone clasts and siderite nodules. These lags are usually succeeded by cross-stratified sandstones. As shown in Fig. 11B, weakly tangential set bases and micaceous/organic drapes on fore-set lamina can be seen in some individual sets, but these are not pervasive through the facies units. Finer grained sandstone capping the fining upwards units (Fig. 11C) show predominant sub-horizontal lamination with current ripples and are occasionally bioturbated, although not very intensively. The associated fine-grained strata are invariably grey and comprise laminated mudrocks, thin-bedded sandstone and thin coal beds with preserved carbonaceous rootlets (Fig. 11D).

In well 7121/4-F-2H the upper part (app. 35 m) of the Tubåen Formation (above ~2720 m in Fig. 10) shows a distinct coarsening upwards (CU) grain size motif. The lower part of this CU-unit is in particular characterised by very fine-grained to fine-grained micaceous sandstones with abundant cm to dm scale wave-ripple lamination (e.g. Fig. 11E) and discrete event beds with internal low-angle cross-stratification (Fig. 11F). These strata pass vertically into coarser grained sandstones and eventually to a sharply based fining upwards unit dominated by dm-scale cross-stratification (Fig. 10).

As annotated in Fig. 10, the cored section can be related to deposition in a variety of terrestrial to marginal marine environments, however with increasing marine impact towards the top. Similar to other interpreted sections (Figs 8 and 9), the major fining upwards sandstone bodies are related to deposition within fluvial channels, however with the observed bioturbation and cross-stratification (tangential set bases and possible drapes) indicating some more marine impact to the west, with possible tidal modulation of bedforms within the channels. The interbedded fine-grained material and thin coal beds are related to deposition and peat accretion (coal) in lakes and swamps of the surrounding delta plain. A minor sandy fining upwards unit is also related to deposition in a possible crevasse channel (Fig. 10). Channel sandstone deposits in the lower to middle part of the Tubåen Formation account for approximately 78% of the terrestrial interval in well 7121/4-F-2H. The relatively high channelised deposits content is generally indicative of good lateral connectivity between sandstone bodies (cf. Bridge & Mackey, 1993).

Fig. 11. Detailed core photographs, Tubåen Formation, well 7121/4-F-2H. (A) sharp-based sandstone bed with abundant coaly debris; (B) medium-grained sandstone with dm-scale cross-beds; (C) fine-grained, micaceous sandstone with small-scale ripple lamination and weak bioturbation; (D) rooted, silty sandstone below thin coal bed; (E) Wave-generated ripple lamination; (F) Wave-generated ripple lamination with low-angle cross-lamination (micro-hummocky). Core width is app. 12 cm.

The sedimentary features in the CU-unit capping the Tubåen Formation seem to record the establishment of a wave-dominated embayment. The observed wave-ripple lamination (Fig. 11E and F) is indicative of relatively strong and persistent wave activity in the lower (deeper) part of the embayment. Furthermore, the observed 35 m thickness of the CU-unit indicates an initial water depth exceeding some tens of metres, thus pointing towards a relatively large and deep embayment. The upper part of the CU-unit seems to record a gradual shallowing upwards succession of facies. Hence a major mouth bar and distributary channel interpretation is favoured, as was also suggested by Gjelberg *et al.* (1987) for the upper part of the Tubåen Formation in the central Hammerfest Basin.

Nature of the Tubåen fluvial system

Together, the interpreted channels and floodplain deposits of the Tubåen Formation constitute a channel-dominated delta plain environment with some weak marine impact towards the west. A braided channel interpretation was invoked for the very sandstone-dominated succession observed to the east, in wells 7124/3-1 and 7226/11-1 (Figs 8 and 9). The very coarse-grained nature of the strata in well 7226/11-1, with quartzite pebbles larger than 10 cm, is clearly pointing at very strong traction currents affecting the channels and also at deposition in a proximal position relative to the source area. This may well accord with a braided type of channel, although sediment calibre alone is not conclusive as regards the morphology of the fluvial channels (e.g. Miall, 1992).

The better defined fining upwards units seen in the Hammerfest Basin, coupled with the sporadic evidence for marine impacts (Fig. 11), may be interpreted in favour of a more meandering fluvial style. Commonly, interpreted meandering river systems have been related to well-defined fining upwards channel fill successions separated by fine-grained intervals (Miall, 1992) and the weak marine influence in the Hammerfest Basin also suggests that deposition took place in the lower part of the delta plain, where fluvial channels typically show a stronger tendency towards meandering patterns (e.g. Collinson, 1996).

From the stratigraphic age assignments (Figs 8 to 10) it appears that fluvial conditions persisted in the Hammerfest Basin during Rhaetian through Hettangian times, followed by more marine conditions, with establishment of a wave-dominated bay and mouth bar systems during the Sinemurian stage (Fig. 10). However, further to the east (i.e. in well 7226/11-1, Fig. 9), fluvial conditions seem to have persisted into the Sinemurian and it can also be demonstrated from other wells that even Pliensbachian strata are predominantly fluvial; for instance in the Nordkapp Basin (e.g. Henriksen *et al.*, 2011). Accordingly, the fluvial system seems to have been most widespread during the Hettangian, so that the lower part of the Tubåen Formation represents the regressive maximum of the system. Subsequently, the fluvial environment seems to have retreated gradually eastwards during Sinemurian to Pliensbachian times.

DEPOSITIONAL TRENDS ACROSS THE TRIASSIC TO JURASSIC BOUNDARY

Changes in depositional trends across the Triassic to Jurassic boundary can be discussed in terms of temporal variations in sedimentary architecture and development of accommodation space available for deposition, in terms of changing provenance areas and hinterland rejuvenation, and from both local and regional climatic variations, as inferred from the Carnian to Sinemurian strata.

Temporal variation in accommodation space development

In the present case, where fluvial-dominated delta plain successions are recorded from the upper Snadd Formation, from the middle and upper parts of the Fruholmen Formation and from the Tubåen Formation, an investigation of alluvial architecture and in particular the stacking density of channel deposits, can provide a reference for temporal changes in basin subsidence; or accommodation space development (e.g. Allen, 1978; Bridge & Leeder, 1979; Bridge & Mackey, 1993; Martinsen *et al*., 1999). Thematically, abundance of channels can be related to slow rates of accommodation space generation, whereas fewer channels and more abundant preservation of floodplain fines reflect a higher rate of accommodation space generation (Bridge & Mackey, 1993). Accommodation space measured over millions of years, as in the present case, is probably governed by tectonic subsidence rates (e.g. Sadler, 1981; Sadler & Strauss, 1990), but also includes space generation related to, for instance, sea-level rise. These are not treated explicitly here and it is assumed that sediment accumulation rates are indeed controlled by tectonic subsidence.

The architectural changes inferred from the Late Triassic to Early Jurassic succession is illustrated in the correlation panel shown in Fig. 12. In the Snadd Formation, the correlated strata comprise more than 200 m of delta plain deposits capped by shallow marine sandstones (as demonstrated from core data in Fig. 8). Despite the large vertical thickness (>200 m) of the delta plain association (Snadd Formation), only one major channel unit seems to have been penetrated in the three wells; an approximately 30 m thick sandstone body in well 7124/3-1. The remaining strata are entirely dominated by fine-grained lithologies, including thin coal beds as inferred from density/porosity wire-line log data, however with some scattered thinner channel sandstones indicated. In summary, the proportion of channel deposits in the upper part of the Snadd Formation is apparently very low. This is similar to well 7222/11-1 on the Loppa High, where only a single thick sandstone body (Fig. 5) was encountered in more than 500 m of delta plain deposits. Based on regional data it seems unlikely that the proportion of channelised sandstone in the upper Snadd Formation is larger than approximately 10% to 20% in the south-west Barents Sea. Some large channels are identified clearly from wells and seismic data, but these appear to be isolated within the surrounding fine-grained material.

In the Fruholmen Formation, pro-deltaic mudrocks and overlying mouth bar sandstones seem to be correlated readily between the wells (Fig. 12), although they expand in thickness to the west into the Hammerfest Basin, along with the rest of the formation. The large channelised sandstone unit observed in well 7124/3-1 (Fig. 8) does not correlate with similar bodies in the studied wells, whilst the overlying lower delta plain association is widespread. A small distributary channel sandstone is recognised in well 7124/3-1 (Fig. 8) and some possible thin, channelised, sandstones are also recognised in the other two wells. The actual proportion of channel deposits in the Fruholmen delta plain association may exceed that of the Snadd Formation but would nevertheless be limited. Accordingly, the delta plain association in the Fruholmen Formation also appears to be dominated by fine-grained material and the few channels present are likely to be isolated within the surrounding fines.

The architecture of the Tubåen Formation is markedly different. Between wells 7226/11-1 and 7124/3-1 the formation seems to form a multilateral, 20 m to 34 m thick, sheet of amalgamated channel sandstones, more or less devoid of fine-grained material. The formation thickens westwards toward well 7121/5-1 (65 m), where some thin, fine-grained and coal-bearing units are preserved between the interpreted channel sandstones. Despite a slightly more heterolithic character to the west, the Tubåen Formation maintains a high content of channel sandstone in

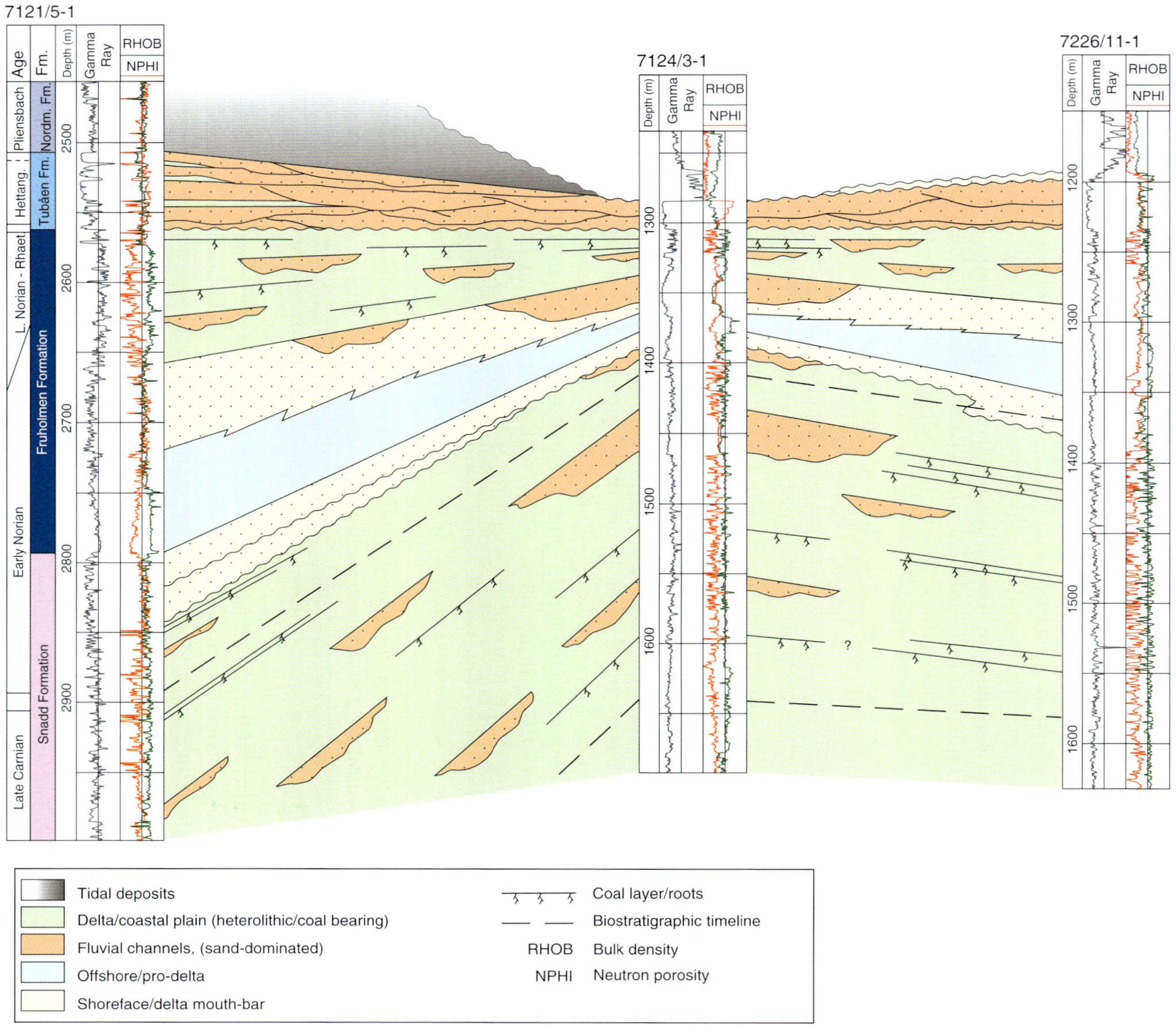

Fig. 12. Correlation of Late Triassic to Early Jurassic strata in the south-west Barents Sea. The upper part of the Snadd Formation and upper part of the Fruholmen Formation are characterised by channelised sandstone units encased within the surrounding floodplain/delta plain deposits. The overlying Tubåen Formation forms a multilateral/multi-storey sandstone body of nested fluvial channel sandstones. For well locations, see Fig. 2B.

the Hammerfest Basin. As noted for well 7121/4-F-2H, the terrestrial section (Fig. 10) comprises approximately 78% channel fill deposits. On average, more than 60% of channelised deposits are observed from the alluvial part of the Tubåen Formation in the Hammerfest Basin.

With reference to process-based alluvial stratigraphy modelling (e.g. Allen, 1978; Bridge & Leeder, 1979; Bridge & Mackey, 1993), the very limited channel deposit proportions seen in the upper part of the Snadd Formation, and also in the delta plain association of the Fruholmen Formation, would be characteristic of a depositional regime dominated by a relatively high subsidence rate and plentiful sediment supply. Under such conditions, which are also referred to as high accommodation settings by other workers (e.g. Martinsen *et al.*, 1999), fine-grained material deposited outside the main channels have a high potential to be preserved. Furthermore, scattered thick channel sandstones will probably also form due to persisting vertical aggradation within channels and channel belts, but would be more or less encased within the surrounding fine-grained material, particularly if the actual proportion of channels is lower than approximately 40% (Bridge & Mackey, 1993).

By contrast, the multilateral stacking of channel sandstone bodies observed in the Tubåen Formation (Fig. 12), coupled with the observed low content of fine-grained material in the vertical sections (Figs 8 to 10), can, in particular, be related to a slower subsidence rate. Under such conditions of low accommodation (e.g. Martinsen *et al.*, 1999), persistent fluvial reworking may lead to predominant preservation of the coarsest (deeper) parts of the channel fills, as inferred for the section in well 7226/11-1 (Fig. 9). Some spatial variation in subsidence rates can also be inferred for the Tubåen Formation. As shown in Figs 10 and 12, fine-grained material is better preserved between the channelised sandstones towards the west, where the formation is thickest. This tendency can be related to less effective fluvial reworking in areas undergoing more rapid subsidence (e.g. Allen, 1978).

The presented architectural data are essentially indicative of a regional reduction in tectonic subsidence rate across the Triassic to Jurassic boundary in the south-west Barents Sea. Delta plain deposits of Late Carnian and Early to Late Norian age are characterised by abundant fine-grained material and isolated channel sandstone deposits, whereas the Early Jurassic fluvial system (Tubåen Formation) is completely dominated by channel sandstone deposits and with little preserved fine-grained material, even though some limited lateral variation is observed.

Considerations of architecture versus accumulation rates, based upon absolute time scale, yield similar results. As noted above, the entire Snadd Formation can reach a thickness of more than 1400 m (e.g. in well 7222/11-1; Table 1) and is commonly more than 500 m thick. These thicknesses include pro-deltaic and marginal marine strata of Late Ladinian to Early Carnian age, in addition to the Late Carnian to Early Norian delta plain deposits described here. However, if a total duration of approximately 20 Ma to 25 Ma can be inferred for the accumulation of the Snadd Formation (Late Ladinian to Early Norian; app. 233.3 Ma to 207.6 Ma; cf. Ogg *et al.*, 2008), accumulation rates (ignoring effects of compaction) should amount to several tens (20 to 70) of metres per million years during the Late Triassic.

For the Tubåen Formation, one can firstly consider the section in well 7226/11-1 (Fig. 9). Here, the available biostratigraphy yields a general Hettangian to Sinemurian age for the entire formation. Accordingly, the 34 m thick alluvial succession appears to be stratigraphically condensed and spanning a possible time interval of up to 10 Myr (189.6 Ma to 199.6 Ma, cf. Fig. 4 and Ogg *et al.*, 2008), yielding an accumulation rate of only 3 m to 4 m per million years. If the lateral thickness variation is included, alluvial successions of about 50 m to 100 m in thickness, deposited mainly during the Hettangian to Sinemurian time interval, can be considered. Given duration of approximately 3 Ma to 10 Ma, accumulation rates would be in the order of 5 m to 30 m per million years, which is still lower than those inferred for the Late Triassic strata. Such considerations serve to illustrate that net accumulation rates were higher during the Late Triassic than they were during the earliest Jurassic. Thus, for fluvial and/or channelised depositional settings, one should expect to see a denser stacking of channel fill deposits in the Hettangian to Sinemurian succession than in the underlying units, as is clearly pictured in Fig. 12.

Changing provenance areas and hinterland rejuvenation

Mineralogical differences between the sandstones of the Snadd Formation and the overlying units have been addressed briefly above. Bergan & Knarud (1993) provided a detailed mineralogical

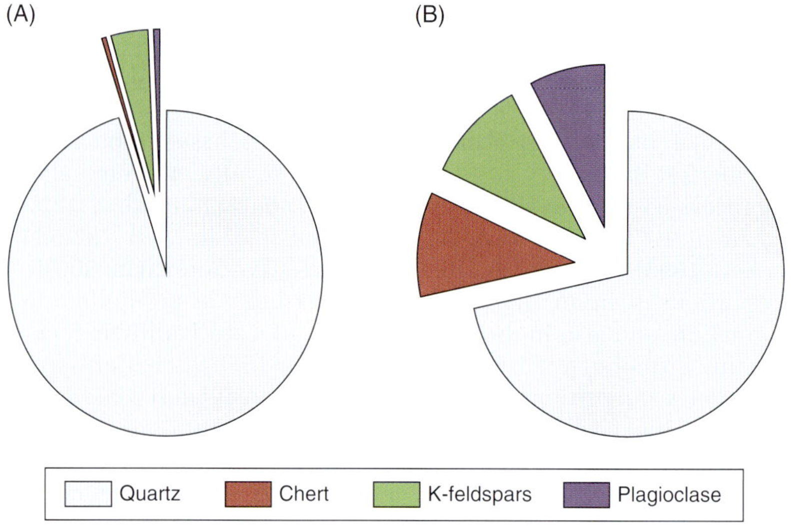

Fig. 13. Relative mineralogical composition of framework (detrital) grains in fluvial sandstones from the Tubåen Formation (A) and Snadd Formation (B). The data are obtained from petrographic thin sections from a well on the Bjarmeland Platform (Fig. 2). Modal data are courtesy of L. Sæbø (Statoil).

account of sandstones in well 7124/3-1 (Fig. 8). They observed a significant change between the Snadd Formation, which in particular is characterised by abundant plagioclase feldspars, and the overlying Fruholmen and Tubåen formations, which contain coarser and more mature, quartz-rich arenites depleted in feldspars and with K-feldspars being more common than plagioclases.

A comparable dataset is shown in Fig. 13. The relative percentages of detrital framework grains (quartz, chert, K-feldspars and plagioclases) in channelised sandstones from the Tubåen and Snadd formations are shown from a well on the Bjarmeland Platform. Here, sandstones in the Tubåen Formation are dominated by quartz grains (95%), with minor amounts of K-feldspar (app. 4%) and with only traces (<1%) of chert grains and plagioclases. In comparison, sandstones in the Snadd Formation contain relatively abundant chert grains (10%), K-feldspars (10%) and plagioclases (app. 8%), in addition to quartz. Following Bergan & Knarud (1993) and Mørk (1999), this mineralogical change is the most distinct and regionally important petrographic turn-around in the Arctic region. The change occurs at the Early Norian marine flooding surface, defining the top of the Snadd Formation, and can be attributed to several factors:

- Change to a more humid type of climate which would lead to more pronounced leaching of feldspars;
- A change or redistribution in provenance areas following the Early Norian transgression;
- Temporal change in the composition of erosion products from the east (Uralides) to a more quartz-rich composition in the Late Triassic to Early Jurassic.

Given a predominantly humid climate in the Barents Sea area from the Late Carnian (Hochuli & Vigran, 2010), a significant climatic change seems to pre-date the observed mineralogical turn-around and it therefore seems unlikely that change to a more humid type of climate alone can account for the temporal variation in sandstone composition. Nevertheless, enhanced tropical weathering from Late Carnian times will dissolve feldspars more effectively and play a part in controlling sandstone composition.

A detailed account of possible provenance areas is beyond the scope of this paper. However, it is noted that the Baltic shield and Caledonides of north-west Scandinavia are the most probable sources for quartz-rich sandstones, particularly in the south-west Barents Sea (Mørk, 1999). Accordingly, the denudation history of northern Scandinavia may provide a direct relationship with the offshore sedimentation patterns, as has been indicated above. Hendriks & Andriessen (2002) inferred rapid Triassic to Jurassic uplift and denudation along the onshore Atlantic rift margin of Troms County from fission track data. They also related this denudation phase to rift shoulder uplift concurrent with subsidence and deposition in the offshore areas including the Hammerfest Basin. Subsequently, Hendriks (2003) presented apatite fission track analyses indicative of a contemporary phase of Late Triassic to Early Jurassic uplift in the onshore regions from eastern Finnmark to the Kola Peninsula, with estimated 2.5 km to 3 km of denudation.

This reported evidence for Late Triassic to Early Jurassic uplift and erosion in onshore areas to the south of the Barents Sea basins may link to the transition to more quartz-rich sandstones above the Snadd Formation. Such a phase of hinterland rejuvenation may also have played its part in introducing much coarser clastic material into the depositional basins, as recorded from the Tubåen Formation, in particular. It is also tempting to relate the inferred reduction in basin subsidence rate (Fig. 12) to concurrent uplift/denudation in the onshore area to the south; although a clear genetic (tectonic) linkage cannot be demonstrated in this study.

With reference to the North Sea area, it is also noteworthy that Paul *et al.* (2009) concluded that rapid Late Triassic uplift in the Caledonides of southern Norway accounted for provenance variations in the Keuper sandstones in Germany. Similar to Hendriks & Andriessen (2002), Paul *et al.* (2009) referred to rift shoulder uplift along the Atlantic rift system to account for the Late Triassic to Early Jurassic increase in denudation. These interpretations support a conclusion that hinterland rejuvenation played an important role in changing depositional trends at the Triassic to Jurassic boundary on the Norwegian shelf; from the North Sea to the Barents Sea areas.

Furthermore, the Early Jurassic palaeogeographic reconstructions of Smelror *et al.* (2009) suggest that the western part of the Barents shelf became shielded from sediment input from the Uralides and Novaya Zemlya by a major marine embayment developing in the South Barents and North Barents basins (Fig. 2), in marked contrast to the dominant Triassic setting with continuous sediment transport across these basins (Riis *et al.* 2008; Smelror *et al.* 2009). Accordingly, it appears that the Early Norian marine incursion may eventually have terminated the supply of sediment from the south-east across the Barents shelf, thus further enhancing the petrographical changes induced by hinterland rejuvenations in Fennoscandia.

Finally, shortening of fluvial transport distances between the Snadd and Tubåen formations can also be considered. If the primary source for sandy sediment in the Snadd Formation is located in the Uralides to the south-east, sediment transport over approximately 500 km to 1000 km is required to supply sediment, for instance, to the Bjarmeland Platform (Fig. 2; see also Riis *et al.*, 2008). In contrast, direct supply of quartz-rich sediment from the Caledonides of northern Norway into the Hammerfest and Nordkapp basins would require transport distances of approximately 100 km to 200 km or less. As river sediment typically becomes finer grained downstream, increasing sediment supply form Fennoscandia across the Triassic to Jurassic boundary may have also contributed to an increase in the sandiness and overall grain size of the Tubåen Formation, as compared to the Snadd Formation. If the Fennoscandian hinterland was subject to additional rejuvenation during Late Triassic to Early Jurassic times, these effects (increased sandstone content and coarser grain sizes) would be amplified even more.

Geographic and temporal climatic variations

Late Triassic climates in the south-west Barents Sea have been addressed by Hochuli & Vigran (2010), who examined climatic variations from palynological records, e.g. in the Nordkapp Basin. These workers found that the entire post-Early Carnian interval is characterised by a very warm climate. Furthermore, an abrupt change from dry to more humid conditions is recorded at the Early to Late Carnian boundary (i.e. within the Snadd Formation). Regional occurrence of coal-bearing strata in the Late Carnian to Early Norian upper part of the Snadd Formation, as well as the dominant grey colour of the floodplain deposits, is in good accordance with a humid climate, as coals (peat) require water-saturated conditions and high groundwater level to form (e.g. McCabe, 1984).

Furthermore, the sedimentary successions described here from the Fruholmen and Tubåen formations (coal-bearing grey beds in the delta plain) accords with a humid climate throughout the Norian to Hettangian stages, thus it may seem that Late Triassic and Early Jurassic climates were relatively similar in terms of overall humidity. However, Bugge *et al.* (2002) have described red beds from the upper part of the Snadd Formation in the Nordkapp Basin; and sporadic red/grey mottled palaeosols also occur in the Fruholmen Formation in some cored sections (e.g. well 7226/11-1; Fig. 9). Such red colouration would indicate some water deficiency in the parent soil and also oxidising conditions. Accordingly, some evidence exists for slightly drier climatic conditions in the Late Triassic than in the Early Jurassic successions in the Barents Sea.

The Late Triassic climate further to the south is apparently very different. As described above, palaeographic reconstructions generally assume that north-west Europe, including the North Sea, was positioned at a palaeolatitude of approximately 40° N during the Late Triassic, approximately 12° south of the Barents shelf (Golonka, 2007). Widespread occurrence of red beds with pedogenic carbonate (calcrete) throughout the Triassic strata in the north west European basins, coupled with climatic circulation models, generally point towards very hot and arid conditions, with a short wet season (e.g. Sellwood & Valdes, 2006) and fluvial systems entering the basins are thought to drain into terminal playa lake systems (McKie & Williams, 2009).

However, following Nystuen *et al.* (this volume), a gradual increase in humidity can be recorded from Late Norian to Rhaetian strata (Lunde Formation) in the North Sea area, reflected by significant changes in clay mineralogy and by a change from mainly ephemeral stream deposits and associated red beds in the lower Lunde Formation, to more perennial fluvial systems in its the upper parts. Nystuen *et al.* (this volume) related this gradual change in climate mainly to continuous continental drift, with a steady northwards shift of e.g. the Viking Graben (Fig. 1) to approximately 50°N in the Early Jurassic. Other studies have also indicated that marine incursion from the south may have contributed to a more humid type of climate during the latest Triassic (Rhaetian) in the North Sea area (Ahlberg *et al.* 2002).

For the Late Triassic on the Halten Terrace, Müller *et al.* (2005) stated that the Carnian halite units (Fig. 4) were precipitated from marine waters in a warm and arid climate. They also inferred a semi-arid setting for overlying Norian red beds, very similar to the inferred climatic condition in the North Sea area. This contrast between arid climates to the south and more humid climates to the north (Barents Sea) during the Late Triassic seems to relate to palaeogeographic variations, particularly in palaeolatitude, with a humid climatic zone to the north of a prevailingly dry zone to the south.

Humid conditions and high ground-water levels in the Late Triassic strata in the Barents Sea area can also be related to deposition in closer proximity to marine environments. The sedimentary facies of the Snadd Formation are indicative of frequent marine incursions (e.g. Fig. 6) in the delta plain and palaeogeographic models (e.g. Riis *et al.* 2008; Glørstad-Clark *et al.* 2010) show the existence of marine systems immediately to the north and west of the interpreted delta plain of the Snadd Formation. General proximity to the Boreal ocean, as opposed to inland desert conditions to the south, should clearly be favourable for a more humid climate on the Barents shelf than, for instance, in the North Sea area, which apparently was located well away from marine waters and within inland basins of a major continent. However, the desert basins to the south may have been partly connected to the Boral Ocean to the north through the Atlantic rift, particularly during deposition of the Carnian halite on the Halten Terrace (Fig. 1), which has been related to marine incursions from the north (Müller *et al.*, 2005).

In contrast to the inferred climatic zonation in the Late Triassic, the Early Jurassic climate seems to be humid and rather equable between the North Sea basins and the Barents Sea area. In the south-west Barents Sea, the fine-grained part of the Tubåen Formation seems to be totally dominated by grey beds and associated coal, which is indicative of completely water saturated and predominantly reducing conditions. In the North Sea, coal-bearing grey beds are frequently observed in the upper part of the Statfjord Formation (Steel & Ryseth, 1990; Ryseth, 2001). Likewise, common coal-bearing strata in the time-equivalent (Hettangian) lower part of the Åre Formation on the Halten Terrace (Morris *et al.* 2009) also accords with a humid climate.

This widespread change to a wetter climate across the Triassic to Jurassic boundary on the Norwegian shelf may relate to continuous northwards continental drift (e.g. Frostick *et al.* 1992; Nystuen *et al.* this volume). However, the turnaround to more humid conditions is apparently common to many European basins. For instance, Hesselbo *et al.* (2007) noted that marine platform carbonates in Iberia (Spain) record a change from arid to humid climate at the Triassic to Jurassic boundary. Based on plate reconstructions placing this part of the western Tethys in the equatorial belt south of 30° N during Rhaetian to Hettangian times (Golonka, 2007), it seems unlikely that continental drift alone can explain the climatic change.

Götz *et al.* (2011) examined coal-bearing strata of Rhaetian to Hettangian age in southern Hungary and concluded, from microfloral data, that there was a change to more humid conditions during Late Rhaetian to Hettangian times, more or less concurrent with the increased humidity inferred for the Lunde to Statfjord transition in the North Sea area. Götz *et al.* (2011) also inferred that the climatic change was sudden and argued that the increase in humidity could be related to the onset of intensive volcanism during the initial break-up of Pangaea. As noted above, this volcanic phase (Central Atlantic Magmatic Province; e.g. Hesselbo *et al.* 2007) is associated with global mass extinctions and increased atmospheric CO_2 concentration, hence any climatic change induced by this volcanism should clearly be widespread enough to affect the climate of the entire Atlantic rift system, including the strata and basins referred to in this study. For the Barents Sea area the significantly coarser sandstones in the Tubåen Formation, than in the underlying strata, can be related to larger fluvial discharge and increased capacity of rivers to transport coarser material into the basins. This can clearly reflect a more humid situation and higher annual precipitation of rain during the Hettangian stage than, for example, during the Carnian to Late Norian stages.

The data presented in the current study do not provide any evidence for the actual causes of the Late Triassic to Early Jurassic climatic changes. Clearly, a gradual increase in humidity, as inferred for the Late Norian to Rhaetian Lunde Formation in the North Sea, can relate to continued northwards plate-tectonic drift of the sedimentary basins, particularly if combined with increasing proximity to marine environments. However, a regionally consistent change towards more humid climatic conditions at the Triassic to Jurassic boundary, stretching from Iberia in the south to the Barents Sea basins to the north, is probably more logically tied to global climatic change (volcanism) than to plate tectonic drift, which only places part of this area to the north of 30° N and within the more humid climatic zone.

CONCLUSIONS

Based on the discussions above, it appears that changing depositional styles across the Triassic to Jurassic boundary in the south-west Barents Sea can, on a sub-regional scale, be related to a combination of changing subsidence rates, hinterland rejuvenation and redistribution of sediment transport routes from the hinterlands. These changes are not synchronous, but may eventually have led to more favourable conditions for the development of coarser-grained and more sand-dominated fluvial systems in the Early Jurassic Tubåen Formation than in underlying Late Triassic strata:

1. Contrasting alluvial architectures in Late Triassic versus Early Jurassic strata indicate that the generation of accommodation space was significantly reduced at the Triassic to Jurassic boundary, thus favouring selective preservation of sandy, channelised deposits.
2. Comparison of rock accumulation rates from absolute age assignments accords with limited deposition and stratigraphic condensation of Early Jurassic strata relative to the underlying Late Triassic successions, again indicating that fluvial reworking and preservation of channelised deposits should be more typical for the Early Jurassic strata.
3. Late Triassic to Early Jurassic rejuvenation of hinterlands to the south of the Barents shelf can be inferred from other studies and may have played an important role in increasing the grain size of sediment delivered to the basin areas during deposition of the Early Jurassic unit (Tubåen Formation).
4. A re-organisation of transport routes with possible termination of sediment flux from the Uralides may be related to the Early Norian marine flooding and concurrent uplift (rejuvenation) to the south, thus changing the mineral composition of sandy sediment.

Comparison of strata in the Barents Sea area, with age-equivalent units from the North Sea basins and the mid-Norway shelf provide additional knowledge regarding more regional climatic controls on deposition.

5. Whereas warm and humid climatic conditions persisted in the Barents Sea area during the Late Triassic (Late Carnian to Norian), warm and generally dry desert conditions prevailed to the south. Accordingly, a strong climatic zonation between a dry desert to the south and more humid, nearshore, setting to the north seems to have persisted until the Rhaetian.
6. A significant climatic change to more humid conditions can be inferred for the Early Jurassic, with establishment of a seemingly more equable and wet climate from the North Sea area to

the Barents Sea basins. This climatic change is apparently very widespread and is referred to from basins as far to the south as Spain and Hungary.

Whereas sub-regional factors such as changing subsidence rates and rejuvenation/reconfiguration of hinterlands can explain the very sandy and coarse-grained nature of the Tubåen Formation, as opposed to (for example) the generally finer-grained Snadd Formation with its more sporadic occurrence of channel sandstones, these effects would clearly be amplified by a wetter climate. In relative terms, the coarser sandstones in the Tubåen Formation would require more powerful streams to form than the finer grained sandstones in the Snadd Formation. This can clearly be accomplished if the river discharge increased due to higher annual precipitation in the rejuvenated hinterland.

Although not entirely conclusive, the data and interpretations presented here accord with other studies indicating increased climatic humidity across the Triassic to Jurassic boundary in many European basins. A direct climatic relationship with the Triassic to Jurassic volcanism to the south can only be speculated on, although the very regional nature of the change to a more humid climate in the Early Jurassic supports a global (volcanic) cause.

ACKNOWLEDGEMENTS

I wish to thank Statoil for the permission to publish this paper. Also, the partners in the Snøhvit Production unit, Petoro AS, Total E&P Norge AS, RWE DEA Norge AS and GDF SUEZ E&P Norge AS, are thanked for the permission to publish data from well 7121/4-F-2H. Reviewers William Helland-Hansen and Atle Folkestad provided many useful comments to the manuscript. Geir Samuelsen is thanked for his careful drafting and many good ideas to improve the graphic work.

REFERENCES

Ahlberg, A., Arndorff, L. and **Guy-Ohlson, D.** (2002) Onshore climatic change during the Late Triassic marine inundation of the Central European Basin. *Terra Nova*, **14**, 241–248.

Allen, J.R.L. (1978) Studies in fluviatile sedimentation: an exploratory quantitative model for the architecture of avulsion-controlled alluvial suites. *Sed. Geol.*, **21**, 129–147.

Badley, M.E., Price, J.D., Rambech Dahl, C. and **Agdestein, T.** (1988) The structural evolution of the northern Viking Graben and its bearing upon extensional modes of basin formation. *J. Geol. Soc. London*, **145**, 455–472.

Bergan, M. and **Knarud, R.** (1993) Apparent changes in clastic mineralogy of the Triassic to Jurassic succession, Norwegian Barents Sea: possible implications for paleodrainage and subsidence. In: *Arctic Geology and Petroleum Potential* (Eds T.O Vorren, E. Bergsager, Ø.A. Dahl-Stamnes, E. Holter, B. Johansen, E. Lie and T.B. Lund), *Norw. Petrol. Soc. Spec. Publ.*, **2**, Elsevier, Amsterdam, 481–493.

Berglund, L.T., Augustson, J., Færseth, R., Gjelberg, J. and **Ramberg-Moe, H.** (1986) The evolution of the Hammerfest Basin. In: *Habitat of Hydrocarbons on the Norwegian Continental Shelf* (Eds A.M. Spencer, C.J. Campbell, S.H. Hanslien, P.H. Nelson, E. Nysaether and E.G. Ormaasen), Proc. Norw. Petrol. Soc., Graham and Trotman, London, 319–338.

Besly, B.M. and **Fielding, C.R.** (1989) Palaeosols in Westphalian coal-bearing and red-bed sequences, central and northern England. *Palaeogeogr. Palaeoclimatol. Palaeoecol.*, **70**, 303–330.

Blystad, P., Brekke, H., Færseth, R., Larsen, B.T., Skogseid, J. and **Tørudbakken, B.** (1995) Structural elements of the Norwegian continental shelf. Part II: The Norwegian Sea region. *Norw. Petrol. Direct. Bull.*, **8**, 1–45.

Breivik, A.J., Faleide, J.I. and **Gudlaugsson, S.T.** (1998) South-western Barents Sea margin: late Mesozoic sedimentary basins and crustal extension. *Tectonophysics*, **293**, 21–44.

Bridge, J.S. and **Leeder, M.R.** (1979) A simulation model of alluvial stratigraphy. *Sedimentology*, **26**, 617–644.

Bridge, J.S. and **Mackey, S.D.** (1993) A revised alluvial stratigraphy model. In: *Alluvial Sedimentation* (Eds C. Puigdefabregas and M. Marzo), *Int. Assoc. Sedimentol. Spec. Publ.*, **17**, 319–336.

Bugge, T., Elvebakk, G., Fanavoll, S., Mangerud, G., Smelror, M., Weiss, H.M., Gjelberg, J., Kristesen, S.E. and **Nilsen, K.** (2002) Shallow stratigraphic drilling applied in hydrocarbon exploration of the Nordkapp Basin, Barents Sea. *Mar. Petrol. Geol.*, **19**, 13–37.

Charnock, M., Kristiansen, I.L, Ryseth, A. and **Fenton, J.** (2001) Sequence stratigraphy of the Dunlin Group, northern North Sea. In: *Sedimentary Environments offshore Norway: Palaeozoic to Recent* (Eds O.J. Martinsen and T. Dreyer), *Norw. Petrol. Soc. Spec. Publ.*, **10**, Elsevier, Amsterdam, 145–174.

Collinson, J.D. (1996) Alluvial sediments. In: *Sedimentary Environments: Processes, Facies and Stratigraphy* (Ed. H.G Reading), pp 37–82. Blackwell.

Dalland, A., Worsley, D. and **Ofstad, K.** (1988) A lithostratigraphical scheme for the Mesozoic and Cenozoic succession offshore Mid- and Northern Norway. *Norw. Petrol. Direct. Bull.*, **4**, 1–65.

Dallmann, W.K. (Ed) (1999) *Lithostratigraphic Lexicon of Svalbard: Upper Paleozoic to Quaternary Bedrock. Review and Recommendation for Nomenclature Use.* Norwegian Polar Institute, Tromsø, 318 pp.

Doré, A.G. (1992) Synoptic palaeogeography of the Northeast Atlantic Seaway: late Permian–Cretaceous. In: *Basins on the Atlantic Seaboard: Petroleum Geology, Sedimentology and Basin Evolution* (Ed. J. Parnell), *Geol. Soc. London Spec. Publ.*, **62**, 421–446.

Doré, A.G., Lundin, E.R., Birkeland, Ø., Eliassen, P.E. and **Jensen, L.N.** (1997) The NE Atlantic Margin: implications of late Mesozoic and Cenozoic events for hydrocarbon prospectivity. *Petroleum Geoscience*, **3**, 117–131.

Færseth, R.B. (1996) Interaction of Permo-Triassic and Jurassic extensional fault blocks during the development of the northern North Sea. *J. Geol. Soc. London*, **153**, 931–944.

Frostick, L., Linsey, T.K. and **Read, I.** (1992) Tectonic and climatic control of Triassic sedimentation in the Beryl Basin, northern North Sea. *J. Geol. Soc. London*, **149**, 13–26.

Gabrielsen, R.H., Færseth, R.B., Jensen, L.N., Kalheim, J.E. and **Riis, F.** (1990) Structural elements of the Norwegian continental shelf. Part I: The Barents Sea region. *Norw. Petrol. Direct. Bull.*, **6**, 1–33.

Gee, D.G., Fossen, H., Henriksen, N. and **Higgins, A.K.** (2008) From the early Paleozoic platforms of Baltica and Laurentia to the Caledonide orogen of Scandinavia and Greenland. *Episodes*, **31**, 44–51.

Gjelberg, J., Dreyer, T., Høie, A., Tjelland, T. and **Lilleng, T.** (1987) Late Triassic to Middle Jurassic sandbody development on the Barents and Mid-Norwegian shelf. In: *Petroleum Geology of North West Europe* (Eds J. Brooks and K. Glennie), pp 1105–1129. Geological Society, London, Graham and Trotman.

Glørstad-Clark, E., Faleide, J.I., Lundschien, B.A. and **Nystuen, J.P.** (2010) Triassic seismic sequence stratigraphy of the western Barents Sea area. *Mar. Petrol. Geol.*, **27**, 1448–1475.

Golonka, J. (2007) Late Triassic and Early Jurassic palaeogeography of the world. *Palaeogeogr. Palaeoclimatol. Palaeoecol.*, **244**, 297–307.

Götz, A., Ruckwied, K. and **Barbacka, M.** (2011) Palaeoenvironment of the Late Triassic (Rhaetian) and Early Jurassic (Hettangian) Mecsek Coal Formation (south Hungary): implications from macro- and microfloral assemblages. *Palaeobiol. Palaeoenviron.*, **91**, 75–88.

Gudlaugsson, S.T., Faleide, J.I., Johansen, S.E. and **Breivik, A.J.** (1998) Late Palaeozoic structural development of the South-western Barents Sea. *Mar. Petrol. Geol.*, **15**, 73–102.

Hallam, A. (1985) A review of Mesozoic climates. *J. Geol. Soc. London*, **142**, 433–445.

Hendriks, B.W.H. (2003) *Cooling and Denudation of the Norwegian and Barents Sea Margins, Northern Scandinavia, Constrained by Apatite Fission Tracks and (U-Th)He Thermochronology.* Unpublished Ph.D thesis. University of Amsterdam, 1–177.

Hendriks, B.W.H. and **Andriessen, P.A.M.** (2002) Pattern and timing of post-Caledonian denudation of northern Scandinavia constrained by apatite fission-track thermochronology. In: *Exhumation of the North Atlantic Margin; Timing, Mechanisms and Implications for Petroleum Exploration* (Eds A.G. Doré, J.A. Cartwright, M.S Stoker, J.P. Turner and N. White), *Geol. Soc. London Spec. Publ.*, **196**, 117–137.

Henriksen, E., Ryseth, A.E., Larssen, G.B., Heide, T., Rønning, K., Sollid, K. and **Stoupakova, A.V.** (2011) Tectonostratigraphy of the greater Barents Sea: Implications for petroleum systems. In: *Arctic Petroleum Geology* (Eds A. Spencer, A. Embry, D. Gautier, A.V. Stoupakova and K. Sørensen), *Geol. Soc. London Mem.*, **35**, 163–195.

Hesselbo, S.P., Robinson, S.A., Surlyk, F. and **Piasecki, S.** (2002) Terrestrial and marine extinction at the Triassic–Jurassic boundary synchronized with major carbon-cycle perturbation: A link to the initiation of massive volcanism. *Geology*, **30**, 251–254.

Hesselbo, S.P., McRoberts, C.A. and **Palfy, J.** (2007) Triassic–Jurassic boundary events: Problems, progress, possibilities. *Palaeogeogr. Palaeoclimatol. Palaeoecol.*, **244**, 1–10.

Hochuli, P.A. and **Vigran, J.O.** (2010) Climate variations in Boreal Triassic–inferred from palynological records from the Barents Sea. *Palaeogeogr. Palaeoclimatol. Palaeoecol*, **290**, 20–42.

Jacobsen, V.W. and van **Veen, P.** (1984) The Triassic North of 62°N. In: *Petroleum Geology of the North European Margin* (Eds A.M. Spencer *et al.*), pp 317–329. Norwegian Petroleum Society. Graham and Trotman, London.

Johansen, S.E., Ostisky, B.K., Birkeland, Ø., Fedrovsky, Y.F., Martirosjan,V.N., Bruun Christensen, O., Cheredeev, S.I., Ignatenko, E.A. and **Margulis, L.S.** (1993) Hydrocarbon potential in the Barents Sea region: play distribution and potential. In: *Arctic Geology and Petroleum Potential* (Eds T.O Vorren, E. Bergsager, Ø.A. Dahl-Stamnes, E. Holter, B. Johansen, E. Lie, and T.B. Lund), *Norw. Petrol. Soc. Spec. Publ.*, **2**, Elsevier, Amsterdam, 273–320.

Johnson, S.Y. (1984) Cyclic fluvial sedimentation in a rapidly subsiding basin, North-west Washington. *Sed. Geol.*, **38**, 361–391.

Lervik, K.S. (2006) Triassic lithostratigraphy of the Northern North Sea Basin. *Norw. J. Geol.*, **86**, 93–116.

Lervik, K.S., Spencer, A.M. and **Warrington, G.** (1989) Outline of Triassic stratigraphy and structure in the central and northern North Sea. In: *Correlation in Hydrocarbon Exploration* (Ed. J.D. Collinson), 173–189. Norwegian Petroleum Society, Graham and Trotman, London.

Marsh, N., Imber, J., Holdsworth, R.E., Brockbank, P. and **Ringrose, P.** (2009) The structural evolution of the Halten Terrace, offshore Mid-Norway: extensional fault growth and strain localisation in a multi-layer brittle-ductile system. *Basin Res.*, **46**, 1–20.

Martinsen, O.J., Ryseth, A., Helland-Hansen, W., Flesche, H., Torkildsen, G. and **Idil, S.** (1999) Stratigraphic base level and fluvial architecture: Ericson Sandstone (Campanian), Rock Springs Uplift, SW Wyoming, USA. *Sedimentology*, **46**, 235–259.

McCabe, P.J. (1984) Depositional environments of coal and coal-bearing strata. In: *Sedimentology of Coal and Coal-bearing Sequences* (Eds R.A. Rahmani and R.M. Flores), *Int. Assoc. Sedimentol. Spec. Publ.*, **7**, 13–42.

McKie, T. and **Williams, B.** (2009) Triassic palaeogeography and fluvial dispersal across the north western European Basins. *Geol. J.*, **44**, 711–741.

Miall, A.D. (1992) Alluvial settings. In: *Facies Models: Response to Sea Level Change* (Eds R.G. Walker and N.P. James), pp 119–142. Geological Association of Canada.

Morad, S., de Ros, L.F., Nystuen, J.P. and **Bergan, M.** (1998) Carbonate diagenesis and porosity evolution in sheet-flood sandstones: evidence from the Middle and Lower Lunde members (Triassic) in the Snorre Field, Norwegian North Sea. *Int. Assoc. Sedimentol. Spec. Publ.*, **26**, 53–85.

Morris, P.H., Cullum, A., Pearce, M. and **Batten, D.J.** (2009) Megaspore assemblages from Åre Formation (Rhaetian–Pliensbachian) offshore mid-Norway and their value as field and regional stratigraphic markers. *J. Micropalaeontol.*, **28**, 161–181.

Müller, R., Nystuen, J.P., Eide, F. and **Lie, H.** (2005) Late Permian to Triassic basin infill history and palaeogeography of the Mid-Norwegian shelf–East Greenland region. In: *Onshore-Offshore Relationships on the North Atlantic Margin* (Eds B. Wandas, J.P. Nystuen, E. Eide and F.M. Gradstein), *Norw. Petrol. Soc. Spec. Publ.*, **12**, Elsevier, Amsterdam, 165–189.

Mørk, M.B.E. (1999) Compositional variations and provenance of Triassic sandstones from the Barents Shelf. *J. Sed. Res.*, **69**, 690–710.

Mørk, A., Dallmann, W.K., Dypvik, H., Johannessen, E.P., Larssen, G.B., Nagy, J., Nøttvedt, A., Olaussen, S., Pcelina, T.M. and **Worsley, D.** (1999) Mesozoic Lithostratigraphy. In: *Lithostratigraphic Lexicon of Svalbard: Upper Paleozoic to Quaternary Bedrock Review and Recommendation for Nomenclature Use* (Ed. W.K. Dallmann), pp 127–214. Norwegian Polar Institute, Tromsø.

Nilsen, K.T., Vendeville, B.C. and **Johansen, J-T.** (1995) Influence of regional tectonics on halokinesis in the Nordkapp Basin, Barents Sea. In: *Salt Tectonics: A Global Perspective* (Eds M.P.A. Jackson, D.G. Roberts and S. Snelson), *AAPG Mem.*, **65**, 413–436.

Nøttvedt, A., Cecchi, M., Gjelberg, J.G., Kristensen, S.E., Lønøy, A., Rasmussen, A., Rasmussen, E., Skott, P.H. and van **Veen, P.M.** (1993) Svalbard–Barents Sea correlation: a short review. In: *Arctic Geology and Petroleum Potential* (Eds T.O Vorren, E. Bergsager, Ø.A. Dahl-Stamnes, E. Holter, B. Johansen, E. Lie, and T.B. Lund), *Norw. Petrol. Soc. Spec. Publ.* **2**, Elsevier, Amsterdam, 363–375.

Nystuen, J.P., Knarud, R., Jorde, K. and **Stanley, K.O.** (1989) Correlation of Triassic to Lower Jurassic sequences, Snorre Field and adjacent areas, northern North Sea. In: *Correlation in Hydrocarbon Exploration* (Ed. J.D. Collinson), pp 273–289. Norwegian Petroleum Society, Graham and Trotman, London.

Nystuen, J.P., Kjemperud, A., Müller, R., Adnestål, V. and **Shomacker, E.** (2014) Late Triassic–Early Jurassic climatic change, northern North Sea region: impact on alluvial architecture, palaeosols and clay mineralogy. *Int. Assoc. Sedimentol. Spec. Publ.*, **46**, (this volume).

Ogg, J.G., Ogg, G. and **Gradstein, F.** (2008) *The Concise Geologic Time Scale.* Cambridge University Press. 1–150.

O'Leary, N., White, N., Tull, S., Bashilov, V., Kuprin, V., Natapov, L. and **MacDonald, D.** (2004) Evolution of the Timan-Pechora and South Barents basins. *Geol. Mag.*, **141**, 141–160.

Paul, J., Wemmer, K. and **Wetzel, F.** (2009) Keuper (Late Triassic) sediments in Germany–indicators of rapid uplift of Caledonian rocks in southern Norway. *Norw. J. Geol.*, **89**, 193–202.

Reading, H.G. and **Collinson, J.D.** (1996) Clastic coasts. In: *Sedimentary Environments: Processes, Facies and Stratigraphy* (Ed. H.G Reading), pp 154–231. Blackwell.

Riis, F., Lundschien, B.A., Høy, T., Mørk, A. and **Mørk, M.B.** (2008) Evolution of the Triassic shelf in the northern Barents Sea region. *Polar Res.*, **27**, 318–338.

Ritzmann, O. and **Faleide, J.I.** (2009) The crust and mantle lithosphere in the Barents Sea/Kara Sea region. *Tectonophysics*, **470**, 89–104.

Ryseth, A. (2001) Sedimentology and palaeogeography of the Statfjord Formation (Rhaetian - Sinemurian), North Sea. In: *Sedimentary Environments offshore Norway – Palaeozoic to Recent* (Eds O.J. Martinsen and T. Dreyer), *Norw. Petrol. Soc. Spec. Publ.*, **10**, Elsevier, Amsterdam, 67–85.

Ryseth, A., Augustson, J.H., Charnock, M., Haugerud, O., Knutsen, S.-M., Midbøe, P.S., Opsal, J.G. and **Sundsbø, G.** (2003) Cenozoic stratigraphy and evolution of the Sørvestsnaget Basin, south-west Barents Sea. *Norw. J. Geol.*, **83**, 107–130.

Sadler, P.M. (1981) Sediment accumulation rates and the completeness of stratigraphic sections. *J. Geol.*, **89**, 569–584.

Sadler, P.M. and **Strauss, D.J.** (1990) Estimation of completeness of stratigraphical sections using empirical data and theoretical models. *J. Geol. Soc. London*, **147**, 471–485.

Schoene, B., Guex, J., Bartoloni, A., Schaltegger, U. and **Blackburn, T.J.** (2010) Correlating the end-Triassic mass extinction and flood basalt volcanism at the 100 ka level. *Geology*, **38**, 387–390.

Sellwood, B.W. and **Valdes, P.J.** (2006) Mesozoic climates: General circulation models and the rock record. *Sed. Geol.*, **190**, 269–287.

Smelror, M., Petrov, O.V., Larssen, G.B. and **Werner, S.** (Eds) (2009) Atlas. Geological History of the Barents Sea. *Geol. Surv. Norw.* Trondheim, 134 pp.

Spencer, A.M., Briskeby, P.I., Christensen, L.D., Foyn, R., Kjølleberg, M., Kvadsheim, E., Knight, S., Rye-Larsen, M. and **Williams, J.** (2008) Petroleum geoscience in Norden–exploration, production and organization. *Episodes*, **31**, 115–124.

Steel, R., and **Ryseth, A.** (1990) The Triassic - Early Jurassic succession in the northern North Sea: megasequence stratigraphy and intra-Triassic tectonics. In: *Tectonic Events Responsible for Britain's Oil and Gas Reserves* (Eds R.F.P. Hardman and J. Brooks), *Geol. Soc. London Spec. Publ.*, **55**, 139–168.

Steel, R.J. and **Worsley, D.** (1984) Svalbard's post-Caledonian strata - an atlas of sedimentational patterns and palaeogeographic evolution. In: *Petroleum Geology of the North European Margin* (Ed. A.M. Spencer), pp 109–135. Norwegian Petroleum Society, Graham and Trotman, London.

Svela, K.E. (2001) Sedimentary facies in the fluvial-dominated Åre Formation as seen in the Åre-1 member in the Heidrun Field. In: *Sedimentary Environments offshore Norway - Palaeozoic to Recent* (Eds O.J. Martinsen and T. Dreyer), *Norw. Petrol. Soc. Spec. Publ.*, **10**, Elsevier, Amsterdam, 87–102.

Vollset, J. and **Doré, A.M.** (1984) A revised Triassic and Jurassic lithostratigraphic nomenclature for the Norwegian North Sea. *Norw. Petrol. Direct. Bull.*, **3**, 1–53.

Worsley, D. (2008) The post-Caledonian development of Svalbard and the western Barents Sea. *Polar Research*, **27**, 298–317.

Ziegler, P.A. (1988) Evolution of the Arctic–North Atlantic and the Western Tethys. *AAPG Mem.*, **46**, 198 pp.

Int. Assoc. Sedimentol. Spec. Publ. (2014) **46**, 215–252

Halten Terrace Lower and Middle Jurassic inter-rift megasequence analysis: megasequence structure, sedimentary architecture and controlling parameters

RODMAR RAVNÅS*§, KARI BERGE*, HEATHER CAMPBELL*, CRAIG HARVEY*† and MIKE J. NORTON*‡

**AS Norske Shell, PO Box 40, 4098 Tananger, Norway*
§Department of Petroleum Engineering, University of Stavanger, Stavanger, Norway
†Qatar Shell Service Company W.L.L, 1st Floor, Al Mirqab Tower, Corniche Road, West Bay, PO Box 3747, Doha, Qatar
‡Premier Oil, 23 Lower Belgrave Street, London

ABSTRACT

The Lower to Middle Jurassic of the Halten Terrace, that is the Åre (uppermost parts), Tilje, Tofte-Ror, Ile and Garn Formations, comprises three megasequences: the Tilje, the Tofte-Ile and the Garn megasequences. These megasequences have a common architectural motif and stratigraphic structure with a basal coarsening or 'sandying' upwards motif, a central sand-rich unit and an upper part of more varied stratigraphy. The lower progradational and parts of the central aggradational segments of these megasequences represent infilling and progradation of predominantly basin-marginal clastic wedges sourced from the Fennoscandian hinterlands. These segments formed during an interval of slowly rising through still-stand to falling and renewed rise of base-level, representing highstand normal through forced to lowstand normal regressions. The upper aggradational and retrogradational segments show a varied architectural motif that formed in response to increased basinal subsidence rates and mild structuring of the area. They are characterised by back-stepping of the rift-marginal clastic wedges and the presence of locally developed clastic systems formed around tectonically rejuvenated, intra-basinal highs. These upper parts or segments of the megasequences portray a sedimentary architecture mimicking mixed non-marine-shallow-marine syn-rift infills. The Lower to Middle Jurassic megasequences represent a series of depositional systems of varied size, origin and provenance. Larger hinterland-sourced fluviodeltaic systems dominated during progradational and aggradational stages, whereas more localised systems become increasingly important during the retrogradational stages. Repeated low-order relative sea-level fall, attributed to broad upwarpings of the rift-margin and its hinterland, forced the basin-marginal fluviodeltaic systems basinward to produce the repeated forced regressive to lowstand wedges, resulting in the formation of high-quality sheet-like reservoir bodies in the core of the megasequences. The uplift of rift-margins appears to be linked to and to represent the initial stages of basin-scale structuring, preceding and heralding subsequent minor rifting phases.

Keywords: megasequences, Halten Terrace, Lower and Middle Jurassic, deltaic to shallow marine, sedimentary architectures.

INTRODUCTION

Rift-basins often have a cyclic infill pattern defined by a series of stacked clastic wedges (Fig. 1). These clastic wedges are defined by alternations of sandier and muddier lithologies, related to the advance and retreat of basin-marginal, fluvio-deltaic depositional systems. Such infill patterns have been argued to form a common motif within both syn-rift infills (Ravnås *et al.*, 2000) as well as the infill of prolonged periods between successive rift-episodes (Steel, 1993); the so-called inter-rift

From Depositional Systems to Sedimentary Successions on the Norwegian Continental Margin, First Edition.
Edited by A. W. Martinius, R. Ravnås, J. A. Howell, R. J. Steel, and J. P. Wonham.

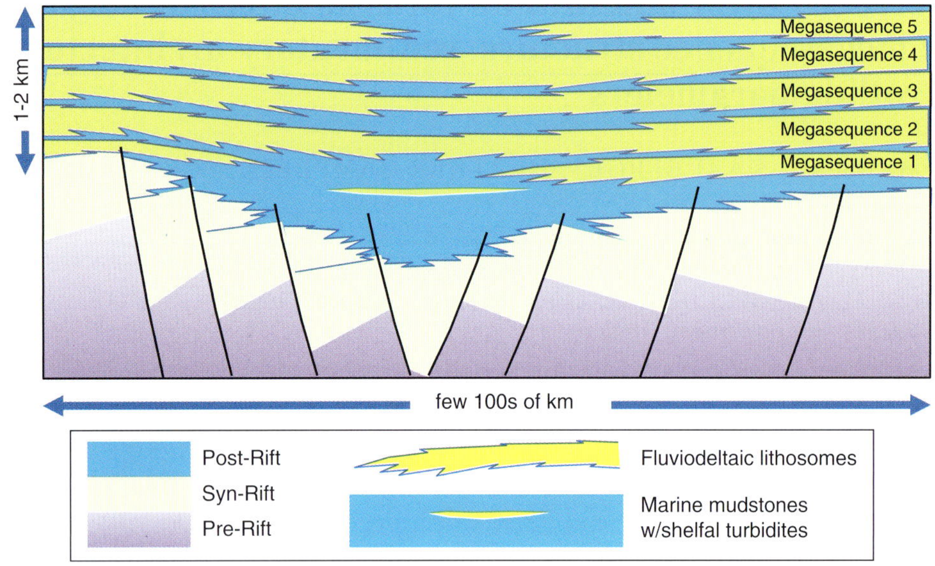

Fig. 1. Conceptual illustration showing characteristics of post-rift or inter-rift succession of rift-basin fills. Note stacking of clastic wedges in a forestepping to backstepping manner. Modified from Steel (1993), Ravnås & Steel (1998) and Ravnås *et al.* (2000). Post-rift strata are shown with a background colour of blue (illustrating marine mudstones) with component large scale (low order) fluviodeltaic lithosomes representing individual megasequences shown in yellow.

infill (Ravnås *et al.*, 2000). The clastic wedges have a regressive to transgressive structure, are bounded at their base and tops by mudstones representing peak transgressions (in marine successions) or peak backstepping (in non-marine successions). They normally represent a time-span of a few (i.e. 3 to 4) to more than eight million years. As such, they represent second order sequences and have been awarded the name 'megasequences' (Steel, 1993; Ravnås *et al.*, 2000).

The inter-rift megasequences (referred to as post-rift by Steel, 1993) differ from their syn-rift equivalents by their simpler sequence structure and component sedimentary architectures, as well as representing a longer time-span. This reflects a less complex basin and sediment infill development during the inter-rift period. Syn-rift megasequences, in contrast, often have a much more complex sedimentary architecture (Ravnås *et al.*, 2000), largely related to the fundamental control that tectonics exert on accommodation creation, sequence development and the resultant stratigraphic architecture (e.g. Ravnås & Steel, 1998; Gawthorpe & Leeder, 2000). However, a growing number of studies support the notion that minor or subdued extensional tectonics persisted throughout the inter-rift periods (Ravnås *et al.*, 2000 and references therein). Thus tectonics, at least partly, influence the stratigraphic architecture of the inter-rift megasequences and some similarities with proper syn-rift megasequences should be expected.

This contribution focuses on the development of the non-marine through marginal-marine to shallow-marine megasequences in the Lower to Middle Jurassic on the Halten Terrace, offshore Mid-Norway (Figs 2, 3 and 4). These megasequences were deposited during an interval of slow stretching, resulting in minor extensional tectonics during the inter-rift period separating the Permo-Triassic and Late Jurassic rift-episodes. The various non-marine to shallow-marine Lower to Middle Jurassic megasequences of the Halten Terrace have a similar progradational-aggradational-retrogradational gross structure. Minor departures from the 'norm' or differences between the stacked megasequences are attributed to temporal changes in basinal conditions. These are predominantly due to variations in sediment dispersal but are also caused by some variability in gross basinal subsidence and sediment supply rates between the various megasequences. The Lower to Middle Jurassic of the Halten Terrace thus provides an opportunity, firstly, for the identification and differentiation of the fundamental control(s) on the megasequence development. Secondly, a comparable study of these megasequences offers the opportunity to demonstrate how variations in

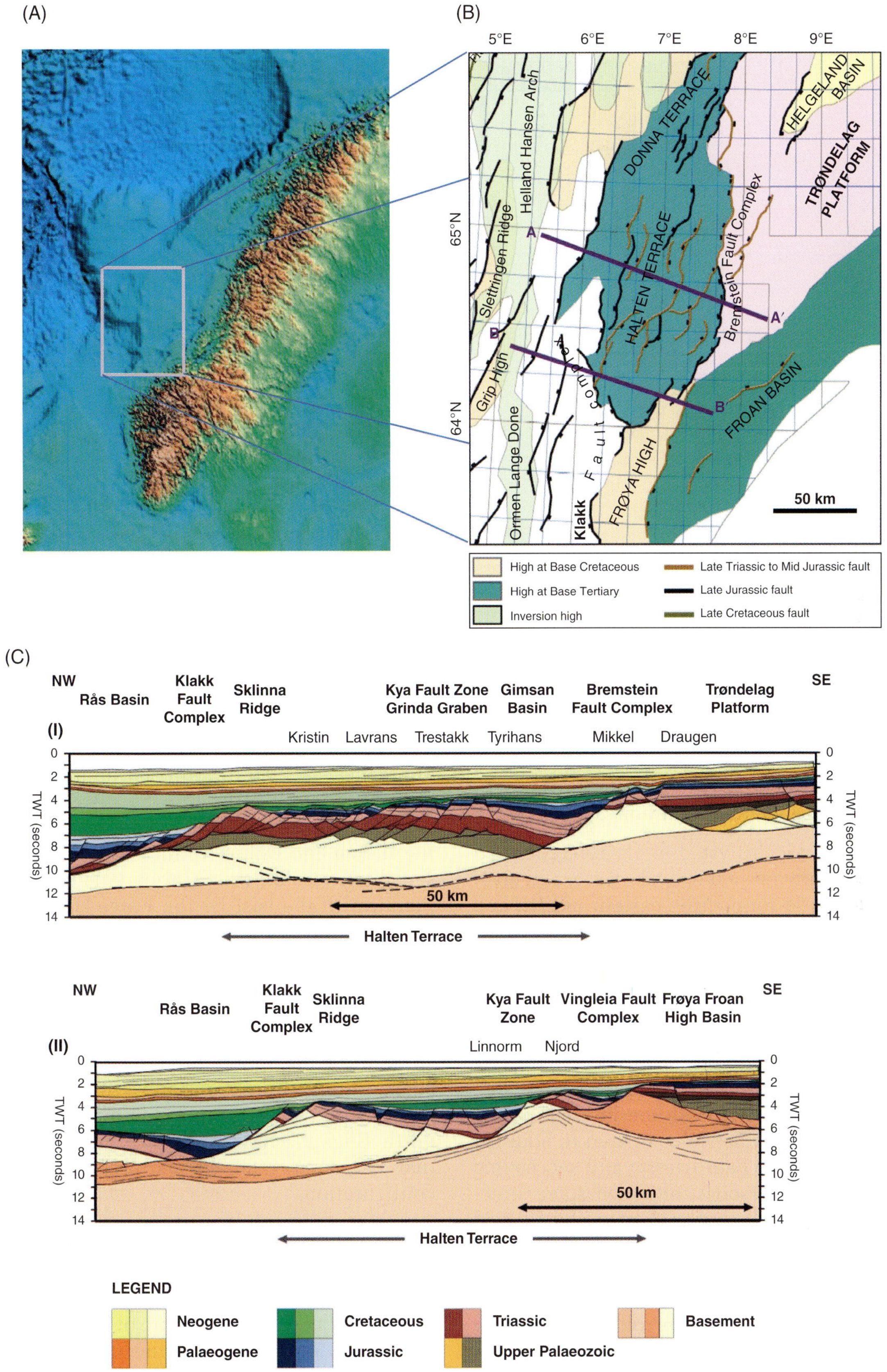

Fig. 2. Location map (A) and structural elements map (B) of the Halten Terrace area, mid Norway. (C) Regional geosection across the Halten Terrace (modified from Osmundsen *et al.*, 2002).

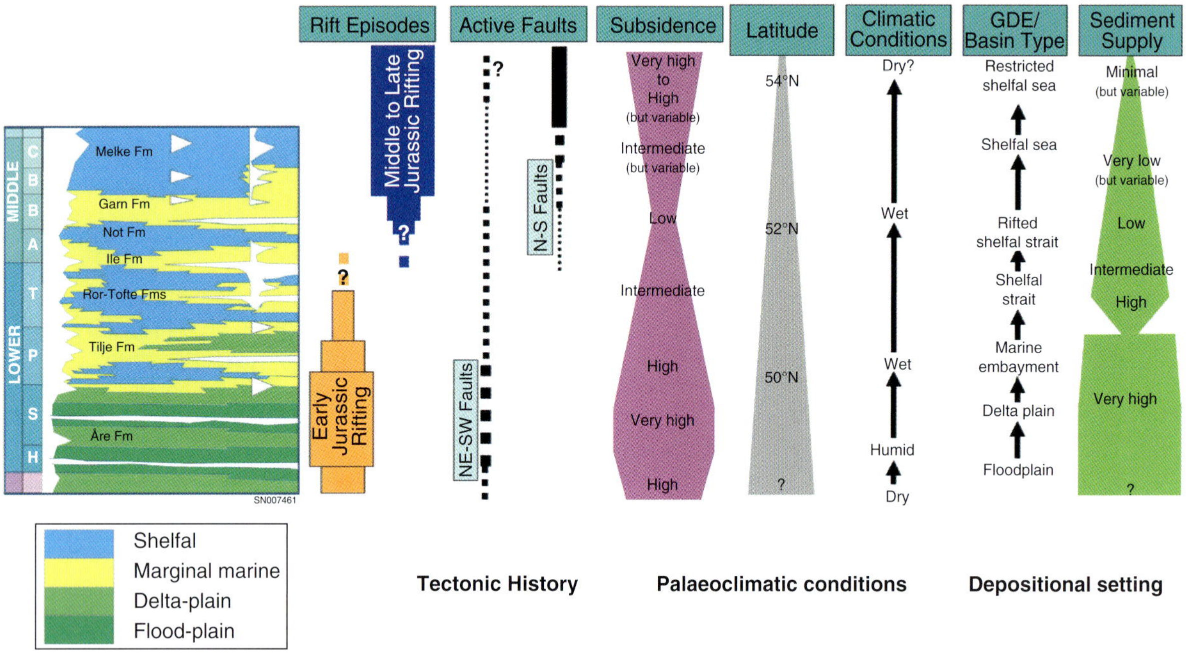

Fig. 3. Tectonostratigraphic summary chart for the Jurassic of the Halten Terrace area, mid Norway.

these controlling factors impact the resultant sequence architecture.

Geological setting & Early to Middle Jurassic semi-regional geological evolution

The Halten Terrace (Blystad *et al.*, 1995), located on the Mid-Norway shelf between 64°N and 65°N and 6°E and 8°E, formed the eastern margin and sub-platform area of the Mesozoic Norwegian-Greenland Sea Rift-basin (Fig. 2). The Halten Terrace together with its northward extension, the Dønna Terrace, both formed in response to two rifting episodes; during the Permo-Triassic and the Late Jurassic (Roberts *et al.*, 1999; Brekke, 2000; Færseth & Lien, 2002; Mosar *et al.*, 2002). The intervening Early to Middle Jurassic period was less tectonically active compared to both the preceding Middle to Late Triassic and succeeding Late Jurassic rift episodes, although some structuring also prevailed during this period (Fig. 3; Marsh *et al.*, 2010).

During the Early and Middle Jurassic, the Norwegian-Greenland Sea rift evolved from a subsiding alluvial-fluvial, south-westward-draining basin to a narrow shelfal strait or epicontinental seaway, connecting the Boreal Sea to the north with the Tethys Sea to the south (Figs 3, 4 and 5; Nøttvedt *et al.*, 2008). The East Greenland and Mid-Norway shelf areas formed the two opposing, or conjugate, margins of this seaway. The Halten and Dønna Terraces then combined with the Trøndelag Platform to form a wide platform area along the eastern rift margin.

During the late Middle to Late Jurassic rifting, the Halten and Dønna Terraces were structurally separated from the rift-marginal Trøndelag Platform to the east by the Frøya High and the Nordland Ridge (Withjack *et al.*, 1989; Brekke, 2000; Færseth & Lien, 2002; Mosar *et al.*, 2002). The Halten and Dønna Terraces were variably tilted eastwards and separated from the deep Late Jurassic basins to the west by a series of structural highs along the Klakk Fault Complex, such as the Sklinna Ridge and a series of smaller, downfaulted terraces and sub-platforms to the north (Fig. 2). It was during this rift episode that the area attained its present structural configuration.

Early to Middle Jurassic structuring & basin setting

The Lower to Middle Jurassic of the Norwegian-Greenland Sea rift-basin forms the post-rift unit to the underlying Permo-Triassic and the pre-rift strata to the overlying Upper Jurassic syn-rift succession (Færseth & Lien, 2002). Previous

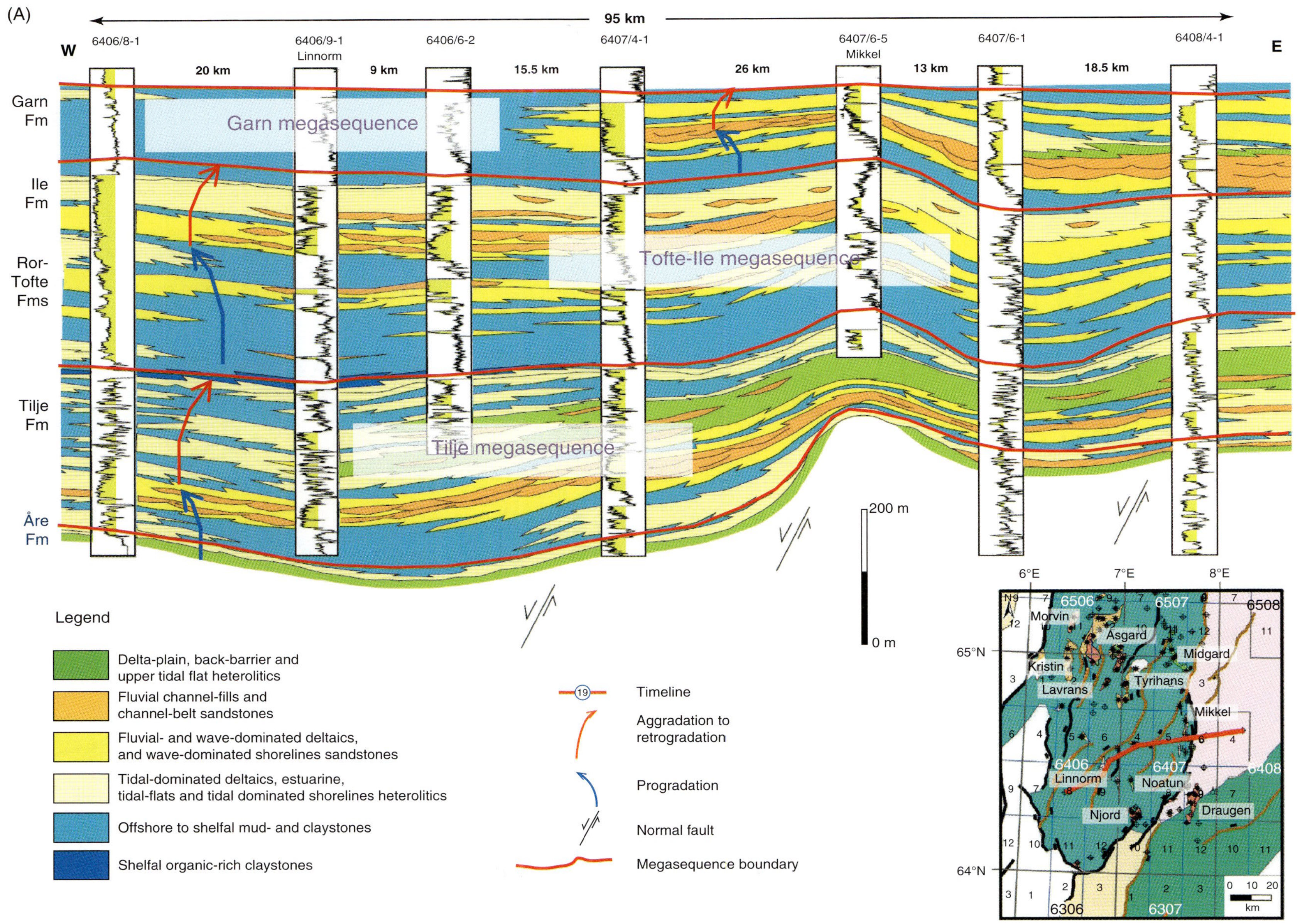

Fig. 4. (A) West-East oriented well correlation of the Lower and Middle Jurassic megasequences on the Halten Terrace area, mid Norway.

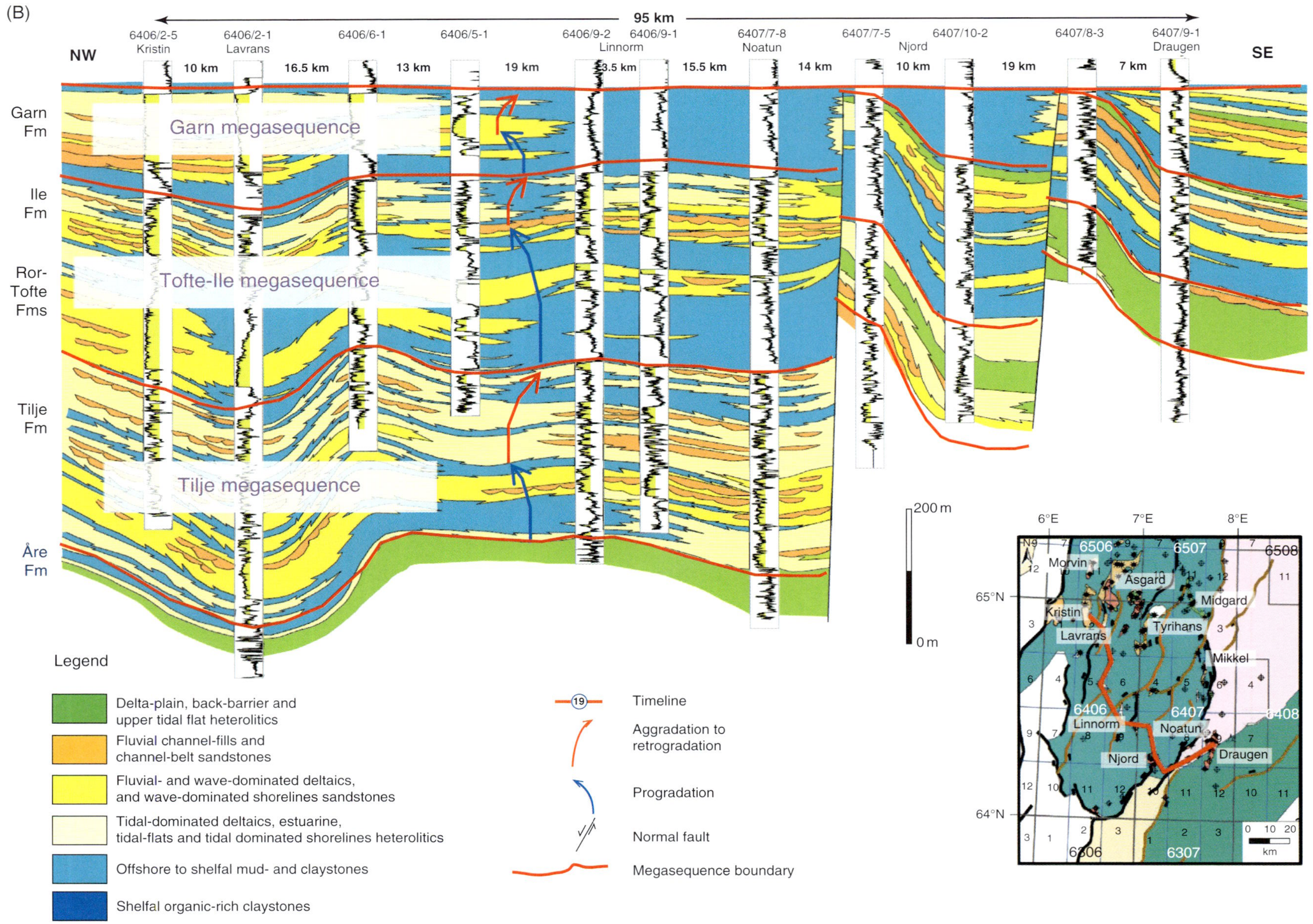

Fig. 4. (*Continued*) (B) North-South oriented well correlation of the Lower and Middle Jurassic megasequences on the Halten Terrace area, mid Norway. Note that section lengths (ca. 95 km) are shorter than the sum of individual inter-well distances due to the zigzag pattern of the well correlations. Log-curve shown is GammaRay (GR).

studies have documented variable structural influence on the infill patterns of the Lower to Middle Jurassic succession (e.g. Gjelberg *et al.*, 1987; Corfield & Sharp, 2000; Corfield *et al.*, 2001; Martinius *et al.*, 2001), testifying to the prolonged but varying rates of subsidence and rifting within this region. Structural styles related to Early to Late Jurassic rifting were controlled by the presence of Upper Triassic salt, which on the Halten Terrace alternates with mudstones. This interbedded salt-mudstone package forms a detachment interval (up to a few hundred metres thick), which commonly results in decoupling of the shorter wave-length Jurassic structures from the deeper-seated basement structures (Withjack *et al.*, 1989; Pascoe *et al.*, 1999; Marsh *et al.*, 2010).

Within the Halten Terrace area, seismically mapable extensional normal faults which trend NE–SW were active in the Early to Middle Jurassic, whereas N–S trending faults were active from the Middle Jurassic and onwards. Gross subsidence rates of the terrace area were relatively high in the Early Jurassic, gradually decreased through the early Middle Jurassic but increased again during the late Middle Jurassic due to renewed rifting (Fig. 3). These variations in subsidence rates reflect a decrease in gross basinal rifting rates, in combination with a westward shift and narrowing of the area of active rifting to the deeper basins west of the Halten Terrace. The westward shift in rift activity is herein related to a narrowing of the rift-basin proper; it resulted in the formation of a broad platform area comprising the Halten-Dønna Terraces and the Trøndelag Platform, along the eastern margin of the Norwegian-Greenland Sea rift-basin.

During the Early to Middle Jurassic, the Halten Terrace was located at ca. 50° to 52° northern palaeolatitude (Fig. 3; see also Dore, 1991) but within a relatively warm and humid palaeoclimate setting (Hallam, 1994). Sediment supply was very high during the Early Jurassic but gradually waned throughout the Middle Jurassic. As a result, a large-scale, overall transgression occurred during the Early to Middle Jurassic, despite the fact that the area experienced a gradual decrease in gross subsidence rates. However, both rifting and sediment supply rates fluctuated significantly (Fig. 3), which during certain intervals show marked departures from the overall gradual decreasing trend.

Lower to Middle Jurassic stratigraphy and palaeogeographies

On the Halten and Dønna Terraces, the Lower to Middle Jurassic forms an overall retrogradational, layered package of alluvial, fluviodeltaic, shallow marine and shelfal strata (Figs 3, 4 and 5). This contains the Rhaetian to Sinemurian, alluvial to fluviodeltaic Åre Formation and the Pliensbachian, fluviodeltaic and estuarine to shallow-marine Tilje Formation. The Toarcian to Bathonian Tofte, Ile and Garn Formations form three sandy fluviodeltaic, paralic, estuarine and shallow marine units that are interbedded with the marine (offshore to shelfal) mudstones of the Ror, Not and Melke Formations (Dalland *et al.*, 1988).

The Lower to Middle Jurassic can be organised into four clastic wedges, which grossly corresponds to the Åre, the Tilje, the Ror-Tofte-Ile and the Not-Garn Formations, respectively (Figs 3 and 4). These basin-marginal clastic wedges are separated by relatively thick offshore mudstones within uppermost Åre and basal Tilje, the lowermost Ror and the basal Not Formations, which represent pronounced marine flooding intervals. The clastic wedges are herein referred to as megasequences (following the terminology of Steel, 1993). They all have a common architectural motif comprising a basal coarsening or 'sandying' upwards motif, a central sand-rich unit and an upper part of more varied stratigraphy. The lower and central part of these megasequences are interpreted as progradation of basin-marginal alluvial to fluviodeltaic depositional systems during an interval of slowly rising through still-stand to falling base-level, typical of normal to forced regressive situations. The upper part, in contrast, shows more varied architectural motifs including aggradation and back-stepping of the rift-marginal deltas and locally developed clastic systems around local intra-basinal highs, producing a relatively thick transgressive succession.

Sediment provenance is inconclusive for the Lower to Middle Jurassic of the Halten Terrace and has been ascribed to easterly (e.g. Ziegler, 1988), westerly (e.g. Gjelberg *et al.*, 1987) or even northerly (e.g. Dore, 1991) sources. Recent studies have proposed repeated changes in provenance through time (e.g. Morton *et al.*, 2009), while other workers (e.g. Martinius *et al.*, 2001; Corfield *et al.*, 2001; Elfenbein *et al.*, 2005) support an overall Norwegian hinterland together with local, intrabasinal sediment sources (Fig. 5).

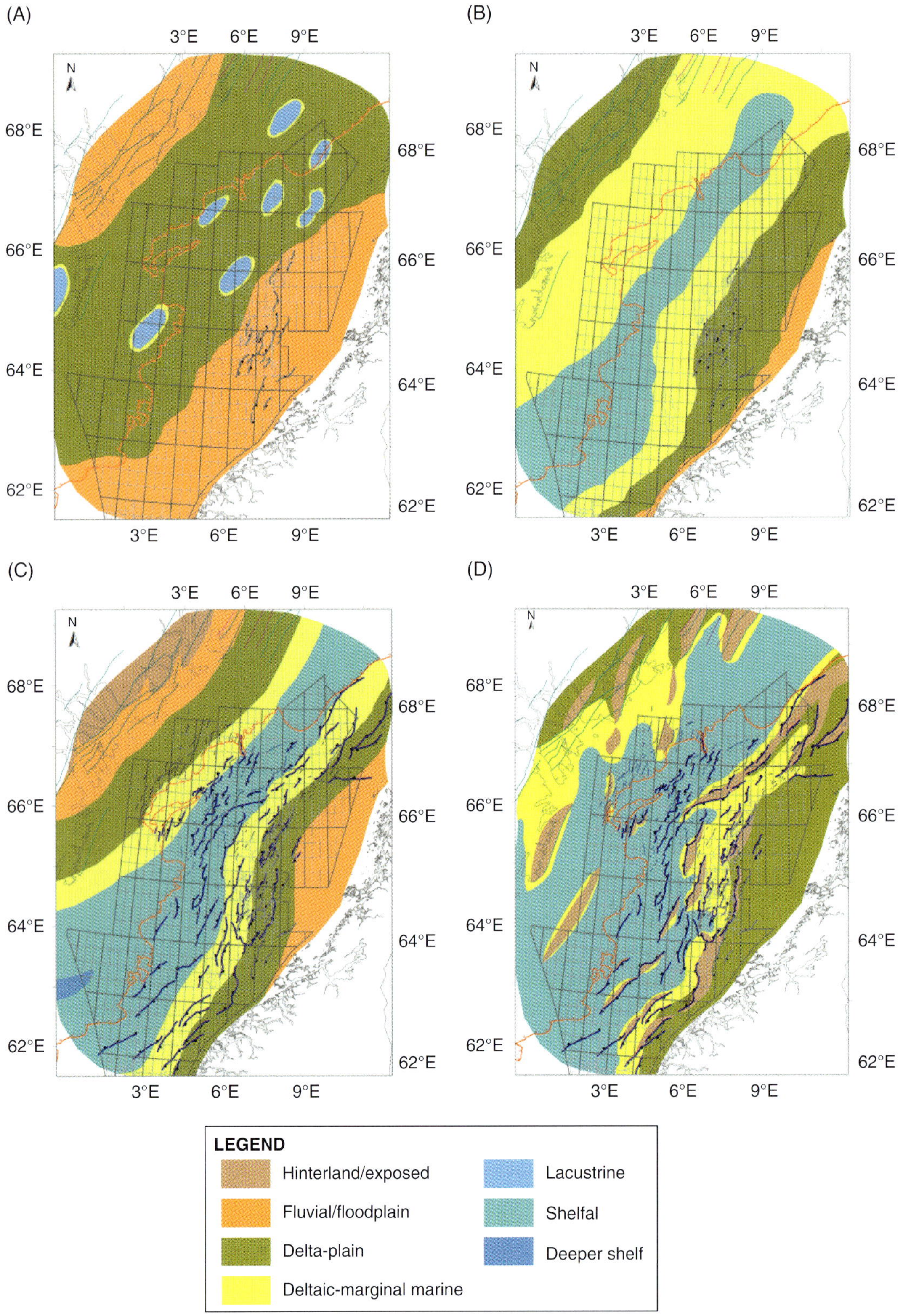

Fig. 5. Early and Middle Jurassic palaeogeographies of the Norwegian-Greenland Sea rift-system. (A) Hettangian to Sinemurian (Åre Formation) fluvial plain; (B) Pliensbachian (uppermost Åre Formation & Tilje Formations) embayment; (C) Toarcian to Aalenian (Ror-Tofte and Ile Formations) shelfal strait; and (D) Bajocian to Bathonian (Not-Garn Formations) rift-structured shelfal strait. Note however that the basin changed back and forth between an unstructured (as shown in B and C) to a moderately structured (as shown in D) basin during the Toarcian to Bathonian. Modified from Nøttvedt *et al.* (2008).

During the Early and Middle Jurassic there was a complex interplay between depositional styles, sediment supply and accommodation creation related to minor structuring or rifting events which occurred intermittently during this inter-rift period. The minor structuring events sustained the background subsidence which, in combination with eustatic sea-level changes, resulted in a gradual drowning of the Norwegian-Greenland Sea rift-system (Fig. 5). As a result, the gross depositional setting changed from a fluviodeltaic, elongated embayment in the earliest Jurassic, through a narrow shelfal strait and into an epicontinental seaway in the earliest Middle Jurassic. This change in depositional setting was paralleled by a gradual increase in basin topography and bathymetry, as a result of the continuous background structuring and a change to an increasingly under-filled basin during the Middle Jurassic.

Lower to Middle Jurassic megasequence analysis

The large number and geographic spread of well penetrations, together with a series of well-defined biostratigraphic time-lines readily identified and correlateable within these marginal marine units, has allowed a detailed semi-regional correlation of these Lower to Middle Jurassic units. However, the significantly fewer well penetrations and the poorer biostratigraphic resolution prevents a similar analysis of the underlying non-marine Åre Formation (Svela, 2001).

This contribution focuses on the evolution of the three mixed non-marine to marginal-marine Tilje, Tofte-Ile and Garn megasequences. The sequential development and stratigraphy of these megasequences are described separately below, with the aim to determine the controls on the variability in their infill trends and resultant stratigraphic architecture.

The analysis is based on all released and proprietary exploration and appraisal wells within the area, regional 2D seismic data as well as a semi-regional merged 3D dataset consisting of 17 individual 3D surveys. Core data have been selected to cover all stratigraphic intervals from all parts of the Halten Terrace, totalling about 5500 m of cores, split by about 1600 m (30%) over the Tilje megasequence, 2200 m (40%) over the Tofte-Ile megasequence and another 1600 m (30%) over the Garn megasequence (Appendices 1, 2 and 3, respectively).

Subsidence and sedimentation rates (Table 1) are based on preserved compacted thicknesses, as interpreted from seismic data and calibrated by wells. The Tilje and Ile-Tofte megasequences consist of highly variable lithologies with locally high proportions of mudstones, sometimes with interbedded coals. Decompacted thicknesses thus should be significantly thicker, suggesting that

Table 1. Calculated sedimentation and subsidence rates for the Tilje, Tofte-Ile and Garn megasequences, based on compacted thicknesses. Note that sedimentation and subsidence rates are nearly the same but with minor differences: The Tilje megasequence is an example of an overfilled basin, with calculated sedimentation rate equal to subsidence rates. Sediment supply rates was however larger and resulted at times in sediment bypass across the Halten Terrace during the formation of this megasequence. Sediment supply rates for the Garn megasequence vary from that equal to the subsidence rates, e.g. in areas where the basin was completely infilled, to less than the subsidence rates in areas where the basin remained under-filled (or submerged) during the formation of the megasequence. For all megasequences, both subsidence/uplift and sedimentation rates varied spatially and through time, sometimes also involving non-deposition or erosion, in the latter situation locally resulting in complete removal of the megasequence.

	Thickness (m)				Sedimentation rate		Subsidence rate	
Mega-sequence	Maximum	Range	Duration/ timespan Myr	Basin infill state	Maximum	Range	Maximum	Range
Garn	185	2 to 185	7 (7+)	Under-filled	<0.026 mm a^{-1}	0.0003* to 0.026 mm a^{-1}	0.026 mm a^{-1}	0.0003* to 0.026 mm a^{-1}
Tofte-Ile	355	20 to 355	9 (8 to 9)	Balanced	0.039 mm a^{-1}	0.002 to 0.039 mm a^{-1}	0.039 mm a^{-1}	0.002 to 0.039 mm a^{-1}
Tilje	330	0 to 330	8 (8+)	Over-filled	>0.041 mm a^{-1}	0.000* to 0.041 mm a^{-1}	0.041 mm a^{-1}	0.000* to 0.041 mm a^{-1}

*Missing stratigraphy. Reflects intervals of erosion and non-deposition. Includes intervals with uplift.

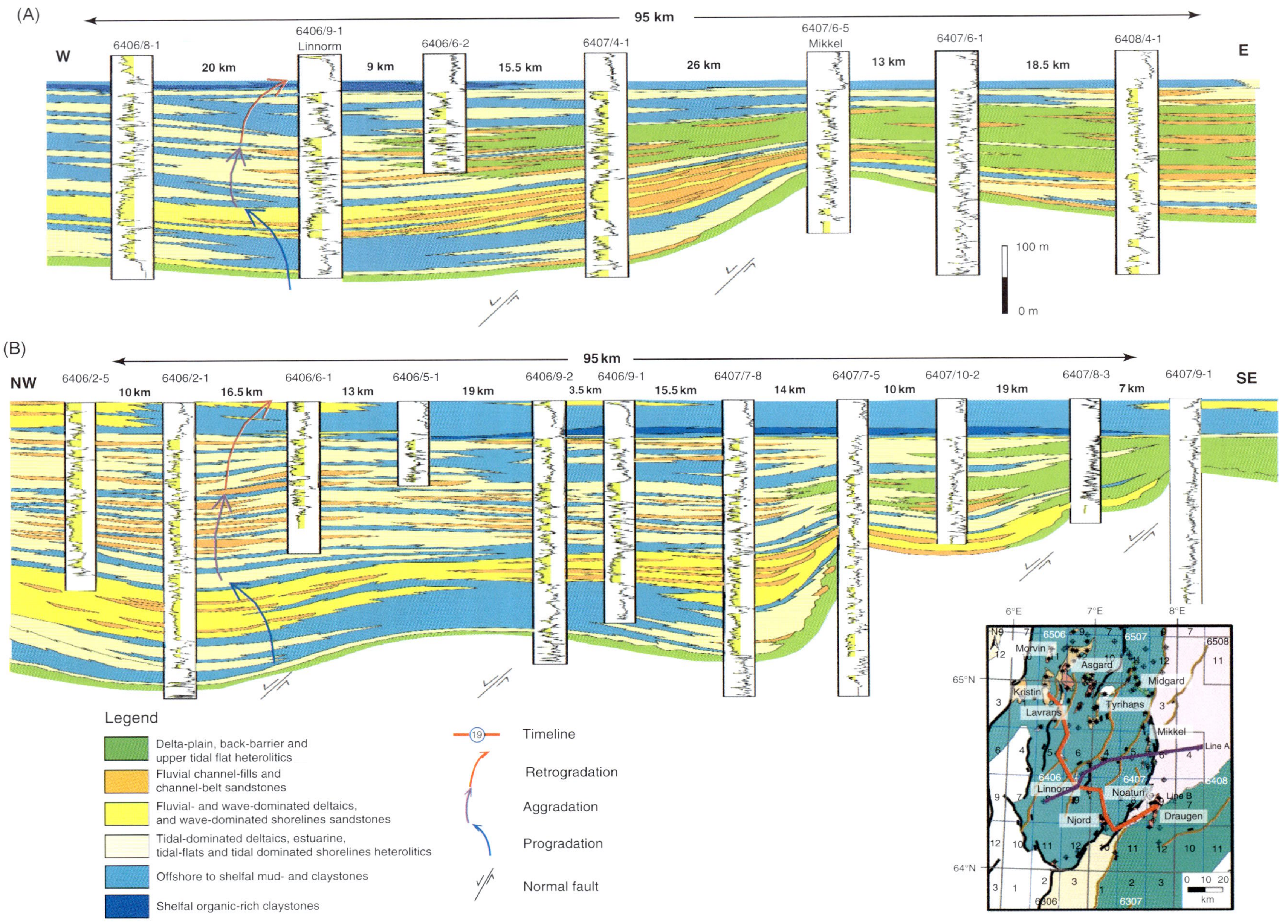

Fig. 6. (A) West-East and (B) North-South oriented well correlations of the Upper Sinemurian to Pliensbachian Tilje megasequence on the Halten Terrace area, mid Norway. Note that section lengths (ca. 95 km) are shorter than the sum of individual inter-well distances due to the zigzag pattern of the correlations. Log-curve shown is GammaRay (GR).

calculated sedimentation rates represent low estimates. The sandier nature of the Garn megasequence indicates that the difference between compacted and decompacted thicknesses should be less. Subsidence rates based on compacted thicknesses are less erroneous as accommodation space creation also involves early compaction of the substrate. The calculated rates thus are only relative estimates, but allow for comparison between the various megasequences.

The Tilje Megasequence

The Pliensbachian Tilje Formation (Fig. 2) together with the upper part of the underlying Åre Formation forms an up to 350 m thick megasequence (Figs 4 and 6), representing a timespan of some 8 + million years. It consists of a series of fluviodeltaic-estuarine lithosomes or higher-order sequences, which stack to form an overall progradational-aggradational-retrogradational megasequence architecture. Delta types range from sandy, low gradient, shoal-water, mixed tide-wave influenced river-deltas transitional into braidplain-deltas (Dalland *et al.*, 1988; Martinius *et al.*, 2001; Ichaso & Dalrymple, this volume) to mixed sandy-muddy, tide-influenced bay-head or protected embayment deltas (see also Kjærefjord, 1999). Estuaries range from smaller-scale inactive distributaries to larger, sub-basin wide, wave-dominated estuaries (Fig. 6, see also Dreyer 1992, 1993, 1999; Saigal *et al.*, 1999; Martinus *et al.*, 2003, 2005; LaFont *et al.*, 2010).

The basal boundary is defined by a relatively thick marine claystone unit within the uppermost Åre to basal Tilje Formations, which represents a major marine flooding. The upper boundary is defined by organic-rich claystones in the lowermost Ror Formation, representing a prolonged interval of basin-marginal delta retreats and basinal sediment starvation.

The progradational (regressive) segment (the upper Åre and the lower Tilje members) of the Tilje megasequence is relatively thick (>100m) and consists of an irregular but overall forward stepping stack of wave-dominated fluviodeltaic and shoreline lithosomes, each some 5 m to 15 m thick. These higher order lithosomes are interpreted as mixed tide- and wave-influenced deltas and shorelines, that most likely infilled a broad embayment (Fig. 7A,B1; Kjærefjord, 1999). The lithosomes represent a mixture of larger basin margin derived and smaller locally derived or intrabasinal sourced deltaic units. The locally sourced deltaic units are present around large intrabasinal highs such as the Sklinna Ridge (Ichaso & Dalrymple, this volume) and Frøya High (e.g. Dreyer, 1992, 1999), that was tectonically rejuvenated during a precursor Sinemurian rifting event.

The forward stepping, normal regressive lithosome stack is capped by an erosively based, sand-rich and laterally extensive fluviodeltaic to estuarine sandstone sheet, informally referred to as the lower Tilje member. On the Halten Terrace the sandstone sheet is composed of amalgamated, mixed wave and tidally-influenced, coarser-grained fluviodeltaics, distributary channel complexes and capping marginal marine (i.e. estuarine and shoreline) sandy lithosomes (Fig. 7A,B2). On the Trøndelag Platform to the east, it is backed by lower delta-plain heterolithics with isolate channel-fills and channel complexes, the latter representing candidate incised valley-fills. It is interpreted to represent a forced regressive to lowstand normal regressive lithosomes stack. 'Intraformational' unconformities and bypass surfaces are locally present across structural highs within the Halten Terrace (e.g. the SB2 of Martinius *et al.*, 2001; Ichaso & Dalrymple, this volume), as well as on the Trøndelag platform to the east.

The aggradational segment (the middle Tilje member) is separated from the underlying progradational segment by a bioturbated mudstone unit representing inner shelf to protected embayment heteroliths, which forms a higher-order flooding surface. The segment has an overall aggradational character, evidenced by a near-vertical superposition of facies belts within the stacked deltaic to estuarine lithosomes, though with a thin forward stepping basal part. It consists of stacked, mixed wave and tide-influenced, mud-prone and heterolithic fluviodeltaic lithosomes (Dreyer 1992, 1993, 1999; Martinius *et al.*, 2001; Ichaso & Dalrymple, this volume), representing the outbuilding of mixed lithology or muddy deltas into protected embayments or bays (Fig. 7A,B3). Based on distribution and orientation of facies belts, a mixed sourcing from basin-marginal and intrabasinal sources is envisaged (see also Ichaso & Dalrymple, this volume).

The retrogradational (transgressive) segment (or the upper Tilje member) comprises a heterolithic unit with significant lateral variability in infill patterns. Across the central part of the Halten Terrace it consists of mixed wave-structured and tidally-structured estuary fills, representing the infill of wave-dominated estuaries and offset backstepping tidally-influenced, deltaic lithosomes, representing

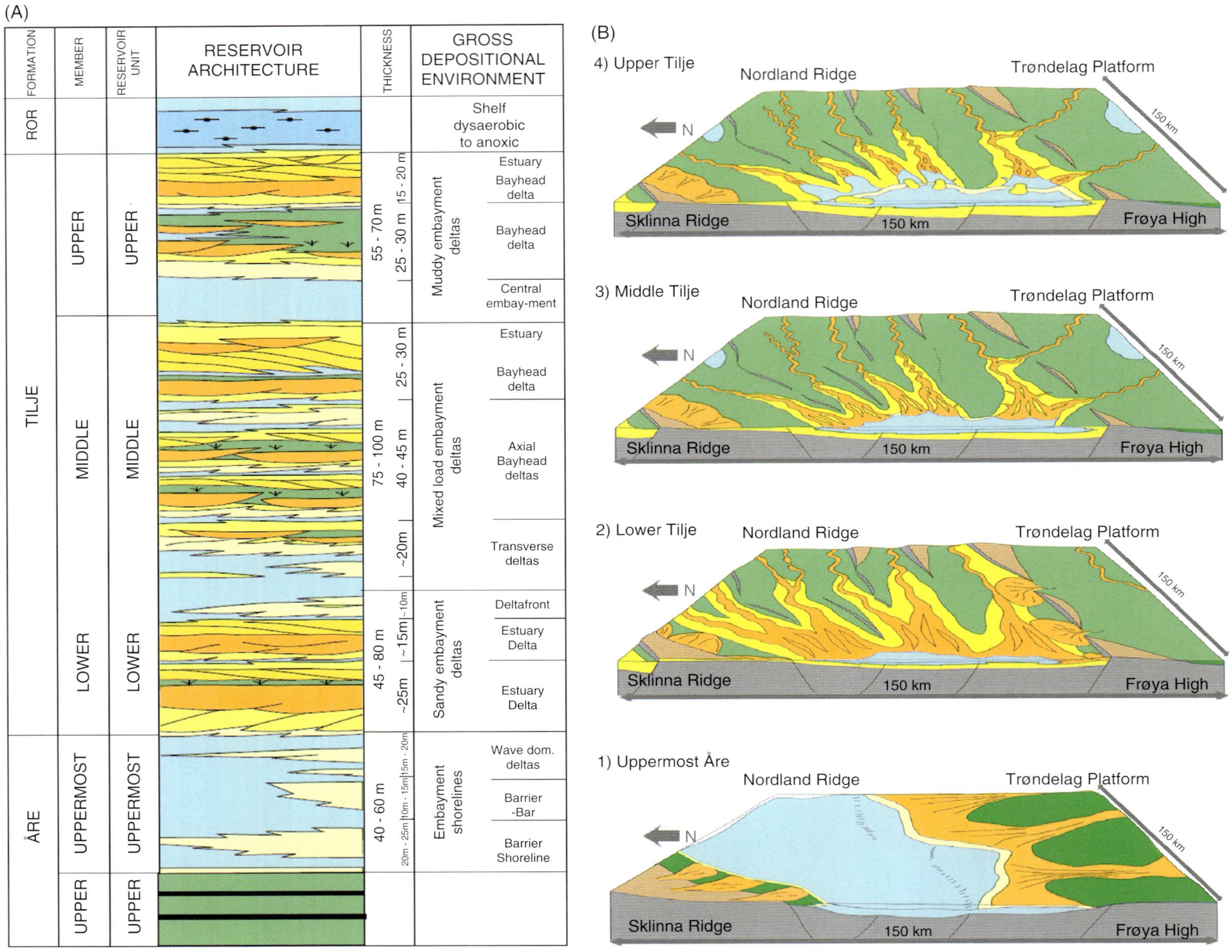

Fig. 7. (A) Stratigraphy, Gross Depositional Environments (GDEs) and Reservoir Architectures of the Tilje Formation in the south-western part of the Halten Terrace area, mid Norway. (B) Conceptualised GDEs for the Tilje megasequence on the Halten Terrace. (B1) Basal normal regressive part; (B2) overlying forced regressive part; (B3) middle aggradational part; and (B4) upper transgressive part. The transition from aggradation to retrogradation (B3 to B4) was paralleled by a change from dominantly deltaic to dominantly estuarine settings. Moreover, the transition was associated with increased sediment supply from intra-basinal provenances, resulting in a gradual backstep of basin-marginal depositional systems whereas the intra-basinal sourced depositional systems shows local progradation (e.g. around the Sklinna Ridge). Similar colour legend as for the well correlation, see Fig. 6.

a series of advancing and retreating bay-head deltaic units (Fig. 7A,B4; see also Dreyer 1992, 1993, 1999; Saigal *et al.*, 1999; Martinius *et al.*, 2001; Ichaso & Dalrymple, this volume). Alluvial fan to fan-deltaic and braidplain-deltaic lithosomes were re-established around structural highs, such as the Sklinna Ridge (Ichaso & Dalrymple, this volume) and downdip of relay-ramps along masterfault-bounded highs, such as the Frøya High. These coarser-grained lithosomes have an overall forestepping structure, although with considerable local variations in their gross sedimentary structure and internal architecture.

Synthesis: The Tilje megasequence is an example of infill during an interval with high rates of subsidence and sediment supply (Fig. 3). The lower part of the progradational segment, comprised by the uppermost Åre Formation, formed during normal regression with high sediment supply and rising relative sea-level. Accordingly, it represents a highstand systems tract. The stratigraphic variations in the dominant reservoir architectures probably represent changing basinal energy responses during infilling of the broad late Sinemurian to early Pliensbachian or 'late Åre embayments'. The capping sandstone sheet of the lower Tilje member is, based on the nature of the basal surface, component facies and coarser lithologies, interpreted to represent a marked basinward shift in facies tracts. It accordingly represents a forced regression and thus a falling stage to early lowstand systems tract. The aggradational segment is interpreted to represent the late lowstand systems tract, with the overlying retrogradational segment representing the transgressive systems tract. Sediments were supplied continuously from intra-basinal and basin marginal sources. Changes in provenance and sediment dispersal during deposition of the upper Tilje member is interpreted to reflect tectonic rejuvenation (e.g. Dreyer 1999; Martinius *et al.*, 2001; Ichaso & Dalrymple, this volume) related to minor rifting (Corfield & Sharp, 2000; Marsh *et al.*, 2010). This is also reflected in the increased spatial variability in infill patterns between and within the various sub-basins on the Halten Terrace (Figs 6 and 7).

Infill was from a mixture of large basin-marginal, longitudinal (NE to SW oriented) and lateral (SE to NW oriented) as well as a series of smaller, basin-interior sourced depositional systems (Fig. 7). During the main part of the megasequence development, these depositional systems linked to form an extensive, very low-relief, delta-plain to coastal (tidal) flat with a predominant longitudinally (i.e. NE to SW directed) sediment dispersal.

The Tofte-Ile Megasequence

The Toarcian to Aalenian Ror-Tofte and Ile Formations (Fig. 3) form another megasequence, more than 350 metres thick, (Figs 4 and 8), which represents a timespan of some 8 Myr to 9 Myr. The Ror-Tofte Formations consists of a series of stacked deltaic units, ranging from shelfal, deeper-water to shoal-water, mixed wave and river-/fluvial-influenced deltas, to braidplain deltas along the south-eastern part as well as braidplain-deltas and fan-deltas fringing highs along the western part of the Halten Terrace (Fig. 9; see also Gjelberg *et al.*, 1987; Dalland *et al.*, 1988; Ehrenberg *et al.*, 1992). Seismically identifiable, lozenge-shaped sandstones embedded in offshore mudstones, interpreted as shelfal sand-ridges, are locally present. A shelfal sand-ridge origin is supported by well- and core-calibration of seismic amplitudes in combination with palaeocurrent data derived from re-oriented cores and FMI-logs. The Ile Formation has traditionally been interpreted as a succession of tide-dominated deltas and estuaries (Gjelberg *et al.*, 1987; Dalland *et al.*, 1988; Harris, 1989; Pedersen *et al.*, 1989; Ehrenberg *et al.*, 1992; Saigal *et al.*, 1999; McIlroy, 2004; Martinius *et al.*, 2005), though braidplain deltas are present locally towards the base of this unit as well. The Tofte-Ile megasequence has an overall progradational-to-retrogradational character in the central and south-eastern part of the Halten Terrace, changing to a simple aggradational or highly irregular gross internal stacking pattern along the north-western part of the Halten Terrace (Fig. 8).

The basal boundary is defined by a relatively thick marine, shelfal claystone unit representing a major flooding interval within the lowermost Ror Formation. The upper boundary is represented by marine claystones in the basal part of the Not Formation, which represent a major landward flooding across the upper Ile tidal flats.

The progradational (regressive) segment (the Ror-Tofte Formations and parts of lower Ile member) consists, in the south-eastern and central part of the Halten Terrace, of a series of intermediate-scale coarsening-upwards motifs (up to >25 m thick), representing the prodelta to delta-top and inner shelf to paralic (back-barrier) deposits of wave-influenced and storm-influenced delta and shoreline systems (Fig. 9A,B1–2). The basal shelfal

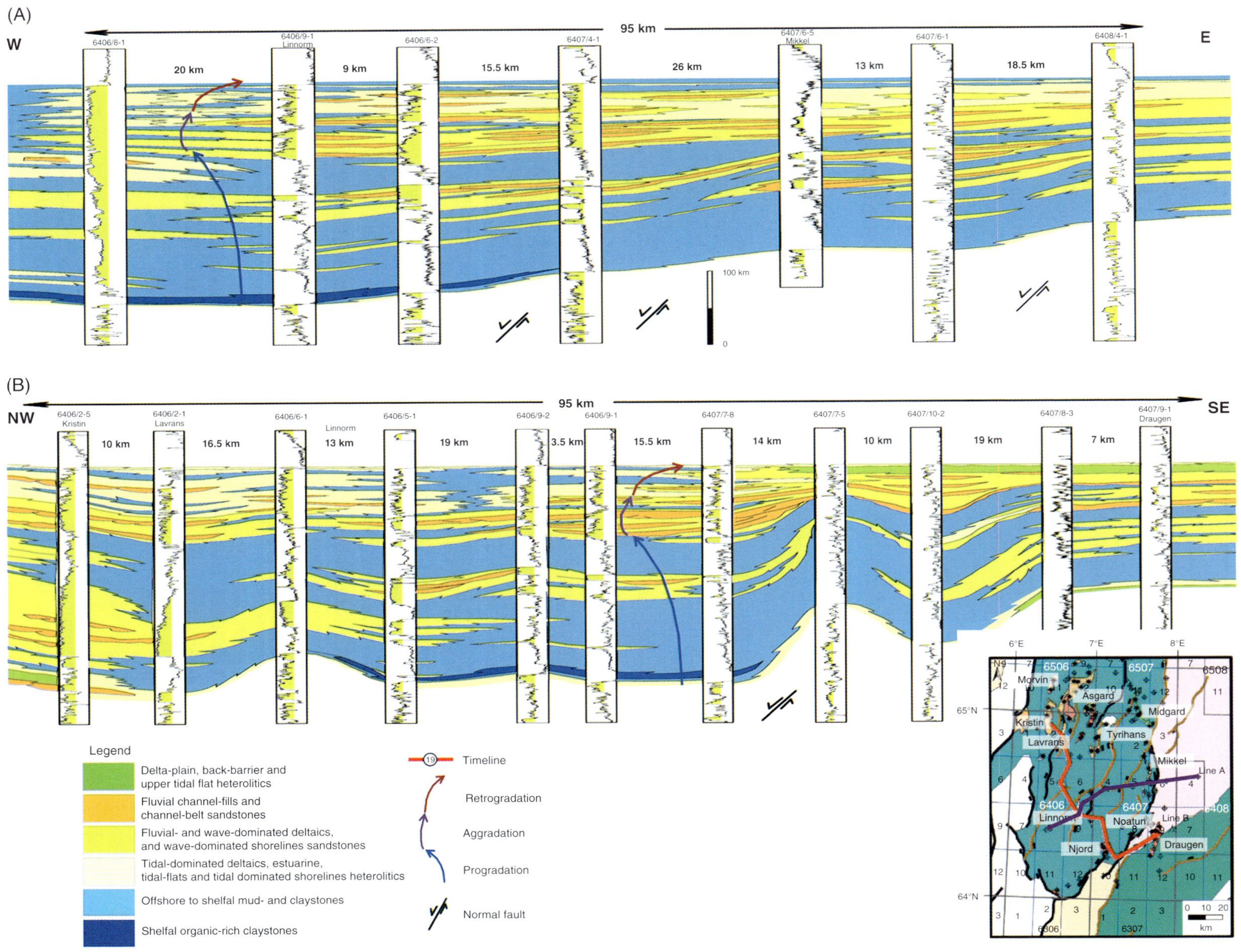

Fig. 8. (A) West-East and (B) North-South oriented well correlations of the Upper Toarcian to Aalenian Tofte-Ile megasequence on the Halten Terrace area, mid Norway. Note that section lengths (ca. 95 km) are shorter than the sum of individual inter-well distances due to the zigzag pattern of the well correlations. Log-curve shown is GammaRay (GR).

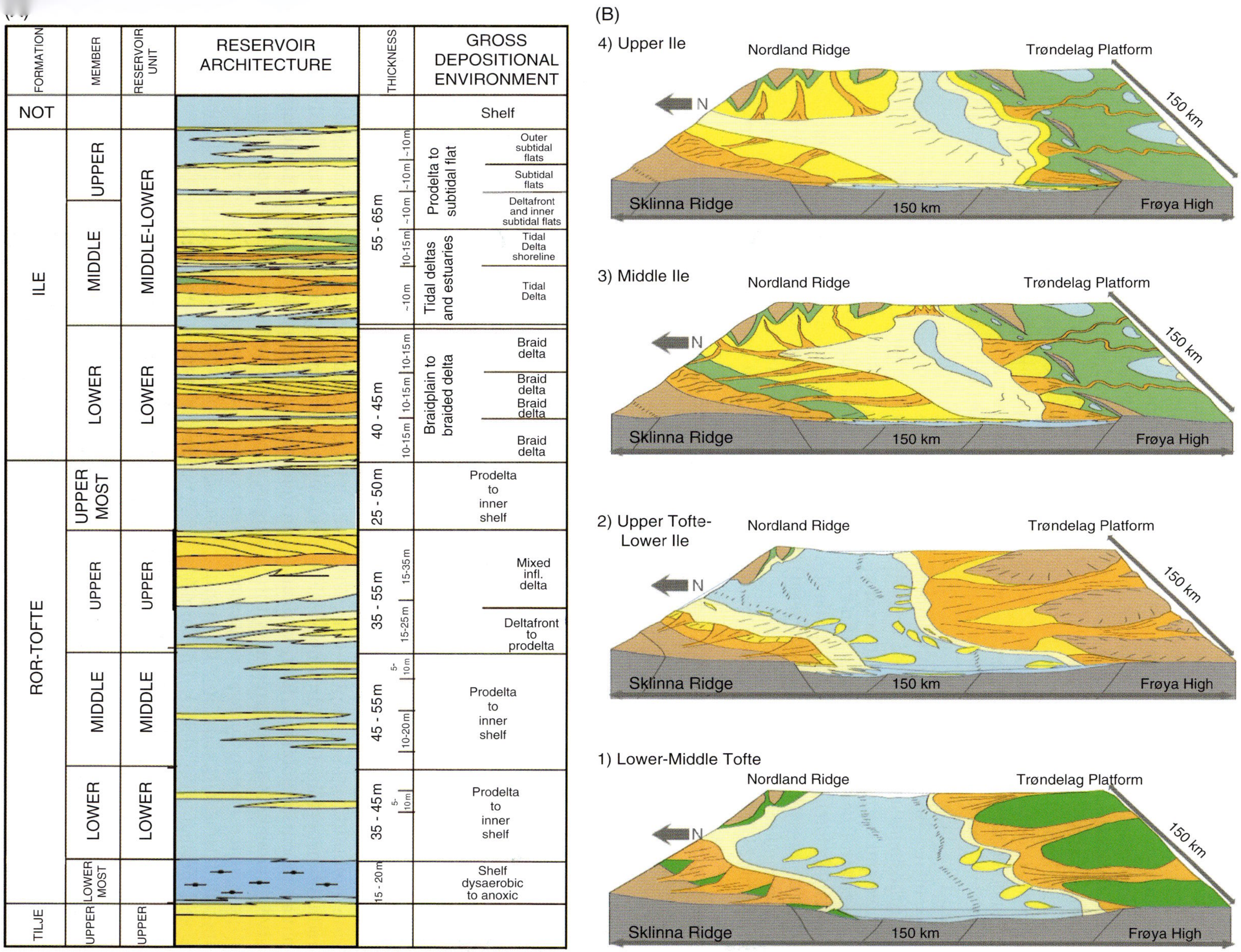

Fig. 9. (A) Stratigraphy, Gross Depositional Environments (GDEs) and Reservoir Architectures of the Ror, Tofte and Ile Formations in the south-western part of the Halten Terrace area, mid Norway. (B) Conceptualised GDEs for the Tofte-Ile megasequence on the Halten Terrace. (B1) Basal regressive-to-transgressive turnaround of local, fault-block sourced systems and peak flooding of basin-marginal clastic wedges to early normal regressive part; (B2) overlying forced regressive part; (B3) aggradational segment and (B4) upper transgressive part or segment. The transition from progradation via aggradation to retrogradation (B2 through B3 to B4) was paralleled by a change from dominantly deltaic to dominantly estuarine settings. The retrogradation is only evident from the basin-marginal depositional systems, i.e. those along the eastern half of the figures. The uplift and increased sediment supply from intra-basinal provenances resulted in a marked outbuilding of intra-basinal sourced depositional systems (e.g. around both the Sklinna and Nordland Ridges). Similar colour legend as for the well correlations, see Fig. 8.

claystones record the drowning of the Tilje Formation deltas and estuaries across the larger part of the Halten Terrace. The coarsening upward motifs within the lower part of the Ror Formation define a forward stepping or normal regressive stack of basin-margin derived deltaic units. The basin-marginal forestepping lithosome stack is capped by two erosively based, sand-rich fluvio-deltaic sandstone sheets, the upper Tofte and the basal part of the lower Ile members. These two sandstone sheets are developed only partially over the southern and central part of the Halten Terrace (Fig. 9A,B2). They mark abrupt and significant basinward shift in facies tracts and represent two detached mixed wave- and tide-influenced fluviodeltaic lithosomes and form a forced regressive to late lowstand normal regressive lithosome stack. On the western Trøndelag Platform, i.e. in the Draugen Field, these forced regressive to lowstand systems tracts are represented by an erosively based (?incised) fluvial sandstone sheet.

Along structural highs within the western part of the Halten Terrace, the Ror-Tofte Formations are developed as a stack of blocky or coarsening-upwards to coarsening-to-fining upwards motifs. These motifs change from aggradational to highly irregular stacks of sandy, coarse-grained fluviodeltaic lithosomes, with or without basal, intervening and capping marine mudstones. Often there is a lower progradational unit overlain by a thicker irregular and thinner retrogradational unit. These locally developed, coarse-grained fluviodeltaic lithosomes represent a stack of braidplain-deltaic and fan-deltaic units (Gjelberg *et al.*, 1987; Dalland *et al.*, 1988), sometimes with a tide and current-reworked delta-front and prodelta area. The locally developed deltaic units represent the outbuilding and stepwise retreat towards gradually peneplaned and drowned local or intrabasinal precursor footwall islands. The two extensive fluviodeltaic sheets, present within the central part of the Halten Terrace, represent outbuilding of basin-marginal, low-gradient, braidplain and normal river deltas, the former often with a steep, gravity-flow dominated delta-front area. However, outbuilding occurred during fluctuating relative sea-level, with high order falls and rises, as evidenced by interbedding of shelfal mudstones.

The aggradational segment (lower (parts) middle (parts) Ile member) comprises an overall aggradational stack of wave-influenced and tide-influenced fluviodeltaic lithosomes ranging in thickness from < 5 m to 15 m. The basal boundary is defined by a bioturbated inner shelf to shoreface mudstone-sandstone unit representing a higher-order flooding surface within the lower Ile member (Fig. 9A).

The component higher-order fluviodeltaic lithosomes show large spatial variability in characteristics and sedimentary architectures. In the south-eastern half of the Halten Terrace, the lower part of the segment is comprised of a relatively sand-rich, laterally extensive fluviodeltaic to estuarine sandstone lithosomes, representing the bulk of the lower Ile member. This package is interpreted to represent the continued outbuilding of the braidplain system, but now with a change to a series of mixed influence deltas bordered laterally by extensive tidal flats and shorelines (Fig. 9A,B3). In other areas, the lower Ile is composed of stacked, variably river-influenced, wave-influenced and tide-influenced fluviodeltaic and marginal marine, i.e. estuarine and protected shoreline, sandy lithosomes, which become increasingly tide-dominated towards the east. On the Trøndelag Platform further to the east, (palaeolandward), the segment is characterised by erosively based, fluvial channel-fills, interpreted to represented candidate incised valley-fills.

Thicker, locally developed, braidplain-deltaic, coarsening-to-fining-upwards sandstone units are present in the north-western and western Halten Terrace, as well as along the Nordland Ridge (e.g. Harris, 1989; Pedersen *et al.*, 1989). These represent a renewed phase of intra-basinal sourced delta progradation along the downfaulted dip-slope of basin-bounding highs, such as the Sklinna and Nordland Ridges, in the western and northern part of the Halten Terrace (Fig. 9A,B3). The progradational structure of these localised, basin-interior sourced fluviodeltaic lithosomes is related to renewed structuring and uplift of intra-basinal provenances. Other parts of these intrabasinal highs, such as the western flank of the Nordland Ridge, were surrounded by shorelines; again formed in response to renewed uplift and subaerial emergence of these structures.

For both the basin-marginal and intrabasinal sourced depositional systems, deltaic facies tracts alternate laterally and vertically with estuarine and shoreline units. This suggests a lower degree of incision and amalgamation of fluviodeltaic units and a larger variability in depositional sub-environments compared to the underlying regressive segment. Co-occurrence of deltaic, shoreface and estuarine facies tracts, suggest along depositional strike translations from deltas to estuaries with intervening shorelines. These were probably situated within a broad tectonic embayment that formed across the wider Halten Terrace (Fig. 9A,B3). Extensive tidal flats developed across the eastern part of the Halten Terrace and represent the inner or landward portions of the embayment. As a corollary, most estuary-fills recorded in the western part of the Halten Terrace probably represent abandoned distributaries within broader, deltaic settings.

The retrogradational (transgressive) segment (the middle (parts) and upper Ile member), is highly variable across the Halten Terrace (Fig. 9A,B4): The south-eastern part is dominated by an offset backstepping stack of tidally structured, fluviodeltaic lithosomes each up to 10 m to 15 m thick. These consist of tide-influenced deltaics, tidal shoreline and tidal flat heteroliths. The north-western part of the terrace is dominated by similar facies (Ehrenberg *et al.*, 1992; Saigal *et al.*, 1999; McIlroy, 2004; Martinius *et al.*, 2005), although forming an asymmetrical forestepping (thinner) to backstepping (thicker) stack of offset backstepping heterolithic lithosomes, that pass landward, i.e. toward structural highs to the west and north-west, into fluvial strata. The central part of the Halten Terrace is dominated by tidally-structured brackish-water heteroliths, representing protected bay or 'central estuary basin' deposits, whereas laterally extensive tidal flat heteroliths continued to be deposited along the eastern Halten Terrace and the western Trøndelag Platform. Sandy to heterolithic tidal bar and barrier/shoreline deposits are present across the south-western part of the Halten Terrace, representing the outer part of an alternating tide- and wave-dominated estuary, respectively.

The overall transgressive character of the segment is mostly evident by the backstepping of the basin-margin attached depositional systems present along the south-eastern part of the Halten Terrace. The locally sourced depositional systems around the intrabasinal highs such as the Sklinna (Ehrenberg *et al.*, 1992; Saigal *et al.*, 1999; McIlroy, 2004; Martinius *et al.*, 2005) and Nordland Ridges (Harris, 1989; Pedersen *et al.*, 1989), in contrast, shows a progradational to retrogradational structure suggesting local, tectonically induced controls on accommodation creation and sediment supply. Moreover, these locally sourced systems developed out of phase, with a diachronous build-out and retreat towards the Nordland Ridge, compared to that around the Sklinna Ridge. This is attributed to local variations in timing and style of structuring across the area.

Synthesis: The Tofte-Ile megasequence represents an interval with high rates of subsidence and intermediate to high sediment supply (Fig. 3) with highly variable but gradually decreasing sediment supply during development of the transgressive segment (Fig. 9). In the south-eastern and central part of the Halten Terrace, the progradational segment represents a normal regression in response to high but overall muddy sediment supply during continuously rising relative sea-level. Preservation is mainly as shelfal to delta-front and shoreface facies tracts. The capping sandstone sheets of the upper Tofte and lower Ile members are interpreted as detached, lowstand deltas and shorelines.

Within the regressive segment, prograding basin-margin attached fluviodeltaic lithosomes were paralleled by prograding to retrograding, coarse-grained braidplain-deltaic to fan-deltaic units along basin-interior highs within the western and north-western parts of the Halten Terrace. There is normally a non-correlative and diachronous sequence development between the basin-marginal and basin-interior, as well as between the various basin-interior sourced depositional systems. This is interpreted to reflect spatial variations in sediment supply and basinal subsidence rates.

The variable sedimentary architecture of the transgressive segment is indicative of rapid flooding across and sediment shut-off, to the former basin marginal lowstand shoreline. This was paralleled by a phase of uplift of basin-bounding highs with rejuvenation of local topography and renewed progradation of local, fault-block sourced depositional systems. As such, there was a marked change in uplift-subsidence patterns and sediment supply from the regressive to the transgressive segment, induced by renewed structuring of the greater Halten Terrace area. The highly variable infill character and lateral changes in depositional subenvironments, contained architectural elements and component stratigraphic surfaces present in the transgressive segment, is characteristic for basins with lateral changes in subsidence-uplift patterns and sediment supply. This suggests a return to a basin with an increase in the rate of tectonic activity, with the entire Halten Terrace forming a large funnel-shaped embayment bordered by structural highs and underlain by a series of half-graben sub-basins.

The presence of shelfal sand-ridges in the uppermost part of the regressive segment is related to

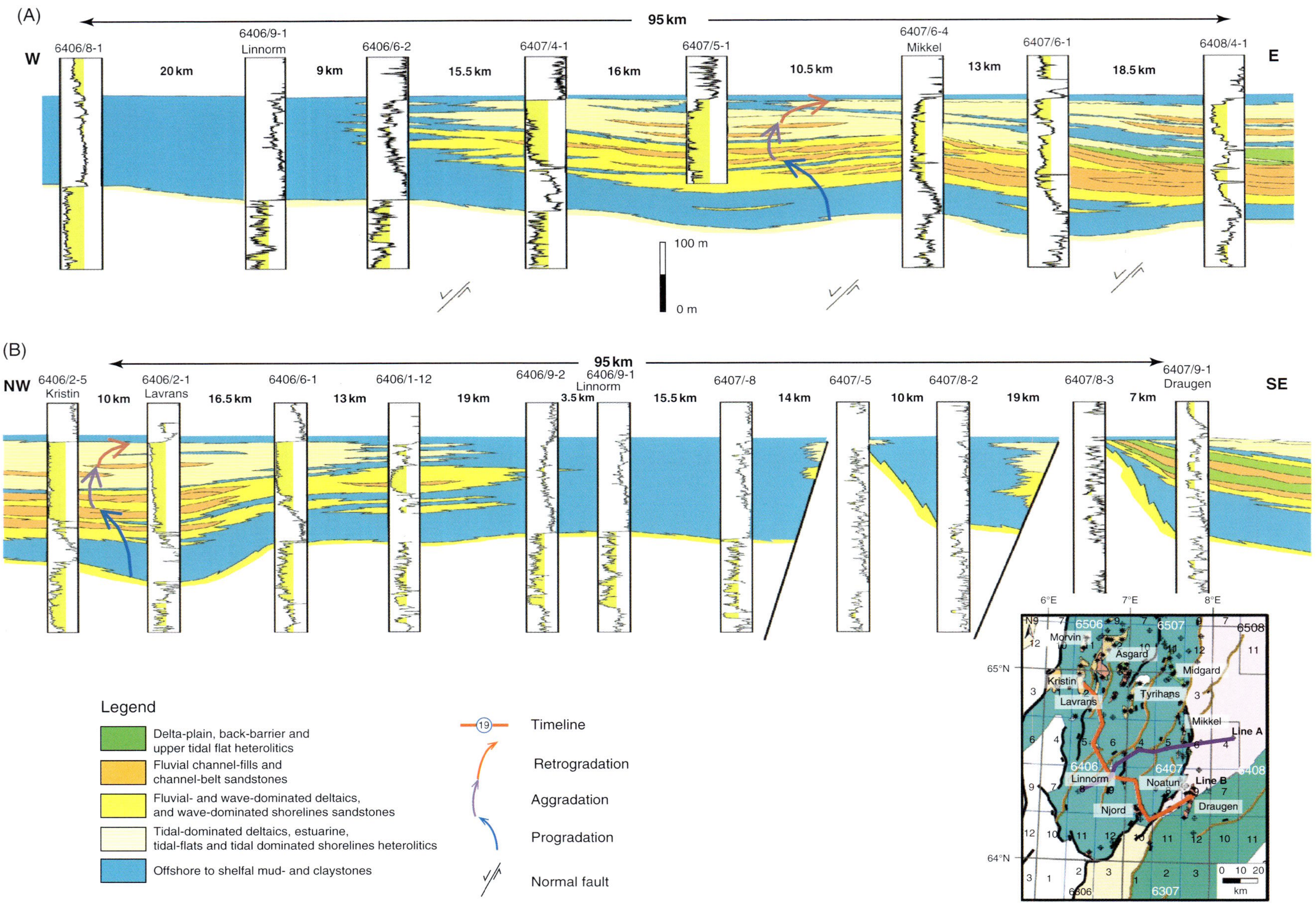

Fig. 10. (A) West-East and (B) North-South oriented well correlations of the Upper Aalenian to Bajocian Garn megasequence on the Halten Terrace area, mid Norway. Note that section lengths (ca. 95) km is shorter than the sum of individual inter-well distances due to the zigzag pattern of the well correlations. Log-curve shown is GammaRay (GR).

the narrowing of the structurally defined seaways that existed during Tofte depositional times. Basin-margin attached deltaic systems were extended across the central Halten Terrace, whereas the renewed uplift of the intra-basinal highs (such as the Sklinna and Nordland Ridges) promoted progradation of the local, rift-interior sourced depositional systems. This created a series of local straits across the western Halten Terrace connecting the various sub-basins, such as the northern and western parts of the Halten Terrace. These straits formed between growing structural highs and the detached, lowstand deltaic shoreline bodies (Fig. 9B2). In these narrow straits, strong, commonly unidirectional currents reworked the former deltaic sands into elongated, strait parallel sand ridges (or compound sand bars), probably in a similar manner as described by Reynaud *et al.* (2006) for Miocene seaways of SE France and by Anastas *et al.* (2006) for Oligo-Miocene seaways of New Zealand. Transgressive stage sand-ridges now appear fully embedded in shelfal mudstones, whereas sand ridges formed during high-order regressive stages appear to be shoreface attached.

The Garn Megasequence

The Garn Formation together with the upper part of the underlying Not Formation and the basal part of the overlying Melke Formation (Fig. 3) form a relatively thinner (up to 200 m thick) megasequence (Figs 4 and 10). The Garn megasequence represents a timespan of at least 7 Myr and has an overall progradational-aggradational-retrogradational character. The Garn Formation consists of braidplain and fluvial strata that are fronted to the south-west by mixed tide-wave-influenced, shoal-water braidplain deltas and intervening wave-dominated shorelines, as well as wave-dominated, sandy estuaries (Fig. 10; see also Gjelberg *et al.*, 1987; Harris, 1989; Pedersen *et al.*, 1989; Ehrenberg *et al.*, 1992; Corfield *et al.*, 2001; Elfenbein *et al.*, 2005). The lower delta-front, outer estuary and shoreface areas were dominated locally by stacked mixed wave- and tide-reworked sand-ridges (Sæther *et al.*, 1999; Messina *et al.*, this volume), which pass distally into the shelfal muddy sandstones and mudstones of the Not and Melke Formations (equivalents to the lower and upper parts of the Garn Formation, respectively; see also Dalland *et al.*, 1988).

The basal boundary is defined by a relatively thick marine, outer shelf mudstone unit in the basal part of the Not Formation. The upper boundary is represented by the Bathonian Melke Formation mudstones, which represent the shut-off of Garn sands to the Halten Terrace and the western Trøndelag Platform areas.

The progradational (regressive) segment (the Not Formation and lower Garn member) consists of a forward stepping stack of wave-structured fluviodeltaic lithosomes some 10 m+ to 35 m thick, representing the outbuilding of wave-influenced deltas and shorelines, most probably as the infill of a broad embayment (Fig. 11A, B1). The forward stepping stack is capped by an erosively based (in places with a thin coarsening upwards basal part), sand-rich and laterally extensive fluviodeltaic to estuarine sandstone sheet (the lower Garn member). The lower Garn member is comprised by amalgamated, mixed wave and tide-influenced fluviodeltaic and marginal marine (estuarine and shoreline-barrier bar), sandy (sand content normally more than 90%) facies tracts. The sandy nature and relatively coarse grain size suggests that the fluviodeltaic strata represent outbuilding of low-gradient, shoal-water braidplain deltas (Fig. 11A,B2; see also Harris, 1989; Ehrenberg *et al.*, 1992), interpreted to represent a forced regressive to late lowstand normal regressive lithosome stack. In narrower and semi-protected sub-basins, the delta-front and outer estuary sandstones were reworked into a series of stacked sand ridges (Messina *et al.*, this volume), whereas the river-structured and wave reworked mouth-bars were preserved in broader and less sheltered sub-basins.

The aggradational segment (the middle Garn member) is separated from the underlying progradational segment by a bioturbated, muddy siltstone unit, which represents a higher-order, minor flooding surface. The remainder of the succession comprises an aggradational stack of higher order fluviodeltaic lithosomes, including a thin forward-stepping basal part. It comprises mixed wave-influenced, tide-influenced and river-influenced sandstones (see also Ehrenberg *et al.*, 1992; Corfield *et al.*, 2001; Messina *et al.*, this volume), forming a sandstone sheet which displays thickness variations across faults and within tilted sub-basins. The unit is interpreted to represent alternations of sandy, mixed energy braidplain deltas and inner to outer parts of sandy, wave-dominated, structurally formed estuaries. These deltas and estuaries translated laterally into a wave dominated prodelta and inner shelf areas, respectively (Fig. 11A,B3).

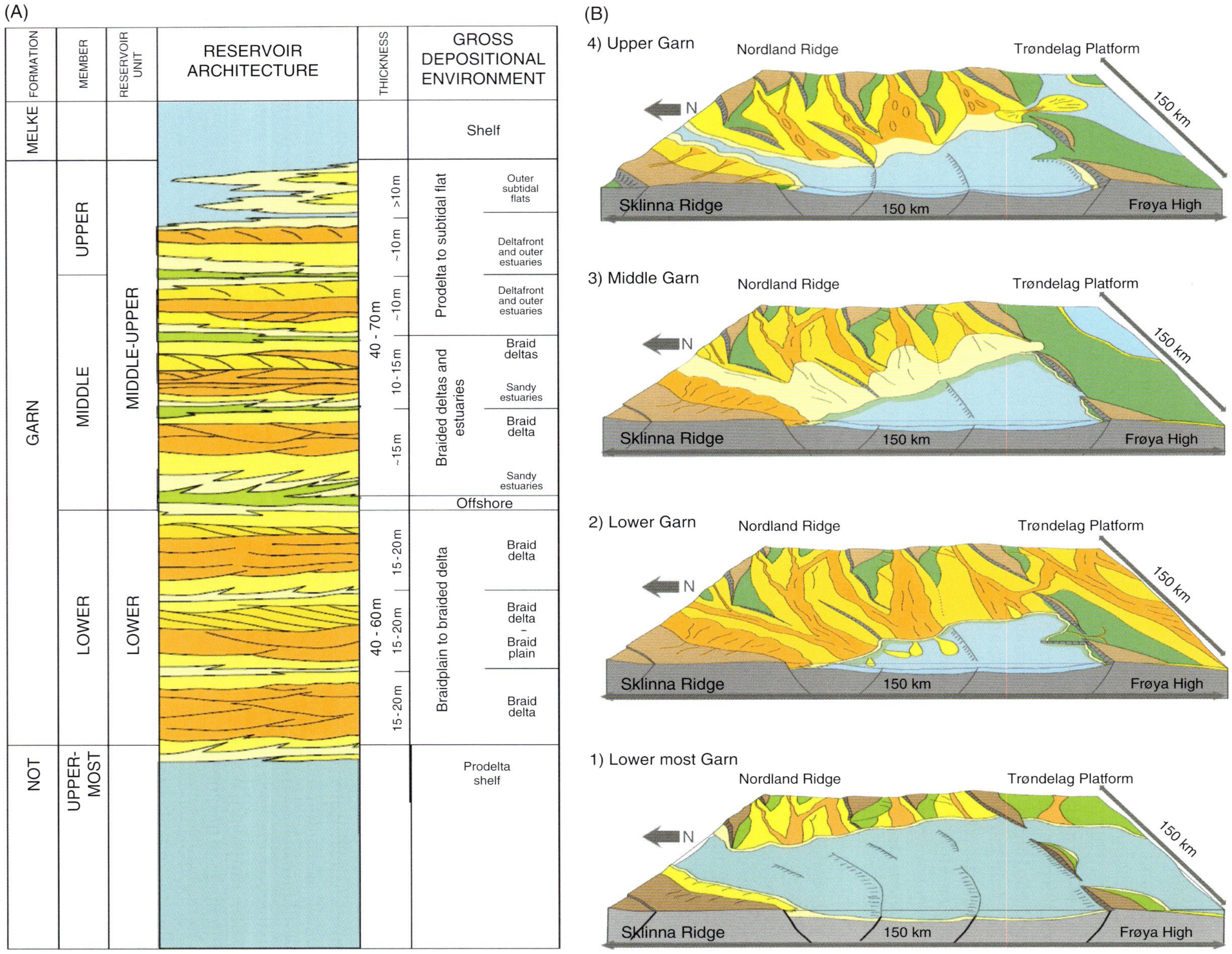

Fig. 11. (A) Stratigraphy, Gross Depositional Environments (GDEs) and Reservoir Architectures of the Garn Formation in the south-western part of the Halten Terrace area, mid Norway. (B) Conceptualised GDEs for the Garn megasequence on the Halten Terrace. (B1) Basal normal regressive part; (B2) overlying forced regressive part; (B3) aggradational part with braidplain deltas and estuaries; and (B4) upper transgressive part or segment. Similar colour legend as for the well correlations, see Fig. 10.

The retrogradational (transgressive) segment (the upper Garn member) is a sandy unit with marked lateral variability in infill patterns and sedimentary architectures (Fig. 11A,B4; see also Ehrenberg, 1991; Corfield *et al.*, 2001). The unit is represented by an unconformity across most structural highs, including local fault-block crests, sometimes with a thin transgressive lag resting atop the unconformity. Sub-basins within the central part of the Halten Terrace are dominated by mixed wave-structured and tidally-structured estuary fills, representing the infill of wave-dominated estuaries, as well as offset backstepping tide-structured, deltaic lithosomes, representing advancing and retreating bay-head deltas. Braidplain-deltaic lithosomes are locally present along structural highs, such as the Sklinna and Nordland Ridges (e.g. Harris, 1989). Along the western Halten Terrace narrow straits formed in swales between structural highs and formed sites of active tidal sand ridge migration (Messina *et al.*, this volume).

Synthesis: The Garn megasequence illustrates infill during low rates of both subsidence and sediment supply (Fig. 3). The lower part of the progradational segment (the Not Formation) represents normal regression of a highstand systems tract. The capping sandstone sheet of the lower Garn member is interpreted as a forced regression and thus a falling stage to early lowstand systems tract. The aggradational segment is interpreted to represent the late lowstand systems tract, with the overlying retrogradational segment representing the transgressive systems tract. The transgressive systems tract has marked lateral stratigraphic variability suggesting large spatial variations in depositional sub-environments and infill trends. Its stratigraphic architecture is characteristic of basins with lateral changes in subsidence-uplift patterns and sediment supply, such as tectonically active basins. However, subsidence rates appear to have remained low throughout the formation of the megasequence, suggesting that structural activity remained relatively low and that any structurally formed topography was subdued and removed by erosion and infilling.

Infill was initially from larger basin-marginal, longitudinal (NE to SW oriented) and a series of smaller, basin-interior sourced depositional systems, although with a predominant longitudinally (i.e. NE-SW directed) sediment dispersal (Fig. 11). With time, sediment supply from basin-marginal systems gradually diminished and there was a relative increase in the supply from basin-interior sources (e.g. the Sklinna and Nordland Ridges). This coincided with a change from dominantly deltaic (i.e. progradational) to estuarine (i.e. aggradational to retrogradational) development of the megasequence.

The sandy nature of the aggradational to retrogradational segments of the Garn megasequence is related to 1) sandy sediment supply during overall low rates of accommodation creation with fines continuously partitioned offshore and 2) uplift and erosion of older lower Garn sands to form added sand-rich sources during the aggradational and retrogradational stages.

Megasequence structure and sedimentary architecture

The Tilje, Tofte-Ile and Garn megasequences display similar gross stratigraphic structure, external geometries and internal stacking patterns (Fig. 12A), despite variability in subsidence, basin physiography, sediment supply and dispersal pattern, depositional systems and sediment calibre. The duration of megasequence (some 7 Myr to 9 + Myr) suggests that they represent candidate second order sequences.

Generic stratigraphic characteristics of, as well as variations in, sedimentary architectures between the various segments that comprise the megasequences are described below. Then a discussion of the candidate controlling mechanism(s) is presented.

Sedimentary architecture of the progradational (regressive) segment

The regressive segments are dominated by a progradational stack of higher-order, mixed wave- and river-/fluvial-influenced fluviodeltaic lithosomes. These represent delta advances and short-term retreats or avulsions along open to semi-protected (embayed) wave-dominated coasts, sometimes with a subordinate tidal influence (Figs 12B and 13).

Normal regressive segment: The early stages of delta-shoreline progradation were apparently slow and during rising relative sea-level, with relatively thick, gradually coarsening-upward, offshore-shoreface or prodelta-delta front-distributary channel fill motifs. These motifs, with their gradual facies tract transitions, represent highstand normal regressions (Van Wagoner *et al.*, 1990; Helland-Hansen & Gjelberg, 1994; Van Wagoner, 1995; Helland-Hansen & Martinsen, 1996) of basin-margin

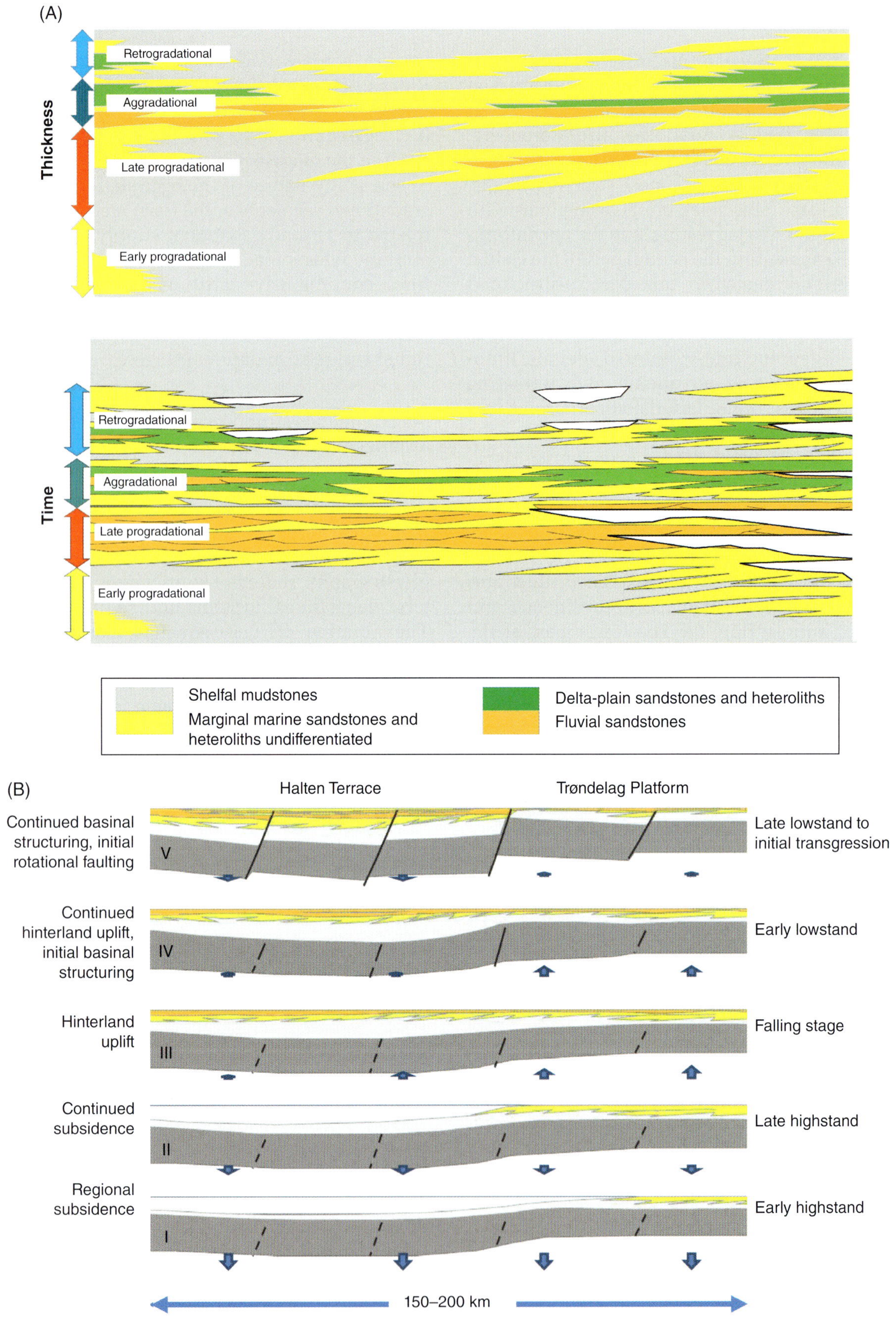

Fig. 12. Conceptualised sequence structure and internal sedimentary architecture in the Halten Terrace Lower and Middle Jurassic mixed, non-marine to shallow-marine megasequences (A). Conceptualised basin development and infill response during the progradational segments (B) and the aggradational to retrogradational segments (C).

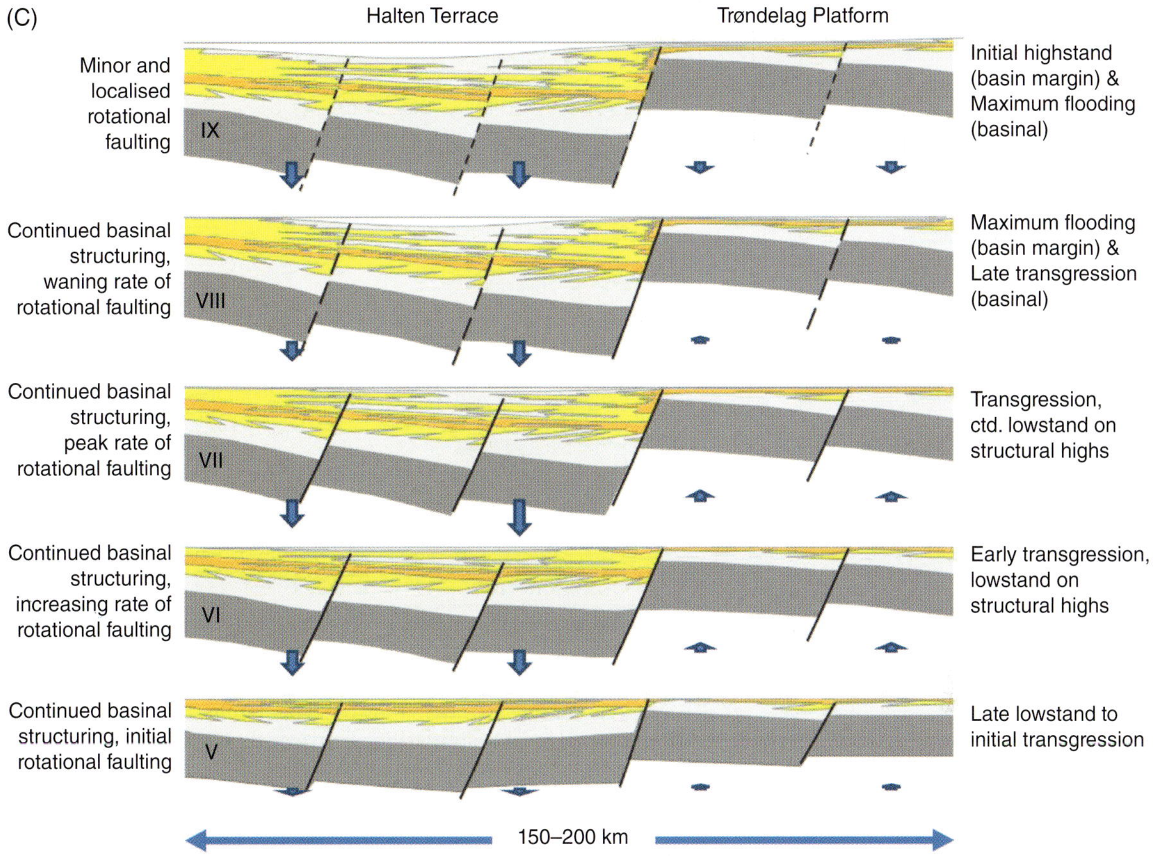

Fig. 12. (*Continued*) Sediment supply from the left part is conceptually envisaged to represent intrabasinal provenances. Deeper basins with departing subsidence and sediment supply histories may be present in the western (left) end of the profile. The profile is an idealised section that represents trends observed along both the N–S and W–E profiles illustrated by the well correlations in Figs 6, 8 and 10. Dark grey colour shading in B and C represent an older megasequence; light grey is marine mudstones, yellow is marginal marine and deltaics undifferentiated and orange is fluvial sheet sandstones.

attached depositional systems, either as transverse or combined transverse-longitudinal deltas and shorelines. The fronting shelfal to offshore area was dominated by alternating muddy inner shelf and gravity flow-dominated prodelta areas. The early progradational phase of the basin-marginal systems was often paralleled by the retreat (or aggradation) of any intrabasinal sourced depositional system.

Forced regressive segment: The upper part of the regressive segment is dominated by thinner coarsening-upwards motif(s), again representing offshore-shoreface or prodelta-delta front-distributary channel fill successions, often with marked and abrupt facies tract boundaries. These motif(s) are capped by erosively based, often detached, fluviodeltaic lithosomes or multi-storey-multilateral fluviodeltaic sheet sands. These detached fluviodeltaic wedges or sheet sands are, in turn, overlain by another forward-stepping stack of rapidly coarsening-upwards motifs of sandy fluviodeltaic lithosomes. This represents a change to coastal and delta outbuilding during falling relative sea-level to lowstand situations and early stages of rising relative sea-level or forced regressive or the falling stage (Hunt & Tucker, 1992; Posamentier *et al.*, 1992), through early to late lowstand systems tract, normal regressive (Van Wagoner *et al.*, 1990; Helland-Hansen & Gjelberg, 1994; Van Wagoner, 1995; Helland-Hansen & Martinsen, 1996) situations (Fig. 12B).

The forced regressive and lowstand units sometimes continued the outbuilding patterns from the underlying highstand situation, i.e. as transverse or combined transverse-longitudinal delta-shoreline systems, such as in the Tofte-Ile and Garn megasequences, often with the additional influx of sediment from local, intrabasinal

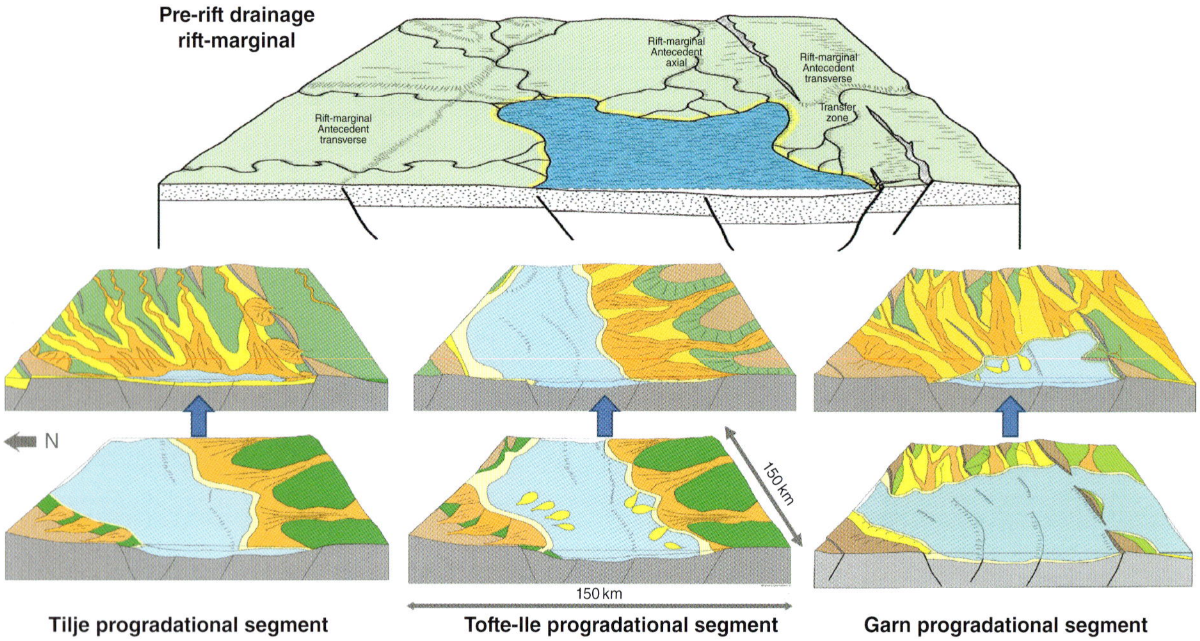

Fig. 13. Conceptualised Gross Depositional Environments (GDEs) for the regressive segments of the Halten Terrace Lower and Middle Jurassic megasequences. Note (1) the inherent contrasts in GDEs for the Tilje, Tofte, Ile and Garn megasequences, despite overall gross similar sequence development and (2) similarities between the gross overall infill style, despite variations in lithologies (e.g. Tilje and Garn megasequences) and basin physiographies and depositional systems (see discussion in text). Blue colouring represents shelfal environments and central parts of embayments, light yellow represents inner shelf/offshore transition zone and embayment margins, yellow represents shorelines and estuaries, orange represents deltaic distributaries, estuarine and fluvial channels, green represents delta-plain and floodplain environments, whereas lighter brown represents alluvial fans, braidplains and fan-deltas.

source areas. In other cases there was a change into a predominantly longitudinal drainage, sediment dispersal and outbuilding character, e.g. as inferred for the Tilje megasequence (Fig. 13). In parallel there was, commonly, a change in basin physiography from more open-coast to semi-protected embayments or straits, often with an increased current or mixed tide-wave influence on depositional facies. These stratigraphic changes in infill patterns probably represent an increasing rate of structuring of the receiving basin or depositional area (Fig. 12B; see also Ravnås & Steel, 1998; Ravnås *et al.*, 2000). As a corollary, the postulated relative sea falls responsible for the marked basinward shift in facies belts were probably tectonic in origin and related to tectonic uplift of the basin margin and hinterlands (cf. Steel, 1993) as well as initial, renewed structuring of the basin itself (cf. Ravnås *et al.*, 2000).

There are also notable differences between the various regressive segments, both in terms of overall sedimentary architecture, outbuilding styles and directions, as well as in their contained architectural elements (Figs 2 and 13). These differences are predominantly related to differences in basin topography and physiography during deposition. The much thicker normal regressive part in the Tilje and Tofte-Ile megasequences suggests outbuilding into a deeper basin with higher subsidence rates. Predominance of longitudinal or axial infill appears to reflect changes from early stage, open and sediment-under-filled to late stage, increasingly confined and sediment-balanced conditions, often reflecting transitions from normal to forced regressive situations. The basin energy regime also seems to follow similar trends; wave influence was predominant during the normal regressive state, whereas tide influence gradually increased and was locally the main reworking process during the late stage normal regressive and the forced regressive to lowstand states. In addition, reworking by continuous and strong shelfal currents appear to have been common during peak regressive state (e.g. during the Tofte-Ile and locally also the Garn megasequences). This is attributed to the gross depositional setting (Fig. 4); initial uplift or broadening of the intra-basinal highs during peak progradation resulted in narrowing of the tectonically silled sub-basins, then connected by a network of narrow and shallow straits. This repeated basinal topography rejuvenation caused changes into semi-protected bays and tectonic estuaries where wave energy was dampened and current activity enhanced (see review in Reynaud & Dalrymple (2012) and references therein).

Sedimentary architecture of the aggradational segment

The aggradational segment forms a relatively thick stack of mixed wave- and tide-influenced fluviodeltaic-estuarine lithosomes. These formed during repeated advances of the basin-marginal deltas and shorelines supplemented by a steadily increasing supply from intrabasinal sourced depositional systems (Fig. 12C) and resulted in infilling of the broader, tidally modified embayments. During higher-order delta retreats or avulsions, the former site of deltaic deposition changed into estuaries, either as a change from a distributary channel to a 'mini-estuary' on a larger delta top or a change of the structurally defined sub-basins into a 'tectonic estuary' or bay. In either case, the prevalent wave influence suggests a predominance of wave-dominated rather than proper tide-dominated estuaries.

The aggradational stacking is attributed to increased rates of basinal subsidence rates. In parallel there was a renewed or increased sediment supply from local intrabasinal sources such as uplifted structural highs, i.e. horst and footwalls along basin-bounding and sub-basin-bounding masterfaults, to form locally developed prograding depositional systems (Figs 12C & 14). These formed a series of lateral (transverse), point-sourced systems which in some megasequences (Tilje and Garn) linked with the larger axial or longitudinal systems, whereas in others (Tofte-Ile) joined laterally to form an elongate coastal/delta plain surrounding the emerging highs. Entry points into basinal areas were at cross-cutting faults and downdip of relay zones (e.g. Quin *et al.*, 2010). Unconformities are present across the most prominent structural highs, where they form isolate unconformity strands or merge to form a composite unconformity. In consort, this suggests a return to a more structured basin subject to spatial variations in tectonically induced subsidence and uplift rates.

Variations in subsidence rates and sediment supply/calibre has resulted in clear differences between the megasequences; the aggradational segment of the Tilje and Tofte-Ile megasequences is comprised by a distinct and layered aggradational stack of mixed-lithology (heterolithic) fluviodeltaics and estuarine facies tracts. The aggradational segment of the Garn megasequence, in contrast, forms a sand-rich unit of amalgamated sandy fluviodeltaic and estuarine facies tracts. However, also in this megasequence, the stratigraphic facies partitioning allows distinction of superposed or stacked higher-order sandy fluviodeltaic-to-estuarine units.

Sedimentary architecture of the transgressive segment

The transgressive segment commonly shows larger infill variability, both between the various megasequences, as well as laterally and stratigraphically within a single megasequence (Figs 12C and 14). Basin-marginal depositional systems form an overall retrogradational or offset backstepping stacking pattern, again composed of stacked, mixed tide- and wave-structured fluviodeltaic lithosomes but now with a clearer tidal signature (compared to the underlying aggradational part). These represent bay-fills with bay-head deltas and infills of wave-dominated to tide-dominated tectonic estuaries defined by individual sub-basins or half-grabens (Tilje and Garn Formations) or a large (basin-wide), partly segmented estuary or embayment (Ile Formation) (Fig. 14).

Sediment supply from local intrabasinal sources, i.e. structural highs along basin-bounding and sub-basin-bounding masterfaults, continued but now as more locally developed depositional systems (Fig. 12C). Braidplains formed along larger and higher-relief intrabasinal highs and sourced smaller and isolate deltas downdip on the structures, whereas short-lived local shorelines formed along smaller and lower relief emergent islands, before these were drowned to form submerged highs. Along the more prominent intrabasinal highs there is often spatial variability in delta types dependent on 1) sediment calibre, resulting in the formation of normal river to braidplain deltas to fan deltas and 2) the local basinal energy regime, with the resultant delta types varying from river to wave to tidally influenced or

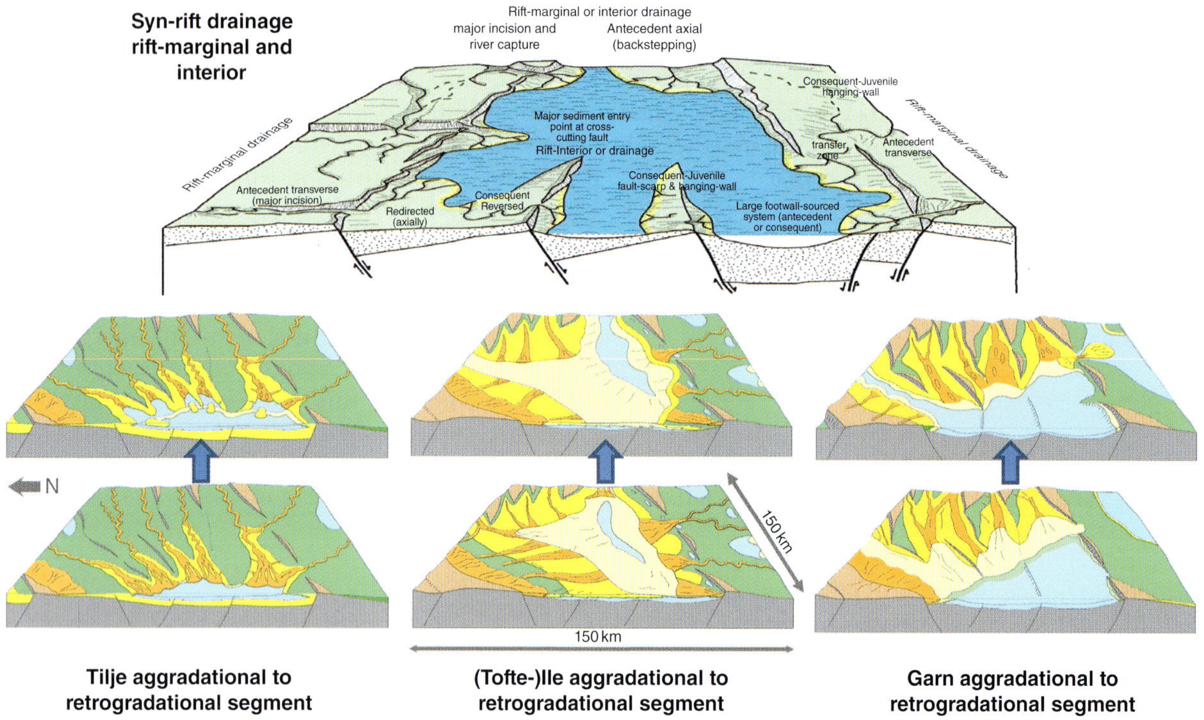

Fig. 14. Conceptualised Gross Depositional Environments (GDEs) for the transgressive segments of the Halten Terrace Lower and Middle Jurassic megasequences. Note the inherent contrasts in GDEs for the Tilje, Tofte, Ile and Garn megasequences, despite overall gross similar sequence development. Note the large variability in component depositional systems for the various megasequences, caused by mainly differences in basin structuring, sediment supply and calibre and basin physiographies. The same colour representations have been used as in Fig. 13.

dominated. In other units, such as in the Tofte-Ile megasequence, there was a dominance of tide-influenced to tide-dominated river deltas and estuaries, sourced by candidate braided fluvial systems.

In combination, this suggests increased rates of structuring and topographic rejuvenation, of both the wider basin areas, individual sub-basins and bordering highs during deposition of the transgressive segments (Fig. 12C). This is typical of basins subject to mild rifting (see Ravnås & Steel, 1998; Ravnås *et al.*, 2000). The presence and thickness of Triassic salt influenced the resultant structural styles strongly, as evidenced by local synclinal sub-basins as well as shorter wave-length fault-blocks/half-grabens and localised horst-and-graben structures (e.g. Pascoe *et al.*, 1999; Corfield & Sharp, 2000; Corfield *et al.*, 2001; Marsh *et al.*, 2010). These lateral changes in style of structuring and thereby rift topography, resulted in rapid to more gradual lateral variations from depositional sinks to bordering highs in terms of topography and basin physiography (Fig. 14). This is reflected in differences and contrasting developments of the transgressive segments between sub-basins in terms of their contained facies tract development and partitioning, depositional (sub-)systems, infill style and stratigraphic architectures.

Megasequence structure

The megasequences, with a lower progradational and an upper aggradational to retrogradational part, can all be divided into systems tracts, including a basal highstand normal regressive systems tract, a falling stage, forced regressive to late lowstand normal regressive systems tract, an aggradational or early transgressive systems tract and a retrogradational or late transgressive systems tract. These can be identified purely on stratal stacking patterns (e.g. see Catuneanu *et al.*, 2009) or via inferred shoreline trajectory patterns (e.g. Helland-Hansen & Gjelberg, 1994; Helland-Hansen & Hamspon, 2009) inferred from well (and core) data.

The megasequences are defined as second order genetic sequences (Galloway, 1989) with the full series of systems tracts (Brown & Fisher, 1977; Van Wagoner *et al.*, 1988, 1990; Posamentier *et al.*, 1988; Posamentier & Allen, 1999). Their component systems tracts define higher-order (candidate third order) genetic sequences, composed in turn by a lithosome stack of quite different architecture, i.e. from normal through forced regressive through aggradational to retrogradational. The lithosome units nested within the various systems tracts define even higher order sequences, which again show variations in component facies tracts/depositional systems and strata architecture. Neglecting that variability may arise when comparing sequences at different hierarchical scales, from contrasting basin settings or variable situations of accommodation creation may have added to the of the variety and misalignment of the alternative approaches to sequence stratigraphy (see Catuneanu *et al.*, 2009, 2010, 2011).

The aggradational segment, herein attributed to the early transgressive systems tract, appears to be a less common architectural element compared to other systems tracts (Catuneanu *et al.*, 2009, 2011; though see Van Wagoner *et al.*, 1990). This may potentially reflect geologic setting; thick aggradational segments are often well-developed in sedimentary cycles (megasequences) defining rift-basin fills and especially those formed during late pre-rift and inter-rift/early post-rift situations (Steel, 1993; Ravnås & Steel, 1998; Ravnås *et al.*, 2000). During such conditions both subsidence and sediment supply rates are often high and potentially near-balanced.

Major controls on gross sequence architecture

The Lower to Middle Jurassic megasequences represent a series of depositional systems of varied size, origin and provenance. Also, the basinal subsidence and stretching (rifting) rates varied spatially and temporarily during deposition of a single megasequence and exerted an additional fundamental control on the megasequence structure and sedimentary architecture (Fig. 12B to C), as well as on basin physiography (see also discussion in Yoshida *et al.*, 2007). In addition, the larger-scale variations in background subsidence and sediment supply (Fig. 3), with an overall reduction throughout the Early and Middle Jurassic, coupled with temporal changes in sediment calibre, had an additional impact on the resultant 'syn-rift' infill. These longer and shorter term temporal changes in basinal subsidence and sediment supply rates resulted in the differences in internal sedimentary architecture between the Lower and Middle Jurassic megasequences. Eustatic signals seem subordinate due to the non-uniform relative sea-level fluctuations and sedimentary response across the basin area.

The temporal and spatial changes in depositional styles, sediment supply and accommodation creation reflect continuous minor and variable structuring during the Early to Middle Jurassic. Repeated rifting sustained the background subsidence and, combined with eustatic sea-level changes, resulted in a gradual drowning of the Norwegian-Greenland Sea rift-system (Fig. 5), as well as repeated topography rejuvenation and thereby changing basin configurations and subsidence patterns of the Halten Terrace area.

Repeated structuring of basin areas was argued by Ravnås and co-workers (2000) as the main controlling factor on the megasequence structure in their study of time-equivalent marine rift-basin infill of the northern North Sea. Steel (1993) favoured hinterland uplift as the main controlling factor on the formation of the sand-rich sandstones forming the core of his 'post-rift megasequences'. The same tectonostratigraphic controls are also invoked for the Lower & Middle Jurassic infill of the Mid-Norway 'inter-rift' or mild syn-rift conditions. As such, these 'megasequences' mirror those argued to be typical of combined 'inter-rift' to early syn-rift origin in the northern North Sea (Steel, 1993; Ravnås *et al.*, 2000).

The falling-stage to lowstand system tract forming the central sand-rich core of the megasequences (Fig. 12A,B) reflects sea-level falls attributed to repeated broad upwarping of the rift-margin and its hinterland, similar to that proposed by Steel (1993) for the northern North Sea equivalents. Thus hinterland structuring is favoured as the major controlling factor on the formation of the high-quality, sheet-like or detached lowstand reservoir bodies, whereas variations in the style and rate of basinal structuring exerted the fundamental control on the sedimentary architecture of the aggradational and transgressive segments. The predictability of the 'transgressive' systems tract or segment development and its stratigraphic architectures remain challenging; also within the mild syn-rift infill intervals (see also Ravnås & Steel, 1998; Gawthorpe & Leeder, 2000; Ravnås *et al.*, 2000). However, some general rules still apply, but careful calibration and integration with the basin's structural history, basin physiography and sediment yield potential of intrabasinal highs is a pre-requisite in order to properly assess and predict the reservoir architecture and potential within these segments.

The stratigraphic occurrence of the lowstand bodies immediately underneath the more typical syn-rift packages, with some indications that initial rifting occurred concomitant with lowstand deposition (12B to C) suggest a linkage with hinterland uplift to the immediately succeeding rifting. It is therefore speculated that the uplift of rift-margins should be linked to, and is a manifestation of, basin-scale structuring preceding and heralding the subsequent mild rifting. However, the structuring of the area was gentle and it is uncertain whether the deeper seated faulting actually resulted in a faulted topography at surface or if drape folding of the supra-salt sediment pile was the dominant structural style during the Early and Middle Jurassic. Proper wedge shaped packages are only locally developed in the hanging-wall of some major basin-bounding faults and only within certain stratigraphic intervals, such as at the Åre to Tilje and Tilje to Ror-Tofte transitions. In such settings a faulted topography probably existed. More commonly, a gradual thinning into the faults is observed, favouring a more 'rounded' topography related to drape folding (e.g. see Sharp *et al.*, 2000; see also Withjack *et al.*, 1989; Corfield & Sharp, 2000 Corfield *et al.*, 2001; Marsh *et al.*, 2010). Moreover, the megaseqeunces and their component segments have an overall tabular appearance within a single sub-basin, with thickness differences observed only across major faults. In turn this suggests predominantly fault-related differential subsidence of individual sub-basins. It is therefore problematic to ascribe a proper syn-rift status to these Lower and Middle Jurassic megasequences.

CONCLUSIONS

On the Halten Terrace, the late Early to early Middle Jurassic (late Sinemurian to Bajocian) was a period of relative tectonic quiescence interrupted by repeated intervals of mild rifting. Three second order, regressive to transgressive, sequences, herein referred to as megasequences, form the basin-fill and display the following features: 1) basal and capping marine flooding intervals form the bounding surfaces and 2) progradational to retrogradational stratigraphic architecture, normally with a middle aggradational segment, representing the outbuilding and retreat of predominantly basin-marginal deltaic and estuarine systems.

Repeated outbuilding and retreat of basin-marginal clastic wedges was primarily in response to the interplay between sediment supply and the background, tectonically induced, basin subsidence and the linked hinterland-basin structuring. The regressive segment consists of a lower highstand normal regressive part, a forced regressive part and a lowstand normal regressive part. The lowstand normal regressive part often thickens up to form a thick forestepping aggradational segment. Outbuilding was initially during an interval of tectonic quiescence (the highstand normal regressive part) followed by an interval of hinterland uplift and initial basinal rifting, which resulted in the forced regressive and lowstand normal regressive part.

The transgressive segments display greater variability in stratigraphic structure and internal sedimentary architecture, formed in response to increased rates of basinal subsidence and structuring during relatively short-lived intervals with mild rifting. The component higher-order sequences nested within the transgressive segments, also show large spatial variability in their (higher-order) sequence development and structure, as well as in stacking patterns.

Spatial and stratigraphic variability resulting from the different combinations of subsidence/uplift rates, sediment supply and calibre, basin physiographies, component depositional systems and their sediment dispersal are the rule, for both a single as well as between the various megasequences. This has resulted in relatively large variability in sedimentary architecture of the megasequences, despite their similar gross stratigraphic structure. Improved reservoir prediction potential, both in terms of reservoir presence, geometries, connectivities as well as reservoir qualities, thus require a detailed knowledge of the wide range of controlling factors responsible for the resultant megasequence development.

Finally, the repeated occurrence and close temporal linkage between hinterland uplift and basinal rifting suggests a casual mechanism that should be further investigated in refined structural models for rift-basin evolution.

ACKNOWLEDGEMENTS

This study reflects the work of a team of Shell Exploration geologists and geophysicists working over a period of several years on the Lower and Middle Jurassic in Mid-Norway. The authors thank colleagues and partners in joint ventures for fruitful discussions over this period. We would also like to thank R.J. Steel, H.D. Johnson and A. Martinus for reviewing and suggesting improvements to this paper.

Appendix 1. Core database for the Tilje megasequence.

Tilje Megasequence Stratigraphic Segment		Well	Core	Location: Geographical	Location: Field/Discovery	Stratigraphic Unit
Retrogradational		6406/2-3	#10 (27 m)	Northwestern Halten Terrace	Lavrans	Upper Tilje mb
		6406/2-1	#11–15 (77 m)	Northwestern Halten Terrace	Lavrans	Upper Tilje mb
		6506/12-6	#9–10 (33.5 m)	Northern Halten Terrace	Smørbukk (Åsgard)	Upper Tilje mb
		6506/12-1	#10–11 (20 m)	Northern Halten Terrace	Smørbukk (Åsgard)	Upper Tilje mb
		6506/12-5	#23–27 (54 m)	Northern Halten Terrace	Smørbukk S (Åsgard)	Upper Tilje mb
		6506/12-3	#11–12 (41 m)	Northern Halten Terrace	Smørbukk S (Åsgard)	Upper Tilje mb
		6406/3-2	#12–14 (40 m)	Northern Halten Terrace	Trestakk	Upper Tilje mb
		6507/11-1	#1–5 (30.5 m)	Northeastern Halten Terrace		Upper Tilje mb
		6507/11-3	#7–9 (29.5 m)	Northeastern Halten Terrace	Midgard	Upper Tilje mb
		6407/2-3	#9 (22 m)	Northeastern Halten Terrace	Midgard	Upper Tilje mb
		6407/4-1	#9 (5.5 m)	Central Halten Terrace		Upper Tilje mb
		6407/7-8	#3–4 (34 m)	Central Halten Terrace	Noatun	Upper Tilje mb
		6407/7-A6H	#1 (25 m)	Southern Halten Terrace	Njord	Upper Tilje mb
		6407/7-5	#4 (24 m)	Southern Halten Terrace	Njord	Upper Tilje mb
		6407/7-4	#2–3 (53 m)	Southern Halten Terrace	Njord	Upper Tilje mb
		6407/7-2	#4–5 (22 m)	Southern Halten Terrace	Njord	Upper Tilje mb
		6407/7-1S	#6–7 (30 m)	Southern Halten Terrace	Njord	Upper Tilje mb
		6407/10-1	#5 (28 m)	Southern Halten Terrace	Njord	Upper Tilje mb
Core Length	596 m					
Aggradational		6406/2-1	#15–20 (103 m)	Northwestern Halten Terrace	Lavrans	Middle Tilje mb
		6506/12-6	#11–12 (47 m)	Northern Halten Terrace	Smørbukk (Åsgard)	Middle Tilje mb
		6506/12-1	#11–14 (42 m)	Northern Halten Terrace	Smørbukk (Åsgard)	Middle Tilje mb
		6506/12-5	#27 (24 m)	Northern Halten Terrace	Smørbukk S (Åsgard)	Middle Tilje mb
		6506/12-3	#12–15 (73 m)	Northern Halten Terrace	Smørbukk S (Åsgard)	Middle Tilje mb
		6406/3-2	#14–17 (46.5 m)	Northern Halten Terrace	Trestakk	Middle Tilje mb
		6406/9-2	#2–3 (35.5 m)	Central Halten Terrace	Linnorm	Middle Tilje mb
		6407/4-1	#10–14 (46 m)	Central Halten Terrace	Linnorm	Middle Tilje mb
		6407/7-8	#4–6 (38 m)	Central Halten Terrace	Noatun	Middle Tilje mb
		6407/7-A6H	#2–3 (39 m)	Southern Halten Terrace	Njord	Middle Tilje mb
		6407/7-4	#3–7 (45 m)	Southern Halten Terrace	Njord	Middle Tilje mb
		6407/7-2	#5–8 (36 m)	Southern Halten Terrace	Njord	Middle Tilje mb
		6407/7-1S	#5–9 (42 m)	Southern Halten Terrace	Njord	Middle Tilje mb
Core Length	617 m					

(*Continued*)

Appendix 1. (*Continued*)

Tilje Megasequence Stratigraphic Segment		Well	Core	Location		Stratigraphic Unit
				Geographical	Field/Discovery	
Progradational Forced Regressive to Lowstand Normal Regressive		6406/2-1	#20–23 (78 m)	Northwestern Halten Terrace	Lavrans	Lower Tilje mb
		6506/12-6	#12–13 (34.5 m)	Northern Halten Terrace	Smørbukk (Åsgard)	Lower Tilje mb
		6406/9-2	#4 (27 m)	Central Halten Terrace	Linnorm	Lower Tilje mb
		6407/7-8	#6–7 (47 m)	Central Halten Terrace	Noatun	Lower Tilje mb
		6407/7-A6H	#3–4 (36 m)	Southern Halten Terrace	Njord	Lower Tilje mb
		6407/7-5	#5 (27 m)	Southern Halten Terrace	Njord	Lower Tilje mb
		6407/7-4	#7–10 (49 m)	Southern Halten Terrace	Njord	Lower Tilje mb
		6407/7-2	#7–9 (30 m)	Southern Halten Terrace	Njord	Lower Tilje mb
		6407/7-1S	#9–11 (52 m)	Southern Halten Terrace	Njord	Lower Tilje mb
Core Length	280.5 m					
Normal Regressive		6407/7-A6H	#5–6 (42 m)	Southern Halten Terrace	Njord	Uppermost Åre Fm
		6406/2-1	#24–26 (48.5 m)	Northwestern Halten Terrace	Lavrans	Uppermost Åre Fm
		6407/7-4	#10 (4 m)	Southern Halten Terrace	Njord	Uppermost Åre Fm
		6407/7-1S	#12–13 (24 m)	Southern Halten Terrace	Njord	Uppermost Åre Fm
Core Length	119.5 m					
Core Length Tilje Megasequence	**1613 m**					

Appendix 2. Core database for the Tofte-Ile megasequence.

Tofte-Ile Megasequence Stratigraphic Segment		Well	Core	Location: Geographical	Location: Field/Discovery	Stratigraphic Unit
Retrogradational		6406/2-5	#7 (16 m)	North-western Halten Terrace	Kristin	Upper Ile mb
		6406/2-3	#7 (12.5 m)	North-western Halten Terrace	Kristin	Upper Ile mb
		6406/2-1	#2 (15 m)	North-western Halten Terrace	Lavrans	Upper Ile mb
		6506/12-7	#4–7 (30.5 m)	Northern Halten Terrace	Smørbukk (Åsgard)	Upper Ile mb
		6506/12-6	#5 (7.5 m)	Northern Halten Terrace	Smørbukk (Åsgard)	Upper Ile mb
		6506/12-5	#14–15 (35 m)	Northern Halten Terrace	Smørbukk S (Åsgard)	Upper Ile mb
		6506/12-3	#7 (14 m)	Northern Halten Terrace	Smørbukk S (Åsgard)	Upper Ile mb
		6506/12-1	#4 (17 m)	Northern Halten Terrace	Smørbukk (Åsgard)	Upper Ile mb
		6507/11-3	#4 (22 m)	North-eastern Halten Terrace	Midgard	Upper Ile mb
		6407/2-2	#7–8 (20 m)	North-eastern Halten Terrace	Midgard	Upper Ile mb
		6407/2-3	#4–5 (32 m)	North-eastern Halten Terrace	Midgard	Upper Ile mb
		6406/8-1	#1–2 (41 m)	Western Halten Terrace		Upper Ile mb
		6406/2-3	#9 (25)	Central Halten Terrace	Trestakk	Upper Ile mb
		6407/1-3	#7 (10 m)	Central Halten Terrace	Tyrihans N	Upper Ile mb
		6406/5-1	#2 (10 m)	Central Halten Terrace		Upper Ile mb
		6406/9-1	#1 (4 m)	Central Halten Terrace	Linnorm	Upper Ile mb
		6407/4-1	#7 (21 m)	Central Halten Terrace		Upper Ile mb
		6407/7-8	#1 (12 m)	Central Halten Terrace	Noatun	Upper Ile mb
		6406/11-1	#1 (1.5 m)	Southern Halten Terrace		Upper Ile mb
Core Length	346 m					
Aggradational		6406/2-5	#7–9 (47 m)	North-western Halten Terrace	Kristin	Middle Ile mb
		6406/2-3	#7–8 (42 m)	North-western Halten Terrace	Kristin	Middle Ile mb
		6406/2-1	#2–3 (40 m)	North-western Halten Terrace	Lavrans	Middle Ile mb
		6506/12-6	#6 (13 m)	Northern Halten Terrace	Smørbukk (Åsgard)	Middle Ile mb
		6506/12-5	#7–8 (40 m)	Northern Halten Terrace	Smørbukk S (Åsgard)	Middle Ile mb
		6506/12-3	#15–17 (28)	Northern Halten Terrace	Smørbukk S (Åsgard)	Middle Ile mb
		6506/12-1	#4–5 (27 m)	Northern Halten Terrace	Smørbukk (Åsgard)	Middle Ile mb
		6406/8-1	#2–3 (33.5 m)	Western Halten Terrace		Middle Ile mb
		6406/3-2	#9–11 (47)	Central Halten Terrace	Trestakk	Middle Ile mb
		6406/5-1	#2–3 (26.5 m)	Central Halten Terrace		Middle Ile mb
		6406/9-1	#1 (11.5 m)	Central Halten Terrace	Linnorm	Middle Ile mb
		6407/4-1	#7–8 (34 m)	Central Halten Terrace		Middle Ile mb
		6507/11-3	#4–5 (28 m)	North-eastern Halten Terrace	Midgard	Middle Ile mb
		6407/2-2	#9–11 (23 m)	North-eastern Halten Terrace	Midgard	Middle Ile mb
		6407/2-3	#6–7 (32.5 m)	North-eastern Halten Terrace	Midgard	Middle Ile mb
		6407/6-5	#1 (30 m)	Eastern Halten Terrace	Mikkel	Middle Ile mb
		6406/11-1	#2–3 (19 m)	Southern Halten Terrace		Middle Ile mb
		6407/7-5	#2 (18 m)	Southern Halten Terrace	Njord	Middle Ile mb
Core Length	539.5 m					

(*Continued*)

Appendix 2. (*Continued*)

Tofte-Ile Megasequence Stratigraphic Segment		Well	Core	Location		Stratigraphic Unit
				Geographical	Field/Discovery	
Progradational Forced Regressive to Lowstand Normal Regressive		6406/2-5	#9–10 (47 m)	North-western Halten Terrace	Kristin	Lower Ile mb
		6406/2-3	#8 (3 m)	North-western Halten Terrace	Kristin	Lower Ile mb-umst Ror mb
		6406/2-1	#3–5 (59 m)	North-western Halten Terrace	Lavrans	Lower Ile mb-umst Ror mb
		6506/12-5	#17–18 (9 m)	Northern Halten Terrace	Smørbukk S (Åsgard)	Lower Ile mb
		6506/12-3	#9 (14 m)	Northern Halten Terrace	Smørbukk S (Åsgard)	Lower Ile mb-umst Ror mb
		6506/12-1	#5–6 (28 m)	Northern Halten Terrace	Smørbukk (Åsgard)	Lower Ile mb-umst Ror mb
		6406/8-1	#4–5 (44.5 m)	Western Halten Terrace		Lower Ile mb
		6406/3-2	#11 (7,5)	Central Halten Terrace	Trestakk	Lower Ile mb
		6406/5-1	#3 (8 m)	Central Halten Terrace		Lower Ile mb
		6406/9-1	#2 (28 m)	Central Halten Terrace	Linnorm	Lower Ile mb
		6406/9-2	#1 (29 m)	Central Halten Terrace	Linnorm	Lower Ile mb
		6407/4-1	#8 (4 m)	Central Halten Terrace		Lower Ile mb
		6507/11-3	#5–6 (19 m)	North-eastern Halten Terrace	Midgard	Lower Ile umst Ror-Tofte mbs
		6407/6-5	#2 (54 m)	Eastern Halten Terrace	Mikkel	Lower Ile umst Ror-Tofte mbs
		6407/6-3	#5 (14 m)	Eastern Halten Terrace	Mikkel	Lower Ile umst Ror-Tofte mbs
		6406/11-1	#4-7 (40 m)	Southern Halten Terrace		Lower Ile umst Ror-Tofte mbs
		6407/7-5	#2-3 (28.5 m)	Southern Halten Terrace	Njord	Lower Ile mb
		6407/7-4	#1 (19 m)	Southern Halten Terrace	Njord	Lower Ile mb
		6406/5-1	#4-5 (25 m)	Central Halten Terrace		Upper Tofte mb
		6406/9-1	#3 (26 m)	Central Halten Terrace	Linnorm	Upper Tofte mb
		6406/11-1	#7-10 (35 m)	Southern Halten Terrace		Upper Tofte mb
		6407/2-1	#4 (15 m)	North-eastern Halten Terrace		Upper Tofte mb
		6407/6-5	#2 (22 m)	Eastern Halten Terrace	Mikkel	Upper Tofte mb
		6407/6-3	#6 (10 m)	Eastern Halten Terrace	Mikkel	Upper Tofte mb
Core Length	613.5 m					
Normal Regressive		6406/11-1	#11 (15 m)	Southern Halten Terrace		Upper Ror-Tofte mbs
		6406/2-5	#10–11 (36.5 m)	North-western Halten Terrace	Kristin	Middle-Upper Ror mbs
		6506/12-1	#6–7 (35 m)	Northern Halten Terrace	Smørbukk (Åsgard)	Middle-Upper Ror mbs
		6506/12-5	#18–23 (99 m)	Northern Halten Terrace	Smørbukk S (Åsgard)	Lower-Upper Ror mbs
		6406/2-1	#5–6 (82 m)	North-western Halten Terrace	Lavrans	Middle-Upper Ror mbs
		6406/2-5	#11–13 (32.5 m)	North-western Halten Terrace	Kristin	Middle Tofte mb
		6406/2-3	#9 (27 m)	North-western Halten Terrace	Kristin	Middle Tofte mb
		6407/7-8	#2 (27 m)	Central Halten Terrace	Noatun	Uppermost Tofte mb
		6407/7-2	#2–3 (35.5 m)	Southern Halten Terrace	Njord	Middle-Upper Ror-Tofte mbs
		6407/7-1	#2–5 (66.5 m)	Southern Halten Terrace	Njord	Lower-Upper Ror-Tofte mbs
		6407/10-1	#3–4 (56 m)	Southern Halten Terrace	Njord	Middle/Upper and uppermost Ror-Tofte mbs
		6406/2-1	#6–9 (38 m)	North-western Halten Terrace	Lavrans	Lower Tofte mb
		6406/2-1	#9–10 (29.5 m)	North-western Halten Terrace	Lavrans	Lower Ror mb
		6506/12-7	#8 (28 m)	Northern Halten Terrace	Smørbukk (Åsgard)	Lower Ror & Tofte mbs
		6506/12-6	#7–9 (74.5 m)	Northern Halten Terrace	Smørbukk (Åsgard)	Lower-Middle Ror-Tofte mbs
		6506/12-1	#8–10 (20 m)	Northern Halten Terrace	Smørbukk (Åsgard)	Lower Ror-Tofte mbs
		6506/12-3	#10 (28.5 m)	Northern Halten Terrace	Smørbukk S (Åsgard)	Lower Ror mb
		6406/8-1	#6 (10.5 m)	Western Halten Terrace		Lower Ror mb
		6406/9-1	#4 (28.5 m)	Central Halten Terrace	Linnorm	Lower Ror mb
		6407/6-3	#7 (28 m)	Eastern Halten Terrace	Mikkel	Lower-Middle Tofte mb
Core Length	772.5 m					
Core Length Tofte-Ile	**2267.5 m**					

Appendix 3. Core database for the Garn megasequence.

Garn Megasequence Stratigraphic Segment		Well	Core	Location		Stratigraphic Unit
				Geographical	Field/Discovery	
Retrogradational		6406/2-5	#1–3 (38.5 m)	Northwestern Halten Terrace	Kristin	Uppermost Garn mb
		6406/2-3	#1 (7 m)	Northwestern Halten Terrace	Kristin	Uppermost Garn mb
		6406/2-1	#1 (8 m)	Northwestern Halten Terrace	Lavrans	Uppermost Garn mb
		6506/12-4	#2 (1.5 m)	Northern Halten Terrace	Smørbukk N (Åsgard)	Uppermost Garn mb
		6506/12-6	#3 (5 m)	Northern Halten Terrace	Smørbukk (Åsgard)	Uppermost Garn mb
		6506/12-5	#7 (22 m)	Northern Halten Terrace	Smørbukk S (Åsgard)	Uppermost Garn mb
		6406/3-3	#1 (3.5 m)	Northern Halten Terrace	Smørbukk S (Åsgard)	Uppermost Garn mb
		6407/2-3	#1 (25 m)	Northeastern Halten Terrace	Midgard	Uppermost Garn mb
		6407/2-2	#1–2 (13 m)	Northeastern Halten Terrace	Midgard	Uppermost Garn mb
		6406/3-2	#1–2 (11 m)	Central Halten Terrace	Trestakk	Uppermost Garn mb
		6406/3-1	#1 (9 m)	Central Halten Terrace		Upper Garn mb
		6406/6-1	#1 (11 m)	Central Halten Terrace		Uppermost Garn mb
		6407/1-2	#1–2 (28.5 m)	Central Halten Terrace	Tyrihans S	Uppermost Garn mb
		6407/5-1	#2 (9 m)	Eastern Halten Terrace		Uppermost Garn mb
		6407/6-4	#1 (19 m)	Eastern Halten Terrace	Mikkel	Uppermost Garn mb
		6407/9-4	#1 (5 m)	Western Trøndelag Platform	Draugen	Uppermost Garn mb
Core Length	215 m					
Aggradational		6406/2-5	#1–3 (38.5 m)	Northwestern Halten Terrace	Kristin	Upper Garn mb
		6406/2-3	#2–5 (42 m)	Northwestern Halten Terrace	Kristin	Upper Garn mb
		6506/12-4	#2–3 (12 m)	Northern Halten Terrace	Smørbukk N (Åsgard)	Upper Garn mb
		6506/12-6	#3–4 (17.5 m)	Northern Halten Terrace	Smørbukk (Åsgard)	Upper Garn mb
		6506/12-1	#1 (13 m)	Northern Halten Terrace	Smørbukk (Åsgard)	Upper Garn mb
		6506/12-5	#7–8 (26 m)	Northern Halten Terrace	Smørbukk S (Åsgard)	Upper Garn mb
		6506/12-3	#1–3 (24 m)	Northern Halten Terrace	Smørbukk S (Åsgard)	Upper Garn mb
		6406/3-3	#1–2 (27 m)	Northern Halten Terrace	Smørbukk S (Åsgard)	Upper Garn mb
		6407/1-3	#1–3 (28 m)	Northern Halten Terrace	Tyrihans N	Upper Garn mb
		6407/1-4	#1 (19.5 m)	Northern Halten Terrace	Tyrihans N	Upper Garn mb
		6407/2-3	#1–2 (14.5 m)	Northeastern Halten Terrace	Midgard	Upper Garn mb
		6407/2-2	#3 (22 m)	Northeastern Halten Terrace	Midgard	Upper Garn mb
		6407/2-1	#2–3 (9.5 m)	Northeastern Halten Terrace		Upper Garn mb
		6406/2-6	#1 (27 m)	Western Halten Terrace		(Middle-) Upper Garn mbs
		6406/3-2	#2–4 (36.5 m)	Central Halten Terrace	Trestakk	Upper Garn mb
		6406/3-1	#1–3 (22.5 m)	Central Halten Terrace		Upper Garn mb
		6406/6-1	#1 (17.5 m)	Central Halten Terrace		Upper Garn mb
		6407/1-2	#3–4 (29 m)	Central Halten Terrace	Tyrihans S	Upper Garn mb
		6407/4-1	#3–4 (29 m)	Central Halten Terrace		Upper Garn mb
		6407/5-1	#2 (6 m)	Eastern Halten Terrace		Upper Garn mb
		6407/6-4	#1 (9 m)	Eastern Halten Terrace	Mikkel	Upper Garn mb
		6407/9-4	#2–5 (27 m)	Western Trøndelag Platform	Draugen	Upper Garn mb
Core Length	495 m					

(*Continued*)

Appendix 3. (*Continued*)

Garn Megasequence Stratigraphic Segment		Well	Core	Location		Stratigraphic Unit
				Geographical	Field/Discovery	
Progradational Forced Regressive to Lowstand Normal Regressive		6406/2-5	#3–5 (37 m)	Northwestern Halten Terrace	Kristin	Lower-Middle Garn mbs
		6406/2-3	#5–6 (27.5 m)	Northwestern Halten Terrace	Kristin	Lower-Middle Garn mbs
		6506/12-4	#3 (16.5 m)	Northern Halten Terrace	Smørbukk N (Åsgard)	Middle Garn mb
		6506/12-7	#2–3 (11 m)	Northern Halten Terrace	Smørbukk (Åsgard)	Lower-Middle Garn mbs
		6506/12-6	#4 (15 m)	Northern Halten Terrace	Smørbukk (Åsgard)	Lower-Middle Garn mbs
		6506/12-1	#1–2 (23 m)	Northern Halten Terrace	Smørbukk (Åsgard)	Lower-Middle Garn mbs
		6506/12-5	#10–11 (39 m)	Northern Halten Terrace	Smørbukk S (Åsgard)	Lower-Middle Garn mbs
		6506/12-3	#3–5 (46.5 m)	Northern Halten Terrace	Smørbukk S (Åsgard)	Lower-Middle Garn mbs
		6407/1-3	#1–3 (28 m)	Northern Halten Terrace	Tyrihans N	Lower-Middle Garn mbs
		6407/1-4	#1–2 (40 m)	Northern Halten Terrace	Tyrihans N	Lower-Middle Garn mbs
		6507/11-2	#2–3 (14 m)	Northeastern Halten Terrace		Lower-Middle Garn mbs
		6407/2-3	#2–3 (20 m)	Northeastern Halten Terrace		Lower-Middle Garn mbs
		6407/2-2	#4–7 (12 m)	Northeastern Halten Terrace	Midgard	Middle Garn mb
		6407/2-1	#3 (5 m)	Northeastern Halten Terrace		Middle Garn mb
		6406/3-2	#4–6 (46 m)	Central Halten Terrace	Trestakk	Lower-Middle Garn mbs
		6406/3-1	#3 (31 m)	Central Halten Terrace		Middle Garn mb
		6406/5-1	#1 (24.5 m)	Central Halten Terrace		Middle Garn mb
		6407/4-1	#4–6 (39 m)	Central Halten Terrace		Middle Garn mb
		6407/6-3	#1–2 (13 m)	Eastern Halten Terrace	Mikkel	Middle Garn mb
		6407/9-4	#5 (8.5 m)	Western Trøndelag Platform	Draugen	Middle Garn mb
Core Length	496.5 m					
Normal Regressive		6406/2-5	#5–6 (47 m)	Northwestern Halten Terrace	Kristin	Lower Garn mb-Not Fm
		6406/2-3	#6 (26.5 m)	Northwestern Halten Terrace	Kristin	Lower Garn mb-Not Fm
		6506/12-5	#11–13 (27 m)	Northern Halten Terrace	Smørbukk S (Åsgard)	Lower Garn mb-Not Fm
		6506/12-3	#5–6 (18.5 m)	Northern Halten Terrace	Smørbukk S (Åsgard)	Lower Garn mb-Not Fm
		6506/12-1	#2–3 (33 m)	Northern Halten Terrace	Smørbukk (Åsgard)	Lower Garn mb-Not Fm
		6407/1-3	#6 (6 m)	Northern Halten Terrace	Tyrihans N	Lower Garn mb-Not Fm
		6507/11-2	#2–3 (14 m)	Northeastern Halten Terrace		Lower Garn mb
		6407/2-2	#3–4 (25.5 m)	Northeastern Halten Terrace	Midgard	Lower Garn mb-Not Fm
		6407/2-2	#7–8 (11 m)	Northeastern Halten Terrace	Midgard	Lower Garn mb-Not Fm
		6406/3-2	#6–8 (31 m)	Central Halten Terrace	Trestakk	Lower Garn mb-Not Fm
		6407/4-1	#6 (22 m)	Central Halten Terrace		Lower Garn mb-Not Fm
		6407/6-1	#1–4 (18 m)	Western Trøndelag Platform		Lower Garn mb
		6407/6-3	#2–4 (50 m)	Eastern Halten Terrace	Mikkel	Lower Garn mb-Not Fm
		6407/10-1	#2 (28 m)	Eastern Halten Terrace	Mikkel	Lower Garn mb-Not Fm
Core Length	357.5 m					
Core Length Tilje Megasequence	**1564 m**					

REFERENCES

Anastas, A., Dalrymple, R.W., James, N.P. and **Nelson, C.S.** (2006) Lithofacies and dynamics of a cool-water carbonate seaway; mid-Tertiary, Te Kuiti Group, New Zealand. In: *Cool-Water Carbonates: Depositional Systems and Palaeoenvironmental Controls* (Eds H.M. Pedley and G. Carannante), *Geol. Soc. Spec. Publ.*, **255**, 245–268.

Blystad, P., Brekke, H., Færseth, R.B., Larsen, B.T., Skogseid, J. and **Tørudbakken, B.** (1995) Structural Elements of the Norwegian Continental Shelf – Part II: the Norwegian Sea Region. *Norw. Petrol. Direct. Bull.*, **8**, 44 pp.

Brekke, H. (2000) The tectonic evolution of the Norwegian Sea continental margin with emphasis on the Vøring and Møre basins. In: *Dynamics of the Norwegian Margin* (Eds A. Nøttvedt), *Geol. Soc. Spec. Publ.*, **136**, 327–378.

Brown, L.F. and **Fisher, W.L.** (1977) Seismic stratigraphic interpretation of depositional systems: examples from Brazilian rift and pull apart basins. In: *Seismic Stratigraphy – Applications to Hydrocarbon Exploration* (Ed. C.E. Payton), *AAPG Mem.*, **26**, 213–248.

Catuneanu, O., Abreu, V., Bhattacharya, J.P., Blum, M.D., Dalrymple, R.W., Eriksson, P.G., Fielding, C.R., Fisher, W.L., Galloway, W.E., Gibling, M.R., Giles, K.A., Holbrook, J.M., Jordan, R., Kendall, C.G.St.C., Macurda, B., Martinsen, O.J., Miall, A.D., Neal, J.E., Nummedal, D., Pomar, L., Posamentier, H.W., Pratt, B.R., Sarg, J.F., Shanley, K.W., Steel, R.J., Strasser, A., Tucker, M.E. and **Winker, C.** (2009) Towards the standardization of sequence stratigraphy. *Earth-Sci. Rev.*, **92**, 1–33.

Catuneanu, O., Bhattacharya, J.P., Blum, M.D., Dalrymple, R.W., Eriksson, P.G., Fielding, C.R., Fisher, W.L., Galloway, W.E., Gianolla, P., Gibling, M.R., Giles, K.A., Holbrook, J.M., Jordan, R., Kendall, C.G.St.C., Macurda, B., Martinsen, O.J., Miall, A.D., Nummedal, D., Posamentier, H.W., Pratt, B.R., Shanley, K.W., Steel, R.J., Strasser, A. and **Tucker, M.E.** (2010) Sequence stratigraphy: common ground after three decades of development. *First Break*, **28**, 21–34.

Catuneanu, O., Galloway, W.E., Gibling, Kendall, C.G.St.C., Miall, A.D., Posamentier, H.W., Strasser, A. and **Tucker, M.E.** (2011) Sequence Stratigraphy: Methodology and Nomenclature. *Newsl. Stratigr. Spec. Issue*, **44/3**, 246 p.

Corfield, S. and **Sharp, I.** (2000) Structural style and stratigraphic architecture of fault-propagation folding in extensional settings: a seismic example from the Smørbukk area, Halten terrace, Mid-Norway. *Basin Res.*, **12**, 329–341.

Corfield, S., Sharp, I., Häger, K.-O., Dreyer, T. and **Underhill, J.** (2001) An integrated study of the Garn and Melke formations (Middle to Upper Jurassic) of the Smørbukk area, Halten terrace, mid-Norway. In: *Sedimentary Environments Offshore Norway – Paleozoic to Recent* (Eds O.J. Martinsen and T. Dreyer), *NPF Spec. Publ.*, **10**, 199–210.

Dalland, A., Worsley, D. and **Ofstad, K.** (1988) A lithostratigraphic scheme for the Mesozoic and Cenozoic succession offshore mid- and northern Norway. *Norw. Petrol. Direct. Bull.*, **4**, pp. 65.

Doré, A.G. (1991) The structural foundation and evolution of Mesozoic seaways between Europe and the Arctic. *Palaeogeogr. Palaeoclimatol. Palaeoecol.*, **87**, 441–492.

Dreyer, T. (1992) Significance of tidal cyclicity for modeling of reservoir heterogeneities in the Lower Jurassic Tilje Formation, mid-Norwegian Shelf. *Nor. geol. tidsskr.*, **72**, 159–170.

Dreyer, T. (1993) Geometry and facies of large-scale flow-units in fluvial-dominated fan-delta-front sequences. In: *Advances in reservoir geology* (Ed. M. Ashton), *Geol. Soc. London Spec. Publ.*, **69**, 135–174.

Dreyer, T. (1999) Sedimentary environments in the cored part of the Tilje Formation, well 6407/7-A6H, Njord Field. In: *Sedimentary environments offshore norway – paleozoic to recent – extended abstracts* (Eds O.J. Martinsen and T. Dreyer), 129–131.

Ehrenberg, S.N. (1991) Kaolinized, potassium-leached zones at the contacts of the GarnFormation, Haltenbanken, mid-Norwegian continental shelf. *Mar. Petrol. Geol.*, **8**, 250–269.

Ehrenberg, S.N., Gjerstad, H.M. and **Hadler-Jacobsen, F.** (1992) Smørbukk Field: A Gas Condensate Fault Trap in the Haltenbanken Province, Offshore Mid-Norway. In: *Giant Oil and Gas Fields of the Decade 1978–1988* (Ed. M.T. Halbouty), *AAPG Mem.*, **54**, 323–348.

Elfenbein, C., Husby, Ø. and **Ringrose, P.** (2005) Geologically based estimation of kv/kh ratios: an example from the Garn Formation, Tyrihans Field, Mid-Norway. In: *Petroleum Geology of Northwest Europe: Proceedings of the 5th Conference* (Eds A.J. Fleet and S.A.R. Boldy), *Geol. Soc. London*, 537–543.

Færseth, R.B. and **Lien, T.** (2002) Cretaceous evolution in the Norwegian Sea – a period characterized by tectonic quiescence. *Mar. Petrol. Geol.*, **19**, 1005–1027.

Galloway, W.E. (1989) Genetic stratigraphic sequences in basin analysis, I. Architecture and genesis of flooding-surface bounded depositional units. *AAPG Bull.*, **73**, 125–142.

Gawthorpe, R.L. and **Leeder, M.R.** (2000) Tectono-sedimentary evolution of active extensional basins. *Basin Res.*, **12**, 195–218.

Gjelberg, J., Dreyer, T., Høie, A., Tjelland, T. and **Lilleng, T.** (1987) Late Triassic to Mid-Jurassic development of the Barents and Mid-Norwegian Shelf. In: *Petroleum Geology of Northwest Europe* (Eds J. Brooks & K. Glennie). Graham & Trotman, London, 1105–1129.

Hallam, A. (1994) Jurassic climates as inferred from the sedimentary and fossil record. In: Palaeoclimates and Their Modelling: With Special Reference to the Mesozoic Era (Eds J.R.L. Allen, B.J. Haskins, B.W. Sellwood, R.A. Spicer and P.J. Valdes), pp 79–88, Chapman & Hall, London.

Harris, N.B. (1989) Reservoir geology of Fangst Group (Middle Jurassic), Heidrun Field, offshore Mid-Norway. *AAPG Bull.*, **73**, 1415–1435.

Helland-Hansen, W. and **Gjelberg, J.G.** (1994) Conceptual basis and variability in sequence stratigraphy: a different perspective. *Sed. Geol.*, **92**, 31–52.

Helland-Hansen, W. and **Hampson, G. J., (**2009) Trajectory analysis: concepts and applications. *Basin Res.*, **21**, 454–483.

Helland-Hansen, W. and **Martinsen, O.J.** (1996) Shoreline trajectories and sequences: description of variable depositional-dip scenarios. *J. Sed. Res.*, **66**, 670–688.

Hunt, D. and **Tucker, M.E.** (1992) Stranded parasequences and the forced regressive wedge systems tract: deposition during base-level fall. *Sed. Geol.*, **81**, 1–9.

Ichaso, A.A. and **Dalrymple, R.W.** (this volume) Eustatic, tectonic and climatic controls on an early syn-rift mixed-energy delta, Tilje Formation (Early Jurassic, Smørbukk Field, offshore Mid-Norway).

Kjærefjord, J.M. (1999) Bayfill successions in the Lower Jurassic Åre Formation, offshore Norway: sedimentology and heterogeneity based on subsurface data from the Heidrun Field and analog data from the Upper Cretaceous Neslen Formation, eastern Book Cliffs. In: *Advanced Reservoir Characterisation for the 21st Century* (Ed. T.F. Hentz), *Gulf Coast Section, SEPM Spec. Publ.*, 19th Annual Research Conference, 149–157.

LaFont, F., Capron, A. and **Schulbaum, L.** (2010) Facies tracts in the Lower-Mid Jurassic (Tilje and Ile) of the Dønna Terrace: Distinguishing sequence stratigraphic patterns from tectonically-forced physiographic changes. In: *From Depositional Systems to Sedimentary Successions on the Norwegian Continental Shelf, NPF Abstract Volume*, 16–19.

Marsh, N., Imber, J., Holdsworth, R.E., Brockbank, P. and **Ringrose, P.** (2010) The structural evolution of the Halten Terrace, offshore Mid-Norway: extensional fault growth and strain localisation in a multi-layer brittle–ductile system. *Basin Res.*, **22**, 195–214.

Martinius, A.W., Kaas, I., Næss, A., Helgesen, G., Kjærefjord, J. and **Leith., D.A.** (2001) Sedimentology of the heterolithic and tide-dominated Tilje Formation (Early Jurassic, Halten Terrace, offshore mid-Norway. In: *Sedimentary Environments Offshore Norway – Paleozoic to Recent* (Eds O.J. Martinsen and T. Dreyer), *NPF Spec. Publ.*, **10**, 103–144.

Martinius, A.W., Ringrose, P.S., Brostrøm, C., Elfenbein, C., Næss, A. and **Ringås, J.E.** (2005) Reservoir challenges of heterolithic tidal sandstone reservoirs in the Halten Terrace, mid-Norway. *Petrol. Geosci.*, **11**, 3–16.

McIlroy, D. (2004) Ichnofabrics and sedimentary facies of a tide-dominated delta: Jurassic Ile Formation of Kristin Field, Haltenbanken Offshore Mid-Norway. In: *The Application of Ichnology to Palaenvironmental and Stratigraphic Analysis* (Ed. D. McIlroy), *Geol. Soc. Spec. Publ.*, **228**, 237–272.

Messina, C., Nemec, W., Martinius, A.W. and **Elfenbein, C.** (this volume) Modelling of a petroleum reservoir composed of large tidal sand ridges encased in shoreface deposits.

Morton, A., Hallsworth, C., Strogen, D., Whitham, A. and **Fanning, C.** (2009) Evolution of provenance in the NE Atlantic rift: The Early–Middle Jurassic succession in the Heidrun Field, Halten Terrace, offshore Mid-Norway. *Mar. Petrol. Geol.*, **26**, 1100–1117.

Mosar, J., Torsvik, T.H. and **the BAT-team** (2002) Opening the Norwegian and Greenland Seas: Plate tectonics in Mid Norway since the Late Permian. In: *BATLAS – Mid Norway Plate Reconstruction Atlas with Global and Atlantic Perspectives* (Ed. E.A. Eide), Geological Survey of Norway, pp. 48–59.

Nøttvedt, A., Johannessen, E.P. and **Surlyk, F.** (2008) The Mesozoic of western Scandinavia and eastern Greenland. *Episodes*, **31**, 59–65.

Osmundsen, P.T., Sommaruga, A., Skilbrei, J.R. and **Olesen, O.** (2002) Deep structure of the Norwegian Sea area, NE Atlantic margin. *Norw. J. Geol.*, **82**, 205–224.

Pascoe, R., Hooper, R., Storhaug, K. and **Harper, H.** (1999) Evolution of extensional styles at the southern termination of the Nordaland Ridge, Mid-Norway: A response to variations in coupling above Triassic salt. In: *Petroleum Geology of Northwest Europe: Proceedings of the 5th Conference* (Eds A.J. Fleet and S.A.R. Boldy), *Geol. Soc. London*, 83–90.

Pedersen, T., Harms, J.C., Harris, N.B., Mitchell, R.W. and **Tooby, K.M.** (1989) The role of correlation in generating the Heidrun Field geological model. In: *Correlation in Hydrocarbon Exploration* (Ed. J.D. Collinson), Graham and Trotman, London, 113–130.

Posamentier, H.W. and **Allen, G.P.** (1999) Siliciclastic sequence stratigraphy: concepts and applications. *SEPM Concepts in Sedimentology and Paleontology*, **7**, 210 pp.

Posamentier, H.W., Jervey, M.T. and **Vail, P.R.** (1988) Eustatic controls on clastic deposition. I. Conceptual framework. In: *Sea Level Changes – An Integrated Approach* (Eds C.K. Wilgus, B.S. Hastings, C.G.St.C. Kendall, H.W. Posamentier, C.A. Ross and J.C. Van Wagoner), *SEPM Spec. Publ.*, **42**, 110–124.

Posamentier, H.W., Allen, G.P., James, D.P. and **Tesson, M.** (1992) Forced regressions in a sequence stratigraphic framework: concepts, examples and exploration significance. *AAPG Bull.*, **76**, 1687–1709.

Quin, J.G., Zweigel, P., Eldholm, E., Hansen, O.R., Christoffersen, K.R. and **Zaostrovski, A.** (2010) Sedimentology and unexpected pressure decline: the HP/HT Kristin Field. In: *Petroleum Geology: From Mature Basins to New Frontiers – Proceedings of the 7th Petroleum Geology Conference* (Eds B.A. Vining and S.C. Pickering), *Geol. Soc. London*, 419–429.

Ravnås, R. and **Steel, R.J.** (1998) Architecture of marine rift basin succession. *AAPG Bull.*, **82**, 110–146.

Ravnås, R., Nøttvedt, A., Steel, R.J. and **Windelstad, J.** (2000) Syn-rift sedimentary architectures in the northern North Sea. In: *Dynamics of the Norwegian Margin* (Eds A. Nøttvedt *et al.*), *Geol. Soc. Spec. Publ.*, **167**, 133–177.

Reynaud, J.Y. and **Dalrymple, R.W.** (2012) Shallow-marine tidal deposits. In: *Principles of Tidal Sedimentology* (Eds R.A. Davis and R.W. Dalrymple), Springer Verlag, 339–359.

Reynaud, J.Y., Dalrymple, R.W., Vennin, E., Parize, O., Besson, D. and **Rubino, J.-L.** (2006) Topographic controls on producing and depositing tidal cool-water carbonates, Uzès basin, SE France. *J. Sed. Res.*, **76**, 117–130.

Roberts, D.G., Thompson, M., Mitchener, B., Hossack, J., Carmichael, S.M.M. and **Bjørnseth, H.M.** (1999) Palaeozoic to Tertiary rift and basin dynamics; mid-Norway to the Bay of Biscay; a new context for hydrocarbon prospectivity in the deep water frontier. In: *Petroleum Geology of Northwest Europe: Proceedings of the 5th Conference* (Eds A.J. Fleet and S.A.R. Boldy), *Geol. Soc. London*, 7–40.

Sæther, T., Bergan, M., Olsen, J., Saigal, G. and **Throndsen, I.** (1999) Wave- and tide-influenced shallow marine facies: Core examples from the Garn Formation. In: *Sedimentary Environments Offshore Norway –*

Paleozoic to Recent – Extended Abstracts (Eds O.J. Martinsen and T. Dreyer), 129–131.

Saigal, G.C., Amundsen, H.E.F., Bergan, M., Knarud, R., Olsen, J. and **Sæther, T.** (1999) Understanding the problems related to tidal depositional environments and their sequence stratigraphic interpretations – a few examples from the Jurassic Tilje and Ile Formations, offshore Norway and our approach. In: *Sedimentary Environments Offshore Norway – Paleozoic to Recent – Extended Abstracts* (Eds O.J. Martinsen and T. Dreyer), 121.

Sharp, I.R., Underhill, J.R., Gawthorpe, R. and **Gupta, S.** (2000) Fault propagation folding in extensional settings: Examples of structural styles and syn-rift sedimentary response from the Suez Rift, Sinai, Egypt. *AAPG Bull.*, **112**, 1877–1899.

Steel, R.J. (1993). Triassic-Jurassic megaesequence stratigraphy in the Northern North Sea: rift to post-rift evolution. In: *Petroleum Geology of Northwest Europe: Proceedings of the 4th Conference* (Ed. R.J. Parker), *Geol. Soc. London*, **2**, 99–315.

Svela, K.E. (2001) Sedimentary facies in the fluvial-dominated Åre Formation as seen in the Åre 1member in the Heidrun Field. In: *Sedimentary Environments Offshore Norway – Paleozoic to Recent* (Eds O.J. Martinsen and T. Dreyer), *NPF Spec. Publ.*, **10**, 87–102.

Van Wagoner, J.C. (1995) Overview of sequence stratigraphy of foreland basin deposits: terminology, summary of papers and glossary of sequence stratigraphy. In: *Sequence Stratigraphy of Foreland Basin Deposits: Outcrop and Subsurface Examples from the Cretaceous of North America* (Eds J.C. Van Wagoner and G.T. Bertram), *AAPG Mem.*, **64**, ix–xxi.

Van Wagoner, J.C., Posamentier, H.W., Mitchum, R.M., Vail, P.R., Sarg, J.F., Loutit, T.S. and **Hardenbol, J.** (1988) An overview of sequence stratigraphy and key definitions. In: *Sea Level Changes – An Integrated Approach* (Eds C.K. Wilgus, B.S. Hastings, C.G.St.C. Kendall, H.W. Posamentier, C.A. Ross and J.C. Van Wagoner), *SEPM Spec. Publ.*, **42**, 39–45.

Van Wagoner, J.C., Mitchum Jr., R.M., Campion, K.M. and **Rahmanian, V.D.** (1990) Siliciclastic sequence stratigraphy in well logs, core and outcrops: concepts for high-resolution correlation of time and facies. *AAPG Methods in Exploration Series*, **7**, 55 pp.

Withjack, M.O., Meisling, K.E.U. and **Russell, L.R.** (1989) Forced folding and basement-detached normal faulting in the Haltenbanken area, offshore Norway. In: *Extensional Tectonics and Stratigraphy of the North Atlantic Margins* (Eds A.J. Tankard and H.R. Balkwill), *AAPG Mem.*, **46**, 567–575.

Yoshida, S., Steel, R.J. and **Dalrymple, R.W.** (2007) Changes in depositional processes – an ingredient in the generation of new sequence-stratigraphic models. *J. Sed. Res.*, **77**, 447–460.

Ziegler, P.A. (1988) Evolution of the Arctic-North Atlantic and the Western Tethys. *AAPG Mem.*, **43**, pp. 198.

Int. Assoc. Sedimentol. Spec. Publ. (2014) **46**, 253–290

Updated depositional and stratigraphic model of the Lower Jurassic Åre Formation, Heidrun Field, Norway

CAMILLA THRANA*, ARVE NÆSS*, SIMON LEARY†, STUART GOWLAND‡, MALI BREKKEN* and ANDREW TAYLOR‡

* *Statoil ASA, Strandveien 4, N-7501 Stjørdal, Norway*
† *Statoil ASA, 2103 CityWest Blvd., Houston, Texas, 77042, USA*
‡ *Ichron Limited, Century House, Gadbrook Business Centre, Northwith, Cheshire CW9 7TL, UK*

ABSTRACT

The Jurassic reservoirs of the Heidrun Field, offshore mid-Norway, have produced in excess of 135 M Sm^3 of oil. The oil recovery factor of the fluviodeltaic (Rhaetian to Early Pliensbachian) Åre Formation is the lowest amongst the reservoir formations, which makes this an important stratigraphic interval when it comes to remaining reserves and IOR potential. In order to address challenges related to stratigraphic correlation and detailed reservoir modelling, a major core and log based reservoir characterisation study of the Åre Formation was performed. The outcome of this study includes a new understanding of the depositional setting and a revised reservoir zonation of the Heidrun Field. The new stratigraphic framework defines seven main zones, which comprise genetically linked facies bounded by field-wide stratigraphic surfaces. The Åre Formation is an overall transgressive succession that progresses upward through: 1) non-marine coastal plain deposits in Åre 1 to 2 reservoir zones, 2) a succession of lower delta plain and brackish-water interdistributary bay deposits in Åre 3 to 4 reservoir zones, 3) mixed wave-influenced and tide-influenced estuarine deposits in Åre 5 to 6 zones; and 4) open marine shoreface deposits in the uppermost Åre 7 reservoir zone. Although the regional sequence stratigraphic significance of surfaces separating these reservoir intervals could not be determined in every case, the stratigraphic division presented here has proven more robust on a field-scale than in previous models. It provides improved stratigraphic control for drilling operations and more accurate predictions of the spatial distribution and geometry of hydrocarbon bearing facies. Long-term benefits will include more precise production forecasts and improved drainage strategies.

Keywords: Fluviodeltaic, estuarine, sequence stratigraphy, industrial application, Jurassic, Norwegian Sea.

INTRODUCTION

The Heidrun Field, located approximately 190 km offshore mid-Norway (Figs 1 and 2; Koenig, 1986; Whitley, 1992), has produced in excess of 135 M Sm^3 of oil (per 2010) since its discovery in 1985. Hydrocarbons are accumulated in the Lower and Middle Jurassic deposits of the Båt Group (Åre and Tilje Formations) and the Fangst Group (Ile, Not and Garn Formations; Fig. 3). Production rate of the Heidrun Field is declining and one of the main challenges today is to target and drain efficiently the remaining oil reserves. A relatively large fraction of remaining reserves is allocated in heterogeneous and low-productivity reservoirs within the Lower Jurassic Åre Formation. Current development plans focus on a revised drainage strategy, utilisation of more complex well solutions and evaluation of small and economically marginal drilling targets. Continued effort to develop improved oil recovery (IOR) measures and enhanced oil recovery (EOR) methods are necessary as significant business opportunities still exist.

Historically, geological understanding and reservoir characterisation of the Åre Formation has faced various challenges: 1) Reservoirs are composed of complex fluvial to marginal-marine deltaic and tide-influenced deposits. Although conceptual depositional models have been in

From Depositional Systems to Sedimentary Successions on the Norwegian Continental Margin, First Edition. Edited by A. W. Martinius, R. Ravnås, J. A. Howell, R. J. Steel, and J. P. Wonham.

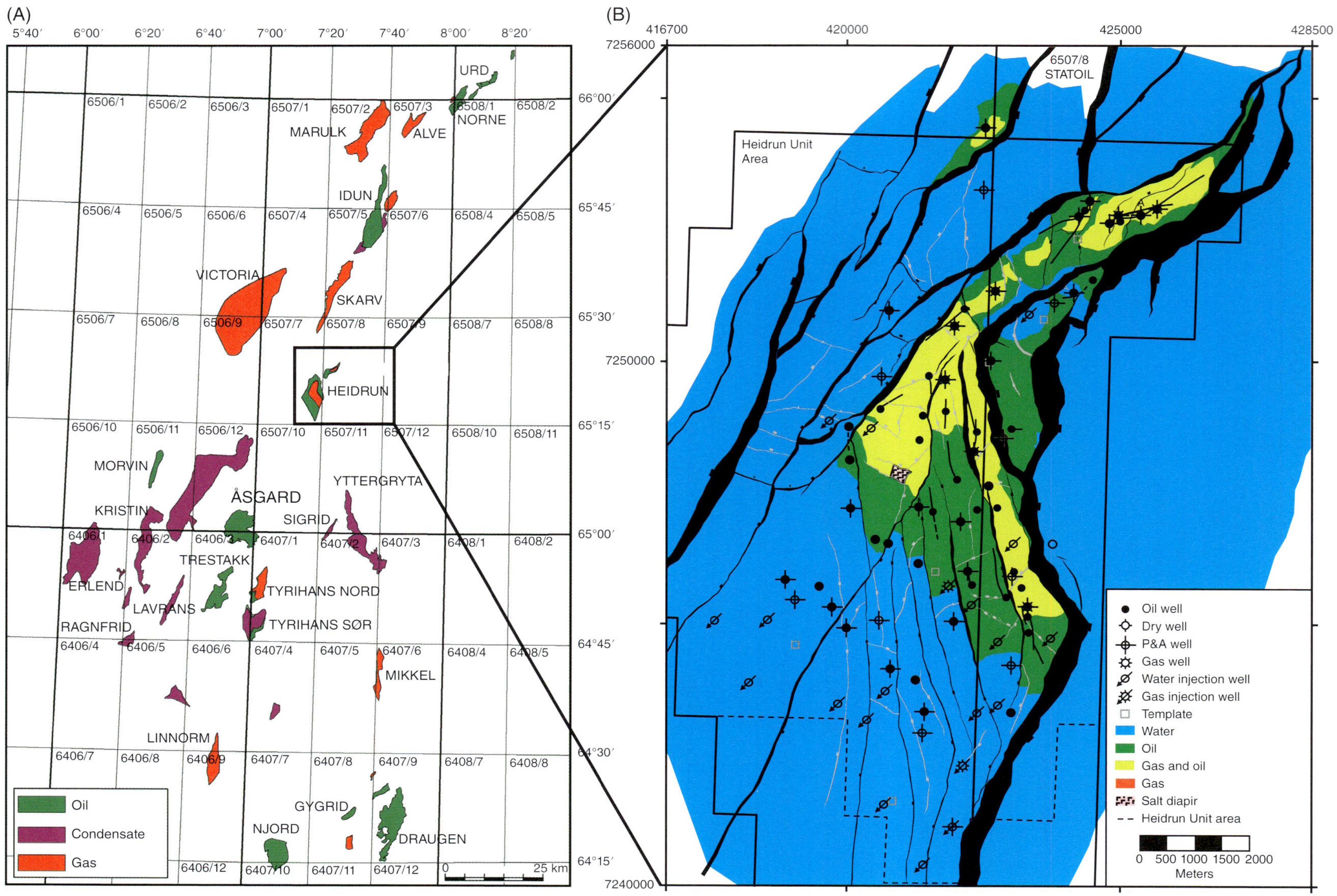

Fig. 1. (A) Location of hydrocarbon fields on the Halten Terrace, offshore Mid-Norway (Modified from Fact Maps, Norwegian Petroleum Directorate, 2009). (B) Detailed map showing the Heidrun Field structure with hydrocarbon contacts for the top of the Åre Formation.

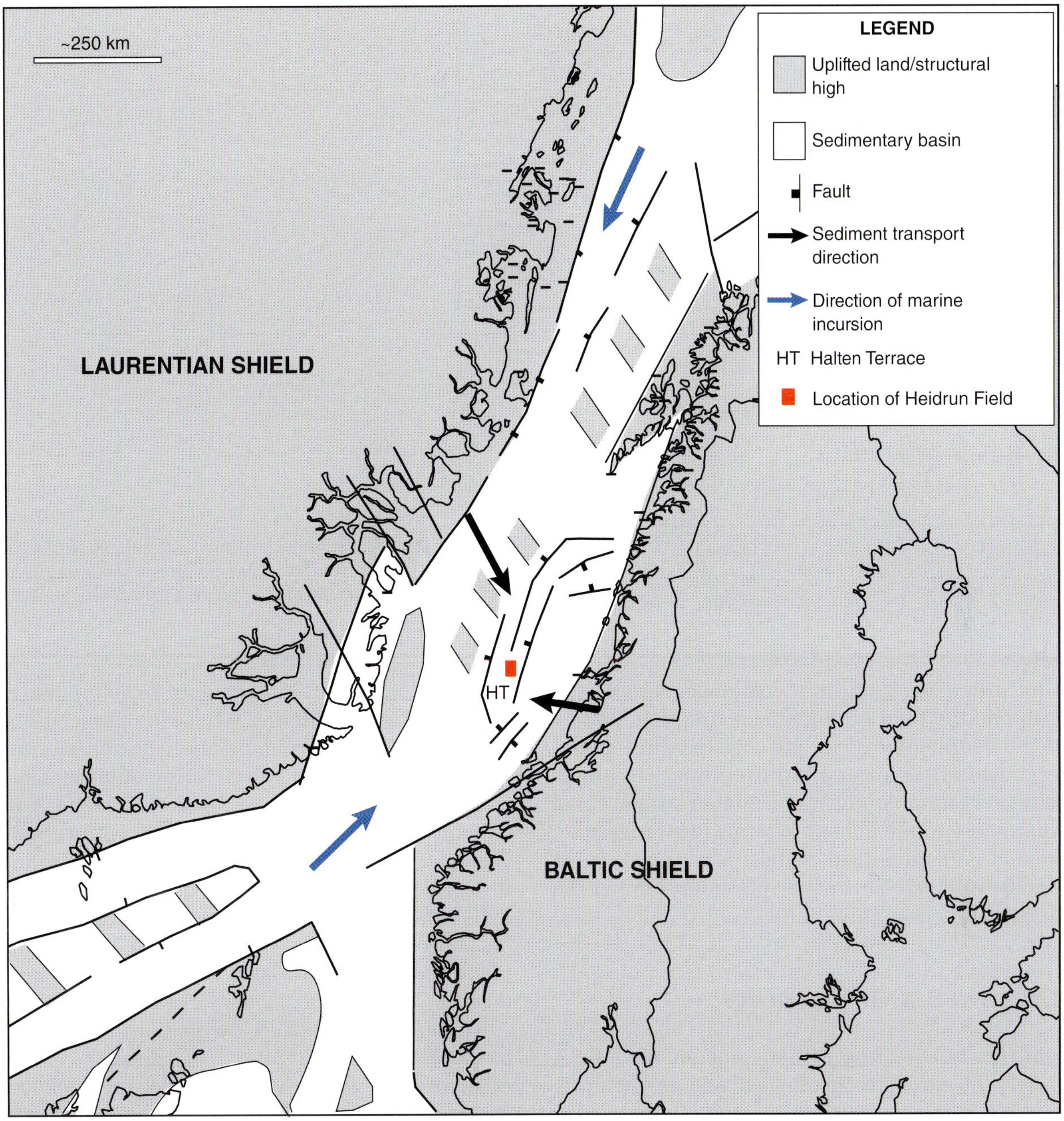

Fig. 2. Palaeogeographic map of the early Jurassic North Atlantic seaway during deposition of the Åre Formation (Modified from Doré, 1991).

place since production start-up, lack of well-defined sequence stratigraphic surfaces in previous models have in many cases prevented prediction of lateral trends and vertical stacking patterns. Furthermore, poor biostratigraphic resolution has hindered construction of a well-defined stratigraphic framework and few marker beds have been recognised and correlated across the field. 2) Structural complexity and the lack of a robust reservoir zonation have created various difficulties in interpretation of well data and major challenges in defining and developing future drilling targets. 3) Previous geological models of the field have become outdated due to inconsistencies between new well results and the former stratigraphic reservoir zonation. The predictive capability of older models is therefore limited.

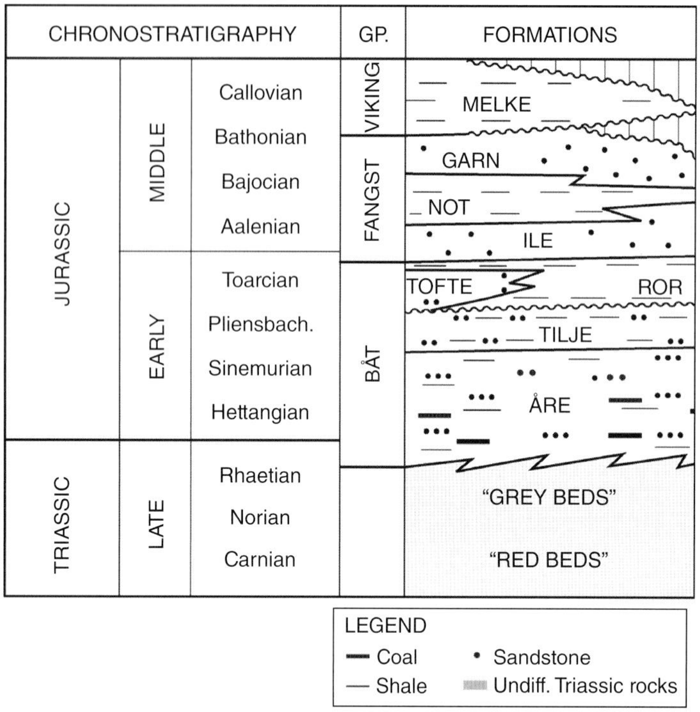

Fig. 3. Generalised time and lithostratigraphic sections from the Triassic–Jurassic succession on the Halten Terrace (modified from Dalland *et al.*, 1988).

In order to meet the challenges associated with late life field production, a complete update of the sedimentological and stratigraphic model was required. A major multidisciplinary reservoir characterisation study was initiated in 2003 and completely implemented in terms of stratigraphic models and numerical realisations of geological models in 2007. This paper presents some of the main results from this study.

DATABASE AND METHODS

Heidrun is a mature oilfield documented by a variety of extensive geological and engineering studies. A major challenge in reservoir characterisation has been integration of the available data to sustain consistency and coherence as new information becomes available and geological concepts advance.

The Heidrun Field covers an area of approximately 16 km × 8 km (Fig. 1B). Although interwell spacing ranges from 150 m to several kilometres, it is typically ~700 m. Although 77 wells penetrate the Åre Formation, only 18 reach the lowermost part of the stratigraphy. Fifteen wells have core taken through the Åre Formation (totalling 1362 m). This study is based on detailed sedimentological and ichnological descriptions of these cores.

A total of 49 lithofacies and 46 ichnofabrics are combined to define 18 facies associations that are indicative of deposition in non-marine, marginal-marine and fully marine environments (Fig. 4). This facies association scheme was extended to the remaining uncored intervals by geological interpretation of conventional wireline logs. Diagnostic wireline log responses (GR/RHOB/NPHI) defined from cored intervals enabled recognition of vertical facies trends and the identification of significant stratigraphic surfaces (Fig. 4).

Although well-to-well correlations over short inter-well distances (150 m to 500 m) are generally relatively straight forward, large lateral thickness variations and complex facies shifts in some stratigraphic intervals add considerable correlation uncertainty. Such complexity and uncertainty is expected in marginal marine environments, where a combination of tidal, wave and fluvial processes can interact to cause significant along-strike variation in deposition.

Analysis of facies correlation and vertical stacking patterns within a framework of key surfaces is used to subdivide the Åre Formation into seven main reservoir zones, with a further 17 subzones (Fig. 4). Detailed facies correlations based on this stratigraphic framework together with sedimentological analysis of dip data from

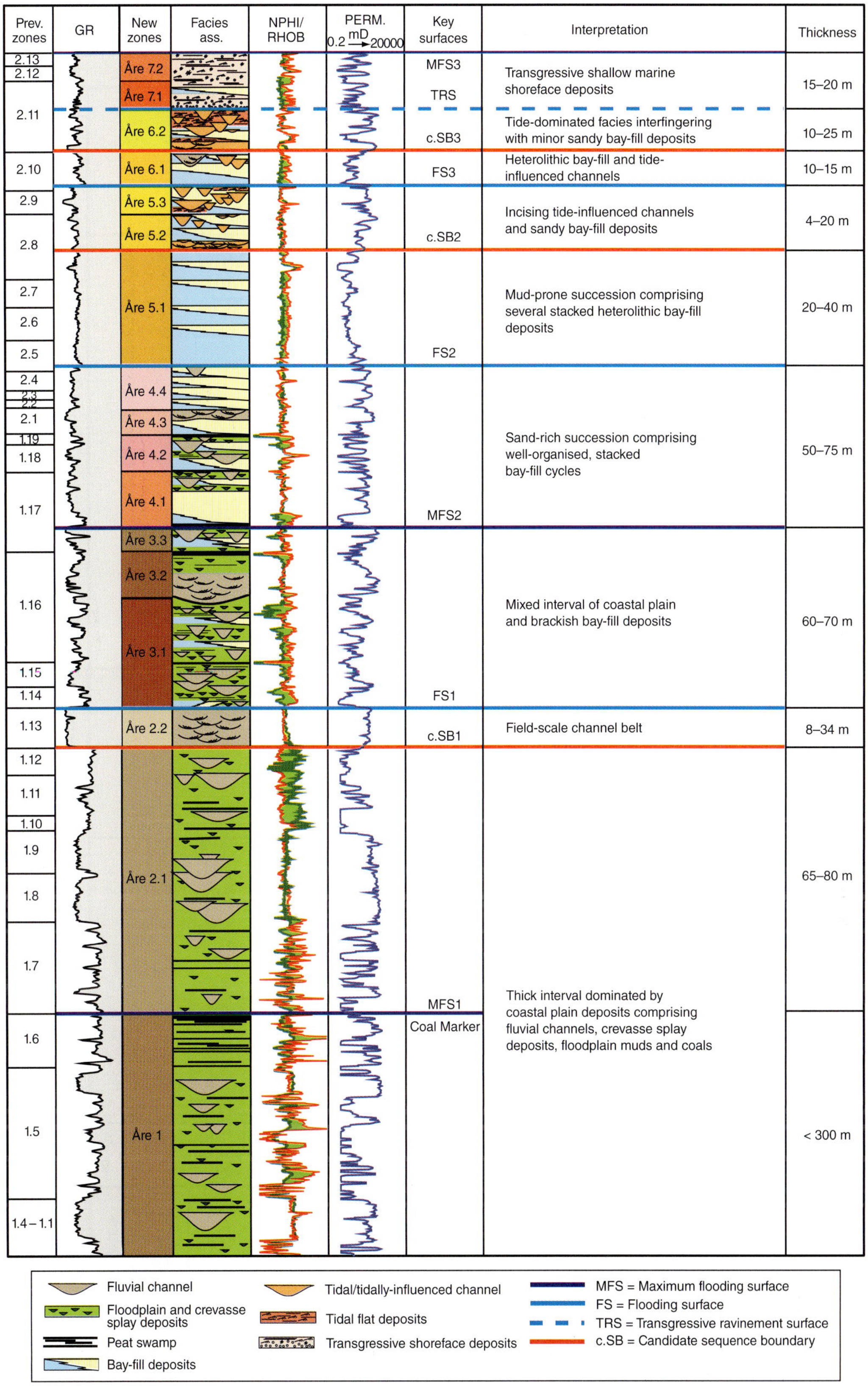

Fig. 4. Reservoir zonation scheme of the Åre Formation, Heidrun Field, showing sedimentary characteristics and associated wireline log signatures of each reservoir zone (based on well 6507/7-A-46). The previous reservoir zonation (left column) was modified from Kjærefjord (1999) and Svela (2001). (GR, NPHI/RHOB, PERM. refer to gamma ray; neutron/density; permeability, respectively).

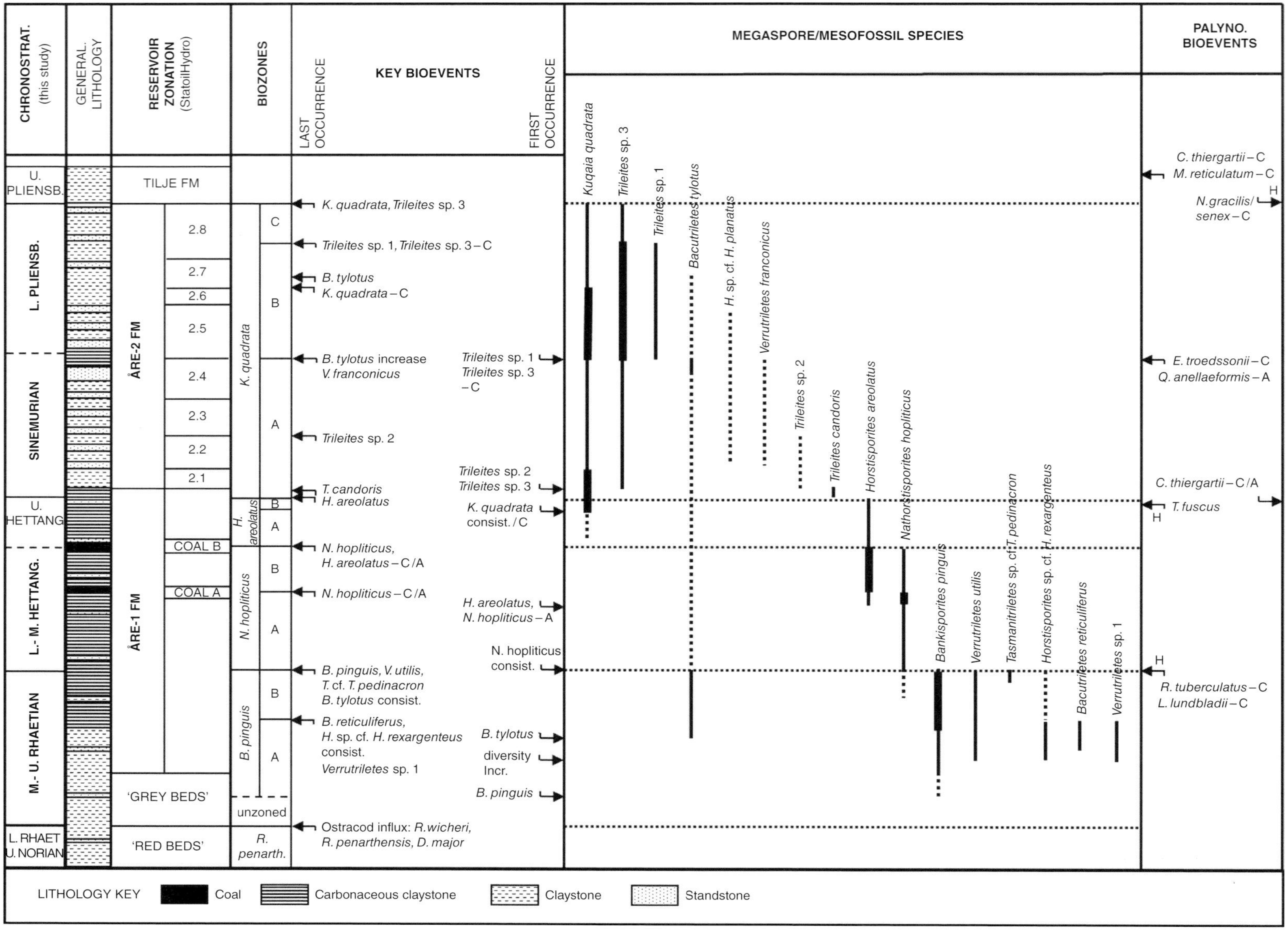

Fig. 5. Megaspore/mesofossil biozonation scheme of the Åre Formation, Halten Terrace region, plotted against palynology and interpreted chronostratigraphy (from Morton, *et al.*, 2009). Note that the entire Åre Formation is divided into two main reservoir zones in the regional scheme (Åre-1 Fm. and Åre-2 Fm.). This scheme cannot be compared straightforwardly with the Åre reservoir zonation scheme from the Heidrun Field but in-house biostratigraphy studies confirm that the regional biozones can be correlated with the Heidrun field reservoir zones. The 'coal A' corresponds to the Mid-Hettangian Top Åre 1 Coal Marker (see Fig. 4); Lowermost Sinemurian 'Åre 2.1' (Base of *K. quadrata* biozone) correlates to the Åre 3 Zone; Lower Pliensbachian Top Åre: 'Top Åre 2.8' corresponds to Top Åre 7 Zone in the Heidrun scheme (Fig. 4).

image logs (FMI; Fullbore Formation MicroImager) are used to derive proximal to distal facies relationships and depositional trends. Facies distribution maps illustrate depositional scenarios and representative 'time-slices' of sedimentary environments for each reservoir zone.

Two biostratigraphic studies validate chronostratigraphic correlation in the Åre Formation. Pedersen *et al.* (1989) and an in-house palynostratigraphic study documented both non-marine acritarchs and non-pollen/spore organic debris and marine palynomorphs (dinoflagellate cysts, acritarchs and prasinophycean algae). However, it has been found that the use of palynological biozones for chronostratigraphic correlation is limited due to their strong facies dependency. A separate megaspore zonation study attempted to improve the stratigraphic resolution of the lowermost, non-marine part of the Åre Formation. Although results from this study are in keeping with the regional megaspore biozonation of the Åre Formation in Mid-Norway (based on Morris *et al.*, 2009; Fig. 5), the reservoir zone resolution within the non-marine succession on Heidrun was not significantly improved by this megaspore zonation.

Some sedimentological and stratigraphic concepts established during this study are tested against dynamic well data (in-house studies). For example, potential and proven barriers and baffles for flow and connectivity of sandbodies identified by petrophysical pressure gradients, well test data and production history have been compared with the geological understanding of the different reservoir zones (see 'Industrial application' chapter for further discussion).

GEOLOGICAL SETTING

The Åre Formation, of the Early Jurassic Båt Group, is approximately 670 m thick and comprises coastal plain to delta plain sandstones, mudstones and coals of Rhaetian to early Pliensbachian age (Figs 3 and 4). Few wells extend through the base of the Åre succession into the underlying alluvial Triassic age 'grey beds'. The Rhaetian age of base Åre is confirmed by both the megaspore *Bankisporites pingus* (in-house study; Fig. 5) and palynological biostratigraphy *Ricciisporites tuberculatus*/*Limbosporites lundbladii* events (Pedersen *et al.*, 1989). The Åre Formation is succeeded by the tide-dominated Tilje Formation of early Pliensbachian to early Toarcian age (Fig. 3). The lithostratigraphic framework of the Mesozoic and Cenozoic succession offshore mid-Norway is defined by Dalland *et al.* (1988). The regional chronostratigraphic framework is, to a large degree, based on palynostatigraphy (Pedersen *et al.*, 1989) and in-house, unpublished studies and reports. The Åre Formation is identified in several fields offshore mid-Norway and northern Norway (cf. Dalland *et al.*, 1988). Although the current stratigraphic subdivision of the Åre Formation differs from field to field, the formation has traditionally been sub-divided into two members: Åre 1 and Åre 2 (Fig. 5). The boundary between these members is defined as the top of the youngest coal-bearing strata (Dalland *et al.*, 1988; Svela, 2001). This boundary separates the dominantly coastal plain deposits of the lower Åre Formation from the marginal-marine dominated deposits of the upper Åre Formation.

The overall transgressive character of the Åre Formation reflects gradual tectonic development of the Halten Terrace Basin during development of the North Atlantic rift system in the early Jurassic. This intracratonic post-rift basin, estimated to be 600 km by 200 km, opened northward to the Tethys Sea (Fig. 2; Doré, 1991; Kjærefjord, 1999). The oldest part of the Åre Formation consists of a succession of fluvial channel sandstones, peat swamp coals and floodplain mudstones, interpreted to correspond to deposition in a low-lying and wet coastal plain environment (Åre 1 and 2 Zones; Rhaetian to Hettangian; Fig. 4 (Dalland *et al.*, 1988). A gradual change into a lower delta plain setting is confirmed by repeated incursions of marginal-marine interdistributary bay, marsh and distributary channel fill facies associations in the middle part of the Åre Formation (Åre 3 and 4 Zones; Hettangian to Sinemurian; Fig. 4; Kjærefjord, 1999; Svela, 2001). A progressive transition into more open marine conditions recorded in the uppermost Åre Formation (Åre 5, 6 and 7 Zones; Sinemurian to early Pliensbachian time; Fig. 4) is sustained throughout deposition of the lowermost Tilje Formation (Kjærefjord, 1999; Martinius *et al.*, 2001). This transition is ascribed to a significant change in the basin geometry related to the establishment of a marine seaway between the Tethys Sea in the south and the Boreal Sea in the north (Kjærefjord, 1999; Martinius *et al.*, 2001).

Provenance

An integrated sediment provenance study using provenance-sensitive heavy mineral ratios, mineral chemistry and U/Pb dating of detrital zircons suggested that the main source area of the Åre Formation shifted between the western basin margin (East Greenland source) towards the eastern basin margin (Norwegian mainland source; Morton *et al.*, 2009). The lowermost, non-marine part of the Åre Formation shows evidence of provenance from the East Greenland margin (Fig. 2). This provenance signal decreases upwards within the Åre Formation and the uppermost part of the Åre and the overlying Tilje Formation are interpreted to have a Norwegian margin source (Martinius *et al.*, 2001). This change in source material is ascribed to restricted sediment supply from East Greenland following gradual sea-level rise. In-house provenance age data derived from Sm/Nd isotope analysis confirm this trend. Published studies from the time-equivalent Statfjord Formation of the Tampen area in the North Sea (~300 km south of Heidrun) by Mearns *et al.* (1989) and Brekke *et al.* (1999) imply a structural high to the north or the west of the basin (Fig. 2) that may have acted as a significant source area for fluvial systems on the Norwegian Shelf during Rhaetian to Sinemurian time (cf. Ryseth, 2001).

DEPOSITIONAL MODEL AND STRATIGRAPHIC FRAMEWORK

A total of 49 lithofacies (Fig. 6B) and 18 facies associations are recognised from core material and interpreted to have formed in a range of coastal plain to marine paralic environments (Figs 4 and 7). The facies associations are grouped into four main depositional settings: 1) fluvial/alluvial coastal plain; 2) lower delta plain/brackish-water interdistributary bay; 3) mixed wave-influenced and tide-influenced estuary and 4) transgressive shoreface (Fig. 4). Significant shifts in depositional environments during deposition of the Åre Formation are recorded in the stratigraphy and define reservoir zone boundaries (Fig. 4). A combination of facies stacking patterns and correlation of key stratigraphic surfaces are used to interpret the sedimentological history and establish a stratigraphic framework.

The previous reservoir zonation of the Åre Formation on Heidrun utilised the Åre 1 and 2 subdivision proposed by Dalland *et al.* (1988). Thirty two reservoir subzones were defined mainly reflecting a separation of flow and no-flow units (Fig. 4). The reservoir zonation presented in this study divides the succession into 7 reservoir zones (Åre 1 to Åre 7 and 17 subzones; Fig. 4).

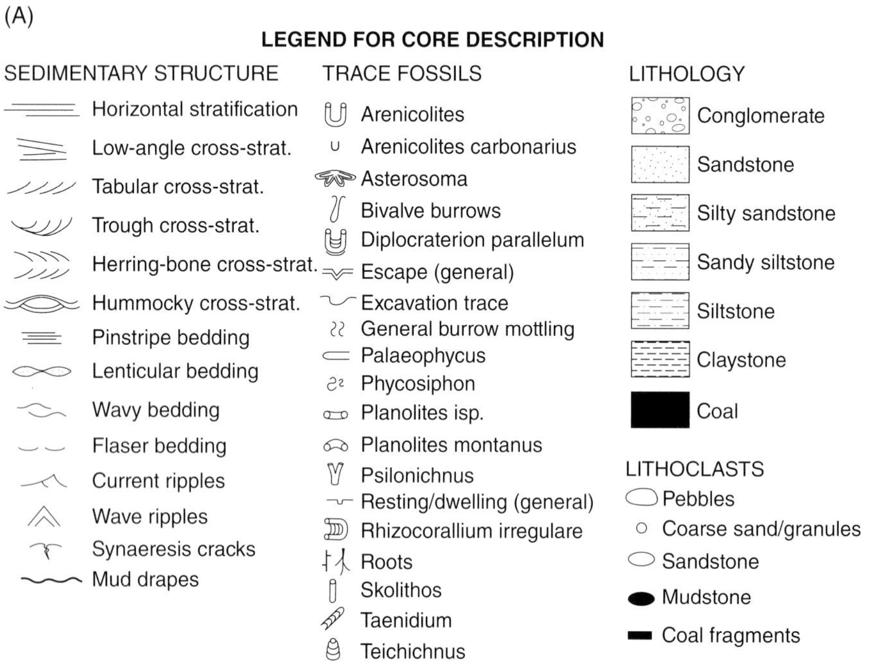

Fig. 6. (A) Legend for the lithofacies and facies association schemes; also core descriptions.

(B)

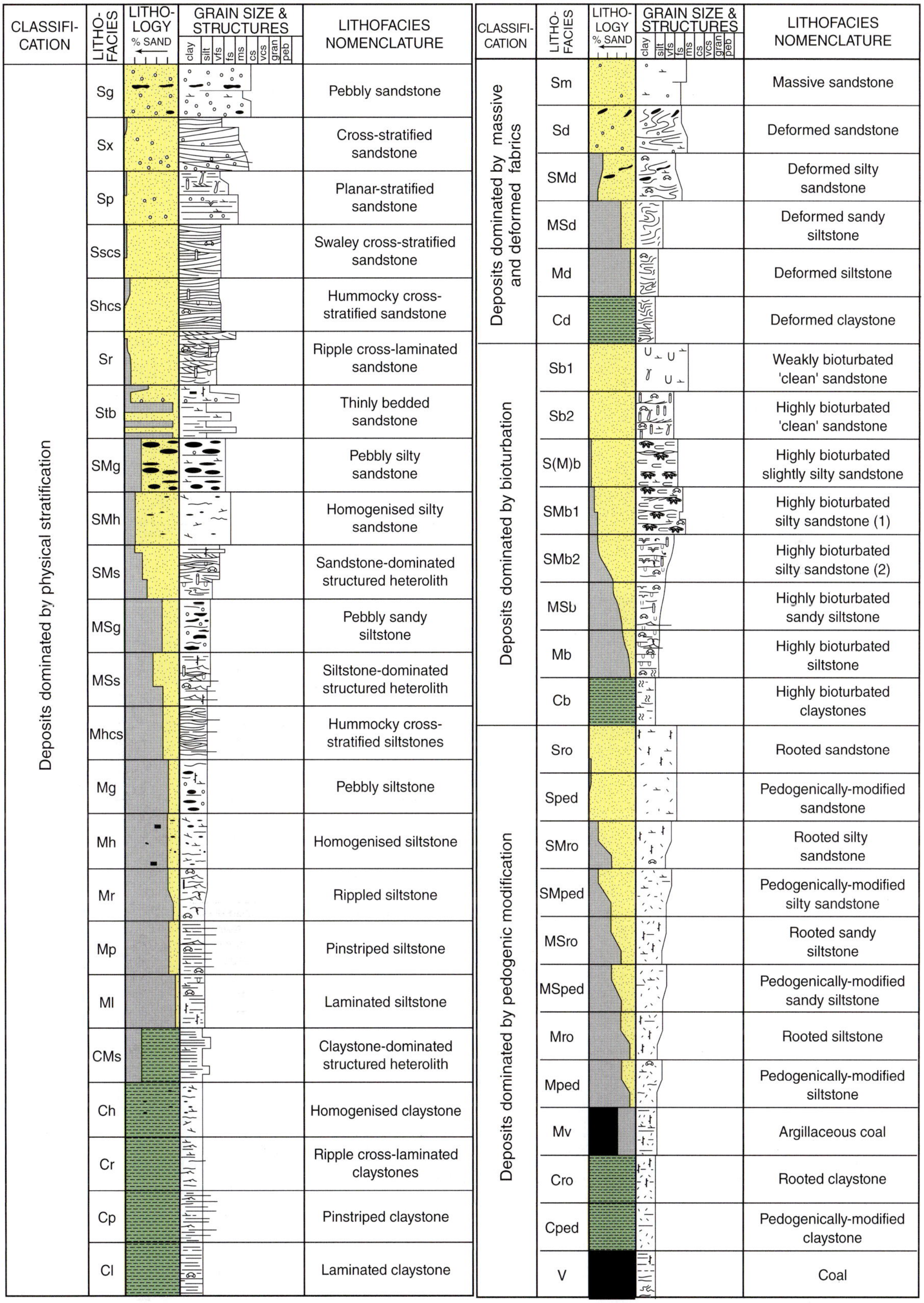

Fig. 6. (*Continued*) (B) Main characteristics of the lithofacies identified in the Åre Formation.

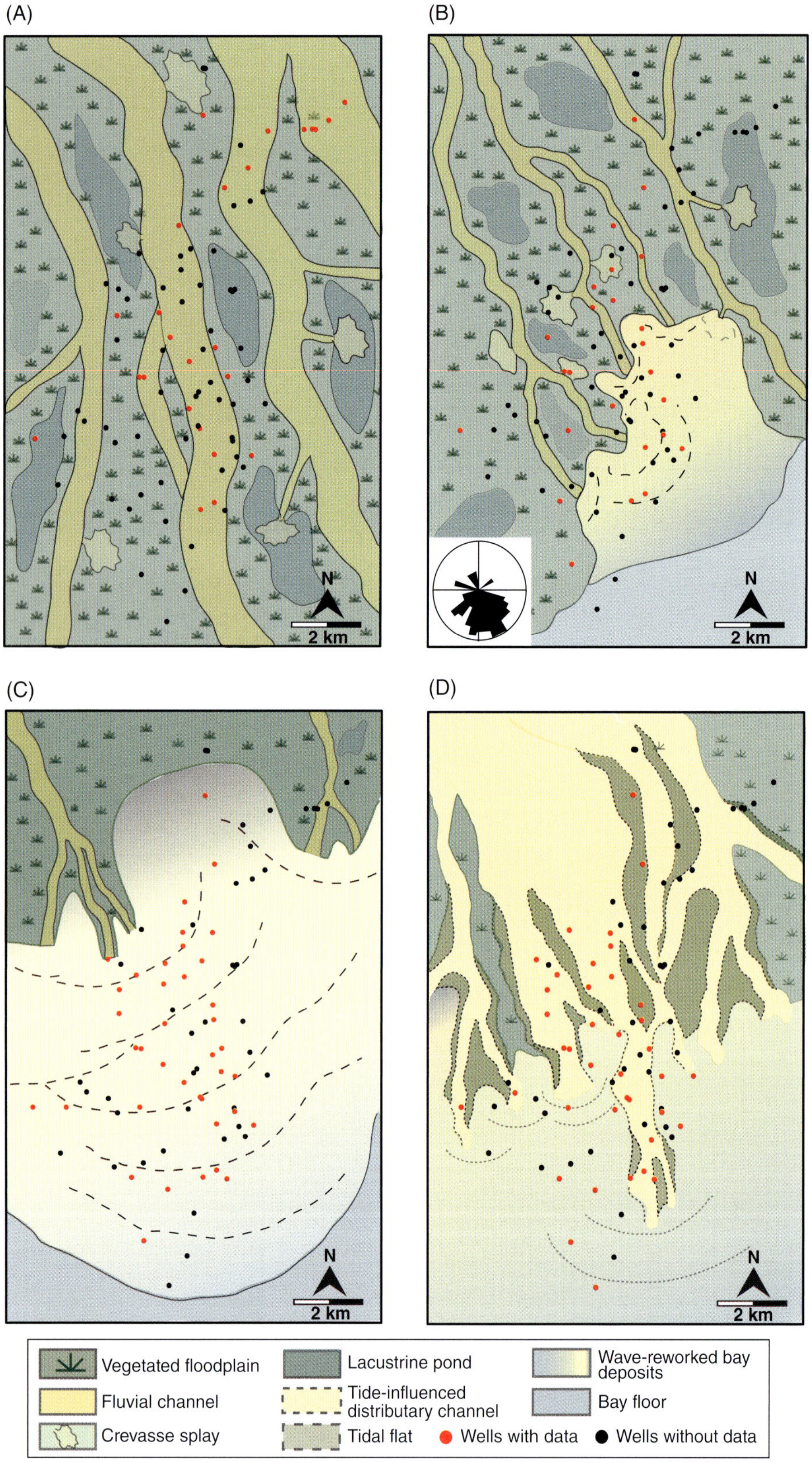

Fig. 7. Palaeogeographic maps showing possible depositional scenarios for selected intervals. The position of sub-environments is conditioned to well-data (red dots) and outside the wells the setting is estimated from general conceptual understanding. (A) Palaeogeographic map illustrating a possible scenario from the fluvial coastal plain setting of the upper Åre 2.1 Sub-zone. (B) Example of a 'time-slice' from the Åre 3.1 Sub-zone, showing the first brackish-water bay development. The rose diagram shows palaeocurrent directions based on available dip data from the entire Åre 3 Zone. (C) Interval within the Åre 4 Zone dominated by laterally amalgamated wave-reworked bay deposits. (D) Palaeogeographic map from the Åre 5.2 Sub-zone. This interval corresponds to the first development of significant tide-influenced facies filling an incised system that drained from the NW towards the SE.

Table 1. Main characteristics of the key stratigraphic surfaces.

Key surface	Seismic marker	Chrono-strat. age	Surface characteristics
Top Åre 7 Zone **MFS3**	No	Early Pliensbachian	1 m to 1.5 m thick bioturbated and calcite cemented mudstone/heterolith. Facies transition: from shoreface deposits of the Åre 7 Zone into storm-influenced prodelta deposits of the Tilje Formation
Top Åre 6 Zone **TRS**	Yes	n/a	Erosive surface characterised by a reworked lag deposit and calcite cementation. Facies change: From tidal to wave-dominated deposits. A significant shift into more marine trace fossil assemblage is seen across this surface
Top Åre 6.1 Sub-zone **c.SB3**	No	n/a	Sharp and erosive surface with no significant relief. Facies change: Marks a shift from wave-influenced bay deposits to tide-dominated deposits
Top Åre 5 Zone **FS3**	No	n/a	0.5 m to 2 m thick mudstone or heterolith. Facies change: A transition from tide-influenced to heterolithic bay-fill deposits
Top Åre 5.1 Sub-zone **c.SB2**	No	n/a	Erosive surface characterised by up to 10 m relief. Associated with basal lag of mud clasts. Facies change: From mud-prone, wave-dominated bay deposits to sandy tide-influenced channel sandstones
Top Åre 4 Zone **FS2**	Local/ segment-scale	n/a	0.5 m to 3 m thick mudstone or heterolith. Facies change: A transition from sand-dominated to mud-prone bay facies
Top Åre 3 Zone **MFS2**	No	Base/Early Sinemurian	2 m to 5 m thick laminated, carbonaceous mudstone and claystone associated with calcite cementation and bioturbation. Facies change: From mixed coastal plain and brackish-water interdistributary bay deposits to extensive sand-dominated bay-fill facies
Top Åre 2 Zone **FS1**	No	n/a	1 m to 3.5 m thick brackish water mudstone, locally alternating with peat swamp deposits. Facies change: from fluvial channel facies to marginal-marine bay deposits
Top Åre 2.1 Sub-zone **c.SB1**	No	n/a	Field-wide down-cutting of a fluvial channel belt. The surface is associated with a pebble lag and variable relief
Top Åre 1 Zone/Coal Marker **MFS1**	Yes	Middle Hettangian	10 m to 46 m thick package of peat swamp coals and lacustrine mudstones. Facies changes: Marks a change in fluvial channel stacking pattern and a shift from wet and immature to better drained floodplain palaeosols

Key stratigraphic markers

The most reliable correlative markers within the Åre Formation are various flooding surfaces which form the basis for the stratigraphic analysis provided here. Within the Åre Formation on the Heidrun Field these are not *sensu stricto* parasequence-bounding flooding surfaces (Van Wagoner *et al.*, 1988), but they are ascribed to relative deepening events marked by condensed deposits and field-scale shifts in depositional style. These transgressive events correspond to either flooding surfaces (FS), maximum flooding surfaces (MFS) or transgressive ravinement surfaces (TRS) (cf. Galloway, 1989; Posamentier & Vail, 1988; Van Wagoner *et al.*, 1988) as specified in Fig. 4 and Table 1. In core, the deepening events are recognised as 0.5 to 5 m thick mudstone-dominated intervals (Fig. 8) characterised by bioturbation (onset of brackish or marine water ichnofabrics) and, in some cases, calcite cementation (e.g. MFS2, TRS and MFS3; Table 1 and Fig. 8). According to Hammer *et al.* (2010) the calcite cemented beds associated with key

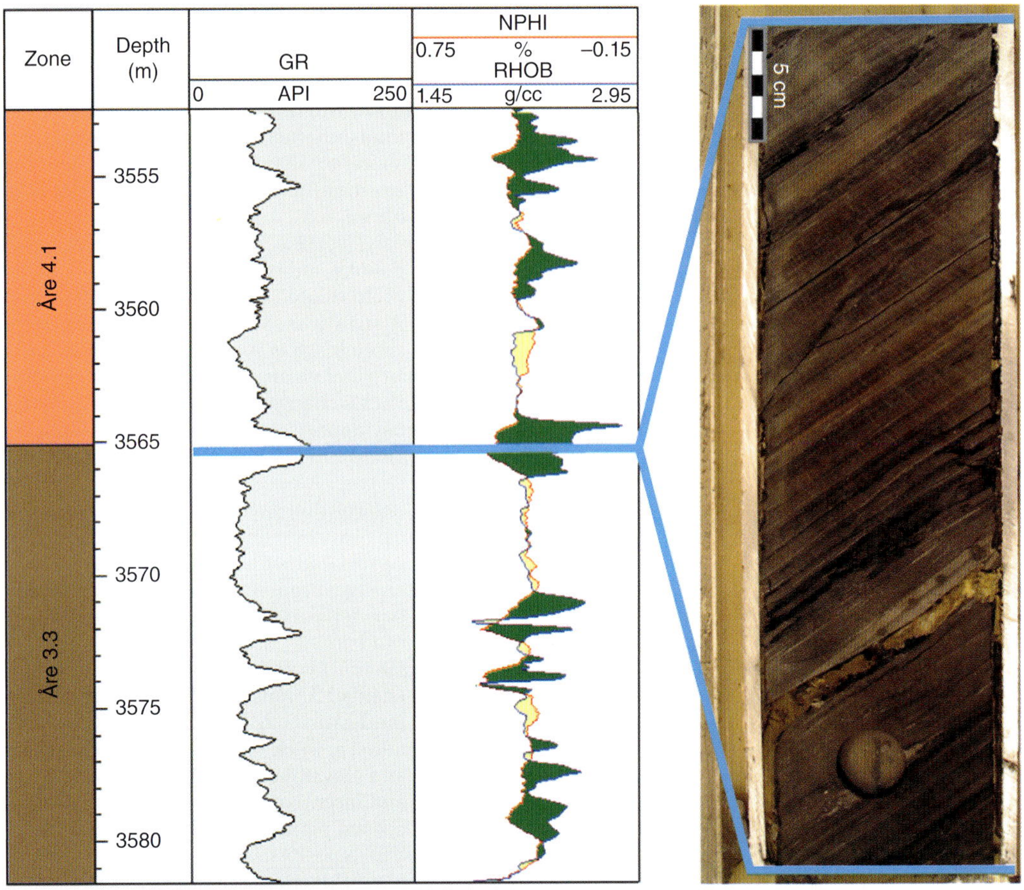

Fig. 8. Characteristic well-log signatures and core expression (well 6507/8-D-4BHT3) of a flooding surface within the Åre Formation (in this case: top Åre 3 Zone). In the core the typical signature is a heterolithic laminated claystone-mudstone, generating a wide shale separation on the NPHI/RHOB logs and a significant increase in the GR log. Note calcite cementation at 3564 m MD as indicated by the peak in the RHOB values. See Fig. 4 for abbreviations.

stratigraphic markers within the Åre Formation, could possibly be related to concentration of fossil fragments. This suggests that the flooding events may have caused sediment starvation and condensation.

Several flooding surfaces identified in this study correspond to biostratigraphic events of chronostratigraphic significance (Table 1). Minor time gaps may also occur at the other defined flooding surfaces; however, the biozone resolution is insufficient to demonstrate this. The lateral extent of the mudstone-rich flooding surfaces within the Åre Formation are, in some cases, confirmed by pressure and production data, although complex fault segmentation often obscures the true extent of these sedimentological barriers and baffles by creating fluid communication windows. Major changes in depositional facies at two reservoir zone boundaries are associated with significant contrasts in acoustic impedance which define distinct seismic horizons (MFS1 and TRS; Table 1).

The reservoir stratigraphy is further divided into subzones (Fig. 4). These are bounded by minor flooding surfaces that are interpreted to reflect local autocyclic shifts within the coastal system and do not correspond to larger-scale landward movements of the palaeocoastline. The term 'candidate' flooding surface has been adapted for these events.

A number of surfaces characterised by significant erosion and marking an abrupt change into sand-dominated channelised deposits have also been identified as correlateable stratigraphic markers and are termed candidate sequence boundaries (c.SB; Table 1) (sensu Sloss *et al.*, 1949). These surfaces have most likely formed in response to base level fall. Candidate sequence boundaries do not commonly delineate reservoir zones, but in some cases define reservoir subzones (Fig. 4).

Åre 1 and 2 Zones: Fluvial/alluvial coastal plain

Fluvial coastal plain deposits of the lower Åre Formation (Åre 1 and 2 reservoir zones; Fig 4) are up to 470 m thick. Facies associations are classified as: fluvial channel (FA 5), crevasse splay (FA 4), palaeosol (FA 3), lacustrine mud (FA 2) and peat swamp (FA 1; Figs 9, 10 and 11). Fluvial channel deposits (FA 5) from the lowermost part of the Åre Formation have been cored extensively and two main types of channel-fill are recognised: 1) thin (3 m to 5 m), sand-rich single-story channel belts and 2) thick (5 m to 35 m), multi-storey sand-dominated channel belts that display frequent internal scour surfaces. Single-story channel belt deposits are characterised by erosive basal boundaries, fining-upward grain size trends and, in their upper parts, inclined heterolithic strata. An upwards decrease in bedform scale (from dune-scale to ripple-scale cross bedding) and grain-size reflects waning current strength due to lateral migration of a channel margin and adjacent point bar. Multi-storey channel belt deposits are both sharp-based and sharp-topped, completely sand-dominated and lack vertical trends in grain size (Fig. 12). This channel-fill style may imply that deposition occurred in downstream accreting bedforms. Interpretation of fluvial channel style based upon vertical profiles from subsurface well dataset is inherently problematic. However, for the purposes of reservoir characterisation, this simple twofold subdivision of channel sandbodies is used to group and model the key elements of the reservoir.

The floodplain deposits within the Åre 1 and 2 Zones are dominated by dark grey-brown claystones and siltstones with coals (Fig. 9). The coals are dm-thick to m-thick beds of vitreous coals or siltstones with high volumes (30% or more) of carbonaceous debris (FA 1; Fig. 10). Pedogenic mottling and rooting is well developed within these peat swamp deposits and they are commonly associated with claystones and siltstones with palaeosol character (FA 3). Locally, the extensively rooted horizons of siltstone and claystone of FA 3 underlie thin coals/carbonaceous siltstones (FA 1) and are therefore regarded as 'seat-earths'.

Lacustrine mudstones (FA 2), a common element of the floodplain deposits (Figs 9 and 10), are dominated by laminated to pinstriped siltstones and claystones. Fine carbonaceous debris

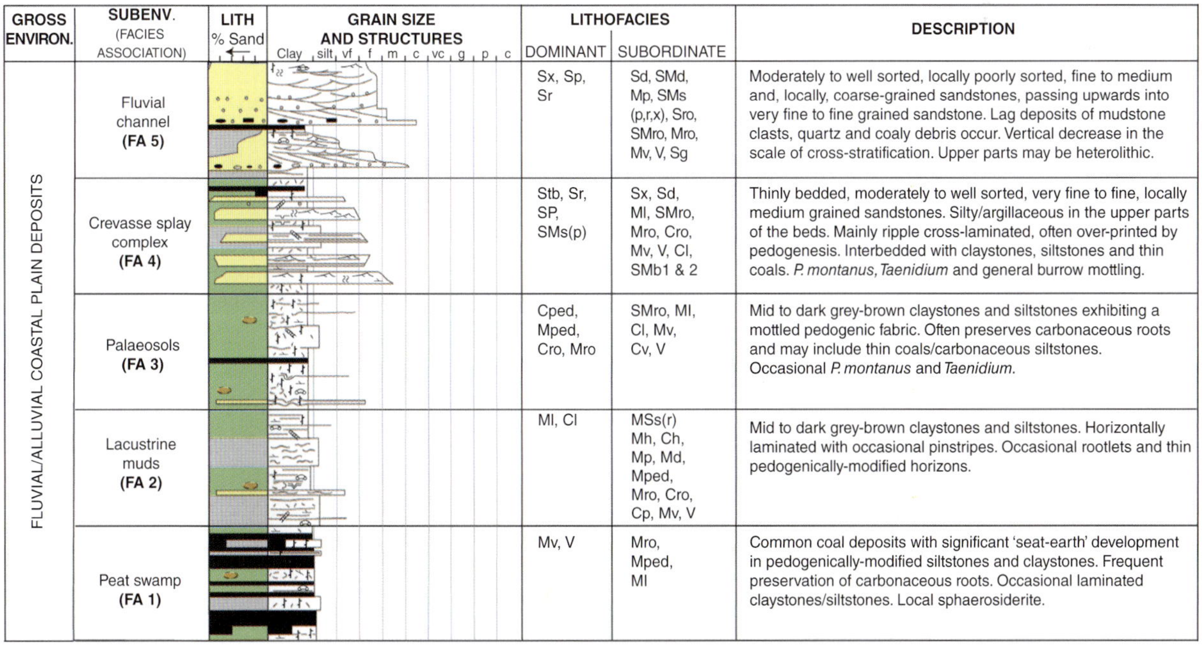

GROSS ENVIRON.	SUBENV. (FACIES ASSOCIATION)	LITHOFACIES DOMINANT	LITHOFACIES SUBORDINATE	DESCRIPTION
FLUVIAL/ALLUVIAL COASTAL PLAIN DEPOSITS	Fluvial channel **(FA 5)**	Sx, Sp, Sr	Sd, SMd, Mp, SMs (p,r,x), Sro, SMro, Mro, Mv, V, Sg	Moderately to well sorted, locally poorly sorted, fine to medium and, locally, coarse-grained sandstones, passing upwards into very fine to fine grained sandstone. Lag deposits of mudstone clasts, quartz and coaly debris occur. Vertical decrease in the scale of cross-stratification. Upper parts may be heterolithic.
	Crevasse splay complex **(FA 4)**	Stb, Sr, SP, SMs(p)	Sx, Sd, Ml, SMro, Mro, Cro, Mv, V, Cl, SMb1 & 2	Thinly bedded, moderately to well sorted, very fine to fine, locally medium grained sandstones. Silty/argillaceous in the upper parts of the beds. Mainly ripple cross-laminated, often over-printed by pedogenesis. Interbedded with claystones, siltstones and thin coals. *P. montanus*, *Taenidium* and general burrow mottling.
	Palaeosols **(FA 3)**	Cped, Mped, Cro, Mro	SMro, Ml, Cl, Mv, Cv, V	Mid to dark grey-brown claystones and siltstones exhibiting a mottled pedogenic fabric. Often preserves carbonaceous roots and may include thin coals/carbonaceous siltstones. Occasional *P. montanus* and *Taenidium*.
	Lacustrine muds **(FA 2)**	Ml, Cl	MSs(r) Mh, Ch, Mp, Md, Mped, Mro, Cro, Cp, Mv, V	Mid to dark grey-brown claystones and siltstones. Horizontally laminated with occasional pinstripes. Occasional rootlets and thin pedogenically-modified horizons.
	Peat swamp **(FA 1)**	Mv, V	Mro, Mped, Ml	Common coal deposits with significant 'seat-earth' development in pedogenically-modified siltstones and claystones. Frequent preservation of carbonaceous roots. Occasional laminated claystones/siltstones. Local sphaerosiderite.

Fig. 9. Facies association scheme of the fluvial/alluvial coastal plain depositional system within the Åre Formation. (See Fig. 6A for legend; and 6B for lithofacies scheme).

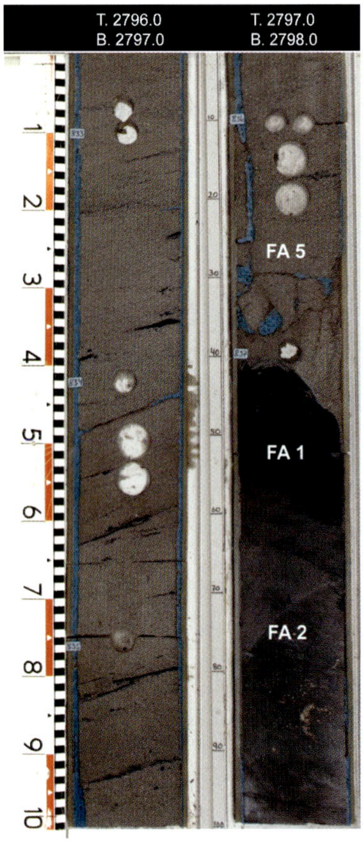

Fig. 10. Core photograph from Åre 1 Zone in well 6507/7-A38. Typical fluvial facies showing fine-laminated lacustrine claystone (FA 2); carbonaceous claystone representing peat swamp deposits (FA 1); and cross-stratified fluvial channel sandstone (FA 5). The ruler is 1.0 m long.

is common and occurs either well dispersed, concentrated into carbonaceous siltstones or even as thin, 'drifted' coals. Rooting and soft-sediment deformation are locally recognised. The type and intensity of bioturbation is variable. Some intervals are non-bioturbated, whilst others show occasional rooting or non-specific burrow mottling. *Planolites montanus* is dominant in low diversity assemblages, whilst *Taenidium* is occasionally present.

Crevasse splay deposits (FA 4) are interbedded with the other types of floodplain deposits (Fig. 9). Individual beds are 0.3 m to 1 m thick sandstones displaying current ripple cross-lamination (with locally climbing current ripples) and planar lamination. Many sandstone beds preserve carbonaceous rootlets, whilst others show a mottled fabric without rootlets, most likely related to pedogenic processes. The trace fossils *Taenidium* and *Planolites montanus* are observed.

These floodplain facies associations provide evidence of an extensive, low-lying, muddy coastal plain and humid climate (Fig. 11). Although much of the floodplain was characterised by extensive shallow lakes, ponds and peat swamps, better drained areas allowed mature palaeosols to develop. Crevasse splay sands were introduced to the floodplain during peak fluvial discharge periods via crevasse channel systems.

The peat swamps and highly vegetated floodplain areas most likely constituted stable substrates

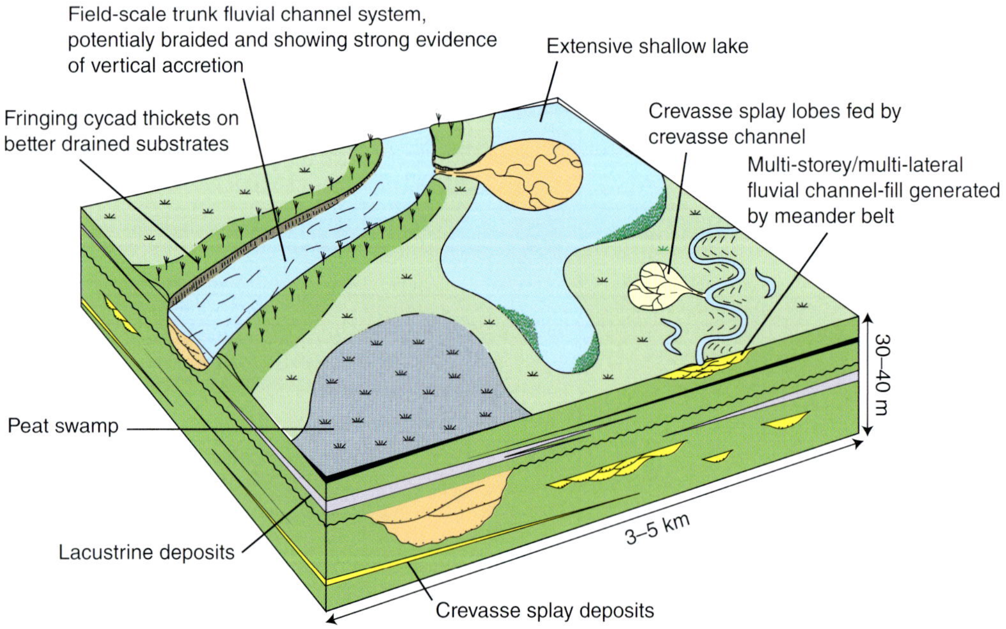

Fig. 11. Conceptual model for the Åre Formation coastal plain setting (reservoir zones Åre 1 and 2).

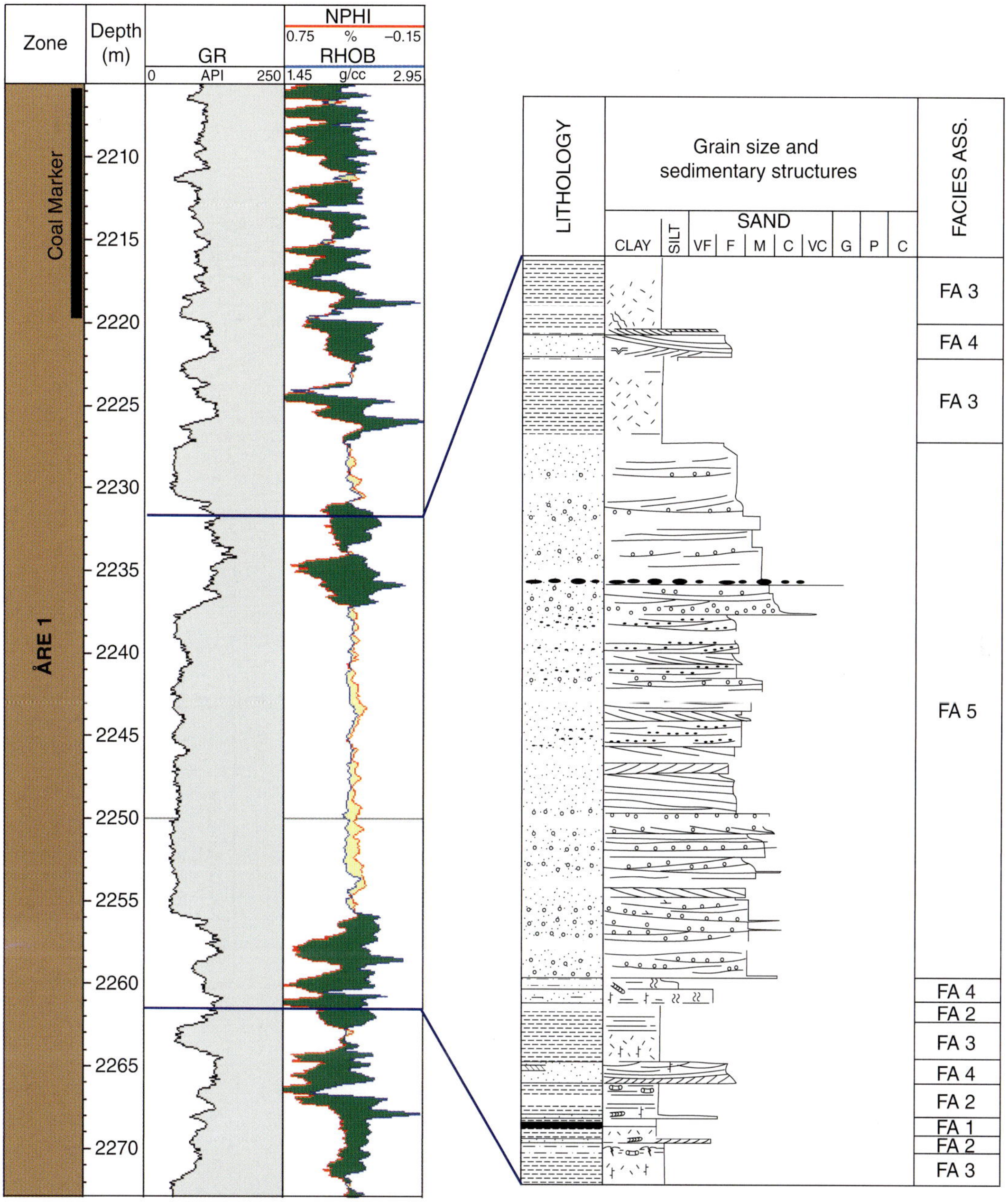

Fig. 12. Wireline log response and core interpretation of typical facies associations within the Åre 1 Zone (see Fig. 9 for facies association descriptions, example from well 6507/8-4). Note the 'blocky' signature on the GR log corresponding to a multi-storey fluvial channel sandstone. A thick package of coals and carbonaceous claystones towards the top of the Åre 1 Zone corresponds to the lower part of the Coal Marker. (See Fig. 6A for core description legend and Fig. 4 for abbreviations).

that restricted the fluvial channel lateral mobility and migration (e.g. Gradzinski *et al.*, 2003). Multi-storey vertical stacking of fluvial channels (e.g. Fig. 12) may also indicate stable channel positions and a tendency of channels to aggrade vertically when the accommodation space allowed for it. Abundant internal scour surfaces suggest multiple phases of cut and fill, pointing towards a braided or anastomosing style of river systems (Gjelberg *et al.*, 1987; Gradzinski *et al.*, 2003; Makaske, 2001; Mjøs & Prestholm, 1993; Pedersen *et al.*, 1989). The channels of the Åre 1 and 2 Zones probably reflect deposition in river reaches characterised by a combination of both

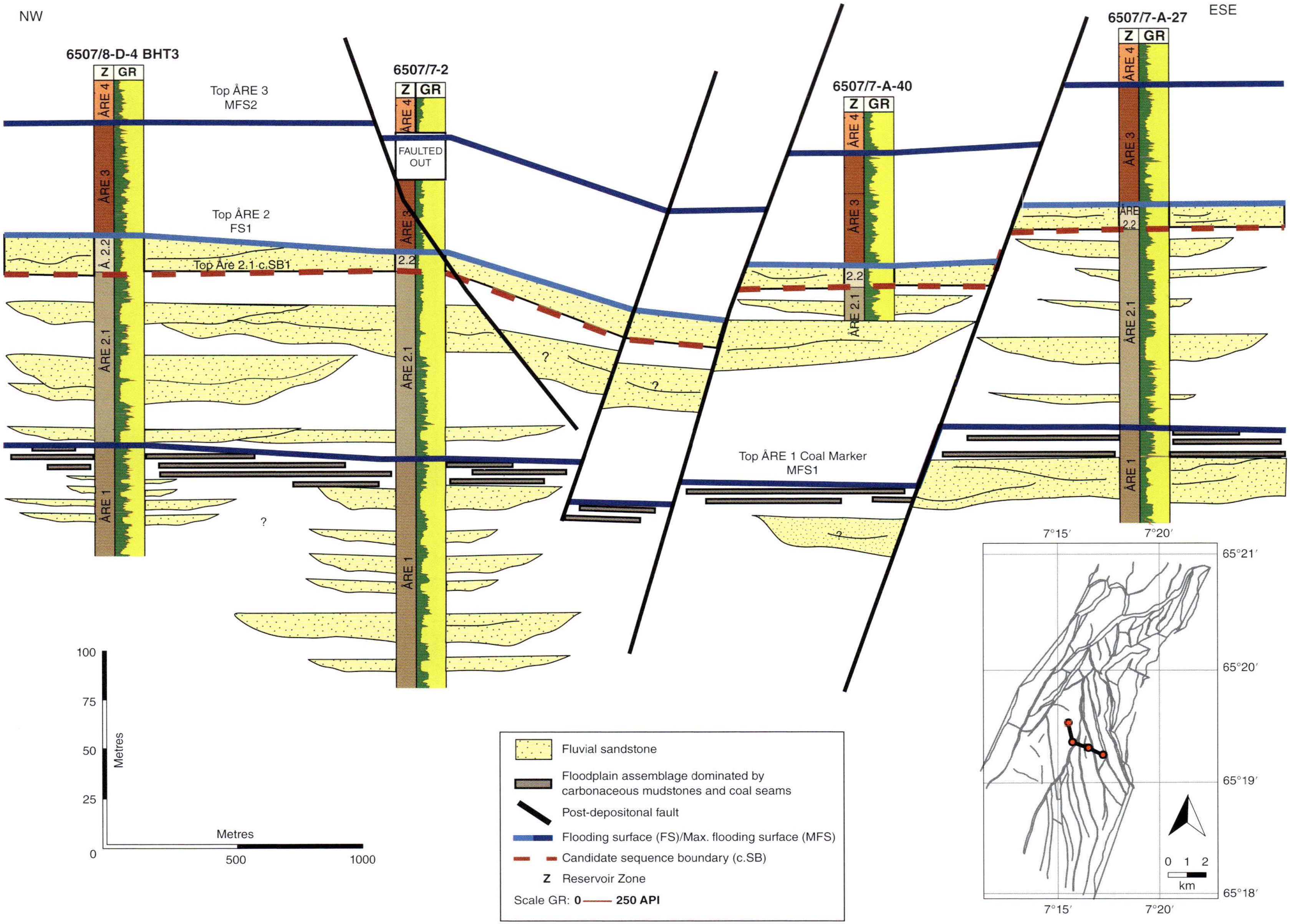

Fig. 13. NW–ESE oriented panel illustrating wells with gamma ray logs (GR) and a possible tie of fluvial sandbodies within the reservoir zones Åre 1, 2.1 and 2.2 (section location is marked on the map). Fault orientation and placement is based on seismic data. The stratigraphic markers within the fluvial-dominated succession are few. Top of reservoir zone Åre 1 Coal Marker/MFS1 represents the most confident stratigraphic surface and is also a seismic marker horizon (see also Table 1 and Fig. 24).

braided and sinuous meandering channels located upstream of the tidal influence. Biostratigraphical samples comprising entirely non-marine fauna and flora confirm a non-marine origin for these facies associations (Morris *et al.*, 2009). Estimated depositional direction of fluvial deposits is partly based on borehole image log dip interpretations of reliable sedimentary features (i.e. dune-scale cross-stratification within a channel fill). A fairly consistent palaeocurrent direction seems to have been from north to south (with variations showing NW to SE and NE to SW-directions; Fig. 7A).

Sequence stratigraphic interpretation of the lowermost Åre Formation is challenging because key stratigraphic surfaces are difficult to recognise in fluvial successions. Variations in water table elevation and accommodation space might be recorded in the way channel sandstones stack and the geometry and distribution of facies belts (Bridge, 2003; Bridge & Tye, 2000; Shanley & McCabe, 1994). Studies such as those referenced have shown that at a large scale, net-to-gross, geometry and connectivity of fluvial channel sandstones are related to responses of the coastal plain to cyclic sea-level change. However, a general lack of frequent stratigraphic marker surfaces has led to a low resolution reservoir zonation within this fluvial-dominated package of the Åre Formation. Based on a field-wide coal marker (see description below of the Top Åre 1 Coal Marker; Table 1 and Fig. 4) and vertical trends in channel architecture, the coastal plain deposits are divided into two reservoir zones; the Åre 1 and 2 Zones. The latter reservoir zone can be further subdivided into sub-zones Åre 2.1 and 2.2 (Figs 4 and 13).

Åre 1 Zone

The Åre 1 Zone (up to 300 m thick) is comprised of coastal plain facies associations as described above. Due to the few and scattered wells penetrating this part of the stratigraphy in the main field, confident well-to-well correlation of fluvial channel sandstones is only possible in a few areas. Although wells are most closely spaced in the northernmost part of the Heidrun Field, the majority of wells are highly deviated and difficult to use for facies correlation. The channel sand proportion vs. floodplain fines is generally less than 30% within this part of the stratigraphy, indicating a limited three-dimensional connectivity of the channel belts (e.g. Mackey & Bridge, 1995). Fluvial sandbodies are dominated by single-storey channel deposits, which are typically less than 5 m thick and inferred to be only 150 m to 200 m wide (based on relationships shown in Bridge & Tye, 2000). Subordinate, thicker, multi-storey channel belts can, in some cases, be correlated for two kilometres perpendicular to the estimated depositional direction (Fig. 7A). The uncertainties regarding correlation of channel sandstones are further increased by lateral differential compaction caused by a high number of peat swamp coals (cf. Hammer, 2010).

Abundance of peat swamp coals, immature palaeosols and lacustrine mudstones suggest wet floodplain conditions prevailed throughout the Åre 1 Zone. The uppermost part of this reservoir zone is characterised by the 'Coal Marker' interval (Table 1, Figs 4, 12 and 13). This is a 10 m to 46 m thick package of floodplain deposits comprising interbedded peat swamp and lacustrine deposits attributed to deposition within a wet floodplain with frequent development of extensive bodies of standing water. Locally, relatively thin fluvial channel sandbodies and/or crevasse splay sands occur within this interval. The Åre 1 Zone top surface is defined by the top of the Coal Marker. Although picking the top of the Coal Marker from wireline logs can be uncertain because of local variations in coal bed thickness and number, this interval defines the most distinct seismic reflector and the most confident stratigraphic horizon within the fluvial-dominated part of the Åre Formation. In-house megaspore biostratigraphy indicates that the Coal Marker interval may be associated with the last occurrence of *Nathorstisporites hopliticus* which occurred during Middle Hettangian ('coal A' in Fig 5; Morris *et al.* 2009). Deposition of the Coal Marker interval was most probably caused by a significant rise in the water table elevation, and presumably increased accommodation space, which allowed extensive growth and accumulation of peat swamps and the development of muddy floodplain lakes. The top of the Åre 1 Zone over the Heidrun Field has been ascribed tentatively to a maximum floodplain aggradation phase corresponding to a maximum flooding surface (MFS1; Table 1 and Fig. 13). An overall rising base level during deposition of the Åre 1 Zone is suggested by indications of a generally wet floodplain system and more isolated channel sandbodies.

Åre 2 Zone

Two sub-zones have been defined within this reservoir zone (Åre 2.1 and 2.2; Figs 4 and 13). The lowermost, the Åre 2.1 Sub-zone (65 m to 80 m thick), is composed of fluvial coastal plain deposits which show an overall increase in the channel to floodplain ratio (>40% sand) compared to the underlying Åre 1 Zone, suggesting improved connectivity of channel belts (e.g. Mackey & Bridge, 1995). Single-storey and multi-storey channel sandstones of limited lateral extent are characteristic for the lowermost part of this sub-zone, but a significant increase in correlateable vertically and laterally accreting channel deposits is observed upwards in the stratigraphy (Fig. 13). In particular, one channel belt is identified in the uppermost part of this reservoir sub-zone and mapped in the seismic data through the northernmost part of the Heidrun Field. Seismic mapping and well-to-well correlations suggest this cut-and-fill-channel belt has a lateral extent of at least 2.5 km perpendicular to the estimated N to S-oriented depositional direction. The frequency of lake and peat swamp deposits is somewhat lower within Åre 2.1 Sub-zone than the Åre 1 Zone, suggesting floodplains were better drained. This change in floodplain facies, together with increased connectivity of channel sandstone bodies, might be attributed to deposition during a phase of highstand progradation.

The succeeding Åre 2.2 Sub-zone comprises a relatively thick interval (typically 10 m to 15 m, maximum 34 m; Fig. 4) of fluvial channel sandstones, with a minor component of crevasse splay sandstones and floodplain mudstones (Fig. 13). This fluvial channel belt is assumed to be laterally extensive over large parts of the field, based upon well correlations, pressure continuity and production data. Evidence from core indicates that the channel sandstones display consistent depositional characteristics (in terms of bedform thickness, presence of internal scour surfaces, grain size and sorting) in most wells. The base of the Åre 2.2 Sub-zone is erosive and typically characterised by a pebble lag. This erosional surface may reflect field-wide down-cutting forming an incised valley filled by braided channel deposits. Deepest incision is seen in the NNW area of the field and the sand-dominated Åre 2.2 Sub-zone thins toward the SE. This thinning may define the margin of the valley. Based on the field-wide extent of the channel belt, the base of Åre 2.2 is considered to be a candidate sequence boundary formed during significant lowering of base level (c.SB1; Table 1 and Fig. 13). The top of the zone shows fairly abrupt transition into overlying floodplain fines and brackish-water bay-floor mudstones. This onset of marginal-marine mud deposition defines a flooding surface (Fig. 4).

Åre 3 and 4 Zones: Lower delta plain and brackish water, interdistributary bay

Reservoir zones Åre 3 and 4 (110 m to 125 m combined thickness; Fig. 4) contain deposits that reflect an interaction of fluvial and marginal marine processes. Like the underlying reservoir zones, these facies associations are interpreted to have formed in the lower coastal plain and include peat swamp (FA 1), lacustrine muds (FA 2) and crevasse splay (FA 4; Fig. 9). However, in these zones, lower coastal plain deposits are intermixed with marginal marine deposits, including those of the interdistributary bay facies associations; tide-influenced distributary channel (FA 6), bay-floor muds (FA 7), wave-influenced bay-fill (FA 8) and bay-margin muds (FA 9; Figs 14, 15 and 16).

The brackish-water interdistributary bay deposits are dominated by 2 m to 10 m thick, upward-coarsening/cleaning units (Figs 14 and 16) grading from: 1) wave-rippled and pinstriped mudstones and siltstones (FA 7), to 2) mud-prone to sand-prone heteroliths showing lenticular bedding and small-scale to medium-scale hummocky cross-lamination; and 3) very fine-grained to fine-grained sandstones displaying medium-scale cross beds, swaley lamina sets, planar stratification and local current ripple cross-lamination (FA 8) (Fig. 16). Synaeresis cracks are abundant within these deposits. Bioturbation is weak to locally high and the trace fossil assemblages are dominated by *Planolites montanus*, *Skolithos* and *Arenicolites carbonarious* (Fig. 14). Carbonaceous, muddy deposits of FA 7 are interpreted to have formed in a low energy bay floor setting swept by occasional wave action. Dominance of wave-generated structures within the heterolithic and sand-rich parts of these bay-fills (FA 8) is consistent with a wave-dominated 'shoreface' development (Figs 14 and 15). Low-diversity, stressed trace fossil assemblages, together with abundant synaeresis cracks, confirm deposition in a brackish water environment and the vertical thickness of complete bay-fills suggests depositional water-depths of less than 10 m.

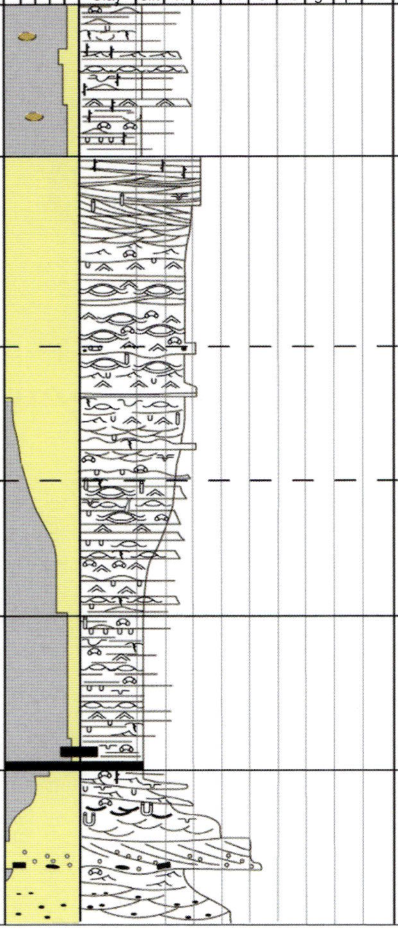

GROSS ENVIRON.	SUBENV. (FACIES ASSOCIATION)	LITHOFACIES DOMINANT	LITHOFACIES SUBORDINATE	DESCRIPTION	
BRACKISH-WATER INTERDISTRIBUTARY BAY	Bay-margin muds **(FA 9)**	Ml, Mp, Mr, MSs(b), Mro	MSs(r)	Dark grey laminated micaceous siltstones, locally with pinstripes, lenses and thin beds of well sorted coarse silt to very fine sand. Current ripples, wave ripples and occasional micro-hummocky cross-lamination. Occasional synaeresis cracks. Prominent rooting and carbonaceous debris, otherwise weak bioturbation, mainly by *P. montanus* and *A. carbonarius*.	
	Wave-influenced bay-fill **(FA 8)** – 'Upper shoreface'	Sp, Sx, Sscs, Shcs, Sr	SMs(r), Ml	Moderately to well sorted, very fine to fine grained sandstones. Medium-scale (dune) cross-stratification and swaley laminae-sets, plus planar stratification, wave and current ripple cross-lamination and hummocky laminae-sets. Sporadic bioturbation by *Skolithos*, *P. montanus*, *A. carbonarius* and escape traces. May be rooted at top.	Typically form upward coarsening/cleaning-units. Parts of the association may occur as single beds. Wave-generated structures dominate, although current-structures occur in the sand-prone beds.
	'Lower shoreface'	Shcs, Sr, Sp, Sb1	Sx, SMs Sg, Sro, SMb1, SMb2, MSs(r)	Moderately to well sorted micaceous very fine to fine grained sandstones/sand-prone heteroliths dominated by hummocky laminae-sets and wave ripple cross-lamination. Occasional synaeresis cracks. Generally weak bioturbation with *P. montanus*, *Skolithos* and *A. carbonarius*, plus escape and excavation traces.	
	'Transition zone'	SMs (r,b,l,hcs), MSs (r,b,l,hcs)	Shcs, Sr, Mr, Ml, Mp, Stb	Mud-prone heterolithic interval with a micaceous quartzose 'sand' component of coarse silt/very fine sand. Dominated by wave ripple and micro-hummocky cross-lamination, although larger scale hummocky cross-stratification occurs. Silt-dominated intervals often show lenticular bedding. Ichnotaxa: *P. montanus* and *A. carbonarius*.	
	Bay-floor muds **(FA 7)**	Ml, Mp, Mr, Mhcs, MSs(l,r)	MSs(r), MSb, Mv, V, CMs(l), Cl	Dark grey-brown micaceous siltstones with thin pinstripes, lenticules and wave-rippled horizons of pale quartz silt/very fine sand. Occasional micro-hummocky lamination. Synaeresis cracks may be common and flooding surfaces may exhibit thin coals/carbonaceous siltstones. Ichnotaxa: *P. montanus*, *A. carbonarius* prominent. Also *Skolithos*, resting and excavation traces.	
	Tidally-influenced distributary channel **(FA 6)**	Sx, Sp, Sm, Sr, Stb, SMs (r,b,p)	SMh, SMb1, SMb2, MSs, Mp, Ml, Mv, Cl, CSs(b), Cp	Broad fining-upward/muddying-upward units. Moderately to locally poorly sorted medium to coarse grained sandstones in the lower part grade up into moderately to well sorted very fine to fine grained sandstones and sand-dominated heteroliths. Lower to middle parts exhibit medium-scale cross-stratification and current ripple cross-lamination; upper parts are rippled heteroliths if preserved. Main ichnotaxa: *P. montanus*, *A. carbonarius*, resting and excavation traces.	

Fig. 14. Facies association scheme of the brackish-water interdistributary bay depositional system within the Åre Formation. (See Figs. 6A for legend and 6B for the lithofacies scheme).

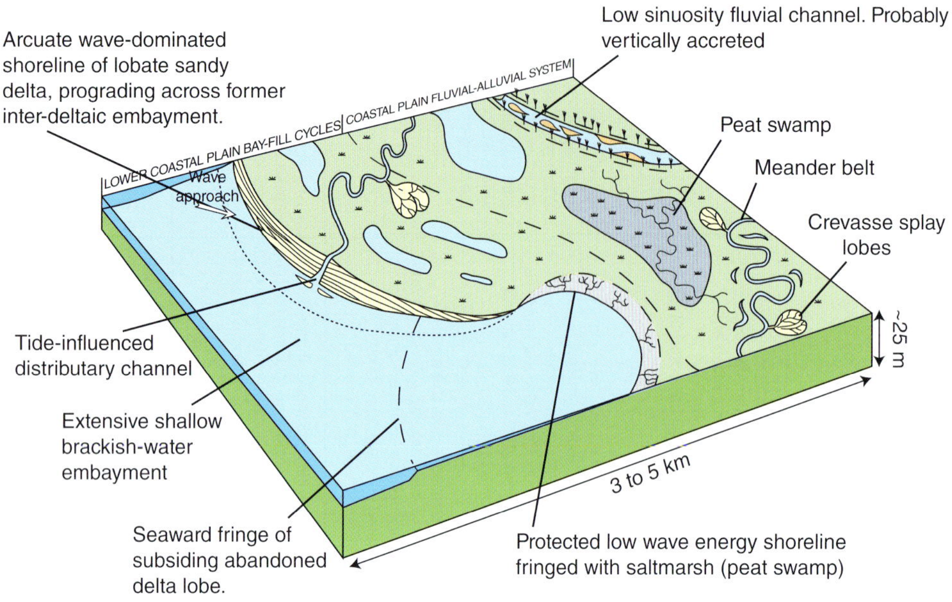

Fig. 15. Conceptual model for the lower delta plain and interdistributary bay depositional setting of the Åre Formation.

Top of bay-fills are locally truncated and succeeded by fining-upward, medium-grained to coarse-grained, cross-stratified distributary channel sandstones (FA 6; Figs 14 and 16). Channel sandstones display possible tidal influence in the form of silty drapes on foresets and troughs within current ripple sets (e.g. Dalrymple & Choi, 2007). Presence of both *Planolites montanus* and *Arenicolites carbonarius* indicates brackish water within distributary channels. The upper boundary of bay-fill deposits may also show rooting and the preservation of coastal plain peat swamp coals (FA1).

Wave-influenced bay-fill deposits and distributary channel sandstones are typically interbedded with organic-rich mudstones characterised by rootlets, wave and current ripples and synaeresis cracks (FA 9; Figs 14 and 16). The general depositional setting for this facies association is interpreted to have been a sheltered, low-energy bay margin or sub-bay (Fig. 15; cf. Kjærefjord, 1999), which was shallow enough to allow colonisation by salt-tolerant vegetation (halophytes).

Bay-fill deposits probably accumulated in muddy, brackish interdistributary embayments, which were locally fringed with wave-reworked prograding lobate deltas alongshore (Fig. 15) (cf. Coleman & Gagliano, 1964; Elliot, 1974).

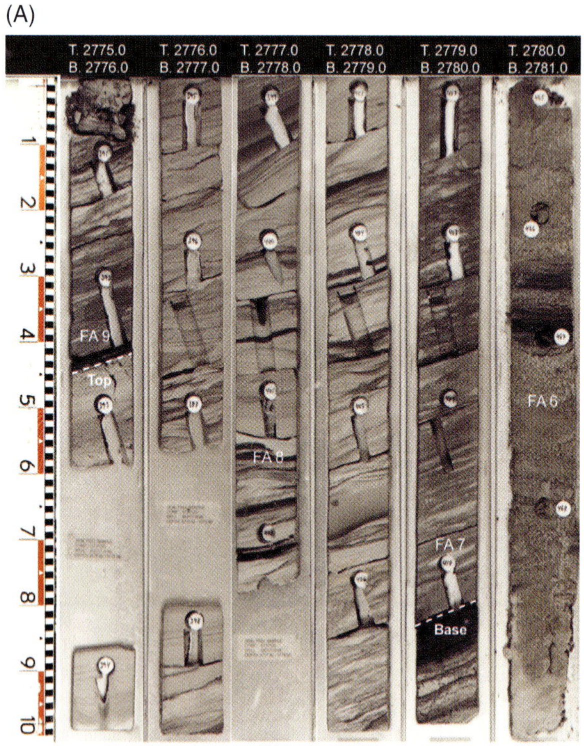

Fig. 16. (A) Core photograph showing typical brackish-water interdistributary bay facies (example from the Åre 4.4 Sub-zone in well 6507/7-A-46). The ruler is 1 m long).

(B)

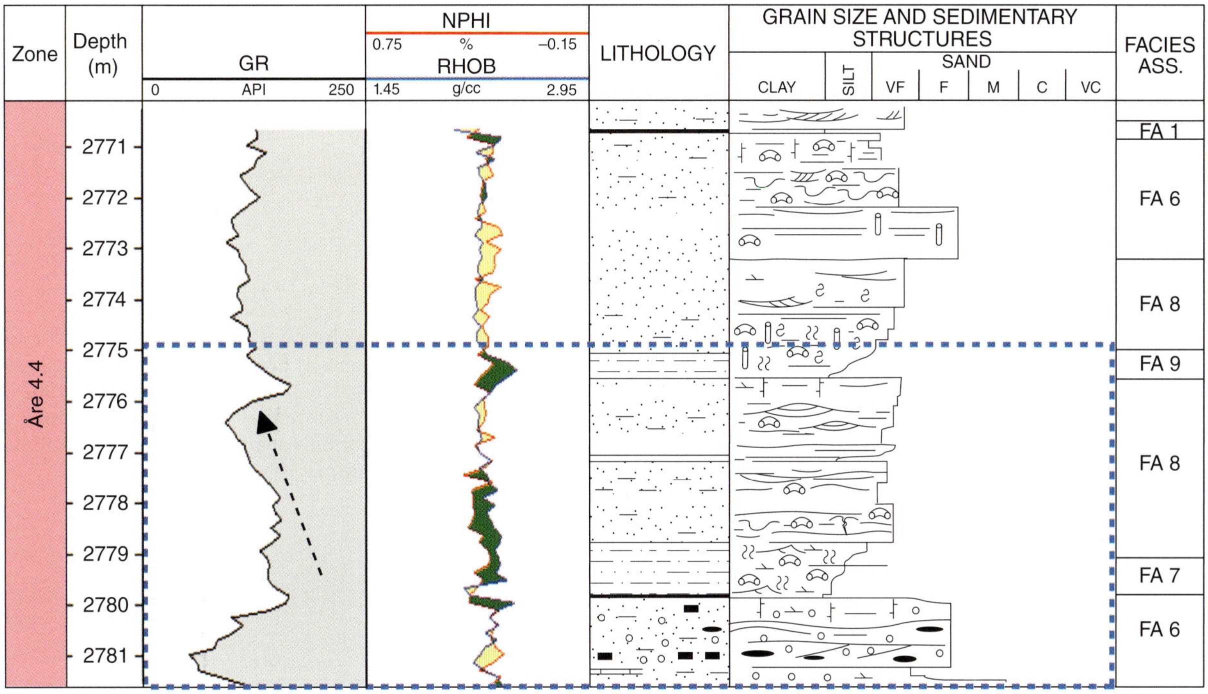

Fig. 16. (*Continued*) (B) Typical log signatures of brackish-water interdistributary bay facies (from well 6507/7-A-46). The dashed blue box corresponds to the cored interval in Fig. 16A. Note the upward-coarsening/cleaning trend of FA 8, reflecting a bay-fill unit (top and base are indicated on the core photo). The base is characterised by carbonaceous bay-floor muds (FA 7) which grade upwards into a heterolithic 'transition zone' and, finally, a sand-dominated wave-reworked upper part (FA 8). The top boundary of this bay-fill unit shows rooting and carbonaceous siltstones corresponding to bay margin muds (FA 9). See Fig. 14 and text for further descriptions of the different facies associations (see Fig. 6A for core description legend and Fig. 4 for abbreviations).

Upward coarsening/cleaning bay-fills thus reflect progradation of sandy delta lobes across the muddy bay floor following autocyclic abandonment and subsidence of adjacent delta lobes along the coast (Fig. 17). Sedimentation in a system of this kind is mainly due to crevasse splay processes, represented either by crevasse channels cutting the levee of distributary channels and forming sheet floods into the adjoining bays (e.g. Mississippi delta; cf. Roberts, 1997; Roberts & Coleman, 1996), or by avulsion of the distributary channel itself into the adjacent bay (Elliot, 1974). A dominance of distributary channel facies relative to crevasse facies may indicate that the latter process have been the most important one during deposition of the Åre reservoir zones 3 and 4. Depending upon location with respect to sediment input points and incident wave activity, the margins of the embayments might comprise clean, wave-dominated sands of shoreface origin in one location and low-energy heteroliths and muddy deposits in another. Despite this complex facies distribution, well-to-well correlations of individual bay-fill cycles can be made laterally in areas with closely spaced wells.

Repeated marine incursions onto the low-relief coastal plain during deposition of Åre 3 and 4 Zones resulted in characteristic brackish water bay-floor mud deposits, typically seen as 0.5 m to 1 m thick beds in the stratigraphy. These mudstones serve to define candidate flooding surfaces bounding reservoir sub-zones. The Åre 3 and 4 Zones are separated by a field-wide flooding surface associated with a singificant shift in the depostional style (MFS2; Fig. 18).

Åre 3 Zone

This reservoir zone (60 m to 70 m thick) marks a transition from a fluvial-dominated setting in the underlying Åre 2 Zone to a system displaying significant brackish-water influence and the first development of shallow-water bays (Fig. 7B).

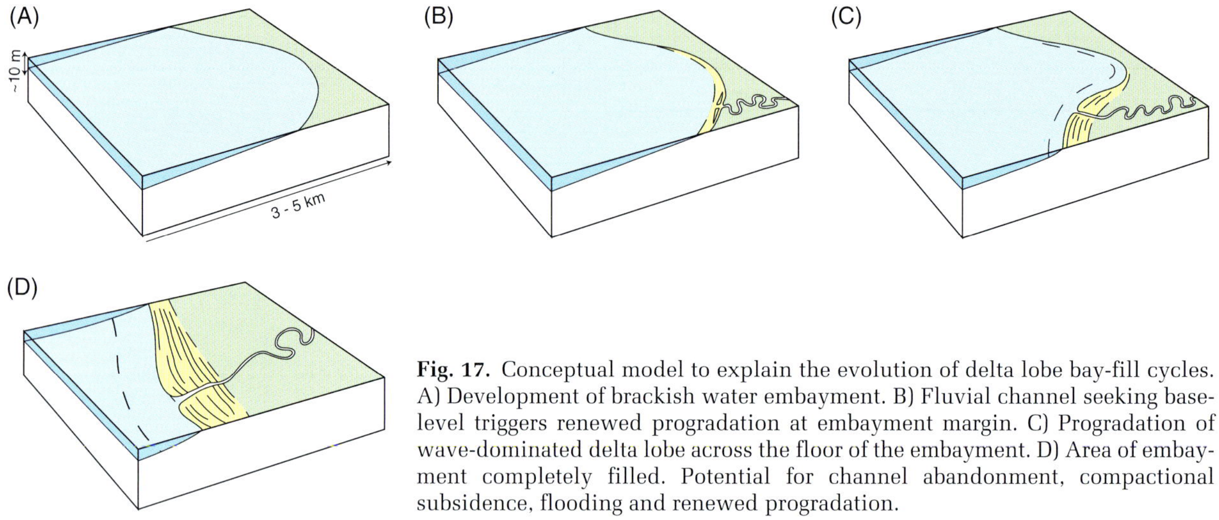

Fig. 17. Conceptual model to explain the evolution of delta lobe bay-fill cycles. A) Development of brackish water embayment. B) Fluvial channel seeking base-level triggers renewed progradation at embayment margin. C) Progradation of wave-dominated delta lobe across the floor of the embayment. D) Area of embayment completely filled. Potential for channel abandonment, compactional subsidence, flooding and renewed progradation.

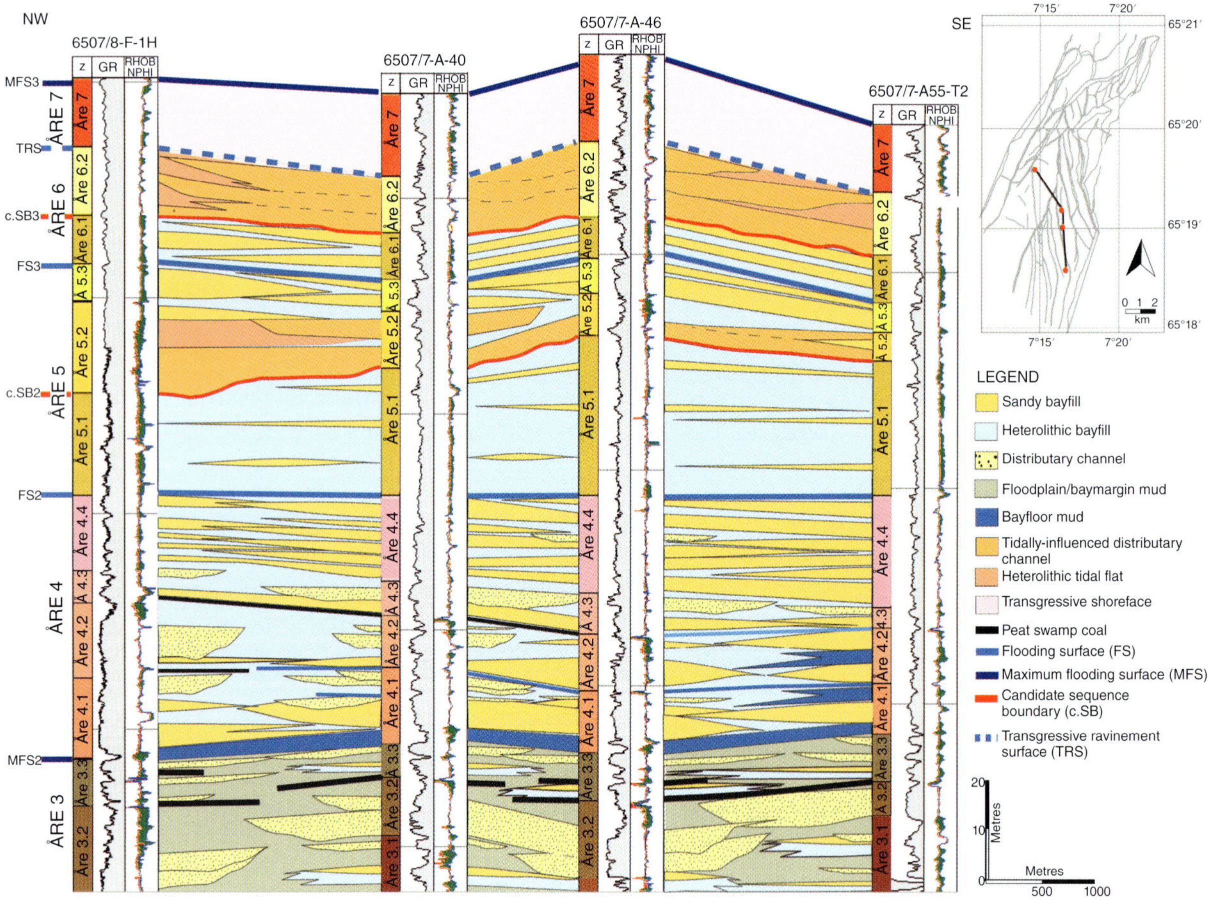

Fig. 18. NW–SE oriented cross-section showing vertical and lateral distribution of facies associations within the Åre 3 to 6 Zones and the interpreted stratigraphic surfaces. Note the down-cutting of the tide-influenced distributary channel deposits at the base of Åre 5.2 to the NW. See Fig. 4 for abbreviations.

The Åre 3 Zone is characterised by rapid lateral and vertical shifts in coastal plain to marginal-marine bay facies belts (cf. Machado, 2009) and the base boundary of the zone is defined by a 1 m to 3.5 m thick mudstone, containing the first evidence of brackish-water conditions, such as synaeresis cracks and trace fossils (e.g. *Arenicolites carbonarius*). In the northernmost part of the field, this marine incursion and associated rise in water table is associated to alternating brackish water mudstones and locally developed peat swamp deposits. The distribution of facies associations indicates that the geometry of the palaeocoastline was probably quite complex during deposition of this reservoir zone. Correlation of facies across the Heidrun Field indicates that the first, rather poorly developed and mud-prone, wave-reworked bay-fill deposits developed initially in the SE, with succeeding subsequent gradual expansion along a SW to NE oriented coastline (the Åre 3.1 Sub-zone; Figs 4 and 7B). There is an overall landward shift of facies belts upwards within the Åre 3 Zone and compared to the Åre 3.1 Sub-zone deposits, the bay-fill units in the uppermost part of the reservoir zone (Åre 3.3 Sub-zone; Fig. 4) are better developed, sandier and more correlateable. In this part of the stratigraphy the bay-fill units can be traced for 2 km to 5 km in a NNE to SSW direction and 1 km to 3 km in a W to E direction.

The top of the Åre 3 reservoir zone is defined at the base of a 2 to 5 m thick interval comprising well-laminated carbonaceous claystone, mudstone and muddy siltstone (Fig. 8). Pervasive calcite cementation is in most cases associated with the uppermost part of this mudstone interval (cf. Hammer *et al.*, 2010). An upward increase in bioturbation (*Arenicolites carbonarious* and *Planolites montanus*) within the mudstones and claystones may suggest increased salinities and more tolerable living conditions in combination with reduced sedimentation rate during deposition. The characteristic log expression of this interval, high GR values and a wide negative separation on the RHOB/NPHI logs together with high RHOB values indicating the calcite cementation, facilitates identification across the field (Figs 8 and 18). This mudstone interval is the most consistent and reliable correlation marker within the Åre Formation and the bay-fill cycle above it (the Åre 4.1 Sub-zone) is the first of these units to have prograded over a significant swathe of the southern Heidrun Field. As such, this surface is considered a maximum flooding event (MFS2; Table 1, Figs 4 and 18). Results from the Heidrun Field megaspore study (in-house) confirm that the uppermost part of the Åre 3 Zone and locally lowermost part of the Åre 4 Zone corresponds to the first occurrence of mesofossils from the *Kuqaia quadrata* Zone. This biozone can be correlated regionally across the Halten Terrace (Morris *et al.*, 2009) and, according to Morris *et al.* (2009), it calibrates with the first occurrence of abundant *Cerebropollenites thiergartii* (palynological bioevent) which defines the base of the Sinemurian stage (Fig. 5; Pedersen *et al.*, 1989). Although, it is difficult to pin the base of the megaspore biozone precisely to the top of the lithostratigraphically defined Åre 3 reservoir zone, the base Sinemurian megaspore assemblage seems to broadly correspond to the MFS2 mudstone interval.

Åre 4 Zone

Although the deposits of the Åre 4 Zone (50 m to 75 m thick) display a similar composition of facies as the Åre 3 Zone, it is distinguished by an increase in brackish water bay-fill relative to lower coastal plain fluvial/alluvial deposits (Figs 4 and 18). Wave-reworked bay-fill deposits form packages of 4 m to 5 m thick, stacked upward-cleaning heteroliths and clean sandstones. Strong wave-influence is indicated by the presence of hummocky and swaley cross-stratification. Facies correlations indicate that these thicker sandy units are continuous for at least 6 km to 10 km in NE to SW directions. Compared with the Åre 3 Zone, the deposits of the Åre 4 reservoir zone contain larger and better organised bay-fill successions. This, combined with less frequent distributary channel deposits, suggests that larger interdistributary bay systems had developed, with increased wave reworking of the delta front areas (Fig. 7C).

Repeated marine incursions during deposition of the Åre 4 Zone have produced a number of candidate flooding surfaces that can be correlated over wide areas (Fig. 4). Hence, several sub-zones have been identified: Åre 4.1 to 4.4. These surfaces most likely represent autocyclic shifts within the interdistributary bay system where relatively low gradients of the coastal plain combined with shallow water depths may have enhanced the extent of apparent palaeocoastline migrations.

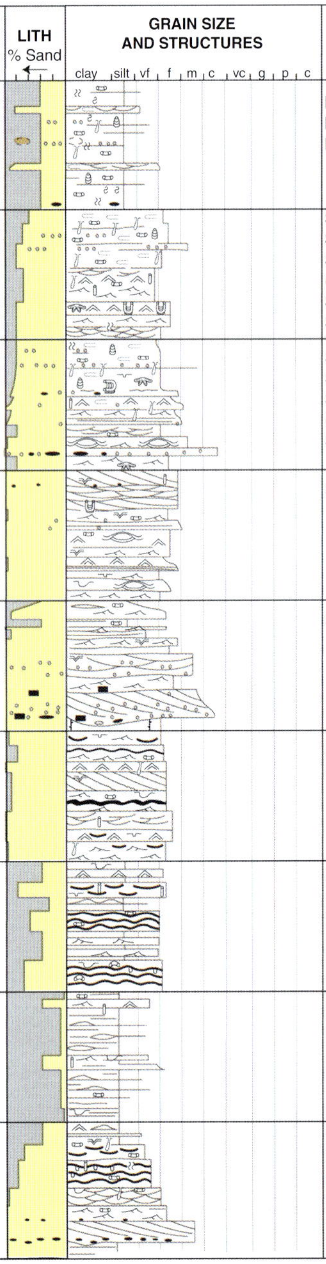

GROSS ENVIRON.	SUBENV. (FACIES ASSOCIATION)	LITHOFACIES DOMINANT	LITHOFACIES SUBORDINATE	DESCRIPTION
TRANSGRESSIVE SHOREFACE	Distal transition zone **(FA 18)**	MSb, MSs(b), Mb	SMb2	Dark grey sandy siltstones with local cm-scale event beds of moderately to well sorted very fine to fine sandstone. The sandstone may be rippled or highly bioturbated. Occasional coarse grained lag deposits, sometimes with sideritised mudstone clasts. *Planolites*, *Teichichnus*, *Phycosiphon*, *Palaeophycus* and bivalve burrows.
TRANSGRESSIVE SHOREFACE	Proximal transition zone **(FA 17)**	Sx, Sr, Shcs, SMs (b,l,p,r), SMb1, SMb2	Sg, Sp, MSs(p), MSs(l), MSb	Moderate to well sorted v.f. to f.-grained silty sandstones with wave- and current-generated ripple scale structures. Highly bioturbated versions which may be poorly sorted where lag deposits have been biogenically reworked. *Planolites* and *Palaeophycus* are the main ichnotaxa, *Asterosoma*, *Teichichnus*, *Skolithos*, bivalve burrows and *Diplocraterion* are also present.
TRANSGRESSIVE SHOREFACE	Lower shoreface **(FA 16)**	Sx, Sp, Sr, Stb, SMs (b,l,p,r), SMb1	Sm, MSs(r)	Moderate to well sorted silty v.f. to f.-grained sandstones. Small- to medium-scale cross-stratification generated by wave activity. Occasional event beds. Locally high levels of bioturbation, with *Palaeophycus* and bivalve burrows prominent, but also including *Asterosoma*, *Rhizocorallium* and *Skolithos*.
TRANSGRESSIVE SHOREFACE	Upper shoreface **(FA 15)**	Sx, Sr	Sp, SMs(r)	Moderate to well sorted f.-grained sandstones exhibiting medium scale cross-stratification, hummocky/swaley laminae-sets and wave and current ripple cross-lamination. Sporadically bioturbated, with occasional resting, escape and excavation traces, plus *Planolites*, *Diplocraterion* and *Skolithos*.
ESTUARINE-FILL	Tide-influenced distributary channel **(FA 14)**	Sx, Sr	SMs(r), MSs(r)	Fining-upward profile with moderately to poorly sorted, locally bimodally sorted, m. to c.-grained sandstones. Lags of mudstone clasts, quartz and coaly debris are present. Higher beds are moderately to well sorted, finer grained and locally heterolithic. They mainly show current ripple cross-lamination, along with sporadic bioturbation in the form of *Planolites* and escape traces.
ESTUARINE-FILL	Sandflat (subtidal) **(FA 13)**	Sx, Sr, SMs(r,b)	Stb, Sp	Moderately to well sorted fine grained sandstones. Micaceous, with up to around 20% mud. Current and wave ripple cross-lamination. Occasional bipolar bedding and paired mud drapes. Generally weak bioturbation with *Planolites*, excavation, resting and escape traces, plus occasional bivalve burrows and *Skolithos*.
ESTUARINE-FILL	'Mixed flat' (subtidal) **(FA 12)**	SMs (b,l,r), MSs(b,l)	Sp, Sx, Ml, Mr	Heterolithic sandstone-mudstone deposits containing ~30-80% moderately to well sorted v.f to f. sandstone. Micaceous and showing an intricate fabric of wave- and current-generated structures. Prominent paired mud drapes and local bipolar foresets. Generally weak bioturbation with *Planolites P.montanus*, *Arenicolites carbonarius*, *Skolithos*, excavation and resting traces.
ESTUARINE-FILL	Mudflat (subtidal) **(FA 11)**	Ml, Mp, MSs(l)	SMs(r)	Dark grey micaceous siltstones that are well laminated, with pin-stripes and lenticules of coarse silt/very fine sandstone. Both wave- and current-generated small-scale structures. Prominent fine carbonaceous debris. Weak bioturbation with mainly *Planolites*, *Skolithos* and excavation traces.
ESTUARINE-FILL	Tidal channel **(FA 10)**	Sx, SMs(r,b), MSs(r,b), Stb, Ml	MSs(p), Ml, MSb, SMb2, SMped, Mped, Mv, V	Moderately sorted m.-grained sandstones pass upwards into to v.f. to f.-grained sandstones and heteroliths. Medium-scale cross-stratification grades vertically to current-rippled fabrics which may exhibit paired mud drapes. Upward increase in bioturbation intensity with Planolites dominant, but may also include bivalve burrows and *Skolithos*, plus some *A. carbonarious* and excavation traces.

Fig. 19. Facies association scheme of the estuarine-fill and transgressive shoreface depositional system (see Figs. 6A for legend, and 6B for the lithofacies scheme).

Åre 5 and 6 Zones: Mixed wave-influenced and tide-influenced estuary

The Åre 5 and 6 reservoir zones (70 m to 115 m in combined thickness; Fig. 4) comprise a diverse mix of wave-influenced and tide-influenced deposits including the following facies association types: bay-floor mud (FA 7); wave-influenced bay-fill (FA 8); bay-margin mud (FA 9); tidal channel (FA 10); mudflat (FA 11); mixed flat (FA 12); sandflat (FA 13) and tide-influenced distributary channel (FA 14) (Figs 9 and 19). Typical tide-influenced facies associations recognised within the Åre 5 and 6 reservoir zones are composed of bimodally sorted, medium-grained to coarse-grained, broadly upward-fining sandstone units (1.5 m to 6 m thick) showing frequent single and paired mud drapes and mud clast lags (FA 14; Figs 19 and 20).

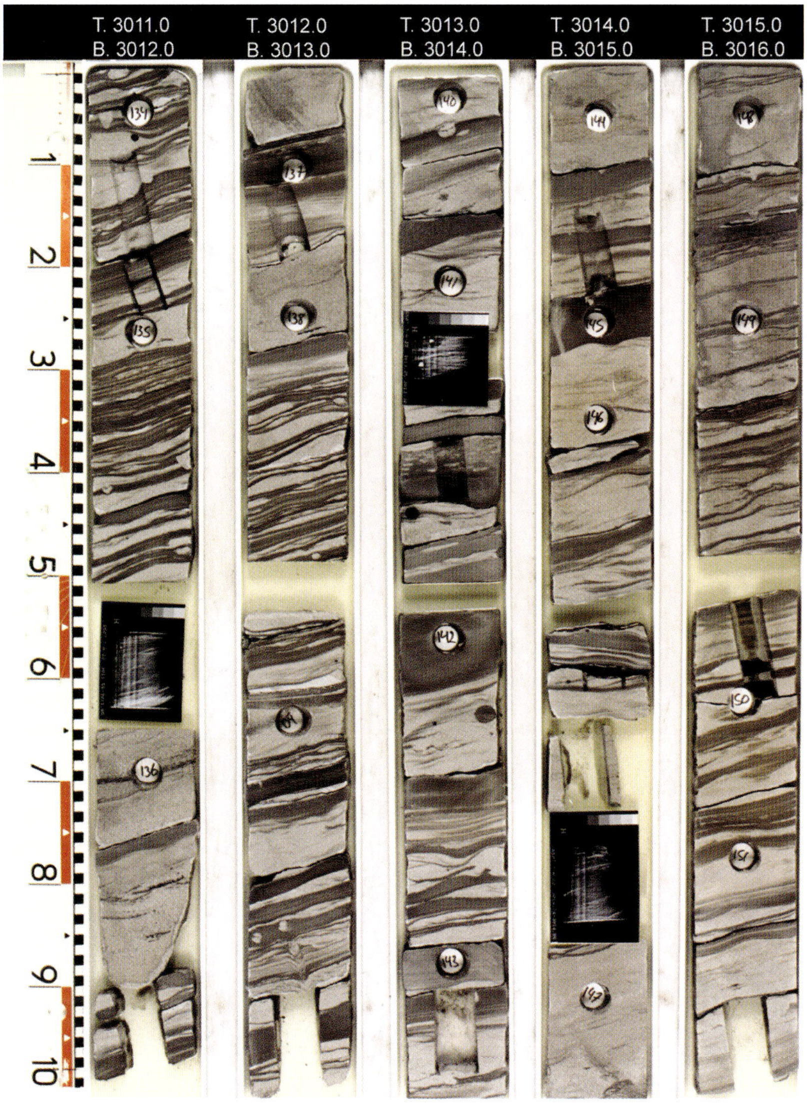

Fig. 20 Core expression of the heterolithic tidal channel facies (FA 10; Fig. 19) characteristic for the Åre 6.2 Sub-zone. The sandstone comprises both wave and current ripple cross-lamination showing single and paired mud drapes. (Example from well 6507/7-A-22. Ruler is 1.0 m long).

Decimetre-scale cross-strata characterise the sandiest deposits, whereas current ripple cross-lamination, flaser and lenticular bedding is present in the muddier and more heterolithic parts. Various sandy to mud-prone heterolithic facies occur, characterised by mixed wave-generated and current-generated ripple cross lamination, bipolar-oriented ripples and paired mud drapes (FA 10, 11, 12 and 13; Fig. 19). Bioturbation intensity is variable; the sandier beds tend to be characterised by occasional *Planolites* burrows, while heterolithic units display a larger range of ichnofauna and increase in bioturbation intensity. Burrow-mottling, excavation traces, occasional *Skolithos*, *Chondrites*, *Phycosiphon*, *Arenicolites* *carbonarius* and bivalve burrows are also observed.

Frequent abrupt changes in sand-silt ratios and grain-size together with paired silt and clay drapes indicate fluctuating energy conditions characteristic of tidal processes. Fining-upward cross-stratified units are interpreted to result from the migration of tidal barforms within tide-influenced distributary channels (Fig. 21). The more heterolithic associations (FA 11, 12 and 13) are inferred to have formed in sub-environments where flow was less confined, i.e. areas with decelerating current velocities in a localised zone of low energy, for example in subtidal flats between the tidal bars within the distributary

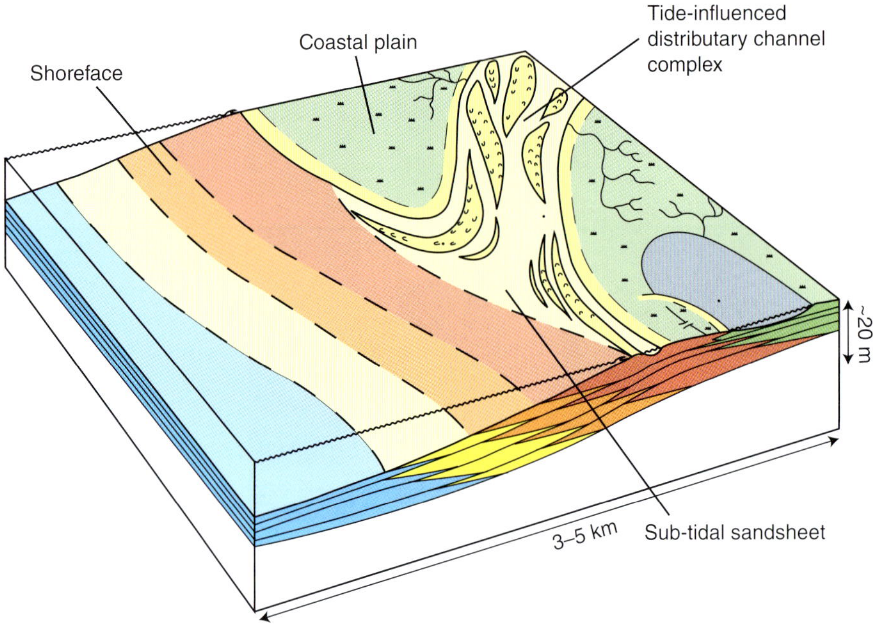

Fig. 21. Conceptual model for the estuarine to transgressive shoreface depositional setting of the Åre Formation.

channels. The tidal deposits alternate with and grade laterally into marginal-marine interdistributary bay deposits (FA 7, 8 and 9). The spatial relationship between the wave-influenced and tide-influenced deposits is locally challenging to resolve but the two sub-environments seem to have been coeval during deposition of some parts of the succession; suggesting that the wide mouths of the tide-influenced channels were closely associated with wave-influenced bay-margin deltas and more muddy sheltered bay areas (Figs 7D and 21).

Åre 5 Zone

The Åre 5 Zone (50 m to 65 m thick) is divided into three sub-zones based on field-scale facies changes and correlateable surfaces (Figs 4 and 18). At the base, the Åre 5 reservoir zone is separated from the underlying Åre 4 Zone by a field-wide mudstone (0.5 m to 3 m thick) and an associated change in facies from sand-dominated bay-fill deposits below into extensive muddy bay deposits above (Åre 5.1 Sub-zone; Fig. 18). This lowermost sub-zone (Åre 5.1; 20 m to 40 m thick) is composed of heterolithic to muddy bay-fill facies displaying crude upward coarsening cleaning cycles. Compared to the sandy bay system of the Åre 4 Zone, these facies reflect a more distal expression of the interdistributary bay environment. The base of the Åre 5 Zone is thus ascribed to a flooding surface marking a field-wide transgressive shift to a larger-scale, mud-prone interdistributary bay setting (FS2; Table 1; Figs 4 and 18).

The Åre 5.2 Sub-zone is defined by prominent change in facies into a field-wide belt of tide-influenced distributary channel sandstones (Figs 18 and 22). Thickness mapping shows that the base of this channel belt is associated with significant down-cutting and a ~ 10 m relief into the underlying Åre 5.1 Sub-zone has been identified in the north-western part of the Heidrun Field (Figs 18 and 23). This erosional surface is characterised with a basal lag of sub-angular to rounded intraformational mud clasts overlain by laterally amalgamated tide-influenced channel and subtidal flat facies filling in the lowermost part of the incision (Fig. 22). Amalgamated channel deposits can be traced for ~ 7 km in E to W direction and 3km to 5 km in an N to S direction. In the south-eastern part of the field, the tide-influenced channel facies grade laterally into wave-influenced sandy bay-fill deposits (Fig. 18). Vertically, there is a field-wide facies change from tide-influenced distributary channel deposits succeeded by an interval displaying a complex mixture of tide-influenced and wave-reworked

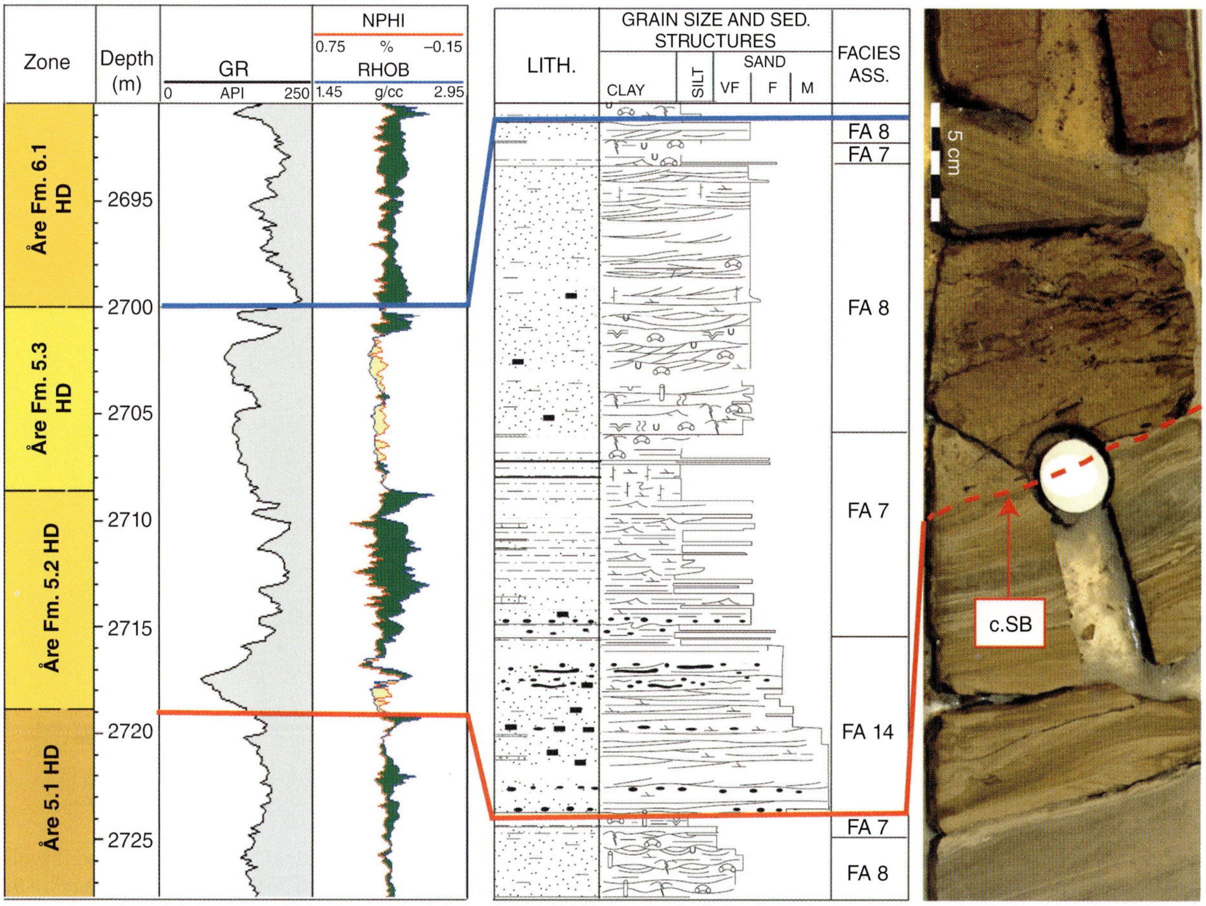

Fig. 22. Example of wireline log signature and core description from sub-zones Åre 5.2 and 5.3 in well 6507/7-A-46. The top of Åre 5.1 Sub-zone is cut by a tide-influenced distributary channel. This erosive surface is characterised by a lag of mud-clasts, and is interpreted as a sequence boundary (c. SB on core photograph). (See Fig. 19 and text for further description of the facies associations, Fig. 6A for the core description legend and Fig. 4 for abbreviations.)

bay facies and eventually a gradual dominance of distal bay deposits (upper part of Åre 5.2 Sub-zone and Åre 5.3 Sub-zone).

Based on its incisive nature and associated basinward shift in facies, base of the Åre 5.2 Sub-zone channel belt is interpreted as a candidate sequence boundary recording a pronounced fall in relative sea-level (c.SB2; Table 1, Figs 4 and 18). The succeeding infill, displaying a gradual landward change in facies, resemble the transgressive evolution of the distal reaches of a wave-dominated estuary, in keeping with the facies descriptions proposed by Dalrymple *et al.* (1992). This facies model describes an estuary with a bay-head delta in the landward end. However, based on the available data, the presence of an estuarine bay-head delta setting cannot be demonstrated conclusively for the deposits within the Åre 5.2 and 5.3 Sub-zones.

The Åre 5.2 Sub-zone and the succeeding Åre 5.3 interval comprise several criteria which are characteristic to incised valley estuaries (cf. Boyd *et al.*, 2006) such as: 1) a basal sequence boundary with erosional relief; 2) basinward shift in facies and 3) transgressive infill. However, there is, based on the facies recognised in the current data, a lack of evidence for lateral interfluves and/or fluvial lowstand deposits within the valley, which are important, but not essential, criteria for identification of incised valley estuaries.

An alternative interpretation of this down-cutting tide-influenced channel belt is that its basal surface is formed by increased tidal energy and resultant scouring due to basin reconfiguration. Similar channelised incisions formed by tidal

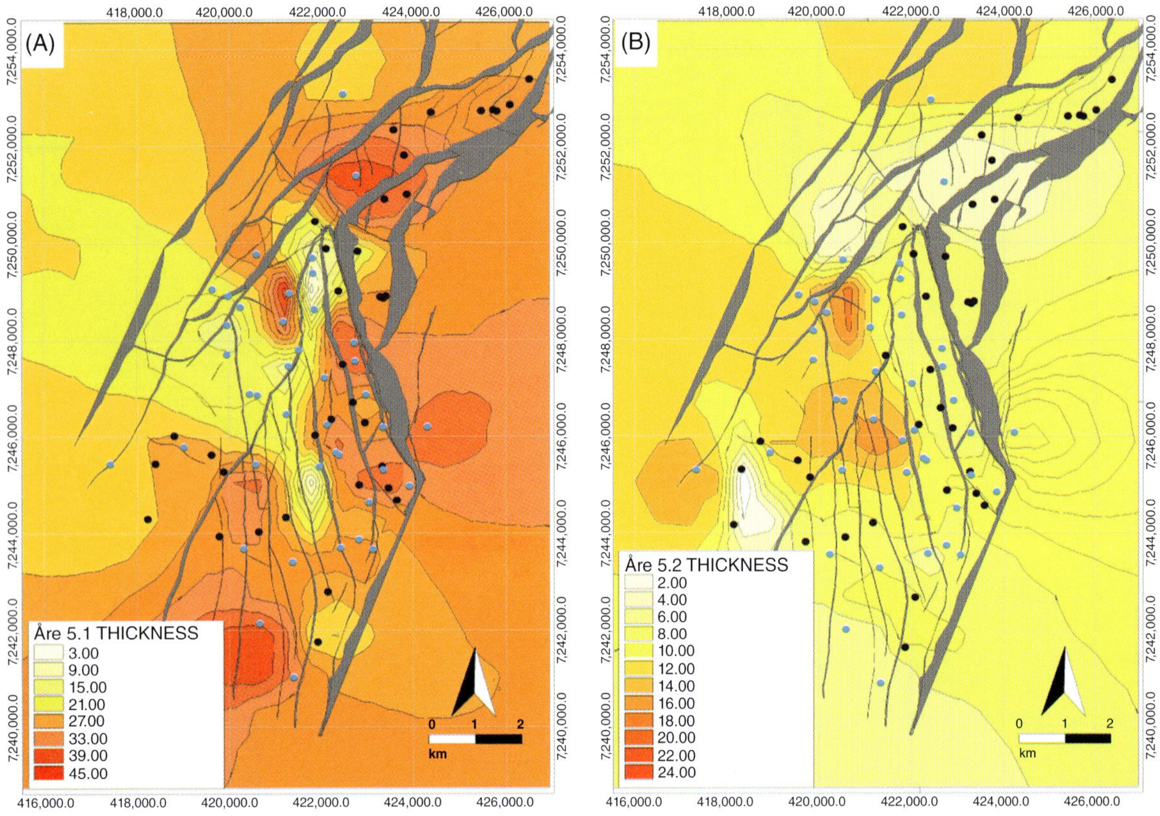

Fig. 23. Thickness maps of the 5.1 (A) and 5.2 (B) Sub-zones (thickness indicated in metres). Blue dots represent wells with data on which the maps are based. The Åre 5.1 Sub-zone shows a thinning towards the NW and the overlying Åre 5.2 Sub-zone displays a corresponding thickening in this area. This trend is related to the down-cutting of tide-influenced distributary channels at the base of the Åre 5.2 Sub-zone.

resonance are described from the Cretaceous Sego Sandstone Member, central Utah, by Willis & Gabel (2001; 2003). However, this interpretation does not fully explain the transgressive facies development succeeding the incision.

Åre 6 Zone

Coeval existence of estuarine facies and wave-influenced bay deposits also characterises the Åre 6 Zone (20 m to 35 m thick). The base of this reservoir zone is characterised by an interval of bay-floor mudstone and heteroliths (0.5 m to 2 m thick) overlying sand-rich deposits of the Åre 5.3 Sub-zone. This surface can be correlated across the field and is interpreted as a flooding event (FS3; Table 11, Figs 4 and 18).

The Åre 6.1 Sub-zone is dominated by mud-prone, wave-influenced bay-fill deposits, with a minor component of sandy tide-influenced channels and tidal flats (Figs 4 and 18). A significant shift into tide-dominated facies marks the transition into the overlying Åre 6.2 Sub-zone. This interval comprises a laterally persistent belt of heterolithic tide-dominated, channel sandbodies (Fig. 20) and subtidal flats, with subordinate brackish-water bay-fill deposits (<10% of the reservoir sub-zone). The onset of channelling in the Åre 6.2 Sub-zone can be identified in a majority of wells, although wireline log-based facies correlations within this reservoir zone are generally uncertain due to the non-characteristic log response of these heterolithic facies associations. Nonetheless, integration with production data confirms that the tide-dominated heterolithic channel bodies are laterally connected for considerable distances (>2 km).

The field-wide channel belts of the Åre 6.2 Sub-zone show strong similarities to the tidal deposits filling-in the incision at the base of the 5.2 Sub-zone, except from two characteristics: 1) the estuarine facies display stronger influence of tide processes; and 2) although the base boundary is sharp, it appears to have low relief (Fig. 18).

The facies change from muddy bay deposits in Åre 6.1 to tide-dominated channelling in Åre 6.2 Sub-Zone is interpreted as a basinward shift in deposition and the base boundary may thus be ascribed to another candidate sequence boundary (c.SB3; Table 1, Fig. 18). However, the low relief characteristics of the surface add some uncertainties to this interpretation. Alternatively, the field-wide belt of tidal channel deposits may be interpreted as a progradational phase of meandering tidal channels containing inclined heteroliths strata as a part of the late stage of estuary filling (e.g. Dalrymple *et al.*, 1992). The multiple cycles of relative sea-level rise and fall identified within the Åre 5 and 6 Zones are proposed to represent compound estuarine valley fill, similar to those described from the overlying Tilje Formation on the Heidrun Field (Martinius *et al.*, 2001).

Åre 7 Zone: Transgressive shoreface

The Åre 7 Zone (15m to 20m) comprises a heterolithic, but nevertheless sandstone-dominated succession (Fig. 4). The deposits display a combination of wave-dominated and current-dominated sedimentary structures and have been divided into upper shoreface (FA 15), lower shoreface (FA 16), proximal transition zone (FA 17) and distal transition zone facies associations (FA 18; Fig. 19). Bioturbation intensity varies. High bioturbation intensity is often associated with moderate to high ichnodiversity and trace fossil assemblages displaying strong marine affinities: *Planolites, Palaeophycus, Teichichnus, Asterosoma, Chondrites, Diplocraterion*, bivalve burrows, plus subordinate *Rhizocorallium* and *Skolithos*. Where bioturbation is less abundant, current ripples, wave ripples, combined flow ripples and both small-scale and larger-scale hummocky cross-stratification are observed in close juxtaposition. Although the sedimentary structures are dominantly formed by wave-generated processes, silt drapes associated with current-ripples suggest that tidal currents may have occurred periodically. Local poorly sorted sandy interbeds within this package of shoreface-dominated facies associations reflect the reworking of the underlying Åre 6.2 Sub-zone deposits.

The basal surface of the Åre 7 Zone erodes the underlying estuarine deposits (Åre 6.2 Sub-zone) and is locally overlain by a lag of reworked pebble-sized intraformational clasts and granular quartz. This lag is commonly associated with a bed of upper shoreface sandstone (FA 15) showing calcite cementation (Fig. 24). The entire Åre 7 Zone contains extensive calcite cemented horizons that results in this interval being a mapable seismic marker across the Heidrun Field. The change in ichnological assemblage seen across the Åre 6 and 7 Zone boundary, from low-diversity, brackish water ichnotaxa to a relatively diverse assemblage indicative of typical marine palaeosalinities, confirms that an important change took place in the depositional environment. Significantly higher palaeosalinity than in the underlying Åre 6.2 Sub-zone is also confirmed by an influx of dinoflagellate cysts and acritarchs, such as prasinophycean algae. Based on these characteristics, the erosive surface marking the base of the Åre 7 Zone is identified as a transgressive ravinement surface (TRS; Table 1, Figs 18 and 24).

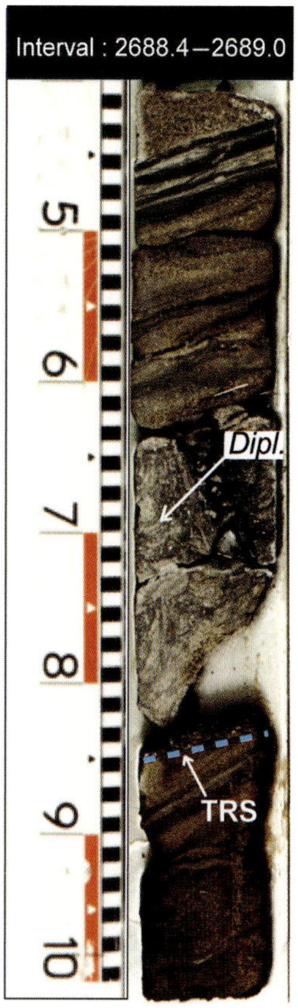

Fig. 24. Core expression (6507/7-A-46) of the transgressive ravinement surface (TRS) separating the Åre 6 and 7 Zones. The surface (blue, dashed line) is overlain by a granule lag and calcite cemented sandstone. A *Diplocraterion* (*Dipl.*) trace is indicated 15 cm above the TRS. (Red bars on the ruler are 10 cm long).

The top surface of the Åre 7 Zone is defined as the base of a calcite cemented mudstone interval (1 m to 1.5 m thick) characterised by intense bioturbation and marking a facies change into prograding storm-influenced prodelta facies of the Tilje Formation (cf. Martinius *et al.*, 2001). This surface is interpreted as a condensed maximum flooding surface (MFS3; Table 1 and Fig. 4; cf. Martinius *et al.*, 2001) Biostratigraphic correlation confirms that this MFS is associated with a regional shift in the composition of megaspore/mesofossil assemblages and is attributed to the top Early Pliensbachian age, defining the onset of Tilje Formation (Morris *et al.*, 2009; Pedersen *et al.*, 1989).

STRATIGRAPHIC SYNTHESIS

The stratigraphic framework presented here is based upon a standard procedure of stratigraphic analysis in terms of recognising genetically linked packages of strata bounded by key surfaces (e.g. Van Wagoner *et al.*, 1990). On a field-scale, the sequence stratigraphic scheme appears to be robust; however, the chronostratigraphic significance of many of the marker horizons remains uncertain. Autocyclic processes have played a major role in the development of the Åre Formation stratigraphy, forming surfaces that look as if they could have sequence stratigraphic significance. Frequent autocyclic channel and delta lobe switching in the lower delta plain and interdistributary bay depositional setting, in addition to local subsidence variations (cf. Hammer, 2010) and related rapid shifts in relative sea-level along the low-gradient shoreline, may have produced local regressions and transgressions (e.g. Demko & Gastaldo, 1996; Read, 1995). The exact nature of the marine incursions and erosional surfaces, whether they are of autocyclic origin or represent regional events, has in many cases not been possible to determine. This has been hampered by the fact that few published sequence stratigraphic studies of the Åre Formation exist. An on-going study of the Åre Formation from the Åsgard Field, 30 km SE of Heidrun, together with in-house data from the Norne and Urd Fields (~90 km NNE of Heidrun) and the Njord Field (~110 km south of Heidrun; Fig. 1A), confirms a regional extent for several of the key marker horizons and the development of similar stratigraphic stacking patterns. The most reliable sequence stratigraphic surfaces within the formation, which are associated with bio-chronostratigraphic events and correlateable across the Halten Terrace Basin, are the maximum flooding surfaces: Top Åre 1 MFS1; Top Åre 3 MFS2 and Top Åre 7 MFS3 (Table 1). The Top Åre 7 MFS3 and the Top Åre 1 Coal Marker MFS1 represent important regional seismic horizons and the latter is the only sequence stratigraphic surface which has been recognised with confidence within the non-marine part of the succession.

Tectonic controls on stratigraphic development

The overall depositional trend within the Åre Formation is clearly transgressive, passing upwards from fluvial to brackish water bay-fills, marginal marine and ultimately open marine deposits (Martinius *et al.*, 2001; Svela, 2001). This trend also coincides with the regional tectono-eustatic development of the Early Jurassic North Atlantic seaway (first-order framework; Hallam, 1988; Hallam, 2001; Haq *et al.*, 1987; Surlyk, 1990).

The overall transgressive character of the Åre Formation reflects regional tectonic subsidence and long-term sea-level rise. Sequence development within the succession is attributed tentatively to second order cyclicity. The regionally correlateable maximum flooding surfaces (MFS1, MFS2 and MFS3; Table 1), related to relative sea-level rise, were probably the result of periods of more rapid tectonic subsidence within the Halten Terrace Basin. According to biostratigraphic dating and correlations with the time-scale of Gradstein *et al.* (2004) the time span between the maximum flooding surfaces; MFS1 (top Åre 1 Coal Marker; middle Hettangian); MFS2 (top Åre 3; lowermost Sinemurian); and MFS3 (top Åre 7; top of early Pliensbachian), could possibly imply sequences of 4 +/−1 Ma to 3 Ma duration. Other studies of Jurassic stratigraphic cyclicity within the Atlantic region (cf. Hallam, 2001) have identified similar second order cycles interpreted to result from sea-level change. Regional early Sinemurian and early Pliensbachian transgressions documented by Hallam (2001) are attributed to plate tectonic processes, such as the opening of rift basins. Glacio-eustatic influence on Jurassic sea-level variations are considered unlikely due to lack of evidence for significant quantities of polar ice during Jurassic time (Hallam, 1988; Hallam, 2001).

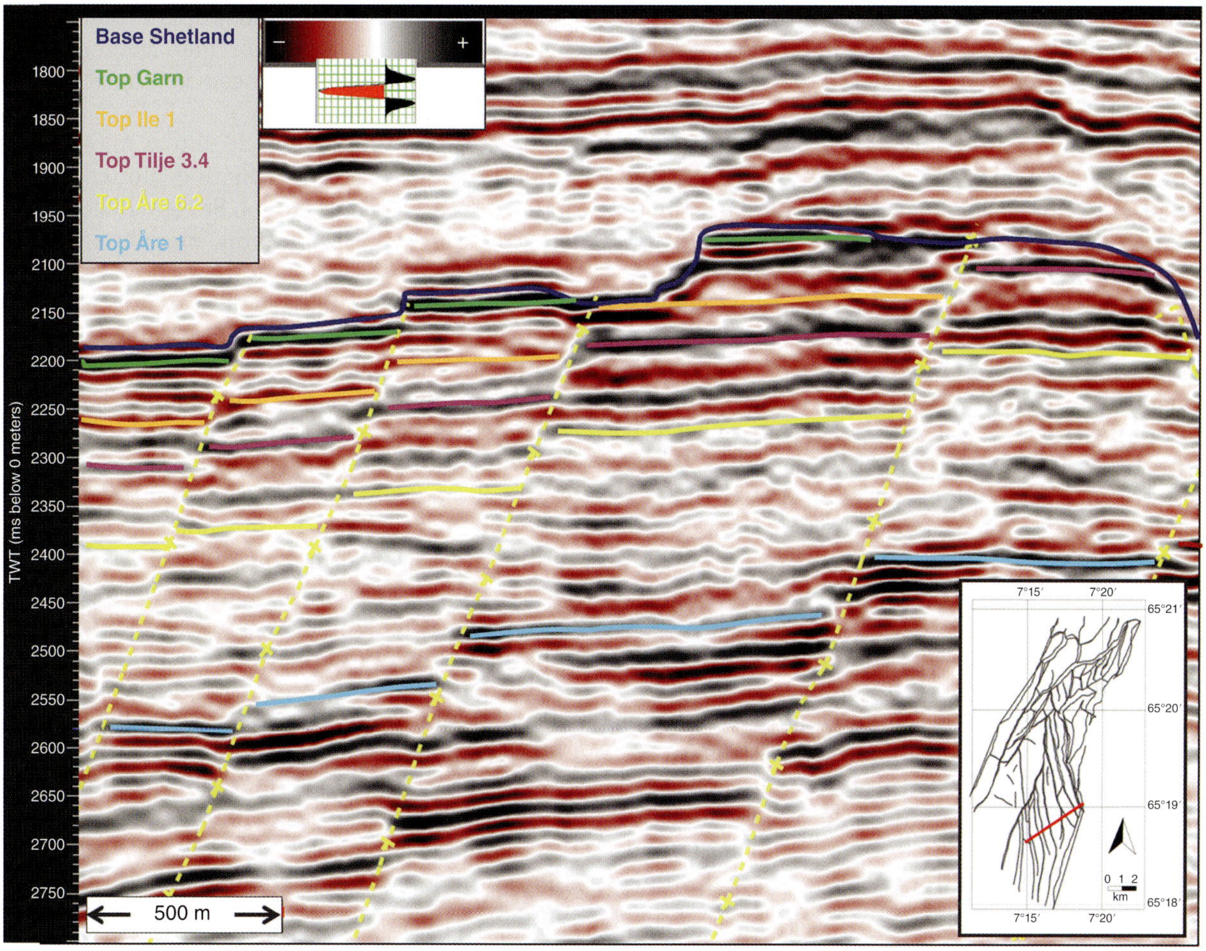

Fig. 25. Seismic section from the southern part of the Heidrun Field. Yellow dashed lines indicate segment bounding faults. Field wide seismic horizons are interpreted at the top of the Åre 1 Zone (Top Åre 1) and at the top of the Åre 6 Zone (Top Are 6.2). Vertical scale is given in ms TWT (milliseconds Two Way Travel-time).

Deposition of the Åre Formation within the Halten Terrace region took place during a rift initiation phase characterised by local faulting and associated variable subsidence (Marsh *et al.*, 2010). These faults are likely to have had a significant control on accommodation space and sediment supply within the basin, as indicated by a seismic study from the Åsgard Field which shows internal thickness variation within the Åre Formation (Marsh *et al.*, 2010). Here, the synsedimentary tectonic control is indicated by, for example, wedge-shaped and thicker stratigraphic intervals in the hangingwalls of active faults relative to the footwall (Marsh *et al.*, 2010). Seismic data from the Heidrun Field, on the other hand, do not show any convincing evidence of thickness variation within the Åre Formation across faults (Fig. 25). This is also confirmed by isochoring of reservoir zones based on well analysis. Thickness maps from main reservoir zones defined by flooding surfaces indicate fairly uniform trends within the Åre Formation across the field. However, significant thickness variations are observed within the reservoir sub-zones which are associated with sequence boundaries (c.SB1, c.SB2 and c.SB3; Table 1). The seismic reflectors within the Åre Formation on Heidrun are approximately 200 m apart (i.e. MFS1 and TRS) and the seismic data quality are insufficient to resolve internal thickness variations on a reservoir sub-zones scale (for example the 10 m relief at the c.SB2 at the Åre 5.2 Sub-zone; Fig. 18). Lowstand fluvial incision and subsequent estuary development during marine transgression is, in most cases, related to eustasy and/or tectonic uplift (Dalrymple *et al.*, 1994). Tectonically induced relative sea-level changes thus cannot be ruled out as a potential control on sequence boundary formation and associated basinward shift in facies as seen within the Åre 5.2 and 6.2 reservoir Subzones. Furthermore, regional change in rift basin geomorphology may have been responsible for amplifying tidal resonance within the basin (e.g. Carr *et al.*, 2003; Yoshida *et al.*, 2007) and causing tidal current dominance in the younger Åre Formation deposits (Åre 5 and 6 Zones). A significant tectonic influence during deposition of the succession is also confirmed by the change in provenance from East Greenland during deposition of the fluvial Åre 1 and 2 Zones to an eastern source for the tide-influenced Åre 5 and 6 Zones (cf. Morton *et al.*, 2009).

INDUSTRIAL APPLICATIONS

The shift of focus from producing from high quality Fangst reservoirs to the low productivity Åre Formation formed one of the main drivers for undertaking a complete update of the Åre reservoir stratigraphy in the Heidrun Field. Lack of robustness in existing stratigraphic models was the other main driver. However, in a large and mature field like Heidrun, the update of the stratigraphic model represents a major investment that affects all subsurface disciplines. In this section, a few examples of industrial applications that motivated this investment are presented.

General impact of a revised stratigraphic model

Within the oil industry, the reservoir zonation is one of the very few elements of reservoir characterisation that is passed on throughout the entire value chain from one discipline to another. It is applied in seismic interpretation, reservoir modelling, production forecast (reservoir simulation), well planning, drilling operations, measures for increased oil recovery and allocation of produced volumes. All disciplines have a close relationship to the field's reservoir zonation and a well-defined stratigraphic model has a significant intrinsic value in all interdisciplinary communication.

In the Heidrun Field's expanding well database, the previous reservoir zonation had become unreliable (Fig. 4). Well-to-well correlations could not be achieved in a consistent manner and maintaining a geologically predictable reservoir zonation became challenging. Both daily operational activities and long term reservoir management decisions were severely influenced by the lack of a reliable stratigraphic model.

The cross-section in Fig. 26 shows an example of the previous reservoir zonation plotted on top of the current updated model. Several distinctions between the two stratigraphic models are illustrated by the well correlations; 1) Some key surfaces were not recognised in the previous reservoir zonation (e.g. the surface cSB2 (Fig. 18) is located within the former Åre 2.8 Sub-zone); 2) Some of the former sub-zones were defined on not-recognisable geological criteria (e.g. inclusion of sub-zones Åre 2.6, Åre 2.7 and partly Åre 2.8 within the current Åre 5.1 Sub-zone) and; 3) the top of the former Åre 2.6 Sub-zone cuts across the key FS-marker at top of the current Åre 4 Zone. These distinctions are primarily caused by lack of documented geological criteria of the stratigraphic framework in the previous model. Several other examples of such distinctions can be inferred from Fig. 4. Inconsistencies, such as those described above, were a major concern and development of a simpler stratigraphic model was a main objective. Despite the underlying complexity in the sedimentology of the Åre Formation, the current update has enabled an intuitive and predictive grouping of facies associations into genetically related packages. The simplicity in this model construction is also easy for non-geologists to adapt.

Impact on reservoir models and calculation of hydrocarbon volumes

Updates of in place hydrocarbon volumes and extractable reserves are primarily dependent on petrophysical evaluation models and the choices made when implementing geological reservoir models into numerical representations. Some of the most important outcomes of the revised stratigraphic framework in the Heidrun Field are improved robustness in volume calculation. This has, in turn, resulted in updates of both hydrocarbon resources and reserves.

Reservoir modelling strategies commonly represent a choice between modelling details in facies distributions and inherent uncertainties in spatial arrangement and more general representations of average reservoir properties disconnected from facies analysis (i.e. data-driven observations of lateral and vertical trends in reservoir quality).

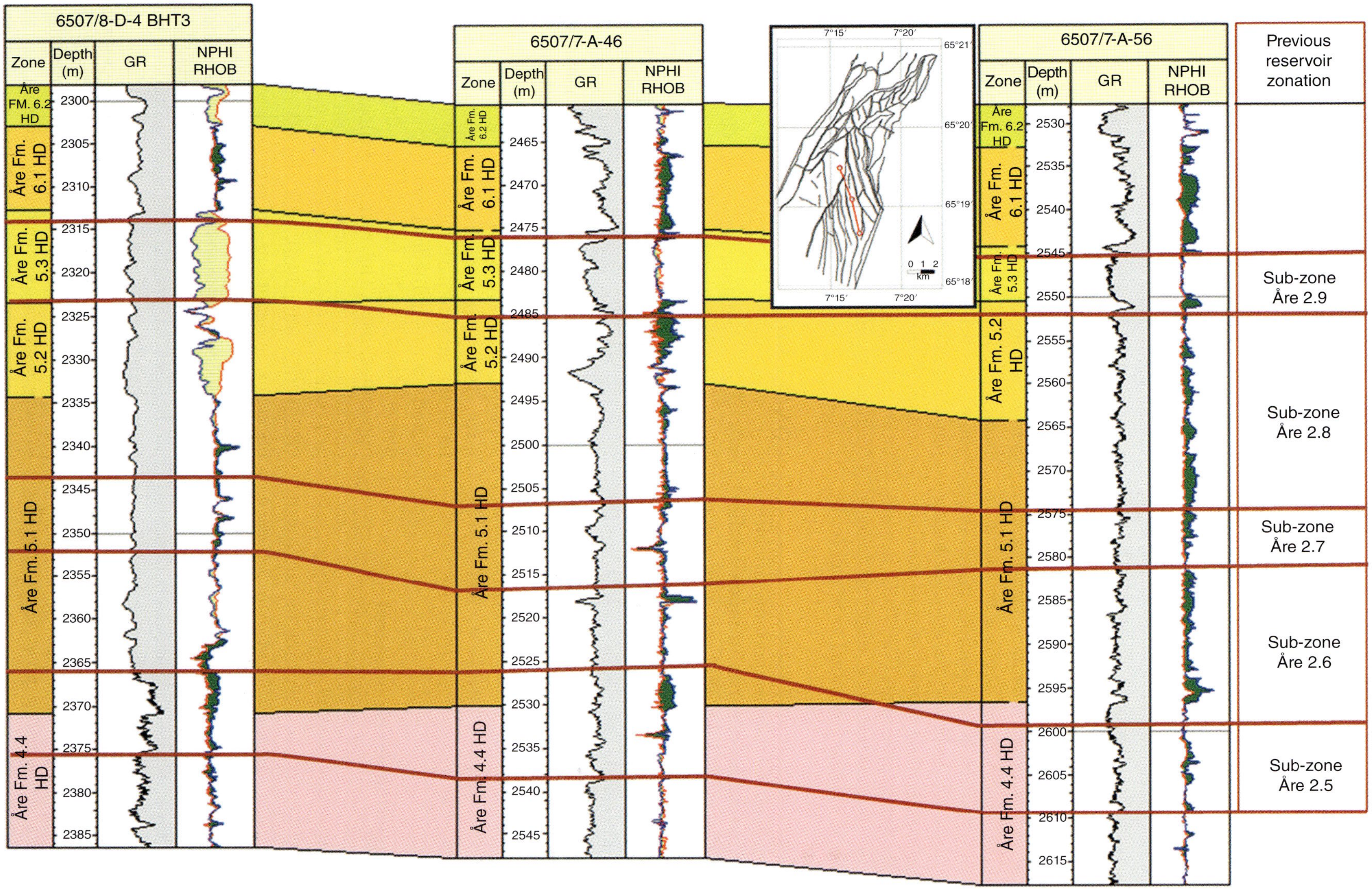

Fig. 26. NNW–SSE correlation panel demonstrating distinctions between the previous and current Åre reservoir zonations. Wells are located approximately 2 km apart. See text for further discussion and Fig. 4 for abbreviations.

Since the Åre Formation spans such a wide variety of depositional systems, the challenges in geological modelling and reservoir forecasting are not similar throughout the succession. In heterogeneous reservoir intervals characterised by facies associations with large contrast in reservoir properties, the challenges related to reservoir modelling are associated commonly with realistic representation of dimensions, distribution and orientation of various facies bodies (Jones *et al.*, 1995). The Åre fluvial successions (Åre Zone 1 and 2) comprise a range of such heterogeneous deposits (see Fig 13). The seismic data from the Heidrun Field cannot resolve individual fluvial channels (see Fig. 25) and a very important model input is provided by the reconstructed palaeogeographic maps (e.g. Fig. 7A). Also, for reservoir zones with higher degree of expected lateral connectivity of sandbodies, the palaeogeographic maps (Fig. 7B/C) in combination with cross-sections (e.g. reservoir zones 4 and 5 in Fig. 18) has been applied as model input parameters defining dimensions and proximal-distal trends of reservoir sandbodies.

In zones where the most important variability in reservoir properties occurs at a much smaller scale than the modelling grid cell dimension, upscaling to effective cell values becomes the main challenge (Kjønsvik *et al.*, 1994; Nordahl & Ringrose, 2008). Examples of such zones in the Heidrun Field include (but are not restricted to) the tidal influenced sediments in the Åre Subzone 5.2 and 6.2. Identifying and constraining these sediments to certain reservoir zones is another significant and important improvement of the current stratigraphic model, as compared to previous versions.

Impact on opportunities for increased oil recovery

Further development of the Heidrun Field is heavily dependent of various measures for increased oil recovery. The recognition of Åre 5.1 as a distinct reservoir zone (see Fig. 18) serves as an illustrative example of how the current updates of the sedimentological and stratigraphic models provide new opportunities for increased oil recovery. Previously, the reservoir sub-zone Åre 5.1, as defined in this study, had not been recognised as a distinct reservoir zone and was not implemented in the reservoir models. The Åre 5.1 Sub-zone is characterised as a low net to gross reservoir interval with low expected production performance. Nevertheless, the sub-zone contains large in place volumes and represents valuable additional reserves. However, development of these volumes is dependent on a robust reservoir characterisation and a reliable forecast of well production. Several cross-sections, similar to the one illustrated in Fig. 18, form important input to the decision processes regarding choice of well solutions and drainage strategies for the Åre 5.1 Sub-zone (e.g. evaluation of long horizontal zone-dedicated producers vs. co-mingled production from several zones). Essential to these decisions are consistent definition of bounding surfaces (i.e. top Åre 5.1 Sub-zone and top Åre 4 Zone) as inclusion of any highly productive intervals in a dedicated horizontal Åre 5.1-well will potentially have a high impact on well performance. Both revision of the stratigraphic concepts and consistent implementation was required in order to exploit this business opportunity.

CONCLUSIONS

The Åre Formation reservoir characterisation study has resulted in an updated and more detailed conceptual understanding of the Rhaetian to Early Pliensbachian-aged depositional systems in the Heidrun area. A total of 18 facies associations are defined and form the basis for a new stratigraphic framework comprising seven main reservoir zones. The Åre Formation is transgressive with non-marine coastal plain deposits at the base, passing into a succession of lower delta plain and interdistributary bay deposits in the middle part, followed by an interval of mixed wave-influenced and tide-influenced deposits and, finally, open marine deposits in the uppermost part. The revised geological characterisation of the Åre Formation permits definition of a framework with correlateable key surfaces. Within this framework, the vertical distribution of facies associations is well-constrained and packaged into genetically related units. Subsequently, well-to-well correlations can be undertaken more confidently and the uncertainty in predictions of spatial facies distribution have been significantly reduced. Results from this study have been applied in updated reservoir modelling strategies and in construction of geological models of the Heidrun Field. In addition to a necessary improvement in decision basis for daily operational activities, these models have also been applied to updates of in place hydrocarbon volumes and extractable reserves. The revised stratigraphic framework has also formed the basis for adding new and previously unidentified opportunities for increased oil recovery from the Åre Formation. In a regional context, the revised Åre stratigraphy developed in the Heidrun Field should also provide a common basis in upcoming sequence stratigraphic studies in the neighbouring fields.

ACKNOWLEDGEMENTS

The Heidrun Licence owners (Statoil ASA, ConocoPhillips Skandinavia AS, Eni Norge AS and Petoro AS) are acknowledged for providing access to necessary data in order to prepare this publication. Furthermore, the Heidrun Field operator, Statoil ASA, is acknowledged for permission to publish the synthesis and analytical presentation of the Åre Formation sedimentology and stratigraphy. The authors also would like to extend the acknowledgements to Allard W. Martinius for initiating the process of a formal presentation of the scientific results from this study and to his continuous effort of following up the process such that the manuscript could be presented. Contributions from three reviewers, Jonathan Wonham, Brian J. Willis and Donatella Mellere, are highly appreciated and have significantly improved the quality of the final manuscript.

REFERENCES

Boyd, R., **Dalrymple, R.W.** and **Zaitlin, B.A.** (2006) Estuary and incised valley facies models. In: *Facies Models Revisited* (Eds H.W. Posamentier and R.G. Walker), *SEPM Spec. Publ.*, **84**, 171–234.

Brekke, H., **Dahlgren, S.**, **Nyland, B.** and **Magnus, C.** (1999) The prospectivity of the Vøring and Møre basins on the Norwegian continental margin. In: *Petroleum Geology of Northwest Europe: Proceedings of the 5th Conference* (Eds A.J. Fleet and S.A.R. Boldy), pp. 261–274. Geological Society London.

Bridge, J.S. (2003) *Rivers and Floodplains – Forms, processes and Sedimentary Record.* Blackwell Publishing, Oxford, 491 pp.

Bridge, J.S. and **Tye, R.S.** (2000) Interpreting the dimensions of ancient fluvial channel bars, channels and channel belts from wireline-logs and cores. *AAPG Bull.*, **84**, 1205–1228.

Carr, I.D., **Gawthorpe, R.L.**, **Jackson, C.A.L.**, **Sharp, I.R.** and **Sadek, A.** (2003) Sedimentology and sequence stratigraphy of early syn-rift tidal sediments: The Nukhul Formation, Suez Rift, Egypt. *J. Sed. Res.*, **73**, 407–420.

Coleman, J.M. and **Gagliano, S.M.** (1964) Cyclic sedimentation in the Mississippi River deltaic plain. *Trans. Gulf Coast Assoc. Geol. Soc.*, **14**, 67–80.

Dalland, A., **Worsley, D.** and **Ofstad, K.** (1988) A lithostratigraphic scheme for the Mesozoic and Cenozoic succession offshore mid- and northern Norway. *Norw. Pet. Direct. Bull.*, **4**, 1–65.

Dalrymple, R.W., **Zaitlin, B.A.** and **Boyd, R.** (1992) Estuarine facies models: conceptual basis and stratigraphic implications. *J. Sed. Petrol.*, **62**, 1130–1146.

Dalrymple, R.W., **Boyd, R.** and **Zaitlin, B.A.** (1994) *Incised-valley Systems: Origin and Sedimentary Sequences*, Tulsa, 399 pp.

Dalrymple, R.W. and **Choi, K.** (2007) Morphologic and facies trends through the fluvial - marine transition in tide-dominated depositional systems: A schematic framework for environmental and sequence stratigraphic interpretation. *Earth-Sci. Rev.*, **81**, 135–174.

Demko, T.M. and **Gastaldo, R.A.** (1996) Eustatic and autocyclic influences on deposition of the lower Pennsylvanian Mary Lee coal zone, Warrior Basin, Alabama. *Int. J. Coal Geol.*, **31**, 3–19.

Doré, A.G. (1991) The structural foundation and evolution of Mesozoic seaways between Europe and the Arctic. *Palaeogeogr. Palaeoclimatol. Palaeoecol.*, **87**, 441–492.

Elliot, T. (1974) Interdistributary bay sequences and their genesis. *Sedimentology*, **21**, 611–622.

Galloway, W.E. (1989) Genetic stratigraphic sequences in basin analysis I: Architecture and genesis of flooding-surface bounded depositional units. *AAPG Bull.*, **73**, 125–142.

Gjelberg, J., **Dreyer, T.**, **Høie, A.**, **Tjelland, T.** and **Lilleng, T.** (1987) Late Triassic to Mid-Jurassic sandbody development on the Barents and Mid-Norwegian shelf. In: *Petroleum Geology of North West Europe* (Eds J. Brooks and K. Glennie), pp. 1105–1129. Graham & Trotman, London.

Gradstein, F.M., **Ogg, J.G.** and **Smith, A.G.** (2004) *A Geologic Time Scale 2004.* Cambridge University Press, 610 pp.

Gradzinski, Baryla, J., **Doktor, M.**, **Gmur, D.**, **Gradzinski, M.**, **Kedzior, A.**, **Paszkowski, M.**, **Soja, R.**, **Zielinski, T.** and **Zurek, S.** (2003) Vegetation-controlled modern anastomosing system of the upper Narew River (NE Poland) and its sediments. *Sed. Geol.*, **157**, 253–276.

Hallam, A. (1988) A reevaluation of Jurassic eustasy in the light of new data and the revised Exxon curve. *Spec. Publ. Soc. Econ. Paleontol. Mineral.*, **42**, 261–273.

Hallam, A. (2001) A review of the broad pattern of Jurassic sea-level changes and their possible causes in the light of current knowledge. *Palaeogeogr. Palaeoclimatol. Palaeoecol.*, **167**, 23–37.

Hammer, E. (2010) *Sedimentological Correlation of Heterogeneous Reservoir Rocks: Effects of Lithology, Differential Compaction and Diagenetic Processes.* PhD thesis, University of Science and Technology, Trondheim, Norway.

Hammer, E., **Mørk, M.B.E.** and **Næss, A.** (2010) Facies controls on the distribution of diagenesis and compaction in fluvial-deltaic deposits. *Mar. Petrol. Geol.*, **27**, 1737–1751.

Haq, B.U., **Hardenbol, J.** and **Vail, P.R.** (1987) Chronology of fluctuating sea-level since the Triassic. *Science*, **235**, 1156–1167.

Jones, A., Doyle, J., Jacobsen, T. and **Kjønsvik, D.** (1995) Which sub-seismic heterogeneities influence waterflood performance? In: *Geol. Soc. Spec. Publ.* (Ed. H.J. de Haan), **84**, pp. 5–18.

Kjærefjord, J.M. (1999) Bayfill successions in the Lower Jurassic Åre Formation, Offshore Norway: Sedimentology and heterogenity based on subsurface data from the Heidrun field and analogue data from the Upper Cretaceous Neslen Formation, Eastern Book Cliffs, Utah. *GCSSEPM Foundation 19th Annual Research Conference*, 149–158.

Kjønsvik, D., Doyle, J., Jacobsen, T. and **Jones, A.** (1994) The effects of sedimentary heterogeneities on production from a shallow marine reservoir; what really matters? In: *European Petroleum Conference*, **1**, pp. 27–40. Soc. Petrol. Eng.

Koch, J. and **Heum, O.** (1995) Exploration trends of the Halten Terrace. In: *Petroleum Exploration in Norway* (Ed. S. Hanslien), *NPF Sp. Publ.*, **4**, pp. 235–251.

Koenig, R.H. (1986) Oil discovery in 6507/7; an initial look at the Heidrun Field. In: *Habitat of Hydrocarbons on the Norwegian Continental Shelf* (Ed. A.M. Spencer), pp. 307–311. Graham & Trotman, London & Stavanger, Norway.

Machado, H.B. (2009) *Reservoir Characterisation of the Fluviodeltaic Succession of the Early Jurassic Åre Formation, Heidrun Field, Halten Terrace.* MSc thesis, Imperial College, London, UK, 118 pp.

Mackey, S.D. and **Bridge, J.S.** (1995) Three dimensional model of alluvial stratigraphy: theory and application. *J. Sed. Res*, **B65**, 7–31.

Makaske, B. (2001) Anastomosing rivers: a review of their classification, origin and sedimentary products. *Earth-Sci. Rev.*, **53**, 149–196.

Marsh, N., Imber, J., Holdsworth, R.E., Brockbank, P. and **Ringrose, P.** (2010) The structural evolution of the Halten Terrace, offshore Mid-Norway: extensional fault growth and strain localisation in a multi-layer brittle-ductile system. *Bas. Res.*, **22**, 195–214.

Martinius, A.W., Kaas, I., Næss, A., Helgesen, G., Kjærefjord, J.M. and **Leith, D.A.** (2001) Sedimentology of the heterolithic and tide-dominated Tilje Formation (Early Jurassic, Halten Terrace, offshore mid-Norway). In: *Sedimentary Environments Offshore Norway – Palaeozoic to Recent* (Eds O.J. Martinsen and T. Dreyer), **10**, pp. 103–144. Norsk Petroleumsforening Spec. Publ., Elsevier.

Mearns, E.W., Knarud, R., Ræstad, N., Stanley, K.O. and **Stockbridge, C.P.** (1989) Samarium-neodymium isotope stratigraphy of the Lunde and Statfjord Formations of Snorre Oil Field, northern North Sea. *J. Geol. Soc. London*, **146**, 217–228.

Mjøs, R. and **Prestholm, E.** (1993) The geometry and organization of fluviodeltaic channel sandstones in the Jurassic Saltwick Formation, Yorkshire, England. *Sedimentology*, **40**, 919–935.

Morris, P.H., Cullum, A., Pearce, M.A. and **Batten, D.J.** (2009) Megaspore assemblages from the Åre Formation (Rhaetian–Pliensbachian) offshore mid-Norway and their value as field and regional stratigraphical markers. *J. Micropalaeontol.*, **28**, 161–181.

Morton, A., Hallsworth, C., Strogen, D., Whitham, A. and **Fanning, M.** (2009) Evolution of provenance in the NE Atlantic rift: The Early-Middle Jurassic succession in the Heidrun Field, Halten Terrace, offshore Mid-Norway. *Mar. Petrol. Geol.*, **26**, 1100–1117.

Nordahl, K. and **Ringrose, P.** (2008) Identifying the representative elementary volume of permeability in heterolithic deposits using numerical rock models. *Math. Geosci.*, **40**, 753–771.

Norwegian Petroleum Directorate (2009) *Fact maps.* URL: https://npdmap1.npd.no/website/NPDGIS/viewer.htm.

Pedersen, T., Harms, J.C., Harris, N.B., Mitchell, R.W. and **Tooby, K.M.** (1989) The role of correlation in generating the Heidrun Field geological model. In: *Correlation in Hydrocarbon Exploration* (Ed. J.D. Collinson), pp. 327–338. Norwegian Petroleum Society.

Posamentier, H.W. and **Vail, P.R.** (1988) Eustatic controls on clastic deposition II - sequence and systems tracts models. In: *Sea-level Changes: An Integrated Approach* (Eds C.K. Wilgus, B.S. Hastings, C.G. Kendall, H.W. Posamentier, C.A. Ross and J.C. Van Wagoner), *SEPM Spec. Publ.*, **42**, pp. 110–124.

Read, W.A. (1995) Sequence stratigraphy and lithofacies geometry in an early Namurian coal-bearing succession in central Scotland. In: *European Coal Geology* (Eds M.K.G. Whateley and D.A. Spears), **82**, pp. 285–297. Geological Society, London, Special Publications.

Roberts, H.H. (1997) Dynamic changes of the Holocene Mississippi River delta plain: The delta cycle. *J. Coastal Res.*, **13**, 605–627.

Roberts, H.H. and **Coleman, J.M.** (1996) Holocene evolution of the deltaic plain: a perspective — from Fisk to present. *Eng. Geol.*, **45**, 113–138.

Ryseth, A. (2001) Sedimentology and palaeogeography of the Statfjord Formation (Rhaetian - Sinemurian), North Sea. In: *Sedimentary Environments Offshore Norway - Palaeozoic to Recent* (Eds O.J. Martinsen and T. Dreyer), pp. 67–85. Norsk Petroleumsforening Spec.Publ., **10**, Elsevier.

Shanley, K.W. and **McCabe, P.J.** (1994) Perspectives on the sequence stratigraphy of continental strata. *AAPG Bull.*, **78**, 544–568.

Sloss, L.L., Krumbein, W.C. and **Dapples, E.C.** (1949) Integrated facies analysis. In: *Sedimentary Facies in Geological History* (Ed. C.R. Longwell), *Geol. Soc. Am. Mem.*, **39**, pp. 91–124.

Surlyk, F. (1990) A Jurassic sea-level curve for East Greenland. *Palaeogeogr. Palaeoclimatol. Palaeoecol.*, **78**, 71–85.

Svela, K.E. (2001) Sedimentary facies in the fluvial-dominated Åre Formation as seen in the Åre 1 member in the Heidrun Field. In: *Sedimentary Environments Offshore Norway – Palaeozoic to Recent* (Eds O.J. Martinsen and T. Dreyer), **10**, pp. 87–102. Norsk Petroleumsforening Spec. Publ., Elsevier.

Van Wagoner, J.C., Posamentier, H.W., Mitchum, R.M., Vail, P.R., Sarg, J.F., Loutit, T.S. and **Hardenbol, J.** (1988) An overview of the fundamentals of sequence stratigraphy and key definitions. In: *Sea-level Changes: An Integrated Approach* (Eds C.K. Wilgus,

B.S. Hastings, C.G. Kendall, H.W. Posamentier, C.A. Ross and J.C. Van Wagoner), *SEPM Spec. Publ.*, **42**, pp. 39–45. Tulsa.

Van Wagoner, J.C., Mitchum, R.M., Campion, K.M. and **Rahmanian, V.D.** (1990) *Siliciclastic Sequence Stratigraphy in Well Logs, Cores and Outcrops; Concepts for High-Resolution Correlation of Time and Facies.* Tulsa, AAPG, 55 pp.

Whitley, P.K. (1992) The geology of Heidrun; a giant oil and gas field on the mid-Norwegian shelf. *AAPG Mem.*, **54**, 383–406.

Willis, B.J. and **Gabel, S.L.** (2001) Sharp-based, tide-dominated deltas of the Sego Sandstone, Book Cliffs, Utah, USA. *Sedimentology*, **48**, 479–506.

Willis, B.J. and **Gabel, S.L.** (2003) Formation of deep incisions into tide-dominated river deltas: Implications for the stratigraphy of the Sego Sandstone, Book Cliffs, Utah, U.S.A. *J. Sed. Res.*, **73**, 246–263.

Yoshida, S., Steel, R.J. and **Dalrymple, R.W.** (2007) Changes in depositional processes - An ingredient in a new generation of sequence-stratigraphic models. *J. Sed. Res.*, **77**, 447–460.

Int. Assoc. Sedimentol. Spec. Publ. (2014) **46**, 291–338

Depositional dynamics and sequence development in a tidally influenced marginal marine basin: Early Jurassic Neill Klinter Group, Jameson Land Basin, East Greenland

JUHA M. AHOKAS*, JOHAN P. NYSTUEN* and ALLARD W. MARTINIUS†

**Department of Geosciences, University of Oslo, P.O. Box 1047, 0316 Oslo, Norway (E-mail: juhaho@detnor.no)*
†Statoil Research Centre, N-7005 Trondheim, Norway

ABSTRACT

During the Early Jurassic, contemporaneous marginal marine successions were deposited on both sides of the approximately 400 km-wide proto-Norwegian–Greenland Sea in the present NE Atlantic domain. The seaway was connected to the Boreal Sea in the north and the Tethys Sea in the south. On East Greenland, deposition occurred in the Jameson Land Basin, a three-way closed embayment that developed along the East Greenland rift zone. Offshore mid-Norway, in the Halten Terrace area, deposition occurred within an open but relatively narrow NE–SW elongated basin. The latest Sinemurian-Aalenian Neill Klinter Group in the Jameson Land Basin comprises an up to 300 m thick mixed-energy, mainly tide- and wave-influenced deltaic to estuarine sandstone succession and up to 110 m thick marine mudstone succession. The succession has been divided into six transgressive-regressive (T/R) sequences separated by subaerial unconformities (SU) and transgressive marine ravinement surfaces (TS), commonly coinciding and forming combined SU/TS sequence boundaries. The transgressive and regressive systems tracts are defined by seven genetically related architectural elements composed of twenty-one facies associations within eight mappable allostratigraphic units. Incised valleys were formed at three stratigraphic levels. The valley infill successions are characterised by regressive-transgressive estuarine successions revealing fluvial activity during the falling stage and earliest lowstand, wave and tidal activity during the main stage of rising sea-level and back to wave-dominance in the late stage of sea-level rise. Transitions in the energy system were triggered by changes in the estuary morphology, from narrow at the early stage to wide and open at the late stage, during rises in relative sea-level. Paralic deposits outside the incised valleys are also characterised by highly variable lateral, as well as vertical, facies changes. The Neill Klinter Group of the Jameson Land Basin can be used as an analogue for the Early Jurassic Båt Group on the Halten Terrace; not by a direct lithostratigraphic or sequence stratigraphic correlation but by comparison of depositional environments, external and internal factors controlling depositional architecture, processes and resulting 3D heterogeneity properties of sandstone bodies.

Keywords: paralic, incised valleys, estuaries, tide-influenced, lateral facies transition, sequence stratigraphy, East Greenland.

INTRODUCTION

The paralic depositional environment includes coastal plain, incised valleys, deltas, estuaries and shoreline to shelf systems, all with a series of sub-environments representing potential sites for reservoir sandstones to accumulate (e.g. Reading & Collinson, 1996; Emery & Myers, 1996). The tide-influenced depositional systems are highly dynamic due to the interplay between fluvial, wave and tidal energy, topography, variation in rate of sediment input and changes in relative sea-level. The different co-existing energy systems and environments result in three-dimensional lithological heterogeneity, which is reflected at all scales of the depositional architecture. In order to

From Depositional Systems to Sedimentary Successions on the Norwegian Continental Margin, First Edition.
Edited by A. W. Martinius, R. Ravnås, J. A. Howell, R. J. Steel, and J. P. Wonham.

identify genetically related depositional units, it is therefore important to distinguish key bounding surfaces in the depositional succession, formed as subaerial unconformities, transgressive surfaces, surfaces that are combinations of subaerial exposure and subsequent transgression and maximum flooding surfaces. In this respect, the change in dominance of energy systems from fluvial to tidal and wave processes during a cycle of fall and rise in sea-level is thought to be of particular significance to understanding of the 3D facies distribution of paralic deposits and their reservoir properties (Emery & Myers, 1996; Posamentier & Allen, 1999; Catuneanu, 2006).

In tectonically stable basins with overall aggrading depositional architecture, tide-influenced sedimentation may occur over one or several orders of sea-level cycles (e.g. Dam & Surlyk, 1998; Martinius *et al.*, 2001; Midtkandal & Nystuen, 2009) and be comprised of nearly identical but complex depositional architecture (Allen, 1990; Choi *et al.*, 2004; Choi & Dalrymple, 2004; Dalrymple *et al.*, 1990, 1991, 2003, 2012; Gastaldo *et al.*, 1995; Gingras *et al.*, 1999; Martinius & Van den Berg, 2011; Tessier, 2012).

The primary aim of this study is to record multi-scaled facies variations, depositional architecture and sequence development as a function of basin configuration and variation in relative sea-level in the Early Jurassic a marginal marine Neill Klinter Group (Rosenkrantz, 1934; Surlyk *et al.*, 1973; Dam & Surlyk, 1998) of the Jameson Land Basin, East Greenland (Figs 1 and 2).

The secondary aim of the study has been to provide qualitative and quantitative data for application in reservoir modelling of analogous reservoir sandstones on the Halten Terrace on the Norwegian continental shelf. Conceptual understanding of the three-dimensional distribution of depositional processes in the formation of tide-influenced sandstone reservoirs is crucial in modelling their multi-scale complex heterogeneity (Martinius *et al.*, 2005). Analogue data from the overall progradational to retrogradational tide-influenced to wave-dominated Neill Klinter Group (Fig. 2) are considered to (1) increase understanding of the range of natural variability in heterolithic tide-influenced paralic depositional systems, (2) contribute with data for characterisation and construction of static and dynamic reservoir models, (3) reduce uncertainties in the choice and range of values of modelling parameters and their effect on geo-model and fluid-flow properties and (4) improve the current understanding of the dynamic behaviour of these reservoir rocks to assist in development and production decisions.

REGIONAL SETTING AND STRATIGRAPHY

The coastal zone of East Greenland is cut by dominantly N–S-trending faults that controlled basin development and sediment infill in the Palaeozoic and the Mesozoic East Greenland rift system (Surlyk, 1990a). Younger NW–SE trending cross-faults are postulated along the present-day fjords (Bütler, 1948; Surlyk, 1977, page 326; Dam *et al.*, 1995; Surlyk, 2003). The approximately 100 km-wide and 250 km-long Jameson Land Basin is located in the southernmost segment of the rift system and forms a slightly southward-plunging (1 to 2°) ~ 17 km thick synclinal structure (Larsen & Marcussen, 1992; Mathiesen *et al.*, 2000). The Jameson Land Basin hosts a ~ 13 km thick continental and marine Late Palaeozoic rift-basin succession overlain by a ~ 4 km thick continental to marine post-rift succession of a latest Palaeozoic to Mesozoic sag-basin (Larsen & Marcussen, 1992; Mathiesen *et al.*, 2000; Surlyk, 2003).

The latest Triassic to Early Jurassic succession along the eastern margin of the Jameson Land Basin is well exposed in response to crustal uplift (2 to 3 km) in connection with the opening of the Norwegian-Greenland Sea during the Cenozoic (Christiansen *et al.*, 1992; Hansen, 2000; Johnson & Gallagher, 2000; Mathiesen *et al.*, 2000). The semi-continuous cliff-faces are 200 to 400m-high with 7° stratal dip towards the west and are aligned sub-parallel to the depositional strike with limited 3D control (Fig. 3). Palaeogene (~ 55 Ma) intrusive dolerite sills are common in the southern Jameson Land, where these follow mudstone units (Fig. 3). These intrusives are rare in the northern area.

The paralic Pliensbachian-Toarcian Neill Klinter Group (*sensu* Dam & Surlyk, 1998; Surlyk, 2003) in the Jameson Land Basin was established by marine inundation of the Rhaetian-Sinemurian lacustrine basin, represented by the Kap Stewart Group (Dam & Surlyk, 1992, 1993; Dam *et al.*, 1995; Surlyk, 2003), from the south. The younger deposits unconformably overlie the older sediments and onlap basement along the basin margins. The basin received sediments from erosion

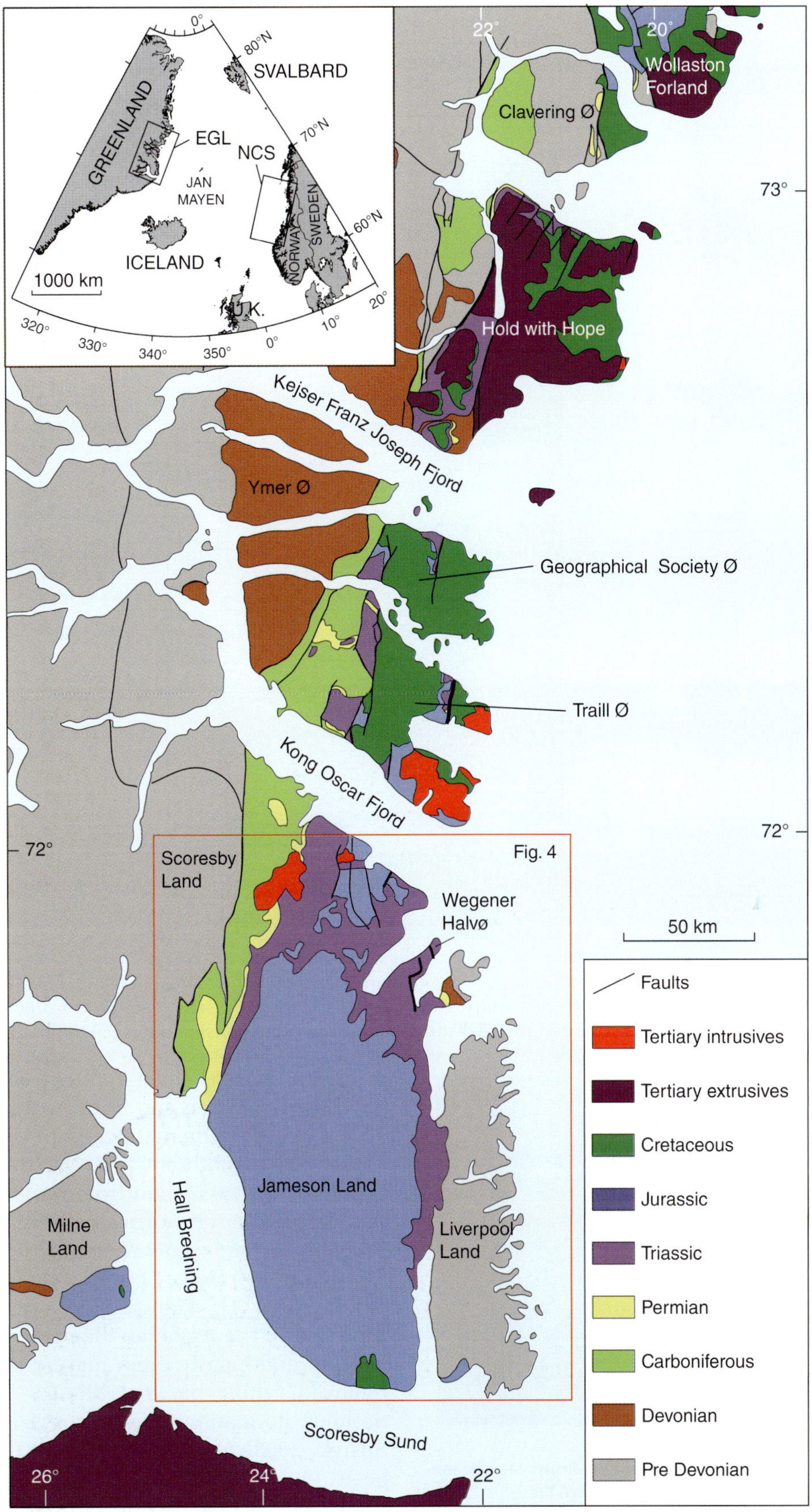

Fig. 1. Geological map of East Greenland. Small map shows present-day North Atlantic (referred to as the Norwegian-Greenland Sea in the text) and discussed areas of the East Greenland (EGL) and the Norwegian Continental Shelf (NCS).

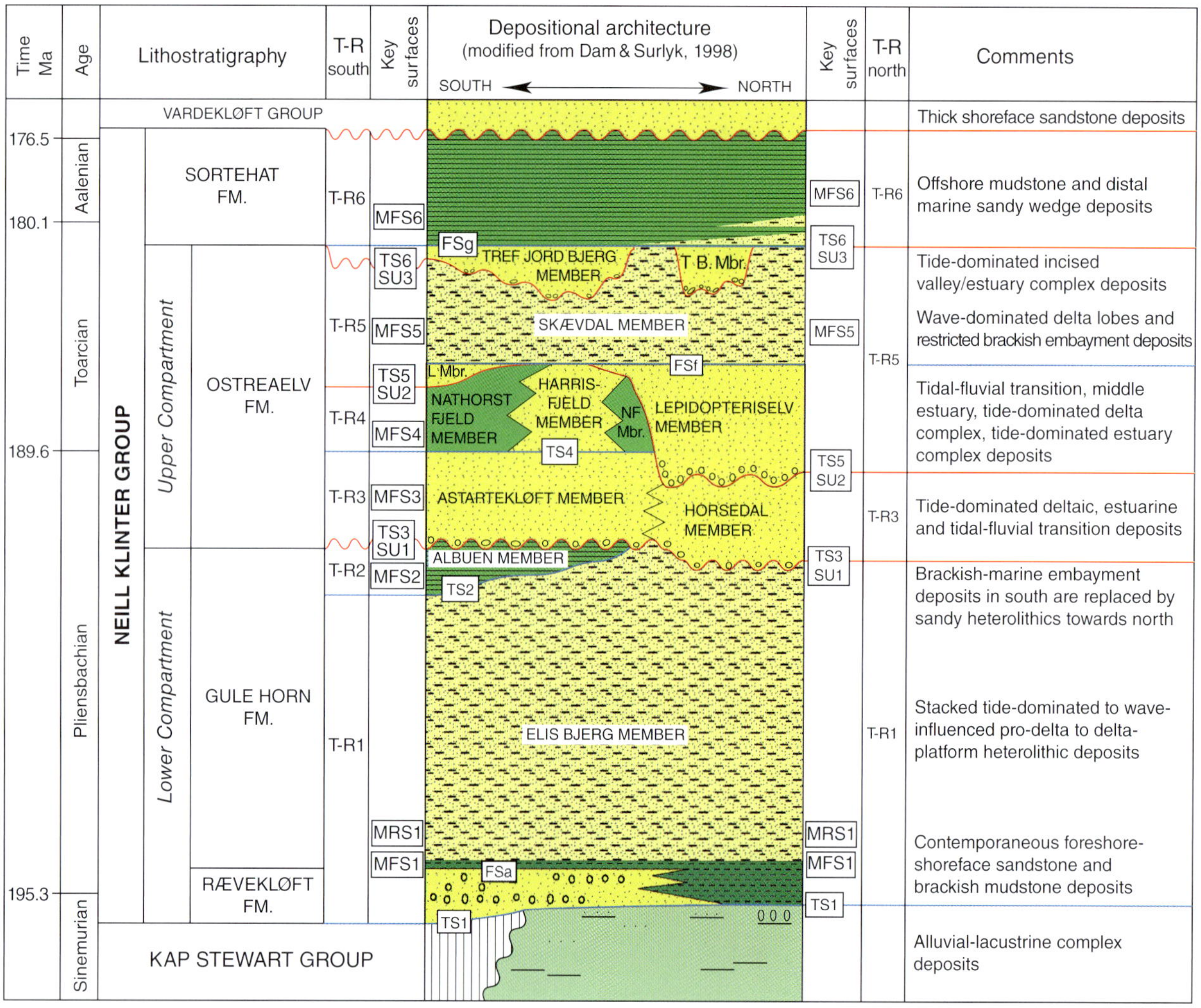

Fig. 2. Lithostratigraphic sub-division of the latest Sinemurian/Pliensbachian – Toarcian/Aalenian Neill Klinter Group, T-R sequences and key surfaces defined in this study (modified from Dam & Surlyk, 1998; Surlyk, 2003 and references therein). Red lines mark subaerial unconformities (SU) and blue lines selected transgressive (TS) or flooding surfaces (FS).

Fig. 3. The well exposed Neill Klinter Group succession along the western side of Hurry Inlet in eastern Jameson Land, East Greenland (a view towards north of locality S4) (photo A. Martinius). The dark stratigraphic levels are Palaeogene dolerite intrusives. For position of locality see Fig. 4.

of crystalline basement rocks from the present-day Liverpool Land and from an area further east; a land probably then attached to the present Jan Mayen microcontinent (e.g. Mjelde *et al.*, 2008; Breivik *et al.*, 2012) and from hinterland areas in the west and north-north-west (Fig. 1). The basin probably extends further south of the Scoresby Sund, but is presently covered by early Cenozoic basalts (Fig. 1). The paralic environment was drowned in the Aalenian–Bajocian and the basin became dominantly open marine at the onset of renewed rifting in the Middle Jurassic. The Sortehat Formation (Koppelhus & Hansen, 2003; Surlyk, 2003) of the Neill Klinter Group has not been included in the present study, as this Formation is considered to represent an open-marine environment, a depositional setting very different from the rest of the Neill Klinter Group.

The Early Jurassic marginal marine Jameson Land Basin was to the south connected to a relatively narrow and long basin system of the proto-Norwegian-Greenland Sea. This basin system is today represented by the tide-influenced and wave-influenced marine Jurassic succession on the mid-Norwegian continental shelf (Gjelberg *et al.*, 1987; Dalland *et al.*, 1988; Skogseid *et al.*, 2000; Brekke *et al.*, 2001; Nøttvedt *et al.*, 2008; Mitchell *et al.*, 2011).

METHODOLOGY AND TERMINOLOGY

The study was carried out in two outcrop areas along the eastern coast of Jameson Land (Fig. 1). The primary dataset consists of 32 measured sections produced in 8 localities along the over 100 km N–S oriented transect (Fig. 4). In addition, selected sections from Dam & Surlyk (1998) were consulted. The measured sections are 30 to 300 m thick and record grain size, sedimentary structures, palaeocurrent directions (measured from dune foresets), nature of bedding, bedding contacts, lateral variability, trace fossil assemblages and degree of bioturbation.

In this study, the Neill Klinter Group is divided into two stratigraphical compartments (Fig. 2) composed in total of eight allostratigraphic units (North American Commission on Stratigraphic Nomenclature, 1983, p. 865; Bhattacharya, 2001). Allostratigraphic units are defined on the basis of observed bounding discontinuities representing changes in relative sea-level, or changes in the ratio between rate in creation of accommodation space *versus* rate of sediment supply (A/S). Each allostratigraphic unit is formed by a certain set of architectural elements (described below) that in a specific manner characterise particular stages in the depositional history of the Neill Klinter Group. Each architectural element consists of one or more facies associations and is bounded by key stratal surfaces reflecting basinward or landward shifts of the palaeo-shoreline.

Key stratigraphic surfaces, architectural elements and their organisation in allostratigraphic units form the descriptive database for a conceptual depositional model and for a transgressive-regressive (T-R) sequence stratigraphic framework of the Neill Klinter Group. A 105 km-long correlation profile was formed by combining new field data, published data and photo-mosaics to illustrate the T-R sequence stratigraphic framework for the Neill Klinter Group (discussed below). Ground and aerial photo-mosaics were used to map large-scale lateral and vertical changes in texture and structure and the geometry of depositional units.

DEPOSITIONAL ARCHITECTURE

Key stratigraphic surfaces in the Neill Klinter Group

The lowest frequency (i.e. higher hierarchy) key surfaces include three subaerial unconformity surfaces (SU), six transgressive surfaces (TS) and five maximum flooding surfaces (MFS). The SU surfaces are extensive, irregular to subhorizontal erosive surfaces overlain by cm-thick to dm-thick matrix-supported or clast-supported conglomeratic beds containing extra-basinal quartz pebbles and/or granules, quartzite and gneiss clasts, coalified wood and plant debris and marine body-fossil fragments (Fig. 5A–C). The TS surfaces are extensive, irregular to subhorizontal surfaces across which a facies of more proximal character is sharply overlain by more basinward distal facies (Fig. 5A and D). Transgressive surfaces are interpreted as marine ravinement surfaces created by tidal scouring and coastal onlap. The TS surfaces mostly coincide with the preceding subaerial unconformities and these two types of surface together form combined SU/TS surfaces (Fig. 5A to D). The maximum flooding surfaces (MFS) are defined by up to several metres-thick condensed zones of the most distal lithofacies. Although the maximum flooding surfaces record relatively slow transitions from landward to basinward sedimentation, they have been applied as correlative surfaces, in accordance with general principles of sequence stratigraphic correlation (Van Wagoner *et al.*, 1988; 1990; Posamentier *et al.*, 1988; Posamentier & Allen, 1999; Catuneanu, 2006).

The high frequency (i.e. lower hierarchy) key surfaces comprise four marine ravinement surfaces (MRS) (regressive or transgressive) and seven flooding surfaces (FS). These surfaces define local basinward or landward shifts in facies belts (Fig. 5A, E and F). Correlatability of key surfaces varies from a few kilometres to more than several tens of kilometres.

Facies and facies associations

A total of 14 lithofacies (labelled F1 to F11) have been defined in the Neill Klinter Group (Table 1). The main lithofacies are sand-dominated and

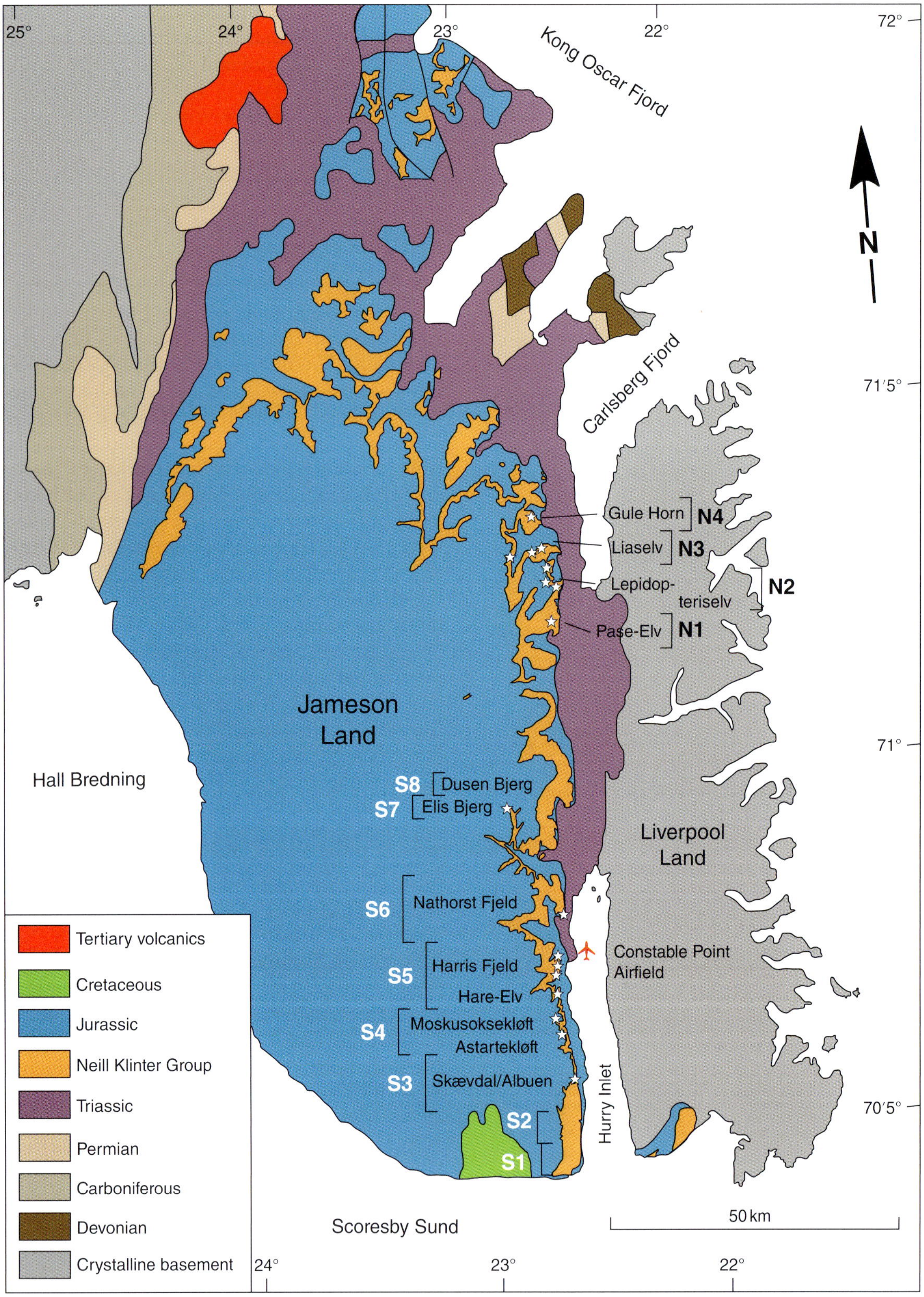

Fig. 4. Geological map of the Jameson Land region showing the outcrop of the Neill Klinter Group (in orange). The investigated localities in this study are S3 to 6 in the southern area and N1 to 4 in the northern area. Sections in localities S1 to 2 and S7 to 8 area are redrawn from Dam & Surlyk, 1998.

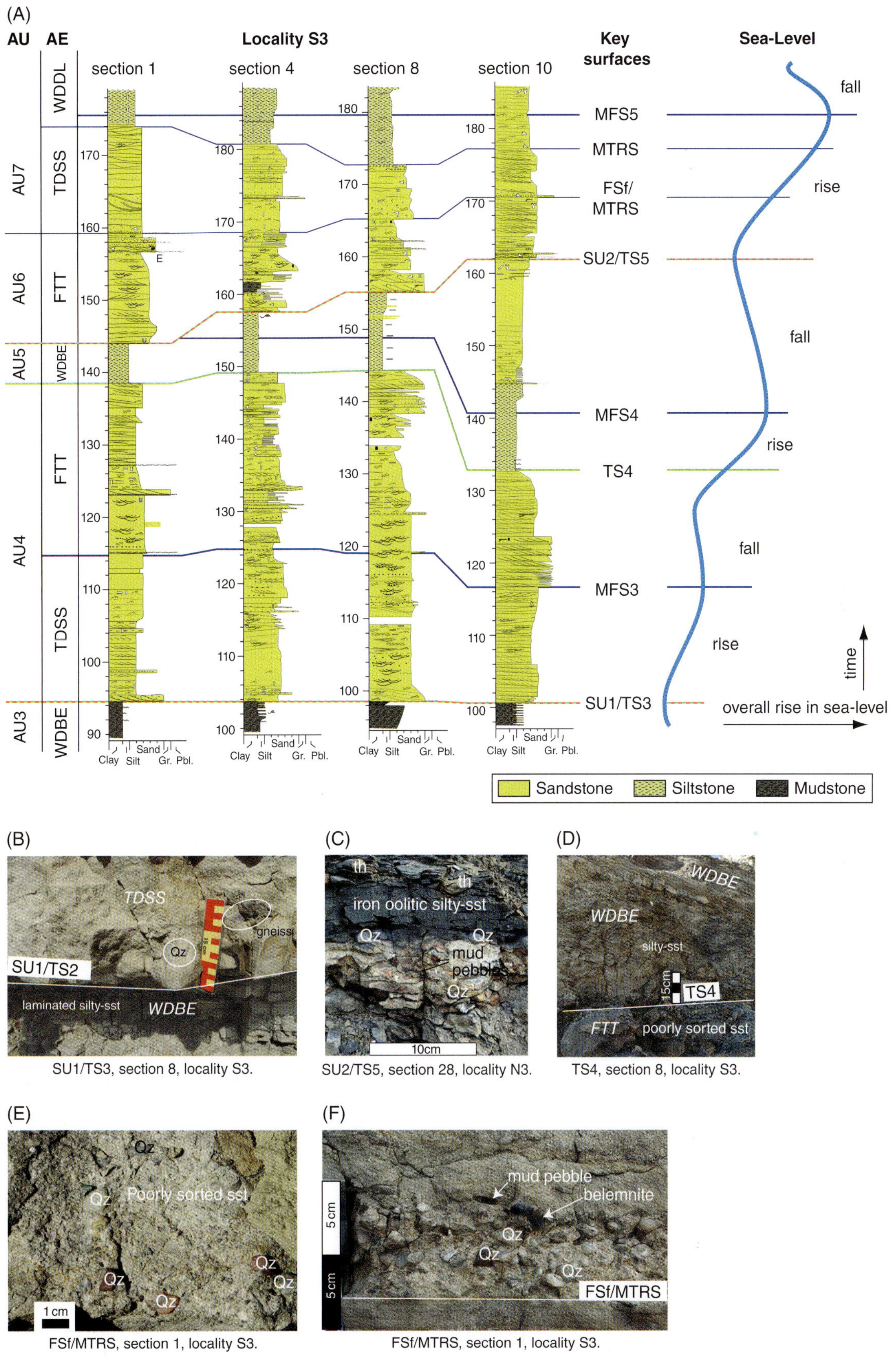

Fig. 5. (A) Selected detailed sedimentary logs (depths in m) through the upper stratigrahical compartment showing sedimentology, key surfaces and changes in relative sea-level (locality S3). Photos of mixed regressive and transgressive lithofacies overlying selected key surfaces at localities S3 and N3; (B) A matrix-supported conglomeratic bed overlying the SU1/TS2 unconformity (measured section 8); (C) Clast-supported and matrix-supported conglomerate beds overlying the SU2/TS5 unconformity (measured section 28, locality N3); (D) Silty structureless sandstone overlying the TS4 unconformity (measured section 8); (E) A matrix-supported conglomerate bed overlying the FS/MTRS (measured section 1); and (F) A mixed fluvial and marine conglomerate bed (measured section 1). (Photos J.M. Ahokas).

Table 1. Summary of facies recorded in the Neill Klinter Group, Jameson Land Basin, East Greenland. Intensity of bioturbation is described using the bioturbation index scheme of Taylor & Goldring (1993).

Facies	Sedimentology	Bedding boundaries and bed conditions	Ichnology (index 0-6)	Interpretation	Appearances in logs, not to scale	Facies Association
1a Lithic conglomerate	Non-stratified to weakly stratified or graded; very poorly sorted extra-basinal lithic pebbles up to clasts 9 cm in diameter, rounded to sub-rounded; intra-basinal tree trunks, mudstone chips, marine body fossils	Erosional lower and sub-horizontal sharp or gradational upper boundaries; bed thickness 5 to 30 cm, lateral extension from < 10 m up to several km	Unbioturbated (0); Belemnite rostra[1]; Occasional vertical burrows from overlying substrate: *Di, ?Mo, Sk*; low intensity (1 to 2)	Deposited either as fluvially derived conglomerate; marine ravinement lag; gravitational sheet flow or by traction in pro-deltaic, delta front, offshore environment	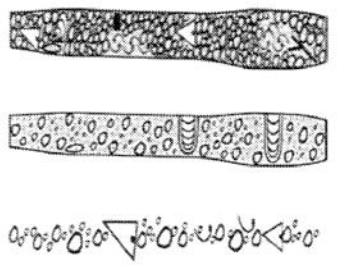	1A of fluvial-tidal transition 1B of fluvial-tidal transition 1C of fluvial-tidal transition 2A† of tidal inclined heterolithic strata 3A† of tidal channel infill 3D† of tidal channel infill 5B† of mixed tide- and wave-influenced heterolithic delta 6A* of wave-dominated brackish-marine embayment mudstone and siltstone
1b Mud clast conglomerate	Matrix-supported; poorly sorted; mudstone clasts dark red-brown, well rounded, elongated, weakly imbricated or trough cross-stratified, cm to dm in diameter; common angular to rounded extra-basinal lithic clasts, organic debris	Erosional lower boundary; upper boundary sharp and sub-horizontal or erosional (relief < 5 to 7 cm); bed thickness 5 cm to 1 m, lateral extent m to 10s of m	Unbioturbated (0)	Mud rip-up conglomerate deposited as shallow tidal channel-fill on tidal delta platform/intra-tidal and/or supra-tidal mudflats	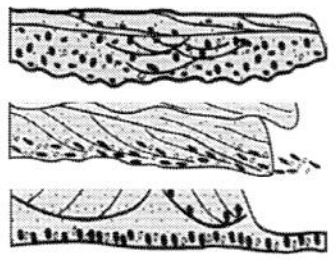	1B of fluvial-tidal transition 3E of tidal channel infill 2A† of tidal inclined heterolithic strata 2B† of tidal inclined heterolithic strata 3B† of tidal channel infill 3D† of tidal channel infill
2 Cross-stratified sandstone	Immature; poorly to well sorted; fine-grained to very coarse-grained; distinct normal grading trend; trough-bedded and planar cross-bedded; intra-basinal debris, angular to subrounded quartz pebbles, feldspar, lithic grains in thin beds, organic debris, angular to rounded, cylindrical cm to dm size mudstone rip-up clasts; poorly to well cemented; floating marine body fossils	Commonly erosive base; upper boundary gradational fining-up; common reactivation surfaces; bed thickness 10 cm < 3 m; trough width > 2 m; bedset thickness 2 to 5 m, lateral extent 100 m up to km	Unbioturbated (0); Rare *Ar, Sk, Ph, Tae, Pl, Di*; horizontal locomotion traces; low intensity (0 to 2); lacustrine fauna and terrestrial flora generally dominate over marine fauna	Deposited by oblique or traverse migration of 2D and 3D dunes and bars in braided fluvial and/or tide-influenced sinuous to straight channels; occasional mud drapes, rip-ups and fluid mud layers suggest some tidal influence	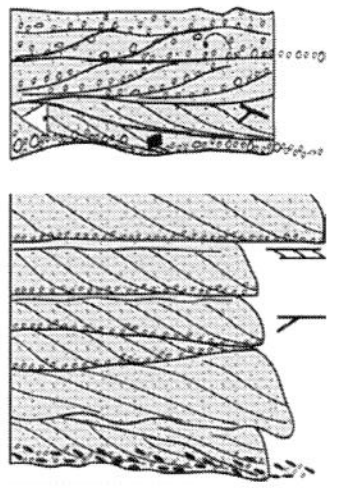	1A of fluvial-tidal transition 1B of fluvial-tidal transition 1C of fluvial-tidal transition 3E of tidal channel infill 4B of wave-dominated shoreface 2B of tidal inclined heterolithic strata

3 Low-angle cross-stratified sandstone	Texturally mature; clean; very well sorted quartz-rich fine-grained to medium-grained; incipient continuous mudstone, silt or heavy-mineral drapes	Sharp lower and upper boundaries, set boundaries distorted; bed thickness > 2.5 m, lateral extent < 100 m, bedset thickness < 10 m	Unbioturbated (0); rare local high intensity (5)	Deposited by waves and/or storms in shoreface; possible swaley cross-stratification	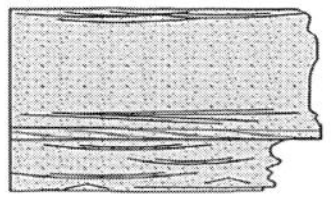	3B of tidal channel infill 4A of wave-dominated shoreface 4B of wave-dominated shoreface
4 Cross-stratified sandstone with mudstone drapes	Fine-grained to very coarse; well to poorly sorted; continuous rhythmic single and/or paired mudstone or siltstone drapes along normally graded tabular to tangential and sigmoidal foresets; mudstone/siltstone/plant debris, drape thickness and abundance vary rhythmically; opposite oriented flow cross-laminae on bottomsets; mudstone rip-up flakes in toesets oriented parallel with bottomsets; rare mudstone rip-up clasts and floating clastic granules; glauconite grains	Erosional lower boundary; upper boundary either gradational and upward fining or sharp and erosive; abundant to rare set bounding accretion surfaces; abundant internal reactivation surfaces; bed thickness 0.1 m up to > 4.2 m, lateral extent several 10s to several 100 m; tabular to lensoid geometries	Coarse fraction: sparse *Ar, Di, Op, Sk*, rare to common Belemnite rostra[1]; low intensity (0 to 2). Fine fraction: *Pl* and worm traces, oyster shell fragments; moderate intensity (3 to 4); Unbioturbated (0)	Deposited by migration of small to giant single and complex fluvial to sub-tidal 2D and 3D dunes, bars in wide high energy tidal or tide-dominated fluvial channels; from mid-channel/thalweg bars to heterolithic foreset portion of sub-tidal compound dunes and distributary mouthbars; Deposited by migration of small complex 3D tidal-fluvial dunes on top of sub-aqueous or exposed mouth bars or on top of tidal-fluvial side attached bars	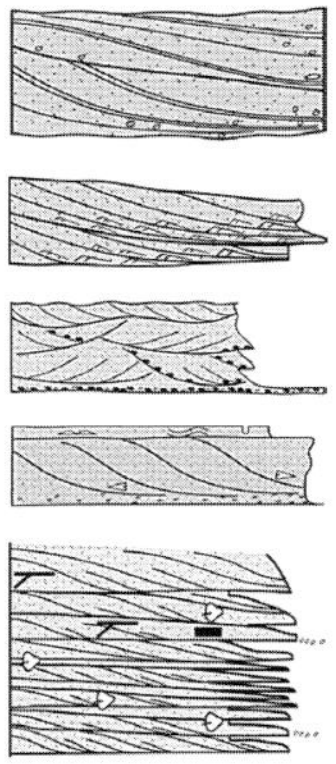	1B of fluvial-tidal transition 1C of fluvial-tidal transition 2A of tidal inclined heterolithic strata 3A of tidal channel infill 3C of tidal channel infill 3D of tidal channel infill 4A of wave-dominated shoreface 2B† of tidal inclined heterolithic strata 2C† of tidal inclined heterolithic strata 5C† of mixed tide- and wave-influenced heterolithic delta 7A of wave-dominated open-marine
5 Compound cross-stratified sandstone with mudstone drapes	Fine-grained to medium-grained; systematic double mudstone drapes; drape thickness and abundance alternate rhythmically; low-angle to high-angle cross-stratification; rare mudstone clasts parallel on foresets; ripple laminae on bottom sets; rare iron ooid-rich beds	Erosive upper and lower bounding accretion surfaces; abundant accretion surfaces, reactivation surfaces separate individual sets; bed thickness 0.2 to 1.0 m	Heterolithic member: unidentified burrow traces, *Th*; moderate intensity (2 to 3). Sandy member: low intensity (0 to 1).	Deposited by migration of complex dunes in relatively deep high energy tidal channels with diurnal bipolarity of current; iron ooids eroded on land and transported in the sand fraction	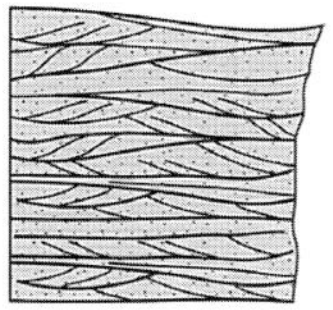	2A of tidal inclined heterolithic strata

(*Continued*)

Table 1. (*Continued*)

Facies	Sedimentology	Bedding boundaries and bed conditions	Ichnology (index 0-6)	Interpretation	Appearances in logs, not to scale	Facies Association
6a Fine-grained wave ripple cross-laminated sandstone	Moderately to well sorted; homogenous; fine-grained; wave-ripples and wave-modified current ripples; absent very thin discontinuous mudstone and siltstone drapes; cleaning-up beds	Sharp lower boundary; gradational coarsening-up trend; bed thickness 5 to 30 cm, bedset-scale lateral extent > 10s of km	Local sparse *Th, Tei*, Belemnite rostra[1]; intensity (0 to 1) to high diversity dominated by: *As, ?Ch, Cy, Op, Pa-h, Pl, Sk, Tei, Ter, Th*; intensity (4 to 5).	Deposition of dominating sand as ripples during dominant wave and sub-dominant current activity and subsequent deposition of mud from suspension during slack-water stage	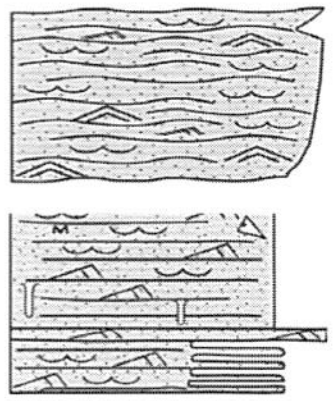	2A of tidal inclined heterolithic strata 2B of tidal inclined heterolithic strata 2C of tidal inclined heterolithic strata 3D of tidal channel infill 5A of mixed tide- and wave-influenced heterolithic delta 7A of wave-dominated open-marine 7B (rare) of wave-dominated open-marine 5B[†] of mixed tide- and wave-influenced heterolithic delta
6b Heterolithic mudstone and sandstone with lenticular bedding	Sharply inter-laminated; very-fine to medium-grained; laterally and vertically isolated ripple-laminae in laminated/non-laminated silt-stone and mudstone	Sandstone bed thickness 10 to 30 cm; muddy siltstone beds 5 to 20 cm thick, bedset-scale lateral extent several km	Low to high intensity (0 to 6). Dominant ichnofabrics: *Ter, Pl, Ph, Ch, Th.*	Deposition of sand as incomplete ripples during current activity on a muddy substratum; some wave modification	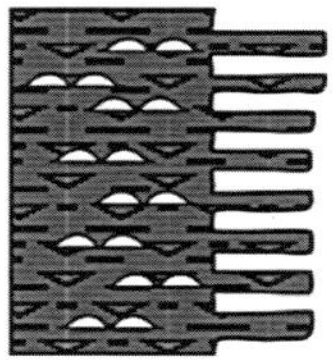	5A of mixed tide- and wave-influenced heterolithic delta 5B of mixed tide- and wave-influenced heterolithic delta 5C of mixed tide- and wave-influenced heterolithic deltaic deposits 7B of wave-dominated open-marine 3D[†] of tidal channel infill

6c Heterolithic ripple cross-laminated sandstone with wavy-bedding	Sharply inter-laminated mudstone and sandstone; very-fine to very coarse-grained; continuous dominant unidirectional and sub-dominant bidirectional ripple-laminae draped by laminated/ structureless siltstone and mudstones; distinct overall coarsening-up trend	Sandstone bed thickness < 30 cm; siltstone and mudstone beds < 5 mm, lateral extent from 100 m up to 3 km	Fine member: moderate to high intensity (3 to 6). Dominant ichnofabrics: *Tei, Ph, Ro, Op, Pl, Pa, Th, Sk, Ar, ?As.* Coarse member: low to moderate intensity (1 to 3). Dominant ichnofabrics: *Ph, Pl, Pa.*	Deposition of sand as ripples during repeating current activity and subsequent deposition of mud from suspension during slack water stage; secondary wave modification	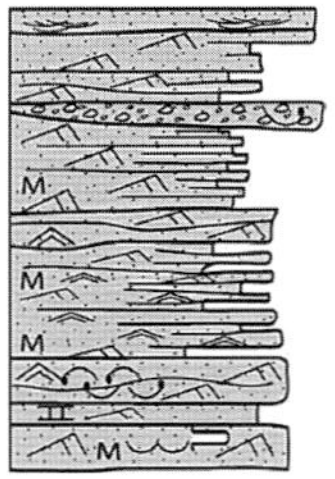	1A of fluvial-tidal transition 2A of tidal inclined heterolithic strata 2B of tidal inclined heterolithic strata 3A of tidal channel infill 3B of tidal channel infill 3C of tidal channel infill 3D of tidal channel infill 5A of mixed tide- and wave-influenced heterolithic delta 5C of mixed tide- and wave-influenced heterolithic delta 7A of wave-dominated open-marine 7B (rare) of wave-dominated open-marine 1B† of fluvial-tidal transition 1C† of fluvial-tidal transition
7 Hummocky and swaley cross-stratified sandstone	Clean; well sorted; very fine-grained to fine-grained; low-angle curving cross-lamination; wavelength 10 to 20 cm	Sharp lower and upper boundaries; bed thickness 5 to 30 cm, lateral extent > 100 m	Unbioturbated (0 to 1); occasional secondary *Pl* and/or *Co*; low to moderate intensity (1 to 3)	Deposition by storms in lower shoreface and pro-delta environment		2C of tidal inclined heterolithic strata 4A† of wave-dominated shoreface 4B† of wave-dominated shoreface 5B† of mixed tide- and wave-influenced heterolithic delta 6A† of brackish-marine embayment 7B† of wave-dominated open-marine

(*Continued*)

Table 1. (*Continued*)

Facies	Sedimentology	Bedding boundaries and bed conditions	Ichnology (index 0-6)	Interpretation	Appearances in logs, not to scale	Facies Association
8 Structureless sandstone	Unstratified; very well to very poorly sorted; medium-grained to coarse-grained; rare floating terrestrial and marine body fossils; lithic pebbles; rare green to black secondary colouring; chlorite-coated sub-angular to sub-rounded quartz grains; glauconite grains; cm to dm-size sideritic concretions	Sharp to gradational lower and upper boundaries; bed thickness 0.2 to 1.5 m, lateral extent < 100 m	Unbioturbated (0 to 1); locally high intensity beds (6) – unspecified; thick walled oysters (6)	Deposition by waves in shoreface; rapid deposition from gravity flows or collapse. Deposition by tides and waves as rare dark green or black beds of iron-rich reducing iron-ooid horizons; as structureless chlorite-coated and glauconitic quartz sands from iron-rich source areas	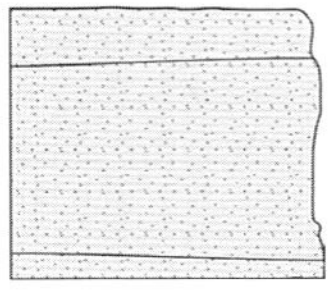	4A of wave-dominated shoreface 6B of brackish-marine embayment 3A† of tidal channel infill 3B† of tidal channel infill 3C† of tidal channel infill 3D† of tidal channel infill 5B† of mixed tide- and wave-influenced heterolithic delta 5C† of mixed tide- and wave-influenced heterolithic delta
9 Structureless muddy siltstone and sandstone	Consolidated or loose; coarsening-up; occasional weak wave ripple laminae or trough cross-stratified; rare flow ripples; rare carbonaceous cone-in-cone structures, floating organic debris and lithic clasts	Sharp unconformable to gradational lower and upper boundaries; bed thickness 5 cm up 12 m, lateral extent from few m up to few km	Consolidated fraction: high intensity (3 to 5) *Ter, Pl, Ph, Ch, Th*; common floating marine body fossils and fully marine fauna. Loose fraction: unbioturbated (0); dominant terrestrial fauna and flora	Deposition from suspension in well oxygenated fully marine offshore to distal pro-delta environment; discontinuous sedimentation rates; intense biological activity. Deposition from suspension in low-energy environment with carbonaceous debris	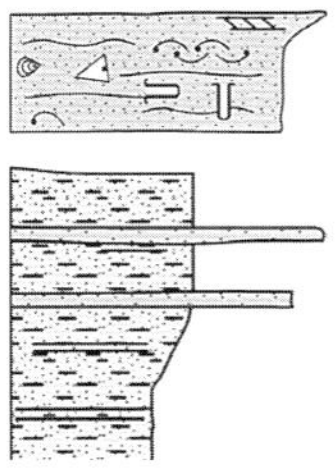	6A of brackish-marine embayment 6B of brackish-marine embayment 1C† of fluvial-tidal transition 4B† of wave-dominated shoreface 5B† of mixed tide- and wave-influenced heterolithic delta 5C† of mixed tide- and wave-influenced heterolithic delta 7A† of wave-dominated open-marine

10 Structureless mudstone	Non-laminated; well cemented red to grey coloured mudstone	Sharp lower and upper boundaries; bed thickness 2 mm up to < 10 cm, laterally discontinuous from cm to few metres	Unbioturbated (0-1)	Deposition as gravitational non-cohesive fluid muds		3E† of tidal channel infill 3B† of tidal channel infill 2A† of tidal inclined heterolithic strata 1C† of fluvial-tidal transition 1B† of fluvial-tidal transition
11 Laminated mudstone	Dark-grey to black, laminated mudstone, paper shale and siltstone; content of rare petrified tree trunks and carbonaceous cone-in-cone structures	Sharp; erosional lithological lower and upper boundaries against varying sandy facies	?Low intensity (0 to 2); brackish to fully marine fauna; mixed flora; floating Ammonites, Belemnite rostra[1]	Deposition from suspension in wave-influenced restricted or wide fully marine or brackish water embayment or lagoon with source for carbonaceous debris	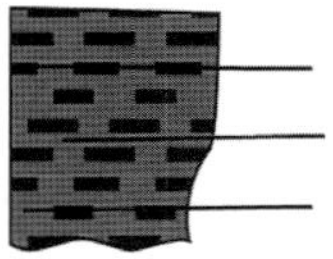	5B of mixed tide- and wave-influenced heterolithic delta 7B of wave-dominated open-marine 6C of brackish-marine embayment 5C† of mixed tide- and wave-influenced heterolithic delta

[Abbreviations: *Arenicolites (Ar); Asterosoma (As); Belemnite (Bel); Chondrites (Ch); Conichnus (Co); Cylindrichnus (Cy); Diplocraterion (Di); Monocraterion (Mo); Ophiomorpha (Op); Paleophycus (Pa) Palaeophycus herbertii (Pa-h); Phycosiphon (Ph); Planolites (Pl); Rhizocorallium (Rh); Rosselia (Ro); Skolithos (Sk); Taenidium (Ta); Teicichnus (Tei); Terebellina (Ter); Thalassinoides (Th)*].

[[1] *Bel rostra* = invertebrate skeleton (Belemnite). The term rostrum (or guard) designates the non-septate posterior portion of the belemnite phragmocone].

†Rare or uncommon occurrence.

Fig. 6. Petrographic thin-section photos show (A) chlorite-coated quartz grains; (B) glauconite between clastic quartz grains; and (C) iron-ooids in quartz-rich sandstone in the northern study area. (D) Photo and detailed sedimentary log show the stratigraphical occurrence of beds rich in iron-ooids at locality N4 (measured section 29). (E) A sedimentary log and photo of glauconite-rich sandstone facies and its stratigraphical occurrence at locality N1 (measured section 18). For localities see Fig. 4. (Photos J.M. Ahokas).

heterolithic, characterised by a very strong bimodal grain-size distribution. Sandstones in the studied succession are quartz-dominated. The feldspar content is generally low, but estimated up to about 20% in the most feldspar-rich sandstone beds. Mica is generally abundant. Chlorite and glauconite are also abundant at certain stratigraphical levels (Fig. 6). The quartz and feldspar grains are generally subrounded to rounded, thus indicating extensive reworking. Some of the conglomeratic beds include abundant angular quartz and quartzite grains, probably delivered directly as first generation detritus from basement rocks. Examples of distinct fluvial sedimentary structures are sparse due to overprinting by tidal reworking.

The genetically related lithofacies are subdivided into seven facies association groups (Table 2). These facies associations groups are the building blocks of architectural elements (Table 3) and genetic successions in the Neill Klinter Group. The depositional architecture of the Neill Klinter Group is summarised in Fig. 7 and localities in Fig. 4.

Bounding surfaces, allostratigraphic units, architectural elements

Lower stratigraphical compartment of the Neill Klinter Group

At all localities, the lower compartment is bounded at its base by a transgressive surface (TS1) and topped by a combined subaerial unconformity and transgressive surface (SU1/TS3; Fig. 7). The compartment is divided into i) retrogradational, ii) progradational to aggradational and iii) retrogradational to progradational architectural elements which comprise allostratigraphic units 1 to 3.

Allostratigraphic unit 1

Description: Retrogradational architectural elements

The base of the Neill Klinter Group was defined by Dam & Surlyk (1998) as a thin lag conglomerate bed with fragmented body fossils and quartzite pebbles. In this study the same boundary is referred to TS1, defining the base of the allostratigraphic unit 1.

Along the localities S1 to 8 (Fig. 7) the TS1 boundary marks a lithofacies change from alternating fine-grained to coarse-grained, wave ripple-laminated to cross-stratified sandstones and black laminated mudstones rich in macroplant fossils (Dam & Surlyk, 1992; this study) into structureless siltstones and cross-stratified sandstones containing marine body fossils, micro fossils and trace fossils (for example, *Diplocraterion parallelum*; Fig. 8A, i-iii). In light of these marine indicators, the terrestrial palynomorph assemblage is suggested to be allochtonous (Koppelhus & Dam, 2003). The up to 15 to 20 m thick upward-fining sandy succession above the TS1 is composed of lower shoreface-foreshore sandstone facies association FA4b (Table 2) and sharply overlain by thinly laminated silty shales. The sandstones of facies association FA4b are thinning northward and are not present at localities N1 to 4, where the corresponding stratigraphic level is an up to 25 m thick black shale unit rich in terrestrial and brackish-water palynomorphs and microfossils with minor amounts of marine specimens (this study). The shale unit is composed of brackish-marine embayment facies associations (FA6a and 6c; Table 2) and sharply overlies an up to 10 m thick light-coloured and poorly sorted quartz-rich sandstone unit. The top of the shale unit is marked by a few metres-thick upward-coarsening facies transition into sandy heterolithic strata (Fig. 8B).

Interpretation: Allostratigraphic unit 1

Based on the change in facies and fossil content, the basal unconformity of allostratigraphic unit 1 is interpreted as a basinwide marine transgressive ravinement surface, TS1, separating alluvial plain and lacustrine strata of the Kap Stewart Group from overlying shallow marine strata of the Neill Klinter Group (Figs 2 and 7). Above TS1, fossiliferous marine sandstones of the lower shoreface to foreshore sandstone facies association (FA4b) define the retrogradational architectural element shoreface-foreshore sandstone SHFS (Fig. 8A; Table 3) in the south. In the north, the shales of brackish-marine embayment facies associations (FA6a and 6c) constitute the contemporaneous wave-influenced brackish embayment architectural element, WIBE (Fig. 8B; Table 3). The along-strike shift in facies reflects basinward (south and west) increasing water depths, while the change in palynomorph fauna suggests a northward decrease in salinity. Combined, the south to north transition from SHFS to WIBE architectural element is interpreted to record an overall retrogradational depositional architecture.

The flooding surface FSa (Fig. 7) on top of the allostratigraphic unit 1 marks the drowning of the

Table 2. Summary of facies associations, facies, occurrence, architectural elements and allostratigraphic units of the Neill Klinter Group, Jameson Land Basin, East Greenland. († = rare occurrence).

FA	Facies	Locality	AE	AU1-8
1: FLUVIAL-TIDAL TRANSITION				
1A: tide-influenced fluvial channel-fill	2 Cross-stratified sandstone 1A Lithic conglomerate 6C Heterolithic ripple cross-laminated sandstone with wavy-bedding	S3-4 N1-4	FTT TDDP	AU4 AU6 AU2
1B: tide-dominated fluvial channel-fill	2 Cross-stratified sandstone 4 Cross-stratified sandstone with mud drapes 1A Lithic conglomerate 1B Mud clast conglomerate 6C Heterolithic ripple cross-laminated sandstone with wavy-bedding† 10 Structureless mudstone†	S3-4 N2-4 N1†	TDDP FTT	AU2 AU4 AU6
1C: tide-influenced fluvial mouthbar and bayhead delta / tidal channel complex	2 Cross-stratified sandstone 4 Cross-stratified sandstone with mud drapes 1A Lithic conglomerate 6C Heterolithic ripple cross-laminated sandstone with wavy-bedding† 9 structureless muddy siltstone and sandstone† 10 Structureless mudstone†	S3-4	FTT	AU4 AU6
2: TIDAL INCLINED HETEROLITHIC STRATA				
2A: sub-tidal inclined heterolithic strata	4 Cross-stratified sandstone with mud drapes 6A Fine-grained wave ripple cross-laminated sandstone 6C Heterolithic ripple cross-laminated sandstone with wavy-bedding 5 Compound cross-stratified sandstone with drapes 1A Lithic conglomerate† 1B Mud clast conglomerate† 10 Structureless mudstone†	S2-N4	TDDP TDSS TDEC†	AU2 AU4 AU8†
2B: mudstone-dominated to sandstone-dominated inclined heterolithic strata	2 Cross-stratified sandstone 4 Cross-stratified sandstone with mud drapes 6A Fine-grained wave ripple cross-laminated sandstone 6C Heterolithic ripple cross-laminated sandstone with wavy-bedding 1B Mud clast conglomerate†	S2-N4	FTT TDDP TDSS TDEC	AU4 AU2 AU7 AU8
2C: wave-dominated sandy heterolithic strata	7 Hummocky and swaley cross-stratified sandstone 6A Fine-grained wave ripple cross-laminated sandstone 4 Cross-stratified sandstone with mud drapes†	N2S2-8	TDSS	AU8 AU4
3: TIDAL CHANNEL INFILL				
3A: compound sub-tidal channel thalweg deposits	4 Cross-stratified sandstone with mud drapes 6C Heterolithic ripple cross-laminated sandstone with wavy-bedding 8 Structureless sandstone† 1A Lithic conglomerate†	S3-4 S6 N2-4	TDSS TDEC	AU4 AU7 AU6 AU8
3B: compound sandy 3D sub-tidal dunes	6C Heterolithic ripple cross-laminated sandstone with wavy-bedding 3 Low-angle cross-stratified sandstone 8 Structureless sandstone† 10 Structureless mudstone† 1B Mud clast conglomerate†	S1-N4	TDSS TDEC	AU2 AU4 AU7 AU6 AU8
3C: compound heterolithic sub-tidal dunes	4 Cross-stratified sandstone with mud drapes 6C Heterolithic ripple cross-laminated sandstone with wavy-bedding 8 Structureless sandstone†	S2-N4	TDSS TDEC TDDP	AU2 AU4 AU7 AU6 AU8

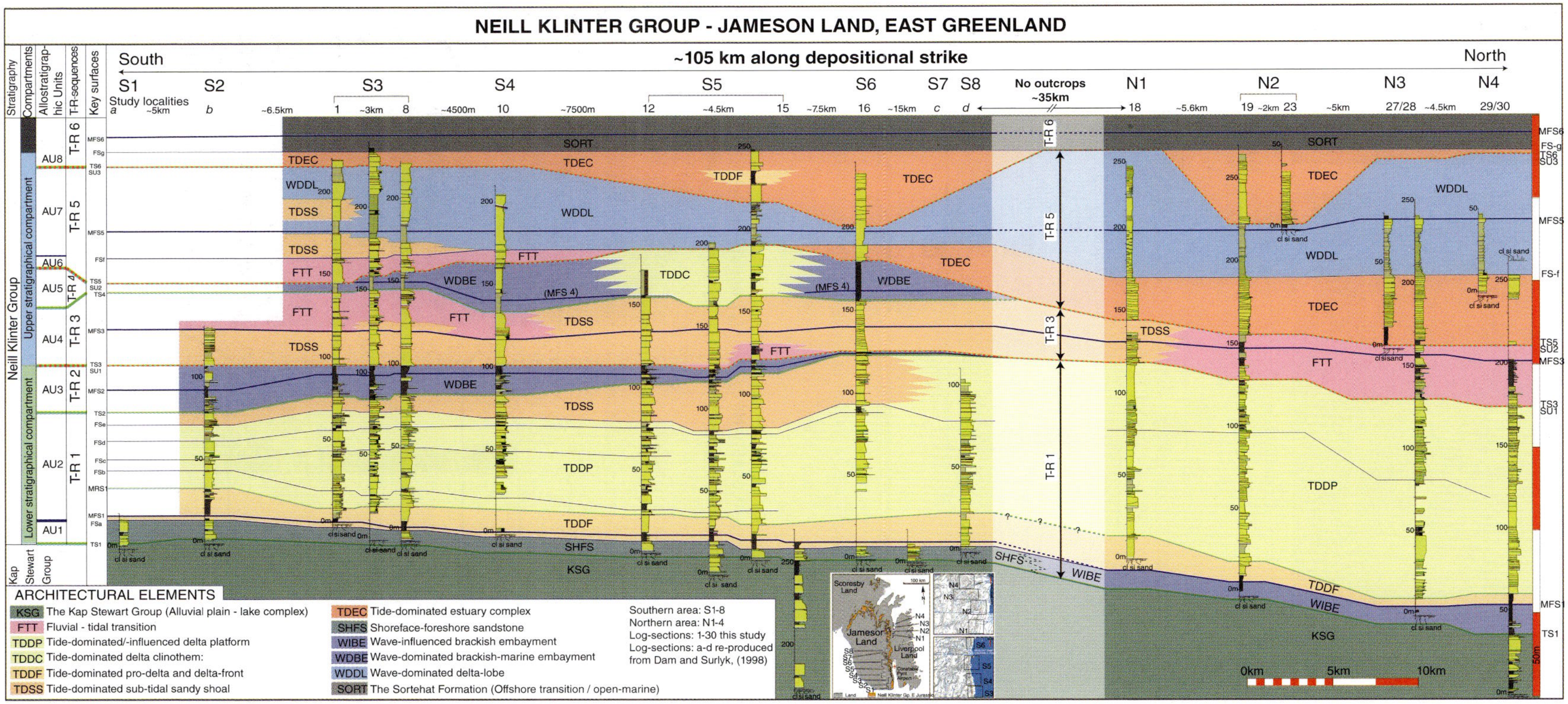

Fig. 7. The depositional architecture of the Neill Klinter Group summarised with lateral and vertical relationships of the eight allostratigraphic units (AU1 to 8) and ten architectural elements (AE) within the two major stratigraphical compartments of this study. The major unconformity separating the upper compartment from the Sortehat Formation, or the stratigraphically next identifiable key stratal surface below, is used as a datum for correlation of measured sections. For study localities see Fig. 4.

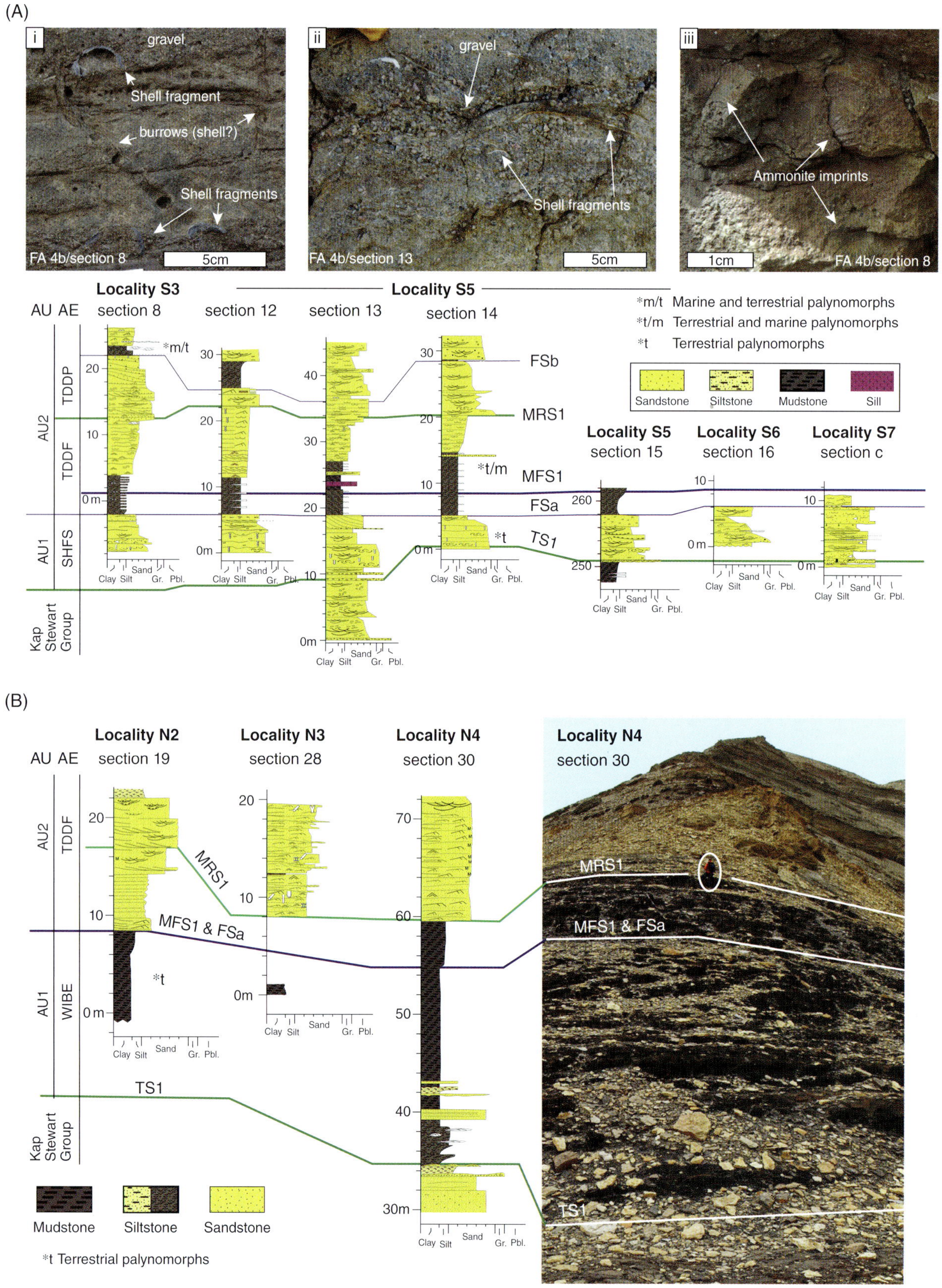

Fig. 8. (A) Selected detailed sedimentary logs show the correlatable poorly sorted and upward-fining facies character of the SHFS architectural element and the AU1 in the southern study area. The overlying TDDF and TDDP elements show upward coarsening facies trends and belong to the AU2. Photos of SHFS sandstone lithofacies show; i) diverse marine bioturbation; ii) co-existing shell fragments and gravel; and iii) marine fossil imprints. (B) Selected sedimentary logs show the mudstone-dominated WIBE architectural element replacing the SHFS within the AU1 in the northern study area. The AU2 show similar character as in the southern area. Photo of the WIBE element at locality N4 (measured section 30). (Photos J.M. Ahokas).

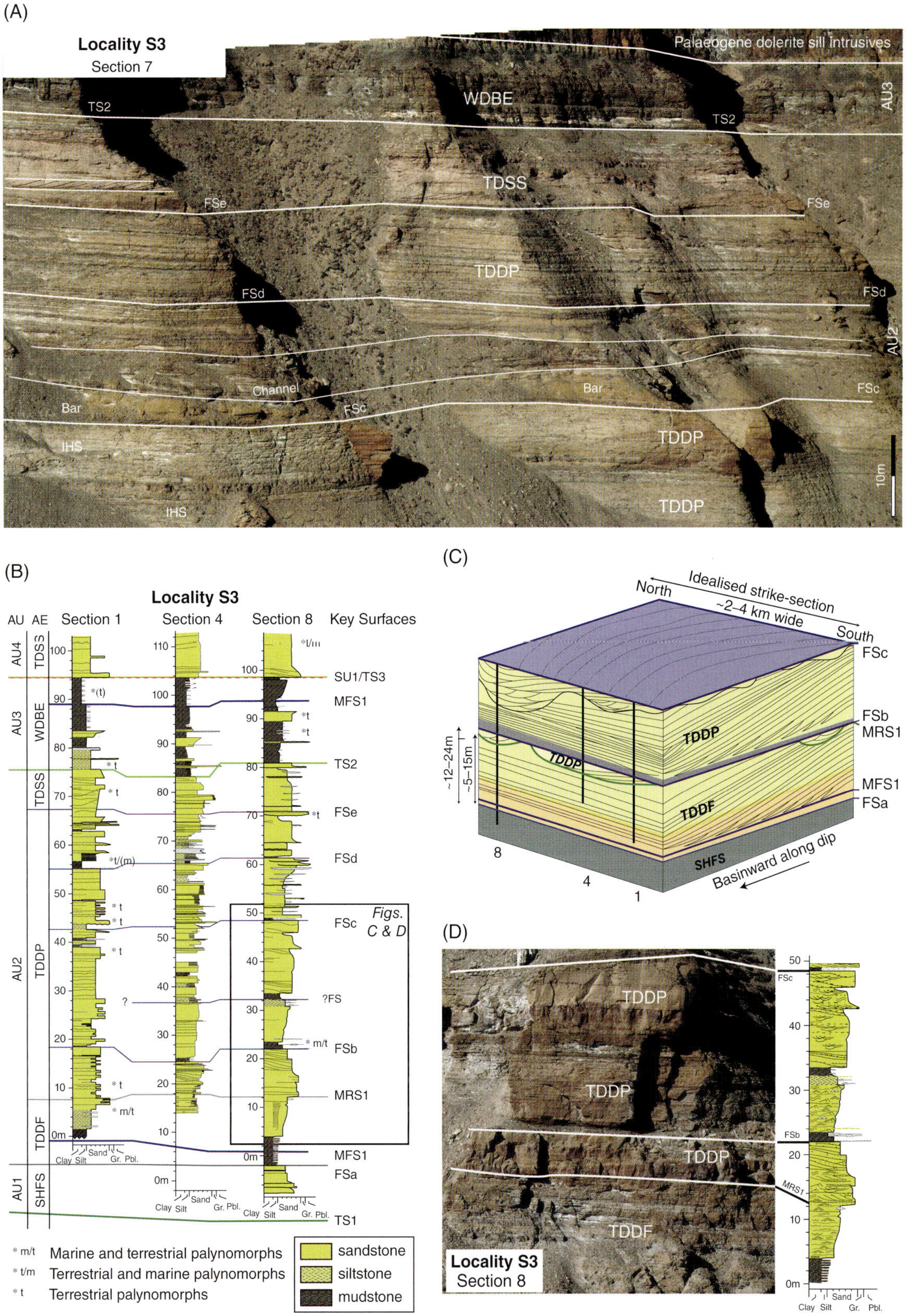

Fig. 9. (A) The sandstone-dominated architectural elements TDDP and TDSS are composed of traceable heterolithic upward-coarsening or upward-fining units (UCU and UFU) within the lower stratigraphical compartment. The mudstone-dominated heterolithic architectural element WDBE overlying the TDSS element in the southern area is commonly distorted by Palaeogene dolerite sill intrusives. The hummocky-stratified beds are seen as light-colored beds towards the base of the WDBE architectural element. (B) Traceability of the flooding surfaces within allostratigraphic unit 2 vary between the measured sections at locality S3. (C) A schematic presentation of the tide-dominated deltaic depositional system at locality S3. (D) The flooding surfaces and/or marine ravinement surfaces separate the TDDF and TDDP elements at locality S3 (measured section 8). (Photos J.M. Ahokas).

amount of plant debris, mixed siderite-cemented mudstone and clastic pebble-filled shallow channel-formed sandstone beds increase significantly towards the top of allostratigraphic unit 2.

Along the southern area (S2 to 6), the muddy to sandy basal part of the allostratigraphic unit 2 is 5 to 15 m thick and composed of tide-influenced heterolithic prodelta mudstone facies association (FA5b) overlain by tide-influenced sandy heterolithic delta front facies association (FA5a), forming together a single upward coarsening unit (Fig. 9B to D; Table 2). The prodelta silty mudstones of FA5b directly above the FSa contain small bivalves, marine and terrestrial shell debris and brackish to open marine microfossils. The upward-coarsening unit is topped by an erosive unconformity lined with thin, discontinuous and gravelly conglomerate along the base of a tide-dominated fluvial channel-fill facies association (FA1b; Table 2). This, up to 9 m thick, upward-fining sandstone unit is topped by a flooding surface (FSb).

Above the FSb in the southern area, the rest of the sandy heterolithic allostratigraphic unit 2 records laterally shifting upward-fining subtidal inclined heterolithic strata facies association (FA2a), upward-coarsening mudstone dominated to sandstone dominated inclined heterolithic strata facies association (FA2b), tide-influenced fluvial channel-fill facies association (FA1a) and sandy heterolithic tidal channel infill facies associations (FA3c, 3d and 3e) (Fig. 9B and D; Table 2). At localities S2 to 6, the 8 to 32 m thick uppermost part of allostratigraphic unit 2 thickens northward and is composed of up to 10 m thick upward-fining sandy tidal channel infill facies associations (FA3b and 3c) and sand-dominated inclined heterolithic strata facies association (FA2b) (Figs 7 and 9A; Table 2). Marine trace fossils (for example, *Diplocraterion parallelum*) are observed in the uppermost part of the allostratigraphic unit 2. The top of the allostratigraphic unit 2 is defined at the base of an up to 2 m thick structureless sandy facies transition into shale-dominated heterolithic beds.

In the northern area, the allostratigraphic unit 2 is thicker than in the southern area. Similarly to the southern area the lower part of allostratigraphic unit 2 is defined by a single upward-coarsening sandstone unit whereas the remaining part of the allostratigraphic unit 2 records superimposed upward-coarsening and upward-fining facies associations bounded by flooding surfaces. Dominance of terrestrial palynomorphs increases upwards in allostratigraphic unit 2 (Koppelhus & Dam, 2003; Ahokas *et al.*, 2014; this study). The up to 9 m thick upward-coarsening units are composed of mixed tide-influenced and wave-influenced heterolithic delta front to prodelta facies associations (FA5a and 5b, respectively) and tidal inclined heterolithic strata facies associations (FA2a and 2b; Table 2). The up to 6 m thick upward-fining units are composed of fluvial-tidal transition facies associations (FA1a and 1b; Table 2).

Several unconsolidated intervals containing chlorite-coated grains (Fig. 6A) are observed within the allostratigraphic unit 2 at locality N4. These 2 to 3 m thick greenish coloured intervals rarely include recognisable sedimentary structures. At all the northern localities, allostratigraphic unit 2 is terminated above by an erosive conglomeratic bed marking the shift from sandy heterolithic facies associations into locally shale-dominated facies associations with brackish-water palynomorphs (Ahokas *et al.*, 2014).

Interpretation: Allostratigraphic unit 2

At all localities, the basal upward-coarsening unit of the allostratigraphic unit 2, directly above the FSa, is interpreted as a tide-dominated prodelta and delta front (TDDF) architectural element (Figs 7 and 9; Table 3). The morphology of the bivalve shells above the FSa suggests that they lived in relatively shallow water, burrowing in sediments close to the sea bed where light conditions and amount of sediment in suspension have been sufficiently high. At all the southern localities, the first maximum flooding surface (MFS1) is picked within a condensed interval of silty mudstones of the TDDF element (Fig. 7). At the northern localities, the FSa flooding surface amalgamates with the MFS1; and this surface is picked at the transition from the brackish architectural element WIBE into the sandy heterolithic and mixed tide-influenced and wave-influenced deltaic strata of the element TDDF (Figs 7 and 8B).

The superimposed upward-coarsening and upward-fining units forming the bulk of the sandy heterolithic allostratigraphic unit 2 are interpreted as the tide-dominated delta platform (TDDP) and tide-dominated subtidal sandy shoal (TDSS) architectural elements (Fig. 9; Table 3). The erosive surface separating the distal TDDF from the proximal TDDP element is interpreted as a regressive marine ravinement surface (MRS1) based on the locally developed gravel lag and its erosive character along the base of tide-influenced channel-fill facies association. The mineralogical immaturity of the coarse fraction suggests short transport distances.

The four flooding surfaces in the TDDP element are interpreted as local features due to their laterally restricted extent, thin overlying shale beds and presence of marine palynomorphs. The surfaces are thought to reflect rapid local changes in relative sea-level in a tide-dominated deltaic depositional system that included several minor distributaries along a highly sinuous coastline. Although similar flooding surfaces in the northern area cannot be physically traced from one locality to the next, these may correspond to those in the southern area.

Allostratigraphic unit 2, overlying FSa in the southern area and MFS1 in the northern area, is interpreted to show an overall progradational to aggradational depositional style. Partly based on the more terrestrial signature of its palynomorph content, the northern area is interpreted to represent a more proximal deltaic depositional setting than corresponding deltaic architectural elements in the southern area. The progradational to aggradational depositional trend determined from the organisation of architectural elements reflects continuously higher rates of sediment input compared to the rate of creation of accommodation space ($A < S$). Laterally shifting channel infill facies associations and inclined heterolithic strata facies associations accompanied by the most proximal facies in the upper part of the TDDP element are interpreted as the most proximal deltaic deposits within allostratigraphic unit 2. The TDSS element recorded in the southern area represents deeper water deposits than the underlying TDDP element and indicates a landward shift in facies and thus a turnaround towards a rise in relative sea-level. The TDSS element is topped by the transgressive surface TS2 that defines the drowning of the distal tide-dominated deltaic depositional system in the southern area. In the northern area the uppermost delta platform deposits are eroded by the combined SU1/TS3 surface (Fig. 7).

Allostratigraphic unit 3

Description: Retrogradational to progradational architectural elements

The muddy heterolithic allostratigraphic unit 3 is up to 26 m thick and restricted to the southern area where it overlies the transgressive surface TS2 (Fig. 7). Allostratigraphic unit 3 is composed of offshore marine mudstone (FA7b) and wave-dominated brackish-marine embayment mudstone and siltstone (FA6a) facies associations (Fig. 9A and B; Table 2). Brackish and marine palynomorphs have been recorded in allostratigraphic unit 3 (Koppelhus & Dam, 2003). The top of allostratigraphic unit 3 is well defined by a sharp facies boundary with occasional rounded quartz pebbles along the base of overlying sandstones (Fig. 5B). The bounding surfaces amalgamate and allostratigraphic unit 3 pinches out north of locality S6 (Fig. 7).

The base of allostratigraphic unit 3 consists of an up to 2 m thick, fine-grained structureless or weakly cross-stratified sandstone bed. This bed passes northwards into multiple thinner, strongly bioturbated (*Diplocraterion parallelum*) and poorly sorted gravelly and pebbly beds (Figs 7 and 9B). Several hummocky-stratified sandstone beds, as well as hundreds of metres wide and few decimetres thick matrix-supported conglomeratic beds composed of well-rounded quartz and quartzite pebbles and elongated cobbles (<6 cm in diameter) are observed within the muddy-heterolithic lower 10 m of the allostratigraphic unit 3 (Fig. 9A and B). The matrix in the conglomeratic beds is the same silty mudstone to fine-grained sandstone as below and above these beds. The bed contacts are sharp, but no clear concave or convex lateral shape has been recorded. The upper half of allostratigraphic unit 3 is better sorted and more sand-rich and shows a slight upward-coarsening trend. Allostratigraphic unit 3 is also distorted by up to 15 m thick horizontal dolerite sills (Fig. 3) and thinner vertical dykes of Palaeogene age (c.f. Larsen & Marcussen, 1992).

Interpretation: Allostratigraphic unit 3

The basal surface of allostratigraphic unit 3 is interpreted as a marine transgressive ravinement surface (TS2) representing a shift from aggradational to retrogradational depositional trend. The mudstone-dominated facies associations in the allostratigraphic unit 3 are interpreted to compose the wave-dominated brackish-marine embayment (WDBE) architectural element (Fig. 9A and B; Table 3). The northward increase in abundance of coarse-grained beds in the lowermost part of the allostratigraphic unit 3 may be due coastal instabilities and along-strike facies variation in a northward shallowing depositional environment. The mudstone-dominated sediments that comprise the lower half of allostratigraphic unit 3 are interpreted to have accumulated in a retrograding prodelta depositional environments. The several laterally restricted pebbly debris flow beds and associated tempestite sandstone beds within the lower half of the allostratigraphic unit 3 are interpreted as storm induced deposits (cf. Dam & Surlyk, 1998) due to the size of beds and the lack of lensoidal development and because no connection to a fluvial source is observed.

Due to heating by Palaeogene dolerite intrusives, palynomorphs are scarce and dominated by resistant terrestrial forms. Despite this, based on facies and its limited occurrence the depositional environment is interpreted as brackish-marine (cf. Koppelhus & Dam, 2003) because the thinner-walled marine palynomorphs are assumed destroyed by the heating, whereas the thicker-walled terrestrial palynomorphs survived the high temperatures.

The maximum flooding surface MFS2 is picked within the most distal facies of the WDBE element and reflects a shift in depositional trend from retrogradational into progradational (Fig. 7). Erosion due to a subsequent fall in relative sea-level, as indicated by the overlying unconformity SU1/TS3, is interpreted to have removed any contemporaneous deposits of allostratigraphic unit 3 (WDBE element) and the TS2 and MFS2 surfaces from the northern area.

Upper stratigraphical compartment of the Neill Klinter Group

At all localities, the upper compartment of the Neill Klinter Group is bounded at its base by a combined SU1/TS3 unconformity and topped by flooding surface (FSg) which equals the base of the Sortehat Formation (Fig. 7). The compartment is subdivided into five retrogradational to progradational allostratigraphic units (AU4 to 8), including three incised valley infill successions.

Subaerial unconformity SU1 and transgressive surface TS3

Description

In the southern area, the bounding surface between mudstones of allostratigraphic unit 3 and the overlying sandstones of allostratigraphic unit 4 is characterised by discontinuous gravel and pebble beds with quartz, quartzite and gneiss clasts (Fig. 5B). The boundary is subhorizontal at localities S1 to 4 and irregular between S5 and S6 (Fig. 7). In the northern area, the unconformity is picked along a patchy conglomeratic bed that separates sandstones of the TDDP element from overlying sandy and muddy heterolithic deposits.

Interpretation: Subaerial unconformity SU1 and transgressive surface TS3

The basinwide unconformity is interpreted as the combined subaerial unconformity and transgressive surface SU1/TS3. The surface is interpreted as having formed by a fall in relative sea-level, leaving the north-eastern part of the Jameson Land Basin exposed to river incision and fluvial transport and deposition of coarse-grained extrabasinal lithic debris. The presence of extrabasinal pebbles on the SU1 is critical in this interpretation as these clasts must have been transported into the basin by rivers. The succeeding rise in relative sea-level and increase in accommodation space led to tidal reworking of the fluvial deposits and intratidal to subtidal deposition. The fall and rise in relative sea-level took place along a subhorizontal surface in the southern area, whereas the dip and relief of the same surface in the northern area is interpreted to reflect the base of an up to several tens of kilometres wide incised valley which has cut through MFS2 and TS2.

Allostratigraphic unit 4

Description: Retrogradational to progradational architectural elements – incised valley fill system 1

The sandstone-dominated allostratigraphic unit 4 is 24 to 45 m thick, bounded from below by the combined SU1/TS3 surface and on top by either TS4 or the combined SU2/TS5 surface (Fig. 7). In the southern area allostratigraphic unit 4 is composed of tidal inclined heterolithic strata facies associations (FA2a and 2b), wave-dominated sandy heterolithic strata facies association (FA2c) and upward-fining and heterolithic tidal channel infill facies associations (FA3a, 3b and 3d), with occasional conglomeratic base (Figs 10A to B, 11A, 11E to H and 12A to E; Table 2).

At localities S2 to 4, the stacked and laterally interfingering upward-fining and upward-coarsening heterolithic units are 6 to 18 m thick; whereas at localities S5 to 6, stacked sandy tidal channel infill facies associations (FA3a, 3b and 3d) attain thicknesses of up to 40 m in places (Fig. 12A and B). At locality S3, the sandy allostratigraphic unit 4 includes locally greenish coloured and up to 5 to 6 m thick beds which rarely include recognisable sedimentary structures. These beds are unconsolidated, probably due to the effect of chlorite coating of detrital quartz grains, observed in thin sections (Figs 6A and lowermost part of Fig. 10A; cf. Ehrenberg, 1993). Other unconsolidated stratigraphic levels without chlorite are also observed. The top part of allostratigraphic unit 4 is defined by poorly sorted fluvial-tidal

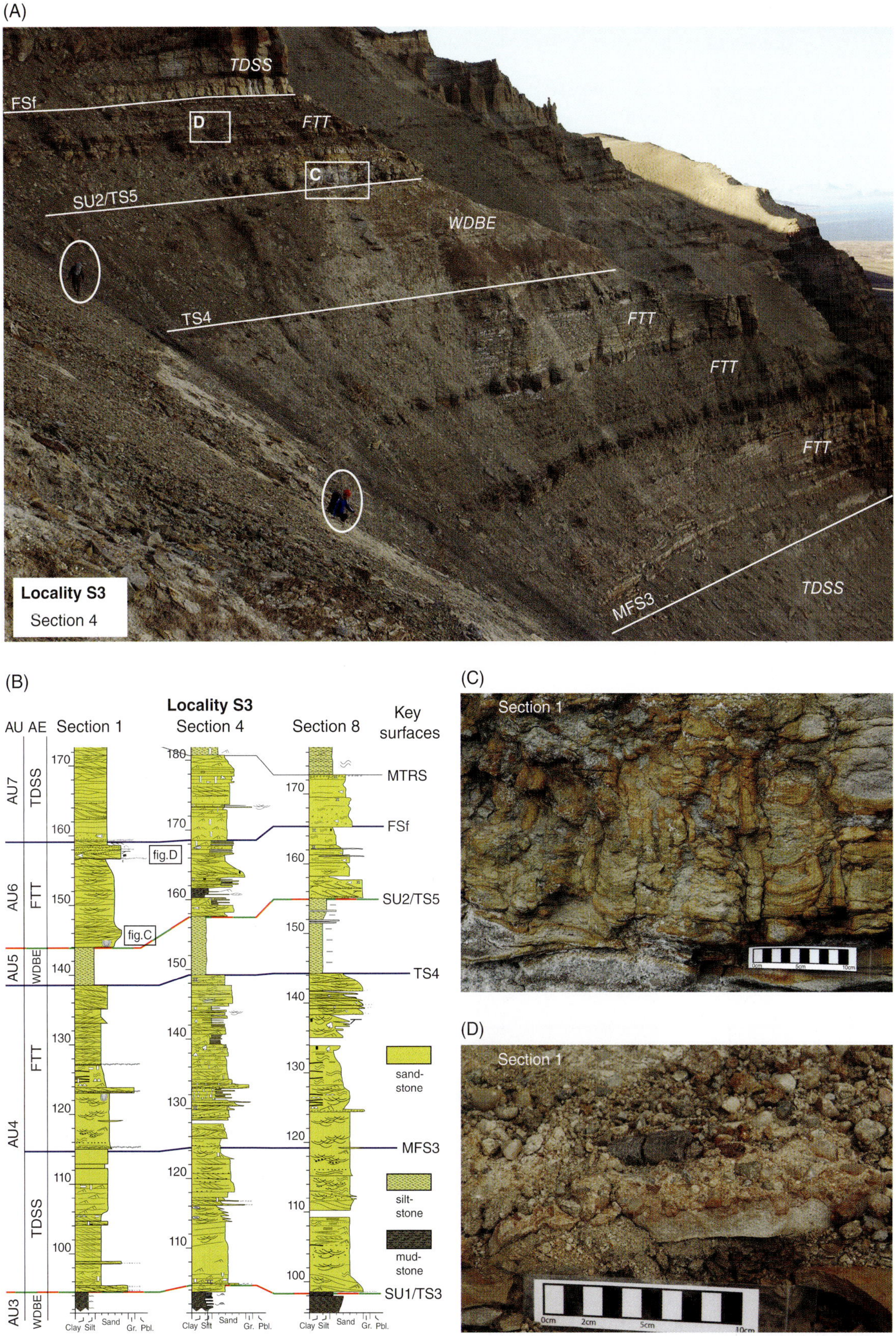

Fig. 10. (A) At locality S3 the architectural elements TDSS, FTT and WDBE are stacked in laterally identifiable successions within the upper stratigraphical compartment (measured section 4, persons circled for scale). (B) Detailed sedimentary logs through the AU4 to AU7 show sedimentology, architectural elements and key surfaces at locality S3. (C) Photo of abundant *Diplocraterion parallelum* burrows along base of FTT element above the SU2/TS5 surface at locality S3 (measured section 1). (D) Photo of a marine transgressive ravinement surface with belemnite fragment (<2 cm in Ø) and quartz and quartzite gravel and pebbles the FTT element at locality S3 (measured section 4). (Photos J.M. Ahokas).

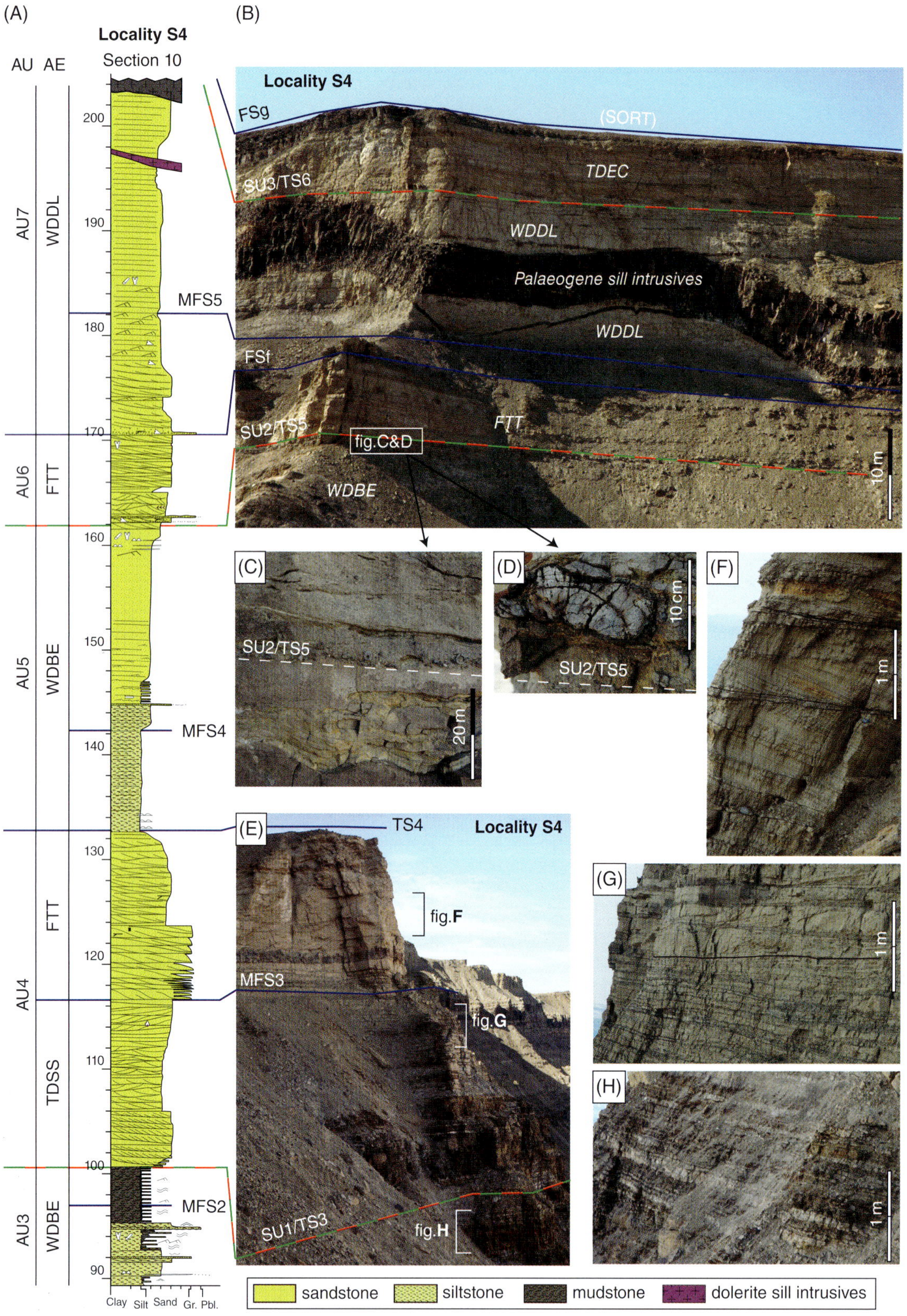

Fig. 11. (A) A detailed sedimentary log through AU3 to AU7 showing sedimentology, architectural elements and key surfaces within the upper stratigraphical compartment at locality S4 (measured section 10). The compartment is here sandstone-dominated and composed of upward-coarsening and -fining units. (B) Photo of the locality S4. In the uppermost part of the compartment, the sandstone-dominated WDDL architectural element is split by an irregular 10 to 15 m thick Palaeogene dolerite sill intrusion. Photographs of lithofacies above the SU2/TS5 unconformity contain; (C) mixed belemnites and gravel; and (D) rare silicified tree trunks. (E) At locality S4, the lower part of the upper stratigraphical compartment is defined by a cleaning-upward depositional trend from muddy heterolithic to clean sandstone deposits. (F–H) Photos of heterolithic lithofacies within the allostratigraphic unit AU4. Upward increase in bed thickness and sand content indicate higher rate of sediment input than rate creation of accommodation space. (Photos J.M. Ahokas).

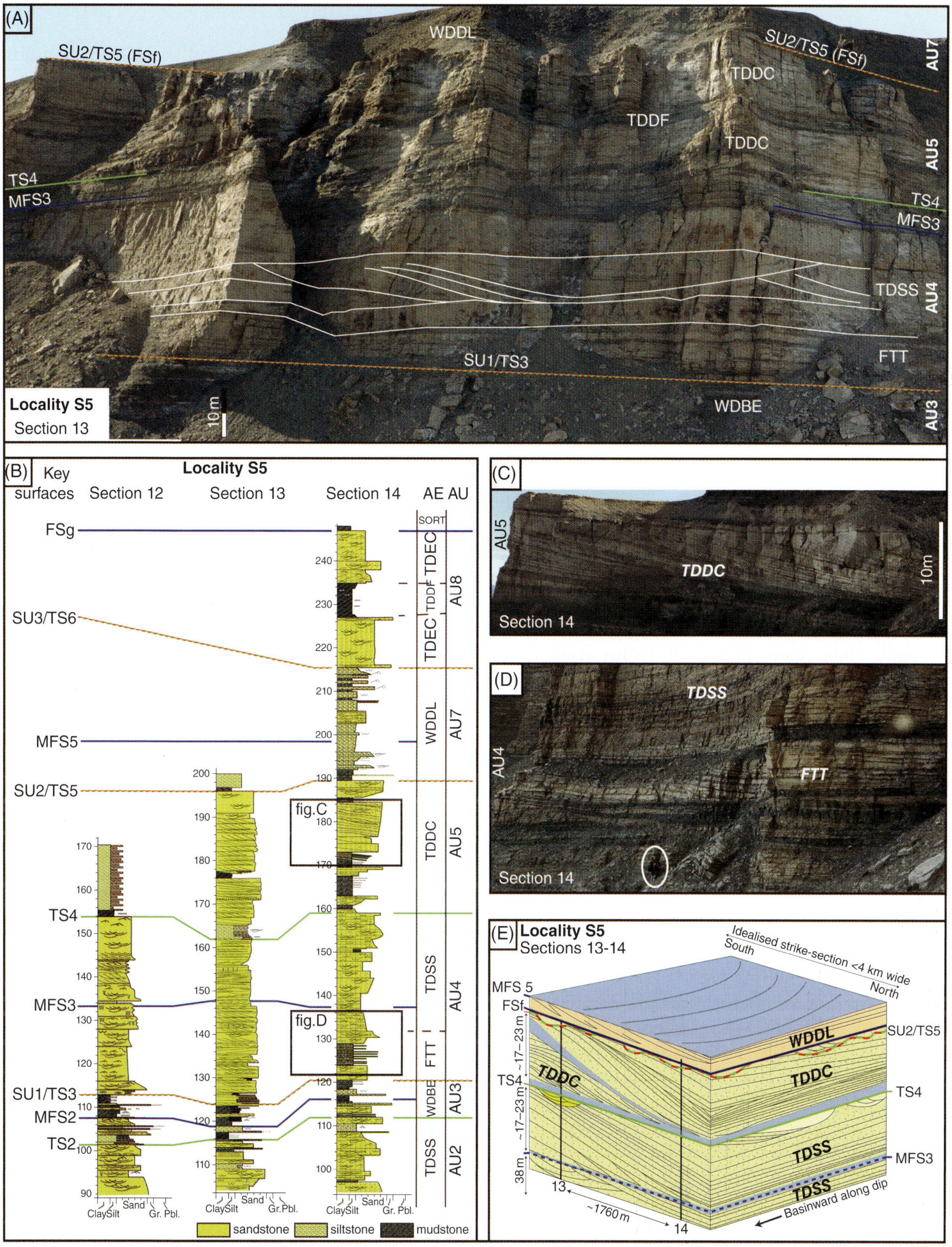

Fig. 12. (A) Photo of locality S5 (measured section 13) show the thick sandstone-dominated TDSS and FTT architectural elements indicating high rate of sediment input and rate of creation of accommodation space. The AU4 is interpreted as a major distributary system (10 to 15 m thick and tens of metres-wide) at this locality. The overlying sandy heterolithic deltaic TDDC and muddy heterolithic TDDF elements define the local deltaic system within an overall retrogradational AU5. Note that the allostratigraphic unit AU6 is absent in this locality. (B) Detailed sedimentary logs show sedimentology, architectural elements and key surfaces at locality S5 (measured sections 12 to 14). (C) Photo of a 12 m thick upward-coarsening deltaic architectural element TDDC in section 14. (D) Photo of laterally accreting heterolithic deposits of architectural element FTT and overlying stacked heterolithic deposits of TDSS element in section 14 (person circled for scale). E) A schematic presentation depositional architecture of TDDS and TDDC elements across sections 13 to 14 at locality S5. (Photos J.M. Ahokas).

transition facies associations (FA1a, 1b and 1c; Table 2) with abundant marine body fossils and trace fossils (similar to Fig. 10C). These facies associations pass laterally into the tidal channel infill facies associations (FA3a and 3b) north of locality S4 (Fig. 7; Table 2).

In the northern area, allostratigraphic unit 4 is initiated by the SU1/TS3 surface that separates either poorly cemented silty sandstones of the TDDP element of allostratigraphic unit 2 from overlying sandy heterolithic deposits of allostratigraphic unit 4 (locality N1), or inclined heterolithic deposits of the TDDP element of allostratigraphic unit 2 from overlying muddy heterolithic deposits of allostratigraphic unit 4 (localities N2 to 4; Fig. 7). At locality N1, the 24 m thick allostratigraphic unit 4 is composed of sandy heterolithic tidal channel infill facies associations (FA3c, 3d, 3e; Table 2). These facies associations are coarse-grained, poorly sorted and rich in mudstone pebbles and beds are thinner than at localities S5 to 6. At localities N2 to 4, the 25 to 38 m thick allostratigraphic unit 4 is composed of sandy fluvial-tidal transition facies associations (FA1a and 1b) and the wave-influenced brackish embayment mudstone facies association (FA6c) with both terrestrial and brackish palynomorphs (Tables 1 and 2) (Ahokas *et al.*, 2014).

Allostratigraphic unit 4 is topped in the southern area by a sharp facies shift into brackish-marine silty shale, whereas in the northern area the corresponding unconformity is defined by a change into tidal sandstones.

Interpretation: Allostratigraphic unit 4

Based on tide-influenced shallow marine to fluvial facies associations and variable palynomorph assemblage (Koppelhus & Dam, 2003; this study), the allostratigraphic unit 4 is interpreted to be composed of the tide-dominated subtidal sandy shoal (TDSS) and the fluvial-tidal transition (FTT) architectural elements (Figs 7, 10A and B, 11A, 11E to G and 12; Table 3). Collectively, allostratigraphic unit 4 records an overall retrogradational to progradational depositional architecture formed subsequent to drowning of the original subaerial unconformity SU1 (Fig. 7). During marine flooding and transgression, the fluvial deposits on top of the subaerial unconformity were reworked by tidal currents. Deposition took place in contemporaneous tide-dominated estuarine and more open wave-influenced environments with constant high rate of sediment input and variable water depths. At this stage, a deltaic distributary environment probably existed around locality S5. The changing depositional architecture along the combined surface SU1/TS3 from south to north suggests that the initial basinward shift in facies (i.e. towards west) was greater in the south than in the north. This resulted in deposits of more proximal character and more subtle shift in facies in the north. Above the SU1/TS3 unconformity, the tide-dominated subtidal sandy shoal (TDSS) and fluvial-tidal transition (FTT) elements appear to fill in a relatively low relief depression cut by that initial subaerial erosion (Fig. 7). This erosional feature is interpreted to be part of a wide topographic depression, itself interpreted as an incised valley system.

Maximum flooding surface MFS3 is difficult to pick because allostratigraphic unit 4 is sand-rich and without any apparent marine shale unit. MFS3 is therefore placed at the base of the FTT element in the southern area. In the northern area, MFS3 is put right above the TDSS element and within the FTT element where marine palynomorphs become dominant. The top bounding surface of allostratigraphic unit 4 is interpreted as the next transgressive surface (TS4). In the northern area this surface is cut by a major unconformity (SU2/TS5) (Fig. 7).

Allostratigraphic unit 5

Description: Retrogradational to progradational architectural elements

The muddy and sandy heterolithic allostratigraphic unit 5 overlies the transgressive surface TS4 and is restricted to the southern area (Figs 5D and 7). At localities S3 to 4 and S6 the TS4 unconformity is characterised by a cm-thick bed of angular quartz and quartzite gravel. The boundary is associated with a sharp facies change from underlying sandstones of allostratigraphic unit 4 into wave-dominated brackish marine embayment mudstone and siltstone and offshore marine mudstone facies associations (FA6a and FA7b, respectively) of allostratigraphic unit 5 (Figs 7 and 10A to B; Table 2). The mudstone facies associations are rich in mixed marine palynomorphs and terrestrial debris (Koppelhus & Dam, 2003; this study).

At locality S5, the up to 35 m thick allostratigraphic unit 5 is composed of heterolithic tidal deltaic clinoform facies association (FA5c) (Fig. 12A to C and 12E; Table 2). The stacked individual heterolithic clinothems are upward-coarsening and up to 21 m thick; their geometry and orientation are relatively uniform with strata dipping towards west-north-west. To the north and south of locality

S5, the clinothems are inferred to grade into WDBE silty mudstones, as no associated incision surface has been recorded.

The allostratigraphic unit 5 is truncated beneath the SU2/TS5 surface both north and south of locality S5. The scale of truncation is up to 20 m to the south and up to 40 m to the north (Fig. 7).

Interpretation: Allostratigraphic unit 5

Based on the change in facies and palynomorph surface TS4 beneath, allostratigraphic unit 5 is interpreted as having formed by marine ravinement during regional transgression. Allostratigraphic unit 5 is unique in the Neill Klinter Group because it contains contemporaneous deposits of i) silty mudstone units formed in a wave-dominated brackish-marine embayment (WDBE architectural element; Figs 7 and 10A to B; Table 3) and ii) stacked units of tide-dominated delta clinothems (TDDC architectural element; Figs 7, 12A to C and 12E; Table 3). The association of these two architectural elements is restricted to the southern area and can be explained by the formation of different adjacent paralic subenvironments in response to coast-parallel differences in sediment input subsequent to transgression across TS4.

A major tide-dominated delta system, defined by the TDDC element, was established on top of the underlying distributary system (allostratigraphic unit 4, locality S5) during the rise in relative sea-level. The contemporaneous embayments with low sedimentation rates hosted the brackish to normal marine silt-rich mud (WDBE element). The TDDC element shows lateral sediment migration, as suggested earlier by Dam & Surlyk (1998). The mud fraction of the sediment load was distributed basinward and into bays adjacent to distributary mouths by suspension plumes, whereas coarser sediments were deposited as prograding delta lobe clinothems or mouth bars by bottom currents.

Picking the maximum flooding surface MFS4 is difficult due to a lack of well-defined distal facies. The MFS4 surface is placed within the lower part of the WDBE element. In the area of locality S5, MFS4 is cut by the lower boundary of the clinothem element TDDC and is therefore inferred to have been located in a more distal position on the downlap surface of the lowermost progradational clinothem. Only the silty shales of the WDBE element were eroded along the succeeding subaerial unconformity SU2 (see below) south and north of locality S5 that formed during a fall in relative sea-level.

Subaerial unconformity SU2 and transgressive surface TS5

Description

South of locality S5, the bounding surface between silty mudstones of allostratigraphic unit 5 and overlying sandstones of allostratigraphic unit 6 is characterised by a few decimetres-thick conglomerate bed with quartz and quartzite pebbles, marine body-fossil debris (belemnites) and decimetre-scale wood fragments (Fig. 11C and D). In places, the bed is extensively bioturbated by *Diplocraterion parallelum* (Fig. 10C). Mapping of the surface underneath the conglomerate bed defines up to 20 m-deep incision into underlying WDBE element strata (allostratigraphic unit 5; Figs 5A and 7). From locality S5 to S6, the same unconformity cuts 9 m into underlying WDBE element strata (allostratigraphic unit 5) and is draped by a centimetre-thick, gravelly sandstone bed with abundant shell imprints and oyster fragments. In the northern area, the unconformity is defined by the overlying i) 10 cm thick lithic conglomerate bed with locally abundant marine body fossils (*belemnite rostra*) and/or ii) mud pebble beds, or iii) an up to 1 m thick matrix-supported conglomerate bed with lithic pebbles and cobbles (up to 9 cm in diameter) composed of quartz, quartzite and gneiss (similar to locality N3, section 28; Figs 5C, 13A and D). Along the northern area, the erosional surface records an incision up to 40 m deep.

Interpretation: Subaerial unconformity SU2 and transgressive surface TS5

This basinwide unconformity is interpreted as a combined subaerial unconformity and transgressive surface SU2/TS5. The subaerial unconformity surface is interpreted to have formed by fluvial erosion and marine regressive ravinement during a fall in relative sea-level. Fluvial gravel was subsequently deposited onto the SU2. During the following rise in relative sea-level, marine transgressive ravinement and sedimentation obliterated much of the previous fluvial erosional and depositional features formed along the unconformity. The strata above the transgressive surface TS5 thus record a change from progradational depositional architecture, formed during the regressive stage, into retrogradational architecture, formed during the transgressive stage. Lateral variation in incision was controlled by the interplay between rate of fall and rise in relative sea-level, slope gradient variation along the coast, rate of sediment supply, distance

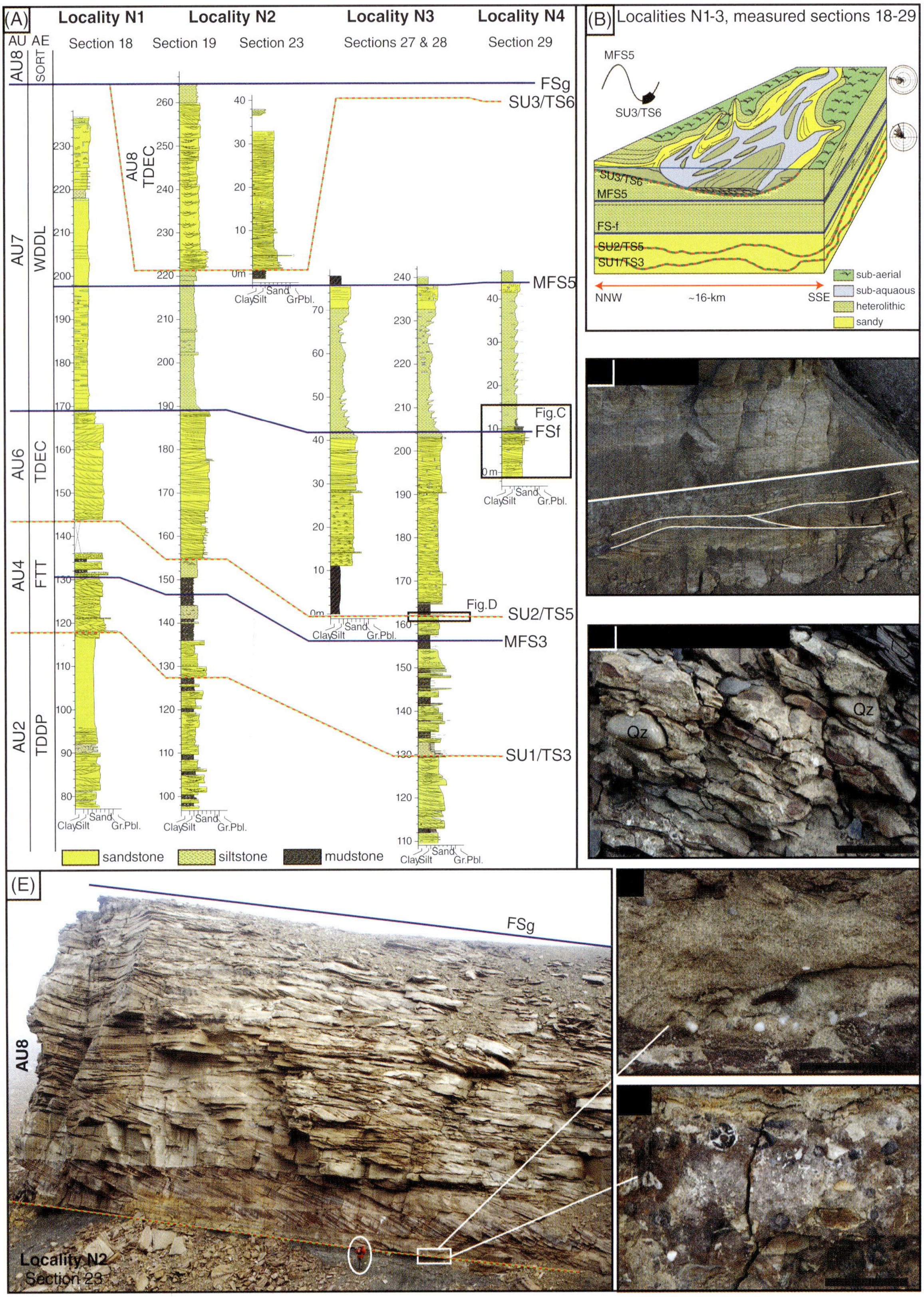

Fig. 13. (A) Detailed sedimentary logs (depths in m) showing sedimentology, architectural element and key surfaces in the northern area (localities N1 to 4). (B) A schematic presentation of a possible depositional scenario for the AU8. Note the relatively similar palaeocurrent directions. (C) Photo of upward facies transition across the flooding surface FSf that separates the sandy heterolithic TDEC and muddy heterolithic WDDL elements at locality N4 (measured section 29). (D) Photo of matrix-supported conglomerate bed directly overlying the combined SU2/TS5 unconformity surface at locality N3 (measured section 28). (E) Photo of the AU8 succession at locality N2 (measured section 23). Photos of mixed conglomerate bed directly overlying the SU3/TS6 unconformity surface along base of TDEC element (measured section 23); (F) quartz clasts; (G) mixed lithic clasts (quartz and quartzite), belemnite fragments and mud clasts. (Photos J.M. Ahokas).

to sediment source, underlying substrate and the relative impact of erosion and deposition by fluvial and tidal currents and waves along the shoreline.

Allostratigraphic unit 6

Description: Retrogradational architecture element – Incised valley fill system 2

At localities S3 to 4, northward thinning, 8 to 16 m thick sandstone-dominated allostratigraphic unit 6 overlies the SU2/TS5 surface. This unit has not been recorded at locality S5, but from locality S6 it thickens northwards from 9 to 44 m (Fig. 7). Allostratigraphic unit 6 is draped by the flooding surface FSf at all locations.

At localities S3 to 4, allostratigraphic unit 6 is composed of one to three upward-fining sandstone units of fluvial-tidal transition facies associations (FA1a, 1b and 1c; Figs 11A to D; Table 2). These facies associations are coarse-grained and poorly sorted with abundant lithic conglomerate beds and marine body fossils (Fig. 10D). At locality S3, the top of allostratigraphic unit 6 is a sharp sandstone-on-sandstone contact against overlying gravelly structureless bed of allostratigraphic unit 7. The boundary records an increase in floating marine body fossils and an abundance of trace fossils. At locality S4, allostratigraphic unit 6 is overlain by cross-stratified sandstone beds of allostratigraphic unit 7.

At localities S6 to N4, allostratigraphic unit 6 is composed of sandy tidal channel infill facies associations (FA3a and 3b) or heterolithic tidal channel infill facies associations (FA3c and 3d), wave-dominated middle to upper shoreface sandstone facies association (FA4a) and rare units of wave-influenced brackish embayment mudstone facies association (FA6c) (Fig. 13A and 13C to D; Table 2). The sandy strata within allostratigraphic unit 6 are relatively homogenous in grain size and organised in both upward-fining and upward-coarsening units. The beds are characterised by their content of marine palynomorphs (Ahokas *et al.*, 2014) but display a low degree of bioturbation of mostly by *Ophiomorpha* and *Skolithos*. The thickness of cross-stratified bed sets varies between 1 and 4 m. At localities N1 and 2, the upper 10 m of the allostratigraphic unit 6 is dominated by coarse-grained trough cross-stratified sandstone (of FA3a and 3b) with a distinct green colour due to abundant glauconite grains (Fig. 6B and E; Table 2). These bed sets are composed solely of quartz and glauconite. At localities N3 and 4 the same stratigraphic level is defined by iron-ooid concentrated beds (Fig. 6C and D).

The lithologically sharp surface on top of allostratigraphic unit 6 in the northern area is subhorizontal and separates cliff-forming sandstones from commonly loose sand-shale heterolithic lithologies above (Fig. 13A and C).

Interpretation: Allostratigraphic unit 6

Based on the facies and mixed palynomorph assemblage (Koppelhus & Dam, 2003) deposits within allostratigraphic unit 6 in the southern area are interpreted as the fluvial-tidal transition (FTT) architectural element (Figs 10A and B, 11A and B; Table 3). Facies associations within allostratigraphic unit 6 in the nortern area are collectively interpreted as the tide-dominated estuary complex (TDEC) architectural element (Fig. 13A and 13C; Table 3). These architectural elements, together with the SU2/TS5 unconformity, define the base of a southern and a northern incised valley system.

The southern valley-fill succession is interpreted to record the fluvial-tidal transitional architectural element FTT deposited during an overall rise in relative sea-level within the inner part of a flooded incised valley. Here, the repeated high-frequency regressive and transgressive sea-level fluctuations caused a dynamic mixture of fluvial and marine processes and created the marine ravinement surfaces characterised by originally fluvial conglomeratic beds mixed with marine body fossils. In this estuary bay-head delta environment, sedimentation was influenced by tidal currents in fluvial distributaries and mouthbars, whereas in the mouth of the estuary sediments were dominantly reworked by tides. The retrogradational architecture of the southern valley fill succession is demonstrated by beds of progressively distal strata capping the fluvial-tidal transitional FTT element advancing in an eastward and landward direction.

The northern valley-fill is interpreted to record the tide-dominated estuary complex architectural element TDEC deposited during transgression. Sediments are inferred to have been derived from the landward side and deposited in an at least 30 km-wide incised valley, as judged from the thickness and width of the observed depositional package as well as from field mapping (Figs 7, 13A). During rise in relative sea-level and continued infilling of the incised valley, the sediments were exposed to tidal reworking, obscuring all fluvial sedimentary structures. The grain size, sedimentary structures and fewer beds, indicating multiple high frequency changes in relative sea-level, are more in favour of a deeper estuary depositional setting than in the southern incised valley system. Both incised valley fill successions are draped on top by the same flooding surface, FSf (Figs 10A and B, 11A and B, 13A and B) and represents the stage when the sea also flooded inter-valley areas at all localities (Fig. 7).

The over 50 km-wide poorly exposed area between localities S6 and N1 may conceal another valley-fill system since the widths of the two observed incised valley systems are in the order of some tens of kilometres. Alternatively, an inter-valley land area or a shallow marine embayment may be present within the unexposed area.

Allostratigraphic unit 7

Description: Retrogradational to progradational architectural elements

At all localities, the sandstone-dominated to sandy mudstone-dominated allostratigraphic unit 7 is 12 to 75 m thick and overlies the flooding surface FSf (Fig. 7). Allostratigraphic unit 7 is dominated by sandy heterolithic wave-dominated open marine facies associations (FA7a and 7b) but includes also brackish-marine embayment facies associations (FA6a and 6b) (Figs 11A and B, 12A, B and E and 13A; Table 2). At all the localities except for N1, allostratigraphic unit 7 is terminated on top by the SU3/TS6 unconformity (Fig. 7).

At locality S3, up to 14 m of basal allostratigraphic unit 7 is defined by southward-thickening, upward-coarsening, mudstone-dominated to sandstone-dominated inclined heterolithic strata facies association (FA2b) and heterolithic tidal channel infill facies associations (FA3a, 3b and 3c) with dominantly marine palynomorphs (Figs 7, 10A and B; Table 2). The base of these cross-stratified sandstone facies associations of allostratigraphic unit 7 are either an up to 2 m thick, greenish upward-fining bed with gravelly base (Fig. 5F), or well sorted light coloured structureless sandstone bed (Fig. 10A). These facies associations also record slight upward increase in amount of body fossils and intensity of trace fossils (Fig. 5A and 10B). The upper boundary of the sandstone-dominated facies associations of allostratigraphic unit 7 is an erosive surface locally covered by a thin conglomerate bed composed of quartz and quartzite pebbles and marine body fossil fragments (Fig. 5E). Across this erosive surface, the diversity and amount of trace fossils increase abruptly in the heterolithic low-angle delta-lobe facies association (FA7a).

At localities S3 to 6, the open marine facies associations of allostratigraphic unit 7 are composed of 10 to 15 m thick, heterolithic upward-coarsening units capped by up to 1.5 m thick and several hundred metres-wide well-cemented fossiliferous sandstone beds (Fig. 11A and B). These fossiliferous beds contain abundant oyster shells with a diameter of up to 10 to 15 cm and shell walls up to 1 cm thick, belemnites, shell remains and local concentrations of abundant small shells (<2 cm in diameter). The degree of bioturbation and species diversity is generally high. Locally, upward-coarsening units may also be capped by few metres-thick bedsets composed of up to 1 m thick cross-stratified sandstone beds of low-angle delta-lobe facies association (FA7a; Table 2).

At locality N1, the well-cemented beds are restricted to the lower half of allostratigraphic unit 7, whereas the upper half of the unit is chiefly wave-dominated and bioturbated heterolithic sandstones of low-angle delta lobe facies association (FA7a) with floating belemnites and high degree and diversity bioturbation compared to other allostartigraphic units. At locality N2, no sandstone beds are observed in allostratigraphic unit 7 which is here composed of poorly consolidated silty brackish-marine embayment facies associations (FA6a and 6b; Table 2). At localities N3 and 4, a single bed of fossiliferous hardground facies association (FA6b) is observed within stacked metre-thick, upward-coarsening and heavily bioturbated muddy heterolithic beds of low-angle delta-lobe facies association (FA7a) (Fig. 13A; Table 2). Biostratigraphy samples collected from the northern area include brackish palynomorphs at locality N2 (Ahokas *et al.*, 2014).

Interpretation: Allostratigraphic unit 7

Based on the facies and high degree of marine micro fossils, trace fossils and body fossils, the dominant sandy heterolithic facies associations of allostratigraphic unit 7 overlying the FSf, are interpreted as wave-dominated delta-lobe (WDDL) architectural element (Figs 7, 11A and B, 12B, 13A; Table 3). At locality S3, the FSf is interpreted as a diachronous drowning and marine transgressive ravinement surface (MTRS) along the base of the retrogradational deltaic deposits of the tide-dominated sub-tidal sandy shoal (TDSS) arhitectural element (Figs 5A, 10A and B; Table 3). Based on the facies architecture allostratigraphic unit 7 is interpreted to have formed by stacking of distal offshore delta lobes. These stacked prodelta and offshore sediments were reworked by waves and longshore currents in an open water setting, but wave energy must have been low enough to allow for a wide diversity of organisms to thrive. The appearance of mixed brackish-marine palynomorphs (this study) at locality N2 may indicate locally lower salinity or lower sedimentation rates in the northern area. The thinner bedded and extensively bioturbated muddy sandstone strata in most of the northern localities suggest generally lower sedimentation rates than along the southern localities.

In the southern area, the maximum flooding surface MFS5 is represented by a condenced section formed by the finest-grained distal marine facies below upward-coarsening units (Fig. 5A). In the northern area the MFS5 is defined by the top of the well-cemented bed of fossiliferous hardground (FA6b; Figs 7 and 13A). The WDDL element is truncated below the SU3/TS6 surface, except at locality N1 where the WDDL element facies appears to pass directly into offshore marine mudstone (FA7b) of the Sortehat Formation (architectural element SORT; Fig. 7). The area of locality N1 is therefore considered to be located between two incised valleys.

Subaerial unconformity SU3 and transgressive surface TS6

Description

The sharp unconformity on top of the shaly WDDL element sandstones of allostratigraphic unit 7 is overlain by thin beds of conglomerates containing extrabasinal pebble-sized quartz clasts, few centimetres-long belemnites and marine shell fragments. At localities S3 to 6, field mapping revealed that the unconformity defines an up to 50 m deep, irregular and northward-deepening incision (Fig. 7). The unconformity is not identified at locality N1, whereas at N2 it is overlain by an up to 10 cm thick conglomerate bed with angular rip-up mud clasts, floating rounded gravel to pebble size quartz clasts and belemnite rostra fragments (Fig. 13E to G), similar to the conglomerate bed at localities S3 to 6. At locality N2, the incision measures up to 45 m, decreasing to a few metres along the localities N2 to 4 and disappearing north of N4 (Figs 7 and 13A).

Interpretation: Subaerial unconformity SU3 and transgressive surface TS6

It is interpreted that the incision of SU3 took place during a fall in relative sea-level that caused fluvial erosion into marine deposits and deposition of coarse sediments on the subaerial unconformity. This process was accompanied basinward by regressive ravinement scouring into the underlying marine strata (Fig. 13B and E). During the subsequent rise in relative sea-level, transgressive ravinement and deposition took place along the same surface, forming the transgressive ravinement surface (TS6). Hence, the deposits above the unconformity surface records retrogradational depositional architecture. In general, the incision is considered to have been controlled by 1) rate of decrease in accommodation space; 2) rate of sedimentation; 3) distance from the sediment source; 4) type of underlying substrate; and 5) the energy system present (fluvial, tides, waves). The combined SU3/TS6 surface defines the lower boundary of the overlying allostratigraphic unit 8, which also is typified by the presence of incised valleys (see next section).

Allostratigraphic unit 8

Description: Retrogradational architectural elements succession – Incised valley system 3 and the Sortehat Formation

At localities S3 to 6, the sandstone-dominated allostratigraphic unit 8 is 10 to 50 m thick, at least 30 km wide and overlies the SU3/TS6 surface (Fig. 7). At locality S3, the allostratigraphic unit 8 is composed of an up to 10 m thick single unit of middle to upper shoreface sandstone facies association (FA4a) with abundant cm-size to dm-size sideritic concretions and sandy tidal channel infill facies associations (FA3a, 3b and 3c) (Table 2). North of locality S3, the allostratigraphic unit 8 is composed of laterally shifting sandy tidal channel infill facies associations (FA3a, 3b and 3c) and large-scale tidal inclined heterolithic strata facies associations (FA2a and 2b) (Table 2). At locality S5, the allostratigraphic unit 8 is dominated by tide-influenced sandy heterolithic delta front facies association (FA5a), compound heterolithic subtidal dune facies association (FA3c) and a single laterally continuous bed of tide-influenced heterolithic prodelta mudstone facies association (FA5b) (Figs 7 and 12B; Table 2). At locality S6, the allostratigraphic unit 8 is dominated by sandy tidal channel infill facies associations (FA3a, 3b and 3c; Table 2).

No sandstone-dominated allostratigraphic unit 8 is observed at locality N1. At localities N2 to 4, the up to 45 m thick and less than 16 km-wide sandstone-dominated allostratigraphic unit 8 is composed of sandy heterolithic tidal channel infill facies associations (FA3a, 3b, 3c and 3d) (Figs 7 and 13; Table 2). In one section at locality N3, the allostratigraphic unit 8 consists mainly of the wave-dominated shoreface facies association (FA4a) and the wave-dominated sandy heterolithic strata facies associations (FA2c). Outcrop mapping and aerial photomosaics show that the allostratigraphic unit 8 is geometrically relatively symmetrical in shape in the northern area (Fig. 7).

At all the localities except N1, the sandstone succession of the allostratigraphic unit 8 is topped by a regional lithological shift into a succession of offshore marine mudstone facies association (FA7b) (Fig. 7; Table 2). The boundary may locally be draped with patches of quartz pebbles (Dam & Surlyk, 1998) and is here defined as the flooding surface FSg. The overlying 60 to 100 m thick Sortehat Formation succession is composed of shaly heterolithic black to dark grey mudstones containing bivalves, local carbonate concretions, plant debris and wood fragments (Surlyk, 2003; Koppelhus & Dam, 2003; Ahokas *et al.*, 2014; this study).

Interpretation: Allostratigraphic unit 8

The allostratigraphic unit 8 is interpreted as a succession of incised valley infill between the SU3/TS6 surface and the marine flooding surface FSg (Fig. 7). The valley infill is composed of tide-dominated estuary-complex (TDEC) and rare tide-dominated prodelta and delta front (TDDF) architectural elements (Figs 7, 11B, 12B and 13A; Table 3) and interpreted as deposited in a variety of estuarine sub-environments. Similarly to allostratigraphic unit 6, two separate but contemporaneous incised valley systems are interpreted along the N-S outcrop belt (Fig. 7).

The southern incised valley system is the deepest, attaining 50 m at locality S6 and is interpreted to have been at least 40 kilometres wide (Fig. 7). The observed cross-section of the valley is asymmetric. The northern incised valley system is deepest (45 m) at locality N2 and areally is about half the size of the southern system and more symmetrical in shape. The field observations and facies within suggest that the present outcrops cut these two incised valley systems in different proximal-distal positions. The northern system records an approximate position basinward of the central basin and close to the mouth of the paleo-estuary with large sand influx from the seaward side (Fig. 13B). The southern valley system records an approximate position landward of central basin and with a higher sediment supply from the landward side. The differences in probable contemporaneous facies development between the two incised valley infills are interpreted to have been caused by variations in hydraulic energy (e.g. tidal contra fluvial) and different sets of physical parameters (e.g. width:depth) as these valley infills were deposited in different proximal-distal estuarine settings.

The dominantly massive southern estuary succession at locality S3 suggests a locally high degree of reworking by bioturbation (Dam & Surlyk, 1998), whereas siderite concretions indicate variation in oxygen content of the bottom water. Northward thickening along the depositional strike, associated with an increasing abundance of several metres-thick tidal cross-stratification within the TDEC element (Fig. 7), shows that in addition to the fluvial source the estuary was supplied with large amounts of sand by alongshore currents. Also, interfingering of sandy and heterolithic tidal deposits (sandy TDEC and muddy TDDF elements; Fig. 7) indicates that various estuarine sub-environments were present, including wave-dominated shoreface (locality S3). Several tide-influenced bay-head deltaic distributaries and tidal bars formed by oblique accretion (localities S4 to 5 and S6) and lower-energy environments associated with the bay-head delta or central basin (locality S5). Therefore the southern incised valley system of the allostratigraphic unit 8 is interpreted as upper estuary succession exposed slightly obliquely to the outcrop belt.

Based on the size and facies the narrower northern incised valley infill is interpreted as a tide-dominated and more confined estuarine depositional system than in the southern one. Fluvial sediments at the valley base were exposed to extensive marine transgressive ravinement by tidal currents when the incised valley was flooded. The outcrops indicate a middle estuary to estuary mouth setting where large bars migrated along the bottom of the valley during early stages of the valley infill (Fig. 13E). These large tidal bars were dissected by highly active transverse tidal channels and smaller swatchways (Fig. 13B). The tidal channels migrated laterally, eroding into the shoal on one side and constructing the shoal on the other side, similar to the modern Oosterschelde estuary in the Netherlands (Martinius & Van den Berg, 2011). The lateral evolution is documented by the lower energy shoal deposits on top of larger basal bar forms.

As the rise in relative sea-level continued, the upper part of the valley widened, tidal dominance decreased and wave-influenced deposits successively became more abundant. The change from tidal energy system to wave dominance is well marked across the regional flooding surface (FSg) previously interpreted as a transgressive erosional ravinement surface (Surlyk, 1990a, b; Hansen & Surlyk, 1994; Koppelhus & Hansen, 2003) (Fig. 7). At this stage, the incised valley of the allostratigraphic unit 8 was filled up.

The heterolithic to muddy deposits overlying the flooding surface (FSg) record continued north-west and north-east retrogradation. The biostratigraphy sample collected approximately 15 m above the FSg (locality N2) is dominated by brackish marine dinoflagellate cysts (*Nannoceratopsis gracilis/senex*), brackish or freshwater planktonic alga (*Botryococcus* spp.) and relatively abundant pollen (*Perinopollenites elatoides, Callialasporites* spp. and *Cerebropollenites* spp.). This indicates mixing of fresh waters and marine waters during the early stages of marine transgression and is in accordance with the interpretation of Koppelhus & Hansen (2003). Petrified wood fragments sampled during this study, approximately 20 m above the FSg, indicate proximity to land areas with suitable living conditions for rather large trees.

The MFS6 surface has not been identified in the measured sections but is inferred to be located within the middle to upper part of the Sortehat Formation. The occurrence of the Vardekløft Group (e.g. Surlyk, 2003) shoreface sandstones above the Sortehat Formation (SORT architectural element; Table 3) defines the next major retrogradational to progradational depositional cycle.

OVERALL TRENDS: SEQUENCE STRATIGRAPHY

The sequence stratigraphical analysis of the Neill Klinter Group is based on key stratal surfaces and the spatial occurrence of architectural elements within the above described allostratigraphic units 1 to 8. In this context, palynostratigraphic data, both published (Koppelhus & Dam, 2003; Koppelhus & Hansen, 2003; Surlyk, 2003) and collected for this study have been of significant importance.

The two stratigraphic compartments of the Neill Klinter Group define an overall transgressive Early Jurassic (Pliensbachian-Aalenian) succession from sea-level lowstand to highstand, confirming earlier reports (Surlyk, 2003 and references therein). This first order transgressive systems tract (cf. Haq *et al.*, 1987; Hallam, 1988) includes six higher frequency transgressive-regressive sequences (*sensu* Embry 1989, 1993, 1995) listed in Table 4. These T-R sequences in the Neill Klinter Group are each composed of one or more of the allostratigraphic units (Table 4) and are summarised in Figs 7 and 14.

T-R sequences

T-R 1 is composed of the allostratigraphic units 1 and 2 (Table 4) and is bounded below by the transgressive marine ravinement surface (TS1) on top of the continental lowstand deposits of the Kap Stewart Group (Figs 7 and 14). The transgressive systems tract TST1 corresponds to the up to 25 m thick allostratigraphic unit 1 and reflects diachronous northwards marine flooding of the Jameson Land Basin during the Early Pliensbachian. The flooding surface FSa marks the drowning of initial transgressive deposits in the southern area. The FSa has not been identified in the northern area, but may here coincide with the MFS1.

MFS1 separates the allostratigraphic units 1 and 2 and hence also TST1 from the regressive systems tract RST1. The up to 127 m thick RST1 records at least four high-frequency cycles of fall and rise in relative sea-level, bounded by flooding surfaces FSb to FSe (Figs 7 and 14). Based on facies, these cycles took place during a period of overall upward shallowing, likely caused by an increase in rate of sedimentation relative to rate in creation of accommodation space. Palynomorphs indicate an upward increase of terrestrial/brackish dominance, thus supporting the interpretation of a basinwide regression (Koppelhus & Dam, 2003; this study).

Strata of the T-R1 sequence show a total thickening towards the north by 60 m over a distance of 50 km (Figs 7 and 14). Surlyk (2003) interpreted this as caused by increased subsidence on the northern side of deep-seated faults between the southern and northern areas. In this study the analysis of onshore seismic data from Grønlands Geologiske Undersøgelse and Arctic Richfield Exploration Company (unpublished seismic data 1986 to 1989, available on loan to academic institutions from the Geological Survey of Denmark http://www.geus.dk/geuspage-uk.htm) do not support the hypothesis of deep-seated faults.

Table 4. Summary of relationship between sequence stratigraphic framework, key surfaces and systems tracts, allostratigraphic units, architectural elements, facies associations and maximum flooding surface description of the Neill Klinter Group, Jameson Land Basin, East Greenland.

T-R sequence 1-6	TST/RST	Stratigraphic range	Occurrence	Allostratigraphic unit	Architectural element	Facies association	MFS characteristic
T-R sequence 6	**RST6 (HST)**	MFS6 to Base Pelion Fm	S1-N4	Sortehat Fm.	SORT	7B	MFS6:
	TST6	SU3/TS6 to MFS6	S1-8, N2-4	AU8 & basal Sortehat Fm.	TDEC	3A 3B 3C 2A 2B 4A 3D 2C	Within the Sortehat Fm above basal backstepping silty sandstones.
					TDDF	5B 5A	
T-R sequence 5	**RST5**	MFS5 to SU3/TS6	S1-N4	AU7	WDDL	7A 6B 6A 7B	MFS5:
					TDSS	2B 3A 3B 3C	Top of correlative hardground sandstone bed with abrupt shift into silty-shale.
	TST5	SU2/TS5 to MFS5	S1-4, S6-N4	AU6 & basal part of AU7	TDEC	3A 3B 3C 3D 4A 6C	
					FTT	1A 1B 1C	
T-R sequence 4	**RST4**	MFS4 to SU2/TS5	S3-4, S6 (S8)	AU5	WDBE	6A 7B	MFS4:
	TST4	TS4 to SU2/TS5	S5		TDDC	5C	Unclear or marked by interval of thin cone-in-cone cemented beds within the basal 10 m of WDBE, eroded within TDDC.
		TS4 to MFS4	S3-4, S6 (S8)		WDBE	6A 7B	
					WDBE		
T-R sequence 3	**RST3**	MFS3 to SU2/TS5	N1-4	AU4	FTT	1A 1B 6C	MFS3:
					TDSS	3C 3D 3E	Within sandy interval, no clear condensed section.
		MFS3 to TS4	S3-6 (-8?)		TDSS	2A 2B 2C 3A 3B 3D	
					FTT	1A 1B 1C	
	TST3	SU1/TS3 to MFS3	S1-N4		TDSS	2A 2B 2C 3A 3B 3D 3C 3E	
					FTT	1A 1B 6C 2A 2B	
T-R sequence 2	**RST2**	MFS2 to SU1/TS3	S1-6	AU 3	WDBE	6A	MFS2:
	TST2	TS2 to MFS2	S1-8		WDBE	7B 6A	Condensed section within WDBE silty mudstone, not observed north of S6.
T-R sequence 1	**RST1**	MFS1 to SU1/TS3	N1-4	AU2	TDDP	2A 2B 1A 1B	MFS1:
					TDDF	5B 5A	Condensed section within basal TDDF mudstone, few metres above SHFS.
		MFS1 to TS2	S1-S6 (S8)		TDSS	3B 3C 2B	
					TDDP	1B 2A 2B 1A 3C 3D 3E	
					TDDF	5B 5A	
	TST1	TS1 to MFS1	S1-N4	AU1	SHFS	4B	
					WIBE	6C 6A	

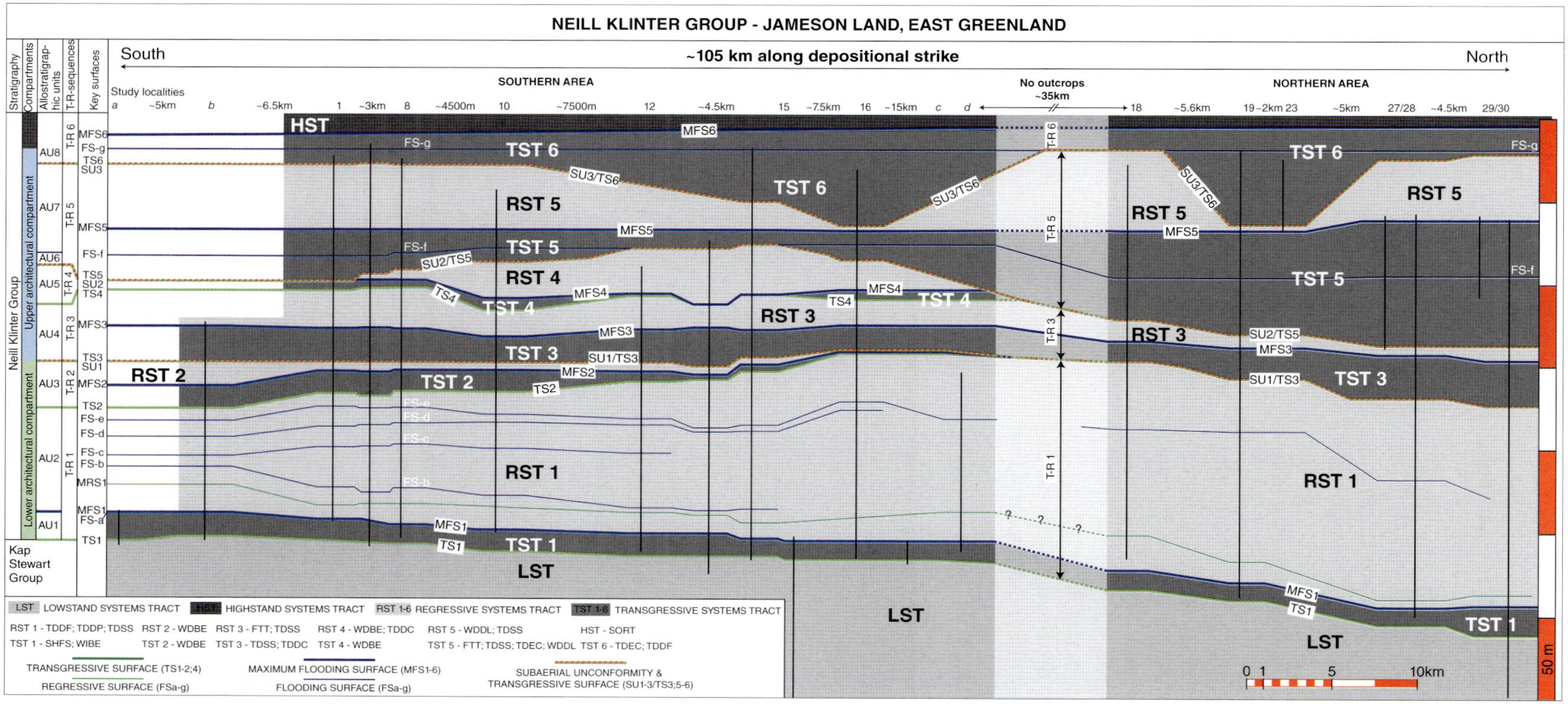

Fig. 14. The Neill Klinter Group divided into six transgressive-regressive sequences (T-R 1 to 6), each composed of a transgressive systems tract (TST) and a regressive systems tract (RST). The Kap Stewart Group defines a lowstand systems tract (LST), whereas the Sortehat Formation defines a highstand systems tract (HST). Transgressive and regressive systems tracts are composed of one or more allostratigraphic units (AU1 to 8), which in turn are composed of certain sets of architectural elements as described in the text and Table 3.

Increased accommodation in the northern area may instead have been created by a higher post-rift crustal cooling and subsidence of this particular part of the basin.

The T-R 2 sequence corresponds with allostratigraphic unit 3 (Table 4) and is restricted to the southern area (Fig. 14). The southward thickening T-R 2 sequence was initiated by the transgressive surface TS2. The overlying transgressive systems tract TST2 displays a semi-regional flooding of the underlying regressive systems tract RST1. MFS2 marks the initiation of the regressive systems tract RST2 (Fig. 14). The northward amalgamation of TS2 with the MFS2 is indicative of a wide transgression at this stage. Proximal depositional environments prevailed towards the north, partly shallow-marine and partly emerged, whereas the southern area was entirely drowned. In the northern area, T-R 2 deposits were either not deposited or thin and eroded during the fall in relative sea-level that caused the establishment of the combined SU1/TS3 surface that terminates the T-R 2 sequence.

The T-R 3 sequence corresponds with the allostratigraphic unit 4 (Table 4). The basal unconformity surface SU1/TS3 reflects a basin-wide fall in relative sea-level and formation of low relief incised valleys on top of the T-R 2 and T-R 1 sequences in the southern and northern areas, respectively. The total amount of subaerial erosion beneath the SU1/TS3 is uncertain. An overall aggradational architecture of the T-R 3 sequence implies that during the rise in relative sea-level, a close balance existed between rate of creation of accommodation space and rate of sediment supply (A/S ratio ~ 1). Due to this, the position of MSF3 is not immediately obvious and the regressive systems tract RST3 resembles much of the TST3. The most proximal deposits of the T-R 3 sequence in the southern area are topped by the TS4 surface marking the succeeding transgressive ravinement and flooding event. This explains the observed mixing of terrestrial and marine debris along TS4 (see above).

The T-R 4 sequence corresponds with the allostratigraphic unit 5 (Table 4) and is restricted to the southern area (Figs 7 and 14). The T-R 4 sequence is bounded below by TS4, which also may have been present in the northern area prior to the erosion that formed the incised valley along the SU2/TS5 surface (Fig. 14). The separation of TST4 and RST4 in the stratal architecture is not obvious. However, the MFS4 is placed at the base of the faint upward-coarsening regressive systems tract RST4 (Fig. 7). The upper surface of the T-R 4 sequence (SU2/TS5) records up to 16 m deep incision south of locality S5, whereas northwards the entire sequence has been removed (Figs 7 and 14).

The T-R 5 sequence is composed of the allostratigraphic units 6 and 7 (Table 4) and is underlain by the unconformity surface SU2/TS5. This surface reflects a basin-wide fall in relative sea-level and formation of a new set of incised valleys. The transgressive systems tract TST5 reflects drowning and infilling of the incised valleys along the FSf that represents a diachronous and abrupt increase in accommodation space. A condensed section with hardground sandstone defines the maximum surface MFS5 and the shift from TST5 into progradational and aggradational RST5 in the southern and northern areas, respectively (Fig. 14). In the regressive systems tract RST5, shell-beds record pauses or stagnation in sediment input. Sequence T-R 5 was terminated by a new fall in relative sea-level and formation of new incised valleys along SU3/TS6. Approximately 40 m of RST5 was locally eroded during this stage (Fig. 14).

The T-R 6 sequence corresponds with the allostratigraphic unit 8 (Table 4). The basal unconformity surface SU3/TS6 reflects a fall and rise in relative sea-level and formation of another set of incised valleys. The transgressive systems tract TST6 reflects infilling and drowning of the incised valleys along the marine FSg that represents a diachronous and abrupt increase in accommodation space and water depth. Across this flooding surface, the Jameson Land Basin turned into an open-marine environment defined by the Sortehat Formation. MFS6, located within the Sortehat Formation, is diachronous towards the north (Dam & Surlyk, 1998; Koppelhus & Dam, 2003; Surlyk, 2003). The strata above FSg suggest regional backstepping of all facies belts and regional transgression towards the north, west and east of the Jameson Land Basin. The unconformity between the Sortehat Formation and the overlying Vardekløft Group deposits marks the top of the T-R 6 sequence and a major Middle Jurassic (Bajocian-Bathonian) regression in East Greenland.

DISCUSSION

Sediment infill dynamics of the Neill Klinter Group

The Neill Klinter Group in Jameson Land, East Greenland, is an excellent field laboratory for the multi-scale stratal complexity resulting from the dynamic interplay between numerous paralic subenvironments during deposition. The sedimentary architecture reflects both the basinal setting and the changes in relative sea-level within the large shallow marine embayment that existed in Jameson Land in the Early Jurassic.

The structural behaviour of the Jameson Land Basin during Early Jurassic was of critical importance for the sediment infill of the basin. The Devonian to Late Jurassic sedimentary strata in Jameson Land represent a complex stack of superimposed basin infill successions representing successive evolving structural conditions. As emphasised above, no rift-faults are recognised to be associated with the structural and depositional architecture of the Neill Klinter Group. The onlap of Jurassic strata on Precambrian crystalline basement in Liverpool Land to the east supports the view that the Neill Klinter Group was deposited in a sag basin. The sag basin may have formed in response to thermal cooling subsequent to the Triassic rifting, similar to the Early Jurassic basins located on the Norwegian continental shelf (Roberts *et al.*, 2009).

Between phases of crustal extension, the northern Pangean plate experienced regional changes in sea-level, as documented by T-R sequences in Triassic to Early Cretaceous rocks of the Canadian and Fennoscandian Arctic (Embry, 1989; Johannessen & Embry, 1989; Mørk *et al.*, 1989; Riis *et al.*, 2008; Midtkandal & Nystuen, 2009; Glørstad-Clark *et al.*, 2010). The relative sea-level changes that controlled the sequence stratigraphic development of the Neill Klinter Group were probably related to regional vertical lithosphere plate movements associated with processes in the lower crust or/and the mantle causing epeirogenic crustal movements and may have been combined with local differential movements along old structural lineaments within Jameson Land.

Changes in relative sea-level are reflected in the Neill Klinter Group by basinward-stepping and landward-stepping depositional architectural elements comprising the lower and upper stratigraphic compartments. The lower stratigraphic compartment of the Neill Klinter Group formed as an overall prograding wave- to tide-influenced deltaic depositional system and the upper stratigraphic compartment as an overall retrograding series of stacked tide-dominated incised valley successions and wave-dominated delta depositional systems (Figs 7, 14 and 15).

Incised valleys and their bounding unconformities of the Neill Klinter Group

Due to the general high sand content in the Neill Klinter Group, identification of key stratal surfaces depends on the interpretation of depositional facies and environment. This study has benefitted much from the previous studies by Dam & Surlyk (1995, 1998), in particular their discussion of criteria that can be used to define key sequence stratigraphic surfaces. However, Dam & Surlyk (1995, p. 489) concluded that "major unconformities associated with significant valley incision have not been observed".

One main, and new, result of this study is the identification of three major incised valley systems. The base of the valley systems is represented by the combined subaerial unconformities and transgressive surfaces (SU/TS) that separate regressive elements, proximal or distal, from overlying transgressive elements. The lithological and stratigraphical character of conglomerate deposits resting on these unconformities has been of critical significance for their recognition. The mixture of extrabasinal clasts with marine body fossils was formed when clasts introduced to the basin by fluvial or/and debris-flow transport were later reworked by tidal currents that supplied the marine fossil fragments. Marine body fossils may also have originated from the substrate incised by the rivers.

From our database, it is not possible to draw any definite conclusion about the lateral extent of the subaerial erosional unconformities beneath the incised valleys. However, the cross-sectional dimensions of the unconformities in the eastern Jameson Land outcrop belt indicate that the incisions represent erosional features that may have affected large parts of the Jameson Land Basin. In basinward direction towards the west, the incised valleys, along their respective stratigraphical positions, may have coalesced and formed regional subaerial unconformities on a low-gradient ramp slope, covered by thin gravel and sand sheets. Several examples of such extensive subaerial unconformities covered with fluvial deposits are referred to by Posamentier & Allen (1999, p. 69). The unconformity beneath the Lower Cretaceous Helvetiafjellet Formation in Spitsbergen is another example (Midtkandal & Nystuen, 2009) and similar regional unconformities have been suggested to be present within the marginal marine to open marine Triassic succession in the Barents Sea (Riis *et al.*, 2008; Glørstad-Clark *et al.*, 2010).

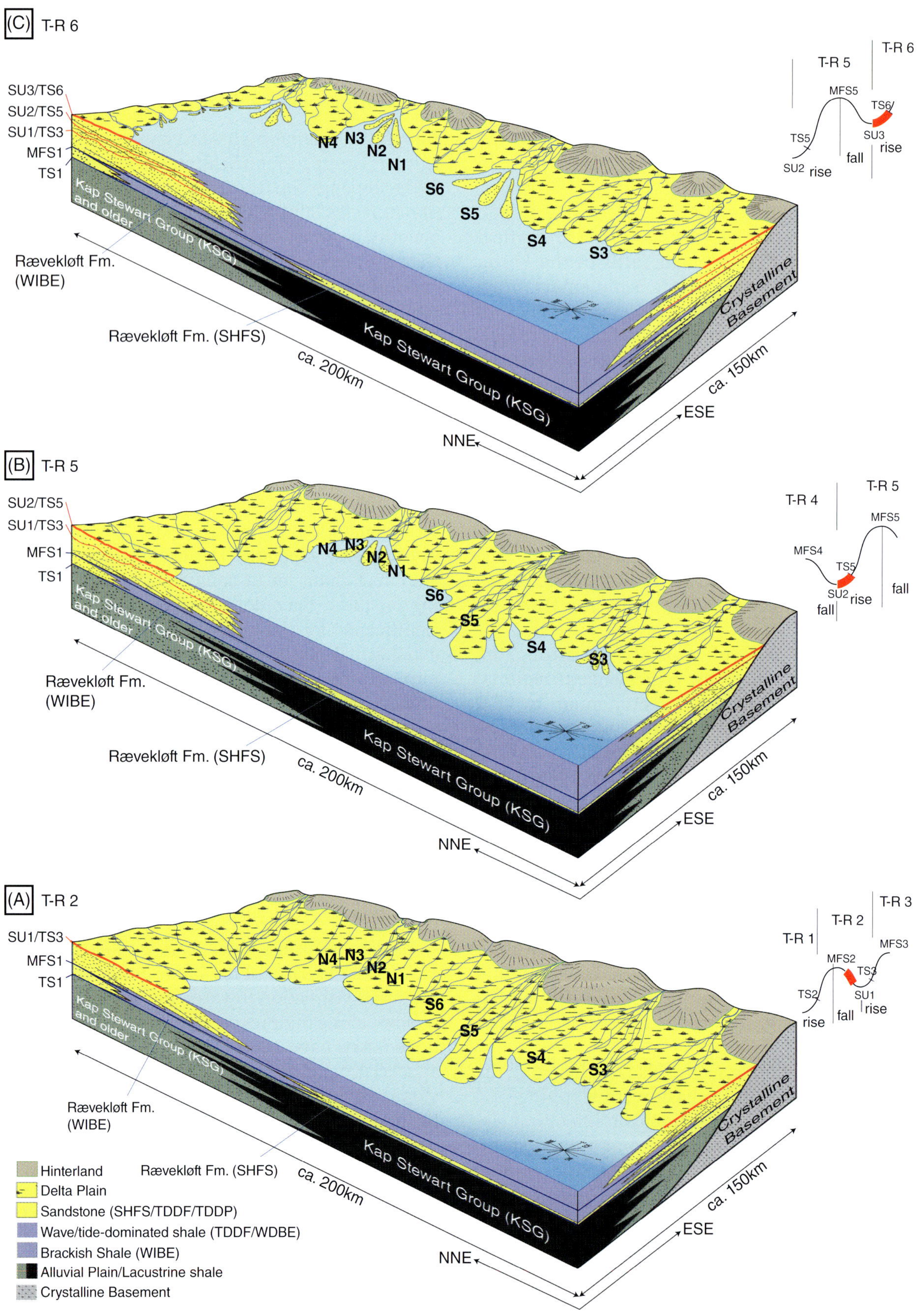

Fig. 15. Successive stages in the development of depositional environments in the eastern Jameson Land Basin. The three block-diagrams correspond to the stratigraphic intervals and systems tracts (Figs 7 and 14). The three sketches illustrate a possible depositional scenario for the different stages of Neill Klinter Group development. See discussion in text for details.

The three incised valley systems were formed during basinward extension of major river systems during events of major fall in relative sea-level. These rivers originated in hinterland areas on present-day Liverpool Land or a larger ancient land area further to the east that may have been the Jan Mayen microcontinent (Figs 15 and 16) (cf. Mjelde *et al.*, 2008; Breivik *et al.*, 2012). The width:depth ratio of the incised valleys and their position and orientation, may have played an important role in creating high-energy tidal systems within the valleys. The estuaries which filled these incised valleys received great amounts of clastic detritus to bay-head deltas and to sediment convergent areas in the middle part of the estuaries. Part of this fluvial sediment input may have bypassed into the shallow marine realm, from where sediments may have been transported back during the infilling of the estuaries. All the estuary infill deposits are considered as tide-influenced to tide-dominated. The predominance of tidal currents during the infill of the estuaries explains the absence of distinct fine-grained central basin deposits and/or wave-dominated barrier deposits. However, the few examples of associated wave-dominated shoreface deposits indicate co-existence of strong tidal and wave energy environments in the outer parts of these estuaries (cf. Dalrymple *et al.* 2012; Tessier, 2012). The increase of wave-formed structures relative to tidal features in the uppermost part of the incised valley infill successions was probably the result of widening of the marine embayments by rising sea-level, exposing them to increasing wave activity.

Sequence development in the Neill Klinter Group

The six T-R sequences in the Neill Klinter Group are inferred to be basin-wide. They represent an overall rise in relative sea-level from lowstand in the continental Kap Stewart Group to highstand in the basin-wide offshore marine Sortehat Formation. The T-R sequences 1 and 2 were formed during the initial flooding stage of the basin and were terminated by a fall in relative sea-level and the establishment of the first SU and incised valley system. Lateral distribution of depositional environments (Fig. 7) within T-R sequences 1 and 2 was controlled mainly by the intrabasinal distribution of fluvial-dominated, tidal-dominated and wave-dominated energy systems. The marked asymmetry of the transgressive systems tract TST1 in the T-R sequence 1, with MFS1 located close to the lower transgressive boundary, indicates a rapid rise in relative sea-level and establishment of maximum water depth (accommodation space) shortly after initial transgression. The aggradational pattern of the regressive systems tract RST 1 (Figs 7 and 14) reveals a close balance between creation of accommodation space and rate of sediment supply. This helps explain why the Neill Klinter Group lacks basin-scale clinoform geometry with shoreface deposits that could define distinct shoreline trajectories.

The relatively low relief of the lowermost unconformity surface SU1/TS3 may imply moderate incision, or that the cross-section occupies a rather distal position in the incised valley (Fig. 7). The higher reliefs of the other incised valleys, defined by the surfaces SU2/TS5 and SU3/TS6, indicate that these two sections represent rather proximal positions of the incised valleys. The considerable lateral and vertical variation in paralic subenvironments within the T-R sequences 3 to 6 (Fig. 7) is interpreted to have been formed along coastlines with fluctuating geomorphology. The ratio between creation of accommodation space and rate of sediment supply varied significantly along the coast, as did the dominating energy system, giving rise to a dynamic variation in the sedimentary environments (updip and downdip, laterally along the coastline and stratigraphically); a common feature in tidally influenced paralic settings (Dalrymple & Choi, 2007; Martinius & Van den Berg, 2011). However, an overall aggradational to retrogradational depositional setting prevailed. The distance of sedimentary transport was relatively short from the surrounding cratonic basement areas that yielded large influxes of clastic material to the Jameson Land Basin. This is interpreted to be related to the warm and humid climate that in the Early Jurassic succeeded the arid to semi-arid climate of the Triassic in the NE Atlantic region (Frostick *et al.*, 1992; Clemmensen *et al.*, 1998; Müller *et al.*, 2005; Ruhl & Kürschner, 2011). The warm and humid climate in the Early Triassic–earliest Jurassic gave rise to an increase in chemical weathering. This is reflected by the dominantly quartz-arenitic composition of the sandstones in the Neill Klinter Group and by the abundance of quartz and quartzite pebbles *versus* granitic pebbles in conglomerates. The formation of glauconite, iron-rich ooids and early-diagenetic chlorite in shallow-marine paralic environments also points to a high delivery of Fe^{2+} ions from land, as

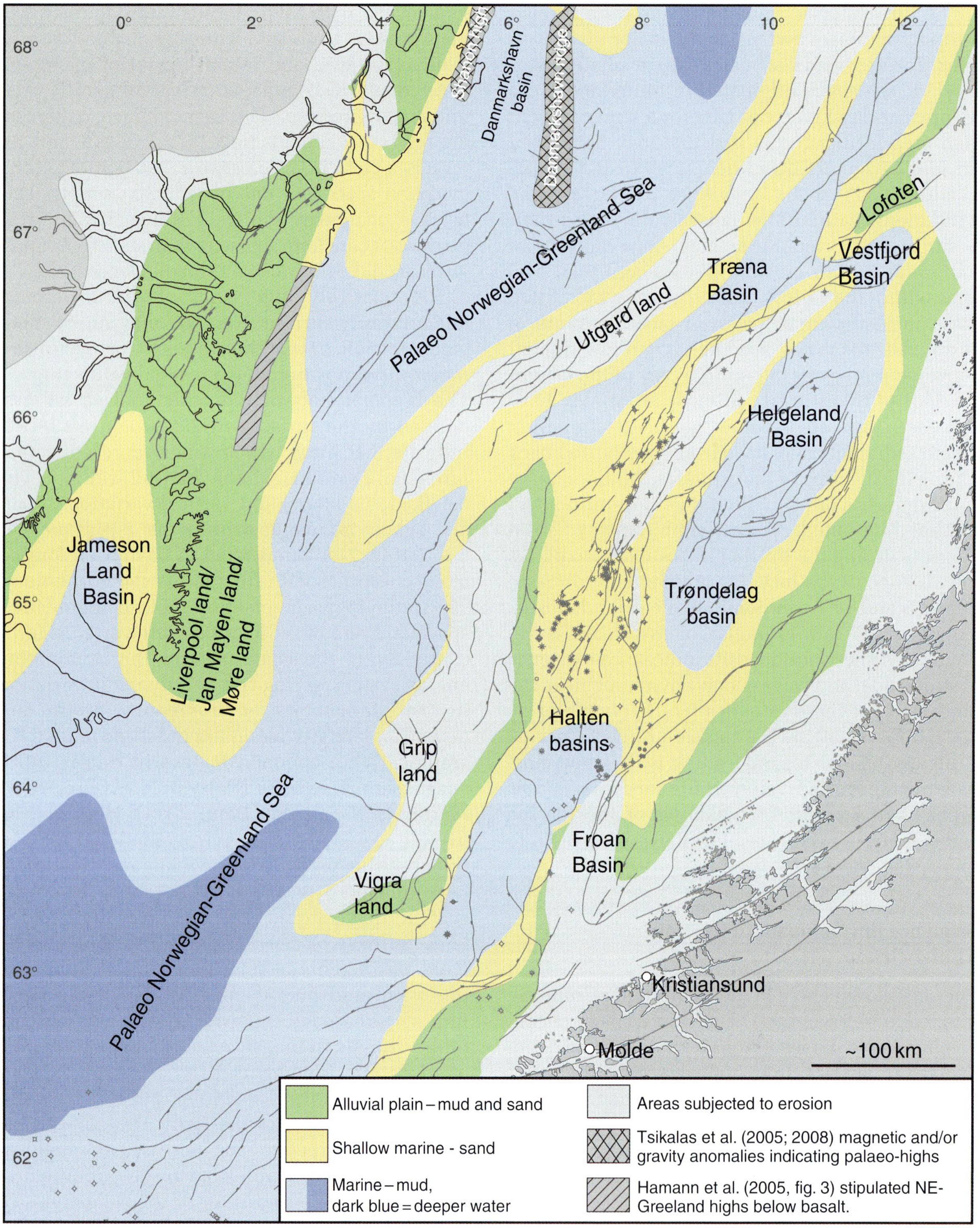

Fig. 16. Early Jurassic palaeo-reconstruction of the Norwegian-Greenland Sea (modified from original figure by J. Gjelberg in Johannessen & Nøttvedt, 2008; palaeo-highs from Tsikalas *et al.*, 2005, 2008; Hamann *et al.*, 2005; Berger & Jokat, 2008). Drainage areas and suggested main depositional systems, as well as present day main faults and wells on the Norwegian offshore, are indicated. Terminology modified from Blystad *et al.* (1995).

a function of chemical weathering. Local variation in precipitation in the hinterland, reflecting variation in altitude, relief and atmospheric circulation pattern, may also have attributed to differences in sediment delivery along the coastline of the Jameson Land Basin.

Neill Klinter Group as an analogue for reservoir sandstones on the Halten Terrace

During the Early Jurassic, very similar paralic and overall heterolithic successions were deposited in shallow marine sag and rift basins on both sides of the proto-Norwegian-Greenland Sea, in East Greenland and the present mid-Norwegian continental shelf, respectively (see above). As indicated by Dam & Surlyk (1995, 1998) and Surlyk (2003) there are several similarities that regard stratigraphy and sedimentological features between the Neill Klinter Group in the Jameson Land Basin and the Båt Group and partly the overlying Fangst Group (Aalenian-Callovian), on the Halten Terrace-Trøndelag Platform.

The similarities between the Neill Klinter and Båt groups include (1) comparable overall regional structural setting along the same seaway on the same lithospheric plate, in the same palaeolatitude and climatic belt; (2) great similarity in depositional history controlled by regional to eustatic variations in sea-level; (3) basal paralic heterolithic and low-gradient mixed-energy delta systems and homogeneous deltaic systems interfingering with offshore fines in the middle part of both groups and (4) comparable style and scale of various types of depositional architectural elements and their internal facies and heterogeneities.

The Neill Klinter and the Båt groups also reveal several differences in setting and properties. These differences comprise (1) structural setting as the result of differences in local pre-rift and syn-rift development on the two sides of the proto-Norwegian-Greenland Sea, including high frequency local base-level changes; (2) basin dimensions; the Båt Group on the Halten Terrace covers a basin area four times larger than that of Neill Klinter Group; (3) localised variation in precipitation, runoff and sediment influx between the two basin areas; (4) magnitude and distribution of autogenic processes (fluvial, tidal, waves) in relation to accommodation space; (5) sediment supply rates and type of sediments.

A direct lithostratigraphic or a one-to-one sequence stratigraphic correlation between the Neill Klinter Group in Jameson Land Basin and the Båt and Fangst groups on the Halten Terrace–Trøndelag platform has been suggested by Dam & Surlyk (1995, 1998) and Gjelberg (in Johannessen & Nøttvedt, 2008). However, lateral facies variation, sequence stratigraphic architecture and presence of incised valleys in the Neill Klinter Group reveal a more composite relationship between the two basin areas rather than a direct correlation. The presence of incised valleys in the Neill Klinter Group demonstrates that there are basic local deviations between the two basinal settings. Despite the differences, important similarities between the Early Jurassic Jameson Land and Halten Terrace basins make the Neill Klinter Group a reference object for petroleum reservoirs on the mid-Norwegian continental shelf. Corresponding incised valleys as those shown to be present in the Neill Klinter Group may also occur on the mid-Norwegian continental shelf in successions of same age. Conceptual depositional models and qualitative and quantitative data concerning dimensions, geometry, facies variation and potential connectivity of sandstone bodies obtained from the Neill Klinter Group can be used to improve static and dynamic reservoir models of tide-influenced heterolithic reservoir sandstones on the Norwegian continental shelf and probably also in other petroleum provinces of the World located in other marginal marine and tide-influenced basins similar to the Early Jurassic Jameson Land Basin.

CONCLUSIONS

1 In contrast to previous descriptions of the Early Jurassic Jameson Land Basin, no eastern basin marginal fault has been shown, nor any intra-basinal NW-SE cross-faults that separate the southern from the northern Jameson Land Basin. The eastern boundary is an onlap contact. Thickness variation within the Neill Klinter Group is related to differential subsidence subsequent to Permo-Triassic rifting.
2 The Neill Klinter Group consists of a lower stratigraphic compartment, formed as an overall prograding wave-influenced to tide-influenced deltaic depositional system and an upper stratigraphic compartment of an overall retrograding series of stacked tide-dominated incised valley successions and wave-dominated delta depositional systems.

3 The stratigraphic compartments are composed of eight allostratigraphic units which comprise six transgressive-regressive sequences (T-R) separated by sequence boundaries formed as combined subaerial unconformities and transgressive surfaces (SU/TS surfaces). The T-R sequences are the result of changes in relative sea-level, probably controlled by vertical movements in the northern Pangean continental plate.

4 The six T-R sequences are internally organised by architectural elements of facies associations formed in adjacent paralic subenvironments, controlled by available accommodation space, sediment supply and relative influence of fluvial, tidal and wave processes.

5 Estuaries formed by flooding of incised valleys are dominated by tidally controlled sand-rich depositional environments, whereas deltas exposed to more open-marine conditions are preferentially wave-dominated.

6 A stratigraphic correlation between the Lower Jurassic of the Jameson Land Basin and of the Halten Terrace–Trøndelag Platform basin on the Norwegian continental shelf is only valid on a large scale, comprising an overall transgressive-regressive development during the Pliensbachian-Toarcian, with deposition of sand-dominated formations, before regional sea-level rise and flooding in the early Aalenian with deposition of open-marine mudstone and shale.

7 The Neill Klinter Group is an analogue for Lower Jurassic sandstone reservoirs in the Båt Group on the Halten Terrace–Trøndelag Platform in terms of (1) improved conceptual understanding of facies, facies associations and architectural elements as function of depositional environment and sequence stratigraphy and (2) modelling and flow simulation by application of parameters on spatial distribution of heterogeneities on various scales of grain-size variation and shale baffles in tide- and wave-influenced paralic sandstone reservoirs.

ACKNOWLEDGEMENTS

The authors would like to thank Statoil ASA for financial support of this Project and permission to publish. Professor Arild Andresen (University of Oslo), Aka Lynge, POLOG and Air Greenland have been extremely helpful in assisting with field work logistics and transportation. Arve Næss, Bjørn Terje Oftedal, Jan Einar Ringås (all Statoil), Torben Olsen (formerly Statoil, now Noreco) and John Suter (ConocoPhillips) are thanked for initiating this Project and for discussions in and off the field. We also would like to thank Ann Christin Aimee Ella Ferrari Holme, Hege Cecilie Nordeide-Felde and Per Inge Myhre who have contributed with collecting field data and/or logistics and a large number of people for accompanying in the field, from Statoil: Øystein Braathen, Scott Bullimore, Nils Haabesland, Marte Høistad, Anita Høiås, Morten Krogh, Simon Leary, Tine Lærdal, Arve Næss, Philip Ringrose, Kristina Rossavik and from ConocoPhillips: Line Haram (now independant consultant) and Hugo Lawrence. Imogen Poole is thanked for wood sample data analysis and Adrian Read language editing. We also thank Brian Willis, Mr. Dixon and one anonumous reviewer for review and comments for improving this manuscript.

REFERENCES

Ahokas, J.M., Nystuen, J.P. and **Martinius, A.W.** (2014) Stratigraphic signatures of punctuated rise in relative sea-level in an estuary-dominated heterolithic succession: Incised valley fills of the Toarcian Ostreaelv Formation, Neill Klinter Group (Jameson Land, East Greenland). *Marine and Petroleum Geology*, **50**, 103–129.

Allen, J.R.L. (1990) The Severn Estuary in southwest Britain: its retreat under marine transgression and fine-sediment regime. *Sed. Geol.*, **66**, 13–28.

Berger, D. and **Jokat, W.** (2008) A seismic study along the East Greenland margin from 72° N to 77° N. *Geophysical Journal International*, **174**, 733–748.

Bhattacharya, J.P. (2001) Allostratigraphy Versus Sequence Stratigraphy. AAPG Hedberg Research Conference, Dallas, Texas, August 26-29, 2001, Abstract. AAPG Search and Discovery Article #90050, 2005.

Blystad, P., Brekke, H., Færseth, R.B., Larsen, B.T., Skogseid, J. and **Tørudbakken, B.** (1995) Structural elemenst of the Norwegian Continental Shelf. Part II: The Norwegian Sea Region. *NPD Bulletin*, **8**, 45 pp.

Breivik, A.J., Mjelde, R., Faleide, J.I. and **Murai, Y.** (2012) The eastern Jan Mayen microcontinent volcanic margin. *Geophysical Journal International*, **188**, 798–818.

Brekke, H., Sjulstad, H.I., Magnus, C. and **Williams, R.W.** (2001) Sedimentary environments offshore Norway. In: *Sedimentary Environments Offshore Norway – Palaeozoic to Recent* (Eds O.J. Martinsen and T. Dreyer), *NPF Special Publication*, **10**, 7–37.

Bütler, H. (1948) Notes on the geological map of Canning Land (East Greenland). *Meddelser om Grønland*, **133**, 99 pp.

Catuneanu, O. (2006*)* Principles of Sequence Stratigraphy. Elsevier, Amsterdam, 386 pp.

Choi, K.S. and **Dalrymple, R.W.** (2004) Recurring tide-dominated sedimentation in Kyonggi Bay (west coast of Korea): similarity of tidal deposits in late Pleistocene and Holocene sequences. *Mar. Geol.*, **212**, 81–96.

Choi, K.S., Dalrymple, R.W., Chun, S.C. and **Kim, S-P.** (2004) Sedimentology of modern, inclined heterolithic stratification (IHS) in the macrotidal Han River Delta, Korea. *J. Sed. Res.*, **74**, 677–689.

Christiansen, F.G., Larsen, H.C., Marcussen, C., Hansen, K., Krabbe, H., Larsen, L.M., Piasecki, S., Stemmerik, L. and **Watt, W.S.** (1992) Uplift study of the Jameson Land Basin, East Greenland. *Nor. Geol. Tidsskr.*, **72**, 291–294.

Clemmensen, L., Kent, D.V. and **Jenkins, F.A.Jr.** (1998) A late Triassic lake system in East-Greenland: facies, depositional cycles and palaeoclimate. *Paleogeogr., Paleoclimatol., Paleoecol.*, **140**, 135–159.

Dalland, A., Worsley, D. and **Ofstad, K.** (1988) A lithostratigraphic scheme for the Mesozoic and Cenozoic succession offshore mid- and northern Norway. *NPD Bull.*, **4**, 65 pp.

Dalrymple, R.W., Makino, Y. and **Zaitlin, B.A.** (1991) Temporal and spatial patterns of rhythmite deposition on mud flats in the macrotidal Cobequid Bay-Salmon River estuary, Bay of Fundy, Canada. In: *Clastic Tidal Sedimentology* (Eds D.G. Smith, G.E. Reinson, B.A. Zaitlin and R.A. Rahmani), *Can. Soc. Petrol. Geol. Mem.*, **16**, 137–160.

Dalrymple, R.W., Knight, R.J., Zaitlin, B.A. and **Middleton, G.V.** (1990) Dynamics and facies model of a macrotidal sand-bar complex, Cobequid Bay-Salmon River Estuary (Bay of Fundy). *Sedimentology*, **37**, 577–612.

Dalrymple, R.W., Baker, E.K., Harris, P.T. and **Hughes, M.G.** (2003) Sedimentology and stratigraphy of a tide-dominated, Foreland-basin delta (Fly River, Papua New Guinea). In: *Tropical Deltas of Southeast Asia – Sedimentology, Stratigraphy and Petroleum Geology* (Eds F.H. Sidi, D. Nummedal, P. Impert, H. Darman and H.W. Posamentier), *SEPM Spec. Publ.*, **76**, 147–173.

Dalrymple, R.W. and **Choi, K.** (2007) Morphologic and facies trends through the fluvial-marine transition in tide-dominated depositional systems: a schematic framework for environmental and sequence-stratigraphic interpretation. *Earth-Sci. Rev.*, **81**, 135–174.

Dalrymple, R.W., Mackay, D.A., Ichaso, A.A. and **Choi, K.S.** (2012) Processes, Morphodynamics and Facies of Tide-dominated Estuaries. In: *Principles of Tidal Sedimentology* (Eds R.A. Davis and R.W. Dalrymple). Springer Science+Business Media B.V. 2012, 109–128.

Dam, G. and **Surlyk, F.** (1992) Forced regressions in a large wave and storm-dominated anoxic lake, Rhaetian–Sinemurian Kap Stewart Formation, East Greenland. *Geology*, **20**, 748–751.

Dam, G. and **Surlyk, F.** (1993) Cyclic sedimentation in a large wave and storm-dominated anoxic lake; Kap Stewart Formation (Rhaetian–Sinemurian), Jameson Land, East Greenland. In: *Sequence stratigraphy and facies associations* (Eds **H**.W. Posamentier, D.P. Summerhays, B.U. Haq and G.P. Allen). *Int. Assoc. Sedimentol. Spec. Publ.*, **18**, 419–448.

Dam, G. and **Surlyk, F.** (1995) Sequence stratigraphic correlation of Lower Jurassic shallow marine and paralic successions across the Greenland–Norway seaway. In: *Sequence stratigraphy on the Northwest European margin* (Eds R.J. Steel, V.L. Felt, E.P. Johannessen and C. Mathieu). *Norwegian Petroleum Society (NPF) Special Publication*, **5**, 483–499.

Dam, G. and **Surlyk, F.** (1998) Stratigraphy of the Neill Klinter Group; a Lower – lower Middle Jurassic tidal embayment succession, Jameson Land, East Greenland. *Geology of Greenland Survey Bulletin*, **175**, 80 pp.

Dam, G., Surlyk, F., Mathiesen, A. and **Christiansen, F.G.** (1995) Exploration significance of lacustrine forced regressions of the Rhaetian – Sinemurian Kap Stewart Formation, Jameson Land, East Greenland. In: *Sequence stratigraphy of the Northwest European margin* (Eds R.J. Steel, V.L. Felt, E.P. Johannessen and C. Mathieu). *Norwegian Petroleum Society, Special Publication*, **5**, pp. 511–527.

Ehrenberg, N.S. (1993) Preservation of anomalously high porosity in deeply buried sandstones by grain-coating chlorite: examples from the Norwegian Continental Shelf. *AAPG Bulletin*, **77**, 1260–1286.

Embry, A.F. (1989) Correlation of Upper Palaeozoic and Mesozoic sequences between Svalbard, Canadian Arctic Archiepelago and northern Alaska. In: *Correlation in Hydrocarbon Exploration* (Ed. J.D. Collinson), pp. 89–98. Norwegian Petroleum Society, Graham & Trotman, London.

Embry, A.F. (1993) Transgressive-regressive (T-R) sequence analysis of the Jurassic succession of the Sverdrup Basin, Canadian Arctic Archipelago. *Can. J. Earth Sci.*, **30**, 301–320.

Embry, A.F. (1995) Sequence boundaries and sequence hierarchies: problems and proposals. In: *Sequence Stratigraphy on the Northwest European Margin* (Eds R.J. Steel, V.L. Felt, E.P. Johannessen and C. Mathieu). *Norwegian Petroleum Society, Special Publication*, **5**, pp. 1–11.

Emery, D. and **Myers, K.J.** (1996) *Sequence Stratigraphy*. Blackwell, Oxford, 297 pp.

Frostick, L.E., Linsey, T.K. and **Reid, I.** (1992) Tectonic and Climatic control of Triassic sedimentation in the Beryl basin, Northern North Sea. *J. Geol. Soc. London*, **149**,13–26.

Gastaldo, R.A., Allen G.P. and **Huc, A-Y.** (1995) The tidal character of fluvial sediments of the modern Mahakam River delta, Kalimantan, Indonesia. In: *Tidal Signatures in Modern and Ancient Sediments* (Eds B.W. Flemming and A. Bartholomä). *Int. Assoc. Sedimentol. Spec. Publ.*, **24**, 171–181.

Gingras, M.K., Pemperton, S.G. and **Saunders, T.** (1999) The ichnology of Modern and Pleistocene Brackish-Water Deposits at Willapa Bay, Washington: Variability in Estuarine Settings. *Palaios*, **14**, 352–374.

Gjelberg, J., Dreyer, T., Høie, A., Tjelland, T. and **Lilleng, T.** (1987) Late Triassic to Mid-Jurassic sandbody development on the Barents and Mid-Norwegian shelf. In: *Petroleum Geology of North West Europe* (Eds J. Brooks and K. Glennie), pp. 1105–1129. Graham and Trotman, London.

Glørstad-Clark, E., Faleide, J.I., Lundschien, B.A. and **Nystuen, J.P.** (2010) Triassic seismic sequence stratigraphy and paleogeography of the western Barents Sea area. *Mar. Petrol. Geol.*, **27**, 1448–1475.

Hallam, A. (1988) A reevaluation of Jurassic eustasy in the light of new data and the revised Exxon curve. *SEPM Spec. Publ.*, **42**, 261–273.

Hamann, N.E., Whittaker, R.C. and **Stemmerik, L.** (2005) Geological development of the Northeast Greenland

Shelf. In: *Petroleum Geology: North-West Europe and Global Perspectives—Proceedings of the 6th Petroleum Geology Conference* (Eds A.G. Dore and B.A. Vining). *Geol. Soc.* London, 887–902.

Hansen, K. (2000) Tracking thermal history in East Greenland: an overview. *Global Planet. Change*, **24**, 303–309.

Hansen, C.F. and **Surlyk, F.** (1994) A major marine flooding surface associated with change from fully to restricted marine conditions in an intracratonic rift basin (Lower?/ Middle Jurassic Sortehat Formation, East Greenland). In: *High resolution sequence stratigraphy: innovations and applications* (Ed. S.D. Johnson), Abstracts, pp. 379–381. Liverpool University.

Haq, B.U., **Hardenbol, J.** and **Vail, P.R.** (1987) Chronology of fluctuating sea-levels since the Triassic. *Science*, **235**, 1156–1167.

Johannessen, E.P. and **Embry, A.F.** (1989) Sequence correlation: Upper Triassic to Lower Jurassic, Canadian and Norwegian Arctic. In: *Correlation in Hydrocarbona Exploration* (Ed. J.D. Collinson), pp. 155–170. Norwegian Petroleum Society, Graham & Trotman, London.

Johannessen, E.P. and **Nøttvedt, A.** (2008) Norway encircled by coastal plain and deltas. Early and Middle Jurassic; 200-161 million years ago. In: *The Making of a Land – Geology of Norway* (Eds I.B. Ramberg, I. Bryhni, A. Nøttvedt and K. Rangnes). The Norwegian Geological Association, Trondheim, Norway, 352–383.

Johnson, C. and **Gallagher, K.** (2000) A preliminary Mesozoic and Cenozoic denudation history of the North East Greenland onshore margin. *Global Planet. Change*, **24**, 261–274.

Koppelhus, E.B. and **Dam, G.** (2003) Palynostratigraphy and palaeoenvironments of the Rævekløft, Gule Horn and Ostreaelv Formations (Lower-Middle Jurassic), Neill Klinter Group, Jameson Land East Greenland. In: *The Jurassic of Denmark and Greenland* (Eds J.R. Ineson and F. Surlyk). *Geological Survey of Denmark and Greenland Bulletin*, **1**, 723–775.

Koppelhus, E.B. and **Hansen, C.F.** (2003) Palynostratigraphy and palaeoenvironment of the Middle Jurassic Sortehat Formation (Neill Klinter Group), Jameson Land, East Greenland. In: *The Jurassic of Denmark and Greenland* (Eds J.R. Ineson and F. Surlyk). Geological Survey of Denmark and Greenland Bulletin, **1**, 777–811.

Larsen, H.C. and **Marcussen, C.** (1992) Sill-intrusion, flood basalt emplacement and deep crustal structure of the Scoresby Sund region, East Greenland. In: *Magmatism and the Causes of Continental Break-up* (Eds B.C. Storey, T. Alabaster and R.J. Pankhurst). *Geol. Soc. Spec. Publ.*, **68**, 365–386.

Martinius, A.W. and **Van den Berg, J.H.** (2011) *Atlas of sedimentary structures in estuarine and tide-influenced river deposits of the Rhine-Meuse-Scheldt system: Their application to the interpretation of analogous outcrop and subsurface depositional systems.* EAGE Publications, The Netherlands, 298 pp.

Martinius, A.W., **Kaas, I.**, **Næss, A.**, **Helgesen, G.**, **Kjærefjord, J.M.** and **Leith, D.A.** (2001) Sedimentology of the heterolithic and tide-dominated Tilje Formation (Early Jurassic, Halten Terrace, offshore mid-Norway). In: *Sedimentary Environments Offshore Norway – Palaeozoic to Recent* (Eds O.J. Martinsen and T. Dreyer). *Norwegian Petroleum Society (NPF) Special Publication*, **10**, 103–144.

Martinius, A.W., **Ringrose, P.S.**, **Brostrøm, C.**, **Elfenbein, C.**, **Næss, A.** and **Ringås, J.E.** (2005) Reservoir challenges of heterolithic tidal sandstone reservoirs in the Halten Terrace, mid-Norway. *Petroleum Geoscience*, **11**, 3–16.

Mathiesen, A., **Bidstrup, T.** and **Christiansen, F.G.** (2000) Denudation and uplift history of the Jameson Land basin, *East Greenland - constrained from maturity and apatite fission track data. Global Planet. Change*, **24**, 275–301.

Midtkandal, I. and **Nystuen, J.P.** (2009) Depositional architecture of a low-gradient ramp shelf in an epicontinental sea: the lower Cretaceous of Svalbard. *Basin Res.*, **21**, 655–675.

Mitchell, A.J., **Allison, P.A.**, **Gorman, G.J.**, **Piggott, M.D.** and **Pain, C.C.** (2011) Tidal circulation Early Jurassic Laurasian epicontinental seaway. *Geology*, **39**, 208–210.

Mjelde, R., **Breivik, A.J.**, **Raum, T.**, **Mittelstaedt, E.**, **Ito, G.** and **Faleide, J.I.** (2008) Magmatic and tectonic evolution of the North Atlantic. *J. Geol. Soc.*, **165**, 31–42.

Mørk, A., **Embry, A.F.** and **Weitschat, W.** (1989) Triassic transgressive-regressive cycles in the Sverdrup Basin, Svalbard and the Barents Shelf. In: *Correlation in Hydrocarbon Exploration* (Ed. J.D. Collinson), pp. 113–130. Norwegian Petroleum Society, Graham & Trotman, London.

Müller, R., **Nystuen, J.P.**, **Eide, F.** and **Lie, H.** (2005) Late Permian to Triassic basin infill history and palaeogeography of the Mid-Norwegian shelf – East Greenland region. In: *Onshore-Offshore Relationships on the North American Margin* (Eds B.T.G. Wandås, J.P. Nystuen, E.A. Eide and F.M. Gradstein), *NPF Special Publication* **12**, 165–189.

North American Commission on Stratigraphic Nomenclature (1983) North American Stratigraphic Code. *AAPG Bull.*, **67**, 841–875.

Nøttvedt, A., **Johannessen, E.P.** and **Surlyk, F.** (2008) The Mesozoic of Western Scandinavia and East Greenland. Episodes vol. 31, pp.

Posamentier, H.W. and **Allen, G.P.** (1999) Siliciclastic Sequence Stratigraphy – Concepts and Applications. *SEPM, Concepts in Sedimentology and Paleontology*, **7**, 210 pp.

Posamentier, H.W., **Jervey, M.T.** and **Vail, P.R.** (1988) Eustatic controls on clastic deposition I – conceptual framework. In: *Sea-Level Changes: An Integrated Approach* (Eds C.K. Wilgus, B.S. Hastings, C.G.St.C. Kendall, H.W. Posamentier, C.A. Ross and C.J. Van Wagoner). *SEPM, Spec. Publ.*, **42**, 109–124.

Reading, H.G and **Collinson, J.D.** (1996) Clastic coasts. In: *Sedimentary Environments: Processes, Facies and Stratigraphy* (Ed. H.G. Reading), 3rd edition, pp. 154–231. Blackwell Science, Oxford.

Riis, F., **Lundschien, B.A.**, **Høy, T.**, **Mørk, A.** and **Mørk, M.B.E.** (2008) Evolution of the Triassic shelf in the northern Barents region. *Polar Res.*, **27**, 318–337.

Roberts, A.M., **Corfield, R.I.**, **Kusznir, N.J.**, **Matthews, S.J.**, **Hansen, E-K.** and **Hooper, R.J.** (2009) Mapping palaeostructure and palaeobathymetry along the Norwegian Atlantic continental margin: Møre and Vøring basins. *Petroleum Geoscience*, **15**, 27–43.

Rosenkrantz, A. (1934) The Lower Jurassic rocks of East Greenland. Part I. *Meddelser om Grønland*, **110**, 150 pp.

Ruhl, M. and **Kürschner, W.M.** (2011) Multiple phases of carbon cycle disurbance from large igneous province formation at the Triassic-Jurassic transition. *Geology*, **39**, 431–434.

Skogseid, J., Planke, S., Faleide, J.I., Pedersen, T., Eldholm, O. and **Neverdal, F.** (2000) NE Atlantic continental rifting and volcanic margin formation. In: Dynamics of the Norwegian Margin (Ed. A. Nøttvedt), *Geol. Soc. London Spec. Publ.*, **167**, pp. 295–326.

Surlyk, F., Callomon, J.H., Bromley, R.G. and **Birkelund, T.** (1973) Stratigraphy of the Jurassic – Lower Cretaceous sediments of Jameson Land and Scoresby Land, East Greenland. *Bull. Grønl. Geol. Unders.*, **105**, 76 pp.

Surlyk, F. (1977) Mesozoic faulting in East Greenland. In: *Fault tectonics in NW Europe* (Eds R.T.C. Frost and A.J. Dikkers). *Geol. Mijnbouw*, **56**, 311–327.

Surlyk, F. (1990a) Timing, style and sedimentary evolution of Late Palaeozoic – Mesozoic extensional basins of East Greenland. In: *Tectonic events responsible for Britain's oil and gas reserves* (Eds R.F.P. Hardman and J. Brooks). *Geol. Soc. London Spec. Publ.*, **55**, 107–125.

Surlyk, F. (1990b) A Jurassic sea-level curve for East Greenland. *Palaeogeogr., Palaeoclimatol. Palaeoecol.*, **78**, 71–85.

Surlyk, F. (2003) The Jurassic of East Greenland: a sedimentary record of thermal subsidence, onset and culmination of rifting. In: *The Jurassic of Denmark and Greenland* (Eds J.R. Ineson and F. Surlyk). *Geological Survey of Denmark and Greenland Bulletin*, **1**, 659–722.

Taylor, A.M. and **Goldring, R.** (1993) Description and analysis of bioturbation and ichnofabric. *Geol. Soc. London. J.*, **150**, 141–148.

Tessier, B. (2012) Stratigraphy of Tide-Dominated Estuaries. In: *Principles of Tidal Sedimentology* (Eds **R.A. Davis** and R.W. Dalrymple). Springer Science+Business Media B.V. 2012, 109–128.

Tsikalas, F., Faleide, J.I., Eldholm, O. and **Wilson, J.** (2005) Late Mesozoic-Cenozoic structural and stratigraphic correlations between the conjugate mid-Norway and NE Greenland continental margins. In: *Petroleum Geology: North-West Europe and Global Perspectives – Proceedings of the 6th Petroleum Geology Conference* (Eds A.G. Doré and B. Vining). Geological Society of London, 785–801.

Tsikalas, F., Faleide, J.I., Kusznir, N.J. (2008) Along-strike variations in rifted margin crustal architecture and lithosphere thinning between northern Vøring and Lofoten margin segments off mid-Norway. *Tectonophysics*, **458**, 68–81.

Van Wagoner, J.C., Posamentier, H.W., Mitchum, R.M., Vail, P.R., Sarg, J.F., Loutit, T.S. and **Hardenbol, J.** (1988) An overview of the fundamentals of sequence stratigraphy and key definitions. In: *Sea-Level Changes: An Intergrated Approach* (Eds C.K. Wilgus, B.S. Hastings, C.G.St.C. Kendall, H.W. Posamentier, C.A. Ross and J.C. Van Wagoner). *SEPM Spec. Publ.*, **42**, 39–45.

Van Wagoner, J.C., Mitchum, R.M., Campion, K.M. and **Rahmanian, V.D.** (1990) Siliciclastic Sequence Stratigraphy in Well Logs, Cores and Outcrops: Concepts for High-Resolution Correlation of Time and Space. *AAPG Methods in Exploration Series*, **7**, 55 pp.

Int. Assoc. Sedimentol. Spec. Publ. (2014) **46**, 339–388

Eustatic, tectonic and climatic controls on an early syn-rift mixed-energy delta, Tilje Formation (Early Jurassic, Smørbukk field, offshore mid-Norway)

AITOR A. ICHASO and ROBERT W. DALRYMPLE

Department of Geological Sciences and Geological Engineering, Queen's University, Kingston, Ontario K7L 3N6, Canada

ABSTRACT

The Early Jurassic Upper Åre and Tilje formations are an important mature hydrocarbon reservoir in the Smørbukk field, offshore Mid-Norway. The Tilje Fm. was deposited in a tide-influenced deltaic setting during an early syn-rift tectonic stage within a NE–SW oriented half-graben sub-basin. The deltaic deposits rest erosively (SB2) over the Upper Åre shoreface units and are arranged vertically in two major sequences that show a diverse range of heterolithic facies preserving tide, combined-flow and wave-generated structures. The first sequence, consisting of the lower-middle Tilje (reservoir zones T1 to T3.1), is composed of stacked, coarsening-upward and sanding-upward distributary mouth bars, delta-front and prodeltaic deposits. This interval shows a complex vertical alternation of more proximal tidal-influenced and medial wave/storm-influenced successions that locally record seasonal variations in river discharge. The lower-middle Tilje facies distribution suggests overall N–NW to SE progradation of 'lobe-shaped' deltas, oblique to the regional structural grain. The second sequence (reservoir zones T3.2 to T6; middle-upper Tilje) rests erosively (SB3) over the delta-front deposits of the underlying sequence. The sandy lower part of the upper sequence is capped by a regional flooding surface that marks the transition from the Tilje Fm to the marine Ror Fm. The middle-upper Tilje is composed of proximal tidal-fluvial distributary-channel and point-bar successions with abundant fluid-mud deposits. The top of the Tilje is mainly composed of medial-distal channelised heterolithic mouth bars and mixed tide-wave delta-front deposits. The propagation of blind normal faults within the study area and the generation of isolated syn-depositional shallow hanging-wall depocentres exerted a subtle control on the sediment-entry points, the thickness of sedimentary successions and the direction of delta progradation. Two main tectonic pulses are recognised, both of which are responsible, in whole or in part, for the two main sequence boundaries recognised in the succession. The first caused an abrupt change from wave-dominated to tide-dominated sedimentation, whereas the second is reflected in an abrupt change in the sediment-transport pathways in the study area.

Keywords: Tilje, Delta, Tide, syn-rift, Early Jurassic, Norway.

INTRODUCTION

In the Mid-Norwegian shelf (Fig. 1), Early Jurassic Sinemurian and Pliensbachian sediments (Brekke *et al.*, 2001; Fig. 2A) were deposited in a seaway/rift-basin setting (Halten-Trøndelag Basin; Fig. 2B) during the opening of the North Atlantic Ocean, at a time of significant eustatic sea-level rise (Hallam, 1988; Surlyk, 1990). These deposits are characterised by a series of tide-influenced and tide-dominated successions (both estuarine and deltaic; Martinius *et al.*, 2005) that were deposited in a mid-palaeolatitudinal location (between 49°N and 53°N; Doré, 1991) and had a warm climate with strong seasonal differences in temperature and rainfall (Hallam, 1994). The mid-upper Åre Formation (Svela, 2001; defined by the first fully marine deposits within the Halten Terrace succession) and the entire Tilje and Ile formations (Fig. 2A), in particular, contain the main intervals

From Depositional Systems to Sedimentary Successions on the Norwegian Continental Margin, First Edition. Edited by A. W. Martinius, R. Ravnås, J. A. Howell, R. J. Steel, and J. P. Wonham.

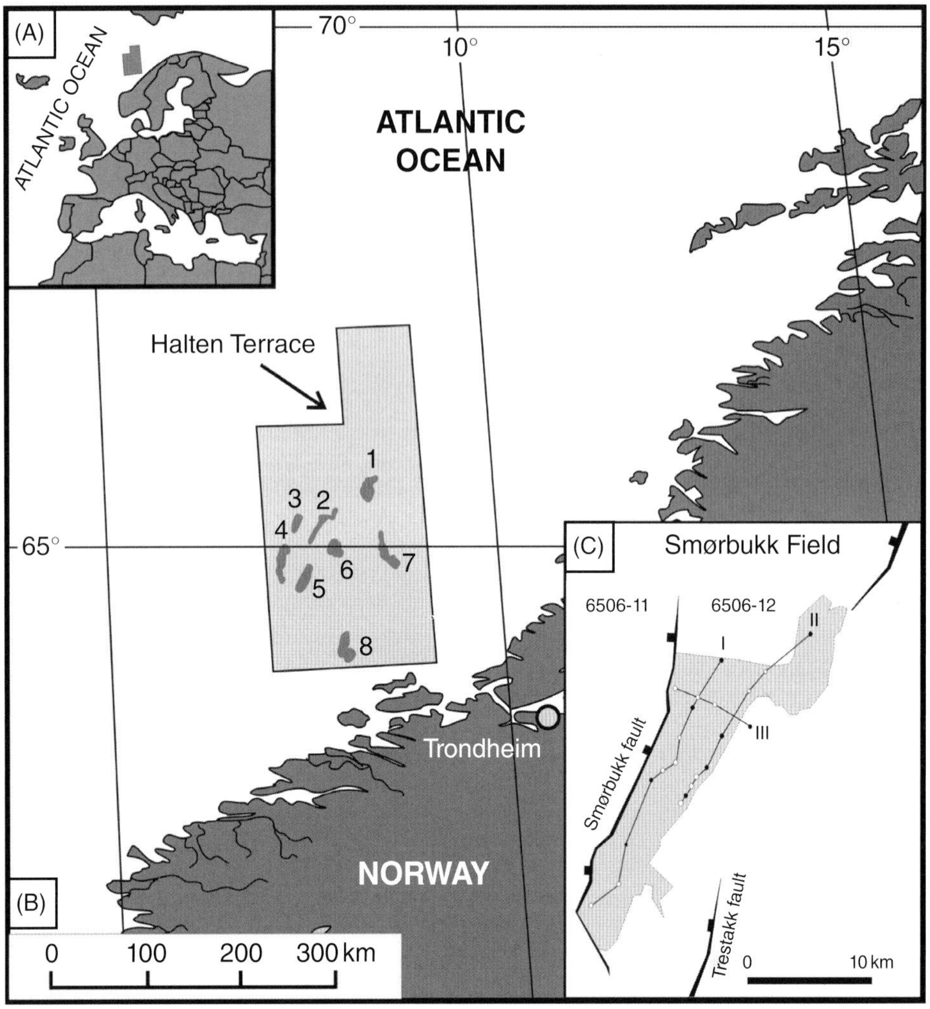

Fig. 1. Area of study. (A) Location map of the Halten Terrace. (B) Map showing the Mid-Norwegian hydrocarbon fields in the Halten Terrace. Main oil fields are: (1) Heidrun, (2) Smørbukk (Åsgard), (3) Morvin, (4) Kristin, (5) Lavrans, (6) Smørbukk Sør (Åsgard), (7) Midgard (Åsgard) and (8) Njord. (C) Map of the Smørbukk field showing main active faults during Tilje deposition and location of cross-section lines (I, II and III) shown in Fig. 19.

of inferred tidal origin (Martinius *et al.*, 2001, 2005). These stratigraphic units form an important, deeply buried reservoir interval and are geographically confined to a NE–SW-oriented belt approximately 470 km long by 150 km wide. The hydrocarbon production from this interval is about 850 BBL (Martinius *et al.*, 2005; Norwegian Petroleum Directorate (NPD) gross production 2010) equivalent per day from several fields on the Halten Terrace (Fig. 1B; Martinius *et al.*, 2001). However, production is adversely affected by the heterolithic nature of the tidal facies (approximately 80% of the total rock volume), deep-burial digenetic features (present-day depth 1 km to 5 km), fracturing and faulting (Martinius *et al.*, 2005) and the complex internal architecture of the reservoir units (Ehrenberg *et al.*, 1992; Martinius *et al.*, 2001). Thus, more detailed sedimentological interpretations, coupled with a better understanding of the stratigraphic architecture and the controls on sedimentation, are required both for individual hydrocarbon fields and at a more regional scale, if production is to be optimised (Martinius *et al.*, 2005).

In this context, the goal of this research is to provide a better sedimentological, palaeoenvironmental, palaeogeographic and sequence-stratigraphic description of the Lower Jurassic upper Åre and Tilje formations. This provides, in turn, an improved understanding of the dimensions and architecture

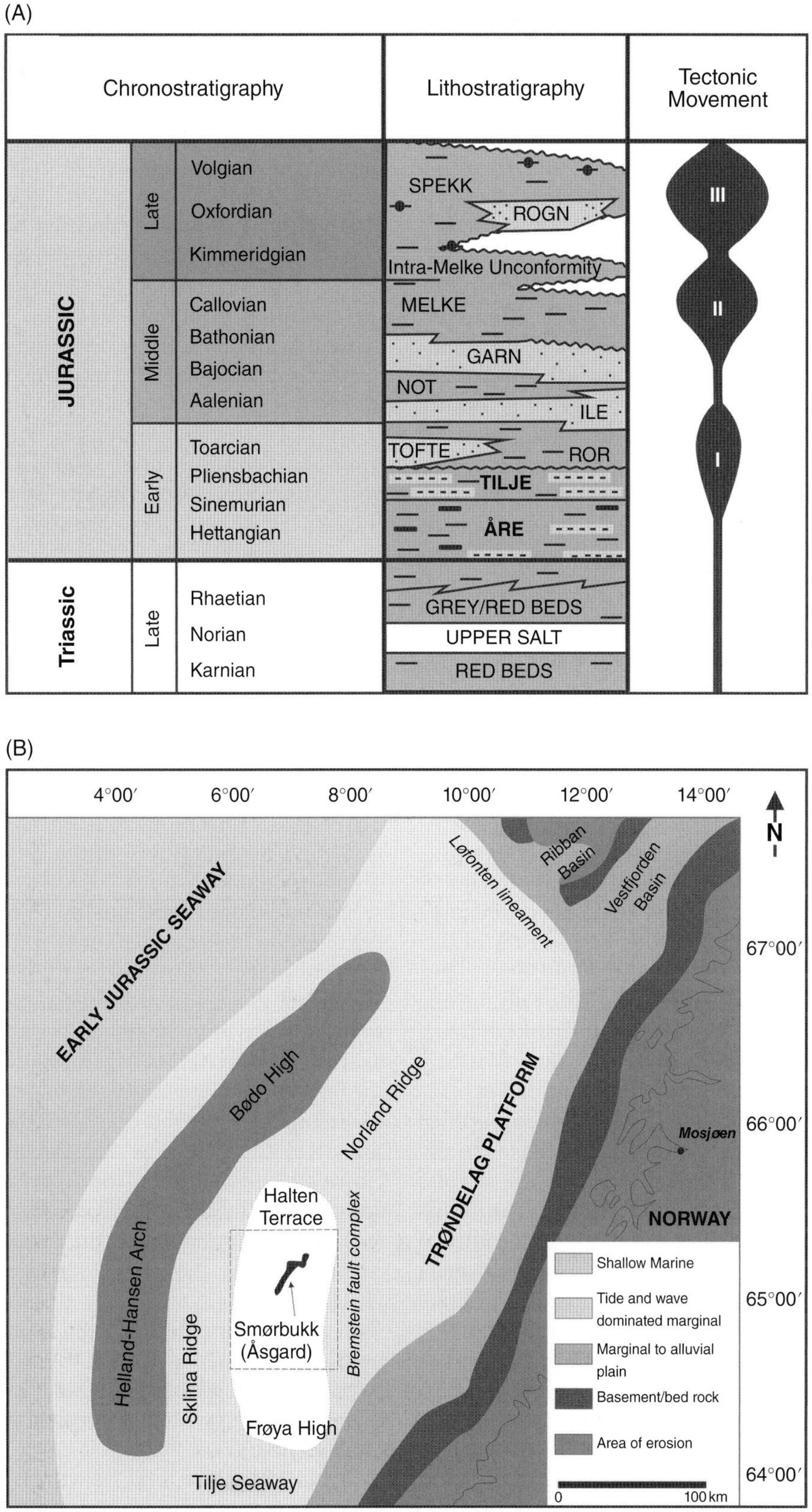

Fig. 2. (A) Summary of Triassic and Jurassic lithostratigraphy and tectonic events in the Halten Terrace area. The tectonic-movement column shows main pulses of rifting that had an impact on the geometry of Jurassic reservoirs: (I) Early Jurassic, (II) Middle Jurassic and (III) Late Jurassic times. (B) Lower Jurassic palaeo-geographic setting of the Halten Terrace and the Early Jurassic seaway elements. Compiled and modified from Doré (1991), Brekke *et al.* (2001).

of sediment bodies within these units and the influence of major eustatic, tectonic and climatic controls on their accumulation. This paper presents a revision of previous sedimentologic and sequence-stratigraphic work on the Tilje Formation carried out by Martinius *et al.* (2001, 2005), building on the recent sedimentological insights on depositional processes, facies distribution, stratigraphic architecture and palaeoenvironmental reconstructions for the Tilje Formation (Ichaso & Dalrymple, 2009; Ichaso, 2012). This paper focuses specifically on the Smørbukk field (Fig. 1C). In more general terms, this study aims to improve our ability to reconstruct the distribution of depositional environments in such complex, ancient tidal, successions, with a particular focus on the interplay between subtle tectonic movements and sedimentation in a rift-basin setting.

Tilje Formation in the Smørbukk field

The Smørbukk field was discovered in 1984. It is situated approximately 230 km west of the Norwegian mainland (mainly in block 6506/12) in water depths between 250 m and 300 m and is considered to be one of the deepest hydrocarbon-bearing structures in the area (Klefstad *et al.*, 2005). Hydrocarbon reserves, mainly gas-rich condensates of mixed marine/terrigenous source, are contained in sandstones of four Lower to Middle Jurassic formations (Åre, Tilje, Ile and Garn). The most common interpretation of the Tilje Formation is that it accumulated in a tide-dominated deltaic environment, during a late pre-rift to early syn-rift stage (Ehrenberg *et al.*, 1992; Martinius *et al.*, 2001; Marsh *et al.*, 2010). This tectono-stratigraphic interpretation is based largely on the absence of basin-bounding fault margins, the absence of volcanic rocks, relative uniformity and continuity of stratigraphic units, the nature of the heterolithic deposits, the large lateral extent of depositional systems, the overall sandier upward succession and the vertical and lateral progradational arrangement of facies (Ehrenberg *et al.*, 1992; Martinius *et al.*, 2001), especially during the lower-middle Tilje stratigraphic interval.

The Tilje Formation (150 m to 200 m thick) was first divided into three major reservoir units (Ehrenberg *et al.*, 1992), mostly based on the inter-well correlation of offshore shales and bioturbated intervals. More recently (2003 to 2004), the operator, Statoil, has divided the Tilje Formation in the Smørbukk field into six reservoir zones (T1 to T6) with several sub-units (e.g. T1.1, T1.2, etc.). This reservoir zonation was based on the integration of sedimentological analysis, regional correlation using ichnofabric analysis, Sm-Nd dating and a sequence-stratigraphic model developed in the Halten Terrace production area (Martinius *et al.*, 2001). As discussed below, most reservoir zones are interpreted here to be parasequence sets that are bounded either by flooding surfaces and/or sequence boundaries. They, therefore, represent approximate chronostratigraphic units that can be correlated throughout the study area.

The sequence-stratigraphic framework interpreted by Martinius *et al.* (2001) for the upper Åre and Tilje formations consists of three sequences (Fig. 3). These sequences are separated by regional second-order hiatal surfaces (SB2 and SB3), with several second-order to third-order, or even higher-order, flooding surfaces (FS) forming parasequences and parasequence sets that are metres to tens of metres thick. Sequence 1 comprises the upper Åre Formation (A3; Fig. 3). Sequence 2 (reservoir zones T1 to T3.1) occupies the lower Tilje Fm., extending from the Åre-Tilje contact (SB2) up to an inferred sequence boundary (SB3) in the mid-Tilje Fm. Sequence 3 (reservoir zones T3.2 to T6) extends from SB3 in the middle of the Tilje Fm. up to the Tilje-Ror contact.

Sm/Nd isotopic data from the Tilje Fm. support the regional and local correlation by indicating a consistent provenance. Martinius *et al.* (2001) suggested a significant change of provenance ages from younger source units within Sequence 2 (1200 Ma to 1400 Ma) to an older source (1500 Ma to 1800 Ma) in the overlying Sequence 3. Previous work had suggested that the main source area for the Tilje Fm. is the Fennoscandian Shield to the east (Fig. 2B; Dalland *et al.*, 1988). However, dip-meter and Sm/Nd isotopic data suggested that the Tilje received sediments from the Ribban Basin structure in the north (dominated by erosion during the entire Jurassic), the Helland-Hansen-Bodø (Utgard) High to the northwest and west and/or from the Nordland ridge (Fig. 2B) in the north-east (Martinius *et al.*, 2001). The area could also have received sediments from local topographic highs associated with the propagation of local normal faults (Marsh *et al.*, 2010).

Structurally, the field lies at the crest of a tilted fault block, bounded on the west by a major normal fault (the Smørbukk fault: Fig. 1C) and on the north by an east-west-trending graben that transects the crest of the fault block (Ehrenberg *et al.*, 1992). Recently, 3D seismic-reflection data were

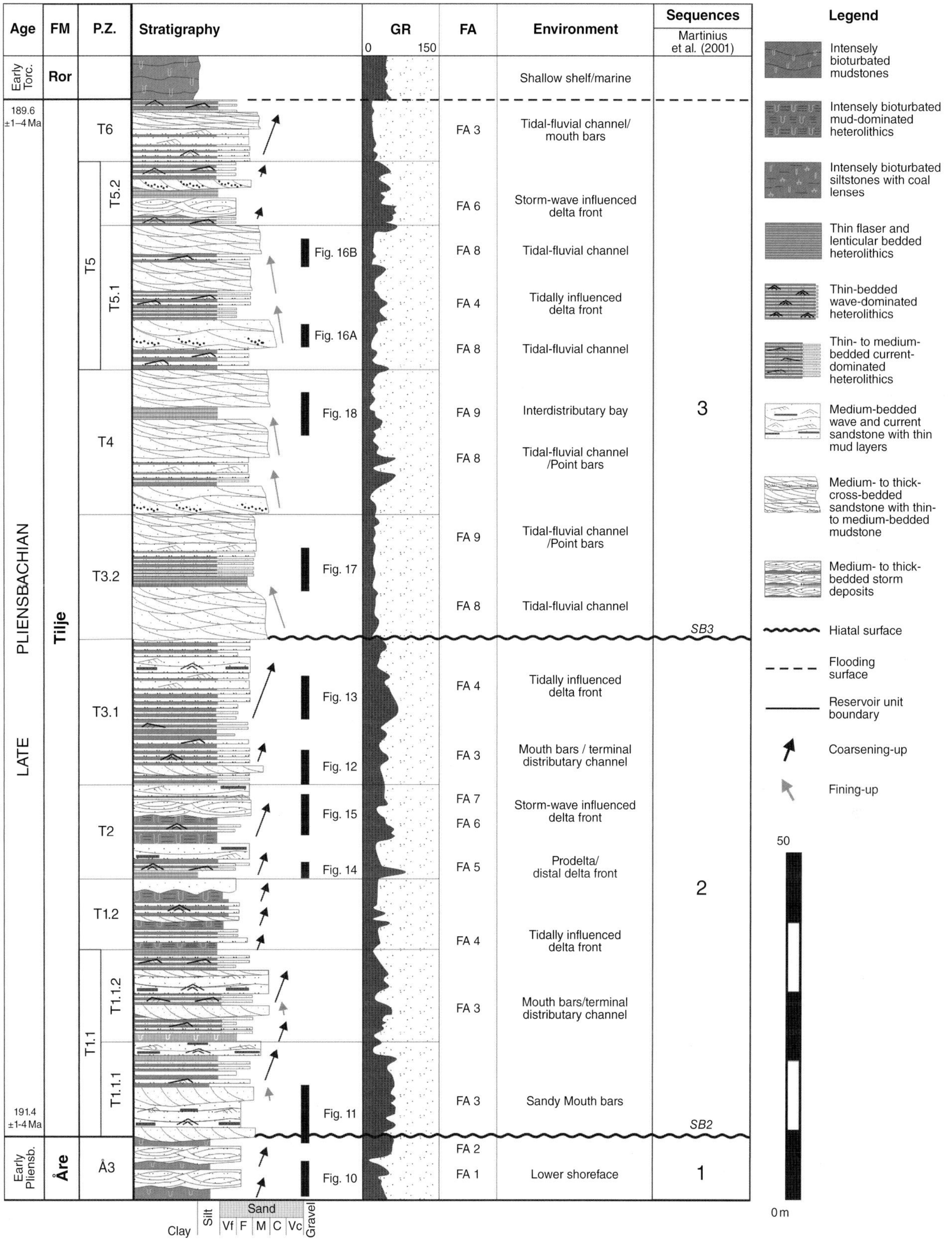

Fig. 3. Composite stratigraphic column for the Smørbukk area, showing production zonation (P.Z.), typical gamma-ray well-log curve, facies associations (FA), depositional environments, interpreted sequences and hiatal surfaces/sequence boundaries (SB). Absolute ages for the base and top of the Tilje Formation after Gradstein *et al.* (1995). The locations of core photos shown in various figures are indicated.

used to constrain the distribution and timing of fault activity during the Early Jurassic to earliest Cretaceous rifting stages in the Åsgard development area. Corfield & Sharp (2000) suggest that rifting was initiated during the early Jurassic (top Åre Formation time) and was characterised by distributed deformation along blind faults (oriented NE–SW) within the basement and by localised deformation along the major Smørbukk and Trestakk normal (westerly dipping) faults within the cover (Fig. 1C). During Tilje deposition, these faults caused slow tectonic subsidence and the formation of local shallow depocentres occurring to the west of the Trestakk Fault and east of the Smørbukk Fault (Marsh *et al.*, 2010). The early syn-rift (rift-initiation stage) Tilje deposition was succeeded by a large-scale rift-climax stage, during which middle-upper Jurassic folds and Early Cretaceous grabens developed at an oblique angle to the main faults (Marsh *et al.*, 2010).

METHODOLOGY

Core material from nine wells that penetrate the upper Åre and Tilje formations in the Smørbukk field (production blocks 6506-11 and 6506-12) was examined and logged in detail (i.e. at the centimetre to decimetre scale), in order to define the different facies (F) and facies associations (FA). The sedimentary facies were distinguished on the basis of lithology, physical sedimentary structures, sand-mud ratio, degree of bioturbation and the specific trace-fossil assemblage present. Heterolithic facies are particularly common in the study interval and are especially challenging to subdivide in a meaningful way. Heterolithic deposits within the Tilje Formation have been described in the past using traditional schemes such as flaser, wavy and lenticular bedding (Dreyer, 1992). Recently, Martinius *et al.* (2001) used a facies classification based on variations of the intralayer characteristics of the heterolithic units such as: grain size and sorting, structures in the coarse-grained layers, trace fossils and ichnofabrics, grain size and thickness variability, frequency of occurrence and stratigraphic position. In this work the heterolithic deposits were classified based on the following characteristics:

1. Sand/Mud ratios: The proportions of sandstone and mudstone are used to create a fivefold subdivision of the heterolithic deposits sandstone (>90 : 10 sandstone : mudstone ratio); (2) sand-dominated heterolithics (60 : 40 to 90 : 10 sandstone : mudstone), (3) mixed sand/mud heterolithics (40 : 60 to 60 : 40), (4) mud-dominated heterolithics (0 : 90 to 40 : 60) and (5) mudstone (<10 : 90).
2. Bed thickness and bedding style: Individual layers were described using the standard bed-thickness classification of Ingram (1954) as laminae (<1 mm thick), very thin-bedded (1 mm to 3 mm), thin-bedded (3 mm to 10 mm), medium-bedded (10 mm to 30 mm), thick-bedded (30 mm to 100 mm) and very thick-bedded (>100 mm). Bedding styles were described following Reineck & Wunderlich's (1968) classification, which includes the classic flaser, wavy, lenticular and pin-stripe types. The terms bed, bedset (group of similar beds), amalgamated (similar beds with no intervening material of different type) and tabular (nearly planar bedding surfaces that are parallel to each other) were used following the definitions in Collinson & Thompson (1982).
3. Grain-size and sedimentary structures: For grain-size descriptions, the standard Wentworth size-class nomenclature was used. The primary sedimentary structures used included those produced by currents, waves and mixed-energy conditions (e.g. Dumas *et al.*, 2005), as well as both deformational and biogenic structures.
4. Degree and type of bioturbation: Descriptive-genetic ichnological classifications (Pemberton *et al.*, 2001) were used for the recognition and interpretation of ichnofacies, with application to the recognition of depositional settings (MacEachern & Bann, 2008). Bioturbation-index values were described following the Bann *et al.* (2008) classification.

Use of all of these criteria has the potential to produce an unwieldy number of facies. Therefore, in this work, classification of heterolithic deposits was focussed at the decimetre to metre scale, whilst paying close attention to details at the mm-scale to cm-scale. This allowed us to have a smaller number of facies than previous authors.

Using the cored wells and the environmental interpretations derived from the facies associations, a sequence-stratigraphic analysis of the succession was undertaken using the standard Exxon-based approach (cf. Catuneanu, 2006; Ichaso, 2012). Correlation of the stratigraphic units defined in this way (parasequences, parasequence sets, systems tracts and sequences) away from the cored wells was done using gamma-ray (GR) and density neutron (DEN/NEU) well-log curves calibrated using the cored wells. The reservoir zones defined by Statoil (Fig. 3) can be correlated directly with these sequence-stratigraphic units. Thus, facies and environmental distribution maps were generated for each reservoir zone using facies percentages and the dominant grain size. Stratigraphic cross-sections (axial and transversal to the field; Fig. 1C) were constructed using the well logs (datum: Tilje-Ror formational contact) and the logged cores. The fault activity-structural maps produced by Marsh *et al.* (2010) for Tilje time were integrated with the facies distribution maps in order to produce evolutionary tectono-stratigraphic interpretations. Together with the stratigraphic cross-sections, the architectural organisation of facies associations and their palaeogeographic distribution in a proximal-distal context are interpreted following Dalrymple & Choi (2007).

FACIES

In general, the upper Åre and Tilje deposits in the Smørbukk and Smørbukk South fields are characterised by the following attributes: (1) a pervasive heterolithic nature; (2) rapid alternation of grain size between claystone or siltstone and fine-grained to medium-grained sandstone; (3) dramatic changes in bedding style and sedimentary structures over short vertical distances, which imparts a high degree of variability to the succession; (4) variable bed thickness with successive beds ranging from thin-bedded to thick-bedded; and (5) variable bioturbation suites composed of mixtures of vertical and horizontal burrows, with bioturbation indexes ranging from sparse or weakly burrowed (BI 0 to 1) to intensely bioturbated (BI 5). Within this heterogeneous succession, seven lithofacies where identified (see Tables 1 and 2 for detailed facies descriptions) on the basis of the characteristics discussed above.

Facies 1: Graded and non-graded conglomerate

The unbioturbated, quartz pebble and granule conglomerates of Facies 1 (Figs 4A & 4B) suggest the presence of high-energy flows. Single and amalgamated, tabular conglomeratic beds (up to 40 cm thick) commonly show normal and inverse grading, with no obvious internal lamination. The normal grading suggests that the conglomerates were deposited rapidly by decelerating turbulent flows, whereas the inverse grading indicates that dispersive pressure operated (Lowe, 1979), or that deposition occurred during waxing flow. Beds that are poorly sorted with a high matrix percentage might indicate deposition from high-concentration sediment flows (Nemec & Steel, 1984), or rapid accumulation in a setting with abundant sand.

Facies 2: Cross-bedded sandstone

The coarse-grained to fine-grained cross-bedded sandstones of Facies 2 (Figs 4C, 4D & 4E), forming fining-upward cosets (up to 2.5 m thick), coupled with the local presence of herringbone stratification and paired mudstone layers, indicate deposition by subaqueous dunes in a setting with moderate to strong currents and a tidal influence (Dalrymple & Choi, 2007). The paired and thick-thin alternations of the mudstone layers in bottomsets of cross-beds are suggestive of a tidal diurnal inequality (Visser, 1980). Unusually thick (1 cm to 3 cm) homogeneous mud layers and rip-up mud clasts derived from such layers (Figs 4D & 4E) occur towards the base of thick cross-bedded successions, suggesting deposition from high-concentration, near-bed suspensions (fluid muds; Ichaso & Dalrymple, 2009). The upward decrease in the thickness of these mud layers indicates a vertical reduction of the suspended-sediment concentration in the water column (cf. Dalrymple *et al.*, 2003). Brackish-water conditions are interpreted from the sparse and low-diversity trace-fossil assemblage, which is mostly composed of large-scale vertical burrows (*Diplocraterion* and *Skolithos*).

Facies 3: Current-rippled sandstone

The fine-grained to very fine-grained sandstones (2 cm to 5 cm thick) of Facies 3, showing current-ripple cross-lamination (Fig. 4F), are interpreted to be deposited by slow currents. Most occurrences show only one direction of ripple migration, but the common presence of true bipolar cross-lamination (as indicated by steep lamina dips, in contrast to the lower dip angles associated with the end-on view ripple troughs) and the alternation with very thin (<2 mm) mudstone laminae imply a tidal influence (Baker *et al.*, 1995; Dalrymple & Choi, 2007; Van den Berg *et al.*, 2007). The occasional occurrence of low-angle climbing-ripple sets indicates rapid deposition from decelerating flows, probably associated with sporadic river floods (Ashley *et al.*, 1982).

Table 1. Upper Åre and Tilje formations lithofacies identified on the basis of their heterolithic nature, grain size, bedding style, sedimentary structures, thickness, and bioturbation style (See Table 2 for heterolithic Facies classification).

Facies	Subfacies	Grain size/description	Thickness	Sedimentary Structures	BI and trace fossils	Interpretation
Facies 1 (F1): Graded and non-graded conglomerate	(i) Normally graded oligomictic quartz-pebble orthoconglomerate. (ii) Normally graded oligomictic quartz-pebble paraconglomerate. (iii) Inversely graded polymictic paraconglomerate. (iv) Non-graded oligomictic paraconglomerate.	Round to sub-rounded quartz pebbles (3 cm to 5 cm) and granules. Siderite and intraformational mudstone clasts are present. By volume the clast fraction is between 60% to 75% and the matrix fraction is between 25% to 40%. Erosive basal contacts.	Beds 10 cm to 30 cm thick, although amalgamated conglomerate beds can be up to 40 cm thick.	Normal and inverse grading	BI 0. Trace fossils absent.	High-energy flows. Normally graded beds deposited by dilute turbulent flows or by waning flows. Inversely graded beds deposited by more concentrated flows in which dispersive pressure operated, or by waxing flows. Poorly sorted beds with a high matrix percentage suggest highly concentrated sediment flows.
Facies 2 (F2): Cross-bedded sandstone	(i) Poorly sorted, coarse-to medium-grained, cross-bedded sandstones. (ii) Moderately sorted, medium-grained to fine-grained, cross-bedded sandstones. (iii) Poorly sorted, medium-grained to fine-grained, low-angle cross-bedded sandstones with angular bioclastic calcite grains.	Sand : mud ratio between 100 : 0 and 90 : 10. Coarse-grained to medium-grained or fine-grained, poorly to moderately sorted grey, to dark brown sandstone. Mica-rich intervals. Thick homogeneous mudstone layers (<3 cm) and mudstone laminae (showing pairs and/or thick-thin alternations) in bottomsets of the cross beds. Sharp and erosive basal contacts.	Fining-upward cosets up to 2.5 m thick.	Large, low-angle to high-angle cross-stratification. Progressively steeper laminae inclinations toward bed tops. Local herring-bone cross-stratification.	BI 0 to 1. Absent to sparse. Low-diversity. Large-scale *Diplocraterion*, and *Skolithos*. Rare *Teichichnus* and *Bergaueria*.	River and tidally influenced, moderate to strong currents. Trace-fossil assemblage suggests brackish-water conditions at times. Thick homogeneous mud layers suggest deposition from high-concentration near-bed suspensions (fluid muds). Local poorly sorted low-angle cross-bedded sandstones product of wave or current energy.
Facies 3 (F3): Current ripple cross-laminated sandstone	No subfacies.	Sand : mud ratio between 100 : 0 and 90 : 10. Fine-grained to very fine-grained, poorly to moderately sorted, light grey to light brown sandstone. Alternation with very thin mudstone laminae. Sharp and erosive basal contacts.	Individual set thicknesses vary between 2 cm and 5 cm. Beds up to 40 cm thick.	Small-scale, low-angle to high-angle, bipolar ripple cross-lamination. Local climbing ripples.	BI 1. Sparse. Low-diversity. Small *Skolithos* and rare *Planolites*	Tidally-influenced slow to moderate currents. The local presence of climbing ripples indicates rapid deposition from decelerating flows, probably associated with river floods.

Facies 4 (F4): Hummocky cross-stratified (HCS) sandstone	(i) Moderately sorted, very-fine grained, micaceous hummocky cross-stratified sandstone. (ii) Well sorted, very-fine grained, hummocky cross-stratified sandstone.	Very-fine grained and moderately to well-sorted grey sandstone. Locally, sandstones are capped by sharp-based, 0.5 cm to 3 cm thick, faintly laminated mudstone. Sharp based and occasionally erosive. At the base, locally preserved residual mud lenses and clasts. Calcite cement is locally present.	Amalgamated bedsets of up to 4 m. Individual bed thickness varies between 15 cm and 1 m.	Planar and/or undulatory lamination, hummocky-cross stratification (HCS) capped by wave ripples, and/or combined-flow ripple cross lamination with asymmetric round-crested formsets. Weakly laminated mudstone on top of HCS.	BI 0 to 1. Absent to sparse. *Skolithos*, *Diplocraterion*, and rare *Planolites*. Bioturbation of the capping mudstone is top-down post-depositional. Locally, vertical to sub-vertical *Glossifungites* towards the top of HCS beds.	Storm-influenced high-energy oscillatory currents. Vertical decrease in wave energy and turbulence (HCS to wave ripples). Wind-driven residual currents during storm events. Superimposed tidal currents. Wave-generated fluid mud caps beds.
Facies 5 (F5): Heterolithics				See Table II		
Facies 6 (F6): Intensely bioturbated sandstone and heterolithics	F6.1. Reddish bioturbated sandstone	Silty fine/medium-grained reddish sandstone. Moderately to poorly sorted. Matrix-supported. Subangular biogenic calcite grains. Homogeneous aspect. Mudstone lenses and interrupted lamination. Burrow mottling, small shell debris and faecal pellets. Iron and calcite cements.	Bedsets up to 50 cm.	Occasional tabular-stratified beds, low-angle lamination and coarsening upward	BI 5 to 6. Abundant to complete. High diversity. *Skolithos*, intergradation with proximal expressions of *Cruziana* Ichnofacies. *Rosselia*, *Planolites*, *Cylindrichnus*, *Skolithos*, *Bergaueria*, *Palaeophycus*, and rare *Thalassinoides*.	Biogenic reworking processes. Large vertical burrows indicative of relatively high levels of wave (below fair-weather wave base but above storm base) or current energy; developed in loose or shifting substrates.
	F6.2. Disturbed sandstone	Fine-grained to medium-grained grey to brown sandstones. Bedding between the sand layers is distinct but, internally, trace fossils tend to cross-cut each other.	20 cm to 40 cm thick	Cross bedding, ripple cross-lamination, and parallel lamination	BI 4 to 5. Common to abundant. Low-diversity. Archetypal *Skolithos* ichnofacies. Large (10 cm to 25 cm long) *Diplocraterion*, small (0.5 cm to 2 cm diameter) *Planolites* and *Palaeophycus*	Moderate to high current energy. Low diversity but abundant U-shaped large burrows as a response to a shifting substrate.

(Continued)

Table 1. (*Continued*)

Facies	Subfacies	Grain size / description	Thickness	Sedimentary Structures	BI and trace fossils	Interpretation
	F6.3. Disturbed Heterolithics	Fine to very-fine sand grain size in mud-dominated and or mixed heterolithics. Trace fossils cross-cut each other. The colour of the beds is dark grey mudstone with light grey sandstone lenses. Associated with F5.2.2 and F5.3.2 (see Table 2).	20 to 40 cm thick	Parallel lamination, synaeresis cracks and ripple cross-lamination are locally visible.	BI 4 to 5. Common to abundant. High-diversity. Proximal expressions of the *Cruziana* Ichnofacies *Planolites*, *Palaeophycus*, *Chondrites*, *Rosselia*, *Rhyzocoralium*, *Diplocraterion*, *Cylindrichnus*, *Teichichnus*, and rare *Thalassinoides* and *Phoebichnus*.	In association with current generated heterolithics 5.2.1. Low energy and relatively slow tidal currents. Moderate intensity trace-fossil suite, and moderate to high diversity suggest low to moderate energy (tidal currents) subjected to chemical and physical variations.
	F6.4. Light grey homogenised sandstone	Silty, Very fine to fine-grained light grey sandstone. Barely visible bedding. Overlap of trace fossils. Discontinuous mudstone and sandstone lenses. Locally nodular with calcite cement.	Up to 30 cm	Discontinuous parallel lamination.	BI 5 to 6. Abundant to complete. High-diversity. Proximal expressions of the *Cruziana* Ichnofacies. Small to large-scale *Planolites*, *Palaeophycus*, *Cylindrichnus*, *Rosselia*, *Ophiomorpha* *Skolithos*, *Thalassinoides*, and less common *Diplocraterion* and *Rhizocorallium*.	Homogenisation and destruction of the original bedding and structures by syn- and/or post-depositional bioturbation. Transitions from suspended sediment settling to shifting substrate conditions under moderate to higher energy.
	F6.5. Rooted sandstone	Very fine-grained to silty sandstone. Barely visible bedding and internal lamination. Small organic lenses locally present toward the top of these units. Mottled aspect.	Up to 40 cm	Parallel lamination locally visible	BI 5 to 6. Abundant to complete. Low diversity. Small plant root-traces showing bifurcated endings, delicate *Skolithos* and *Ophiomorpha*.	Immature palaeo-soil. Low-energy setting with salt water influence.
Facies 7 (F7): Pinstripe-laminated mudstones	No subdivisions	Sand:mud ratios between 10:90 and 0:100. Silt-clay dominated. Sandstones are very fine-grained, sharp based, showing normal grading into siltstone.	Sandstone lenses 0.5 cm to 1.5 cm; mudstone layers 0.5 cm to 3 cm	Pinstripe bedding. Parallel lamination and ripple cross-lamination occur locally, showing bipolar ripple cross-lamination.	BI 0 to 2. Sparse to uncommon. Mm-scale burrows like *Chondrites* and rare *Planolites*.	Low-energy conditions. Low tidal velocities or a scarcity of sand.

Table 2. Heterolithic deposits of the upper Are and Tilje formations. Facies 5 (F5) and sub-facies were classified on the basis of their sand: mud ratios, thickness, sandstone grain size, mudstone characteristics, sedimentary structures, and bioturbation style.

Facies 5 (F5): Heterolithics		Thickness (cm)	Sandstone grain sizes and description	Mudstone description	Sedimentary structures	BI and trace fossils	Interpretation
Facies 5.1 (F5.1) Sand-dominated heterolithics (sand : mud ratios 90 : 10 to 60 : 40)		Sandstone 5 cm to 30 cm; Mudstone 0.2 cm to 2 cm; Fluid-mud layers 3 cm to 15 cm, upward decrease in thickness of fluid-mud layers.	Very fine to medium with occasional gravel and pebble lags. Normal grading or alternations between coarse and fine sandstone. Sharp and erosive basal contacts.	Continuous paired laminae/beds and lenticular geometries. Thicknesses decrease upward into thin laminae. Very thick homogenous mud layers (fluid muds) occur toward the base of intervals of medium and fine-grained sandstone.	Cross-bedding and ripple cross-lamination showing bipolarity. Stylolites occur in the thickest and coarsest sands.	BI 0 to 2. Sparse to uncommon. Vertical burrows in the sandy portions: *Skolithos* and *Diplocraterion*. Small-scale *Planolites* within the laminated mudstones.	Tidally influenced unidirectional currents with moderate speeds. Paired mudstones formed by tidal currents during ebb-flood tidal cycles (i.e. they represent tidal rhythmites in which the neap-spring cyclicity is not well developed). High suspended sediment concentrations (SSC) formed fluid-mud layers. Upward decrease in thickness of fluid-mud layers indicates a vertical reduction of the SSC in the water column (cf. Dalrymple *et al.*, 2003).
Facies 5.2 (F5.2) Mixed sand/mud heterolithics (sand : mud ratios 60 : 40 to 40 : 60)	5.2.1. (F5.2.1) Wavy-bedded current-dominated heterolithics	Sandstone 0.5 cm to 5 cm; Mudstone few mm to 2 cm to 3 cm	Fine to medium with some coarse-grained sand and even granules. Wavy bedding styles. Basal contacts tend to be sharp or erosive.	Wavy bedding styles. Homogeneous (commonly associated with the coarsest sand fraction) or parallel laminated. Presence of paired poorly to more laminated mudstones.	Current-ripple cross lamination showing reverse/bipolar directions.	BI 1 to 3. Sparce to common. *Paleophycus*, *Planolites*, *Chondrites*, *Skolithos*, *Teichichnus*, *Monocraterion and Diplocraterion*. BI 3 heterolithics (associated with fine-grained sandstone) alternate with BI 1 heterolithics containing coarse-grained sandstone.	Tidally influenced unidirectional currents with low to moderate speeds. Alternation of times of low and high fluvial discharge. Homogeneous mudstones associated with the coarsest sands (high discharge) suggest high SSC and the formation of fluid muds.

(*Continued*)

Table 2. (*Continued*)

Facies 5 (F5): Heterolithics		Thickness (cm)	Sandstone grain sizes and description	Mudstone description	Sedimentary structures	BI and trace fossils	Interpretation
	5.2.2. (F5.2.2) Wavy-bedded combined-flow heterolithics	Sandstone 0.5 cm to 4 cm; Mudstone 0.2 cm to 2 cm	Very fine to fine. Wavy bedding styles. Occasionally normal grading into parallel laminated siltstone and homogeneous mudstone. Dispersed coarse sand grains at the base of the sandy layers	Wavy bedding styles. Poorly laminated mudstones alternate in pairs (almost rhythmically stratified) with sandstone.	Parallel lamination, wave, current (bipolar) and small-scale asymmetric (round-crested) combined-flow ripples. Diastasis cracks (cf. Cowen & James, 1992). Local mantle and swirl structures.	BI 1 to 3. Sparse to moderate. *Planolites, Paleophycus, Skolithos, Chondrites, Arenicolites* and large *Diplocraterion, Teichichnus,* and *Cylindrichnus*. BI 3 intervals (with parallel-laminated mudstones and small-scale wave ripples and, rare, current ripples), commonly alternate with BI 1 intervals (with poorly laminated, normally or inversely graded mudstones and current-flow-rippled or combined-flow-rippled sandstones).	Low to moderate-strength oscillatory and unidirectional currents. Deposition might have occurred by means of hyperpycnal flows (cf. Bhattacharya & MacEachern, 2009). Alternation of periods of low river flow or discharge (BI3, wave ripples, fine grained sands) with periods of high river discharge (BI1, current ripples, fine-coarse sands, fluid- mud layers).
Facies 5.3 (F5.3) Mud dominated heterolithics (sand:mud ratios 10 : 90 to 40 : 60)	5.3.1. (F5.3.1) Lenticularly bedded wave-dominated heterolithics	Sandstone 0.5 cm to 3 cm; Mudstone 0.2 cm to 3 cm	Very fine to fine. Sand lenses. Thin beds showing aggradation.	Lenticular bedding. Locally show normal grading. Thick weakly laminated and /or homogeneous mudstones (fluid muds) are also present.	Parallel lamination and symmetric (oscillatory) to round-crested ripples. Small-scale HCS.	BI 1 to 2. Sparse to moderate. Small-scale *Planolites*, *Paleophycus*, and *Chondrites*. Vertical burrows are small (1 cm to 2 cm long) and consist mostly of *Skolithos* and rare larger *Cylindrichnus* (3 cm to 5 cm long).	Weak oscillatory currents. Deposition of parallel laminated mudstones by settling from suspension. Small-scale HCS formed by sporadic increase of oscillatory motion due to storms. Wave generated fluid-mud layers deposited from resuspended mud, which could possibly migrate as hyperpycnal fluid-mud bodies (cf. Mulder *et al.*, 2003).
	5.3.2. (F5.3.2) Lenticularly bedded current-dominated heterolithics	Sandstone 1 cm to 4 cm; Mudstone 0.5 cm to 3 cm	Very fine. Locally upper-medium to coarse. Sand lenses.	Lenticular bedding. No vertical changes in grain size.	Ripple cross-lamination showing local bipolarity. Parallel lamination.	BI 1 to 3. Moderate to common. *Cylindrichnus*, *Skolithos*, *Diplocraterion*, *Planolites*, *Paleophycus*, and *Chondrites*. Locally, *Glossifungites* surfaces.	Weak unidirectional and tidally influenced currents. Deposition of parallel laminated bioturbated mudstones by settling from suspension. Sporadic high river discharges, similar to what is seen in subfacies 5.2.1 and 5.2.2.

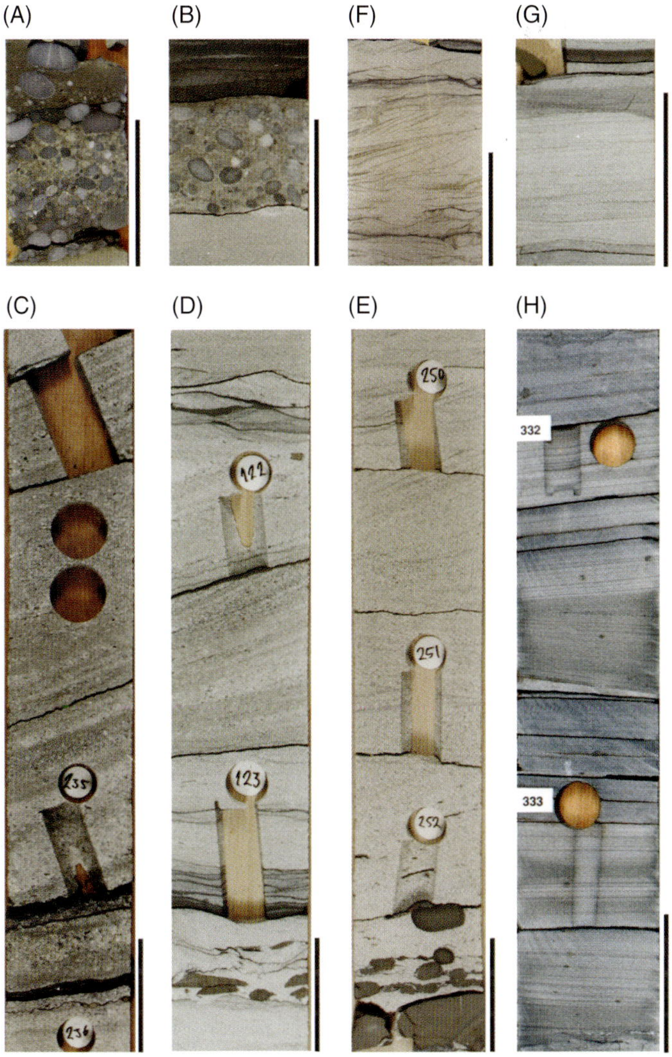

Fig. 4. Conglomerate and sandstone facies in the Smørbukk area. Dark scale bar adjacent to each photo is 10 cm long. Facies 1: (A) Inversely graded paraconglomerate; (B) Matrix supported conglomerate, overlain by a structureless fluid-mud layer. Facies 2: (C) Cross-bedded coarse-grained to medium-grained sandstone showing 2 cm thick organic-rich mud layers in the bottomsets that are interpreted to be fluid-mud deposits; (D) Cross-bedded medium-grained to fine-grained sandstone showing basal flat to rounded rip-up mud clasts and thin mud layers; (E) Cross-bedded medium-grained to fine-grained sandstone showing thick basal fluid-mud deposits and fluid-mud rip-up clasts. Facies 3: (F) Ripple cross-laminated sandstone showing bidirectional palaeocurrent indicators. Facies 4: (G) Low-angle parallel-laminated fine-grained to very fine-grained sandstone (HCS) capped by wave-generated fluid-mud layer. (H) Thick-bedded, low-angle, laminated, very fine grained sandstone (HCS) showing internal truncations.

Rippled sandstone beds (up to 40 cm thick) show a sparse, low-diversity, trace-fossil assemblage mainly composed of small *Skolithos* and, rarely, *Planolites*, suggesting brackish-water conditions.

Facies 4: Hummocky cross-stratified (HCS) sandstone

The amalgamated (up to 4 m thick) fine-grained sandstone beds and bedsets of Facies 4 (Figs 4G & 4H) show pervasive hummocky cross stratification (HCS) and are interpreted to be deposited under high-energy oscillatory currents associated with large storm waves (Hunter & Clifton, 1982; Klein & Marsaglia, 1987; Duke *et al.*, 1991; Yang *et al.*, 2006). Wave ripples occur immediately on top of the HCS beds in many instances but the presence of combined-flow ripples in some cases indicates that there was a unidirectional component to the water motion, due either to wind-driven residual motion (Hill *et al.*, 2003) and/or the presence of superimposed tidal currents (Yang *et al.*, 2005). Locally, 0.5 cm to 3 cm thick wave-generated fluid-mud deposits (cf. Traykovski *et al.*, 2000) are present and commonly cap HCS beds. Fluid-mud layers can be distinguished from those that accumulated by slow settling over long periods of time by the absence of distinct lamination and bioturbation, except for post-depositional, top-down colonisation dominated by a low-diversity trace-fossil assemblage composed of large-scale *Diplocraterion, Skolithos* and rare *Planolites* (Ichaso & Dalrymple, 2009). Less commonly, vertical to sub-vertical *Glossifungites* are present at the top of the amalgamated HCS beds.

Facies 5: Heterolithics

Facies 5 (see also Table 2) consists of heterolithic deposits composed of alternating sandstone and mudstone with sand:mud ratios between 90:10 and 10:90. This heterogeneous facies displays diverse bedding styles showing current-dominated, wave-dominated and combined-flow sedimentary

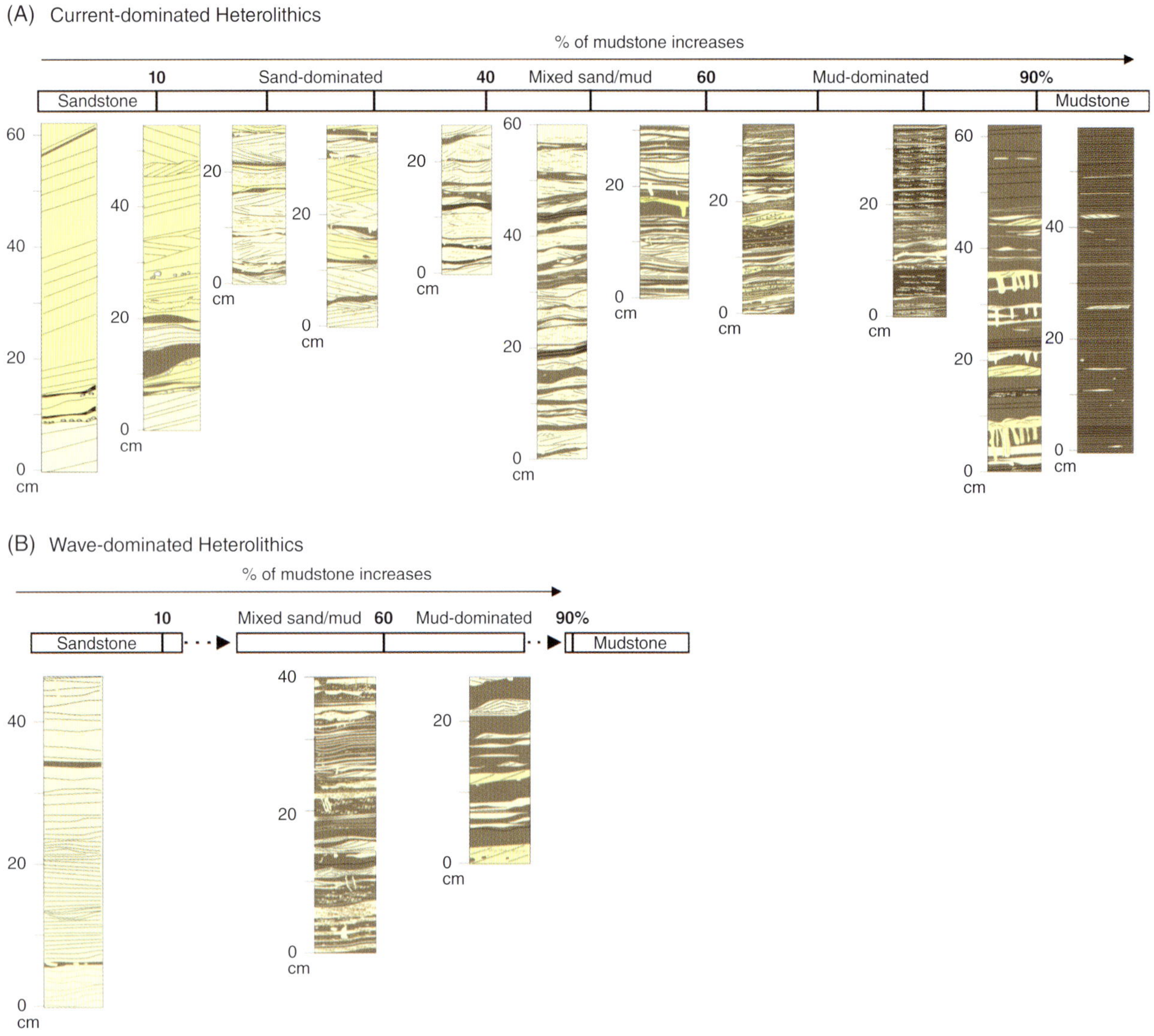

Fig. 5. Heterolithic facies (Facies 5) in the Smørbukk field. All figures represent detailed sketches from core examples. (A) Current-dominated heterolithics showing % of mudstone (dark) vs. sandstone (light). Notice the great variety of facies due the variability of parameters used for classification (e.g. thickness, sand grain size, sedimentary structures, bioturbation). (B) Wave-dominated heterolithics showing % of mudstone.

structures and diverse levels of bioturbation ranging from sparse to abundant. The heterolithic deposits were divided into 3 main groups based on their sand:mud ratio (Fig. 5) and then subdivided further based on their dominant sedimentary structures (current vs. wave influence), bedding style, grain size and bioturbation-index value.

5.1) Sand-dominated heterolithics

The flaser-bedded and heterolithic nature (sandstone:mudstone ratios between 90:10 and 60:40; Fig. 5) of facies 5.1 (Figs 6A, 6B & 6C), coupled with the presence of bidirectional dune cross-bedding, current-ripple cross-lamination and paired, poorly to more laminated mudstones (Figs 6B & 6C), indicate that tidal currents played an important role in its deposition. The presence of sporadic gravel and pebble lags showing normal grading into coarse to fine sands with thick, structureless, mudstone layers that are interpreted as fluid muds (Ichaso & Dalrymple, 2009) suggests that river floods were important: the higher-than-normal discharge is required to introduce the gravel and is also responsible for increasing the suspended-sediment concentration. The sparse and low-diversity bioturbation is

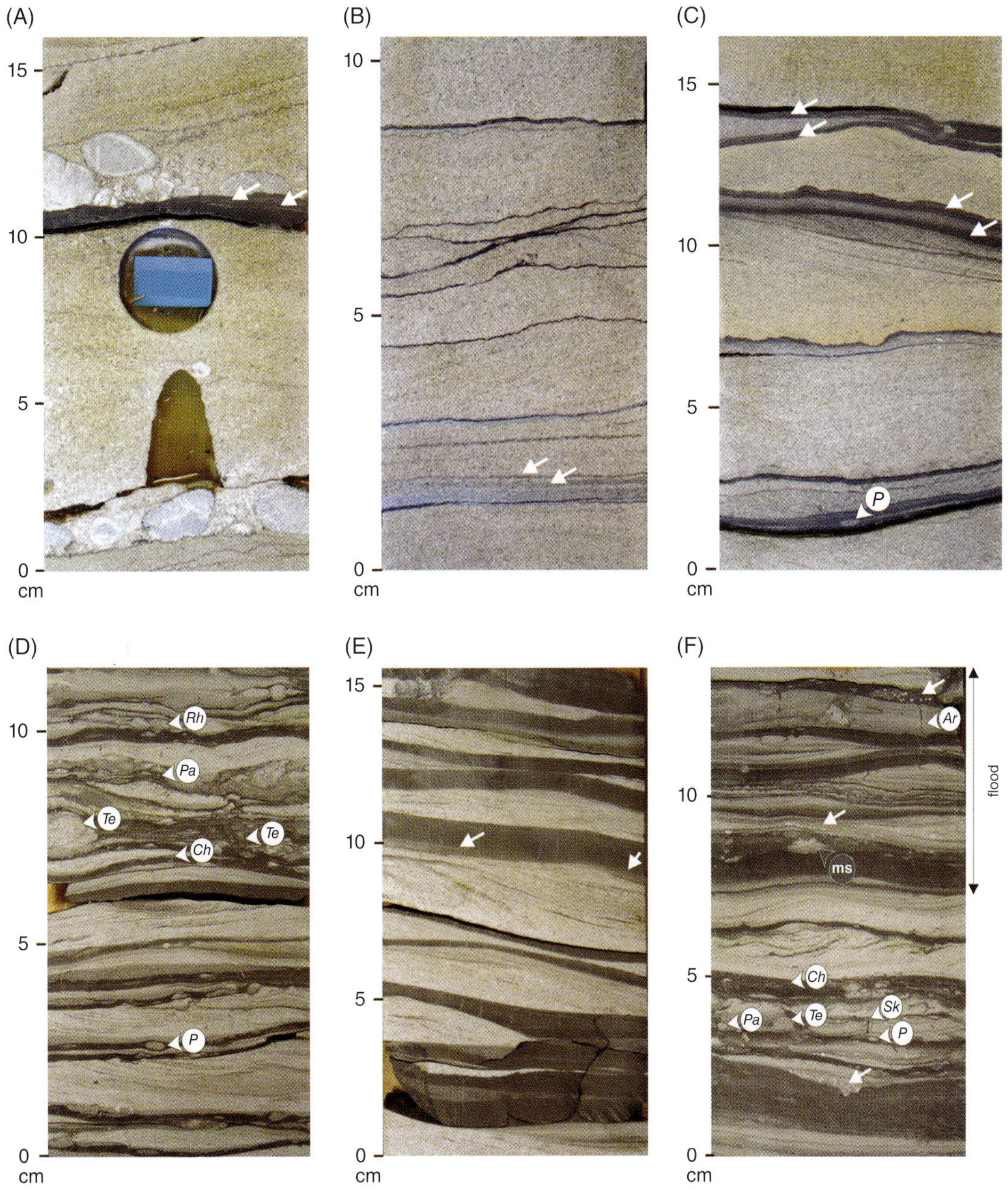

Fig. 6. Sandstone-dominated and mixed sandstone/mudstone heterolithics in the Smørbukk field. Sandstone-dominated heterolithics (Facies 5.1): (A) Thin conglomerate layers interbedded with cross-bedded and ripple cross-laminated sandstones and 2 cm to 3 cm thick homogeneous mudstone layers (white arrows) that are interpreted as fluid-mud deposits; (B) Ripple-cross laminated sandstone showing stylolites, interbedded with thin mudstone laminae showing double-mudstone drapes (white arrows); (C) Ripple cross-laminated sandstone interbedded with 0.2 cm to 0.3 cm thick double mudstone layers (white arrows) with scarce bioturbation mainly dominated by *Planolites* (*P*). Mixed sandstone/mudstone heterolithics (Facies 5.2.1): (D) Ripple cross-laminated sandstone interbedded with 0.5 cm to 1 cm thick mudstone showing sparse (BI 0-1), moderate-diversity bioturbation, predominantly showing *Chondrites* (*Ch*), *Planolites* (*P*), *Palaeophycus* (*Pa*), *Rhizocorallium* (*Rh*) and *Teichichnus* (*Te*); (E) Wavy-bedded heterolithics showing current-ripple and combined-flow ripple cross-laminated sandstone interbedded with 1 cm to 3 cm thick homogeneous mudstone; notice the vertical gradation from the fine-grained ripple cross-laminated sandstone into laminated siltstone to homogeneous unbioturbated mudstone (fluid muds; white arrows). Mixed sand/mud heterolithics (Facies 5.2.2): (F) Round-crest ripples and ripple cross-laminated sandstone with basal coarse sand lenses and/or dispersed coarse sand grains (white arrows) and 0.5 cm to 2 cm thick poorly laminated to homogeneous mudstone showing weakly bioturbated tops (BI 1) suggesting river flood periods, interbedded with moderately to intensely (BI 3-4) bioturbated sandstones and laminated mudstone representing inter-flood deposits. Common trace fossils include *Chondrites* (*Ch*), *Planolites* (*P*), *Palaeophycus* (*Pa*), *Skolithos* (*Sk*), *Teichichnus* (*Te*), *Arenicolites* (*Ar*) and 'mantle and swirl (ms)' structures (cf. Bhattacharya & MacEachern, 2009). Such alternation between weakly bioturbated and moderately bioturbated intervals is interpreted to have a seasonal origin (see text for discussion).

dominated by vertical burrows (*Skolithos* and *Diplocraterion*) and locally (within the laminated mudstones) by small *Planolites*. As with Facies 3, the trace-fossil suite is suggestive of brackish-water conditions.

5.2) Mixed sand/mud heterolithics

These deposits consist of near equal amounts of sandstone and mudstone (sandstone:mudstone ratios between 40:60 and 60:40) and are predominantly wavy-bedded showing current, wave and/or combined-flow structures (Fig. 5). Sand grain size varies between fine and very-fine with occasional quartz granule layers. Commonly, these deposits show sparse to moderate bioturbation (BI 0 to 3). Two subfacies are recognised.

Subfacies 5.2.1: Wavy-bedded current-dominated heterolithics

These mixed sand-mud heterolithitcs show dominantly wavy bedding styles (Figs 6D & 6E), together with bidirectional current-ripple cross lamination and paired poorly to more laminated mudstones, suggesting that they were formed by tidal currents with low to moderate speeds. The alternation between more bioturbated, fine-grained thin-bedded deposits (up to 30cm thick) and weakly bioturbated, coarse-grained, thicker bedded deposits is interpreted to reflect the alternation of times of low and high fluvial discharge (Ichaso & Dalrymple, 2009). At times of high discharge (river floods) the sand fraction is coarser grained and sedimentation rates are higher, creating thicker beds. This, coupled with a reduced salinity, minimises the degree of bioturbation, which is dominated by *Skolithos* and *Diplocraterion*.

Homogeneous mudstones associated with the coarsest sands (high discharge) suggest high suspended-sediment concentrations and the formation of fluid muds, because of the greater supply of fine sediment by the river (SSC values commonly correlate positively with river discharge; Xu, 2002) and the seaward displacement of the turbidity maximum. By contrast, during times of low river discharge, sand grain sizes are finer, sedimentation rates are lower, creating thinner beds and salinities are higher, allowing more intense bioturbation, which is dominated by small *Planolites*, *Palaeophycus*, *Chondrites* and *Teichichnus*, with local *Skolithos* and *Monocraterion*. Such decimetre-scale cycles are unlikely to be generated by neap-spring tidal cycles because the variation in sand-layer grain size is too large and the layer-thickness variations are not sufficiently regular.

Subfacies 5.2.2: Wavy-bedded combined-flow heterolithics

Similarly to subfacies 5.2.1, wavy bedding is the predominant bedding style in subfacies 5.2.2, but with a mixture of wave, bipolar current and small-scale combined-flow ripples. The presence of a spectrum of pervasive small-scale combined-flow structures suggests a variation of the type of energy during deposition (Dumas *et al.*, 2005), ranging from oscillatory currents forming wave (symmetric) ripples to unidirectional currents (river flood or tidal currents) superimposed on wave action to produce asymmetric round-crested, combined-flow ripples (Fig. 6F). Poorly laminated mudstones, associated with sandstones preserving wave ripples and/or combined-flow ripples, show local bottom-up cracks of mechanical origin (diastasis or syneresis cracks; Cowan & James, 1992). However, a tidal influence is indicated by the mud-drape pairs and local bipolar ripple cross lamination. The low to moderate diversity of the mixed *Cruziana-Skolithos* trace-fossil suite is indicative of brackish-water conditions.

The nature of these complex heterolithic deposits showing dominant combined-flow sedimentary structures, local mantle and swirl structures (suggesting soup-ground conditions) and normal and inverse grading in the poorly laminated mudstones suggest that deposition might have occurred by means of hyperpycnal flows (Bhattacharya & MacEachern, 2009). The alternation between moderately bioturbated intervals showing small-scale asymmetric (current) ripples and parallel-laminated mudstones and weakly bioturbated intervals containing abundant current and combined-flow ripples suggest the alternation of periods of low river flow with periods of high river discharge (floods), as was proposed for subfacies 5.2.1 above.

5.3) Mud-dominated heterolithics

This facies has sandstone:mudstone ratios between 10:90 and 40:60, and consists dominantly of lenticularly bedded heterolithics (Fig. 5). Sand grain size varies between fine and very-fine with occasional quartz granule layers. Commonly, these deposits show sparse to moderate bioturbation (BI

1 to 3). Two subfacies can be distinguished: wave-dominated and current-dominated.

Subfacies 5.3.1: Lenticular-bedded wave-dominated heterolithics

The continuous to discontinuous, sandy lenticularly-bedded nature of this subfacies, together with the presence of parallel lamination, symmetric (oscillatory) to round-crested ripples (Figs 7A & 7B) and a dominantly horizontal trace-fossil suite (*Planolites, Palaeophycus and Chondrites*), suggest deposition by weak oscillatory currents. Local small-scale HCS reflects elevated oscillatory motion due to sporadic storms (cf. Dumas *et al.*, 2005). The storm waves appear to have re-suspended mud from the sea floor, forming high-density suspensions that then migrated offshore as a hyperpycnal fluid-mud body (Mulder *et al.*, 2003), depositing some of the thick, homogeneous and/or weakly laminated mudstones. Mudstone showing small-scale horizontal and vertical burrows and continuous and distinct parallel lamination is interpreted to be deposited by settling from suspension in the presence of very weak currents or wave action.

Subfacies 5.3.2: Lenticular-bedded current-dominated heterolithics

As is the case with Subfacies 5.3.1, sandy lenticular bedding is present in subfacies 5.3.2. However, the predominance of asymmetrical ripples and parallel lamination (Figs 7C, 7D & 7E) suggest that Subfacies 5.3.2 was formed by weak unidirectional currents, with tidal influence indicated by local bipolar cross lamination and the mixed *Cruziana-Skolithos* trace-fossil assemblage. Laminated mudstone layers showing moderate and more diverse bioturbation suggest slow sedimentation from suspension during periods of lower stress. The local presence of coarse-grained sandstone lenses, with abundant ripple cross-laminated fine-grained sandstone and poorly laminated mudstone showing low bioturbation indexes by a low diversity assemblage, suggest periods of high river discharge, similar to what was seen in subfacies 5.2.1 and 5.2.2.

Facies 6: Intensely bioturbated sandstone and heterolithics

This facies consists of intensely to completely bioturbated (BI 4 to 6) muddy sandstone (Fig. 8), in which mudstone laminae are barely discernible due to complete post-depositional biogenic homogenisation. Sand grain sizes vary between very-fine and fine. Occasional granules occur. Facies 6 has been subdivided into five subfacies based on the grain size, texture of the deposits, colour and the degree and type of bioturbation.

Subfacies 6.1: Reddish bioturbated sandstone

The homogenised aspect of the reddish silty fine/medium-grained sandstone showing abundant to complete bioturbation (high-diversity mixed *Skolithos-Cruziana* suite; Figs 8A & 8B), together with the presence of sub-angular limestone grains, shell debris and mixing of sand and mud, suggest extensive biogenic reworking at times of slow to negligible sedimentation. The large vertical burrows (*Rosselia*, *Cylindrichnus*, *Skolithos* and *Berguaeria*; Fig. 8A) and occasional tabular-stratified beds, cross-bedding, low-angle lamination and coarsening upward bedsets (up to 1 m thick), are indicative of relatively high levels of wave or current energy. The local presence of horizontal burrows (*Planolites*, *Palaeophycus* and rare *Thalassinoides*; Fig. 8B) indicates intergradation with proximal expressions of the *Cruziana* Ichnofacies suggesting moderate energy levels, probably lying below fair-weather wave base but above storm wave base (MacEachern *et al.*, 2006).

Subfacies 6.2: Disturbed sandstones

The fine-grained to medium-grained sandstones (20 cm to 40 cm thick) showing common to abundant bioturbation (low-diversity, archetypal *Skolithos* Ichnofacies (cf. MacEachern *et al.*, 2006), coupled with locally preserved current-dominated structures, suggest moderate to high current energy. U-shaped large dwellings (10 cm to 25 cm long *Diplocraterion*; Figs 8C & 8D) indicate that the organisms had to respond to an ever shifting substrate, probably related to tidal processes.

Subfacies 6.3: Disturbed heterolithics

These mud-dominated and/or mixed heterolithics (20 cm to 40 cm thick) show common to abundant bioturbation (high-diversity proximal expressions of the *Cruziana* Ichnofacies (MacEachern & Bann, 2008)). The trace-fossil elements commonly tend to cross-cut each other (Figs 8E & 8F); however, the original bedding is still recognisable. The sand : mud ratio and

bioturbation, coupled with the ichnogenera present, suggest an alternation between mainly suspended-sediment settling and shifting-substrate conditions under higher energy.

Subfacies 6.5: Rooted sandstones

Bedding and internal lamination of the very fine-grained to silty sandstone in this subfacies (up to 40 cm thick) are barely visible. The abundant to complete bioturbation by vertical root structures and small *Skolithos* and *Ophiomorpha* (Figs 8I & 8J), together with the local presence of small organic-matter lenses, suggest that Subfacies 6.5 represents an immature palaeo-soil, developed in a relatively low-energy setting with some salt-water influence.

Facies 7: Pinstriped-laminated mudstones

The pinstriped-laminated mudstone (Fig. 7F) of Facies 7 represents the mud-dominated end member (sandstone : mudstone ratios between 10 : 90 and 0 : 100) of the spectrum of heterolithic deposits (Fig. 5). Facies 7 is interpreted to have been deposited under relatively low-energy conditions. Sand lenses (0.5 cm to 1.5 cm thick) showing small-scale bi-directional ripple cross-lamination indicate the presence of tidal currents. The predominance of pinstripe bedding presumably results from the dominance of mud over sand, indicating relatively low tidal velocities or a scarcity of sand. Bioturbation tends to be sparse to uncommon and consists of *Chondrites* and rare *Planolites*, suggesting deposition within highly stressed environments as a result of either a high sedimentation rate or reduced salinity.

FACIES ASSOCIATIONS AND ENVIRONMENTAL INTERPRETATIONS

The common co-occurrence of certain facies groupings, together with vertical changes in the assemblage of deposit characteristics, allowed the combination of closely and genetically related facies into ten facies associations (Table 3) that vary in their abundance through the Upper Åre and Tilje formations. Using these facies associations (FA), it was possible to deduce the combination of processes that occurred and, hence, the different sedimentary environments that existed (Fig. 9).

Facies association 1 (FA1): Lower to middle shoreface

FA1 (Fig. 10) is predominantly sandy. Commonly, this association shows an upward-sanding and locally upward-coarsening, trend up to 5 m thick. At the base, FA1 consists of intensely bioturbated (mixed *Cruziana* and *Skolithos* Ichnofacies) to homogenised silty and very fine-grained sandstones (F6.4) that indicate lengthy periods (many months) of low energy. The bioturbated sandstones are overlain locally by mixed sandstone/mudstone wavy bedded combined-flow heterolithics (F5.2.2). These basal deposits are sharply overlain by low-angle laminated (HCS) very fine-grained to fine-grained amalgamated (up to 3 m thick) sandstones (F4). Erosion surfaces are common within the amalgamated sandy beds. Thick amalgamated HCS beds in FA1 are interpreted as proximal storm beds. Sustained wave motion during storms eroded any fair-weather mud, with the exception of locally preserved residual mud lenses and clasts. Local, weakly laminated to homogeneous mudstones on top of HCS-bearing event beds suggest resuspension of the mud and the generation of high suspended-sediment concentrations forming fluid muds. Alternation of HCS beds and bioturbated sandstones indicates interruption of fair-weather conditions by storms. The nature of the facies (dominance of wave-generated structures) and their sanding-up vertical arrangement, coupled with the mixed trace-fossil assemblage showing a proximal expression of the *Cruziana* Ichnofacies, suggest that deposition of FA1 occurred in a storm-influenced lower to middle shoreface setting (cf. Walker & Plint, 1992; Bhattacharya & Giosan, 2003). The local presence of combined-flow heterolithic deposits might imply a nearby source of mud (i.e. transition into a delta-front setting).

Facies association (FA2): Transgressive shelf lags

FA2 is overall sand-dominated and consists of tabular-stratified beds that are typically < 0.5 m thick, but locally reach more than 1 m in thickness. This FA is erosively based and consists of moderately to poorly sorted fine-grained to medium-grained reddish sandstone (F6.1) that displays a homogeneous aspect due to the complete bioturbation. Vertical burrows are large and diverse, displaying a proximal

Table 3. Facies associations (FA) in the upper Are and Tilje formations used to deduce depositional processes and environments.

(FA) Facies associations	Facies	General Description	Sedimentary structures	Internal organisation	Ichnology		Environmental Interpretation
					Ichnofacies	Trace Fossils	
FA1: Lower-shoreface	F4, F6.4, locally F5.2.2	Light grey very-fine to fine-grained sandstone and intensely bioturbated sandstone. Local sand/mud mixed heterolithics.	HCS, low angle to planar lamination, wave ripples, wavy lamination.	Upward-sanding trend (up to 5 m thick). Basal sandstones moderately to intensely bioturbated alternate locally with wavy-bedded combined-flow heterolithics. Overlain sharply by amalgamated HCS sandstones (up to 3 m thick).	Moderate to intense, high diversity. Mixed *Cruziana* and *Skolithos* assemblage.	*Paleophycus*, *Skolithos*, *Rosselia*, *Planolites*, *Cylindrichnus*, *Thalassinoides*, *Ophiomorpha* *Diplocraterion*, and *Rhizocorallium*.	Storm-influenced lower shoreface. Periods of low energy succeed by high-energy oscillatory and combined-flow conditions during storm events. Interruption of fair-weather by seasonal storms.
FA2: Transgressive shelf lags	F6.1, and F2.	Brown to reddish and/or grey, poorly sorted, silty, fine- to medium-grained sandstones. Locally, coarse-grained. Floating angular bioclastic calcite grains are common.	Low-angle cross-bedding, low-angle lamination, parallel lamination.	Sharp and erosively based tabular beds. Occasionally coarsening- or fining-upward successions (10 cm to 70 cm). Homogeneous aspect.	Abundant, high diversity. Proximal expressions of *Cruziana* Ichnofacies.	*Rosselia*, *Planolites*, *Cylindrichnus*, *Skolithos*, *Berguaeria*, *Paleophycus*, and rare *Thalassinoides*.	Marginal to normal-marine lags. Reworking processes by tidal or wave currents under conditions of slow sedimentation.
FA3: Distributary mouth bars	F2, F3, F5.1, locally F4	Light brown and grey, well sorted, very-fine to fine-grained sandstone. Local homogeneous mudstones.	Cross-beds, current-ripple cross-lamination (locally bipolar), low-angle planar lamination (HCS). Double mud-drapes.	Sharp-based. Cleaning and sandier upward (2 m to 4 m thick). Fluid-mud layers (3 cm thick) and/or angular rip-up mud clasts towards the base of cross-beds. Cross-bedded sandstone alternates with ripple-cross-laminated and double mud- draped sandstones.	Low to moderate, low diversity. Proximal *Skolithos*.	*Skolithos*, *Arenicolites*, *Berguaeria*, *Planolites*, and *Paleophycus*.	Mouth-bar region. Relatively high energy. Tidally influenced currents and sporadic storms.
	F5.1, F5.2.1, F5.5, F2, locally F5.2.2	Mixed sand-mud heterolithics. Bimodal, from coarser granules to fine and/or very-fine grained sand. Laminated to homogeneous mudstones.	Current-ripple cross-lamination (locally bipolar), cross-bedding, wave and asymmetric combined-flow ripples, small-scale HCS, wavy bedding, synaeresis cracks.	Stacked coarsening-upward and thickening-upward successions (2 m to 3 m thick), showing alternation between bioturbated fine-grained heterolithics and less bioturbated heterolithics with bimodal sand grain size. The heterolithics are punctuated by upward-fining sandstones (up to 1 m thick).	Low to intense. Mixed *Cruziana* and *Skolithos* assemblage.	*Paleophycus*, *Skolithos*, *Planolites*, *Cylindrichnus*, *Teichichnus*, and *Diplocraterion*.	Mouth-bar region. Moderate and high energy. Alternation between times of low and high fluvial discharge modulated by tidal currents and waves. Mouth-bar deposits interbedded with terminal distributary channel cross-bedded sandstones.

(*Continued*)

Table 3. (*Continued*)

(FA) Facies associations	Facies	General Description	Sedimentary structures	Internal organisation	Ichnology		Environmental Interpretation
					Ichnofacies	Trace Fossils	
FA4: Proximal to medial tidally influenced delta front	F5.2.1, F5.2.2, and F5.1.	Mixed sand-mud heterolithics. Fine to very-fine grained sandstones. Laminated to homogeneous mudstones.	Parallel lamination, current-ripple cross-lamination (bipolar), combined-flow and locally wave ripples, climbing ripples, and double mud drapes.	Stacked coarsening-upward and thickening-upward successions (3 m to 5 m thick). Current dominated heterolithics alternate with combined-flow-dominated deposits. Heterolithic intervals are overlain by sand-dominated heterolithics.	Low to moderate. Mixed *Cruziana* and *Skolithos*. Deposit-feeder structures.	*Paleophycus, Chondrites, Skolithos, Rosselia, Planolites, Teichichnus, and Rhizocorallium*	Tidally influenced delta front. Mostly in a low-relief, proximal to medial prograding delta-front lobe. More distal than FA3. Fewer seasonal indicators.
FA5: Proximal prodelta	F7, F5.4 and F5.5	Mud dominated heterolithics. Poorly laminated, thin fluid-mud layers and laminated mudstones. Sand lenses are fine to very-fine grained.	Parallel lamination (pinstriped), lenticular bedding, current-ripple cross-lamination, symmetric ripples.	Sanding and coarsening-upward. Pinstripe-laminated mudstones overlain by the interbedding of current and wave-generated, mud-dominated heterolithics (maximum thickness of 4 m).	Low to moderate. Deposit-feeder structures. Impoverished expression of *Cruziana*	*Planolites, Paleophycus, Chondrites, Skolithos*, and locally *Teichichnus*.	Tidal and wave influence proximal prodeltaic setting.
FA6: Storm-influenced delta front	F5.3.1, F5.2.1, F5.2.2, and F4.	Mixed sand-mud heterolithics. Fine to very-fine grained sandstones. Laminated to homogeneous mudstones.	Parallel lamination, symmetric ripples, small-scale and large-scale HCS, combined-flow and current ripples, and lenticular and wavy bedding.	Thickening and sanding upward. Medium to thick beds (30 cm to 60 cm) mud-dominated heterolithics are overlain by mixed sand-mud heterolithics (40 cm to 60 cm thick) that alternate locally with 20 cm to 30 cm thick HCS beds.	Low to moderate. Dominated by deposit-feeders. Mixed *Cruziana* and *Skolithos*.	*Skolithos, Rosselia, Arenicolites, Cylindrichnus, Diplocraterion Planolites, Paleophycus, Chondrites*, and *Teichichnus*.	Storm-dominated mouth bar within the delta-front area. Seasonality between fair-weather and inter-storm periods.
FA7: Distal wave-influenced delta front	F5.2.2, F6.4, F5.1, and F6.3	Mixed sand-mud heterolithics. Fine to very-fine grained sandstones (locally medium grained). Laminated to homogeneous mudstones	Combined-flow and current ripples, chevron lamination, small-scale HCS, parallel lamination, and wavy bedding.	Stacked coarsening-upward (up to 3 m to 5 m thick). Sharp-based wavy-bedded heterolithics overlain and/or alternating with intensely bioturbated intervals. Locally, erosive based cross-bedded sandstones (20 cm to 30 cm thick) alternate with the bioturbated intervals.	Moderate to intense. Mixed *Cruziana* and *Skolithos*.	*Planolites, Paleophycus, Chondrites, Teichichnus, and subordinate Cylindrichnus, Thalassinoides*, and *Phycosiphon*.	Wave-influenced delta front. Moderate to high net energy levels. Ichnology suggests a more open-marine setting compared with FA3 and FA4.

FA8: Tidal-fluvial channels	F1, F2, F3, and F5.1	Sand dominated. Coarse to fine-grained sandstones. Local conglomerates (20 cm to 40 cm thick), compound fluid-mud layers (up to 15 cm thick).	Cross-bedding, current-ripple cross-lamination (bipolar), herringbone cross-stratification, parallel lamination, double mud-drapes.	Stacked fining-upward successions (1 m to 5 m thick). Erosionally based conglomerates overlie by cross-bedded sandstones with basal rip-up mud clasts and/or fluid-mud layers followed by ripple cross-laminated sandstones and sand-dominated heterolithics.	Low to absent. *Skolithos.*	*Skolithos*, large *Diplocraterion*, and locally *Planolites.*	Migration of dunes in tidally influenced fluvial channels. High-energy current-dominated. Episodic brackish-water conditions.
FA9: Heterolithic tidal bars	F5.1, F5.2.1, and F5.5	Mixed sand-mud and sand-dominated heterolithics. Medium to fine-grained sandstone. Laminated to homogeneous mudstone. Inclined mudstones dipping 5 degrees (IHS).	Current-ripple cross-lamination (bipolar), cross-bedding, syneresis cracks, stylolites, double mud drapes, parallel lamination.	Overall successions (up to 5 m thick) displaying upward thinning of the interbedded sandstone and mudstone layers. Locally, sharp-based stacked, upward sandier intervals. Rare upward-coarsening heterolithic successions.	Low to absent. *Skolithos.*	*Skolithos*, large *Diplocraterion*, and locally *Planolites* and *Paleophycus.*	Elongate tidal bars or bank-margin tidal point bars. Moderate to high energy. Stressed brackish-water conditions with high sedimentation rates.
FA10: Interdistributary bay deposits	F6.5, F5.3.1, F5.2.1, F6.3, and F5.1.	Mixed silt to very fine-grained sandstone and mudstone.	Current-ripple cross-lamination (bipolar), parallel lamination, lenticular bedding, wave ripples, wavy bedding. Organic lenses are common.	Upward thickening and sandier successions (up to 3 m thick). Mud-dominated heterolithics passing upward into mixed sand-mud and sand-dominated heterolithics capped by rooted sandstones and siltstones.	Moderate to intense. Mixed *Cruziana* and *Skolithos.* Deposit-feeders and plant roots.	*Planolites, Paleophycus*, small-scale *Skolithos, Diplocraterion*, and *Ophiomorpha.* Rare oblique trace fossils. Plant roots.	Prograding minor mouth-bar or tidally influenced interdistributary bay. Low wave energy. Local aggradation to water level. Local channel deposits by excess discharge from distributary channels.

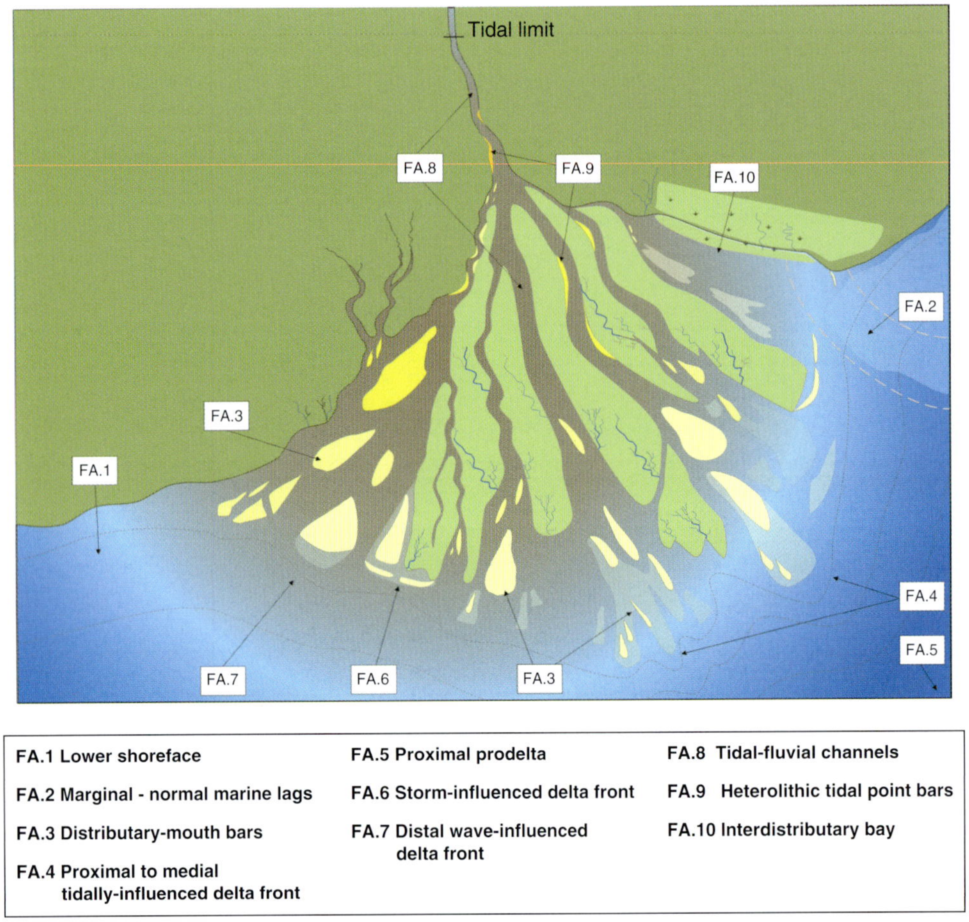

Fig. 9. Map-view representation of a mixed tidally-influenced and wave-influenced lobate deltaic system and adjacent shoreface, showing location of the interpreted facies associations for the upper Åre and Tilje formations in the Smørbukk field.

expression of the *Cruziana* Ichnofacies. Mudstone lenses and interrupted lamination are visible. Locally, angular limestone grains, small shell fragments and faecal pellets are present at the base. These deposits are occasionally overlain by sharp-based, poorly sorted, matrix-supported sandstones that commonly display cross-bedding and floating angular limestone grains (F2). The matrix-supported nature of the sandstones, their homogenised texture and the local abundance of floating angular limestone grains and local shell fragments imply reworking in a setting with slow to negligible sedimentation. Low-angle cross-bedding, coupled with the presence of large vertical burrows, indicate moderate-energy to high-energy levels of wave motion (below fair-weather wave base but above storm base) or tidal currents. FA2 is interpreted to represent sandy lags deposited in a shallow, marginal to open-marine setting. In situations where these lags are more than ca. 1 m to 2 m thick, the possibility exists that it represents a transgressive shelf ridge or shoal, such as those present on the US East Coast or on the outer margin of abandoned and submerged lobes of the Mississippi Delta (Snedden & Dalrymple, 1998).

Facies association 3 (FA3): Distributary-mouth bars

FA3 (Figs 11 and 12) comprises both sandy and mixed sandstone-mudstone deposits. The sandy examples coarsen and become sandier upward in successions 2 m to 4 m thick. The erosively and/or sharp-based intervals of cross-bedded sandstone (F2 and F5.1) contain fluid-mud layers (up to 3 cm thick) and/or unconsolidated (angular) rip-up mud clasts and are commonly overlain by stacked cross-bedded sandstones alternating with bipolar ripple cross-laminated sandstones containing

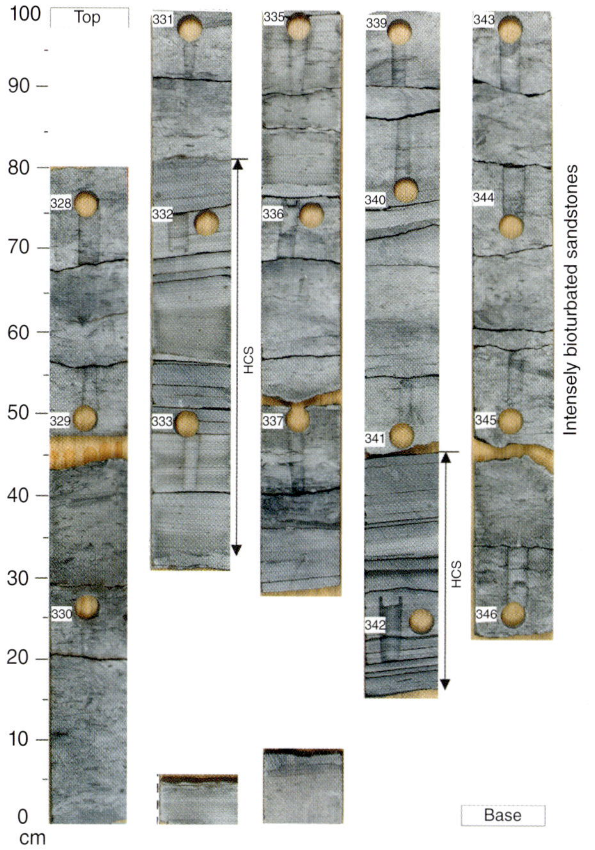

Fig. 10. Facies association FA1. Lower to middle shoreface deposits are predominantly sandy and show intensely bioturbated sandstone overlain by low-angle laminated sandstone interpreted as hummocky cross-stratification (HCS) forming cleaning and sanding upward successions. Example taken from reservoir zone Åre 3 (A3).

double mud drapes (F3). HCS beds (F4) are present locally. The abundance of cross-bedding indicates that currents dominated sedimentation. The low degree and diversity of vertical trace fossils, together with the preserved sedimentary structures, suggest a relatively high-energy environment with moderate to high sedimentation rates, physical reworking and/or salinity variations. The double mud drapes and bipolar ripple cross-lamination indicate tidal influence during deposition and the local presence of HCS suggests sporadic storm/wave action. The cleaning, sandier-upward trends on the higher part of the succession suggest that suspended-sediment values were lower at higher levels in the water column, whereas the stratigraphically lower intervals show thicker mud (fluid muds) layers suggesting deposition in lower levels within the water column (cf. Dalrymple *et al.*, 2003).

The more heterolithic examples of FA3 show a similar coarsening/sandier and thickening upward organisation. Wavy-bedded mixed sand-mud (F5.2.1 and F5.2.2) and mudstone-dominated (F5.5) deposits showing current (bipolar) and combined-flow ripples are commonly overlain by thick fine-grained sand-dominated heterolithics (F5.1). At the 10 cm to 30 cm scale, these heterolithic deposits (F5.2.1 and F.2.2) contain a decimetre-scale alternation between higher-energy deposits (coarse-grained, abundant fluid-muds, bioturbation uncommon, thicker bedded) and lower-energy (fine-grained, more intensely bioturbated and thinner bedded) that is interpreted as the result of high-energy river floods that alternate with lower-energy inter-flood sedimentation. These deposits are commonly punctuated by upward-fining, cross-bedded, medium-grained to fine-grained sandstones (F2) that were deposited by relatively higher energy unidirectional (river) currents.

The stacked coarsening/sandier upward sand and mixed sand-mud dominated successions of FA3, showing persistent current-generated structures containing tidal indicators and mixed *Cruziana-Skolithos* bioturbation assemblages, are interpreted to be deposited in a tidally influenced outer distributary channel to upper delta-front mouth-bar region. The local presence of oscillatory and combined-flow ripples and HCS, suggest that mouth bars were also subjected to episodic wave energy. The overall upward-fining cross-bedded successions that punctuated the most heterolithic mouth-bar deposits are interpreted as terminal distributary channels (Olariu & Bhattacharya, 2006; Ichaso & Dalrymple, 2009), where cross bedding was generated by the migration of dunes and ripples. These deposits will be referred in the following sections as sandy mouth-bars and channelised mouth-bar successions, based on the absence or presence of terminal distributary channels, respectively.

Facies association 4 (FA4): Proximal to medial tidally influenced delta front

FA4 (Fig. 13) is dominated by mixed sandstone-mudstone heterolithics. These deposits are comprised of stacked coarsening-upward and thickening-upward successions (3 m to 5 m thick). Commonly, sharp-based current-dominated heterolithics (F5.2.1) alternate with combined-flow heterolithics (F5.2.2). Heterolithic units show mixed *Cruziana* and *Skolithos* assemblages, mostly showing deposit-feeder structures. These deposits are generally overlain by fine-grained

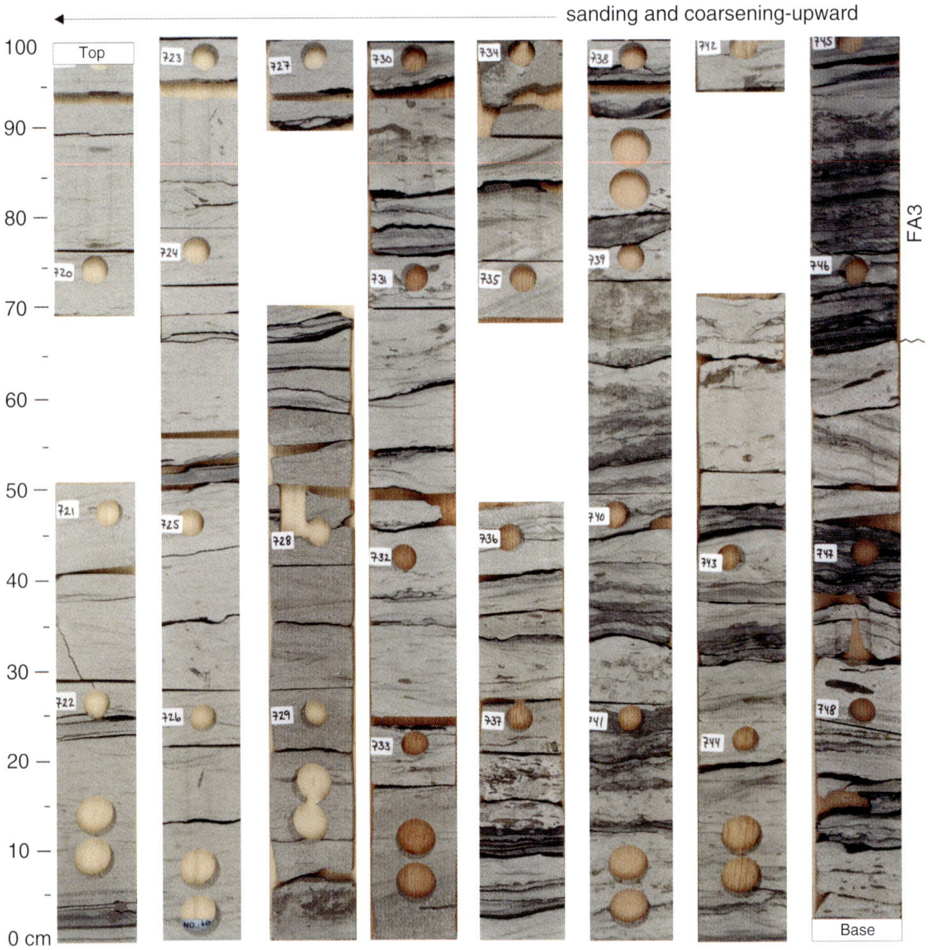

Fig. 11. Facies associations FA2 and FA3: Sanding and coarsening-upward, sand-dominated distributary-mouth bars. Example taken from reservoir zone Tilje 1.1.1 (T1.1.1).

sand-dominated heterolithics with abundant bipolar ripple cross-lamination and double mud drapes (F5.1). Individual sandstone layers tend to increase in thickness vertically in the succession (from 2 cm to 5 cm), alternating with thick (3 cm) fluid-mud layers. Seasonal indicators similar to those seen in FA3 are present, but they are less common in FA4. Combined-flow/wave-generated structures are also present; however, structures indicative of tidal currents (e.g. single and double mud drapes, bipolar ripple cross lamination) are pervasive. The tidal indicators are commonly found associated with fine-grained sandstone layers and thick homogeneous mudstones. These thick mudstone layers formed by deposition from high suspended-sediment concentrations during one slack-water period, or amalgamation of fluid muds formed during several slack-water periods. The high suspended-sediment concentrations and water turbidity formed during river flood events, as indicated by the thicker tidal drapes associated with the coarser sand layers, caused the suppression of suspension feeders from the *Skolithos* ichnofacies (MacEachern *et al.*, 2005). Due to the lack of channelised facies and the overall thickening, sanding and coarsening upward trends, FA4 is interpreted to have formed in low-relief, proximal to medial delta-front prograding lobes, probably in a more distal location than, or laterally adjacent to, FA3.

Facies association 5 (FA5): Proximal prodelta

FA5 (Fig. 14) is overall mud dominated. However, it shows sanding-upward and coarsening-upward successions that attain a maximum thickness of 4 metres. The basal pinstripe-laminated mudstones (F7) commonly alternate or are overlain by lenticularly-bedded wave- and/or current-dominated heterolithics (F5.4

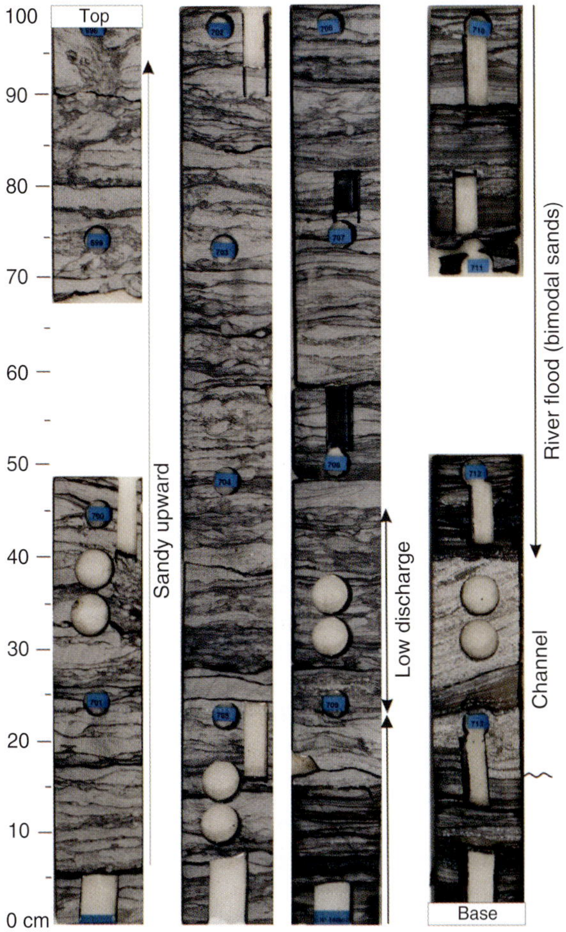

Fig. 12. Facies association FA3: Heterolithic distributary-mouth-bar, punctuated by terminal distributary-channel deposits. Mixed sandstone/mudstone heterolithic deposits showing bimodal sand grain sizes alternating with moderate to intensely bioturbated heterolithics, suggesting river discharge seasonality. Medium-grained cross-bedded sandstone interpreted as terminal distributary channels commonly punctuate the heterolithic mouth-bar deposits. Example taken from reservoir zone Tilje 3.1 (T3.1).

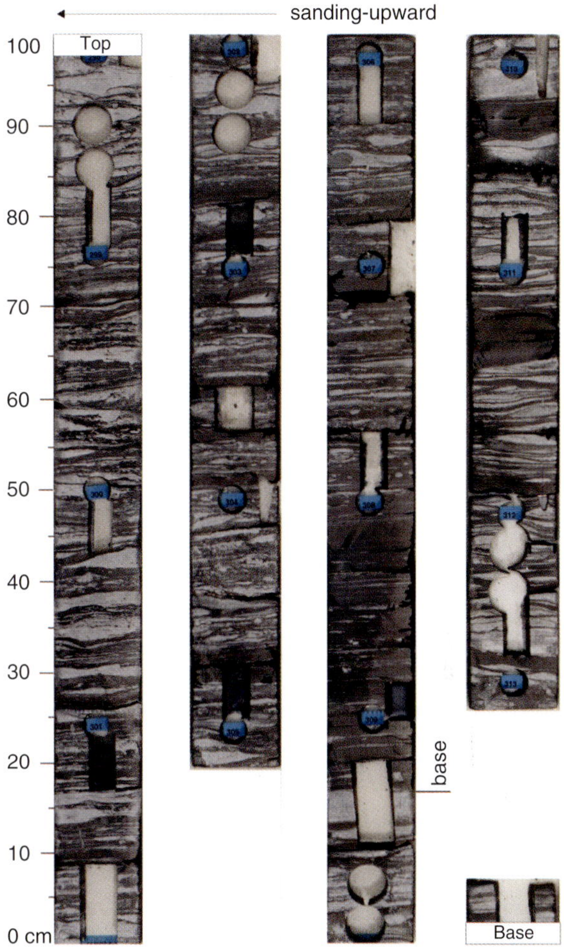

Fig. 13. Facies association FA4: Proximal-medial tidally influenced delta-front deposits showing coarsening-upward and sanding-upward successions. Mudstone-dominated heterolithics change gradationally upward into mixed sandstone/mudstone and sandstone-dominated heterolithics showing ripple cross-laminated sandstone with local bipolar palaeocurrent directions. Thick homogeneous mudstone layers are interpreted as fluid-mud deposits generated during times of high river discharge. Example taken from reservoir zone Tilje 3.1 (T3.1).

and F5.5). These mud-dominated facies showing internal lamination and low to moderate burrowing (dominated by *Cruziana* Ichnofacies deposit feeders) suggest slow rates of sedimentation, largely from suspension. The very-fine grained sand lenses (F5.4 and F5.5) become thicker upward (up to 3 cm). The sand was mostly introduced by low-energy currents during storms (wave-dominated heterolithics) and flood events (current-dominated heterolithics). Mudstone layers associated to flood events are poorly laminated to homogeneous (fluid-mud deposits) and generally unbioturbated. The current-dominated heterolithics showing low intensities of bioturbation indicate the presence of brackish-water conditions. Similarly to FA4, fluid-mud layers preserved in these deposits suggest the formation of substrates that inhibited infaunal colonisation and the suppression of suspension feeders from the *Skolithos* ichnofacies. The overall mudstone-dominated facies and very-fine grained, sanding upward vertical succession, coupled with the impoverished *Cruziana* trace-fossil assemblage, suggest that FA5 was deposited in a low-energy prodeltaic setting.

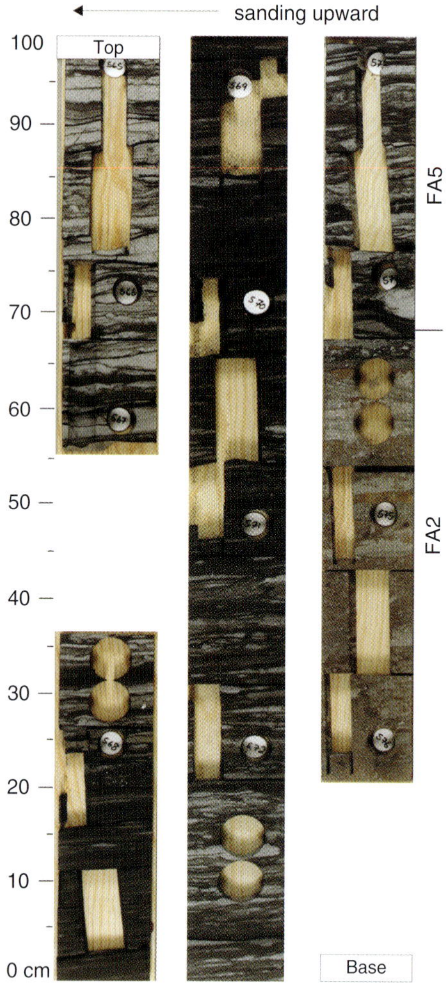

Fig. 14. Facies association FA5: Proximal prodelta deposits composed of thick pinstripe-laminated mudstones and mud-dominated heterolithics that become sandier upward to very fine-grained mixed sandstone-mudstone heterolithics showing ripple cross-lamination (locally bipolar). These deposits locally overlie thick reworked marginal-marine lags (FA2). Example taken from reservoir zone Tilje 2 (T2).

Facies association 6 (FA6): Storm-influenced delta front

FA6 (Fig. 15) is overall mixed sand-mud dominated and bedding generally thickens and becomes sandier upward. Lenticularly-bedded mudstone-dominated heterolithics (F5.3.1) are commonly overlain by mixed sandstone/mudstone heterolithics (F5.2.1 and F5.2.2). The pervasive presence of symmetric and combined-flow ripples and small-scale HCS, indicate moderate levels of oscillatory and combined-flow energy. Locally preserved bipolar current-ripple cross-lamination suggests a tidal influence. Bioturbation is low to moderate and dominated by sediment-feeding structures from transitional to inner-neritic zones of the *Cruziana* Ichnofacies, alternating with a mixed *Cruziana-Skolithos* suite. The heterolithic deposits are commonly overlain by sharp-based very fine to fine-grained sandstone (F4) showing negligible bioturbation and large-scale HCS capped by wave ripples and/or wave-generated fluid-mud deposits (Ichaso & Dalrymple, 2009). These deposits are interpreted as proximal episodic tempestites, indicating temporary high-energy oscillatory energy that deposited the event bed and created the high-density suspension that formed the capping fluid-mud layer (cf. Traykovski *et al.*, 2000). Alternation of moderately bioturbated intervals with weakly bioturbated intervals (associated with HCS beds) suggest fair-weather (inter-storm) periods interrupted by episodic storm events. The wave-dominated heterolithic nature of these deposits and their sandier-upward organisation, coupled with the trace-fossil assemblages and the lack of channelised facies, suggest that FA6 was deposited in the proximal portion of the delta-front area.

Facies association 7 (FA7): Distal wave-influenced delta front

FA7 consists dominantly of mixed sandstone-mudstone heterolithics that commonly are arranged in stacked coarsening-upward and sanding-upward successions. Sharp-based combined-flow heterolithics (F5.2.2) show a dominance of wave-generated structures (e.g. wave ripples and small-scale HCS) and low levels of bioturbation dominated by deposit feeders. These deposits alternate with disturbed heterolithics (F6.3) and are overlain locally by light grey homogenised sandstones (F6.4) showing mixed *Cruziana-Skolithos* suites. Locally, erosively based medium-grained to fine-grained cross-bedded sandstone beds (up to 30 cm thick) alternate with the bioturbated intervals. The dominance of wave-generated structures suggest moderate to high oscillatory-current energy levels. Obvious tidal indicators are absent or are not preserved due to the degree of bioturbation. Bimodal bioturbation degrees (low and intense) and the local presence of cross-bedded sandstones are indicative of seasonality (similar to FA3 and FA4). The overall presence of wave and combined-flow sedimentary structures and the vertical organisation of facies suggest deposition in a wave-influenced delta-front setting.

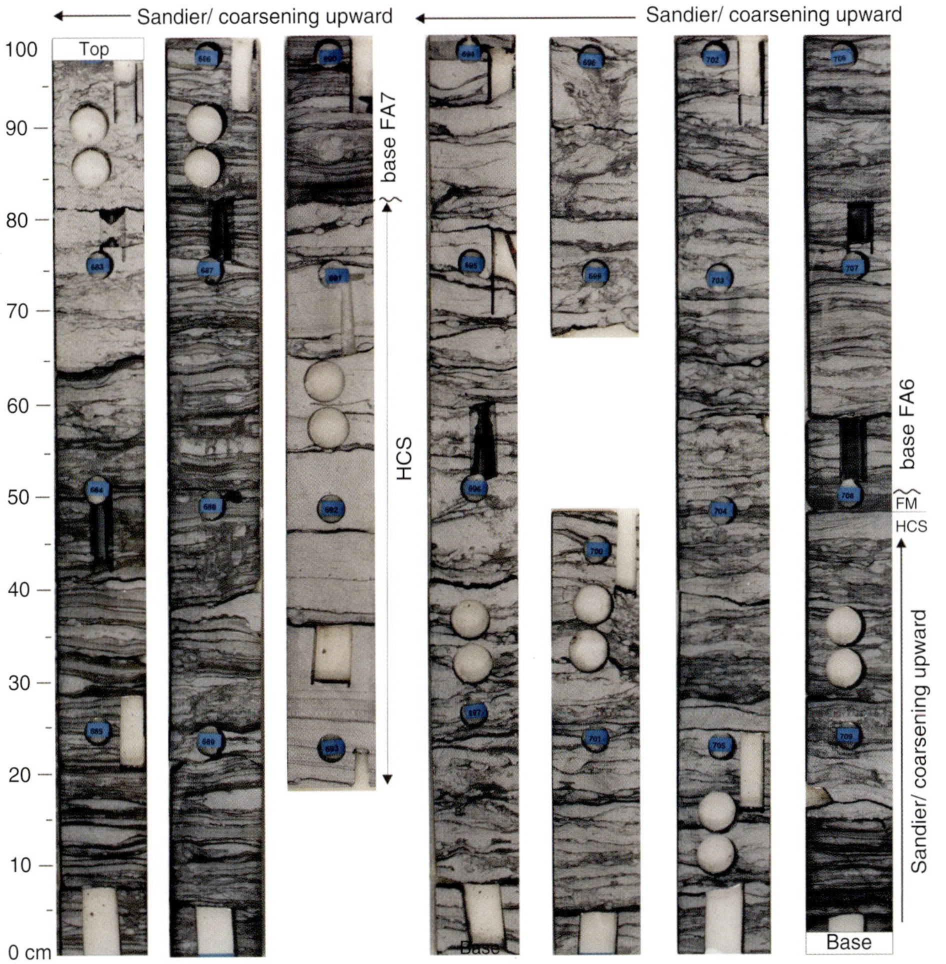

Fig. 15. Facies associations FA6 and FA7: Storm-influenced and wave-influenced delta-front sanding and coarsening-upward successions. In FA6, mudstone and mixed sandstone/mudstone heterolithics are commonly overlain by wave-dominated heterolithics showing overall wavy bedding and local combined-flow (round-crested) and ripple (bidirectional) cross-lamination. These deposits are commonly capped by thick (up to 1 m) HCS beds. The overlying wave-influenced delta-front (FA7) units are overall composed of mudstone-dominated deposits that change upwards into mixed sandstone/mudstone heterolithics showing wave ripples and intensely bioturbated sandstone. FM = fluid muds. Example taken from reservoir zone Tilje 2 (T2).

The ichnological assemblage and sedimentary structures indicate that these deposits formed in a more open and/or distal setting with less fluvial/tidal influence compared to the tidally influenced delta-front deposits (FA4).

Facies association 8 (FA8): Tidal-fluvial channels

FA8 (Fig. 16) consists of sand-dominated, stacked fining-upward successions. Graded and non-graded basal conglomerates (F1) are present locally. The base of each FA8 succession is characterised by erosionally based, coarse-grained cross-bedded sandstones (F2) containing large rip-up mud clasts and/or thick compound fluid-mud layers. These deposits are overlain by medium and fine-grained cross-bedded sandstones (F2) alternating with ripple-cross-laminated sandstones (F3) and/or sand-dominated heterolithics (F5.1). Although the sand in the succession tends to fine upward, the abundance of homogeneous mud layers at the base produces an upward-sanding trend in the lower part of each succession. Overall, the thickness of the mud layers decreases to a few millimetres toward the top of the successions; this thinning reflects an upward decrease of the suspended-sediment concentrations in the water column (Ichaso & Dalrymple,

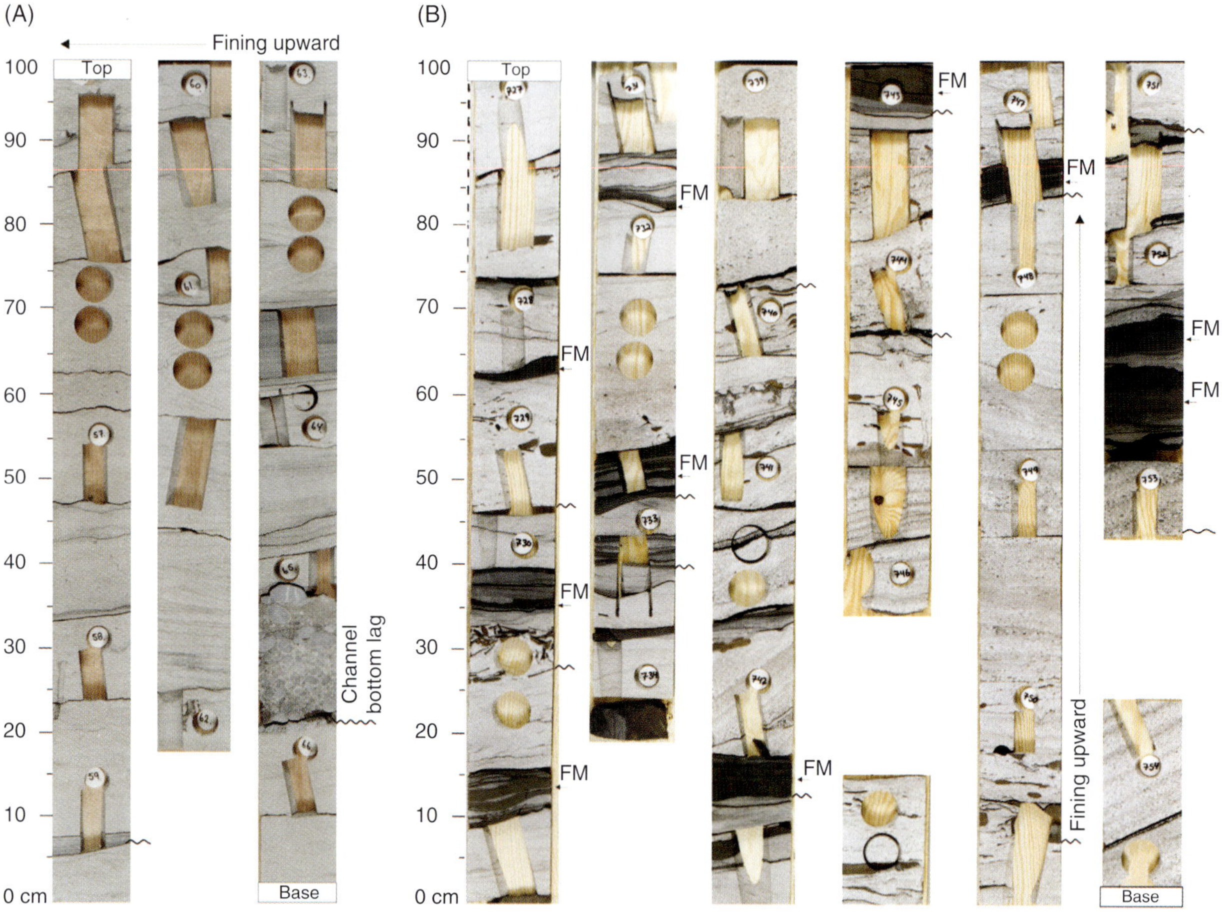

Fig. 16. Facies association FA8: Tidal-fluvial channel deposits. A) Basal conglomerates (channel-bottom lags) overlain by fining-upward and cleaning-upward, cross-bedded, medium-grained to fine-grained sandstone. B) Single and compound fluid-mud layers at the base of tidal-fluvial channel succession. Fluid-muds (FM) are commonly overlain by fining-upward, cross-bedded, medium to fine-grained sandstone showing rip-up mud clasts. Notice the sparse bioturbation. Examples taken from reservoir zone Tilje 5.1 (T5.1).

2009). FA8 shows dominant unidirectional current-generated sedimentary structures produced by the migration of dunes; this, coupled with the presence of conglomerates and coarse-grained sandstones, is suggestive of a high-energy depositional setting. Locally, bipolar ripple-cross lamination and herring-bone cross-stratification and abundant mudstones showing thick-thin alternations are indicative of a tidal influence. The low-diversity trace-fossil assemblage dominated by vertical burrows suggests a stressed environment with brackish-water conditions. All of these characteristics indicate that FA8 was deposited in a tidal-fluvial channel setting. Given the expected downstream fining of bed material in deltaic systems (Dalrymple & Choi, 2007), the local presence of conglomeratic facies suggests that deposition occurred in a relatively proximal location, perhaps near the limit of tidal and brackish water influence during low river flow.

Facies association 9 (FA9): Heterolithic tidal bars

FA9 (Fig. 17) is dominated by mixed sandstone-mudstone heterolithics that are thin to medium bedded. Overall, FA9 successions (up to 4 m to 5 m thick) show an upward thinning of the interbedded sandstone and mudstone layers. Locally, sharp-based, stacked upward-sandier intervals are present. Rare upward-coarsening heterolithic successions (2 m to 3 m thick) are visible. The base of each FA9 succession is characterised by sharp-based, lenticularly-bedded, mudstone-dominated heterolithics (F5.) and

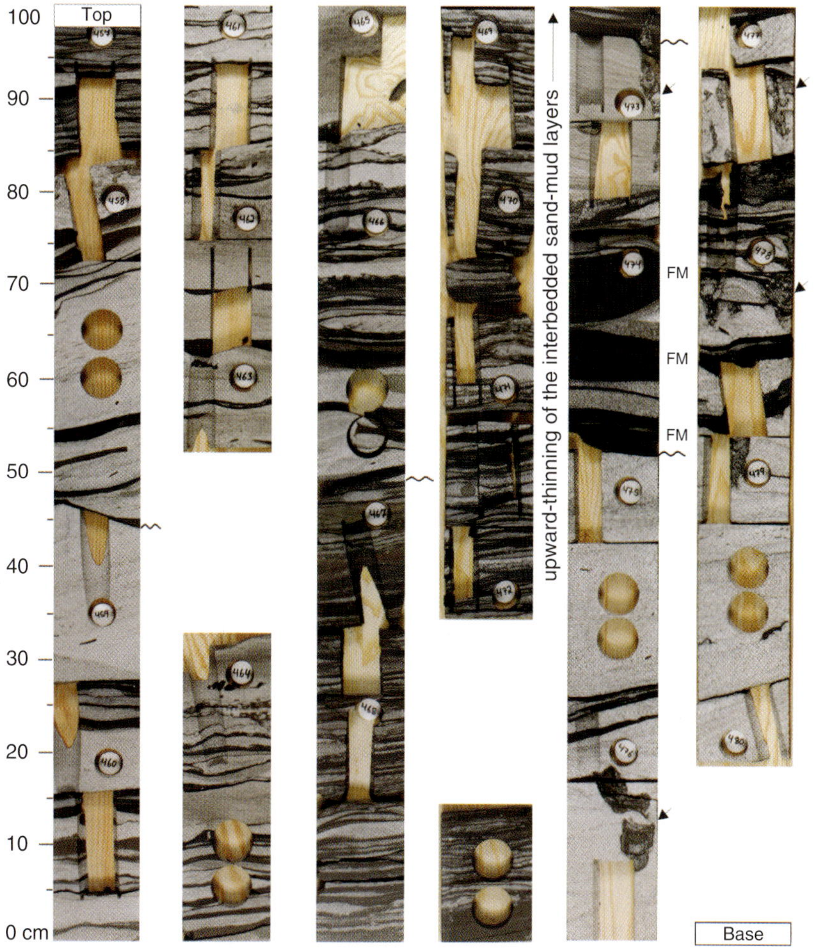

Fig. 17. Facies association FA9. Heterolithic tidal (point) bars displaying upward-thinning of the interbedded sandstone and mudstone layers with local sharp-based, stacked, upward-sandier intervals. Notice the presence of large vertical *Diplocraterion* burrows (black arrow). Example taken from reservoir zone Tilje 3.2 (T3.2).

mixed sandstone-mudstone wavy-bedded heterolithics (F5.2.1). These deposits are transitionally overlain by medium to fine-grained sandstone-dominated heterolithics (F5.1) showing basal cross-bedded sandstones (with rip-up mud clasts) that alternate with inclined and coupled fluid-mud layers. Vertically in the succession, the thicknesses of mudstone-layers decrease to thin laminae. The sedimentary structures within the sandstone layers suggest unidirectional currents. However, reverse palaeocurrent indicators and double mud drapes indicate a tidal influence. The lower stratigraphic intervals show abundant thick homogeneous mudstone layers interpreted to result from high suspended-sediment concentrations (fluid muds), indicating a channel-bottom depositional setting (Ichaso & Dalrymple, 2009). The inclined mudstones (~5°) interbedded with sandstone layers are interpreted as accretion surfaces of inclined heterolithic deposits (IHS), commonly developed near the top of point bars (Thomas *et al.*, 1987; Makaske & Weerts, 2005). Sandstone layers containing rip-up mud clasts are interpreted to represent minor erosional events. The upward thinning of the sandy layers and decrease of mudstone-layer thickness and abundance suggest vertical decrease in the suspended-sediment concentration in the water column. FA9 is interpreted to be deposited on within-channel elongate tidal bars or bank-margin tidal point bars. The low-diversity ichnological assemblage and the reduced level of bioturbation suggest stressed conditions probably due to high rates of sedimentation along the channel margins, together with low salinity levels.

Facies association 10 (FA10): Interdistributary-bay deposits

FA10 (Fig. 18) consists of mixed siltstone, very fine sandstone and mudstone. In general, FA10 successions (3 m to 4 m thick) commonly thicken and become sandier/siltier upward. The base of these deposits is characterised by sharp-based thin-bedded mud-dominated lenticular (F6.5) and disturbed heterolithic (F6.3) deposits (1 m to 2 m thick) showing sandstone lenses with ripple cross-lamination (occasionally bidirectional) and a mixed *Cruziana-Skolithos* trace-fossil suite. Locally, the mudstone-dominated heterolithics pass upward into mixed sandstone-mudstone heterolithics (F5.2.1) showing current-and wave-generated sedimentary structures. Locally, sharp-based to erosively-based sandstone-dominated heterolithic deposits (F5.1) showing cross-bedding and ripple cross-lamination are present. The heterolithic deposits are commonly overlain by medium-bedded very-fine grained to silty, rooted sandstones (F6.5) containing organic-matter lenses.

The nature of the basal muddy heterolithic deposits, the presence of bipolar current to ripple cross-lamination and the trace-fossil assemblage suggest deposition by tidally influenced conditions in a low-energy brackish-water setting. The overlying mixed sandstone-mudstone heterolithics containing both small-scale wave-generated and current-generated structures indicate a relative increase of energy, probably due to the interaction of tides and waves reworking and redistributing the fine suspended sediment. This, coupled with the trace-fossil assemblage, suggests deposition in an embayment (Elliott, 1974). The overlying rooted sandstones and siltstones indicate aggradation to the water level and colonisation by plants, where the total disruption of the original sedimentary structures and the mottled texture of these deposits indicate extensive biogenic activity under low energy and slow rates of sediment accumulation (Turner *et al.*, 1981). Local, sharp to erosively based sandstone-dominated heterolithics are interpreted as tidal-fluvial crevasse channels. The overall coarsening upward of the sandy portion within the heterolithic facies, together with evidence of tidal and wave processes and a moderate to diverse trace-fossil assemblage suggest that FA10 was deposited in a prograding minor mouth-bar/interdistributary-bay or tidal plain area. The local coarser cross-bedded deposits suggest that the excess discharge from the distributary channels was diverted into the bay during flood periods (Elliott, 1974; Brown & Richards, 1989).

Fig. 18. Facies association FA10: Interdistributary-bay deposits. Muddy to very fine-grained bioturbated sandstone showing lenticularly-bedded, wave-dominated and rare current-dominated heterolithic deposits capped by rooted intervals. Organic debris is common locally (black arrow). Example taken from reservoir zone Tilje 4 (T4).

STRATIGRAPHIC MODEL AND PALAEOENVIRONMENTS

In the most general terms, the upper Åre and Tilje interval in the study area shows an overall shallowing upward succession. Internally, the two sequences that are recognised in the Tilje Formation (sequences 2 and 3; Martinius *et al.*, 2001; Ichaso, 2012; Fig. 3) are each composed of several sandier upward trends (i.e. parasequences 2 m to 5 m thick) of shoreface and deltaic origin that are punctuated by smaller-scale (i.e. 0.5 m to 1 m thick) fining-upward channel successions (common in sequence

2 and towards the top of sequence 3) and coarse-grained, sand dominated, fining-upward successions (more common in sequence 3). Both types of succession are themselves organised in parasequence-set-scale (i.e. 10 m to 15 m thick) units. Detailed examination of cores and well logs (GR/RES) by previous workers (Martinius *et al.*, 2001) and in this study shows that such parasequence-set-scale units can be correlated across the Smørbukk field, thereby allowing the studied succession to be subdivided into a number of broadly tabular stratigraphic packages (Statoil reservoir zones A3 and T1 to T6). These reservoir units, which range in thickness from 5 m to 25 m, are bounded by inferred sequence boundaries and/or significant flooding surfaces that can be traced regionally (Ichaso, 2012); thus, each reservoir unit is believed to represent a time slice with minimal diachroneity.

Within each reservoir unit, gradual lateral variations in facies occur. Examination of these facies distributions (Figs 20 to 22) allow the reconstruction of the temporal evolution of the palaeogeography and sediment-input points as a function of eustatic sea-level fluctuations and fault movement. Sediment-input points and dispersal pathways within each time slice were determined using the down-system grain-size fining that is expected to occur in all sedimentary systems (Parker, 1991; Frings, 2008; Dalrymple 2010), together with the observed proximal-distal facies trends (Dalrymple & Choi, 2007). In the determination of these spatial fining trends, comparisons were made using the coarsest grain size present, rather than the bulk grain size of a local succession. This was done to ensure that differential trapping as a result of spatial variation in subsidence did not bias the results (Dalrymple, 2010). Detailed examinations were undertaken to determine whether there was a compensational style of delta-lobe stacking, whereby younger delta lobes occupied the topographic low between older lobes (cf. Straub *et al.*, 2009; Bhattacharya, 2010). Such stacking was not detected. Instead, deltaic units show aggradational and progradational patterns with the finest sediment and most distal facies toward the S-SW and E-NE parts of the field, indicating a simple 'layer-cake' stacking behaviour at the scale of the study area.

Sequence 1: Upper Åre Fm. (A3)

In the Smørbukk field, the uppermost Åre Formation (reservoir unit A3, the only part of sequence 1 examined in this study) is dominated by bioturbated sandstones and heterolithics and storm-dominated lower-shoreface to middle-shoreface deposits (FA1). This unit lacks channel deposits and evidence of fresh-water influence. The moderate to intensely bioturbated, ichnologically diverse intervals are erosively overlain by fine-grained hummocky cross-stratified sandstones forming at least two sandier and coarsening-upward units (recognisable in core material and GR/RES well log curves) suggesting deposition of upward-shoaling successions representing a progradational or regressive coast/shoreline system. Locally, toward the north-east, the lowest sandier-upward succession within A3 is separated from the upper one by a subordinate flooding surface capped by a 5 cm to 10 cm thick erosively based medium-grained to coarse-grained sandstone interpreted as a lag deposit (FA2) that might have been derived from reworked coarse-grained fluvial deposits; elsewhere, this lag is absent. The upper Åre (A3) is approximately 10 m thick, increasing up to 15 m towards the NW (Fig. 19, cross section I). HCS beds are thicker and more common in the central and northern areas (30% to 60% of the interval by thickness; Fig. 20A), whereas the southern areas show a dominance of more intensely bioturbated deposits (up to 75%) and thinner HCS beds (Fig. 20A). This facies distribution suggests that the water deepened toward the south.

Sequence 2: Lower to middle Tilje Fm. (T1 to T3.1)

The shoreface deposits (FA1) of the upper Åre (A3) are overlain abruptly by transgressive shelf lags (FA2) and/or distributary-mouth bar (FA3) deposits of the basal Tilje Formation (T1.1.1). In accordance with previous studies, the stratigraphic contact between the two formations is interpreted as a hiatal surface (SB2 of Martinius *et al.* 2001) based on: 1) the regional extent of the erosive contact, showing low relief and subtle incision towards the south-west and north-east of the field (Fig. 19); 2) the abrupt vertical change of grain size and facies from bioturbated and/or very-fine grained wave/storm-generated sandstones of the upper Åre Fm. to medium-grained and coarse-grained lags and pervasively heterolithic deposits of the Tilje Fm. that show extensive evidence of tidal sedimentation; and 3) widespread development of a coarse-grained lag (FA2) at the contact; and 4) the local presence of *Glossifungites* beneath the lag, all of which indicate a depositional hiatus.

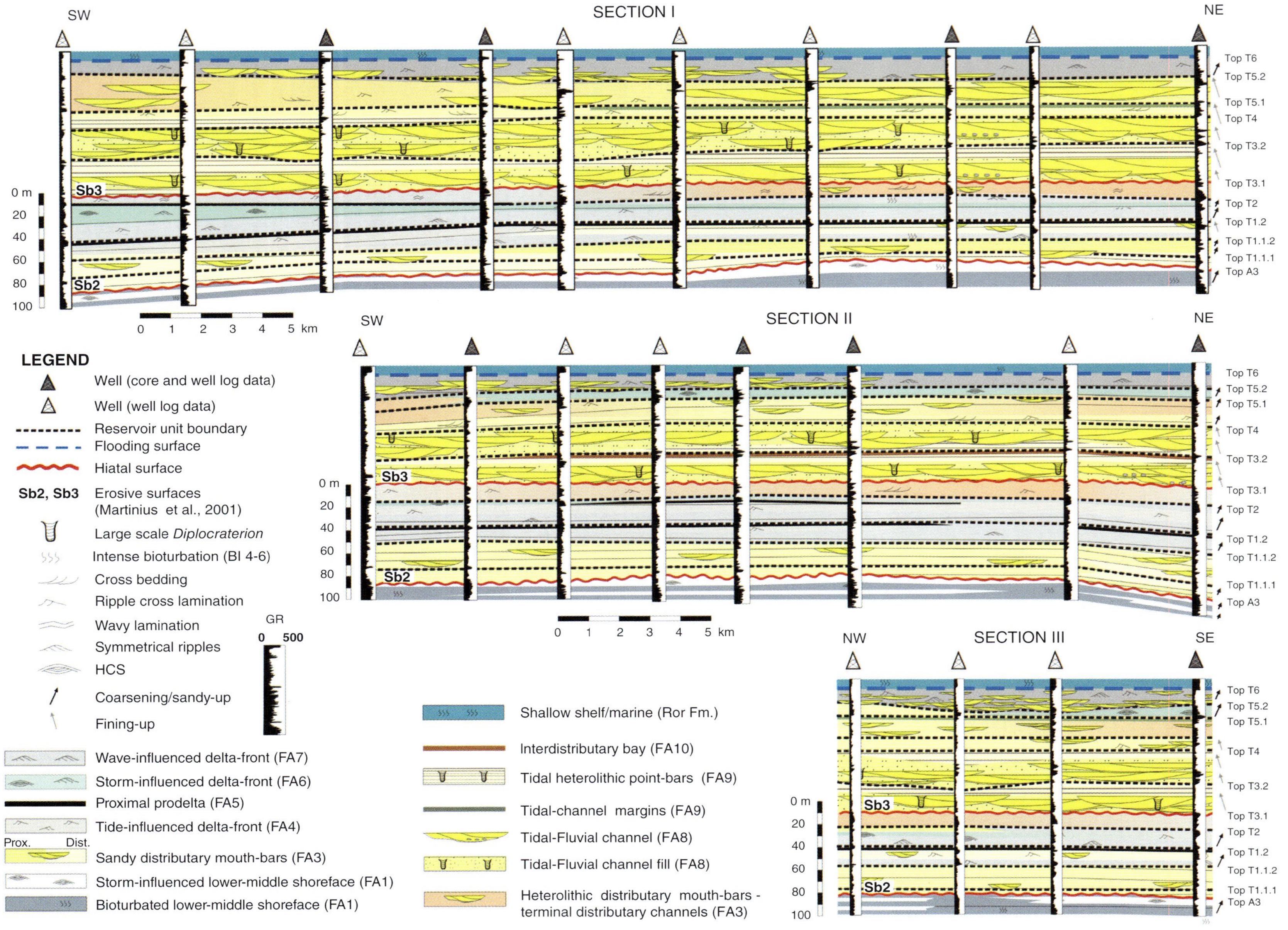

Fig. 19. Longitudinal (top) and transverse cross-sections of the Smørbukk field, developed using core and well-log data. See Fig. 1C for locations.

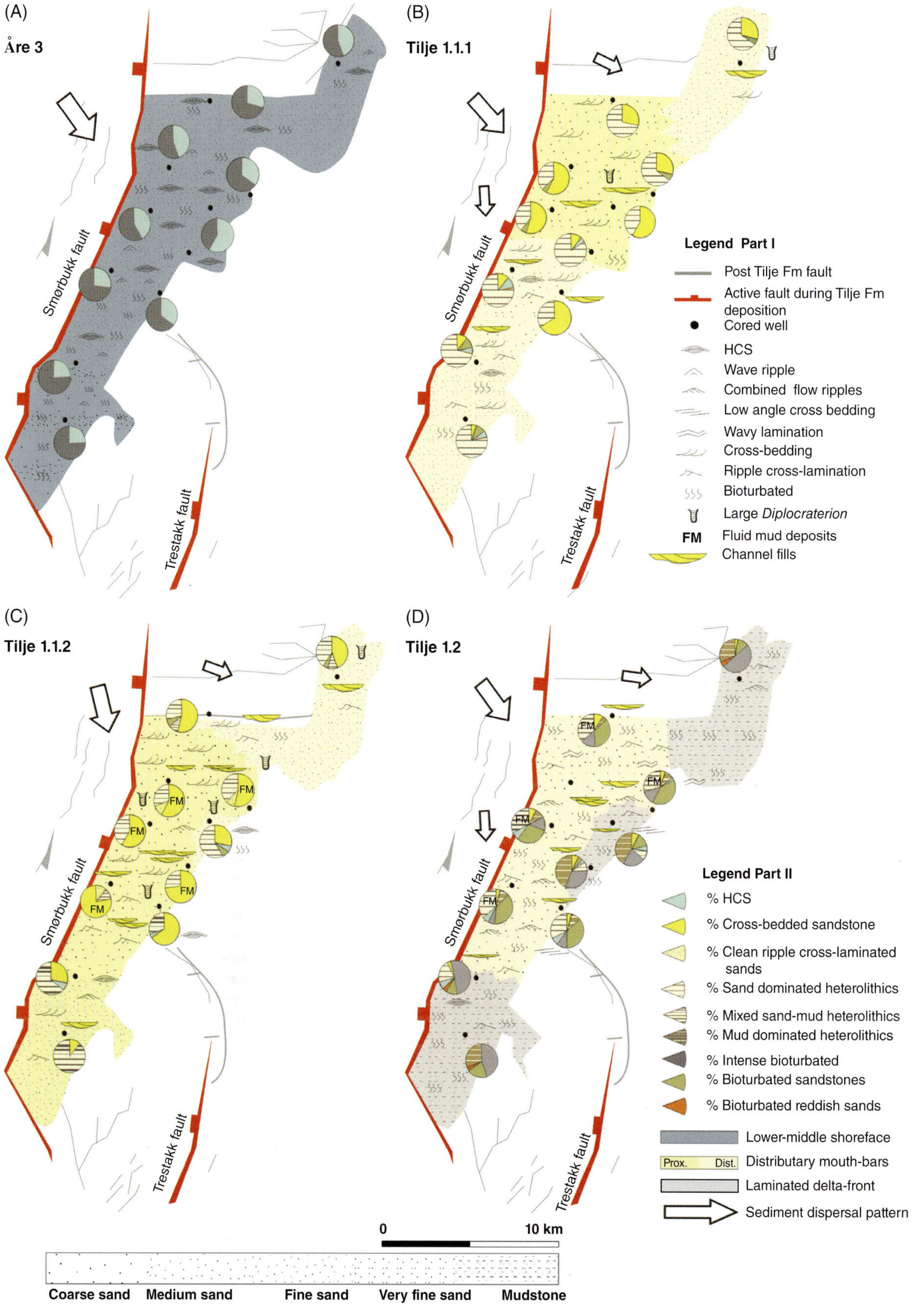

Fig. 20. Facies-distribution and palaeogeographic maps for the Smørbukk field from the upper Åre (A3) and lower-middle Tilje (T1.1–T1.2). The maps integrate the structural background, sedimentary structures (processes) and dominant facies associations (environments) for each reservoir zone. Active normal faults within the cover (after Marsh *et al.*, 2010) are marked in red. Background colour in each map shows the dominant Facies Association (see legend). Background dot-line texture fills represent the dominant grain size from coarse sandstone to mudstone (see grain-size bar). Dominant sedimentary structures are indicated schematically (see legend). Pie graphs show semi-quantitative percentages of the different facies (see part II of the legend).

Environmentally, previous workers have interpreted the lower half of the Tilje as the deposits of an estuarine environment (Martinius *et al.*, 2001). We suggest, instead, that this interval in the Smørbukk field formed in a tidally influenced, open deltaic setting with widespread development of distributary mouth-bar/delta-front sub-environments, with small terminal distributary channels. This interpretation is supported by several lines of evidence including: the pervasive development throughout the basal Tilje of sandier-upward (deltaic) successions alternating with fining-upward (channel) successions that show a predominant palaeocurrent in one direction; the widespread presence of decimetre-scale alternations in sand grain size, the abundance of fluid-mud layers and the degree of bioturbation that together are interpreted as representing seasonal variations in river discharge; and the presence throughout of a low-diversity, mixed *Cruziana-Skolithos* ichnofacies suggestive of a brackish-water assemblage.

Although this assemblage of features represents clearly a deltaic environment, these deltas could have been situated in the bayhead-delta region of a wave-dominated estuary (cf. Dalrymple *et al.*, 1992; Boyd, 2010). Although this cannot be discounted unequivocally on the basis of this local study, we think it is unlikely. The basal part of the Tilje, both in the Smørbukk field (reservoir unit T1.1) and regionally, is aggradational to progradational in character, as is the upper part of sequence 2 (reservoir units T2 and T3.2); the only part of sequence 2 that is retrogradational is reservoir unit T1.2 (Ichaso, 2012). Such a stacking pattern is not consistent with an estuary *sensu* Dalrymple *et al.* (1992). Also, there are no facies that could represent such barrier-related environments as tidal inlets and flood-tidal deltas, the parts of a barrier that have the greatest preservation potential. Therefore, we conclude that the lower part of the Tilje is unlikely to have formed in a closed estuary. We suspect that the widespread presence of a brackish-water trace-fossil assemblage, coupled with confusion over the two definitions of 'estuary' (cf. Dalrymple *et al.*, 2011) contributed to the use of the term 'estuarine' in previous work.

Basal Tilje (T1.1)

The basal Tilje Fm. comprises the reservoir production zone T1.1 (15 m to 25 m thick), which has been subdivided into two sub-units (subzones T1.1.1 and T1.1.2) based on the vertical increase of grain size and thickness of cross-bedded sandstones and the abundance of tidally influenced channel deposits (Fig. 3). These vertical changes are recognised in well logs as a marked shift from high to low GR readings and a change in density-neutron separation from fine mixed mudstone-sandstone to clean sandstone. Tilje 1.1.1 shows local development of a basal transgressive shelf lag (FA2) where the thickest deposits (up to 25 cm thick) occur towards the south and NE of the field. The shelf lags are commonly overlain by distributary mouth bar deposits (FA3) that show a lateral transition from clean medium-grained cross-bedded sandstones in the north-central part of the study area to more fine sand-mud heterolithics to very-fine sand-mud heterolithics towards the S-SW part of the field. The north and central deposits are punctuated by local fining-upward terminal distributary-channel deposits, whereas, toward the south, there is an increase in the proportion of wave-influenced mixed sand-mud heterolithics alternating with HCS beds and bioturbated intervals (Fig. 20B). The presence of the wave-generated structures and abundant bioturbation towards the south indicates less sheltering and deeper-water conditions with salinities that were closer to normal marine.

The overlying sharp-based mouth-bar deposits (FA3) of T1.1.2 show medium-grained, clean cross-bedded sandstones that change laterally into fine-grained sandstones and bioturbated heterolithics toward both the NE and SW (Fig. 20C). Locally, the base of T1.1.2 consists of mudstone-dominated and bioturbated heterolithic deposits, showing a more diverse marine trace-fossil suite; thus, the contact between T1.1.1 and T1.1.2 represents a higher-order flooding surface. Tidally influenced channel fills containing thick fluid-mud deposits are more prominent in the central area. Wave-generated structures are less common than in T1.1.1; however, small-scale HCS occurs towards the E and SW margins of the field. The facies distribution (Figs 20B & 20C) of the basal Tilje, the regional variation of grain sizes within the sandy portion and the southern occurrences of wave-generated structures suggest an overall NW to SE transition from proximal (coarse-medium grained, cross-bedded, abundant fluid-mud layers, sandstone-dominated heterolithics and low bioturbation) to medial (medium-fine grained, cross-bedded, HCS, sand and mixed sandstone-mudstone heterolithics, fewer fluid muds and low to moderate bioturbation) mouth-bar types.(cf. Dalrymple & Choi, 2007; Tanavsuu-Milkeviciene & Plink-Björklund, 2009).

Lower Tilje (T1.2)

The distributary mouth-bar (FA3) deposits of Tilje 1.1 (T1.1.2) are overlain by channelised distributary mouth-bar (FA3) and tidally influenced delta-front deposits (FA4) of reservoir zone T1.2 (Fig. 19). Locally, the E and NE parts of the study area show basal mud-dominated heterolithics with intense bioturbation and a moderately diverse mixed *Cruziana-Skolithos* assemblage, indicative of relatively distal settings compared with the upper deposits of T1.1.2. This indicates that the contact between T1.1 and T1.2 is a significant flooding surface. The T1.2 interval (10 m to 15 m thick) is recognised in GR well logs as two broadly coarsening-up packages that are variably developed (Fig. 3). The channelised mouth-bar deposits (FA3) consist predominantly of fine to very-fine sandstone and are more bioturbated than the basal deposits of T1.1.1. The mouth bars of T1.2 in the north and central areas are composed mainly of mixed sandstone-mudstone heterolithics showing both current and combined-flow structures, locally interbedded with intensely bioturbated sandstones. These deposits change laterally (SW and NE; Fig. 20D) and vertically into tidally influenced delta-front deposits (FA4) that lack channels and show intense bioturbation. Mudstone-dominated heterolithics, showing wave influence, occur locally toward the SW, forming stacked coarsening-upward and thickening-upward successions (3 m to 5 m thick). The trace-fossil suites also show a gradational change from a low-diversity mixed *Cruziana-Skolithos* assemblage in the central area to a more diverse facies-crossing *Cruziana-Skolithos* assemblage (cf. MacEachern & Bann, 2008) toward the NE and SW.

Overall the T1.2 interval shows fewer terminal distributary channels and storm-influenced deposits than the lower stratigraphic intervals. Tidal currents are the dominant depositional process in the delta-front area. Sedimentological observations, facies and trace-fossil distributions within T1.2 suggest the dominance of heterolithic mouth bars with localised terminal distributary-channel deposits towards the NW, whereas the SW and NE areas show tidally influenced proximal to medial delta-front deposits. This distribution (Fig. 20D) suggests a delta-lobe configuration with the delta lobe facing toward the SE. The coarsening-upward trends, coupled with the increase of vertical thickness in southern and north-eastern areas of the Smørbukk field, with interpreted low-angle clinoforms (dipping 2 to 5 degrees) in well-log correlation panels (Fig. 19, cross sections I and II), suggest delta-front progradation toward the SW and NE.

Middle Tilje (T2 to T3.1)

The base of zone T2, which is 16 m to 25 m thick, is a high-order flooding interval that marks a marine incursion followed by renewed progradation. It is characterised by the presence of local marginal-marine to normal-marine lags (FA2), where the thickest deposits (up to 15 cm thick) occur toward the SW and the eastern parts of the field. This basal lag commonly consists of coarse-grained cross-bedded reddish sandstone (similar of the basal T1.1.1 lags). Locally, the thickest marine lags show interbedding with sandstone and intensely bioturbated (high diversity) heterolithic deposits forming a 60 cm thick overall upward-thinning succession. These deposits are commonly overlain by 2 m to 3 m thick, proximal prodelta units (FA5). This distinctive sharp-based mud-dominated deposit has been used by Martinius *et al.* (2001) as a regional stratigraphic marker (Fig. 19) that is characterised by a shift from low to high GR readings (Fig. 3) and, in density-neutron separation, from fine sandstones to mixed mudstone-sandstone. These prodelta deposits (FA5) show sanding-upward and coarsening-upward successions that consist of pinstripe mudstones and lenticularly-bedded, wave-dominated heterolithics, locally preserving evidence of tidal, river flood and storm deposition.

The prodeltaic unit (FA5) is commonly overlain by storm-influenced (FA6) and wave-influenced (FA7) delta-front deposits, which typically show a series of correlatable coarsening-up successions (visible in GR/RS well-log curves; Fig. 3) and commonly thicken toward the SW and NE of the field (Fig. 19 cross sections I and II). These deposits include a variety of heterolithic facies; however, lenticularly wave-dominated and mixed sandstone-mudstone wavy bedded varieties are most common. These heterolithics generally become thicker bedded and sandier upward and locally are interbedded with thick unbioturbated HCS beds. These deposits lack channels. The facies distribution maps (Fig. 21A) indicate that HCS beds are thicker and more common in the central and NW parts of the field; the abundance of HCS

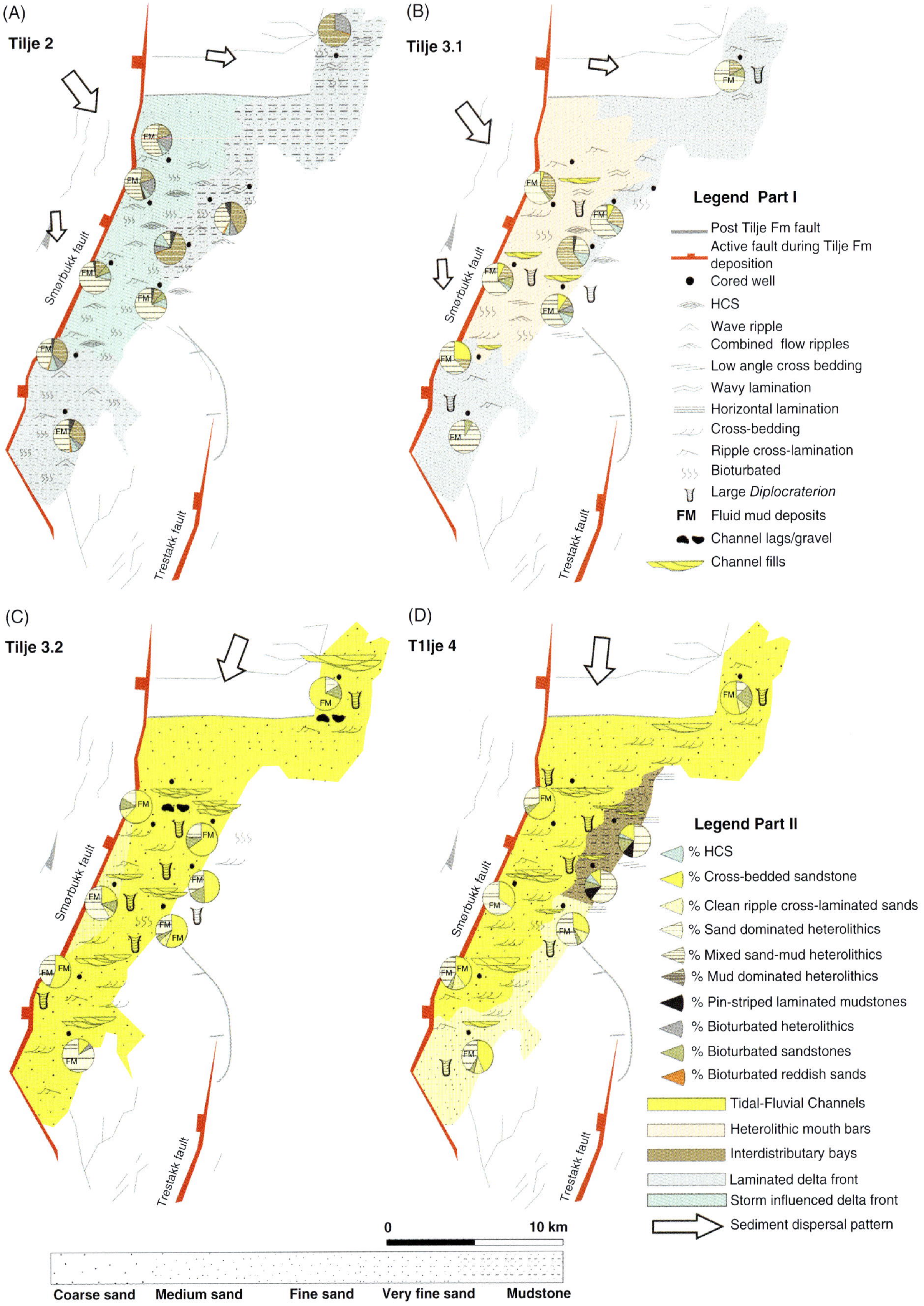

Fig. 21. Facies-distribution and palaeogeographic maps for the Smørbukk field from the lower-middle (T2 to T3.1) and middle-upper (T3.2 to T4) Tilje. See caption of Fig. 20 for explanation of symbols.

diminishes towards the E, NE and SW where wave, combined-flow and current (locally bipolar) ripples are the most common sedimentary structures (FA7). The sandiest intervals in the central area show low to moderate levels of bioturbation mainly composed of a stressed, mixed *Cruziana-Skolithos* assemblage, whereas the muddier successions in the SW and NE show moderate bioturbation composed of small-scale proximal *Cruziana* traces. The distribution of wave and current sedimentary structures and trace-fossil assemblages suggest an overall NW to SW, E and NE transition from medial to a more distal delta-front setting.

The deposits of zone T2 are overlain by zone T3.1 (15 m to 20 m thick). The base of this zone can be recognised in well logs as an inflection point in the GR-log curve with a corresponding increase of mudstone separation on the DEN/NEU logs. This level correlates with the mid-Tilje Formation Nd143/Nd144 marker of Martinius *et al.* (2001) and is another prominent, high-frequency flooding surface. The T3.1 zone displays at least two thick (5 m to 6 m) coarsening-upward successions composed of distributary-mouth bar (FA3) and proximal to medial, tidally influenced delta-front (FA4) deposits. As in the lower Tilje (T1.2), the distributary mouth-bar deposits (FA3) are mainly composed of current-dominated, mixed sandstone-mudstone heterolithics that pass upward into sandstone-dominated heterolithic deposits. The mouth bars are punctuated by fining-upward (1 m to 3 m thick) cross-bedded deposits interpreted as terminal distributary channels that show a decrease in thickness and grain size toward the SW of the field (Fig. 21B). FA3 successions dominate the north and central areas (Fig. 21B) and they change upward and laterally (towards the SW and NE; Fig. 19) into more laminated tidally influenced delta-front deposits (FA4) and more locally into laminated and muddy proximal prodeltaic deposits (FA5). The delta-front units show a variety of current-dominated and wave-dominated heterolithic facies displaying small-scale HCS and intensely bioturbated sandstones that are especially common towards the east side of the field (Fig. 21B). Evidence of seasonal variations in river discharge is present (cf. Ichaso & Dalrymple, 2009). Overall, the middle Tilje succession shows a 'lobate' distribution of facies, similar to the lower stratigraphic interval T1.2.

Sequence 3: Middle-upper Tilje Fm. (T3.2 to T6)

Distributary mouth-bar (FA3), tidally-influenced delta-front (FA4) and prodeltaic deposits (FA5) of T3.1 are overlain by stacked fining-upward tidal-fluvial channel fills (FA8) and heterolithic tidal point-bar deposits (FA9) of reservoir unit T3.2 (20 m to 25 m thick). The stratigraphic contact between T3.1 and T3.2 (SB3) is erosional, with only a small amount of incision in the SW part of the field (Fig. 19, cross section I). Nevertheless, it has been interpreted as a regional hiatal surface (SB3 of Martinius *et al.*, 2001). This interpretation is based on: 1) the abrupt change on grain size from silt, very-fine and fine-grained sand in T3.1 to medium and coarse-grained sandstones and locally conglomerates, in T3.2; 2) the pronounced vertical change of facies associations, which expresses a shift of facies from medial-distal deltaic deposits to the most proximal deposits within the Tilje Formation; 3) the presence of terrestrially derived humic debris within T3.2, suggesting proximity to a fluvial source and freshwater environments in the catchment area; and 4) the change of bioturbation styles from high-diversity trace-fossil assemblages in the underlying T3.1 units, to weak or almost non-existent bioturbation above SB3.

Middle-upper Tilje (T3.2 to T4)

Reservoir unit T3.2 is characterised by erosively based, fining-upward, thick, tidal-fluvial channel deposits (FA8) with GR and DEN/NEU log signatures that show either fining up or blocky patterns. The channel style changes upward within the succession (Fig. 19) from single channels (fining-upward sandstones of FA8 overlain by or replaced laterally by heterolithic tidal point-bar deposits of FA9), to more amalgamated, stacked packages of FA8 sandstones. The FA8 successions are mainly composed of coarse-grained to medium-grained cross-bedded sandstone that fines upward into fine-grained cross-bedded sandstone, bipolar ripple cross-laminated sandstone and sand-dominated heterolithics. Basal conglomeratic lags are more common towards the north of the field, whereas large rip-up mud clasts and anomalously thick (up to 15 cm) fluid-mud deposits (Ichaso & Dalrymple, 2009) are common across the field and within the middle-upper Tilje interval, reflecting periodic high fluvial discharge. These tidal-fluvial channel units commonly change upward or/and towards the W and SW into weakly bioturbated, fine-grained to medium-grained heterolithic tidal point-bar deposits (2 m to 3 m thick) of FA9 (Fig. 21C).

Overall, the mid-Tilje interval (T3.2) shows more-or-less constant thickness across the study area, with a slight lateral increase (up to 25 m) towards the southern and north-eastern parts of the Smørbukk field (Fig. 19, cross sections I and II). The facies distribution within reservoir zone T3.2 (Fig. 21C) shows only slight variability across the field. The percentage of coarse and medium cross-bedded sandstone is, however, higher in the north and central areas, in comparison with the SE, where medium-grained to fine-grained sandstone-dominated heterolithics are more abundant. These subtle thickness and grain size variations suggest a NE to SW proximal-to-distal trend within the channel deposits.

The T3.2 succession is overlain by the stacked tidal-fluvial channels (FA8), point-bar (FA9), interdistributary bay (FA10) and distributary mouth-bar deposits (FA3) of reservoir zone T4 (~25 m thick). In well logs (Fig. 3), the base of T4 is difficult to recognise; however, there is a subtle inflection on the GR logs (from lower to higher readings) and a lithologic separation (from finer to coarser) in the DEN-NEU log, which is more marked towards the east and less pronounced in the central and northern areas. In GR-log curves, the upper part of T4 consists of relatively clean, coarsening-up or blocky sandstone. The top of the unit (T4–T5.1 contact) is marked at the inflection point (high to low GR reading) and is easy to identify in the north but is less prominent towards the SW part of the field.

The erosively based tidal-fluvial channel deposits (FA8) of T4 show similar characteristics to those in zone T3.2. However, vertical burrowing (large-scale *Diplocraterion*) is more abundant and conglomeratic basal lags are rare and mainly localised in the NE and central areas (Fig. 21D). Locally, the tidal-fluvial channels change towards the SW into fine-grained point-bar deposits (FA9) and to the east into 0.5 m to 1 m thick mud and mixed heterolithic interdistributary bay (FA10) deposits (Fig. 21D), showing rooted beds and organic-matter lenses and small-scale wave-generated structures. Distributary mouth-bar deposits (FA3) are also present and they commonly occur towards the south of the field (Fig. 21D).

As in the underlying reservoir zone, the T4 channel deposits are overall thicker and more common in the northern and central parts of the field, with interdistributary-bay deposits occurring toward the east. Coarse-grained to medium-grained channel-fill deposits are replaced towards the south by medium-grained to fine-grained distributary-mouth bars suggesting an NE to SW proximal to distal trend. Grain-size and facies variations are also accompanied by lateral changes in the trace-fossil assemblage, from high-energy, low-diversity (large *Diplocraterion* burrows) within channel fills to more brackish-water assemblages showing low to small-scale *Skolithos* and capping plant roots in the interdistributary bay deposits and to a moderately diverse mixed *Cruziana* and *Skolithos* suite in the southern distributary-mouth bars.

Upper Tilje (T5 to T6)

The T4 tidal-fluvial channel deposits (FA8) are regionally overlain by heterolithic distributary mouth bar (FA3), tidal-fluvial channel fill (FA8) and wave-influenced (FA6) and storm-influenced (FA7) delta-front deposits in reservoir zone T5 (25 m to 30 m thick). The contact between T4 and T5 is interpreted here as a high-order flooding surface based on the increase of trace-fossil diversity at the base of T5 and the upward switch to more wave-influenced and storm-influenced delta-front deposits, indicating a more distal depositional setting compared with the proximal channel-fill deposits of T4. This contact is well defined (in the north and central areas) in GR/RES-log curves as an inflection point from low to high values (Fig. 3), coupled with the change in lithologic separation on the Density-Neutron logs. However, the inflection point in the GR/RES curve and lithological separation are less prominent towards the south of the field.

The thicknesses of T5 increases slightly towards the SW and NE of the field (Fig. 19, cross sections I and II), where clinoforms dipping (2 degrees to 5 degrees) towards these directions have been interpreted from core and well-log data. This reservoir unit is divided in two sub-zones (T5.1: 20 m thick; T5.2: 5 m to 10 m thick) based on the upward change of facies and the change from low to high GR-log readings (Fig. 3).

The base of T5.1 consists of channelised distributary-mouth bar (FA3) and tidal-fluvial channel (FA8) deposits. The mouth bars (FA3) show coarsening-upward and sanding-upward successions composed mostly of sand and mixed sandstone-mudstone, tide-dominated, heterolithic deposits. These deposits are commonly punctuated by metre-scale fining-upward, medium-grained, cross-bedded sandstones, interpreted as terminal distributary channels. Bioturbation within the mouth bars is moderate and wave-generated structures are locally present. They also record seasonal cyclicity in river discharge (cf. Ichaso & Dalrymple, 2009) in the form of regular variations in grain size, bioturbation and mud-layer thickness, as was also present in the middle Tilje (T2 to T3.1). The mouth bars display lateral facies transitions from sand-dominated in the NW-central areas, to mixed sandstone-mudstone deposits toward the SW (Fig. 22A). This variation, coupled with the overall south-westerly decrease of the sand grain size and the increase of the mudstone ratios and bioturbation diversity, suggest an overall N to SW proximal-to-distal transition. The distributary mouth bars commonly alternate or change laterally into thick, medium-grained to fine-grained, stacked, tidal-fluvial channel deposits (FA8), which show similar characteristics to the T3.2 and T4 channel fills. Channel lags are locally present in the central part of the field (Fig. 22A).

These deposits pass upward into the T5.2 reservoir zone, which is composed of tidal-fluvial channels (FA8), heterolithic distributary-mouth bars (FA3) and storm-influenced delta-front (FA6) successions. This upward shift to somewhat more distal deposits indicates the presence of a high-order flooding surface at the base of T5.2. The thickness of this zone varies between 10 m in the central area to 15 m towards the SW, E and NE (Fig. 19). Vertically staked, fining-upward, tidal-fluvial channels (FA8) showing large *Diplocraterion* and *Skolithos* burrows are less common than in the lower stratigraphic intervals. However, they occur locally towards the north and central parts of the field (Fig. 22B), where conglomeratic channel lags and thick fluid-mud layers are present. These deposits change towards the SW into moderatly bioturbated distributary-mouth bars (FA3) and towards the NE and E into storm-influenced delta-front (FA6) deposits (Fig. 22B) showing thick (up to 40 cm) sharp-based large-scale HCS beds.

Locally towards the E, a few coarse channel fills are abruptly overlain by thick, laminated mudstones, bioturbated sandstones and HCS beds, suggesting the abandonment of the channels and filling by low-energy marine processes followed by episodic storms. Distributary channels or inter-bar tidal-fluvial channels appear to be absent in the NE region. In general, this facies distribution suggests that the western margin of the field (Fig. 22B) was dominated by more proximal tidal-fluvial environments, whereas the eastern margin of the field was subjected to more open- to marginal-marine wave and tidal processes.

The uppermost T6 reservoir zone is characterised by the dominance of wave-dominated and tide-dominated delta-front deposits (FA4 and FA7 respectively) and distributary-mouth bars (FA3) showing channelisation. The thickness of this unit varies between 10 m and 20 m, with the greatest thicknesses (20 m) localised in the north-central part of the field (Fig. 19, cross sections I and III). Well and core-log correlation panels suggest that the base of T6 truncates the top of T5.2 locally (Fig. 19, cross section III) in the central and SW parts of the field, implying the filling of subtle incision into T5. A lag (FA2) was not recognised at the base of this unit.

Distributary-mouth bar deposits of T6, showing abundant tidal and combined-flow structures and moderate to intense bioturbation of mixed *Cruziana-Skolithos* ichnofacies, are locally punctuated by sharp and erosively based tidal-fluvial terminal distributary-channel deposits (up to 1 m thick). More proximal, fine and medium-grained, sand-dominated heterolithic mouth bars and the abundance of channel-fill deposits seem to be concentrated in the central part of the field (Fig. 22C). Terminal distributary channels (up to 80 cm thick) are also present towards the SW, where medial mouth bars become muddier and more bioturbated. Tidally influenced mouth bars change laterally towards the E and NE into more distal, laminated and tide-wave influenced delta-front (FA4 and FA7) deposits lacking channels (Fig. 22C). This unit should be regarded as the first indicator of a later flooding event, as it marks transitional or precursor facies to the fully marine facies of the Ror Formation.

Lower Ror Fm.

The top of the Tilje Fm. (top of reservoir zone T6) is defined by the abrupt transition from tide and wave-influenced, channelised mouth-bar (FA3) and delta-front deposits (FA4 and FA7) into thinly laminated and intensely bioturbated marine mudstones and silty bioturbated sandstones of the Ror Fm (Fig. 22D). The contact between the Tilje and Ror formations is defined at the regionally extensive, abrupt transition from low to high readings in the GR/RES-log, coupled with the change in DEN/NEU

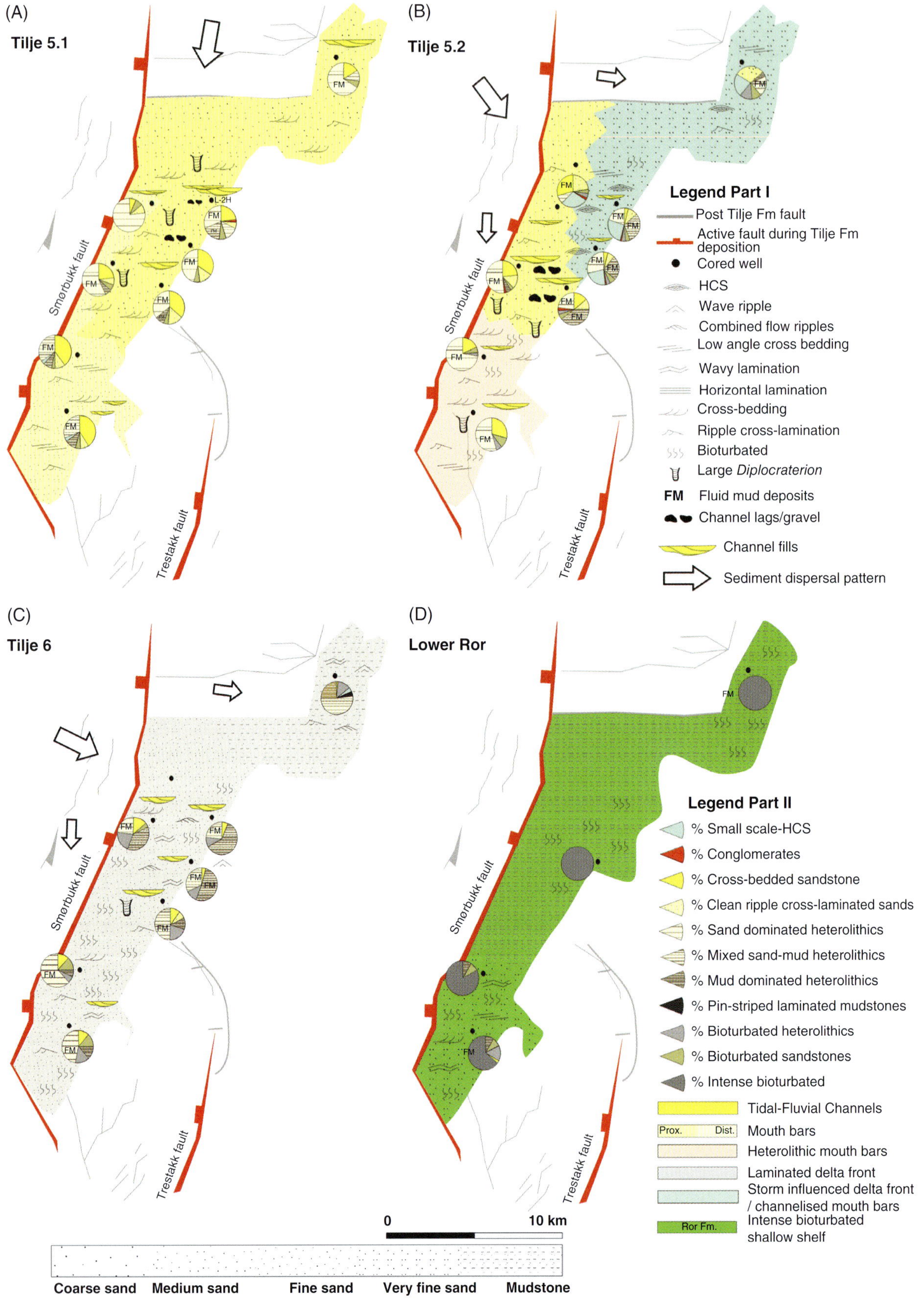

Fig. 22. Facies-distribution and palaeogeographic maps for the Smørbukk field from the upper (T5 to T6) Tilje and lower Ror interval. See caption of Fig. 20 for explanation of symbols.

separation from fine-sand to mud. This surface represents a major flooding event that drowned the underlying deltaic system and was followed by the prolonged deposition of shelf mudstone.

CONTROLS ON FACIES DISTRIBUTION, PALAEOGEOGRAPHY AND SEDIMENT DISPERSAL

Previous regional studies (Ehrenberg *et al.*, 1992; Corfield & Sharp, 2000) have shown that significant crustal extension had occurred in the Halten Terrace area by the time of Tilje deposition, leading to the development of subsidence and the generation of accommodation. However, faulting was not sufficiently active to cause surface rupture and to generate local fault-related topography (Marsh *et al.*, 2010). Thus, the Tilje is generally regarded to be an early syn-rift deposit. The facies distribution patterns reported above (Figs 20 to 22) indicate that sediment input to the Smørbukk study area was predominantly from the north. The environmental interpretation proposed here indicates that the Tilje Fm. accumulated in a deltaic system in which fluvial and tidal processes predominated, but with wave processes influencing and even dominating deposition at certain times and places.

The stratigraphy of syn-rift successions is controlled by a variety of autogenic and allogenic processes which all leave distinctive signatures in the deposits (Bosence, 1998). The stage of rifting (early, main, or late) and the dynamics of its structures affect considerably the depositional patterns by creating sites of uplift and erosion and by controlling the pathways of sediment transport/dispersal (Gawthorpe & Leeder, 2000; Withjack *et al.*, 2002), but more importantly, together with relative eustatic sea-level changes, by defining the accommodation for sediment accumulation (Howell & Flint, 1996). The relationship between space, climate, hydrodynamic processes and sediment supply determine the depositional regime of marginal marine systems in syn-rift settings where a variety of infill patterns can be observed throughout the basin evolution (Ravnås *et al.*, 2000).

Hydrodynamic processes

The relative importance of fluvial, tidal and wave energy varies through the deltaic fluvio-marine transition (Dalrymple & Choi, 2007), as well as with distance away from the river mouth; the complex interaction of these processes in turn exerts a major control on the distribution of the resultant sedimentary structures and deposits (Bhattacharya, 2010). From a proximal-distal perspective, the Tilje channel deposits (T3.2 to T5.1; Figs 21C to 22A) show the coarsest sand fraction (medium to coarse) and are dominated by cross-bedding and current-ripple cross-lamination indicative of unidirectional, strong to moderate currents. The channel deposits also preserve thick fluid-mud beds (Ichaso & Dalrymple, 2009) and other typical tidal signatures such as double mud-drapes and indicators of bidirectional currents. These tidally generated features were produced by the landward increase of tidal energy (in a general northerly direction) because of the compression of flood currents into the smaller cross-sectional area of the distributary channels (Dalrymple & Choi, 2007; Bhattacharya, 2010). The relative decrease in the abundance of tidal indicators in unit T4, coupled with the presence of the coarsest channel deposits, indicates that this unit is the most proximal in the succession and was deposited near the tidal limit. Tidal signatures increase in abundance within the more distal tidal-fluvial channel deposits and proximal mouth-bars of reservoir units T1.1 and T3.1 (Figs 20B & 21B), where deposition was controlled by strong tidal currents in the outer part of the mixed tidal-fluvial transition (Dalrymple & Choi, 2007). The medial to distal distributary-mouth and delta-front deposits of T1.2, T2 and T6 (Figs 20D, 21A & 22C) formed in the marine-dominated deltaic zone and also preserve tidal signatures, but in less abundance than their proximal equivalents. Wave-generated structures become important in these more distal reservoir zones because of the rapid decrease of tidal-current energy away from the river mouth. Similarly, the fine sediments in areas laterally adjacent to the river mouth (e.g. T5.2; Fig. 22B) are commonly moulded by wave and/storm action (cf. Mángano *et al.*, 1994).

As indicated by the above, the Tilje Formation accumulated in a mixed-energy environment in which fluvial, tidal and wave processes where all active. However, based on the vertical distribution and abundance of the preserved sedimentary structures (Fig. 3 and Figs 20 to 22), the Tilje Formation shows vertical changes in the apparent intensity of these three processes: tidal and wave processes are alternatively more intense in sequence 2; tidal-fluvial processes dominate in the lower part of sequence 3; tidal and wave processes are again more important toward the top of sequence 3. These vertical hydrodynamic variations are believed to be controlled by the longitudinal shifting of depositional environments (proximal and distal) as the result of changing accommodation (see eustacy and tectonics discussion below) and by the lateral shifting of entry points from NW (sequence 2) to N-NE (sequence 3). Thus, fluvial and tidal processes are more strongly represented at times when more proximal environments occupied the study area, whereas wave processes are recorded more during episodes when distal and/or flanking environments were present in the study area.

Overall, it is believed that the relative intensity of the three processes did not change significantly throughout Tilje deposition. On the other hand, the dramatic switch from the wave/storm-dominated upper Åre Fm. to the overall tide-influenced Tilje Fm. might be related to major allogenic controls, such as regional structural changes at the rift-basin scale. The abrupt decrease in wave energy, coupled with the increase in the importance of tidal currents can both be explained if structural movements created a more funnel-shaped embayment, which was partially sheltered from wave action and in which tidal currents were enhanced.

Climate

The studied area was situated in a mid-palaeolatitudinal location (Doré, 1991) with a warm climate that experienced strong seasonal differences in temperature and rainfall (Hallam, 1994). This seasonality is reflected in the presence of distinct flood-interflood bedding (see above and also discussion in Ichaso & Dalrymple (2009)) within the distributary-mouth bar deposits of the middle part of sequence 2 (unit T3.1) and the upper sequence 3 (T5.1). In these environments, the seasonal discharge pulses (cf. Day *et al.*, 1995) promoted the formation of current-generated structures and the input of coarser sand grains in locations that during inter-flood periods were commonly dominated by tides and/or waves. The presence of thick fluid-mud layers within the distal mouth-bar and proximal delta-front deposits of T3.1 and T5.1 indicates that river floods caused the expulsion of the turbidity maximum from the river mouth (cf. Bhattacharya & MacEachern, 2009), depositing fluid muds in more distal locations than was the case during interflood times. Within the same depositional environment however, these seasonal indicators are more prominent within the middle Tilje T3.1 deposits and less pronounced in T5.1. This suggests that, during the early Tilje deposition, the rainfall and runoff events were somewhat more intense than later in the Tilje.

Accommodation (eustacy and tectonics)

Eustatic sea-level curves indicate the presence of a relative sea-level rise of up to 100 m during the Early Jurassic (Haq *et al.*, 1987; Hallam, 1988; Surlyk, 1990). This favoured the generation of accommodation on a regional scale. However, the simultaneous tectonic subsidence associated with rifting during the opening of the North Atlantic Ocean (Doré, 1991) was probably the most important factor responsible for generating the space in which the 150 m to 220 m thick Tilje Formation accumulated. However, over the entire duration of the studied succession, there was an overall balance between the creation of accommodation and sediment supply, as shown by the fact that depositional environments remained within a few kilometres of the coast throughout the Tilje. Thus, the Smørbukk area represents a sediment-overfilled to balanced basin. This is a common characteristic of the rift-initiation (late pre-rift to early syn-rift) phase (Ravnås & Steel, 1998; Withjack *et al.*, 2002).

In the Smørbukk area, interpretation of seismic data (Marsh *et al.*, 2010) indicates that rift initiation occurred during deposition of the upper Åre Formation (Fig. 23) and consisted on localised, 'slow' down-to-the-west movement on the NE-SW-oriented normal faults, including the blind Testakk and Smørbukk faults, forming NE-SW axially oriented half-graben morphologies (Corfield & Sharp, 2000). In general, the tabular nature of the reservoir units indicates that differential movement on these faults was significantly less than the regional rate of subsidence/sediment accumulation. Through Tilje Fm. time, the Trestakk and Smørbukk faults did, however, propagate towards the north and south respectively (Marsh *et al.*, 2010) creating isolated, syn-depositional, shallow depocentres on the east side of the Smørbukk field, in the hanging-wall of the Smørbukk Fault (Fig. 23). These areas of relatively more rapid subsidence exerted a subtle control on the Tilje succession, causing the thickening of the lower part of sequence 2 (T1.1.1 to T2) and the lower part of sequence 3 (T3.2 to T4)

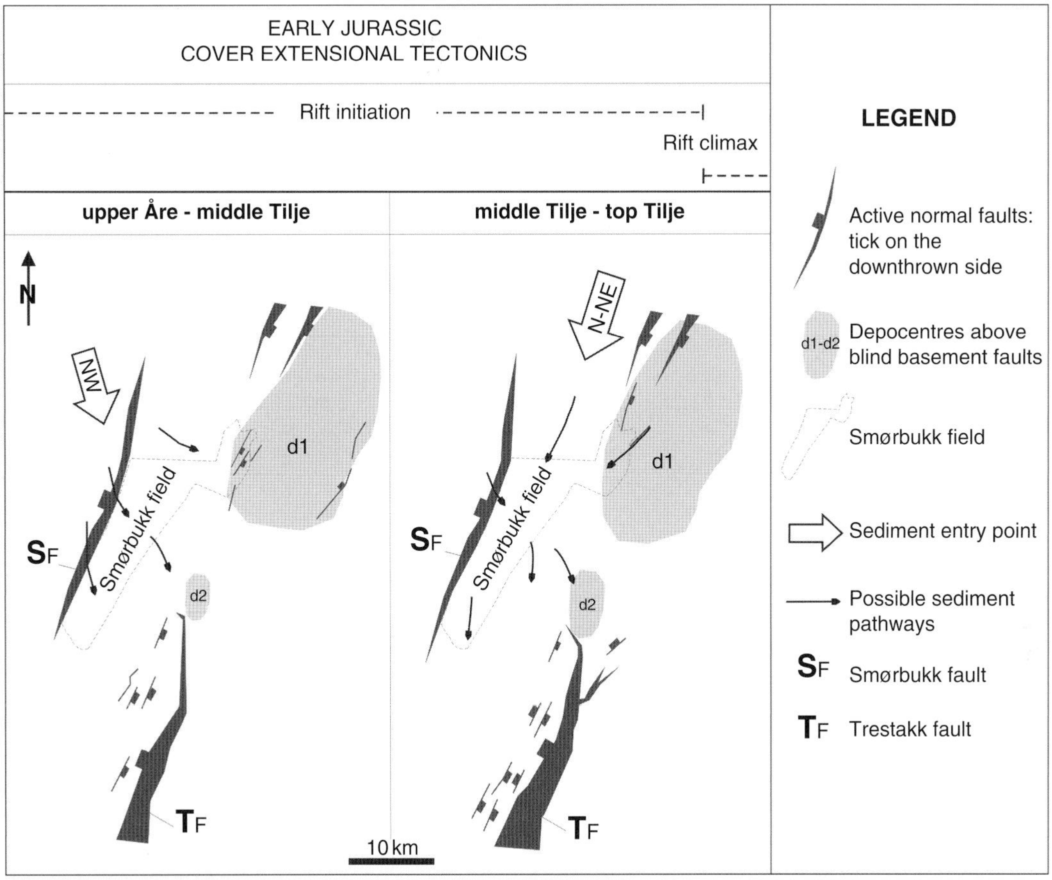

Fig. 23. Schematic map of the Smørbukk field showing the active faults and the inferred sediment-dispersal patterns during deposition of the Åre and Tilje formations. Modified from Marsh *et al.* (2010).

towards the S and E of the field (Fig. 19 cross-sections I and II), by generating areas with greater local accommodation. These localised thicker areas indicate that the generation of accommodation varied temporally throughout the upper Åre and Tilje deposition suggesting that the rate of movement of the Smørbukk and Trestakk faults was not constant. This implies that rift-initiation tectonic pulses or deformation clines (cf. Gawthorpe & Leeder, 2000; Ravnås *et al.*, 2000) were common and produced localised multiple subsidence events in the area of study, mostly intensified during the early and middle Pliensbachian. The presence of clinoforms within reservoir units T1.1.2 and T1.2 (Fig. 19 cross-sections I and II) are associated to the localised thickening and developed into the areas of relatively accelerated subsidence. The height of these clinoforms (up to 15 m thick each) indicates that water depths were approximately 20 m to 25 m, assuming a water depth of 10 m at the topset-foreset break in slope (cf. Hansen & Rasmussen, 2008) and perhaps as much as 40 m if compaction is taken into account (cf. Terwindt & Breusers, 1972; Nordahl, 2004). An apparent acceleration of rifting (Marsh *et al.*, 2010), accompanied by major extensional tectonic activity (cf. Ravnås & Steel, 1998; Gawthorpe & Leeder, 2000), resulted in a rapid regional flooding event and the deposition of the widespread shelf deposits of the Ror and Tofte Formations (Fig. 22D).

The facies distributions presented here are broadly consistent with the interpreted dispersal path models (Fig. 23) of Marsh *et al.* (2010) and indicate an abrupt change of the sediment entry point, involving a shift from a NW entry point in sequence 2 (Fig. 24A) to an entry point in the N-NE in sequence 3 (Fig. 24B). The first of these was oblique to the structural trend at the scale of the study area, whereas the later was broadly parallel to the NE-SW oriented faults and to the grain of any differential subsidence related to their motion. The entry-point switch correlates with the formation of SB3

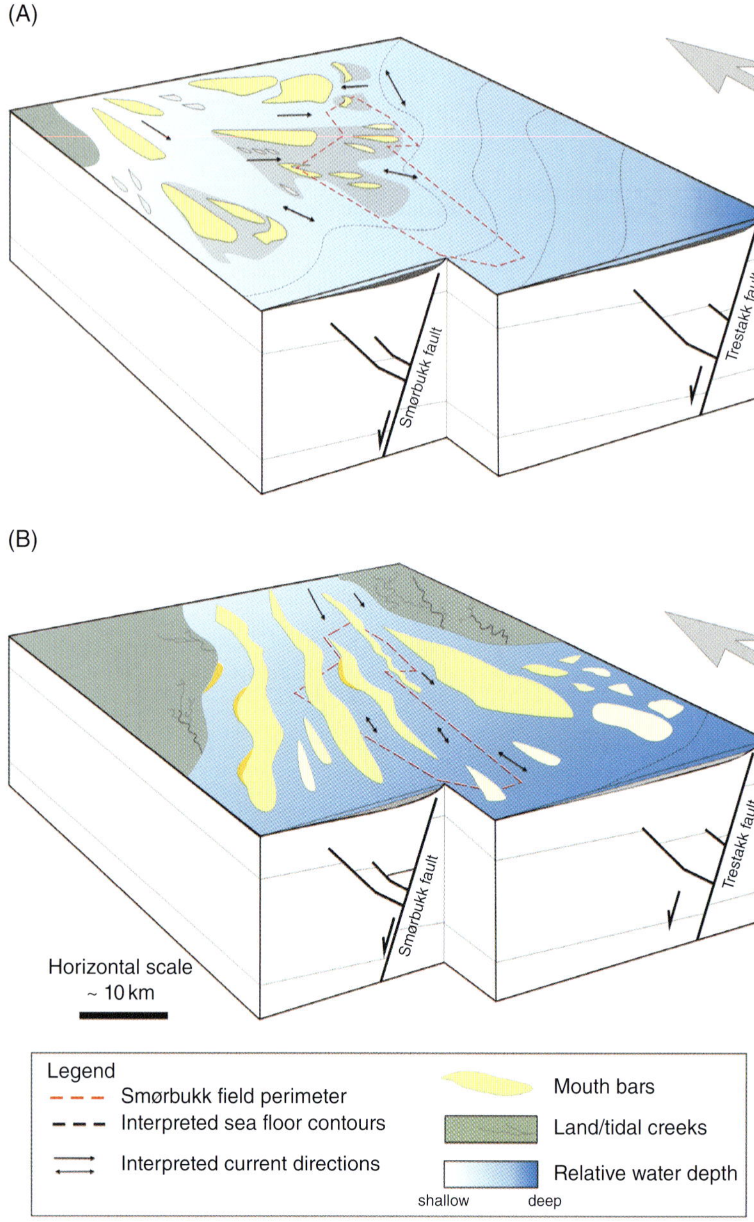

Fig. 24. Schematic block diagrams showing the interpreted palaeogeographic reconstructions for the Tilje Formation in the Smørbukk field (dotted outline), relative to the Smørbukk and Trestakk normal faults, during the lower-middle Tilje (T1.2 to T3.1) and B) the middle-upper Tilje (T3.2 to T4) times.

(Martinius *et al.*, 2001), which coincidentally, occurred during a second-order eustatic sea-level drop in the mid-upper Pliensbachian (Hallam, 1988; Surlyk, 1990; Hallam, 1999). This palaeogeographic change could be caused by two possible processes: 1) river avulsion or 2) the repositioning of the river system by tectonic activity. However, the constant location of the sediment-entry point over a very long period of the early Tilje depositional time (during sequence 2), as indicated by the constant fining trends (Figs 20 to 21) and the absence of compensation-style stratigraphic architectures, suggests that the main fluvial system was probably confined by structural/topographic controls towards the N-NW. Therefore, the abrupt switch during the formation of SB3 implies that a major tectonic movement deflected the river and locked it into a new axially oriented (N-NE to S-SE) course (cf. Bosence, 1998).

CONCLUSIONS

Seven lithofacies and eight subfacies were identified in the upper Åre (A3) to top Tilje (T6) stratigraphic interval in the Smørbukk field, on the basis of their sedimentological attributes. Complex heterolithic facies were classified based on their sandstone/mudstone ratio, bedding style, grain size, sedimentary structures and degree and type of bioturbation. Ten facies associations (FA) were recognised as the product of the co-occurrence of certain facies groups.

The Tilje Fm. (150 m to 220 m thick) is divided in two main overall tabular sequences. Sequence 2 (lower-middle Tilje, T1 to T3.1) is bounded (base and top) by regional hiatal surfaces (SB2 and SB3). Sequence 2 rests erosionally over the lower-middle shoreface deposits of the upper Åre (A3), with only subtle incision towards the south. In general, it shows a N-NW to S-E lobe-shaped proximal-to-distal facies trend, as a result of sediment input from the NW, obliquely across the trend of the Smørbukk fault. The lower part of sequence 2 is composed of tidally influenced, channelised distributary-mouth bars that show a lateral transition into delta-front and prodeltaic deposits. Vertically, these deposits form metres to tens of metres-thick coarsening- and sanding-upward heterolithic parasequences and parasequence sets, which are bounded at their base and top by high-order flooding surfaces, suggesting intermittent periods of eustatic/tectonic sea-level rise. Internally, the sequence shows a complex alternation of more proximal tidal-influenced and more distal wave/storm-influenced delta-front successions. This alternation might be the result of delta-lobe switching during periods of base-level rise and flooding, allowing the wave and storm energy to take over from tidal processes in the delta-front area. Overall, the Tilje Formation shows a dominance of tidal and fluvial influence; however, this study demonstrates that wave and storm energy are also stratigraphically important and should not be ignored. In this context we propose that the Tilje Fm. in the Smørbukk field represents a mixed tide-wave influenced deltaic system, rather than tide-dominated environment, as it was previously interpreted.

Sequence 3 (middle-upper Tilje, T3.2 to T6) rests erosionally (SB3), but with almost no incision, over the mouth-bar and delta-front deposits of T3.1 and is capped by a flooding surface that marks the transition to the shallow-marine deposits of the Ror Fm. The T3.2 to T5.1 interval preserves the most proximal facies of the entire Tilje Fm. It is mainly composed of tidal-fluvial distributary channel and point-bar successions that were deposited by rivers flowing from the N-NE towards the SSW during a relative base-level drop. The upper T5.2 and T6 zones record a flooding event and a major retreat of the tidal-fluvial deltaic system towards the N-NE during the subsequent sea-level rise. The upper Tilje, which is composed mainly of channelised heterolithic distributary-mouth bars and mixed tide-wave delta-front deposits, is here considered to reflect the onset of the regional flooding event that led to the deposition of the Ror Formation in a shallow-shelf setting.

The Tilje Fm. also records seasonal variations in river discharge. High-discharge periods are characterised by weakly bioturbated fine-grained to coarse-grained sandstones and thick fluid-mud layers within the channelised distributary-mouth bar successions. These deposits alternate with moderately to intensely bioturbated, very-fine to fine-grained heterolithic deposits suggestive of periods of low discharge when the current speeds and suspended-sediment concentrations were less and the salinity was higher. This pattern reflects the strong seasonal differences in rainfall that characterised the warm Jurassic climate, a factor that probably exerted major influence on the weathering of sediment sources and sediment supply. Seasonal variations were more prominent during the lower-middle Tilje (sequence 2) depositional times, suggesting that there was a secular change in seasonality.

The integration of models of timing of fault activity and our stratigraphic interpretations indicates that the Tilje formed during an early syn-rift stage within a NE-SW oriented half-graben formed by down-to-the-west fault-block rotation. Short-term tectonic pulses are responsible for causing or enhancing the two major sequence boundaries within the succession (SB2 and SB3), both of which mark the onset of accelerated differential subsidence that created localised, shallow depocentres immediately to the west of the blind Trestakk and Smørbukk normal faults. These depocentres are reflected in localised thickening of reservoir units and had water depths that were up to 25 m to 40 m deeper than the surrounding areas. The first pulse (SB2) caused an abrupt change from wave-dominated to tide-dominated sedimentation, perhaps because of creation of a funnel-shaped fault basin that enhanced tidal activity while limiting the influence of waves, whereas the second pulse (SB3) is apparently responsible for an abrupt change in the sediment distribution pathways.

ACKNOWLEDGMENTS

Research funding to Dr R.W. Dalrymple was granted by Statoil ASA under the regional palaeoenvironmental Lower Jurassic upper Åre and Tilje project. The owners of the Statoil operated Heidrun and Åsgard Units–Statoil ASA, ConocoPhillips Scandinavia, Petoro, Eni Norge AS, Total E&P Norge and ExxonMobil are acknowledged for permissions to undertake sampling of relevant Tilje core material. Special thanks are extended to Allard Martinius for initiating this project, technical support and discussions during its execution. We thank Arve Næss and Morten Krogh for providing access to core material, well-log data and relevant Statoil ASA internal reports and Bjørn Terje Oftedal for discussions and core review during Tilje Formation Workshops in 2007 and 2009. We thank the Statoil-Heidrun operational unit for providing information and technical support.

REFERENCES

Ashley, G.M., Southard, J.B. and **Boothroyd, J.C.** (1982) Deposition of climbing-ripple beds: a flume simulation. *Sedimentology*, **29**, 67–79.

Baker, E.K., Harris, P.T., Keene, J.B. and **Short, S.A.** (1995) Patterns of sedimentation in the macrotidal Fly River delta, Papua New Guinea. In: *Tidal Signatures in Modern and Ancient Sediments* (Eds B.W. Flemming and A. Bartholoma), *Int. Assoc. Sedimentol. Spec. Publ.*, **24**, 193–211.

Bann, K.L., Tye, S.C., MacEachern, J.A., Fielding, C.R. and **Jones, B.G.** (2008) Ichnological and sedimentological signatures of mixed wave- and storm-dominated deltaic deposits: examples from the Early Permian Sydney basin: Australia. In: *Recent Advances in Models of Siliciclastic Shallow-Marine Stratigraphy* (Eds G.J. Hampson, R.J. Steel, P.P. Burges and R.W. Dalrymple), *SEPM Spec. Publ.*, **90**, 293–332.

Bhattacharya, J.P. (2010) Deltas. In: *Facies Models 4* (Eds N.P. James and R.W. Dalrymple), St. John's, Geol. Assoc. Can., 233–264.

Bhattacharya, J.P. and **Giosan, L.** (2003) Wave-influenced deltas: geomorphological implications for facies reconstruction. *Sedimentology*, **50**, 187–210.

Bhattacharya, J.P. and **MacEachern, J.A.** (2009) Hyperpycnal rivers and prodeltaic shelves in the Cretaceous seaway of North America. *J. Sed. Res.*, **79**, 184–209.

Bosence, D.W.J. (1998) Stratigraphic and sedimentological models of rift basins. In: *Sedimentation and Tectonics of Rift Basins: Red Sea-Gulf of Aden* (Eds B.H. Purser and D.W.J. Bosence), Chapman & Hall, London, 9–25.

Boyd, R. (2010) Transgressive wave-dominated coasts. In: *Facies Models 4* (Eds N.P. James and R.W. Dalrymple), St. John's, *Geol. Assoc. Can.*, 265–294.

Brekke, H., Sjulstad, H.I., Magnus, C. and **Williams, R.W.** (2001) Sedimentary environments offshore Norway – an overview. In: *Sedimentary environments offshore Norway – Palaeozoic to recent* (Eds O.J. Martinsen and T. Dreyer), *Nor. Petrol. Soc. Spec. Publ.*, **10**, 7–37.

Brown, S. and **Richards, P.C.** (1989) Facies and development of the Middle Jurassic Brent Delta near the northern limit of its progradation, UK North Sea. In: *Deltas: Sites and Traps for Fossil Fuel* (Eds M.K. Whateley and K.T. Pickering), *Geol. Soc. London Spec. Publ.*, **41**, 253–267.

Catuneanu, O. (2006) *Principles of Sequence Stratigraphy*. Elsevier, Amsterdam.

Collinson, J.D. and **Thompson, D.B.** (1982) *Sedimentary Structures*, 2nd edn, Harper Collins Academic, NY, 208 p.

Corfield, S. and **Sharp, I.R.** (2000) Structural style and stratigraphic architecture of fault propagation folding in extensional settings: a seismic example from the Smørbukk area, Halten Terrace, Mid-Norway. *Basin Res.*, **12**, 329–341.

Cowan, C.A. and **James, N.P.** (1992) Diastasis cracks: mechanically generated syneresis-like cracks in the Upper Cambrian shallow water oolite and ribbon carbonates. *Sedimentology*, **39**, 1101–1118.

Dalland, A., Worsley, D. and **Ofstad, K.** (1988) A lithostratigraphic scheme for the Mesozoic and Cenozoic succession offshore mid-and northern Norway, *Norw. Petrol. Direct. Bull.*, **4**, 87 pp.

Dalrymple, R.W., Zaitlin, B.A. and **Boyd, R.** (1992) Estuarine facies models: conceptual basis and stratigraphic implications. *J. Sed. Petrol.*, **62**, 1130–1146.

Dalrymple, R.W., Baker, E.W., Harris, P.T. and **Hughes, M.G.** (2003) Sediment and stratigraphy of a tide-dominated, foreland basin delta (Fly River, Papua, New Guinea). In: *Tropical Deltas of Southeast Asia–Sedimentology, Stratigraphy and Petroleum Geology* (Eds H. Sidi, D. Nummedal, P. Imbert, H. Darman and H.W. Posamentier), *SEPM Spec. Publ.*, **76**, 147–173.

Dalrymple, R.W. and **Choi, K.** (2007) Morphologic and facies trends through the fluvial–marine transition in tide-dominated depositional systems: a schematic framework for environmental and sequence-stratigraphic interpretation. *Earth-Sci. Rev.*, **81**, 135–174.

Dalrymple, R.W. (2010) Introduction to siliciclastic facies models. In: *Facies Models 4* (Eds N.P. James and R.W. Dalrymple), St. John's, Geol. Assoc. Can., 59–72.

Dalrymple, R.W., Mackay, D.A., Ichaso, A.A. and **Choi, K.S.** (2011) Processes, morphodynamics and facies of tide-dominated estuaries. In: *Principles of Tidal Sedimentology* (Eds R.A. Davis, Jr. and R.W. Dalrymple), New York, Springer, 79–107.

Day, Jr., J.W., Pont, D., Hensel, P.F. and **Ibanez, C.** (1995) Impacts of sea-level rise on deltas in the Gulf of Mexico and the Mediterranean: the importance of pulsing events to sustainability. *Estuaries*, **18**, 636–647.

Doré, A.G. (1991) The structural foundation and evolution of Mesozoic seaways between Europe and the Artic. *Palaeogeogr. Palaeoclimatol. Palaeocol.*, **87**, 441–492.

Dreyer, T. (1992) Significance of tidal cyclicity for modelling of reservoir heterogeneities in the lower Jurassic Tilje Formation, mid-Norwegian shelf. *Nor. Geol. Tidsskr.*, **72**, 159–170.

Duke, W.L., Arnott, R.W.C. and **Cheel, R.J.** (1991) Shelf sandstones and hummocky cross-stratification: New insights on a storm debate. *Geology*, **19**, 625–628.

Dumas, S., Arnott, R.W.C. and **Southard, J.B.** (2005) Experiments on oscilatory-flow and combined-flow bed forms: implications for interpreting parts of the shallow-marine sedimentary record. *J. Sed. Res.*, **75**, 501–513.

Ehrenberg, S.N., Gjerstad, H.M. and **Hadler-Jacobsen, F.** (1992) Smørbukk field: A gas condensate fault trap in the Haltenbanken Province, offshore mid-Norway. In: *Giant Oil and Gas Fields of the Decade 1978–1988* (Ed. M.T. Halbouty), *AAPG Mem.*, **54**, 323–348.

Elliott, T. (1974) Interdistributary bay sequences and their genesis. *Sedimentology*, **21**, 611–622.

Frings, R.M. (2008) Downstream fining in large sand-bed rivers. *Earth-Sci. Rev.*, **87**, 39–60.

Gawthorpe, R.L. and **Leeder, M.R.** (2000) Tectono-sediemntary evolution of active extensional basins. *Basin Res.*, **12**, 195–218.

Gradstein, F.M., Agterberg, F.P., Ogg, J.G., Hardenbol, J., van **Veen, P., Thierry, J.** and **Huang, Z.H.** (1995) A Triassic, Jurassic and Cretaceous time scale. In: *Geochronology, Time Scales and Global Stratigraphy Correlation* (Eds W.A. Berggren, D.V. Kent, M.P. Aubry and J. Hardenbol), *SEPM Spec. Publ.*, **54**, 95–126.

Hallam, A. (1988) A re-evaluation of Jurassic eustasy in the light of new data and the revised Exxon curve. In: *Sea-level Changes: an Integrated Approach* (Eds C.K. Wilgus, B.S. Hastings, C.C. St. C. Kendall, H. Posamentier, C.A. Ross and J. Van Wagoner), *SEPM Spec. Publ.*, **42**, 261–273.

Hallam, A. (1994) Jurassic climates as inferred from the sedimentary and fossil record. In: *Palaeoclimates and Their Modelling: With Special Reference to the Mesozoic Era* (Eds J.R.L. Allen, B.J. Haskins, B.W. Sellwood, R.A. Spicer and P.J. Valdes), 79–88, Chapman & Hall, London.

Hallam, A. (1999) Evidence of sea-level fall in sequence stratigraphy: Examples from the Jurassic. *Geology*, **27**, 343–346.

Hansen, J.P.V. and **Rasmussen, E.S.** (2008) Structural, sedimentologic and sea-level controls on sand distribution in a steep-clinoform asymmetric wave-influenced delta: Miocene Billund sand, Eastern Danish North Sea and Jylland. *J. Sed. Res.*, **78**, 130–146.

Haq, B.U., Handenbol, J. and **Vail, P.R.** (1987) Chronology of fluctuating sea level since the Triassic. *Science*, **235**, 1156–1167.

Hill, P.R., Meulé, S. and **Longuépée, H.** (2003) Combined flow processes and sedimentary structures on the shoreface of the wave-dominated Grande-Rivière-de-la-Baleine-delta. *J. Sed. Res.*, **73**, 217–226.

Howell, J.A. and **Flint, S.S.** (1996) A model for high resolution sequence stratigraphy within extensional basins. In: *High Resolution Sequence Stratigraphy: Innovations and Applications* (Eds J.A. Howell and J.F. Aitken), *Geol. Soc. Lond. Spec. Publ.*, **104**, 129–137.

Hunter, R.E. and **Clifton, H.E.** (1982) Cyclic deposits and hummocky cross-stratification of probable storm origin in Upper Cretaceous rocks of the Cape Sebastian area, southwestern Oregon. *J. Sed. Petrol.*, **52**, 0127–0143.

Ichaso, A.A. (2012) *Spatial and Temporal Controls on the Development of Heterolithic, Lower Jurassic Tidal Deposits (Åre and Tilje Formations), Haltenbanken Area, Offshore Norway.* Queen's University, Unpubl. Ph.D. thesis, 252 pp.

Ichaso, A.A. and **Dalrymple, R.W.** (2009) Tide-and wave-generated fluid mud deposits in the Tilje Formation (Jurassic), offshore Norway. *Geology*, **37**, 539–542.

Ingram, R.L. (1954) Terminology for the thickness of stratification and parting units in sedimentary rocks. *Geol. Soc. Am. Bull.*, **65**, 937–938.

Klefstad L., Kvarsvik, S., Ringås, J.E., Stene, J.J. and **Sundsby, O.** (2005) Characterization of deeply buried heterolithic tidal reservoirs in the Smørbukk Field using inverted post-stack seismic acoustic impedance. *Petrol. Geosci.*, **11**, 47–56.

Klein, G. de V. and **Marsaglia, K.M.** (1987) Hummocky cross-stratification, tropical hurricanes and intense winter storms. *Sedimentology*, **34**, 333–359.

Lowe, D.R. (1979) Sediment gravity flows: their classification and some problems of application to natural flows and deposits. In: *Geology of continental slope* (Eds L.J. Doyle and J.H. Pilkey), *SEPM Spec. Publ.*, **27**, 75–82.

MacEachern, J.A. and **Bann, K.L.** (2008) The role of Ichnology in refining shallow marine facies models. In: *Recent Advances in Models of Siliciclastic Shallow-Marine Stratigraphy* (Eds G.J. Hampson, R.J. Steel, P.P. Burges and R.W. Dalrymple), *SEPM Spec. Publ.*, **90**, 73–116.

MacEachern, J.A., Bann, K.L., Bhattacharya, J.P. and **Howell, C.D.,** Jr. (2005) Ichnology of deltas, organism responses to the dynamic interplay of rivers, waves, storms and tides. In: *River Deltas; Concepts, Models and Examples* (Eds L. Giosan and J.P. Bhattacharya), *SEPM Spec. Publ.*, **83**, 49–85.

MacEachern, J.A., Bann, K.L., Gingras, M.K. and **Pemberton, S.G.** (2006) The ichnofacies paradigm: high-resolution palaeoenvionmental interpretation of the rock record. In: *Applied Ichnology* (Eds J.A. MacEachern, K.L. Bann, M.K. Gingras and S.G. Pemberton), *SEPM Sht. Crs. Notes,* **52**, 26–63.

Makaske, B. and **Weerts, H.J.T.** (2005) Muddy lateral accretion and low stream power in a sub-recent confined channel belt, Rhine-Meuse delta, central Netherlands. *Sedimentology*, **52**, 651–668.

Mángano, M.G., Buatois, L.A., Wu, X., Sun, J. and **Zhang, G.** (1994) Sedimentary facies, depositional processes and climatic controls in a Triassic lake, Tanzhuang Formation, western Henan Province, China. *J. Palaeontol.*, **11**, 41–65.

Marsh, N., Imber, J., Holdsworth, R.E., Brockbank, P. and **Ringrose, P.** (2010) The structural evolution of the Halten Terrace, offshore Mid-Norway: extensional fault growth and strain localisation in a multi-layer brittle-ductile system. *Basin Res.*, **22**, 195–214.

Martinius, A.W., Kaas, I., Næss, A., Helgesen, G., Kjærefjord, J.M. and **Leith, D.A.** (2001) Sedimentology of the heterolithic and tide-dominated Tilje Formation (Early Jurassic, Halten Terrace, offshore mid-Norway). In: *Sedimentary Environments Offshore Norway – Palaeozoic to Recent* (Eds O.J. Martinsen and T. Dreyer), *Nor. Petrol. Soc. Spec. Publ.*, **10**, 103–144.

Martinius, A.W., Ringrose, P.S., Brostrøm, C., Elfenbein, C., Næss, A. and **Ringås, J.E.** (2005) Reservoir challenges of heterolithic tidal sandstone reservoirs in the Halten Terrace, mid-Norway. *Petrol. Geosci.*, **11**, 3–16.

Mulder, T., Syvitski, J.P.M., Migeon, S., Faugères, J.-C. and **Savoye, B.** (2003) Marine hyperpycnal flows: initiation, behaviour and related deposits. A review. *Mar. Petrol. Geol.*, **20**, 861–882.

Nemec, W. and **Steel, R.J.** (1984) Alluvial and coastal conglomerates: Their significant features and some comments on gravelly mass-flow deposits. In: *Sedimentology of gravels and conglomerates* (Eds E.H. Koster and R.J. Steel), *Can. Soc. Petrol. Geol. Mem.*, **10**, 1–31.

Nordahl, K. (2004) A petrophysical evaluation of tidal heterolithic deposits. Doctoral Thesis, 432 p. NTNU, Trondheim.

Olariu, C. and **Bhattacharya, J.P.** (2006) Terminal distributary channels and delta front architecture of river-dominated delta systems. *J. Sed. Res.*, **76**, 212–233.

Parker, G. (1991) Selective sorting and abrasion of river gravel. I: Theory. *J. Hyd. Eng.*, **117**, 131–149.

Pemberton, S.G., Spila, M., Pulham, A.J., Saunders, T., MacEachern, J.A., Robbins, D. and **Sinclair, I.K.** (2001) Ichnology & Sedimentology of shallow to marginal marine systems: Ben Nevis & Avalon reservoirs, Jeane d'Arc Basin. *Geol. Assoc. Can. Sht. Crs. Notes*, **15**, 343 pp.

Ravnås, R. and **Steel, R.J.** (1998) Architechture of marine riftbasin successions. *AAPG Bull.*, **82**, 110–146.

Ravnås, R., Nøttvedt, A., Steel, R.J. and **Windelstad, J.** (2000) Syn-rift sedimentary architechtures in the Northern North Sea. In: *Dynamics of the Norwegian Margin* (Eds A. Nøttvedt, B.T. Larsen, S. Olaussen, B. Torudbakken, J. Skogseid, R.H. Gabrielsen, H. Brekke and O. Birkeland), *Geol. Soc. London Spec. Publ.*, **167**, 133–177.

Reineck, H.E. and **Wunderlich, F.** (1968) Classification and origin of flaser and lenticular bedding. *Sedimentology*, **11**, 99–104.

Snedden, J.W. and **Dalrymple, R.W.** (1998) Modern shelf sand bodies: an integrated hydrodynamic and eolutionary model. In: *Isolated Shallow Marine Sandbodies: Sequence Stratigraphic Analysis and Sedimentologic Interpretation* (Eds K. Bergman and J.W. Snedden, *SEPM Spec. Publ.*, **64**, 13–28.

Straub, K.M., Paola, C., Mohrig, D., Wolinsky, M.A. and **George, T.** (2009) Compensational stacking of channelized sedimentary deposits. *J. Sed. Res.*, **79**, 673–688.

Surlyk, F. (1990) A Jurassic sea-level curve for East Greenland. *Palaeogeogr. Plaeoclimatol. Palaeoecol.*, **78**, 71–85.

Svela, K.E. (2001) Sedimentary facies in the fluvial-dominated Åre Formation as seen in the Åre 1 member in the Heidrun Field. In: *Sedimentary environments offshore Norway – Palaeozoic to recent* (Eds O.J. Martinsen and T. Dreyer), *NPF Spec. Publ.*, **10**, 87–102.

Tanavsuu-Milkeviciene, K. and **Plink-Björklund, P.** (2009) Recognizing tide-dominated versus tide-influenced deltas:Middle Devonian strata of the Baltic Basin. *J. Sed. Res.*, **79**, 887–905.

Terwindt, J.H.J. and **Breusers, H.N.C.** (1972) Experiments on the origin of flaser, lenticular and sand-clay alternating bedding. *Sedimentology*, **19**, 85–98.

Thomas, R.G., Smith, D.G., Wood, J.M., Visser, J., Calverley-Range, E.A. and **Koster, E.H.** (1987) Inclined heterolithic stratification: terminology, description, interpretation and significance. *Sed. Geol.*, **53**, 123–179.

Traykovski, P., Geyer, W.R., Irish, J.D. and **Lynch, J.F.** (2000) The role of wave-induced density-driven fluid mud flows for cross-shelf transport on the Eel River continental shelf. *Cont. Shelf Res.*, **20**, 2113–2140.

Turner, B.R., Stanistreet, I.G. and **Whateley, M.K.G.** (1981) Trace fossils and palaeonvironments in the Ecca Group of the Nongoma Graben, northern Zululand, South Africa. *Palaeogeogr. Palaeoclimatol. Palaeocol.*, **36**, 113–123.

Van den Berg, J.H., Boersma, J.R. and **Van Gelder, A.** (2007) Diagnostic sedimentary structures of the fluvial-tidal transition zone – Evidence from deposits of the Rhine and Meuse. *Geol. Mijnbouw*, **86**, 287–306.

Visser, M.J. (1980) Neap-spring cycles reflected in Holocene subtidal large-scale bedform deposits: a preliminary note. *Geology*, **8**, 543–546.

Walker, R.W. and **Plint, A.G.** (1992) Wave- and storm-dominated shallow marine systems. In: *Facies Models Response to Sea Level Change* (Eds R.G. Walker and N.P. James), St. John's, *Geol. Assoc. Can.*, 219–238.

Withjack, M.O., Schlische, R.W. and **Olsen, P.E.** (2002) Rift-basin structure and its influence on sediment systems. In: *Sedimentation in Continental Rifts* (Eds R.W. Renault and G.M. Ashley), *SEPM Spec. Publ.*, **73**, 57–81.

Xu, J. (2002) Implication of relationships among suspended sediment size, water discharge and suspended sediment concentration: the Yellow River basin, China. *Catena*, **49**, 289–307.

Yang, B.C., Dalrymple, R.W. and **Chun, S.S.** (2005) Sedimentation on a wave-dominated, open-coast tidal flat, south western Korea: summer tidal flat – winter shoreface. *Sedimentology*, **52**, 235–252.

Yang, B.C., Dalrymple, R.W. and **Chun, S.S.** (2006) The significance of hummocky cross-stratification (HCS) wavelenghts: evidence from an open-coast tidal flat, South Korea. *J. Sed. Res.*, **76**, 2–8.

Int. Assoc. Sedimentol. Spec. Publ. (2014) **46**, 389–416

Tectonic influence on the Jurassic sedimentary architecture in the northern North Sea with focus on the Brent Group

ATLE FOLKESTAD*, TORE ODINSEN*, HAAKON FOSSEN[†,a] and MARTIN A. PEARCE[‡]

**Statoil, Bergen, Box 7200, Bergen 5020, Norway (E-mail: atlef@statoil.com)*
[†] Department of Earth Science, University of Bergen, Allégaten 41, 5007 Bergen, Norway
[a] Natural History Collections, University Museum of Bergen, PO Box 7800, 5020 Bergen, Norway
[‡] Statoil Gulf Services LLC, 2103 CityWest Boulevard, Houston, Texas, 77042, USA

ABSTRACT

It is known from previous studies that the Middle Jurassic succession in the northern North Sea has been influenced and controlled by syn-depositional fault activity. In this study from the Gullfaks-Kvitebjørn area, we build upon that knowledge with seismic cross-sections, well-correlations and cored intervals to evaluate features that can be linked to Middle–Late Jurassic rifting in the northern North Sea. A regional east–west transect shows an overall wedge-shaped Jurassic succession in the strike-section between two long-lived Permo–Triassic faults, with a marked asymmetric thickness distribution from the Ness Formation and upward. In a local section across the Kvitebjørn Field the same pattern is identified, but here the thickness differences are more pronounced. We suggest that the sedimentological response to this Middle Jurassic tectonic activity is reflected by the formation of local depocentres with stacked tidal dunes, differences in lithological characteristics along strike within stratal units and facies variability along an irregular coast with enhancement of tidal currents in the funnel-shaped hangingwall areas of rotated fault-blocks. These integrated data suggest that the Middle–Late Jurassic rift phase started in the Early Bajocian (basal Ness Formation) within the Gullfaks to Kvitebjørn transect and with flexing at the crest of the Permo–Triassic mega-block leading to a complex stratigraphic development of the Brent Group at that location.

Keywords: Brent Group, syn-sedimentary tectonics, Middle Jurassic, early to main rift-phase, northern North Sea.

INTRODUCTION

The Jurassic northern North Sea (Fig. 1) is one of the best-studied extensional rift basins in the world, due to extensive hydrocarbon exploration and production, especially within the economically important Middle Jurassic Brent Group. The deposition of clastic Jurassic formations in this basin is directly or indirectly influenced by tectonic activity. Accordingly, it is important to understand how this activity affected the depositional environment. The interaction between tectonics and deposition of the Jurassic succession of the northern North Sea has been discussed in a series of studies, albeit with different focus and interpretations (e.g. Helland-Hansen *et al.*, 1992; Steel, 1993; Johannessen *et al.*, 1995; Færseth, 1996; Fjellanger *et al.*, 1996; Ravnås *et al.*, 2000; Davies *et al.*, 2000; Hampson *et al.*, 2004).

The North Sea Basin is built upon a structural framework of faults and shear zones formed during the Caledonian orogeny and the subsequent extensional collapse of this orogenic belt in the Devonian (e.g. Fossen *et al.*, 2008). The basin was affected by two post-orogenic lithospheric rift episodes, one during the Late Permian–Early Triassic (Beach *et al.*, 1987; Gabrielsen *et al.*, 1990; Færseth *et al.*, 1995a) and one during the

From Depositional Systems to Sedimentary Successions on the Norwegian Continental Margin, First Edition.
Edited by A. W. Martinius, R. Ravnås, J. A. Howell, R. J. Steel, and J. P. Wonham.

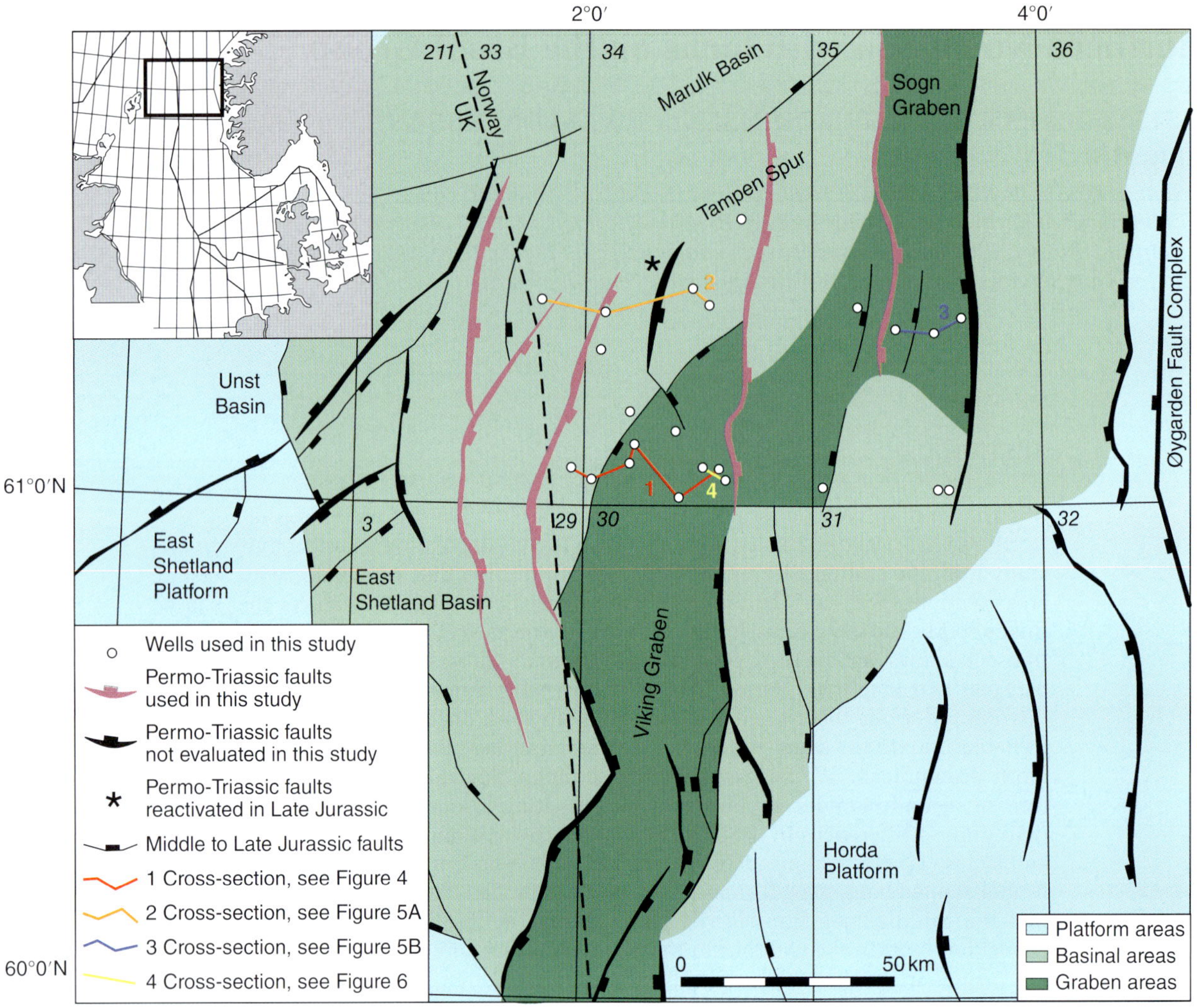

Fig. 1. Main structural elements of the northern North Sea with indication of cross-sections and wells used. Modified from Færseth (1996).

Middle–Late Jurassic (Leeder, 1983; Badley *et al.*, 1988; Rattey & Hayward 1993; Færseth *et al.*, 1997). The details regarding the timing, significance and lateral extent of the Permo–Early Triassic stretching has been a matter of debate for a long time (Giltner 1987; Gabrielsen *et al.*, 1990, Færseth *et al.*, 1995a; Roberts *et al.*, 1995; Færseth, 1996). Large tilted fault-blocks bounded by master-faults with throws of the order of several kilometres formed in a 150 km wide, north–south oriented basin in the Late Palaeozoic. During the thermal subsidence that followed the rifting, faulting occurred on both margins (Steel & Ryseth, 1990) due to the interaction between lateral variations in thermal subsidence, sediment loading, compaction and flexure (Badley *et al.*, 1988).

Precise dating of the initiation of the Jurassic rifting has also been the subject of debate (e.g. Gabrielsen *et al.*, 1990). Evidence of increased subsidence and early fault-block rotation in Bajocian–Bathonian times has led many workers to conclude that rifting was initiated at this time (Badley *et al.*, 1988; Helland-Hansen *et al.*, 1992; Ravnås *et al.*, 1997). Permo–Triassic master-faults were reactivated during the Jurassic rifting and, together with newly-formed Jurassic master-faults, influenced the general structural pattern of the entire basin, promoting segmentation and controlling the subsidence pattern in some areas (Yielding *et al.*, 1992; Færseth, 1996; Odinsen *et al.*, 2000a; Hampson *et al.*, 2004).

The Brent Group (Fig. 2) is the main reservoir unit in the northern North Sea and its development

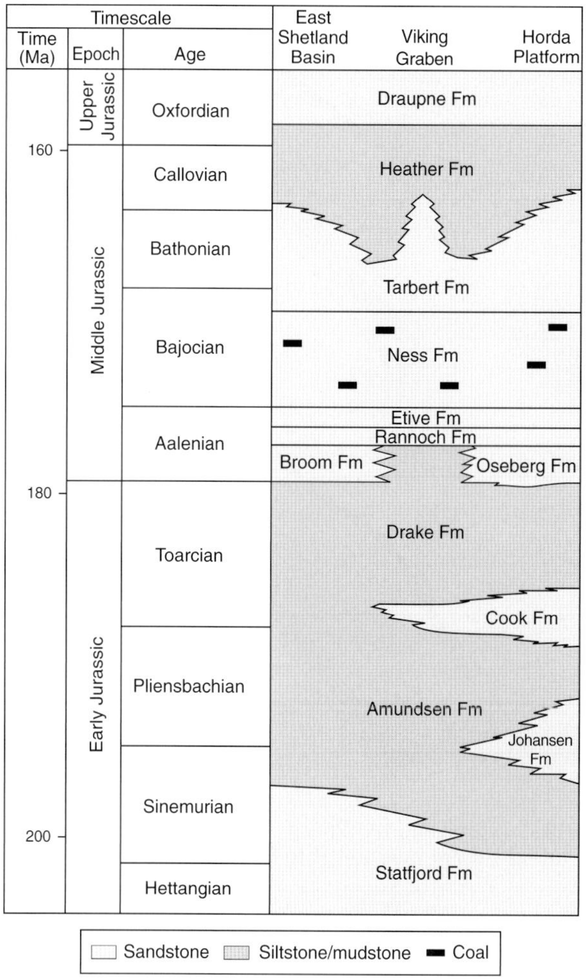

Fig. 2. Stratigraphic column of the Jurassic northern North Sea. Modified from Husmo *et al.* (2003).

is strongly linked to the North Sea dome uplift and to the late pre-rift and early syn-rift tectonic subsidence in the northern North Sea. The 'Brent delta' advanced northward as a consequence of the uplift and erosion of the North Sea dome. It reached as far north as about 62° N, where its advancement was halted by a combination of three factors: (1) increased fault activity and subsidence related to the onset of the Jurassic rift phase (Ravnås *et al.*, 1997); (2) a relative sea-level rise; and (3) exhausted sediment supply due to an over-extended delta front (Helland-Hansen *et al.*, 1992). Several studies have speculated on the exact timing and duration of the Jurassic phase of rifting. Some authors (Jennette & Riley, 1996; Hampson *et al.*, 2004) assigned the rift-phase to the Late Jurassic, which renders the Middle Jurassic Brent Group entirely pre-rift (or post-rift with respect to the preceding Permo–Triassic phase). Others have suggested that rifting started during deposition of the upper Brent Group (Tarbert Formation of Late Bajocian age; Johannessen *et al.*, 1995; Løseth *et al.*, 2009; Davies *et al.*, 2000). A few studies, including Helland-Hansen *et al.* (1992), Fjellanger *et al.* (1996); Færseth (1996) and Ravnås *et al.* (1997) have suggested that syn-rift sedimentation started within the Bajocian Ness Formation (upper) (Fig. 2). Fält *et al.* (1989) pointed to the thick delta plain succession of the Ness Formation as a possible indication of syn-sedimentary fault activity.

As stated by Davies *et al.* (2000), the early phase of Jurassic rifting is poorly understood due to subtle rift-initiation indicators. Importantly, fault activity involved in both the early and main stages of the Jurassic rift phase would have had a pronounced impact on the sedimentary infill style, drainage pattern, facies distribution and shoreline complexity (e.g. Gawthorpe & Leeder, 2000). In order to investigate the initiation of the rifting and the sedimentary response and its potential impact on Brent Group stratigraphy, a framework of generic pre-rift to syn-rift models is presented below.

Conceptual pre-rift to syn-rift models

Nøttvedt *et al.* (1995) emphasised that rift systems are generally described by a three-stage model, where active crustal stretching and faulting (the syn-rift stage) is preceded by a proto-rift (pre-rift; Fig. 3) and followed by a post-rift stage. The proto-rift stage is characterised by gentle basin flexure and minor vertical movements along pre-existing faults (Gabrielsen *et al.*, 1990). The post-rift stage is characterised by sediment infilling of the basin topography inherited by the active stretching stage and where minor fault movements may occur along some of the Jurassic master-faults (White & McKenzie, 1988; Nøttvedt *et al.*, 1995). The rift stages are described below in more detail.

Generic pre-rift to syn-rift structural models

The Jurassic active stretching evolution of the northern North Sea has generally been described as a three-stage structural model, as summarised below (e.g. Nøttvedt *et al.*, 1995; Færseth *et al.*, 1995a; Ravnås *et al.*, 2000):

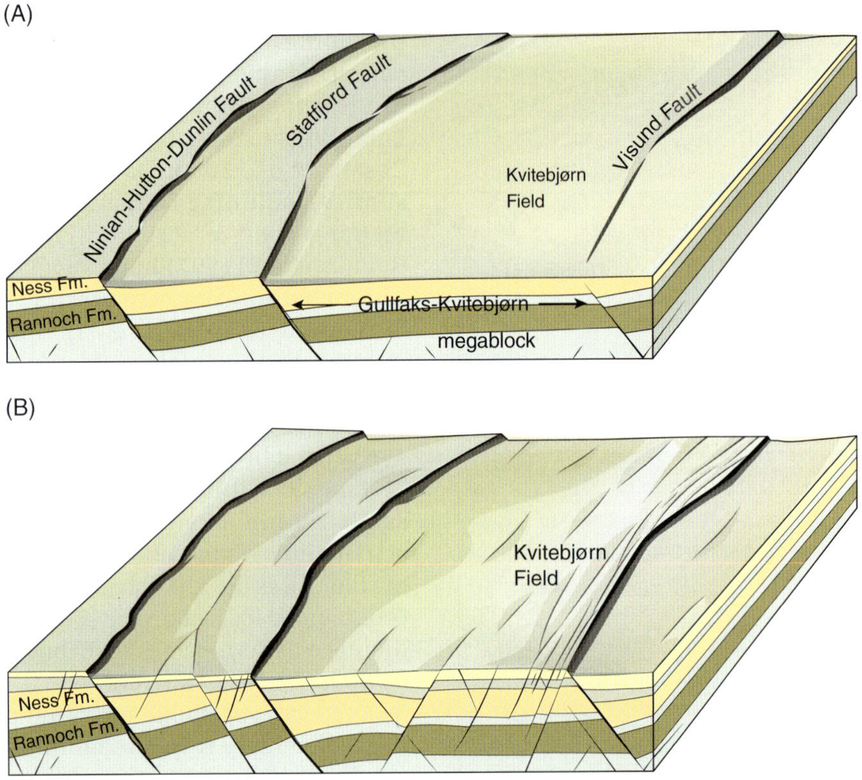

Fig. 3. Fault population and fault linkage trough time during rift development. (A) Pre-rift with Permo–Triassic faults indicated. (B) Initial rift-stage. The isolated small faults of the initial rift are interpreted to be responsible for the local depocentres recorded within the base Ness sandstone.

- Early (initial) rift stage:
  - Initial rift stage is characterised by scattered normal fault population and development of scattered local depressions;
  - Fault propagation with initial fault linkage and early fault death of some of these faults;
  - Initial footwall flexing;
  - Initial fault-block tilting;
- Main rift stage:
  - Transition to rift-climax is correlated with a sharp increase in the rate of basin-wide subsidence early in the rift event (e.g. Steckler *et al.*, 1988). Gupta *et al.* (1998) suggested that the transition from rift-initiation to rift-climax occurs as fault activity became localised onto linked arrays (Fig. 3). With a decline in the number of active faults, rates of fault displacement increase; hence, the rate of tectonic subsidence increases;
  - Increasing degree of fault linkage and fault death;
  - Escalating rotation of fault blocks and potential slumping at footwall fault crests;
  - Episodic fault movement leading to variations in tilt rate and accommodation space;
  - Footwall flexing;
- Late rift (or transition to post-rift) stage:
  - Reduced tilt-rates;
  - Submerged fault-block crests;
  - Initial rotation toward the rift axis of the basin.

Generic pre-rift to syn-rift infill models

To identify the different stages within the transition from pre-rift to syn-rift in a sedimentary basin, a set of sedimentary signatures can be recognised within each stage, as outlined by Yielding *et al.* (1992); Prosser (1993); Nøttvedt *et al.* (1995); Ravnås & Steel (1998) Gawthorpe & Leeder (2000); Sharp *et al.* (2000) and Nøttvedt *et al.* (2000):

- Pre-rift stage characterised by tabular/evenly thick stratal units within each fault-block. The thicknesses of these units may vary from block to block due to differential subsidence.
- Proto-rift stage characterised by near-tabular stratal units due to minor fault movements of pre-existing faults.
- Early rift stage showing the following characteristics:
  - Early-stage scattered local depocentres;
  - Asymmetrical stratal units, typically shaped as wedges (downflank) as a response to the fault-block rotation;
  - Potential erosion, starvation or low deposition-rates at footwall crests, due to fault-block rotation and isostasy-driven footwall uplift;
  - At the coast, the hangingwall areas show a high palaeo-shoreline trajectory with an aggradational style, whereas the footwall areas have a low palaeo-shoreline trajectory that reaches farther basinward;
  - Facies and lithological segregation within rotating fault-blocks. The hangingwall area is prone to show transgressive strata, whereas footwall areas are dominated by progradational events;
  - Occurrence of isolated fluvial channels in the hangingwall area due to axial drainage being steered into the subsiding areas;
  - Development of irregular coastline morphology in terms of spit and embayments, due to contrasting/asymmetric subsidence rates along strike as the shoreline of the footwall areas extends further into the basin compared to the shoreline at the hangingwall areas;
- Main rift stage:
  - Enhanced stratal thickness development in hangingwall and stratal wedge-shaped units toward the footwall crest;
  - Development of footwall islands with erosion of elevated areas;
  - Flooding of fault-blocks and landward retreat of shoreline;
- Late rift (or transition to post-rift) stage:
  - Passive infill with parallel build-up and onlap strata;

The aim of the study

The purpose of this paper is to describe and interpret the stratal architecture of the Jurassic packages and depositional environment variability within the Brent Group with regard to the generic pre-rift to syn-rift structural and infill models, as listed above. This has been achieved by examination of four east–west profiles (Fig. 1), using a combination of seismic, biostratigraphic and well data. We discuss how the early stage of rifting can be interpreted from the sedimentological response to fault population and their influence on coastal morphology.

GEOLOGICAL SETTING AND BASIN HISTORY

Following the Permo–Triassic rift phase, the Early Jurassic northern North Sea experienced thermal subsidence along the inherited rift topography (Gabrielsen *et al.*, 1990; Odinsen *et al.*, 2000a, b). This subsidence led to the transgression (north-directed) of the fluvial-dominated Statfjord Formation and a north–south seaway was established through the Viking Graben at approximately Sinemurian time (Steel, 1993). The lower part of the Dunlin Group (Sinemurian–Pliensbachian) encompasses the Amundsen and Burton formations, which consist of shales and siltstones (Vollset & Doré, 1984) deposited in a shelfal setting (Husmo *et al.*, 2003). The shallow marine Cook Formation built out into this seaway from the Norwegian mainland during Pliensbachian time, in response to basin margin uplift and erosion (Charnock *et al.*, 2001; Folkestad *et al.*, 2012). Subsidence continued in the northern North Sea, as the Cook Formation was draped by offshore mudstones of the Drake Formation. The overall thickness distribution of the Lower Jurassic Statfjord Formation and the Dunlin Group suggests that the future Viking Graben was built on a broader, possibly asymmetric basin (Færseth & Ravnås, 1998); a basin configuration inherited from the Permo–Triassic stretching phase (Færseth, 1996).

The Oseberg and Broom formations have traditionally been included in the Brent Group but they are not genetically linked to the other formations of the group (Steel, 1993). The Broom and the time-equivalent Oseberg formations (Aalenian) represent coarse-grained fan deltas that prograded into the basin from the uplifted margins of the Viking Graben and are restricted to the western (Broom Formation) and eastern

(Oseberg Formation) basin margins due to basin-margin uplift (Ziegler, 1982; Underhill & Partington, 1994). In Aalenian time, the thermal uplift of the 'Mid North Sea Dome' created a regional high at the triple junction between Viking Graben, Moray Firth and Central Graben (Underhill & Partington, 1993; Fjellanger *et al.*, 1996). This elevated area was subjected to erosion and shed sediments into the adjacent basins. As a result, the 'Brent delta' prograded northward over a time-span of only 2 million years (Helland-Hansen *et al.*, 1992) with a pinch-out of the delta at about 62° N (Mjøs, 2009). This axial, southerly-derived sediment input is illustrated by the coastal plain and delta top depositional environment passing into shallow-marine sandstones in the north (Husmo *et al.*, 2003). Additional sediment supply for the 'Brent delta' came from the uplifted flanks on the margins of the northern North Sea Basin (Helland-Hansen *et al.*, 1992; Steel, 1993; Johannessen, *et al.*, 1995; Fjellanger *et al.*, 1996).

The Middle Jurassic northern North Sea Basin was a ramp basin without shelf-edge topography, flanked by the Hordaland and Shetland platforms (Fjellanger *et al.*, 1996) and with the Møre Basin (Doré, 1991; Gabrielsen, 1989) to the north. The Møre Basin had probably started to subside at this time (Brekke, 2000) and that promoted wave-currents which affected the Brent Group. Such a ramp basin promoted very low-gradient clinoforms and was partly the reason for the rapid northward deltaic progradation (Helland-Hansen *et al.*, 1992). In the Tampen area, the lateral thickness distribution of the prograding part appears uniform due to high sedimentation rates, onlaps and thins above the Broom and Oseberg formations in the platform areas (Helland-Hansen *et al.*, 1992; Johannessen *et al.*, 1995; Fjellanger *et al.*, 1996; see Fig. 1). The 'Brent delta' reached its regressive maximum in the northern part of the Tampen Spur area in the Early Bajocian (Helland-Hansen *et al.*, 1992; Løseth *et al.*, 2009).

The progradational part of the 'Brent delta' consists of the Rannoch, Etive and lower Ness formations (see Table 1 for detailed facies interpretations). The Rannoch Formation forms a wave-reworked lower delta front environment overlying and interfingering with the offshore mudstones of the Drake Formation. The overlying Etive Formation represents an upper delta front with shoreface and mouth-bar successions. The Ness Formation constitutes a delta top environment, with coal layers and mudstones in the lower part, followed by fluvial channel, bay-head delta and mouth-bar deposits. At the time of maximum extension, the stacking style of the 'Brent delta' changed from progradational to aggradational, due to eustatic sea-level rise and increased tectonic activity (Helland-Hansen *et al.*, 1992; Johannessen *et al.*, 1995).

The aggradational shoreline trajectory (see Helland-Hansen & Gjelberg, 1994) changed to retrogradational during transgression, resulting in the retreat of the delta plain of the Ness Formation and the formation of shoreface and embayment environments (estuaries, bay and barrier; see Table 1 for detailed facies interpretations) of the Tarbert Formation. The Tarbert and Ness formations retreated southwards along the Viking Graben in an offset backstepping manner (Helland-Hansen *et al.*, 1992, Johannessen *et al.*, 1995; Fjellanger *et al.*, 1996; Ravnås *et al.*, 1997). South of 60°N, the delta front and delta top deposits are named the Hugin and Sleipner formations, respectively (Fält *et al.*, 1989). The Tarbert and Hugin formations are draped by, and interfinger with, the offshore mudstones of the Heather Formation, which is then capped by the organic-rich shales of the Draupne Formation (Kyrkjebø *et al.*, 2001; Kjennerud *et al.*, 2001).

Several studies have pointed out that fault-block rotation started in the Bajocian (Helland-Hansen *et al.*, 1992, Johannessen *et al.*, 1995; Fjellanger *et al.*, 1996; Ravnås *et al.*, 1997) but increased fault-related subsidence may have started earlier, during Late Aalenian–Early Bajocian in parts of the wide, asymmetrical basin. This is probably best seen as an increase in thickness per time unit of the lower Brent Group, compared to the underlying Dunlin Group (Steel, 1993; Færseth & Ravnås, 1998). This tectonic activity and fault-block rotation accelerated through the Middle Jurassic with the development of footwall islands and submerged half grabens in Bathonian (Ravnås *et al.*, 1997), until maximum strain rates were achieved in the Late Jurassic (Yelding *et al.*, 1992). Primarily N–S striking faults controlled the basin geometry in the Middle Jurassic and it is suggested that the Viking Graben retained the asymmetry inherited from the Permo–Triassic block tilting until faults with a more NE–SW orientation became active in the Late Jurassic (Færseth *et al.*, 1997; Færseth and Ravnås, 1998; Gabrielsen *et al.*, 2001).

Table 1. Facies associations of the Rannoch, Etive, Ness and Tarbert formations within the study area: description and interpretation of the facies associations.

Formation	Description	Interpretation	Comments
Tarbert Fm	Coarsening upward units (2 to 7 m) from mudstones to fine-grained sandstones with micro-hummocks, wave and current ripples, planar lamination, trough cross-bedding, lenticular-bedding and flaser -bedding capped by palaeosols. Bioturbation, syneresis cracks and coal clasts occur.	These units are probably bay-head deltas building into bays or lagoons.	See Figs 9A and 12B,C.
	Fining-upward units from medium to very-fine grained sandstones with cross-bedding, sharp/erosive bases, current ripples and coal-clasts.	Crevasse splays.	See Figs 9B and 12D.
	Well-sorted and planar laminated units of fine-grained sandstones generally with bioturbation, wave and current ripples.	Back barrier deposits and washover fans.	See Figs 9A, 12C.
	Up to 20 m thick units of medium-grained to coarse-grained sandstones with erosive lower boundary and cross-bedding with even-spaced mud or organic drapes.	Channel complex with fluvial and tidal interaction in an estuarine setting.	See Fig. 12B. Also recorded by Davies *et al.*, (2000) in block 34/7.
	Medium to very fine-grained sandstones constitute the upper part of the Tarbert Formation with trough cross-stratification, high and low-angle lamination, swaley cross-bedding, wave and current ripples (Fig. 3A).	Beach barrier complex with shoreface at the top.	See Figs 9A and 12A,D.
Ness Fm	Interbedded coal-layers and mudstones (Fig. 4D) with minor amounts of thin silt and very fine-grained sandstones with root-horizons.	Delta plain with overbank deposits and stagnant swamp forming coal-beds.	See Figs 9A,B,C and 12B,C.
	Coarse to fine grained sandstones, 1 to 5 m thick fining upward units with erosive or sharp-based, cross-bedding and current ripples.	Fluvial channels.	See Figs 9A,C and 12C.
	1 to 8 m thick siltstones coarsening-upward to fine-grained sandstones with wave ripples, syneresis cracks, mud-drapes, micro-hummocks and laminated beds capped by palaeosols.	Bay head deltas.	See Figs 9A and 12B.
	Tens of metres (~20 to 40 m) of medium to fine-grained (bimodal) cross-bedded sandstones, mud-drapes (Fig. 4J) and sporadically bioturbation.	Stacked tidal dune deposits in an estuarine setting.	See Figs 9A, C. See also Mjøs (2009) who has similar interpretation of the depositional environment.
Etive Fm	Moderate sorted medium sandstones (coarse to fine) with massive, cross-bedded and planar-laminated units with some coal-clasts and mud-clasts. Sharp or erosive bases, stacked in a general fining-upward trend. Palaeosols are common at the top of these sandstones.	Mouthbars to wave-reworked mouthbars.	See Figs 9A,C and 12D.
Rannoch Fm	Very fine-grained to fine-grained and well-sorted sandstones with horizontal, low-angle lamination, hummocky and swaley cross bedding, wave ripples. Some muddrapes and stylolites.	Shoreface environment in a delta front setting.	See Figs 9A, C and 12D.
	Alternating mudstones, siltstones and very-fine grained sandstones with lenticular and flaser bedding, horizontal lamination, micro-hummocks, wave and current ripples, muddrapes, syneresis-cracks, coal-fragments and clasts. Abundant *Botryococcus* (Batten & Grenfell, 1996).	Pro-delta deposits. The occurrence of abundant *Botryococcus*, syneresis cracks, current ripples and coal-material indicates strong fresh water influx. The bioturbation, wave-ripples and micro-hummocks indicates marine influence. The muddrapes indicated tidal influence.	Fig. 12A, (well 35/4-1) show this facies as a 90 m thick aggrading unit. Palynological data indicate that the lower part was deposited under freshwater condition that became progressively more brackish in the upper part.

DATABASE AND METHODOLOGY

Well data

This study from the Gullfaks–Kvitebjørn area in the central parts of the Northern Viking Graben is based on 31 wells with wireline logs from blocks 33/12; 34/7, 8, 10, 11; 35/4, 10, 11 (Fig. 1). Slabbed core material (2875 m) of the Brent Group has been examined in terms of grain size, internal sedimentary structures and bioturbation and interpreted in terms of depositional environment as identified from wireline logs. Palynological data from consultant biostratigraphic reports have been re-analysed to provide a chronostratigraphic framework for the evolution of the basin and to provide palaeoenvironmental support for the sedimentological interpretations.

Timelines

Unlike in fully-marine shelf environments, where high diversities and the continued basinward transport of well-mixed dinoflagellate cyst assemblages result, characteristically, in the relatively uniform distribution of marker taxa, care must be taken in marginal marine areas, where changing environmental conditions play a more significant role in the distribution of key taxa. The palynological assemblages from the studied wells are dominated by spores and pollen throughout, indicating a clear proximity to land but the presence of a generally moderate to high diversity of dinoflagellate cysts in the Bathonian interval (that broadly includes the lower Heather and upper Tarbert formations), which suggests relatively normal marine conditions. As a result, confidence is relatively high that the uppermost occurrence of biostratigraphically significant dinocyst species in the studied wells represents predictably correlateable stratigraphic events (e.g. their last regional occurrence/extinction); and industry-standard palynological markers for the Upper and Lower Bathonian can be used to pick these horizons.

Despite the marked evolutionary radiation of dinocysts known to occur through the preceding Bajocian, the number of dinocyst taxa in the studied wells from the Bajocian sections is generally much lower than expected, suggesting a significantly more stressed, marginal marine environment than during the Bathonian. Although confidence is inherently reduced due to a reduction in number of marker taxa, picks for the Late and Early Bajocian are almost consistently picked on distribution characteristics of the dinoflagellate cyst *Nannoceratopsis gracilis* and/or *N. deflandrei senex* (hereafter *Nannoceratopsis gracilis/senex*). Species of *Nannoceratopsis* are considered to have been derived from dinoflagellates that were well-adapted to euryhaline conditions (Riding, 1983, 2006) and particularly to marginal marine (and even estuarine) settings. As a result, it is reasonable to assume that the presence of *Nannoceratopsis gracilis/senex* is correlateable in time across the study area since they tolerate a wide salinity range and are palaeo-environmentally *in situ*. In other words, basing stratigraphic events on the distribution of species that are derived from a more offshore setting will clearly result in correlations with a palaeoenvironmental overprint driven by sea-level change.

The highest stratigraphic occurrence of *Nannoceratopsis gracilis/senex* is used to place the position if the Bajocian/Bathonian boundary, whilst the highest consistent occurrence (or significant increase) in the relative number of these taxa is traditionally employed as a marker for the end of the Early Bajocian. Occasionally, where *Nannoceratopsis gracilis/senex* are particularly sporadic, the first down-hole occurrence of *Evansia granulata*, down-hole occurrence of common *Batiacasphaera* spp (including *B. rudis*) and the last down-hole occurrence of common *Dissiliodinium willei* were used in conjunction with *Nannoceratopsis gracilis/senex* to pick the Bajocian/Bathonian boundary.

The significant increase in relative abundance of *Nannoceratopsis gracilis/senex* (as a common to abundant constituent of the marine assemblage) is used to position the Aalenian/Bajocian boundary. Although the placement of this boundary is beyond the scope of the present study, this relative abundance characteristic serves to highlight the principle that a significant increase in number or bloom of a species with respect to the background assemblage, is characteristic of an adaptation to stressful environments (e.g. low salinity), further increasing confidence in the biostratigraphic use of *Nannoceratopsis* in deltaic palaeoenvironments of the Brent Group.

Palynological sample density is generally high throughout the Bajocian and Bathonian, typically much better than one sample per ten metres. Much of the Bajocian sections were cored and sample spacing is often on the sub-metre scale, indicating that biostratigraphic precision is potentially high. Over uncored sections, where the majority of

Table 2. Stratigraphic thicknesses of the different formations seen in wells 33/12-6 and 34/11-A-15 in Fig. 4: depth of top formation (TVD = true vertical depth), thicknesses ($m\,Ma^{-1}$), thickness-expansion rates (E-W incr.) and time span of formation.

W					E	
33/12-6					34/11-A-15	
Formation	Top TVDm	Thickness (m)	E-W incr	~Million years	Top TVDm	Thickness (m)
Heather	2645	295 (29.5 m per my)	37x	10	3970	8 (0.8 m per my)
Ness-Tarbert	2940	250 (50 m per my)	3.5x	5	3978	72 (14.5 m per my)
Rannoch-Etive	3190	65 (26 m per my)	1.2x	2.5	4050	55 (22 m per my)
Drake	3255	115 (10.5 m per my)	1.9x	11	4105	60 (5.5 m per my)
Cook	3370	100 (33 m per my)	1.3x	3	4165	75 (25 m per my)
Amundsen-Burton	3470	220 (36.6 m per my)	1.2x	6	4240	185 (30.8 m per my)
Statfjord	3690				4240	

Million years estimates from Husmo *et al.* (2003).

biostratigraphic samples were derived from cuttings, stratigraphic tops are favoured over bases, due to the inherent problem of caving.

Seismic data

The three reprocessed 3D seismic surveys ST11M12, ST11M07 and ST09M01 (Figs 4 and 6) have been used for reflection seismic interpretation. They are partly overlapping and have a total areal extent of ~ 1730 km^2, with a 585 km^2 final migration size. In-line direction is east–west with a line spacing of 12.5 m and cross-line direction is north–south with a line-spacing of 25 m. The vertical seismic resolution (reflection separation) is depth-dependent and varies throughout the area from ~ 25 ms Two-Way-Time to ~ 50 ms TWT at a depth of 3500 ms TWT. Lateral resolution is dependent on line-spacing and depth and would typically be 200 m to 300 m at 3500 ms TWT.

The seismic lines have not been depth-converted since they are used to show relative lateral thickness variations of the sedimentary strata (see also Table 2 for thickness-variations of the formations). However, velocity well data have been utilised for both time and depth migration of the seismic surveys and show that the velocity field is generally uniform laterally for regional (km-scale) interpretation or depth conversion purposes. Hence, the thickness variations in time reflect true stratigraphic thickness variations. Time-to-depth relationships are derived from well data and show that the equivalent true vertical depth (TVD) at 3500 ms TWT is approximately 4000 metres.

Interpretation of the seismic data was based on correlation with the well data. Chronostratigraphically significant and lithostratigraphically significant surfaces identified from the interpretation of core samples, electrical logs and biostratigraphic information were tied to the seismic volumes using velocity log data.

PRE-RIFT TO SYN-RIFT SIGNATURES

Geological cross-section across the Gullfaks–Kvitebjørn area

Fig. 4 shows a regional geological cross section that runs from the Kvitebjørn Field via the Valemon Field and across the southern part of the Gullfaks Field toward the Statfjord Field (Fig. 1, cross-section 1). The wells along this correlation have a complete section of the Jurassic stratal package. Fig. 4 shows the east–west stratal thickness trends for the Dunlin, Brent and Viking groups. The Dunlin Group shows a tendency to thicken westward with a gentle (~1.3 times, see Table 2) thickness increase. The Brent Group shows a more pronounced (~2 times) westward increase in thickness towards a Permo–Triassic master-fault (See Fig. 1) but with significant internal variations. The Rannoch and Etive formations have a near tabular thickness distribution, whereas the Ness and Tarbert formations together show clearer wedge-shaped geometries that involves a pronounced (~3.5 times; Table 2) westward thickness increase. Internally, the sedimentary architecture of the lower Ness–Tarbert formations show near tabular strata while the upper part show a gradual angular variation upward in the form of westward bed expansion. An apparently isolated and several tens of metres thick sandstone unit occurs locally at the base of Ness Formation (see well 34/10-23 in Fig. 4).

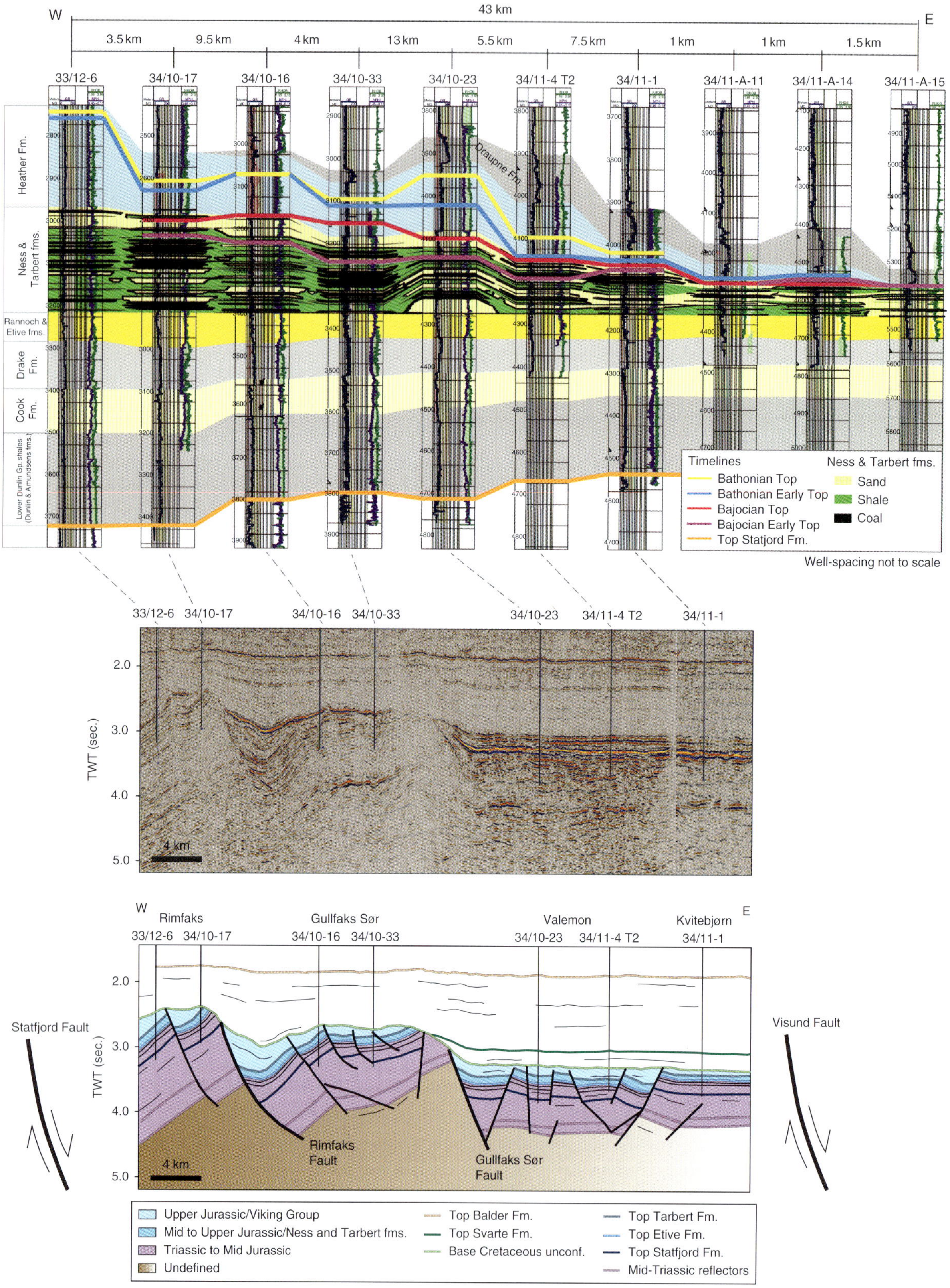

Fig. 4. A regional east-west cross-section of the Jurassic succession from the top Statfjord Formation (orange surface at the base) of the Tampen area (Gullfaks-Kvitebjørn); see Fig. 1 for location. The seismic section (in time) shows fault and horizon interpretation (lower) with well tie. The section displays the general Triassic-Jurassic thinning from west to east. An apparently isolated and several tens of metres thick sandstone unit occurs at the base of the Ness Formation (see well 34/10-23).

In terms of net-to-gross ratio (sand content *vs.* shale content) within the Ness-Tarbert formations, the western part shows lower ratios with low sand-connectivity and thicker and more abundant coal layers. In the eastern part of the well correlation panel (Fig. 4), there is a higher net-to-gross ratio with only a few thin coal layers. In the overlying Viking Group, the Heather Formation shows a pronounced (37 times) westward thickness increase (Table 2), whereas the Draupne Formation portrays a more irregular thickness distribution.

The seismic cross-section (Fig. 4) is tied to formation-picks in the wells. The quality of the seismic resolution at this depth is not sufficient to replicate the thickness trends within the Brent Group directly, as seen in the wells. However, a general westward-thickening trend of the combined Triassic and Jurassic package is seen on the seismic cross-section, with the wedge-shaped unit of the Late Jurassic syn-rift strata being the most obvious. Three major fault-blocks are seen in the seismic cross-section, which progressively steps down eastward, i.e. into the northern Viking Graben. The western Rimfaks fault-block is the shallowest and the Kvitebjørn fault-block to the east is the deepest. The smaller normal faults are predominately limited to the Jurassic or older strata and tend to terminate within the Upper Jurassic syn-rift strata. Only a few of these are seen to offset the Base Cretaceous.

Fig. 5 shows two well correlations across Permo–Triassic faults (according to Færseth,

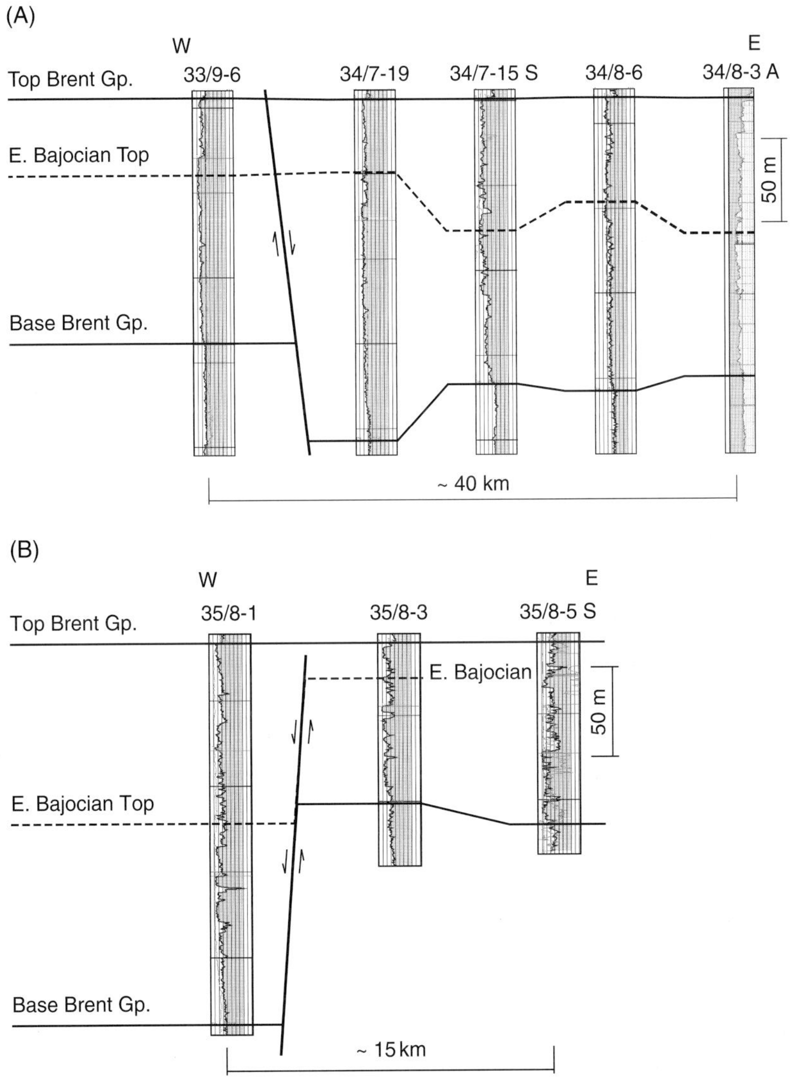

Fig. 5. (A) West-east well correlation across the Visund (34/8), Snorre (34/7) and the Statfjord Field (33/9), showing westward thickening of the Brent Group toward the Statfjord fault. (B) West-east well correlation across the Vega Field (35/8) showing increased Brent Group thickness across a Jurassic fault (see Fig. 1).

1996) in the western and eastern part of the North Viking Graben (Fig. 1). Both correlations show marked thickening of the Brent Group into the hangingwall of the Permo–Triassic faults. The same thickness variation is observed in the well-correlation in Fig. 4 between two deep-rooted Permo–Triassic faults, the Statfjord Fault in the west and the Visund Fault to the east. These faults are identified as Permo–Triassic faults by Færseth *et al.* (1996; see also Fig. 1) and define Permo–Triassic fault-blocks (mega-blocks) with widths of up to about 50 km (Færseth, 1996, fig. 2). Although the seismic data have limited resolution, it shows in combination with well data that this extensive fault-block, the Gullfaks–Kvitebjørn/Snorre-Visund mega-block (Figs 1 and 3) experienced block-rotation throughout Jurassic time. The displacement of the Statfjord Fault during the Jurassic seems to be of the same order as the movement of the Ninian-Hutton-Dunlin Fault reported by Hampson *et al.* (2004) in the same time-interval. Thus, it appears that these deep-rooted faults were active during the Jurassic. The well correlations in Figs 4 and 5 indicate similar marked thickness differences of the Brent Group across these Permo–Triassic faults. In summary, these observations suggest that the Permo–Triassic faults had a significant impact on the Brent Group in terms of subsidence and facies variation in the North Viking Graben. This interpretation is in agreement with observations made by several previous authors (Fält *et al.*, 1989; Fjellanger *et al.*, 1996; Ravnås *et al.*, 1997) and will be discussed further in the sections below.

On a finer scale, the basal Dunlin Group show a slightly westward thickness increase, which is probably due to minor movement of the Statfjord Fault during the Early Jurassic (Table 2). The slight wedge shape of this unit is negligible considering the timespan of this unit, rendering it as near-tabular. Therefore, a general proto-rift stage is interpreted for this unit. The Rannoch and Etive formations of the basal Brent Group are also near-tabular, which assigns them to the proto-rift stage. The Ness-Tarbert formations (~5 Myr; Table 2) form a pronounced east–west stratal wedge-shaped unit, which suggests that this interval was deposited during an interval with increased rates of rotational faulting. The transition from the proto-rift to an early syn-rift stage is therefore placed within the lower Ness Formation. The overlying wedge-shaped Heather Formation (~10 Myr; Table 2) shows a dramatic thickness expansion in the western part and this formation, together with the bulk of the Draupne Formation, represents the main stage of rifting with high rotation rates (Færseth *et al.*, 1995b; Færseth & Ravnås, 1998). The timelines in Fig. 4 suggest that the western part was transgressed earlier by the Heather Formation than the eastern part of the cross-section.

The exact timing of rift-initiation is difficult to constrain from the stratal architecture since early syn-rift beds would be near-tabular to slightly wedge-shaped. However, the thick (30 m) isolated sandstone unit at the basal Ness Formation in Fig. 4 (Well 34/10-23) may represent a local depocentre formed during rift-initiation as outlined above. This feature will be treated in more detail below.

The east–west lithological distribution of the Ness–Tarbert formations in Fig. 4 suggests low net-to-gross ratios towards the hangingwall area (western part) as compared to the area towards footwall (eastern part). This difference can be explained by increased rate of accommodation space generation and resulting in preferential preservation of strata in the western part through the trapping of river-supplied sediments such as sands and muds, resulting in isolated sandstone units (channels) and common formation of swamps (coal layers). In contrast, the eastern part shows higher net-to-gross ratios with significantly fewer coal layers and shale units, indicating lower rates of accommodation creation. Such a difference in net-to-gross as observed in the east–west lithological distribution of the Ness-Tarbert formations is another indication that rifting created variable accommodation space. An alternative interpretation for these east–west trends of net-to-gross ratio can be explained by differential compaction along the transect (Fig. 4). This would imply significant sea-bed topography prior to deposition of the Brent Group. The near tabular nature of Rannoch and Etive formations supports the first interpretation.

Geological cross-section across the Kvitebjørn Field

The Early–Middle Jurassic geological evolution of the Kvitebjørn area is illustrated in Fig. 6 with the Rannoch, Etive, Ness and Tarbert formations with wells tied to the seismic cross-section. The Cook, Drake, Rannoch and Etive formations show a tabular thickness distribution whereas the Ness and

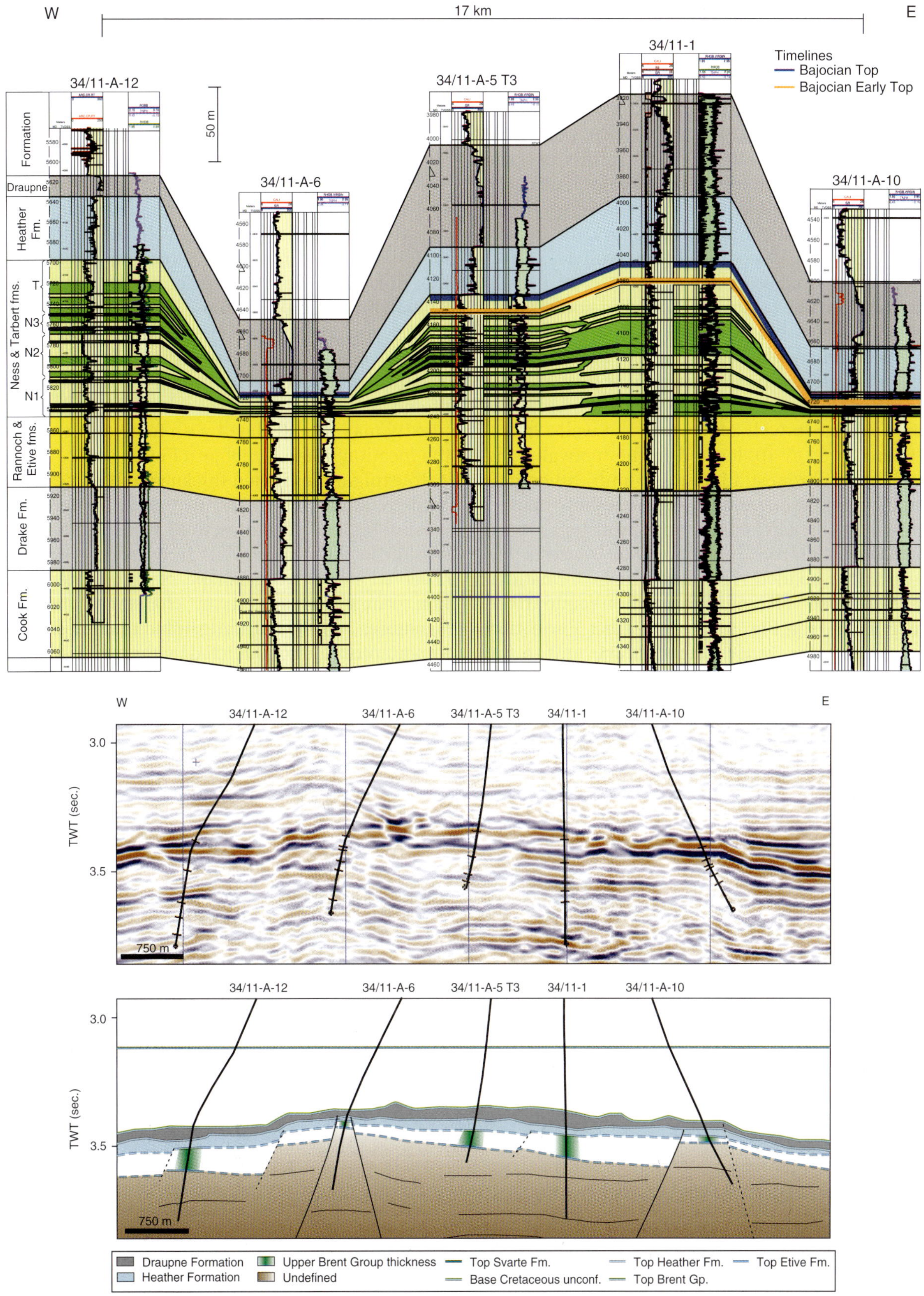

Fig. 6. East-west cross section of the Kvitebjørn Field, showing the Cook and Drake formations and the Brent and the Viking groups. See Fig. 4 for a colour legend for the well-correlation. The seismic cross-section is shown in time with interpretation based on stratigraphic well picks. The interpretation at the bottom shows fault-delimited thickness variations of the Ness and Tarbert formations across the field. Green-coloured intervals are well-defined Ness-Tarbert formation thicknesses, whereas the white intervals are interpreted thicknesses with uncertainties. Note the termination of faults in the Heather Formation (Upper Jurassic).

Tarbert formations show overall wedge-shaped stratal geometry from a very thin condensed section in well 34/11-A-6 through a thicker section in well 34/11-A5-T3, to an even thicker section in well 34/11-1. Internally, the beds in this package show a near-tabular appearance in the lower part with a more pronounced wedge-shaped geometry up-section. Further eastward, a dramatic reduction in thickness is observed from well 34/11-1 to a very condensed sequence in 34/11-A-10. This is associated with a westward-throwing normal fault (see the lower cross-section of Fig. 6) located between the two wells. Well 34/11-A-12 in the western part of the cross-section, shows another expanded thickness of the Ness and Tarbert formations, similar to well 34/11-1. In the wells 34/11-A-6 and 34/11-A-10, located in the footwalls, the combined chronostratigraphic and lithostratigraphic correlations and log-patterns show stacked thinly-bedded sandstones and mudstones with streaks of coal layers. In contrast, this suggests that these thinly-bedded Ness and Tarbert formations are condensed versions of the section found in the other wells (see Fig. 6). The Heather Formation shows a similar wedge-shaped stratal unit to the Ness and Tarbert formations in the same wells.

The tabular nature of the Cook, Drake, Rannoch and Etive formations (Fig. 6) is consistent with a pre-rift stage. The observed stratal wedge geometry of the Ness and Tarbert formations suggests that this unit represents the early-rift stage, possibly entering the main-rift stage as the internal stratal wedge architecture show a gradual angular variation up-section (Fig. 7; note the similarity to the wedge-shaped Ness-Tarbert unit in Fig. 4). The complex stratigraphic development of the Ness and Tarbert formations in the Kvitebjørn Field can be put into a more regional context, as this area is located at the footwall crest of the Kvitebjørn–Gullfaks Permo–Triassic mega-block (Figs 1 and 4). The westerly dipping faults, indicated in Fig. 7 with the stratal wedge in between, are antithetic when seen in relation to the deep-rooted Statfjord Fault (Figs 1 and 4). This complexity may be associated with footwall flexure at the location of the Kvitebjørn Field.

To explain the limited thickness of the Ness and Tarbert formations in the two wells in terms of erosion (or faulted out section) would also imply erosion in the nearby wells in order to create the observed stratal wedge shape. Such an interpretation seems unlikely, as it challenges the log-signatures and the biostratigraphic timelines. More importantly, it is contradicted by the wedge-shaped geometry of individual members of the Ness and Tarbert formations. Moreover, it seems unlikely that a scenario of faulted-out section would apply to all wells with anomalous Ness–Tarbert thicknesses.

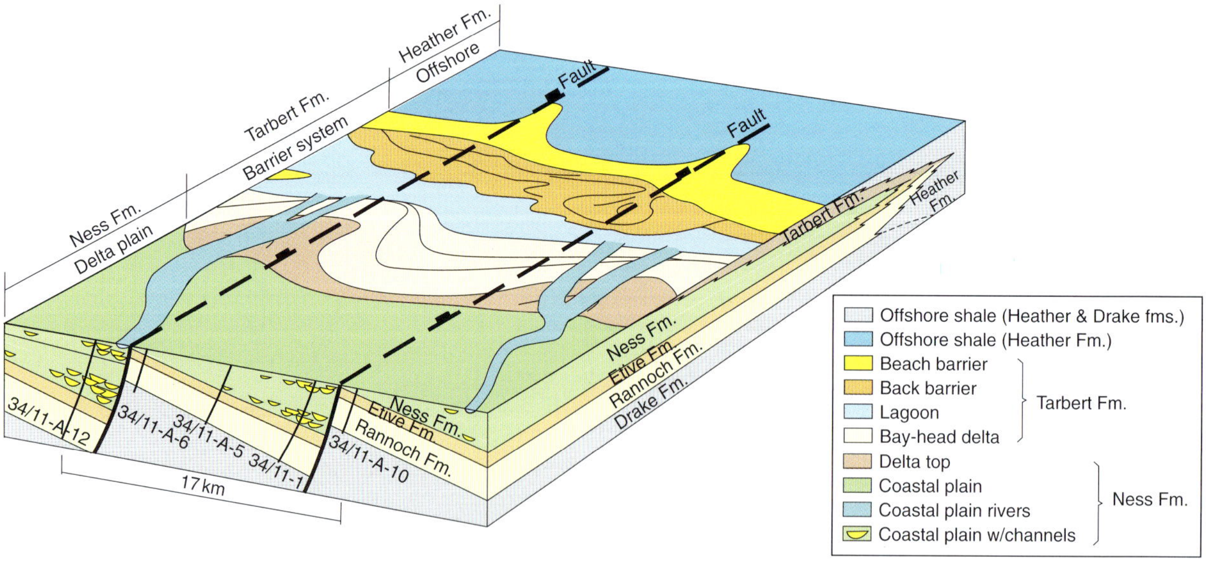

Fig. 7. Schematic 3D illustration of the Brent Group on the Kvitebjørn Field. Syn-sedimentary tectonic development, producing a wedge-shaped geometry for the Ness-Tarbert package across the Kvitebjørn Field contrasts the tabular Rannoch-Etive formations (below) and is taken to reflect early-stage rifting. The coastline undulations are largely formed by the fault crests (see Fig. 11).

Local depocentres

The stacked (tens of metres thick) sandstones at the basal Ness Formation in 34/10-23 (Fig. 4), are conspicuously anomalous in the correlation panel. Similar stacked sandstones at the basal Ness Formation are seen in several other wells (Figs 8, 9 and 10). These sand-units are geographically isolated and are not observed in the surrounding wells (see Fig. 8). Where these sand-units are cored (Fig. 9A and C), they display cross-bedding with mudstone-drapes. Dipmeter data for such a unit in 34/11-A-15, Fig. 10D) show landward-directed palaeo-currents (i.e. to the south). Palynomorphs are continuously recorded in all samples, particularly spores and pollen; however, marine palynomorphs including the dinoflagellate cyst *Nannoceratopsis gracilis* and sphaeromorphs acritarchs are sporadically present indicating episodic marine incursions. These sandstones are interpreted as stacked tidal dunes (see also Mjøs, 2009 with regard to 35/11-7, Fig. 9C), forming thick aggrading units in a nearshore position (above the Etive Formation) displaying similarities to estuarine depositional environments (e.g. Dalrymple & Choi, 2007). The interval from 35/11-7 is dominated by spores and pollen and, although no *in situ* dinoflagellate cysts have been observed, it contains sporadic acritarchs and a significant proportion of *Botryococcus* spp., especially towards the top of the unit. These data also suggest some influence of marine waters and the known tolerance of modern day *Botryococcus* spp. of 0-4ppt salinity (see Qin, 2005; Rao *et al.*, 2007) further suggests a brackish marine influence.

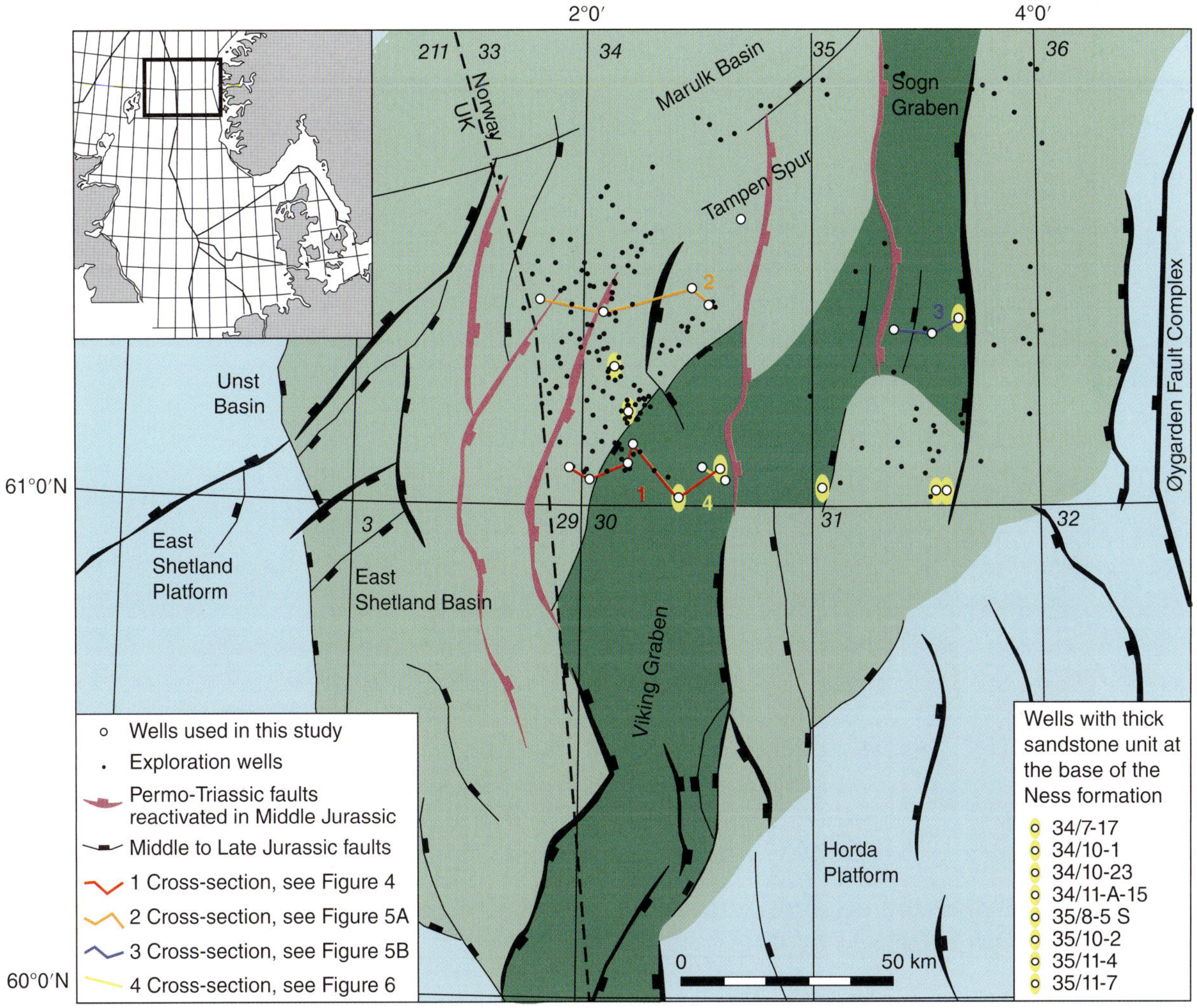

Fig. 8. Geographical location of wells at the base Ness sandstone level in the northern North Sea.

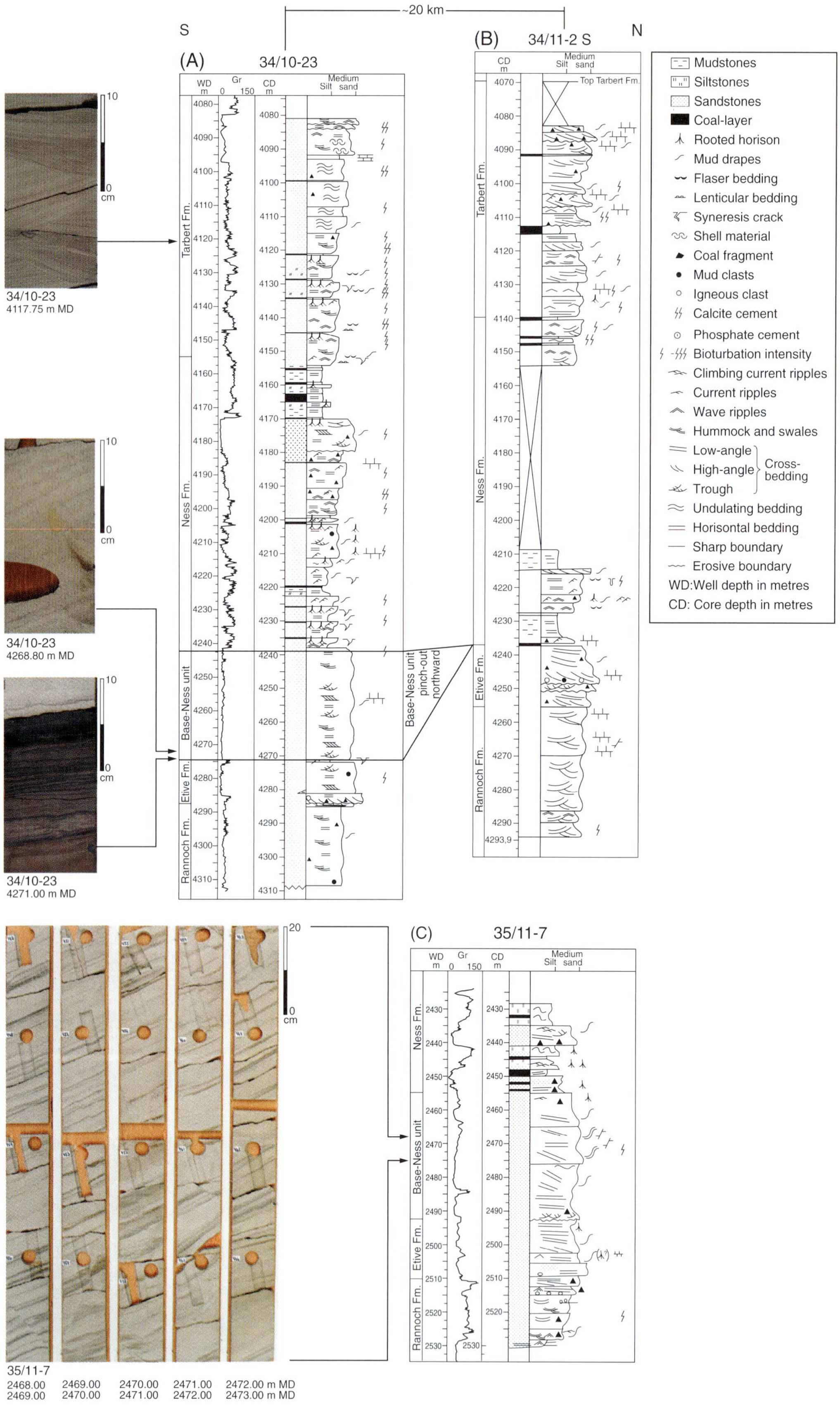

Fig. 9. (A) Log of the Brent Group in well 34/10-23 with the base Ness sandstone indicated. The palaeosol at the top of the Etive Fm is illustrated in the photo from 4271 m. The photo at 4268.8 m shows weak but rhythmic mudstone-drapes of the base Ness sandstone unit. A reactivation surface is also observed in the lower part of the photo. (B) Log of the Brent Group in well 34/11-2S (about 20 km north or basinward of 34/10-23; see Fig. 1) with the base Ness sandstone unit absent. (C) Log of the lower half of the Brent Group of well 35/11-7, including the base Ness sandstone unit. The core picture shows the mudstone-draped cross-bedded sandstones interpreted as stacked tidal dunes.

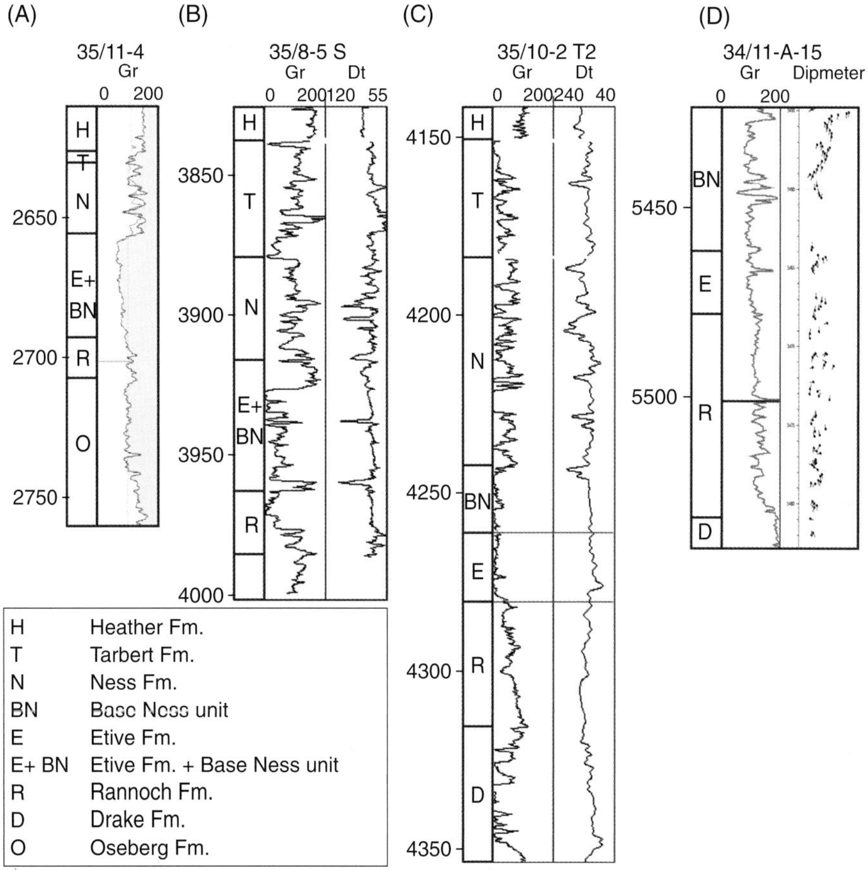

Fig. 10. Base Ness sandstone unit shown in selected well logs as a supplement to Fig. 9. Note that well 34/11-A-15 also shows dipmeter data from the Formation Micro Image log with the tadpoles directed generally southward (down-page) in the base Ness sandstone unit.

The local occurrence of these units (Fig. 8) with a thick stack of tidal dunes favours interpretation as local depocentres that experienced considerable accommodation space generation, formed during the initial stages of the early rift-phase. The initial rift-stage is characterised by fault population evolving into fault-linkage and fault-death (Gawthorpe & Leeder, 2000; Fig. 3). The stratigraphic occurrence of these local depocentres at the base of the wedge-shaped Ness and Tarbert formations (Figs 4 and 8) suggests that rift-initiation occurred at this stratigraphic level in the study area. A tidal influence has been recorded in the Rannoch Formation (Table 2) but tidal processes appear to have been suppressed by wave activity. The recognition of tidal influence in the deltaic intervals makes the suggested estuarine interpretation of these units probable, as tidal currents may have been amplified landward and not suppressed by other processes.

An alternative, non-tectonic interpretation of these thick sandstone units is to view them as distributary channels that were subjected to rapid sea-level rise. Such an interpretation is possible (as dunes are common within channels) but the strong tidal character of the sandstones and the occurrence of palynomorphs would be unusual within fluvial distributary channels due to the winnowing nature of these uni-directional high-energy environments. Interpreting these units as an aggrading backshore unit would contradict their local occurrence as shown in Figs 4 and 8. The stratigraphic occurrence at the base of the stratal wedge makes a tectonic influence during deposition a more likely scenario.

Coastal morphology – progradational part

The pinch-out morphology of the 'Brent delta' in Early Bajocian time has usually been interpreted to map as a linear E–W coastline (Olsen & Steel, 1995; Løseth *et al.*, 2009; Mjøs *et al.*, 2009). However, as indicated in Figs 4 and 5, the Permo–Triassic faults seem to have been active at that time. Active faults will affect the palaeogeography of the coastline in the sense that hangingwall areas generally show a transition of progradational to more aggradational shorelines due to increased accommodation. By contrast, the shoreline reaches further basinward at the footwall highs and are characterised by progradation with low-accommodation facies as wave-dominated strata, as indicated in Fig. 11 (see Gawthorpe & Leeder, 2000). This explains the observed lateral facies variability of the progradational part of the Brent Group (Fig. 12) in the pinch-out area within such a fault-influenced context.

Well 35/4-1 is located in a hangingwall position relative to a Permo–Triassic fault (Fig. 11) and shows heterolithic strata (low net to gross ratio of sand) displaying fluvial, tidal and wave influence (Fig. 12A). Biostratigraphic analysis of the Brent Group in this well indicates a gradual change from freshwater influence (abundant *Botryococcus*; see Batten & Grenfell, 1996) in the lower part of well 35/4-1 (see Table 2) to brackish-water to marine conditions towards the upper part (see Fig. 12A).

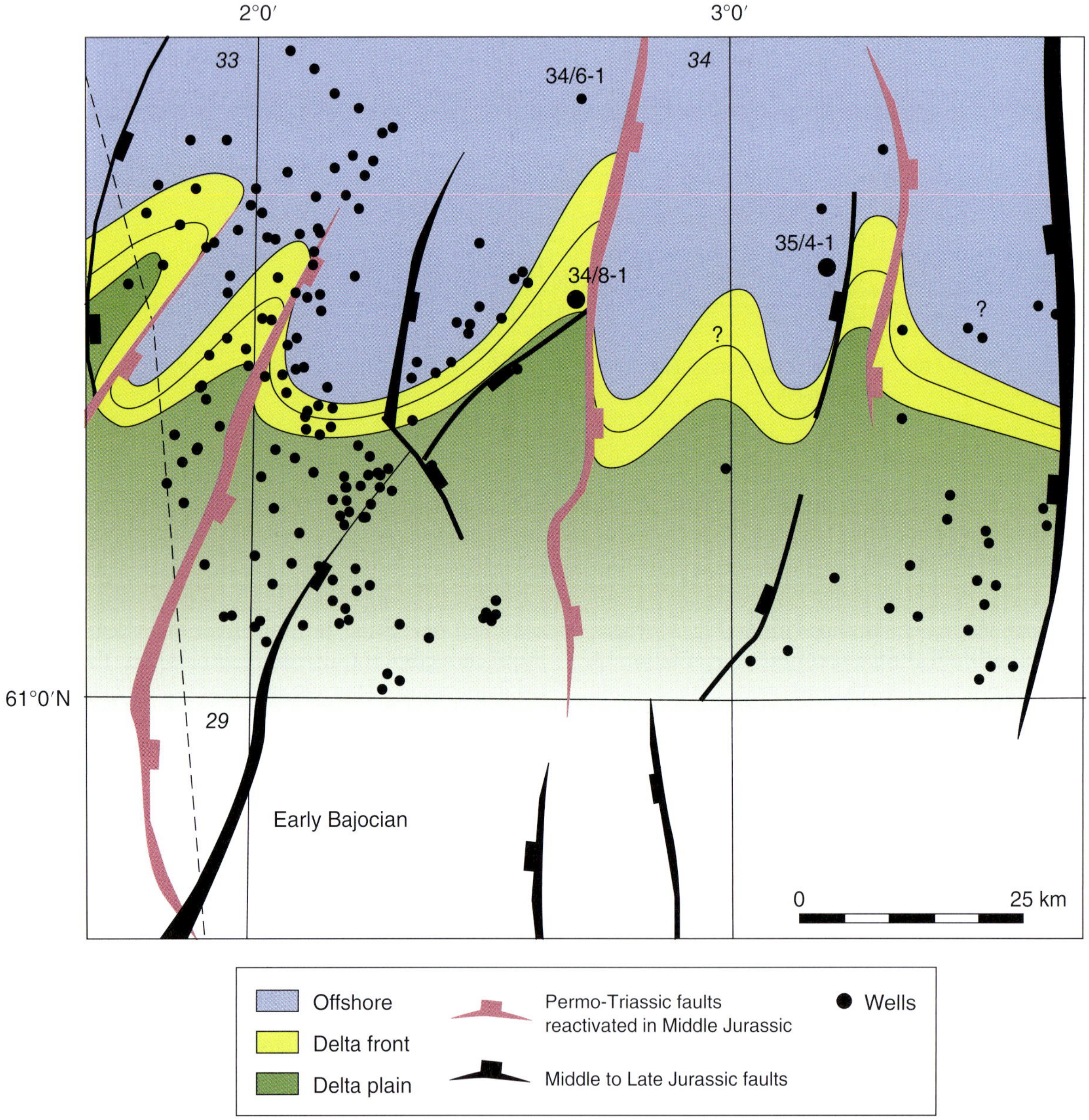

Fig. 11. Schematic illustration of the prograding Brent Group (not progradational maximum), portrayed as an undulating coastline where the footwall areas promote basinward deltaic extension with a relatively low shoreline trajectory and the hangingwall areas show aggrading shoreline trajectories with a pinch-out somewhat more southwards than the footwall areas.

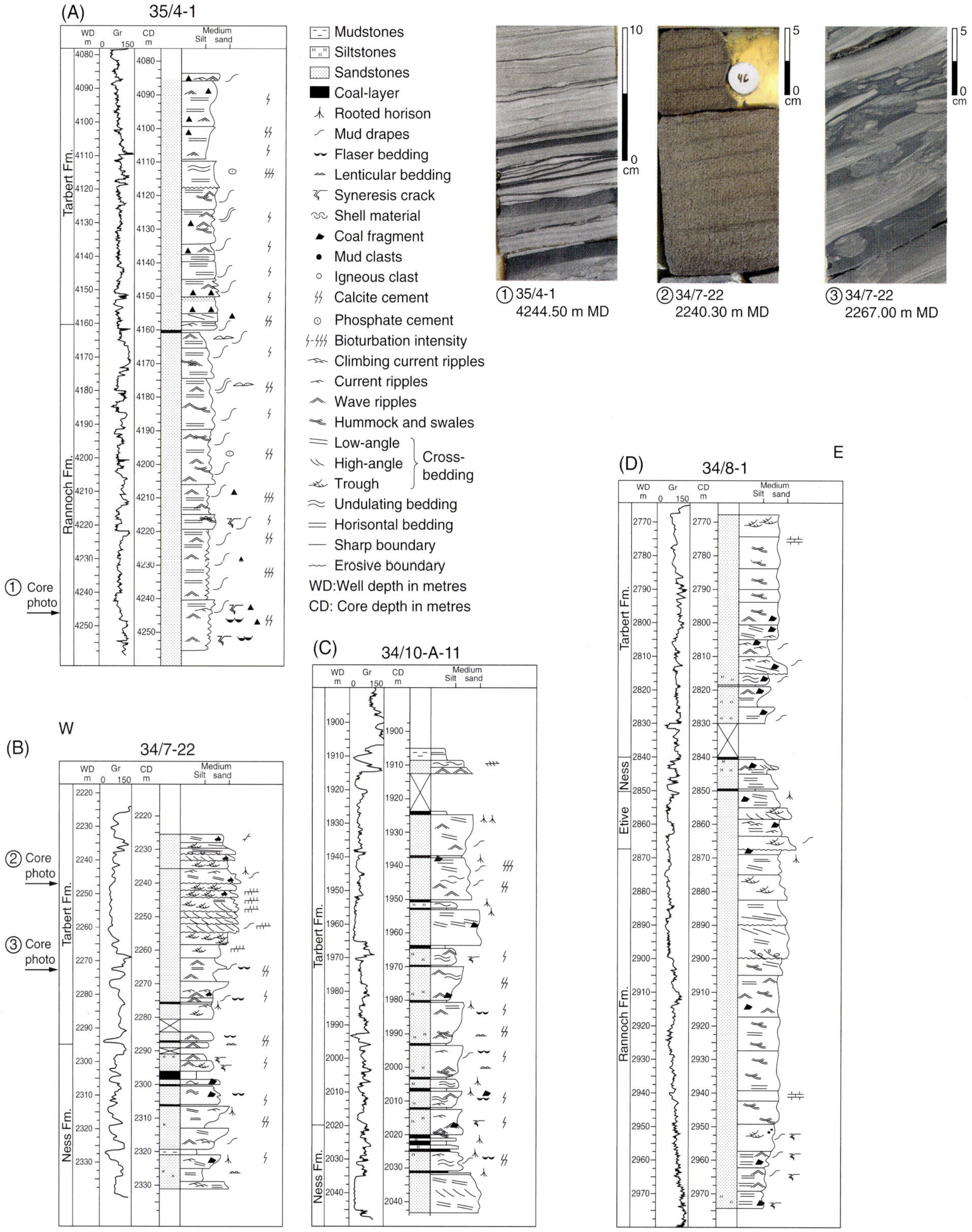

Fig. 12. Log of the Brent Group in (A) well 35/4-1, note the 90 m thick aggrading unit Rannoch Formation. Palynological data indicate a trend upwards from freshwater condition becoming progressively more brackish in the upper part; see Fig. 11 for location; (B) well 34/7-22. Note the cross-bedded sandstone interval in the middle part of the Tarbert Formation; (C) well 34/10-A-11; (D) well 34/8-1, the location of this well is important in Figs 11 and 13.

Well 34/8-1 is located at the footwall high of a Permo–Triassic fault (Fig. 11) and shows predominantly wave-dominated strata in the Brent Group. Here, the progradational portion extends as far into the basin as well 34/6-1. An alternative interpretation of the facies differences in the prograding 'Brent delta' would be to regard them as an effect of lateral deltaic variability owing to the distance between the feeder systems. However, in the context of the fault-block rotations at this time (Figs 4 and 5), the related fault locations (Figs 1 and 11) and the manner in which the coastline evolved later (see below) leads up to a strong impression that these faults had an impact on the shape of the coastline.

Coastal morphology – retrogradational part

The Tarbert Formation represents the retreating part of the 'Brent delta' and shows a variation in depositional environments along the coast (east–west; Fig. 12B, C and D) in the Middle–Late Bajocian. The Tarbert Formation shows wave reworking in well 34/8-1 as spit environment (see also Ravnås *et al.*, 1997; Figs 12D and 13), transforming to stacked bay-head deltas capped by beach-barrier complexes in 34/10-A-11 (Figs 12C and 13) that pass laterally into a wave-dominated estuary in 34/7-22 (Figs 12B and 13; expressed as fluvial-tidal channel; cf. Davis *et al.*, 2000). In light of the argued influence of fault activity in the Ness and Tarbert formations, the facies distribution pattern described above for the Tarbert Formation can best be explained by rotation of the Permo–Triassic fault-blocks giving an asymmetric subsidence pattern along the coastline. In this way the estuary development represents high-subsidence and accommodation space creation in the hangingwall part (well 34/7-22) and the wave-reworked area as the lesser subsidence (with low accommodation-space creation) in the footwall part of the Permo–Triassic fault block (34/8-1; Fig. 13). It is typical for rift basins to show this kind of variability in sediment infill patterns and facies types along the coast (Gawthorpe & Leeder, 2000), as observed here for the Tarbert Formation. Rift or fault activity within the Tarbert Formation, with some impact on the lateral facies variation, has been reported by several authors (Fält *et al.*, 1989; Fjellanger *et al.*, 1996; Ravnås *et al.*, 1997; Davies *et al.*, 2000; Ravnås *et al.*, 2000).

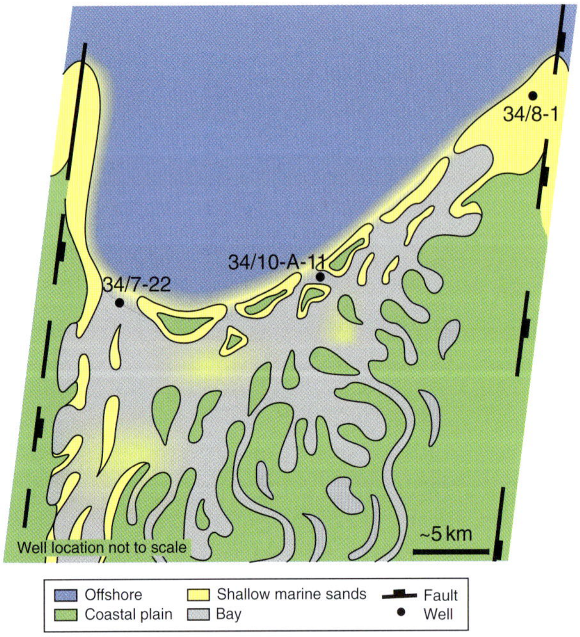

Fig. 13. Schematic illustration of the variation in depositional environments of the Ness and Tarbert formations (during transgressive) in wells 34/7-22, 34/10-A-11 and 34/8-1 and their relationship with the faults. Well position not to scale. Here, the shoreline at the footwall area extends further into the basin, whereas the hangingwall area experienced transgression (see Gawthorpe & Leeder, 2000).

DISCUSSION

This study has demonstrated that Early Bajocian fault-activation led to wedge-shaped stratal development in the Ness and Tarbert formations in the Tampen area. In terms of displacement and strain, these fault movements are subordinate but sufficiently large to have influenced the distribution and lithological partitioning (net to gross ratio) of Middle Jurassic strata at the scale of 100s of metres. In this section, we discuss how our findings add to the understanding of Jurassic rift-initiation in this part of the North Sea rift system.

Rift initiation

The period between the Late Permian–Middle Triassic and Middle–Late Jurassic rift episodes is traditionally characterised as a post-rift period (i.e. a period of tectonic quiescence). By contrast, there is evidence of fault activity at various stages during this time interval that suggests this period is better described as an inter-rift period during which minor fault-slip accumulated due to subsidence and compaction. It has also been speculated that fault-slippage during this period was caused by pulses of mild extension (Ravnås *et al.*, 2000). For the most part, these inter-rift fault movements occurred on Permo–Triassic faults that propagated upward during deposition of the inter-rift sedimentary sequence. Such faults include the Statfjord East fault (Fig. 4) and the Ninian-Hutton-Dunlin fault, both of which are reported to have been active throughout the Jurassic period (Hampson, *et al.*, 2004).

There are at least two end-member models that can explain such fault activity during the inter-rift period. One is that extension continued more or less continuously at a very low rate, activating faults at different locations in the basin at different times. The other is that the basin experienced a number of short-lived (some millions of years) pulses of extension, as advocated by Ravnås *et al.* (2000). At present the resolution of available data makes it difficult to discriminate between these two models.

The suggested time of initiation for the main Middle–Late Jurassic rift phase varies from study to study, ranging from Late Bajocian (Helland-Hansen *et al.*, 1992; Mitchener *et al.*, 1992, Johannessen *et al.*, 1995) to Kimmeridgian (Badley *et al.*, 1988). This wide range may indicate that rift-initiation was not synchronous throughout the basin (Ravnås & Bondevik, 1997) because wedge-shaped stratal packages (with internal stratal thinning or expansion) in a depositional strike section are taken to indicate syn-depositional fault-block rotation (i.e. syn-rift deposition; e.g. Yielding *et al.*, 1992; Ravnås & Steel, 1998). Considering the data presented in this study, we suggest that the Jurassic rift episodes of the northern North Sea rift system initiated in the Early Bajocian within the central part of the basin, marked by the base of the wedge-shaped Ness and Tarbert formations in the Gullfaks-Kvitebjørn area (Figs 4 and 6).This would correspond approximately at the base of the Early Bajocian. Work by Graue *et al.* (1987) and Fält *et al.* (1989) suggests that the Ness Formation shows evidence of syn-sedimentary faulting along major structural lineaments that most likely represent Permo–Triassic faults. Rifting first occurred in the proto-Viking Graben and the adjacent areas and then expanded to affect the lateral platform-areas at a later stage to cause the syn-rift activity reflected in the Tarbert Formation in the East Shetland Platform (Davies *et al.*, 2000) and the Horda Platform (Ravnås *et al.*, 1997).

The initial rift stage (Early Bajocian) was probably characterised by a combination of increased movement of the Permo–Triassic faults as well as the development of a new Jurassic fault-population at the base of the large-scale Ness and Tarbert stratal wedge (Fig. 4) that created scattered local depocentres (see also Fig. 3; Gawthorpe & Leeder, 2000; Sharp *et al.*, 2000; Davies *et al.*, 2000). We refer to the anomalously thick sandstone units, which formed in the fault-induced local depocentres, as the 'base Ness sandstone unit' (Figs 8, 9A, 9C and 10D). Davies *et al.* (2000) found fault-induced local depocentres at the rift-initiation level within the Tarbert Formation on the East Shetland Platform which appears to be analogous to the 'base Ness sandstone unit'. In other locations with limited sand supply either further landward or offshore, these local depocentres may have had a different sedimentological expression than seen in Fig. 9A and C.

The base Ness sandstone unit in well 34/10-23 (Fig. 9A) has earlier been interpreted as the result of rapid sea-level rise, with a step-up of the shoreline with cross-bedded sandstones representing shoreface (Bullimore & Helland-Hansen, 2009). Such a rapid sea-level rise is not indicated in any of the laterally adjacent wells (see Figs 4 and 8). In the UK sector, a similar unit with tens of metres of cross-bedded sandstones has been interpreted as incised-valley fills (Morris *et al.*, 2003; Hampson *et al.*, 2004). This latter interpretation explains the aggradational style and isolated occurrence of the sandstones but is at odds with the lack of a time-equivalent seaward shoreline detached sedimentary wedge, at least in the Norwegian sector of the 'Brent delta' (Graue *et al.*, 1987; Helland-Hansen *et al.*, 1992; Fjellanger *et al.*, 1996; Bullimore & Helland-Hansen, 2009; Mjøs, 2009).

The transition from the rift-initiation stage to a phase of increased fault movement has been correlated to a sharp increase in the rate of basin-wide subsidence early in the rift event (e.g. Steckler *et al.*, 1988). Gupta *et al.* (1998) suggested that the transition from rift-initiation to syn-rift occurs as fault activity becomes localised into linked arrays (Table 2 and Fig. 3). With a decline in the number of active faults, the rate of fault displacement increased; hence, the rate of tectonic subsidence increased. Observations from the East Shetland Basin (Dawers & Underhill, 2000; McLeod *et al.*, 2000; Hampson *et al.*, 2004), the Tampen Spur area (Bruaset *et al.*, 1999; Ravnås *et al.*, 2000) and the Horda

Platform (Fjellanger *et al.*, 1996; Færseth *et al.*, 1997; Ravnås *et al.*, 1997) are in general agreement with the predictions of the model described by Gupta *et al.* (1998). For the Gullfaks–Kvitebjørn mega-block, we have demonstrated an earlier rift-initiation than has been reported from the basin flanks. At least two possible explanations can account for this. Either the position of the Kvitebjørn area, as part of a major step-over of the Viking Graben (ramp and horst, according to Fossen *et al.*, 2010), had reduced subsidence rates in that location, or increased subsidence rates occurred more or less synchronously across the Gullfaks–Kvitebjørn Permo–Triassic mega-block.

The asymmetrical Ness and Tarbert wedge, recognised here from the Kvitebjørn area, places the onset of initial rifting within the Ness Formation (Fig. 6). The incipient faulting that influenced the depositional patterns of the Ness Formation indicates that the footwall part of the Gullfaks–Kvitebjørn mega-block (Fig. 4) underwent extension at this time. Ravnås and Steel (1998) described a similar setting with segmentation of the footwall collapse and Sharp *et al.* (2000) showed footwall flexing with development of half-grabens during block-rotation. These two examples illustrate similarities to the development of the faulted Kvitebjørn Field from the Ness Formation (Early Bajocian) and onwards in terms of stratigraphy.

As the population of small faults within the Gullfaks–Kvitebjørn mega-block accumulated strain, they propagated and linked, with some developing into major faults through fault linkage processes (e.g. see Cowie, 1998; Dawers & Underhill, 2000). At this stage, other Permo–Triassic faults were active, including the major Snorre and Gullfaks faults (Fig. 1). The well correlations of the Brent Group in this study show that this occurred at the last stages of Brent Group deposition or during submergence of the Brent Group. The large fault-blocks outlined by a few Permo–Triassic faults were active in the Early–Middle Jurassic (Fig. 3), a collection of minor faults during deposition of the Ness–Tarbert formations (Fig. 3) that gradually developed into the well-known Late Jurassic fault population. The exact reason for the selective reactivation of Permo–Triassic faults is unclear but it is likely to relate to variations in frictional strength, geometric restrictions, fault interaction and/or orientation relative to the regional stress field.

Early syn-rift

The Ness Formation shows a general aggradational stacking of facies, with a balance between sediment supply and accommodation space creation (Fig. 4). The transition from the progradational style of the advancing 'Brent delta' (Rannoch, Etive and lower Ness formations) to an aggradational style was explained by Ravnås *et al.* (1997) to be the result of increased rate of fault-related subsidence and thus added rate of accommodation space creation – a model that is supported by the findings of the present study of the Gullfaks–Kvitebjørn mega-block. A 'Mid-Ness-Shale' unit within the wedge-shaped Ness–Tarbert is described by Fjellanger *et al.* (1996) in the western part of block 34/10 (north of Fig. 4) as an organic-rich lagoonal mudstone. This observation adds to the picture of a westward rotation of the Kvitebjørn-Gullfaks Permo–Triassic mega-block, where the formation of this thick lagoonal mudstone unit is indicative of rapid subsidence. Hampson *et al.* (2004) also reports the occurrence of a wedge-shaped 'Mid-Ness-Shale' unit (lagoonal mudstone) on the UK side, which thickens toward the Ninian-Hutton fault. These 'Mid-Ness-Shale' units are associated with the hanging wall depocentres of Permo–Triassic faults and, together with the lithologic distribution within the Ness Formation (Fig. 4), show a net to gross reduction westward, with preservation of thicker mudstone intervals between the sandstone units as the river-supplied muds became trapped and preserved in the high-accommodation setting of the hangingwall area. The hangingwall area probably experienced transgressive events (see timelines in Fig. 4) that reached farther inland than the footwall area and with more immature and laterally restricted coal-layer accumulations in the hangingwall than the footwall area. The occurrence of isolated fluvial channels in the hangingwall area is the result of axial drainage being attracted into the subsiding areas.

The Brent Group forms a continuum from the rapidly advancing deltaic Rannoch-Etive formations, without recordable tectonic influence, to the delta pinch-out in Early Bajocian time where the initiation of fault movements created an irregular delta front (Fig. 11). This can be seen as the precursor to the retreating Tarbert Formation (Rønning & Steel, 1987). The timelines in Fig. 4 demonstrate that the hangingwall area was transgressed earlier than the footwall area of the Gullfaks–Kvitebjørn Permo–Triassic mega-block,

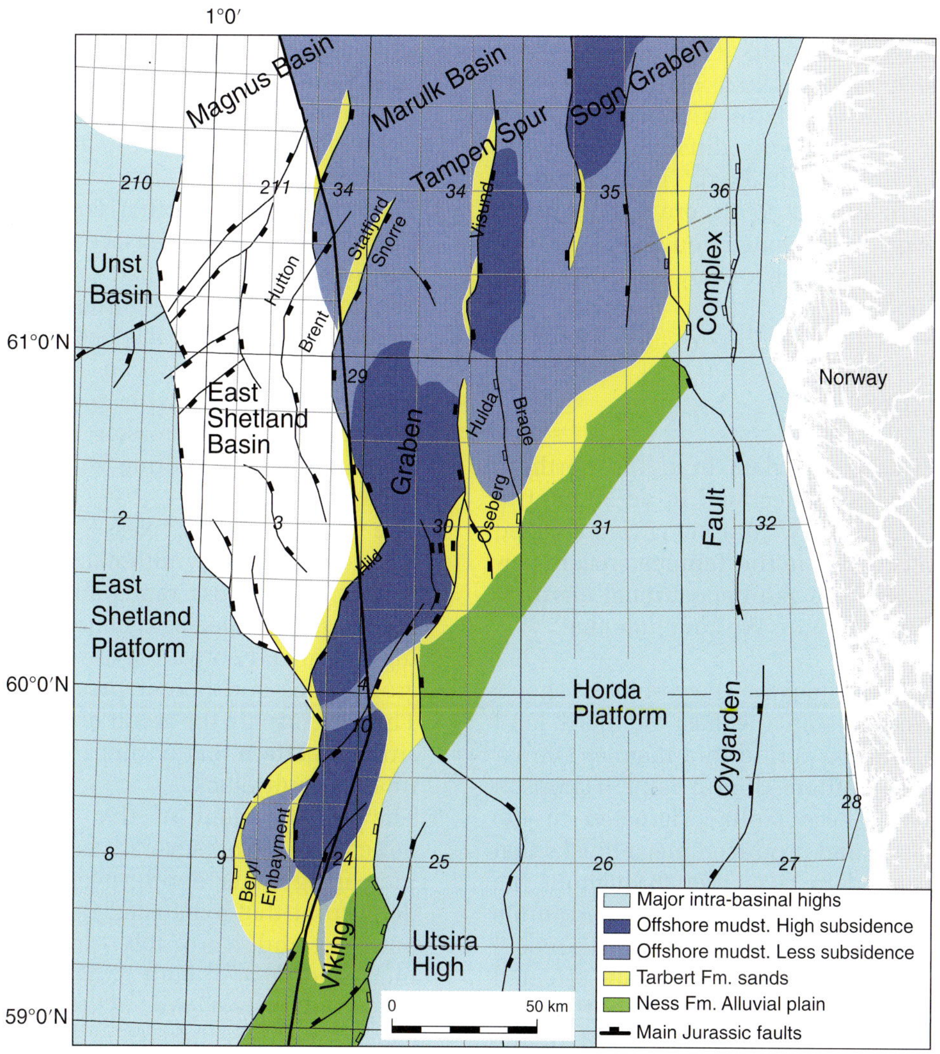

Fig. 14. Schematic illustration of the transgressed northern Viking Graben with the footwall islands in the early part of the Late Jurassic Epoch.

which made the footwall area of this block a (temporary) footwall island. The ongoing block rotations created estuaries with amplified tidal currents in the hangingwall and spits/shoreface at the footwall highs (Figs 13 and 14).

Under such transgressive conditions the net sediment transport is directed landward (van der Molen & van Dijck, 2000), where barrier-inlets (as shown in Fig. 13) will act as grain size-filters (Oost, 1995). Finer-grained sediments pass through the inlets whilst coarser-grained sediments will be kept within or outside the inlets. These observations can explain why the back-barrier area of the Tarbert Formation has finer-grained sediments (Figs 9A, 9B, 12B and C) and may also explain the origin of the enigmatic 'silk sand' (Bruaset *et al.*, 1999; named after its textural properties) found in the Gullfaks Field. The 'silk sand' unit is structureless and trough cross-bedded, medium grained, 30 m to 54 m thick and about 1.5 km wide (E–W), tidally influenced sandstone with high permeability (Bruaset *et al.*, 1999). Its tidal character and the uniform grain-size trend, typical of the back-barrier area (Oost, 1995), indicate that the 'silk sand' may represent a tidal-bar complex or a flood-tidal delta. A full analysis is beyond the scope of the present study but it is interesting to note its presence in our study area as a consequence of the transgression caused by the rotation of the Permo–Triassic megablock.

The Late Bajocian transgression of the Brent Group with block-faulting along the pre-existing

Permo–Triassic fault lineaments has been documented by several authors (Fält *et al.*, 1989; Helland-Hansen *et al.*, 1992; Fjellanger *et al.*, 1996). This phase correlates directly to rifting, convincingly, as the syn-sedimentary tectonic signs stand out (on seismic data). The earlier stage of rifting is more difficult to discern (Davies *et al.*, 2000). However, as demonstrated in this study, the sedimentary responses to early stages of rifting can be identified.

CONCLUSIONS

Seismic, well and core data from the northern North Sea have been integrated in an attempt to constrain when and how Jurassic rifting affected the Jurassic succession in the Tampen area; in particular with respect to the depositional environments of the Brent Group. We conclude that:

1 The Gullfaks–Kvitebjørn mega-block shows fault-displacement on the East Statfjord fault during the Jurassic period in a manner similar to the Ninian–Hutton–Dunlin fault (Hampson *et al.*, 2004). These faults originated in the Permo–Triassic and produced mega-blocks that either continued to be active throughout the Triassic and Jurassic or were reactivated in the Late Triassic–Early Jurassic. We suggest a model wherein a few weak and favourably oriented Permo–Triassic faults remained active throughout the Jurassic and where the onset of the Middle–Late Jurassic rift phase populated the early mega-blocks with smaller faults that developed into mature Late Jurassic fault populations over time.

2 The onset of Middle–Late Jurassic rifting in the northern Viking Graben is reflected by the wedge-shaped Ness–Tarbert formations by means of regional well-correlations (Fig. 4) across the Kvitebjørn Field (Fig. 6), where rifting appears to have initiated during deposition of the lowermost part of the Ness Formation at that location. This interpretation agrees with a model where 'base Ness sandstone units' were deposited within local depocentres formed by normal faults in the early rift phase. This pattern of normal faults and their associated local depocentres developed into a more mature fault population, as some of the small faults linked up to form larger faults that took over much of the fault activity. The early fault population caused the added accommodation space which, in sum, created the wedge-shaped Ness and Tarbert formations, in accordance with general models of rift-initiation (Gawthorpe & Leeder, 2000; Sharp *et al.*, 2000; Davies *et al.*, 2000).

3 These data indicate that fault activity in the Gullfaks-Kvitebjørn transect area (Figs. 4 and 6) was initiated earlier than faulting in the East Shetland Platform and Horda Platform, where a similar development has been reported for the Tarbert Formation. The location of the Kvitebjørn footwall with respect to the Viking Graben step-over (Fossen *et al.* 2010) can explain the early faulting (footwall flexure and collapse).

4 The wedge-shaped Ness-Tarbert formations (Figs 4 and 6) show internal trends in the form of low net-to-gross ratios and the trapping of landward-supplied sediments toward the hangingwall area. Footwall areas have higher net to gross ratios and poorer preservation of strata. These trends are the result of the variable depositional pattern that occurs along the strike of a rotating fault-block.

5 The advancement of the 'Brent delta' was halted due to several factors: northward increasing water depth; over-extension of the delta-front that exhausted its sediment supply budget; and, probably most importantly, increased rates of fault movements, leading to increased subsidence. This fault initiation caused irregular delta-front morphology at the northern extension of the 'Brent delta' (Fig. 11). Footwall areas were prone to progradational facies-stacking patterns while the hangingwall areas developed more aggradational stacking patterns. This fault-influenced irregular delta front became increasingly irregular, with increased tectonic topography, as the faults moved and the coastline retreated southwards. This led to the development of an estuarine environment in the hangingwall area and spit/shoreface-development in the footwall area (Fig. 14). Originally weak tidal currents operating at the delta front were enhanced into strong tidal currents due to the funnel-shaped coastal morphology created by the flooded hangingwall of the fault-blocks. This motif of a complex shoreline morphology shifted southwards with time into the South Viking Graben due to the southward extension of the rift-activity in the Late Jurassic.

ACKNOWLEDGEMENTS

We thank the editor Rodmar Ravnås and the reviewers Lars-Magnus Fält, Gary Hampson and James MacEachern for helpful and instructive comments that improved an earlier version of the manuscript. Statoil is thanked for allowing us to publish these data. Statoil's graphic department is thanked for assistance with many of the figures.

REFERENCES

Badley, M.E., Price, J.D., Rambech-Dahl, C. and **Agdestein, T.** (1988) The structural evolution of the northern Viking Graben and its bearing upon extensional modes of basin formation. *J. Geol. Soc. London*, **145**, 455–472.

Batten, D.J. and **Grenfell, H.R.** (1996) Botryococcus. In: *Palynology: Principles and Applications* (Eds J. Jansonius and D.C. McGregor), *American Association of Stratigraphic Palynologists Foundation*, **1**, 205–214.

Beach, A., Bird, T. and **Gibbs, A.** (1987) Extensional tectonics and crustal structure: deep seismic reflection data from the northern North Sea Viking Graben. In: *Continental Extensional Tectonics* (Eds M.P. Coward, J.F. Dewey and P.L. Hancock) *Geol. Soc. Spec. Publ.*, **28**, 467–476.

Brekke, H. (2000) The tectonic evolution of the Norwegian Sea Continental Margin with emphasis on the Vøring and Møre Basins. In: *Dynamics of the Norwegian margin* (Eds A. Nøttvedt, B.T. Larsen, S. Olaussen, B. Tørudbakken, J. Skogseid, R.H. Garbrielsen, H. Brekke and Ø. Birkeland), *Geol. Soc. Spec. Publ.*, **167**, 327–378.

Bruaset, V., Båtevik, A., Jakobsen, K.G. and **Helland-Hansen, W.** (1999) An anomalous thick, elongated sandbody, Tarbert Formation, Gullfaks Field: High accommodation fluvial channel stacking, incised valley deposition or both? Palaeozoic to Recent sedimentary environments, offshore Norway, Oslo, *Norw. Petrol. Soc.* (abstract), 101–104.

Bullimore, S.A. and **Helland-Hansen, W.** (2009) Trajectory analysis of the lower Brent Group (Jurassic), Northern North Sea: contrasting depositional patterns during the advance of a major deltaic system, *Basin Res.*, **21**, 559–572.

Charnock, M.A., Kristiansen, I.L., Ryseth, A. and **Fenton, J.P.G.** (2001) Sequence stratigraphy of the Lower Jurassic Dunlin Group, northern North Sea. In: *Sedimentary Environments Offshore Norway – Palaeozoic to Recent* (Eds O.J. Martinsen and T. Dreyer,). *Norw. Petrol. Soc. Spec. Publ.*, **10**, 145–174.

Cowie, P.A. (1998) A healing-reloading feedback control on the growth rate of seismogenic faults. *J. Struct. Geol.*, **20**, 1075–1087.

Dalrymple, R.W. and **Choi, K.** (2007) Morphologic and facies trends through the fluvial-marine transition in tide-dominated depositional systems; a schematic framework for environmental and sequence stratigraphic interpretation. *Earth-Sci. Rev.*, **81**, 135–174.

Davies, S.J., Dawers, N.H., McLeod, A.E. and **Underhill, J.R.** (2000) The structural and sedimentological evolution of early syn-rift successions: the Middle Jurassic Tarbert Formation, North Sea. *Basin Res.*, **12**, 343–365.

Dawers, N.H. and **Underhill, J.R.** (2000) The role of fault interaction and linkage in controlling synrift stratigraphic sequences: late Jurassic, Statfjord East area, northern North Sea. *AAPG Bull.*, **84**, 45–64.

Doré, A.G. (1991) The structural foundation and evolution of Mesozoic seaways between Eurpoe and the Artic. *Palaeogeogr. Palaeoclimatol. Palaeoecol.*, **87**, 441–492.

Færseth, R.B. (1996) Interaction of Permo–Triassic and Jurassic extensional fault-blocks during the development of the northern North Sea. *J. Geol. Soc.*, **153**, 931–944.

Færseth, R.B., Gabrielsen, R.H. and **Hurich, C.A.** (1995a) Influence of basement in structuring of the North Sea Basin offshore southwest Norway. *Norsk Geologisk Tidsskrift*, **75**, 105–119.

Færseth, R.B., Sjøblom, T.S., Steel, R.J, Lijedahl, T., Sauer, B.E. and **Tjelland, T.** (1995b) Tectonic controls on Bathonian-Volgian syn-rift successions on the Visund fault block, northern North Sea. In: *Sequence stratigraphy of the northwest European margin* (Eds R. Steel, V.L. Felt, E.P. Johannessen and C. Mathieu), *Norw. Petrol. Soc. Spec. Publ.*, **5**, 325–346.

Færseth, R.B., Knudsen, B.-E., Liljedahl, T., Midbøe, P.S. and **Søderstrøm, B.** (1997) Oblique rifting and sequential faulting in the Jurassic development of the northern North Sea. *J. Struct. Geol.*, **19**, 1285–1302.

Færseth, R.B. and **Ravnås, R.** (1998) Evolution of the Oseberg fault-block in context of the northern North Sea structural framework. *Mar. Petrol. Geol.*, **15**, 467–490.

Fält, L.M., Helland, R., Jacobsen, R.V.W. and **Renshaw, D.** (1989) Correlation of transgressive-regressive depositional sequences in the Middle Jurassic Brent Vestland megacycle, Viking Graben, Norwegian North Sea. In: *Correlation in hydrocarbon exploration* (Ed. J.D. Collinson), *Norw. Petrol. Soc.*, Graham and Trotman, London, 191–200.

Fjellanger, E., Olsen T.R. and **Rubino, J.L.** (1996) Sequence stratigraphy and palaeogeography of the Middle Jurassic Brent and Vestland deltaic systems, Northern North Sea. *Norsk Geol. Tidsskr.*, **76**, 75–106.

Folkestad, A., Veselovsky, Z. and **Roberts, P.** (2012) Utilising borehole image logs to interpret delta to estuarine system: A case study of the subsurface Lower Jurassic Cook Formation in the Norwegian northern North Sea. *Mar. Petrol. Geol.*, **29**, 255–275.

Fossen, H., Dallman, W. and **Andersen, T.B.** (2008) The mountain chain rebounds and founders. The Caledonides are worn down; 405–359 million years. In: *The Making of a Land. Geology of* Norway (Eds I.B. Ramberg, I. Bryhni, A. Nøttvedt and K. Rangnes), *Norsk Geologisk Forening* (The Norwegian Geological Association), 232–259.

Fossen, H., Schultz, R.A., Rundhovde, E., Rotevatn, A. and **Buckley, S.J.** (2010) Fault linkage and graben stepovers in the Canyonlands (Utah) and the North Sea Viking Graben, with implications for hydrocarbon migration and accumulation. *AAPG Bull.*, **94**, 597–613

Gabrielsen, R.H. (1989) Reactivation of faults on the Norwegian continental shelf and its implications for earthquake occurrence. In: *Causes and Effects of Earthquakes at Passive Margins and in Areas with Postglacial Rebound on both Sides of the North Atlantic* (Eds S. Gregersen and P. Basham), Elsevier, Amsterdam, 69–92.

Gabrielsen, R.H., Farseth, R.B., Steel, R.J., Idil, S. and **Klovjan, O.S.** (1990) Architectural styles of basin fill in the northern Viking Graben. In: *Tectonic Evolution of*

North Sea Rifts (Eds D.J. Blundell and A.D. Gibbs), Oxford University Press, pp 158–179.

Gabrielsen, R.H., **Odinsen, T.** and **Grunnaleite, I.** (1999) Structuring of the Northern Viking Graben and the Møre Basin; the influence of basement structural grain and the particular role of the Møre-Trøndelag Fault Complex. *Mar. Petrol. Geol.* **16**, 443–465.

Gabrielsen, R.H., **Kyrkjebo, R.**, **Faleide, J.I.**, **Fjeldskaar, W.** and **Kjennerud, T.** (2001) The Cretaceous post-rift basin configuration of the northern North Sea. *Petroleum Geoscience*, **7**, 137–154.

Gawthorpe, R.L. and **Leeder, M.R.** (2000) Tectono-sedimentary evolution of active extensional basins. *Basin Res.*, **12**, 195–218.

Giltner, J.P. (1987) Application of extensional models to the northern Viking Graben. *Norsk Geol. Tidsskr.*, **67**, 339–352.

Graue, E., **Helland-Hansen, W.**, **Johnsen, J.**, **Lømo, L.**, **Nøttvedt, A.**, **Rønning, K.**, **Ryseth, A.** and **Steel, R.** (1987) Advance and retreat of Brent Delta system, Norwegian North Sea. In: *Petroleum Geology of North West Europe* (Eds J. Brooks and K. Glennie), Graham and Trotman, 915–937.

Gupta, S., **Cowie, P.A.**, **Dawers, N.H.** and **Underhill, J.R.** (1998) A mechanism to explain rift basin subsidence and stratigraphic patterns through fault array evolution. *Geology*, **26**, 595–598.

Hampson, G.J., **Sixsmith, P.J.** and **Johnson, H.D.** (2004) A sedimentological approach to refining reservoir architecture in a mature hydrocarbon province: the Brent Province, UK North Sea. *Mar. Petrol. Geol.*, **21**, 457–484.

Helland-Hansen, W., **Ashton, M.**, **Lomo, L.** and **Steel, R.J.** (1992) Advance and retreat of the Brent Delta: recent contributions to the depositional model. In: *Geology of the Brent Group* (Eds A.C. Morton, R.S. Haszeldine, M.R. Giles and S. Brown), *Geol. Soc. Spec. Publ.*, **61**, 109–127.

Helland-Hansen, W. and **Gjelberg, J.G.** (1994) Conceptual basis and variability in sequence stratigraphy: a different perspective. *Sed. Geol.*, **92**, 31–52.

Husmo, T., **Hamar, G.P.**, **Høiland, O.**, **Johannessen, E.P.**, **Rømuld, A.**, **Spencer, A.M.** and **Titterton, R.** (2003) Lower and Middle Jurassic. In: *The Millennium atlas: petroleum geology of the central and northern North Sea* (Eds D. Evans, C. Graham, A. Armour, and P. Bathurst), *Geol. Soc.*, 129–155.

Jennette, D.C. and **Riley, C.O.** (1996) Influence of relative sea-level on facies and reservoir geometry of the Middle Jurassic lower Brent Group, UK, North Viking Graben. In: *High Resolution Sequence Stratigraphy Innovations and Applications* (Eds J.A. Howell and J.F. Aitken), *Geol. Soc. Spec. Publ.*, **104**, 87–113.

Johannessen, E.P., **Mjøs, R.**, **Renshaw, D.**, **Dalland, A.** and **Jacobsen, T.** (1995) Northern limit of the 'Brent Delta' at the Tampen Spur - a sequence stratigraphic approach for sandstone prediction. In: Sequence stratigraphy of the northwest European margin (Eds R. Steel, V.L. Felt, E.P. Johannessen and C. Mathieu), *Norw. Petrol. Soc. Spec. Publ.*, **5**, 213–256.

Kjennerud, T., **Faleide, J.I.**, **Gabrielsen, R.H.**, **Gillmore.**, **G.K.**, **Kyrkjebo, R.**, **Lippard, S.J.** and **Løseth, H.** (2001) Structural restoration of Cretaceous-Cenozoic (post-rift) palaeobathymetry. In: *Sedimentary Environments Offshore Norway – Palaeozoic to Recent* (Eds O. Martinsen and T. Dreyer), *Norw. Petrol. Soc. Spec. Publ.*, **9**, 347–364.

Kyrkjebø, R., **Kjennerud, T.**, **Gillmore, G.K.**, **Faleide, J.I.** and **Gabrielsen, R.H.** (2001) Cretaceous – Tertiary palaeobathymetry in the northern North Sea; integration of palaeo-water depth estimates obtained by structural restoration and micropalaentological analysis. In: *Sedimentary Environments Offshore Norway – Palaeozoic to Recent* (Eds O. Martinsen and T. Dreyer), *Norw. Petrol. Soc. Spec. Publ.*, **9**, 321–345.

Leeder, M.R. (1983) Lithospheric stretching and North Sea Jurassic clastic source lands. *Nature*, **305**, 510–514.

Løseth, T.M., **Ryseth, A.E.** and **Young, M.** (2009) Sedimentology and sequence stratigraphy of the middle Jurassic Tarbert Formation, Oseberg South area (northern North Sea). *Basin Res.*, **21**, 597–619.

McLeod, A.E., **Dawers, N.H.** and **Underhill, J.R.** (2000) The propagation and linkage of normal faults: insights from the Strathspey-Brent-Statfjord fault array, northern North Sea. *Basin Res.*, **12**, 263–284.

Mitchener, B.C., **Lawrence, D.A.**, **Partington, M.A.**, **Bowman, M.B.J.** and **Gluyas, J.** (1992) Brent Group: sequence stratigraphy and regional implications. In: *Geology of the Brent Group* (Eds A.C. Morton, R.S. Haszeldine, M.R. Giles and S. Brown), *Geol. Soc. Spec. Publ.*, **61**, 45–80.

Mjøs, R. (2009) Anatomy of the seaward steps and seaward termination of the Brent clastic wedge. *Basin Res.*, **21**, 573–596.

Morris, J.E., **Hampson, G.J.** and **Maxwell, G.** (2003) Controls on facies architecture in the Brent Group, Strathspey Field, UK North Sea: Implications for reservoir characterization. *Petroleum Geoscience*, **9**, 209–220.

Nøttvedt, A., **Gabrielsen. R.H.** and **Steel, R.J.** (1995) Tectonostratigraphy and sedimentary architecture of rift basins with reference to the northern North Sea. *Mar. Petrol. Geol.*, **12**, 881–901.

Nøttvedt, A., **Berge, A.M.**, **Dawers, N.H.**, **Færseth, R.B.**, **Hager, K.O.**, **Mangerud, G.** and **Puigdefabregas, C.** (2000) Syn-rift evolution and resulting play models in the Snorre-H area, northern North Sea. In: *Dynamics of the Norwegian Margin.* (Ed. A. Nøttvedt), *Geol. Soc. Spec. Publ.*, **167**, 179–218.

Odinsen, T., **Christiansson, P.**, **Gabrielsen, R.H.**, **Faleide, J.I.** and **Berge, A.M.** (2000a) The geometries and deep structure of the northern North Sea rift system. In: *Dynamics of the Norwegian Margin.* (Ed. A. Nøttvedt), *Geol. Soc. Spec. Publ.*, **167**, 41–57.

Odinsen, T., **Reemst, P.**, **van der Beek, P.**, **Faleide, J.I.** and **Gabrielsen, R.H.** (2000b) Permo–Triassic and Jurassic extension in the northern North Sea: results from tectonostratigraphic forward modelling. In: *Dynamics of the Norwegian Margin.* (Ed. A. Nøttvedt), *Geol. Soc. Spec. Publ.*, **167**, 83–103.

Olsen, T.R. and **Steel, R.J.** (1995) Shoreface pinch-out style on the front of the 'Brent delta' in the easterly Tampen Spur. In: *Sequence Stratigraphy on the Northwest European Margin* (Eds R.J. Steel, V.L. Felt, E.P. Johannessen and C. Mathieu), *Norw. Petrol. Soc. Spec. Publ.*, **5**, 273–289. Elsevier Science B.V., Amsterdam.

Oost, A.P. (1995) Dynamics and Sedimentary Development of the Dutch Wadden Sea with Emphasis on the Frisian

Inlet. A Study of Barrier Islands, Ebb-Tidal Deltas, Inlets and Drainage Basins, Geologica Ultraiectina. PhD thesis, Utrecht University, Utrecht, pp. 454.

Prosser, S. (1993) Rift-related linked depositional systems and their seismic expression. In: *Tectonics and seismic sequence stratigraphy* (Eds G.D. Williams and A. Dobb), *Geol. Soc. Spec. Publ.*, **71**, 35–66.

Qin, J. (2005) Bio-Hydrocarbons from Algae: Impacts of temperature, light and salinity on algal growth. *Rural Industries Research and Development Corporation*, report 05/025, 18 pp.

Rao, A.O., Davananda, C., Sarada, R., Shamala, T.R and **Ravishankar, G.A.** (2007) Effect of salinity on growth of green alga *Botryococcus braunii* and its constituents. *Bioresource Technology*, **98**, 560–564.

Rattey, R.P. and **Hayward, A.B.** (1993) Sequence stratigraphy of a failed rift system: the Middle Jurassic to Early Cretaceous basin evolution of the Central and Northern North Sea. In: *Petroleum Geology of Northwest Europe. Proceedings of the 4th Conference.* (Ed. J.R. Parker) *Geol. Soc.* London, 215–249.

Ravnås, R. and **Bondevik, K.** (1997) Architecture and controls on Bathonian Kimmeridgian shallowmarine synrift wedges of the Oseberg-Brage area, northern North Sea. *Basin Res.*, **9**, 197–226.

Ravnås, R. and **Steel, R.J.** (1998) Architecture of marine rift-basin successions. *AAPG Bull.*, **82**, 110–146.

Ravnås, R., Bondevik, K., Helland-Hansen, W., Lømø, Leif., Ryseth, A. and **Steel, R.J.** (1997) Sedimentation history as an indicator of rift-initiation and development: the Late Bajocian – Bathonian evolution of the Oseberg – Brage area, northern North Sea. *Norsk Geol. Tidskr.*, **77**, 205–232.

Ravnås, R., Nøttvedt, A., Steel, R.J. and **Windelstad, J.** (2000) Syn-rift sedimentary architectures in the Northern North Sea. In: *Dynamics of the Norwegian margin* (Eds A. Nøttvedt, B.T. Larsen, S. Olaussen., B. Tørudbakken, J. Skogseid, R.H. Garbrielsen., H. Brekke and Ø. Birkeland), *Geol. Soc. Spec. Publ.*, **167**, 133–177.

Riding, J.B. (1983) The palynology of the Aalenian (Middle Jurassic) sediments of Jackdaw Quarry, Gloucestershire, *England. Mercian Geol.*, **9**, 111–120.

Riding, J.B. (2006) A palynological investigation of the Brent Group from UK North Sea wells 211/28-H15 and 211/28-H1 (Hutton Field). *Brit. Geol. Surv. Internal Report* **IR/06/100**, 12 pp.

Roberts, A.M., Yielding, G. and **Badley, M.E.** (1995) A kinematic model for the orthogonal opening of the Late Jurassic North Sea Rift System, Denmark-Mid Norway. In: *Tectonic Evolution of the North Sea Rifts* (Eds D.J. Blundell and A.D. Gibbs) Clarendon, Oxford, 180–199.

Rønning, K. and **Steel, R.J.** (1987) Depositional sequences within a "transgressive" reservoir sandstone unit: the Middle Jurassic Tarbert Formation, Hild area, northern North Sea. In: *North Sea Oil and Gas Reservoirs* (Eds J. Kleppe, E.W. Berg, T.A. Buller, O. Hjelmeland and O. Torsæter), Graham and Trotdman, London, 169–176.

Sharp, I.R. Gawthorpe, R.L., Underhill, J.R. and **Gupta, S.** (2000) Fault-Propagation Folding in Extensional Settings: Examples of Structural Style and Synrift Sedimentary Response from the Suez Rift, Sinai, Egypt. GSA Bull., **112**, 1877–1899.

Steckler, M.J., Berthelot, F., Lyberis, N. and **LePichon, X.** (1988) Subsidence in the Gulf of Suez: Implications for rifting and plate kinematics. *Tectonophysics*, **153**, 249–270.

Steel, R.J. (1993) Triassic-Jurassic megasequence stratigraphy in the Northern North Sea: rift to post-rift evolution. In: *Petroleum Geology of Northwest Europe. Proceedings of the 4th Conference.* (Ed. J.R. Parker), *Geol. Soc.* London, 299–315.

Steel, R. and **Ryseth, A.** (1990) The Triasssic - early Jurassic succession in the northern North Sea: megasequence stratigraphy and intra-Triassic tectonics. In: *Tectonic Events Responsible for Britain's Oil and Gas Reserves.* (Eds R.P.F. Hardman and J. Brooks), *Geol. Soc. Spec. Publ.*, **55**, 139–168.

Underhill, J.R. and **Partington, M.A.** (1993) Jurassic thermal doming and deflation of the North Sea: implications of the sequence stratigraphic evidence. In: *Petroleum Geology of Northwest Europe. Proceedings of the 4th Conference.* (Ed. J.R. Parker), *Geol. Soc.* London, 337–346.

Underhill, J.R. and **Partington, M.A.** (1994) Use of maximum flooding surfaces in determining a regional control on the intra-Aalenian (Mid-Cimmerian) sequence boundary: implications for the North Sea basin development and Exxon's sea-level chart. In: *Recent Advances in Siliciclastic Sequence Stratigraphy* (Eds H.W. Posamentier and P.J. Wiemer), *AAPG Mem.*, **58**, 449–484.

Van der Molen, J. and **Van Dijck, B.** (2000) The evolution of the Dutch and Belgian coasts and the role of sand supply from the North Sea. *Global Planet. Change*, **27**, 223–244.

Vollset, J. and **Doré, A.G.** (1984) A revised Triassic and Jurassic lithostratigraphic nomenclature for the Norwegian North Sea. *Norw. Petrol. Direct. Bull.*, **3**, 2–53.

White, N. and **McKenzie, D.** (1988) Formation of the steer's head geometry of sedimentary basins by different stretching of the crust and mantle. *Geology*, **16**, 250–253.

Yielding, G., Badley, M.E. and **Roberts, A.M.** (1992) The structural evolution of the Brent Province. In: *Geology of the Brent Group* (Eds A.C. Morton, R.S. Haszeldine, M.R. Giles and S. Brown). *Geol. Soc. Spec. Pub.*, **61**, 27–43.

Ziegler, P.A. (1982) Geological Atlas of Western and Central Europe. The Hague: Shell Internationale Petroleum, Elsevier Scientific.

Int. Assoc. Sedimentol. Spec. Publ. (2014) **46**, 417–452

Mesozoic and Cenozoic basin configurations in the North Sea

ERLEND MORISBAK JARSVE, JAN INGE FALEIDE, ROY HELGE GABRIELSEN and JOHAN PETTER NYSTUEN

Department of Geology, University of Oslo, P.O. Box 1047, Blindern, N-0316 Oslo, Norway (E-mail: e.m.jarsve@geo.uio.no)

ABSTRACT

The North Sea Basin system is the result of repeated episodes of extension and differential subsidence. In order to inspect changes in style, geometry and basin infill dynamics of basin development in relation to source-to-sink relationships, a series of 2D seismic sections have been studied. This study shows the spatial distribution of source and sink areas in the North Sea area throughout the Late Mesozoic and Cenozoic and, hence, periods of uplift and denudation of the hinterland. The results suggest that the structural framework in the North Sea Basin can be divided into four main basin configurations in Late Mesozoic and Cenozoic times. The tectonic episodes were associated with asymmetrical uplift and erosion of the basin flanks, resulting in temporal, asymmetric, basin configuration. The later basin configurations of the Jurassic graben formation were characterised by a complex system of rotated fault blocks and sub-basins, with infill sourced by sediments derived from the Norwegian mainland. Parts of the sediments were trapped in structural lows at the Horda Platform, Stord Basin and the Egersund Basin, whereas a considerable volume bypassed the platform area. The Jurassic extension was followed by a typical post-rift development where the earliest Cretaceous sedimentation was partly sourced from local, rotated, fault blocks and uplifted areas at the margins of the basin system. This suggests that the basin configuration, which had developed during the syn-rift stage in Late Jurassic, prevailed throughout the early post-rift phase in Early Cretaceous. During the Late Cretaceous the hitherto subaerially exposed margins eventually became drowned. Thus, the subsidence pattern associated with the post-rift stage of the Jurassic to Cretaceous North Sea Basin system, although interrupted by periods of uplift affecting semi-regional and local parts of the basin system, can be traced throughout the Cretaceous. Also, truncation of lower Cretaceous siliciclastics is observed within graben structures developed during the Late Jurassic syn-rift stage, which indicates denudation of a substantial part of the North Sea area. However, from early Palaeogene times, the basin system was characterised by a much wider subsidence area and also by the development of migrating depocentres and uplift of source areas. A western source area dominated in the Palaeogene, switching to a main eastern source area in the Late Palaeogene and Neogene, concurrently with diachronous development of accommodation space across the North Sea area, which continued to develop throughout Neogene times. This indicates lithospheric processes acting on both the hinterland and the basin part of the north-western European continental plate; a structurally coupled source-to-sink system.

Keywords: North Sea, basin configurations, differential subsidence, sequence stratigraphy, structural geology.

From Depositional Systems to Sedimentary Successions on the Norwegian Continental Margin, First Edition.
Edited by A. W. Martinius, R. Ravnås, J. A. Howell, R. J. Steel, and J. P. Wonham.

INTRODUCTION

The tectono-sedimentological development of the North Sea Basin system has been extensively studied during more than 50 years of hydrocarbon exploration. Despite this, many aspects of the development of the North Sea Basin remain less well understood, such as the structural relationship between source-and-sink, or hinterland and basin, basin subsidence and inversion mechanisms, the interlinked effects of tectonics and climate in sediment production and deposition and the eustatic control *versus* tectonic control on creation and destruction of accommodation space.

The North Sea Basin developed as a continental basin in post-Caledonian (Devonian) time, becoming a lowland region of north-west Europe, underlain by the European continental plate, which included Caledonian rocks accreted to the Precambrian crust of Europe. Since the major marine transgression in the Early Jurassic (e.g. Evans *et al.*, 2003), the North Sea Basin system has been, dominantly, a marine epicontinental basin. Sub-basins developed during late Palaeozoic and Mesozoic time through several events of crustal extension and rifting, fault block rotation and basin inversion and, in the Cenozoic, by fault tectonics, halokinesis and differential subsidence, driven by various tectonic mechanisms, as well as differential sediment compaction. Sediment composition, grain size and routing were controlled by bedrock, relief, slope gradient and orientation, development of drainage systems and climate in the hinterlands of the North Sea Basin. Since the early Jurassic, potential sediment source areas to the North Sea Basin system were located to the west, east and south. These source areas were locally supplemented by intermittent intrabasinal highs formed by remnant flanks of rotated fault blocks and basin inversions (Gabrielsen *et al.*, 2001; Kyrkjebø *et al.*, 2001). The marine basin opened to the Borealic Basin in the north and was linked by several shallow-marine seaways to the proto-Tethys and Tethys in the south (e.g. Ziegler, 1982, 1990; Evans *et al.*, 2003; Anell *et al.*, 2012).

The bulk sediment infill was controlled by the evolution through time of the ratio between rate of creation of accommodation space (A) and rate of sediment influx (S). The A/S ratio varied in time and space as a function of intrabasinal and hinterland tectonics, rate and geometry of basin subsidence, climate and eustasy. The complex interplay of all of these factors is reflected in the basin by sediment thickness, textural and mineralogical composition, depositional geometry and sequence stratigraphy.

The scope of this paper is to obtain an improved understanding of some of the complexities in the development of the North Sea Basin system, by studying the geometry of basin infill sequences and their bounding surfaces. This includes the transition between late Jurassic to early Cretaceous syn-rift and post-rift basin stages (Nøttvedt *et al.*, 1995; Gabrielsen *et al.*, 2001), the interplay between basin subsidence and uplift and erosion of basin margins and, not least, the denudation of basin hinterlands, the sediment transport systems and the sediment deposition in basin realms (e.g. Kyrkjebø *et al.*, 2000; Gabrielsen *et al.*, 2001, 2005, 2010a; Faleide *et al.*, 2002). For the Cenozoic, the relative influence of tectonics *versus* climate on sediment yield from the hinterland has become an important issue of discussion (Lindstrom, 1993; Japsen & Chalmers, 2000; Huuse *et al.*, 2002; Gabrielsen *et al.*, 2005, 2010a,b; Doré *et al.*, 2008; Anell *et al.*, 2009, 2010, 2012; Nielsen *et al.*, 2009, 2010a,b; Chalmers *et al.*, 2010; Goledowski *et al.*, 2012; Goledowski *et al.*, 2014; Rasmussen & Dybkjær, 2014).

Our hypothesis is that the North Sea Basin and its hinterland areas, being located on the same continental plate, operated as a coupled structural system and not as a decoupled system. In a structurally coupled source-to-sink system, crustal processes controlling basin subsidence are balanced or linked with uplifts in adjacent hinterland areas; in a totally decoupled system, basin and hinterland areas behave tectonically independently of each other. The study focuses on the Late Mesozoic and Cenozoic depositional systems, particularly emphasising transitions in basin configuration resulting from shifting tectono-thermal stages of development, plate margin tectonics and far field tectonic influences, as well as climate. Based on the variations in sequence development, the study aims also at establishing criteria which can be used to differentiate various basin configurations.

The main study area is situated between 1°E to 8°E and 56°N to 63°N (Fig. 1). However, for the discussion of Cenozoic sediment distribution, the development of the surrounding areas has also been taken into account.

Database and methodology

The present seismic interpretation utilised 60,000 km of long-offset regional 2D reflection seismic data (NSR survey), acquired by Fugro Multi Client Services and TGS Nopec between

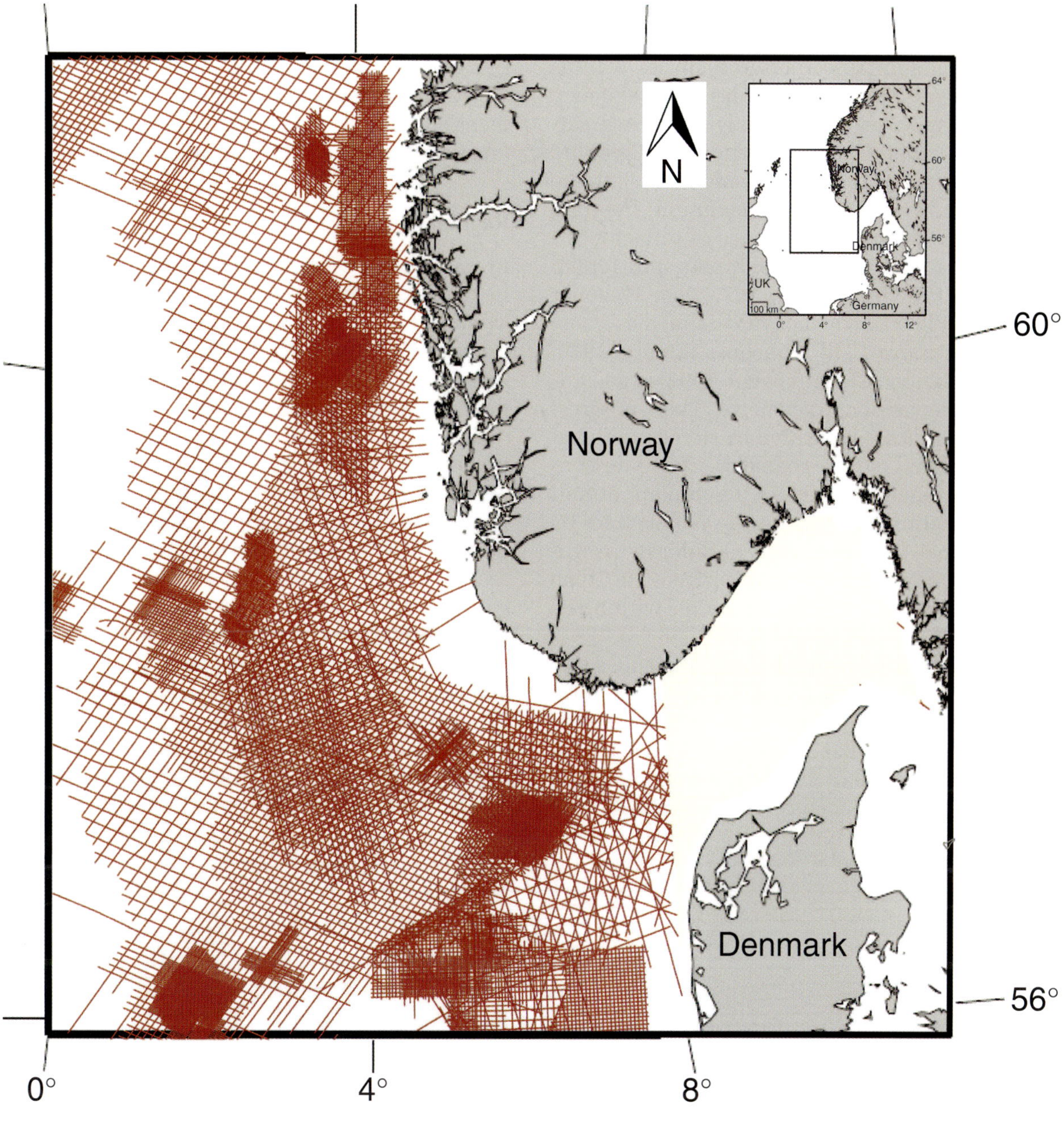

Fig. 1. Location of the study area and the reflection seismic data applied in the present study.

2003 and 2009. Furthermore, several additional 2D seismic surveys were utilised, including the surveys DCS, DK1, DK2, DT97, SP82 and UCGE97, mainly in the Danish sector (Fig. 1). The NSR survey is a regional, state-of-the-art reflection seismic survey, which offers excellent seismic resolution of the Mesozoic, as well as the Cenozoic successions. This survey covers the Norwegian part of the study area. Combined with the other 2D seismic surveys used in this study, the data coverage exceeds that of previous investigations, particularly so for the southern parts of the Norwegian Sector of the North Sea Basin. Thus, the present database allows a more complete interpretation of the Late Mesozoic (Late Jurassic to Cretaceous) and Cenozoic successions.

The seismic data have been tied to a large number of wells which provide biostratigraphic and lithostratigraphic information and are used to calibrate the seismic stratigraphic framework.

The seismic mapping and interpretations are based on seismic stratigraphic principles, such as the identification and mapping of units with seismic stratigraphic patterns indicative of progradation, aggradation and transgression, as well as variations in reflection amplitudes and frequency spectra. Seismic sequence stratigraphy was carried out by identifying sequence stratigraphic surfaces like subaerial unconformities (SU), transgressive surfaces (TS) and maximum flooding surfaces (MFS) from the presence of reflector terminations as downlap, onlap, toplap and toplap truncations (Mitchum *et al.*, 1977; Van Wagoner *et al.*, 1988; Fig. 2). Sequences are defined as transgressive-regressive (T-R) sequences (Embry, 1988, 1993, 1995; Embry & Johannessen, 1992) in accordance with depositional geometry. The transgressive systems tract is identified by retrogradational or aggradational strata geometry above unconformities defined by underlying toplap or/and toplap truncation and regressive systems tract by progradational geometry with clinoforms downlapping on maximum flooding surfaces. In some sequences SU, TS and MFS coincide within the present seismic resolution (Fig. 2).

Each of the seismic stratigraphic sequences was mapped and the results are presented in the form of time-thickness maps and examples from the seismic lines. Particular attention was given to the identification and mapping of positions of maximum thickness (depocentres) for each unit. The sequences are described and discussed in terms of their geometrical characteristics following compaction and for some sequences, also formed by post-depositional erosion. Therefore, time–thickness maps and clinoform heights referred to in the text must be considered as minimum values.

Regional geology

In this section we summarise the regional geology in the North Sea Basin, with emphasis on the Mesozoic and Cenozoic periods, as the depositional patterns presented in this study were highly influenced by the regional tectonics throughout the Late Mesozoic and Cenozoic.

The North Sea Basin configuration reflects two major post-Caledonian basin-forming events, namely the Permo-Triassic and the Late Jurassic to earliest Cretaceous. The Permo-Triassic development affected the entire south-west Norwegian realm including the Skagerrak to Oslo Graben system, whereas the Late Jurassic to earliest Cretaceous phase mainly affected the (present) continental shelf, including the northern, central and eastern North Sea (Badley *et al.*, 1988; Ziegler & van Hoorn, 1989; Gabrielsen *et al.*, 1990; Roberts *et al.*, 1990; Ziegler, 1990, 1992; Yielding & Roberts, 1992; Milton, 1993; Rattey & Hayward, 1993; Nøttvedt *et al.*, 1995; Færseth, 1996;

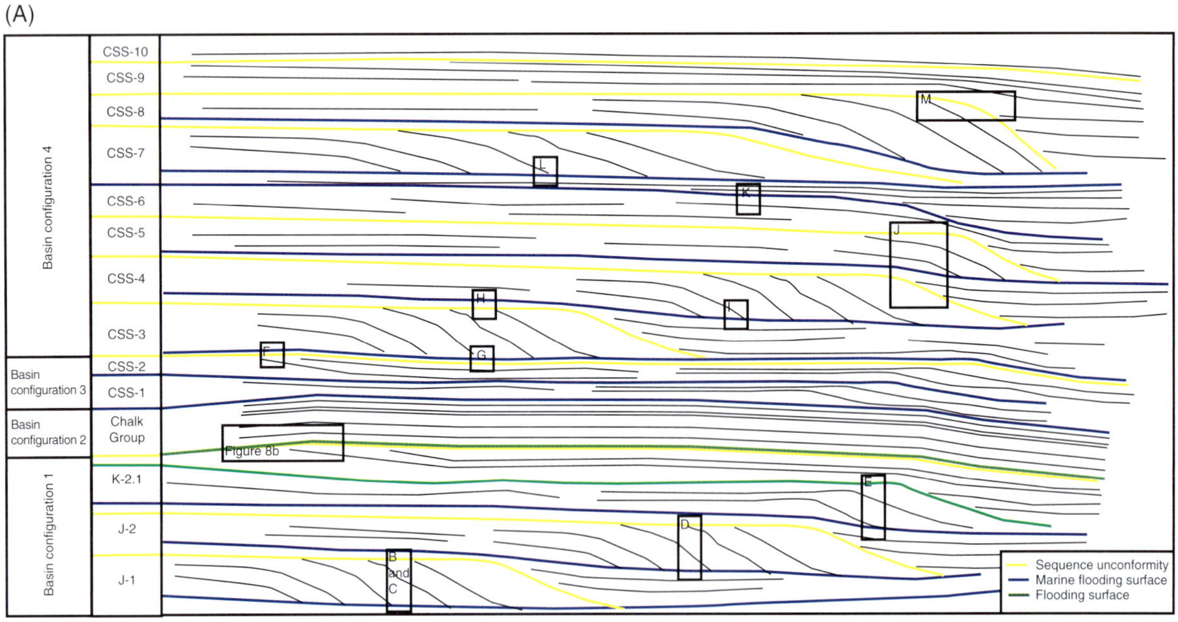

Fig. 2. (A) Schematic figure showing the various sequences recorded in this study, with emphasis on the sequence boundaries.

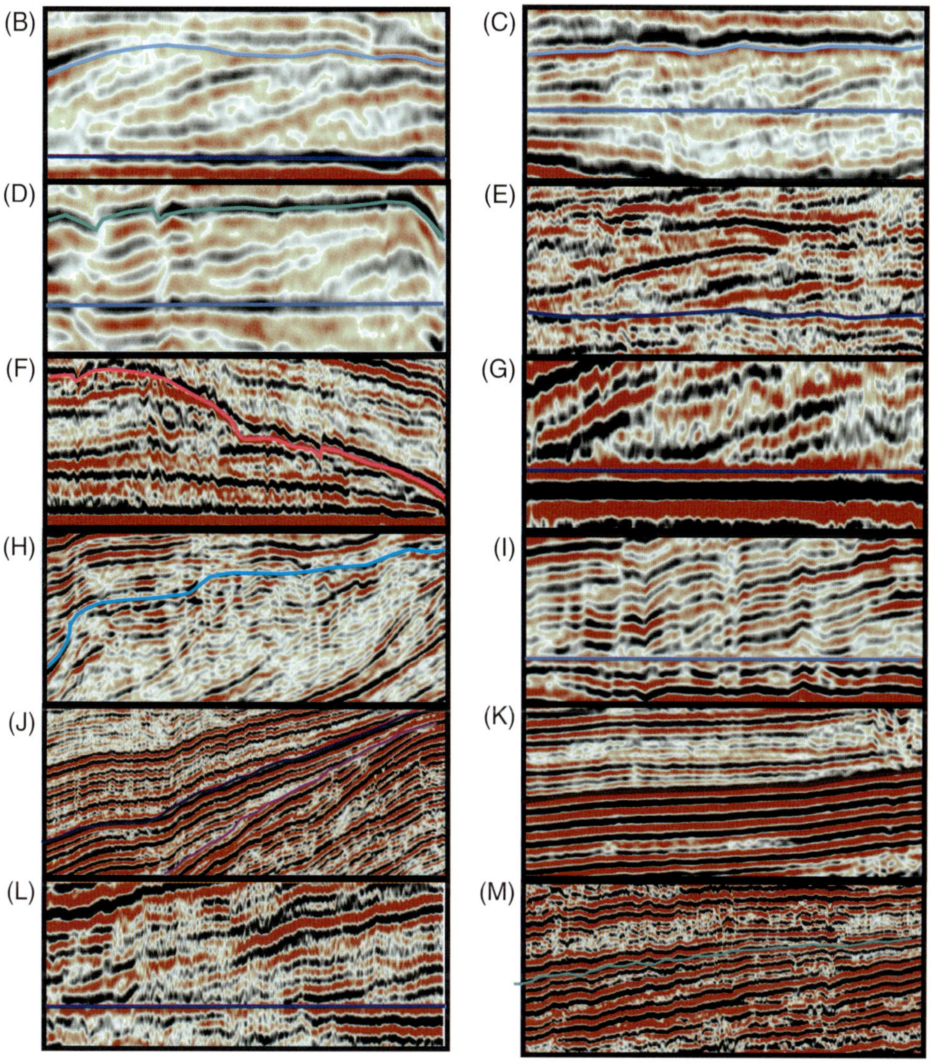

Fig. 2. (*Continued*) (B) Downlap and toplap truncation related to sub-sequence J-2.1 in the Stord Basin/Åsta Graben (flattened on the downlap surface). (C) Downlap and toplap truncation related to sub-sequence J-2.1 in the Egersund Basin (flattened on the downlap surface). (D) Downlap and toplap truncation related to sub-sequence J-2.2 in the Egersund Basin (flattened on the downlap surface). (E) Downlap of sub-sequence K-2.1 in the Egersund Basin. (F) Truncation of Eocene strata in the Norwegian-Danish Basin (flattened on top sequence CSS-1). (G) Downlap of sequence CSS-3 in the Norwegian-Danish Basin (flattened on the downlap surface). (H) Toplap truncation CSS-3 NDB. (I) Downlap sequence CSS-4. (J) Onlap CSS-5 and CSS-6. Truncation at top CSS-5. (K) Toplap and downlap of sub-sequence J-2.1 in the Åsta Grabe/Stord Basin. (L) Downlap CSS-7 onto MFS in lower part of the same sequence (flattened on the downlap surface). (M) Toplap truncation at top sequence CSS-8. Onlap onto the SU from sequence CSS-9.

Færseth *et al.*, 1997; Odinsen *et al.*, 2000; Faleide *et al.*, 2002).

During the Late Carboniferous to Early Permian, the northern and southern Permian basins with their axes oriented E-W to NE-SW, dominated the palaeogeography in the southern and central North Sea area (e.g. Ziegler, 1982, 1987, 1992; Heeremans *et al.*, 2004). In detail, the eastern part of the Northern Permian Basin, which coincides with the Norwegian-Danish Basin, is bounded by the Central Graben and the Coffee-Soil Fault Complex in the west and south and by the Sorgenfrei-Tornquist Zone to the north. This basin is NW-SE striking and contains Palaeozoic, Mesozoic and Cenozoic sedimentary successions (Ziegler & Van Hoorn, 1989; Frederiksen *et al.*, 2001). To the north (Viking Graben area) and in the north-east (Skagerrak-Oslo Graben) N-S-striking rift systems developed throughout the latest Carboniferous to Permian (Ziegler, 1990;

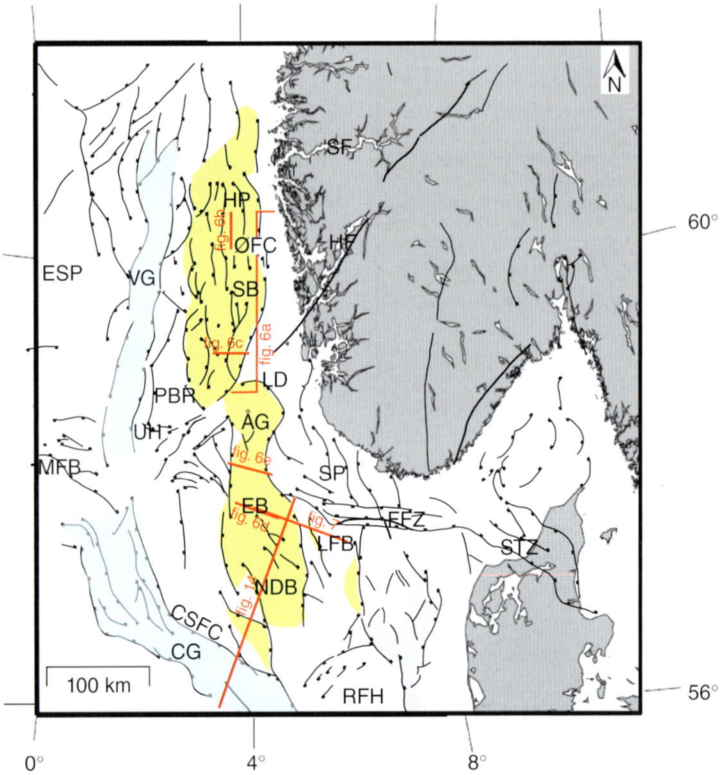

Fig. 3. Main structural elements in the study area (main Permian and Late Jurassic rift basins marked in yellow and blue respectively): Central Graben (CG), Coffee-Soil Fault Complex (CSFC), Egersund Basin (EB), East Shetland Platform (ESP), Fjerrislev Fault Zone (FFZ), Hardangerfjord (HF), Horda Platform (HP), Ling Depression (LD), Lista Fault Blocks (LFB), Moray Firth Basin (MFB), Norwegian Danish Basin (NDB), Patchbank Ridge (PBR), Ringkøbing Fyn High (RFH), Stord Basin (SB), Sognefjord (SF), Stavanger Platform (SP), Sorgenfrei Torquist Zone (STZ), Utsira High (UH), Viking Graben (VG), Øygarden Fault Complex (ØFC) and Åsta Grabe (ÅG).

Gabrielsen *et al.*, 1990; Heeremans & Faleide, 2004). The active stretching stage of this rift event is considered to have ceased in earliest Triassic times. However, the post-rift subsidence stretched well into the Late Triassic (Gabrielsen *et al.*, 1990; Odinsen *et al.*, 2000).

The Late Jurassic to earliest Cretaceous rift phase mainly affected the (present) continental shelf, including the northern, central and eastern North Sea (Badley *et al.*, 1988; Ziegler & van Hoorn, 1989; Gabrielsen *et al.*, 1990; Roberts *et al.*, 1990; Ziegler, 1990, 1992; Yielding & Roberts, 1992; Milton, 1993; Rattey & Hayward, 1993; Steel, 1993; Nøttvedt *et al.*, 1995; Færseth, 1996; Færseth *et al.*, 1997; Odinsen *et al.*, 2000; Faleide *et al.*, 2002). During the Late Jurassic to earliest Cretaceous syn-rift phase, NW to SE trending structures like the Central Graben dominated the central North Sea, whereas structures with NNE to SSW orientation were dominant in the northern North Sea (e.g. Viking Graben: Ziegler, 1987; Ziegler & Van Hoorn, 1989; Fig. 3). During the syn-rift stage, extensive hinterland areas, like the south Norwegian mainland and the East Shetland Platform, sourced the basin. However, several more local source areas, like the uplifted margins of the basins and emergent crests of rotated fault blocks, supplied sediments to the basin system as well (e.g. Gabrielsen *et al.*, 2001, 2010a).

During the Late Cretaceous to Early Palaeocene, the Chalk Group, representing dominantly shallow marine depositional environments, accumulated in the Egersund Basin and Åsta Graben areas to a thickness of up to 900 metres and in the Central Graben area up to 1500 metres (Surlyk *et al.*, 2003). It should be emphasised that chalk sediments have later been completely eroded in areas east and north-east of the Norwegian Channel during the Cenozoic.

During the Late Palaeocene to Eocene, break-up of the NE Atlantic influenced the northern North Sea by the uplift and eastward tilting of the East

Shetland Platform and rapid, regional subsidence in the northern North Sea area (Ziegler, 1982; Jordt *et al.*, 1995; Faleide *et al.*, 2002). The break-up and initial opening generated explosive volcanism, which is reflected by a marked seismic amplitude anomaly of earliest Eocene age, representing volcanic tuff in the study area (Knox & Morton, 1988; Eldholm & Thomas, 1993).

The Palaeogene sediments of the North Sea were mainly derived from the East Shetland Platform in the west (Jordt *et al.*, 1995, 2000; Michelsen *et al.*, 1995; Faleide *et al.*, 2002; Anell *et al.*, 2009, 2012). These authors noted a marked change in the Neogene, where the main source area appears to have switched from west to become situated to the east and north-east. This change is traditionally considered to reflect the onset of a renewed stage of vertical uplift of Scandinavia. However, there is an ongoing debate as to whether the change in source area, from west in the early Palaeogene to east and north-east in the Oligocene and the Neogene, is a response to a renewed vertical uplift of the south Norwegian dome in central, western South Norway (e.g. Jordt *et al.*, 1995, 2000; Michelsen, 1995; Rohrman *et al.*, 1995; Martinsen *et al.*, 1999; Japsen & Chalmers, 2000; Japsen *et al.*, 2002; Faleide *et al.*, 2002; Japsen *et al.*, 2007; Anell *et al.*, 2009, 2012; Gabrielsen *et al.*, 2010a,b), or whether this dome is just a topographic feature inherited from the Silurian-Devonian Caledonian mountain chain and lowered through time by erosion controlled by eustasy and climate changes (e.g. Huuse *et al.*, 2001; Huuse, 2002; Nielsen *et al.*, 2009). The first interpretation is a structurally coupled source-to-sink model, whereas the latter interpretation implies a structurally decoupled source-to-sink model.

Seismic stratigraphic sequences and lithostratigraphy

The principal lithostratigraphic scheme for the northern and eastern North Sea was presented by Deegan & Scull (1977), later to be modified by Morton (1982), Stewart (1987) and Mudge & Copestake (1991). A revised nomenclature for the Triassic and Jurassic lithostratigraphy for the Norwegian North Sea was given by Vollset & Doré (1984) and of the Cretaceous and the Tertiary by Isaksen and Tonstad (1989). The seismic stratigraphic sequence subdivision of the Cretaceous succession, as applied in the present paper, builds on work by Gabrielsen *et al.* (2001) and for the Cenozoic succession on the work by Jordt *et al.* (1995, 2000), Michelsen & Danielsen (1996) and Faleide *et al.* (2002) (Fig. 4). The present work includes a similar seismic stratigraphic subdivision of the Upper Jurassic succession (Oxfordian to Tithonian) (Fig. 4). However, some adjustments concerning the subdivision of the Cenozoic succession are suggested. This modification mainly concerns the Oligocene of the eastern North Sea Basin.

The Upper Jurassic is divided into two seismic stratigraphic sequences, namely the J-1 and J-2, of Oxfordian and Volgian age respectively. Here, sequence J-1 is equivalent with sequence A; and sequence J-2 is time-equivalent with sequences B to E, in Fraser *et al.* (2002). The sequence J-2 is further subdivided into two sub-sequences (J-2.1 and J-2.2) where subsequence J-2.1 includes the Flekkefjord Formation and subsequence J-2.2 includes the Sauda Formation. The Cretaceous succession is divided into sequences K-1 to K-6, where sequences K-1 and K-2 represent the Early Cretaceous (Berriasian to Albian) and sequences K-3 to K-6 the Late Cretaceous to early Palaeocene (Cenomanian–Danian; Gabrielsen *et al.*, 2001).

The Cretaceous seismic stratigraphic sequences are dated based on more than 50 released wells, mainly in the Norwegian northern North Sea (Gabrielsen *et al.*, 2001). The seismic stratigraphic scheme of the Cenozoic succession is generally in accordance with that suggested by Jordt *et al.* (1995). Hence, the Cenozoic succession is divided into sequences CSS-1 to CSS-10, where sequences CSS-1 and CSS-2 represent the latest Palaeocene to earliest Eocene and the early to late Eocene, respectively. The upper surface of CSS-1 is time-equivalent with the top Balder Formation. Sequences CSS-3 and CSS-4 are early and late Oligocene in age, whereas sequences CSS-5 to CSS-7 are early, middle and late Miocene. Sequence CSS-8 is Pliocene of age and sequences CSS-9 and CSS-10 are dated to the early and late Pleistocene (Fig. 4). Jordt *et al.* (1995, 2000) dated the Cenozoic sequences based on biostratigraphic zonation of calcareous nannofossils, dinoflagellates and foraminifera (Steurbaut *et al.*, 1991; Eidvin & Riis, 1992; Gradstein & Bäckström, 1996). However, in this work we also include dating and correlation of the Cenozoic sequences in the eastern North Sea (Danish sector: Rasmussen & Dybkjær, 2005; Dybkjær & Rasmussen, 2007, Schiøler *et al.*, 2007; Dybkjær *et al.*, 2012).

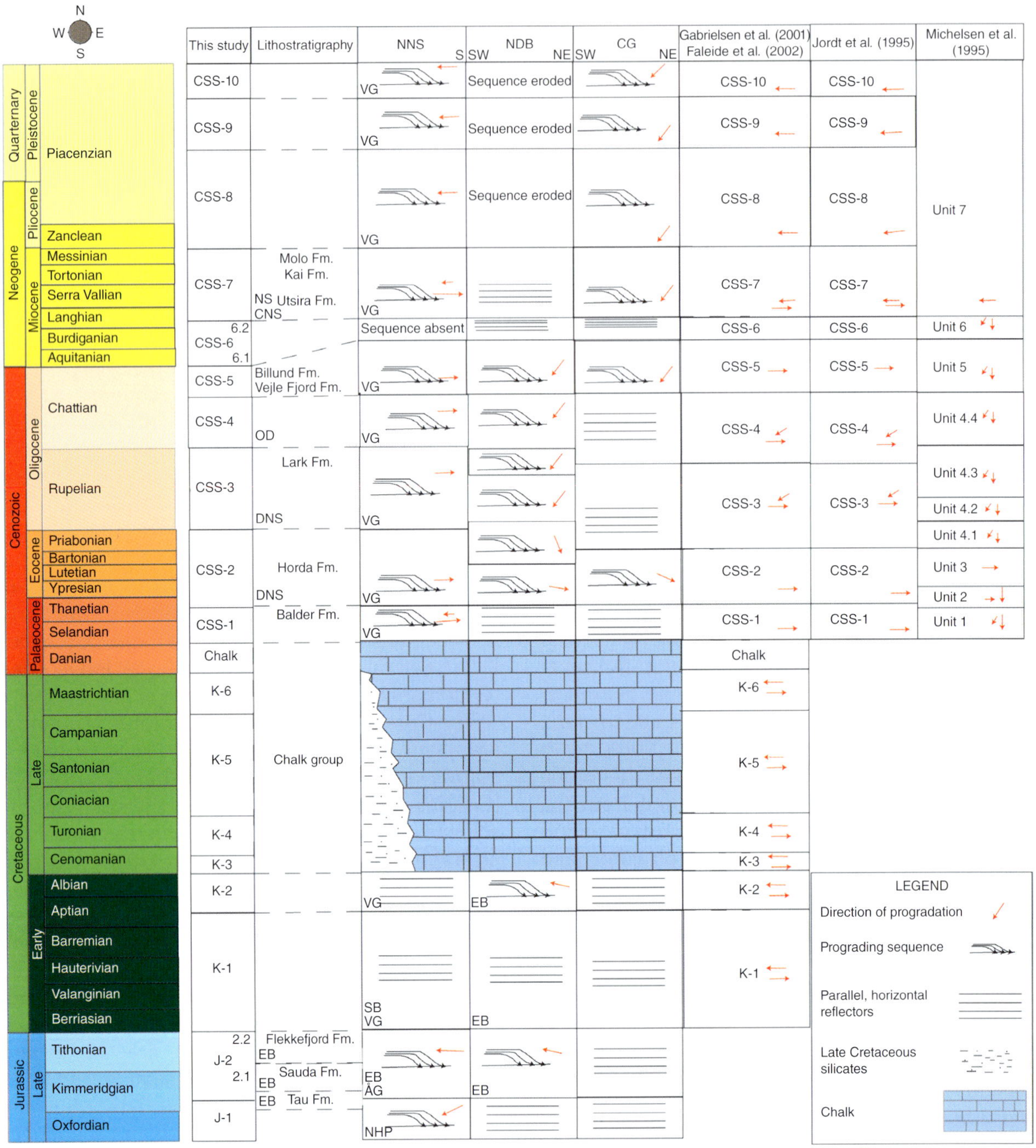

Fig. 4. Lithostratigraphic and seismic stratigraphic scheme in the northern North Sea, Norwegian-Danish Basin and Central Graben. The seismic sequences are correlated to previous work. Arrows show direction of progradation and aggradation. VG = Viking Graben, ÅG = Åsta Graben, EB = Egersund Basin, NHP = northern Horda Platform, DNS = Danish North Sea, OD = onshore Denmark, NS = Norwegian Sea, CNS = Central North Sea. Lithostratigraphy is from Deegan & Scull (1977), Eidvin *et al.* (2007), Schiøler *et al.* (2007) and Rasmussen *et al.* (2010).

In this work we propose that the study area has experienced four major changes in basin configuration in late Mesozoic and Cenozoic time. The four basin configurations are based on different basin infill patterns and basin geometry, including width, depth and orientation. These are the Late Jurassic to Early Cretaceous (basin configuration 1), Late Cretaceous (basin configuration 2), Early Palaeogene (basin configuration 3) and Late Palaeogene to Pleistocene (basin configuration 4).

Accordingly, the seismic sequences described in this paper are divided according to the various basin configurations and this subdivision is further discussed in the discussion chapter.

Seismic sequences

In this section, we present all the seismic sequences which reflect the major changes in basin configuration in the study area. The sequences are presented according to the basin configurations that they are proposed to be related to. These include (see Fig. 4) basin configuration 1, with sediments deposited in the Late Jurassic and Early Cretaceous sequences (J-1 to K-2, Fig. 4), basin configuration 2, reflected by the Late Cretaceous sequences (K-3 to K-6; Chalk Group, Fig. 4), basin configuration 3, represented by the Palaeocene and Eocene sequences (CSS-1 and CSS-2; Fig. 4) and basin configuration 4 which reflects the Late Palaeogene to Pleistocene sequences (CSS-3 to CSS-10; Fig. 4). Criteria for distinguishing these basin configurations are further proposed and discussed in the discussion chapter.

Sequences of the latest syn-rift and early post-rift stages: Basin configuration 1

In reflection seismic data, the Upper Jurassic and Lower Cretaceous sequences are well imaged across the entire study area. The structuring is dominated by moderate faulting in the platform areas and more intense deformation and development of rotated fault blocks in the sub-platform settings (Kyrkjebø *et al.*, 2001), typical of the syn-rift and early post-rift stages. In the northern North Sea, part of the sediment was trapped in structural lows on the northern Horda Platform (including the Troll Field) whereas a considerable volume bypassed the platform area (Gabrielsen *et al.*, 1990, 2001).

Outside the Viking Graben, four main progradational systems are recorded (Fig. 5A), namely in the northern Horda Platform area (J-1), the Åsta Graben and the Stord Basin areas (J-2.1) and in the Egersund Basin (J-2.1 and J-2.2). The relationship between sequences J-1 and J-2.1 is illustrated in Fig. 6A. In the Norwegian-Danish Basin, the Upper Jurassic and Lower Cretaceous successions

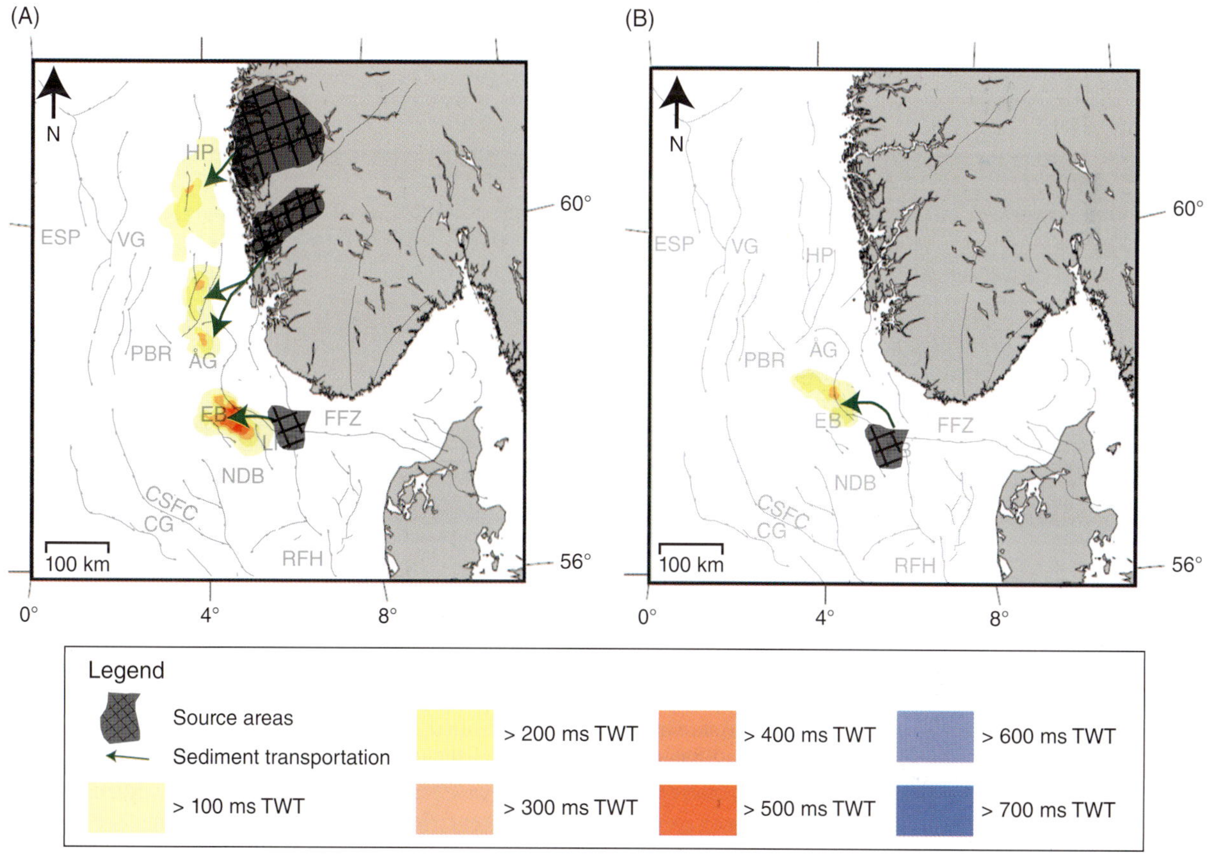

Fig. 5. Time–thickness maps, stage 1. Sequences J-1 and J-2 in (A) and sequence K-2.1 in (B). Abbreviations see Fig. 3.

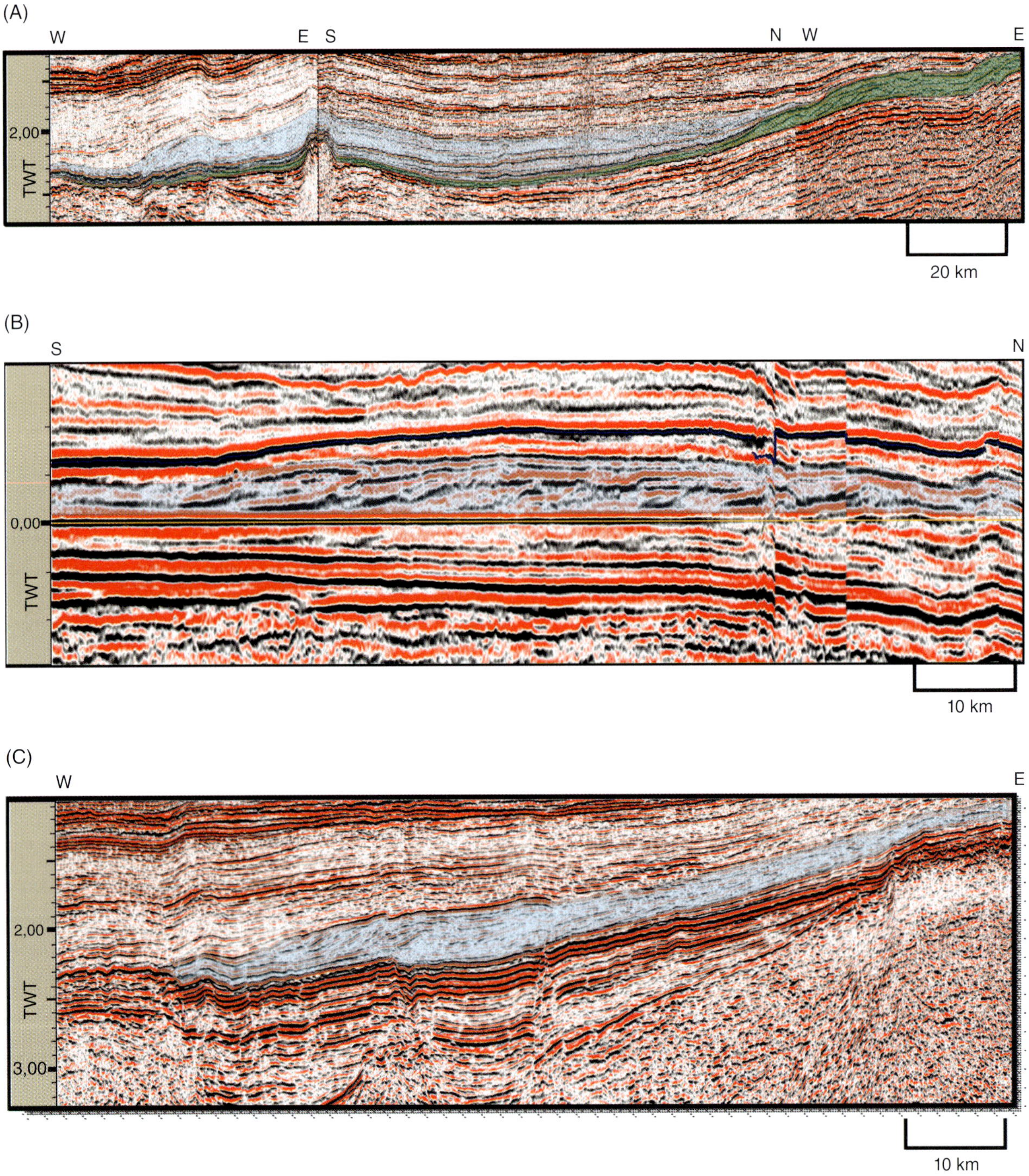

Fig. 6. Seismic lines (see Fig. 5 for location). (A) Arbitrary line from J-1 on the northern Horda Platform (marked in green) and J-2.1 in the Stord Basin/Åsta Graben (marked in blue). (B) Clinoforms belonging to sequence J-1 on the Horda Platform stacked in a prograding pattern (marked in blue). Note truncation of the topset. The seismic section is flattened on the downlap surface. (C) J-2.1 in the Stord Basin/Åsta Graben (marked in blue). Note the progradation of the clinoforms. (D) Progradation of J-2.1 and J-2.2 in the Egersund Basin. (E) K-2.1 in the Egersund Basin (marked in blue).

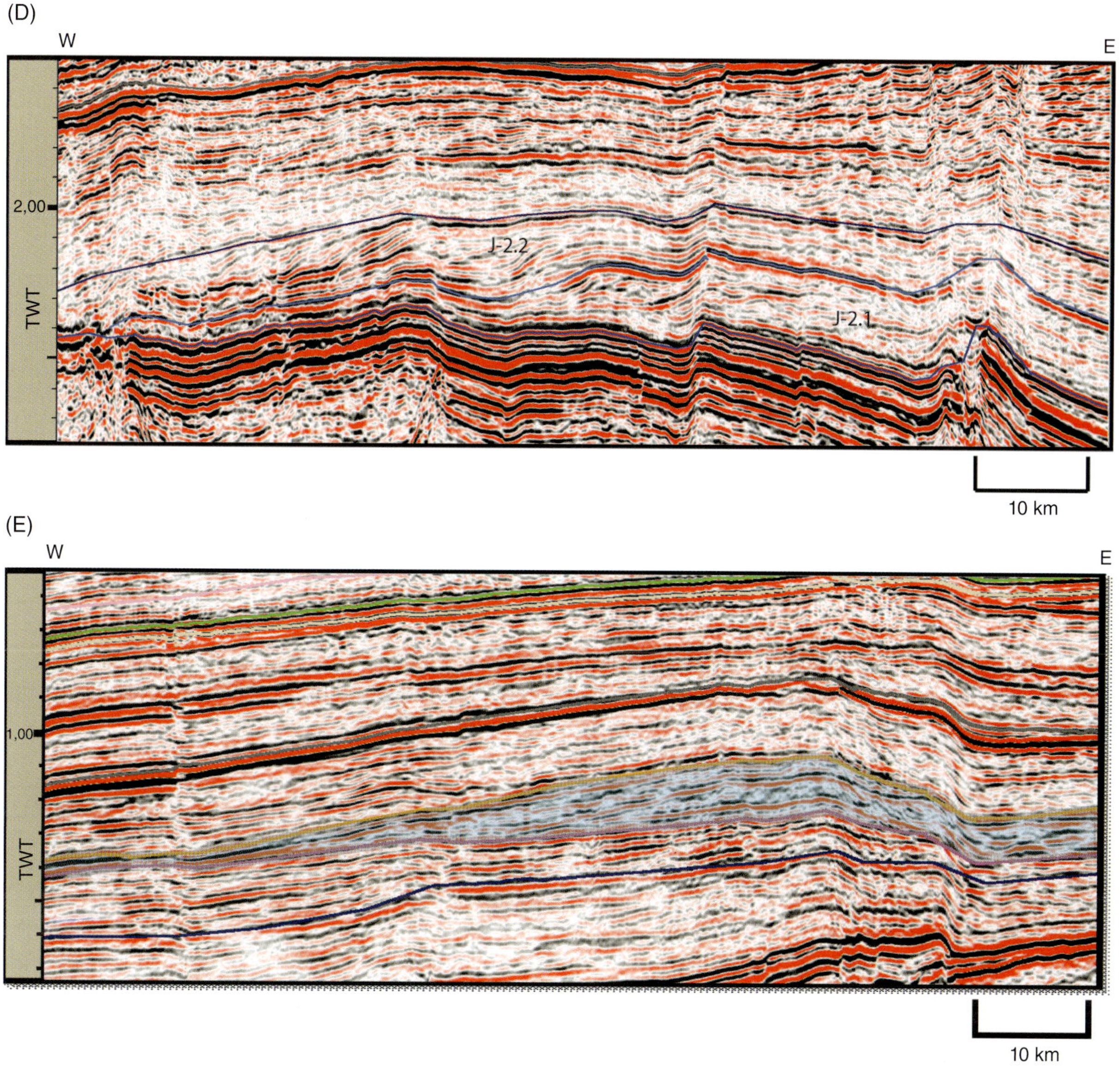

Fig. 6. (*Continued*)

are thin (<200 m) and only minor faulting occurred during this period.

Oxfordian (sequence J-1)

Description: The sequence J-1 is mainly found in the northern Horda Platform area. It displays good data quality in the southern part and south of the present outlet of Sognefjord, whereas poor data quality is typical seen across the Horda Platform north of the present outlet of Sognefjord (Fig. 5A).

The maximum thickness of sequence J-1 is located south-west of Sognefjord, where it reaches 300 ms TWT (Fig. 5A). The sequence thins southwards to less than 100 ms TWT in the Åsta Graben area; and also westwards (Fig. 5A). The bottom bounding surface of sequence J-1 is characterised by downlap on top of a high seismic reflector which represents sediments draping an erosional surface (Fig. 6A and B). The upper surface is characterised by toplap truncation, as seen in the area of maximum thickness (Figs 5A and 6B).

The sequence incorporates SW dipping, oblique clinoforms, where the topset is truncated below the offlap break (Figs 5A and 6B). The clinoform heights are approximately 200 ms TWT.

Interpretation: The lower bounding surface is interpreted as a combined sequence unconformity (SU) and maximum flooding surface (MFS), based on the erosional surface and downlap pattern from sequence J-1. The oblique shape of the clinoforms indicates that the sediments were deposited in a high-energy environment, with mainly coarse-grained material trapped in the clinothems (e.g. Mitchum *et al.*, 1977). This is in agreement with observations by Gibbons *et al.*, (1993) and Goldsmith (2000), who concluded that the Upper Jurassic succession in the northern Horda Platform area consists of shoreface to proximal shelf sandstones interbedded with lagoonal to marine embayment facies. The SW direction of progradation indicates the existence of a source area to the north-east. The accommodation space created during the syn-rift stage was completely infilled, as indicated by the truncation of the topsets (Fig. 6B).

Volgian (sequence J-2)

Description: For sequence J-2, three independent prograding systems have been mapped in the study area, namely in the Åsta Graben-Stord Basin area (J-2.1) and in the Egersund Basin (J-2.1 and J-2.2) (Fig. 5A). The top of the J-2.1 sub-sequence is equivalent to the top Sauda Formation, whereas the top of J-2.2 sub-sequence is equivalent to the top Flekkefjord Formation (Fig. 4). The data quality of both sub-sequences is excellent (Figs 6C and D).

In the Åsta Graben-Stord Basin, the bottom bounding surface of sub-sequence J-2.1 is characterised by onlap against the prograding J-1 sequences offshore Sognefjord, whereas the upper sequence boundary is characterised by toplap truncation, as seen in the Åsta Graben area (Figs 2 and 6C). A high and continuous seismic reflector is recorded internally in sequence J-2.1, which also acts as a downlap surface in the Åsta Graben area (Fig. 6C). Similar boundary conditions are recorded for sub-sequences J-2.1 and J-2.2 in the Egersund Basin (Figs 2 and 6D). In the Åsta Graben-Stord Basin, the maximum thickness of sequence J-2 is 300 ms TWT and the sub-sequence pinches out towards the Utsira High in the west. In the Egersund Basin, sequence J-2 reveal thicknesses up to 500 ms TWT (Fig. 5A). Outside these basinal areas, sequence J-2 has a maximum thickness of less than 100 ms TWT (Fig. 5A). Both sub-sequences J-2.1 and J-2.2 in the Egersund Basin pinch out towards the west and south-west. On the Utsira High and the Jæren High they are absent or have an accumulated thickness below that of the seismic resolution.

In the Åsta Graben-Stord Basin, sub-sequence J-2.1 displays westerly to SW dipping clinoforms (Fig. 6C) which have a tangential, oblique geometry. The height range of the clinoforms is 150 ms TWT to 350 ms TWT.

In the Egersund Basin, the sub-sequence J-2.1 represents a W-SW prograding system, which incorporates W-SW dipping clinoforms (Fig. 6D). Here, sub-sequence J-2.1 is characterised by truncation below the offlap break of the topset (Fig. 7). The maximum height of the clinoforms is approximately 250 ms TWT and their dip is approximately 4°. They have a parallel oblique geometry (Fig. 6D).

The J-2.2 sub-sequence, which is only recorded in the Egersund Basin, shares the same characteristics

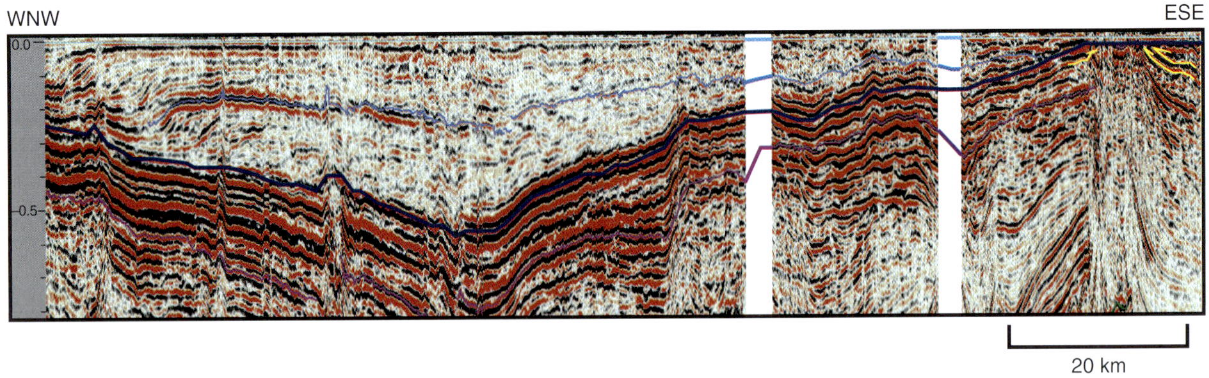

Fig. 7. Correlation of sequence J-2 in the Egersund Basin and the erosional surface on the Lista Fault Blocks. The section is flattened on the top J-2.2 sub-sequence. Note the truncation on the Lista Fault Blocks and onlap patterns alongside the structure.

as sub-sequence J-2.1 and is characterised by a prograding pattern with dip-direction of the clinoforms towards W-SW (Fig. 6D). The clinoforms are downlapping onto the sub-sequence J-2.1 with a parallel oblique geometry. The maximum height of the clinoforms is approximately 400 ms TWT and they dip between 3° and 4°.

The J-2 sequence, as seen in the Egersund Basin, onlaps a structure at the Lista Fault Blocks in the east, which is also associated with an erosional surface in the Upper Jurassic (Fig. 7).

Interpretation: The downlap surfaces within both sub-sequences are interpreted as maximum flooding surface (MFS) based on their downlap patterns, both in the Åsta Graben-Stord Basin and in the Egersund Basin. In the Egersund Basin, sub-sequences J-2.1 and J-2.2 are divided by the upper sequence unconformity of sub-sequence J-2.1. The sediments of sub-sequence J-2 were sourced from the east, into the northern Horda Platform area, indicating a continued depositional history from the sub-sequence J-1. The W-SW progradation of sub-sequence J-2.1 in the Åsta Graben-Stord Basin may reflect a source area to the E-NE. The sub-sequence J-2.2 in the Egersund Basin was deposited to the south-west of sub-sequence J-2.1, which is taken to indicate that the shoreline was displaced farther into the basin during deposition of sub-sequence J-2.2. The source area may have been the Lista Fault Blocks, based on the erosional surface, which is correlated to the Jurassic sequence J-2 (Fig. 7). The water depths appear to have been approximately the same during deposition of sub-sequences J-2.1 and J-2.2 in the Egersund Basin, as shown by the clinoform heights, although exposure of the shorefaces is indicated by the upper sequence boundaries of both sub-sequences. The oblique geometry of the prograding systems belonging to sequence J-2 is typical for coarse-grained systems (Mitchum *et al.*, 1977; Veeken, 2007).

Aptian-Albian (sub-sequence K-2.1)

Description: In the study area, one prograding sub-sequence (K-2.1), which is incorporated in the K-2 sequence of Gabrielsen *et al.* (2001; Aptian to Albian age), is recorded in the area west of the Stavanger Platform and the Lista Fault Blocks. The data quality of the Cretaceous succession is good, although seismic mapping is challenged by low contrasts in acoustic impedances. However, mapping the sub-sequence K-2.1 was unproblematic.

The lower bounding surface of sub-sequence K-2.1 corresponds with the upper SU of sequence J-2 in the Egersund Basin, whereas the upper bounding surface is characterised by onlap from younger Lower Cretaceous strata. A downlap surface is recorded internally in sub-sequence K-2.1 (Fig. 6E). The sub-sequence K-2.1 has a maximum thickness of 400 ms (TWT) (Fig. 6E) and it pinches out towards the Utsira High in the west and against Sele High and the southern boundary of the Egersund Basin in the south-west and south, respectively. Below the downlap surface, sub-sequence K-2.1 consists of parallel, sub-horizontal seismic reflectors whereas SW dipping clinoforms are seen above the downlap surface (Fig. 6E). The clinoforms included in sub-sequence K-2.1 have a sigmoidal geometry. The height of the clinoforms is approximately 250 ms TWT.

Interpretation: The sigmoidal geometry of the prograding clinoforms indicates that sub-sequence K-2.1 consists mainly of fine-grained material (e.g. Bullimore *et al.*, 2005). The prograding K-2.1 probably represents shelf deposits, based on the clinoform heights. This is in accordance with Wood & Barton (1983) and White & Latin (1993) who state that the Lower Cretaceous succession in the Norwegian-Danish Basin is characterised by a package of sediments characterised by horizontal reflections representing inner shelf deposits. The pinching out of K-2.1 towards the Utsira High, the Sele High and the southern boundary of the Egersund Basin indicates that in the Early Cretaceous the Egersund Basin was separated from the Viking Graben in the west by the Utsira High and the Sele High, like it was in the Late Jurassic. The sediment transport may have been through the Hardangerfjord, similar as for sub-sequence J-2.1 in the Stord Basin-Åsta Graben (Fig. 5B).

Sequences of the late post-rift stage: Basin configuration 2

Basin configuration 2 includes the Late Cretaceous to Danian Chalk Group. At the time of deposition, most of the structures which acted as local source areas during basin configuration 1 had been eroded to base level and subsequently transgressed. This gave rise to a continuous basin that extended across the entire study area (e.g. Gabrielsen *et al.*, 2001; Kyrkjebø *et al.*, 2001; Surlyk *et al.*, 2003), which is the most prominent feature of basin configuration 2.

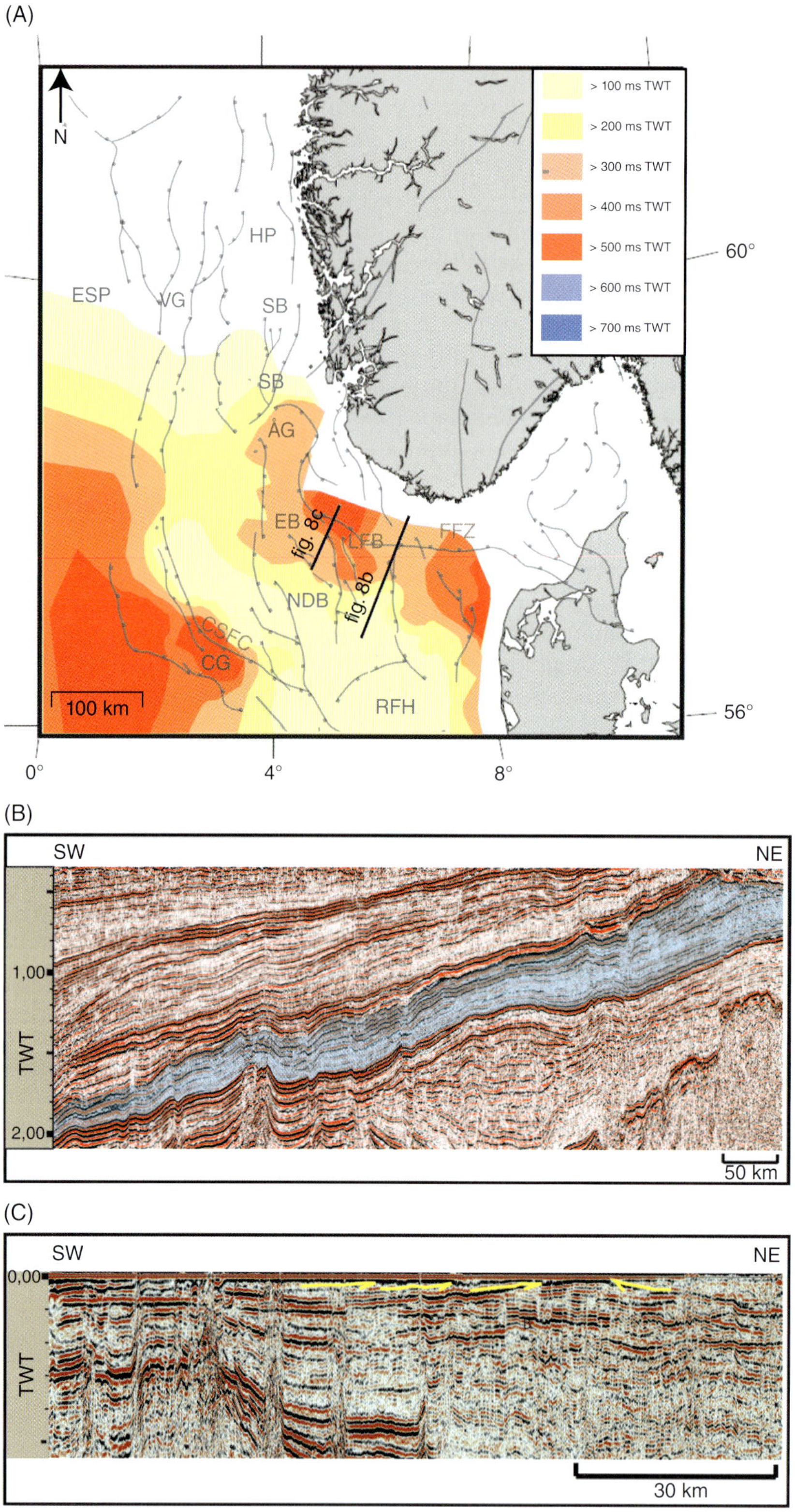

Fig. 8. (A) Time–thickness map of the Late Cretaceous Chalk Group. (B) Seismic section with the chalk sequences marked in blue. Notice the increase in thickness towards north-east. For location of the seismic line, see A. (C) The seismic line is flattened at base of Chalk Group. Note the truncation of the lower Cretaceous strata, indicated by yellow arrows.

Cenomanian–Danian (sequences K-3–K-6)

Description: The Chalk Group consists of pelagic chalk and is present in the study area with thicknesses up to 500 m to 800 m, as seen in the Norwegian-Danish Basin (Clausen & Huuse, 1999), and more than 1500 m in the Central Graben (Surlyk *et al.*, 2003). The Chalk Group was deposited in a quite setting in an epicontinental sea, although locally affected by ocean bottom currents (e.g. Esmerode *et al.*, 2008). Sediments belonging to the basin configuration 2 are seen across the whole study area. However, the thickness varies greatly from south to north (Fig. 8). The data quality is very good across the entire study area, revealing internal seismic reflectors belonging to the Chalk Group.

Both the lower and the upper boundaries of the Chalk Group are marked by high-amplitude reflections. The internal seismic reflection pattern is parallel to sub-parallel. At some structural features, as in the area of the Utsira High and in the area at the Øygarden Fault Complex, the Chalk Group rests on an angular unconformity. An erosional surface is also indicated beneath the Chalk Group in the Egersund Basin, where the group rests unconformable on Lower Cretaceous strata (Fig. 8).

The sediments of the Chalk Group are less than 100 ms TWT thick in the south-western parts of the Norwegian-Danish Basin, across the Ringkøbing-Fyn High and along the Coffee-Soil Fault Complex. In contrast to this the Chalk Group thickens gradually towards mainland southern Norway in the north-eastern parts of the Norwegian-Danish Basin (>500 ms TWT; Figs 8 and 9). In the Norwegian Channel, where it is preserved beneath an erosional unconformity at the base of the Pleistocene, the Chalk Group thins rapidly due to Plio-Pleistocene glacial erosion. The Chalk Group thins northwards in the northern North Sea Basin until it becomes absent in parts of the northern Viking Graben area.

Interpretation: The angular unconformity recorded beneath the Chalk Group on the Utsira High and along the Øygarden Fault Zone indicates subaerial exposure of these structural features in Early to Middle Cretaceous time. This was followed by transgression on top of these SUs and with deposition of the Chalk Group during rising sea-level and highstand. The Egersund Basin was probably infilled during Early Cretaceous with subsequent exposure and later transgression. The recorded increase in thickness of the Chalk Group towards mainland Norway is in this respect of particular interest and will be discussed further below.

Deposits influenced by rifting and initial opening of the NE Atlantic: Basin configuration 3

In the late Palaeocene and Eocene times, deposition of sequences CSS-1 and CSS-2 corresponded with uplift and eastward tilting of the East Shetland Platform and rapid regional subsidence in the northern North Sea area, which was not related to faulting (Ziegler, 1990; Milton *et al.*, 1990; Jordt *et al.*, 1995, 2000; Faleide *et al.*, 2002; Anell *et al.*, 2011).

Late Palaeocene to earliest Eocene (sequence CSS-1)

Description: Sequence CSS-1 conformably overlies the Upper Cretaceous to lowest Palaeocene Chalk Group and the upper boundary is equivalent to the top of the Balder Formation. Both boundaries are characterised by and mapped along strong amplitude anomalies in the Central North Sea area, which conformably overlie the Chalk Group and underlie the Eocene sequence, respectively (Fig. 9). The data quality of sequence CSS-1 is poor throughout the study area, although upper and lower boundaries are easily traced along the amplitude anomalies. Outside this area, the identification of seismic reflections inside the section is hampered by low acoustic impedance contrast, although the lower and upper seismic reflections are easily mapped along their amplitude anomalies. One depocentre of sequence CSS-1, dated to late Palaeocene to early Eocene, is located in the southern Viking Graben area (≥700 ms TWT). Sequence CSS-1 in this sub-basin thins northward along the axis of the Viking Graben structure. In the northern North Sea, another depocentre of the same age is documented offshore Sognefjord and a third is located in the Ling Depression offshore Hardangerfjord (>300 ms TWT; Fig. 10A). In the south-eastern parts of the study area, in the Norwegian-Danish Basin, sequence CSS-1 reveals one minor depocentre (>200 ms TWT; Fig. 10A). The CSS-1 sequence, including the depocentre offshore Sognefjord, is cut by an erosional unconformity closest to the Norwegian mainland.

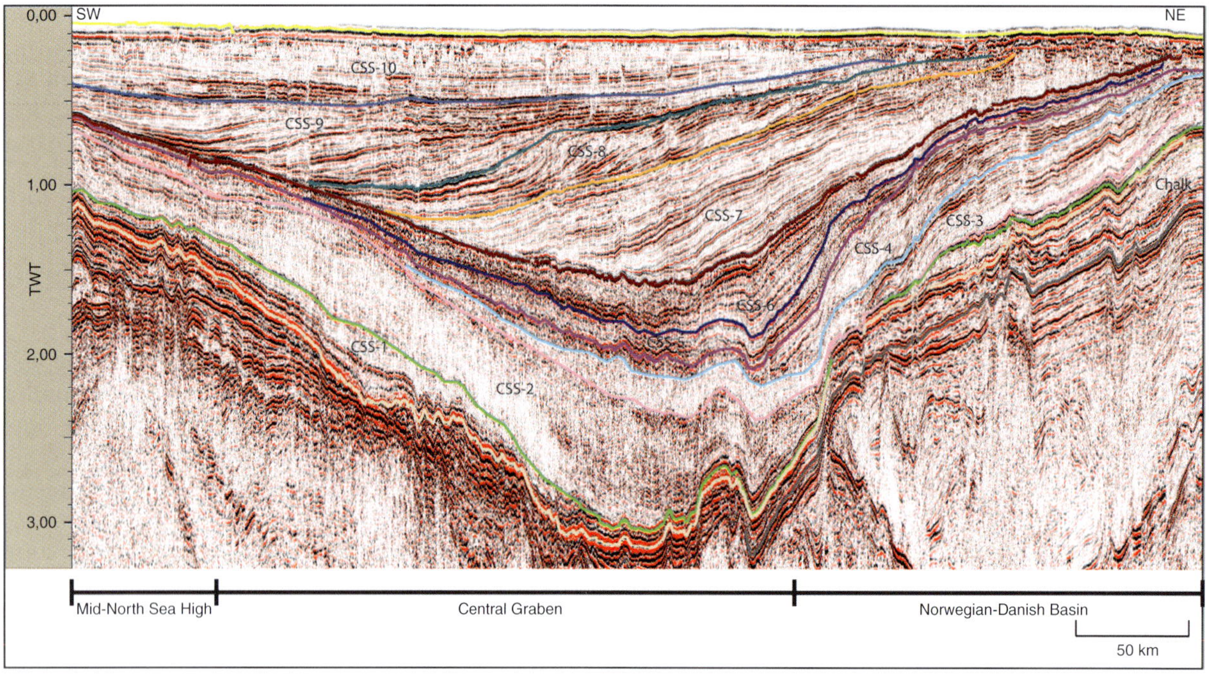

Fig. 9. Seismic line across the Central Graben, illustrating the Cenozoic sequences. Note the major input of Oligocene CSS-3 and CSS-4 sequences in the Norwegian-Danish Basin, sourced from the northeast, and the major sediment input in Miocene time, particularily sequence CSS-7, in the Central Graben, also sourced from the northeast. Also, note the Pliocene CSS-8 and early Pleistocene CSS-9 sequences, prograding towards the southwest. Late Pleistocene sequence CSS-10 is sub-parallel with the seabed.

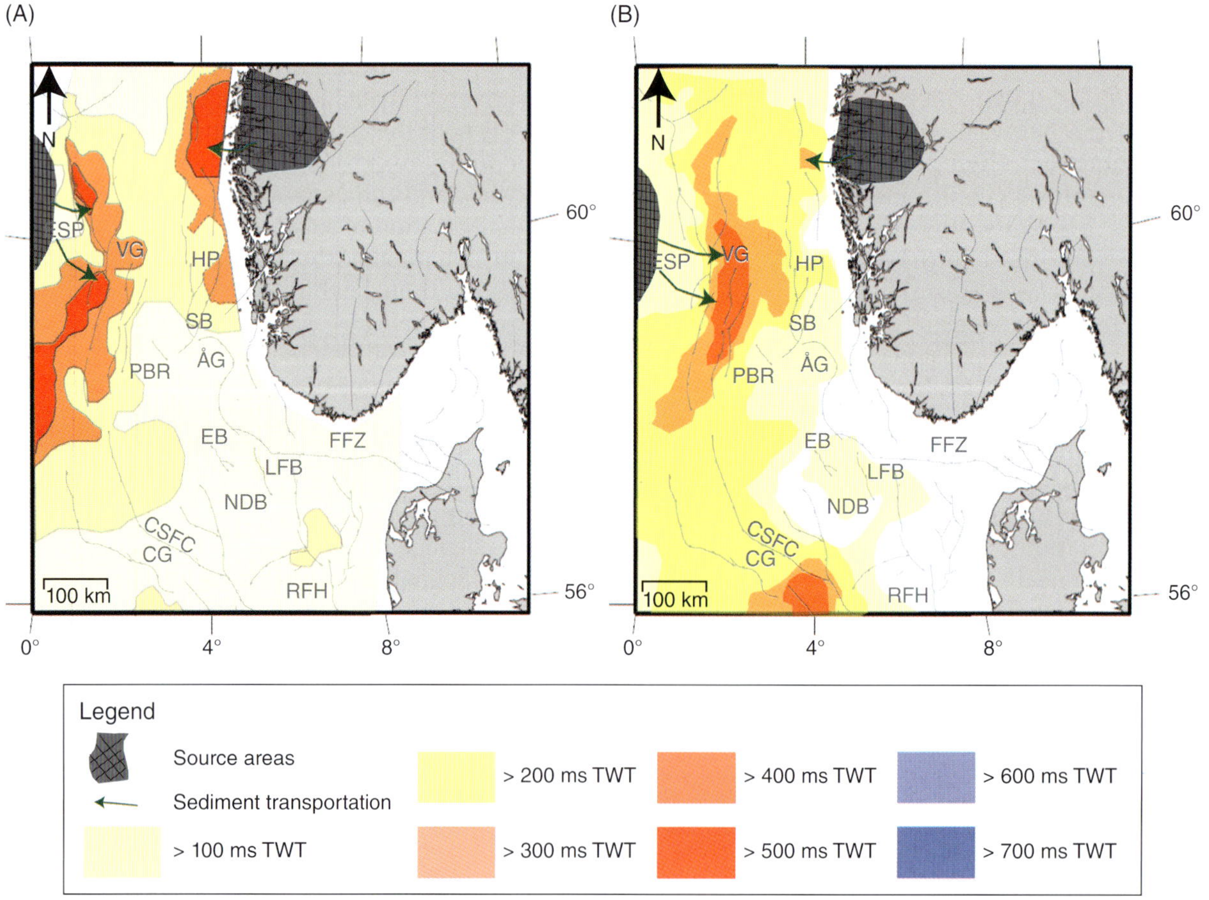

Fig. 10. Time–thickness maps of CSS-1 (A) and CSS-2 (B).

Easterly-dipping clinoforms are observed at the western margin of the Viking Graben segment.

Interpretation: The amplitude anomaly in the reflection seismic data, picked as the upper bounding seismic reflector of sequence CSS-1, coincides with a bentonite bed in the Balder Formation (early Eocene; Spjeldnæs, 1975; Thyberg *et al.*, 2000). This correlates to the time interval when continental break-up took place in the North Atlantic (e.g. Eldholm *et al.*, 2002; Faleide *et al.*, 2008). Eastward-dipping clinoforms in the western part of the Viking Graben area are in favour of a sediment source area located at the uplifted East Shetland Platform. The depocentre offshore Sognefjord had a local source area in the north-western part of mainland southern Norway (e.g. Jordt *et al.*, 1995, 2000; Faleide *et al.*, 2002). The Ling Depression depocentre was probably sourced from the north-western part of southern Norway as well, although sediment transport took place through a route along the present Hardangerfjord. However, deposits in both depocentres became truncated by glacial erosion in Plio-Pleistocene time, forming the erosional unconformity bounding the sequence CSS-1 in these areas.

Eocene (sequence CSS-2)

Description: The sequence CSS-2 is recorded across the entire study area. It is characterised by seismic reflections downlapping onto sequence CSS-1 in the western parts of the study area. In the Central Graben area, the data quality is too poor to reveal whether the sequence CSS-2 reflections are conformable with sequence CSS-1, or not. Still, the CSS-2 is easily distinguished from sequence CSS-1 and the overlying sequence CSS-3 due to amplitude anomalies at both lower and upper boundaries. In the Norwegian-Danish Basin, the upper boundary of the Eocene sequence CSS-2 is distinguished from the Lower Oligocene sequence CSS-3 by an erosional surface. The CSS-2 sequence is time-equivalent with the deep marine Søvind Marl Formation onshore Denmark (Michelsen, 1994).

The CSS-2 sequence has its main depocentres in the south-eastern Central Graben area (>700 ms TWT) and southern Viking Graben area (>700 ms TWT), east of the main depocentre of sequence CSS-1 (Fig. 10B). Clinoforms of the sequence dip towards the E-SE along the East Shetland Platform.

A minor depocentre (>200 ms TWT) of Eocene age is recorded in the Norwegian-Danish Basin and was separated from the Central Graben area by the Coffee-Soil Fault Complex/Ringkøbing Fyn High (Fig. 10B). No truncation is observed within the Cenozoic sedimentary record at the Coffee-Soil Fault Complex, although the basin segment along the fault complex may have been an area of positive relief compared to the Central Graben area and the Norwegian-Danish Basin. The depocentre is characterised by parallel seismic reflectors, although these are truncated by an angular unconformity in the southern and south-eastern margins of the depocentre (Fig. 2 F).

Sandstone is indicated in the Eocene succession in the Norwegian well 11/9-1 (Strass, 1982) which penetrates the depocentre in the Norwegian Danish Basin.

Interpretation: The lower bounding surface, characterised by downlap from easterly dipping clinoforms in the western parts of the study area, is interpreted as a maximum flooding surface. The prograding CSS-2 sequence here indicates that significant accommodation space was created during this transgression. The eastward dipping clinoforms within sequence CSS-2 indicate, like the clinoforms of CSS-1 in the western Viking Graben area, sediment transport and progradation from the Shetland Platform into the North Sea Basin. Sequence CSS-2 has prograded farther into the North Sea Basin, as a result of an eastward movement of the deepest part of the depocentre. A minor source area is also indicated along parts of present southern Norway, from the minor depocentre in the Norwegian-Danish Basin.

Deposits affected by renewed uplift of southern Norway: Basin configuration 4

In Early Oligocene time, Scandinavia was subjected to an increasing supply of coarse-clastic material, either this was caused by renewed uplift, climatic change or a combination (see later discussion). This gave rise to a different basin infill pattern than that prevailing during deposition of the CSS-1 and CSS-2 sequences. The main sediment transport direction changed, from eastwards during basin configuration 3 to westwards during basin configuration 4. Basin configuration 4 includes seven sequences illustrating various phases during uplift and denudation of the Scandinavian landmasses and subsidence in the basinal areas (see later discussion).

Early Oligocene (sequence CSS-3) and Late Oligocene (sequence CSS-4)

Description: The Early Oligocene sequence CSS-3 corresponds to the units 4.1, 4.2 and lower parts of 4.3 of Michelsen & Danielsen (1996). The Late Oligocene CSS-4 sequence is equivalent to the upper part of unit 4.3 and the unit 4.4 of the same authors (Fig. 4). Sequence CSS-3 is observed to have accumulated across the entire study area, although with enhanced thickness in two separate depocentres, one located in the Viking Graben area (>300 ms TWT) and the other in the eastern Norwegian-Danish Basin (>500 ms TWT; Fig. 11A). The Late Oligocene sequence CSS-4 is also present throughout the entire study area. Similar to sequence CSS-3, sequence CSS-4 displays two main depocentres, one in the Viking Graben area (>400 ms TWT) and one in the eastern Norwegian-Danish Basin (>600 ms TWT; Fig. 11B).

The lower boundary of sequence CSS-3 is characterised by an amplitude anomaly across extensive parts of the study area. Also, in the eastern parts of the Norwegian-Danish Basin sequence CSS-3 is downlapping an erosional unconformity terminating the underlying CSS-2 sequence (Fig. 2F and G). Termination of sequence CSS-2 is also seen at the Patchbank Ridge, offshore Hardangerfjord (Fig. 12). The boundary separating sequences CSS-3 and CSS-4 is characterised by an amplitude anomaly in most of the northern North Sea and the Central Graben area. In the eastern North Sea, sequence CSS-4 onlaps the sequence boundary on top of sequence CSS-3. The upper boundary of sequence CSS-4 is characterised by toplap truncation (Fig. 2).

Easterly dipping clinoforms are observed within the CSS-3 sequence in the Viking Graben area. In the Norwegian-Danish Basin, sequence CSS-3 downlaps onto sequence CSS-2, with clinoforms

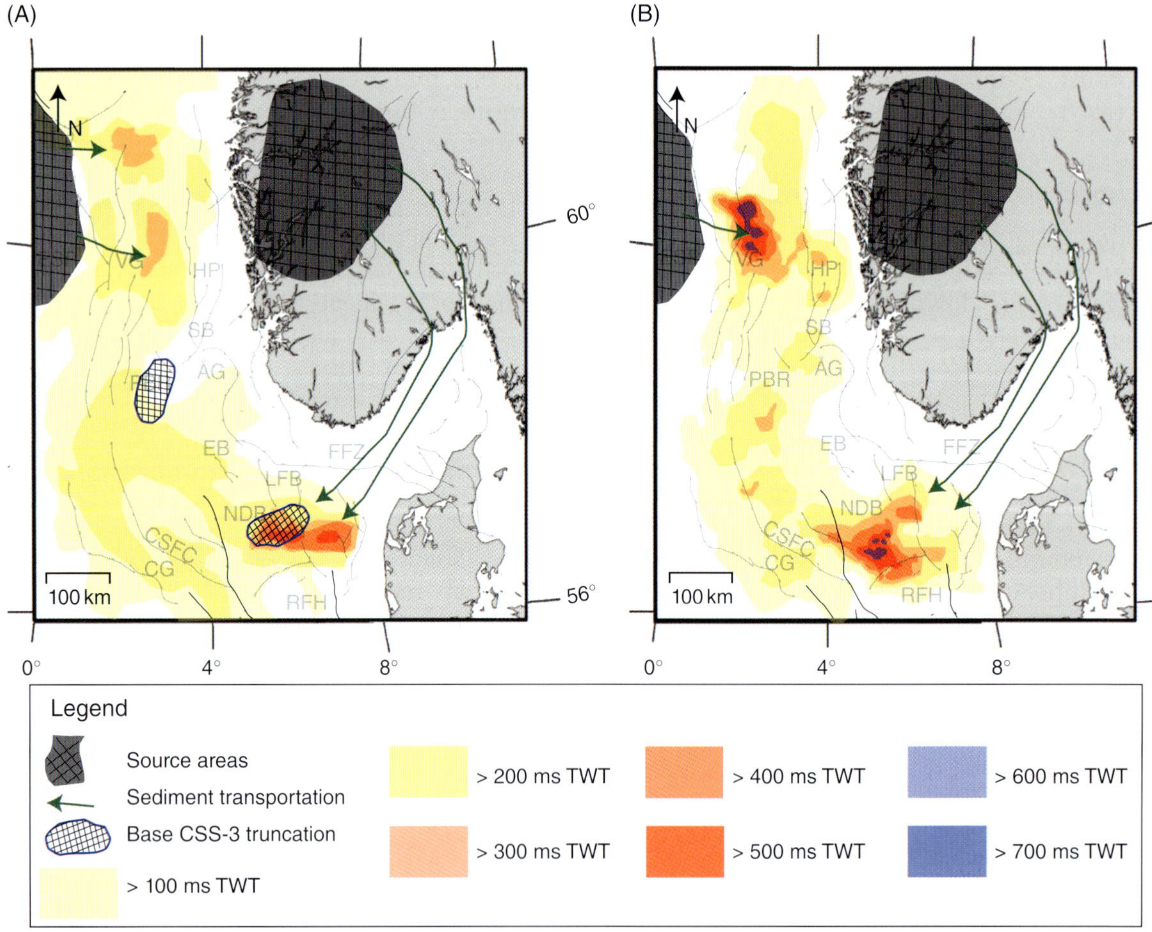

Fig. 11. Time–thickness maps of the Oligocene sequences CSS-3 (A) and CSS-4 (B).

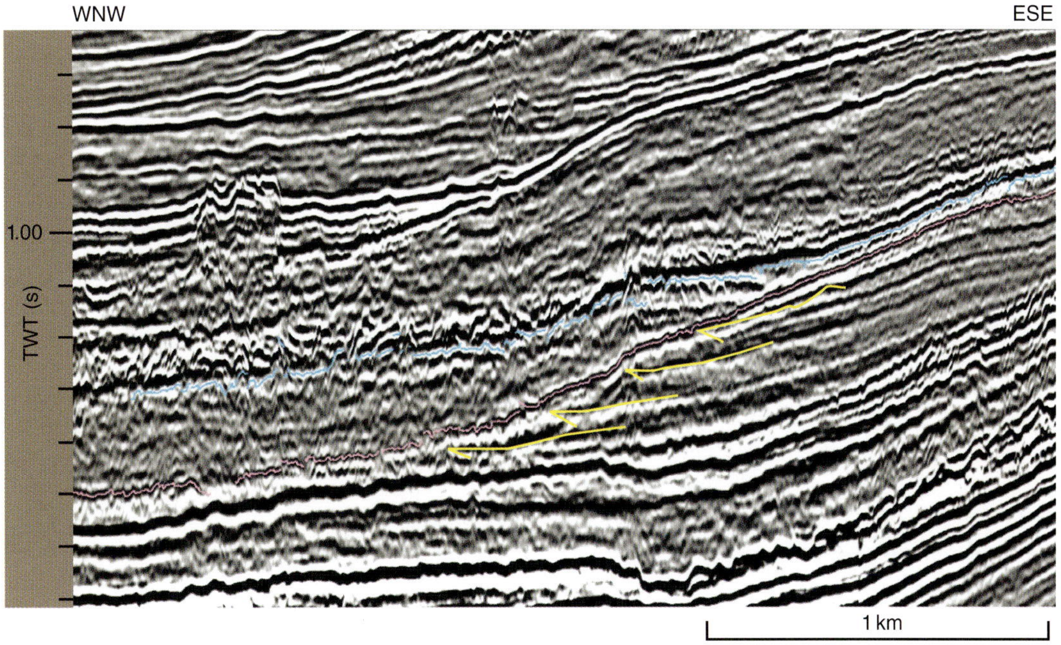

Fig. 12. Seismic section, showing truncation of Eocene strata at the offshore elongation of Hardangerfjorden, in the Patchbank Ridge area. Truncation of Eocene strata is highlighted by the yellow arrows.

dipping towards S-SW (Fig. 9; Michelsen & Danielsen, 1996). The clinoforms belonging to sequence CSS-3 have an oblique geometry in the Norwegian Danish Basin, with clinoform heights up to 250 ms TWT.

In the Viking Graben area, the CSS-4 depocentre is located south-east of the sequence CSS-3 depocentre, whereas in the Norwegian-Danish Basin it is located south-west of the depocentre of CSS-3 (Fig. 11). Progradation continued in an eastward direction in the northern North Sea. In the eastern North Sea, progradation shifted to a more SW direction compared to sequence CSS-3. In the Norwegian Danish Basin, the clinoforms belonging to sequence CSS-4 have a sigmoidal geometry with clinoform heights up to 300 ms TWT.

Interpretation: The lower SU at the base of sequence CSS-3 in the eastern North Sea and on the Patchbank Ridge area indicates lowering of the relative sea-level at the Eocene-Oligocene transition. This is also indicated in the mid-Oligocene, based on the SU and truncation of clinoform topsets in the eastern North Sea at sequence CSS-3 top. In the eastern North Sea, base CSS-3 represents a combined SU/TS/MFS, indicated by the downlapping pattern directly onto the SU. Such combined surfaces imply rapid creation of accommodation space subsequent to erosion and potential exposure. The regional extent, thickness variation and internal geometry of sequences CSS-3 and CSS-4 reflect a major sediment source in the west on the East Shetland Platform, filling in a depocentre in the Viking Graben area by eastward prograding sedimentary systems. This depocentre migrated eastwards from Early to Late Oligocene. The other depocentre system, in the eastern part of the Norwegian-Danish Basin, was shifted towards the south-west during the Oligocene and was probably sourced with sediments from mainland Norway and filled by progradation as seen from internal clinoform orientation. The depocentres in the eastern North Sea, revealing S-SW direction of progradation, indicate exposure and development of a source area in the north-north-east, as compared to during deposition of the Chalk Group (basin configuration 2) and the sequences CSS-1 and CSS-2 (basin configuration 3).

Based on the clinoform geometries, the CSS-3 sequence may have been more sand-rich compared to the upper CSS-4 sequence (Bullimore *et al.*, 2005). However, the oblique clinoform geometry may also indicate limited accommodation space during deposition and longshore drift of sediments at the platform margin (e.g. Emery & Myers, 1996).

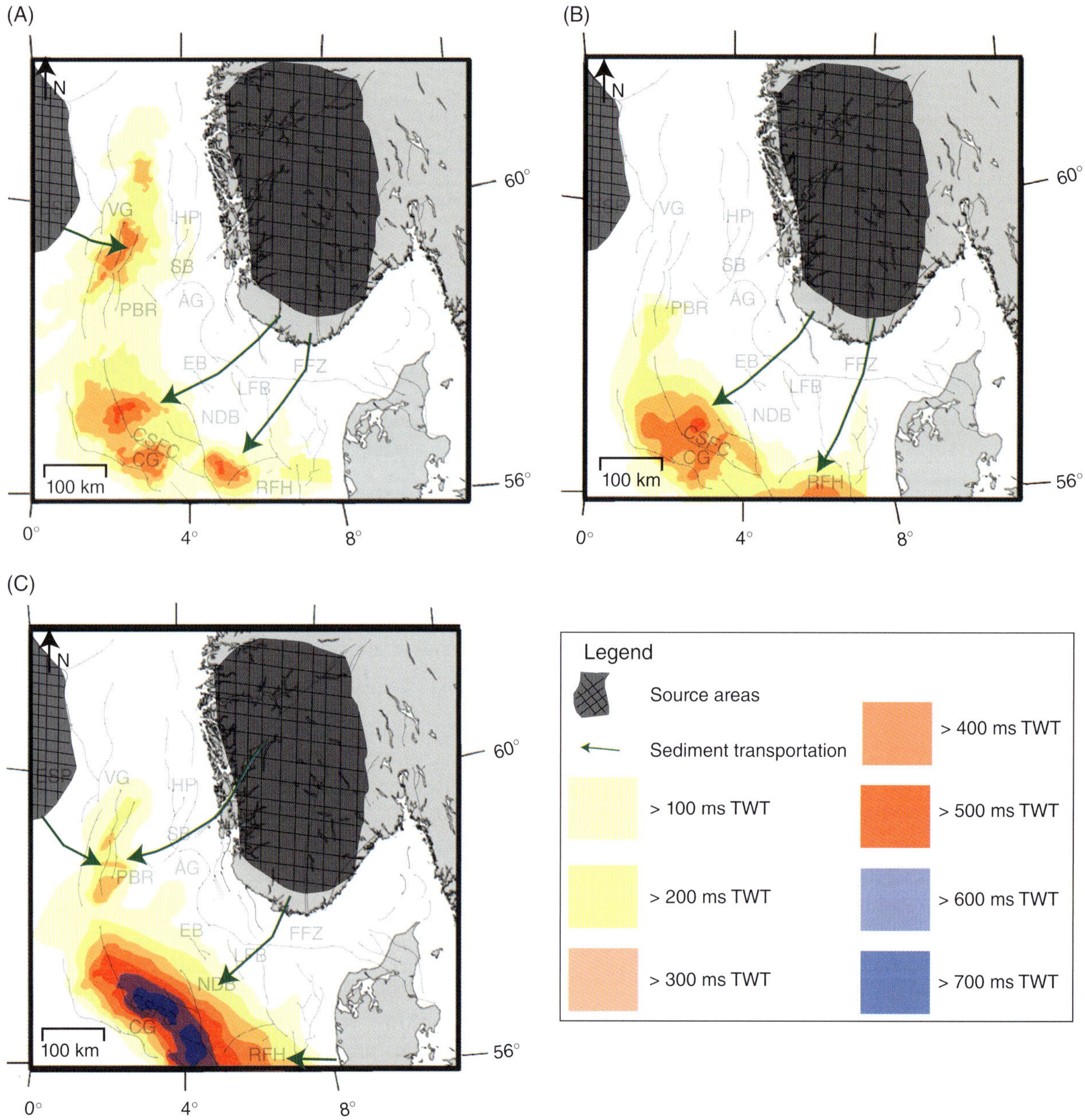

Fig. 13. Time–thickness maps of the Miocene (A) CSS-5, (B) CSS-6, and (C) CSS-7 sequences.

Early, mid and late Miocene (sequences CSS-5, CSS-6 and CSS-7)

Description: The CSS-5 sequence is present throughout most parts of the study area with thickened sequences in three main depocentres (Fig. 13A). The sequence CSS-5 thins towards north-east, across the Coffee-Soil Fault Complex and the Ringkøbing-Fyn High and becomes truncated in the north-eastern part of the Norwegian-Danish Basin. The sequence is therefore absent across the northern part of the Norwegian-Danish Basin. The data quality of the sequence is excellent.

The lower bounding surface of sequence CSS-5 is characterised by onlapping against the SU at the top of the CSS-4 sequence in the eastern North Sea area (Figs 2 & 9). The upper bounding surface is characterised by an erosional surface in the Norwegian Danish Basin and onlap from the overlying CSS-6 sequence.

The main depocentres belonging to sequence CSS-5 are in the southern Central Graben area (with thickness > 500 ms TWT), in the southern Viking Graben area (>500 ms TWT) and in the southern part of the Norwegian-Danish Basin (>500 ms TWT).

In the southern Viking Graben area sequence CSS-5 reveals internal seismic reflections dipping to the east. Sequence CSS-5 is characterised by a parallel to sub-parallel seismic reflection pattern in the southern Central Graben area, however, with onlap relations against the Mid-North Sea High.

The depocentre in the Norwegian-Danish Basin is located south-west of the CSS-4 depocentre, including a basinward shift in onlap onto the CSS-4 sequence. Here, CSS-5 is characterised by SW dipping clinoforms, located south-west of the Oligocene sequences. Correlation of sequence CSS-5 to onshore Denmark reveals that it is time-equivalent with the Billund and Vejle Fjord formations (Rasmussen *et al.*, 2010).

Sequence CSS-6 is mainly recorded in the southern parts of the study area (Fig. 13), although it thins to less than 100 ms TWT across the Coffee-Soil Complex and the Norwegian-Danish Basin. In the north-eastern part of the Norwegian-Danish Basin CSS-6 is truncated by glacial erosion in Pliocene and Pleistocene times. The sequence thins to below seismic resolution in the northern North Sea and for this reason is not recorded within the Viking Graben area.

Sequence CSS-6 can be subdivided into two sub-sequences, namely CSS-6.1 and CSS-6.2. These are separated by an onlap surface, where the sub-sequence CSS-6.2 onlaps sub-sequence CSS-6.1 in the Coffee-Soil Fault Complex area. The upper bounding surface of sequence CSS-6 is characterised by an amplitude anomaly across the study area, with onlap relations from the overlying CSS-7 sequence in the south-western parts of the Norwegian-Danish Basin.

The Central Graben area was the main depocentre during deposition of the sub-sequences CSS-6.1 and CSS-6.2 (Fig. 13B). The CSS-6.1 sub-sequence has its maximum thickness in the north-eastern part of the Central Graben area and pinches out towards south-west. The CSS-6.2 sub-sequence is of constant thickness throughout the Central Graben area but thins towards the NE and SW, across the Coffee-Soil Fault Complex and the Mid-North Sea High (Figs 9 and 13).

The CSS-7 sequence is time-equivalent to the Utsira Formation in the central and northern North Sea and the Molo and Kai formations in the Norwegian Sea (e.g. Eidvin *et al.*, 2007). The base of the CSS-7 sequence corresponds to the mid-Miocene unconformity (e.g. Huuse & Clausen, 2001), which is also present onshore Denmark (base sequence E in Rasmussen, 2004). Onlap patterns are recorded in the south-western part of the Norwegian-Danish Basin, against top sequence CSS-6 (Fig. 2 K). An amplitude anomaly is recorded within the lower part of sequence CSS-7, which acts as a downlap surface for the upper part of the same sequence in the Central Graben area. Two main depocentres of mid-late Miocene age are observed in the study area: in the southern Viking Graben area (>300 ms TWT) and in the Central Graben area (>1000 ms TWT; Fig. 13C). These are separated by the Jæren High.

In the Viking Graben area, sequence CSS-7 is represented by the Utsira Formation. The seismic reflectors onlap against the East Shetland Platform in the west and downlap the CSS-6 sequence in the east, at the Utsira High area. In the Central Graben area, sequence CSS-7 downlaps onto the underlying CSS-6 sequence, thins across the Ringkøbing-Fyn High and Coffee-Soil Fault Complex and finally becomes truncated in the southern part of the Norwegian-Danish Basin by glacial erosion in Pliocene and Pleistocene time. The sequence onlaps the Mid-North Sea High at the south-western margin of the Central Graben area. Within the Central Graben area, sequence CSS-7 is characterised by SW dipping clinoforms (Fig. 9).

Interpretation: The Miocene sequences reveal a further development of the Oligocene trend regarding the infilling of the North Sea Basin. A lowering of the relative sea-level at the transition between sequence CSS-4 and CSS-5 is indicated by the SU at top CSS-4. A subsequent transgression is indicated from the onlapping patterns in the same area (Fig. 2). Sequence CSS-5 in the depocentre in the southern Viking Graben area was sourced from the East Shetland Platform in the west. Within the depocentre in the Norwegian-Danish Basin, the clinoform orientation reflects a continuing progradation from South Norway. The lower part of sequence CSS-6, sub-sequence CSS-6.1, was transported towards the SW from its source area in mainland Norway. In contrast sequence CSS-7, the Utsira Formation, was mainly sourced from the west (East Shetland Platform), although in the north-eastern parts it also received sediments from southern Norway (e.g. Gregersen *et al.*, 1997; Eidvin *et al.*, 2007; Gregersen &

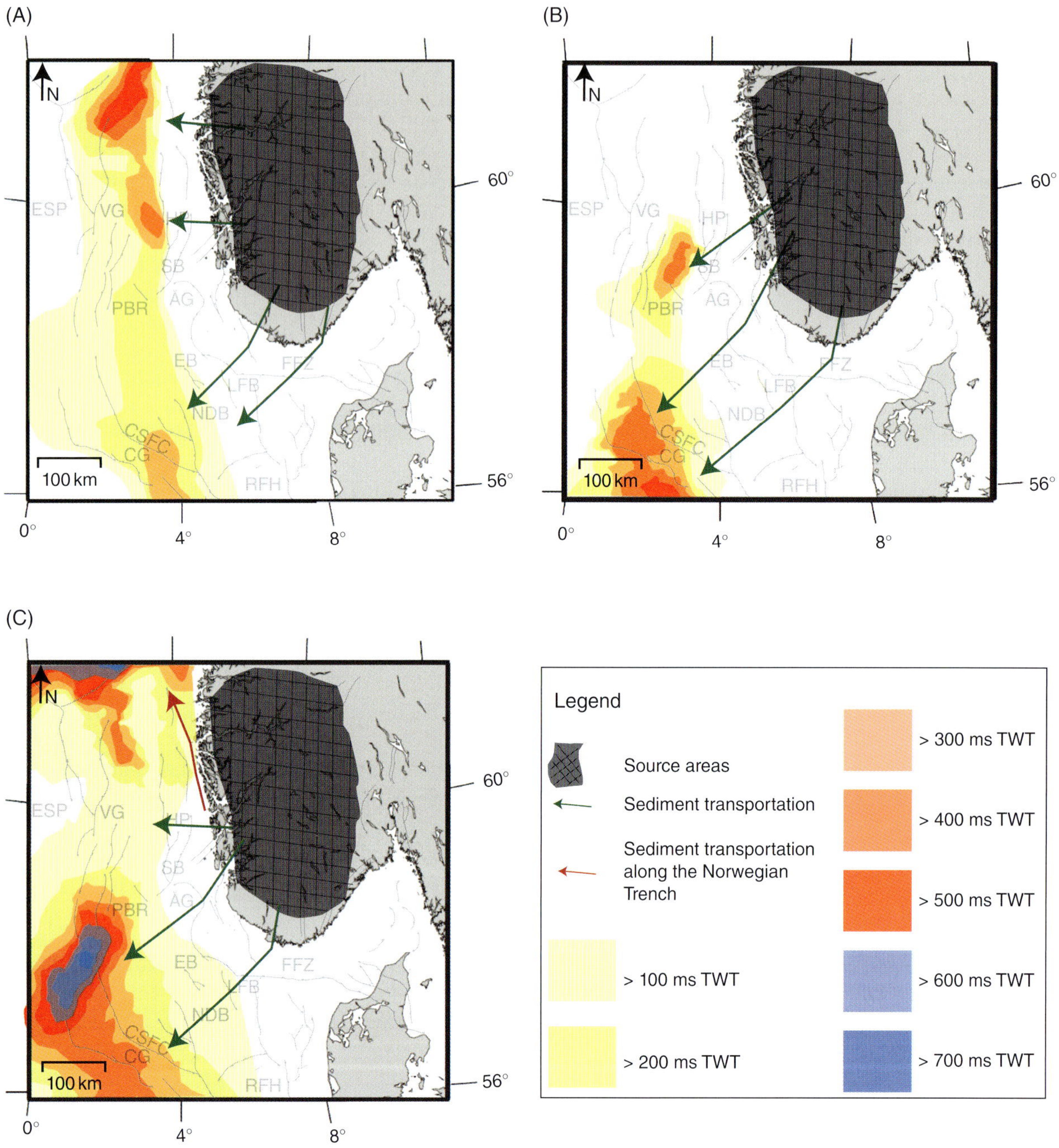

Fig. 14. Time–thickness maps of sequences (A) CSS-8, (B) CSS-9 and (C) CSS-10.

Johannessen, 2007). In the Central Graben area, downlap directions and thickness distribution of CSS-7 indicate a source area situated to the north and east (Fig. 9). Although no unconformity is documented on the Mid-North Sea High, the onlap patterns of the Miocene sequences alongside the structure indicate a positive topography here in Miocene time.

Pliocene (sequence CSS-8)

Description: The Pliocene CSS-8 sequence is recorded in the northern North Sea and the Central Graben area (Fig. 14A), where the seismic resolution is excellent, with the identification of internal seismic reflections. The base of sequence CSS-8 is characterised by downlap onto the underlying upper Miocene CSS-7 sequence (Fig. 9). The upper boundary of the CSS-8 sequence reveals toplap truncation in the south-western parts of the Norwegian-Danish Basin, with onlap from the following sequence CSS-9 (Fig. 2). The Pliocene sequence is truncated in the Norwegian-Danish Basin as far south as to the Ringkøbing-Fyn High and the Stavanger Platform. In the Northern North Sea, sequence CSS-8 is truncated along the Øygarden Fault Zone in the east and as far west as to the Utsira High.

Two main depocentres of Pliocene age are observed along the axis of the Central Graben (>300 ms TWT) and in the northern North Sea (>400 ms TWT; Fig. 14A). In the Central Graben area the sequence is characterised by SW dipping clinoforms (Fig. 9), whereas, in the northern North Sea Basin, sequence CSS-8 is characterised by westerly dipping clinoforms.

Interpretation: Sequence CSS-8 was probably deposited during a fall in relative sea-level, which may have corresponded to the growth of the ice-sheets across northern Europe and North America (Buchard, 1978; Haq *et al.*, 1987) in Mid-Pliocene time 2.5 Ma (Kiltgord & Schouten, 1986). That sequence CSS-8 downlaps onto the CSS-7 sequence in the Central Graben area indicates that this surface is a combined TS/MFS and hence, reflecting a rapid rise in sea-level. Toplap truncation in the Norwegian-Danish Basin indicates that the shoreline of the southern Norwegian landmasses stretched as far out as to the north-eastern margin of the Central Graben area in Late Pliocene time. The W-SW direction of progradation of sequence CSS-8 indicates that in the east and south-east of southern Norway acted as the source area. The truncation of the Pliocene sequence in the Norwegian-Danish Basin as far south as to the Ringkøbing-Fyn High and the Stavanger Platform appears to coincide with the maximum extent of the Plio-Pleistocene glaciers in the Central North Sea area (Fig. 9).

Pleistocene (CSS-9 and CSS-10)

Description: The Pleistocene CSS-9 and CSS-10 sequences are well imaged in the reflection seismic data in the study area. Sequence CSS-9 is truncated at the Coffee-Soil Fault Complex and the Sele High in the Central North Sea and at the Øygarden Fault Complex and the Utsira High in the northern North Sea.

The lowermost boundary of sequence CSS-9 is characterised by regional downlap onto the CSS-8 sequence in the Norwegian-Danish Basin (Fig. 2) and onlaps the CSS-6 sequence at the mid-North Sea High (Fig. 9). Sequences CSS-9 and CSS-10 are separated by an erosional surface, recorded in the north-eastern and northern parts of the study area parallel to the present coast line of southern Norway. This erosional surface truncates the underlying Cenozoic succession, as these are inclined upwards towards north-east and east. The top of the CSS-10 sequence corresponds with the present day seabed. In the northern North Sea area, sequence CSS-9 is seen as a westward progradational unit (>400 ms TWT), whereas in the Central Graben area, a more than 300 ms TWT thick prograding wedge built out towards SW, onlapping the Mid-North Sea High (Figs 9 and 14B). Also, easterly dipping prograding clinoforms belonging to sequence CSS-9 are recorded in the western part of the study area, in the Moray Firth Basin area.

The CSS-10 sequence infilled the North Sea Basin in the Central Graben area (>400 ms TWT) and pinches out northwards across the Utsira High area and the Patchbank Ridge (Fig. 14C). The CSS-10 sequence filled the remaining accommodation space in the Central Graben area; seismic reflections, being sub-parallel to the present-day sea floor (Fig. 9) are in favour of a vertical aggradation mechanism for the sediment infill succession.

Interpretation: The lowermost boundary is interpreted as a combined TS/MFS and hence a non-accretionary TST. The westerly and SW progradation of sediments belonging to sequence CSS-9 indicates that the main sediment source area was located to the east and north-east. However, erosion of landmasses in the UK is also indicated from the easterly dipping clinoforms in the Moray Firth Basin area. The parallel to sub-parallel nature of the seismic reflections of sequence CSS-10 in the Norwegian Trench indicates that the sediments were deposited after the main tilting of the Mesozoic and Cenozoic successions.

DISCUSSION

The present study reviews basinal configurations in the North Sea area throughout the Late Mesozoic and Cenozoic. The two main tectonic phases which affected the study area, were the Permo-Triassic and Late Jurassic to earliest Cretaceous rifting (e.g. Ziegler, 1975, 1989, 1990; Gabrielsen *et al.*, 1990;

Underhill and Partington, 1993; Nøttvedt *et al.*, 1995; Færseth *et al.*, 1997; Fraser *et al.*, 2002). However, this study starts with the Late Jurassic to earliest Cretaceous rift phase. Following the Late Jurassic to earliest Cretaceous basin formation, the period from earliest Cretaceous to Early Palaeocene was ruled by the post-rift development (Gabrielsen *et al.*, 2001). At this time, sedimentary strata became gradually inclined towards the basin axis during post-rift (Nøttvedt *et al.*, 1995). Hence, the syn-rift to post-rift transition is well established (Gabrielsen *et al.*, 2001; Kyrkjebø *et al.*, 2001). However, the North Sea basin configuration underwent transformations that cannot be accounted for by post-rift related processes alone in the mid-Cretaceous, when the intra-basinal highs were transgressed (Kyrkjebø *et al.*, 2001). Noting these differences, Nøttvedt *et al.*, (1995) subdivided the post-rift development into two separate phases; namely early and late post-rift. This subdivision was based on the sediment infilling of the graben system, where early post-rift sedimentation took place in areas of subsidence separated by rotated fault blocks developed during syn-rift stage. Thus, late post-rift sediments draped the entire graben system, including the platform areas. In this study, we emphasise this subdivision. In doing so we incorporate the early post-rift phase together with the late syn-rift phase in basin configuration 1, because the early post-rift phase to a great extent was governed by the late syn-rift relief. In contrast, we see that the late post-rift phase, represented by basin configuration 2, to a greater extent was dominated by thermal cooling and associated processes (differential sediment loading and compaction). This is further discussed below.

Following basin configuration 2, the subsequent basin configuration was characterised by the following patterns; (1) initiation of sediment transportation and deposition from west in the Late Palaeocene (basin configuration 3) and (2) a switch from a mainly western source area to a dominantly easterly to north-easterly source area in the Oligocene (basin configuration 4).

Following is a discussion of the dynamics of the changing basin configurations during the Late Mesozoic and Cenozoic.

Basin configuration 1

During basin configuration 1, the North Sea Basin was characterised by structural highs and lows related to the extensional regime which prevailed in the Late Jurassic to earliest Cretaceous. When compared to this period of extension, the Horda Platform represents a stable platform (e.g. Gabrielsen *et al.*, 1990; Fraser *et al.*, 2002). The deepest parts of the rift basin in the northern North Sea were clustered along the central axis of the Viking Graben, although accommodation space was also created along rotated fault blocks and graben structures on the platform areas (e.g. Nøttvedt *et al.*, 1995; Odinsen *et al.*, 2000; Gabrielsen *et al.*, 2001).

Basin configuration 1 includes the late syn-rift and early post-rift phases. Although the syn-rift to post-rift transition is defined by the switch from lithospheric heating to cooling (McKenzie, 1978; Nøttvedt *et al.*, 1995), a change in the basin configuration is delayed due to the strong structural impact of the late syn-rift stage prevailing into the early post-rift (Gabrielsen *et al.*, 2001, Kyrkjebø *et al.*, 2001). Thus, during the late syn-rift and early post-rift phase (basin configuration 1), the North Sea area was characterised by several emerged structural highs (such as the Utsira High and the Ringkøbing-Fyn High) and regional and local lows with water depths up to 1000 metres, such as the axis of the Viking Graben and rotated fault blocks along its margins (e.g. Ziegler, 1987; Nøttvedt *et al.*, 1995; 2002; Gabrielsen *et al.*, 2001; Kjennerud *et al.*, 2001; Fraser *et al.*, 2002; Martinsen *et al.*, 2005). We therefore suggest that basin configuration 1 includes both the late syn-rift and early post-rift phases, incorporating the sequences J-1, J-2 and K-2.1, as described in the present study.

Basin development and sediment supply

As the sediments belonging to the J-1 sequences in the study area were sourced from the east, it is probable that the terrigenous detritus was transported through sediment routes along a valley system located at the present Sognefjord (e.g. Gabrielsen *et al.*, 2010a). Simultaneously, the point sources for the sub-sequence J-2.1 of the Stord Basin-Åsta Graben were located at the mouth of a river system aligned along the present Hardangerfjord (see also Gabrielsen *et al.*, 2010a). In contrast and in accordance with the dip-direction of the internal clinoforms of sub-sequences J-2.1 and J-2.2 in the Egersund Basin, these prograded from east to west. The source area has probably been the Lista Fault Blocks, based on the erosional surface truncating these structural heights and which corresponds to the basal bounding surface of the J-2 sequence and onlapping patterns alongside the structure (Fig. 7). This suggests that the feeder system migrated southwards, from Sognefjord at the time of deposition of sequence J-1 to Hardangerfjord and the Lista Fault Blocks further south, during deposition of sequence J-2. However, we cannot rule out that these feeder systems acted at the same time. The truncation of the topsets of each of the prograding sequences indicates that the available accommodation space became overfilled at the end of the depositional period for sequences J-1 and J-2, suggesting that by the Late Jurassic the shoreline had advanced basinward and was located in the Egersund Basin, Stord Basin-Åsta Graben and the northern Horda Platform.

According to Mitchum *et al.* (1977), obliquely oriented clinoforms are common in sequences which have been deposited in a higher-energy environment, indicating coarse sediments and short transport distances. Accordingly, the presence of oblique clinoforms within sequences J-1 and J-2 are interpreted to reflect increased coarse-clastic sediment supply in the Stord Basin, Åsta Graben and the Egersund Basin (Fig. 15A), probably due to renewed exposure and uplift of parts of southern Norway. Such a structural relationship between basin and hinterland is in accordance with expected tectonic configuration in a late syn-rift setting (e.g. Nøttvedt *et al.*, 1995). An elevated Jurassic topography in southern Norway is supported by Apatite Fission Track studies (Rohrman *et al.*, 1995, 1996; Hendriks & Andriessen, 2002; Hendriks *et al.*, 2007) and morphological considerations (Mosar, 2003; Gabrielsen *et al.*, 2010a). However, uncertainties still prevail as to whether the whole of southern Norway was exposed or if only local source areas were available in latest Jurassic time. Either way, parts of the same landmass continued to deliver sediments in the Early Cretaceous, as indicated by the prograding sub-sequence K-2.1. This is also supported by sand distribution patterns in the Lower Cretaceous, which indicate that the Norwegian mainland was supplying sediments to the North Sea Basin (Thomson *et al.*, 1999; Gabrielsen *et al.*, 2001). That those topographic highs continued as source areas in Early Cretaceous time is also supported by uplift of intra-basinal topographic highs, such as the Utsira High (Gabrielsen *et al.*, 2001; Kyrkjebø *et al.*, 2001). This uplift probably increased the time duration for basin configuration 1 to throughout Early Cretaceous; hence, a close structural relationship between basin formation and hinterland tectonic uplifts, sediment supply and infill dynamics seems to be characteristic for basin configuration 1.

Basin configuration 2

At the onset of basin configuration 2, the remaining topography became drowned, including the platform areas, which is characteristic for a typical late post-rift palaeogeographic setting (Gabrielsen *et al.*, 1990, 2001, Kyrkjebø *et al.*, 2001). Siliciclastic sedimentation dominated in the northern Viking Graben in a deep marine setting (e.g. Gabrielsen *et al.*, 2001; Kjennerud *et al.*, 2001), with a gradual increase of chalk deposition southwards (e.g. Surlyk *et al.*, 2003).

As the intra-basinal lows were filled with sediment and the structural highs were transgressed (Gabrielsen *et al.*, 2001) in mid-Cretaceous time, this gave rise to a wider and more open regional marine basin, compared to basin configuration 1 with its sub-basins and local source areas. We therefore suggest that the Late Cretaceous to Danian represents a period with a different basin configuration compared to basin configuration 1. This also agrees with the findings of Nøttvedt *et al.* (1995), who divided the early-rift and late post-rift phases on the same criteria.

Basin development and sediment supply

The recorded thickness distribution of the Chalk Group found in the present study (Fig. 8) is in accordance with previous studies, where thicknesses up to 500 m to 800 m are seen in the Norwegian-Danish Basin and more than 1500 m in the Central Graben (Surlyk *et al.*, 2003). This may be taken as an indication that a shallow epicontinental sea persisted for a considerable time interval as the North Sea Basin in Late Cretaceous to Danian time. The aggradational style of chalk facies demonstrates that the rate of pelagic carbonate production and sedimentation kept pace with the rate of subsidence. The thinning of the Chalk Group across the Coffee-Soil Fault Complex and at the Ringkøbing-Fyn High was probably related to tectonic inversion of the basin succession along these structures, related to Alpine compression, in the Late Cretaceous (e.g. Liboriussen *et al.*, 1987; Handy *et al.*, 2010).

The presence of the Chalk Group in the Norwegian-Danish Basin itself indicates that no

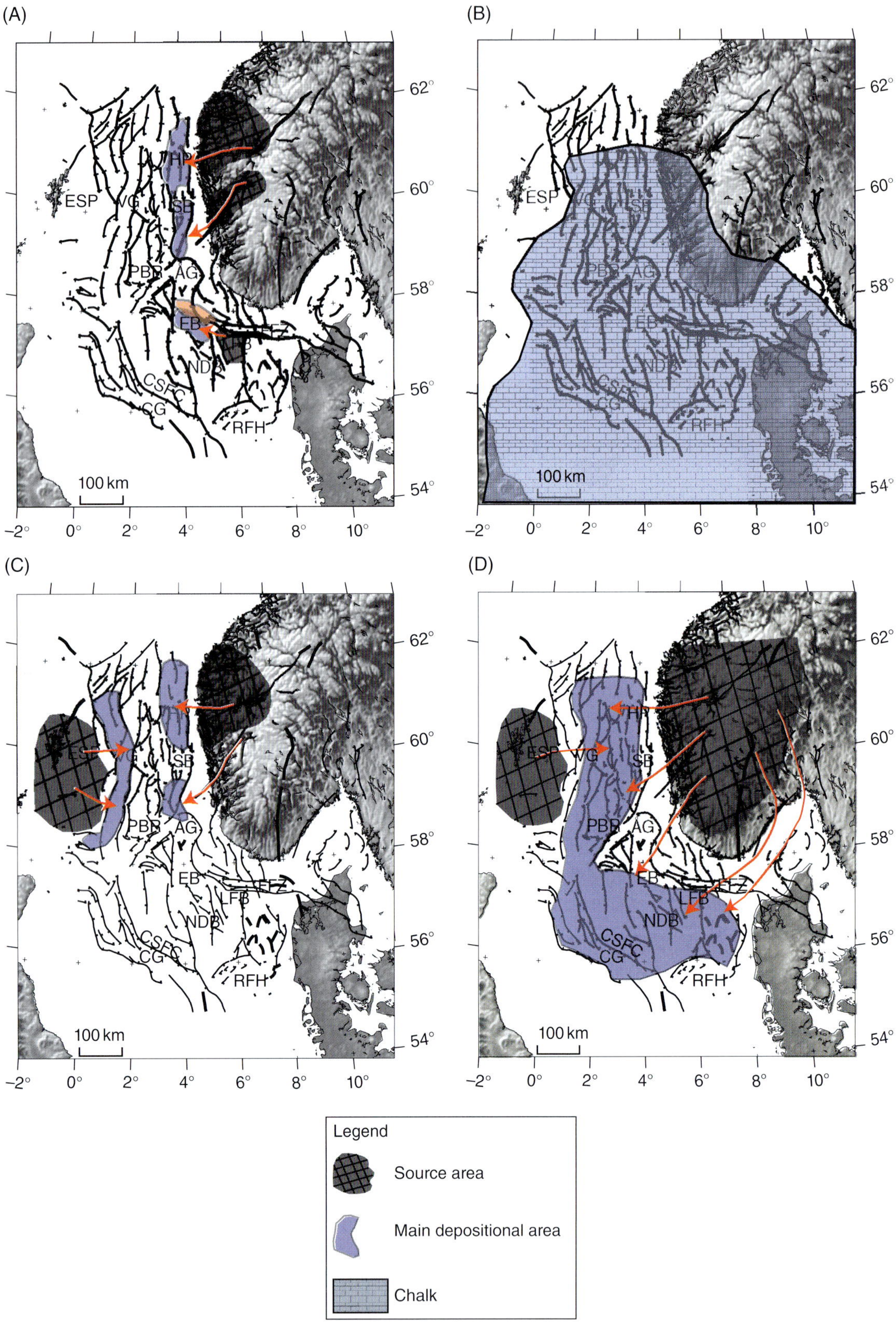

Fig. 15. Proposed stages in basin development. (A) Stage 1. Areas in southern Norway which are suggested to have acted as source areas during stage 1. The prograding sequences J-1, J-2 (dark blue) and K-2.1 (red) are marked with proposed drainage patterns. (B) Stage 2. The carbonates covered most of the North Sea area, including parts of southern Norway in Late Cretaceous time. However, the extent of the carbonates across southern Norway is uncertain. Note that base of the Chalk Group thins towards the northern part of the northern North Sea. Southern and western limit of the Chalk Group is adapted from Surlyk *et al.* (2003). (C) Stage 3. Vertical uplift of parts of southern Norway and the East Shetland Platform gave rise to basin stage 3. Sediments where probably sourced via drainage systems located at the same positions as the Sognefjord and possibly Hardangerfjord. (D) Stage 4. Renewed uplift of southern Norway. The sediments were probably sourced via old lineaments, indicated in the figure. See Fig. 3 for abbreviations.

topographic highs were available in southern Norway at this time to supply siliciclastic sediments towards the S-SW. Advanced chemical weathering in the warm and humid climate of northern Europe in the Late Cretaceous (Huber *et al.*, 1995, 2002; Norris *et al.*, 2002; Friedrich *et al.*, 2012) might have altered most feldspar and other silicates into clay minerals. However, there should still have been a lot of quartz left for deposition in adjacent basins, provided sufficient relief existed for sediment transport from a source area in southern Norway. The north-eastward thickening of the Chalk Group in the Norwegian-Danish Basin (i.e. thickening towards southern Norway; Fig. 8) is here interpreted to reflect that the sea probably covered substantial parts of South Norway in the Late Cretaceous during the first order eustatic highstand at this time (Haq *et al.*, 1987; Fig. 15B). This interpretation is supported by the following arguments: (1) the Chalk Group rests on an angular unconformity that cuts sedimentary successions of basin configuration 1 on intrabasinal highs, as seen at the Utsira High and along the Øygarden Fault Complex, as well as in depocentres, implying (2) that the formation of this regional unconformity also affected mainland Norway with denudation and potential peneplanation, thus (3) allowing the sea to transgress across parts of South Norway, (4) preventing large amounts of siliciclastic sediments to be transported to the southern part of the North Sea Basin system.

A Late Cretaceous depositional area in parts of southern Norway is in accordance with observations by Thyberg *et al.* (2000), who recorded chalk clasts, suggested to be of Late Cretaceous age, in prograding sequences of Pliocene age in the northern North Sea. However, the siliciclastic sedimentary input in the northern North Sea area indicates exposed landmasses in the surrounding areas. Brekke (2000) suggested a western or eastern source for the Late Cretaceous siliciclastic succession in the Møre Basin, north of the northern North Sea, based on onlap patterns and thinning of Upper Cretaceous strata against basin flanks in the east and the west. This is also in harmony with Hancock (1990), who suggested a northern source area for the siliciclastics in the northern North Sea. Due to the proximity, the same source area probably delivered sediments into the northern North Sea in Late Cretaceous time as well and indicates that parts of mid-Norway and northern Norway may have been exposed in Late Cretaceous time. However, the siliciclastic input ceased during Maastrichtian time, with subsequent chalk deposition also in the northern North Sea (Hancock, 1990).

Basin configuration 3

At the onset of basin configuration 3, uplift of the East Shetland Platform took place, with the subsequent deposition of sequences CSS-1 and CSS-2 in a rapidly deepening North Sea basin (Fig. 8; e.g. Jordt *et al.*, 1995, 2000; Martinsen *et al.*, 1999; Kjennerud *et al.*, 2001; Faleide *et al.*, 2002; Gabrielsen *et al.*, 2005, 2010a; Anell *et al.*, 2011). This uplift corresponds in time with major igneous activity in the North Atlantic Igneous Province, north-west of the study area (e.g. Knox & Morton, 1988; White & Lovell, 1997) and an overall change from shallow marine conditions during Late Cretaceous to deep marine during Early Palaeogene (Michelsen, 1994; Kyrkjebø *et al.*, 2001).

We propose that the onset of deposition of the CSS-1 sequence in Late Palaeocene time represents a transition to a third basin configuration within the study area. Deposition of the sequences CSS-1 and CSS-2 within rapidly subsiding depocentres indicates new tectono-thermal conditions, with more localised depocentres and the development of structurally coupled source areas in the hinterlands. This indicates that the lithospheric properties changed, as compared to during basin stage 2, where the study area was mainly affected by thermal cooling and sediment loading (Nøttvedt *et al.*, 1995).

Basin development and sediment supply

The outline of the sequences trapped in depocentres and the eastward dipping clinoforms in the Viking Graben area (CSS-1 and CSS-2) and in the Central Graben area (CSS-2) indicate eastward progradation, suggesting a westerly source in Late Palaeocene and Eocene times. This implies a tectonic uplift of the East Shetland Platform (Fig. 15C), as the vast chalk sediment deposition indicate that no source area was situated here during basin configuration 2. At the same time, accumulations of sequence CSS-1 offshore Sognefjord and Hardangerfjord indicates uplift of north-western parts of southern Norway. A source area in the east in Palaeocene time is also suggested by Dmitrieva *et al.* (2012), based on recordings of proximal basin floor fan systems in the eastern Viking Graben area and by Jordt *et al.* (1995, 2000) which suggested an eastern source for sediments which infilled the depocentre offshore Sognefjord.

As this occurred simultaneously with volcanic activity in the North Atlantic Igneous Province, the uplift and subsidence observed in Late Palaeocene and Eocene was probably linked to a common set of processes, which led to rifting and later break-up of the NE Atlantic.

Because the Eocene depocentre within the Norwegian-Danish Basin contains sandstone (Strass, 1982), this accommodation space was probably situated close to a source area. Erosional channels are recorded, cut into the top strata of the Chalk Group at the Sorgenfrei Tornquist Zone and the Ringkøbing Fyn High (Huuse, 1999), reflecting that these structures became inverted, probably due to intraplate stresses related to Alpine compression in southern Europe. These or similar channels may have acted as pathways for siliciclastic material from the Fennoscandian Shield.

Basin configuration 4

At the Eocene-Oligocene boundary, vast quantities of sediment were transported into the eastern North Sea area, with S-SW progradation, indicating the existence of a new source area in southern Norway. This depositional pattern continued throughout the Neogene. As glacio-isostatic rebound in Plio-Pleistocene times (Riis & Fjeldskaar, 1992) cannot explain the Neogene tilting of the sedimentary layers itself (Riis, 1996), an additional vertical uplift of Scandinavia has been suggested as an explanation (Jordt *et al.*, 1995, 2000; Michelsen, 1995; Rohrman *et al.*, 1995; Martinsen *et al.*, 1999; Japsen & Chalmers, 2000; Japsen *et al.*, 2002, 2007; Faleide *et al.*, 2002; Anell *et al.*, 2009, 2010, 2012; Chalmers *et al.*, 2010; Gabrielsen *et al.*, 2010a, b). However, increased sediment input in late Palaeogene and Neogene times was also controlled by climate changes and eustasy (Huuse, 2002; Nielsen *et al.*, 2009; Goledowski *et al.*, 2012).

The observed change in infill pattern and the diachronous creation of accommodation space in the study area, concurrently with uplift of southern Norway, is here suggested to reflect a new basin configuration as compared to basin configuration 3. We therefore propose that this change in the main direction of sediment transport, contemporaneous with changes in geometry in the North Sea Basin system, indicates that hinterland and basinal area were structurally coupled by some sets of common tectonic processes in the European crustal plate. This is further discussed below.

Basin development and sediment supply

A major shift in sediment provenance occurred at the onset of the Oligocene sequence CSS-3, as indicated by the W-SW progradation in the eastern North Sea area at that time (Fig. 9). The main depocentres of Oligocene age in the Norwegian-Danish Basin (Fig. 11) are here suggested to have been sourced from the NNE (Fig. 15D). The presence of sandstones in the Lower Oligocene (Danielsen *et al.*, 1997) supports the hypothesis of shoreline progradation towards a southern direction in CSS-3 time, becoming situated close to the source area. This is in accordance with Michelsen *et al.* (1995) and Danielsen *et al.* (1997), who found that the Oligocene sequences consist of sandstone bodies representing near-shore marine sediments deposited during a S-SW progradation of the shoreline. The abrupt switch from a mainly western source to a dominant eastern/north-eastern source in Early Oligocene is a strong indication that renewed uplift and exposure of landmasses occurred in earliest Oligocene time.

A sudden decrease in ocean temperature in early Late Oligocene (Molnar & England, 1990; Sliwinska *et al.*, 2010) is used to explain the sudden input of erosional material in the eastern North Sea (Goledowski *et al.*, 2012) and hence, that changing depositional patterns are controlled by a climatic change. However, temperature drops cannot explain the rapid creation of accommodation space in the eastern North Sea, which followed a period of truncation at the Eocene to Oligocene boundary (Fig. 2). Also, truncation of Eocene strata and the absence in some places of Lower Oligocene strata on the Patchbank Ridge (Fig. 12) indicates that Early Oligocene uplift occurred within the basin, in addition to affecting extensive areas of southern Norway. Therefore, although climatic changes probably contributed to increased sediment production in mainland Norway in Oligocene time (e.g. Huuse *et al.*, 2001; Huuse, 2002; Nielsen *et al.*, 2009), differential tectonic uplift is, nevertheless, also probably exposed landmasses for increased erosion due to rejuvenation of the relief. This is also supported by observations of Upper Oligocene (CSS-4) NW-SE trending normal faults, which reflect NE-SW gravitationally driven extension on the Horda Platform (Clausen *et al.*, 1999). These faults were active during tilting of the Horda Platform, forced by vertical uplift of southern Norway. The tilting of the basin strata also explains the pinch-out of the Early Oligocene CSS-3 sequence along the eastern margin in the northern part of the North Sea Basin (Faleide *et al.*, 2002). The uplift is here suggested to have been most pronounced in western parts of South Norway, thus explaining major sediment contribution to the eastern and central North Sea Basin and very limited sediments in the northernmost North Sea area. The sediments were probably transported along established topographic lineaments and faults (Fig. 15D), as has also been proposed by Eidvin *et al.* (2010).

Phases of tectonic uplift of the hinterland continued throughout the Oligocene, indicated by the presence of coarse-grained sediments in the Upper Oligocene succession, deposited simultaneously with major subsidence in the Norwegian-Danish Basin (Jarsve *et al.*, submitted).

The progradational Miocene CSS-5 to CSS-7 sequences in the central North Sea reflect a prolonged existence of a source area in Scandinavia. This is substantiated by observations of truncation at the Fjerrislev Fault Zone of Upper Oligocene strata in Late Oligocene to Early Miocene time (e.g. Rasmussen, 2009) and hence indicates renewed phases of uplift of the hinterland. The thinning of the Miocene CSS-5, CSS-6 and CSS-7 sequences across the Norwegian-Danish Basin indicates that the accommodation space here was filled during Oligocene time. Incision of Miocene channels into Oligocene strata in the northern North Sea has been allocated to a drop in relative sea-level in Early Miocene time (Rundberg *et al.*, 1995; Gregersen, 1998; Faleide *et al.*, 2002). However, the presence of sequence CSS-5 depocentre in the southern parts of the study area indicates uplift of parts of the northern North Sea as well. If this erosion had been caused by a eustatic drop in sea-level, then it is also probable that central parts of the North Sea would have been exposed for erosion.

The major Late Miocene sediment accumulation (CSS-7) in the Central Graben area was deposited during rapid subsidence of the basin. Although this phenomenon occurred simultaneously with increased sediment input from the Baltic river system in the east and south-east (Bijlsma, 1981; Gibbard, 1988; Cameron *et al.*, 1993) and a potential global cooling (Zachos *et al.*, 2001), it is difficult by these processes alone to explain the subsidence which was created to accommodate for the vast quantities of sediments deposited during Late Miocene time. Therefore, tectonic subsidence, although non-fault related, is suggested here to have occurred in Miocene time. This was also strengthened from observation of a regional unconformity at the base of Pliocene in the eastern North Sea by Japsen *et al.* (2007), who proposed that the Norwegian-Danish Basin was exhumed in Late Miocene and Early Pliocene times. This indicates that sediments belonging to sequence CSS-7 in the Central Graben area include re-worked sediments.

The truncation of the Pliocene and Early Pleistocene sequences (CSS-8 and CSS-9) across the Norwegian-Danish Basin to the Ringkøbing Fyn High and the Stavanger Platform indicates that glaciers reached the north-eastern margin of the Central Graben area at the glacial maximum. This was also proposed by Sejrup *et al.* (2005). This is also observed in the Norwegian Sea (Stuevold & Eldholm, 1996), where more extensive glaciers of Pleistocene age reached the shoreline and eroded the earlier Mesozoic and Cenozoic sedimentary layers (Jansen & Sjøholm, 1991).

The sub-parallel configuration of the seismic reflections of sequence CSS-10 relative to seabed indicates that the sequence was deposited after most of the uplift of southern Norway, both tectonic uplift and uplift related to isostatic rebound from glacial erosion of the hinterland. The main tilting therefore occurred during deposition of sequences CSS-8 and CSS-9, due to isostatic rebound from glacial erosion and tectonic uplift of southern Norway.

Coupled source-to-sink dynamics

As described above, there appears to be a direct relationship between creation of accommodation space in the North Sea area and influx of siliciclastic debris from hinterlands at the basin margins. We suggest that this can be best explained by a geodynamic model in which crustal uplifts in the hinterlands are linked to basin formation and subsidence in the North Sea area.

A coupled source-to-sink model is established for basin configuration 1; the prograding sequences filled the accommodation space that developed during the active stretching phase in the Late Jurassic to earliest Cretaceous. The creation of the accommodation space by basin subsidence can thus be linked to uplift of parts of southern Norway and increased rates of erosion and production of sediments belonging to sequences J-1 and J-2 and sub-sequence K-2.1.

During basin configuration 2, both the North Sea area and the hinterlands were probably exposed to only minor tectonic activity, as indicated by the prolonged existence of the Chalk Sea, gradually extending northwards in the northern North Sea throughout Late Cretaceous (Surlyk *et al.*, 2003). However, we propose that the coupled source-to-sink relationship as seen during basin configuration 1 also functioned during basin configuration 2. This interpretation is based on the assumption that the tectonic influence in the North Sea area was negligible during the Late Cretaceous, where subsidence were mainly caused by sediment loading and thermal cooling (e.g. Gabrielsen *et al.*, 2001; Kyrkjebø *et al.*, 2001). As this occurred without any simultaneous uplift of the basin margins, although subsequent to denudation and transgression of parts of the hinterlands, we suggest that the North Sea area and present adjacent landmasses were stable during basin configuration 2.

A coupled source-to-sink relationship is also suggested for basin configuration 3. This is evident from the rapid subsidence of the North Sea Basin, with palaeo-water-depths up to 600 metres, as recorded by Jordt *et al.* (1995), Kjennerud *et al.* (2001) and Kyrkjebø *et al.* (2001), simultaneously with uplift of the East Shetland Platform and the north-western parts of southern Norway (Faleide *et al.*, 2002). The uplifted hinterlands acted as source areas for sediments belonging to sequences CSS-1 and CSS-2. Several uplift mechanisms have been proposed to explain this tectonodynamic process: transient thermal uplift due to buoyancy of the lithosphere when heated by the Icelandic Plume (e.g. Sleep, 1990; Clift & Turner, 1998), uplift due to mantle fluid flow (e.g. Nadin *et al.*, 1995, 1997) and underplating, giving rise to a permanent isostatic uplift (e.g. Brodie & White, 1994; Clift, 1999). However, although the coupled source-to-sink relationship is well established, further work is necessary to constrain the processes which acted on the basin configuration during Late Palaeocene to Eocene time.

A vast increase in sediment input, sourced from southern Norway in Early Oligocene, combined with development of diachronous accommodation space in the North Sea area, is a strong indication of coupled source-to-sink dynamics also for basin configuration 4. Although a climatic change at the Eocene-Oligocene boundary triggered an increase in erosion and run-off in the hinterland (Huuse *et al.*, 2001), this cannot explain the formation of contemporaneous sub-basins and migration of their depocentres throughout the Oligocene and Neogene, as discussed above. The basin subsidence, as recorded throughout the Cenozoic, probably was related to lithospheric flexuring, as the southern Norwegian landmasses were uplifted. However, as mentioned above, the mechanism of such coupled crustal processes is still not constrained. The reader is referred to Faleide *et al.* (2002) and Anell *et al.* (2009) for a review of the discussion.

CONCLUSIONS

As illustrated in this study, sedimentation in the study area was related to four main different basin configurations when basin configuration and uplift of the hinterlands during Late Mesozoic and Cenozoic times are concerned, namely the Late Jurassic to Early Cretaceous (basin configuration 1), the Late Cretaceous to Early Palaeocene (basin configuration 2), the configuration influenced by break-up and initial opening of the NE Atlantic (Late Palaeocene-Eocene; basin configuration 3) and, finally, the last phase of uplift of South Norway, which was initiated in the Oligocene (basin configuration 4).

1. This way of separating the basin configurations from each other implies that the latest syn-rift and the early post-rift phases are incorporated in basin configuration 1 and that basin configuration 2 coincides with the late post-rift phase.
2. A coupled source-to-sink dynamic is revealed from the close relationship between diachronous development of accommodation space in the North Sea area and uplift of hinterlands throughout the Late Mesozoic and Cenozoic.
3. The late syn-rift to early post-rift phases (basin configuration 1), is characterised by terrigenous erosional material that was trapped in depocentres located between topographical highs, both in the northern and eastern North Sea basinal areas.
4. The eastern topographic highs were located in parts of southern Norway (Fig. 14A) and at the Lista Fault Blocks south of the Sorgenfrei

Tornquist Zone, acting as source areas for the Late Jurassic J-1 and J-2 sequences located in the northern Horda Platform area, Stord Basin-Åsta Graben and Egersund Basin and for the Lower Cretaceous K-2.1 sub-sequence.

5 Sediment infill patterns and source areas were similar during the late syn-rift and early post-rift phases, implying that the basin configuration which developed during late syn-rift phase prevailed throughout the early post-rift phase.

6 Southern Norway continued as a sediment source area throughout the early post-rift phase in Early Cretaceous. Denudation prevailed and a regional unconformity was formed in extensive areas in the North Sea basin and probably in parts of southern Norway to mid-Cretaceous time with subsequent deposition of the Chalk Group in the late post-rift phase (basin configuration 2).

7 During basin configuration 2, vast areas of southern Norway were part of a depositional area covered by a shallow sea.

8 Shallow sea conditions prevailed to Late Palaeocene, when topography on the East Shetland Platform and in north-western parts of southern Norway caused increased erosion and sediment supply into the North Sea Basin. This created a different basin configuration, as compared to basin configuration 2, as uplift of these areas exposed new source areas and this occurred simultaneously with rapid subsidence in the basin, particularly in the Viking Graben area.

9 This basin configuration continued until creation of a source area in the north-east and development of diachronous accommodation space in the eastern North Sea area, initiated in the Early Oligocene.

10 This uplift probably started in the western part of southern Norway, with later uplift of more extensive areas in Late Oligocene and Miocene.

11 Although climatic changes probably occurred during basin configuration 4, these cannot explain the observed basin configuration and sediment infill style alone; climatic change towards increased precipitation and run-off acted as an external factor simultaneously with uplift of the hinterland. Potentially, the climatic change was influenced by the uplift itself.

ACKNOWLEDGEMENTS

We want to thank Fugro Multi Client Services and TGS Nopec for the seismic data, which increased the quality of the seismic interpretation. Also, thank you to Brit Inger Thyberg for inspiring discussions during this work and Olav Antonio Blaich for technical help with the maps. The authors also want to thank Mike Young, Erik S. Rasmussen and Allard Martinius for valuable comments on the manuscript. Also, we want to thank Adrian Read for proofreading the manuscript.

REFERENCES

Anell, I., Thybo, H. and **Artemieva, I.M.** (2009) Cenozoic uplift and subsidence in the North Atlantic region: Geological evidence revisited. *Tectonophysics*, **474**, 78–105.

Anell, I., Thybo, H. and **Stratford, W.** (2010) Relating Cenozoic North Sea sediments to topography in southern Norway: The interplay between tectonics and climate. *Earth Planet. Sci. Lett.*, **300**, 19–32.

Anell, I., Thybo, H. and **Rasmussen, E.** (2012) A synthesis of Cenozoic sedimentation in the North Sea. *Basin Res.*, **23**, 1–26.

Badley, M.E., Price, J.D., Rambech Dahl, C. and **Agdestein, T.** (1988) The structural evolution of the northern Viking Graben and its bearing upon extensional modes of basin formation. *J. Geol. Soc. London*, **145**, 455–472.

Bijlsma, S. (1981) Fluvial sedimentation from the Fennoscandian area into the north-west European Basin during the Late Cenozoic. *Geol. Mijnb.*, **60**, 337–345.

Brekke, H. (2000) The tectonic evolution of the Norwegian Sea Continental Margin with emphasis on the Vøring and Møre Basins. In: *Dynamics of the Norwegian Margin* (Eds A. Nøttvedt and B.T. Larsen). *Geol. Soc. London Spec. Publ.*, **167**, 327–378.

Brodie, J. and **White, N.** (1994) Sedimentary basin inversion caused by igneous underplating: North-west European continental shelf. *Geology.* **22**, 147–150.

Buchard, B. (1978) Oxygen isotope palaeotemperatures from the Tertiary period in the North Sea area. *Nature*, **275**, 121–123.

Bullimore, S.A., Henriksen, S., Liestøl, F.M. and **Helland-Hansen, W.** (2005) Clinoform stacking patterns, shelf-edge trajectories and facies associations in Tertiary coastal deltas, offshore Norway: implications for the prediction of lithology in prograding systems. *Norwegian Journal of Geology*, **85**, 169–187.

Cameron. T.D.J., Bulat, J. and **Mesdag, C.S.** (1993) High resolution seismic profile through a Late Cenozoic delta complex in the southern North Sea. *Mar. Petrol. Geol.*, **9**, 591–599.

Chalmers, J.A., Green, P., Japsen, P. and **Rasmussen, E.S.** (2010) The Scandinavian mountains have not persisted since the Caledonian orogeny. A comment on Nielsen *et al.* (2009). *Journal of Geodynamics*, **50**, 94–101.

Clausen, O.R. and **Huuse, M.** (1999) Topography of the top chalk surface on- and offshore Denmark. *Mar. Petrol. Geol.*, **16**, 677–691.

Clausen, O.R., Gabrielsen, R.H., Reksnes, P.A. and **Nysæther, E.** (1999) Development of intraformational (Oligocene–Miocene) faults in the northern North Sea: influence of remote stresses and doming of Fennoscandia. *J. Struct. Geol.*, **21**, 1457–1475.

Clift, P.D. (1999) The thermal impact of Palaeocene magmatic underplating in the Faeroe-Shetland- Rockall region. In: *Petroleum Geology of North-west Europe; Proceedings of the 5th Conference* (Eds A.J. Fleet and S.A.R. Bold), *Geol. Soc. London*, 585–593.

Clift, P.D. and **Turner, J.** (1998) Palaeogene igneous underplating and subsidence anomalies in the Rockall-Faeroe-Shetland area. *Mar. Petrol. Geol.*, **15**, 223–243.

Danielsen, M., Michelsen, O. and **Clausen, O.R.** (1997) Oligocene sequence stratigraphy and basin development in the North Sea sector based on log interpretations. *Mar. Petrol. Geol.*, **14**, 931–950.

Deegan, C.E. and **Scrull, B.J.** (1977) A standard lithostratigraphic nomenclature for the Central and Northern North Sea. *Institute of Geological Sciences*, **1**, 36 pp.

Dmitrieva, E., Jackson, C. A-L., Huuse, M. and **McCarthy, A.** (2012) Palaeocene deep-water depositional systems in the North Sea Basin: a 3D seismic and well data case study, offshore Norway. *Petroleum Geoscience*, **18**, 97–114.

Doré, A.G., Lundin, E.R., Kusznir, N.J. and **Pascal, C.** (2008) Potential mechanism for the genesis of Cenozoic domal structures on the NE Atlantic margin: pros, cons and some new ideas. In: *The Nature and Origin of compression in Passive Margins* (Eds H. Johnson, A.G. Doré, R.W. Gatliff, R. Holdsworth, E.R. Lundin and J.D. Ritchie), *Geol. Soc. London*, **306**, 1–26.

Dybkjær, K. and **Rasmussen, E.S.** (2007) Organic-walled dinoflagellate cyst stratigraphy in an expanded Oligocene – Miocene boundary section in the eastern North Sea Basin (Frida-1 well, Denmark) and correlation from basinal to marginal areas. *Journal of Micropalaeontology*, **26**, 1–17.

Dybkjær, K., King, C. and **Sheldon, E.** (2012) Identification and characterisation of the Oligocene-Miocene boundary (base Neogene) in the eastern North Sea Basin – based on dinocyst stratigraphy, micropalaeontology and δ^{13} C-isotope data. *Palaeogr. Plaeoclimatol. Palaeoecol.*, http://dx.doi.org/10.1016/j.palaeo.2012.08.007.

Eidvin, T., Bugge, T. and **Smelror, M.** (2007) The Molo Formation, deposited by coastal progradation on the inner Mid-Norwegian continental shelf, coeval with the Kai Formation to the west and Utsira Formation in the North Sea. *Norwegian Journal of Geology*, **87**, 75–142.

Eidvin. T. and **Riis. F.** (1992) En biostratigrafisk og seismostratigrafisk analyse av tertiære sedimenter i nordlige deler av Norskerenna med hovedvekt på øvre pliocene vifteavsetninger. *Norwegian Petroleum Directorate Contribution*, **32**. 40 pp.

Eidvin, T., Skovbjerg Rasmussen, E., Riis, F. and **Rundberg, Y.** (2010) Oligocene to Lower Pliocene deposits of the Norwegian continental shelf, with correlation to the Norwegian Sea, Greenland, Svalbard, Denmark and their relation to the uplift of Fennoscandia. In: *29th Nordic Geological Winter Meeting* (Eds H.A. Nakrem, A.O. Harstad and G. Haukdal), *Abstract and proceedings of the Geological Society of Norway, Oslo*, **1**, 43–44.

Eldholm, O. and **Thomas, E.** (1993) Environmental impact of volcanic margin formation. *Earth Planet. Sci. Lett.*, **117**, 319–329.

Eldholm, O., Tsikalas, F. and **Faleide, J.I.** (2002) Continental Margin off Norway 62–75°N: Palaeogene tectonomagmatic segmentation and sedimentation. In: *The North Atlantic Igneous Province: Stratigraphy, Tectonic, Volcanic and Magmatic Processes* (Eds D.W. Jolley and B.R. Bell), *Geol. Soc. London Spec. Publ.*, **197**, 39–68.

Embry, A. (1988) Triassic sea-level changes: evidence from Canadian Arctic Archipelago. In: *Sea-level Changes – an integrated approach* (Eds C.K. Wilgus, B.S. Hastings, H. Posamentier, J.V. Wagoner, C.A. Ross and C. Kendall), *SEPM Spec. Publ.*, **42**, 249–259.

Embry, A. (1993) Transgressive-regressive (T-R) sequence analysis of the Jurassic succession of the Sverdrup Basin, Canadian Arctic Archipelago. *Can. J. Earth Sci.*, **30**, 301–320.

Embry, A. (1995) Sequence boundaries and sequence hierarchies: problems and proposals. In: *Sequence Stratigraphy on the North-west European Margin* (Eds R. Steel, V.L. Felt, E.P. Johannessen and C. Mathieu) *Norwegian Petroleum Society Special Publication*, **5**, 1–11.

Embry, A. and **Johannessen, E.** (1992) T-R sequence stratigraphy, facies analysis reservoir distribution in the uppermost Triassic-Lower Jurassic succession, western Sverdrup Basin, Arctic Canada. In: *Arctic Geology and Petroleum Potential* (Eds T. Vorren, E. Bergsager, Ø.A. Dahl-Stamnes, E. Holter, B. Johansen, E. Lie and T.B. Lund), *Norwegian Petroleum Society Special Publication*, **2**, 121–146.

Emery, D. and **Myers, K.J.** (1997) *Sequence Stratigraphy*. Blackwell Science, Oxford, 297 pp.

Esmerode, E.V., Lykke-Andersen, H. and **Surlyk, F.** (2008) Interaction between bottom currents and slope failure in the Late Cretaceous of the southern Danish Central Graben, North Sea. *J. Geol. Soc. London*, **165**, 55–72.

Evans, D., Graham, C., Armour, A. and **Bathurst, P.** (2003) The Millennium Atlas: Petroleum geology of the Central and Northern North Sea. *Geol. Soc. London*, pp 989.

Færseth, R.B. (1996) Interaction between Permo-Triassic and Jurassic extensional fault-blocks during the development of the northern North Sea. *J. Geol. Soc. London*, **153**, 931–944.

Færseth, R.B., Knudsen, B.E., Liljedahl, T., Midboe, P.S. and **Soderstrom, B.** (1997) Oblique rifting and sequential faulting in the Jurassic development of the northern North Sea. *J. Struct. Geol.*, **19**, 1285–1302.

Faleide, J.I., Kyrkjebø, R., Kjennerud, T., Gabrielsen, R.H., Jordt, H., Fanavoll, S. and **Bjerke, M.D.** (2002) Tectonic impact on sedimentary processes during Cenozoic evolution of the northern North Sea and surrounding areas. In: *Exhumation of the North Atlantic Margin: Mechanisms and Implications for Petroleum Exploration* (Eds A.G. Doré, J.A. Cartwright, M.S. Stoker, J.P. Turner & N. White) *Geol. Soc. London Spec. Publ.*, **196**, 235–269.

Faleide, J.I., Tsikalas, F., Breivik, A.J., Mjelde, R., Ritzmann, O., Engen, Ø., Wilson, J. and **Eldholm, O.** (2008) Structure and evolution of the continental margin off Norway and the Barents Sea, *Episodes*, **31**, 82–91.

Fraser, S.I., Robinson, A.M., Johnson, H.D., Underhill, J.R., Kadolsky, D.G.A., Connell, R., Johannessen, P. and **Ravnås, R.** (2002) Upper Jurassic. In: *The Millenium Atlas: petroleum geology of the central and northern North Sea* (Eds D. Evans, C. Graham, A. Armour and P. Bathurst). *Geol. Soc.* London, 372–439.

Frederiksen, S., Nielsen, S.B. and **Balling, S.** (2001) A numerical dynamic model for the Norwegian-Danish Basin. *Tectonophysics*, **343**, 165–183.

Friedrich, O., Norris, R.D. and **Erbacher, J.** (2012) Evolution of middle to Late Cretaceous oceans – A 55 m.y. record of Earth's temperature and carbon cycle. *Geology*, **40**, 107–110.

Gabrielsen, R.H., Færseth, R.J., Steel, R.J., Idil, S. and **Kløvjan, O.S.** (1990) Architectural styles of basin fill in the northern Viking Graben. In: *Tectonic evolution of the North Sea Rift* (Eds D.J. Blundell and A.D. Gibbs), Oxford University Press, Oxford. 158–179.

Gabrielsen, R.H., Kyrkjebø, R., Faleide, J.I., Fjeldskaar, W and **Kjennerud, T.** (2001) The Cretaceous post-rift basin configuration of the northern North Sea. *Petroleum Geoscience*, **7**, 137–154.

Gabrielsen, R.H., Braathen, A., Olesen, O., Faleide, J. I., Kyrkjebø, R. and **Redfield, T.F.** (2005) Vertical movements in south-western Fennoscandia: a discussion of regions and processes from the present to the Devonian. In: *Onshore-Offshore Relationships on the North Atlantic Margin* (Eds B.T.G. Wandas, J.P. Nystuen, E. Eide and F. Gradstein), *Norwegian Petroleum Society Special Publication*, **12**, 1–28.

Gabrielsen, R.H., Faleide, J.I., Pascal, C., Braathen, A., Nystuen, J.P., Etzelmuller, B and **O'Donnell, S.** (2010a) Latest Caledonian to Present tectonomorphological development of southern Norway. *Mar. Petrol. Geol.*, **27**, 709–723.

Gabrielsen, R.H., Faleide, J.I., Pascal, C., Braathen, A., Nystuen, J.P., Etzelmuller, B. and **O'Donnell, S.** (2010b) Reply to discussion of Gabrielsen *et al.* (2010) by Nielsen *et al.* Latest Caledonian to present tectonomorphological development of southern Norway. *Mar. Petrol. Geol.*, 27, 1290–1295.

Gibbard, P.L. (1988) The history of the great north-west European rivers during the past three million years. *Phil. Trans. Roy. Soc. London*, **318**, 559–602.

Gibbons, K., Hellem, T, Nio, S.D. and **Vebenstad, K.** (1993) Sequence architecture, facies development and carbonate cemented horizons in the Troll Field reservoir, offshore Norway. In: *Advances in Reservoir Geology* (Ed. **M. Ashton**), *Geol. Soc. London Spec. Publ.*, **69**, 1–31.

Goldsmith, P.J. (2000) Exploration potential east of the Troll Field, offshore Norway, after dry well 32/4-1. In: *Improving the Exploration Process by Learning from the Past* (Eds K. Ofstad, J.E. Kittilsen and P. Alexander-Marrack), *Norwegian Petroleum Society Special Publication*, **9**, 65–97.

Goledowski, B., Nielsen, S.B. and **Clausen, O.R.** (2012) Patterns of Cenozoic sediment flux from western Scandinavia. *Basin Res.*, **23**, 1–24.

Goledowski, B., Clausen, O.R. and **Nielsen, S.B.** (2014) Reply to comment on E.S. Rasmussen and K.Dybkjær on Patterns of Cenozoic sediment flux from western Scandinavia, by B.Goledowski, S.B.Nielsen and O.R.Clausen, Basin Research (2012), 24(4), 377–400. *Basin Res.*, **26**, 347–350.

Gradstein. F. and **Bäckström. S.** (1996) Cenozoic biostratigraphy and palaeobathymetry, northern North Sea and Haltenbanken. *Norsk Geol. Tidsskr.*, **76**. 3–32.

Gregersen, U. (1998) Upper Cenozoic channels and fans on 3D seismic data in the northern Norwegian North Sea. *Petroleum Geoscience*, **4**, 67–80.

Gregersen, U. and **Johannesen, P.N.** (2007) Distribution of the Neogene Utsira Sand and the succeeding deposits in the Viking Graben area, North Sea. *Mar. Petrol. Geol.*, **24**, 591–606.

Gregersen, U., Michelsen, O. and **Sørensen, J.C.** (1997) Stratigraphy and facies distribution of the Utsira Formation and the Pliocene sequences in the northern North Sea. *Mar. Petrol. Geol.*, **14**, 893–914.

Hancock, J.M. (1990) Cretaceous. In: *Introduction to the petroleum geology in the North Sea* (Ed. **K.W. Glennie**), Oxford, Blackwell Scientific Publications, 255–272.

Handy, M.R., Schmid, S.M., Bousquet, R., Kissling, E. and **Bernoulli, D.** (2010) Recoiling plate-tectonic reconstructions of Alpine Tethys with the geological-geophysical record of spreading and subduction in the Alps. *Earth-Sci. Rev.*, **102**, 121–158.

Haq, B.U., Hardenbol, J. and **Vail, P.R.** (1987) Chronology of fluctuating sea-levels since the Triassic. *Science*, **235**, 1156–1166.

Heeremans, M. and **Faleide, J.I.** (2004) Late Carboniferous-Permian tectonics and magmatic activity in the Skagerrak, Kattegat and the North Sea. In: *Permo-Carboniferous magmatism and Rifting in Europe* (Eds M. Wilson, E.R. Neumann, G.R. Davies, M.J. Timmerman, M. Heeremans and B.T. Larsen). *Geol. Soc. London Spec. Publ.*, **223**, 157–176.

Heeremans, M., Faleide, J.I. and **Larsen, B.T.** (2004) Late Carboniferous – Permian of NW Europe: an introduction to a new regional map. In: *Permo-Carboniferous Magmatism and Rifting in Europe* (Eds M. Wilson, E.R. Neumann, G.R. Davies, M.J. Timmerman, M. Heeremans and B.T. Larsen). *Geol. Soc. London Spec. Publ.*, **223**, 75–88.

Heilmann-Clausen, C., Nielsen, O.B. and **Gersner, F.** (1985) Lithostratigraphy and depositional environment in the Upper Palaeocene and Eocene of Denmark. *Bull. Geol. Soc. Denmark*, **33**, 287–323.

Hendriks, B. and **Andriessen, P.** (2002) Pattern and timing of the post-Caledonian denudation of northern Scandinavia constrained by apatite fission-track thermochronology. In: *Exhumation of the North Atlantic Margin: Timing, Mechanisms and Implications for Petroleum Exploration* (Eds A.G. Doré, J.A. Cartwright, M.S. Stoker, J.P. Turner and N. White) *Geol. Soc. London*, **196**, 117–137.

Hendriks, B., Andriessen, P., Huigen, Y., Leighton, C., Redfield, T., Murrell, G., Gallagher, K. and **Nielsen, S.B.** (2007) A fission track data compilation for Fennoscandia. *Norwegian Journal of Geology*, **87**, 143–155.

Huber, B.T., Hodell, D.A. and **Hamilton, C.P.** (1995) Middle–Late Cretaceous climate of the southern high latitudes: stable isotopic evidence for minimal equator-to-pole thermal gradients. *Geol. Soc. Am. Bull.*, **107**, 1164–1191.

Huber, B.T., Norris, R.D. and **MacLeod, K.G.** (2002) Deep-sea paleotemperature record of extreme warmth during the Cretaceous. *Geology*, **30**, 123–126.

Huuse, M. (1999) Detailed Morphology of the Top Chalk Surface in the Eastern Danish North Sea. *Petroleum Geoscience*, **5**, 303–314.

Huuse, M. (2002) Cenozoic uplift and denudation of southern Norway: insights from the North Sea Basin. In: *Exhumation of the North Atlantic Margin: Timing, Mechanisms and Implications for the Petroleum Exploration* (Eds A.G. Doré, J.A. Cartwright, M.S. Stoker, J.P. Turner and N. White) *Geol. Soc. London Spec. Publ.*, **196**, 209–233.

Huuse, M. and **Clausen, O.R.** (2001) Morphology and origin of major Cenozoic sequence boundaries in the eastern North Sea basin: top Eocene, near-top Oligocene and the mid-Miocene unconformity. *Basin Res.*, **13**, 17–41.

Huuse, M., **Lykke-Andersen, H.** and **Michelsen, O.** (2001) Cenozoic evolution of the eastern Danish North Sea. *Mar. Geol.*, **177**, 243–269.

Isaksen, D. and **Tonstad, K.** (1989) A revised Cretaceous and Tertiary lithostratigraphic nomenclature for the Norwegian North Sea. *NPD Bulletin* **5**, 59 pp.

Jansen, E. and **Sjøholm, J.** (1991) Reconstruction of glaciation over the past 6 Myr from ice-borne deposits in the Norwegian Sea. *Nature*, **349**, 600–603.

Japsen, P., **Bidstrup, T.** and **Lidmar-Bergstrøm, K.** (2002) Neogene uplift and erosion of southern Scandinavia induced by the rise of the South Swedish Dome. In: *Exhumation of the North Atlantic Margin: Timing, Mechanisms and Implications for Petroleum Exploration* (Eds A.G. Doré, J.A. Cartwright, M.S. Stoker, J.P. Turner and N. White). *Geol. Soc. London Spec. Publ.*, **196**, 183–207.

Japsen, P. and **Chalmers, J.A.** (2000) Neogene uplift and tectonics around the North Atlantic: overview. *Global Planet. Change*, **24**, 165–173.

Japsen, P., **Green, P.F.**, **Nielse, L.H.**, **Rasmussen, E.S.** and **Bidstrup, T.** (2007) Mesozoic-Cenozoic exhumation events in the eastern North Sea Basin: a multi-disciplinary study based on palaeothermal, palaeoburial, stratigraphic and seismic data. *Basin Res.*, **19**, 451–490.

Jordt, H., **Faleide, J.I.**, **Bjorlykke, K.** and **Ibrahim, M.T.** (1995) Cenozoic sequence stratigraphy of the central northern North Sea Basin: tectonic development, sediment distribution and provenance areas. *Mar. Petrol. Geol.*, **12**, 845–879.

Jordt, H., **Thyberg, B.I.** and **Nottvedt, A.** (2000) Cenozoic evolution of the central and northern North Sea with focus on differential vertical movements of the basin floor and surrounding clastic source areas. In: *Dynamics of the Norwegian Margin* (Eds A. Nottvedt, B.T. Larsen, R.H. Gabrielsen, S. Olaussen, H. Brekke, B. Torudbakken, O. Birkelnad and J. Skogseid), *Geol. Soc. London Spec. Publ.*, **167**, 219–243.

Kiltgord, K.D. and **Schouten, H.** (1986) Plate kinematics and the Central Atlantic. In: *The Geology of North America Margin, the Western North Atlantic Region* (Eds P.R. Vogt and B.E. Tucholke), *Geol. Soc. Am.*, 351–378.

Kjennerud, T., **Faleide, J.I.**, **Gabrielsen, R.H.**, **Gillmore, G.K.**, **Kyrkjebø, R.**, **Lippard, S.J.** and **Løseth, H.** (2001) Structural restoration of Cretaceous – Cenozoic palaeobathymetry in the northern North Sea. In: *Sedimentary Environments offshore Norway – Palaeozoic to recent* (Eds O.J. Martinsen and T. Dreyer) *Norwegian Petroleum Society*, **10**, 347–364.

Knox, R.W.O. and **Morton, A.C.** (1988) The record of early-Tertiary N Atlantic volcanism in sediments of the North Sea Basin. In: *Early Tertiary Volcanism and the Opening of the NE Atlantic* (Eds A.C. Morton and L.M. Parson), *Geol. Soc. London Spec. Publ.*, **39**. 407–419.

Kyrkjebø, R., **Hamborg, M.**, **Faleide, J.I.**, **Jordt, H.** and **Christiansson, P.** (2000) Cenozoic tectonic subsidence from 2D depositional simulations of a regional transect in the northern North Sea basin. In: *Dynamics of the Norwegian Margin* (Ed. A. Nøttvedt), *Geol. Soc. London Spec. Publ.*, **167**, 273–294.

Kyrkjebø, R., **Kjennerud, T.**, **Gillmore, G.K.**, **Faleide, J.I.** and **Gabrielsen, R.H.** (2001) Cretaceous – Tertiary palaeo-bathymetry in the northern North Sea; integration of palaeo-water depth estimates obtained by structural restoration and micropalaeontological analysis. In: *Sedimentary Environments Offshore Norway – Palaeozoic to Recent* (Ed. O. Marthinsen), *Norwegian Petroleum Society Special Publication*, **9**, 321–345.

Liboriussen, J., **Ashton, P.** **and Tygesen, T.** (1987) The Tectonic evolution of the Fennoscandian Border Zone in Denmark. *Tectonophysics*, **137**, 21–29.

Lindstrom, K. (1993) Denudation surfaces and tectonics in the southernmost part of the Baltic shield. *Precambrian Res.*, **64**, 337–345.

Martinsen, O.J., **Boen, F.**, **Charnock, M.A.**, **Mangerut, G.** and **Nøttvedt, A.** (1999) Cenozoic development of the Norwegian margin 60-64°N: sequences and sedimentary response to variable basin physiography and tectonic setting. In: *Petroleum Geology of NW Europe, Proceedings of the 5th Conference* (Ed. A.J. Fleet and S.A.R. Boldy). *Geol. Soc. London*, 293–304.

Martinsen, O.J., **Lien, T.** and **Jackson, C.** (2005) Cretaceous and Palaeogene turbidite systems in the North Sea and Norwegian Sea Basins: source, staging area and basin physiography controls on reservoir development. In: *Petroleum Geology North-west Europe and Global Perspectives – Proceedings of the 6th Petroleum Geology Conference* (Eds A.G. Doré and B.A. Vinning) *Geol. Soc. London*, 1147–1164.

McKenzie, D. (1978) Some remarks on the development of sedimentary basins. *Earth Planet. Sci. Lett.*, **40**, 25–32.

Michelsen, O. (1994) Stratigraphic correlation of the Danish onshore and offshore Tertiary successions based on sequence stratigraphy. *Bull. Geol. Soc. Denmark*, **41**, 145–161.

Michelsen, O., **Danielsen, M.**, **Heilmann-Clausen, C.**, **Jordt, H.**, **Laursen, G.V.** and **Thomsen, E.** (1995) Occurrence of major sequence stratigraphic boundaries in relation to basin development in Cenozoic deposits of the south-eastern North Sea. In: *Sequence Stratigraphy on the North-west European Margin* (Eds R.J. Steel, V.L. Felt, E.P. Johannessen and C. Mathieu), *Norwegian Petroleum Society*, **5**, 415–427.

Michelsen, O. and **Danielsen, M.** (1996) Sequence and systems tract interpretation of the epicontinental Oligocene deposits in the Danish North Sea. In: *Geology of Siliciclastic Shelf Seas* (Eds M. De Batist and P. Jacobsen), *Geol. Soc. Spec. Publ.*, **117**, 1–13.

Milton, N.J. (1993) Evolving depositional geometries in the North Sea Jurassic rift. In: *Petroleum geology of North-west Europe* (Ed. J.R. Parker), *Proceedings of the 4th Conference*, 425–442.

Milton, N.J., Bertram, G.T. and **Vann, I.R.** (1990) Early Palaeogene tectonics and sedimentation in the Central North Sea. In: *Tectonic Events Responsible for Britain's Oil and Gas Reserves* (Eds R.F.P. Hardman and J. Brooks) *Geol. Soc. Spec. Publ.*, **55**, 339–351.

Mitchum, R.M.J., Vail, P.R. and **Thompson, S.I.** (1977) Seismic stratigraphy and global changes in sea-level, part 2: the depositional sequence as the basic unit for stratigraphic analysis. In: *Seismic Stratigraphy: Application to Hydrocarbon Exploration* (Ed. C. Payton), *AAPG Mem.*, **26**, 53–62.

Molnar, P. and **England, P.** (1990) Late Cenozoic uplift of mountain ranges and global climate change: chicken or egg. *Nature*, **346**, 29–34.

Morton, A.C. (1982) Lower Tertiary Sand Development in Viking Graben, North Sea. *AAPG Bull.*, **66**, 1542–1559.

Mosar, J. (2003) Scandinavia's North Atlantic passive margin. *J. Geophys. Res.*, **108**, 1–18.

Mudge, D.C. and **Copestake, P.** (1991) Revised lower Palaeogene lithostratigraphy for the Outer Moray Firth, North Sea. *Mar. Petrol. Geol.*, **9**, 53–69.

Nadin, P.A., Kusznir, N.J. and **Toth, J.** (1995) Transient regional uplift in the Early Tertiary of the northern North Sea and the development of the Iceland Plume. *J. Geol. Soc.*, **152**, 953–958.

Nadin, E.A., Kusznir, N.J. and **Cheadle, M.J.** (1997) Early Tertiary plume uplift of the North Sea and Faeroe-Shetland basins. *Earth Planet. Sci. Lett.*, **148**, 109–127.

Nielsen, S.B., Gallagher, K., Leighton, C., Balling, N., Svenningsen, L., Holm Jacobsen, B., Thomsen, E., Nielsen, O.B., Heilmann-Clausen, C., Egholm, D.L., Summerfield, M.A., Rønø Clausen, O., Piotrowski, J.A., Thorsen, M.R., Huuse, M., Abrahamsen, N., King, C. and **Lykke-Andersen, H.** (2009) The evolution of western Scandinavian topography: A review of Neogene uplift versus the ICE (isostacy – climate – erosion) hypothesis. *Journal of Geodynamics*, **47**, 72–95.

Nielsen, S.B., Clausen, O.R., Pedersen, V.K., Leseman, J.E., Goledowski, B., Huuse, M., Gallagher, K. and **Summerfield, M.A.** (2010a) Discussion of Gabrielsen *et al.* (2010): Latest Caledonian to present tectonomorphological development of southern Norway. *Mar. Petrol. Geol.*, **27**, 1285–1289.

Nielsen, S.B., Clausen, O.R., Jacobsen, B.H., Thomsen, E., Huuse, M., Gallagher, K., Balling, N. and **Egholm, D.** (2010b) The ICE hypothesis stands: How the dogma of late Cenozoic *tectonic* uplift can no longer be sustained in the light of data and physical laws. *Journal of Geodynamics*, **50**, 102–111.

Norris, R.D., Bice, K.L., Magno, E.A. and **Wilson, P.A.** (2002) Jiggling the tropical thermostat in the Cretaceous hothouse. *Geology*, **30**, 299–302.

Nøttvedt, A., Gabrielsen, R.H. and **Steel, R. J.** (1995) Tectonostratigraphy and sedimentary architecture of rift basins, with reference to the northern North Sea. *Mar. Petrol. Geol.*, **12**, 881–901.

Odinsen, T., Reemst, P., van der Beek, P., Faleide, J.I. and **Gabrielsen, R.H.** (2000) Permo-Triassic and Jurassic extension in the northern North Sea: results from tectonostratigraphic forward modelling. In: *Dynamics of the Norwegian Margin* (Ed. A. Nottvedt). *Geol. Soc. London Spec. Publ.*, **167**, 83–103.

Rasmussen, E.S. (2004) The interplay between true eustatic sea-level changes, tectonics and climatic changes: what is the dominating factor in sequence formation of the Upper Oligocene–Miocene succession in the eastern North Sea Basin, Denmark? *Global and Planetary Change*, **41**, 15–30.

Rasmussen, E.S. (2009) Neogene inversion of the Central Graben and Ringkøbing-Fyn High, Denmark. *Tectonophysics*, **465**, 84–97.

Rasmussen, E.S. and **Dybkjær, K.** (2005) Sequence stratigraphy of the Upper Oligocene – Lower Miocene of eastern Jylland, Denmark: role of structural relief and variable sediment supply in controlling sequence development. *Sedimentology*, **52**, 25–63.

Rasmussen, E.S., Dybkjær, K. and **Piasecki, S.** (2010) Lithostratigraphy of the Upper Oligocene – Miocene succession of Denmark. *Geological Survey of Denmark and Greenland Bulletin*, **22**, 100 pp.

Rasmussen, E.S. and Dybkjær, K. (2014) Patterns of Cenozoic sediment flux from western Scandinavia: discussion. *Basin Res.*, **26**, 338–346.

Rattey, R.P. and **Haward, A.B.** (1993) Sequence stratigraphy of a failed rift system: the Middle Jurassic to Late Cretaceous basin evolution of the Central and Northern North Sea. In: *Petroleum Geology of North-west Europe.* (Ed. J.R. Parker), *Proceedings of the 4th Conference. Geol. Soc. London Spec. Publ.*, 215–251.

Riis, F. (1996) Quantification of Cenozoic vertical movements of Scandinavia by correletaion of morphological surfaces with offshore data. *Global Planet. Change*, **12**, 331–357.

Riis, F. and **Fjeldskaar, W.** (1992) On the magnitude of the Late Tertiary and Quaternary erosion and its significance for the uplift of Scandinavia and the Barents Sea. In: *Structural and Tectonic Modelling and its Application to Petroleum Geology* (Ed. R.M. Larsen), *Norwegian Petroleum Society Special Publication*, **1**, 163–185.

Roberts, A.M., Yielding, G. and **Badley, M.E.** (1990) A kinematic model for the orthogonal opening of the Late Jurassic North Sea Rift System, Denmark-Mid Norway. In: *Tectonic evolution of the North Sea Rifts* (Eds D.J. Blundell and A.D. Gibbs), Oxford University Press, Oxford.

Rohrman, M., Van der Beek, P., Andriessen, P. and **Cloetingh, S.** (1995) Meso-Cenozoic morphotectonic evolution of southern Norway: Neogene domal uplift inferred from apatite fission track thermochronology. *Tectonics*, **14**, 704–718.

Rohrman, M., Andriessen, P. and **Van der Beek, P.** (1996) The relationship between basin and margin thermal evolution assessed by fission track thermochronology: an application to offshore southern Norway. *Basin Res.*, **8**, 45–63.

Rohrman, M. and **Van der Beek, P.** (1996) Cenozoic post-rift domal uplift of North Atlantic margins: An asthenospheric diapirism model. *Geology*, **24**, 901–904.

Rundberg, Y., Olaussen, S. and **Gradstein, F.** (1995) Incision of Oligocene strata; evidence for northern North Sea Miocene uplift and key to the formation of the Utsira sands. *Geonytt*, **22**, (abstract).

Schiøler, P. andsbjerg, J., Clausen, O.R., Dam, G., Dybkjær, K., Hamber, L., Heilmann-Clausen, C., Johannessen, E.P., Kristensen, L.E., Prince, I. and **Rasmussen, J.A.** (2007) Litostratigraphy of the Palaeogene – Lower Neogene succession of the Danish North Sea. *Geological Survey of Denmark and Greenland*, **12**, 77 pp.

Sejrup, H.P., Hjelstuen, B.O., Torbjørn Dahlgren, K.I., Haflidason, H., Kuijpers, A., Nygård, A., Praeg, D., Stoker, M.S. and **Vorren, T.O.** (2005) Pleistocene glacial history of the NW European continental margin. *Mar. Petrol. Geol.*, **22**, 1111–1129.

Sleep, N.H. (1990) Hotspots and mantle plumes; some phenomenology. *J. Geophys. Res. (B)*, **95**, 6715–6736.

Sliwinska, K.K., Clausen, O.R. and **Heilmann-Clausen, C.** (2010) A mid-Oligocene cooling (Oi-2b) reflected in the dinoflagellate record and in depositional sequence architecture. An integrated study from the eastern North Sea basin. *Mar. Petrol. Geol.*, **27**, 1424–1430.

Spjeldnæs, N. (1975) Palaeogeography and facies distribution in the Tertiary of Denmark and surrounding areas. In: *Petroleum Geology and Geology of the North Sea and North-west Atlantic Continental Margin* (Eds A. Whiteman, D. Roberts and M.A Sellevoll), *Nor. Geol. Unders. Bull.*, **29**, 289–311.

Steel, R.J. (1993) Triassic – Jurassic megasequence stratigraphy in the northern North Sea: rift to post-rift evolution. In: *Petroleum Geology of North-west Europe: Proceedings of the 4th Conference* (Ed. J.R. Parker), *Geol. Soc. London*, 299–315.

Steurbaut, E., Spiegler, D., Weinelt, M. and **Thiede, J.** (1991) *Cenozoic Erosion and Sedimentation on the North-west European Continental Margin.* Geomar, Kiel.

Stewart, I.J. (1987) A revised stratigraphic interpretation of the Early Palaeogene of the Central North Sea. In: *Petroleum Geology of North West Europe* (Eds J. Brooks and K.W. Glennie), *London*: *Graham & Trotman*, 557–576.

Strass, I.F. (1982) The Norwegian-Danish Basin, *Norwegian Petroleum Directorate*, **31**, 79 pp.

Stuevold, L.M. and **Eldholm, O.** (1996) Cenozoic Uplift of Fennoscandia inferred from a study of the mid-Norwegian Shelf. *Global Planet. Change*, **12**, 359–386.

Surlyk, F., Dons, T., Clausen, C.K. and **Higham, J.** (2003) Upper Cretaceous. In: *The Millenium Atlas: Petroleum Geology of the Central and Northern North Sea* (Eds D. Evans, C. Graham, A. Armour and P. Bathurst). *Geol. Soc. London*, 489–547.

Thomson, K., Underhill, J.R., Green, P.F., Bray, R.J. and **Gibson, H.J.** (1999) Evidence from apatite fission track analysis for the post-Devonian burial and exhumation history of the northern Highlands, Scotland. *Mar. Petrol. Geol.*, **16**, 27–39.

Thyberg, B.I., Jordt, H., Bjørlykke, K. and **Faleide, J.I.**, (2000) Relationships between sequence stratigraphy, mineralogy and geochemistry in Cenozoic sediments of the northern North Sea. *Geol. Soc. London Spec. Publ.*, **167**, 245–272.

Underhill, J.R. and **Partington, M.A.** (1993) Jurassic thermal doming and deflation in the North Sea: implications of the sequence stratigraphic evidence. In: *Petroleum Geology of North-west Europe: Proceeding of the 4th Conference* (Ed. **J.R. Parker**), *Geol. Soc. London*, 337–345.

Van Wagoner, J.C., Posamentier, H.W., Mitchum, R.M., Vail, P. R., Sarg, J.F., Loutit, T.S. and **Hardenbol, J.** (1988) An overview of sequence stratigraphy and key definitions. In: *Sea-level Changes – an integrated approach* (Eds C.K. Wilgus, B.S. Hastings, H. Posamentier, J.V. Wagoner, C.A. Ross and C. Kendall), *SEPM Spec. Publ.*, **42**, 39–45.

Veeken, P.C.H. (2007) Seismic Stratigraphy, Basin Analysis and Reservoir Characterisation. Elsevier, Amsterdam. pp 453.

Vollset, J. and **Doré, A.G.** (1984) A revised Triassic and Jurassic lithostratigraphic nomenclature for the Norwegian North Sea. *Nor. Pet. Direct. Bull., Stavanger*, **3**, 33 pp.

White, N and **Latin, D**. (1993) Subsidence analysis from the North Sea 'triple-junction'. *J. Geol. Soc. London*, **150**, 473–488.

White, N. and **Lovell, B.** (1997) Measuring the pulse of a plume with the sedimentary record. *Nature*, **387**, 888–891.

Wood, R. and **Barton, P.** (1983) Crustal thinning and subsidence in the North Sea. *Nature*, **302**, 134–136.

Yielding, G. and **Roberts, A.** (1992) Footwall uplift during normal faulting – implications for structural geometries in the North Sea. In: *Structural Modelling and Its Application to Petroleum Geology* (Eds R.M. Larsen, H. Brekke, B.T. Larsen and E. Talleraas), *Norwegian Petroleum Society Special Publication*, **1**, 289–304.

Zachos, J., Pagani, M., Sloan, L., Thomas, E. and **Billups, K.** (2001) Trends, rhythms and aberrations in global climate 65 Ma to present. *Science*, **292**, 686–693.

Ziegler, P.A. (1975) Geologic evolution of North Sea and its tectonic framework. *AAPG*, **59**, 1073–1097.

Ziegler, P.A. (1982) *Geological Atlas of Western and Central Europe*. Elsevier, Amsterdam, pp 130.

Ziegler, P.A. (1987) Late Cretaceous and Cenozoic intra-plate compressional deformations in the Alpine foreland – a geodynamic model. *Tectonophysics*, **137**, 389–420.

Ziegler, P.A. (1990) *Geological Atlas of Western and Central Europe 1990.* Shell Internationale Petroleum Maatschappij B.V, 239 pp.

Ziegler, P.A. (1992) North Sea Rift System. *Tectonophysics*, **208**, 55–75.

Ziegler, P.A. and **Van Hoorn, B.** (1989) Evolution of the North Sea Rift system. In: *Extensional Tectonics and Stratigraphy of the North Atlantic Margins* (Eds A.J. Tankard and H. Balkwill), *AAPG. Mem.*, **46**, 471–500.

Int. Assoc. Sedimentol. Spec. Publ. (2014) **46**, 453–472

3D forward modelling of the impact of sediment transport and base level cycles on continental margins and incised valleys

DIDIER GRANJEON

Department of Geology-Geophysics-Geochemistry, IFP Energies nouvelles, 1 à 4 avenue de Bois, Préau, 92852 Rueil-Malmaison, France

ABSTRACT

This paper describes a 3D stratigraphic forward model, Dionisos, intended to reconstruct the stratigraphic architecture of sedimentary basins at a regional scale. This model accounts for basin deformation, clastic and carbonate supplies and sediment transport, both in continental and marine environments. Sediment is defined as a mixture of several grain-size fractions; and the transport of each grain-size fraction is controlled by a slow creeping law and a faster linear slope-driven and non-linear water-driven diffusion law. Attention is focussed on sediment transport laws. Passive-margin and source-to-sink theoretical models were built using independent data from the XES 02 flume experiment. A few quantitative metrics were defined to analyse the numerical experiments, such as the number of channel per strike sections, the location of the centroid and erosion and sedimentation rates in the source or transitions areas. These numerical experiments demonstrate that a non-linear water-driven diffusion equation could be used to predict the evolution of sedimentary systems under extreme forcing conditions such as extreme sea-level variations. These applications at a regional scale are very encouraging, although more studies are required to calibrate transport parameters. The long-term and short-term evolution of the shoreline and centroid of sedimentary systems could be investigated. The detailed internal architecture of sedimentary units could be simulated; such as progradation and downlap of foresets, backstepping and fluvial onlap, subaerial truncation and incised valleys. These applications illustrate that a 3D stratigraphic forward model based on a simple non-linear water-driven diffusion approach could be used at a basin scale to better understand evolution of depositional systems and thus reduce uncertainties in exploration and reservoir risk assessment.

Keywords: stratigraphic forward model, diffusion equation, passive margin, incised valley, source-to-sink.

INTRODUCTION

Frontier exploration areas such as deep-water to ultra-deep-water environments, arctic basins or subsalt sedimentary formations have the least geological information recorded about them as they are remote basins and difficult to image or drill. Exploration and development of such complex areas requires accurate prediction of sedimentary system characteristics and petroleum potential. Sedimentary architecture is the result of the complex interplay between accommodation space created in the basin, sediment supply brought into or produced in the basin and transport of sediment within this basin (Swift & Thorne, 1991; Helland-Hansen & Martinsen, 1996; Catuneanu *et al.*, 2009).

Geologists have followed different approaches to understand the evolution of sedimentary systems at a basin scale. Outcrops and subsurface data (e.g. Porebski & Steel, 2003; Uroza & Steel, 2008), modern systems and natural experiments (e.g. Wright & Nittrouer, 1995; Tucker, 2009) and flume experiments (e.g. Heller *et al.*, 2001; Muto & Steel, 2001, 2004; Postma *et al.*, 2008; Strong & Paola, 2008; Martin *et al.*, 2009; Paola *et al.*, 2009) have been used to propose process-response models of continental margins (Porebski & Steel, 2003; Catuneanu *et al.*, 2009; Helland-Hansen, 2009). These huge amounts of data and concepts have

From Depositional Systems to Sedimentary Successions on the Norwegian Continental Margin, First Edition.
Edited by A. W. Martinius, R. Ravnås, J. A. Howell, R. J. Steel, and J. P. Wonham.

allowed researchers to develop and validate three-dimensional (3D) numerical process-based stratigraphic forward models. Numerical models provide an ideal method to test geological and physical concepts and causal relationships between stratigraphic parameters such as basin deformation or sediment transport and evolution of sedimentary basins. These numerical models have been around since the early 1970s (Harbaugh & Bonham-Carter, 1970; Merriam, 1972; Schwarzacher, 1975; Tetzlaff & Harbaugh, 1989; Cross & Harbaugh, 1989; Waltham, 1992; Slingerland *et al.*, 1994). The models have become increasingly sophisticated and have been designed to represent the dynamics of sedimentary systems and simulate the filling of sedimentary basins. They can be classified roughly into two classes, although the boundary between these classes is less and less noticeable. The first class corresponds to physical processes-based models, which describe water and sediment flow in two dimensions over a short time interval using simplified or empirical laws with the aim of analysing the complex interactions between physical processes (Tetzlaff & Harbaugh, 1989; Martinez & Harbaugh, 1993; Griffiths *et al.*, 2001; Dalman & Weltje, 2008; Hutton & Syvitsky, 2008). The second class corresponds to the diffusion process-based models, which describe the behaviour of sedimentary systems at larger spatial and temporal scales with the aim of describing the full continental margin from source, or at least from the lowest part of the catchment system, to sink (Kenyon & Turcotte, 1985; Rivenaes, 1988; Granjeon, 1996; Granjeon & Joseph, 1999; Meijer, 2002; Clevis *et al.*, 2003; Meijer *et al.*, 2008). The former approach is appropriate for constructing small-scale alluvial, deltaic or turbiditic reservoir models by modelling the fundamental processes that formed them (Storm & Swift, 2003; Dalman & Weltje, 2008, 2012; Miller *et al.*, 2008). However, many of these models are dedicated to a single depositional environment. Numerous sedimentary environments, from alluvial and deltaic to deep-water settings, have to be included in the model when studying sedimentary systems at a basin scale.

The main objective of this study was to improve and test a diffusion process-based model, called Dionisos (Granjeon, 1996; Granjeon *et al.*, 1999), able to simulate sediment routing systems at a regional scale. Classical applications of this model are to test sequence-stratigraphy concepts (Burgess *et al.*, 2006; Somme *et al.*, 2009a) and to evaluate geological scenarios based on regional seismic data and exploration wells (Rabineau *et al.*, 2005; Burgess & Steel, 2008; Ku Shafie & Madon, 2008; Alzaga-Ruiz *et al.*, 2009; Csato *et al.*, 2012). In this paper, the main principles of Dionisos are presented and their performances on passive margin settings are illustrated. The focus of attention is on the transport of sediment, looking at the fundamental timing and styles of incised valleys and fluvial to shelf deposition.

MODEL DESCRIPTION

Dionisos is a 3D stratigraphic forward model which aims to simulate the geometry and facies of sedimentary units over regional spatial scales from several tens of kilometres to hundreds of kilometres and over geological time scales from tens of thousands of years to hundreds of millions of years, in a sequence of time steps (Granjeon, 1996; Granjeon *et al.*, 1999). Spatial discretisation is usually set at 1 km to 10 km and time steps at around a few thousand years to hundreds of thousands of years. At each time step, the numerical model accounts for accommodation, sediment supply and sediment transport.

Accommodation is calculated by defining a eustatic sea-level curve and subsidence or uplift maps. Internal deformation of stratigraphy due to salt and shale diapirs, or growth faults, thrust and decollement levels, could also be defined (Alzaga-Ruiz *et al.*, 2009). Sedimentary units above decollement surfaces are moved to account for horizontal displacement of sedimentary rocks, using the horizontal deformation rate imposed by users and assuming vertical shear deformation.

Sediment supply may correspond either to boundary conditions (fluvial inflow of water and sediment), basement erosion (controlled by the transport equations described below), carbonate production or evaporite precipitation. Carbonate production is controlled by setting a maximum growth rate curve for each producer, such as corals, algals or ooids. This maximum rate could be modulated by local parameters (water depth, wave energy, ecological rules etc.).

This paper is focused on the simulation of the third process to be accounted for in stratigraphic modelling: the transport of sediment. Since the early work of Culling (1960), the diffusion equation has been used in geomorphology and geology to represent large-scale, spatially averaged transport of sediment by creep, overland flow and channel flow processes. This diffusion law has been used in various forms to study evolution of scarps (Colman & Watson, 1983; Hanks *et al.*, 1984; Begin, 1988, 1993), hill slopes (Culling, 1960; Carson & Kirkby, 1972; Hirano, 1975), alluvial fans, rivers and floodplains (Gessler, 1971; Begin *et al.*, 1981; Green, 1987; Pizzuto, 1987; Begin, 1988; Parker *et al.*, 1998; Roering *et al.*, 1999; Pelletier *et al.*, 2006), mountains and foreland basins (Moretti & Turcotte, 1985; Flemings & Jordan, 1989; Jordan & Flemings, 1991; Sinclair *et al.*, 1991; Paola *et al.*, 1992; Willgoose *et al.*, 1992; Tucker & Slingerland, 1994; Chalaron *et al.*, 1996; Davy & Crave, 2000; Clevis *et al.*, 2003; Carretier & Lucazeau, 2005), or deltas and continental margins (Kenyon & Turcotte, 1985; Rivenaes, 1988; Flemings & Grotzinger, 1996). Recent studies based on flume experiments have also demonstrated that diffusion laws could be used to simulate time-averaged transport of sediment (Paola *et al.* 1992; Postma *et al.*, 2008), although pointing out that insufficient data are available to verify values of diffusion laws for real-world prototypes (Postma & Van den Berg van Saparoea, 2007).

In our model, following the classical approach used in landscape evolution models (Willgoose *et al.*, 1991 a, b; Tucker & Slingerland, 1994), all processes leading to the movement of sediment grains are lumped into two large-scale diffusion processes: a slow slope-driven creeping and a fast water-driven and slope-driven diffusion transport (Granjeon, 1996; Granjeon & Joseph, 1999). The creeping of grains is assumed to be proportional to the local slope of landscape, leading to a non-linear slope-driven diffusion equation, mainly active along hill slopes and continental margin slopes.

$$Q_{sc,k} = K_{c,k}\, c_k\, S^{m_c} \qquad (1)$$

Equation 1: definition of the creeping flux, $Q_{sc,k}$ (m^2s^{-1}), of the grain-size fraction *k*, where c_k is the surface concentration of this grain-size fraction (–); S is the local landscape slope (–); $K_{c,k}$ is the creeping diffusion coefficient of the grain-size fraction k, defined as a function of water depth; m_c is a constant, usually between 1 and 2 (Carson & Kirkby, 1972).

The second process is the more efficient of the two as it allows a quick transport of sediment grains from sources to sinks. At the regional spatial scale and over geological time, upscaling of local-scale and short-term transport processes is assumed to lead to a large-scale non-linear water-driven and slope-driven diffusion equation.

$$Q_{sw,k} = K_{w,k}\, c_k\, \bar{Q}_w^n\, S^{m_w} \text{ with } \bar{Q}_w = (Q_w / Q_{wo}) \qquad (2)$$

Equation 2: definition of the non-linear water-driven and slope-driven flux, $Q_{sw,k}$ [m^2s^{-1}], of the grain-size fraction k, where c_k is the surface concentration (–), Q_w the local water discharge (m^3s^{-1}), $\bar{Q}_w$, is the local dimensionless water discharge, Q_{wo}=1 m^3s^{-1} is the reference water discharge; S is the basin slope (–), $K_{w,k}$ is the water-driven diffusion coefficient of the grain-size fraction k, defined as a function of water depth; *n* and m_w are two constants, usually between 1 and 2 (Tucker & Slingerland, 1994).

In classical landscape evolution models, fluvial water flow is routed across the simulated area using a single-direction routing method in which water flows along the steepest slope (Willgoose *et al.*, 1991a, b; Tucker & Slingerland, 1994; Tucker, 2004). In our model, a multiple-direction method is used: water is routed to all the local lower neighbours of a given cell, according to their slope ratios. This multiple-direction method handles diverging flow better (Freeman, 1991; Tucker & Hancock, 2010) and is well adapted to simulate overland sheet flow and braided systems (Murray & Paola, 1994, 1997; Coulthard *et al.*, 1996).

These two non-linear diffusion equations define the transport capacity of the system. Availability of grains is constrained by weathering and incision rate. This rate is highly variable and is a function of climate, topographic elevation and slope (Willgoose *et al.*, 1992; Tucker & Slingerland, 1994; Coulthard, 2001; Roering *et al.*, 2007). A maximum erosion rate is defined as a function of the excess of shear stress; that is the difference between basal shear stress induced by fluvial water flow and the critical threshold stress below which no incision can occur. The actual transport rate is finally defined as the minimum of the transport capacity and the grain-size fraction availability. Sedimentation and erosion rate of each grain-size fraction at each point of the basin is computed from the mass conservation equation and the actual sediment flux. Slope stability of the depositional profile is then checked assuming that the angle of the depositional slope predicted by the long-term diffusion equation cannot exceed a static angle of repose that depends on grain-size fractions. At any point of the basin, if the local slope is higher than the static angle of repose, then all sediments above the safety plane move downslope until the new depositional slope is lower than the dynamic slope angle. Finally, compaction of sediments is computed using burial depth-porosity laws and elastic flexure due to water, while sediment loading is calculated assuming a 3D elastic lithosphere. The solution of this numerical model is obtained using an implicit finite volume scheme (Eymard *et al.*, 2004; Gervais, 2004; Gervais & Masson, 2004).

METHODOLOGY

The main objective of this paper is to present and illustrate the use of non-linear diffusion equations to simulate transport of sediment at a regional scale. Direct numerical simulation of a full sedimentary system remains largely empirical (Paola *et al.*, 2009), in particular at a regional scale, as in Dionisos. The study of a sedimentary basin where all controlling parameters, boundary conditions and sedimentary architecture are known would be the best natural experiment to validate transport equations. Due to the lack of fully documented basin systems, stratigraphic flume experiments are candidates for such a study (Postma *et al.*, 2008).

Passive margin model

A theoretical passive margin was defined from the well-documented XES02 flume experiment (Strong & Paola, 2008; Martin *et al.*, 2009, 2010), conducted in the Experimental EarthScape (XES) system and often referred to as 'Jurassic Tank' (Paola, 2000; Paola *et al.*, 2001). The XES facility is a large experimental basin (3 m x 6 m) where scaled down experiments on natural systems can be tested. The XES 02 experiment was designed to examine the effects of slow and rapid base-level cycles on the stratigraphy of a passive margin. It is a perfect candidate to explore and analyse the behaviour of sediment transport at a regional scale and thus to test the capacity of our numerical model to reproduce important morphologic characteristics such as erosional unconformities, valley incisions and preservation of stratigraphy. The parameters of the XES 02 experiment were upscaled to construct a theoretical basin-scale passive margin using simple similarities rules.

The characteristics of height, time, slope and sediment concentration of the XES02 experiments are respectively 110 mm, 20 h, 35° and 30 g l^{-1} (Table 1). Scaling parameters were defined to produce a theoretical passive margin setting looking like a real work example. XES02 base level cycles (period ~ 20 hours, amplitude = 110 mm) were supposed to be similar to extreme glacial sea-level cycles (period ~ 20 ka, amplitude = 110 m). The time ratio was around 10^9 and the height ratio is 10^3. The angle of the XES 02 slope is around 35° (or 700 m km^{-1}), whilst classical continental margin slope is around a few degrees (e.g. Adams *et al.*, 1998; Adams & Schlager, 2000; O'Grady *et al.*, 2000; Kertznus & Kneller, 2009; Covault *et al.*, 2011). The slope ratio was thus defined as equal to 0.1, which led to a margin slope equal to 4° and a length ratio equal to 10^4. The length and height ratios defined the upscaled sediment volume and supply, around 1820 km^3 My^{-1}.

Table 1. Key parameters used in the XES 02 experiment (Strong & Paola, 2008); upscaled parameters used in Dionisos and scaling ratios.

	XES 02 experiment	Dionisos model	Scaling ratio
Duration of the short-term eustatic cycle	20 h	20 ka	$t = 3.2\ 10^8$
Time interval	320 h	320 ka	$t = 3.2\ 10^8$
Amplitude of sea-level cycles	110 mm	110 m	$h = 10^3$
Margin slope	35° (700 m km^{-1})	4° (70 m km^{-1})	$h/l = 0.1$
Fluvial slope	2° (35 m km^{-1})	0.2° (3.5 m km^{-1})	$h/l = 0.1$
Basin length	6 m	60 km	$l = 10^4$
Basin area	6 × 3 = 18 m^2	1800 km^2	$l^2 = 10^8$
Subsidence	3.71 mm h^{-1}	3710 m My^{-1}	$h/t = 3.2\ 10^{-6}$
Sediment supply	0.303 l min^{-1}	1820 km^3 My^{-1}	$h.l^2/t = 320$
Sediment concentration	30 g l^{-1}	0.6 g l^{-1}	$c = 0.02$
Water supply	25 l min^{-1}	240 $m^3 s^{-1}$	$h.l^2/c.t = 16{,}000$

This value corresponds to the yield of small modern rivers such as the Wanganui River, New Zealand, or the Homathko and Klinaklini Rivers, Canada (Syvitski & Milliman, 2007). The yearly-average sediment concentration in these rivers is around 0.5 $g\,l^{-1}$ and 0.7 $g\,l^{-1}$, whilst the sediment concentration in the XES 02 inflow is 30 $g\,l^{-1}$. The concentration ratio was set to 0.02, which allowed definition of the upscaled water discharge; equal to 240 $m^3 s^{-1}$.

Using these scaling parameters, a passive margin model was built. Its characteristic parameters were defined as 60 km, 20 ka, 4° and 0.6 $g\,l^{-1}$. The simulated area was a rectangular box measuring 60 × 30 km^2 and the spatial resolution was 300 m. The total duration of the simulation was around 300 ka and the numerical time step was 5 ka. The initial bathymetry was a plane with a gentle slope of 0.01 m km^{-1} (Fig. 1A). The accommodation was created by a downstream-increasing subsidence and sea-level cycles, as in the XES 02 experiments. The sea-level curve (Fig. 2) was divided into two stages: stage 1, with a first smooth long-term cycle (duration = 108 ka) and a second smooth short-term sea-level cycle (duration = 18 ka) and stage 2, with superimposed long and short-term cycles. Water, 240 $m^3 s^{-1}$, and sediment supply, 1820 km^3 My^{-1}, were held constant and fed into the basin via a point source at its western boundary (Fig. 1A).

Source-to-sink model

A second model was defined to test the influence of the constant boundary condition imposed in the flume experiments on the fluvial to sink areas and in particular on the water flow structure observed and analysed previously. The passive-margin model was extended to include a 60 km-long upper catchment area (Fig. 1C). The initial morphology of the extended basin was a gentle slope, as previously. The subsidence (Fig. 1C) was defined using a tilted plane with uplift of the upper catchment area and subsidence of the lower catchment and shelf areas. Two triangle plateaus were added around the hinge line to force water to focus on a single point. The downstream part of the extended basin was identical to the initial passive margin model. Rainfall was set constant over the full subaerial environments. Its value was defined using a trial-and-error manual inversion used to get nearly the same final physiography and total sediment volume, 1820 km^3 My^{-1}, in the lower catchment area as in the initial passive margin model. This constant rain fall was 4000 mm y^{-1}.

Transport parameters

The diffusion coefficients used in both experiments were defined using a simple geometrical rule (see equation 3), assuming that (1) sediment transport is ruled by a linear slope-driven, non-linear water-driven diffusion law ($m_w = 1.0$, $n = 1.5$) and (2) the fluvial equilibrium slopes, S_{eq}, is 0.3 m km^{-1} (~0.02°) for sand grains and 0.1 m km^{-1} (~0.006°) for fine grains.

$$K_w = \frac{Q_{s,o}}{\bar{Q}^n_{w,o}\, S^{m_w}_{eq}} \qquad (3)$$

Equation 3: estimation of the diffusion coefficients K_w ($m^2 s^{-1}$), where $Q_{s,o}$ is the sediment inflow ($m^2 s^{-1}$); $\bar{Q}_{w,o}$ is the dimensionless water inflow (–); S_{eq} is the equilibrium slope (–); n and m_w are the two exponents controlling the non-linear behaviour of the transport equation.

The diffusion coefficients obtained from this geometrical rule were 400 $m^2 y^{-1}$ for the sand grains and 1200 $m^2 y^{-1}$ for the fine grains. Slow creeping coefficients are usually in the range 10^{-4} $m^2 y^{-1}$ to 10^{-2} $m^2 y^{-1}$, whilst fast fluvial transport coefficients are in the range 10^2 $m^2 y^{-1}$ to 10^4 $m^2 y^{-1}$ (Flemings & Jordan, 1989; Avouac & Burov, 1996). Our estimated coefficients are thus in the lower part of the fluvial transport coefficient range.

CONTROLS

A direct comparison between the flume tank and numerical experiments is quite complex due to the problematic upscaling and inherently chaotic results in such experiments. Over basic snapshots of the geomorphology and stratigraphy, quantitative metrics were used to analyse these numerical experiments and measure the dispersion of the water flow and the sediment distribution.

The water flow is the most important parameter in the experiments as it controls the transport of sediment. To get an automatic measure of the dispersion of the water flow, the number of channels crossing each strike plane of the simulated domain was counted. To keep this measure as simple as possible, the maximum water flow crossing each strike plane was counted and then a channel was defined as a cell in which the local water flow exceeds 10% of this maximum value. In the case of a uniform water flow, the number of channels is

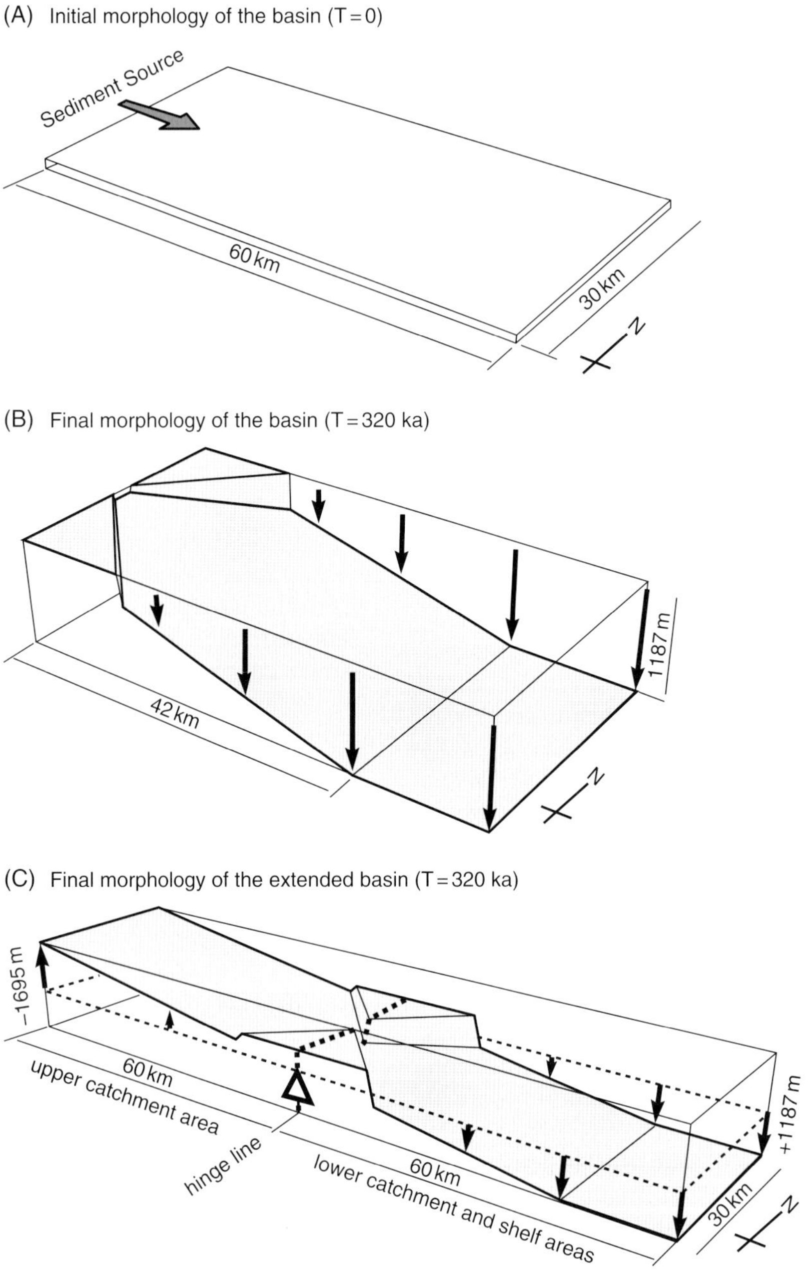

Fig. 1. Schematic diagrams of (A) the initial morphology, (B) the final morphology of the passive-margin model (these parameters are defined from the parameters of the XES 02 experiments, Strong & Paola, 2008) and (C) the final morphology of the extended source-to-sink model.

equal to the total number of cells in the strike direction. On the opposite, in the case of a focused flow, for example inside an incised valley, the number of channels is very small and close to one.

The location of the centroid was also used (Martin *et al.*, 2009), that is the depositional centre of mass, to measure planview dispersion of sediment in the experiment. This work has improved a little upon the definition proposed by Martin *et al.*, 2009. The centroid position, (X,Y), of a sedimentary layer is measured at the time of its deposition and not at the end of the simulation. This accounts for the spatial distribution of erosion and deposition.

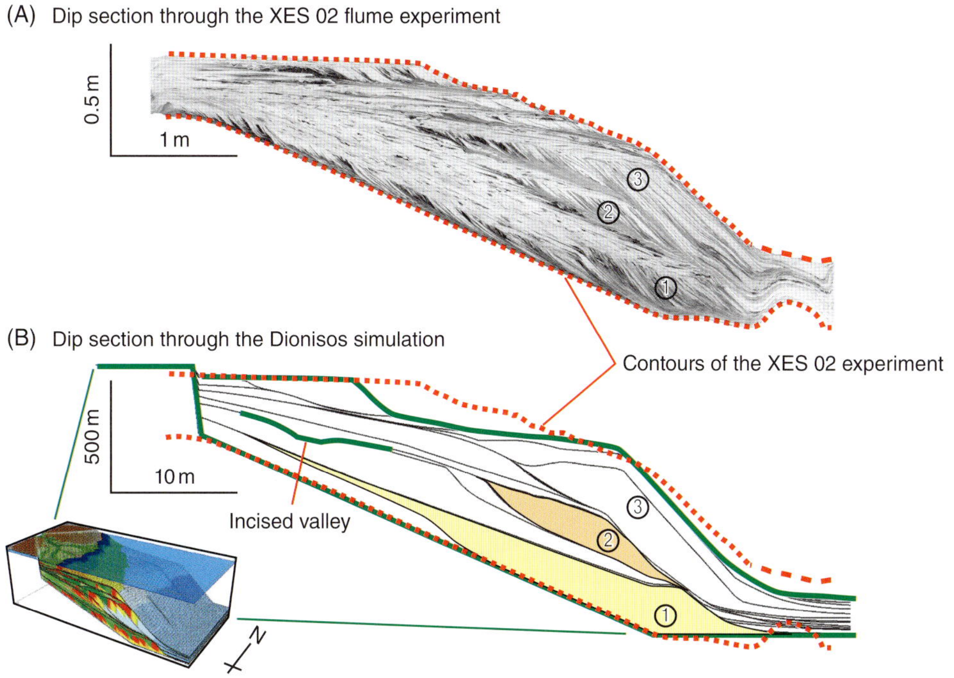

Fig. 2. Comparison of (A) a dip section through the XES 02 flume experiment (from Strong & Paola, 2008) and (B) a dip section through the 3D numerical passive-margin model.

$$H = \frac{1}{A}\iint_{\Omega} \frac{\partial h}{\partial t}(x,y)\, dxdy\ \delta t \qquad (4)$$

$$X = \frac{1}{AH}\iint_{\Omega} x\,\frac{\partial h}{\partial t}(x,y)\, dxdy\ \delta t$$

$$Y = \frac{1}{AH}\iint_{\Omega} y\,\frac{\partial h}{\partial t}(x,y)\, dxdy\ \delta t$$

Equation 4: definition of the average thickness, H, of a sedimentary layer and the centroid location, (X,Y) (m), where A is the simulated area (m^2) and Ω the simulated domain; ∂h/∂t is the local sedimentation rate, which could be positive (deposition of sediment), or negative (erosion), measured at the time of the deposition (m); δt is the duration of the sedimentary layer.

RESULTS

Passive margin model

The first numerical simulation was performed to simulate the evolution of a passive margin. The simulated geometry and stratigraphy are in good agreement with the flume experiment results (Fig. 2). The numerical model reproduced the large-scale architecture of this XES 02 experiment. A first, huge, sedimentary unit was deposited during the slow cycle (unit 1 on Fig. 2). A second, smaller, unit exhibited erosional surface, toplap truncation and steep downlap onto unit 1 (unit 2 on Fig. 2). A composite and complex third unit was composed of a series of smaller units similar to the second unit (unit 3 on Fig. 2). In more detail, during the initial steady phase (0 ka to 26 ka, Fig. 3A-1), despite all parameters being constant and accommodation rate being higher than sediment supply, a long-term progradation of the sedimentary system was recorded. This initial progradation corresponds to a transitory phase characterised by a widespread fluvial plain and numerous channels (Fig. 3A-1), a fast downward movement of the shoreline and centroid (Fig. 4, 0 ka to 26 ka). Sedimentary systems progressively reached equilibrium during this transitory phase. During the slow sea-level cycle (26 ka to 78 ka, Fig. 3A-2), subsidence was lower than sea-level fall on the west side of the basin. Accommodation was decreasing and a small incised valley was formed. The falling rate was not fast enough to maintain channel at the same location and the small incised valley gave birth to multiple fluvial channels (Fig. 4). Despite a nearly constant accommodation and supply, autogenic cycles were characterised by channel

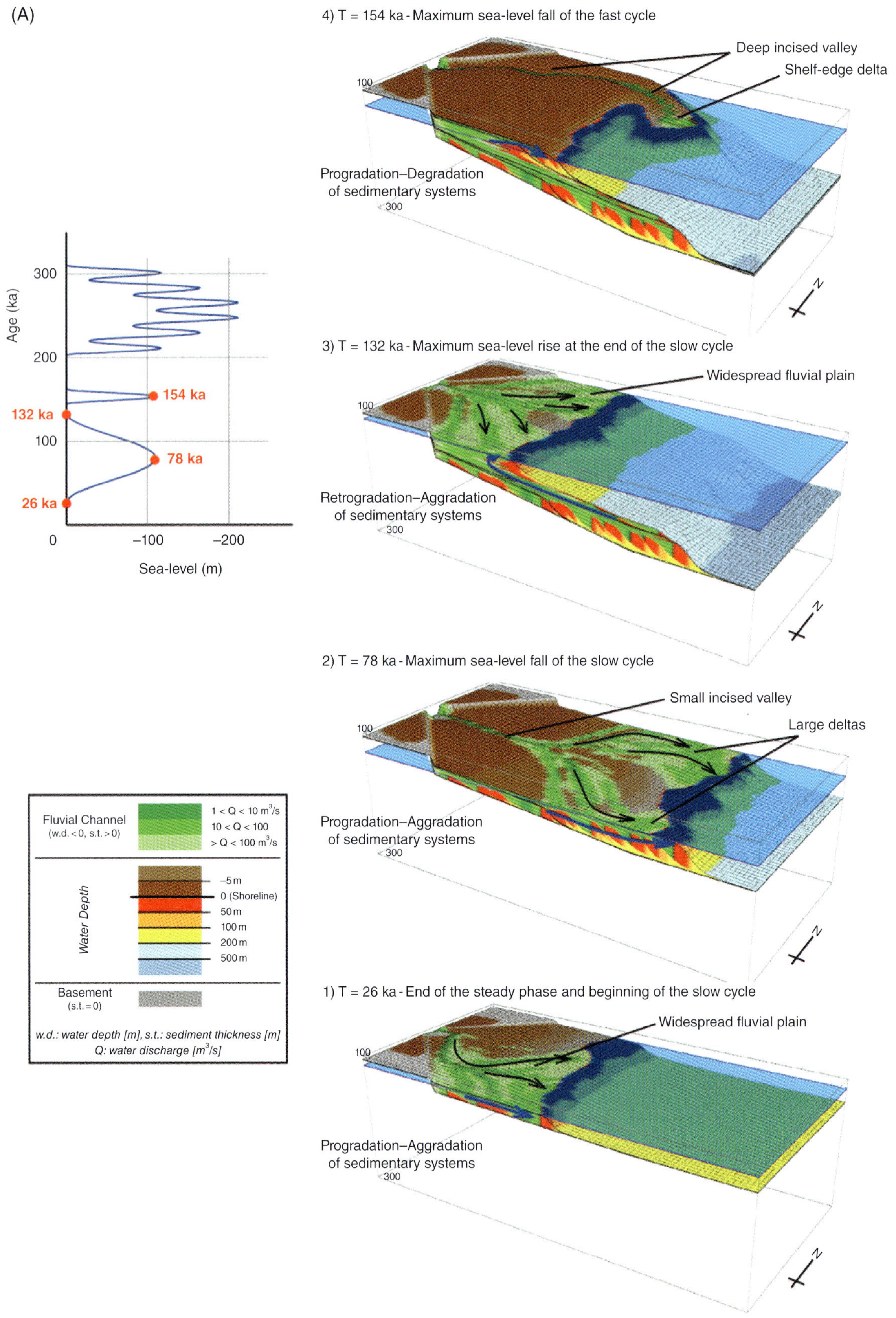

Fig. 3. Analysis of the evolution of sedimentary environments in response to sea-level cycles in the case of the passive-margin model.

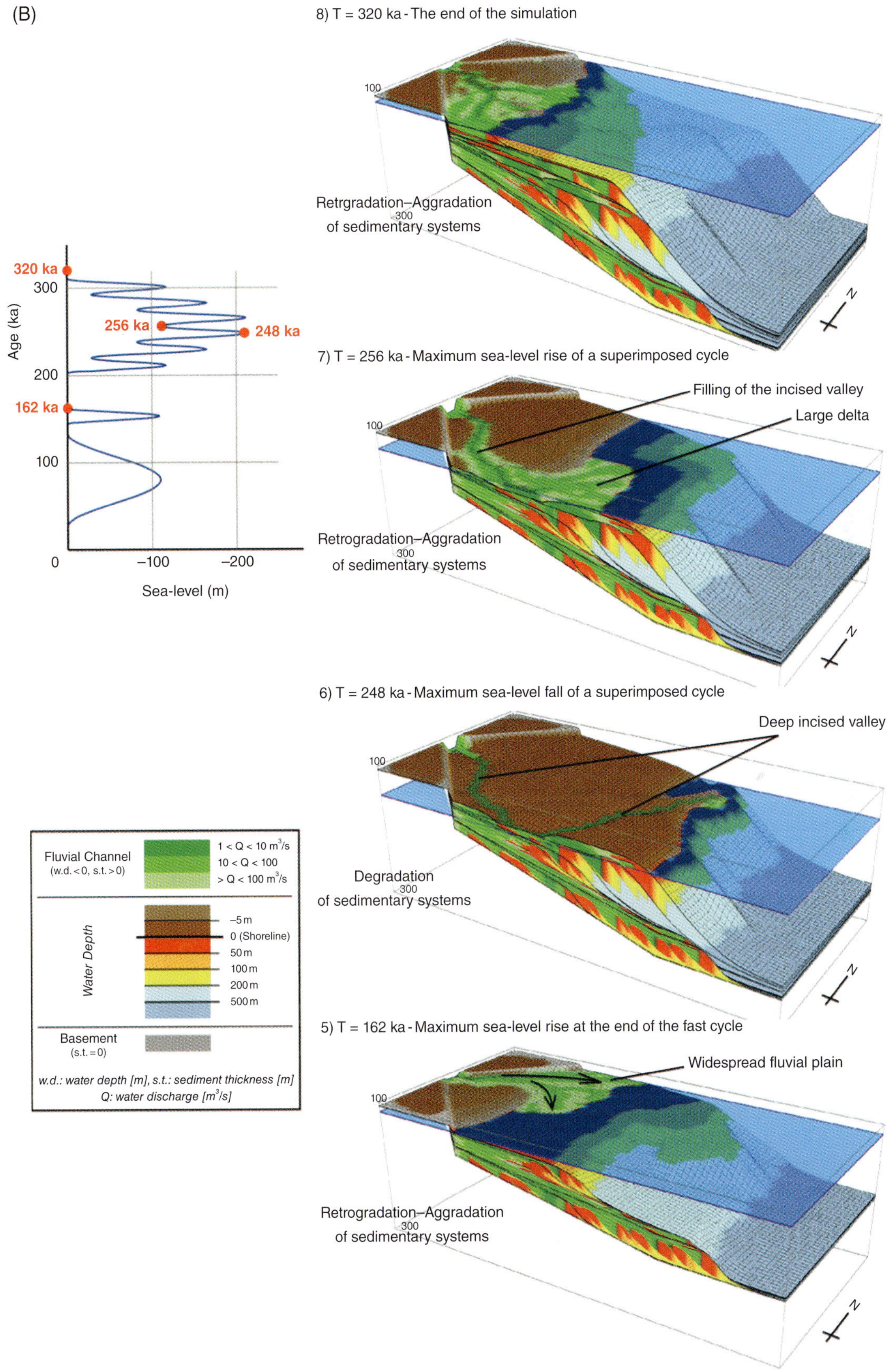

Fig. 3. (*Continued*)

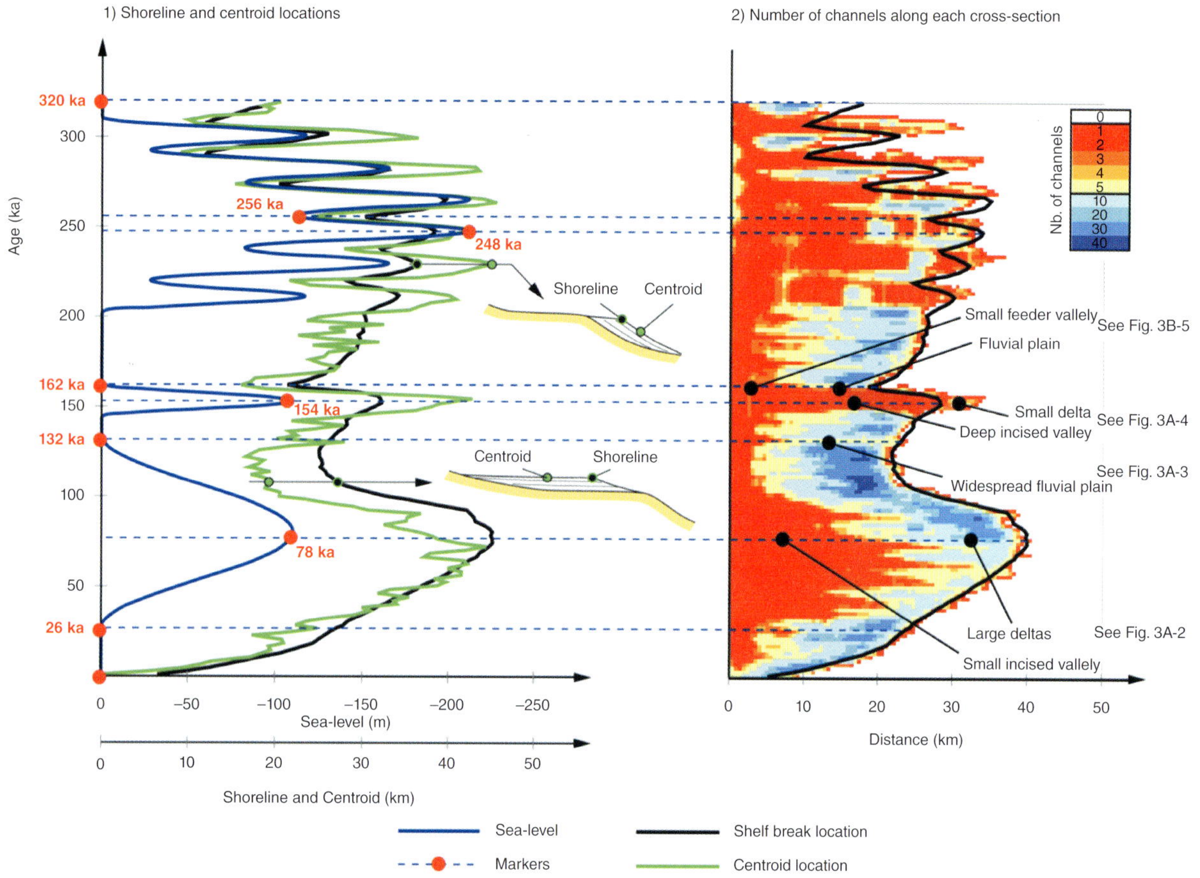

Fig. 4. (1) Comparison of sea-level variations and migration of the shoreline and centroid of the passive margin, (2) analysis of the number of channel at each age along each strike section of the numerical model.

incision, filling and avulsion. On the contrary, on the east side of the basin, subsidence was higher than the eustatic sea-level fall. Channel fed downslope large deltas. This normal regression was characterised by a broad and composite erosional surface westward and aggradation and progradation of the margin eastward. Shoreline and centroid moved eastward more or less at the same rate (Fig. 4). This regression was followed by a retrogradation then aggradation of the depositional systems, characterised by a widespread fluvial plain (78 ka to 144 ka, Figs 3A-3 and Fig. 4). Fluvial onlaps onto flat-lying fluvial strata could be observed both along dip section (Fig. 5-1) and on the statistical analysis of centroid and shoreline location. During this regression, the centroid was moving back towards the west side of the basin and was around 20 km to 30 km upstream of the shoreline (Fig. 4).

During the fast sea-level cycle (Fig. 3A-4, 132 ka to 154 ka and Fig. 4), the eustatic sea-level fall was now faster than the subsidence rate everywhere. It was so fast westward that the feeder channel was forced to incise and had to stay in place. A deep incised-valley was formed (Figs 4 and 5), which fed downslope a small lobate shelf-edge delta (Fig. 3A-4). This accretionary forced regression was characterised by a major subaerial unconformity eastward, truncated toplaps westward and steep marine foresets which downlap the previous maximum flooding surface (Fig. 5-2). The centroid was now moving very fast and was nearly at 50 km downstream the shoreline (Fig. 4). This forced regression was followed by a rapid flooding and transgression of depositional systems (Fig. 3B-5), characterised by the infill of the deep incised valley by backstepping units and onlaps onto the subaerial exposure surface (Fig. 5-3). During the second stage (200 ka to 320 ka), with superimposed slow and fast sea-level cycles, the same evolution of the depositional systems was recorded, with deep incision of fluvial valleys during sea-level fall (Fig. 3B-6) and filling during sea-level rise (Figs 3B-7 and 8).

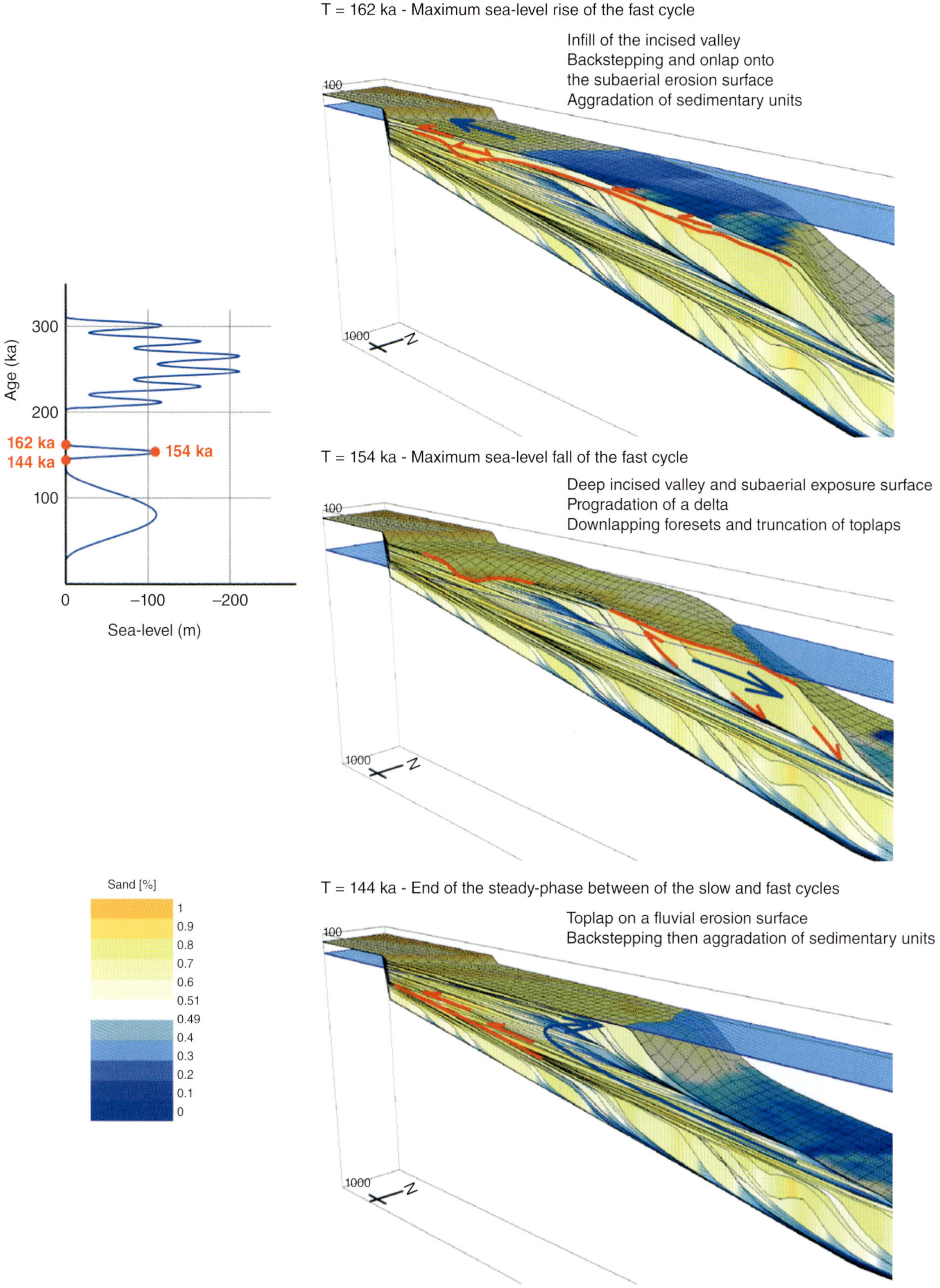

Fig. 5. Detailed internal architecture of sedimentary units deposited during the fast sea-level cycle (T = 144 ka to 162 ka).

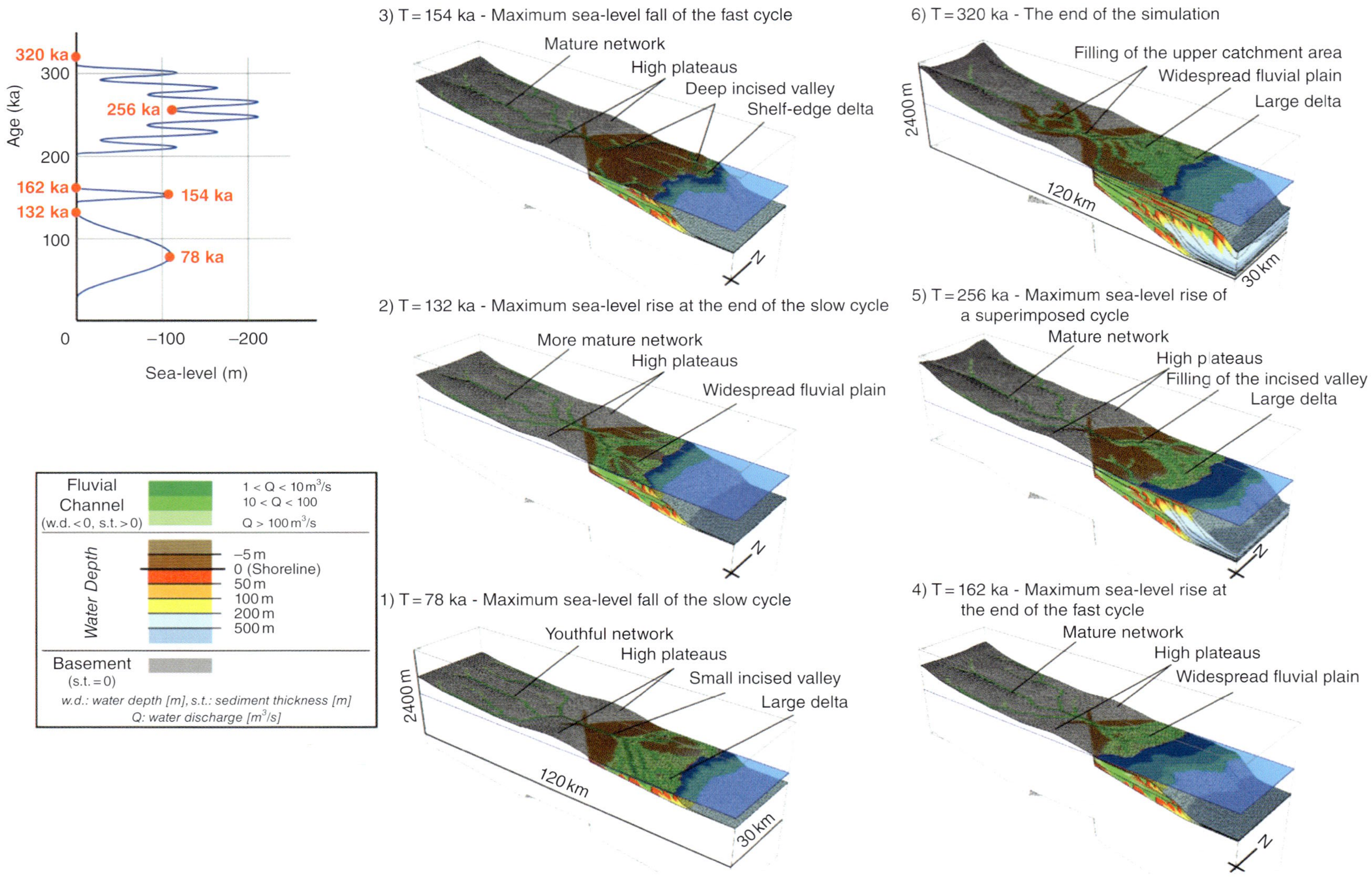

Fig. 6. Analysis of the evolution of sedimentary environments in response to sea-level cycles in the case of the source-to-sink model.

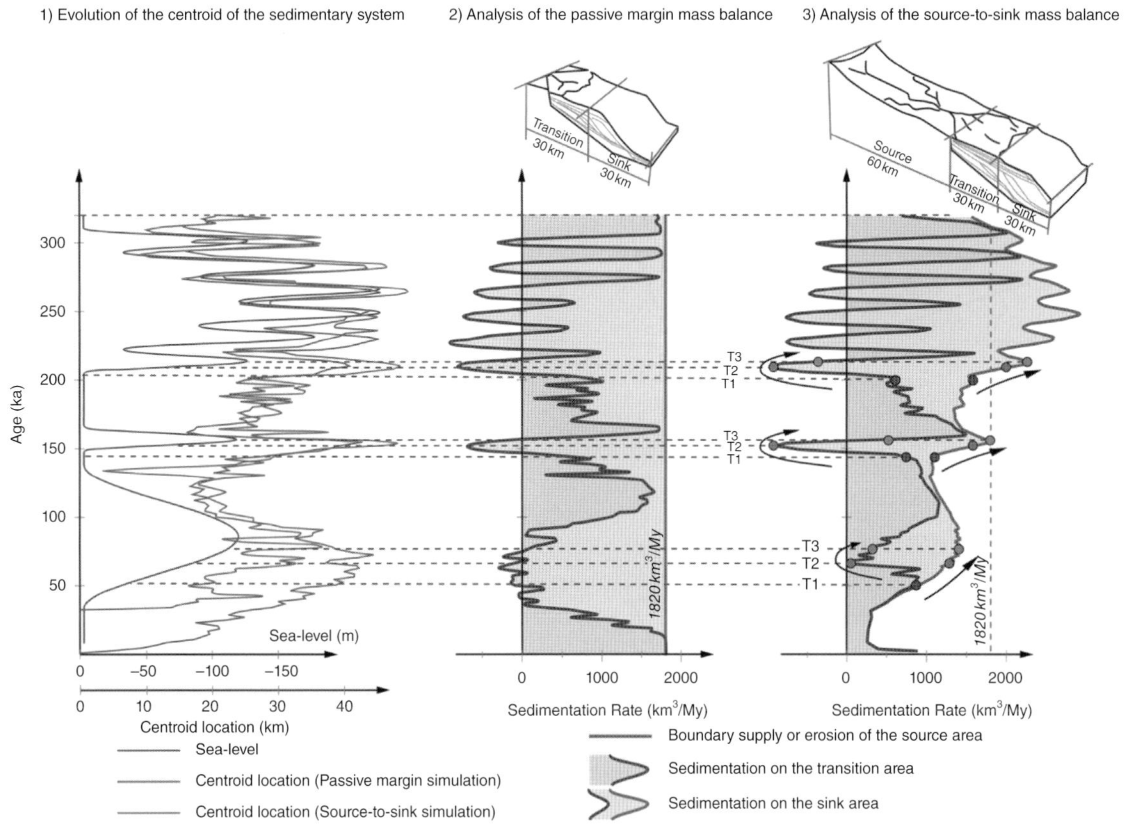

Fig. 7. (1) Comparison of sea-level variations and migration of the centroid of the passive-margin model and of the source-to-sink model, (2) analysis of the sedimentation and erosion rates in the passive margin model and (3) analysis of the sedimentation and erosion rates in the source-to-sink model.

Source-to-sink model

The second numerical simulation was performed to simulate the evolution of a full source-to-sink system, which included an uplifted drainage area and a passive margin, separated by two plateaus. The evolution of this source-to-sink model was quite similar to that of the passive margin model (Figs 6 and 7). The drainage networks grew progressively upstream of the two high plateaus while a fluvial plain prograded and aggraded downstream during the steady phase and slow sea-level fall (Fig 6-1). This regression was followed by a backstepping then progradation of a widespread fluvial plain during the slow sea-level rise (Fig. 6-2). An increasingly mature network fed a deep incised valley during the fast sea-level fall (Fig. 7-3) and once again a widespread fluvial plain backstepped and aggraded during the fast sea-level rise (Fig. 7-4). During the second stage (200 ka to 320 ka), with superimposed slow and fast sea-level cycles, fluvial valleys were deeply incised during sea-level fall and filled during the sea-level rise up to the two plateaus (Fig. 7-5) and finally even inside the drainage network itself (Fig. 7-6).

DISCUSSION

Natural vs. flume experiments

This study explores the use of a non-linear diffusion equation to simulate transport of sediments at a regional scale. A passive-margin model and a source-to-sink model were built using data from a flume experiment. Debates are still very active about the utility and application of experimental geomorphology and stratigraphy, as these experiments may be viewed as not representative of real field-scale systems and as a scaling of all parameters is not possible (Peakall *et al.*, 1996; Paola *et al.*, 2009). Analogue flume modelling presents well-known scaling problems as all the dimensionless scaling parameters such as the Froude or

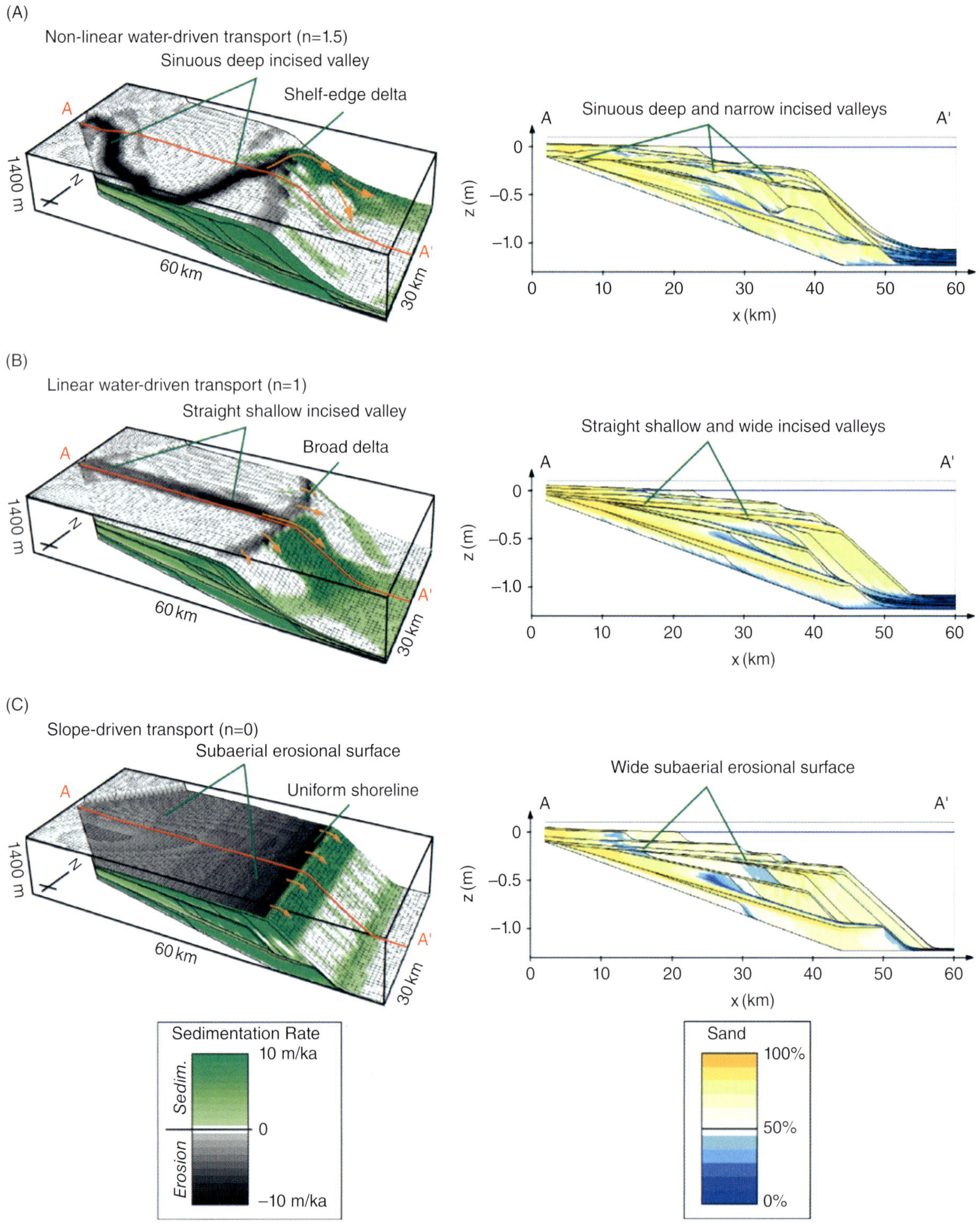

Fig. 8. Comparison of the evolution of the passive margin model obtained using (A) a non-linear water driven diffusion equation (non-linear exponent n = 1,5), (B) a linear water-driven diffusion equation (n = 1) and (C) a simple slope-driven diffusion equation (n = 0).

criticial Shields numbers cannot be respected at the same time as is usual in aerodynamics or fluid flow modelling (Peakall *et al.*, 1996; Paola, 2000; Postma *et al.*, 2008; Martin *et al.*, 2009). As classical full-scale-modelling procedures are too restrictive and unsuited to geomorphologic and stratigraphic experiments, Hooke (1968) proposed to focus on the similarity of processes rather than the true and full dynamic similarity. A geomorphic experiment should be considered as a small natural system in its own right and not just as a small model. Flume experiments have been used over the last decades to understand the qualitative behaviour of natural systems and have been proven to yield stratigraphic responses that are consistent to natural systems, opening up new ways to calibrate stratigraphic forward models (Whipple *et al.*, 1998; Heller *et al.* 2001; Muto & Steel, 2001; Van Heijst *et al.*, 2001a, b; Strong & Paola, 2008; Postma *et al.* 2008; Van den Berg van Saparoea & Postma, 2008; Martin *et al.*, 2009; Van Dijk *et al.*, 2009; Martin *et al.*, 2010). The passive-margin and source-to-sink models built from the flume experiments are thus extreme models. In particular, sea-level cycles (110 m in 20 ka) correspond to really extreme glacial cycles. Although they may be not common in real world systems, these extreme values have the benefit to enhance the role of sediment transport on the stratigraphic architecture, both in the flume and numerical experiments.

Linear vs. non-linear fluvial transport

Numerical models provide an ideal method to test assumptions on transport laws and causal relationships between parameters and stratigraphy. The application of the linear slope-driven, non-linear water-driven diffusion model demonstrates that the simulation results are consistent with the flume experiment. Not only the large-scale behaviour of the flume experiment was reproduced, but the complex stratigraphic details were also simulated: the broad and composite erosional surface during the slow long-term eustatic sea-level fall, the narrow and deep incised valley during the fast short term eustatic sea-level fall. To control the influence of the non-linearity, two simulations were performed using a linear-water driven and slope-driven transport equation ($Q_{sw} = K_w\ Q_w\ S$, Fig. 8B) and using only the classical linear slope-driven diffusion equation ($Q_{sw} = K_w\ S$, Fig. 8C). Transport coefficients were modified to obtain similar fluvial slope in each simulation (Table 2). Sediment transport in marine environments was simulated using the same avalanching code: all sediment grains above the static angle of repose move downslope until the depositional slope angle is lower than this dynamic angle of repose.

Table 2. Transport coefficients used in the non-linear water-driven (Fig. 6A), linear water-driven (Fig. 6B) and linear slope-driven simulations (Fig. 6C).

	Case a: non-linear water-driven transport	Case b: linear water-driven transport	Case c: slope-driven transport
n (–)	1.5	1.0	0.0
Transport equation	$Q_{sm} = K_w\ \overline{Q}_w^{1.5}\ S$	$Q_{sw} = K_w\ \overline{Q}_w\ S$	$Q_{sw} = K_w\ S$
K_{sand} (m^2y^{-1})	400	1100	9000
K_{silt} (m^2y^{-1})	1200	3300	27,000

The basin-scale evolution of the three simulated passive margins is very similar, both in terms of total sediment volume distribution and in terms of sand/silt distribution (Fig. 8). This large-scale architecture is controlled by sediment transport in marine environments, which acts as a vacuum cleaner at the mouth of the fluvial systems. The large-scale internal stratigraphic architecture is also very similar. The three main sedimentary units controlled by the extreme sea-level cycles were reproduced in each simulation. A first, thick, sedimentary unit was deposited during the slow cycle (unit 1 on Fig. 2). A second and smaller unit exhibited erosional surface, toplap truncation and steep downlap onto unit 1 (unit 2 on Fig. 2). A composite and complex third unit was composed of a series of smaller units similar to the second unit (unit 3 on Fig. 2).

Despite the large-scale similitude of these three simulations, the detailed internal stratigraphic architectures were very different. The classical slope-driven diffusion transport equation assumes that sediment transport is only controlled by the local slope. During each sea-level fall, this slope-driven diffusion induced widespread subaerial erosion from the source point downslope to a uniform shoreline parallel to the strike direction (Fig. 8C). The water-driven diffusion equation assumes that a local slope, but also a water flow, are required to transport sediment. As soon as a valley was created in the simulated domain, water flow locally converged towards this valley. As the sediment transport was linearly proportional to the water discharge, the sediment transport efficiency along this valley increased whilst almost no transport was possible along the sides of this valley. This convergence of the water flow was more sensitive during sea-level fall. Each sea-level fall induced a fluvial base level fall. The drainage network could not escape this valley and a narrow shallow incised valley was formed (Fig. 6B).

As the transport equation was linear and as the simulated domain was symmetric, the incised valley was straight and fed a symmetric broad delta. The non-linear water driven diffusion equation behaves in a similar way but has a more explosive behaviour due to the non-linearity. Basin-scale water flow structure has a major influence on the sediment transport. In the case of channel confluences in the simulated domain, water flow increased leading to local erosion to satisfy the increased sediment transport capacity. On the contrary, in the case of bifurcations or diffluences, water flow decreased leading to fast sedimentation due to the loss of transport capacity. It should be noted here that no avulsion or confluence rules were used in our numerical model. This chaotic behaviour arose from the water flow routing and the non-linear equation itself. It was amplified during extreme sea-level fall such as during the fast cycle (Figs 3A-4, 5 & 6 and Fig. 6A): water flow focused in a single sinuous channel, which deeply incised the previously deposited sedimentary units.

This analysis of the linear versus non-linear water-driven diffusion equation demonstrates that the three sediment transport equations tested here gave similar results along a 2D dip-section. The detailed fluvial architecture and, in particular, the water flow structure calculated by these equations were completely different. These simulations were characterised by a uniform shoreline parallel to the strike direction without any delta (slope-driven transport, Fig. 6C) or by a broad delta fed by a straight and shallow channel (linear water-driven transport, Fig. 6B) or a small shelf-edge delta fed by a sinuous narrow and deeply incised channel (non-linear water-driven transport, Fig. 6A). The linear slope-driven, non-linear water-driven diffusion equation led to realistic fluvial evolution with sinuous channels, but more studies of real-world systems are still required to calibrate all the transport parameters such as the exponent or diffusion coefficients.

Passive margin vs. source-to-sink model

The above analysis of the non-linear water-driven transport equation is focused on the passive-margin model that is on the lower catchment to sink areas. It has been recently recognised that these areas are not alone but part of a full source-to-sink system (Martinsen & Helland-Hansen, 2009; Somme *et al.*, 2009b; Martinsen *et al.*, 2010). The second simulation performed at a source-to-sink scale allowed exploration of the importance of the drainage area that fed the passive margin. The evolution of the sedimentary systems of the source-to-sink model (Fig. 6) was quite similar to that of the passive margin model, as illustrated by the evolution of the centroid (Fig. 8-1).

A long transitory phase (0 ka to 80 ka) was observed at the beginning of the simulation. During this transitory phase, the upper catchment area evolved progressively from an immature to a mature landscape. To better explore the distribution of sediment, the total erosion of the source area and the sedimentation rate in the transition area were calculated. These erosion and sedimentation rates were compared to the boundary condition defined in the passive margin model (1820 km^3 My^{-1}). The first-order evolution of the erosion and sedimentation rates was quite similar in the two simulations and mimicked the sea-level cycles.

In detail, the evolution of the source-to-sink model was far more complex than the passive margin model. In particular, each sea-level fall was characterised by two phases (Fig. 8-3). During the first phase (T1-T2), sea-level fall steepened the fluvial base level. This induced an increase of the erosion rate of the source area. In the transition area, this fall induced either a decrease of the sedimentation rate in the transition area when sea-level fall was slow, or a major erosion of this transition area when sea-level fall was very fast. During the second phase (T2-T3), sea-level continued to fall up to a minimum value but at a slower rate than during the first phase. Erosion of the catchment area continued to increase, while in the transition area erosion decreases or even sedimentation starts again. These first results on a source-to-sink system illustrate the importance of simulating the full sedimentary system. A constant boundary condition hides high-frequency variations of the erosion along the drainage area and thus of the sediment inflow into the sink area.

These simulations also illustrate the diachronicity of erosional surfaces, as already noticed in flume experiments (Strong & Paola, 2008). These results are promising but have been obtained with extreme parameters, such as sea-level cycle or rainfall. More studies of real-world systems are required to calibrate erosion and transport parameters and quantify the evolution of catchment, fluvial, shelf and sink areas in response to sea-level; also climate and tectonics.

CONCLUSIONS

It has been demonstrated that a 3D stratigraphic forward model can successfully reproduce the evolution of a passive margin either at a regional or source-to-sink scale. The overall geometry and stratigraphic architecture of flume and numerical experiments were analogous. The detailed internal architecture could not be compared directly due to the chaotic behaviour of the natural system and the non-linear response of the numerical model, but the statistical characteristics such as the shoreline or centroid evolutions correspond. The sea-level-forced units were reproduced correctly with a first, thick, sedimentary unit deposited during a sea-level slow cycle and characterised by a major progradation and aggradation of fluvial systems and small shallow incised valleys. A second, smaller, unit during a fast sea-level cycle exhibited subaerial erosional surface and sinuous narrow and deep incised valley, toplap truncation and steep downlap onto the first unit. Finally a composite and complex third unit was composed of a series of smaller units similar to the second unit.

The 3D numerical model uses a linear slope-driven, non-linear water-driven diffusion equation to simulate transport of sediment. A simple slope-driven diffusion equation induced very smooth landscapes without any fluvial rivers or deltas. A linear water-driven diffusion equation created more complex landscape. Only the non-linear water-driven diffusion equation was able to reproduce realistic water flow structure characterised by sinuous channels and incised valleys, without using any explicit rules to force confluences or diffluences of rivers.

Finally, it has been demonstrated that the non-linear water-driven diffusion equation could be applied both at a passive margin scale and at a full source-to-sink system scale. The results of the source-to-sink model are similar to the local passive margin model but also provide some new insights into the evolution of catchment and fluvial areas in response to sea-level variations. In particular, the detailed timing of the erosion and sedimentation processes, or the diachronicity of erosional surfaces, could be investigated at a regional scale.

This study illustrates that simulating sediment routing systems at a regional scale from source, or at least from fluvial system or shelf, to sink gives a better understanding of depositional systems. This numerical modelling may be used either as a scientific tool to quantify the interaction between stratigraphic parameters or as an industrial guide to reduce uncertainties in exploration and reservoir risk assessment (Roure *et al.*, 2009). More studies of real-world systems (e.g. Csato *et al.*, 2012) are required to calibrate transport parameters and to explore the consequences of these non-linear fluvial laws on shelf and deep-water turbiditic systems.

ACKNOWLEDGEMENTS

We thank, gratefully, the Dionisos Consortium (Chevron, ConocoPhillips, ENI-Agip, ExxonMobil, GdF-Suez, Petrobras, Respol, Shell, Statoil and Total). Cornel Olariu and Rory Dalman are acknowledged for providing highly valuable critical reviews.

REFERENCES

Adams, E.W. and **Schlager, W.** (2000) Basic types of submarine slope curvatures. *J. Sed. Res.*, **70**, 814–828.

Adams, E.W., Schlager, W. and **Wattel, E.** (1998) Submarine slopes with an exponential curvatures. *Sed. Geol.*, **117**, 135–141.

Alzaga-Ruiz, H., Granjeon, D., Lopez, M., Seranne, M. and **Roure, F.** (2009) Gravitational collapse and Neogene sediment transfer across the western margin of the Gulf of Mexico: Insights from numerical models. *Tectonophysics*, **470**, 1–2, 21–41.

Avouac, J.P. and **Burov, E.B.** (1996) Erosion as a driving mechanism of intracontinental mountain growth. *J. Geophys. Res.*, **101**, 17747–17769.

Begin, Z.B. (1988) Application of a diffusion-erosion model to alluvial channels which degrade due to base-level lowering. *Earth Surf. Proc. Land.*, **13**, 487–500.

Begin, Z.B. (1993) Application of quantitative morphologic dating to paleo-seismicity of the northwestern Negev, Israël. *Isr. J. Earth Sci.*, **41**, 95–103.

Begin, Z.B., Meyer, D.F. and **Schumm, S.A.** (1981) Development of longitudinal profiles of alluvial channels in response to base-level lowering. *Earth Surf. Proc. Land.*, **6**, 49–68.

Burgess, P.M. and **Steel, R.** (2008) Stratigraphic forward modelling of basin margin clinoform systems: Implications for controls on topset and shelf width and timing of formation of shelf-edge delta. *SEPM Spec. Publ.*, **90**, 35–45.

Burgess, P.M., **Lammers, H.**, **Van Oosterhout, C.** and **Granjeon, D.** (2006) Multivariate sequence stratigraphy: tackling complexity and uncertainty with stratigraphic forward modelling, multiple scenarios and conditional frequency maps. *AAPG Bull.*, **90**, 1883–1901.

Carretier, S. and **Lucazeau, F.** (2005) How does alluvial sedimentation at range fronts modify the erosional dynamics of mountain catchments? *Basin Res.*, **17**, 361–381.

Carson, M.A. and **Kirkby, M.J.** (1972) Hillslope form and process. Cambridge Univ. Press, 475 p.

Catuneanu, O., **Abreu, V.**, **Bhattacharya, J.P.**, **Blum, M.D.**, **Dalrymple, R.W.**, **Eriksson, P.G.**, **Fielding, C.R.**, **Fisher, W.L.**, **Galloway, W.E.**, **Gibling, M.R.**, **Giles, K.A.**, **Holbrook, J.M.**, **Jordan, R.**, **Kendall, C.G.St.C.**, **Macurda, B.**, **Martinsen, O.J.**, **Miall, A.D.**, **Neal, J.E.**, **Nummedal, D.**, **Pomar, L.**, **Posamentier, H.W.**, **Pratt, B.R.**, **Sarg, J.F.**, **Shanley, K.W.**, **Steel, R.J.**, **Strasser, A.**, **Tucker, M.E.** and **Winker, C.** (2009) Towards the standardization of sequence stratigraphy. *Earth-Sci. Rev.*, **92**, 1–33.

Chalaron, E., **Mugnier, J.L.**, **Sassi, W.** and **Mascle, G.** (1996) Tectonics, erosion and sedimentation in an overthrust system: a numerical model. *Comput. Geosci.*, **22**, 2, 117–138.

Clevis, Q., **De Boer, P.L.** and **Wachter. M.** (2003) Numerical modelling of drainage basin evolution and three-dimensional alluvial fan stratigraphy. *Sed. Geol.*, **163**, 85–110.

Colman, S.M. and **Watson, K.** (1983) Ages estimated from a diffusion equation model for scarp degradation. *Science*, **22**, 263, 265.

Coulthard, T.J. (2001) Landscape evolution models: a software review. *Hydrol. Process.*, **15**, 165–173.

Coulthard, T.J., **Kirkby, M.J.** and **Macklin, M.** (1996) A cellular automaton fluvial and slope model of landscape evolution. In: *The First International Conference on GeoComputation*. University of Leeds, 168–185.

Covault, J.A., **Fildani, A.**, **Romans, B.W.** and **McHargue, T.** (2011) The natural range of submarine canyon- and channel longitudinal profiles. *Geosphere*, **7**, 2, 313–332.

Cross, T.A. and **Harbaugh, J.W.** (1989) Quantitative dynamic stratigraphy: A workshop, a philosophy, a methodology. In: *Quantitative Dynamic Stratigraphy* (Ed. T.A. Cross), pp 3–20. Prentice-Hall, Englewood Cliffs, NJ.

Csato, I., **Granjeon, D.**, **Catuneanu, O.** and **Baum, G.R.** (2012) A three-dimensional stratigraphic model for the Messinian crisis in the Pannonian Basin, eastern Hungary. *Basin Res.*, 0, 1–18, doi: 10.1111/j.1365-2117.2012.00553.x

Culling, W.E.H. (1960) Analytical theory of erosion. *J. Geol.*, **68**, 3, 336–344.

Dalman, R.A.F. and **Weltje, G.J.** (2008) Sub-grid parameterisation of fluvio-deltaic 368 processes and architecture in a basin-scale stratigraphic model. *Comput. Geosci.*, **34**, 1370–1380.

Dalman, R.A.F. and **Weltje, G.J.** (2012) SimClast: an aggregated forward stratigraphic model of continental shelves. *Comput. Geosci.*, **38**, 115–126.

Davy, P. and **Crave, A.** (2000) Upscaling local-scale transport processes in large-scale relief dynamics. *Phys. Chem. Earth (A)*, **25**, 6–7, 533–541.

Eymard, R., **Gallouet, T.**, **Granjeon, D.**, **Masson, R.** and **Tran, Q.H.** (2004) Multi-lithology stratigraphic model under maximum erosion rate constraint. *Int. J. Numer. Methods Ingrg.*, **60**, 527–548.

Flemings, P.B. and **Grotzinger, J.P.** (1996) STRATA: Freeware for Solving Classic Stratigraphic Problems. *GSA Today*, v. **6**, no. **12**, pp. 1–7.

Flemings, P.B. and **Jordan, T.E.** (1989) A synthetic stratigraphic model of foreland basin development. *J. Geophys. Res.*, **94**, B4, 3851–3866.

Freeman, T.G. (1991) Calculating catchment area with divergent flow based on a regular grid. *Comput. Geosci.*, **17**, 3, 413–422.

Gervais, V. (2004) Etude et simulation d'un modele stratigraphique multi-lithologique sous contrainte de taux d'erosion maximal, These de doctorat, Universite de Marne-la-Vallee.

Gervais, V. and **Masson, R.** (2004) *Mathematical and Numerical Analysis of a Stratigraphic Model*. M2AN 38, **4**, 585–611.

Gessler, L. (1971) Critical shear stress for sediment mixtures. In: *Proceedings International Association for Hydraulic Research Congress, Changes in alluvial beds composed of non-uniform material*, **3**, 1–8.

Granjeon, D. (1997) Modelisation stratigraphique deterministe: conception et applications d'un modele diffusif 3D multilithologique, These de Doctorat. Universite de Rennes 1.

Granjeon, D. and **Joseph, P.** (1999) Concepts and applications of a 3-D multiple lithology, diffusive model in stratigraphic modelling. In: *Numerical Experiments in Stratigraphy: Recent Advances in Stratigraphic and Sedimentologic Computer Simulations* (Eds J.W. Harbaugh, W.L. Watney, E.C. Rankey, R. Slingerland, R.H. Goldstein and E.K. Franseen), *SEPM Spec. Publ.*, **470**, 62, 197–210.

Granjeon, D., **Joseph, P.**, **Assier-Rzadkiewicz, S.**, **Bassant, P.**, **Brieuc, O.**, **Hugot, A.** and **Moreira, J.L.P.** (1999) Application of 3D fluvial and turbiditic sediment transport laws in stratigraphic modelling of siliciclastic and carbonate formations. In: *Proceedings of the 5th Annual Conference of the International Association for Mathematical Geology* (Eds S.J. Lippard, A. Naess and R. Sinding-Larsen), Trondheim, 515–520.

Green, T. (1987) The importance of double diffusion to the settling of suspended material. *Sedimentology*, **34**, 319–331.

Griffiths, C.M., **Dyt, C.**, **Paraschivoiu, E.** and **Liu, K.** (2001) Sedsim in hydrocarbon exploration. In: *Geologic Modelling and Simulation* (Eds D. Merriam and J.C. Davis), pp 71–97. Kluwer Academic, New York.

Hanks, T.C., **Bucknam, R.C**, **Lajoie, K.R.** and **Wallace, R.E.** (1984) Modification of wave cut and faulting controlled landforms. *J. Geophys. Res*, **89**, 7, 5771–5790.

Harbaugh, J.W. and **Bonham-Carter, G.** (1970) Computer Simulation in Geology: Wiley-Interscience, New York, 575 pp.

Helland-Hansen, W. (2009) Towards the standardization of sequence stratigraphy. *Earth-Sci. Rev.*, **94**, 1–4, 95–7.

Helland-Hansen, W. and **Martinsen, O.J.** (1996) Shoreline trajectories and sequences: a description of variable depositional-dip scenarios. *J. Sed. Res.*, **66**, 4, 670–688.

Heller, P.L., **Paola, C.**, **Hwang, I.-G.**, **John, B.** and **Steel, R.J.** (2001) Geomorphology and sequence stratigraphy due to slow and rapid base-level changes in an experimental subsiding basin, XES 96-1. *AAPG Bull.*, **85**, 5, 817–838.

Hirano, M. (1975) Simulation of development process of interfluvial slopes with reference to graded form. *J. Geol.*, **83**, 113–123.

Hooke, R.L. (1968) Model geology: prototype and laboratory streams: discussion. *Geol. Soc. Amer. Bull.*, **79**, 391–394.

Hutton, E.W.H. and **Syvitski, J.P.M.** (2008) Sedflux 2.0: an advanced process-rsponse model that generates three-dimensional stratigraphy. *Comput. Geosci.*, **34**, 10, 1319–1337.

Jordan, T.E. and **Flemings, P.B.** (1991) Large-scale architecture, eustatic variation and unsteady tectonism: a theoretical evaluation. *J. Geophys. Res.*, **96**, B4, 6681–6699.

Kenyon, P.M. and **Turcotte, D.L.** (1985) Morphology of a delta prograding by bulk sediment transport. *Geol. Soc. Am. Bull.*, **96**, 1457–1465.

Kertznus, V. and **Kneller, B.** (2009) Clinoform quantification for assessing the effects of external forcing on continental margin development. *Basin Res.*, **21**, 5, 738–758.

Ku Shafie K.R., and **Madon, M.** (2008) A review of stratigraphic simulation techniques and their applications in sequence stratigraphy and basin analysis. *Bulletin of the Geological Society of Malaysia*, **54**, 81–89.

Martin, J., **Paola, C.**, **Abreu, V.**, **Neal, B.** and **Sheets, B.** (2009) Sequence stratigraphy of experimental strata under known conditions of differential subsidence and variable base level. *AAPG Bull.*, **93**, 4, 503–533.

Martin, J.L., **Abreu, V.**, **Neal, J.E.** and **Pratson, L.** (2010) Sequence stratigraphy of siliciclastic systems – the ExxonMobil methodology. In: *SEPM Concepts in Sedimentology and Paleontology* (Eds V. Abreu, J.E. Neak, K.M. Bohacs and J.L. Kalbas), **9**, 199–208.

Martinez, P.A. and **Harbaugh, J.W.** (1993) *Simulating Nearshore Environments*. Pergamon, New York, 265 p.

Martinsen, O.J. and **Helland-Hansen, W.** (2009) Source-to-Sink: Principles, Methods and Applications. *AAPG Annual Convention & Exhibition, Abstracts.*, p.138.

Martinsen, O.J, **Somme, T.O.**, **Thurmond, J.B.**, **Helland-Hansen, W.** and **Lunt, I.** (2010) Source-to-sink systems on passive margins: theory and practice with an example from the Norwegian continental margin. In: *Petroleum Geology: From Mature Basins to New Frontiers – Proceedings of the 7th Petroleum Geology Conference* (Eds B.A. Vining and S.C. Pickering), GSL, 1244 pp.

Meijer, X.D. (2002) Modelling the drainage evolution of a river-shelf system forced by Quaternary glacio-eustasy. *Basin Res.*, **14**, 361–377.

Meijer, X.D., **Postma, G.**, **Burrough, P.A.** and **de Boer, P.L.** (2008) Modelling the preservation of sedimentary deposits on passive continental margins during glacial-interglacial cycles. In: *Analogue and numerical forward modelling of sedimentary systems; from understanding to prediction* (Eds P.L. de Boer, G. Postma, C.J. van der Zwan, P.M. Burgess and P. Kukla), *Int. Assoc. Sedimentol. Spec. Publ.*, **40**, 223–238.

Merriam, D.F. (1972) *Mathematical Models of Sedimentary Processes; an international symposium*. Plenum Publ., New York, 271 p.

Miller, J., **Sun, T.**, **Li, H.**, **Stewart, J.**, **Genty, C.**, **Dachang, L.** and **Lyttle, C.** (2008) Direct Modeling of Reservoirs Through Forward Process-Based Models: Can We Get There? DOI 10.2523/12729-MS.

Moretti, I. and **Turcotte, D.L.** (1985) A model for erosion, sedimentation and flexure with application to New Caledonia. *J. Geodynamics*, **3**, 155–168.

Murray, A.B. and **Paola, C.** (1994) A cellular model of braided rivers. *Nature*, **371**, 54–57.

Murray, A.B. and **Paola, C.** (1997) Properties of a cellular braided stream model. *Earth Surf. Proc. Land.*, **22**, 1001–1025.

Muto, T. and **Steel, R.J.** (2001) Autostepping during the transgressive growth of deltas: Results from flume experiments: *Geology*, **29**, 771–774.

Muto, T. and **Steel, R.J.** (2004) Autogenic response of fluvial deltas to steady sea-level fall: Implications from flumetank experiments. *Geology*, **32**, 401–404.

O'Grady, D., **Syvitski, J.P.M.**, **Pratson, L.F.** and **Sarg, J.F.** (2000) Categorizing the morphologic variability of siliciclastic passive continental margins. *Geology*, **28**, 207–210.

Paola, C. (2000) Quantitative models of sedimentary basin filling. *Sedimentology*, **47**, suppl. 1, 121– 178.

Paola, C., **Heller, P.L.** and **Angevine, C.L.** (1992) The large-scale dynamics of grain-size variation in alluvial basins, 1: theory. *Basin Res.*, **4**, 73–90.

Paola, C., **Mullin, J. Ellis, C.**, **Mohrig, D.C.**, **Swenson, J.B.**, **Parker, G.**, **Hickson, T.**, **Heller, P.L.**, **Pratson, L.**, **Syvitski, J.**, **Sheets, B.** and **Strong, N.** (2001) Experimental stratigraphy. *GSA Today*, **117**, 4–9.

Paola, C., **Straub, K.**, **Mohrig, D.** and **Reinhard, L.** (2009) The "unreasonable effectiveness" of stratigraphic and geomorphic experiments. *Earth-Sci. Rev.*, **97**, 1–43.

Parker, G., **Poala, C.**, **Whipple, K.X.** and **Mohrig, D.** (1998) Alluvial fans formed by channelized fluvial and sheet flow: I: theory. *J.Hydraul. Eng.*, **124**, 10, 985–995.

Peakall, J., **Ashworth, P.J.** and **Best, J.L.** (1996) Physical modelling in fluvial geomorphology: principles, applications and unresolved issues. In: *The scientific nature of geomorphology* (Eds B.L. Rhoads and C.E. Thorn), *Proceedings of the 27th Binghamton Symposium on Geomorphology*, pp 221–254. John Wiley and Sons, Chichester.

Pelletier, J.D., **DeLong, S.B.**, **Al-Suwaidi, A.H.**, **Cline, M.**, **Lewis, Y.**, **Psillas, J.L.** and **Yanites, B.** (2006) Evolution of the Bonneville shoreline scarp in west-central Utah: Comparison of scarp-analysis methods and implications for the diffusion model of hillslope evolution. *Geomorphology*, **74**, 257–270.

Pizzuto, J.E. (1987) Sediment diffusion during overbank flows. *Sedimentology*, **34**, 301–317.

Porebski, S.J. and **Steel, R.J.** (2003) Shelf-margin deltas: their stratigraphic significance and relation to deepwater sands. *Earth-Sci. Rev.*, **62**, 283–326.

Postma, G. and **Van den Berg van Saparoea, A.-P.** (2007) Flume modelling of river-delta systems at geological relevant time scales: Templates for sea-level induced flux. In: *Analogue and numerical forward modelling of sedimentary systems; from understanding to prediction* (Eds P.L. de Boer, G. Postma, C.J. van der Zwan, P.M. Burgess and P. Kukla), *Int. Assoc. Sedimentol. Spec. Publ.*, **40**, 191–206.

Postma, G., **Kleinhans, M.G.**, **Meijer, P.T.** and **Eggenhuisen, J.T.** (2008) Sediment transport in analogue flume models compared with real-world sedimentary systems: a new look at scaling evolution of sedimentary systems in a flume. *Sedimentology*, **55**, 1541–1557.

Rabineau, M., Berne, S., Aslanian, D., Olivet, J.-L., Joseph, P., Guillocheau, F., Bourillet, J.-F., Ledrezen, E. and **Granjeon, D.** (2005) Sedimentary sequences in the Gulf of Lion: A record of 100,000 years climatic cycles. *Mar. Petrol. Geol.*, **22**, 775–804.

Rivenaes, J.C. (1988) Application of a dual-lithology, depth dependent diffusion equation in stratigraphic simulation. *Basin Res.*, **4**, 133–146.

Roering, J.J., Kirchner, J.W. and **Dietrich, W.E.** (1999) Evidence for nonlinear, diffusive sediment transport and implications for landscape morphology. *Water Resour. Res.*, **35**, 853–870.

Roering, J.J., Perron J.T. and **Kirchner, J.W.** (2007) Functional relationships between denudation and hillslope form and relief. *Earth Planet. Sci. Lett.*, **264**, 245–258.

Roure, F., Alzaga-Ruiz, H., Callot, J.-P., Ferket, H., Granjeon, D., Gonzalez-Mercado, G.E., Guilhaumou, N., Lopez, M., Mougin, P., Ortuno-Arzate, S. and **Seranne, M.** (2009) Long lasting interactions between tectonic loading, unroofing, post-rift thermal subsidence and sedimentary transfers along the western margin of the Gulf of Mexico: Some insights from integrated studies. *Tectonophysics*, **475**, 169–189.

Schwarzacher, W. (1975) Sedimentation models and quantitative stratigraphy. *Developments in Sedimentology*, **19**, Elsevier Scientific Publishing, New York. ISBN 0 444 41302 2.

Sinclair, H.D., Coakley, B.J., Allen, P.A. and **Watts, A.B.** (1991) Simulation of foreland basin stratigraphy using a diffusion model of mountain belt uplift and erosion: an example from the Central Alps, Switzerland. *Tectonics*, **10**, 3, 599–620.

Slingerland, R., Harbaugh, J.W. and **Furlong, K.** (1994) *Simulating Clastic Sedimentary Basins: Physical Principles and Computer Programs for Creating Dynamic Systems*. Prentice-Hall, Englewood Cliffs, 220 pp.

Somme, T.O., Helland-Hansen, W. and **Granjeon, D.** (2009a) Impact of eustatic amplitude variations on shelf morphology, sediment dispersal and sequence stratigraphic interpretation: Icehouse versus greenhouse systems. *Geology*, **37**, 7, 587–590.

Somme, T.O., Helland-Hansen, W., Martinsen, O.J. and **Thurmond, J.B.** (2009b) Relationships between morphological and sedimentological parameters in source-to-sink systems: a basis for predicting semi-quantitative characteristics in subsurface systems. *Basin Res.*, **21**, 4, 361–387.

Storm, J.E.A. and **Swift, D.J.P.** (2003) Shallow-marine sequences as the building blocks of stratigraphy: insights from numerical modelling. *Basin Res.*, **15**, 287–303.

Strong, N. and **Paola, C.** (2008) Valleys that never were: time surfaces versus stratigraphic surfaces. *J. Sed. Res.*, **78**, 579–593.

Swift, D.J.P. and **Thorne, J.A.** (1991) Sedimentation on continental margin, I: a general model for shelf sedimentation. In: *Shelf sands and sandstone bodies* (Eds D.J.P. Swift, G.F. Oertel, R.W. Tilman and J.A. Thorne), *Int. Assoc. Sedimentol. Spec. Publ.* **14**, 3–31.

Syvitski, J.P.M. and **Milliman, J.D.** (2007) Geology, geography and humans battle for dominance over the delivery of fluvial sediment to the coastal ocean. *J. Geol.*, **115**, 1–19.

Tetzlaff, D.M. and **Harbaugh, J.W.** (1989) *Simulating Clastic Sedimentation*. Reinhold Van Nostrand, New York. 202 pp.

Tucker, G.E. (2004) Drainage basin sensitivity to tectonic and climatic forcing: implications of a stochastic model for the role of entrainment and erosion thresholds. *Earth Surf. Proc. Land.*, **29**, 401–422.

Tucker, G.E. (2009) Natural experiments in landscape evolution. *Earth Surf. Proc. Land.*, **34**, 1450–1460.

Tucker, G.E. and **Hancock, G.R.** (2010) Modelling landscape evolution. *Earth Surf. Proc. Land.*, **35**, 28–50.

Tucker, G.E. and **Slingerland, R.** (1994) Erosional dynamics, flexural isostasy and long-lived escarpments: a numerical modelling study. *J. Geophys. Res.*, **10**, 12, 229–243.

Uroza, C.A. and **Steel, R.J.** (2008) A highstand shelf-margin delta system from the Eocene of West Spitsbergen, Norway. *Sed. Geol.*, **203**, 3–4, 229–245.

Van den Berg van Saparoea, A.-P. and **Postma, G.** (2008) Control of climate change on the yield of river systems. In: *Recent advances in models of siliciclastic shallow-marine stratigraphy* (Eds G.J. Hampson, R.J. Steel, P.M. Burgess and R.W. Dalrymple), *SEPM Spec.* Publ. **90**, 15–33.

Van Dijk, M., Postma, G. and **Kleinhans, M.G.** (2009) Autocyclic behaviour of fan deltas: an analogue experimental study. *Sedimentology*, **56**, 1569–1589.

Van Heijst, M.I.W.M., Postma, G., Meijer, X.D., Snow, J.N. and **Anderson, J.A.** (2001a) Analogue flume-model study of the Late Quaternary Colorado/Brazos shelf. *Basin Res.*, **13**, 243–268.

Van Heijst, M.I.W.M. and **Postma, G.** (2001b) Fluvial response to sea-level changes: a quantitative analogue, experimental approach. *Basin Res.*, **13**, 269–292.

Waltham, D. (1992) Mathematical modelling of sedimentary basin processes. *Mar. Petrol. Geol.*, **9**, 265–273.

Whipple, K.X., Parker, G., Paola, C. and **Mohrig, D.** (1998) Channel dynamics, sediment transport and the slope of alluvial fans: experimental study. *J. Geol.*, **106**, 677–693.

Willgoose, G.R., Bras, R.L. and **Rodriguez-Iturbe** (1991a) A physically based coupled network growth and hillslope evolution model, 1, theory. *Water Resour. Res.*, **27**, 1671–1684.

Willgoose, G.R., Bras, R.L. and **Rodriguez-Iturbe** (1991b) A physically based coupled network growth and hillslope evolution model, 2, applications. *Water Resour. Res.*, **27**, 1685–1696.

Willgoose, G., Bras, R.L. and **Rodriguez-Iturbe, I.** (1992) The relationship between catchment and hillslope properties: implications of a catchment evolution model. *Geomorphology*, **5**, 1–2, 21–37.

Wright, L.D. and **Nittrouer, C.A.** (1995) Dispersal of river sediments in coastal seas: six contrasting cases. *Estuaries*, **18**, 3, 494–508.

Int. Assoc. Sedimentol. Spec. Publ. (2014) **46**, 473–512

Tectonic control on sedimentation, erosion and redeposition of Upper Jurassic sandstones, Central Graben, North Sea

JONATHAN P. WONHAM*, IAN RODWELL†, TORE LEIN-MATHISEN* and MICHEL THOMAS*

* *Total E&P Norge, Finnestadveien 44, Dusavik, P.O. Box 168, N-4001 Stavanger, Norway*
† *Total E&P UK, Crawpeel Road, Altens, Aberdeen, Scotland*

ABSTRACT

The distribution of Upper Jurassic sandstones in the Central North Sea Basin was influenced by three structural mechanisms: (1) the contemporaneously evolving Central Graben rift system that formed the basin and created fault block highs and basinal grabens; (2) the movement of Zechstein salt (late Permian age) which can be locally very thick and (3) bathymetric features controlled by the Jurassic deformation of Triassic pods (salt minibasins). Rift development involved movement on deep seated, normal faults with offset seen at Rotliegend level (Permian). Zechstein salt decouples this fault displacement from the deformation of younger strata. As a result of the individual or combined effects of footwall uplift, salt movement or grounding of pods on the underlying Rotliegend, areas of former Triassic or Upper Jurassic deposition were subsequently reworked erosively and redistributed. It has been found that spiculitic sandstone layers within Upper Jurassic Fulmar Formation sandstones serve as useful horizons for correlation and help highlight intra-Jurassic erosive truncation of shoreface sandstones. Erosion of structurally controlled highs can result in reworking of earlier deposited shoreface sandstones, the products of this erosion being sometimes re-deposited as turbidites that are transported short distances before being ponded in depositional lows such as the hangingwall of major faults. The development of significant erosion surfaces within Upper Jurassic shallow marine sandstones is linked to tectonic mechanisms. It is probable that these surfaces are not basin wide sequence boundaries; rather they are local surfaces reflecting the dynamic and geographically varied evolution of the basin. Previous proposals that many phases of incised valley development occurred during the Late Jurassic generating 3rd order sequence boundaries do not appear to be backed by sedimentary evidence. While turbidite sandstones are evidently developed at various times in the basin history, their relationship to proximal fluvial systems and incised valleys is unproven. A preferred interpretation is that local tectonic uplift and shoreface erosion are the main mechanisms for redistributing Jurassic and older sediments.

Keywords: North Sea Central Graben, Upper Jurassic, tectonics, turbidite, shoreface, sequence stratigraphy.

INTRODUCTION

In the UK Central North Sea and the adjacent Norwegian Southern North Sea sectors, Upper Jurassic sandstones form prolific, good quality, reservoir rocks in large fields such as Gyda and Ula (Stewart, 1993) in the Norwegian Sector and Elgin-Franklin (Lasocki *et al.*, 1997; 1999), Shearwater (Jeremiah & Nicholson, 1999, Gilham *et al.*, 2005) and Fulmar (Johnson *et al.*, 1986; Spaak *et al.*, 1999; Kuhn *et al.*, 2003) in the UK sector. These fields contain, or have produced, economically important hydrocarbon reserves. Elgin-Franklin, for example, is the world's largest high pressure, high temperature, development and has supplied up to 7% of the UK's natural gas requirement.

Several of these fields sit within the high pressure and high temperature (HPHT) play of the

From Depositional Systems to Sedimentary Successions on the Norwegian Continental Margin, First Edition.
Edited by A. W. Martinius, R. Ravnås, J. A. Howell, R. J. Steel, and J. P. Wonham.

North Sea Central Graben. This region has been explored for thirty years (Erratt *et al.*, 2005) and tests ever deeper and hotter targets approaching 200°C and beyond 1100 bar pressure. The cost of exploration and development in such an environment means that significant efforts have been made to understand the sedimentology and reservoir architecture of Upper Jurassic sandstones. These sandstones developed throughout the Central Graben and are known by a variety of formation names, although commonly referred to as Fulmar Formation in the UK sector and Ula Formation in the Norwegian sector. They range from Oxfordian to Volgian in age and overlie a varied subcrop stratigraphy ranging from Middle Jurassic fluvial deposits with coals (Pentland Formation in the UK sector, Bryne Formation in the Norwegian sector) to Triassic continental/lacustrine deposits (Skagerrak and Smith Bank Formations) and Permian Rotliegend Group or Zechstein salt. The reason for this variety is their transgressive character, progressively overlying older rocks that had been exposed following the Middle Jurassic Mid-Cimmerian uplift. This uplift defined a regional thermal doming focused on the North Sea triple rift junction (Underhill & Partington, 1993) and strongly influenced the subsequent rift-related structuring of the basin (Errat *et al.*, 1999). Greater erosion on the unconformity surface occurs towards the centre of the dome. Detailed study by Davies *et al.* (1999) based on regional mapping across the triple rift junction area identified this Mid-Cimmerian Unconformity as intra-Aalenian in age, although intra-Callovian and intra-middle Oxfordian age unconformities were also recognised. The intra-Callovian unconformity is developed between the Pentland Formation and the overlying Heather Shales or equivalent age Fulmar Formation deposits.

Broadly, the Upper Jurassic sandstones of the Central North Sea can be characterised as belonging to one of two depositional types: (1) shoreface deposits and (2) turbidite deposits. The criteria for recognising these different types of deposit are discussed later. Both types of sandstone can pass laterally into offshore shales. Coastal plain coal-bearing horizons and lagoonal shales are locally identified, but are generally rare.

In the Central North Sea basin, Upper Jurassic shoreface sandstones can be locally very thick (up to 300 m or more) while the turbidite sandstones are generally thinner, forming units where the gross thickness is rarely more than 100 m. For both sandstone types, prediction of reservoir presence, thickness and quality often proves difficult because of the complex underlying structural controls on accommodation and facies development. These structural controls have a variety of origins and will be expanded upon in more detail in later sections. Reservoir thickness may also reflect variable preservation due to intraformational or post-depositional truncation. In much of the region discussed, the sandstones are at a depth of 4 km to 5 km so that seismic resolution is commonly poor and such truncations are difficult to observe seismically.

This paper has three main objectives: (1) to show examples of the tectonic context for converting areas of former deposition into areas of erosion and to propose a simple scheme for classifying three recognised types of tectono-sedimentary setting in which this occurs; (2) to examine the basis on which major erosion surfaces can be identified in sandstones of shallow marine Oxfordian age by focusing on the importance of *Rhaxella* sponge spicule-bearing sandstones as correlateable units; and (3) to examine the products of erosion in the form of turbidite sandstones.

STRUCTURAL FRAMEWORK

The stratigraphic complexity of the Upper Jurassic sandstones in the Central Graben is well known and primarily a function of the varied structural controls on their deposition. The structural evolution of the Central North Sea Graben is the subject of an extensive literature with many studies based on 3D seismic data presented in Parker (1993) and Fleet & Boldy (1999). The Central Graben can be regarded as a series of N–S oriented sub-basins, progressively offset along Tornquist basement lineaments (Errat *et al.* 1999, Fraser *et al.*, 2002). The development of extension along normal fault systems as seen in the Viking Graben area (e.g. Rattey & Hayward, 1993) also occurred in the Central North Sea. However, the variety and complexity of tectonic structures was greater due to superposition of the Central North Sea Basin above the older North Permian Basin, which displays a thick development of evaporitic halite deposits, originally up to 2 km thick in the basin centre (Stewart & Clark, 1999).

A variety of views exist on the structural importance of salt in the area. Eggink *et al.* (1996) consider that salt withdrawal played a role in the structural development of the area but argue that the movement of the salt was, in many cases, triggered by reactivation of deeper fault zones and, as such, was passively deformed during the structural development. Other authors (e.g. Erratt, 1993; Helgeson, 1999) were more interested in demonstrating the importance of Zechstein salt in the mechanical decoupling between basement Rotliegend fault blocks and the overlying Triassic. They showed examples of the relationship between faulting and halokinesis that focused on the interaction of basement tectonics and pod/interpod complexes. In the view of these authors, a pure extensional setting existed in the Central Graben with structures in the Triassic-Jurassic carapace influenced by extension and halokinesis. An alternative interpretation is that an oblique-slip regime had existed in the Central Graben (Bartholomew *et al.*, 1993; Sears *et al.*, 1993) with the local development of transpressional and transtensional domains.

The presence of two clear extensional fault trends in the Central Graben aligned NNW–SSE and NW–SE (Fig. 1) is interpreted by Erratt (1999) and Davies *et al.* (1999) to reflect a polyphase Jurassic extension history in which the kinematic axes were aligned with an initial W–E extension during the Bajocian and Bathonian and a later NW–SE extension during the Kimmeridgian to Volgian. The oblique-slip model places more emphasis on the role of basement structures and their continued influence on the structural style of the Jurassic to explain the same trends (Bartholomew, 1993).

Where the Zechstein salt is thick, salt tectonics and Triassic minibasin development can very

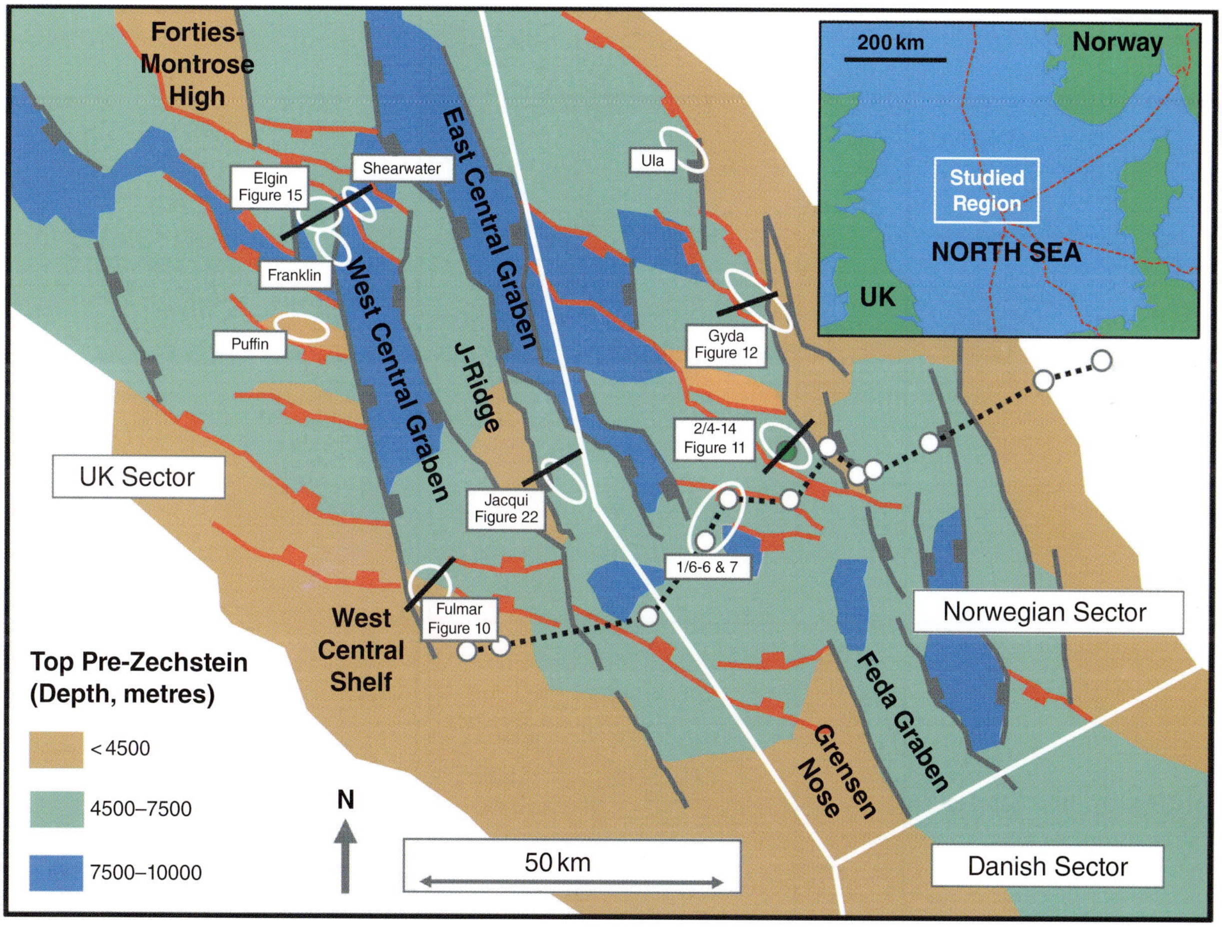

Fig. 1. Top Pre-Zechstein simplified depth map showing location of major faults that have been reactivated by Upper Jurassic rifting. Early developed NNW–SSE aligned extensional faults are shown in black whilst later developed NW–SE aligned extensional faults are shown in red. The location of a geological structural section shown in Fig. 2 is indicated by the black dotted line between white well locations. The locations of several geological sections discussed in this paper are also indicated on the map with their associated figure numbers.

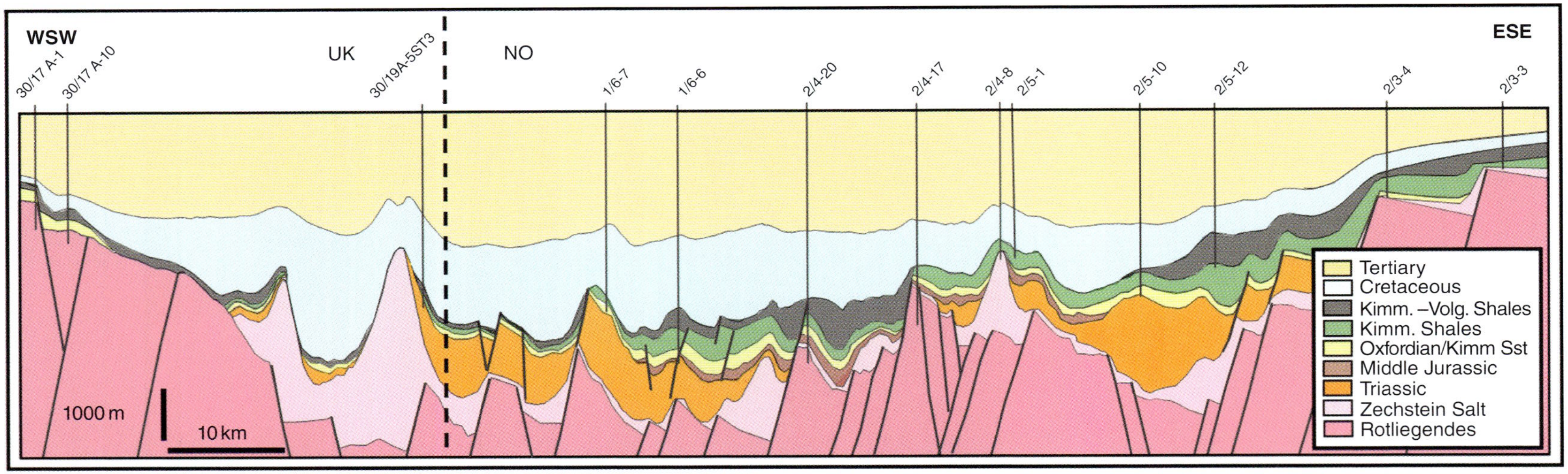

Fig. 2. Regional structural section across the Central Graben showing some areas where salt is the dominant control on later structural development and sedimentation (e.g. near the border region) and other areas where the salt is thinner and Upper Jurassic sedimentation is directly influenced by the movement of Rotliegend faults (e.g. 2/4-20 well area). The geographical location of this geological section is indicated on Fig. 1.

often become an important controlling factor on Upper Jurassic deposition (Vendeville & Jackson, 1992a/b; Vendeville & Guerin, 1998; Guerin *et al.*, 1998), notably over salt highs or in areas of inter-pod salt wall collapse (Fig. 2). Zechstein salt, where present, had two main influences: (1) it controlled the deposition of the Triassic with the formation of Triassic salt minibasin or 'pod' structures (Smith *et al.*, 1993; Stewart *et al.*, 1999; Stewart & Clark, 1999; Stewart, 2007; Jackson *et al.*, 2010); and (2) it decoupled the movement of basement Rotliegend faults from that of overlying faults. Thus, the controls on Upper Jurassic deposition are a hybrid between normal extensional tectonism as seen in the Northern North Sea and halokinesis.

STRATIGRAPHIC FRAMEWORK

The lithostratigraphic nomenclature of the Upper Jurassic is quite complex and has not been used in a consistent manner across the basin. There are several reasons for this: (1) the transgressive and diachronous character of the Oxfordian-Volgian sandstones; (2) the relative scarcity/unreliability of the biostratigraphic dating, commonly made more difficult by HPHT conditions which degrade the microfossils; (3) the naming of sandstones according to the field in which they were drilled, rather than to a regional reference and (4) the geographical location of the Central Graben basin straddling the UK/Norway/Danish borders, leading to three sets of nomenclature for the same formations.

In this study, we prefer to make use of the genetic stratigraphy for the Jurassic introduced by Partington *et al.* (1993a, b) which is well known and has been used in the description of several major fields such as Fulmar (Kuhn *et al.*, 2003), Erskine (Coward, 2003), Gyda (Partington *et al.*, 1993a) and Elgin-Franklin (Lasocki *et al.*, 1999). The lithostratigraphic naming is placed within the context of the biostratigraphic division (Fig. 3) in the same manner as Partington *et al.* (1993b).

Fig. 3 provides a summary of the presence of shoreface and turbidite sandstones in various fields throughout the basin and biostratigraphically dated maximum flooding surfaces (MFS) within the shales above and below these sandstones. Certain MFSs are described as 'Tectonically Enhanced' (TEMFS), for example the Eudoxus J63 TEMFS and the Fittoni J71 TEMFS. These TEMFSs are developed following the drowning of fault block footwalls and are said to have caused a switching off of deposition of coarse clastics (i.e. shoreface sandstone deposits) in both the basin centre and marginal areas due to major reorganisation of both basinal and shelfal palaeogeographies (Partington *et al.*, 1993a). Regional synthesis work on sequence definition such as that carried out by Fraser *et al.* (2002) rely on earlier biostratigraphic studies (e.g. Partington *et al.*, 1993a; Duxbury *et al.*, 1999) to define genetic, tectonostratigraphic sequences bounded by MFSs that are used as a basis for Upper Jurassic palaeogeographic mapping.

The aim of biostratigraphic dating is to identify regionally developed sedimentary packages of coeval deposits which can be grouped into genetically related depositional sequences (Partington *et al.*, 1993a, b; Jeremiah & Nicholson, 1999). The conceptual framework of sequence stratigraphy (Van Wagoner *et al.*, 1988; Van Wagoner *et al.*, 1990) aims to explain the development of these depositional sequences in terms of relative sea-level changes. If such sea-level fluctuations were entirely of eustatic origin, high-stand and low-stand events would be very precise time markers. However, rifting and salt movement were both very active in the Central Graben and this has led to both regional and local tectonic overprints, the effects of which cannot be separated from eustatic variations.

More recently, the concept of synchronous sequence stratigraphic surface development around the basin has been a contentious issue. The relatively comforting statement of Fraser *et al.* (2002) that: "Whatever their main controls, observed sequence boundaries and maximum flooding surfaces (MFSs) are still probably isochronous within the relatively small area represented by the North Sea Basin" are challenged by the work of biostratigraphers who consider that "rifting-induced tectonic events, produced coeval flooding and uplift" (Jeremiah & Nicholson, 1999). A number of studies have placed sequence stratigraphic development within a context of varying accommodation space creation due to extension of normal faults and deal with some of the challenges of applying sequence stratigraphy in a tectonically active setting (Howell & Flint, 1996; Ravnås & Steel, 1998; Ravnås *et al.*, 2000).

Biostratigraphic dating of the more deeply buried Upper Jurassic sandstones has a poor resolution. The sandstones of the Elgin and Franklin

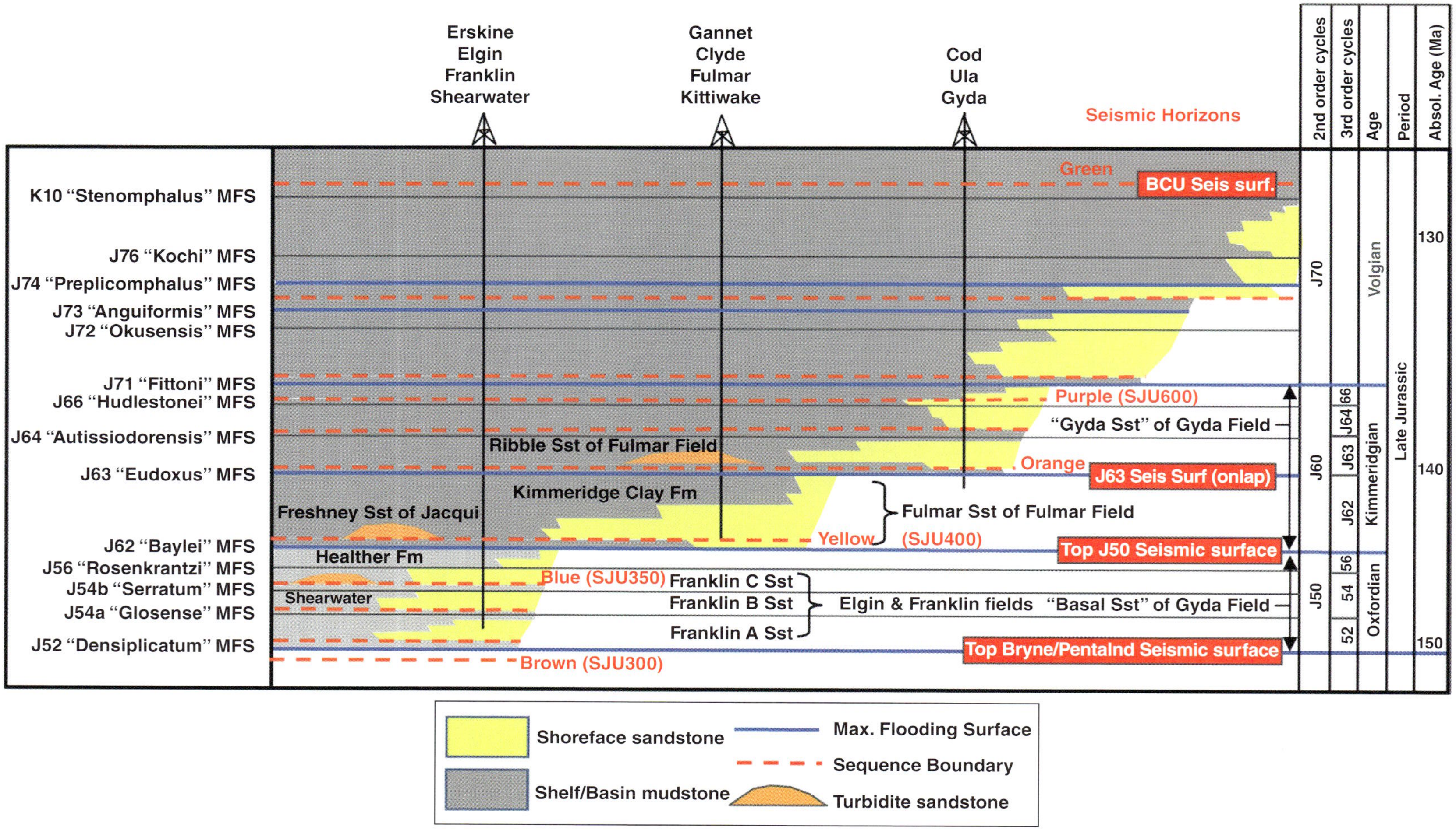

Fig. 3. Central North Sea biostratigraphic scheme showing schematic long term retrogradational stratigraphy of sequence sets J50 to J70. This figure has been adapted after Partington *et al.* (1993b), retaining only data from the Central Graben. Sequence boundary locations proposed by Partington *et al.* are indicated with names based on the colour scheme of Donovan *et al.* (1993) and the SJU sequence boundary naming scheme of Veldkamp *et al.* (1996). Relationship of named sequence boundaries to named MFS follows that given by Veldkamp *et al.* (1996). Brown, Yellow and Green sequence boundaries were not shown on the original figure of Partington *et al.* from which this figure is adapted. Dating of the Fulmar Sst of the Fulmar Field is from Kuhn *et al.* (2003). Dating of the Gyda Sandstone of the Fulmar Field is from Partington *et al.* (1993a) and dating of the Franklin Sst is from Lasocki *et al.* (1999) and subsequent studies. Seismic horizons interpreted by this study are also indicated. The turbidite sandstone marked 'Shearwater' is the turbidite sandstone lying immediately above shoreface sandstones in the 22/30b-11 well (as described by Jeremiah and Nicholson, 1999).

fields, for example, have a considerable thickness (up to 400 m) and many marine flooding events deposited intervals of offshore claystones. However, the latter are poorly constrained biostratigraphically, despite considerable core acquisition and micropalaeontological analysis of cutting samples from many wells. The reasons for the uncertainty on stratigraphic dating are well explained by Price *et al.* (1993). The main problem is that below around 4300 m, the dinocysts (palynomorphs) on which Upper Jurassic stratigraphy is typically based are destroyed by thermal maturation (at temperatures of 150 °C to 200 °C). The micropalaeontological scheme in this setting therefore relies on foraminifera (agglutinating and calcareous benthonic types), radiolaria and sponge spicules, but this is relatively undeveloped and facies dependent. MFSs have typically been named after the ammonite zone in which they are assumed to occur (Partington *et al.*, 1993a; b). However, ammonite remains are only rarely recovered in the subsurface and some uncertainty about the precise tie of dinocyst and ammonite biozonations still exists. Some studies have proposed that MFSs should be named after dinocyst zonal species (Duxbury *et al.*, 1999; Jeremiah & Nicholson, 1999). In this study, we follow the naming approach of Partington *et al.* (1993a, b).

Central to discussion of stratigraphic complexity and structural control on deposition of Upper Jurassic sandstones is the development of erosion surfaces of variable extent. Whilst the scheme of Partington *et al.* (1993b) is said to represent a genetic stratigraphy based on the recognition of maximum flooding surfaces (Galloway, 1989), it also makes reference to the work of Van Wagoner *et al.* (1990) and includes eight sequence boundaries with associated incision in the Oxfordian to Volgian age succession (Fig. 3). The evidence for these incisions is not presented or discussed by Partington *et al.* (1993b) and appears to be conceptual. Other studies of the same period (Price *et al.*, 1993) emphasise that, while they would have liked to apply a sequence stratigraphic approach in which the sequence subdivision depends on the identification of subaerial erosion surfaces traced basinwards, lowstands systems tracts are not readily identifiable and no unconformities related to sequence boundaries are described apart from the basal Mid-Cimmerian unconformity.

Donovan *et al.* (1993) in their study of UK Central Graben Quad 21, paid less attention to the dating of MFSs and more to the identification of seven sequence boundaries produced by relative falls in sea-level within the Late Jurassic marine succession. The most significant among these sequence boundaries was the Yellow SB (142 Ma, Kimmeridgian age). This sequence boundary, observed as a burrowed firm ground in core, was interpreted to result in major erosional truncation of previously deposited shoreface sequences (shown schematically as SB2 in Fig. 4). Howell & Flint (1996) and Carruthers *et al.* (1996) developed other criteria for sequence boundary recognition including the interpretation of an incised valley and tidal deposit development. Studies by Veldkamp *et al.* (1996) and Jeremiah & Nicholson (1999) continued to highlight the importance of depositional sequence boundaries, adapting and translating the naming system of Donovan *et al.* (1993) into a SJU nomenclature (Upper Jurassic sequences SJU350, SJU400 etc.). The relationship of named sequence boundaries to the named MFSs of Partington *et al.* (1993b) and the colour code scheme of Donovan *et al.* (1993) is given by Veldkamp *et al.* (1996) and has been included for reference in Fig. 3.

A recent study of Sansom (2010) presents a wide ranging review of the Upper Jurassic in the UK sector of the Central Graben. The sequence stratigraphic approach used follows the nomenclature scheme of Partington *et al.* (1993a, b) rather than that of Veldkamp *et al.* (1996) and recognises Transgressive-Regressive cycles bounded by

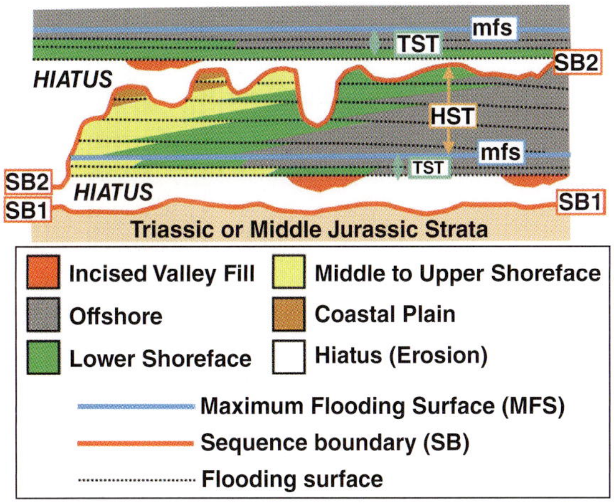

Fig. 4. Stylistic representation of intraformational unconformities developed within the Late Jurassic coastal to marine succession of UK Quad 21 as pictured by Donovan *et al.* (1993). TST: Transgressive Systems Tract; HST: Highstand Systems Tract.

sequence boundaries or maximum regression surfaces (MRS). The study focuses particularly on the Base Volgian-Ryazanian Unconformity (BVRU), an erosive unconformity developed on structural highs and considered to have been generated by a major pulse of rifting, Triassic pod rotation and salt redistribution. The BVRU separates an underlying succession (Callovian–Kimmeridgian age megasequence; sequences J44 to J63) characterised by conformable deposition of widespread shoreface sandstone and relatively thick packages of coeval turbidite mass flow sandstones from an overlying succession (Volgian–Ryazanian age megasequence; sequences J64 to K10) of more localised shoreface deposition, often condensed or truncated and with thin coeval turbidite mass flow sandstones. Sansom (2010) consider that, at the time the BVRU was developed, a number of

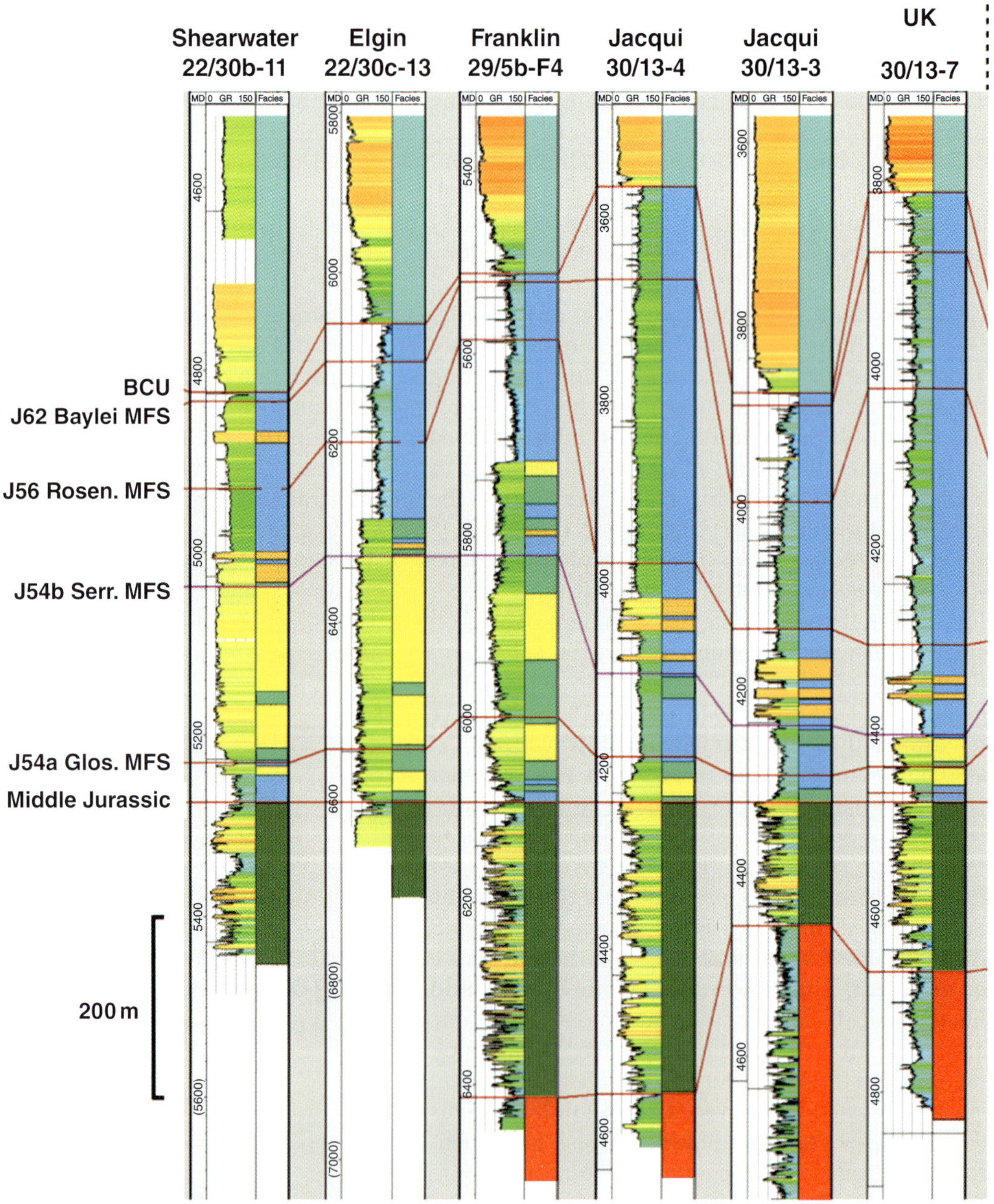

Fig. 5. Regional correlation showing wells referred to by this study. Gamma Ray log colour display according to values: warm colours (orange) more sandy, cool colours (blue) more shaley. The geographical location of this correlation section is indicated on Fig. 1. MD: Measured Depth (m); GR: Gamma Ray log.

basinal highs may have become prominent: e.g. Forties-Montrose High and the J-Ridge, converting areas of former shallow marine deposition to exposed highs that served as sediment source areas for adjacent basins.

This current study considers that a genetic stratigraphic approach is most appropriate for describing the Upper Jurassic of the Central Graben and follows the J-Sequence system developed by Partington *et al.* (1993a, b). Many biostratigraphic studies have been reviewed from throughout the UK Central Graben and locally, chemostratigraphic analysis based on cuttings has also been tested. A regional view of stratigraphic correlation using MFSs is given in Fig. 5. This figure shows how considerably the thickness of

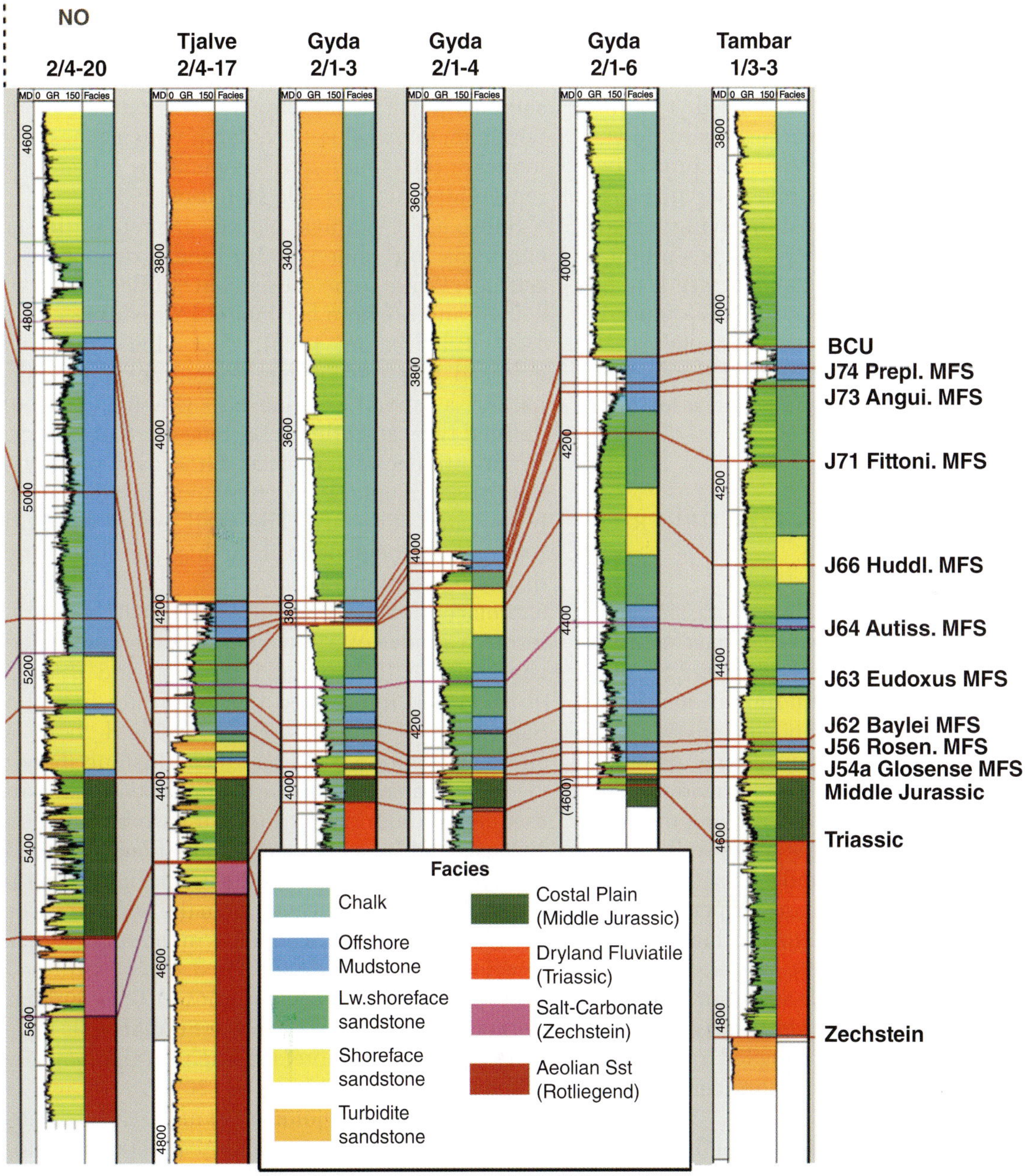

Fig. 5. (*Continued*)

major depositional successions varies across the basin, reflecting the changing geographical localisation of structurally-controlled depocentres through time. At the same time, this study relies on geological and geophysical integration, for example: identifying sections in wells that have been removed by faults and correctly calibrating seismic data to geological interpretations at wells.

Locally significant erosive surfaces, such as that developed across the Elgin-Franklin Field (see later discussion) are considered to be controlled by local tectonics rather than eustasy. This implies that regional correlation of such erosive surfaces throughout the basin should be resisted. The aim of recognising such erosion surfaces (and down-dip basinward shifts of facies) is to be predictive about the possibility of sandstone reservoirs produced by erosion and re-deposition lying further down depositional dip. If the erosion is a product of local tectonism, this predictability will also be purely local. Thus the emphasis of this paper is placed on recognising erosion surfaces where they exist within or above the shoreface sandstones and considering their development in a tectono-sedimentary context.

DEPOSITIONAL MODELS FOR UPPER JURASSIC SANDSTONES

A large number of studies of Upper Jurassic sandstones in the Central North Sea basin have contributed to the understanding of these deposits. Studies of wells and cores have their limitations when reconstructing depositional models and resulting sand body geometries or depositional patterns. Outcrop analogues for the Fulmar Formation have been sought, one of the most cited being the well exposed Liassic outcrop of the Wessex Basin in southern England. These sandstones appear as massive tabular units with many cemented shell beds of large lateral extent (up to 2 km) in cliff outcrops (Kantorowicz *et al.*, 1987). Low angle, metre-scale undulating bedforms have also been described (Pickering, 1995). More recent study has shown the development of large-scale, shingled clinoforms observed in 3D seismic data from the nearby Wytch Farm oilfield (Morris *et al.*, 2006).

Detailed sedimentological studies of the Fulmar Formation are mainly based on the analysis of closely spaced wells in large fields such as Fulmar (Johnson *et al.*, 1986), Ula (Spencer *et al*, 1986), Gannet (Armstrong *et al.*, 1987) or Elgin-Franklin (Lasocki *et al.*, 1999). These studies have sometimes been equivocal about the depositional origin of the sandstones, pointing to the fact that primary sedimentary structures, which might have indicated a depositional process origin, have almost always been entirely obliterated by bioturbation. Johnson *et al.* (1986) considered that storm-induced currents were responsible for depositing a large part of the succession, while gravity-driven turbidity flows deposited a minor component (roughly 10%). These processes were interpreted to have operated in either of two different depositional settings: (1) a shoreface setting and (2) a shelf setting. The authors were not explicit about which model was preferred or at which times one model or the other might be operating; rather they represented two possible scenarios.

Subsequent studies placed greater or lesser emphasis on shoreface or shelf depositional settings. While some subsequent studies preferred a shelf and offshore bar model (Armstrong *et al.*, 1987; Price *et al.*, 1993), a growing consensus emerged that the Fulmar Formation was dominantly a shoreface deposit. Gowland (1996) defined two categories of shoreface: a high energy, storm-influenced shoreface with some preservation of sedimentary structures and a low-energy, bioturbated shoreface. The shelf model was still retained, although deposits of this type show little difference sedimentologically to the low energy shoreface model. Donovan *et al.* (1993) also preferred a model of a low-energy wave-dominated shoreline for the Fulmar Formation and Lasocki *et al.* (1999) interpreted the Oxfordian Franklin Sandstone of the Elgin-Franklin Field to be a shoreface deposit. The Franklin Sandstone shows: (1) clear coarsening-upwards, progradational parasequences bounded at top and base by flooding surfaces and (2) a clear and gradual facies change from north to south across the Franklin field passing from shoreface succession in the north towards interfingering offshore deposits in the south where a major, salt withdrawal-related, depocentre had been identified. Recent studies of the Fulmar Formation type sections in the Fulmar field have firmly ascribed a shoreface origin to the majority of the sandstones (Kuhn *et al.*, 2003).

The suggestion of Johnson *et al.* (1986) that turbidity current processes might be responsible for some Oxfordian age sandstones of the Fulmar Field has been followed by other reports of turbidite deposits from elsewhere in the Central Graben, notably studies by Robinson (1990), Howell &

Flint (1996), Howell *et al.* (1996) and Jeremiah & Nicholson (1999). Howell & Flint (1996) focus on the role of tectonism in a sequence stratigraphic context and identify a variety of depositional responses to eustatically controlled sea-level fall when comparing deposition within a single chronostratigraphic interval in the hangingwall and footwall succession. In particular, they interpret the development of an incised valley (Curlew Member) with a tidally-influenced sedimentary fill, incision over an uplifted footwall crest (Puffin fault block, location Fig. 1) and deposition of contemporaneous turbidite sandstones successions (up to 50 m thick) in the adjacent footwall basin.

Shoreface depositional model

Thick, Oxfordian age, sandstones of the Central Graben are now most commonly referred to as shoreface, lower shoreface and offshore deposits. Deposition is interpreted to have occurred by the progradation of shoreface sandstone facies into a mud-dominated basin. Fulmar Formation deposits typically occur in the form of large-scale coarsening-upward successions (10 m to 50 m thick) that evolve from claystone or siltstone into fine-grained to medium-grained sandstone. Such a succession can define a flooding surface bounded parasequence that reflects shoreline progradation during a period of fairly constant (or gently rising) relative sea-level. The coarsening upwards character of parasequences reflects increasing wave energy when passing from the offshore to the nearshore depositional setting (Fig. 6).

The shoreface deposits of the Central Graben typically comprise highly bioturbated sandstones and argillaceous sandstones. The sandstones are normally highly reworked by burrowing organisms that produced a wide range of ichnofabric types (Taylor & Gawthorpe, 1993; Fig. 6) and a sedimentary fabric to the rock in which it is usually difficult to identify primary sedimentary structures (Fig. 7). A number of facies types (here termed S1 to S5) are typically identified within prograding shoreface successions:

Facies S1: Mudstone – consisting of silty claystone or argillaceous siltstone, sometimes calcareous, ranging from well laminated to highly bioturbated. Bioturbation when present consists of *Terebellina*, *Chondrites* and *Helminthopsis horizontalis*. This facies is interpreted as an offshore shelf deposit.

Facies S2: Strongly bioturbated sandy mudstone to very muddy sandstones. This facies is strongly bioturbated and exhibits *Chondrites*, *Helminthopsis*, *Phycosiphon*, *Planolites* and *Terebellina* trace fossils. This facies is interpreted to reflect the shoreface to offshore transition zone.

Facies S3: Bioturbated very fine grained sandstones with abundant horizontal burrows. This facies exhibits *Palaeophycus*, *Teichichnus*, *Planolites*, *Terebellina*, *Asterosoma* and *Macaronichnus* traces. This facies is interpreted to reflect the lower shoreface depositional setting.

Facies S4: Strongly bioturbated fine-grained to medium-grained sandstones, with occasional relict stratification. Low angle discontinuities and scour-and-fill structures are occasionally observed. This facies exhibits *Ophiomorpha* and *Skolithos* traces and is interpreted to be deposited in an upper shoreface setting.

Facies S5: Stratified sandstone – consisting of sandstones with cross-stratification and planar-stratification and carbonaceous-rich mudstone, ripple stratified sandstones and thin coal beds. Bioturbation is uncommon and consists of *Ophiomorpha* traces. This facies has a low preservation potential in the Central Graben, probably due to transgressive shoreface erosion, and it is rarely observed. It is interpreted as a foreshore to coastal plain deposit.

Sandstone deposits produced by stacked phases of parasequence progradation generate thick sandstone units that may often exceed 300 m in thickness (e.g. Stockbridge & Gray, 1991; Lasocki *et al.*, 1999). Parasequences may be stacked progradationally, aggradationally or retrogradationally, thereby indicating the interplay of sediment supply and relative sea-level change. Sandbody development at the basin margins occurred simultaneously with deposition of more basinal sandy turbidite facies (see later) and siltstone or mudstone of the Heather and Kimmeridge Clay formations.

In a regional study of the Fulmar Formation, Gowland (1996) made a further refinement of the shoreface model by defining 12 facies that were grouped into 6 facies associations and used to construct three broad depositional models: (1) Model 1: storm-influenced shoreface; (2) Model 2: bioturbated shoreface and (3) Model 3: a speculative bioturbated shelf model. Model 1 shorefaces are said to have well-preserved event beds that occur seaward of the upper shoreface, whereas

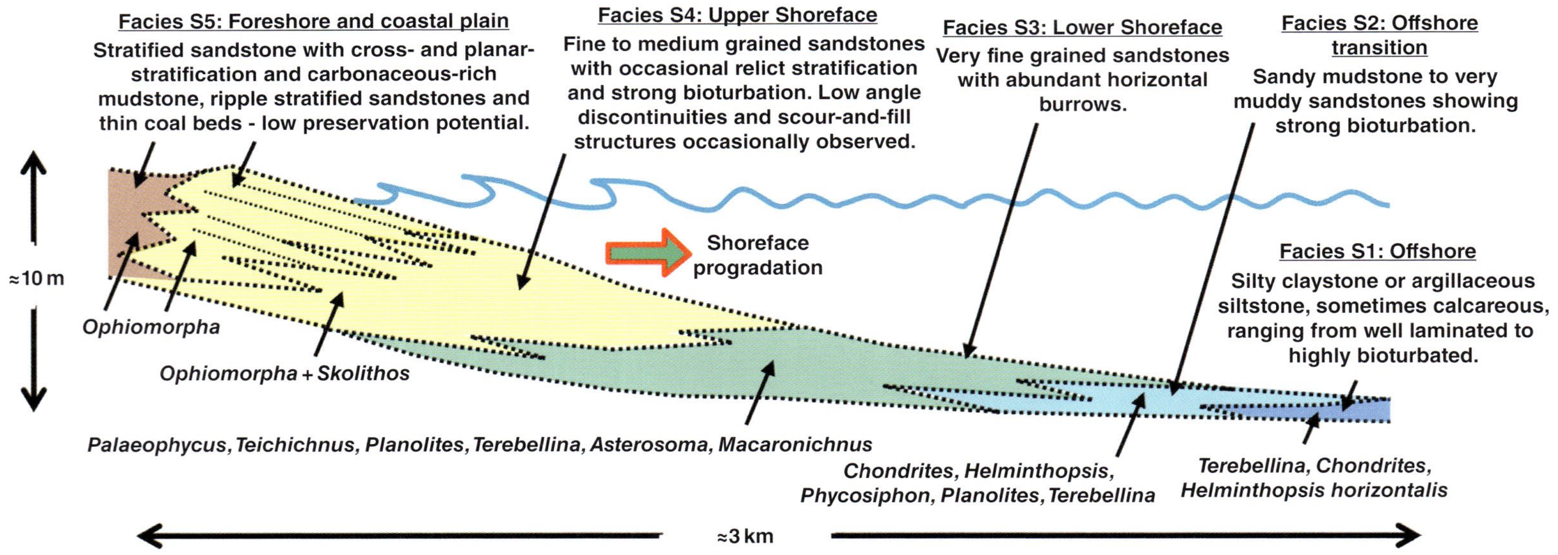

Fig. 6. Shoreface progradation model (after Gowland, 1996).

Fig. 7. Comparative examples of shoreface sandstone and turbidite sandstone from the Upper Jurassic of the UK Central North Sea: (A) Facies S3: bioturbated fine grained sandstone, 22/30b-15Z, 5710.7 mRKB; (B) Facies T4: planar laminated sandstone, 22/30b-15Z, 5652.8 m RKB; (C) Facies T3: fine grained sandstone and mudstone heterolithic couplets, 30/16A-31, 4006.4 mRKB; (D) Facies T4: fine grained sandstone displaying both planar stratification and climbing ripple cross-lamination (marked 'c'), Franklin Sands, 29/4d-4, 5700.85 mRKB; (E) Facies T2: pin-striped micaceous siltstone developed below a massive fine grained sandstone unit exhibiting loading at the base, 22/30b-11, 5017.8 mRKB.

Model 2 shorefaces are characterised by intense bioturbation of the lower shoreface and adjacent shelf. Model 3 shelves display no clear evidence of shoreline connection and exhibit abundant siliceous sponge spicules together with other open marine indicators such as belemnites and rare ammonites. The bulk of deposition is considered to have occurred in water depths of less than 20 m.

Turbidite depositional model

Turbidite facies are considered by this study to have special significance for understanding Upper Jurassic depositional history in the Central Graben due to: (1) their relative rarity and (2) their poorly understood mechanism of genesis. The possibility of turbidite development was first mentioned by Johnson *et al.* (1986) in relation to deposition of the Kimmeridgian Ribble Sandstone Member of the Fulmar Field for which there exists an extensive well and core database. Several studies (Robinson, 1990; Clelland *et al.*, 1993; Kuhn *et al.*, 2003) have documented the sedimentology and stratigraphy of the Ribble Member based on extensive well and core data. Robinson (1990) made a detailed study and described five turbidite facies types, here referred to as T1 to T5:

Facies T1: Fissile laminated, non-bioturbated mudrocks – consisting of uniform, dark grey to black laminated mudrocks (Fig. 7F) containing carbonaceous material, pyrite and macrofossils (ammonites and bivalves). A high gamma ray response of up to 140 API units is characteristic. The lack of bioturbation suggests a hemipelagic origin under anoxic conditions. This is the characteristic Kimmeridge Shale facies.

Facies T2: Silt laminated mudrocks – consisting of dark, laminated mudrocks with regular thin (mm-scale to cm-scale) silt laminae (Fig. 7E) that correspond well to the T_1 to T_4 fine grained turbidite divisions of Stow & Shanmugam (1980). Laminae are sometimes disrupted by bioturbation. This facies is interpreted as a distal or lobe fringing turbidite deposit.

Facies T3: Heterolithic sand-mud couplets – consisting of fine to medium-grained sandstones with a wide-range of bed thickness (cm to dm thickness). The sandstones are often heterolithically interbedded with T2 facies and can form fining upwards packages up to 2 m thick. A hierarchy of structures resembling Bouma's B-C-D-E divisions (Bouma, 1962) is commonly developed; however, laminae are sometimes disrupted by bioturbation. Sandstone beds, often with scattered siltstone clasts, show a sharp loaded basal contact with underlying mudstones. Sandstones are massive and structureless or with dish and pillar or flame structures suggesting dewatering. Indistinct horizontal lamination is present and thin beds (<3 cm thick) display low amplitude ripples with mud drape development (Fig. 7C). This facies is interpreted as the deposit of mixed-load turbidity currents. Gowland (1996) postulates an origin either by storm events or moderate density turbidity currents for this facies type.

Facies T4: Massive amalgamated sandstone – consisting of largely massive, medium-grained sandstone forming units typically 1 m to 10 m thick. Graded bedding is mainly absent. Carbonaceous material is scattered throughout and becomes concentrated into horizontal laminae towards the top of sandstone beds (Fig. 7B). Climbing ripples are sometimes observed (Fig. 7D). The top of sandstone units can be bioturbated by *Chondrites*. The sandstones are laterally continuous and mapping shows a lateral extension of up to 6 km^2. This facies is interpreted as the deposit of high-density turbidity currents. A similar interpretation for this type of facies is postulated by Gowland (1996). Mapping suggests that sandstone units of this type were deposited as lobes or sand sheets in depositional lows.

Facies T5: Bioturbated silty sandstone (called 'Clyde Member') – consisting of massive, fine to medium-grained, poorly sorted and extensively bioturbated grey-green sandstones. Thin graded sandstone beds occur occasionally and clay-lined *Zoophycus* burrows are present throughout. This facies is interpreted as a shallow shelf deposit with occasional storm influence. Biostratigraphic dating suggests that this facies is laterally equivalent to Facies T4 over a distance of approx. 400 m.

Gowland (1996) describes one further facies type which might be attributed to mass flow processes. This is a matrix or clast-supported pebble sandstones with intraclasts of reworked Triassic, Middle Jurassic or phosphatic material. This deposit is attributed to either: (a) storm-generated flows and wave current winnowing associated with lowstand shoreline development and transgressive surfaces, or (b) high density turbidity currents or debris flows compatible with a deeper marine setting.

The major control on the morphology of the Ribble Sandstone is structural causing the thickness to range from 60 m to 2.5 m over less than 1 km distance (Robinson, 1990). This is a common feature of turbidite deposition throughout the Central Graben as confirmed by mapping of local and regional tectonic elements. Within the Ribble Sandstone Member and elsewhere in the Fulmar Formation of Fulmar Field, soft sediment deformation structures are widespread. Robinson (1990) considers it probable that the observed dewatering, fracturing and mudslides (within the Avon Shale) are the result of external shocks rather than sediment overloading. In addition to remobilising the turbidites, faulting, salt movement and periodic earthquakes are considered a possible trigger mechanism for turbidite flows to be initiated. In other areas, sand remobilisation features such as injectite beds and dykes can be observed. These sand injectites erode and transport claystone clasts that have been prised and ripped from the shales in which they are encased (e.g. Kimmeridgian Farsund Formation core of well NO-2/4-18R; Fig. 7F; well situated 1 km to the NE of well 2/4-14 shown on Fig. 1).

Other authors have considered these, or similar facies, in different ways. Jeremiah & Nicholson (1999), for example, identified sandstones of turbidite origin in several Central Graben wells (22/30b-11, 23/26b-15, 30/13-4, from the Erskine, Shearwater and Elgin-Franklin area (located at the four corners area of Quad 22, 23, 29 and 30). The sandstones were interpreted as being deposited by high density turbidites to form stacked amalgamated sandstone beds with planar lamination, occasional current ripples or symmetrical wave ripples and no bioturbation. Age equivalent sandstones of Franklin Field were equivocally interpreted as either turbidite sandstones or high energy foreshore/shoreface deposits (Lasocki *et al.*, 1997). The Late Oxfordian/early Kimmeridgian Freshney Sandstone Member in the Jacqui area (Fraser *et al.*, 2002) is a further example of turbidite sandstone units.

Previous studies have concluded that turbidite deposits such as the Ribble Member are the deposits of turbidite lobes directly sourced from adjacent structural highs (Robinson, 1990). Other studies have tried to show evidence that these lobes were sourced during lowstands by fluvial systems that delivered sediment through incised valleys (e.g. Howell & Flint, 1996; Carruthers *et al.*, 1996).

An alternative model previously proposed is that hyperpycnal flows (Mulder & Syvitski, 1995; Mulder *et al.*, 2003) brought sand-rich sediment from high sediment-load rivers. Forster (2005) used this model to explain the thick, sandstone-dominated succession (Oxfordian-Volgian age) of the Buzzard Field in the Outer Moray Firth (Doré & Robbins, 2005). These deposits consist of parallel laminated or massive sandstones similar to the Ribble Sandstone Member T4 facies and are interpreted to be sourced from an updip braided river system.

TECTONO-SEDIMENTARY SETTINGS A, B AND C

A schematic map (Fig. 8) indicates the global distribution of Oxfordian shoreface to basinal sandstones within the Central Graben. The extent of the coastal plain is largely inferred since most of these deposits were reworked by shoreface erosion during subsequent transgression. The map is adapted from that of Erratt (1999) and Fraser *et al.* (2002), with local updates. The broad shoreface

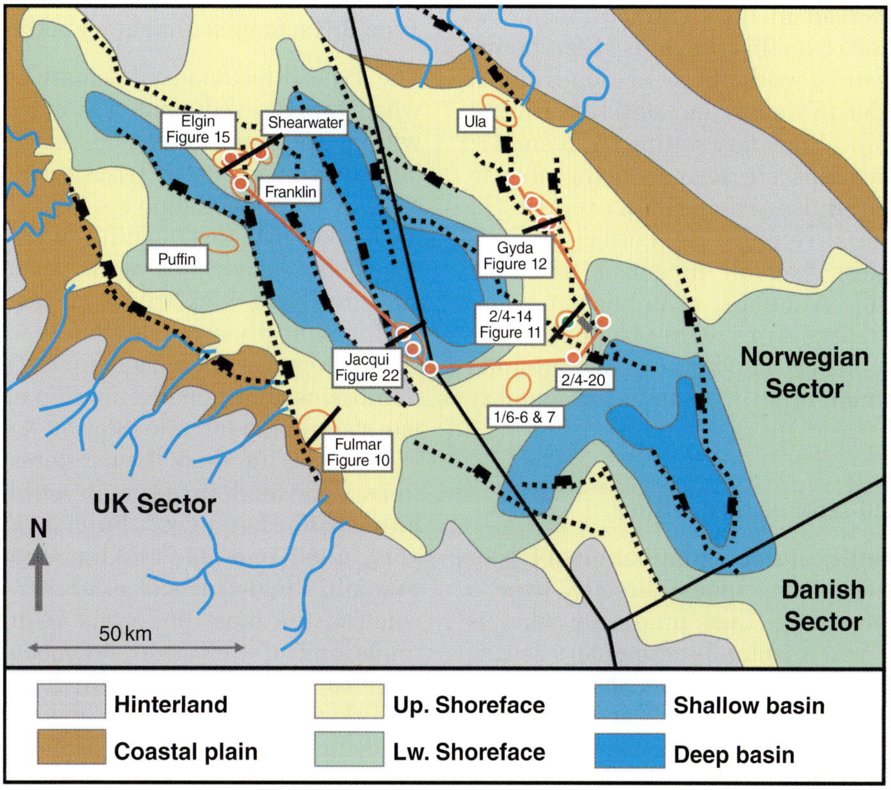

Fig. 8. Central Graben depositional setting during the Oxfordian. Facies belts are based on generalised models of Fraser *et al.* (2002) and Erratt *et al.* (1999) with local adaptions. Top Pre-Zechstein faults (as shown in Fig. 1) are superimposed to indicate the manner in which these faults are perceived to influence depositional trends. The figure is at the same scale as Fig. 1 with which it is intended to be compared. The location of a correlation section shown in Fig. 5 is indicated by red well locations, as are the location of several geological sections that are indicated on the map with their associated figure number.

Table 1. Type A, B and C tectono-sedimentary settings with a number of field examples.

Tectono-Sedimentary Setting	Example Fields / Discoveries
Type A: Hanging Wall Trap Door Setting	Flora Field (Hayward *et al.*, 2003) Angus Field (Hall, 1992; Spathopoulos *et al.*, 2000) NO-2/4-14 Discovery (Mjelde, 1991; Jones *et al.*, 2013)
Type B: Reactivated Pod Setting	Elgin-Franklin Field (Lasocki *et al.*, 1999) Fulmar Field (Johnson *et al.*, 1986) Clyde Field (Stevens & Wallis, 1991) Jacqui / Jessica (UK 30/13-3, UK 30/13-7)
Type C: Interpod Setting	Kittiwake Field (Wakefield *et al.*, 1993) Durward Field, UK 21/16 (Stewart *et al.*, 1999) Dauntless Field, UK 21/11 (Stewart *et al.*, 1999) Ula Field (Brown *et al.*, 1992) Gyda Field (Spencer *et al.*, 1986) Nemo discovery, NO-7/7-2

trends are supported by well data, whereas the extent of deep basin deposits remains sometimes conceptual because these regions are less calibrated by wells. Turbidite deposition is largely absent, apart from the shallow basin area around Jacqui and east of Puffin. As an introduction to the complex relationships between structure and sedimentation observed in the Central Graben, the following sections combine previous observations of structural setting with those of depositional controls in order to define and describe several types of tectono-sedimentary setting (A, B and C). Practical applications are seen in being able to identify different styles of deposition occurring in different structural settings, e.g. correct application of analogies; reservoir prognosis based on structural setting. A number of published examples of fields or discoveries within Upper Jurassic sandstones are listed in Table 1 and grouped according to their type of tectono-sedimentary setting.

Type A: Tectono-sedimentary setting

Where major faults caused accumulation of Upper Jurassic shoreface sandstones directly over a hangingwall trap door, they produced what is here termed a Type A tectono-sedimentary setting (Fig. 9). The approximate locations of a few major Rotliegend-offsetting faults are shown in Fig. 8, since they influenced Upper Jurassic deposition. The movement of major Rotliegend fault blocks, and salt migration, played a fundamental role in developing the landscape over which Middle Jurassic fluvial systems and transgressive Upper Jurassic shoreface sandstones were deposited. Pure trap door settings of this type are abundant and well documented in the northern Viking Graben area (e.g. Ravnås *et al.*, 2000), but are less common in the Central Graben where the presence of Zechstein salt and Triassic minibasin development generally creates a more complex structural setting.

Type B: Tectono-sedimentary setting

The Type B tectono-sedimentary setting occurred when Jurassic sediment was able to accumulate within the same minibasin structures that have already accommodated Triassic deposition. In such settings, accommodation space creation for Jurassic deposition remained significant and thick Jurassic sandstone successions were developed immediately overlying thick Triassic deposits. This type of setting typically occurs within the Central Graben rift where salt was typically up to 2 km thick. Here, minibasins continued to develop and ground at a late stage in the Jurassic (Stewart & Clark, 1999).

Examples of thick Upper Jurassic accumulations deposited in Type B setting include the Fulmar Formation of Fulmar Field (Johnson *et al.*, 1986) and the Franklin Sandstone of Elgin-Franklin Field (Lasocki *et al.*, 1999) which host some of the most important hydrocarbon accumulations of the Central Graben. The Fulmar (Fig. 10) and Elgin-Franklin fields also display reworking of previously deposited shoreface sandstones as turbidite sandstone to areas around the margins of the collapsing pod, as demonstrated in correlations of the Fulmar Field made by Kuhn *et al.* (2003). It should be noted that a Type B setting can evolve into a Type A setting. This is the case for the Fulmar Field where there exist elements of both settings.

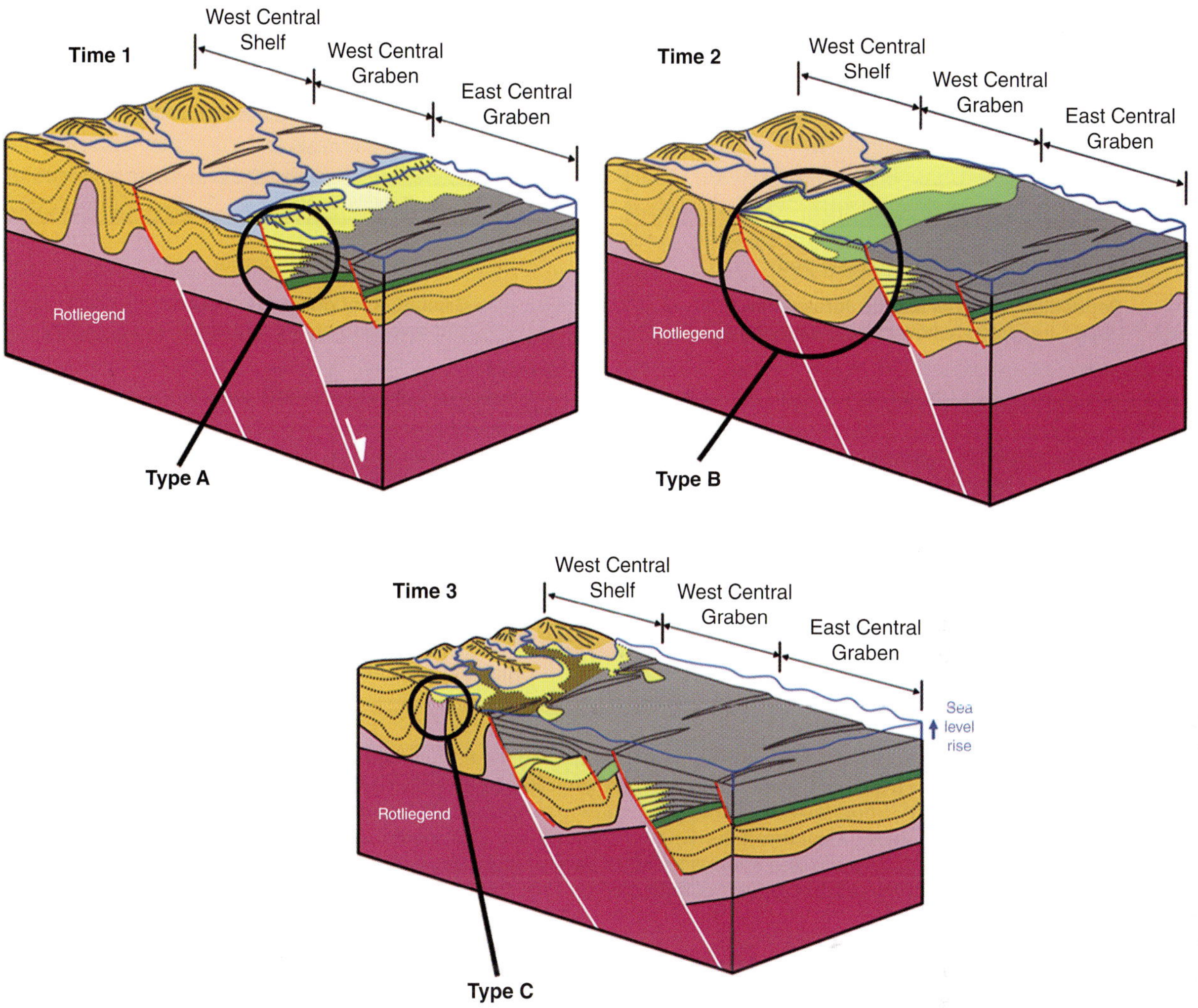

Fig. 9. Successive stages are illustrated in the development of Upper Jurassic sandstones highlighting successive tectono-sedimentary settings. At Time 1, Oxfordian rift related deposition occurs in the East Central Graben against the foot wall producing a 'Type A – Hanging Wall Trap Door Setting' for shallow marine and turbidite sandstone accumulation. At Time 2, the graben broadens as a result of ongoing extension and marine transgression causing a pod structure initiated in the Triassic to become reactivated with a fill of Upper Jurassic shoreface deposits, namely a 'Type B – Reactivated Pod Setting'. At Time 3, contemporaneous transgressive sandstone deposition takes place within drowned salt cored valleys producing a 'Type C – Interpod Setting' for shallow marine sandstone accumulation (adapted after Fraser *et al.*, 2002).

Type C: Tectono-sedimentary setting

A Type C tectono-sedimentary setting resulted from the marine flooding of former salt highs (swells, walls and diapirs) at the margins of Triassic minibasins (Goldsmith *et al.*, 2003; Vendeville & Jackson, 1992a, 1992b; Stewart & Clark, 1999). These small, Upper Jurassic, interpod basins often became infilled with littoral sandstones (Fig. 9). The basins originated as valleys where salt deflation has occurred (dissolution or withdrawal of salt) and can be subsequently enlarged by continued salt withdrawal or sediment loading. Supra-salt minibasins of this kind can form networks surrounding highs formed by grounded and inverted Triassic minibasins. The Upper Jurassic sandstone trends shown on Fig. 8 are influenced by the location of Triassic minibasins that are known to have served as sediment source terrains on both margins of the Central Graben (Stewart *et al.*, 1999).

Type C tectono-sedimentary settings typically occur over structurally high areas that correspond to relatively thin original Zechstein isopachs (Stewart & Clark, 1999). In such regions,

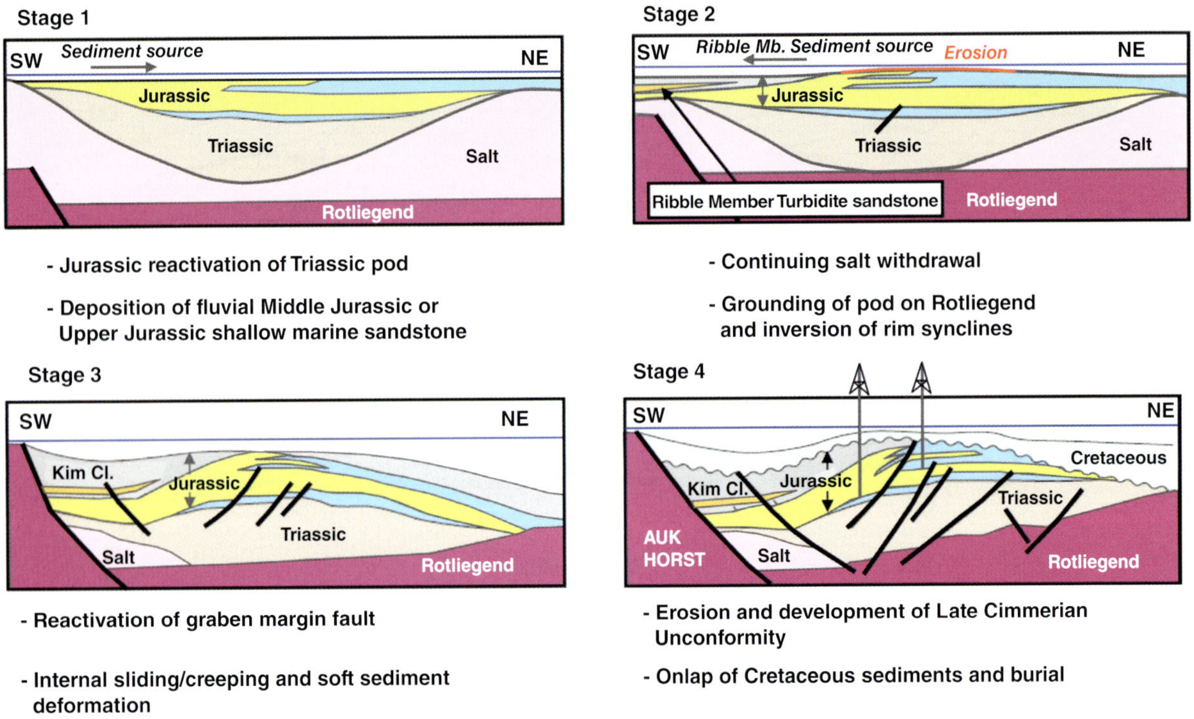

Fig. 10. Time step progression showing the development of a tectono-sedimentary setting of 'Type B – Reactivated Pod Setting'. The example comes from the Fulmar Field (adapted after Johnson *et al.*, 1986). Marine shales and transition zones siltstones are shown in blue. Shoreface sandstones are shown in yellow. Reworking of previously deposited shoreface sandstones can occur on top of the structure following pod grounding. This model is based on examination of correlations made by Kuhn *et al.* (2003). The reworked shoreface sandstones were re-deposited to the west as turbidites of the Ribble Sandstone Member. The location of this section is indicated on Figs 1 and 8.

the grounding of Triassic pods occurred relatively early. On the West Central Shelf for example (Fig. 1), grounding occurred before the Mid-Triassic (Fraser *et al.* 2002). The grounded pods were surrounded by salt highs which were differentially eroded leading to the development of salt cored valleys. These valleys were flooded during the Upper Jurassic after a long period of exposure, in the order of 80 Ma according to Rattey & Hayward (1993) and they became commonly infilled with Upper Jurassic littoral sandstones (Fig. 9). Examples of Type C tectono-sedimentary setting have been described from the Durward and Dauntless fields (Stewart *et al.*, 1999).

Regional structural controls on deposition

In addition to local structural controls such as those described above, there are also regional changes to consider due to the overall development of the Central Graben rift. Upper Jurassic extension took place on basement faults that may show local displacements of up to 3 km (Stewart, 2007). This implies rapid rates of extension and subsidence during the Late Jurassic and consequent important evolution of the location of sediment sinks. Rifting occurred initially in the most central parts of the basin such as the East Central Graben and the Feda Graben and extended gradually towards the margins of the basin. This led to major shifts in shoreface sandstone depocentres both in the UK sector (Fig. 9) and the Norwegian sector (e.g. stratigraphic evolution from well 2/4-20 to well 2/1-4 shown in Fig. 5).

CASE STUDIES

Case Study 1: NO-2/4-14 well area: Type A – Hangingwall trapdoor setting

Examples of the Type A hangingwall trapdoor tectono-sedimentary setting are abundant in

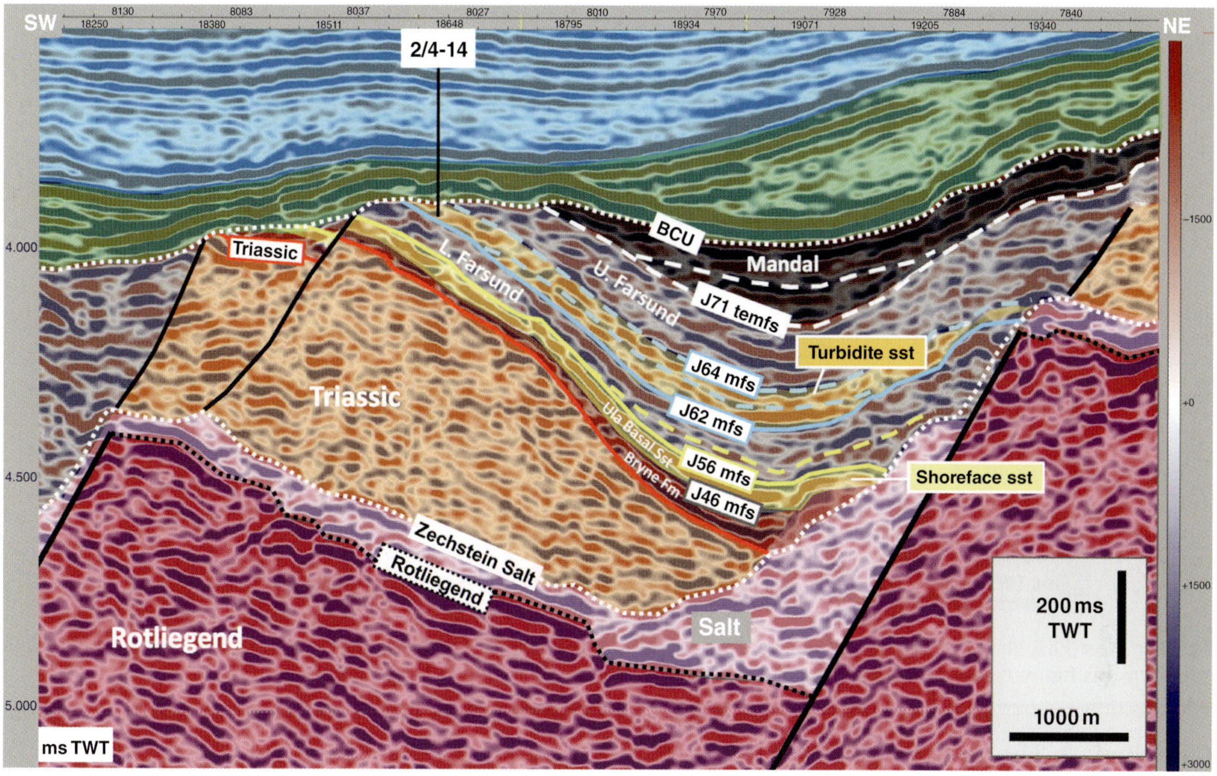

Fig. 11. Genetic sequences showing onlap associated with MFS in the vicinity of the 2/4-14 well. Onlapping strata mark renewed clastic influx in an overall shaley succession. The location of this section is indicated on Figs 1 and 8.

the Viking Graben area but less common in the Central Graben where the presence of Zechstein salt and Triassic pod development generally creates a more complex structural setting. However, where salt thickness is limited and Triassic pod grounding occurred relatively early (i.e. prior to Jurassic), the Jurassic deposits form a depositional wedge with a geometry that is primarily influenced by the movement of Rotliegend fault blocks (Fig. 11).

In Type A settings, Middle and Upper Jurassic deposits thin towards the crest of the fault block and genetic sequences defined by MFSs show stratal onlap on seismic sections. Such tectonically controlled onlaps have now been widely documented form the North Sea rift basin (e.g. Underhill, 1991). Within the Kimmeridgian succession, onlapping strata mark renewed clastic influx in an overall shaley succession. Onlaps are often associated with turbidite 'ponding' in structural lows. The J62 MFS in the example shown in Fig. 11, the King Lear discovery (Jones *et al.*, 2013) is overlain by hydrocarbon-bearing turbidite sandstones that are calibrated by wells NO-2/4-14 and NO-2/4-18R (Mjelde, 1991; Landrø, 2011). Dipmeter data from the 2/4-18R well suggest that these turbidite sandstones were sourced from the North and deposited within a basin depocentre oriented NW–SE, following the structural trend imposed by the underlying major Rotliegend faults. The turbidite sandstones bodies (T4/T2 facies; 5 m to 20 m thick) are of varying quality and occur as a multi-story succession up to 120 m thick. Lateral continuity is variable, but some thicker sandstones extend for at least 1 km.

This type of tectono-sedimentary setting produces stratigraphic series that are relatively layer cake in geometry but wedging away from major sub-basin boundary faults. Hangingwall onlap related to phases of rapid fault movement is a characteristic feature and an important control on the geometry and geographical localisation of sandstone bodies. Erosion on the crest of fault blocks can be anticipated, which may cannibalise older stratigraphy. However, late erosion at Base Cretaceous Unconformity (BCU) level may mean that any stratal evidence for earlier phases of erosion is removed.

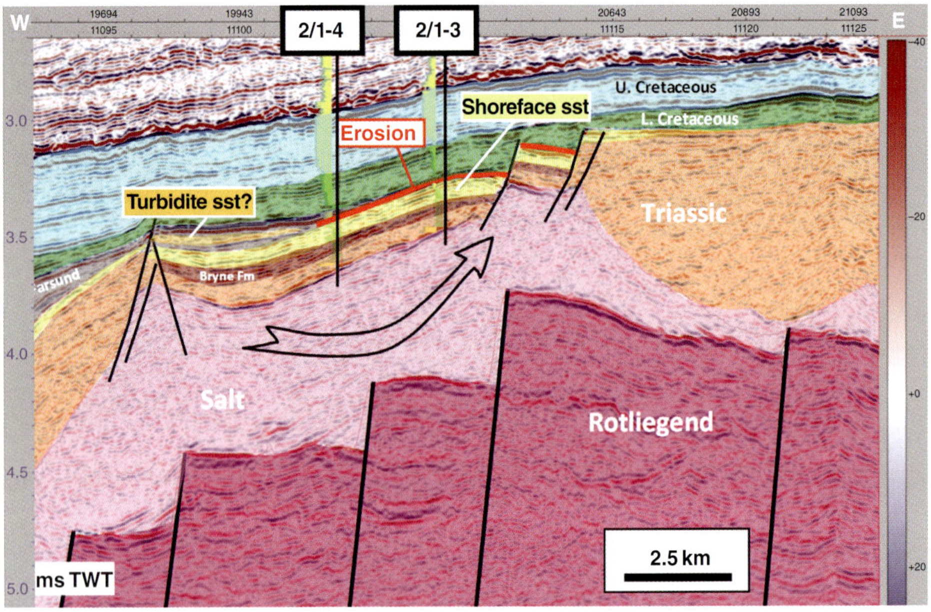

Fig. 12. West-East structural cross-section over Gyda Field. The location of an area of interpod crestal erosion has been added to this figure based on the correlation shown in Fig. 14. The location of this section is indicated on Figs 1 and 8.

Case Study 2: The Gyda Field: Type C – Interpod setting

The Gyda Field forms part of the Ula Trend, a petroleum province located on the eastern margin of the Central Graben (Spencer *et al.*, 1986; Brown *et al*, 1992; Bjørnseth & Gluyas, 1995; Fig. 1). Fields within this trend include the Ula Field, which has produced more than 500 MMboe and the Gyda Field which has produced more than 300 MMboe (Norwegian Petroleum Directorate figures).

The Gyda Field is located in an area of the basin where thick deposits of Zechstein salt were developed. In this area, late Jurassic rifting took place through the reactivation of older Permian faults. The Ula-Gyda fault zone is defined by deep Rotliegend faulting and marks the easternmost boundary of the Jurassic graben. West of this zone, Triassic pods were rotated forming terraces that step down into the graben (Bjørnseth & Gluyas, 1995). These terraces contain the Gyda Field developed over a thick (~1000 m) Zechstein salt layer as an interpod structure (Fig. 12). The characteristic features of this Type C interpod setting are the development of Jurassic sandstones over a former salt high which deflated by salt dissolution or salt withdrawal during the late Jurassic, thereby providing accommodation space for deposition of the Middle Jurassic Bryne Formation and the overlying Gyda Sandstone.

The sandstones of the Gyda Field (Fig. 5) are dominated by shoreface (S4) to lower shoreface (S3) facies and are of late Kimmeridgian age, i.e. relatively young compared to other shallow marine sandstones of the Central North Sea. Nonetheless, their sedimentary characteristics are much the same as elsewhere. Studies by Aase *et al.* (1996) have shown the occurrence of *Rhaxella* sponge spicules in certain layers, as high-lighted by the localised occurrence of secondary porosity development resulting from sponge spicule dissolution (Fig. 13). The spicule-rich zone belongs predominantly to the retrogradational part of the shoreface progradational-retrogradational cycles and is useful for sequence stratigraphic correlation.

Evidence for uplift, erosion and redeposition of the Gyda Sandstone comes from the integration of well correlations with seismic data (Fig. 13). Erosion of the upper parts of the J66 Sequence can be identified by extrapolating the spicule zone identified by Aase *et al.* (1996) using the characteristic low porosity signature on neutron-density logs across the field towards the 2/1-4 and 2/1-3 wells where this zone is interpreted to be progressively eroded (Fig. 14). This erosion surface on the eastern part of the Gyda Field (Fig. 12) suggests that the Gyda interpod structure was tilted and uplifted by salt movement post-J71 TEMFS.

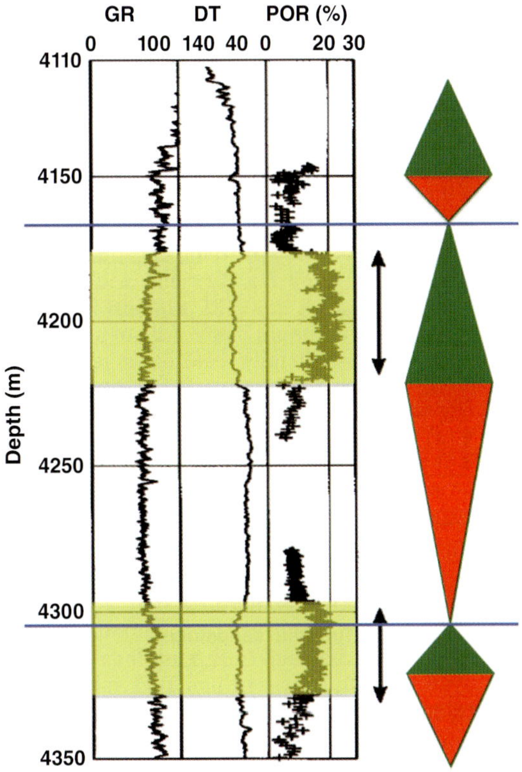

Fig. 13. Enhanced porosity due to secondary porosity development (yellow shading) in zones where sponge spicule occurrence has been demonstrated by Aase *et al.* (1996). Extensive dissolution of sponge spicules can generate secondary porosity as shown in Fig. 18. In this example (and several others) it is observed that sponge spicules are much more abundant during phases of shoreface retrogradation. Red triangle: long term progradational trend; green triangle: long term retrogradational trend. GR: Gamma Ray log; DT: Delta Time (Sonic) log; POR: total porosity (%) measured on core.

Wells situated on the flanks of the Gyda structure, for example the 1/3-3 Tambar well (Fig. 14), have several thin beds (20 cm to 2 m thick) of sandstone that are present in the basal 50m of the shale-dominated J71 Sequence (Farsund Formation) immediately overlying the Ula Formation. The best examples of these sandstones are not cored in the 1/3-3 well, but those which are show highly bioturbated, fine-grained sandstone beds corresponding to previously described Facies T3. It is suggested that these sandstone units are turbidite deposits resulting from erosion on the crest of the Gyda structure. This erosion would have reworked the pre-J71 TEMFS shoreface sandstones and redistributed them as thin sandstone stringers within the overlying shales.

Case Study 3. The Elgin-Franklin Field: Type B – Reactivated pod setting

The Elgin and Franklin fields are structural traps, the main reservoir being the shallow marine Franklin Sandstone (equivalent to the Puffin Formation of Price *et al.*, 1993). The sandstones are sealed by the overlying Heather and Kimmeridge shales and, ultimately, by the Cromer Knoll Group of Lower Cretaceous age. Minor production also comes from the underlying Pentland Formation, a fluvial reservoir of Middle Jurassic age.

The structural context of the Elgin Field can be demonstrated with a W–E geoseismic section across the area (Fig. 15). The deepest observable seismic event is the Top Rotliegend, which is at approximately 7 km depth (over 5 s two-way-time). The predominant basement-linked faults trend NNW–SSE with secondary faults trending ENE–WSW. The Top Zechstein salt pick indicates that there is a marked detachment between the sub-salt and supra-salt successions. Salt movement has created sediment 'pods' infilled by Triassic and Jurassic strata. Both salt movement (swelling and withdrawal) and grounding of pods on the underlying Rotliegend fault blocks have influenced the distribution of Upper Jurassic reservoir sandstone thickness. Hence, the Elgin-Franklin Field is an example of a Type B tectono-sedimentary setting. The structure also reflects a phase of late Jurassic extension, when thick Kimmeridge Clay was deposited in the hangingwall basins.

The Franklin Sands net reservoir thickness averages 200m and consists of a succession of fine- to medium-grained sandstones that are mainly strongly bioturbated. The succession is divided into three major lithostratigraphic units separated by minor shale intervals: Franklin A, B and C Sandstone (Lasocki *et al.*, 1999). Each sandstone unit can be more finely divided into minor progradational-retrogradational sequences on the basis of changes in sandstone clay content detected by the GR log (Fig. 16).

Reduced vertical seismic resolution at more than 5km depth (18Hz to 20Hz dominant frequency) makes intra-reservoir surfaces difficult to pick in a consistent manner. In addition, the homogenous character of the reservoir has led to a number of different sedimentology-based correlations, each with significant uncertainty. A genetic sequence stratigraphic approach was used to recognise multiple phases of progradation and retrogradation within the Franklin Sands (Lasocki

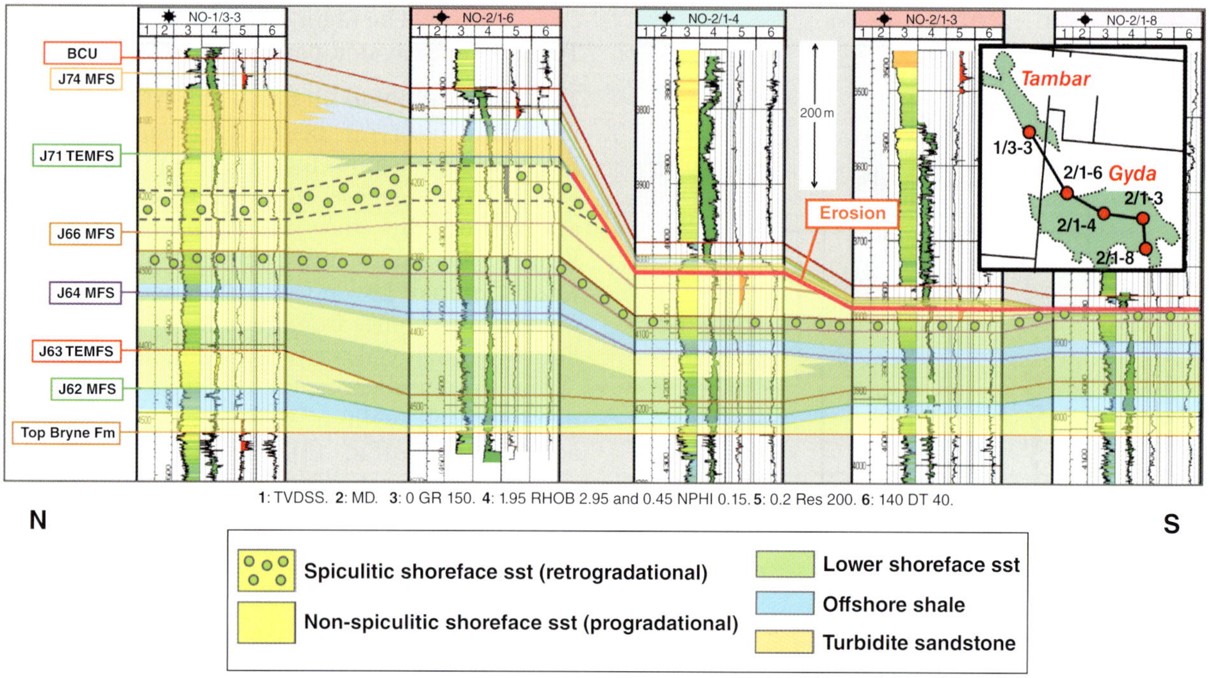

Fig. 14. N–S aligned correlation panel from Tambar Field (left well) to Gyda Field (4 wells to right) showing truncation of the Upper Jurassic sandstone in a crestal location (strong red line), removing spiculite sandstone (yellow with green spots) and linked to downdip stratigraphic expansion dated to the J71 Fittoni TEMFS. Spiculite sandstone definition follows that given by the study of Aase *et al.* (1996). TVDSS: True Vertical Depth Sub Sea (m); MD: Measured Depth (m); GR: Gamma Ray log; DT: Delta Time (Sonic) log; RHOB: Bulk Density; NPHI: Neutron Porosity; Res: Resistivity.

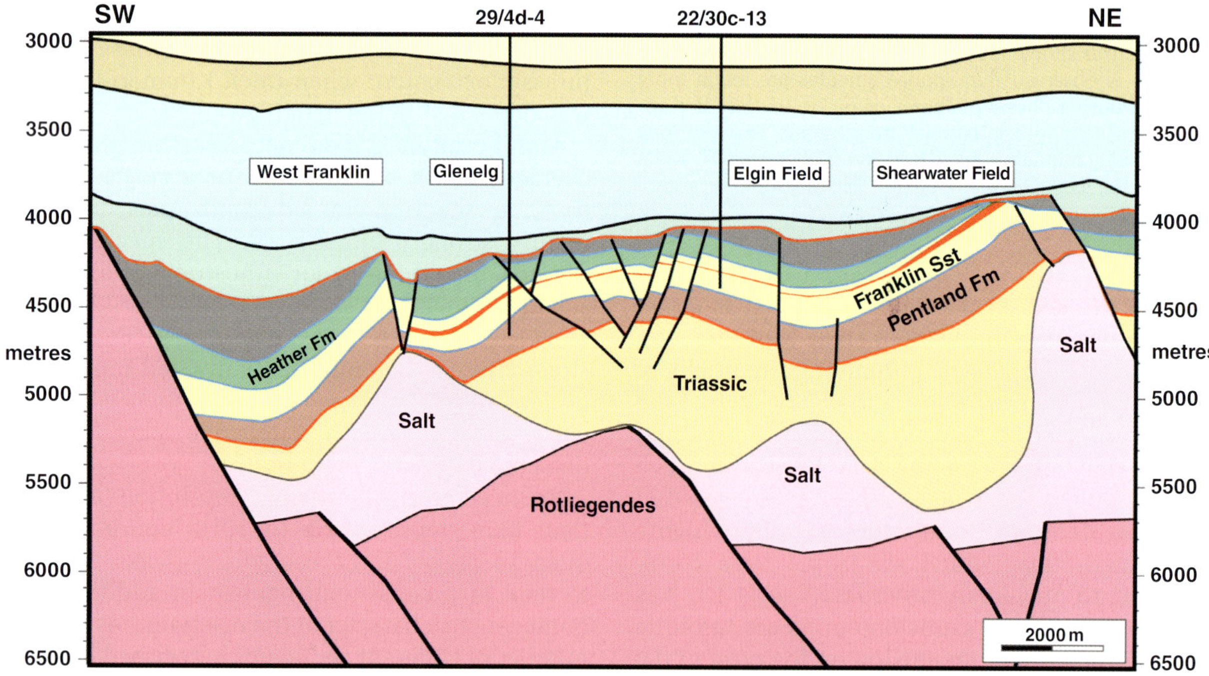

Fig. 15. Geoseismic section showing the structural development of the West Franklin, Glenelg, Elgin and Shearwater fields. A red unit towards the top of the Franklin Sandstone schematically indicates a thickening of the Top B Sands turbidite facies interval towards Shearwater and Glenelg, away from the crest of Elgin Field which is supported on an underlying Rotliegend fault block crest (figure adapted from Suiter *et al.*, 2005). The location of this section is indicated on Figs 1 and 8.

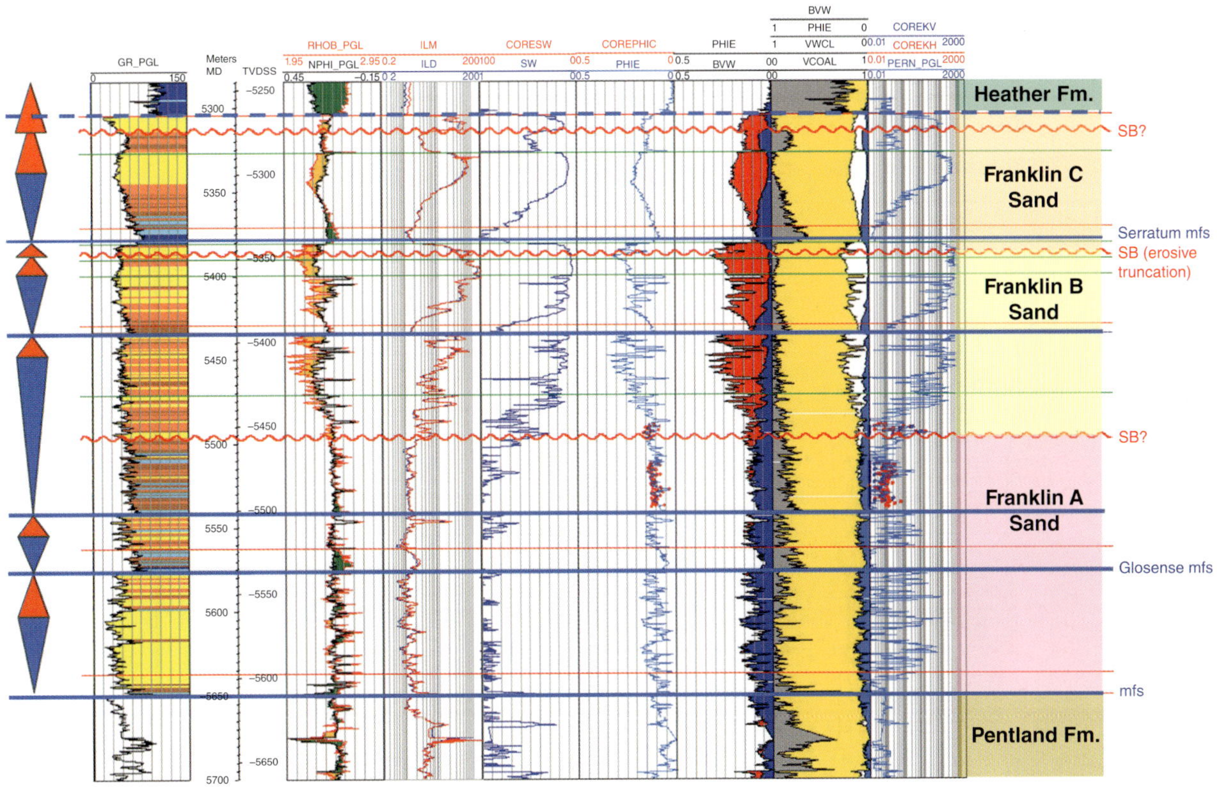

Fig. 16. Progradational (blue triangle) and retrogradational (red triangle) stacking pattern development in the Franklin Sandstone (Elgin Field well 22/30c-8). GR_PGL: Normalised Gamma Ray; MD: Measured Depth (m); TVDSS: True Vertical Depth Sub Sea (m); RHOB_PGL: Bulk Density; NPHI_PGL: Neutron Porosity; ILM/ILD: Induction Log Medium/Deep; CORESW: Water Saturation Core; SW: Water Saturation Log; COREPHIC: Corrected Total Porosity Core; PHIE: Effective Porosity Log; PHIB: Density Porosity; BVW: Bulk Volume Water; VWCL: Volume Wet Clay; VCOAL: Volume Coal; COREKV: Vertical Permeability Core; COREKH: Horizontal Permeability Core; PERM_PGL: Calculated Permeability.

et al., 1999). However, due to the lack of grain size differentiation and the low shale content, cycle correlation is uncertain, particularly where wells are spaced more than 500m apart. The work of Lasocki *et al.* (1999) was superseded by that of Wonham *et al.*, (2002) following the integration of numerous studies of core data across the field (~1300m) and an improved intra-reservoir correlation was achieved using a combined analysis of wireline logs, core sedimentology (especially trace fossil analysis), core chemostratigraphy, spectral gamma ray logs and point-counted petrographic data. These studies highlighted an important intra-reservoir erosive event (Fig. 17; Wonham *et al.*, 2002).

The erosion surface is observed in core in several wells (Elgin well 22/30c-13; Franklin well 29/5b-6 and Glenelg well 29/4d-4; Fig. 18). In well 29/5b-6 it is overlain by a sharp-based lag (~1m thick) consisting of matrix-supported mud clasts, pebbles and oyster coquina material. The lag overlies a thick interval of bioturbated, slightly argillaceous sandstones (Franklin B Sandstone) interpreted as lower shoreface to transition zone deposits. Above the erosive base is a clean sandstone (~9 m thick) consisting of a basal lag overlain by structureless, clean, fine to very fine sandstone (~2 m) that grades upwards into laminated or massive sandstone (~6 m thick). The upper 1 m of the laminated sandstone shows an increase in bioturbation, marking a return to burrowed offshore transition zone facies, similar to the facies that underlies the erosive base. This ~ 9 m succession forms the basal part of the Franklin C Sandstone.

The base Franklin C Sandstone erosion surface removes variable proportions of underlying reservoir sandstone (e.g. the Upper Franklin B Sandstone shown with spotted yellow pattern in Fig. 17). This sandstone has distinctive porosity

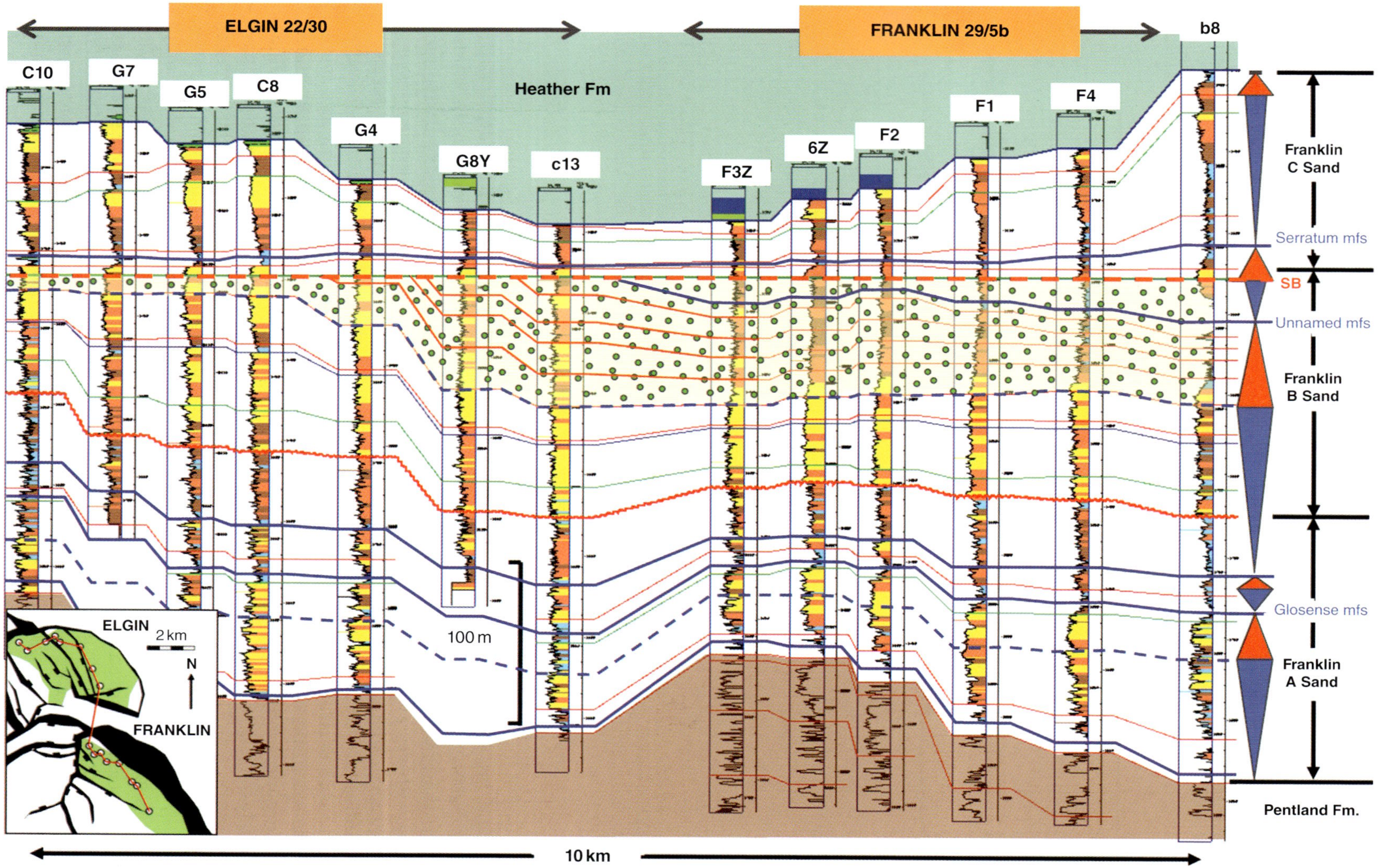

Fig. 17. Correlation of the Elgin and Franklin fields (from Wonham *et al.*, 2002). Comparison can be made with an earlier correlation of Lasocki *et al.* (1999) which proposes a conformable succession. In this new correlation, the upper part of the Franklin B Sands can be seen to be significantly truncated in the northern part of Elgin Field, removing the spicule-bearing sandstones in this area (yellow unit with green spots).

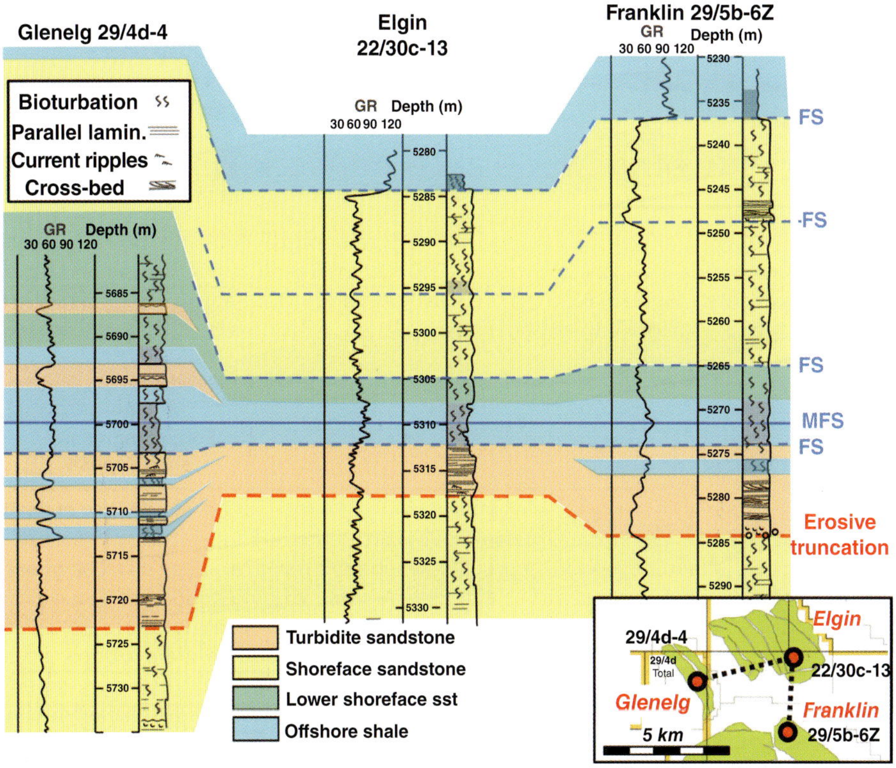

Fig. 18. Correlation between cored intervals of Glenelg, Elgin and Franklin fields showing the preservation of parallel laminated sandstone of turbidite origin overlying a surface of truncation with local pebble lag development (Franklin Field well 29/5b-6). GR: Gamma Ray log.

characteristics related to the presence of *Rhaxella* sponge spicule material (see later section on diagenesis and influence of *Rhaxella* spicule).

The Franklin B Sandstone represents the thickest and most important reservoir in the field. The Upper Franklin B Sandstone was deposited during a long term retrogradational phase while the lower part of the Franklin B Sandstone was deposited during the preceding progradational phase (Fig. 17). However, the erosive unconformity surface at the base of the Franklin C Sandstone has produced pronounced thickness variations of the Franklin B Sandstone. Passing from south to north across the structure, the uppermost interval of the Franklin B Sandstone is erosionally thinned from 90 m to only 10 m thick. In addition, this thinning takes place in the same direction in which the sandstone was initially thickening (as indicated by thickness trends of parasequences in the erosional remnants, see Fig. 17). Thus, a preferential depocentre was effectively inverted into a preferential site of erosion. The variably eroded interval is also notable for containing large quantities of *Rhaxella* sponge spicule debris, which has resulted in high secondary porosity content.

Lasocki *et al.* (1997) noted that the lag and overlying parallel laminated sandstone were "open to several possible interpretations with major consequences on reservoir correlation and trends". The sandstone overlying the erosive surface was initially interpreted as a mass flow deposit (Lasocki *et al.*, 1997). Subsequently, it was reinterpreted with the base of the unit taken to represent a planar regression surface at which an unknown amount of erosion had taken place. The overlying structureless to parallel laminated sands above the erosional surface were interpreted to have been deposited in a high-energy foreshore or shoreface environment. Later, Jeremiah & Nicholson (1999) placed the same interval of the 29/5b-6 well in the context of adjacent wells such as the 22/30b-11 well of Shearwater Field. In the 22/30b-11 well, a thick interval of parallel laminated sandstones were interpreted as turbiditic mass flow sandstone deposits developed above a major erosion surface identified as SJU310. Several other wells with

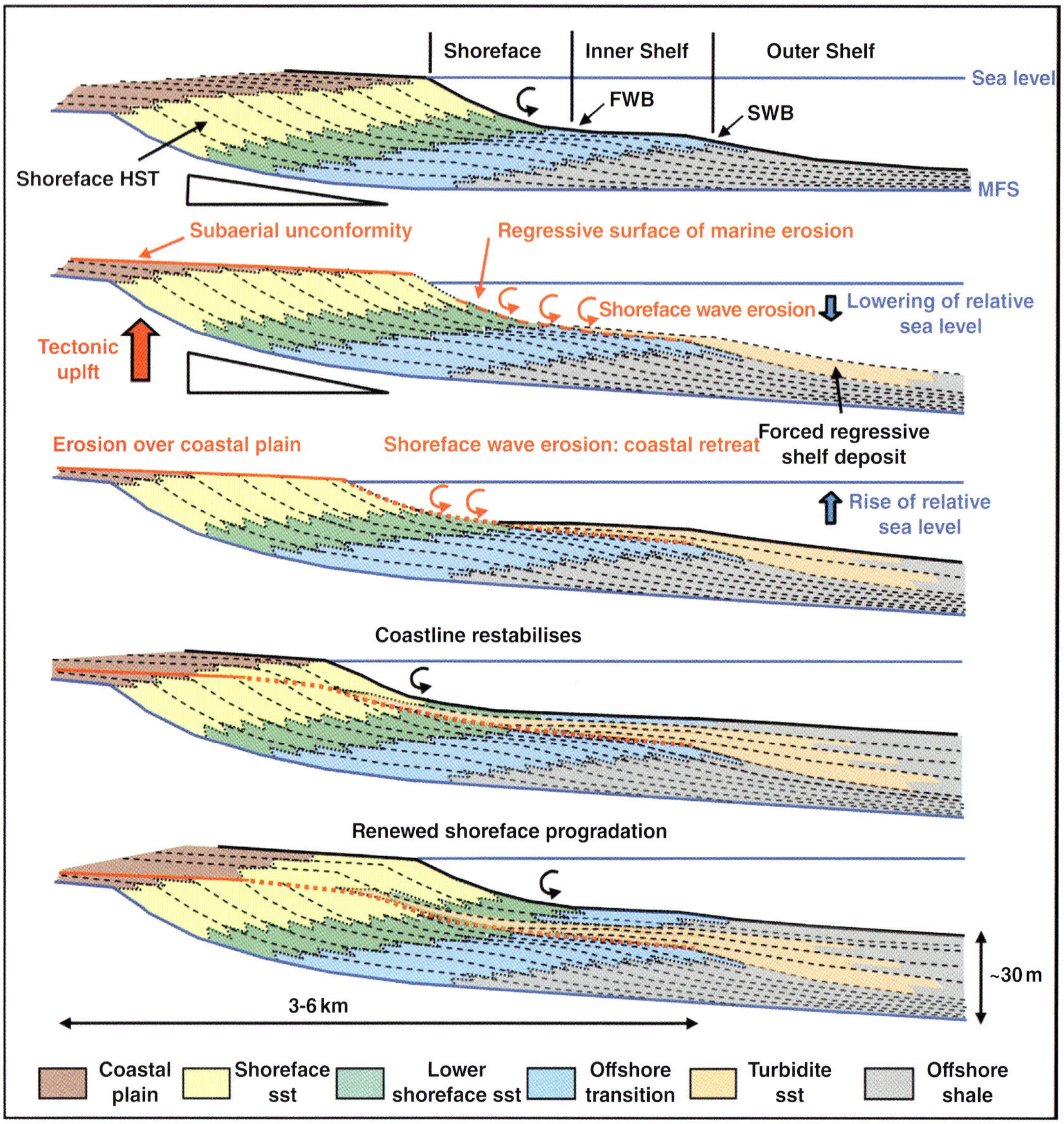

Fig. 19. Offshore turbidite generation from reworking of earlier shoreface deposits in a tectonically active context. Shoreface facies characteristics based on Gowland (1996). Wave reworking of shoreface following relative sea-level fall and forced regression based on Figure 4.23 of Catuneanu (2006). HST - highstand systems tract; MFS maximum flooding surface; FWB - fairweather wave base; SWB - storm wave base.

core in the same sandstone indicate that this parallel laminated sandstone sometimes occurs as discrete dm-thick beds separated by siltstones (e.g. Glenelg well 29/4d-4; Fig. 18).

The parallel-laminated sandstone is here interpreted as a widespread, turbiditic, mass flow deposit and is considered to overlie a major erosion surface related to transgressive shoreface reworking (Fig. 19) which possibly also includes a component of earlier subaerial erosion. During the reworking, storm events could have been responsible for redistributing sand over the sea floor as tempestite deposits. While the sandstone is widespread around the zone of erosion, it may be also locally discontinuous. A number of authors have argued in the past for tempestite or turbidity current generation on storm-dominated shelves, notably Walker (1985a/b) who argued that turbidity currents could be generated by earthquakes, sediment failures or storm surge relaxation. In the Elgin-Franklin area, fault development related to pod grounding could have led to both earthquakes and sediment failures. However, both tectonically-influenced effects and tropical storms could have contributed to generating shallow water turbidity current in the Central Graben. Similar interactions are inferred for other wave-dominated shelves (Pattison *et al.*, 2007).

With the benefit of more wells, it has been demonstrated that the base Franklin C Sandstone erosive unconformity cuts down around 90 m into the underlying Franklin B Sandstone succession in the northern part of the Elgin Field (Fig. 17). The location of this erosion reflects the dynamic interaction of salt movement, sediment loading and, most significantly, grounding of the Elgin pod on the underlying Rotliegend (Fig. 15). The Franklin C Sandstone, which lies above the unconformity surface, shows different thickness trends to the Franklin B Sandstone below the unconformity due to a change from pod filling to pod collapse. Thickness trends in the Franklin C Sandstone are the same as those in the overlying Heather Shale Formation and markedly different to changes in parasequence thickness recognised through correlation of the Franklin B Sandstone (shown in Fig. 17). Hence, deposition was controlled by the interaction between tectonism and the change from shoreface to turbidite deposition (Fig. 19).

Updated correlation and mapping of the major intra-reservoir erosion surface described above allows a more rigorous correlation of the parasequences in the sandstones above and below the erosion surface and the improved mapping of reservoir quality based on well log and core-based facies and petrophysics. Most importantly, the new stratigraphic scheme better captures proximal-distal facies trends, which are the main control on reservoir quality. Together with changes to the structural model, these are the two most significant factors which probably influence field volumetric calculations.

Case Study 4: The Fulmar Field: Type B – Reactivated pod setting

Within the Fulmar Field, the Ribble Sandstone Member represents the last phase of Fulmar Sandstone deposition (Johnson *et al.*, 1986; Robinson, 1990; Stockbridge & Gray, 1991; Clelland *et al.*, 1993; Kuhn *et al.*, 2003; Figs 3 and 20). The

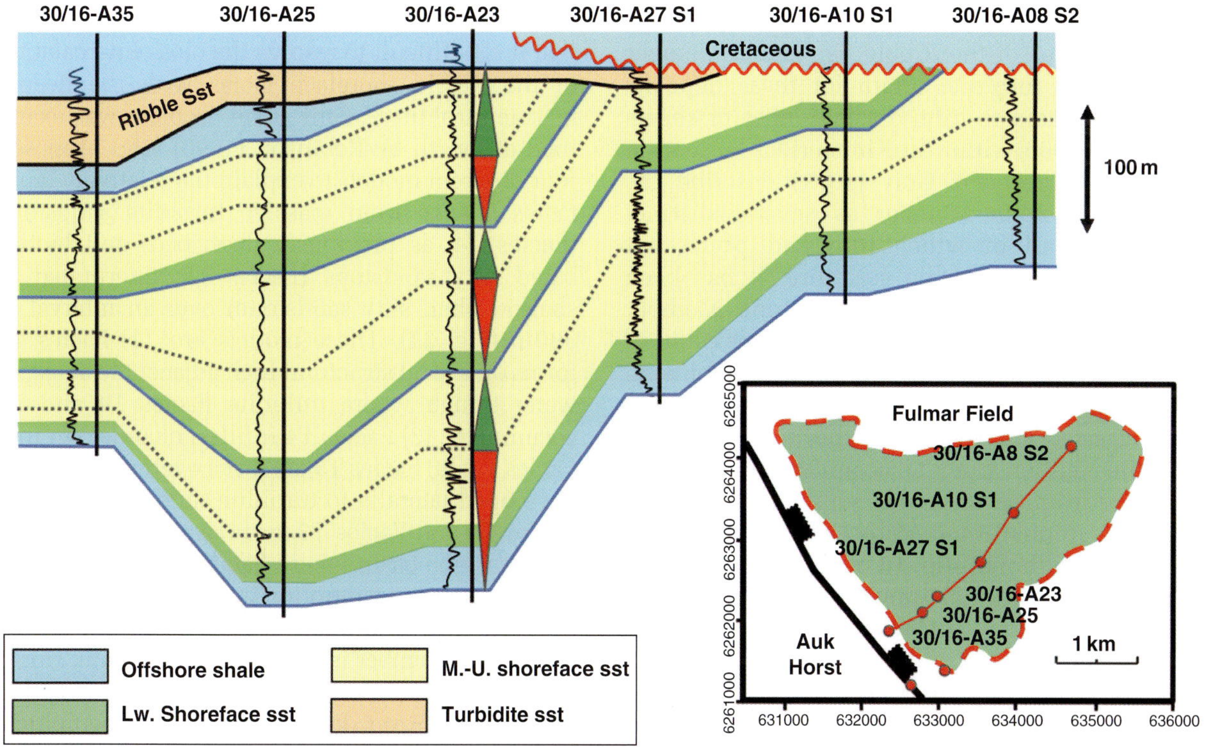

Fig. 20. Fulmar Field correlation showing truncation of the Fulmar Formation shoreface sandstones and reworking as Ribble Member turbidite sandstone (after Kuhn *et al.*, 2003). Gamma ray logs indicate changes in clay content.

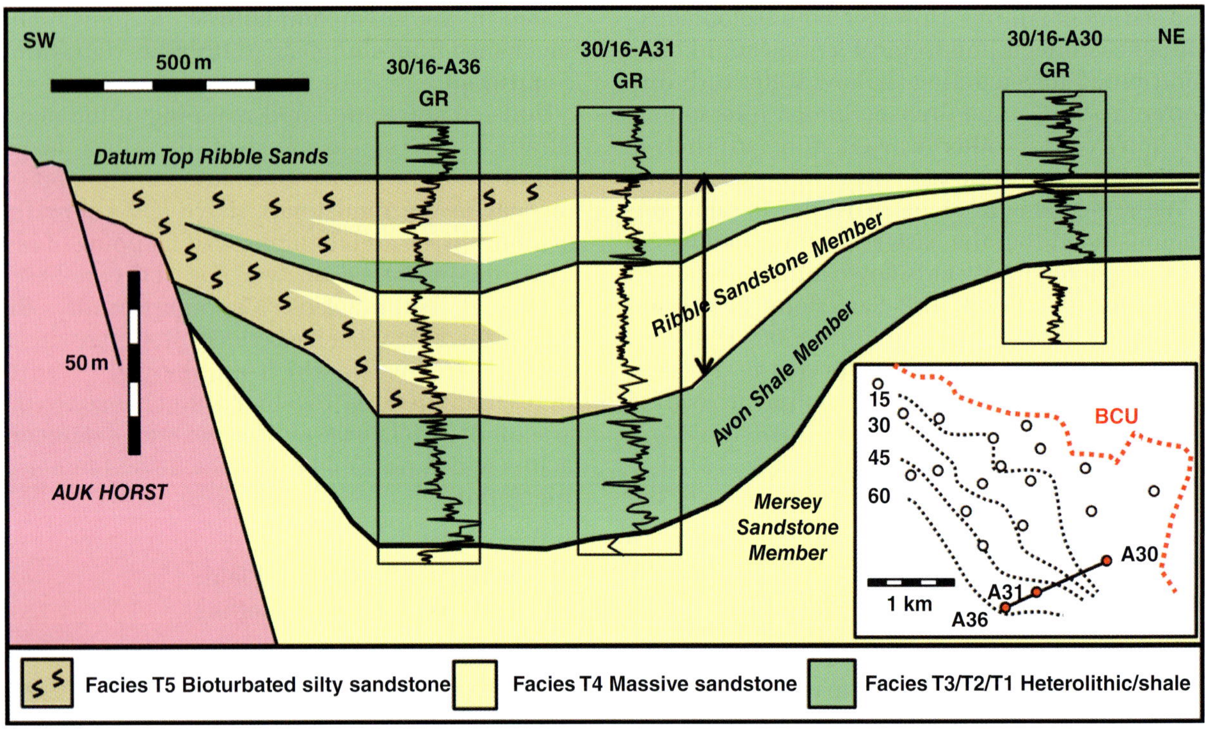

Fig. 21. Rapid lateral facies changes in the Ribble Sandstone Member (adapted after Robinson, 1990). Facies are described in more detail in the text. Inset map shows the location of wells and the thickness of the unit with contours in metres. BCU: Base Cretaceous Unconformity; GR: Gamma Ray log.

earliest studies of the Ribble Sandstone Member recognised a turbidite origin for these deposits which were said to form discrete lobes or sheets (Robinson, 1990) within the Kimmeridge Formation shale. Stockbridge & Gray (1991) describe the Ribble Sandstone Member as several distinct sandstones, all of which thin and pinch out towards the crest of the structure. It has been postulated that the Ribble Sandstone Member might have been derived from the erosion of the Auk Horst (Clelland *et al.*, 1993; Robinson, 1990). Subsequent study by Kuhn *et al.* (2003) has suggested a derivation of the Ribble Member turbidites either from adjacent shoreface sandstones or from an unknown source to the north-west. A depositional direction towards the south-east is shown (fig. 10 of Kuhn *et al.*, 2003) with the depositional transport direction being influenced by active palaeofaults with a NW–SE strike trend.

The same study indicates that the Ribble Sandstone Member comprises an interfingering of well bioturbated distal shoreface deposits and turbiditic units. Kuhn *et al.* (2003) made a study of the stratigraphy of the Fulmar Field (Fig. 20) and state that it is "difficult to resolve the close intercalation of the Ribble shoreface facies with the deeper marine turbidites". This same problem was examined at length by Robinson (1990) who proposed four different models to explain the relatively rapid facies change from massive turbidite sandstone (Facies T4, as previously described) and bioturbated silty sandstone (Facies T5, interpreted as shallow shelf silty sandstone) over an interval of 380 m (Fig. 21). These models are: (1) Facies T5 forms on a local structural high adjacent to a deeper water trough where turbidite Facies T4 and T3 accumulate; (2) Facies T4 infills a channel cut into Facies T5; (3) Facies T5 are levee overbank deposits developed laterally to turbidite sandstone channel Facies T4 and Facies T3, or (4) a salt ridge formed an intraslope basin where Facies T5 accumulated. There is a basic assumption in all these models that the sand grade sediment that comprises the Ribble Sandstone Member is sourced from the Auk Horst. Kuhn *et al.* (2003) supported the idea of deposition of Facies T5 on fault-controlled topographical highs, but also mentioned the possibility that the Ribble Member could have been sourced by reworking of older Upper Jurassic sandstones.

A number of the models proposed by Robinson appear to be unsupported, based on the knowledge that there is little evidence to suggest structural compartmentalisation between the two wells concerned (30/16-A36 and 30/16-A31). In addition, there is no strong reason to suppose that the Ribble Sandstone was sourced directly from the Auk Horst. Mineralogically, the sands of the Ribble Sandstone Member are arkosic or subarkosic and quite similar in composition to the sandstones in the shoreface sandstone members of the Fulmar Formation below. The Ribble Sandstone might just as easily have been sourced from the underlying shoreface sandstone uplifted to the north-east of the Ribble Sandstone depositional area, as from the Auk Horst.

It is suggested here that the observed facies change could reflect a rapid loss of depositional energy within turbidite sandstones whose flows had a very short run out. A lower energy setting and an increase in mud content of the sediment could have been sufficient for this facies to be attractive for bioturbating organisms giving Facies T5 a very different textural appearance to the massive or parallel laminated sandstones of Facies T4. A facies change of this kind is interpreted to have resulted from a depositional model such as that shown in Fig. 19 in which shoreface sandstone is reworked to produce turbidite mass flow sands by the combined influence of local tectonic uplift and transgressive shoreface erosion and retreat. Correlations of Kuhn *et al.* (2003) suggest that shoreface sandstones underlying the Ribble Sandstone Member are eroded on the crest of the Fulmar Field structure (Fig. 20) and it is here suggested that this eroded shoreface sandstone material could be the main source of the Ribble Sandstone Member, rather than the Auk Horst to the west. This idea is supported by the observed facies change between Well 30/16-A31 and well 30/16-A36 which suggest a lower energy system to the south-west than to the north-east (Fig. 21).

Elgin-Franklin and Fulmar fields share a common tectono-stratigraphic setting (Type B reactivated pod setting) and also appear to show a similar stratigraphic and sedimentological development, notably: (1) initial shoreface deposition within reactivated pod structure and (2) turbidite sandstone deposition over erosive unconformity following pod grounding onto the Rotliegend. Faulting during pod collapse may have assisted and focused the distribution of reworked shoreface sandstones as turbidite deposits away from the pod crest as illustrated in Fig. 10. For both fields, a structural inversion and repositioning of sediment source and sink may have occurred.

Case Study 5: Jacqui turbidites: Type B – Reactivated pod setting

The Freshney Sandstone Member of the Jacqui area presents a further example of turbidite sandstone deposition, based on the results of wells 30c/13-3, 30c/13-4, 30c/13-6 and 30c/13-7). These wells have proven a transition from sandy shoreface to distal silty offshore deposits passing from the south-east towards the north-west (Fig. 5). However, the overlying turbidite sandstones show a different trend and are scattered broadly across the area. Correlation of the turbidite sandstone units is not straightforward and indicates either the lobate geometry of sandstone units or, perhaps more probably, the influence on their deposition of marked tectonic uplift on the Joanne Horst to the west. Seismic lines (e.g. Fig. 22) show significant uplift and steepening of strata over the underlying Rotliegend footwall in such a way that previously deposited shoreface sandstones and Triassic sediments could have been eroded and transported the short distance to the site where turbidite deposition occurred. These turbidite sandstones could, therefore, have been the result of deposition from high density but low energy turbiditic flows that rapidly deposited their sediment load, a scenario similar to that envisaged for the Elgin-Franklin-Shearwater and Fulmar Field turbidite deposits. Unlike these fields, pod grounding in the Jacqui area appears to have occurred very late so that the turbidites are ponded within the pod, rather than shedding to the sides of the pod as seen in the other examples.

OCCURRENCE OF *RHAXELLA* SPONGE SPICULES AND THEIR INFLUENCE ON RESERVOIR QUALITY

The Fulmar sandstones of Elgin and Franklin fields are typically subarkosic arenites which show: (1) field-wide average porosity/permeability of 17%, 25 mD and (2) local maximum porosity/permeability of 30% and several Darcies. Secondary porosity evolution has been enhanced through the dissolution of feldspar and sponge spicules. Early compactional effects have been lessened as a result of the growth of early authigenic minerals (microquartz and dolomite) which

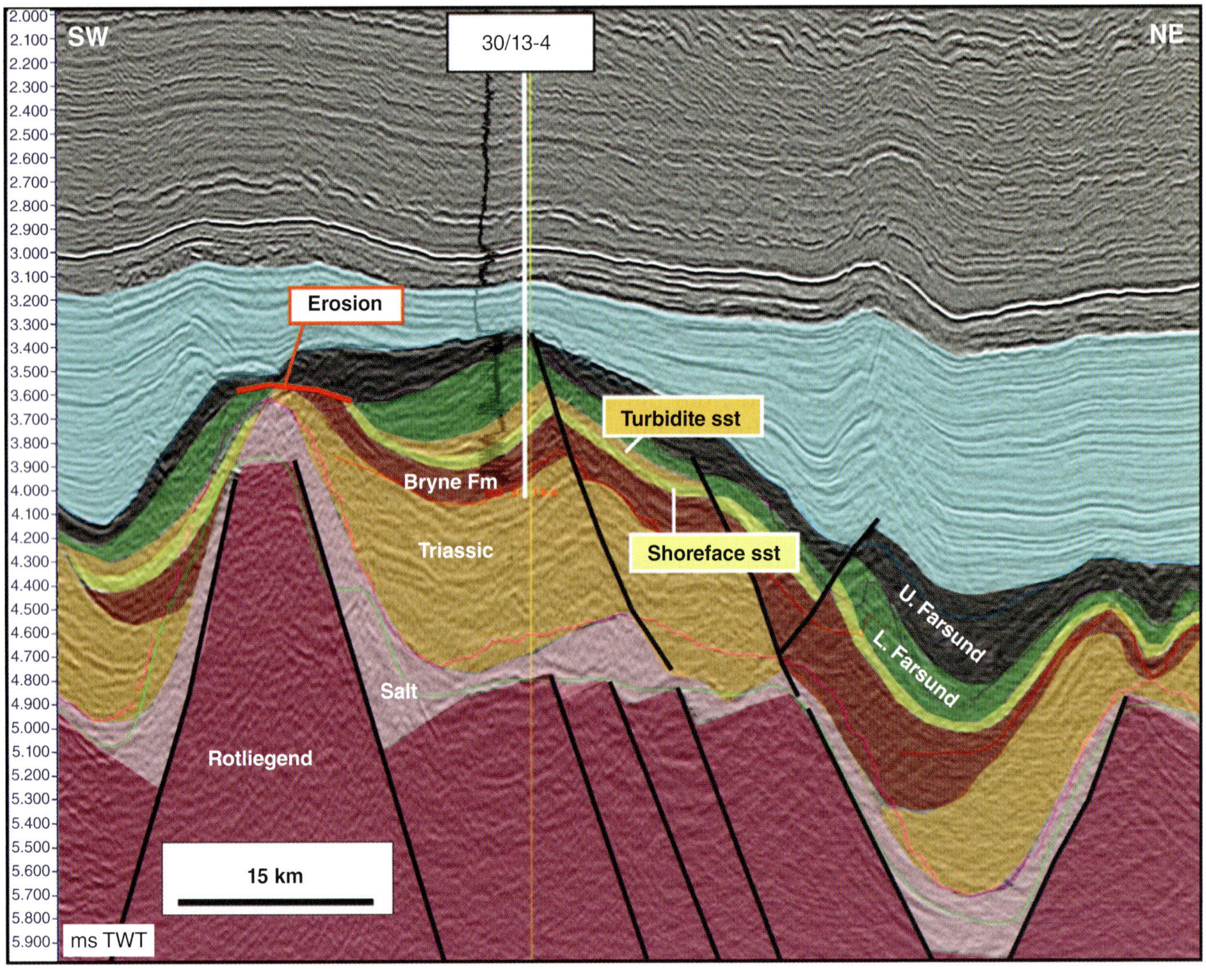

Fig. 22. Uplift of shoreface sandstone on the Joanne Horst to the west of the 30/13-4 well giving the potential for reworking of previously deposited shoreface sandstones (yellow) and subsequent re-deposition as Freshney Member turbidites (orange). The location of this section is indicated on Figs 1 and 8.

helped provide a rigid grain framework. During burial, the development of increasingly high overpressure within the reservoir helped to minimize further compactional effects, as indicated by the general absence of significant pressure solution effects at grain contacts. Compaction tends to be higher in the muddier reservoir facies where secondary porosity predominates.

The most significant controls on primary porosity within these dominantly fine-grained sandstones have been summarised by Lasocki *et al.*, 1999 as follows: (1) disseminated clay which occludes porosity (up to 37% clay content) and is related to shoreface proximal-distal trends and the role of bioturbating organisms which re-distribute the clay; (2) carbonate cementation as concretions, which are locally abundant, localised as overgrowths on winnowed shell beds (e.g. well 22/30c-13) and composed mainly of dolomite and ankerite (Hendry *et al.*, 2000a/b; Burns *et al.*, 2005); (3) mesoquartz cementation and (4) late pyrobitumen formation that occludes around 2% of the porosity.

In addition, a variety of factors influence secondary porosity development including: (1) secondary porosity creation through dissolution of sponge spicules and associated microquartz coating of pore spaces (Walgenwitz & Wonham, 2003); (2) feldspar dissolution (Wilkinson *et al.*, 1997; 2006) and (3) minor carbonate (shell debris) dissolution. The role hydrocarbon charging history played in inhibiting quartz cementation and the preservation of porosity is debatable (Osborne & Swarbrick, 1997; Wilkinson *et al.*, 2006; Taylor

Fig. 23. Secondary porosity produced by *Rhaxella* sponge spicule dissolution and lining of pore spaces with microcrystalline quartz which inhibits later quartz overgrowth.

et al., 2010). However, spicule dissolution occurred relatively early in the burial process, before hydrocarbon charging and makes the resulting secondary porosity an important contribution to the overall porosity. Spicule dissolution porosity, due to its secondary nature and limited pore connectivity, does not enhance permeability. Hence, hydrocarbon-bearing sandstones with spicule porosity tend to contribute more significantly to production in gas condensate fields than oil fields.

Rhaxella sponge spicules have been documented in many fields, for example the Gyda Field (Aase *et al.*, 1996; discussed in more detail later), the Clyde Field (Stevens & Wallis, 1991) and the Fulmar Field (Johnson *et al.*, 1986; Kuhn *et al.*, 2003) where it occurs in two distinct heterogeneous intervals (Lower Usk and Clyde Sands facies of Unit 3 Ribble Sands). Spicules can be found within a large range of different rock types from Sinemurian to Kimmeridgian in age (Vagle *et al.*, 1994). They are found in the Moray Firth in silty or fine-grained sandstones of middle Oxfordian age and are abundant in the Alness Spiculite Formation (Johnson *et al.*, 2005).

Early spicule dissolution and re-precipitation occurs due to the inherent instability of biogenic opal-A silica, of which spicules are formed (Hendry & Trewin, 1995). Together, sponge spicule dissolution and feldspar dissolution can account for up to half the point-counted total porosity. Spicule-derived porosity can be easily observed in petrographic sections due to the circular or oval form of the pores (Fig. 23). Microquartz coating of pore spaces is thought to inhibit the growth of quartz cement (Bloch *et al.*, 2002; Walgenwitz & Wonham, 2003; Taylor *et al.*, 2010) and microquartz crystals may resist pressure solution by solidifying contacts between quartz grains (Osborne & Swarbrick, 1999).

Significantly, secondary porosity created by sponge spicule dissolution is mainly developed in the upper, transgressive unit characterised by retrogradational stacking of shoreface parasequences (Wonham *et al.*, 2002; Fig. 24). The significance of such observations for mapping reservoir quality can be seen on phi-K cross-plots comparing the characteristics of (1) retrogradational, spicule-bearing Upper Franklin B Sandstone and (2) progradational, Lower Franklin B Sandstone with no spicules (Fig. 25). This cross-plot emphasises a marked change in the relationship of porosity and permeability between the Lower and Upper Franklin B Sandstone.

Abundant spicules within the Upper Franklin B Sandstone may reflect slower rates of sedimentation primarily within Facies S3 lower shoreface deposits during the retrogradational phase of deposition. Ajdukiewicz *et al.* (2001) concluded that spicules are most common in the relatively distal reaches of transgressive systems tracts, where clear water and limited volumes of detrital fines provided favourable environmental conditions. The development of lower permeability within the retrogradational succession has been recognised earlier (Cannon & Gowland, 1996) but the link to spicule presence was not made. Gowland (1996) speculatively associated abundant spicule development to a shelf depositional environment. However, the observation of the retrogradational nature of the spicule-bearing sandstones is more useful for predictive purposes. Gowland (1996) also suggested that palynofacies data demonstrated that spicule-rich

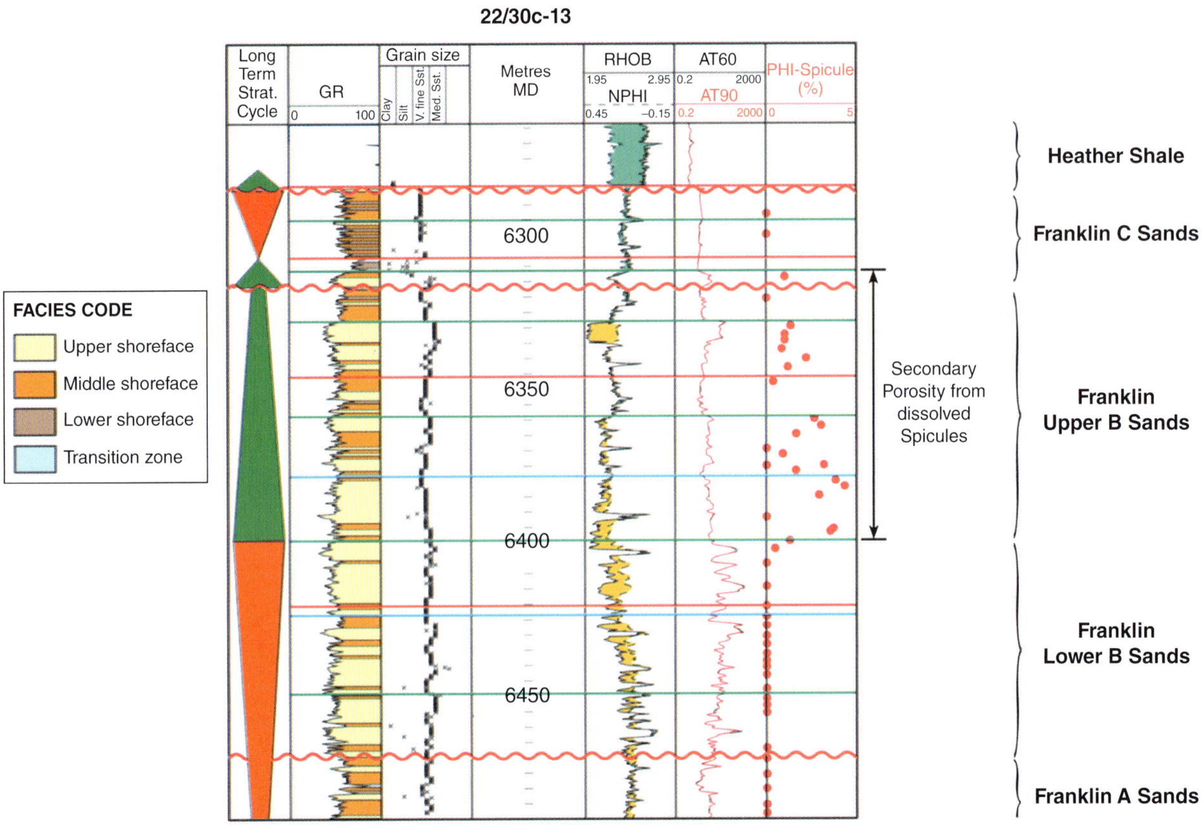

Fig. 24. Secondary porosity development related to sponge spicule dissolution in the upper part of the Franklin B Sands. GR: Gamma Ray Log; MD: Measured Depth (m); RHOB: Bulk Density; NPHI: Neutron Porosity; AT60: Medium Resistivity; AT90: Deep Resistivity; PHI-Spicule: petrographically estimated core secondary porosity related to spicule dissolution.

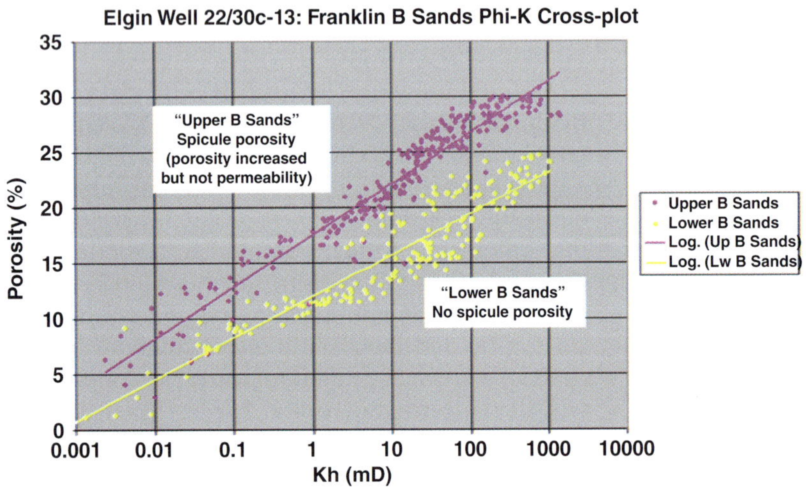

Fig. 25. Comparison of porosity-permeability relationships for the progradational Lower B Sands vs. the retrogradational Upper B Sands (of Fig. 17) where secondary porosity related to sponge spicule dissolution is developed.

Fulmar deposits display an increased marine dinocyst component and relatively more abundant saccate pollen. This could be attributed to the more condensed character of the retrogradational succession.

DISCUSSION

Key findings of this study challenge a number of concepts concerning stratigraphy and sedimentation in the Central Graben, most notably:

(1) sequence boundaries or their correlative conformities can be traced as synchronous surfaces around the basin; (2) turbidite deposits are introduced into the basin during synchronous, eustatically controlled lowstand periods and (3) turbidite deposits indicate a deep-water setting. These conclusions are typically based on regional studies but are not generally supported by the findings of local case studies, such as those reviewed in this paper, where a more complex relationship between shoreface and turbidite mass flow sandstones is documented.

In previous studies, sequence boundaries have been used for regional correlation (e.g. Donovan *et al.*, 1993); possibly with as many as eight sequence boundaries indicated to be associated to lowstand incision phases in the Upper Jurassic section (Partington *et al.*, 1993b). However, it is considered here that the erosion surfaces previously postulated to be related to incision events were more probably produced at different times due to local tectonic changes, as supported by a number of seismic-scale examples. In addition, it is considered that in a strongly wave-influenced setting, such as the Central Graben during the Oxfordian, preservation potential for sequence boundaries may be very limited due to later shoreface erosion during transgression. This may also contribute to the lack of preservation of time-equivalent continental deposits.

The current study extends the stratigraphic subdivision of the Upper Jurassic in the North Sea Central Graben using four main concepts: (1) sandstone depositional environment; (2) parasequence stacking trend analysis; (3) maximum flooding surface correlation and (4) identification of significant erosion surfaces. Detailed studies based on numerous wells in the Elgin and Franklin fields show that there is a close relationship between local tectonic changes related to grounding of Triassic pods, development of a significant erosion surface over the grounded pod crest and the synchronous/subsequent deposition of turbidite deposits in areas adjacent to the pod. A similar relationship may be present over the Fulmar Field where grounding of a Triassic-Jurassic pod is interpreted to have resulted in erosive reworking of shoreface sandstone at the crest of the structure (inverted high) and deposition of turbidite sandstones on the flanks of the structure.

Turbidite sandstones are observed in a number of areas to immediately overlie the shoreface (Shearwater Field UK-22/30b-11 well) or shoreface-shelf transition deposits (Jacqui UK-30/13-3 well). Turbidite sandstone development appears to occur widely post J54b Serratum MFS. While local structural mechanisms may have created local sources for these sandstones, it remains possible that turbidite deposition was linked to an increase in the rate of rifting. This is also consistent with the regional translation of shoreface deposition from the central parts of the basin (e.g. NO-2/4-20 well of Fig. 5) to the margins of the basin (NO-2/1-4 well of Fig. 5). This translation of the shoreface occurs above the widely recognised J56 Rosenkrantzi MFS.

The link between increased turbidite deposition and increased rifting may have occurred via a sequence of related events: (1) increased rates of extensional fault movement; (2) increased flow of salt up major faults and local differential salt movement over fault block crests and (3) subsequent increased rates of pod grounding and erosive reworking of shoreface sandstone as turbidite sandstones. During the period between the Serratum and Rosenkrantzi MFSs, the normal shoreface depositional style became increasingly disrupted by local tectonic controls. Factors that enhanced the rapid deposition of shallow water turbidites became increasingly common. The most important of these might have been pod grounding and associated erosion of previously deposited shoreface deposits.

While a deepening of the basinal setting is expected, this may not have been very substantial. Equally, the distances over which the turbidite deposits were transported may not have been large, perhaps only 1 km to 5 km (Fig. 26). Turbidite sandstone deposition is interpreted to have taken place over a shelfal area with rapid sediment supply due to a combination of initial tectonic uplift and subsequent transgressive shoreface erosion. The presence of lowstand fluvial systems supplying turbidites to the shelf via hyperpycnal flows cannot be ruled out, but there is no evidence for such systems in the examples studied. Rather, these studies show that local sourcing of sediments eroded from the crests of grounded structures was probably a more important source of turbidite or tempestite sandstones.

There has been, over the last few years, an ongoing reassessment of a number of littoral deposits that show signs of turbidite style deposition. Pattison *et al.* (2007) reviewed a number of such

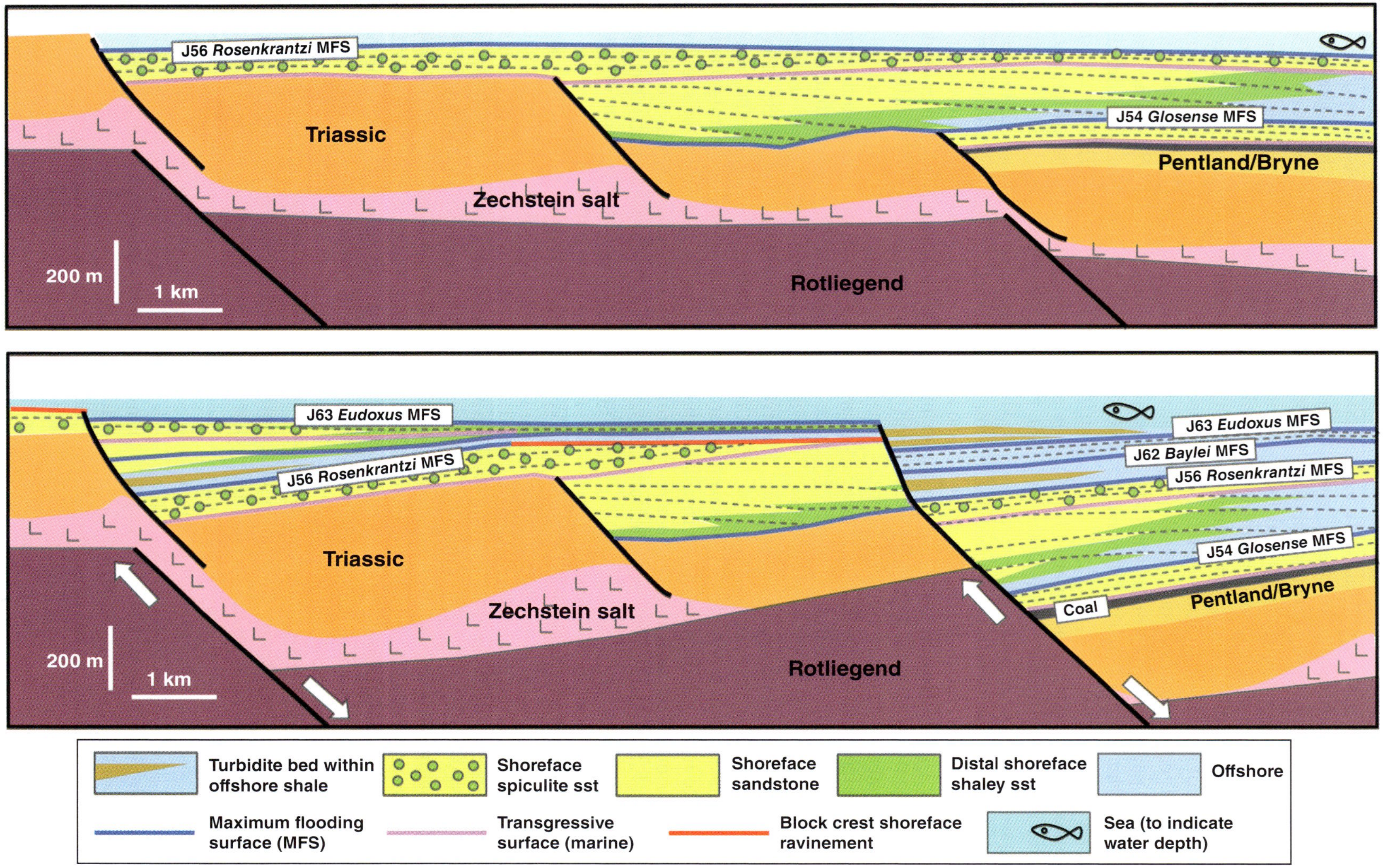

Fig. 26. Schematic tectono-sedimentary model showing the mechanism for reworking of shoreface sandstones from fault block crests and consequent truncation of spicule-bearing sandstones of the transgressive, retrogradational trend.

studies and concluded, based on their own observations of the well exposed Campanian shoreface to shelf strata of Book Cliffs, Utah that shoreface-to-shelf facies models should be revised to incorporate turbidite-rich shelf deposits in some cases. Other Western interior studies have also revised earlier turbidite interpretations to a shallow, littoral interpretation, e.g. Cardium Formation of the Ricinus Field (Walker 1985c; Walker, 1995). Whilst the Cretaceous of the US Western Interior foreland basin is a very different tectonic setting, it indicates that the processes for generating turbidite style deposits in shallow water are being reinvestigated, particularly in the light of new knowledge about modern depositional settings (also reviewed by Pattison *et al.*, 2007). These studies add confidence to the inference that the recognition of turbidite type facies does not necessarily imply a major increase in water depth. The observation of interfingering between shoreface and turbidite deposition on both Elgin-Franklin and Fulmar fields tends to suggest that a linked mechanism of tectonically-enhanced regressive erosion followed by transgressive ravinement is a more probable mechanism to explain the combination of significant erosion and turbidite mass flow deposition observed in these areas (Fig. 26).

To summarise, the four local structural scenarios explaining the deposition of four of the best known Oxfordian to Early Kimmeridgian age turbidite deposits of the Central Graben casts some doubt on whether there is evidence in the Central Graben for a global eustatic sea-level mechanism that could have introduced a widespread and dateable phase of turbidite deposition by lowstand incised valley formation around the basin margin. This conclusion compares well with that of Underhill (1991) in a study of the Late Jurassic seismic stratigraphy of the Inner Moray Firth which concluded that active tectonics in extensional basins probably overprint any regional eustatic signal due to a variety of local tectonic and sedimentary processes documented by seismic onlaps and downlaps.

CONCLUSIONS

1 Upper Jurassic sandstone deposits of the North Sea Central Graben can be characterised according to three tectono-sedimentary models termed A, B and C which depend on the evolving geometry of the rift, geographical location (rift margin or rift centre) and the timing of Triassic pod grounding on the Rotliegend due to displacement of Zechstein salt.

2 The development of significant erosion surfaces within Upper Jurassic age shallow marine sandstones was linked to tectonic mechanisms. It is also probable that these surfaces are not basin wide sequence boundaries, rather they are local surfaces reflecting the dynamic and geographically varied evolution of the basin. Maximum flooding surfaces provide a more reliable method of correlation throughout the basin.

3 Identification of erosive truncation has improved the understanding of the relationships between tectonics and sedimentation. Intra-Jurassic erosion of structurally controlled highs, such as (1) fault block crests, or (2) grounded pods, can result in uplift and subaerial erosion or submarine shoreface reworking of previously deposited Oxfordian age shoreface sandstones.

4 Tectonic uplift has resulted in localised forced regression with associated shoreface erosion. The products of this erosion, produced especially during storm events, may be re-deposited as turbidites. These turbidites are transported short distances (a few kilometres) and are ponded in depositional lows such as the hangingwall of major faults. Five case studies have been presented which suggest that local structural controls are able to explain: (1) the development of local unconformities and (2) the local deposition of Oxfordian to Early Kimmeridgian age turbidite deposits.

5 It is suggested that some aspects of Upper Jurassic sandstone deposition as previously presented require revision. The proposal that many phases of incised valley development occurred during the Upper Jurassic generating 3rd order sequence boundaries (Partington *et al.*, 1993b) does not appear to be backed by sedimentary evidence. While turbidite sandstones are evidently developed at various times in the basin history, their relationship to proximal fluvial systems and incised valleys is unproven. Such turbidites may have been triggered by either: (1) local tectonic uplift promoting fault footwall failure, or (2) storm wave erosion of shoreface deposits due to relative sea-level fall.

6 Transgressive, retrogradationally stacked shoreface sandstones are commonly spiculite rich. This is related to low energy shorefaces with low

sediment supply that characterise retrogradationally stacked shoreface parasequence sets. Spiculitic sandstone layers can also serve as useful horizons for correlation and help to highlight significant, intra-Jurassic erosive truncation of otherwise massive shoreface sandstones.

ACKNOWLEDGEMENTS

This joint Total E&P Norge – Total E&P UK study reflects the work of a team of geologists and geophysicists working over a period of several years on a database that includes more than 1000 wells. The authors thank co-venturers in the UK Elgin-Franklin project (Eni, BG International, Ruhrgas, Esso, Chevron, Dyas and Oranje-Nassau) for permission to publish studies of the Elgin-Franklin correlation update completed in 2002. Parts of the current study were first presented at the AAPG annual meeting in Houston of the same year. We also thank partners in Norwegian licences PL018, PL044 and PL146 for permission to publish data from these areas: Statoil in PL146/333 and Statoil, ConocoPhillips, Eni Norge and Petoro in PL018/044. Many Total E & P geoscientists provided support to this study, notably: J. Suiter, F. Walgenwitz, C. Gout, A. Jourdan, M. Dougherty, B. Magnier, D. Michoux, P. Stinson, A. Braghiroli and C. Mathieu. We thank all concerned for their contributions. We would also like to thank T. McKie, H.D. Johnson and R. Ravnås for reviewing and suggesting improvements to this paper.

REFERENCES

Aase, N.E., **Bjorkum, P.A.** and **Nadeau, P.H.** (1996) The effect of grain-coating microquartz on preservation of reservoir porosity, *AAPG Bull.*, **80**, 1654–1673.

Ajdukiewicz, J.M., **Paxton, S.T.**, **Jennette, D.C.**, **Nicholson, P.H.** and **Erratt, D.** (2001) *Reservoir Quality Controls in the Deep Fulmar Formation, Central North Sea (Abstract).* AAPG Annual Convention, Denver, Colorado.

Armstrong, L.A., **Ten Have, A.** and **Johnson, H.D.** (1987) The geology of the Gannet Fields, Central North Sea, UK Sector. In: *Petroleum Geology of North West Europe*, (Eds J. Brooks and K.W. Glennie), pp 533–548. Graham and Trotman, London.

Bartholomew, I.D. (1993) Structural styles and their evolution in the North Sea area Introduction and review. In: *Petroleum Geology of Northwest Europe: Proceedings of the 4th Conference* (Ed. J.R. Parker), pp 1081–1082. Geological Society, London.

Bartholomew, I.D., **Peters, J. M.** and **Powell, C. M.** (1993) Regional structural evolution of the North Sea: oblique slip and the reactivation of basement lineaments. In: *Petroleum Geology of Northwest Europe: Proceedings of the 4th Conference* (Ed. J.R. Parker), pp 1109–1122. Geological Society, London.

Bjørnseth, H.M. and **Gluyas, J.** (1995) Petroleum exploration in the Ula Trend. In: *Petroleum Exploration and Exploitation in Norway* (Ed. S. Hanslien), *NPF Spec. Publ.*, **4**, 85–96. Elsevier, Amsterdam.

Bloch, S., **Lander, R.H.** and **Bonnell, L.** (2002) Anomalously high porosity and permeability in deeply buried sandstone reservoirs: origin and predictability. *AAPG Bull.*, **86**, 301–328.

Bouma, A.H. (1962) *Sedimentology of Some Flysch Deposits: a Graphic Approach to Facies Interpretation.* Elsevier, Amsterdam.

Brown, A., **Mitchell, A.W.**, **Nilssen, I.R.**, **Stewart, I.J.** and **Svela, R.T.** (1992) Ula Field: relationship between structure and hydrocarbon distribution. In: *Structural and Tectonic Modelling and its Application to Petroleum Geology* (Eds R.M. Larsen, H. Brekke, B.T. Larsen and E. Talleraas), *NPF Spec. Publ.*, **1**, 409–420. Elsevier, Amsterdam.

Burns, F.E., **Burley, S.D.**, **Gawthorpe, R.L.** and **Pollard, J.E.** (2005) Diagenetic signatures of stratal surfaces in the Upper Jurassic Fulmar Formation, Central North Sea, UKCS, *Sedimentology*, **52**, 1155–1185.

Cannon, S.J.C. and **Gowland S.** (1996) Facies controls on reservoir quality in the Late Jurassic Fulmar Formation, Quadrant 21, UKCS. In: *Geology of the Humber Group: Central Graben and Moray Firth, UKCS* (Eds A. Hurst, H.D. Johnson, S.D. Burley, A.C. Canham and D.S. MacKertich) *Geol. Soc. London Spec. Publ.*, **114**, 215–233.

Carruthers, A., **McKie, T.**, **Price, J.**, **Dyer, R.**, **Williams, G.** and **Watson, P.** (1996) The application of sequence stratigraphy to the understanding of Late Jurassic turbidite plays in the Central North Sea, UKCS. In: *Geology of the Humber Group: Central Graben and Moray Firth, UKCS* (Eds A. Hurst, H.D. Johnson, S.D. Burley, A.C. Canham and D.S. MacKertich), *Geol. Soc. London Spec. Publ.*, **114**, 29–46.

Catuneanu, O. (2006) *Principles of Sequence Stratigraphy.* Elsevier, Amsterdam.

Clelland, W.D., **Kantorowicz J.D.** and **Fens, T.W.** (1993) Quantitative analysis of pore structure and its effect on reservoir behaviour: Upper Jurassic Ribble Member sandstones, Fulmar Field, UK North Sea. In: *Advances in Reservoir Geology* (Ed. **M. Ashton**), *Geol. Soc. London Spec. Publ.*, **69**, 57–79.

Coward, R.N. (2003) The Erskine Field, Block 23/26, UK North Sea. In: *United Kingdom Oil and Gas Fields: Commemorative Millennium Volume* (Eds J.G. Gluyas and H.M. Hichens), *Geol. Soc. London Mem.*, **20**, 523–536.

Davies, R.J., **O'Donnell, D.**, **Bentham, P.N.**, **Gibson, J.P.C.**, **Curry, M.R.**, **Dunay, R.E.** and **Maynard, J.R.** (1999) The origin and genesis of major Jurassic unconformities within the triple junction area of the North Sea, UK. In: *Petroleum Geology of Northwest Europe: Proceedings of the 5th Conference* (Eds A.J. Fleet and S.A.R. Boldy), pp 117–131. Geological Society, London.

Donovan, A.D., **Djakic, A.W.**, **Ioannides, N.S.**, **Garfield, T.R.** and **Jones, C.R.** (1993) Sequence stratigraphic control on Middle and Upper Jurassic reservoir distribution within the UK Central North Sea. In: *Petroleum Geology of Northwest Europe: Proceedings of the 4th Conference* (Ed. J.R. Parker), pp 251–269. Geological Society, London.

Doré, G. and **Robbins, J.** (2005) The Buzzard Field. In: *Petroleum Geology: North-West Europe and Global Perspectives - Proceedings of the 6th Petroleum Geology Conference* (Eds A.G. Doré and B.A. Vining), pp 241–252. Geological Society, London.

Duxbury, S., Kadolsky, D. and **Johansen, S.** (1999) Sequence stratigraphic subdivision of the Humber Group in the Outer Moray Firth Area (UKCS, North Sea). In: *Biostratigraphy in Production and Development Geology* (Eds R. Jones and M. Simmons), *Geol. Soc. London Spec. Publ.*, **152**, 23–54.

Eggink, J.W., Riegstra, D.E. and **Suzanne, P.** (1996) Using 3D seismic to understand the structural evolution of the UK Central North Sea. *Petrol. Geosci.*, **2**, 83–96.

Erratt, D. (1993) Relationships between basement faulting, salt withdrawal and Late Jurassic rifting, UK Central North Sea. In: *Petroleum Geology of Northwest Europe: Proceedings of the 4th Conference* (Ed. J.R. Parker), pp 1211–1220. Geological Society, London.

Erratt, D., Thomas, G.M. and **Wall, G.R.T.** (1999) The evolution of the Central North Sea Rift. In: *Petroleum Geology of Northwest Europe: Proceedings of the 5th Conference* (Ed. J.R. Parker), pp 63–82. Geological Society, London.

Erratt, D., Nicholson, P. H., Winefield, P., Milton-Worssell, R. J., Cayley, G. T. and **Arter, G.** (2005) Exploration history of the high-pressure, high-temperature plays: UK Central North Sea. In: *Petroleum Geology: North-West Europe and Global Perspectives, Proceedings of the 6th Petroleum Geology Conference* (Eds A.G. Doré and B.A. Vining), pp 253–267. Geological Society, London.

Fleet, A.J. and **Boldy, S.A.R.** (1999) *Petroleum Geology of Northwest Europe: Proceedings of the 5th Conference.* Geological Society, London, 1398 pp.

Fraser, S.I., Robinson, A.M., Johnson, H.D., Underhill, J.R., Kadolsky, D.G.A, Connell, R. Johannessen, P. and **Ravnås, R.** (2002) Upper Jurassic. In: *The Millenium Atlas: Petroleum Geology of the Central and Northern North Sea* (Eds D. Evans, C. Graham, A. Armour and P. Bathurst), pp 157–189. Geological Society, London.

Galloway, W.E. (1989) Genetic stratigraphic sequences in basin analysis I: Architecture and genesis of flooding surface bounded depositional units. *AAPG Bull.*, **69**, 125–142.

Gilham, R., Hercus, C., Evans, A. and de **Haas, W.** (2005) Shearwater (UK Block 22/30b): managing changing uncertainties through field life. In: *Petroleum Geology: North-West Europe and Global Perspectives—Proceedings of the 6th Petroleum Geology Conference*, (Eds A.G. Doré and B.A. Vining), pp 663–673. Geological Society, London.

Goldsmith, P.J., Hudson, G. and **Van Veen, P.** (2003) Triassic. In: *The Millennium Atlas: Petroleum Geology of the Central and Northern North Sea*, (Eds D. Evans, C. Graham, A. Armour and P. Bathurst), pp 105–127. Geological Society, London.

Gowland, S. (1996) Facies characteristics and depositional models of highly bioturbated shallow marine siliciclastic strata: an example from the Fulmar Formation (Late Jurassic), UK Central North Sea. In: *Geology of the Humber Group: Central Graben and Moray Firth, UKCS* (Eds A. Hurst, H.D. Johnson, S.D. Burley, A.C. Canham and D.S. Mackertich), *Geol. Soc. London Spec. Publ.*, **114**, 185–214.

Guerin, G., Vendeville, B.C. and **Raillard, S.** (1998) Triassic structural distribution of the Zechstein salt in southern North Sea and consequences for further tectonic history (abstract), *Annales Geophysicae*, **16**, p. C 76.

Hall, S.A. (1992) The Angus Field, a subtle trap. In: *Exploration Britain - Geological Insights for the Next Decade* (Ed. R.F.P. Hardman), *Geol. Soc. London Spec. Publ.*, **67**, p. 151–185.

Hayward, R.D., Martin, C.A.L., Harrison, D., Van Dort, G., Guthrie, S. & Padget, N. (2003) The Flora Field, Blocks 31/26a, 31/26c, UK North Sea. In: *United Kingdom Oil and Gas Fields, Commemorative Millenium Volume (Eds J.G. Gluyas and H.M. Hichens) Geol. Soc. London Mem.*, **20**, 549–555.

Helgeson, D.E. (1999) Structural development and trap formation in the Central North Sea HP/HT play. In: *Petroleum Geology of Northwest Europe: Proceedings of the 5th Conference* (Eds A.J. Fleet and S.A.R Boldy), pp 1029–1034. Geological Society, London.

Hendry, J.P. and **Trewin, N.H.** (1995) Authigenic quartz microfabrics in Cretaceous turbidites: evidence for silica transformation processes in sandstones. *J. Sed. Res.*, **65A**, 380–392.

Hendry, J.P., Wilkinson, M., Fallick, A.E. and **Haszeldine, R.S.** (2000a) Ankerite cementation in deeply buried Jurassic sandstone reservoirs of the Central North Sea. *J. Sed. Res.*, **70**, 1, 227–239.

Hendry, J.P., Wilkinson, M. and **Fallick, A.E.** (2000b) Disseminated 'jigsaw piece' dolomite in Upper Jurassic shelf sandstones, Central North Sea: an example of cement growth during bioturbation? *Sedimentology*, **47**, 631–644.

Howell, J.A. and **Flint, S.S.** (1996) A model for high resolution sequence stratigraphy within extensional basins. In: *High Resolution Sequence Stratigraphy: Innovations and Applications* (Eds J.A. Howell and J.F. Aitken), *Geol. Soc. London Spec. Publ.*, **104**, 129–137.

Howell, J.A., Flint, S.S. and **Hunt, C.** (1996) Sedimentological aspects of the Humber Group (Upper Jurassic) of the South Central Graben, UK North Sea. *Sedimentology*, **43**, 89–114.

Jackson, C.A.-L., Kane, K.E. and **Larsen, E.** (2010) Structural evolution of minibasins on the Utsira High, northern North Sea; implications for Jurassic sediment dispersal and reservoir distribution. *Petrol. Geosci.*, **16**, 105–120.

Jeremiah, J.M. and **Nicholson, P.H.** (1999) Middle Oxfordian to Volgian sequence stratigraphy of the Greater Shearwater area. In: *Petroleum Geology of Northwest Europe: Proceedings of the 5th Conference* (Eds A.J. Fleet and S.A.R. Boldy), pp 153–170. Geological Society, London.

Johnson, H.D., Mackay, T.A. and **Stewart, D.J.** (1986) The Fulmar Oil Field (Central North Sea): geological aspects of its discovery, appraisal and development. *Mar. Pet. Geol.*, **3**, 99–125.

Johnson, H., Leslie, A.B., Wilson, C.K., Andrews, I.J. and **Cooper, R.M.** (2005) Middle Jurassic, Upper Jurassic and Lower Cretaceous of the UK Central and Northern North Sea. British Geological Survey Research Report, RR/03/001, 42 pp.

Jones, P., Wonham, J., Sharp, D., Nordfjord, S., Ekroll, M. and **Haugen, J.E.** (2013) King Lear: Rewriting the Play (Abstract), *Exploration Revived 2013: The 5th NPF Biennial Petroleum Geology Conference.* Published by

Norsk Petroleumsforening (NPF) in their conference proceedings of March 19, 2013.

Kantorowicz, J.D, Bryant, I.D. and **Dawans, J.M.** (1987) Controls on the geometry and distribution of carbonate cements in Jurassic sandstones: Bridport Sands, Southern England and Viking Group, Troll Field, Norway. In: *Diagenesis of Sedimentary Sequences* (Ed. J.D. Marshall), *Geol. Soc. London Spec. Publ.*, **36**, 103–118.

Kuhn, O., Smith, S.W., Van Noort, K. and **Loiseau, B.** (2003) The Fulmar Field, Blocks 30/16, 30/11b, UK North Sea. In: *United Kingdom Oil and Gas Fields, Commemorative Millenium Volume* (Eds J.G. Gluyas and H.M. Hichens), *Geol. Soc. London Mem.*, **20**, 563–585.

Landrø, M. (2011) Seismic monitoring of an old underground blowout – 20 years later. *First Break*, **29**, 39–48.

Lasocki, J., Lejay, A., Russell, K. and **Thickpenny, A.** (1997) Problematic event beds in the Franklin Sands, Well 29/5b-6, 6Z. In: *Cores from the Northwest European Hydrocarbon Province: An Illustration of Geological Applications from Exploration to Development* (Eds C.D. Oakman, J.H. Martin and P.W.M. Corbett), pp 205–214. Geological Society, London.

Lasocki, J., Guemene, J. M., Hedayati, A., Legorjus, C. and **Page, W. M.** (1999) The Elgin and Franklin fields: UK Blocks 22/30c, 22/30b and 29/5b. In: *Petroleum Geology of Northwest Europe: Proceedings of the 5th Conference* (Eds A.J. Fleet and S.A.R. Boldy), pp 1007–1020. Geological Society, London.

Mjelde, Ø. (1991) Geology, Pressure Prognosis and Status of Knowledge. Rock mechanics. In: *2/4-14 Experience Transfer Seminar*, Stavanger, January 16-17, 1991, Saga Petroleum AS and Rogalands-Forskning, Stavanger, 1–14.

Morris, J.E., Hampson, G.J. and **Johnson, H.D.** (2006) A sequence stratigraphic model for an intensely bioturbated shallow-marine sandstone: the Bridport Sand Formation, Wessex Basin, UK, *Sedimentology*, **53**, 1229–1263.

Mulder, T. and **Syvitski. J.P.M.** (1995) Turbidity currents generated at river mouths during exceptional discharges to the world oceans. *J. Geol.*, **103**, 285–299.

Mulder. T, Syvitski, J.P.M., Migeon, S., Faugeres, I.E. and **Savoye, B.**, (2003). Marine hyperpycnal flows: initiation, behaviour and related deposits. A review. *Mar. Petr. Geol.*, **20**, 861–882.

Osborne, M.J. and **Swarbrick, R.E.** (1997) Mechanisms for generating overpressure in sedimentary basins: a reevaluation. *AAPG Bull.*, **81**, 1023–1041.

Osborne, M.J. and **Swarbrick, R.E.** (1999) Diagenesis in North Sea HPHT clastic reservoirs – consequences for porosity and overpressure prediction. *Mar. Petr. Geol.*, **16**, 337–353.

Parker, J.R. (1993) *Petroleum Geology of Northwest Europe: Proceedings of the 4th Conference*, Geological Society, London.

Partington, M.A., Mitchener, B.C., Milton, N.J. and **Fraser, A.J.** (1993a) Genetic sequence stratigraphy for the North Sea Late Jurassic and Early Cretaceous. In: *Petroleum Geology of Northwest Europe: Proceedings of the 4th Conference* (Ed. J.R. Parker), pp 347–370. Geological Society, London.

Partington, M.A., Mitchener, B.C. and **Underhill, J.R** (1993b) Biostratigraphic calibration of genetic stratigraphic sequences in the Jurassic - lowermost Cretaceous (Hettangian-Ryazanian) of the North Sea and surrounding areas. In: *Petroleum Geology of Northwest Europe: Proceedings of the 4th Conference* (Ed. J.R. Parker), pp 371–386. Geological Society, London.

Pattison, S.A.J., Ainsworth, R.B. and **Hoffman, T.A.** (2007) Evidence of across-shelf transport of fine-grained shelf channels in the Campanian Aberdeen Member, Book Cliffs, Utah, USA. *Sedimentology*, **24**, 1033–1063.

Pickering, K.T. (1995) Are enigmatic erosional sandy wave-like bedforms in Jurassic Bridport Sands, Dorset, due to standing waves? *J. Geol. Soc. London*, **152**, 481–485.

Price, J., Dyer, R., Goodall, I., McKie, T., Watson, P. and **Williams, G.** (1993) Effective stratigraphical subdivision of the Humber Group and the Late Jurassic evolution of the UK Central Graben. In: *Petroleum Geology of Northwest Europe: Proceedings of the 4th Conference* (Ed. J.R. Parker), pp 443–458. Geological Society, London.

Rattey, R.P. and Hayward, A.B. (1993) Sequence stratigraphy of a failed rift system: the Middle Jurassic to Early Cretaceous basin evolution of the Central and Northern North Sea. In: *Petroleum Geology of Northwest Europe: Proceedings of the 4th Conference* (Ed. J.R. Parker), pp 215–249. Geological Society, London.

Ravnås, R. and **Steel, R.J.** (1998) Architecture of marine rift-basin successions. *AAPG Bull.*, **82**, 110–146.

Ravnås, R., Nøttvedt, A., Steel, R.J. and **Windelstad, J.** (2000) Syn-rift sedimentary architectures in the Northern North Sea. In: Dynamics of the Norwegian Margin (Ed. A. Nøttvedt), *Geol. Soc. London Spec. Publ.*, **167**, 133–177.

Robinson, A.J. (1990) *The Sedimentology and Stratigraphy of the Ribble Sandstone Member and Related Deposits, Fulmar Field, Central North Sea.* M.Sc. Thesis, University of Aberdeen.

Sansom, P.J. (2010) Reappraisal of the sequence stratigraphy of the Humber Group of the UK Central Graben. In: *Petroleum Geology: From Mature Basins to New Frontiers – Proceedings of the 7th Petroleum Geology Conference* (Eds B.A. Vining and S.C. Pickering), pp 177–211. Geological Society, London.

Sears, R.A., Harbury, A.R., Protoy, A.J.G. and **Stewart, D. J.** (1993) Structural styles from the Central Graben in the UK and Norway. In: *Petroleum Geology of Northwest Europe: Proceedings of the 4th Conference* (Ed J.R. Parker), pp 1231–1244. Geological Society, London.

Smith, R.I., Hodgson, N. and **Fulton, M.** (1993) Salt control on Triassic reservoir distribution, UKCS Central North Sea. In: *Petroleum Geology of Northwest Europe: Proceedings of the 4th Conference* (Ed. J.R. Parker), pp 547–557. Geological Society, London.

Spaak, P., Almond, J., Salahudin, S., Mohd Salleh, Z. and **Tosun, O.** (1999) Fulmar: a mature field revisited. In: *Petroleum Geology of Northwest Europe: Proceedings of the 5th Conference* (Eds A.J. Fleet and S.A.R. Boldy), pp 1089–1100. Geological Society, London.

Spathopoulos F., Doubleday P.A. and Hallsworth C. (2000) Structural and depositional controls on the distribution of the Upper Jurassic shallow marine sandstones in the Fife and Angus fields area, Quadrants 31 & 39, UK Central North Sea. *Marine and Petroleum Geology*, **17**, 1053–1082.

Spencer, A.M., Home, P.C. and **Wiik, V.** (1986). Habitat of hydrocarbons in the Jurassic Ula Trend, Central Graben, Norway. In: *The Habitat of Hydrocarbons in the Norwegian Continental Shelf* (Eds A.M. Spencer *et al.*), pp 111–127, Graham and Trotman, London.

Stevens, D.A. and **Wallis, R.J.** (1991) The Clyde Field, Block 30/17b, UK North Sea. In: *United Kingdom Oil and Gas Fields, 25 Years Commemorative Volume.* (Ed. I.L. Abbotts), *Geol. Soc. London Mem.*, **14**, 309–316.

Stewart, I.J. (1993) Structural control on the late Jurassic age shelf system, Ula Trend, Norwegian North Sea. In: *Petroleum Geology of Northwest Europe: Proceedings of the 4th Conference* (Ed. J. R. Parker), pp 469–483. Geological Society, London.

Stewart, S.A. (2007) Salt tectonics in the North Sea Basin: a structural style template for seismic interpreters. In: *Deformation of the Continental Crust: The Legacy of Mike Coward* (Eds A.C. Ries, R.W.H. Butler and R.H. Graham), *Geol. Soc. London Spec. Publ.*, **272**, 361–396.

Stewart, S.A. and **Clark, J.A.** (1999) Impact of salt on the structure of the Central North Sea hydrocarbon fairways. In: *Petroleum Geology of Northwest Europe: proceedings of the 5th Conference* (Eds A.J. Fleet and S.A.R. Boldy), pp 179–200. Geological Society, London.

Stewart, S.A., Fraser, S.I., Cartwright, J.A., Clark, J.A. and **Johnson, H.D.** (1999) Controls on Upper Jurassic sediment distribution in the Durward-Dauntless area, UK Blocks 21/11, 21/16. In: *Petroleum Geology of Northwest Europe: proceedings of the 5th Conference* (Eds A.J. Fleet and S.A.R. Boldy), pp 879–896. Geological Society, London.

Stockbridge, C.P. and **Gray, D.I.** (1991) The Fulmar Field, blocks 30/16 & 30/1 lb. UK North Sea. In: *United Kingdom Oil and Gas Fields, 25 Years Commemorative Volume.* (Ed. I.L. Abbotts), *Geol. Soc. London Mem.*, **14**, 309–316.

Stow, D.A.V. and **Shanmugam, G.** (1980) Sequence of structures in fine grained turbidites: comparison of recent deep-sea and ancient flysch sediments. *Sediment. Geol.*, **25**, 23–42.

Suiter, J., Romani, R., Arnaud, J., Hollingworth, S. and **Hawkins, K.** (2005) Reducing structural uncertainties through anisotropic pre-stack depth imaging: examples from the Elgin/Franklin/Glenelg HP/HT fields area, Central North Sea. In: *Petroleum Geology: North-West Europe and Global Perspectives - Proceedings of the 6th Petroleum Geology Conference*, (Eds A.G. Doré and B.A. Vining), pp. 1435–1448. Geological Society, London.

Taylor, A.M. and **Gawthorpe, R.L.** (1993) Application of sequence stratigraphy and trace fossil analysis to reservoir description: examples from the Jurassic of the North Sea. In: *Petroleum Geology of Northwest Europe: Proceedings of the 4th Conference* (Ed. J.R. Parker), pp. 317–335. Geological Society, London.

Taylor, T.R., Giles, M.R., Lori, A.H., Diggs, T.N., Braunsdorf, N.R., Birbiglia, G.V., Kittridge, M.G., Macaulay, C. and **Espejo, I.S.** (2010) Sandstone diagenesis and reservoir quality prediction: Models, myths and reality. *AAPG Bull.*, **94**, 8, 1093–1132.

Underhill, J.R. (1991) Controls on Late Jurassic seismic sequences, Inner Moray Firth, UK North Sea: a critical test of a key segment of Exxon's original global cycle chart. *Basin Res.*, **3**, 79–98.

Underhill, J.R. and **Partington, M.A.** (1993) Use of genetic sequence stratigraphy in defining and determining a regional tectonic control on the "Mid-Cimmerian Unconformity" - implications for North Sea Basin development and the Global Sea Level Chart. In: *Siliciclastic Sequence Stratigraphy: Recent Developments and Applications* (Eds H.W. Posamentier and P. Wiemer), *AAPG Mem.*, **58**, 449–484.

Vagle, G. B., Hurst, A. and **Dypvik, H.** (1994) Origin of quartz cements in some sandstones from the Jurassic of the Inner Moray Firth (UK). *Sedimentology*, **41**, 363–377.

Van Wagoner, J.C, Posamentier, H.W., Mitchum, R.M., Vail, P.R., Sarg, J.F., Loutit, T.S. and **Hardenbol, J.** (1988) An overview of the fundamentals of sequence stratigraphy and key definitions. In: *Sea-Level Changes: an Integrated Approach.* (Eds C.K. Wilgus, B.S. Hastings, H. Posamentier, J. Van Wagoner, C.A. Ross and C.G. St. C. Kendall), *SEPM Spec. Publ.*, **42**, 39–45.

Van Wagoner, J.C., Mitchum, R.M., Campion, K.M. and **Rahmanian, V.D.** (1990) *Siliciclastic Sequence Stratigraphy in Well Logs, Cores and Outcrops: Concepts for High-Resolution Correlation of Time and Facies*, pp. 55. AAPG, Tulsa.

Veldkamp, J.J., Gaillard, M.G., Jonkers, H.A. and **Levell, B.K.** (1996) A Kimmeridgian time-slice through the Humber Group of the central North Sea: a test of sequence stratigraphic methods. In: *Geology of the Humber Group: Central Graben and Moray Firth, UKCS* (Eds. A. Hurst, H.D. Johnson and S.D. Burley), *Geol. Soc. London Spec. Publ.*, **114**, 1–28.

Vendeville, B.C. and **Guerin, G.** (1998) Structural decoupling by the Zechstein salt during multiphase tectonics in the southern Norwegian North Sea (ext. abs.). In: *AAPG International Conference and Exhibition Extended Abstracts Volume* (Eds M.R. Mello and P.O. Yilmaz), Rio de Janeiro, 972–973.

Vendeville, B.C. and **Jackson, M.P.A.** (1992a). The rise of diapirs during thin-skinned extension. *Mar. Petr. Geol.*, **9**, 331–353.

Vendeville, B.C. and **Jackson, M.P.A.** (1992b) Fall of diapirs during thin-skinned extension. *Mar. Petr. Geol.*, **9**, 354–371.

Wakefield, L.L., Droste, H., Giles, M.R. and **Janssen, R.** (1993) Late Jurassic plays along the western margin of the Central Graben. In: *Petroleum Geology of Northwest Europe: Proceedings of the 4th Conference* (Ed. J.R. Parker), 459–468. Geological Society, London.

Walgenwitz, F. and **Wonham, J.P.** (2003) Causes for the Preservation of High Reservoir Quality at Great Depth in the HP-HT Reservoirs of the Elgin-Franklin fields, Central North Sea Graben, UK (Abstract) AAPG International Conference (September 21–24, 2003), Barcelona, Spain.

Walker, R.G. (1985a) Geological evidence for storm transportation and deposition on ancient shelves. In: *Shelf Sands and Sandstone Reservoirs* (Eds R.W. Tillman, D.J.P. Swift and R.G. Walker), *SEPM Short Course Notes*, **13**, 243–302.

Walker, R.G. (1985b) Comparison of shelf environments and deep basin turbidite systems. In: *Shelf Sands and Sandstone Reservoirs* (Eds R.W. Tillman, D.J.P. Swift and R.G. Walker), *SEPM Short Course Notes*, **13**, 465–502.

Walker, R.G. (1985c) Cardium Formation at Ricinus Field, Alberta: a channel cut and filled by turbidity currents in Cretaceous Western Interior Seaway. *AAPG Bull.*, **69**, 1963–1981.

Walker, R.G. (1995) An incised valley in the Cardium Formation at Ricinus, Alberta: re-interpretation as an estuary fill. In: *Sedimentary facies analysis : a tribute to the research and teaching of Harold G. Reading* (Ed. A.G. Plint). *Int. Assoc. Sedimentol. Spec. Publ.*, **22**, 47–74.

Wilkinson, M., Darby, D., Haszeldine, R.S. and **Couples, G.D.** (1997) Secondary porosity generation during deep burial associated with overpressure leak-off; Fulmar Formation, United Kingdom Central Graben. *AAPG Bull.*, **81**, 5, 803–813.

Wilkinson, M., Haszeldine, R.S and **Fallick, A.E.** (2006) Hydrocarbon filling and leakage history of a deep geopressured sandstone, Fulmar Formation, United Kingdom North Sea. *AAPG Bull.*, **90**, 12, 1945–1961.

Wonham, J., Suiter, J., Gerard, J. and **Walgenwitz, F.** (2002) *Reappraisal of Fulmar Formation Correlation and Impact on Reservoir Quality Distribution: Elgin and Franklin fields, UK Central North Sea (Abstract)*, AAPG Annual Meeting, Houston, Texas.

Int. Assoc. Sedimentol. Spec. Publ. (2014) **46**, 513–550

The Garn Formation (Bajocian-Bathonian) in the Kristin Field, Halten Terrace: its origin, facies architecture and primary heterogeneity model

CARLO MESSINA*‡, WOJCIECH NEMEC†, ALLARD W. MARTINIUS* and CARSTEN ELFENBEIN*[1]

**Statoil Research Centre, Rotvoll, 7005 Trondheim, Norway*
†Department of Earth Science, University of Bergen, 5007 Bergen, Norway
‡Present address: Statoil Gulf Services LLC, Houston 77042 Texas, USA (E-mail: cmes@statoil.com)
[1]Present address: Det Norske Oljeselskap ASA, Føniks, 7011 Trondheim, Norway.

ABSTRACT

The Garn Formation of Halten Terrace, Norwegian Continental Shelf, has long been recognised as a gas-bearing and condensate-bearing succession of shallow-marine arenitic sandstones deposited by early syntectonic sedimentation during Jurassic rifting. However, the origin and spatial architecture of this formation remained unclear and rather controversial, which also led to simplistic models of reservoir heterogeneity. The present study from the Kristin Field indicates that the sand was deposited in an actively subsiding, incipient graben within the narrow Jurassic seaway that linked the Boreal and Tethys open seas. Deposition involved repetitive cycles of tide-dominated to wave-dominated sedimentation, attributed to subsidence-driven relative sea-level changes. Sedimentological analysis indicates that sedimentation involved the development of tidal sand ridges, with episodic storm influence and a concurrent accumulation of sand in inter-ridge swales. Fairweather wave action with ridge erosion and sand bypass prevailed once accommodation became filled by sea floor aggradation, until a new increase in accommodation occurred due to tectonic subsidence. The transgressive-regressive cycles formed an overall transgressive parasequence set ~100 m thick, culminating in major marine flooding due to the regional climax of the rifting phase with associated structural collapse of the shelf. The suggested facies anatomy of the formation has important implications for reservoir heterogeneity, with the tidal sandstone ridges as a main architectural element. These semi-isolated sandstone bodies are more permeable than the surrounding sandstone facies and consist of dune cross-strata sets that form highly anisotropic reservoir mini-compartments. The two types of anisotropy are considered to be the main source of reservoir heterogeneity and a probable cause of the rapid pressure decline and low gas recovery in the Kristin Field. Considerable amounts of gas condensate may potentially be entrapped in the tidal sandstone ridges as the relative water saturation rises and its capillary effect rapidly increases. As an improvement for reservoir model, the dimensions of the sandstone ridges are estimated and a statistical approach is used to estimate the frequency distribution of cross-set volumes from the measured cross-set thicknesses. The estimates allow these sandstone bodies in the reservoir model to be populated with realistic percentages of cross-set volumes. The new heterogeneity model may facilitate more reliable simulation of condensate flow in the reservoir and a better assessment of gas recovery efficiency.

Keywords: Dune cross-sets, facies analysis, Norwegian Continental Shelf, parasequences, sedimentology, reservoir anisotropy, statistical analysis, tidal sand ridges.

From Depositional Systems to Sedimentary Successions on the Norwegian Continental Margin, First Edition.
Edited by A. W. Martinius, R. Ravnås, J. A. Howell, R. J. Steel, and J. P. Wonham.

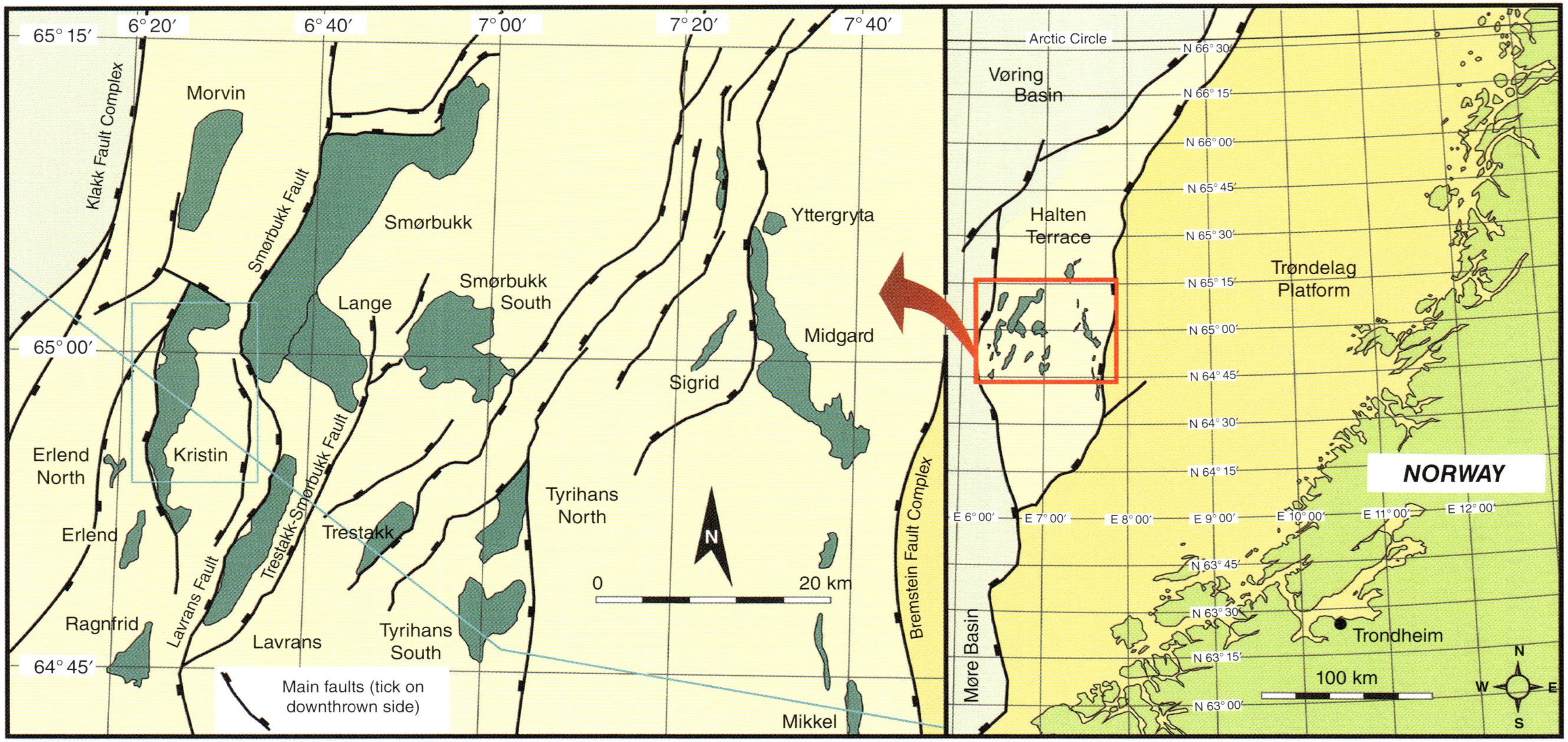

Fig. 1. The location of the Halten Terrace and adjacent structural units of the Norwegian Continental Shelf (right) and the location of the Kristin Field among the terrace's petroleum fields (left). The blue line in the left-hand map indicates the cross-section shown in Fig. 2B and the blue frame indicates the study area shown in Fig. 4.

INTRODUCTION

The Halten Terrace (Fig. 1) is the most prolific hydrocarbon province in offshore mid-Norway, hosting more than a dozen fields with gas, condensate and light to medium oil in the Lower to Middle Jurassic siliciclastic reservoirs (Spencer *et al.*, 1993; Koch & Heum, 1995). The hydrocarbon-bearing formations are mainly tidal and deltaic deposits of the Early Jurassic Båt Group and the Middle Jurassic Fangst Group (Gjelberg *et al.*, 1987; Dalland *et al.*, 1988). Large parts of the reservoirs consist of heterolithic deposits comprising irregularly interlayered sandstones, siltstones and mudstones. The extreme heterogeneity of these deposits renders reservoir characterisation and hydrocarbon recovery difficult and has been addressed by numerous studies (Dreyer, 1992; Wen *et al.*, 1998; Jackson *et al.*, 1999; Kjærefjord, 1999; McIlroy *et al.*, 1999; Martinius *et al.*, 2001, 2005; McIlroy, 2004; Brandsæter *et al.*, 2005; Klefstad *et al.*, 2005; Nordahl *et al.*, 2005; Ringrose *et al.*, 2005; Rivenæs *et al.*, 2005). Little research has thus far been done on the heterogeneity of the sandy and relatively monolithic uppermost unit of the Fangst Group, the Garn Formation (Dalland *et al.*, 1988), although the type, scale and spatial configuration of sedimentary structures in sandstone reservoirs are generally considered to be important to the fluid flow and hydrocarbon recovery (Emmett *et al.*, 1971; Weber, 1982, 1986; Weber & Van Geuns, 1990; Zweigel, 1998; Jackson *et al.*, 1999; Elfenbein *et al.*, 2005).

The present study of the Garn Formation focuses on its development in the Kristin Field (Fig. 1), where this sandstone succession is ~100 m thick and its sedimentological characteristics suggest strong tidal influence accompanied by wave action. Tidal currents are thought to have been enhanced in the area by fault-related topography combined with a relative sea-level fall in the early Bajocian. Sedimentological analysis of well cores is used to analyse the facies anatomy of the Garn Formation and thereby provide a better understanding of its origin and sedimentary heterogeneity. Dune cross-strata sets appear to be an important element of the primary heterogeneity and a statistical analysis of their thickness frequency distribution is used to predict the frequency distribution of their volumes as an important component of an improved reservoir model.

The study also contributes to the sparse existing knowledge concerning the accumulation patterns of tidal deposits in sand-prone rift basins and the record of relative sea-level changes in thick subtidal sandstone successions.

THE HALTEN TERRACE

Tectonics and palaeogeography

The tectonic structure of the Norwegian Continental Shelf evolved through the Permo-Triassic, Late Jurassic, Middle Cretaceous and Palaeocene rifting phases (Bukovics *et al.*, 1984; Doré, 1992; Doré *et al.*, 1999; Brekke *et al.*, 2001). The Cretaceous phase had the greatest magnitude of extension (Pascoe *et al.*, 1999; Corfield *et al.*, 2001) and the Palaeocene phase eventually separated the Fennoscandian and Greenland cratons and opened the North Atlantic. However, it was probably not until the early Pliocene that the hydrocarbons were generated and reservoir overpressure developed in the Halten Terrace (Skar *et al.*, 1999).

The rifting with the influence of Triassic salt resulted in listric detachment faults and deep planar faults at the continental margin (Jackson & Hastings, 1986; Withjack *et al.*, 1989; Pascoe *et al.*, 1999; Corfield & Sharp, 2000; Marsh *et al.*, 2010). The Halten Terrace structure is a stairway of tilted fault blocks striking SW–NE and down-thrown towards the NW (Fig. 2A), with internal horsts, grabens and half-grabens (Fig. 2B). The Jurassic physiography of the Halten Terrace is not reflected in its present-day structure but can be reconstructed from sedimentary facies distribution and structural restoration (Corfield *et al.*, 2001).

The base of Jurassic syn-rift deposits in the Halten Terrace was originally placed between the shallow-marine sandstones of the Fangst Group and the neritic mudstones of the Viking Group (Fig. 3; Dalland *et al.*, 1988; Ehrenberg *et al.*, 1992; Koch & Heum, 1995) but subsequent studies have indicated that the rifting phase commenced earlier and extended into the earliest Cretaceous. Syndepositional growth faults and fault-propagation flexures formed in the Early to Middle Jurassic (Blystad *et al.*, 1995; Corfield & Sharp, 2000; Corfield *et al.*, 2001; Marsh *et al.*, 2010) and this deformation had an important effect on basin physiography, on the location of hanging-wall depocentres and on the development of sediment fairways and accommodation for the Garn Formation (Corfield *et al.*, 2001).

(A)

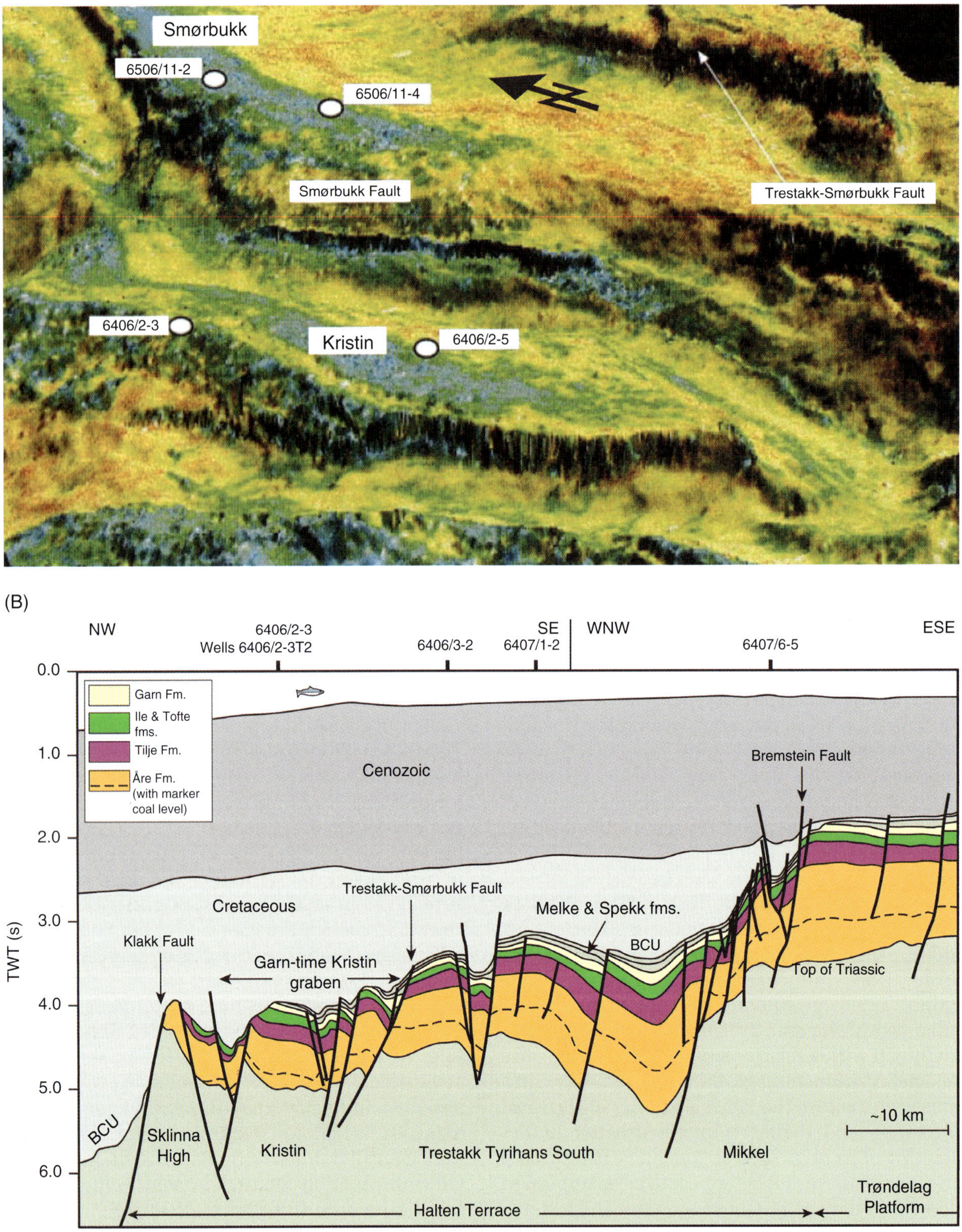

Fig. 2. (A) An eastward perspective view of the Halten Terrace structural image draped with a seismic-reflection intensity map of 48-ms window below the base-Cretaceous reflector, sampling the Upper Jurassic. The zones of low intensity (blue-green colours) are attributed to the erosion of the Upper to Middle Jurassic (Corfield & Sharp, 2000). (B) Geoseismic cross-section through the Halten Terrace along the traverse line shown in Fig. 1 (left-hand map), with the BCU symbol denoting the base-Cretaceous regional unconformity (modified from Martinius *et al.*, 2005).

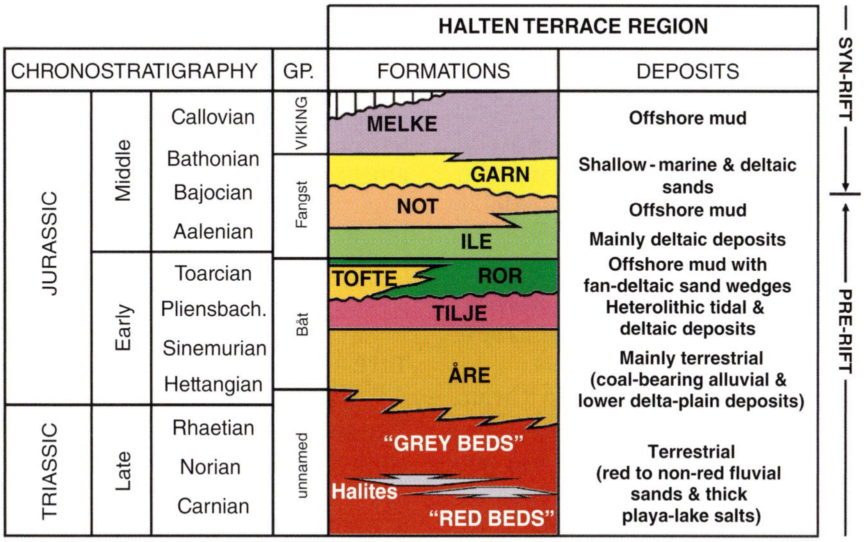

Fig. 3. The Upper Triassic to Middle Jurassic stratigraphy of the Halten Terrace (modified from Martinius *et al.*, 2005; based on the original proposal of Dalland *et al.*, 1988).

The Jurassic latitude of the Halten Terrace was between 49° N and 53° N (Smith *et al.*, 1994) and the regional climate was warm and probably seasonal (Hallam *et al.*, 1994). The elevated area to the west of the Kristin Field, referred to as the Grip High (Johannessen & Nøttvedt, 2008), was subject to uplift and subaerial erosion in the Early to Middle Jurassic, before collapsing in the Cretaceous (Doré, 1992; Brekke, 2000; Brekke *et al.*, 2001). Sediment deposition occurred in the Halten Terrace and adjacent Trøndelag Platform to the east (Figs 1 and 2B). This south-trending basin had a fault-related physiography and acted as a narrow seaway, ~250 km wide and > 1500 km long, linking the Boreal Sea with the Tethys Sea to the south and hosting wave-dominated to tide-dominated sedimentation (Gjelberg *et al.*, 1987; Doré, 1992; Brekke *et al.*, 2001; Martinius *et al.*, 2005; Quin *et al.*, 2010). Sand was delivered from the Grip High shoreline (Jongepier *et al.*, 1996; Ehrenberg *et al.* 1998; Brekke *et al.*, 2001; Johannessen & Nøttvedt, 2008), which probably included small river or fan deltas (Gjelberg *et al.*, 1987; Dalland *et al.*, 1988). The Scandinavian shoreline was located > 200 km to the east (Smith *et al.*, 1994; Doré, 1992; Brekke *et al.*, 2001, fig. 18) and the adjoining Trøndelag Platform was a mud-dominated shelf area.

Evidence from the Jameson Land of East Greenland indicates a narrower parallel branch of the Jurassic seaway on the western side of the Grip High (Doré, 1992; Brekke *et al.*, 2001; Quin *et al.*, 2010, fig. 1b), where southward marine transport with deposition of tidal sand bodies has been recognised (Surlyk, 1991, 2003).

Jurassic stratigraphy

The Mesozoic stratigraphy of the Halten Terrace (Fig. 3) was established by Gjelberg *et al.* (1987) and Dalland *et al.* (1988). Salt-bearing terrestrial Triassic deposits are overlain by the Åre Formation (Hetangian to Sinemurian) composed of coal-bearing alluvial and tidally influenced delta-plain deposits. The overlying Tilje Formation (Pliensbachian) consists of heterolithic deposits representing tide-dominated coastal plains, deltas and bays or estuaries. The transgressive succession culminates in the Ror Formation (latest Pliensbachian to Toarcian) composed of neritic offshore mudstones intercalated with the westerly deltaic sandstone wedges of the Tofte Formation (Fig. 3). The Ile Formation (late Toarcian to Aalenian) comprises heterolithic deposits of tide-dominated deltas overlain by the bioturbated offshore mudstones of the Not Formation (Aalenian to earliest Bajocian), which is another transgressive succession, although passing into sand-rich heterolithic tidal deposits near the top.

The overlying Garn Formation (Bajocian to Bathonian) is limited in its extent to the Halten Terrace (Brekke *et al.*, 2001) and is a time-stratigraphic equivalent of the Brent Group in the northern North Sea (Helland-Hansen *et al.*, 1992).

Its base is sharp and mainly erosional and the formation comprises littoral sandstones with a deltaic or foreshore gravelly component to the west. The Melke Formation (Bajocian to early Oxfordian) consists of heterolithic sublittoral deposits and bioturbated neritic mudstones. In the Trøndelag Platform to the east, the Melke Formation is a time-stratigraphic equivalent of the Garn Formation (Brekke *et al.*, 2001) but it is time-transgressive, onlapping the latter formation westwards in the Halten Terrace (Gjelberg *et al.*, 1987; Corfield *et al.*, 2001).

This muddy lower unit of the Viking Group (Fig. 3) marked the incipient tectonic collapse and inundation of the western hinterland in the late Middle Jurassic. The Late Jurassic rifting pulse and eustatic sea-level rise then dramatically increased this palaeogeographic change, with neritic sedimentation proceeding in the Trøndelag Platform and an anoxic, deep neritic to bathyal environment extending from the Halten Terrace westwards to the evolving Vøring Basin (Spekk Formation; Brekke *et al.*, 2001). Local submarine erosion of the uplifted footwalls of rotated fault blocks (Fig. 2) occurred until the Kimmeridgian.

The Kristin Field

The siliciclastic sandstones of the Garn and Ile formations (Fig. 3) are the main hydrocarbon reservoirs in the Kristin Field (Helgesen *et al.*, 2000; Martinius *et al.*, 2005). The hydrocarbon trap is structural, formed by a north-trending horst block tilted towards the east (Fig. 2), as is characteristic of the Halten Terrace fields. The hydrocarbon source rocks are the coals and coaly shales of the underlying Åre Formation (Helgesen *et al.*, 2000; Martinius *et al.*, 2005; Quin *et al.*, 2010). The field's estimated original in-place hydrocarbon volume includes ~ 100 GSm^3 of gas and ~ 100 MSm^3 of condensate, with a dew point at ~ 400 bar (Quin *et al.*, 2010). The reservoirs are the depth of 4600 m to 5600 m, with high temperature (170 °C) and high initial fluid pressure (911 bars). The field was discovered by Saga Petroleum in 1996 and production commenced in November 2005, with Statoil as the operator.

The porosity of the deeply buried reservoirs in the Halten Terrace generally tends to be reduced by quartz cementation (Bjørlykke *et al.*, 1986; Ehrenberg, 1990; Walderhaug, 1994, 1996). However, the sandstones of the Garn Formation in the Kristin Field at the depth of 4500 m to 5000 m show a helium-based porosity of 18% to 20%, which is twice greater than expected from the general porosity/depth trend in the region (Ehrenberg, 1993; Chuhan *et al.*, 2001). The relatively high preservation of primary porosity is attributed to the early diagenetic coating of sandstone grains with illite and mixed illite-chlorite (Ehrenberg, 1993; Chuhan *et al.*, 2001; Storvoll *et al.*, 2002).

THE GARN FORMATION

The Garn Formation is 95 m to 120 m thick in the studied wells in the Kristin Field but thins gradually to the north and south (Quin *et al.*, 2010) and is also considerably thinner or locally eroded on the Halten Terrace horsts (e.g. on the Sklinna High at the western margin of the terrace, Fig. 2B). The formation consists mainly of fine-grained to medium-grained subarkosic arenites (Chuhan *et al.*, 2001) with coarse to very coarse sandstone intercalations. Its base is a regional erosional unconformity of earliest Bajocian age, recording fault-block tilting accompanied by a relative sea-level fall (Corfield *et al.*, 2001). Biostratigraphic data from the adjacent Smørbukk and Smørbukk South fields indicate that the base, although a subtle angular unconformity with onlap in seismic sections, can be regarded as approximately isochronous. Erosional truncation of the underlying Not Formation was very limited and probably did not exceed 5 m to 10 m even on structural highs (Corfield *et al.*, 2001).

The most voluminous lower sandstones of the Garn Formation were deposited in synclinal troughs, mainly incipient grabens and half-grabens bounded by fault ruptures and/or blind-fault flexures. The uppermost sandstones, in contrast, were deposited as progradational wedges back-stepping onto adjoining structural highs and interfingering with the heterolithic to muddy deposits of the Melke Formation (Corfield *et al.*, 2001). The littoral zones of sand deposition progressively migrated from the depressions onto the adjoining highs, giving way to the neritic environment of the Melke Formation.

The marked regional diachroneity of the upper boundary of the Garn Formation was first recognised by Gjelberg *et al.* (1987), who suggested that the formation was interfingering with the Melke Formation while shrinking in lateral extent. This interpretation remained largely overlooked by researchers until Corfield *et al.* (2001) had documented it from seismic sections, well logs and biostratigraphic data in the Smørbukk area, demonstrating also strong diachroneity on a lateral scale of a few kilometres. This evidence has important implications for the palaeogeography and stratigraphic development of the Garn Formation.

Corfield *et al.* (2001) suggested a three-stage depositional scenario for the Garn Formation: (1) an early Bajocian stage of differential erosion due to syntectonic forced regression, followed by lowstand accumulation of voluminous sand in belt-shaped depocentres; (2) a transgressive stage of 'pulsed progradation', when the littoral zones stepped back onto structural highs as normal-regressive wedges; and (3) a stage of final drowning, when the littoral sand was generally covered by neritic mud in the early Bathonian.

Despite these important insights, the depositional environment of the Garn Formation remained poorly understood. Gjelberg *et al.* (1987) originally suggested a shoreface environment with tidal currents and high-energy waves, passing westwards into mouth bars of a fan-deltaic system, whereas Dalland *et al.* (1988) postulated east-advancing lobes of braided-river deltas dominated by fluvial and wave processes. Subsequent interpretations generally invoked a high-energy shallow-marine environment with deltaic, coastal spit, shoreface and shelfal sand accumulations (Doré, 1992; Brekke *et al.*, 2001; Corfield *et al.*, 2001; Elfenbein *et al.*, 2005; Quin *et al.*, 2010). However, no fluvial or other terrestrial deposits are documented. Corfield *et al.* (2001) recognised elongate sandstone bodies with subtle positive relief and suggested that the fault-related physiography and regional sea-level fall probably focused tidal currents in belt-like sand depocentres parallel to the main faults. Tidal dunes were recognised (Elfenbein *et al.*, 2005) and possible tidal sand ridges were suggested (Corfield *et al.*, 2001) but no attempt was made to recognise these sandstone bodies in cores and take them into account in reservoir models. Quin *et al.* (2010) attributed deposition of the Garn Formation to a structurally controlled, eastward-advancing upper shoreface environment or possibly tidal deltas, but noted that the abundance of relatively coarse-grained, cross-stratified sandstones in the central zone of the Kristin Field, many kilometres away from the Grip High palaeoshoreline, was difficult to explain in this scenario.

The earlier stratigraphic interpretations generally considered the Garn Formation to be a uniform, tabular sandstone unit deposited during a pre-rifting period of 'nearly total tectonic quiescence' (e.g. Ehrenberg *et al.*, 1992; Doré, 1992). On this stratigraphic premise was also developed the original layer-cake zonation model for the Garn Formation reservoir in the Kristin Field, extended further to adjacent fields to the north and east (Elfenbein *et al.*, 2005; Quin *et al.*, 2010, fig. 2). However, it has been pointed out by Corfield *et al.* (2001) that a simple layered reservoir model for the Garn Formation, as used in the Smørbukk, Smørbukk South, Heidrun, Kristin, Trestakk and Tyrihans fields, is probably unreasonable.

METHODS AND TERMINOLOGY

The present study is based on cores and geophysical data from six wells located in the axial zone of the Kristin Field, with lateral well spacing of 1.9 km to 5.1 km (Fig. 4). The Garn Formation was almost fully cored in these wells; and a total of ~ 700 m of cores has been logged for this study.

Conventional sedimentological logging was used, with a bed-thickness measurement accuracy of 0.5 cm. In the measurement of bed thicknesses and identification of sedimentary structures, special care was taken to account for the regional structural dip derived from resistivity imaging logs and the inclination of some of the wells. Wells 6406/2-5AT2, 6406/2-R-4H and 6506/11-N-3H (Fig. 4) are non-vertical and the apparent thickness of the Garn Formation in these wells was corrected accordingly (i.e. multiplied by a correction factor of 0.87, 0.72 and 0.62, respectively). The top contact of the Garn Formation with the overlying Melke Formation was not cored in the three other wells (6406/2-5, 6406/2-3 T3 and 6506/11-6, Fig. 4). The basal contact with the underlying Not Formation was cored in all wells. The occurrence of large-scale cross-stratification was verified and its dip direction assessed on the basis of dip-picking data using resistivity imaging logs from two wells.

The descriptive sedimentological terminology used in this study is after Collinson & Thompson (1982) and Harms *et al.* (1982). The term 'sedimentary facies' refers to the basic types of sedimentary deposits distinguished on the basis of their bulk macroscopic characteristics and attributed to different modes of sediment deposition

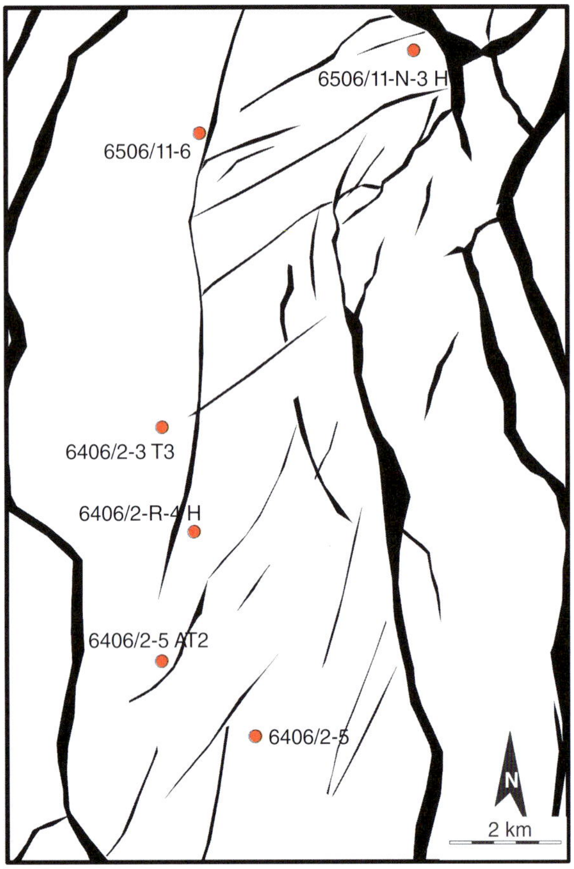

Fig. 4. The location of main wells in the Kristin Field (see study area indicated by the blue frame in the left-hand map in Fig. 1). Cores and wireline logs from the six wells are the basis of the present study.

(Harms *et al.*, 1975; Walker, 1984). The term 'facies association' denotes an assemblage of spatially and genetically related facies representing a particular depositional subenvironment.

The statistical notions and methods employed in this study are from Middleton *et al.* (1995), Drummond & Wilkinson (1996) and Davis (2002). Some crucial geometric-statistical concepts, derived from Malinverno (1997), are reviewed in the Appendix.

FACIES ANALYSIS OF THE GARN FORMATION

The base of the Garn Formation is a regional erosional unconformity (Corfield *et al.*, 2001). It is generally sharp, although the facies change across this boundary is not necessarily profound. In the Kristin Field, the neritic mudstones of the underlying Not Formation grade upwards into bioturbated heterolithic tidalites, which are sharply overlain by the littoral sandstones of the Garn Formation (Fig. 5). The shallowing-upward facies succession of the uppermost Not Formation suggests a normal regression, which was interrupted by a phase of erosion and followed by littoral sandy sedimentation. A similar stratigraphic relationship has been reported from the adjacent Smørbukk area, with the formation boundary attributed to an episode of forced regression (Corfield *et al.*, 2001).

The similarity of facies below and above the formation erosional boundary across the study area (Fig. 5) suggests that the basal palaeorelief of the Garn Formation along the line of wells studied in the Kristin Field (Fig. 4) was probably negligible.

Sedimentary facies

The sandstone succession of the Garn Formation is nearly devoid of mudstone drapes or thicker interlayers and is considered to be of shallow-marine origin. Sporadic trace fossils include *Trichichnus*, *Skolithos*, *Arenicolites*, *Ophiomorpha*, *Cylindrichnus*, *Planolites*, *Palaeophycus*, *Chondrites*, bivalve burrows and resting traces (Quin *et al.*, 2010), which support the notion of a shallow-marine environment (Bromley, 1996). Six sedimentary facies have been distinguished on the basis of sedimentological core description (Table 1 & Fig. 6):

Facies S_{PS}: sandstones with planar parallel stratification — attributed to the action of waves with high near-bottom orbital velocities (Komar & Miller, 1975) or possibly unidirectional currents combined with waves (Arnott, 1993) or boosted to the upper flow regime (Harms *et al.*, 1975) by storms and topographic confinement (Amos *et al.*, 1995).

Facies S_{SS}: sandstones with swaley cross-stratification — attributed to an episodic scour-and-fill action of storm waves combined with unidirectional currents under a low net rate of sediment accretion (Arnott & Southard, 1990; Dumas *et al.*, 2005; Dumas & Arnott, 2006).

Facies S_{HS}: sandstones with hummocky cross-stratification — attributed to a similarly episodic action of storm waves combined with unidirectional currents under a high net rate of sediment accretion (Arnott & Southard, 1990; Dumas & Arnott, 2006).

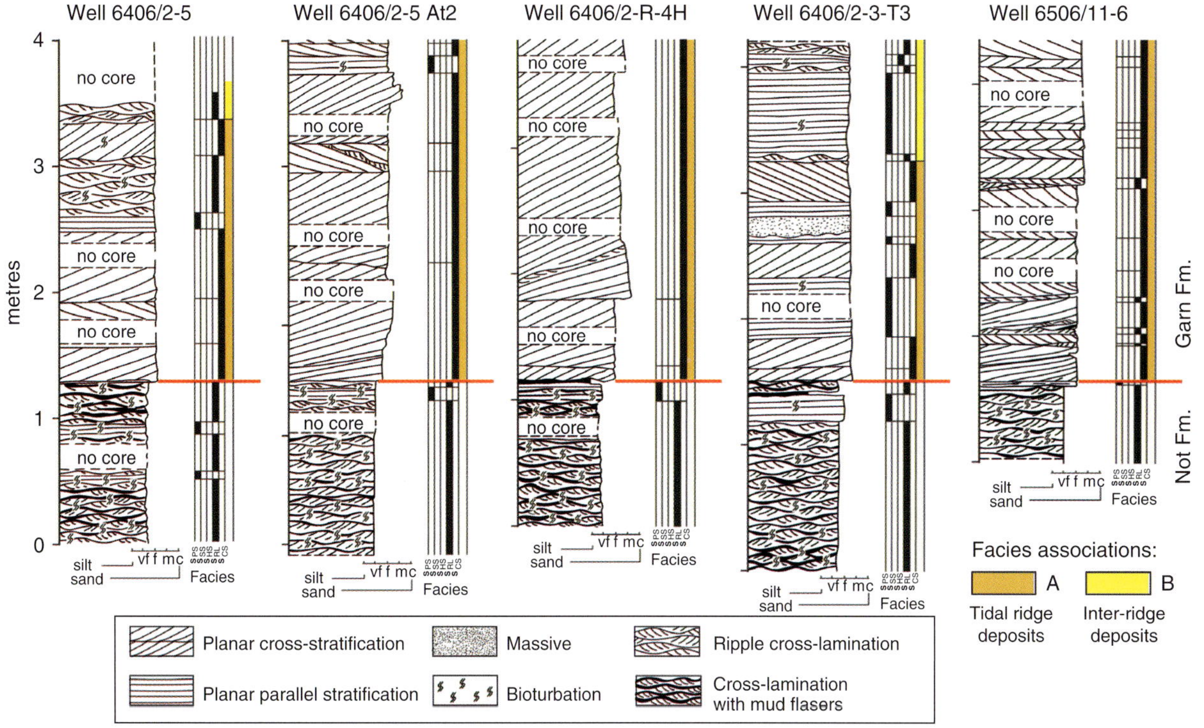

Fig. 5. The basal contact of the Garn Fm. with the underlying Not Fm. (Fig. 3) in cores from the Kristin Field (for the location of wells see Fig. 4). The sharp unconformable base of the formation, considered to be a wave-cut erosional surface, is used as a datum.

Facies S_{RL}: sandstones with ripple cross-lamination — attributed to the action of waves with low near-bottom orbital velocities (Komar & Miller, 1975) or to weak unidirectional currents in the lower range of lower flow regime (Harms *et al.*, 1975).

Facies S_{CS}: sandstones with planar cross-stratification — attributed to the action of unidirectional currents in a higher range of lower flow regime, forming long-crested subaqueous 2D dunes (Harms *et al.*, 1982).

Facies S_M: massive sandstones — attributed to sporadic dune collapses or local sea floor liquefaction due to earthquakes (Seed, 1979; Field *et al.*, 1982) or cyclic loading by storm waves (Seed & Rahman, 1978; Okusa & Uchida, 1980).

The descriptive characteristics of the individual facies are summarised in Table 1 and their relative thickness percentages in the studied cores are given in Table 2. The most abundant are facies S_{CS} (28% to 60%) and S_{PS} (13% to 60%). Facies S_M is rare (≤0.1%) and its only significant occurrence (6.6%) in well 6506/11-N-3H suggests local liquefaction, perhaps due to proximity to a seismically active fault (cf. Fig. 4).

The spectrum of facies indicates a current-dominated and wave-dominated depositional setting with sporadic storm events, characterised by a sufficiently high hydraulic energy to transport sediment of up to granule size and keep mud perennially in suspension. Such conditions typify many inner shelves, particularly the shoreface zone and other littoral shoals (Johnson & Baldwin, 1996; Molgat & Arnott, 2001), as well as sand-prone, narrow epicontinental seaways with enhanced tidal currents (Keene & Harris, 1995; Mellere & Steel, 1996; Slingerland *et al.*, 1996; Carr *et al.*, 2003; Longhitano & Nemec, 2005).

Facies associations

On the basis of the stratigraphic grouping of sedimentary facies in the well logs and their tentative lateral correlation, the Garn Formation in the study area is recognised to consist of three main facies associations (Figs 7 & 8). They are described briefly and interpreted below. Volumetric estimates are one-dimensional, based on the thicknesses of facies and their assemblages in the individual wells (Table 2).

Table 1. A summary of the descriptive characteristics and genetic interpretation of the sedimentary facies distinguished in the studied cores of the Garn Fm. in Kristin Field.

Facies code	Facies name and description	Genetic interpretation
S_{PS} 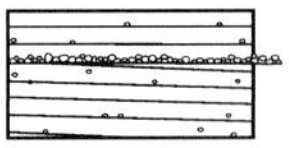	**Sandstones with planar parallel stratification**: These arenitic sandstones are very fine-grained to very coarse-grained, occasionally bearing granules. The units of this facies are solitary sets or cosets of planar parallel strata and range in thickness from < 1 cm to 640.5 cm, averaging 26 cm. Strata sets are up to 25 cm thick, separated by slightly inclined (≤5°) planar truncation surfaces in cosets.	This facies represents tractional deposition of sand by waves with high near-bottom orbital velocities (Komar & Miller, 1975) or by the action of marine currents, either combined with waves (Arnott, 1993) or boosted to the upper flow regime (Harms *et al.*, 1975) by storms and topographic confinement (Amos *et al.*, 1995). The flat truncation surfaces, commonly with very coarse sand and/or granules, may represent erosive storm events (Clifton & Dingler, 1984; Arnott, 1993).
S_{SS} 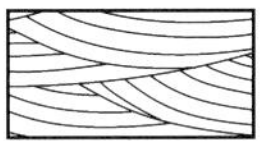	**Sandstones with swaley stratification**: These arenitic sandstones range from very fine/fine-grained to medium/coarse-grained and form units 7 cm to 39.5 cm thick, averaging 14.5 cm. The units have erosional bases and are isolated sets or sporadically cosets of concave-upward and flattening-upward parallel strata recognised as swaley stratification.	This facies represents tractional deposition of sand by an episodic scour-and-fill action of storm waves combined with unidirectional current under a low net rate of sediment accretion (Arnott & Southard, 1990; Dumas *et al.*, 2005; Dumas & Arnott, 2006).
S_{HS} 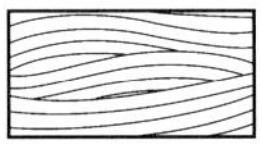	**Sandstones with hummocky stratification**: These arenitic sandstones range from fine-grained to medium/coarse-grained and form units 7 cm to 29 cm thick, averaging 12 cm. The units have erosional bases and are sets of nearly flat or slightly concave-upward to convex-upward parallel strata recognised as hummocky stratification.	This facies represents tractional deposition of sand by an episodic action of storm waves combined with unidirectional current under a high net rate of sediment accretion (Arnott & Southard, 1990; Dumas & Arnott, 2006).
S_{RL} 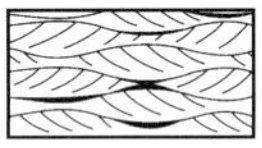	**Sandstones with ripple cross-lamination**: These sandstones are very fine-grained to coarse-grained and form units 0.5 cm to 186 cm thick, averaging 16 cm and ranging from solitary sets to cosets of ripple-scale cross-laminae. They vary from arenitic to slightly wackey, with occasional intra-set mud flasers 0.1 cm to 0.3 cm thick and inter-set mud drapes 0.1 cm to 0.7 cm thick, sporadically up to 7.5 cm. Cross-lamination patterns indicate ripples ranging from symmetrical to asymmetrical, including mud-draped bidirectional bedforms.	This facies represents tractional deposition of sand by the action of waves with low near-bottom orbital veloc-ities (Komar & Miller, 1975) or by sea currents in the lower range of lower flow regime (Harms *et al.*, 1975). Bidirectional ripples indicate a reversing flow and the occurrence of mud drapes/flaser indicates flow slackening episodes, as is generally characteristic of tidal currents (Reineck & Singh, 1980).
S_{CS} 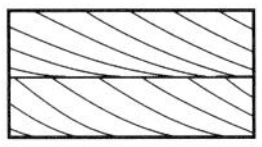	**Sandstones with planar cross-stratification**: These arenitic sandstones are fine-grained to very coarse-grained, forming planar cross-strata sets with mainly tangential basal parts. The cross-sets are 7 cm to 250 cm thick, averaging 31 cm, and range from solitary to multiple, stacked upon one another as cosets up to 12 m in thickness. Many thin cross-sets are erosional relics (bottomsets) of thicker ones. Some cross-sets show interstratal mud drapes, intrasets of backflow ripple cross-lamination and/or reactivation surfaces. Dipmeter data indicate that the cross-strata are dipping mainly towards the SSW, less commonly towards the NNE.	This facies represents tractional deposition of sand by marine currents in the higher range of lower flow regime, forming straight-crested or sinuous-crested subaqueous 2D dunes (Harms *et al.*, 1982). The evidence of flow reversals (opposedly dipping cross-strata sets, backflow ripples, reactivation surfaces) and brief pauses in sand transport (interstratal mud drapes) indicate tidal currents (Reineck & Singh, 1980).

Table 1. (*Continued*)

Facies code		Facies name and description	Genetic interpretation
S_M	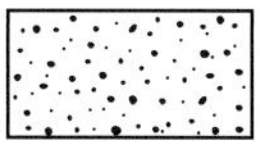	**Massive sandstones**: These non-stratified arenitic sandstones range from fine-grained to medium/coarse-grained, occasionally with an admixture of very coarse sand grains. This facies is rare, forming isolated units 10 cm to 40 cm thick, with sharp tops and mainly diffuse bases. Its only significant occurrence is in well 6506/11-N-3H (Table 2), where it forms a unit ~5 m thick, with local traces of planar parallel strata and a diffuse basal transition to the underlying unit of facies SPS.	Depending on its stratigraphic context, the origin of this facies can be attributed to sporadic dune collapses on tidal-bar slopes or to local sea floor liquefaction triggered by earthquakes (Seed, 1979; Field *et al.*, 1982) or by the cyclic loading by storm waves (Seed & Rahman, 1978; Okusa & Uchida, 1980). The relatively thick unit of this facies in well 6506/11-N-3H suggests major liquefaction and may indicate proximity to a seismically active fault.

Facies association A consists of facies S_{CS} (66 vol.% to 92 vol.%) intercalated with subordinate thin units of facies S_{RL} (3 vol.% to 23 vol.%) and S_{PS} (1 vol.% to 15 vol.%) and sporadic isolated beds of facies S_M (≤8 vol.%), S_{SS} (≤2.5 vol.%) and/or S_{HS} (≤0.5 vol.%). The correlation of sedimentological logs (Fig. 8) suggests that this facies assemblage forms mound-shaped, broadly lenticular sandstone bodies elongated in the north-south direction, with measured in-well thicknesses of up to 12 m and estimated lengths of up to ~20 km. They may correspond to the elongate sandstone bodies with subtle positive relief recognised by Corfield *et al.* (2001). The sandstone bodies of facies association A constitute 30 vol.% to 68 vol.% of the Garn Formation and are interpreted to have been deposited as wave-modified complexes of vertically-stacked 2D dunes forming tidal sand bars. The lack of large-scale foresets with comparably great thicknesses indicates flow-aligned sand ridges (Fig. 9), rather than flow-transverse sand waves with avalanching lee-side slopes (Off, 1963). Linear sand ridges of this type, subject to frontal growth and oblique lateral migration, have been reported from continental shelves, narrow epicontinental seaways and straits, coastal embayments and large estuaries (Caston, 1972; Swift, 1975; Swift & Freeland, 1978; Reineck & Singh, 1980; Swift & Field, 1981; Huthnance, 1982; Cameron *et al.*, 1992; Harris *et al.*, 1992; Lanckneus & De Moor, 1995; Goff *et al.*, 2005; Liu *et al.*, 2007). The evidence from cores combined with dip-picking data suggests that the tidal system was asymmetrical, with the majority of dunes migrating southwards and subordinate ones formed by the reversing current. The stratigraphic stacking of semi-isolated sandstone ridges (Fig. 8) resembles the imbricate pattern described by Evans (1970).

Facies association B consists of facies S_{PS} (7 vol.% to 83 vol.%) and facies S_{RL} (12 vol.% to 70 vol.%, mainly current ripple cross-laminated), which alternate with each other and are intercalated with single or double cross-sets of facies S_{CS} (5 vol.% to 21 vol.%) as well as isolated beds of facies S_M (≤7 vol.%), S_{HS} (≤2 vol.%) and/or S_{SS} (≤2 vol.%). This facies assemblage constitutes 12 vol.% to 65 vol.% of the Garn Formation and laterally interfingers with the sandstone bodies of facies association A (Figs 8 and 10). Facies association B is interpreted to have been deposited by tidal currents with wave influence in inter-ridge swales (Swift & Field, 1981; Dyer & Huntley, 1999; Liu *et al.*, 2007). The swales are thought to have funnelled the near-bottom water flow, boosting it to the upper flow regime at tidal current peaks and thereby causing plane-bed transport and considerable downflow bypass of sand. The swale-confined, helicoidal flow of the tidal current swept sand as dunes onto the adjoining flanks of sand ridges, whereas transport of sand as current ripples predominated at the weakest flow stages.

Little or no mud will be deposited in the tidal system if wave-generated ambient turbulence persists (Montenat *et al.*, 1987; Keene & Harris, 1995; Longhitano & Nemec, 2005). Sand tends to be swept by waves from the tidal ridges and redistributed further by tidal currents in the swales. Depending on local hydraulic conditions, the inter-ridge swales may thus accumulate tide-worked and wave-worked heterolithic deposits (Hein, 1987) or mud-devoid sand (e.g. see the Nantucket sand-ridge field in Swift, 1975, fig. 19; Houthuys & Gullentrops,

Fig. 6. Core intervals showing sedimentary facies of the Garn Fm., Kristin Field. The facies letter code is as in Table 1 and in the text; other symbols: RS – reactivation surface; MD – mud drape; BFR – backflow ripples; B – burrows. For the location of wells, see Fig. 4. The cores samples are from: (A) well 6506/11-6, depth 4763 – 4764 m; (B) well 6506/11-6, depth 4762 to 4763 m; (C) well 6506/11-6, depth 4743 to 4744 m; (D) well 6506/11-6, depth 4742 to 4743 m; (E) well 6506/11-6, depth 4741 to 4742 m;

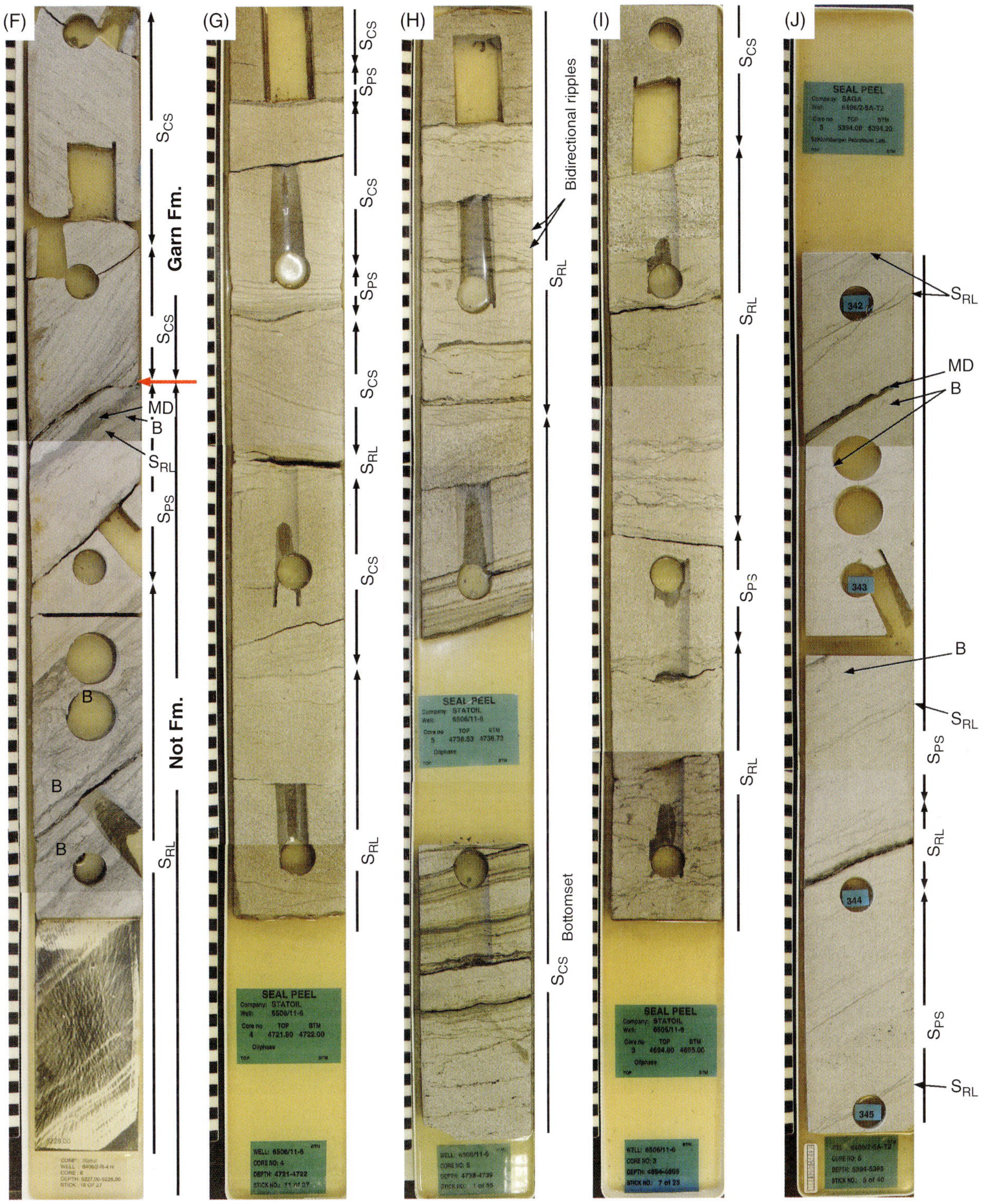

Fig. 6. (*Continued*) (F) well 6406/2-R-4H, depth 5227 to 5228 m; (G) well 6506/11-6, depth 4721 to 4722 m; (H) well 6506/11-6, depth 4738 to 4739 m; (I) well 6506/11-6, depth 4694 to 4695 m; and (J) well 6506/2-5 AT2, depth 5394 to 5395 m. Scale bars are in cm.

Table 2. The thickness percentages of the individual facies and their assemblages in the studied cores. These thickness data may serve as one-dimensional first-order estimates of the corresponding volumes.

Facies and associations	WELLS					
	6406/2-5	6406/2-5 AT2	6406-2-R-4H	6406/2-3 T3	6506/11-6	6506/11-N-3H
Facies S_{PS}	26.7%	60.5%	35.0%	42.9%	12.6%	34.0%
Facies S_{SS}	0.6%	0.4%	0.0%	0.3%	0.8%	2.0%
Facies S_{HS}	0.7%	0.0%	0.2%	1.3%	2.3%	0.5%
Facies S_{RL}	44.3%	10.2%	6.2%	23.3%	24.1%	23.3%
Facies S_{CS}	27.7%	28.9%	58.5%	32.1%	60.2%	33.6%
Facies S_{M}	0.0%	0.0%	0.1%	0.1%	0.0%	6.6%
FA A	32.0%	30.0%	60.0%	33.0%	68.0%	45.0%
FA B	64.0%	65.0%	14.0%	63.0%	12.0%	43.0%
FA C	4.0%	5.0%	26.0%	4.0%	20.0%	12.0%

1988), or may possibly be subject to erosional sand bypass (Le Bot *et al.*, 2005).

Facies association C consists of facies S_{PS} (54 vol.% to 95 vol.%) and facies S_{RL} (3 vol.% to 32 vol.%, mainly wave-ripple cross-laminated), which alternate with each other and are interspersed with the isolated beds of facies S_{HS} (≤11 vol.%) or S_{SS} (≤6 vol.%) and solitary small cross-sets of facies S_{CS} (≤8 vol.%). This facies assemblage constitutes 4 vol.% to 26 vol.% of the Garn Formation, occurs repeatedly as a depositional cover of the previous two facies associations (Fig. 10) and virtually predominates in the uppermost part of the formation in two wells (Fig. 8). Facies association C is interpreted to represent episodes of wave-dominated and tidally-influenced sedimentation, similar to that characterising shorefaces and other littoral shoals on many shelves (Grant & Madfen, 1979; Swift *et al.*, 1979; Bourgeois, 1980; Clifton & Dingler, 1984; Swift & Thorne, 1991; Cacchione *et al.*, 1994). These deposits in the present case are thought to signify shallower-water conditions than those of facies associations A and B, because of the evidence of a perennial wave action (facies S_{PS} and S_{RL}) punctuated by storm events (facies S_{HS} and S_{SS}) and the scarcity or poor preservation of tidal dunes (wave-truncated facies S_{CS}).

Depositional model

Regional palaeogeographical reconstructions suggest that the Early to Middle Jurassic sedimentation in the Halten Terrace area occurred within the northern reaches of a narrow seaway extending southwards from the open Boreal Sea, bounded by the sand-supplying Grip High to the west and passing into a mud-prone Trøndelag Platform to the east (Gjelberg *et al.* 1987; Doré 1992; Brekke *et al.* 2001; Johannessen & Nøttvedt, 2008). The platform was a part of the seaway, but was broad and receiving little sand from the Scandinavian shoreline. It received sand from the west, spread out by storms and tidal currents whenever the accommodation for sand storage on the adjoining Halten Terrace reached its minimum. The deeper axial part of the seaway would have amplified tidal currents (Montenat *et al.*, 1987; Sztanó & De Boer, 1995; Corfield *et al.*, 2001), which might have been circulatory and inherently asymmetrical or whose southward flow could have been enhanced by Boreal storms (see McBride, 2003). Southward palaeocurrents and tidal sandstone bodies were also recognised in the adjacent East Greenland seaway (Surlyk, 1991, 2003).

The Halten seaway probably comprised a series of incipient shallow grabens and half-grabens formed by extensional faults or fault-propagation flexures (Blystad *et al.*, 1995; Corfield & Sharp, 2000; Corfield *et al.*, 2001; Marsh *et al.*, 2010). The Garn Formation in the Kristin Field consists of shallow-marine sandstones and has a considerable thickness. The thickness increases towards the east and abruptly decreases across the Trestakk-Smørbukk Fault (Fig. 2B), which suggests deposition in an actively subsiding shallow graben (Fig. 11). The occurrence of conglomeratic facies at the western extremity of the field (well 6506/11-3) may correspond to the western margin of the hypothetical graben and supports the notion of westerly sediment supply (Fig. 11). The Grip High horst at the time was probably an elongate island or a belt of islands (Doré 1992; Quin *et al.*, 2010) and would have had small water catchments, unlikely to form large river deltas. Its palaeoshoreline is not preserved but was probably wave-dominated and might have included small gravelly river deltas or fan deltas

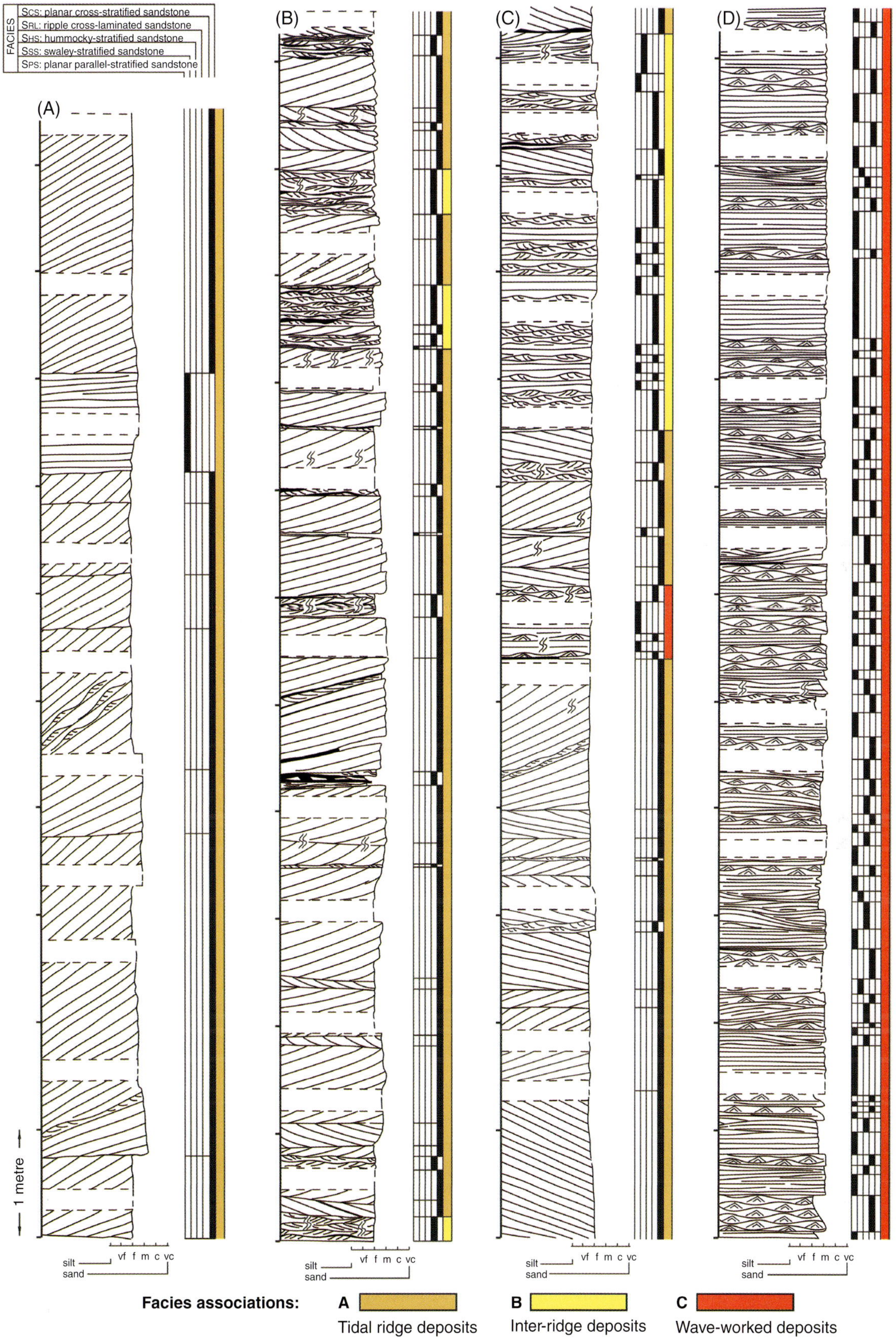

Fig. 7. Example intervals of core logs showing the three main facies associations of the Garn Fm. in Kristin Field. The log portions are from: (A) well 6406/2-R-4H, depth 5216.80 to 5227.25 m; (B) well 6506/11-6, depth 4736.50 to 4747.95 m; (C) well 6406/2-5 AT2, depth 5392.30 to 5403.75 m; and (D) well 6506/11-6, depth 4659.30 to 4670.75 m.

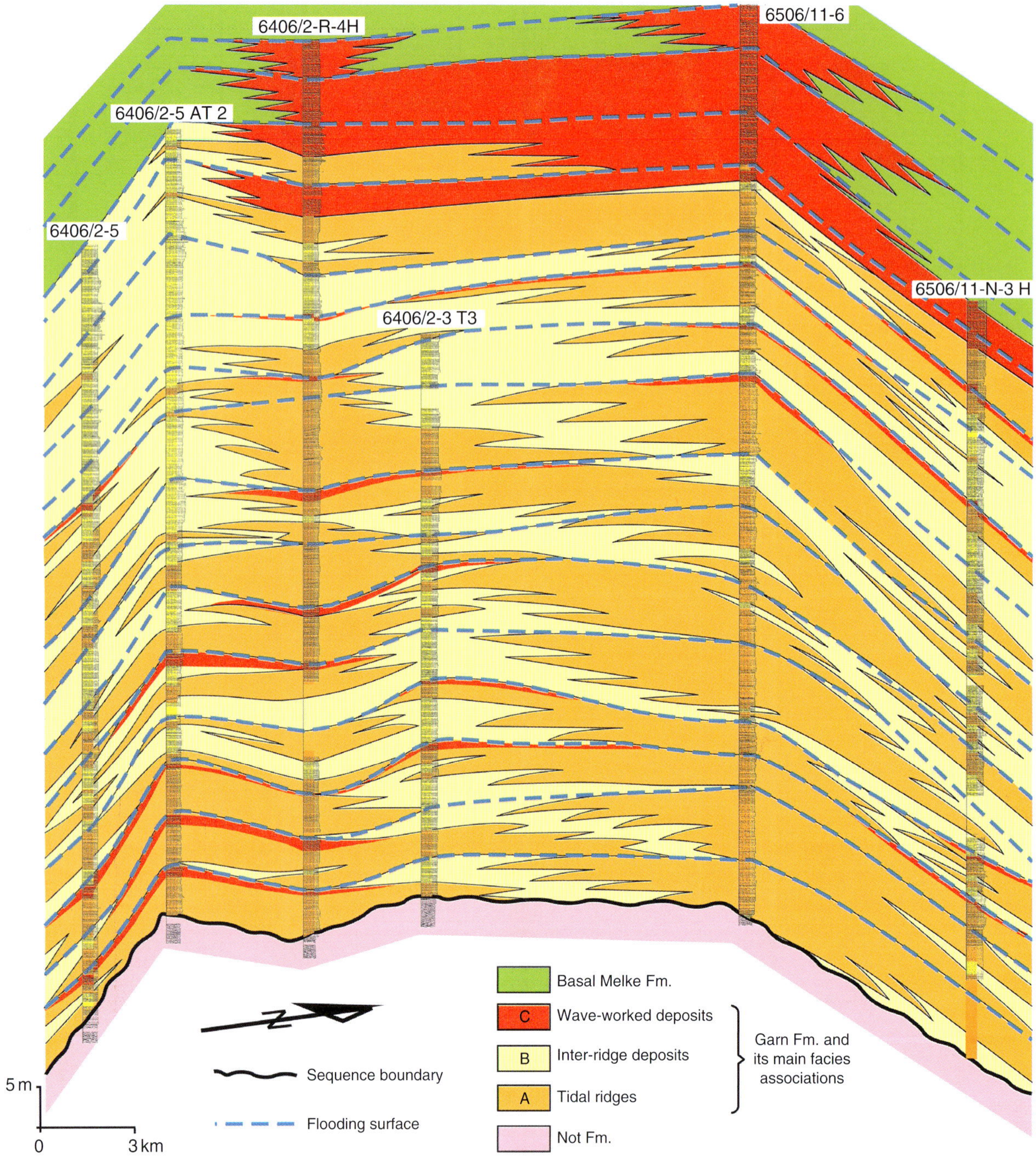

Fig. 8. Core-log correlation panel with an interpreted stratigraphic distribution of the main facies associations of the Garn Fm. (for the location of wells, see Fig. 4). The unconformable base of the formation, used as a datum, is considered to be a wave-cut erosional surface with negligible primary relief in the N–S direction (cf. Fig. 5).

(Gjelberg *et al.*, 1987; Dalland *et al.*, 1988). The depositional system is considered to have been a sandy shoreface that extended eastwards across an active shallow graben where tidal currents were amplified, reworking the entrapped sand into longitudinal bars. Subaqueous dune complexes forming tidal sand bars have been reported from many modern and ancient marine grabens (e.g. Montenat *et al.*, 1987; Santisteban & Taberner, 1988; Surlyk, 1991, 2003; Keene & Harris, 1995; Mellere & Steel, 1996; Carr *et al.*, 2003; Longhitano & Nemec, 2005) and have also been suggested for the Garn Formation in

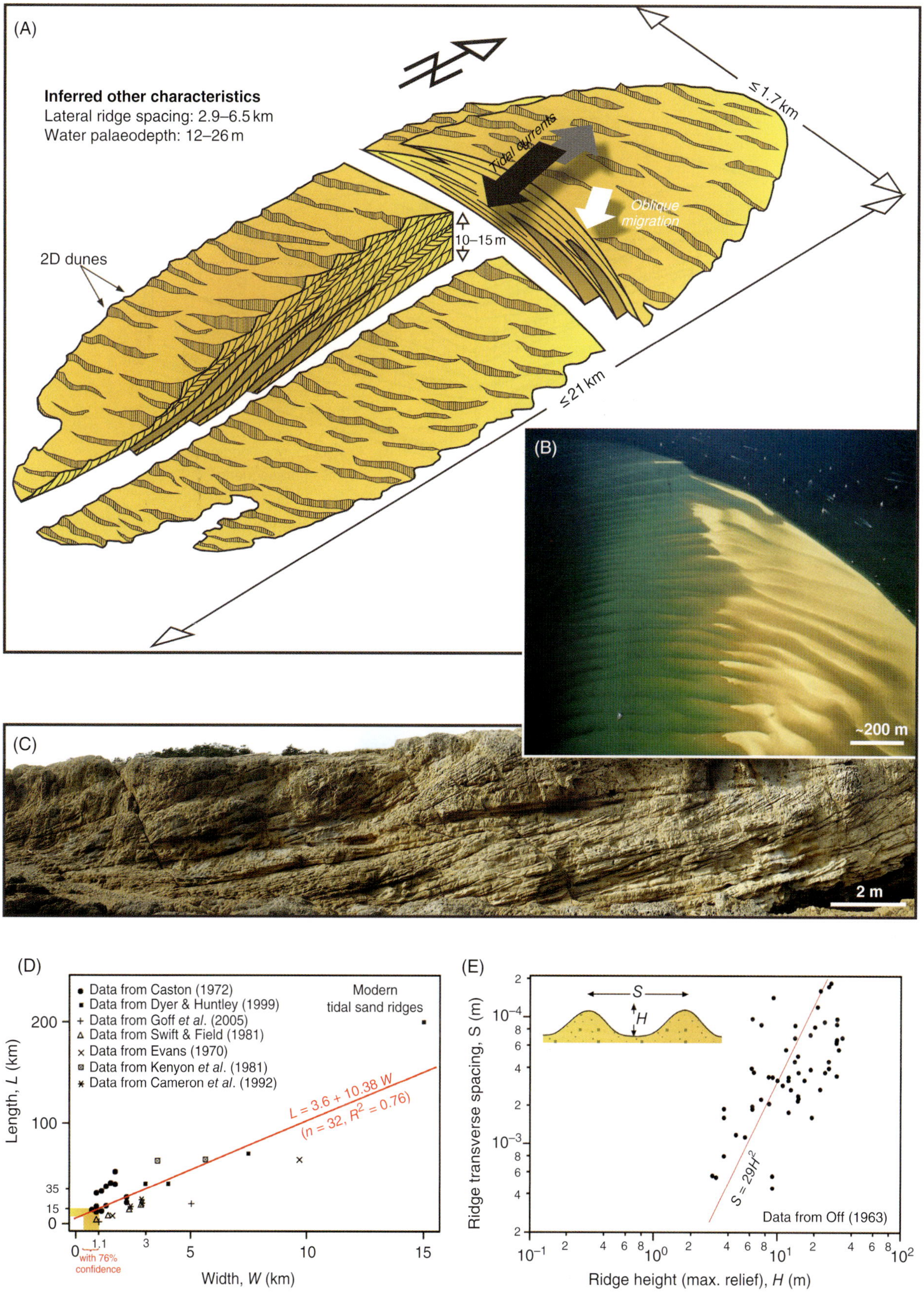

Fig. 9. (A) The inferred shape, dimensions, lateral spacing and palaeobathymetry of the tidal sand ridges in the Garn Fm., Kristin Field. (B) Aerial photograph of a modern tidal sand ridge, Banc d'Arquin in the English Channel, covered with only a few metres of water in the crestal zone at ebb tide; the white spots on the dark sea surface are the wake bores of bypassing motorboats. (C) Longitudinal outcrop section of an Eocene tidal sandstone ridge, Rab island, Croatia; the dominant palaeocurrent direction is to the left. (D) Statistical linear relationship between the width and length of modern tidal sand ridges; n – number of data, R^2 – coefficient of determination (76%) of the ridge length by its width. (E) Statistical linear relationship between the height and lateral spacing of modern tidal sand ridges (after Off, 1963; Allen, 1982).

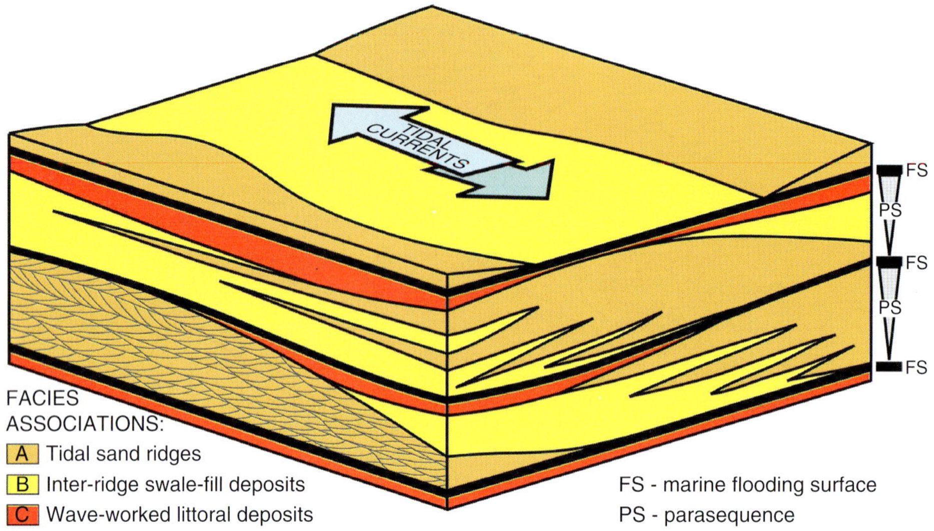

Fig. 10. Schematic diagram showing the interpreted spatial relationships and stratigraphic organisation of the main facies assemblages of the Garn Fm. in the Kristin Field (cf. Fig. 8); the vertical scale is grossly exaggerated for graphical purposes. The diagram represents an approximate area of 2 km × 2 km and a stratigraphic thickness of 10 m to 15 m. Note that each parasequence consists of the tide-dominated deposits A and B, covered with the wave-worked littoral deposits C or truncated by a correlative equivalent surface of wave erosion.

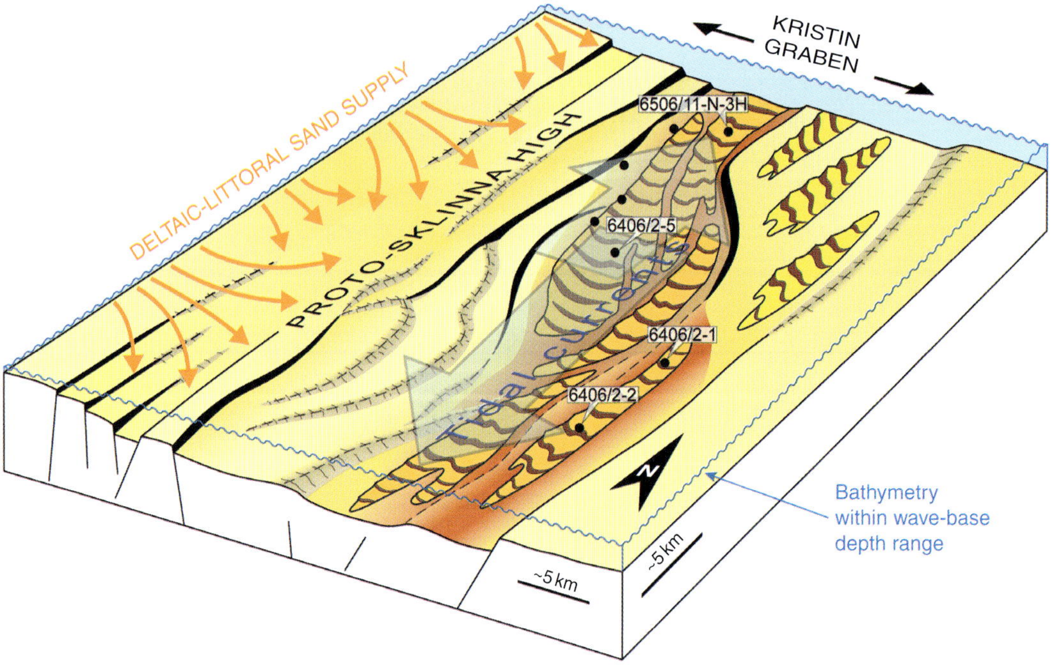

Fig. 11. Hypothetical reconstruction of the incipient Kristin graben and its zones of greatest subsidence during deposition of the Garn Fm. The elevated fault blocks west of the proto-Sklinna High subsequently collapsed during the climax of Late Jurassic rifting (cf. Fig. 2B).

the nearby Smørbukk Field to the north (Corfield *et al.*, 2001).

Isopach maps of the Garn Formation indicate sand depocentres in the central zone of the Kristin Field, along the eastern margin of the hypothetical graben (Fig. 11). The sandstone succession reaches its thickness maxima at different stratigraphic levels in the individual wells before passing upwards into the neritic mudstones of the Melke Formation (Fig. 8). This relationship implies short-distance facies diachroneity and indicates an uneven sea floor topography evolving along the graben's western margin during deposition of the Garn Formation. An uneven sea floor topography related to faults might explain both non-uniform eastward advances of the sandy shoreface environment and its non-uniform overall retreat towards the west (Fig. 8). The graben margin was probably defined by antithetic faults with local relay ramps and/or relay-breach transverse anticlines, on which littoral sedimentation under transgression would persist longer than in adjacent lower-lying areas (Fig. 11).

Subtidal sand ridges (Fig. 9) are considered to have been an important morphodynamic element of the depositional system, apparently related to topography-enhanced tidal currents and relative sea-level rise. The formation of sand ridges is thought to be favoured by transgression on continental shelves characterised by the following four conditions (Snedden & Dalrymple, 1999): (1) pre-existing irregularities in sea floor topography; (2) sufficient supply of sand; (3) sand-transporting tidal and/or storm-driven currents; and (4) sufficient time for the sand to be moulded into a ridge or ridge field. The reported velocities of ridge-forming currents are in the range of 0.5 m/s to 2.5 m/s, which may mean a tidal prism in excess of 3 m for an open continental shelf (McBride, 2003), but possibly little more than 0.5 m for some narrow marine straits (Montenat *et al.*, 1987). The morphology and spatial configuration of sand ridges result from the system's evolution towards equilibrium between tidal current dynamics, sea floor topography and sediment supply (Dyer & Huntley, 1999).

Modern case studies (Caston, 1972; Swift, 1975; Kenyon *et al.*, 1981; Swift & Field, 1981; Cameron *et al.*, 1992; Dyer & Huntley, 1999; Goff *et al.*, 2005; Liu *et al.*, 2007) indicate that the height (H) of sand ridges may reach 40 m and their length (L) and width (W) are in the range of 5 km to 120 km and 0.5 km to 8 km, respectively (McBride, 2003), with the following approximate relationship (Fig. 9D):

$$L \approx 3.6 + 10.4W \tag{1}$$

Sand ridges commonly form fields with a ridge-and-swale topography (Swift, 1975; Hein, 1987; Houthuys & Gullentrops, 1988; Le Bot *et al.*, 2005) and with the lateral spacing (S) of the ridges scaling both with their height (Fig. 9E):

$$S \approx 29H^2 \tag{2}$$

and with the water depth (d_w) (McBride, 2003):

$$S \approx 250d_w \tag{3}$$

The correlation of core logs (Fig. 8) indicates that the sandstone ridges of facies association A are up to 12 m thick and 21 km in length, which suggests ridge widths of up to 1.7 km (eq. 1). Since the log-derived greatest thickness is unlikely to be the true H_{max} and since not all ridges are likely to have attained the same height, we assume a conservative maximum thickness range of 10 m to 15 m for the sandstone ridges. Their inferred lateral spacing would then be in the range of 2.9 km to 6.5 km (eq. 2), suggesting a water palaeodepth range of 12 m to 26 m (eq. 3). A fairweather wave base at, say, ~5 m depth might set the upper limit for ridge-top aggradation.

As the subtidal sand ridges grew in size and aggraded, the vertical stacking of dunes was increasingly interrupted by storm events and the accommodation space became gradually filled, leading to dune truncation, ridge-top erosion and a temporal predominance of littoral wave action with sand transfer both southwards and farther to the east. The coeval tidal deposits of facies associations A and B were thus covered with the wave-worked deposits of facies association C, which itself tended to be locally eroded by waves and become laterally discontinuous at the accommodation limit (Fig. 10). A significant rise in relative sea-level then created new accommodation space and promoted development of a new generation of sand ridges, some of them superimposed directly on the wave-truncated previous ones. The stratigraphic organisation of facies assemblages (Fig. 8) indicates a repetitive pattern of relative sea-level changes. The transgressive-regressive cycles are attributed to pulses of tectonic subsidence and their record defines a series of sixteen recognisable parasequences bounded by marine-flooding surfaces (Figs 8 and 10). The Garn

Formation is bounded by an erosional surface of forced regression at the base (Fig. 5) and consists of a parasequence set (Figs 8 and 10) representing the transgressive systems tract of a major regional sequence (Johannessen & Nøttvedt, 2008). The development of tidal sand ridges was controlled by bathymetry and recurred only for so long as the tectonically-rejuvenated graben topography was sufficient to boost tidal currents and accommodate vertical stacking of dunes.

The thickness of the individual parasequences varies laterally (Fig. 8), reflecting the combined effect of pre-flooding depositional topography and post-flooding sediment accumulation, often compensational in terms of sea floor relief. Therefore, the use of parasequence thicknesses for recognition of lower-order sea-level changes (Eriksson & Simpson, 1990) may be quite unreliable in such a sedimentary succession, formed by cyclic aggradation of a subtidal system with ridge-and-swale topography.

RESERVOIR HETEROGENEITY MODEL

The Garn Formation, despite its deep burial, has retained primary porosities of ~13% and shows high permeability variation recognisably related to diverse sandstone facies (Quin *et al.*, 2010), which suggests that the formation's primary heterogeneity may be of crucial importance to the reservoir characterisation and geological model. The primary heterogeneity of a sedimentary succession stems from the depositional environment and is defined by the succession's architectural elements and their internal characteristics.

A conceptual hierarchy of the architectural elements of the Garn Formation in the study area, including their spatial organisation and relative volumetric percentages, is shown in Fig. 12. The sediment strata of variable texture form strata sets, whose geometrical varieties and volumes define the basic 'building blocks' (sedimentary facies, Table 1). The spatial distribution and grouping of

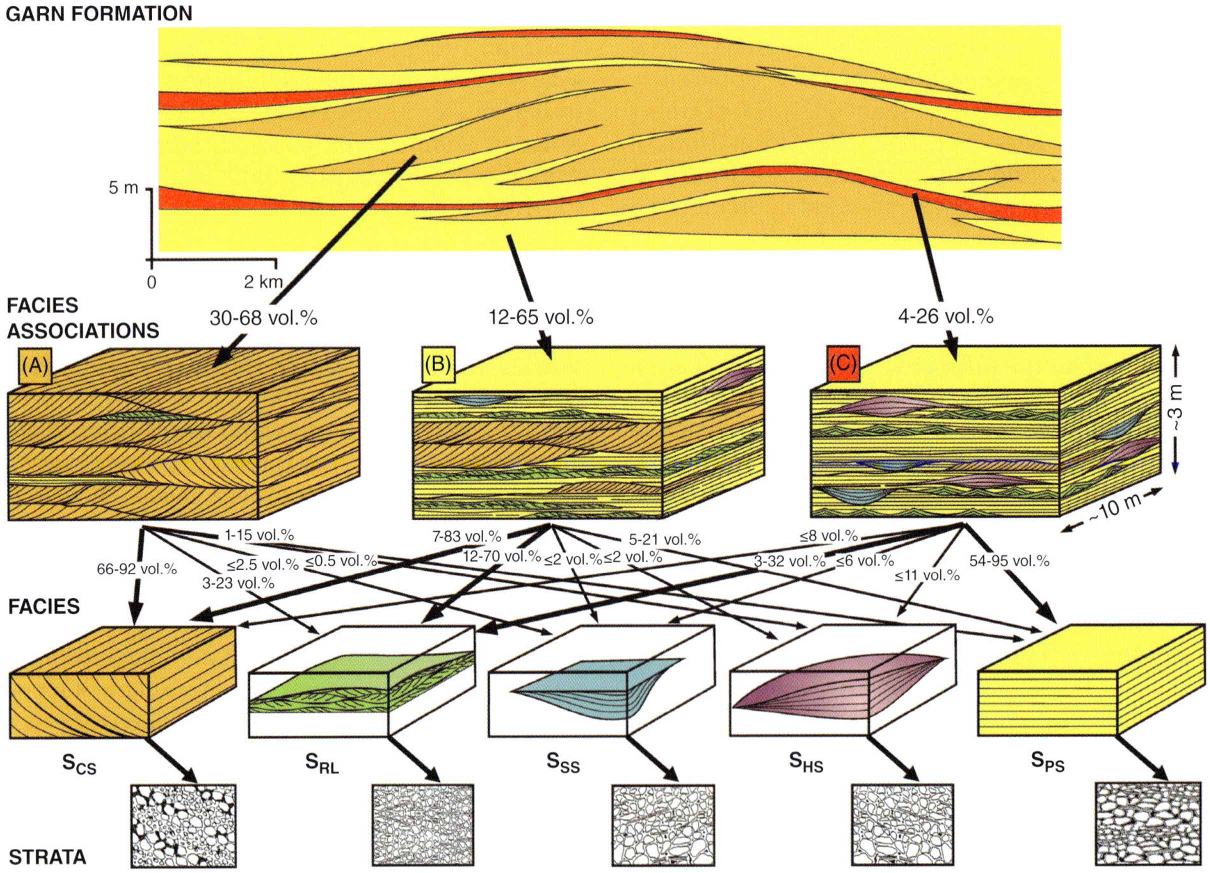

Fig. 12. The hierarchy and simplified model of heterogeneities for the Garn Fm. reservoir in the Kristin Field. The volumetric percentages are one-dimensional estimates based on measured thicknesses.

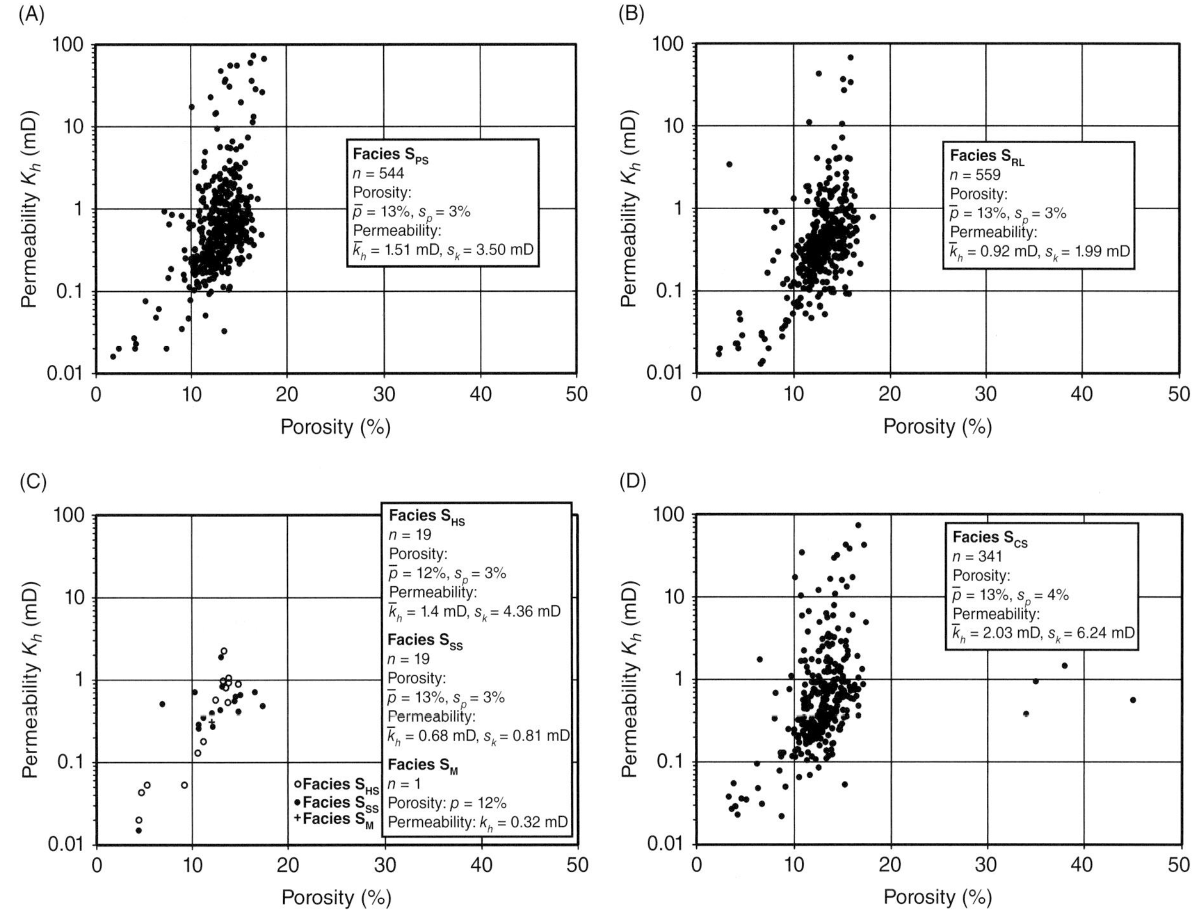

Fig. 13. Porosity and horizontal permeability of the Garn Formation sandstones of: (A) facies S_{HS}; (B) facies S_{RL}; (C) facies S_{HS}, S_{SS} and S_{MA}; and (D) facies S_{CS}. Core-plug data from wells 6406/2-5, 6406-2-5AT2, 6406-2-R-4H, 6406-2-3-T3 and 6406-11-N-H3 in the Kristin Field. Letter symbols: n – number of data; $\bar{p}$ – mean porosity; s_p – porosity standard deviation; $\bar{k}_h$ – mean horizontal permeability; and s_k – permeability standard deviation.

facies define, in turn, three main types of 'megablocks' (facies associations). The megablocks of type A consist chiefly of facies S_{CS} (dune cross-strata sets), whereas the megablocks of type B and C are dominated by facies S_{PS} (planar-parallel strata sets). The other facies are volumetrically subordinate (facies S_{RL}) or minor components (facies S_{HS} and S_{SS}), although some of the units of ripple cross-laminated sandstone contain mudstone drapes or flasers and may be extensive; possibly from tens to hundreds of metres in lateral extent. Facies S_M is of local occurrence and volumetrically negligible and hence is disregarded in the model.

An upscaled model for reservoir flow simulation usually consists of a few hundred thousand grid cells (Fanchi, 2006; Keogh *et al.*, 2007) and hence the reservoir heterogeneity model here is developed at the 'megablock' scale of facies associations (Fig. 12) in order to meet this practical limitation. Each megablock type is characterised in terms of the relative percentage (Table 2) and internal heterogeneity (Fig. 13) of its component facies. The net effect of facies heterogeneities and their spatial configuration should be assessed further for each type of megablock through flow simulation, as there may be an interdependence of the effects of different anisotropies (Weber, 1986).

Facies heterogeneity

The sandstone facies (Table 1) differ markedly in their internal characteristics, which bears directly on their permeability variation and anisotropy. This issue is reviewed below on the basis of detailed laboratory studies of analogous shallow-marine sandstone facies (Rosvoll, 1989; Corbett,

1992), with reference to the core porosity and permeability data (Fig. 13).

Sandstone porosities are known to be normally distributed (Krumbein & Graybill, 1965). The arithmetic mean and standard deviation of porosity for the individual facies have thus been calculated directly from the laboratory datasets. Sandstone permeabilities, in contrast, are log-normally distributed (Krumbein & Graybill, 1965; Rosvoll, 1989) and their corresponding statistics have been calculated from log-transformed, normalised datasets (see formulae in Wilks, 1995, p. 95). Assuming that the standard systematic core-plug measurements of porosity and permeability represent adequately variation of the two parameters in the sandstone facies (Fig. 13), the general meaning of the calculated statistics for a reservoir model would be as follows: ~70 vol.% of a particular facies has the parameter values in the range defined by the mean ±1 standard deviation; ~95 vol.% has values in the range defined by the mean ±2 standard deviations; and ~100 vol.% has values in the range defined by the mean ±3 standard deviations. The frequency distribution of each parameter for a given facies volume would thus be fully described by the two calculated statistics (Fig. 13).

Planar parallel stratification (facies S_{PS}) is not considered to be a major source of reservoir heterogeneity, as the strata are not micaceous and generally lack strong textural variability. The sandstones of facies S_{PS} have a mean porosity of 13%, with a standard deviation of 3% and a mean horizontal permeability of 1.51 mD with a standard deviation of 3.50 mD (Fig. 13A). Permeability in parallel-stratified sandstones varies on the centimetre scale, from one thin bundle of strata to another (Rosvoll, 1989) and tends to be significantly higher in bed-parallel than in bed-normal direction. The ratio of vertical to horizontal permeability is mainly in the range of 0.6 to 0.8. The main permeability baffles are the truncation surfaces separating strata sets, because these surfaces — formed by episodic storms — are broad subtle angular unconformities accompanied by abrupt textural changes. There may be little difference in the capillary properties of sandstone pores in the direction parallel and perpendicular to strata but the displacement of liquid hydrocarbon will generally be more effective by bed-parallel flow (Corbett, 1992). The arithmetic mean may be adequate for both bed-parallel and bed-normal permeability. The use of a harmonic mean, as conventionally applied for bed-normal flow, leads to a considerable overestimation of fractional oil recovery (Corbett, 1992) and thus may also overestimate the flow of condensate liquid.

Ripple cross-lamination (facies S_{RL}) generally involves finer-grained sand and thinner strata, which renders permeability low, with mud flasers and drapes acting as small-scale major baffles. The porosity of facies S_{RL} is similar to that of the previous facies but its permeability is considerably lower, with a mean of 0.92 mD and a standard deviation of 1.99 mD (Fig. 13B). The ratio of vertical to horizontal permeability is mainly in the range of 0.4 to 0.7. ANOVA tests of permeability probe datasets from similar sandstone facies indicate that most variation (65% to 68%) occurs within, rather than between, cross-lamina sets and that the percentage of cross-set internal variance may be as high as 82% to 88% in the presence of mud drapes and flasers (Rosvoll, 1989). Most of the pore gas will be recovered but the amount of condensate initially entrapped by water capillarity may be significant due to the volumetric percentage of this sandstone facies (Fig. 12, Table 2).

Hummocky and swaley stratifications (facies S_{HS} and S_{SS}) are expected to have a relatively high degree of internal heterogeneity, as they form sets of fining-upwards strata that tend to be slightly micaceous near the top and are often capped with wave-ripple cross-lamination. The hummocky and swaley strata sets have scoop-shaped erosional bases and these angular unconformities are commonly marked by an abrupt textural change. The mean porosities of facies S_{HS} and S_{SS} are 12% and 13%, respectively, both with a standard deviation of 3% (Fig. 13C). The mean horizontal permeability in facies S_{HS} is 1.40 mD, with a standard deviation of 4.36 mD, but is only 0.68 mD in facies S_{SS}, with a low standard deviation of 0.81 mD (Fig. 13C). The permeability in either type of strata set is expected to decrease upwards, which may have little effect on the recovery of gas but may affect the flow of liquid (see oil-recovery tests in Corbett, 1992). However, the two facies are volumetrically subordinate (Fig. 12, Table 2) and hence of minor significance to the bulk recovery.

Massive sandstones (facies S_M) are volumetrically insignificant and represented by only one core-plug sample in the dataset, with a porosity of 12% and a horizontal permeability of only 0.32 mD (Fig. 13C). The parameters are lower than in the underlying sandstones of similar grain size (facies S_{CS} or S_{PS}), which supports the notion of sand liquefaction for the origin of facies S_M. An escape of pore water would increase grain packing and redistribute pore-clogging fines (Allen, 1982; Nago *et al.*, 1993; Vaid & Thomas, 1995), thereby reducing both primary porosity and permeability.

Planar cross-stratification (facies S_{CS}) is an important source of heterogeneity in sandstone reservoirs (Weber, 1986; Weber & Van Geuns, 1990) and the permeability anisotropy of cross-strata sets differs markedly from that of the other sandstone facies (Fig. 14; Weber, 1982). This aspect is reflected in the high variance of core-plug data. The mean porosity of facies S_{CS} is 13%,

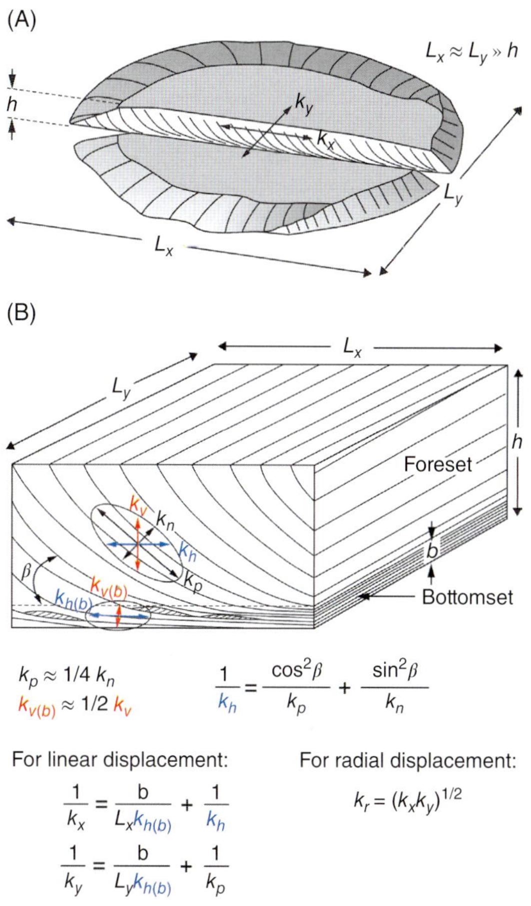

Fig. 14. (A) Definition sketch of dune cross-set dimensions, with k denoting permeability. The horizontal permeability of dune cross-set is considerably different in the direction parallel and transverse to the strike of cross-strata, with $k_x < k_y$. (B) The inherent high permeability anisotropy renders tidal dune cross-sets (facies S_{CS}) the main element of Garn Fm. reservoir heterogeneity. Relationships compiled from Weber (1982), with the symbols denoting: k_h and k_v – horizontal and vertical foreset permeability; $k_{h(b)}$ and $k_{v(b)}$ – horizontal and vertical bottomset permeability; k_p and k_n – permeability parallel and normal to foreset cross-strata.

with a standard deviation of 4% and the mean horizontal permeability is 2.03 mD with a standard deviation of 6.24 mD (Fig. 13D). The ratio of vertical to horizontal permeability is mainly in the range of 0.4 to 0.7, but the permeability parallel to foreset strata varies and may be up to four times higher than the permeability normal to the strata (Emmett *et al.*, 1971). Importantly, the semi-isolated sandstone ridges composed of facies S_{CS} (Fig. 12) are surrounded by less permeable facies and are internally quite heterogeneous, which may hinder the release of condensate from these reservoir compartments and considerably delay spatial equalisation of reservoir pressure.

The sand grain size of foreset strata in facies S_{CS} varies between very fine/fine and medium, or between fine/medium and coarse to very coarse in some cross-sets. The bundles of coarser-grained and finer-grained strata may differ in permeability by nearly an order of magnitude (Weber, 1986; Rosvoll, 1989). This heterogeneity and anisotropy of sandstone cross-strata sets will be of minor importance for the recovery of gas from a permeable reservoir but will be of major importance in an oil reservoir (Emmett *et al.*, 1971; Weber, 1986) and may also play a significant role in the flow and retention of fluids in a less permeable reservoir containing gas and condensate (Elfenbein *et al.*, 2005; Quin *et al.*, 2010).

Approximately 20% of the cross-strata sets in facies S_{CS} include bottomsets, 5 cm to 25 cm thick (e.g. Fig. 6H), which are finer-grained, poorer sorted, richer in mud drapes and containing backflow ripple intrasets. Although the bottomsets are likely to be laterally discontinuous and of limited extent (De Raaf & Boersma, 1971; Nio *et al.*, 1983), they may act as significant permeability baffles (Fig. 14).

In a reservoir, the cross-strata sets act as anisotropic mini-compartments separated by high-angle angular unconformities and textural permeability baffles. The cross-strata sets form sandstone ridges that constitute a major part of the Garn Formation (Fig. 8) and hence the volumetric aspect of cross-sets is of crucial importance to the reservoir characterisation. The measured cross-set thicknesses show great variation and a similarly wide variation can be expected for the unknown cross-set volumes. In a subsequent section, we demonstrate how the frequency distribution of measured cross-set thicknesses can be used on a statistical basis to estimate the frequency distribution of cross-set volumes.

Spatial variation in cross-set thicknesses

The Student *t*-test and the non-parametric RAM and RUD tests and Spearman rank-correlation test (Davis, 2002; Longhitano & Nemec, 2005) failed to reveal any significant spatial trends in cross-set thicknesses. The differences in the cross-set thickness variances for the sandstone ridges, both within and between the individual wells, also appear to be insignificant at ≥90% confidence level. This means a lack of significant differences in cross-set thicknesses between the sandstone ridges and also between their central and lateral parts. The thickest sections of sandstone ridges encountered by the wells show some irregular upward thinning of dune cross-sets, which probably reflects an increase in the erosional impact of waves in an aggrading tidal sand ridge. There is little textural difference between the thicker and thinner cross-sets, which supports the notion of erosional truncation.

The tidal sand ridges were laterally extensive and had a gentle relief, as the inter-ridge sediment accumulation kept pace with the ridge growth (Fig. 10). The reported mean side slopes of modern sand ridges are mainly <1° (McBride, 2003). The lateral bathymetric differences were reduced at aggradation maxima, with the impact of waves increasing and becoming more uniform. This subtidal sea floor physiography differs from that described, for example, by Keene & Harris (1995) from the Torres Strait of north-east Australia and Longhitano & Nemec (2005) from the Amantea Basin of southern Italy. In the latter case, the strait-wide mounded complex of tidal dunes formed at a greater water depth and developed considerable lateral relief, with no wave influence and with the dune cross-set thicknesses reaching 20 m and showing recognisable lateral variation. A pronounced sea floor relief and lack of wave impact may thus be critical for a significant lateral differentiation of dune heights within their subaqueous fields such as tidal sand ridges.

Estimation of cross-set volumes

The dune cross-strata sets form natural depositional units, or beds, and are considered further as such here. Their thickness measurements from the individual cores (Fig. 8) have been analysed by statistical methods, with a special focus on the thickness frequency distribution and its implications for reservoir characterisation.

Cross-set thickness distribution — The statistical character of a dataset frequency distribution is most easily revealed by an exceedence frequency plot (Drummond & Wilkinson, 1996). The exceedence frequency, $EF(x_i)$, of a particular value x_i of the measured variable (x) is defined as the number of data (n_i) with values greater than x_i, divided by the total number of data (n):

$$EF(x_i) = \frac{n_i(x > x_i)}{n} \tag{4}$$

In contrast to a conventional cumulative plot, the exceedence frequency plot is 'decumulative' – made by the back-stripping of data frequency for successively higher *x*-values instead of piling them up by accumulation. An exceedence frequency

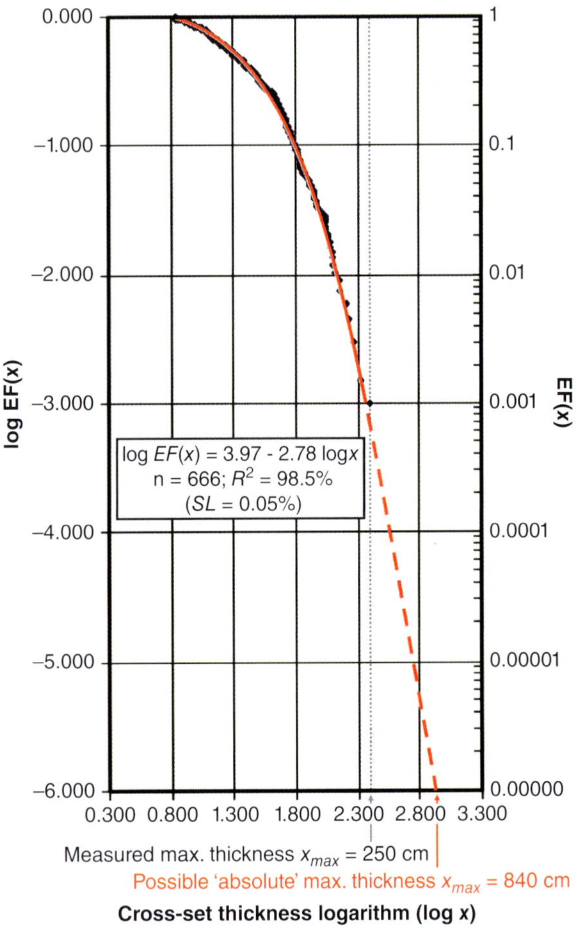

Fig. 15. The exceedence frequency plot of measured dune cross-set thicknesses (bulk dataset from all six well cores) matched by a lognormal distribution curve. Letter symbols: R^2 – coefficient of determination (regression curve goodness of fit); n – number of data; $SL=0.05\%$ is the significance level and means 99.95% confidence for the lognormal trend. An extension of the lognormal curve to the lower frequency limit suggests the possible occurrence of rare cross-sets up to 840 cm thick.

plot of the measured cross-set thicknesses, using their logarithm values (log$x$) and logarithmic frequency, log$EF(x)$, appears to be best matched by a lognormal distribution curve (Fig. 15), with a goodness-of-fit slightly in excess of 98%. However, the dataset distribution can be almost equally well approximated by two straight regression lines, each with a goodness-of-fit coefficient of ~98% (Fig. 16). The same bipartition with a break-point thickness value $x_b \approx 43$ cm characterises data subsets from the individual wells, which implies a homogeneous data population with a statistical property that is stationary (i.e. laterally persistent). A similar property of tidal dune cross-sets has been recognised by Longhitano & Nemec (2005).

The linear trend means that the dataset can be approximated by a two-tier power-law distribution and regarded as a bimodal fractal population (Turcotte, 1992; Middleton *et al.*, 1995). The fractal dimension, estimated as the trend-line gradient, is $D_1 = 0.78$ for the thickness range of 7 cm to 43 cm and $D_2 = 2.78$ for the thicker cross-sets in the measured range of 43 cm to 250 cm (Fig. 16). The minimum cross-set thickness has been taken as $x_{min} = 7$ cm, which is the minimum dune height (Ashley, 1990). The maximum cross-set thickness measured in cores is $x_{max} = 250$ cm, although an extension of the lognormal curve to the lower frequency limit (Fig. 15) suggests that one cannot preclude possible rare cross-sets up to 840 cm thick, not encountered in the studied cores.

The fractal approximation has important implications for the spatial heterogeneity of the sedimentary succession (Stølum, 1991) and gives two major advantages in the present case. Firstly, it allows the frequency distribution of the cross-set thicknesses in their two ranges to be described fully by a single statistic, the fractal dimension (D). The use of Gaussian statistics, such as an arithmetic mean and standard deviation, would be meaningless for the raw dataset, because its frequency distribution is not normal. Second, it allows the frequency distribution of cross-set volumes to be estimated on the basis of Malinverno's (1997) statistical theory.

Cross-set volume distribution — The geometrical relationships used here have been adopted from Malinverno (1997) and are reviewed in the Appendix. Although Malinverno's theory was originally used for turbidite beds, it is by no means generic and accounts for the geometry and spatial stacking pattern of beds irrespective of their origin. Arguably, some of the theory's assumptions are better suited for dune cross-sets (Longhitano & Nemec, 2005) than for lobate turbidite beds tapering with distance.

Malinverno (1997) derived a theoretical relationship between the frequency distribution of bed thicknesses and the frequency distribution of bed volumes, assuming that the former can be approximated by a power-law distribution:

$$EF(x_i) \approx x_i^{-D} \quad (5)$$

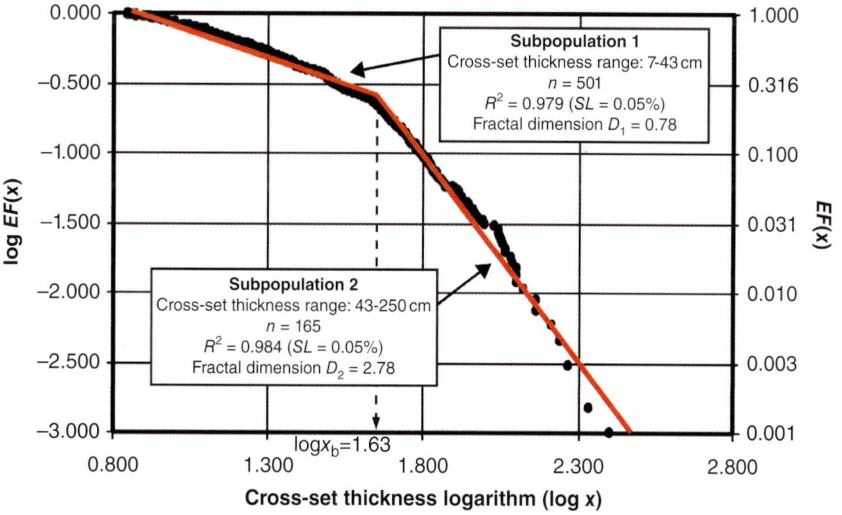

Fig. 16. The exceedence frequency distribution of measured dune cross-set thicknesses approximated as a two-tier power-law distribution. Letter symbols: n – number of data in subpopulation; R^2 – coefficient of determination (regression line goodness of fit); $SL = 0.05\%$ is the significance level and means 99.95% confidence for the linear trend; D – subpopulation fractal dimension (regression line gradient).

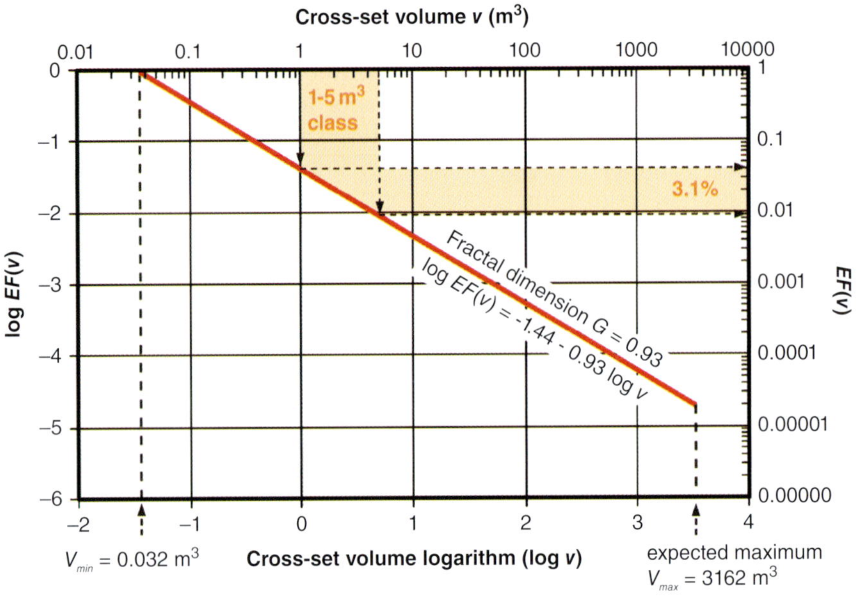

Fig. 17. The exceedence frequency distribution of cross-set volumes, *EF*(*v*), derived from the distribution of measured cross-set thicknesses approximated as a two-tier power-law distribution (Fig. 16). Note that the distribution line allows the number frequency of cross-sets of any particular volume, or volume class, to be predicted. For example, the relative frequency of beds in a volume class of 1 m^3 to 5 m^3 is 3.1%.

where x_i are the measured bed thicknesses and D is the distribution's characteristic exponent, referred to as the fractal dimension. He demonstrated that the corresponding bed-volume distribution will similarly follow a power-law distribution:

$$EF(v_i) \approx v_i^{-G} \tag{6}$$

where $EF(v_i)$ is the frequency of beds with a volume greater than v_i and G is an unknown characteristic exponent (fractal dimension) to be calculated.

The calculation procedure is given in the Appendix. The G-value is derived directly from the values of exponents D_1 and D_2 (Fig. 16) and the resulting equation in the present case is:

$$EF(v_i) = -1.44 v_i^{-0.93} \tag{7}$$

which gives a straight line:

$$\log EF(v_i) = -1.44 - 0.93 \log v_i \tag{8}$$

This line describes completely the expected exceedence frequency distribution of bed volumes (Fig. 17), with the negative G-value defining the line's gradient and with the line's highest point constrained by the frequency value $\log EF(v_{min}) = \log EF(x_{min})$ and the value $\log v_{min}$ for estimated $v_{min} = 0.032\,\text{m}^3$ (see Appendix). The dimensionless coefficient −1.44 is the value of $\log EF(v_i)$ for $\log v_i = 0$. The right-hand limit of the distribution line indicates the expected maximum bed-thickness volume, estimated as $v_{max} \approx 3200\,\text{m}^3$. On this basis, the relative frequency of beds of any particular volume, or volume class, can directly be estimated. For example, the relative frequency of cross-set beds with volumes of 1 m^3 to 5 m^3 is expected to be 3.1% (Fig. 17). In this way, the tidal-ridge sandstone bodies in a reservoir heterogeneity model for the Kristin Field (Fig. 12) can readily be populated with a realistic frequency of cross-set volumes.

Comparative remarks on the reservoir model

As pointed out by Quin *et al.* (2010), the initial interpretation of the Garn Formation in the Halten Terrace was somewhat confused by the tectono-stratigraphic situation where the sediment accumulation occurred in the footwall zone of a major relay structure and the primary supply of sand was from the hanging wall; a configuration opposite to that generally expected in rift basins. It was also not realised that a secondary internal faulting within the footwall zone might have had an important effect on the sedimentation pattern and spatial facies distribution.

The original facies model for the Garn Formation in the Kristin Field was based on the discovery well 6406/2-3, located in the volumetrically important central part of the field. The good-quality reservoir sandstones encountered there were interpreted as coast-proximal shoreface deposits, with a prediction of similar quality sandstones directly to the west and to the east. Structural control upon the shoreface sedimentation was not considered and the Garn Formation as an extremely high net/gross reservoir was not expected to be significantly heterogeneous.

The spatial distribution of facies in subsequent wells was unexpected (see discussion by Quin *et al.*, 2010) and also the reservoir pressure declined much more rapidly than predicted by the original model at the appraisal stage. These subsequent wells indicated that the potential reservoir heterogeneity has been poorly predicted and that the overall sandstone permeability was overestimated by extrapolation from the discovery well (Quin *et al.*, 2010). Seismic surveys allowed detailed mapping of the Kristin Field structure but gave little insight into the reservoir's facies anatomy and spatial heterogeneity.

The reservoir model was subsequently modified, still applying the same assumption of a structurally undisturbed sandy shoreface palaeoenvironment extending eastwards from the Grip High. The Garn Formation in the Kristin Field was subdivided into six stratigraphic zones defined as parasequences. Four zones of this layer-cake model were extended northwards to the Smørbukk and Smørbukk South fields and eastwards to the Tyrrihans North and South fields (Fig. 1; see Elfenbein *et al.*, 2005, fig. 2; Quin *et al.*, 2010, fig. 2). The layer-cake model suggested laterally uniform sandstone units (parasequences) characterised by a gradual eastward decrease in thickness and reservoir quality and with the main heterogeneity on a field scale limited to facies changes between, rather than within, these units.

This sequence-stratigraphic interpretation of the Garn Formation was highly arbitrary, based on recognition of vertical grain-size changes and sporadic mudstone drapes considered to represent marine-flooding surfaces. The coarsening-upwards and 'cleaning-upwards' (i.e. better sorted upwards) textural trends, reportedly discernible and correlative in the adjacent fields to the north and east, are hardly recognisable in the Kristin Field (as also noted by Quin *et al.*, 2010, p. 422).

The present study of the Garn Formation concurs with the notion of a 'structurally controlled' sandy shoreface extending eastwards, influenced by waves and tides and affected by tectonically-driven changes in relative sea-level (Quin *et al.*, 2010). The new depositional model for the Garn Formation in the Kristin Field (Fig. 11) postulates that the shoreface environment terminated by morphological entrapment of sand in an incipient shallow graben, where tidal currents were enhanced by topographic confinement (Sztanó & De Boer, 1995) and repeatedly formed sand ridges. Wave action and eastward sand transfer prevailed each time the graben's accommodation space was temporarily exhausted (Figs 8 and 10). The sandstone succession in the Kristin Field comprises sixteen recognisable shoaling-upwards parasequences and its main heterogeneity is considered to be due to facies variation within, rather than between, the parasequences. The parasequence boundaries are commonly erosional surfaces of sediment bypass, separating similar superimposed facies assemblages (Fig. 8). A crucial aspect of reservoir heterogeneity is the occurrence of tidal sandstone ridges, which are volumetrically important (Fig. 8, Table 2), composed of anisotropic dune cross-sets (Fig. 14) and surrounded by less permeable sandstone facies (Figs 12 and 13).

The new depositional model for the Garn Formation in the Kristin Field explains the 'unusual' distribution of sandstone facies, with 'anomalously abundant' coarser-grained and most permeable cross-stratified sandstones in the central zone of the Kristin Field (Quin *et al.*, 2010), where also the formation is thickest. The notion of semi-isolated sandstone ridges explains the local occurrences of this best-quality reservoir facies assemblage (Quin *et al.*, 2010), which dominates in some wells but is sparse or nearly absent in adjacent ones (Figs 8 and 12, top). The depositional model also bears the following main implications for the development of the Garn Formation in the adjacent fields: (1) the shoreline of the Grip High was probably highly uneven due to fault relay ramps, relay-breach anticlines and local coarse-grained deltas, whereby the eastward advances of its sandy littoral zone were non-uniform; (2) incipient fault-related depressions similar to the inferred Kristin graben (Fig. 11) may have affected sand dispersal and enhanced tidal current in other segments of the Halten seaway (e.g. tidal sandstone ridges were suggested for the

Smørbukk Field by Corfield *et al.*, 2001); (3) the individual grabens and half-grabens of the seaway would likely differ in their internal pattern of subsidence and hence the parasequence stratigraphy of the Garn Formation may not be regionally correlative; and (4) the eastward transfer of sand across these tidal current-dominated depressions depended on their accommodation space and was probably highly non-uniform on regional scale.

The revised model for the primary heterogeneity and permeability compartmentalisation of the Garn Formation in the Kristin Field (Fig. 12) may explain why the reservoir's initial pressure has so rapidly declined (Quin *et al.*, 2010) and the recovery factor has been considerably lower than in other gas-bearing and condensate-bearing fields. The reservoir's overall permeability is low. The release of fluids from the more permeable reservoir compartments (tidal sandstone ridges) is probably hindered by their high degree of internal heterogeneity and baffled by the surrounding less permeable facies, which delays the spatial transfer of fluids and equalisation of reservoir pressure. Considerable amounts of condensate liquid may be entrapped by capillary forces in these compartments at the early stage of production, as the background pressure decreases and pore-water saturation rises. This aspect of the recovery efficiency should thus be simulated and carefully assessed with the aid of the quantitative model provided by the present study. In addition, a possible role of faults (Fig. 4) in secondary compartmentalisation of reservoir should also be taken into account.

CONCLUSIONS

Analysis of the Garn Formation in the Kristin Field indicates that sand deposition occurred in a subsiding, shallow incipient graben within a narrow Jurassic seaway linking the Boreal and Tethys seas. Deposition involved repetitive cycles of tide-dominated to wave-dominated sedimentation, attributed to tectonically-driven changes in relative sea-level. The transgressive-regressive cycles formed a transgressive parasequence set ~ 100 m thick, culminating in a major marine flooding and the deposition of the muddy Melke Formation as a result of the climax of Jurassic rifting, which caused a structural collapse of the shelf.

Facies analysis indicates that the sedimentation involved a vertical stacking of tidal dunes into sand ridges, accompanied by episodic storm action and concurrent sand accumulation in inter-ridge swales. Fairweather wave action predominated as the accommodation was gradually filled up by sea floor aggradation, until a new increase in accommodation occurred due to tectonic subsidence. Littoral wave-dominated sedimentation eventually prevailed, but soon retreated onto the graben flank as the regional rise in relative sea-level caused stepwise encroachment of muddy neritic environment and a general marine flooding.

The depositional setting and facies anatomy of the Garn Formation revealed by the study explain the spatial distribution of sandstone facies and have important implications for reservoir heterogeneity. Tidal sandstone ridges are a major architectural element of the Garn reservoir in the Kristin Field. These semi-isolated, more permeable sandstone bodies constitute locally up to ~ 70% of the bulk formation thickness and are estimated to be up to ~ 20 km long, 1.7 km wide and 10 m to 15 m thick, with a lateral spacing of 2.9 km to 6.5 km.

The tidal sandstone ridges are encased in less permeable sandstone facies and consist of dune cross-strata sets that act as highly anisotropic permeability mini-containers. The combination of the two scales of reservoir heterogeneity is considered to be responsible for the rapid decline of reservoir pressure and low recovery factor. Large amounts of liquid condensate may potentially be withheld in these sandstone bodies at the early stages of production, to be released only later when eventually turning into gas.

A statistical method has been used to estimate the cross-set volume frequency distribution from the frequency distribution of the cross-set thicknesses measured in cores. Such quantitative estimates are valuable for the characterisation and modelling of petroleum reservoirs, where only bed thicknesses can be measured from the cores. On this basis, the sandstone ridge bodies in a reservoir model for the Garn Formation in the Kristin Field can readily be populated with realistic cross-set volumes and their anisotropy be accounted for to simulate the flow of fluids in the reservoir and assess better the recovery.

The study indicates that a high net/gross and seemingly homogeneous sandstone reservoir with low overall permeability may in reality be highly heterogeneous and show a production performance suggesting strong compartmentalisation, even though the reservoir contains only high-pressure gas and condensate.

ACKNOWLEDGEMENTS

The paper stems from the first author's research project sponsored by Statoil A.S.A. and conducted through a post-doctoral fellowship at the University of Bergen in 2006. The licence holders of the Kristin Field (Statoil, Petoro, Exxon Mobil, Eni Norge and Total Norge) are thanked for their permission to publish the data. We thank also the Statoil's Kristin Field operation unit for providing additional information and technical support and Jamie Quin for discussions. The manuscript was critically reviewed by Jonathan Wonham, Gary Hampson and Snorre Olausen, whose helpful comments are much appreciated by the authors.

APPENDIX

The tidal ridge bodies in the Garn Formation consist of sandstone beds that are dune cross-sets, stacked directly upon one another or separated by beds of other sandstones facies. The cross-set beds have a common origin and can be regarded as small architectural elements of a sandstone ridge, scattered in its 3D space. Let us consider that the thicknesses of these beds are measured along a vertical sampling line (Fig. A-1), such as an outcrop or well-core log. The following review of Malinverno's (1997) statistical theory explains how the bed-volume frequency distribution can be estimated from the frequency distribution of the measured bed thicknesses.

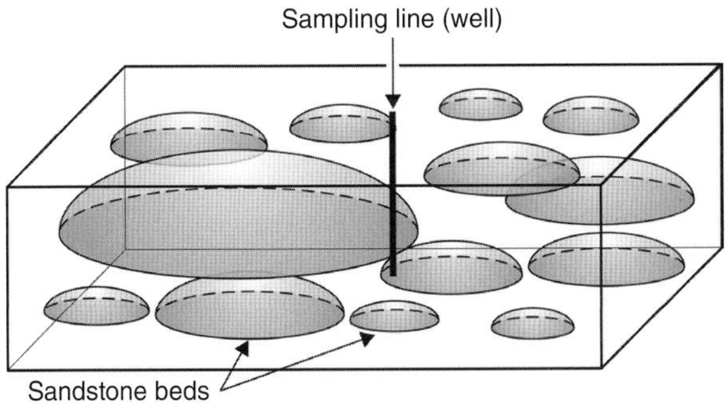

Fig. A-1. Schematic illustration of the notion of the thicknesses of sandstone beds, such as dune cross-sets within a tidal sandstone ridge, measured along a vertical sampling line such as a well core. Diagram slightly modified from Malinverno (1997, Fig. 4).

Definition of bed geometry

As an approximation of dune cross-set shape (Fig. A-2A), the bed geometry is assumed to have a roughly circular base and the maximum thickness (h) in the central part (Fig. A-2B). The diameter of the bed base, or bed breadth, is $w = ah^{\gamma}$, where a is a constant and the exponent $\gamma \geq 0$ serves to scale the bed breadth with its central thickness (Fig. A-3A). If $\gamma = 0$, all beds have the same breadth $w = a$ irrespective of their central thicknesses. If $\gamma = 1$, the bed breadth scales linearly with the central thickness (i.e. a bed that is twice as thick will also have a lateral extent that is twice as great).

As argued by Malinverno (1997), the bed thickness (x) away from the bed centre is only a function of the lateral distance r and of the thickness h in the central part (Fig. A-2B):

$$x(h,r) = h[1-(2r/ah^{\gamma})^{\alpha}] \text{ for } 0 \leq r \leq 0.5ah^{\gamma} \quad \text{(A1)}$$

where $\alpha > 0$ is an exponent controlling the bulk bed shape (Fig. A-3B). If $0 < \alpha < 1$, the bed thickness decreases laterally most rapidly near the centre. If $\alpha = 1$, the thickness decreases uniformly and the bed is broadly conical in shape. If $\alpha > 1$, there is little or no lateral thickness decrease near the bed centre but rapid decrease at the margins, such that the bed top is flat or slightly convex upwards and the bed shape resembles an inverted

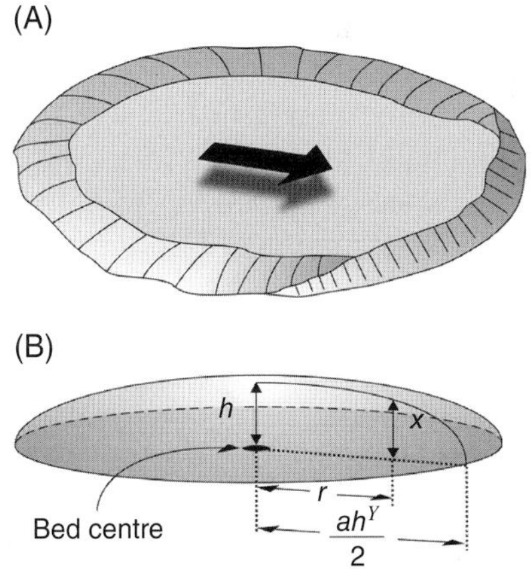

Fig. A-2. (A) Schematic geometry of a long-crested subaqueous dune that has migrated slightly in excess of twice its wavelength (based on Allen, 1982). The height (thickness) of dunes may range from ~7 cm to ≥20 m and scales with their wavelength (Ashley, 1990). (B) A simplified 'inverted-saucer' geometry assumed to represent a dune cross-set bed. The bed has an approximately circular base, a maximum thickness h in the central part and a thickness x at a distance r from the centre (modified from Malinverno, 1997, Fig. 5).

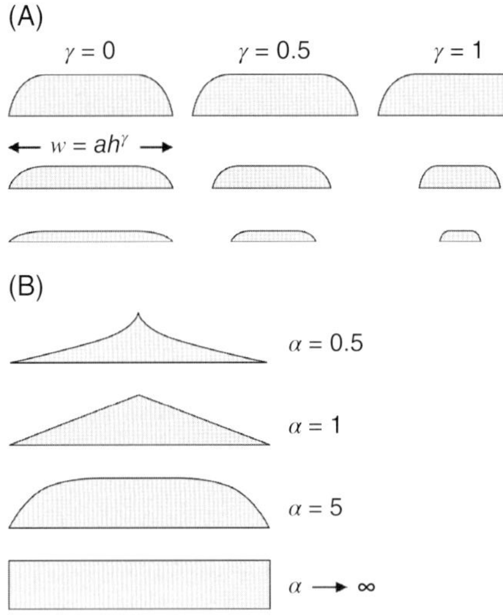

Fig. A-3. The scaling of bed geometry defined by equation A1 in the text. (A) The exponent γ controls how the bed breadth w scales with the bed central thickness h. (B) The exponent α controls how the bed thickness varies laterally with distance from the bed centre. Compiled and slightly modified from Malinverno (1997, Figs 6 & 7).

saucer. And as $\alpha \to \infty$, the bed becomes cylindrical in shape.

This formulation of bed geometry is obviously a simplification, since the lateral change in bed thickness may be irregular and also the bed base may be elliptical, rather than circular. However, it may be a reasonable approximation for long-crested tidal dunes that have migrated over distances comparable to twice their wavelength (Fig. A-2A; see Allen, 1982; Bridge & Demico, 2008). Also, the notion of central maximum thickness (h) can be replaced conceptually with the notion of a representative prevalent thickness (Fig. A-4A). As pointed out by Malinverno (1997), one could define a bed base of any desirable shape, but this would render the mathematical formulation considerably more complicated. It should be noted that, by allowing γ to take any positive value, the geometrical formulation allows for a wide range of possible scaling of the bed breadth with its thickness, whereas the value of α, though possible to estimate, appears to be unimportant because this exponent does not enter the actual relationship between the frequency distributions of the bed thickness and bed volume.

Malinverno (1997) derived the bed volume from equation A1 as follows:

$$v(h) = \int_0^{0.5ah^\gamma} \eta(h,r) 2\pi r \cdot dr = [(a2\pi\alpha)/4(2+\alpha)]h^{(2\gamma+1)} \qquad \text{(A2)}$$

where a is an integration constant (see Fig. A-2B). By combining the latter equation with the power-law relationship for bed volumes:

$$EF(v_i) \approx v_i^{-G} \qquad \text{(A3)}$$

he obtained the frequency distribution of bed thicknesses for the bed central parts as:

$$EF(h_i) \approx h_i^{-G(2\gamma+1)} \qquad \text{(A4)}$$

For example: If $\gamma = 0$, all beds have the same breadth, their volume depends linearly on h and the exponent of this frequency distribution is G, same as of the bed-volume frequency distribution. If $\gamma = 1$, the bed breadth scales linearly with h, the bed volume is proportional to h^3 and the exponent of the distribution is $3G$.

Spatial distribution of beds and the sampling-line problem

Although the relationship defined by equation A4 might seem like the desired solution, Malinverno (1997) points out that this would be the case only if we were able to measure the 'central' maximum thickness h of all encountered beds. In reality, we are measuring the bed thickness x at some unknown distances from the bed centre, whereby the measured value may be significantly lower (Fig. A-1). Even if the bed thicknesses were invariable with distance from the bed centre (Fig. A-3B, bottom), there would still be another sampling-line problem: Since the cross-set beds are not ridge-wide, we are quite unlikely to measure the thicknesses of all dune cross-sets stacked within any given tidal ridge.

The sampling problem to be accounted for is shown schematically in Figure A-5. In this hypothetical example, the bed breadth is proportional to the bed central thickness ($\gamma > 0$) and the bed centres are scattered throughout their depositional area. Let us draw a vertical sampling line through the sedimentary succession at some location and measure the thicknesses x of all beds intersected by the sampling line. Beds with greater thicknesses are more likely to be sampled because of their greater lateral extent, which means that the sampling will be biased towards thick beds, whereas the number of thin beds will be underestimated. There are many more thin beds than thick beds in this example (Fig. A-5), yet the sampling line intersects the same number of both. Only if the centres of all beds had the same location in plan view would a vertical sampling line close to this location intersect nearly all beds,

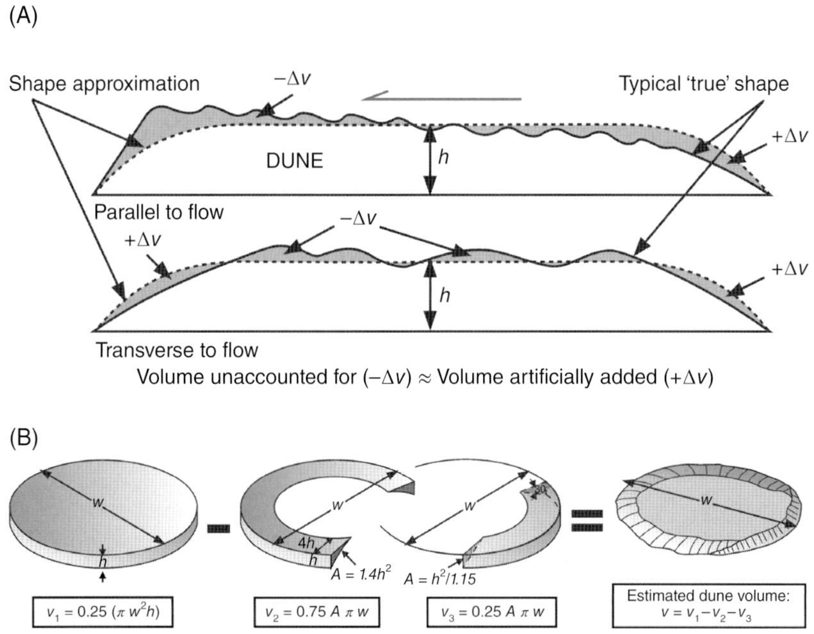

Fig. A-4. (A) In the bed shape considered (Fig. A-2), the dune volume unaccounted for is compensated by the volume artificially added by the 'inverted-saucer' approximation, whereas the notion of a central maximum thickness can be replaced with the notion of a representative prevalent thickness. (B) Diagram explaining how the volume of a dune-enclosing cylinder is geometrically reduced to approximate the idealised shape of a dune that has migrated slightly in excess of its double wavelength.

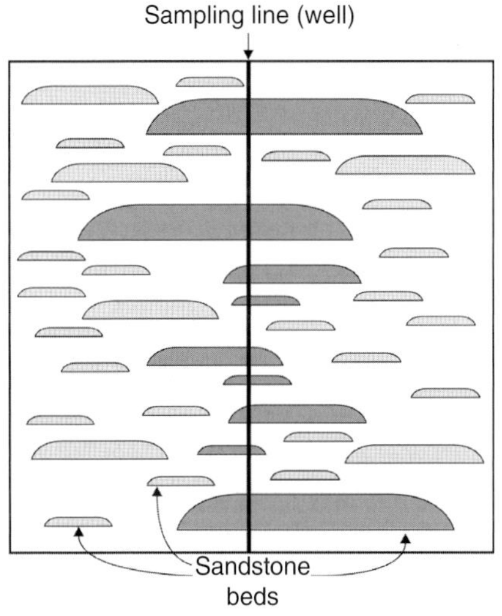

Fig. A-5. Hypothetical cross-section through a stratigraphic succession intersected by vertical sampling line (slightly modified from Malinverno, 1997, Fig. 9). Note that there are many more thin than thick beds in the succession but the sampling line intersects the same number of both because of their different lateral extent.

whereby the distribution exponent for the measured bed thicknesses would be the same as that expected for the bed-centre maximum h-values (eq. A4).

To quantify the spatial distribution of bed centres in a plan view, Malinverno (1997) defined as $M(r_i)$ the number of bed centres whose lateral distance from the sampling line is less than r_i and he assumed that this number is proportional to some power of r_i:

$$M(r_i) \approx r_i^Z \qquad \text{(A5)}$$

where the value of exponent Z is between 0 and 2 (Fig. A-6). If all the bed centres were located at the same point in a horizontal plane and coincided with the location of the sampling line, $M(r)$ would be a constant and $Z=0$ (Fig. A-6A). If the bed centres were located along a line in the horizontal plane, $M(r)$ would be linearly proportional to r and $Z=1$ (Fig. A-6B). If the bed centres were scattered randomly in the horizontal plane, $M(r)$ would be proportional to r^2 and $Z=2$ (Fig. A-6C). In the present case of dunes clustered into elongate sand ridges, an intermediate relationship would be more appropriate, such as $M(r)$ proportional to $r^{1.5}$ and $Z=1.5$ (Fig. A-6D).

The relationship between D and G

According to the preceding conceptual notions, the exponent D of the frequency distribution of bed thicknesses:

$$EF(x_i) \approx x_i^{-D} \qquad \text{(A6)}$$

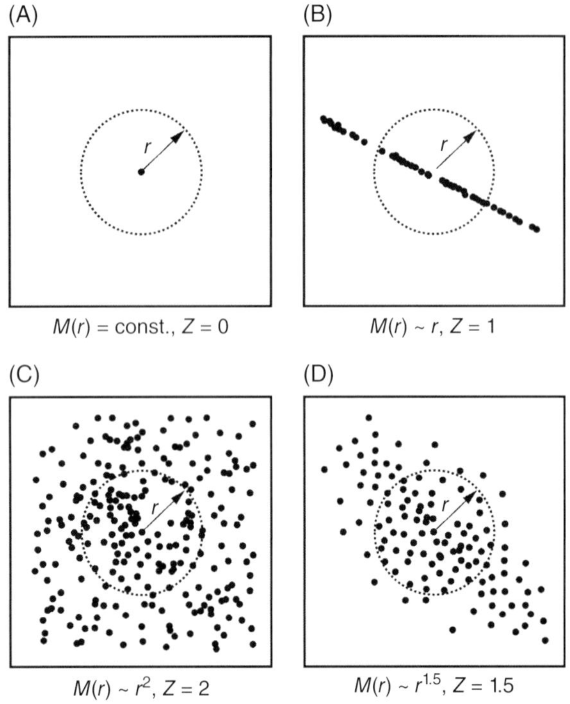

Fig. A-6. The spatial distribution of bed centres (black dots) in plan view; slightly modified from Malinverno (1997, Fig. 8) (A) If all the bed centres have the same location in the horizontal plane, the number of centres $M(r)$ within a circle of radius r is constant and the exponent $Z=0$. (B) If the bed centres are clustered along a line, $M(r) \approx r$ and $Z=1$. (C) If the bed centres are scattered randomly, $M(r) \approx r^2$ and $Z=2$. (D) If the bed centres are clustered within an elliptical area, the parameters will be intermediate, such as $M(r) \approx r^{1.5}$ and $Z=1.5$.

is expected to depend on several factors, each quantified by a different dimensionless exponent: the scaling between the bed breadth and central thickness, quantified by γ ($\gamma \geq 0$); the lateral decrease in bed thickness with distance from the bed centre, quantified by α ($\alpha > 0$); the frequency distribution of the bed volumes, quantified by G ($G > 0$); and the spatial distribution of the bed centres relative to the sampling line, quantified by Z ($0 \leq Z \leq 2$).

Malinverno (1997) has demonstrated that the general formula relating these exponents to the exponent D has the following form:

$$D = G + \gamma(2G - Z) \quad \text{(A7)}$$

This equation shows that the details of bed shape may actually be irrelevant, so long as the shape of all beds is roughly the same and is adequately approximated for some value of $\alpha > 0$ (Fig. A-3B). The latter coefficient is not involved in the equation.

In general, the exponent D will not be equal to the corresponding exponent G, unless $\gamma = 0$ or $2G = Z$. As pointed out by Malinverno (1997), the exponent γ will be zero only if all beds cover the same area irrespective of their thicknesses, which means a peculiar case. Likewise, the equality $2G = Z$ might hold in some unusual case but there is no obvious reason to assume that the scaling of bed breadth with bed thickness is related to the spatial distribution of the bed centres. In short, one cannot infer the frequency distribution of bed volumes from the distribution of bed thicknesses alone; for this purpose, something must be known or assumed about the scaling of bed breadth with bed thickness and about the spatial distribution of bed centres.

The estimation of exponents γ and G

In statistical terms, the vertical stacking of dunes into a tidal sand ridge can be compared to the dropping at random of a large number of 'inverted-saucer' beds (Fig. A-2) of various sizes onto an elongate, elliptical area that is at least as wide as the largest bed breadth (Fig. A-6D). The pertinent question is as to what will be the frequency distribution of the bed thicknesses measured along a vertical sampling line through such a dune complex.

Equation A7 shows that the exponent D of the bed-thickness frequency distribution depends on the exponent G of the bed-volume frequency distribution, on the exponent γ relating bed thickness and breadth (Fig. A-3A) and on the exponent Z of the spatial distribution of the bed centres (Fig. A-6). Malinverno (1997) distinguished two limiting cases for the spatial distribution of the bed centres confined to an area. If the bed breadth is $w \geq ½\, w_{max}$, the bed will always be intersected by a vertical sampling line located near the area centre (which means $Z \approx 0$; Fig. A-6A). If the bed breadth instead is $w < w_{max}$, the bed centre is no longer constrained to be close the area centre (which means $Z \approx 2$; Fig. A-6C). The spatial distribution of bed centres confined to an elongate tidal ridge will represent an intermediate case (Fig. A-6D), where the exponent Z can be any non-integer value between 1 and 2.

As for the two limiting cases, Malinverno (1997) had shown that their transition would occur at the 'break-point' value x_b of bed thickness, defined as:

$$x_b = 0.37^{1/\gamma} x_{\max} \quad \text{(A8)}$$

where x_{max} is the measured maximum bed thickness. In the present case, $x_b = 0.43$ m and

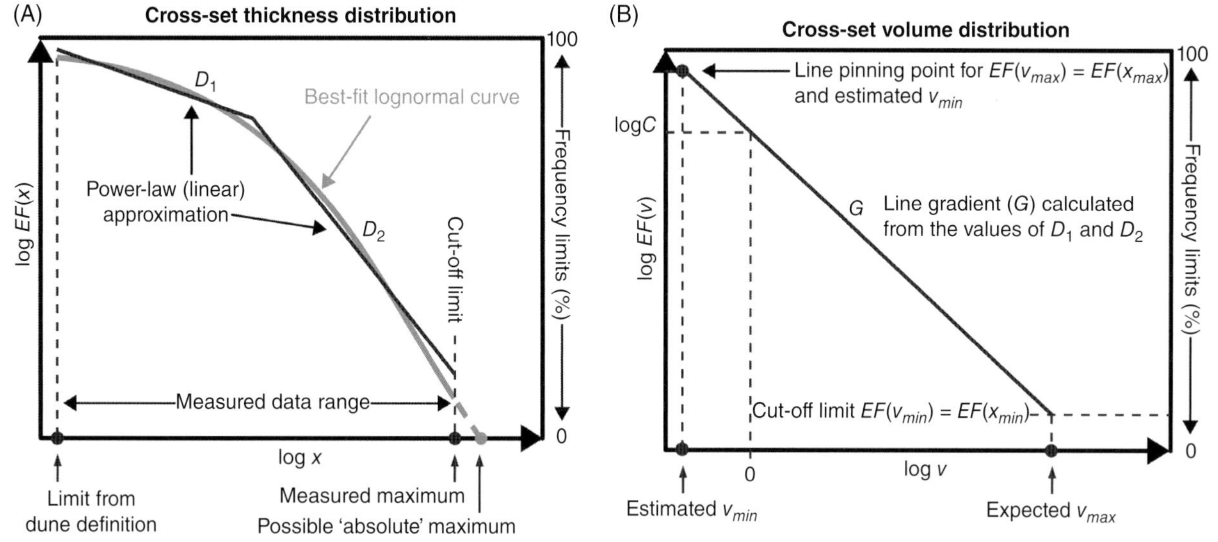

Fig. A-7. Diagrams showing how of a lognormal bed-thickness frequency *EF*(*x*) approximated as a two-tier power-law distribution (A) allows the bed-volume frequency distribution *EF*(*v*) to be determined and numerically constrained (B).

x_{max} = 2.5 m, which gives γ = 0.56. From equation A7, one can expect that the bed-thickness distribution will thus have two exponents:

$$D_1 = G(1+2\gamma) - 2\gamma \text{ for the thinner beds } (x \le x_b)$$
$$D_2 = G(1+2\gamma) \text{ for the thicker beds } (x > x_b) \tag{A9}$$

Malinverno (1997) had tested this observation with a simple numerical experiment and rearranged the two equations to obtain the unknown exponent *G* from the known values of the exponents D_1 and D_2 of bed-thickness distribution (with $D_2 > D_1$, Fig. A-7A):

$$G = \frac{D_2}{1 + D_2 - D_1} \tag{A10}$$

For the bed-thickness dataset in the present case, this exponent is *G* = 0.93 and allows us to determine the frequency distribution of bed volumes, $EF(v_i)$.

The derivation of bed-volume frequency distribution

The exceedence frequency distribution of bed volumes has the following general form:

$$EF(v_i) = Cv_i^{-G} \tag{A11}$$

which, in a log-log plot, defines a straight line (Fig. A-7B):

$$\log EF(v_i) = \log C - G \log v_i \tag{A12}$$

The exponent *G* (negative value) defines the distribution line gradient and the slant line is constrained further as follows (Fig. A-7B):

The frequencies of the volumetrically smallest and largest beds, $EF(v_{min})$ and $EF(v_{min})$, are assumed to be the same as those of the thinnest and thickest beds, $EF(x_{min})$ and $EF(x_{min})$, since the thinnest beds are expected to be the smallest and the thickest beds to be the largest.

The horizontal co-ordinate of the line's left-hand pinning point is the lower cut-off limit $\log v_{min}$, calculated for an estimated v_{min} (see below).

The line's coefficient log*C* is the value of $\log EF(v_i)$ corresponding to $\log v_i = 0$.

The v_{min} value for dune cross-sets can be indirectly estimated by using the Flemming empirical formula (Ashley, 1990, Fig. 9A):

$$h = 0.0677\,L^{0.8098} \tag{A13}$$

to derive the dune wavelength (*L*) for dune minimum height (x_{min}) and take this value to approximate the dune breadth as $w \approx 2L$. The wavelength for *h* = 0.07 m is *L* = 0.96 m and the corresponding dune volume can be calculated from the geometrical equations in Figure A-4B, which yields $v_{min} = 0.16\ m^3$. The right-hand limit of the distribution line simultaneously implies an expected v_{max} value of ~ 3200 m^3. By substituting a with w/h^{γ} (Fig. A-2A) in equation A2, similar bed volumes are obtained for $\alpha = 2.8$, which gives an estimate of the latter exponent (cf. Fig. A-3B).

REFERENCES

Allen, J.R.L. (1982) *Sedimentary Structures: Their Character and Physical Basis*. One-volume Edition, *Dev. Sedimentol.*, **30**, 1256 pp. Elsevier, Amsterdam.

Amos, C.L., Barrie, J.V. and **Judge, J.T.** (1995) Storm-enhanced sand transport in a microtidal setting, Queen Charlotte Islands, British Columbia, Canada. In: *Tidal Signatures in Modern and Ancient Sediments* (Eds B.W. Flemming and A. Bartholomä), *Int. Assoc. Sedimentol. Spec. Publ.*, **24**, 53–68.

Arnott, R.W.C. (1993) Quasi-planar laminated sandstone beds of the Lower Cretaceous Bootlegger Member, north-central Montana: evidence of combined-flow sedimentation. *J. Sed. Petrol.*, **63**, 488–494.

Arnott, R.W.C. and **Southard, J.B.** (1990) Exploratory flow-duct experiments on combined-flow bed configurations and some implications for interpreting storm-event stratification. *J. Sed. Petrol.*, **60**, 211–219.

Ashley, G.H. (1990) Classification of large-scale subaqueous bedforms: a new look at an old problem. *J. Sed. Petrol.*, **60**, 160–172.

Bjørlykke, K., Aagaard, P., Dypvik, H., Hastings, D.S. and **Harper, A.S.** (1986) Diagenesis and reservoir properties of Jurassic sandstones from the Haltenbanken area, offshore mid-Norway. In: *Habitat of Hydrocarbons on the Norwegian Continental Shelf* (Eds A.M. Spencer, C.J. Campbell, S.H. Hanslien, E. Holter, P.H.H. Nelson, E. Nysæther and E.G. Ormaasen), 275–286. *Norw. Petrol. Soc. Proc.*, Graham and Trotman, London.

Blystad, E, Brekke, H., Færseth, R.B., Larsen, B.T., Skogseid, J. and **Tørudbakken, B.** (1995) Structural elements of the Norwegian Continental Shelf. Part II: The Norwegian Sea region. *Norw. Petrol. Direct. Bull.*, **8**, 1–45.

Bourgeois, J. (1980) A transgressive shelf sequence exhibiting hummocky stratification: the Cape Sebastian Sandstone (Upper Cretaceous), southwestern Oregon. *J. Sed. Petrol.*, **50**, 681–702.

Brandsæter, I., McIlroy, D., Lia, O., Ringrose, P. and **Næss, A.** (2005) Reservoir modelling and simulation of Lajas Formation outcrops (Argentina) to constrain tidal reservoirs of the Halten Terrace (Norway). *Petrol. Geosci.*, **11**, 37–46.

Brekke, H. (2000) The tectonic evolution of the Norwegian Sea Continental Margin with emphasis on the Vøring and Møre basins. In: *Dynamics of the Norwegian Margin* (Eds A. Nøttvedt, B.T. Larsen, S. Olaussen, B. Torudbakken, J. Skogseid, R.H. Gabrielsen, H. Brekke and O. Birkeland), *Geol. Soc. London Spec. Publ.*, **167**, 327–378.

Brekke, H., Sjulstad, H.I., Magnus, C. and **Williams, R.W.** (2001) Sedimentary environments offshore Norway — an overview. In: *Sedimentary Environments Offshore Norway – Palaeozoic to Recent* (Eds O.J. Martinsen and T. Dreyer), *Norw. Petrol. Soc. (NPF) Spec. Publ.*, **10**, 7–37.

Bridge, J.S. and **Demico, R.V.** (2008) *Earth Surface Processes, Landforms and Sediment Deposits*. Cambridge University Press, Cambridge (U.K.), 815 pp.

Bromley, R.G. (1996) *Trace Fossils: Biology, Taphonomy and Applications*. 2nd edn., Chapman & Hall, London, 367 pp.

Bukovics, C., Cartier, E.G., Shaw, N.D. and **Ziegler, P.A.** (1984) Structure and development of the mid-Norway continental margin. In: *Petroleum Geology of the North European Margin* (Ed. A.M. Spencer), pp 407–423. Graham and Trotman, London.

Cacchione, D.A., Drake, D.E., Ferreira, J.T. and **Tate, G.B.** (1994) Bottom stress estimates and sand transport on northern California inner continental shelf. *Cont. Shelf Res.*, **14**, 1273–1289.

Cameron, T.D.J., Crosby, A., Balson, P.S., Jeffery, D.H., Lott, G.K., Bulat, J. and **Harrison, D.J.** (1992) *The Geology of the Southern North Sea*. British Geological Survey, London.

Carr, I.D., Gawthorpe, R.L., Jackson, C.A.L., Sharp, I.R. and **Sadek, A.** (2003) Sedimentology and sequence stratigraphy of early syn-rift tital sediments: the Nukhul Formation, Suez Rift, *Egypt. J. Sed. Res.*, **73**, 407–420.

Caston, V.N.D. (1972) Linear sand banks in the southern North Sea. *Sedimentology*, **18**, 63–78.

Chuhan, F.A., Bjørlykke, K. and **Lowrey, C.J.** (2001) Closed-system burial diagenesis in reservoir sandstones; examples from the Garn Formation, Haltenbanken area, offshore Mid-Norway. *J. Sed. Res.*, **71**, 15–26.

Clifton, H.E. and **Dingler, J.R.** (1984) Wave-formed structures and paleoenvironmental reconstruction. *Mar. Geol.*, **60**, 165–198.

Collinson, J.D. and **Thompson, D.B.** (1982) *Sedimentary Structures*. Allen and Unwin, London, 207 pp.

Corbett, P.W.M. (1992) *Geological Description for Reservoir Simulation*. Short Course Lecture Notes, Department of Petroleum Engineering, Heriot-Watt University.

Corfield, S. and **Sharp, I.R.** (2000) Structural style and stratigraphic architecture of fault propagation folding in extensional settings: a seismic example from the Smørbukk area, Halten Terrace, Mid-Norway. *Basin Res.*, **12**, 329–341.

Corfield, S., Sharp, I.R., Håger, K.-O., Dreyer, T. and **Underhill, J.** (2001) An integrated study of the Garn and Melke formations (Middle to Upper Jurassic) of the Smørbukk area, Halten Terrace, mid-Norway. In: *Sedimentary Environments Offshore Norway – Palaeozoic to Recent* (Eds O.J. Martinsen and T. Dreyer), *Norw. Petrol. Soc. (NPF) Spec. Publ.*, **10**, 199–210.

Dalland, A., Worsley, D. and **Ofstad, K.** (1988) A lithostratigraphic scheme for the Mesozoic and Cenozoic succession offshore mid- and northern Norway. *Norw. Petrol. Direct. Bull.*, **4**, 1–65.

Davis, J.C. (2002) *Statistics and Data Analysis in Geology*. 3rd edn., John Wiley and Sons, New York, 638 pp.

De Raaf, J.F.M. and **Boersma, J.R.** (1971) Tidal deposits and their sedimentary structures. *Geol. Mijnbouw.*, **50**, 479–504.

Doré, A.G. (1992) Synoptic palaeogeography of the Northeast Atlantic Seaway: late Permian to Cretaceous. In: *Basins on the Atlantic Seaboard: Petroleum Geology, Sedimentology and Basin Evolution* (Ed. J. Parnell), *Geol. Soc. London Spec. Publ.*, **62**, 421–446.

Doré, A.G., **Lundin, E.R.**, **Jensen, L.N.**, **Birkeland, Ø.**, **Eliassen, RE.** and **Fichler, C.** (1999) Principal tectonic events in the evolution of the northwest European Atlantic margin. In: *Petroleum Geology of Northwest Europe — Proceedings of the 5th Conference* (Eds A.J. Fleet and S.A.R. Boldy), 41–61. Geol. Soc. London.

Dreyer, T. (1992) Significance of tidal cyclicity for modelling of reservoir heterogeneities in the lower Jurassic Tilje Formation, mid-Norwegian shelf. *Nor. Geol. Tidsskr.*, **72**, 159–170.

Drummond, C.N. and **Wilkinson, B.H.** (1996) Stratal thickness frequencies and the prevalence of orderedness in stratigraphic sequences. *J. Geol.*, **104**, 1–18.

Dumas, S. and **Arnott, R.W.C.** (2006) Origin of hummocky and swaley cross-stratification — the controlling influence of unidirectional current strength and aggradation rate. *Geology*, **34**, 1073–1076.

Dumas, S., **Arnott, R.W.C.** and **Southard, J.B.** (2005) Experiments on oscillatory-flow and combined-flow bed forms: implications for interpreting parts of shallow-marine sedimentary record. *J. Sed. Res.*, **75**, 501–513.

Dyer, K.R. and **Huntley, D.A.** (1999) The origin, classification and modeling of sand banks and ridges. *Cont. Shelf Res.*, **19**, 1285–1330.

Ehrenberg, S.N. (1990) Relationship between diagenesis and reservoir quality in sandstones of the Garn Formation, Haltenbanken, mid-Norwegian continental shelf. *AAPG Bull.*, **74**, 1538–1558.

Ehrenberg, S.N. (1993) Preservation of anomalously high porosity in deeply buried sandstones by grain-coating chlorite: Examples from the Norwegian continental shelf. *AAPG Bull.*, **77**, 1260–1286.

Ehrenberg, S.N., **Gjerstad, H.M.** and **Hadler-Jacobsen, E.** (1992) Smørbukk Field: a gas condensate trap in the Haltenbanken province, offshore mid-Norway. In: *Giant Oil and Gas Fields of the Decade 1978–1988* (Ed. M.T. Halbouty), *AAPG Mem.*, **54**, 323–348.

Ehrenberg, S.N., **Dalland, A.**, **Nadeau, P.H.**, **Mearns, E.W.** and **Amundsen, H.E.F.** (1998) Origin of chlorite enrichment and neodymium isotopic anomalies in Haltenbanken sandstones. *Mar. Petrol. Geol.*, **15**, 403–425.

Elfenbein, C., **Husby, Ø.** and **Ringrose, P.S.** (2005) Geologically based estimation of k_v/k_h ratios: an example from the Garn Formation, Tyrihans Field, Mid-Norway. In: *Petroleum Geology: North-West Europe and Global Perspectives — Proceedings of the 6th Petroleum Geology Conference* (Eds A.G. Doré and B.A. Vining), 537–543. Geol. Soc. London.

Emmett, W.R., **Beaver, K.W.** and **McCaleb, J.A.** (1971) Little Buffalo Basin Tensleep heterogeneity — its influence on infill drilling and secondary recovery. *J. Petrol. Technol.*, **23**, 161–168.

Eriksson, K.A. and **Simpson, E.L.** (1990) Recognition of high-frequency sea-level fluctuations in Proterozoic siliciclastic tidal deposits, Mount Isa, Australia. *Geology*, **18**, 474–477.

Evans, W.E. (1970) Imbricate linear sandstone bodies of Viking Formation in Dodsland-Hoosier area of southwestern Saskatchewan, Canada. *AAPG Bull.*, **54**, 469–486.

Fanchi, J.R. (2006) *Principles of Applied Reservoir Simulation*. Elsevier, Amsterdam, 511 pp.

Field, M.E., **Gardner, J.V.**, **Jennings, A.E.** and **Edwards, B.D.** (1982) Earthquake-induced sediment failures on a 0.25° slope, Klamath River delta, California. *Geology*, **10**, 542–546.

Gjelberg, J., **Dreyer, T.**, **Høie, A.**, **Tjelland, T.** and **Lilleng, T.** (1987) Late Triassic to Mid-Jurassic development on the Barents and Mid-Norwegian shelf. In: *Petroleum Geology of North West Europe* (Eds J. Brooks and K. Glennie), pp 1105–1129. Graham and Trotman, London.

Goff, J.R., **Austin, J.A.**, **Jr.**, **Gulick, S.**, **Nordfjord, S.**, **Christensen, B.**, **Sommerfield, C.**, **Olson, H.** and **Alexander, C.** (2005) Recent and modern marine erosion on the New Jersey outer shelf. *Mar. Geol.*, **216**, 275–296.

Grant, W.D. and **Madfen, O.S.** (1979) Combined wave and current interaction with a rough bottom. *J. Geophys. Res.*, **84C**, 1797–1808.

Hallam, A., **Crame, J.A.**, **Mancenido, M. O.**, **Francis, J.** and **Parrish, J.T.** (1994) Jurassic climates as inferred from the sedimentary and fossil record. In: *Palaeoclimates and Their Modelling; With Special Reference to the Mesozoic Era* (Eds J.R.L. Allen, B.J. Hoskins, B.W. Sellwood, R.A. Spicer and P.J. Valdes), pp 79–88. Chapman & Hall, London.

Harms, J.C., **Southard, J.B.**, **Spearing, D.R.** and **Walker, R.G.** (1975) *Depositional Environments as interpreted from Primary Sedimentary Structures and Stratification Sequences*. SEPM Short Course No. 2 Lecture Notes, Society of Economic Paleontologists and Mineralogists, Dallas, 161 pp.

Harms, J.C., **Southard, J.B.** and **Walker, R.G.** (1982) *Structures and Sequences in Clastic Rocks*. SEPM Short Course No. 9 Lecture Notes, Society of Economic Paleontologists and Mineralogists, Tulsa, 250 pp.

Harris, P.T., **Pattiaratchi, C.B.**, **Cole, A.R.** and **Keene, J.B.** (1992) Evolution of subtidal sandbanks in Moreton Bay, Eastern Australia. *Mar. Geol.*, **103**, 225–257.

Hein, F.J. (1987) Tidal/littoral offshore shelf deposits – Lower Cambrian Gog Group, Southern Rocky Mountains, *Canada. Sed. Geol.*, **52**, 155–182.

Helgesen, J., **Magnus, I.**, **Prosser, S.**, **Saigal, G.**, **Aamodt, G.**, **Dolberg, D.** and **Busman, S.** (2000) Comparison of constrained sparse spike and stochastic inversion for porosity prediction at Kristin Field. *Leading Edge*, **19**, 400–407.

Helland-Hansen, W., **Ashton, M.**, **Lømo, L.** and **Steel, R.J.** (1992) Advance and retreat of the Brent Delta: recent contributions to the depositional model. In: *Geology of the Brent Group* (Eds A.C. Morton, R.S. Haszeldine, M.R. Giles and S. Brown), *Geol. Soc. London Spec. Publ.*, **61**, 109–127.

Houthuys, R. and **Gullentops, F.** (1988) The Vlierzele Sands (Eocene, Belgium): a tidal ridge system. In: *Tide-Influenced Sedimentary Environments and Facies* (Eds P.L. de Boer, A. van Gelder and S.D. Nio), 139–152. D. Reidel Publishing Company, Dordrecht.

Huthnance, J.M. (1982) On one mechanism forming linear sand banks. *Estuar. Coast. Shelf Sci.*, **14**, 79–99.

Jackson, J.S. and **Hastings, D.S.** (1986) The role of salt patterns and hydrocarbon play generation. movements in the Tectonic history of Haltenbanken and Traenabanken and its relationship to structural styles. In: *Habitat of Hydrocarbons on the Norwegian Continental Shelf* (Ed. A.M. Spencer), pp 241–257. Graham and Trotman, London.

Jackson, M.D., Yoshida, S., Johnson, H.D., Muggeridge, A.H., Næss, A. and **Ringrose, P.S.** (1999) Three-dimensional reconstruction, flow simulation and upscaling of complex bedform-scale sedimentary structures within tidal sandstone reservoirs. In: *Advanced Reservoir Characterization for the Twenty-First Century* (Ed. T.F. Hentz), pp 169–178. SEPM Gulf Coast Section 19th Annual Research Conference Proceedings, Society of Economic Paleontologists and Mineralogists, Houston.

Johannessen, E.P. and **Nøttvedt, A.** (2008) Norway encircled by coastal plains and deltas. In: *The Making of a Land — Geology of Norway* (Eds I.B. Ramberg, I. Bryhni, A. Nøttvedt and K. Rangnes), pp 356–383. Norwegian Geological Association, Trondheim.

Johnson, H.D. and **Baldwin, C.T.** (1996) Shallow clastic seas. In: *Sedimentary Environments: Processes, Facies and Stratigraphy* (Ed. H.G. Reading), 232–280. Blackwell Science, Oxford.

Jongepier, K., Rui, J.C. and **Grue, K.** (1996) Triassic to Early Cretaceous stratigraphic and structural development of the northeastern Møre Basin margin, off mid-Norway. *Nor. Geol. Tidsskr.*, **76**, 199–214.

Keene, J.B. and **Harris, P.T.** (1995) Submarine cementation in tide-generated bioclastic sand dunes: epicontinental seaway, Torres Strait, north-east Australia. In: *Tidal Signatures in Modern and Ancient Sediments* (Eds B.W. Flemming and A. Bartholomä), *Int. Assoc. Sedimentol. Spec. Publ.*, **24**, 225–236.

Kenyon, N.H., Belderson, R.H., Stride, A.H. and **Johnson M.A.** (1981) Offshore tidal sand banks as indicators of net sand transport and as potential deposits. In: *Holocene Marine Sedimentation in the North Sea Basin* (Eds S.-D. Nio, R.T.E. Shüttenhelm and T.C.E. van Weering), *Int. Assoc. Sedimentol. Spec. Publ.*, **5**, 257–268.

Keogh, K.J., Martinius, A.W. and **Osland, R.** (2007) The development of fluvial stochastic modelling in the Norwegian oil industry: a historical review, subsurface implementation and future directions. *Sed. Geol.*, **202**, 249–268.

Kjærefjord, J.M. (1999) Bayfill successions in the Lower Jurassic Åre Formation, Offshore Norway: Sedimentology and heterogeneity based on subsurface data from the Heidrun Field and analog data from the Upper Cretaceous Neslen Formation, eastern Book Cliffs, Utah. In: *Advanced Reservoir Characterization for the Twenty-First Century* (Ed. T.F. Hentz), pp 149–157. SEPM Gulf Coast Section 19th Annual Research Conference Proceedings, Society of Economic Paleontologists and Mineralogists, Houston.

Klefstad, L., Kvarsvik, S., Ringås, J.E., Stene, J.J. and **Sundsby, O.** (2005) Characterization of deeply buried heterolithic tidal reservoirs in the Smørbukk Field using inverted post-stack seismic acoustic impedance. *Petrol. Geosci.*, **11**, 47–56.

Koch, J.-O. and **Heum, O.R.** (1995) Exploration trends of the Halten Terrace. In: *Petroleum Exploration and Exploitation in Norway* (Ed. S. Hanslien), *Norw. Petrol. Soc. (NPF) Spec. Publ.*, **4**, 235–251.

Komar, P.D. and **Miller, M.C.** (1975) The initiation of oscillatory ripple marks and the development of plane-bed at high shear stresses under waves. *J. Sed. Petrol.*, **45**, 697–703.

Krumbein, W.C. and **Graybill, F.A.** (1965) *An Introduction to Statistical Models in Geology*. McGraw-Hill, New York, 475 pp.

Lanckneus, J. and **De Moor, G.** (1995) Bedforms on the Middelkerke Bank, southern North Sea. In: *Tidal Signatures in Modern and Ancient Sediments* (Eds B.W. Flemming and A. Bartholomä), *Int. Assoc. Sedimentol. Spec. Publ.*, **24**, 33–51.

Le Bot, S., Van Lancker, V., Deleu, S., De Batist, M., Henriet, J.P. and **Haegeman, W.** (2005) Geological characteristics and geotechnical properties of Eocene and Quaternary deposits on the Belgian continental shelf: synthesis in the context of offshore wind farming. *Geol. Mijnbouw.*, **84**, 147–160.

Liu, Z., Berné, S., Saito, Y., Yu, H., Trentesaux, A., Uehara, K., Yin, P., Liu, J.P., Li, C., Hu, G. and **Wang, X.** (2007) Internal architecture and mobility of tidal sand ridges in the East China Sea. *Cont. Shelf Res.*, **27**, 1820–1834.

Longhitano, S.G. and **Nemec, W.** (2005) Statistical analysis of bed-thickness variation in a Tortonian succession of biocalcarenitic tidal dunes, Amantea Basin, Calabria, southern Italy. *Sed. Geol.*, **179**, 195–224.

Malinverno, A. (1997) On the power law size distribution of turbidite beds. *Basin Res.*, **9**, 263–274.

Marsh, N., Imber, J., Holdsworth, R.E., Brockbank, P. and **Ringrose, P.S.** (2010) The structural evolution of the Halten Terrace, offshore Mid-Norway: extensional fault growth and strain localisation in a multi-layer brittle–ductile system. *Basin Res.*, **22**, 195–214.

Martinius, A.W., Kaas, I., Næss, A., Helgesen, G., Kjærefjord, J.M. and **Leith, D.A.** (2001) Sedimentology of the heterolithic and tide-dominated Tilje Formation (Early Jurassic, Halten Terrace, offshore mid-Norway). In: *Sedimentary Environments Offshore Norway — Palaeozoic to Recent* (Eds O.J. Martinsen and T. Dreyer), *Norw. Petrol. Soc. (NPF) Spec. Publ.*, **10**. 103–144.

Martinius, A.W., Ringrose, P.S., Brostrøm, C., Elfenbein, C., Næss, A. and **Ringås, J.E.** (2005) Reservoir challenges of heterolithic tidal sandstone reservoirs in the Halten Terrace, mid-Norway. *Petrol. Geosci.*, **11**, 3–16.

McBride, R.A. (2003) Offshore sand banks and linear sand ridges. In: *Encyclopedia of Sediments and Sedimentary Rocks* (Ed. G.V. Middleton), pp 737–739. Kluiver Academic Publishers, Dordrecht.

McIlroy, D. (2004) Ichnofabrics and sedimentary facies of a tide-dominated delta: Jurassic Ile Formation of Kristin Field, Haltenbanken, Offshore Mid-Norway. In: *The Application of Ichnology to Palaeoenvironmental and Stratigraphic Analysis* (Ed. D. McIlroy), *Geol. Soc. London Spec. Publ.*, **228**, 237–272.

McIlroy, D., Flint, S.S. and **Howell, J.A.** (1999) Applications of high-resolution sequence stratigraphy to reservoir prediction and flow unit definition in aggradational tidal successions. In: *Advanced Reservoir Characterization for the Twenty-First Century* (Ed. T.F. Hentz), pp 121–132. SEPM Gulf Coast Section 19th Annual Research Conference Proceedings, Society of Economic Paleontologists and Mineralogists, Houston.

Middleton, G.V., Plotnick, R.E. and **Rubin, D.M.** (1995) *Nonlinear Dynamics and Fractals: New Numerical Techniques for Sedimentary Data.* SEPM Short Course No. 36 Lecture Notes, Society for Sedimentary Geology, Tulsa, 174 pp.

Mellere, D. and Steel, R.J. (1996) Tidal sedimentation in Inner Hebrides half grabens, Scotland: the Mid-Jurassic Bearreraig Sandstone Formation. In: *Geology of Siliciclastic Shelf Seas* (Eds M. De Batist and P. Jacob). *Geol. Soc. London Spec. Publ.*, **117**, 49–79.

Molgat, M. and Arnott, R.W.C. (2001) Combined tide and wave influence on sedimentation patterns in the Upper Jurassic Swift Formation, south-eastern Alberta. *Sedimentology*, **48**, 1353–1369.

Montenat, C., Barrier, P. and **Di Geronimo, I.** (1987) The Strait of Messina, past and present: a review. *Doc. Trav. IGAL (Paris)*, **11**, 7–13.

Nago, H., Maeno, S., Matsumoto, T. and **Hachiman, Y.** (1993) Liquefaction and densification of loosely deposited sand bed under water pressure variation. In: *Proc. 3rd International Offshore and Polar Engineering Conference*, pp 578–594. International Society of Offshore and Polar Engineers, Singapore.

Nio, S.-D., Siegenthaler, C. and **Yang, C.S.** (1983) Megaripple cross-bedding as a tool for the reconstruction of the palaeohydraulics in a Holocene subtidal environment, S.W. Netherlands. *Geol. Mijnbouw.*, **36**, 499–509.

Nordahl, K., Ringrose, P.S. and **Wen, R.** (2005) Petrophysical characterization of a heterolithic tidal reservoir interval using a process-based modelling tool. *Petrol. Geosci.*, **11**, 17–28.

Off, T. (1963) Rhythmic linear sand bodies caused by tidal currents. *AAPG Bull.*, **47**, 324–341.

Okusa, S. and **Uchida, A.** (1980) Pore-water pressure change in submarine sediments due to waves. *Mar. Geotechnol.*, **4**, 145–161.

Pascoe, R., Hooper, R., Storhaug, K. and **Harper, H.** (1999) Evolution of extensional styles at the southern termination of the Nordland Ridge, Mid-Norway: a response to variations in coupling above Triassic salt. In: *Petroleum Geology of Northwest Europe — Proceedings of the 5th Conference* (Eds A.J. Fleet and S.A.R. Boldy), pp 83–90. Geol. Soc. London.

Quin, J.G., Zweigel, P., Eldholm, E., Hansen, O.R., Christoffersen, K.R. and **Zaostrovski, A.** (2010) Sedimentology and unexpected pressure decline: the HP/HT Kristin Field. In: *Petroleum Geology: From Mature Basins to New Frontiers* (Eds B.A. Vining and S.C. Pickering), *Petroleum Geology Conference Series*, **7**, pp 419–429. Geol. Soc. London.

Reineck, H.-E. and **Singh, I.B.** (1980) *Depositional Sedimentary Environments.* 2nd edn, Springer-Verlag, Berlin, 549 pp.

Ringrose, P.S., Nordahl, K. and **Wen, R.-J.** (2005) Vertical permeability estimation in heterolithic tidal deltaic sandstones. *Petrol. Geosci.*, **11**, 29–36.

Rivenæs, J.C., Otterlei, C., Zachariassen, E., Dart, C. and **Sjøholm, J.** (2005) A 3D stochastic model integrating depth, fault and property uncertainty for planning robust wells, Njord Field, offshore Norway. *Petrol. Geosci.*, **11**, 57–65.

Rosvoll, K.J. (1989) *Små-skala permeabilitetsvariasjon i reservoarsandsteiner: laboratorie-analyse og kvantitativ vurdering.* Cand. Scient. Thesis, University of Bergen, 170 pp. (with Appendix, 310 pp.).

Santisteban, C. and **Taberner, C.** (1988) Geometry, structure and geodynamics of a sand wave complex in the southeast margin of the Eocene Catalan Basin, Spain. In: *Tide-Influenced Sedimentary Environments and Facies* (Eds P.L. de Boer, A. van Gelder and S.D. Nio), pp 123–138. D. Reidel Publishing Company, Dordrecht.

Seed, H.B. (1979) Soil liquefaction and cyclic mobility evaluation for level ground during earthquake. *J. Geotech. Eng.*, **105**, 201–255.

Seed, H.B. and **Rahman, M.S.** (1978) Wave-induced pore pressure in relation to ocean floor stability of cohesionless soils. *Mar. Geotechnol.*, **3**, 123–150.

Skar, T., Van Balen, R.T., Arnesen, L. and **Cloetingh, S.** (1999) Origin of overpressures on the Halten Terrace, offshore mid-Norway: the potential role of mechanical compaction, pressure transfer and stress. In: *Muds and Mudstone: Physical and Fluid-Flow Properties* (Eds A.C. Aplin, A.J. Fleet and J.H.S. Macquaker), *Geol. Soc. London Spec. Publ.*, **158**, 137–156.

Slingerland, R., Kump, L.R., Arthur, M.A., Fawcett, P.J., Sageman, B.B. and **Barron, E.J.** (1996) Estuarine circulation in the Turonian Western Interior Seaway of North America. *Geol. Soc. Am. Bull.*, **108**, 941–952.

Smith, A.G., Smith, D.G. and **Funnel, B.M.** (1994) *Atlas of Mesozoic and Cenozoic Coastlines.* Cambridge University Press, Cambridge, 112 pp.

Snedden, J.W. and **Dalrymple, R.W.** (1999) Modern shelf sand ridges: from historical perspective to a unified hydrodynamic and evolutionary model. In: *Isolated Shallow Marine Sand Bodies: Sequence Stratigraphic Analysis and Sedimentological Perspectives* (Eds K.M. Bergman and J.W. Snedden), *SEPM Spec. Publ.*, **64**, 13–28.

Spencer, A.M., Birkeland, Ø. and **Koch, J.-O.** (1993) Petroleum geology of the proven hydrocarbon basins, offshore Norway. *First Break*, **11**, 161–176.

Stølum, H.H. (1991) Fractal heterogeneity of clastic reservoirs. In: *Reservoir Characterization II* (Eds L.W. Lake, H.B. Carroll and T.C. Wesson), pp 579–612. Academic Press, San Diego.

Storvoll, V., Bjørlykke, K., Karlsen, D. and **Saigal, G.** (2002) Porosity preservation in reservoir sandstones due to grain-coating illite: a study of the Jurassic Garn Formation from the Kristin and Lavrans fields, offshore Mid-Norway. *Mar. Petrol. Geol.*, **19**, 767–781.

Surlyk, F. (1991) Sequence stratigraphy of the Jurassic-lowermost Cretaceous of East Greenland. *AAPG Bull.*, **75**, 1468–1488.

Surlyk, F. (2003) The Jurassic of East Greenland: a sedimentary record of thermal subsidence, onset and culmination of rifting. In: *Jurassic of Denmark and Greenland* (Eds J.R. Ineson and F. Surlyk) *Geol. Surv. Den. Greenl. Bull.*, **1**, 659–722.

Swift, D.J.P. (1975) Tidal sand ridges and shoal-retreat massifs. *Mar. Geol*, **18**, 105–134.

Swift, D.J.P. and **Field, M.E.** (1981) Evolution of a classic sand ridge field: Maryland Sector, North American inner shelf. *Sedimentology*, **28**, 461–482.

Swift, D.J.P. and **Freeland, G.L.** (1978) Mesoscale current lineations on the inner shelf, Middle Atlantic Bight of North America. *J. Sed. Petrol.*, **48**, 1257–1266.

Swift, D.J.P. and **Thorne, J.A.** (1991) Sedimentation on continental margins, I: a general model for shelf sedimentation. In: *Shelf Sand and Sandstone Bodies: Geometry, Facies and Sequence Stratigraphy* (Eds D.J.P. Swift, G.F. Oertel, R.W. Tillman and J.A. Thorne), *Int. Assoc. Sedimentol. Spec. Publ.*, **14**, 3–31.

Swift, D.J.P., Freeland, G.L. & **Young, R.A.** (1979) Time and space distribution of megaripples and associated bedforms, Middle Atlantic Bight, North American Atlantic shelf. *Sedimentology*, **26**, 384–406.

Sztanó, O. and **De Boer, P.L.** (1995) Basin dimensions and morphology as controls on amplification of tidal motions (the Early Miocene North Hungarian Bay). *Sedimentology*, **42**, 665–682.

Turcotte, D.L. (1992) *Fractals and Chaos in Geology and Geophysics*. Cambridge University Press, Cambridge (U.K.), 221 pp.

Vaid, Y.P. and **Thomas, J.** (1995) Liquefaction and postliquefaction behavior of sand. *J. Geotech. Geoenviron. Engineering*, **121**, 163–173.

Walderhaug, O. (1994) Precipitation rates for quartz cement in sandstones determined by fluid-inclusion microthermometry and temperature-history modelling. *J. Sed. Res.*, **64**, 324–333.

Walderhaug, O. (1996) Kinetic modelling of quartz cementation and porosity loss in deeply buried sandstone reservoirs. *AAPG Bull.*, **80**, 731–745.

Walker, R.G. (1984) General introduction: facies, facies sequences and facies models. In: *Facies Models* (Ed. R.G. Walker), 2nd edn, *Geosci. Can. Repr. Ser.*, **1**, 1–9.

Weber, K.J. (1986) How heterogeneity affects oil recovery. In: *Reservoir Characterization* (Eds L.W. Lake and H.B. Carroll, Jr.), pp 487–544. Academic Press, Orlando.

Weber, K.J. (1982) Influence of common sedimentary structures on fluid flow in reservoir models. *J. Petrol. Technol.*, **34**, 665–672.

Weber, K.J. and **Van Geuns, L.C.** (1990) Framework for constructing clastic reservoir simulation models. *J. Petrol. Technol.*, **42**, 1248–1253, 1296–1297.

Wen, R.-J., Martinius, A.W., Næss, A. and **Ringrose, P.S.** (1998) Three-dimensional simulation of small-scale heterogeneity in tidal deposits — a process-based stochastic method. In: *Proceedings 4th Annual Conference of the International Association of Mathematical Geology* (Eds A. Buccianti, G. Nardi and A. Potenza), pp 129–134. De Frede, Naples.

Wilks, D.S. (1995) *Statistical Methods in the Atmospheric Sciences*. Academic Press, San Diego, 467 pp.

Withjack, M.O., Meisling, K.E. and **Russell, L.R.** (1989) Forced folding and basement-detached normal faulting in the Haltenbanken area, offshore Norway. In: *Extensional Tectonics and Stratigraphy of the North Atlantic Margin* (Eds A.J. Tankard and H.R. Balkwill), *AAPG Mem.*, **46**, 567–575.

Zweigel, J. (1998) Reservoir analogue modelling of sandy tidal sediments, Upper Marine Molasse, SW Germany, Alpine foreland basin. In: *Cenozoic Foreland Basins of Western Europe* (Eds A. Mascle, C. Puigdefàbregas, H.P. Luterbacher and M. Fernàndez), *Geol. Soc. London Spec. Publ.*, **134**, 325–337.

Int. Assoc. Sedimentol. Spec. Publ. (2014) **46**, 551–586

Channel development in the chalk of the Tor Formation, North Sea: evidence of bottom current activity

MATTEO GENNARO* and JONATHAN P. WONHAM†

**Department of Earth Science, University of Bergen, 5007, Bergen, Norway*
Corresponding author (E-mail: matteo.gennaro@conocophillips.com)
†Total Norge A.S., Dusavika, 4029, Stavanger, Norway

ABSTRACT

The establishment of a large epeiric sea over NW Europe during the Late Cretaceous sea-level highstand provided the ideal conditions for calcareous coccolithophorid algae to flourish and become the primary components of pelagic sedimentation. The coccoliths, the skeletal remains of the coccolithophorid algae, were deposited as thick successions of calcareous chalk ooze on the sea floor. In tectonically active areas, downslope mass movement of chalk occurred along basin margin slopes, uplifted areas and around halokinetic structures. The resulting allochthonous deposits consist of slides, slumps, debrites and turbidites. Simultaneously, bottom currents swept the sea floor creating an assortment of topographic features. This study focuses on a particular area of the Norwegian North Sea that includes the southern part of the Ekofisk Field and adjacent areas. Seismic and well data are used to examine the characteristics of the Maastrichtian Tor Formation, where a large channel structure has been identified. The channel is shown to be a significant and extensive feature (~30 km long). It is oriented WNW–ESE, following the bathymetric contours of major structural features, notably the Lindesnes Ridge located to the south, and is interpreted to have been formed by bottom currents that originated from the deepest part of the Norwegian Central Graben to the WNW. Log and core data indicate a multiphase history of channel development characterised by variable current energy and interaction with downslope mass flow processes. The channel has similar features to other recently described channels observed in offshore chalk deposits from the North Sea region (Germany, Denmark) that are interpreted to have formed as a result of intensification of bottom current strength during the Late Cretaceous.

Keywords: chalk, contour current, Ekofisk Field, Maastrichtian, North Sea, Tor Formation.

INTRODUCTION

The Maastrichtian Tor Formation in the Norwegian sector of the North Sea is an excellent high porosity carbonate reservoir from which the giant Ekofisk Field has produced for more than 40 years. This formation constitutes part of the Chalk Group (Upper Cretaceous to Danian), a thick carbonate succession deposited in a large epeiric sea that covered most of NW Europe during a global sea-level highstand. The sea was characterised by low terrigenous sediment influx, normal salinity and warm-temperate surface temperatures. These conditions allowed calcareous coccolithophorid algae to flourish and accumulate in thick, laterally continuous, chalk deposits at water depths of hundreds of metres (Surlyk *et al.*, 2003). Chalk is composed mainly of tiny low-Mg calcite plates (coccoliths), the remains of the coccolithophorid algae.

Since the late 1970s, when many hydrocarbon exploration and production wells began to be drilled into the Chalk Group reservoirs of the Norwegian sector and elsewhere in the North Sea, evidence of both pelagic deposition and syndepositional re-sedimentation of chalk has been increasingly identified from the inspection

From Depositional Systems to Sedimentary Successions on the Norwegian Continental Margin, First Edition.
Edited by A. W. Martinius, R. Ravnås, J. A. Howell, R. J. Steel, and J. P. Wonham.

of cores and seismic data (Kennedy 1987a, 1987b; Skirius *et al.*, 1999). Re-sedimented chalks occur in several areas of the Central Graben, notably where inverted or halokinetic structures were active (Perch-Nielsen *et al.*, 1979; Hardman & Kennedy, 1980; Kennedy, 1987a, 1987b; Watts *et al.*, 1980; Hardman, 1982; Brewster & Dangerfield, 1984; Brewster *et al.*, 1986; Hatton, 1986; Nygaard *et al.*, 1993; Bramwell *et al.*, 1999; Farmer & Barkved, 1999; Van der Molen, *et al.*, 2005).

More recent studies of seismic data from the Danish and German offshore sectors, as well as in the Paris Basin, have identified stratal geometries in the Chalk interpreted to have been produced by contourite bottom currents capable of producing both erosional and constructional features (Lykke-Andersen & Surlyk, 2004; Surlyk & Lykke-Andersen, 2007; Esmerode *et al.*, 2007, 2008; Surlyk *et al.*, 2008; Esmerode & Surlyk, 2009). The presence of ridges, furrows and channels in chalk successions has also been described in the extensive cliff exposures of Etretat, northern France (10 m to 60 m deep, <1 km wide channels; Quine & Bosence, 1991; Lasseur, 2007; Lasseur *et al.*, 2009). Channels have in addition been observed in seismic data from the offshore sector of the UK (Evans & Hopson, 2000; Evans *et al.*, 2003), the Netherlands (Van der Molen, 2004) and Denmark (Back *et al.*, 2011). Quine & Bosence (1991) interpreted the channels observed at Etretat to be the product of submarine erosion by tidal or oceanic currents during lowering of relative sea-level. A similar interpretation is favoured for the UK examples, with shallowing suggested to have been induced by tectonic uplift (Evans & Hopson, 2000; Evans *et al.*, 2003). The Danish channel examples are interpreted to have been formed by bottom currents (Esmerode *et al.*, 2007) or alternatively by gravity flows (Back *et al.*, 2011).

The existing evidence from well cores, seismic datasets and outcrop studies indicates that a range of processes operated and interacted in the European Chalk Sea. Pelagic deposition was the principal 'background' process, while re-deposition by mass-flows and reworking or preferential deposition by bottom currents played an important erosional and depositional role due to the creation of sea floor topography during tectonic uplift (Cartwright, 1989; Vejbæk & Andersen, 2002) and the varying spatial pattern of epicontinental seawater circulation. The Chalk Sea occupied Europe for nearly 35 Ma and hence a detailed stratigraphic knowledge is required to reconstruct palaeogeographic evolution. No detailed and regional stratigraphic synthesis at the scale of the North Sea currently exists. Current understanding of palaeogeographic development has not been developed yet and changing internal dynamics therefore relies on a patchwork of local case studies such as this one.

The present study focuses on the Maastrichtian Tor Formation in a central area of the Norwegian North Sea sector, where both pelagic and gravity-driven depositional processes as well as bottom currents are recognised to have occurred. In this area, a large channel (up to 5 km wide, ~30 km long) has been identified within the Tor Formation. Identification of the channel based on well and seismic data was first published by Bramwell *et al.* (1999). The channel has been drilled by the 2/4-12 well in 1975 and by the 2/8-15 well in 1995. The present study interprets the depositional characteristics and evolution of the channel structure utilising a regional 3D seismic dataset, well logs, biostratigraphic data and core material from wells within and close to the channel. A regional merged 3D seismic cube (PGS Megasurvey) with line spacing of 12.5 m for both in-lines and crosslines has been used. The dominant frequency of the seismic dataset at the studied depth is about 26 Hz with vertical resolution of ~ 40 m. The lateral resolution is around 200 m. The seismic data were calibrated with a number of wells (2/7-4, 2/7-14, 2/4-12, 2/4-A-8, 2/4-A-6 and 2/8-15) provided with gamma ray, neutron porosity, sonic and density logs. Cores from wells 2/4-12 and 2/4-A-8 were logged and described. Detailed core descriptions and core photographs for wells 2/7-4, 2/7-14, 2/4-A-6 and 2/8-15 were also available and are presented below.

The methodology used in this study is a seismic stratigraphic analysis integrated with well logs and core description. Key seismic horizons were selected for interpretation based on their usefulness in defining a series of seismic reflection packages marked by truncation or onlap relationships. Amplitude and coherency attribute maps were used to highlight the position and lateral extent of the channel. Once identified on seismic attribute maps, the channel was interpreted on the basis of 3D seismic data as a seismically defined body. The well data were integrated by using synthetic seismograms to calibrate the well logs and formation markers. Dating and lateral correlation of the seismic horizons have been verified by numerous wells with biostratigraphic data. The selected boreholes have been correlated using wireline logs calibrated to core data. Correlations are based on seismic, lithofacies and log responses, allowing the evolution of the channel and adjacent areas to be determined.

GEOLOGICAL BACKGROUND

This paper describes in detail a sector of a large regional 3D seismic study area that covers ~10,000 km^2 of the Central Graben in the southern Norwegian North Sea (Fig. 1). The thickness of the Chalk Group (Upper Cretaceous–Danian) in this region can be >1000 m. The structural configuration of the Central Graben is principally the result of pre-Cretaceous tectonic extensions (Knott *et al.* 1993) followed throughout the Late Cretaceous by post-rift subsidence combined with local inversion probably due to alpine compressional tectonism (Cartwright, 1989; Vejbæk & Andersen, 2002). Locally, sedimentation is influenced by uplift or subsidence related to halokinetic movements of the Zechstein salt (Ziegler, 1990; Gowers *et al.*, 1993; Oakman & Partington, 1998).

During deposition of the Chalk Group, the Central Graben was subsiding along its major boundary faults while the surrounding areas were relatively stable platforms (Cartwright, 1989). This regional post-rift thermal subsidence was

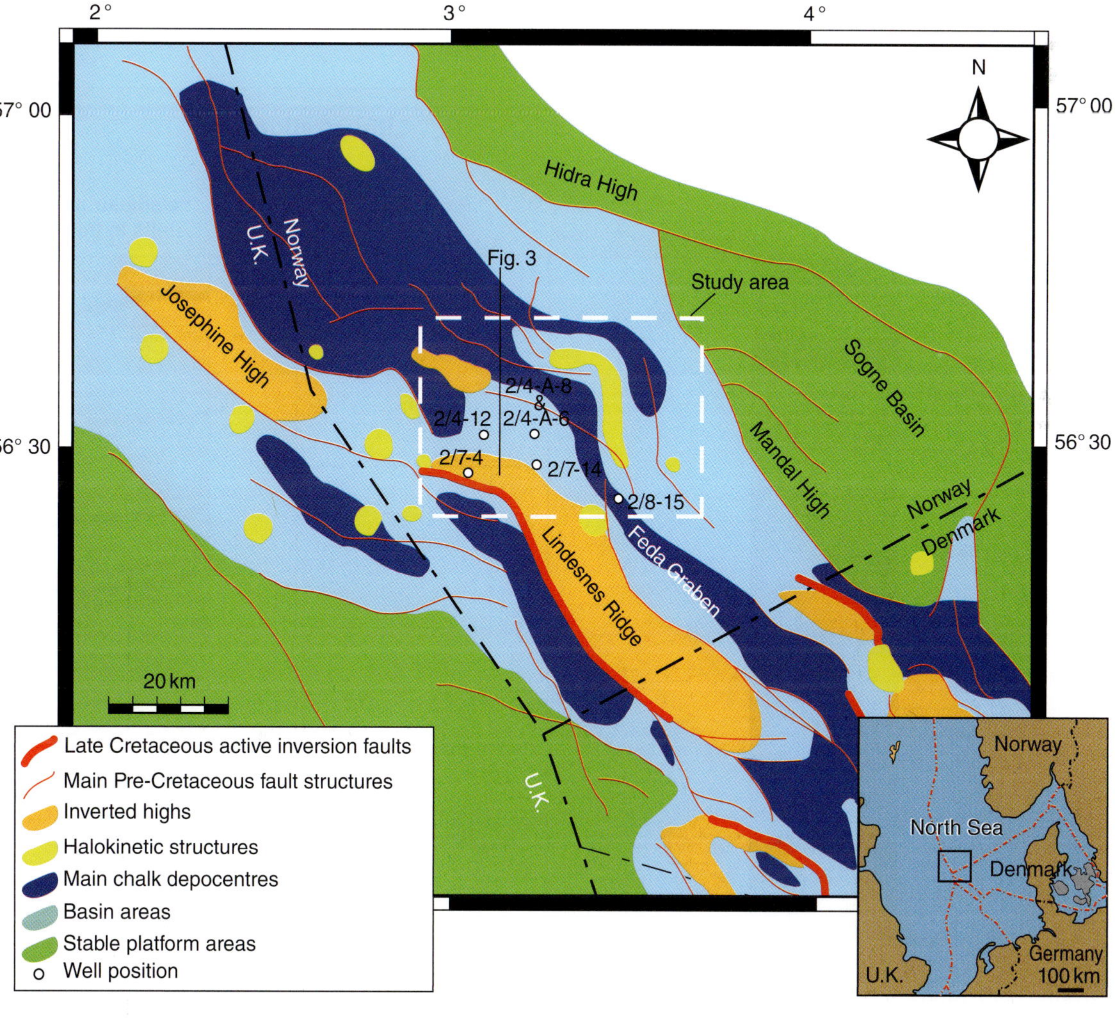

Fig. 1. Structural map of the Norwegian Central Graben, North Sea, showing the main structural elements, inverted highs, halokinetic structures, basin areas and location of the studied wells (modified from Bailey *et al.*, 1999). The inset map in the bottom right corner indicates the position of the study area in the North Sea.

modified locally in many areas by structural inversion resulting from NNE–SSW oriented compression during the Santonian to late Maastrichtian (Cartwright, 1989; Vejbæk & Andersen, 2002). Exact timing of the compressional movements probably differs from one tectonic lineament to another. Inversion occurred primarily along major reactivated extensional faults generating anticlinal fold structures such as the Lindesnes Ridge (Bramwell *et al.*, 1999). The inversion areas commonly developed shallow extensional faults on their flanks and crests as they grew. Such structures became unstable and collapsed due to gravity (e.g. Farmer & Barkved, 1999). Seismic data show that inversion structures were also subject to erosion, particularly during deposition of the Tor Formation (Farmer & Barkved, 1999; Sikora *et al.*, 1999).

Influence of tectonics on deposition

Depositional models for the Chalk Group published during the 1970s to 1980s tended to divide the chalk sedimentary facies into autochthonous and allochthonous types (Kennedy, 1987b). Autochthonous chalk is a result of pelagic sedimentation and trace fossils such as *Planolites*, *Chondrites*, *Zoophycos* and *Thalassinoides* are common (Ekdale & Bromley, 1991). Hiatus horizons and intervals representing low sedimentation rate are expressed by nodular chalk, firmgrounds and hardgrounds (Bromley & Ekdale, 1986; Bromley, 1996).

Allochthonous chalk is a result of re-sedimentation by gravitational processes. The most common cause of chalk re-sedimentation was slope failure due to syndepositional tectonism. Likely tectonic mechanisms include development of growth-fold

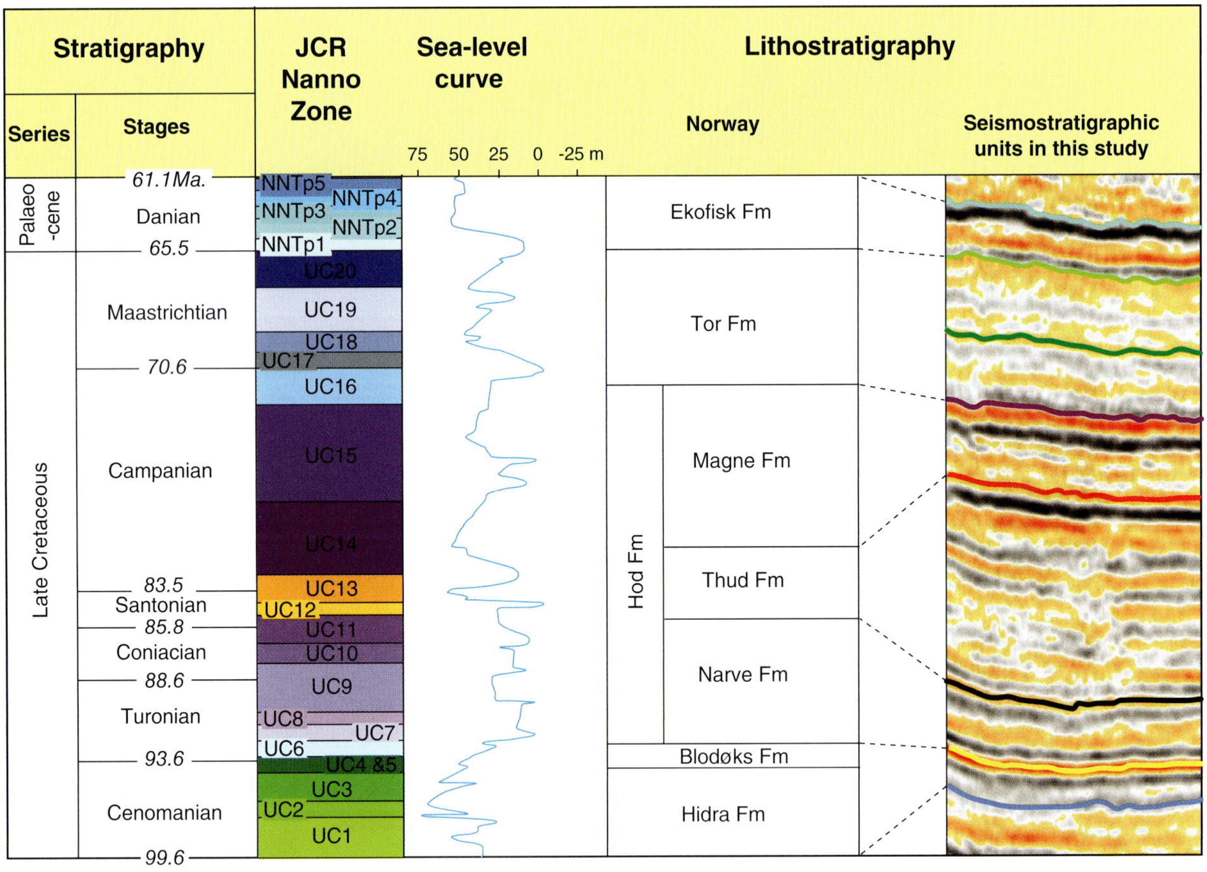

Fig. 2. Lithostratigraphy of the Chalk Group in the Danish sector and southern Norwegian sector of the North Sea and the corresponding seismostratigraphic units used in this study. JCR = Joint Chalk Research Phase V (Bailey *et al.*, 1999). Ages after Gradstein *et al.* (2005) and Ogg *et al.* (2008) and eustatic sea-level curve from Kominz *et al.* (2008). The sea-level estimate of Kominz *et al.* (2008) is the most detailed and updated for the stratigraphic interval of interest in this study and was chosen to evaluate the possible influence of sea-level on the deposition of chalk.

inversion anticlines, halokinetic uplift and basin margins steepening. Tectonically active areas initiated mass movements of chalk, resulting in the deposition of allochthonous facies into the adjacent basins (i.e. Van der Molen *et al.*, 2005). Examples of allochthonous chalk facies have been commonly described from the North Sea region (Watts *et al.*, 1980; Hatton, 1986; Kennedy, 1980, 1987a, 1987b) and many other European localities (see Bromley & Ekdale, 1987 and references therein). These studies have suggested slope failures to be the main cause of re-sedimentation. A study from the Danish offshore sector (Esmerode *et al.*, 2008) has suggested that chalk mass movements could be triggered by bottom currents that destabilised submarine slopes by erosion. The allochthonous facies produced by mass movements include deposits of slides, slumps, debris flows and turbidity currents.

Stratigraphy of the Tor Formation

The stratigraphic nomenclature used by this study (Fig. 2) corresponds to the Chalk Group subdivision of Bailey *et al.* (1999) that comprises, in ascending order, the Hidra, Blodøks, Narve, Thud, Magne, Tor and Ekofisk formations. The bulk of the Tor Formation is of Maastrichtian age. Its thickness in the study area averages ~ 300 m, but locally reaches 500 m. The formation is generally characterised by chalk with low clay content, as indicated by gamma ray logs from wells and petrographic studies (Bramwell *et al.*, 1999).

The lower boundary of the Tor Formation (Fig. 2; violet horizon in Fig.3) is a significant unconformity that is expressed seismically by erosional truncation on uplifted highs and widespread depositional onlap (Bramwell *et al.*, 1999). In the Valhall Field, core data show the unconformity to be developed as extensive hardgrounds, condensed sections and omission surfaces (Kennedy, 1980, 1987a; Farmer & Barkved, 1999; Sikora *et al.*, 1999).

The upper boundary of the Tor Formation with the Ekofisk Formation (Fig. 2; light-green horizon in Fig. 3) is also an unconformity surface with a stratigraphic hiatus identified from nannofossils and foraminifera in several wells in the study area (Bailey *et al.*, 1999; Lottaroli & Catrullo 2000). This hiatus at the base of the Ekofisk Formation is usually represented by hardground and flint passing upwards into a condensed chalk succession (~10 m to 20 m thick) with high clay content (5 wt % to 20 wt %).

This argillaceous interval at the base of the Ekofisk Formation has been informally referred to as Ekofisk dense zone or Ekofisk tight zone (Perch-Nielsen *et al.*, 1979; Brewster *et al.*, 1986; D'Heur 1986; Surlyk *et al.*, 2003). No truncation has been observed in the seismic sections at this level; however local downlap terminations have been noted in structural lows.

As a seismic unit, the Tor Formation (Fig. 3) has a variable thickness in the study area, showing marked thinning towards the basin margins and local inverted structures, such as the Lindesnes Ridge. The reflectivity of this seismic unit is generally good, although the seismic facies are highly varied, ranging from strong parallel reflections to chaotic packages. The Tor Formation can be divided into two seismic sequences separated by the horizon picked on strong negative amplitude trough (dark-green horizon in Fig. 3). This horizon shows features indicating a sequence boundary, such as truncation of the underlying seismic reflections (Fig. 3) and onlap or downlap by reflections of the overlying sequence. This inferred mid-Tor Formation sequence boundary has been dated to the boundary between the lower and upper Maastrichtian and is interpreted to have been caused by a eustatic sea-level fall (Bramwell *et al.*, 1999). The two seismic sequences of the Tor Formation are readily distinguishable in basinal areas where continuous sedimentation formed a relatively thick Maastrichtian succession. Where the Tor Formation thins onto the Lindesnes Ridge, the two sequences are more difficult to recognise due to their reduced thickness and the formation's irregular seismic aspect.

The lower sequence has seldom been cored due to its poor reservoir quality. Therefore, facies interpretation of this sequence is based on well logs and seismic characteristics. The high frequency, lateral continuity and weak amplitude of seismic reflections are interpreted to indicate autochthonous pelagic chalk sediments deposited under tectonically quiet conditions. However, local onlap terminations and chaotic reflections are observed within this sequence and might indicate the presence of gravity flow deposits (cf. Van der Molen *et al.*, 2005).

The reflectivity in the upper sequence varies from moderate to strong and seismic reflections are generally parallel to slightly discontinuous (Fig. 3); however, local mound-like features and chaotic packages are also observed. The channel that is the focus of this study is located in the upper sequence and its top corresponds to the

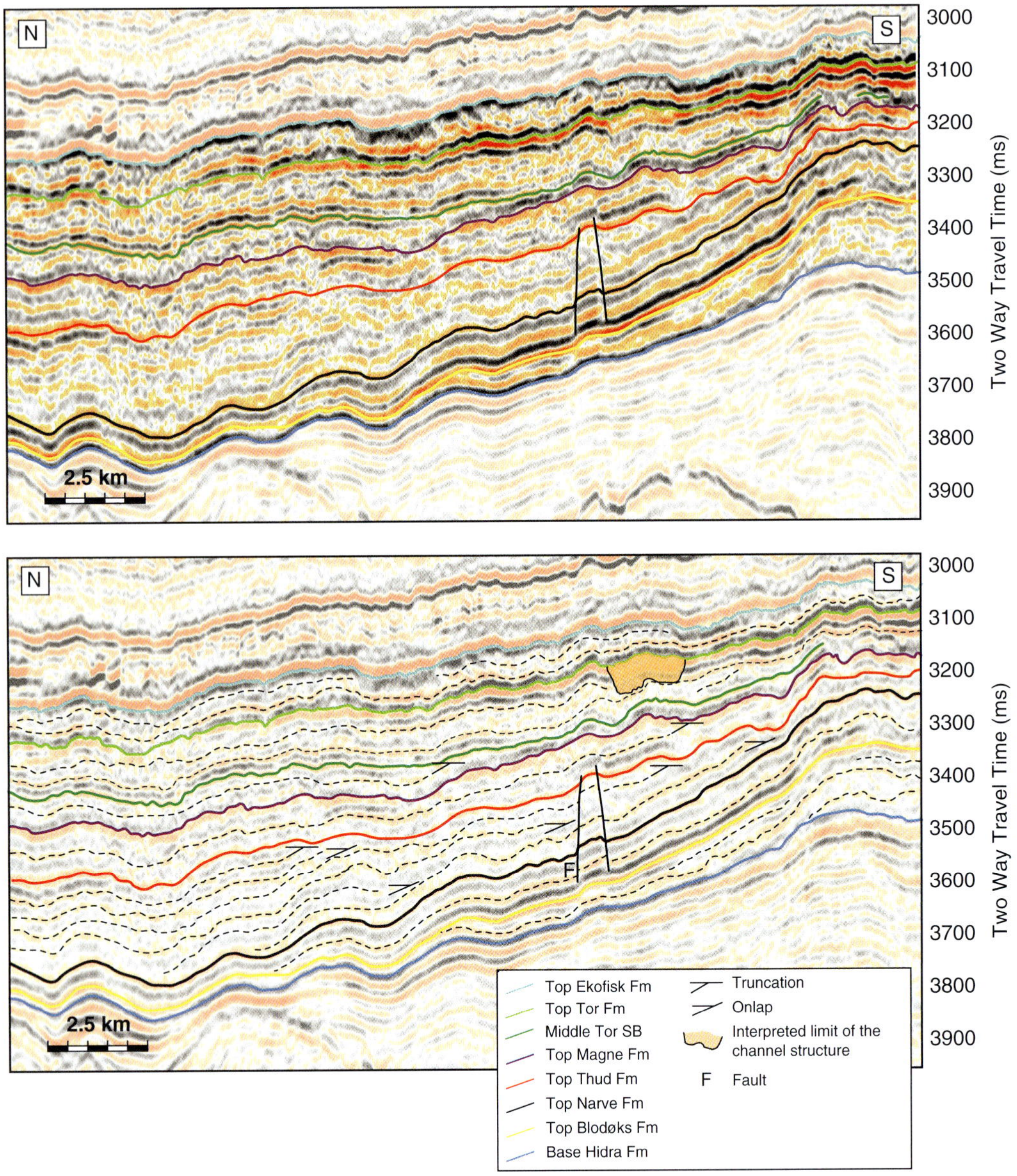

Fig. 3. Reflection configuration and terminations within the seismic units of the Chalk Group (cf. stratigraphy in Fig. 2); note the location of the studied channel.

upper boundary of the Tor Formation (light green horizon of Figs 3, 5, 6 and 7) and the base of the Danian clay-rich Ekofisk dense zone.

Biostratigraphic dating of the Tor Formation is based on nannofossil and microfossil data documented in well reports, from the cored intervals of well 2/4-12 (Martsolf, 1994), wells 2/4-A8 and 2/7-4 (Bailey *et al.*, 1999) and well 2/7-14 (Young & Yang Logan, 1994). The biostratigraphic dates indicate that the upper sequence is upper Maastrichtian, corresponding to the UC19 and UC20 nannofossil zones (Fig. 2) (Bailey *et al.*, 1999). Biofacies are dominated by planktonic and benthic foraminifera. In well 2/7-14, radiolarians have been noted in the intervals 3278 m to 3290 m and 3308 m to 3310 m. Ostracods, *Inoceramus* bivalve fragments and scaphopods are also present but do not provide useful biostratigraphic datings.

Seismic evidence of the channel

Seismic amplitude mapping of the top Tor Formation reflector reveals a distinctive strong negative amplitude in the region adjacent to wells

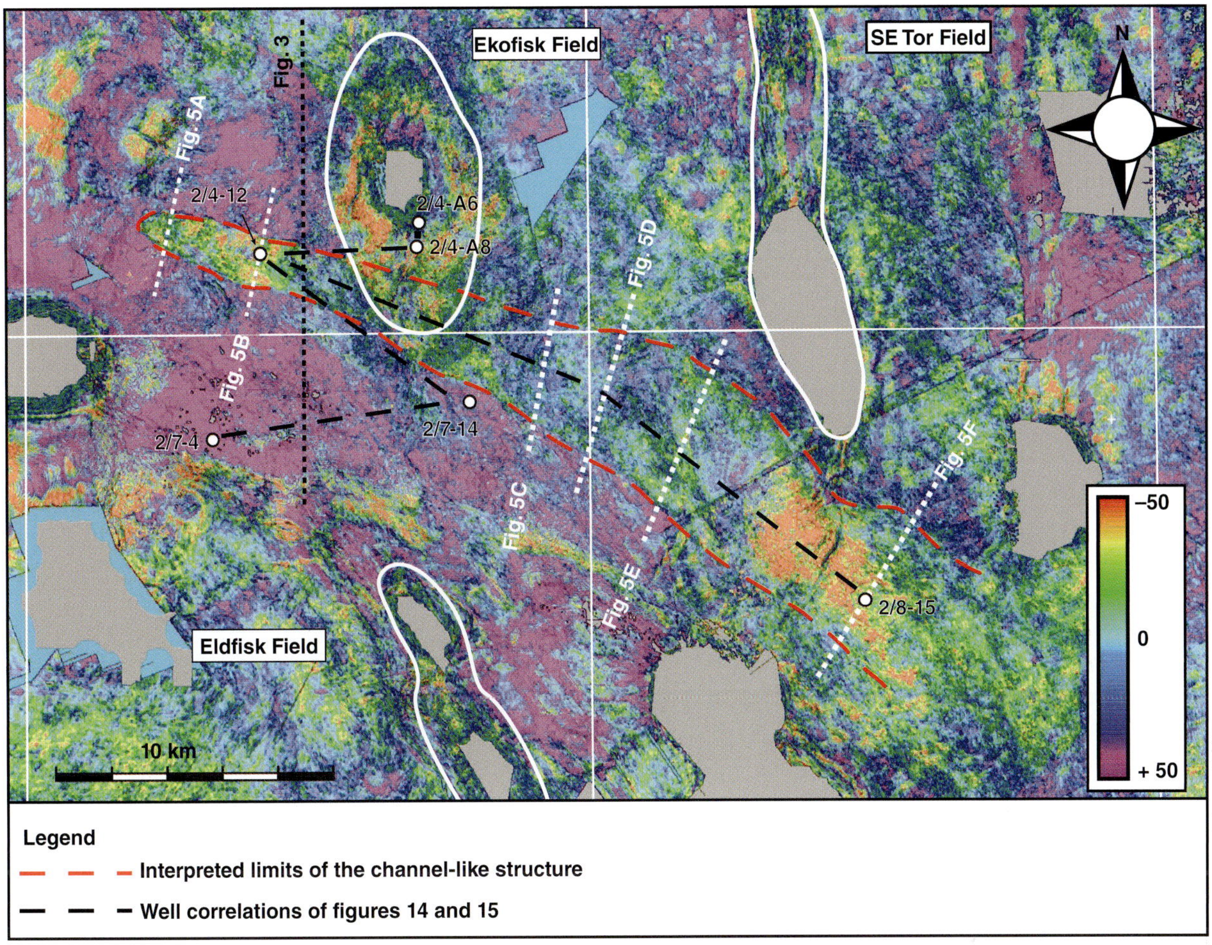

Fig. 4. Maximum signed (positive and negative) amplitude map for a 70 m thick layer below the top of the Tor Formation. Grey areas represent data gaps or the absence of seismic interpretation due to poor seismic imaging. Amplitudes are given as unitless.

2/4-12 and 2/8-15 (Fig. 4). The amplitude anomaly is developed in the south-west part of the Ekofisk Field and extends eastwards to the centre of Block 2/8. Analysis of seismic attribute maps and seismic cross-sections suggests that this anomaly can be interpreted as a channel-like feature trending WNW–ESE, parallel to the northern slope of the Lindesnes Ridge. In map view, the channel is at least 30 km long and has a funnel-shaped geometry widening from 1 km to nearly 5 km towards the ESE. The channel is narrowest in Block 2/4 and widens eastwards across Block 2/7, curving gently towards the ESE before terminating with a wide, splay-like feature in Block 2/8. In some areas, the margins of the channel are less defined and its position is more difficult to map with certainty.

In seismic cross-sections, the channel is recognisable by an erosional incision into fairly parallel seismic reflections that flank this feature and by an expansion of reflections in the upper part of the channel fill, when passing from the channel margins into the channel fill (Fig. 5, Sections A and B). The channel can be subdivided into three segments on the basis of maps and cross-sections: (1) a western segment (Fig. 5, Sections A and B), developed in the southern part of the Ekofisk Field within Block 2/4; (2) a central segment (Fig. 5, Sections C and D), extending south-eastwards from the Ekofisk Field towards the SE Tor structure in the four corners area of Blocks 2/4, 2/5, 2/7 and 2/8 and (3) an eastern segment (Fig. 5, Sections E and F) extending southwards of the SE Tor structure in Block 2/8.

In all three segments, the reflections of the channel fill allow recognition of two units: a lower unit dominated by positive amplitude (light grey

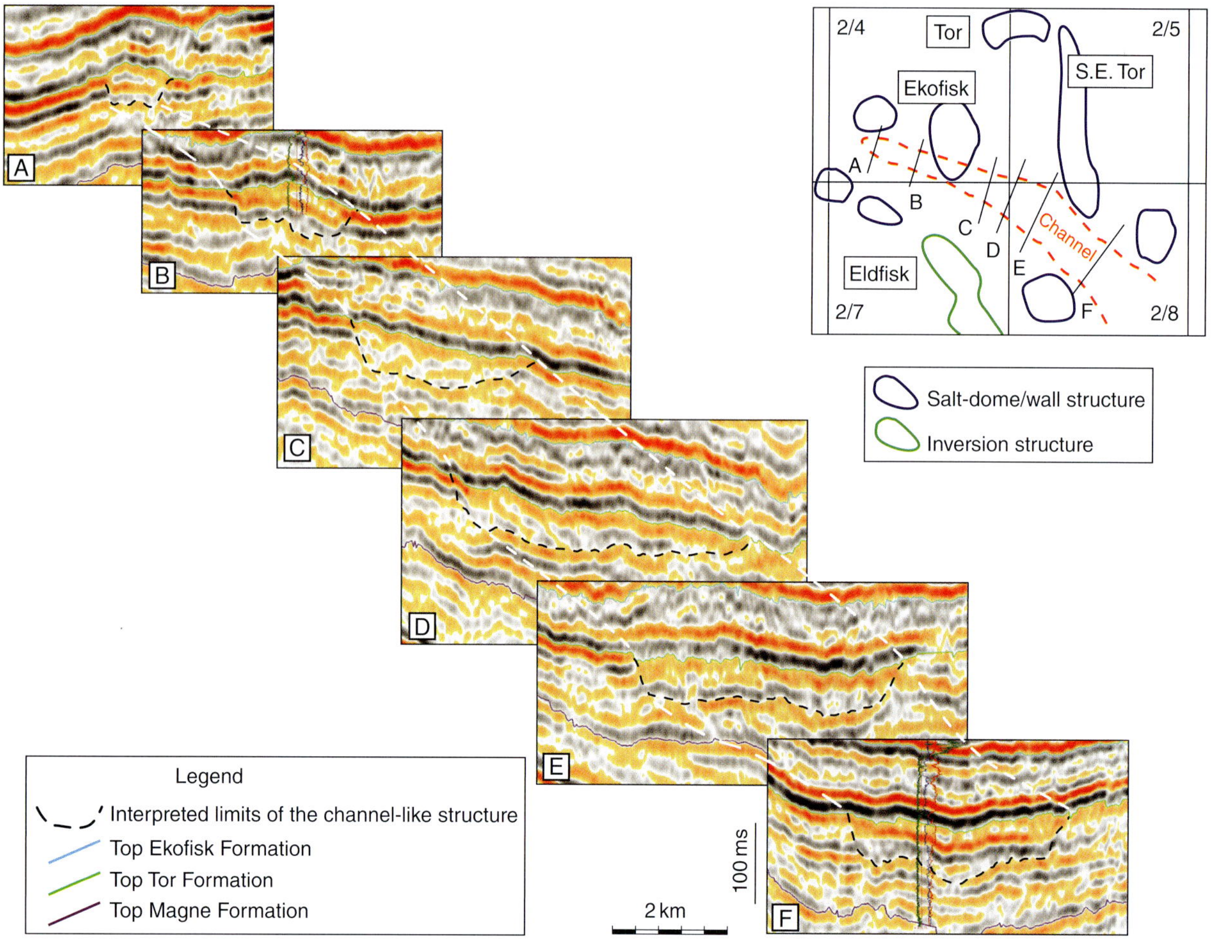

Fig. 5. Series of seismic sections across the channel developed parallel to the northern slope of the Lindesnes Ridge. The inset map shows the location of the seismic lines and the funnel-shaped, plan-view geometry of the palaeochannel extending towards the south-east (Fig. 4).

to black interval in Figs 5, 6 and 7) and an upper unit characterised by negative amplitude (orange to red interval in Figs 5, 6 and 7). The upper unit was penetrated both by wells 2/8-15 and 2/4-12, whereas the lower unit was partially penetrated by well 2/8-15 (Figs 6 and 7).

In the western segment, the channel has a well-defined perimeter and fairly parallel margins (Fig. 4). Close to the Ekofisk Field, the channel's northern margin is slightly irregular, probably because of the local alteration of seismic response by fluids and faults. In seismic cross-sections, the western segment of the channel fill has a tabular geometry with a uniform thickness of 100 m to 150 m and width of 1 km to 2 km. The channel-fill internal reflections have moderate to weak amplitudes (Fig. 5, Sections A and B). The channel base is irregular and incised into the surrounding seismic facies of parallel reflections. The channel-fill top correspond to the boundary between the Tor and Ekofisk formations and coincides on seismic data with a zero phase crossing at the base of a strong positive amplitude reflection correlating to the high gamma ray interval of the Ekofisk dense zone (Figs 5, 6 and 7). In Block 2/4, the Tor-Ekofisk boundary is somewhat irregular and the channel margins appear to be well defined and relatively steep (Fig. 5, Sections A and B). The transition from the western to the central segment of the channel is near seismic Section C (Fig. 5) on the eastern flank of the Ekofisk Field.

The channel's central segment is characterised by a sudden expansion of its northern margin towards the NE in the four-corners area of blocks 2/4, 2/5, 2/7 and 2/8 (Fig. 4), resulting in channel widening from 2 km in its western segment to

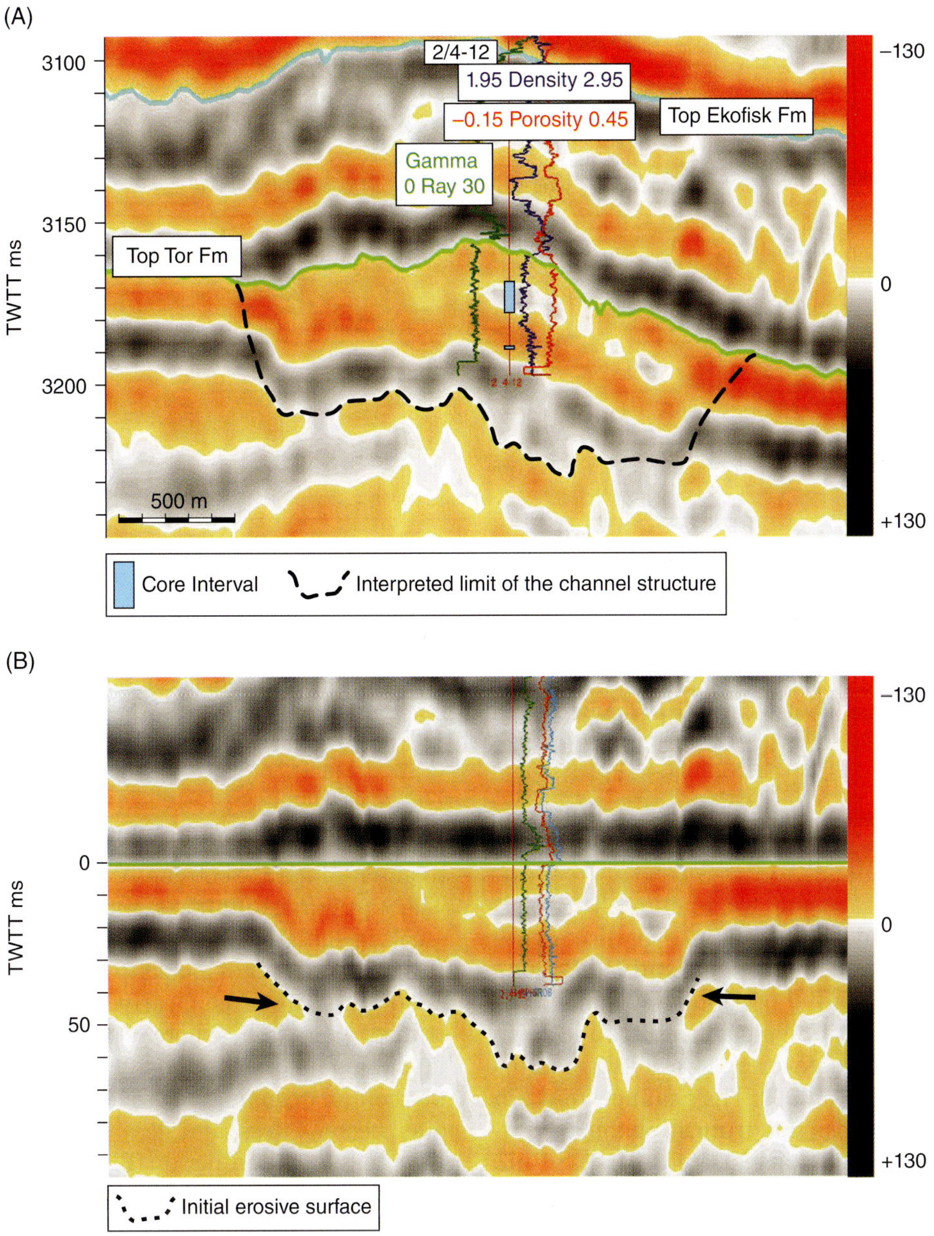

Fig. 6. (A) Seismic Section B of Fig. 5 calibrated to well 2/4-12. Note the positive amplitude (light grey to black colours) and negative amplitude (orange to red colours) reflections defining the lower and the upper channel-fill unit, respectively. The strong positive reflection at the base of the Ekofisk Formation, directly above the top Tor Formation horizon (green line), is generated by the density contrast between the Ekofisk dense zone and underlying porous Tor Formation. This transition from the porous Tor Formation to the overlying Ekofisk dense zone can be seen clearly on the density and porosity logs. The Ekofisk dense zone is characterised by high gamma ray values due to its higher argillaceous content. (B) Seismic Section B of Fig. 5 flattened at Top Tor Formation. Amplitudes are given as unitless.

nearly 5 km in the central segment. Farther towards the ESE, the channel margins are roughly parallel, with the southern margin being fairly straight and the northern margin showing minor irregularities (Fig. 4). The channel-fill seismic facies in the central segment change from flat and uniform to chaotic, convoluted or discontinuous reflections (Fig. 5C). The channel-fill shows NE-inclined reflections, dipping from the southern margin towards the channel axis (Fig. 5D). The

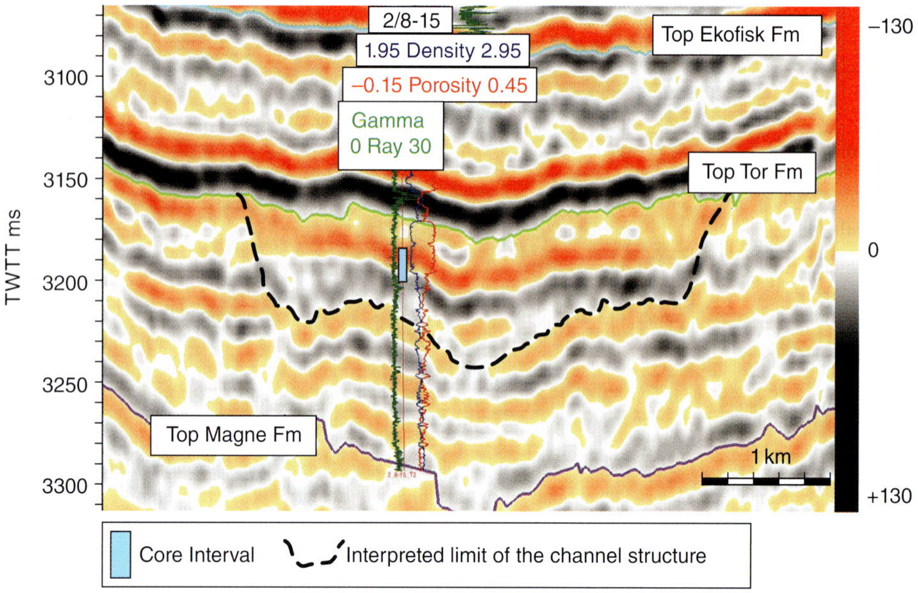

Fig. 7. Seismic Section F of Fig. 5 calibrated to well 2/8-15. Note the correlation between the negative amplitude (orange to red colours) of the upper channel-fill unit to the higher porosity and lower density values on the logs. Similarly, the positive amplitude (light grey to black) of the lower channel-fill unit correspond to lower porosity and higher density values on the logs. Amplitudes are given as unitless.

channel base here is still erosional, but less than in the western segment.

In the central segment, the channel-fill top is uneven and undulating, with an overall concave upward shape (Fig. 5, Sections C and D). Passing from the western to the central segment, the channel cross-section changes from tabular with steep margins to arcuate with more gentle margins (Fig. 5, Sections B, C and D). The channel southern margin is always erosional, truncating the subhorizontal parallel reflections located to the south. The northern channel margin is erosional in the region to the west of the boundary between blocks 2/4 and 2/7. To the east of this boundary the channel margin is no longer clearly erosional and exhibits instead a low gradient pinch-out geometry (Fig. 5, Sections C and D). The transition from the central and the eastern segment is located between the seismic Sections D and E (Fig. 5).

The eastern segment of the channel has a roughly uniform width, but the margins tend to be irregular and locally indistinct, particularly towards the ESE termination of the channel path. At the ESE termination of the channel, bright amplitudes delineate a splay-like geometry. Passing from the central to the eastern segment, the channel base loses its erosional character and becomes only slightly irregular and undulating (compare Fig. 5, Sections A and B with E and F). The uppermost limit is roughly flat with only small irregularities.

Near the border of blocks 2/7 and 2/8, the channel southern margin is clearly erosional and steep walled (Fig. 5E), whereas the northern margin maintains its pinch-out geometry similar to that in the central segment. However, both channel margins become less and less distinct towards the channel's ESE termination (Fig. 5 F). In this terminal area, the channel cross-section is characterised by a gradual decrease in depth and a gradual flattening of the base and margins, until the channel feature flattens out and loses its identity.

WELL CHARACTERISTICS OF THE CHANNEL FILL

Well data overview

Two wells drilled into the channel fill are of particular importance to this study: well 2/4-12 located in the channel western segment and well 2/8-15 located in its eastern segment (Figs 4, 8 and 9). Both these wells were cored and their core samples were calibrated to wireline logs (Figs 8 and 9).

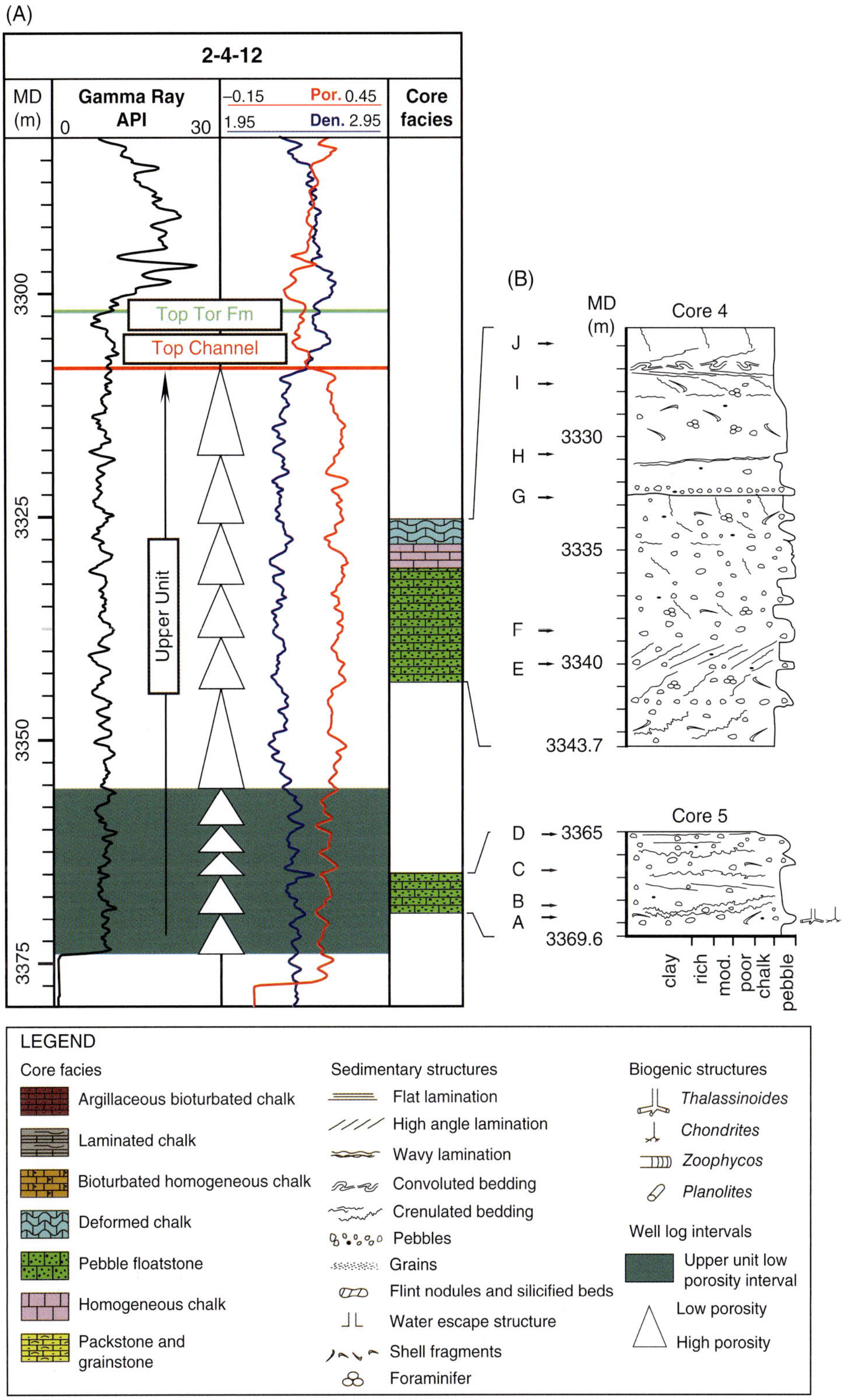

Fig. 8. Composite log of well 2/4-12 (Fig. 6), including (A) wireline logs, core facies and (B) related sedimentological logs. Note the cyclic changes in porosity in the upper unit. MD = measured depth; Den = density log in g cm^{-3}; Por = neutron porosity. API = American Petroleum Institute units. Capital letters at the margin of sedimentological log refer to facies photographs in Fig. 13. Well location in Figs 1 and 4.

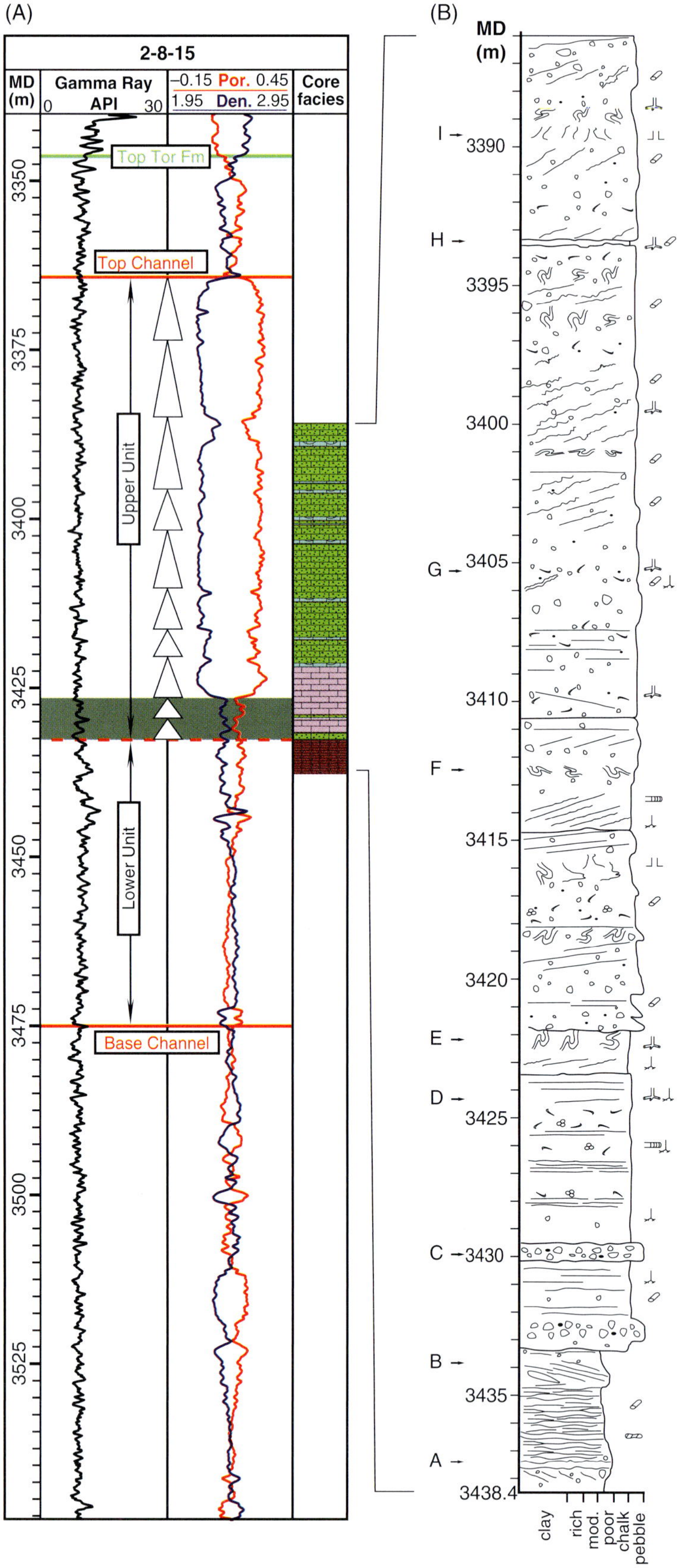

Fig. 9. Composite log of well 2/8-15 (Fig. 7) showing (A) wireline logs, core facies and related (B) sedimentological logs. The bottom and top limits of the channel fill are deduced from the correlation with the seismic line of Fig. 14. MD = Measured depth; Den = density log in g cm^{-3}. Por = neutron porosity in fraction. API = American Petroleum Institute units. Legend as for Fig. 8. Capital letters beside the sedimentological log refer to facies photographs of Fig. 12. Well location in Figs 1 and 4.

Exploration wells 2/4-12 and 2/8-15 were drilled in areas characterised by strong negative amplitude anomalies at the top of the Tor Formation (Fig. 4). Neither well discovered hydrocarbons, as the formation proved to be water-saturated. Calibration of log data to seismic sections (Figs 5, 6 and 7) indicates that the upper channel-fill unit (red amplitude) has a higher porosity than the lower unit (black amplitude) in both wells (Table 1).

Table 1. Overview of the gamma ray, density and neutron porosity values for the channel-fill facies association in wells 2/4-12 and 2/8-15.

	Upper Channel Unit		Lower Channel Unit
Well	2/4-12	2/8-15	2/8-15
Average gamma ray (API)	8	8	12
Density (g cm^{-3})	2.35	2.15	2.40
Neutron porosity (fraction)	26	28	15

API = American Petroleum Institute units.

Detailed core descriptions and sedimentological logs from wells located outside the channel were also investigated (Figs 10 and 11). The condition of the core material is extremely variable; older cores, such as those from wells 2/4-12, 2/4-A8, 2/4-A-6, 2/7-4 and 2/7-14 are in poor condition with numerous rubble intervals and extensive sampling disturbance. The preservation of the core material from more recent wells, such as 2/8-15, is much better with fewer, shorter rubble intervals. Core descriptions and biostratigraphic data in this paper are in log depth. Cores are shifted to log depth on the composite log figures and on the well correlation panels.

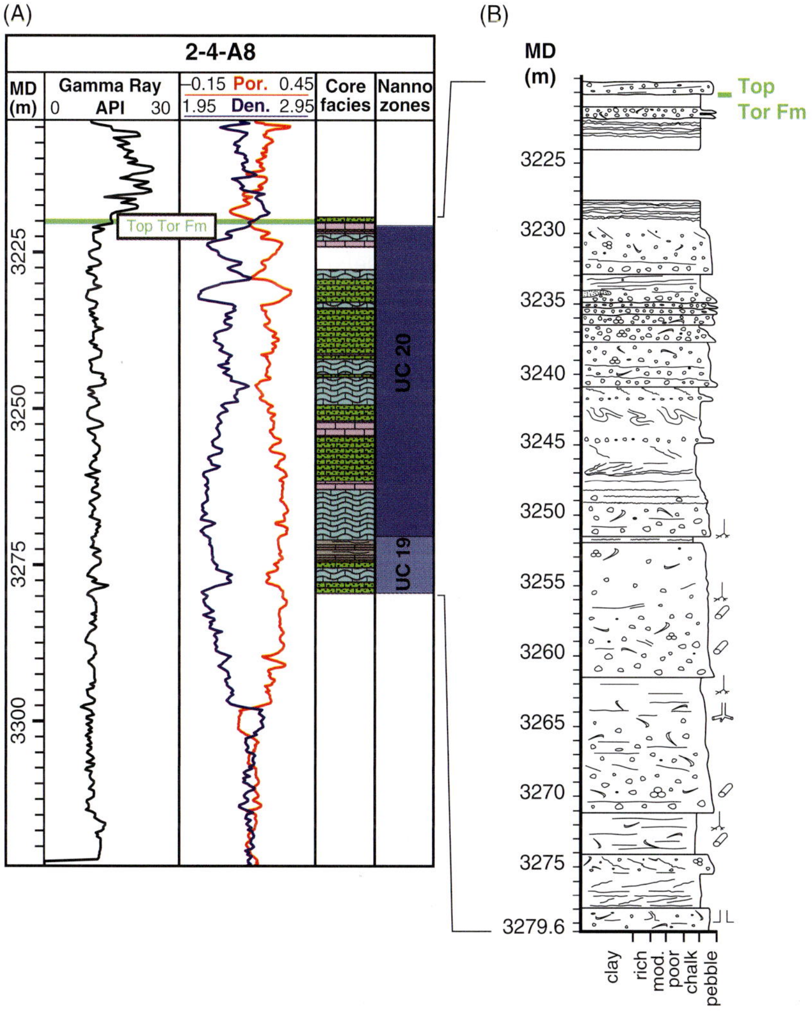

Fig. 10. Composite log of well 2/4-A8, including (A) wireline logs, core facies (legend in Fig. 8) and (B) related sedimentological log. MD = measured depth in m; Den = density log in g cm^{-3}; Por = neutron porosity in fraction. API = American Petroleum Institute units. Well location in Figs 1 and 4.

The margins of the channel have been drilled in its western segment, with two wells at the northern margin (2/4-A8 and 2/4-A6) and one well at the southern margin (2/7-14). Well 2/7-4 located in the Edda Field at 7 km from the channel has also been studied and is included for reference.

Chalk facies

Facies have been distinguished based on depositional texture (Dunham, 1962), diagnostic sedimentary structures and biogenic features. Several facies classifications have earlier been proposed for the chalk (Kennedy, 1987b; Crabtree *et al.*, 1996; Røgen *et al.*, 1999). The classification used in this study (Table 2) is after Bramwell *et al.* (1999) (Table 3). The mode of deposition is interpreted following a similar approach to that of Kennedy (1987b). Grain size is not generally recorded in chalk since this classification of core samples would allow recognition of little more than a fairly monotonous succession dominated by mudstone and wackestone lithologies. To better differentiate these lithologies, an estimation of argillaceous content has been made, indicating the relative amount of carbonate vs. terrigenous material. Pebble floatstone and grain-supported sediments are recorded as pebble-grade deposits in the sedimentological logs. Short descriptions and depositional interpretations of the seven chalk lithofacies that have been recognised are given in Table 2.

Recognition of facies in non-cored well intervals is tentative and based upon relationships between petrophysical log response and core facies. Hatton (1986), Kennedy (1987b) and Campbell & Gravdal (1995) have demonstrated that log responses reflect different chalk facies: (1) Allochthonous facies are generally characterised by intervals with low density, high porosity and low gamma ray values. These facies were deposited by a variety of depositional processes: (a) debris flow, mud flow, low density turbidity currents and mud cloud deposits show distinct low density/high porosity signals, commonly with sharp bases and tops; (b) thick slump and slide deposits show low density/high porosity but with a gradual decrease of porosity and increase in density at the top; (c) coarse turbidites and winnowed lags are distinguished by thin spikes of high porosity and low density response; (2) Autochthonous facies; commonly correspond to broad and less sharply delineated log signals with low porosity and high density; a distinction between clay-rich and clay-poor autochthonous chalks can be based on the gamma ray log responses with clay-rich beds showing high gamma ray values. Although several studies have noted that allochthonous chalk is generally more porous than autochthonous chalk (Hatton, 1986; Kennedy 1987a, 1987b; Taylor & Lapré, 1987; Brasher & Vagle, 1996), other factors such as diagenesis, pore fluid type, sediment texture and composition also undoubtedly influence the degree of porosity preservation. Some studies take the view that there is no straightforward relationship between high porosity development and re-deposition of chalk (Maliva & Dickson, 1992; Fabricius *et al.*, 2007).

The channel fill is subdivided into a lower and upper unit based on seismic characteristics and calibrated well-log signatures (e.g. Figs 6 and 7). The lower unit represents 40% of the channel succession whilst the upper unit makes up the remaining 60%. The following sections describe the facies associations which make up the depositional successions of the lower and upper channel-fill units and the channel margins, respectively.

Facies association of the lower channel-fill unit

Description: The basal part of the lower channel-fill unit, calibrated to seismic section, is characterised by thin, high porosity deposits that are interpreted from wireline logs to be either packstone and grainstone chalk of facies PGC (see Table 2) (e.g. 3 m thick in well 2/8-15, where the channel base is at 3475 m according to the wireline logs and seismic calibration).

No core samples are available from either these basal deposits or from the overlying thick succession of inferred bioturbated homogeneous chalk (facies BHC; Table 2) or laminated chalk (facies LC) that make up ~ 80% of the lower channel-fill unit. Density and porosity logs of this succession show high-frequency facies alternation at decimetre-scale to metre-scale. These deposits pass upwards into a bioturbated argillaceous chalk (facies BAC) that makes up the remaining ~ 20% of the lower channel-fill unit and that is represented by a 5 m core interval in well 2/8-15 (Figs 7, 12A and 12B).

Interpretation: The lowermost 3 m of the lower channel-fill unit (Fig. 9, well 2/8-15 at 3475 m to 3472 m) is a packstone/grainstone (PGC facies, Table 2) chalk deposit formed by erosive reworking

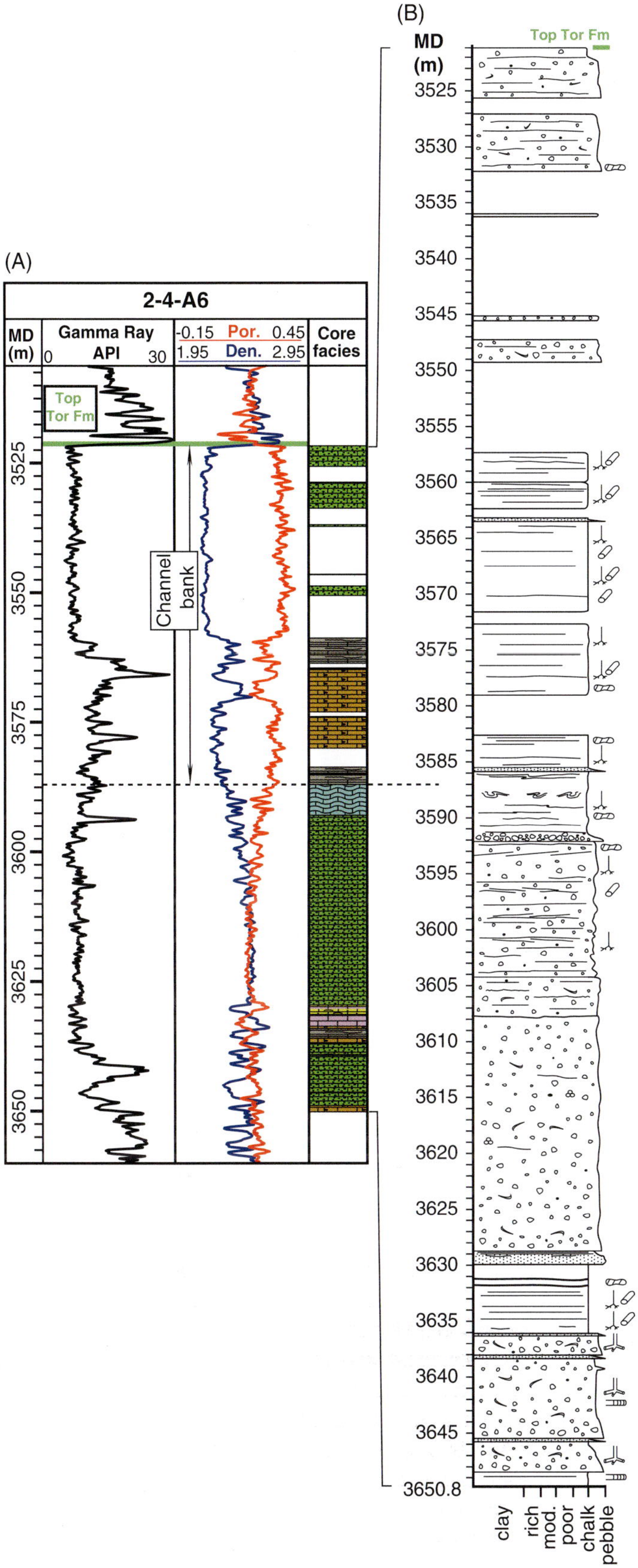

Fig. 11. Composite log of well 2/4-A6, including (A) wireline logs, core facies (legend in Fig. 8) and (B) related sedimentological log. MD = measured depth in m; Den = density log in g cm^{-3}; Por = neutron porosity in fraction. API = American Petroleum Institute units. Well location in Figs 1 and 4.

Table 2. Summary of lithofacies observed in the Tor Formation.

Facies and code	Lithology and texture	Sedimentary and diagenetic features	Biogenic structures	Well log signature	Interpretation
Bioturbated argillaceous chalks (BAC)	Mudstone and wackestone with intermediate to high clay content.	Interbedded clay-poor and clay-rich chalk beds. Clay-rich beds are commonly well laminated, but also show wispy to flasery clay-rich laminations and solutions seams. Clay-poor intervals are intensively bioturbated with numerous horizontal to sub-vertical burrows. Pyrite grains and pyritised shell fragments appear occasionally. Bedding marked by changes in the clay content.	Common to abundant bioturbation induced mottling with indeterminate burrows. Identifiable traces are *Chondrites*, *Planolite*s and *Thalassinoides*.	Broad and poorly delineated signals with low porosity and high density showing high gamma ray response.	Pelagic/autochthonous chalk.
Laminated chalk (LC)	Mudstone and wackestone with low to intermediate clay content.	Weakly to well defined and undisturbed planar-parallel to sometimes wavy-parallel sedimentary laminations with bedding well developed and bed thickness varying from several centimetres to 50 centimetres. Laminations are usually well defined, but occasionally disturbed or cut by bioturbation and bioturbation induce mottling. Common and sparse shell fragments.	Rare bioturbation and bioturbation induced mottling with poorly defined horizontal and vertical burrows. Identifiable traces are *Chondrites*, *Planolite*s and *Zoophycos*.	Broad and poorly delineated signals with low porosity and high density showing intermediate to low gamma ray response.	Pelagic/autochthonous chalk periodically interrupted by deposition of low density turbidity currents.
Bioturbated homogeneous chalk (BHC)	Mudstone and wackestone with low to intermediate clay content.	Homogeneous to indistinct lamination. Bedding poorly to moderately developed with beds varying from 10 cm to 100 cm thick.	Abundant bioturbation induced mottling. Traces include *Chondrites*, *Planolite*s and *Zoophycos*.	Broad and poorly delineated signals with low porosity and high density showing intermediate to low gamma ray response.	Pelagic/autochthonous or biogenically reworked allochthonous chalk if relatively thin (<1 m).
Deformed chalk (DC)	Mudstone and wackestone with variable clay content.	Convoluted, crenulated, folded and shredded steeply dipping laminae with common changes of dip direction. Laminae may be variably argillaceous or cemented. Bedding moderately developed with thickness varying from few decimetres to several metres.	Very rare bioturbation; highly deformed if present.	Broad interval characterised by low density/high porosity but with a gradual decrease of porosity and increase in density towards the top.	Resedimented/ allochthonous chalk: slide, slump and creep.

Pebble floatstone chalk (PFC)	Matrix: Mudstone and wackestone with variable clay content. Clasts: cemented silicified and dolomitised chalk intraclasts.	Matrix supported, highly disorganised, polymictic conglomerate with rounded clasts of coarse sand to pebble size showing poor-moderate sorting and local normal grading. Bedding poorly developed (cm-m thickness). Sparse shell fragments. Local evidence of inclined parallel and crenulated sedimentary laminations occasionally cut or disturbed by bioturbations. Rare cm-thick beds of clast-supported conglomeratic chalk with clasts as above showing local imbrication. Base of the beds is sharp and erosional while the top is marked by gradual disappearance of clasts.	Rare bioturbation usually confined to certain intervals. Identifiable traces are *Chondrites*, *Planolites* and *Thalassinoides*.	Low density/high porosity signals, often with sharp bases and tops.	Resedimented/ allochthonous chalk: cohesive and non-cohesive debris flow.
Homogeneous chalk (HC)	Poor argillaceous mudstone and wackestone.	Uniform and massive with rare laminae or traces of plastic flow. Rare millimetric chalk grains and shell fragments. Poorly developed bedding locally a few decimetres to several metres thick.	Rare to common bioturbation usually confined to certain intervals. Identifiable traces are *Planolite*s and *Chondrites*.	Variable density and porosity signals but commonly similar to PFC facies, commonly with sharp bases and tops.	Pelagic/autochthonous chalk or resedimented/ allochthonous chalk in the form of low density turbidite or turbulent mud clouds.
Packstone and grainstone (PGC)	Poor argillaceous packstone and grainstone.	Uniform and massive. Grainstone may by strongly laminated and with planar laminations and planar cross-laminations. Locally associated with silica cementation. Moderate amount of shell fragments. Base of beds is sharp and locally erosional while the top is marked by the gradual disappearance of clasts.	Rare to common bioturbation. Identifiable traces are *Planolite*s and *Chondrites*.	High porosity and low density response.	Resedimented/ allochthonous chalk: tempestite and turbidite; winnowed lags by bottom current.

Table 3. Summary of lithofacies description from Bramwell *et al.* (1999).

Lithofacies	Texture	Colour	Hardness	Bedding	Sedimentary and biogenic structures	Clay and quartz content	Reservoir quality
Laminated chalk facies	Carbonate mudstones composed of calcareous mud generally devoid of framework material, except for sporadic occurrences of rare, macroscopic skeletal material.	Light grey to yellowish brown in colour.	Well indurated, and hard, although hardness is variable.	Medium bedded with bedding commonly defined by the presence/absence of laminations.	Common, thin, planar-parallel laminations.	<2% quartz in the Tor Formation; 1% to 20% quartz in the Ekofisk Formation. <1% clay minerals.	Laminated chalk seldom exhibit reservoir quality porosity.
Burrowed Laminated chalk facies	Carbonate mudstones composed of calcareous mud and generally devoid of framework material.	Light grey to yellowish brown.	Well indurated, and hard, although hardness is somewhat variable.	Bedding, which is most commonly defined by variability in the degree of burrowing, ranges from thin (<10 centimetre) to very thick (>1 metre).	Planar to somewhat wavy parallel laminations disrupted by rare to common *Planolites* and *Chondrites* burrows.	10% to 35% quartz in the Magne and Ekofisk Formation, <2% quartz in the Tor Formation.	Burrowed Laminated chalk intervals contain occasional, poor reservoir quality porosity.
Homogeneous chalk facies	Calcareous mudstones containing little or no framework material.	Yellowish brown to dark yellowish brown or very rarely light grey.	Well indurated, and very commonly soft to occasionally somewhat hard.	Primary bedding is absent, or at best, difficult to recognise.	Uniform, massive lacking of any internal structure, although occasionally, very rare burrows may occur in some thinner units.	1% to 4% quartz and 1% to 2% clay minerals are common in the Tor Formation; 2% to 15% quartz and 1% to 5% clay minerals are more representative of the Ekofisk Formation.	Commonly contains very good reservoir porosity; commonly the best reservoir porosities that occur in a well are associated with homogeneous chalk.
Burrowed Homogeneous chalk facies	Calcareous mudstones that contain little, if any, framework material.	Yellowish brown to light grey in colour.	Well-indurated, and hard to moderately soft.	If primary bedding can be recognised, it occurs in the medium to thick range (>1 metre).	Fairly uniform and massive chalks, but contain very rare to rare (to occasionally common), *Chondrites* and *Planolites* and burrow mottling.	1% to 3% quartz, <2% clay minerals.	Burrowed Homogeneous chalk exhibit occasional reservoir quality porosity.
Pebble floatstone facies	Calcareous mudstones to wackestones, and very rarely packstones, that contain rare to locally abundant, high-sphericity, round, sand-sized to pebble-sized chalk clasts.	Yellowish brown to light grey in colour.	Well indurated, and soft to hard.	Primary bedding, when recognised, ranges from medium to thick (>1 metre) defined by clast content and texture.	Pebble Floatstones are either massive or exhibit normal graded bedding.	1% to 3% quartz, <1% clay minerals.	Pebble floatstones exhibit reservoir quality porosity.

Argillaceous chalk facies	Argillaceous, carbonate mudstones.	Olive grey to grey.	Hard, well indurated.	Bedding is usually defined by variability in the clay mineral content.	Thin to medium bedded, are commonly thinly interbedded with other chalk lithofacies, and usually exhibit somewhat discontinuous, wavy-parallel laminations and rare to locally common burrowing (primarily *Chondrites* and *Planolites*).	15% to 35% quartz, 5% to 20% clay minerals.	Argillaceous chalk seldom demonstrate significant reservoir quality porosity.
Burrowed argillaceous chalk facies	Argillaceous, carbonate mudstones.	Light olive grey, to yellowish brown, to medium and dark grey.	Well indurated, hard.	Thin (<10 centimetre) to very thick bedded (>1 metre).	Common to abundant burrow structures (*Planolites, Chondrites*,and *Zoophycos*) and relict wavy- to planar-parallel lamination.	15 to 35% quartz, 2 to 12% clay minerals.	Reservoir quality porosities have not been identified
Calcarenite facies	Packstones composed of sand-sized chalk clasts and skeletal debris.	Yellowish brown to grey.	Hard, well-indurated.	Thin to medium bedded (typically occur as single beds less than 30 centimetres thick).	Normal grading and occasional planar-laminations or ripple-laminations.	5% to 35% quartz.	Calcarenites are lacking in reservoir quality porosity.
Conglomeratic chalk facies	Conglomeratic texture of lithified, granule- to pebble-sized lithoclastic material. Clasts tend to be polymictic, larger, more irregularly shaped and more abundant, compared to the small, round, chalk clasts that occur in Pebble Floatstone facies. Carbonate mud composes the matrix portion of the rock.			Thin to medium bedded (typically occur as single beds less than 30 centimetres thick).	No internal structure, with clasts being fairly chaotic in their distribution.		No reservoir quality porosity has been identified in this facies.
Deformed chalk facies		Yellowish brown to light grey.	Hard, and well indurated.	Bedding ranges from thin (<10 centimetres) to thick (>1 metre), when it can be recognised.	Defined by the weak to very strong deformation (contortion) of bedding or internal structures/ fabric.	1% to 4% quartz and < 2% clay minerals in the Tor Formation. In the Ekofisk Formation, 3% to 20% quartz and 1% to 5% clay minerals.	Deformed chalks commonly contain reservoir quality porosity.
Calcareous shale facies	Very calcareous, siliciclastic clayshales and mudshales.	Grey to reddish brown.	Fairly soft and poorly indurated.	Bedding is usually thin to very thin (<10 centimetre).	Discontinuous, wavy lamination and very rare burrowing.	>40% quartz and > 15% clay minerals.	Calcareous shales do not contain reservoir quality porosity.

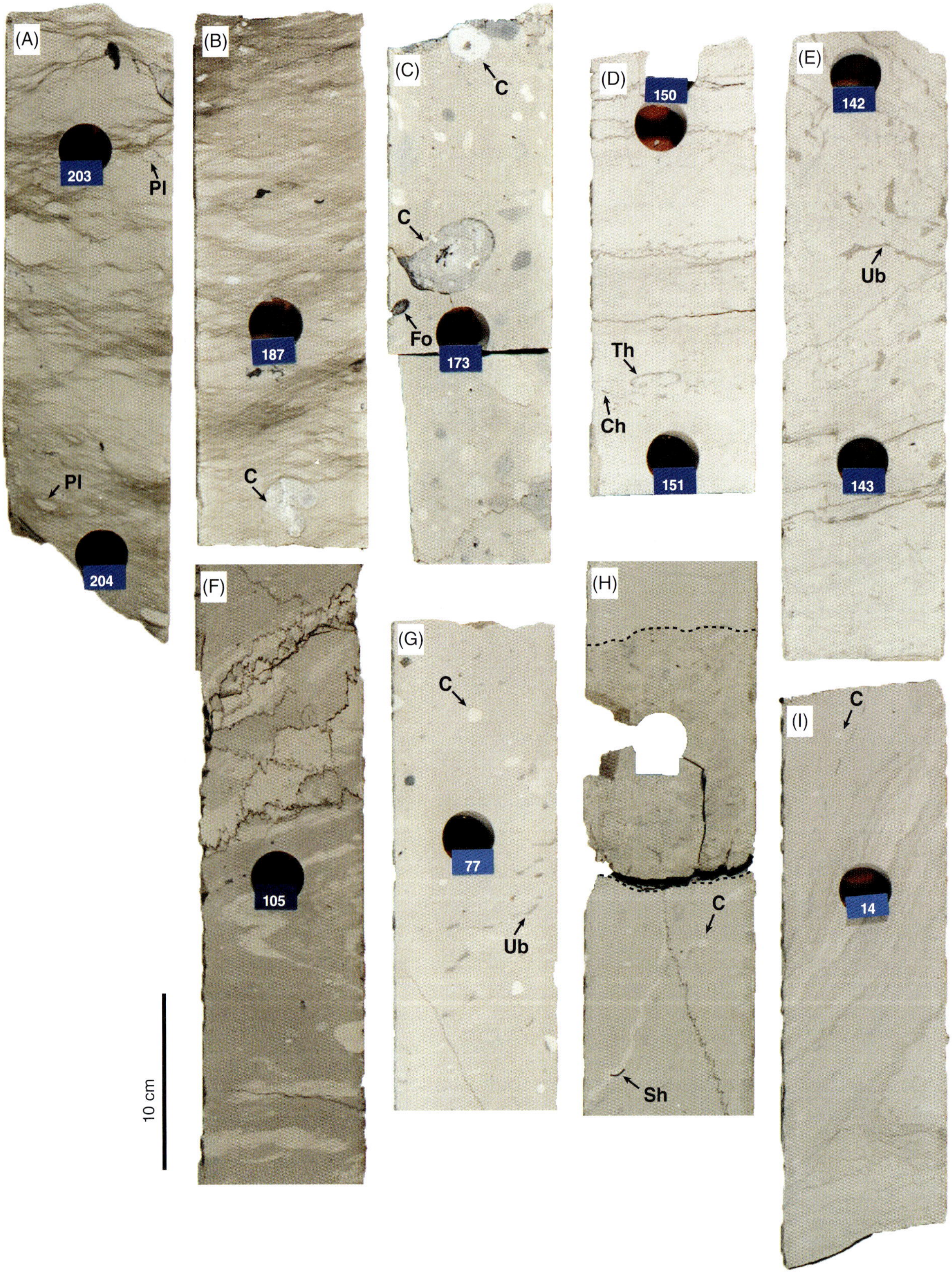

Fig. 12. Facies photographs from the cored part of well 2/8-15 (Figs 7 and 9). (A, B) Facies BAC, bioturbated argillaceous chalk (Table 2). (C) Facies PFC, pebble floatstone chalk with large polymictic clasts. (D) Facies HC, homogeneous chalk with burrows and faint lamination. (E) Facies DC, deformed chalk with burrows. (F) Facies DC, deformed chalk with slump folds. (G) Facies PFC, pebble floatstone with deformed burrows. (H) Facies HC, homogeneous chalk with a strongly bioturbated interval (marked by black dotted lines). (I) Facies DC, deformed chalk with water-escape structures. Letter symbols: C = chalk clast; Sh = shell fragment; Fo = foram fragment; Ch = *Chondrites* burrow; Pl = *Planolites* burrow; Th = *Thalassinoides* burrow; Ub = unidentified burrow. Well location in Figs 1 and 4. For abbreviations see Table 2.

and winnowing of an unconsolidated to partly lithified chalk by a bottom current. This deposit might represent an initial erosive phase of the channel development and might be similar in character to the chalk overlying the base of channels in the Etretat cliffs, France, where conglomeratic, packstone and hardground chalk facies pass abruptly upwards into chalk wackestone and marly chalk (Quine & Bosence, 1991). The thick overlying succession of facies BHC/LC in the lower channel unit reflects vertical aggradation of pelagic chalk within the channel (Table 2). The decimetre-scale to metre-scale variations in density and porosity values observed in the log curves can originate from various mechanisms (see Niebuhr & Prokoph, 1997; Scholle *et al.*, 1998; Bramwell *et al.*, 1999; Stage 1999, 2001; Niebuhr *et al.*, 2001; Damholt & Surlyk, 2004 for detailed discussions). However, in the absence of core data it is impossible to determine the mechanism that caused the cyclicity or indeed to identify the lithological changes that are occurring. The uppermost interval of the lower channel-fill unit is characterised in core by bioturbated argillaceous chalks (facies BAC, Table 2), which suggests sea floor colonisation by infauna and probably represents a period of sediment starvation with relative enrichment of the terrigenous vs. the carbonatic fraction (cf. Bramwell *et al.* 1999) in the channel preceding the deposition of the upper channel-fill unit.

Facies association of the upper channel-fill unit

Description: The upper channel-fill unit sharply overlies the lower channel-fill unit and is dominated by the pebble floatstone chalk of facies PFC (Table 2, Fig. 12C). The chalk here is generally more porous than any of the facies in the lower channel unit and also has lower clay content, as indicated by the gamma ray log (Table 1). Facies PFC forms ~ 80% of this unit and contains well-rounded chalk clasts ranging from coarse sand to pebble in size (Fig. 13C). Shell fragments and foraminifera tests occur throughout this stratigraphic interval, but bioturbation is rare (Fig. 13A). Bedding is sometimes recognisable due to inclined parallel to crenulated primary laminations (Figs 13B, 13C, 13D and 13E) rarely disturbed by bioturbations. However most of samples appear with a homogeneous matrix (Figs 12G and 13F).

Generally, the PFC facies (Table 2) are divided into metre-thick beds showing coarse-tail grading,

Fig. 13. Photographs of allochthonous chalk facies in cores 4 and 5 from well 2/4-12 (Figs 6 and 8). (A) Facies PFC, pebble floatstone chalk with deformed *Thalassinoides* and small dark *Chondrites* burrows. (B–D) Facies PFC, pebble floatstone chalk with crenulated lamination. (E) Facies PFC, pebble floatstone chalk with high-angle lamination. (F) Facies PFC, pebble floatstone chalk with isolated pebbles in fine-grained matrix. (G) Facies DC, deformed chalk with sub-vertical lamination. (H) Facies HC, homogeneous chalk with stylolites and fractures. (I) Facies DC, deformed chalk with high-angle crenulated lamination. (J) Facies DC, deformed chalk with water-escape structures. Letter symbols: C = chalk clast; Sh = shell fragment; Ch = *Chondrites* burrow; Th = *Thalassinoides* burrow. Well location in Figs 1 and 4. For abbreviations see Table 2.

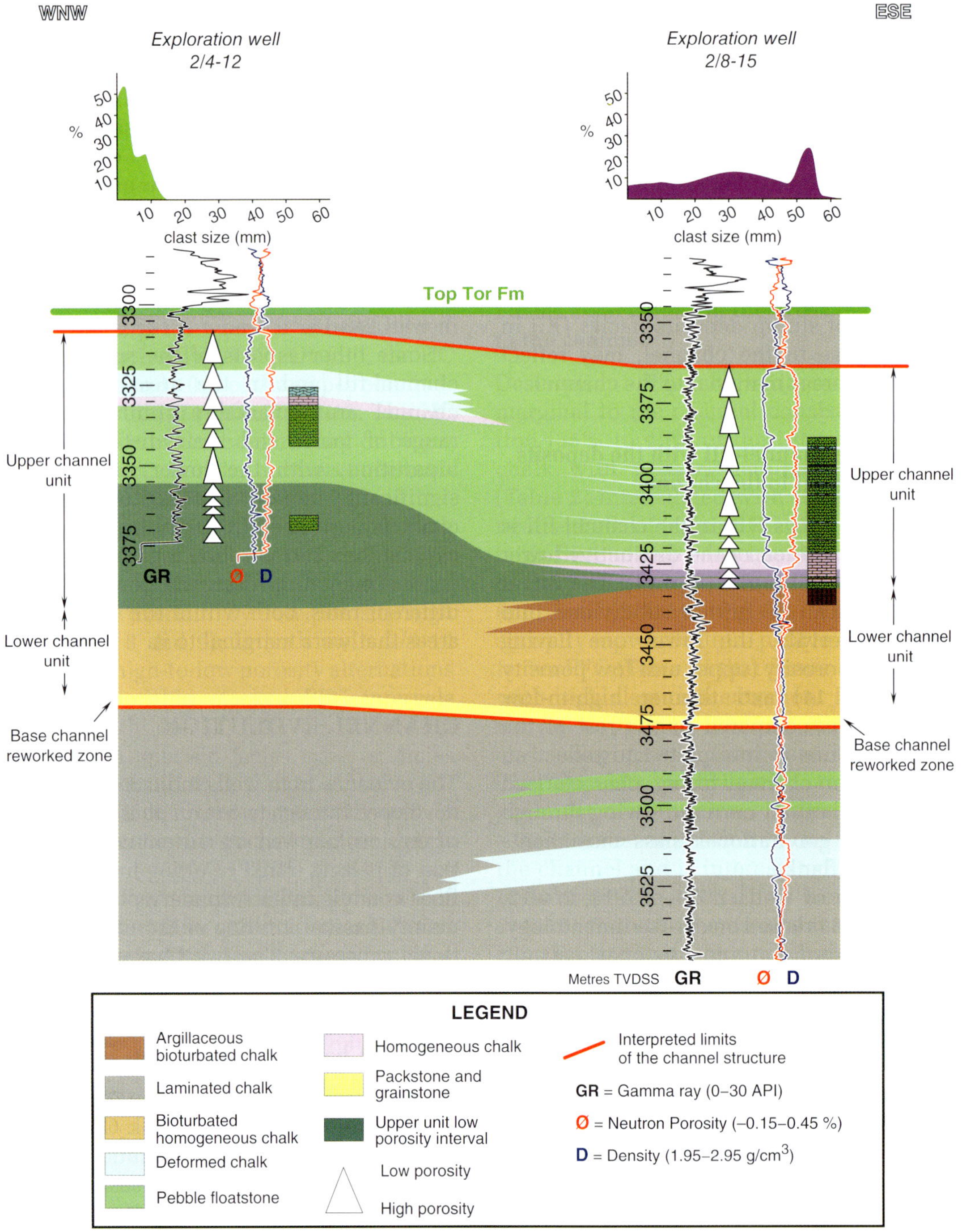

Fig. 14. Well correlation panel parallel to the channel axis, flattened at the top of Tor Formation. The position of the well correlation is indicated in Fig. 4. Depths are given in True Vertical Depth Sub Sea (TVDSS).

feature may be related to both turbidity and bottom currents and that a complex interaction of inversion tectonics, gravity-driven processes and superimposed bottom currents controlled chalk reworking and re-deposition.

The channel of the current study is interpreted to have been developed by contour-parallel bottom currents. It is broadly oriented in a contour parallel rather than a downslope direction, making gravity flow processes a less likely generative mechanism.

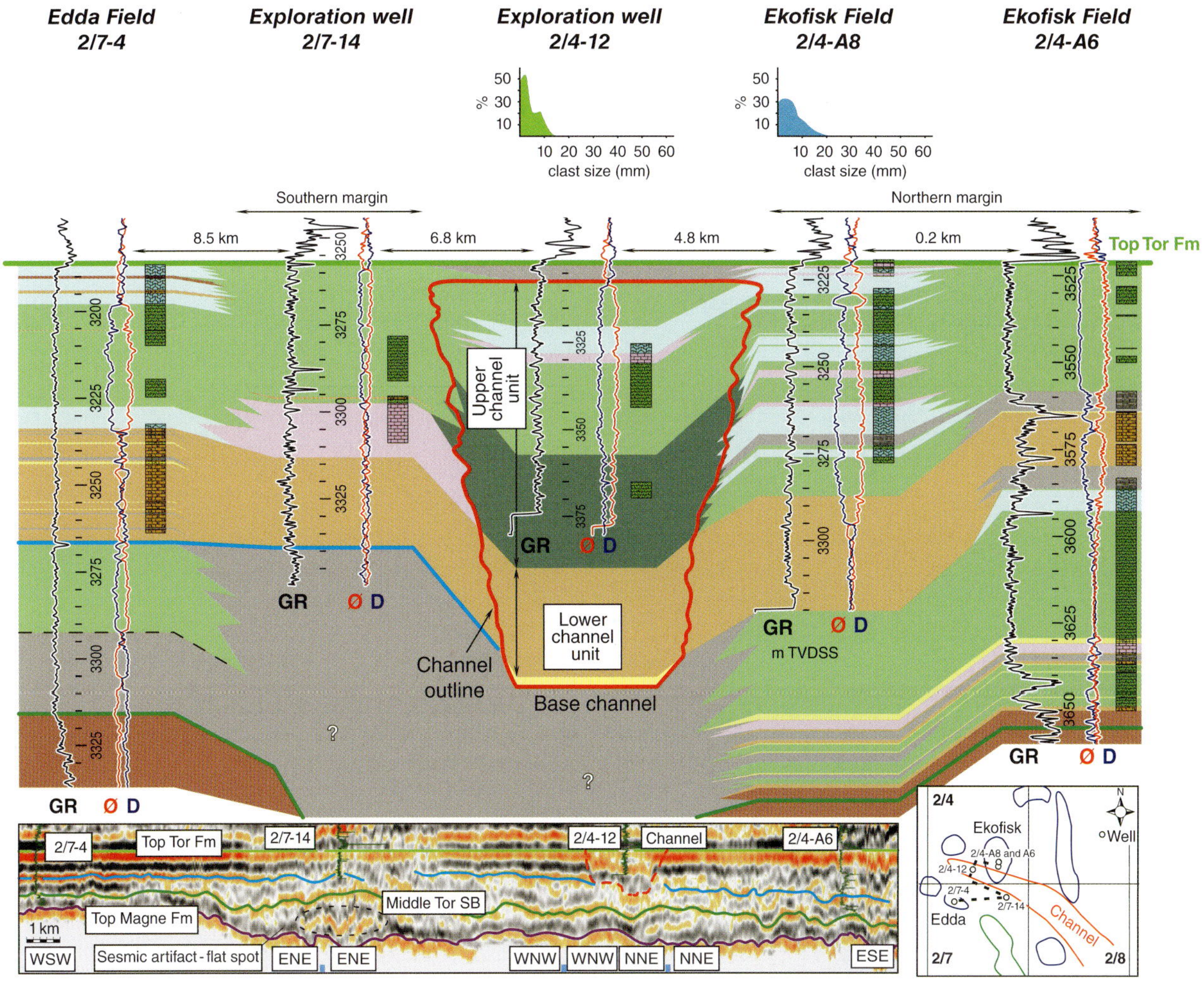

Fig. 15. Well correlation panel across the western segment of the channel. The cross-section is flattened at the top of Tor Formation. The location of wells is shown in Fig. 4 and the legend is as given for Fig. 14. Depths are given in True Vertical Depth Sub Sea (TVDSS).

A number of key factors point towards a contour current origin and are discussed below.

Channel incised base: an incised character is seen at the base of the channel on seismic data. Similar incisions have been seen in the Chalk Group of the Danish Sector and are interpreted to have been produced by bottom currents (Esmerode *et al.*, 2007). The channel is interpreted to have been initiated by a phase of sea floor erosion, as indicated by the incised irregular base (Fig. 5, Sections A and B) and the presence of facies PGC interpreted to have been produced by current winnowing and erosion (Table 2). Sediment erosion would be within the capability of bottom currents. Surlyk & Lykke-Andersen (2007) consider that erosion of non cemented chalk ooze could occur when currents flow at 8 cm s^{-1} to 20 cm s^{-1} at approximately 20 m above the sea floor. Studies of modern bottom currents indicate a wide range of flow strengths, largely independent of the current generative mechanism. For instance, the velocities of thermohaline bottom currents vary from 20 cm s^{-1} (Hollister & Heezen, 1972) to 300 cm s^{-1} (Gonthier *et al.*, 1984) and also the velocity of deep-water tidal currents, recorded down to 4000 m sea depth, can vary from 25 cm s^{-1} to 280 cm s^{-1} (Shepard *et al.*, 1979; Shanmugam, 2008). Wind-driven bottom currents are generally stronger, with velocities of up to 200 cm s^{-1} recorded at water depth of 100 m in the Gulf of Mexico (Cooper *et al.*, 1990).

Phase 1: Incision of the sea floor and channel margin aggradation

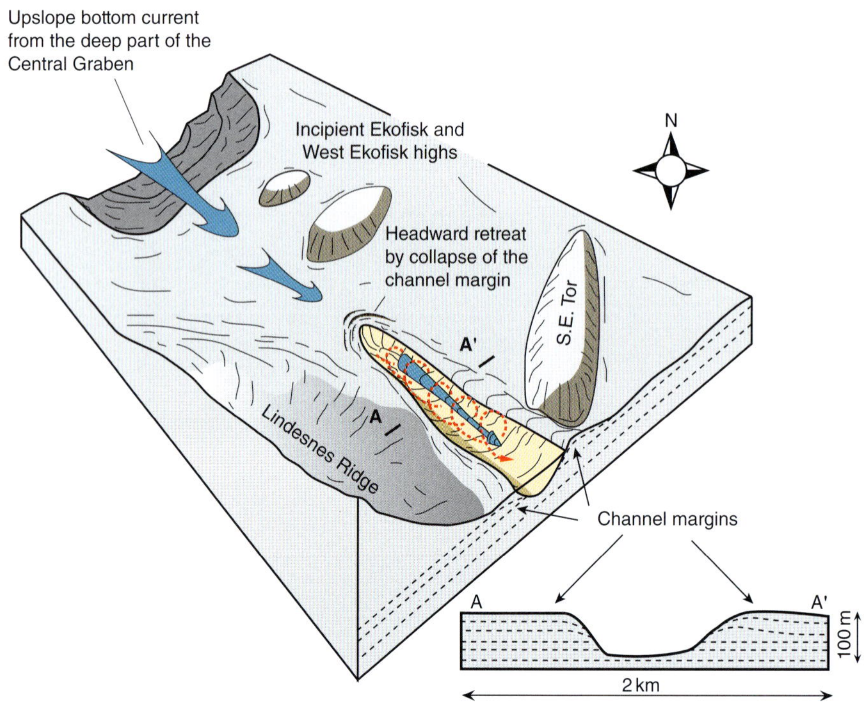

Phase 2: Passive infilling by allochthonous chalk material

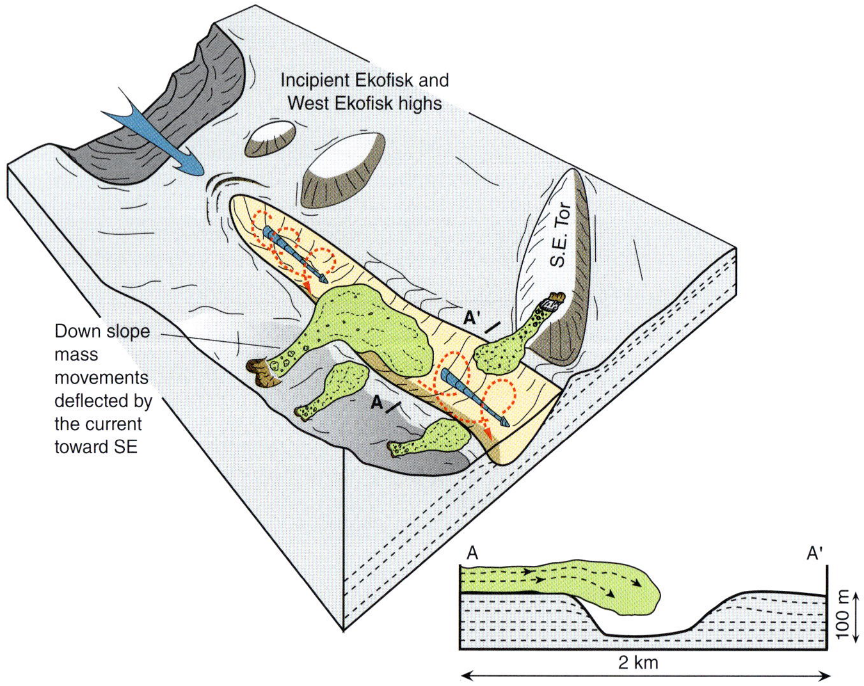

Fig. 16. Block-diagram cartoon showing the inferred two main phases of the channel development, corresponding to the deposition of the lower and upper channel units. The bipartite subdivision of the Tor Formation with pelagic facies below (BHC facies in Figs 14 and 15) and allochthonous material above (e.g. PFC and DC facies in Figs 14 and 15) represent respectively the channel initial channel erosion and aggradation (Phase 1: channel development) and the subsequent infill by gravity flow deposits (Phase 2: channel-fill).

Funnel-shaped channel planform geometry: the funnel-shaped geometry of the channel suggests gradual retrogression by a progressive headward retreat towards the WNW (Fig. 16). Erosion of the chalk ooze gradually produced a scour on the sea floor and failure of the unstable chalk ooze at the channel head. As the ESE-flowing current was funnelled into the channel, it tended to increase its velocity and erosive capability, thus keeping the channel active and open until the widest point at the ESE end of the channel where erosive capability again decreased.

Channel orientation: the channel is orientated parallel to the palaeobathymetric contours of the Lindesnes Ridge, suggesting that it was generated by bottom currents flowing parallel to this major palaeotopographic feature. The broad southwards-trending curve of the channel path towards its ESE end is interpreted to have been controlled by sea floor palaeotopography, notably by the position of the Lindesnes Ridge and the SE Tor anticlinal structure located respectively to the S and NE. These topographic highs were already prominent features on the chalk sea floor during the late Maastrichtian (Bramwell *et al.*, 1999) and most likely constrained the course of the channel during its development.

The asymmetric geometry of the channel cross-section in its central and eastern segments (Fig. 5, Sections C, D and E) is interpreted to be indicative of a current flowing towards the ESE as interpreted in other modern and ancient contourite channels in the northern hemisphere (e.g. Akhmetzhanov *et al.*, 2007; Surlyk *et al.*, 2008).

The asymmetric cross-sectional morphology of this and other contourite channels has been attributed to lateral asymmetry in the flow erosion capability due to deflection of the current by the Coriolis force. This leads to greater erosion on the right side of the current flow and lesser erosion on the left side (northern hemisphere; cf. Surlyk *et al.*, 2008).

In the case of the channel studied, it is interpreted that the Coriolis force gradually deflected the current towards the right side (southern side) of the channel. Lateral and vertical velocity gradients created a helicoidal flow where sedimentation preferentially occurred on the left side of the channel and erosion on the right side.

The channel does not appear to have been formed by erosional processes alone. In seismic cross sections, the channel appears as a scour with sharp margins and without any sort of lateral levees (Fig. 5, Sections A and B). However, if formed solely by erosion, the margins of the channel would be unstable unless the chalks were already lithified as hardgrounds during erosion (cf. Esmerode & Surlyk, 2009). If this would have been the case and erosion was the main mechanism for the formation of the channel, the bottom current would have necessarily been very powerful in order to erode such a deep scour (~100 m deep) into lithified chalk sediments. Alternatively, the steep flanks of the channel may result from a persistent aggradation of the channel margins, while the axial zone was dominated by sediment bypass and erosion alternating with periodic accumulation of pelagic chalk. Stabilisation of the channel flanks during progressive aggradation probably occurred through the development of hardgrounds during periods of non deposition (cf. Quine & Bosence, 1991).

After an initial erosive phase, sediment bypass and periodic pelagic deposition predominated in the channel; currents persisted along the channel axis, whilst aggradation continued at the channel margins. The absence of allochthonous chalk facies in the lower channel-fill unit indicates that the channel margins remained gentle enough to avoid collapsing. The channel was thus a low-relief feature defined by differential sedimentation rate and the growth of its margins.

A similar mode of channel formation has been suggested for the chalk of the Paris Basin (Esmerode & Surlyk, 2009) and the chalk of Etretat, Normandy (Quine & Bosence, 1991). At Etretat, the channels are up to 1 km wide and 60 m deep and have been produced by bottom currents through sediment erosion, winnowing and bypass combined with alternating deposition and cementation of the channel surface represented by the formation of softgrounds and hardgrounds. Those channels resemble the present one in terms of both geometry (width, depth, inclination of margins) and channel-fill succession of autochthonous to allochthonous chalk shared by the channel and its margins (Fig. 17).

Phase 2: Channel infill

The initial erosive and constructive phase was followed by a phase of channel filling with mainly allochthonous chalk, derived probably from the channel flanks and the surrounding topographic highs, such as the Lindesnes Ridge and the SE Tor structure. A dominant southern provenance from the Lindesnes Ridge for the material has been identified by Sikora *et al.* (1999) based on

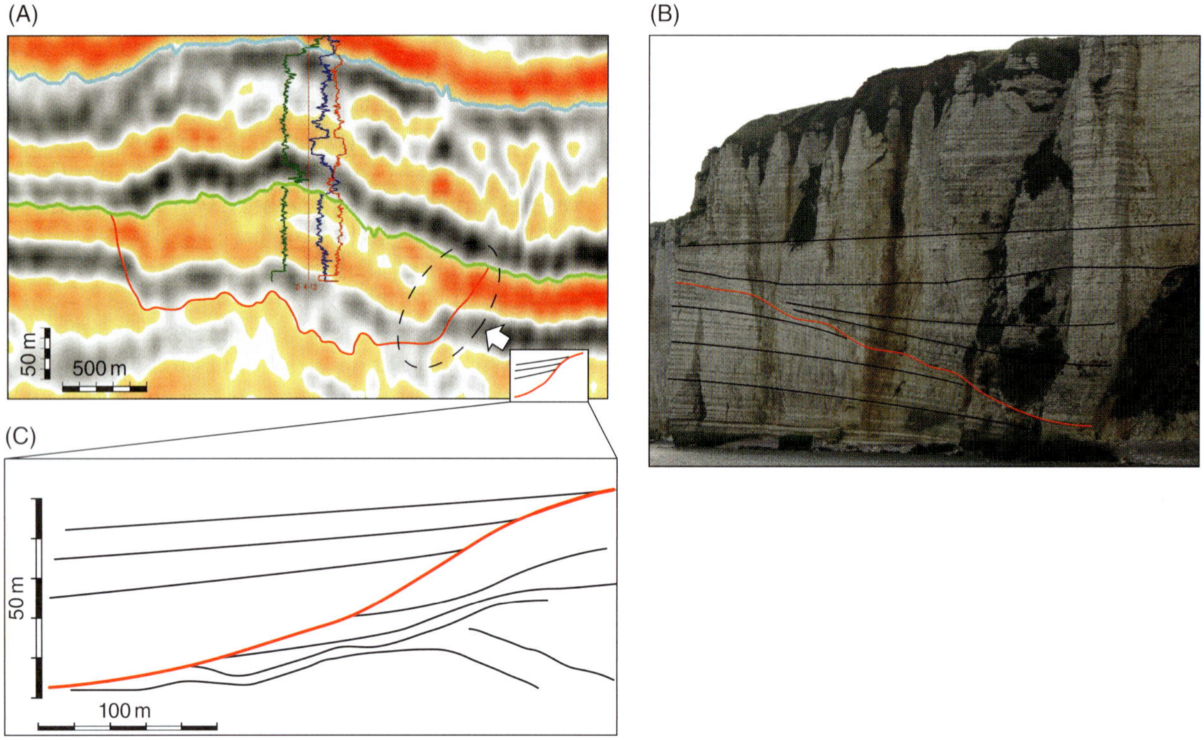

Fig. 17. Comparison between the (A) seismic section of Fig. 5B and (B) channel development as observed at the cliffs of Etretat, Normandy, France. Vertical scale in (A) is approximate and estimated from well 2/4-12. The outcrops sketch in (C) has been rescaled to match approximately to the seismic. The height of the cliffs is 40 m. The photograph shows the northern channel margin, while (C) the 500 m wide correlation sketch (from Lasseur, 2007) shows the southern margin. The total width of this channel in outcrop is 1 km.

biostratigraphic evidence. At this stage, the current may still have been active but apparently was not powerful enough to erode, winnow and re-suspend large parts of the allochthonous chalk deposits. Downslope mass movements originating from the adjacent highs and channel margins consist of chalk material with various degrees of lithification, implying softground or hardground development in the source areas (e.g. Valhall Field on the Lindesnes Ridge; see Kennedy 1980; Sikora *et al.*, 1999). During the initial phase of channel infill (upper channel-fill unit low porosity interval in Fig. 14) the current may have been capable of redistributing part of the allochthonous sediments within the channel, thereby thinning the accumulated debris flow deposits.

In the overlying interval of the upper channel-fill unit, the mass-flow deposits emplaced in the channel, recognised in wireline logs as high-to-low porosity alternations, show a thickening-upward trend. This may reflect a gradual decline in the bottom current strength and its erosive potential. The current was not capable of removing the allochthonous material and this lead to the preservation of thicker debris flow deposits. Analogy can be made with the channels and channel-fill seen at Etretat (cf. Quine & Bosence, 1991) where waning of the current caused gradual infilling of the channel by progressively thicker succession of pelagic and allochthonous chalks associated with a gradual decline of the channel margins inclination.

Accumulation of mass-flow deposits occurred both within the channel axis and on the marginal areas of the channel. Progressive emplacement of gravity-flow deposits gradually filled the channel (Fig. 16). Once filled, gravity-flows were not confined any longer in the channel depression and downslope mass movements could by-pass the earlier channel feature and deposit over a wider area. The morphology of the channel allowed the capture of allochthonous material and hence the channel allochthonous infill (upper channel unit) is relatively thicker compared to coeval intervals in adjacent areas (Fig. 15). The localised bottom-current activity completely ceased above the Tor-Ekofisk Formation boundary at the base of the clay-rich Ekofisk dense zone (Figs 5, 14 and 15) which marks a major basin-scale turnover of the chalk depositional style with the development of hardgrounds and significant input of terrigenous clay/silt grade sediment (Bramwell *et al.*, 1999).

DISCUSSION

Sequence stratigraphic implications

On seismic sections, numerous unconformities have been recognised in the Chalk Group as erosional truncations followed by bedding onlap (Bramwell *et al.*, 1999). In the chalk successions of the Norwegian North Sea, major unconformities have been interpreted to be produced by tectonic uplift and subsequent erosion by gravity-driven processes, e.g. unconformities at Top Narve and Top Thud formations (Bailey *et al.*, 1999; Hampton *et al.*, 2010). In other chalk successions in the North Sea and in the Danish Basin, prominent unconformities have been considered by some authors as evidence of increased bottom-current activity and sea-level fall and may have possibly played an important role in the intensification of bottom currents (Esmerode *et al.*, 2007, 2008; Surlyk & Lykke-Andersen, 2007; Surlyk *et al.*, 2008). This notion would imply that sequence bounding unconformities might have formed in basins sharing the same bottom current system (Faugères *et al.*, 1993). However, the applicability of sequence stratigraphic concepts to relatively deep-water successions with unconformities/erosion surfaces produced by bottom current processes remains uncertain, mainly on account of diachronous variations in current strength and uncertainties in sediment dating (Faugères *et al.*, 1999; Stow *et al.*, 2002).

The North Sea was tectonically active during the Late Cretaceous and hence there are also reasons to believe that tectonic seismicity and subsequent erosion by gravity-driven re-sedimentation caused the formation of at least some of the unconformity surfaces (e.g. Bramwell *et al.*, 1999). Progressive uplift of the inversion zones during the late Maastrichtian associated with lowering of the sea-level may have further focused the flow of bottom currents in addition to have increased gravity-driven erosion and remobilisation of previously deposited chalks.

It is possible that the erosion associated with the base of the channel might reflect a minor relative sea-level fall that narrowed sea gateways and intensified bottom currents. Studies of the Chalk Group in the German North Sea sector have indeed indicated a climax of bottom current activity in the late Maastrichtian (Surlyk *et al.*, 2008), a period of pronounced changes in global ocean temperatures (Pucéat *et al.*, 2003) and significant sea-level falls (Hardenbol *et al.*, 1998; Miller *et al.*, 2004, 2005; Kominz *et al.*, 2008, Surlyk *et al.*, 2008) possibly due to ice-cap growth (Miller *et al.*, 2004; Bornemann *et al.*, 2008). A hypothetical link between the origin of the studied channel and the sequence stratigraphic development of the Tor Formation postulated by Bramwell *et al.* (1999) is suggested in Fig. 18.

Geographical location of the channel

The location and geometry of the channel suggest that it was generated by a bottom current flowing ESE parallel to the bathymetric contours of the Lindesnes Ridge (Fig. 19). Interpretation of flow towards the ESE is based on the funnel shaped geometry of the channel, the proposed mechanism of enlargement by headward retrogression and by erosional versus depositional margins. The channel location also suggests strong influence by other topographic features. For example, the channel appears to have been constrained by the Ekofisk and SE Tor anticline structures that flank the channel along its northern margin (Fig. 19).

The WNW end of the channel can be identified clearly from seismic data (Figs 4 and 5) and is located in a horizontally stratified area to the north of the Eldfisk Field. Strata in this area are gently dipping to the NW and represent a sort of shelfal region when compared to the deeper basin located to the NW. Since it is considered that the current in the channel was flowing from the WNW to the ESE, the bottom current that formed the channel must have originated as a current that followed the elongate, deep basinal areas of the Central Graben. The current is interpreted to have flowed along the eastern margin of the Josephine High before passing between the narrow sea-pathway flanked by the Albuskjell Field in the N and a salt diapir in the south (Fig. 19). This may have forced the current to travel upslope and reduce its flowing section (i.e. confinement), thereby increasing its velocity and erosive potential. Once the current reached the flat area

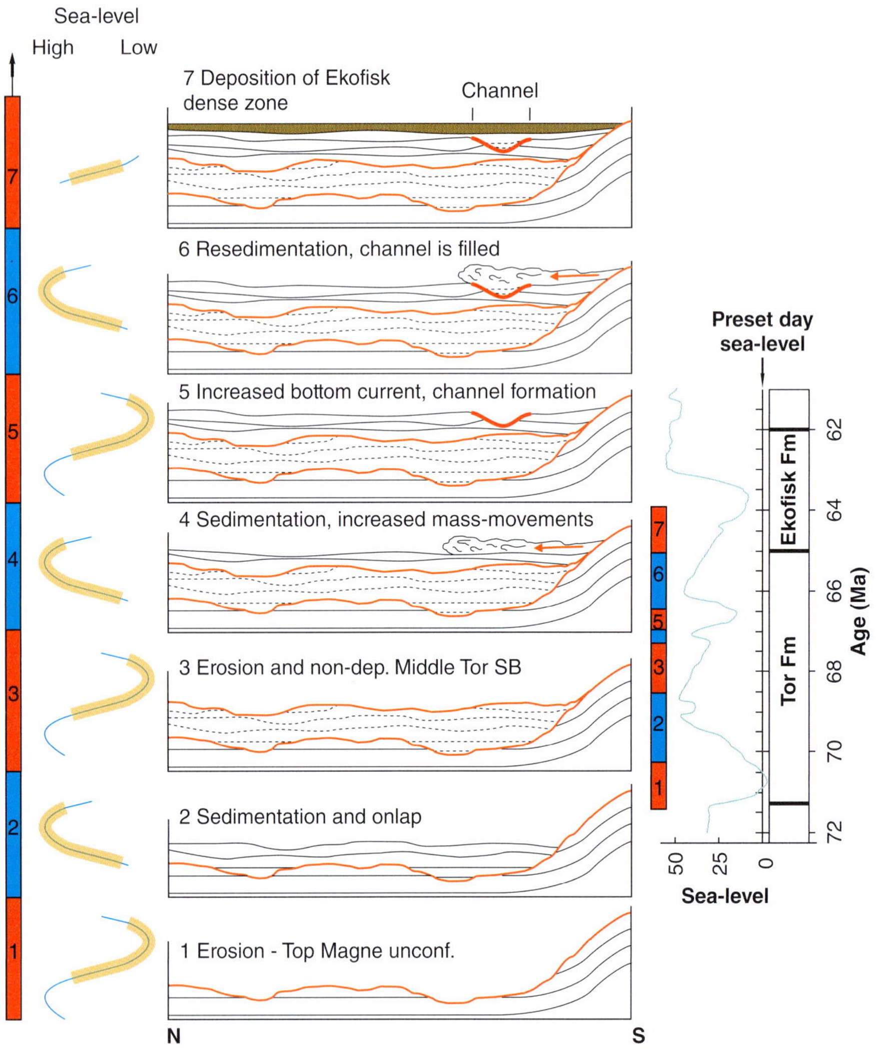

Fig. 18. Simplified interpretation of the N–S seismic cross-section in Fig. 3 illustrating the sequence stratigraphic interpretation of the deposition of the Tor Formation including the development of the channel described in the present study. The successive development phases are as follows: (1) formation of an extensive erosional surface at the top of the Magne Formation; (2) resumption of chalk sedimentation and progressive onlap of the erosional surface; (3) an increase in bottom-current activity and formation of the mid-Tor Formation unconformity; (4) resumption of chalk sedimentation and increase in its reworking; (5) another increase in bottom-current activity and formation of the channel; (6) a decrease in current activity and filling of the channel; and (7) cessation of current activity followed by the deposition of clay-rich chalk and formation of the Ekofisk dense zone. The sea-level curve is after Kominz *et al.* (2008).

on top of the slope it began to scour the poorly consolidated chalk and incise the sea floor. Although Fig. 19 depicts the present day structural configuration of the study area, various studies of chalk thickness maps and seismic cross-sections (e.g. Bramwell *et al.*, 1999) show that this 3D seismic map mirrors strongly the physiography of the study area during the late Maastrichtian. It therefore provides an illustration of the inferred current flow in relation to the coeval topography.

The microfaunal assemblage of well 2/7-14 indicates that the channel margin deposits identified here may reflect deposition under the influence of deep water masses since they are enriched in planktonic foraminifera and nannofossils and radiolarians that could have been transported by bottom currents of deep water provenance. However, changes in the planktonic/benthonic foraminifera ratio may also be due to other factors, such as relative sea-level changes or fluctuating saline/nutrient-rich plumes (e.g.

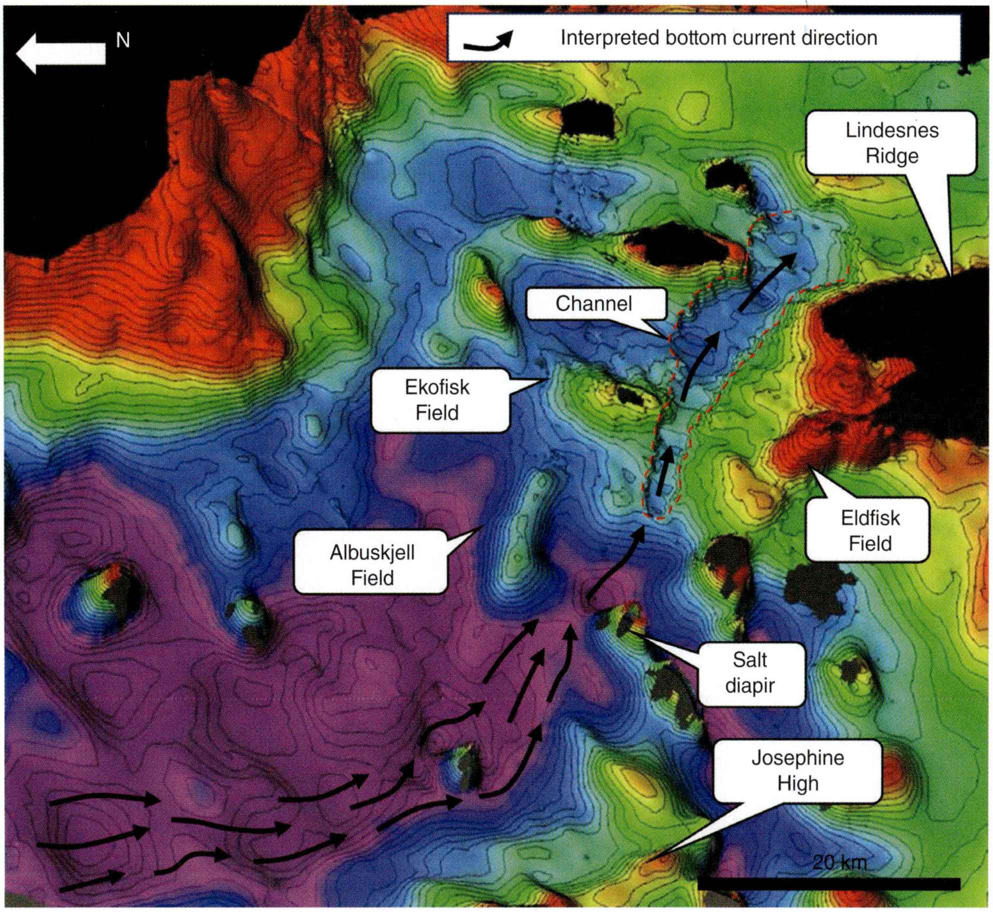

Fig. 19. Seismic map showing the channel extent relative to the main tectonic and halokinetic topographic features. A SSE-moving bottom current is interpreted to have flowed alongside the eastern margin of the Josephine High and then through a narrow topographically defined passage near the Albuskjell Field. The current became more confined and was forced to flow up-slope where it began to erode the channel. The lack of interpretation over certain areas is due to the low quality seismic image. The isopleth contour lines are every 20 ms^{-1}.

Lüning *et al.*, 1998) and preservation, i.e. dissolution of pelagic forms.

This bottom current could be generated by a series of distinct forces, such as differences in temperatures and salinity (thermohaline currents), wind shear stress (wind-driven currents) or gravitation of the moon (deep-water tidal currents). However, the available data do not allow the nature of the current to be specified, apart from indicating that it was topographically confined and its activity was long-lived. Nonetheless, a thermohaline circulation of water masses driven by density variations in response to changes in water temperature and salinity is the most favoured mechanism since this tends to generate regular, long-lived bottom current systems (cf. Surlyk & Lykke-Andersen, 2007).

Late Cretaceous marine circulation in the Chalk Sea

The Late Cretaceous epeiric chalk sea was characterised by strong bottom currents that swept the sea floor creating deep channels, moats and sediment drifts in many areas (Lykke-Andersen & Surlyk, 2004; Surlyk & Lykke-Andersen, 2007; Esmerode *et al.*, 2007, 2008; Surlyk *et al.*, 2008; Esmerode & Surlyk, 2009). These morphogenic features are similar to those formed along the modern continental margins where bottom currents flow parallel to the slope bathymetric contours (Viana & Rebesco, 2007; Rebesco & Camerlenghi, 2008).

Bathymetric interpretations of the chalk sea vary quite considerably. Distefano *et al.* (1980) suggested water depths ranging from 150 m to 600 m for the Chalk Group in the Norwegian Central Graben, while Bramwell *et al.* (1999) postulated that chalk deposition occurred between the effective wave base and 300 m. Based on sedimentological evidence, trace fossil assemblage and vertical difference in the topographic relief of a moat-drift complex, Surlyk & Lykke-Andersen (2007) estimated an average water depth of 300 m to 400 m, with peak of 600 m, for the chalk successions in the Danish Basin. Surlyk *et al.* (2008) pointed out that it is difficult to estimate the depth of chalk deposition, because depositional processes below wave base are similar and practically depth-independent. However, the postulated water depths are within the range estimated by Viana *et al.* (1998) for mid-water (300 m to 2000 m) and shallow-water (50 m to 300 m) bottom current systems.

The current that formed the channel in the present case is inferred to have flowed principally southward, similarly to the current directions identified by Surlyk *et al.* (2008) in the Maastrichtian chalk of the German North Sea sector. A southerly direction is also comparable to the direction of the currents interpreted by Jarvis (1980) and Esmerode *et al.* (2008) in the Santonian and Turonian to Campanian successions of the Paris Basin and Danish Central Graben, respectively. These current directions differ markedly from the NW direction interpreted in the Danish Basin (Fig. 20) (Esmerode *et al.*, 2007; Surlyk & Lykke-Andersen, 2007).

The channel in the present study is thus considered to be a local feature formed within an extensive system of currents (Fig. 20) that drove water circulation in the Chalk Sea since the Cenomanian. In the Norwegian North Sea, this system of bottom current principally flowed southwards along the Central Graben before being funnelled by sea floor topography and forced to flow upslope along the northern flank of the Lindesnes Ridge. The current then passed through the Feda Graben before reaching the Danish and German sectors of the

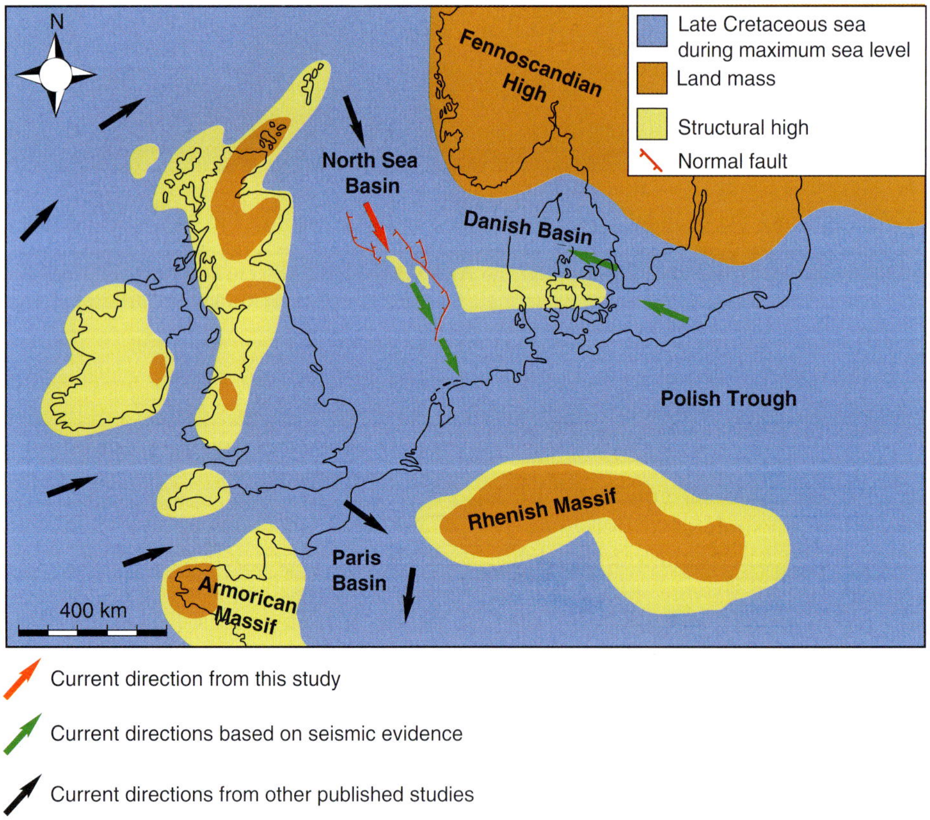

Fig. 20. Palaeogeographical map showing the extent and directions of the Late Cretaceous marine circulation in NW Europe. Simplified from Ziegler (1990), with palaeocurrent directions based on seismic data after Lykke-Andersen & Surlyk (2004), Esmerode *et al.* (2007, 2008), Surlyk & Lykke-Andersen (2007), Surlyk *et al.* (2008) and Esmerode & Surlyk (2009) and other interpretations from Jarvis (1980) and Hart (2007).

southern Central Graben, where it continued its erosional and depositional activity (Surlyk *et al.*, 2008). A northward flow of returning water seems to have occurred through the Danish Basin (Esmerode *et al.*, 2007; Surlyk & Lykke-Andersen, 2007) (Fig. 20). The postulated directions of these palaeocurrents system in the North Sea and Danish Basin are similar to those proposed in recent palaeogeographic modelling (cf. Hart, 2007; Fluteau *et al.*, 2007; Pearce *et al.*, 2009).

CONCLUSIONS

1 Seismic sections of Maastrichtian chalk show a large, WNW–ESE trending channel crossing the southern part of the Ekofisk Field in the Norwegian North Sea. The channel was ~ 100 m deep, 2 km to 5 km wide and ~ 30 km long. The channel has a funnel-shaped planform opening to the SE and relatively steep southern margin, which suggests that it was created by a topographically-guided bottom current that flowed towards the ESE parallel to the slope of the Lindesnes Ridge. The current was probably affected and deviated by the Coriolis force towards its right-hand side leading to erosion or non-deposition on the southern channel margin, while sediment accumulation characterised the northern channel flank.

2 A simple phase of major erosion is discounted as a mechanism for the channel origin. Although initiated by substrate erosion, the channel is considered to have formed by alternating episodes of sediment bypass, deposition and minor erosion that occurred concurrently with the aggradation of the channel margins. Well logs from the channel-fill and channel-margin deposits indicate a two-phase history of channel development: (1) an early phase of erosion and pelagic chalk deposition, with sediment bypass predominating in the channel accompanied by channel margin build-up; and (2) a later phase of allochthonous chalk deposition by sediment gravity processes, which draped both the channel and its margins and eventually filled up the former.

3 Although the channel feature is recognisable in seismic sections, the chalk facies associations of the channel-fill and channel margins are similar and show analogous two-phase accumulation. The channel fill to channel margin stratigraphic correlation and similarity of facies assemblages suggest that the channel formed and continued to develop as a low-relief feature, owing its prolonged existence mainly to a differential aggradational sedimentation along the thalweg zone and the margins.

4 The channel is interpreted to have formed during a phase of sea-level fall that intensified bottom current activity and to have filled during the subsequent sea-level rise and high-stand phase. The correlative bipartition of the Maastrichtian chalk succession in the Norwegian Central Graben, with autochthonous and predominantly allochthonous deposits separated by an extensive submarine erosional surface, has led previous researchers to the recognition of two seismic sequences. The present study indicates another minor unconformity within the upper sequence, the base of which is defined by the erosive channel feature described by this study.

5 The channel is considered to be a local expression of a regional-scale system of bottom currents that characterised the European Chalk Sea reaching a marked activity during the late Maastrichtian sea-level fall. The bottom current flowing southwards along the North Sea grabens probably originated from the Boreal Sea and during its southern course it was driven and constrained by the sea floor topographic relief in the Norwegian Central Graben.

ACKNOWLEDGEMENTS

This paper is dedicated to the memory of Professor Michael R. Talbot, who passed away in November 2009. Mike was an exceptional person and scientist and will be deeply missed. He planned the first author's PhD study on which this paper is based and supervised the research project at its early stages. The study at the University of Bergen has subsequently been supervised jointly by his two colleagues, Gunnar Sælen and Wojciech Nemec. Bruno Caline (Total E&P France), Michel Thomas (Total E&P Norge) and Eric Lasseur (BRGM) are thanked for sharing their sedimentological expertise of the chalk. The comments, corrections and suggestions by reviewers F. Surlyk and A. S. van der Molen significantly improved the early version of the manuscript and their contribution to the final paper is highly appreciated. Total E&P Norge is thanked for sponsoring this research at the

University of Bergen and for providing access to data. Partners in PL018: ConocoPhillips Skandinavia AS, Statoil, Eni Norge and Petoro are thanked for permission to access data and for authorisation to publish this study.

REFERENCES

Akhmetzhanov, A., Kenyon, N. H., Habgood, E., Van der Molen, A.S., Nielsen, T., Ivanov, M. and **Shashkin, P.** (2007) North Atlantic contourite sand channels. In: *Economic and Palaeoceanographic Significance of Contourite Deposits* (Eds A.R. Viana and M. Rebesco), *Geol. Soc. London Spec. Publ.*, **276**, 25–47.

Back, S., Van Gent, H., Reuning, L., Grötsch, J., Niederau, J. and **Kukla, P.** (2011) 3D Seismic Geomorphology and Sedimentology of the Chalk Group, Southern Danish North Sea. *J. Geol. Soc. London*, **168**, 393–405.

Bailey, H., Gallagher, L., Hampton, M., Krabbe, H., Jones, B., Jutson, D., Moe, A., Nielsen, E.B., Petersen, N.W., Riis, F., Sawyer, D., Sellwood, B., Strand, T., Øverli, P.E. and **Øxnevad, I.** (1999) *Joint Chalk Research Phase V: A Joint Chalk Stratigraphic Framework.* Norwegian Petroleum Directorate (NPD), Stavanger.

Bornemann, A., Norris, R.D., Friedrich, O., Beckmann, B., Schouten, S., Damste, J.S., Vogel, J., Hofmann, P. and **Wagner, T.** (2008) Isotopic evidence for glaciation during the Cretaceous supergreenhouse. *Science*, **319**, 189–192.

Bramwell, N.P., Caillet, G., Meciani, L., Judge, N., Green, M. and **Adam, P.** (1999) Chalk exploration, the search for a subtle trap. In: *Petroleum Geology of Northwest Europe: Proceedings of the 5th Conference* (Eds A.J. Fleet and S.A.R. Boldy), pp. 911–937. Geological Society, London.

Brasher, J.E. and **Vagle, K.R.** (1996) Influence of lithofacies and diagenesis on Norwegian North Sea chalk reservoir. *AAPG Bull.*, **80**, 746–769.

Brewster, J. and **Dangerfield, J.A.** (1984) Chalk fields along the Lindesnes Ridge, Eldfisk. *Mar. Petrol. Geol.*, **1**, 239–278.

Brewster, J., Dangerfield, J.A. and **Farrell, H.** (1986) The geology and geophysics of the Ekofisk field waterflood. *Mar. Petrol. Geol.*, **3**, 139–169.

Bromley, R.G. (1996) *Trace Fossils. Biology, Taphonomy and Applications.* Chapman & Hall, London, 361 pp.

Bromley, R.G. and **Ekdale, A.A.** (1986) Composite ichnofabrics and tiering of burrows. *Geol. Mag.*, **123**, 59–65.

Bromley, R.G. and **Ekdale, A.A.** (1987) Mass transport in European Cretaceous chalk; fabric criteria for its recognition. *Sedimentology*, **34**, 1079–1092.

Campbell, S.J.D. and **Gravdal, N.** (1995) The prediction of high porosity chalks in the East Hod Field. *Petrol. Geosci.*, **1**, 57–69.

Cartwright, J.A. (1989) The kinematics of inversion in the Danish Central Graben. In: Inversion tectonics (Eds M.A. Cooper and G.D. Williams), *Geol. Soc. London Spec. Publ.*, **44**, 153–176.

Clayton, C.J. (1986) The chemical environment of flint formation in Upper Cretaceous chalks. In: *The Scientific Study of Flint and Chert* (Eds G. Sieveking and M.B. Hart), pp. 43–54. Cambridge University Press, Cambridge.

Cooper, C., Forristall, G.Z. and **Joyce, T.M.** (1990) Velocity and hydrographic structure of two gulf of Mexico warm-core rings. *J. Geophys. Res.*, **95**, 1663–1679.

Crabtree, B., Fritsen, A., Mandzuich, K., Moe, A., Rasmussen, F., Siemers, T., Soiland, G. and **Tirsgaard, H.** (1996) Description and Classification of Chalks North Sea Central Graben. Joint Chalk Research Phase IV, *Norw. Petrol. Dir. Spec. Publ.*

Damholt, T. and **Surlyk, F.** (2004) Laminated-bioturbated cycles in Maastrichtian chalk of the North Sea: oxygenation fluctuations within the Milankovitch frequency band. *Sedimentology*, **51**, 1323–1342.

D'Heur, M. (1986) The Norwegian Chalk Fields. In: *Habitat of Hydrocarbons on the Norwegian Continental Shelf* (Ed. A.M. Spencer), 77–89. Graham and Trotman, London.

Distefano, M., Feazel, C.T., Park, R.K., Peterson, R.M. and **Wilson, K.M.** (1980) Geological Ekofisk task force, Volume IIa, lithostratigraphy, sedimentology and diagenesis. Internal Phillips Petroleum Company Report.

Dunham, R.J. (1962) Classification of carbonate rocks according to depositional texture. In: *Classification of Carbonate Rocks* (Ed. W.E. Ham), *AAPG Mem.*, **1**, 108–121.

Ekdale, A.A. and **Bromley, R.G.** (1991) Analysis of composite ichnofabrics: an example in uppermost Cretaceous chalk of Denmark. *Palaios*, **6**, 232–249.

Esmerode, E.V. and **Surlyk, F.** (2009) Origin of channel systems in the Upper Cretaceous Chalk Group of the Paris Basin. *Mar. Petrol. Geol.*, **26**, 1338–1349.

Esmerode, E.V., Lykke-Andersen, H. and **Surlyk, F.** (2007) Ridge and valley systems in the Upper Cretaceous chalk of the Danish Basin: contourites in an epeiric sea. In: *Economic and Palaeoceanographic Importance of Contourites* (Eds A.R. Viana and M. Rebesco), *Geol. Soc. London Spec. Publ.*, **276**, 265–282.

Esmerode, E.V., Andersen, H.L. and **Surlyk, F.** (2008) Interaction between bottom currents and slope failure in the late Cretaceous of the southern Danish Central Graben, North Sea. *J. Geol. Soc. London*, **165**, 55–72.

Evans, D.J. and **Hopson, P.M.** (2000) The seismic expression of synsedimentary channel features within the Chalk of southern England. *Proc. Geol. Assoc.* **111**, 219–230.

Evans, D.J., Hopson, P.M., Kirby, G.A. and **Bristow, C.R.** (2003) The development and seismic expression of synsedimentary features within the Chalk of southern England. *J. Geol. Soc. London*, **160**, 797–813.

Fabricius, I.L., Røgen, B. and **Gommesen, L.** (2007) How depositional texture and diagenesis control petrophysical and elastic properties of samples from five North Sea chalk fields. *Petroleum Geoscience*, **13**, 81–95.

Farmer, C.L. and **Barkved, O.I.** (1999) Influence of syndepositional faulting on thickness variations in chalk reservoirs – Valhall and Hod fields. In: *Petroleum Geology of Northwest Europe: Proceedings of the 5th Conference* (Eds A.J. Fleet and S.A.R. Boldy), pp. 949–957. Geological Society, London.

Faugères, J.–C., Mézerais, M.L. and **Stow, D.A.V.** (1993) Contourite drift types and their distribution in the North and South Atlantic Ocean basins. *Sed. Geol.*, **82**, 189–203.

Faugères, J.-C., **Stow, D.A.V.**, **Imbert, P.** and **Viana, A.R.** (1999) Seismic features diagnostic of contourite drifts. *Mar. Geol.*, **162**, 1–38.

Fluteau, F., **Ramstein, G.**, **Besse, J.**, **Guiraud, R.** and **Masse, J.P.** (2007) Impacts of palaeogeography and sea level changes on mid-Cretaceous climate. *Palaeogeogr. Palaeoclimatol. Palaeoecol.* **247**, 357–381.

Gonthier, E., **Faugères, J.-C.** and **Stow, D.A.V.** (1984) Contourite facies of the Faro Drift, Gulf of Cadiz. In: *Fine-Grained Sediments: Deep-Water Processes and Facies* (Eds D.A.V. Stow and D.J.W. Piper), *Geol. Soc. London Spec. Publ.*, **15**, 275–292.

Gowers, M.B., **Holtar, E.** and **Swensson, E.** (1993) The structure of the Norwegian Central Trough (Central Graben area). In: *Petroleum Geology of the Northwest Europe: Proceedings of the 4th Conference* (Ed. J.R. Parker), pp. 1245–1254. Geological Society, London.

Gradstein, F.M., **Ogg, J.G.** and **Smith, A.G.** (2005) *A Geologic Time Scale 2004*. Cambridge University Press, Cambridge.

Hampton, M.J., **Bailey, H.W.** and **Jones, A.D.** (2010) A Holostratigraphic Approach to the Chalk of the North Sea Eldfisk Field, Norway. In: *Petroleum Geology: From Mature Basins to New Frontiers – Proceedings of the 7th Petroleum Geology Conference* (Eds B.A. Vining and S.C. Pickering), *Petroleum Geology Conference Series* **7**, 473–492. Geological Society, London.

Hardenbol, J., **Thierry, J.**, **Farley, M.B.**, **Jacquin, T.**, **DeGraciansky, P.-C.** and **Vail, P.** (1998) Cretaceous sequence chronostratigraphy. In: *Mesozoic and Cenozoic sequence chronostratigraphic framework of European basins* (Eds **P.**-C. De Graciansky, J. Hardenbol, T. Jacquin and P.R. Vail), *SEPM Spec. Publ.* **60**, Chart 4.

Hardman, R.P.F. (1982) Chalk reservoirs of the North Sea. *Bull. Geol. Soc. Denmark*, **30**, 119–137.

Hardman, R.P.F. and **Kennedy, W.J.** (1980) Chalk reservoirs of the Hod Fields, Norway. In: *The Sedimentation of the North Sea Reservoir Rocks. Norsk Petroleums forening*, Elsevier, Amsterdam.

Hart, M.B. (2007) Late Cretaceous climate and foraminiferal distribution. In: *Deep-Time Perspectives on Climate Change: Marrying the Signal from Computer Models and Biological Proxies* (Eds M. Williams, A.M. Haywood, F.J., Gregory, F.J. and D.N. Schmidt) *Geol. Soc. London Spec. Publ.*, **240**, 235–250.

Hatton, I.R. (1986) Geometry of allochthonous Chalk Group members, Central Trough, North Sea. *Mar. Petrol. Geol.*, **3**, 79–99.

Hollister, C.D. and **Heezen, B.C.** (1972) Geological effects of ocean bottom currents: Western North Atlantic. In: *Studies in Physical Oceanography* 2 (Ed. A.L. Gordon), pp. 37–66. Gordon and Breach, New York.

Jarvis, I. (1980) The initiation of phosphatic chalk sedimentation-the Senonian (Cretaceous) of the AngleParis Basin. In: *Marine Phosphorites-Geochemistry, Occurrence, Genesis* (Ed. Y. K. Bentor), *Spec. Publs Soc. Econ. Paleont. Miner.*, **29**, 167–192.

Kennedy, W.J. (1980) Aspects of chalk sedimentation in the southern Norwegian offshore. In: *The Sedimentation of the North Sea Reservoir Rocks*. Geilo, pp 11–14. Norwegian Petroleum Society, **9**. 29 pp.

Kennedy, W.J. (1987a) Late Cretaceous and Early Palaeocene Chalk Group sedimentation in the greater Ekofisk area. North Sea Central Graben. *Bull. Centr. Rech. Explor.-Prod. Elf-Acquitaine*, **11**, 91–126.

Kennedy, W.J. (1987b) Sedimentology of Late Cretaceous-Palaeocene chalk reservoir, North Sea Central Graben. In: *Petroleum Geology of North West Europe* (Eds J. Brooks and K.W. Glennie), pp. 469–481. Graham & Trotman, London.

Knott, S.D., **Burchell, M.T.**, **Jolley, E.J.** and **Fraser, A.J.** (1993) Mesozoic to Cenozoic plate reconstruction of the North Atlantic hydrocarbon plays of the Atlantic margins. In: *Petroleum Geology of Northwest Europe: Proceedings of the 4th Conference* (Ed. J.R. Parker), pp. 953–974, Geological Society, London.

Kominz, M.A., **Browning, J.V.**, **Miller, K.G.**, **Sugarman, P.J.** and **Mizintseva, S.** (2008) Late Cretaceous to Miocene sea level estimates the New Jersey and Delaware coastal plain coreholes: an error analysis. *Basin Res.*, **20**, 211–226.

Lasseur, E. (2007) *La Craie du Bassin de Paris (Cénomanien-Campanien, Crétacé supérieur). Sédimentologie de Faciès, Stratigraphie Séquentielle et Géométrie 3D*. PhD Thesis, Univ. of Rennes 1, 435 pp.

Lasseur, E., **Guillocheau, F.**, **Robin, C.**, **Hanot, F.**, **Vaslet, D.**, **Coueffe, R.** and **Neraudeau, D.** (2009) A Relative Water-Depth Model for the Normandy Chalk (Cenomanian–Middle Coniacian, Paris Basin, France) Based on Facies Patterns of Metre-Scale Cycles. *Sed. Geol.*, **213**, 1–26.

Lottaroli, F. and **Catrullo, D.** (2000) The Calcareous Nannofossil Biostratigraphic Framework of the Late Maastrichtian–Danian North Sea Chalk. *Mar. Micropaleontl.*, **39**, 239–263.

Lüning, S., **Marzouk, A.M.** and **Kuss, J.** (1998) Late Maastrichtian litho- and ecocycles from hemipelagic deposits of eastern Sinai, Egypt. *J. Afr. Earth Sci.*, **27**, 363–395.

Lykke-Andersen, H. and **Surlyk, F.** (2004) The cretaceous-palaeogene boundary at Stevns Klint, Denmark: inversion tectonics or sea floor topography? *J. Geol. Soc. London*, **161**, 343–352.

Maliva, R.G. and **Dickson, J.A.D.** (1992) Microfacies and diagenetic controls of porosity in Cretaceous/Tertiary chalks, Eldfisk Field, Norwegian North Sea. *AAPG Bull.*, **76**, 1825–1838.

Martsolf, R. (1994) Micropalaeontological report of the 2/4-12 Well, Ekofisk Field, Norwegian North Sea, Internal Report, Phillips Petroleum Company.

Miller, K.G., **Sugarman, P.J.**, **Browning, J.V.**, **Kominz, M.A.**, **Olsson, R.K.**, **Feigenson, M.D.** and **Hernandez, J.C.** (2004) Upper Cretaceous sequences and sea level history, New Jersey coastal plain. *Geol. Soc. Am. Bull.*, **116**, 368–393.

Miller, K.G., **Kominz, M.A.**, **Browning, J.V.**, **Wright, J.D.**, **Mountain, G.S.**, **Katz, M.E.**, **Sugarman, P.J.**, **Cramer, B.S.**, **Christie-Blick, N.** and **Pekar, S.F.** (2005) The Phanerozoic Record of Global Sea-Level Change. *Science*, **310**, 1293–1298.

Niebuhr, B. and **Prokoph, A.** (1997) Periodic-cyclic and chaotic successions of the Upper Cretaceous (Cenomanian to Campanian) pelagic sediments in the North German Basin, *Cretaceous Res.*, **18**, 731–750.

Niebuhr, B., **Wiese, F.** and **Wilmsen, F.** (2001). The cored Konrad 101 borehole (Cenomanian–Lower Coniacian, Lower Saxony): calibration of surface and subsurface log data for the lower Upper Cretaceous of northern Germany, *Cretaceous Res.*, **22**, 643–676.

Nygaard, E., **Lieberkind, K.** and **Frykman, P.** (1983) Sedimentology and reservoir parameters of the Chalk Group in the Danish Central Graben. *Geologie Mijnbouw.*, **62**, 177–190.

Oakman, C.D. and **Partington, M.A.** (1998) Cretaceous. In: *Petroleum Geology of the North Sea*, (Ed. K.W. Glennie), pp 294–349. Blackwell Scientific Publications, Oxford.

Ogg, J.G., **Ogg, G.** and **Gradstein, F.M.** (2008) *The Concise Geologic Time Scale*. Cambridge University Press, Cambridge, New York, Melbourne.

Pearce, M.A., **Jarvis, I.** & **Bruce, A.T.** (2009) The Cenomanian–Turonian Boundary Event, OAE2 and Palaeoenvironmental Change in Epicontinental Seas: New Insights from the Dinocyst and Geochemical Records. *Palaeogeogr. Palaeoclimatol. Palaeoecol*, **280**, 207–234.

Perch-Nielsen, K., **Ulleberg, K.** and **Eversen, J.A.** (1979) Comments on the terminal Cretaceous event. In: *Cretaceous–Tertiary Boundary Events Symposium*, **2**, pp. 106–111. Univ. of Copenhagen, Copenhagen.

Pucéat, E., **Lécuyer, C.**, **Sheppard, S.M.F.**, **Dromart, G.**, **Réboulet, S.** and **Grandjean, P.** (2003) Thermal evolution of Cretaceous Tethyan marine waters inferred from oxygen isotope composition of fish tooth enamels. *Paleoceanography*, **18**, 1–11.

Quine, M. and **Bosence, D.** (1991) Stratal geometries, facies and sea floor erosion in Upper Cretaceous chalk, Normandy, France. *Sedimentology*, **38**, 1113–1152.

Rebesco, M. and **Camerlenghi, A.** (2008) *Contourites. Developments in Sedimentology*, **60**. Elsevier, Amsterdam, 663 pp.

Røgen, B., **Gommesen, L.** and **Fabricius, I.L.** (1999) Chalk Rock Catalogue. In: *Joint Chalk Research Phase V: Chalk Rock Catalogue*. Norwegian Petroleum Directorate (NPD), Stavanger, 101 pp.

Scholle, P.A. (1977) Chalk diagenesis and its relation to petroleum exploration: Oil from chalks, a modern miracle? *AAPG Bull.*, **61**, 982–1009.

Scholle, P.A., **Albrechtsen, T.** and **Tirsgaard, H.** (1998) Formation and diagenesis of bedding cycles in uppermost Cretaceous chalks of the Dan Field, Danish North Sea. *Sedimentology*, **45**, 223–243.

Shanmugam, G. (2008) Deep-water bottom currents and their deposits. In: *Contourites* (Eds M. Rebesco and A. Camerlenghi). *Developments in Sedimentology*, **60**, pp. 60–81. Elsevier, Amsterdam.

Shepard, F.P., **Marshall, N.F.**, **McLoughlin, P.A.** and **Sullivan, G.G.** (1979) Currents in submarine canyons and other sea valleys. *AAPG Studies in Geology*, Vol. **8**, American Association Petroleum Geologists, Tulsa, 179 pp.

Sikora, P.J., **Bergen, J.A.** and **Farmer, C.L.** (1999) Chalk palaeoenvironments and depositional model: Valhall and Hod fields, southern Norwegian North Sea. In: *Biostratigraphy in Production and Development Geology* (Eds R.W. Jones and M.D. Simmons), *Geol. Soc. London Spec. Publ.*, **152**, 113–137.

Skirius, C., **Nissen, S.**, **Haskell, N.**, **Marfurt, K.**, **Hadley, S.**, **Ternes, D.**, **Michel, K.**, **Reglar, I.**, **D'Amico, D.**, **Deliencourt, F.**, **Romero, T.**, **D'Angelo, R.** and **Brown, B.** (1999) 3-D seismic attributes applied to carbonates. *Leading Edge*, **3**, 384–393.

Stage, M. (1999) Signal analysis of cyclicity in Maastrichtian pelagic chalks from the Danish North Sea, *Earth Planet. Sci. Lett.*, **173**, 75–90.

Stage, M. (2001) Recognition of cyclicity in the petrophysical properties of a Maastrichtian pelagic chalk oil field reservoir from the Danish North Sea, *AAPG Bull.*, **85**, 2003–15.

Stow, D.A.V., **Faugères, J.–C.**, **Howe, J.A.**, **Pudsey, C.J.** and **Viana, A.R.** (2002) Bottom currents, contourites and deep-sea sediment drifts: current state-of-the-art. In: *Deep-Water Contourites: Modern Drift and Ancient Series, Seismic and Sedimentary Characteristics* (Eds D.A.V. Stow, C.J. Pudsey, J.A. Howe, J.–C. Faugères and A.R. Viana), *Geol. Soc. London Mem.*, **22**, 7–20.

Surlyk, F. and **Lykke-Andersen, H.** (2007) Contourite drifts, moats and channels in the Late Cretaceous chalk of the Danish Basin. *Sedimentology*, **54**, 405–422.

Surlyk, F., **Jensen, S.K.** and **Engkilde, M.** (2008) Deep channels in the Chalk Group of the German North Sea Sector: results of long-lived constructional and erosional bottom currents. *AAPG Bull.*, **92**, 1565–1586.

Surlyk, F., **Dons, T.**, **Clausen, C.K.** and **Highham, J.** (2003) Upper Cretaceous. In: *The Millenium Atlas, Petroleum Geology of the Central and Northern North Sea* (Eds D. Evans, C. Graham, A. Armour and P. Bathurst), pp. 213–233. Geological Society, London.

Taylor, S.R. and **Lapré, J.F.** (1987) North Sea chalk diagenesis: its effect on reservoir location and properties. In: *Petroleum Geology of North West Europe* (Eds J. Brooks and K.W. Glennie), pp 483–495. Graham & Trotman, London.

Van der Molen, A. (2004) Sedimentary development, seismic stratigraphy and burial compaction of the Chalk Group in the Netherlands North Sea area. *Geol. Ultraiect.*, **248**, 1–175.

Van der Molen, A.S., **Dudok van Heel, H.W.** and **Wong, T.E.** (2005) The influence of tectonic regime on chalk deposition: examples of the sedimentary development and 3D-seismic stratigraphy of the Chalk Group in the netherlands offshore area. *Basin Res.*, **17**, 63–81.

Vejbæk, O.V. and **Andersen, C.** (2002) Post mid-Cretaceous inversion tectonics in the Danish Central Graben-regionally synchronous tectonic events? *Bull. Geol. Soc. Denmark*, **49**, 129–144.

Viana, A.R. and **Rebesco, M.** (2007) *Economic and Palaeoceanographic Significance of Contourite Deposits*. Geological Society, London, 350 pp.

Viana, A.R., **Faugères, J.–C.**, **Kowsman, R. O.**, **Lima, J A M.**, **Caddah, L.F.G.**, **Rizzo, J. G.** (1998) Hydrology, morphology and sedimentology of the Campos continental margin, offshore Brazil. *Sed. Geol.*, **115**, 133–158.

Watts, N.L., **Lapré , J.F.**, **Van Schijndel-Goester, F.S.** and **Ford, A.** (1980) Upper Cretaceous and Lower Tertiary chalks of the Albuskjell area, North Sea: deposition in a base of-slope environment. *Geology*, **8**, 217–221.

Young, C.R. and **Yang Logan, L.C.** (1994) Biostratigraphy and palaeoenvironmental interpretation of the Ekofisk 2/7-14 well, Norwegian North Sea, Phillips Petroleum Company.

Ziegler, P.A. (1990) Tectonic and palaeogeographic development of the North Sea rift system. In: *Tectonic Evolution of the North Sea Rifts* (Eds D.J. Blundell and A.D. Gibbs), pp. 1–36. Clarendon Press, Oxford.

Int. Assoc. Sedimentol. Spec. Publ. (2014) **46**, 587–608

MassFLOW-3D™ as a simulation tool for turbidity currents: some preliminary results

RICCARDO BASANI*, MICHAL JANOCKO†, MATTHIEU J.B. CARTIGNY‡, ERNST W.M. HANSEN* and JORIS T. EGGENHUISEN‡

* *Complex Flow Design A.S., P.O. Box 1248, 7462 Trondheim, Norway*
† *Department of Earth Science, University of Bergen, 5007 Bergen, Norway*
‡ *Department of Earth Sciences, Utrecht University, PO Box 80021, 3508 TA Utrecht, Netherlands*

ABSTRACT

Turbidity currents are the most important mechanism for the dispersal and deposition of sand in the deep-sea setting and thus the main phenomenon leading to the formation of oil and gas reservoirs in deep water deposits. The flow characteristics of turbidity currents are difficult to observe and study from the modern environment and their experimental approximations in the laboratory are typically limited by scaling issues, unrealistic flume geometries and short durations. Computational fluid dynamic (CFD) analysis, realised as numerical simulations, has been developed to fill the gap between the small and large scale, integrating data from theory, nature and experiments. CFD can also shed light on flow parameters which are so far impossible to deduce from experimental and field studies, such as detailed density and turbulent kinematic energy distributions. The deterministic process modelling CFD software MassFLOW-3D™ has been developed and used successfully to construct a three-dimensional model for the simulation of turbidity currents. All principal hydraulic properties of the flow (e.g. velocity, density, sediment concentration, apparent viscosity, turbulence intensity and bottom shear stress) and its responses to topography can be monitored continuously in three dimensions over the whole duration of the turbidity current. In this paper, comparisons made between the numerical output of MassFLOW-3D™ and the physical experiments are presented. In addition, the code is used to model the spatial characteristics, velocity structure and deposits of high-density turbidity currents and the flow dynamics of low-density turbidity currents in a sinuous channel. The numerical simulations show close fit to experimental sandy turbidity current dynamics for flows with sediment concentrations up to 27%. However, despite this initial success, on-going customisation and validation of these models, together with implementation of improved subroutines aimed at sediment transport and deposition, is essential in improving the computational code and our understanding of the natural phenomena.

Keywords: computational fluid dynamics, CFD, deterministic process simulations, turbidity currents, sinuous channels.

INTRODUCTION

Sandstone successions of turbiditic origin have long been recognised as important hydrocarbon reservoirs on the Norwegian Continental Shelf (e.g. Hastings, 1986). Despite the fact that these deposits are mainly high-risk targets (complex geometries, uncertain stratigraphic closures and facies heterogeneity) and occur predominantly in deep-water offshore areas where the development poses practical problems, exploration in the near future is bound to move further seaward as many of the hydrocarbon fields on the Norwegian Continental Shelf have now passed their production peaks and face total depletion within the first half of the 21st century (Norwegian Petroleum Directorate, 2009). Similar exploration strategies have already been established in major petroleum provinces around the world, including offshore West Africa, the Gulf of Mexico and offshore

From Depositional Systems to Sedimentary Successions on the Norwegian Continental Margin, First Edition. Edited by A. W. Martinius, R. Ravnås, J. A. Howell, R. J. Steel, and J. P. Wonham.

Brazil. Accordingly, with exploration moving ever further offshore, there is a growing need for deeper understanding of turbidite successions and their depositional processes. Early 3D numerical analyses on Norwegian hydrocarbon fields were undertaken using a previous version of the software described herein (Heimsund, 2007). This approach proved to be a useful tool for gaining a better understanding of the geometry of these turbidite reservoirs. Whilst the conceptual knowledge developed from the analysis of seismic and outcrop data has seen remarkable progress in recent years (e.g. Duller *et al.*, 2010; Dykstra & Kneller, 2009; Hadler-Jacobsen *et al.*, 2007; Mulder & Alexander, 2001; Schwab *et al.*, 2007), the fundamental science of sediment transport and deposition based on controlled experiments and direct observations of natural flows has been hampered by scaling issues and practical problems in measuring the most vital parameters (see Meiburg & Kneller, 2010). Currently, it seems that process-based numerical simulations are the only way to bridge the gap between laboratory experiments and large-scale natural phenomena and, most importantly, to concurrently calculate all flow parameters (velocity, concentration, grain size, apparent viscosity, turbulence intensity and bottom shear stress) for the entire flow field during the whole duration of the flow, which is impossible in physical experiments. However, simulating the flow behaviour, erosion and deposition of turbidity currents in a realistic manner without rendering many parameters negligible or constant has proven highly challenging. More specifically, problems arose in simulating sediment particle transport in the inner region of turbidity current where the concentrations are greatest and particle interactions play an important role. In high-density turbidity currents the boundary layer right above the bed may even show transportation behaviour similar to granular flows (Haughton *et al.*, 2009; Postma *et al.*, 1988); in which case the previously established bed-load models for turbid flows (e.g. Huang *et al.*, 2005; Schmeeckle & Nelson, 2003) cannot be applied. Further difficulties emerge from the fact that most of the established physical principals in fluid dynamics have been designed originally for much larger particle sizes than those which occur in turbidity currents. Numerical models thus need to be improved and correlated with physical experiments to account more realistically for the interaction between the turbidity current and the erodible substrate.

A CFD code that accounts for erosion, deposition and the transport of sediment in high-density and low-density turbidity currents has been implemented recently in MassFLOW-3D™ (Basani & Hansen, 2009), which is a customised version of the commercial software Flow-3D™. To test the capabilities of this computational code, a comprehensive study has been undertaken with a focus on flow dynamics and sedimentation in submarine channels and terminal lobes. The main objective of the study is to carefully compare the output of MassFLOW-3D™ with the results of laboratory experiments and to define empirically some of the theoretically-uncertain input parameters in the simulation model. The aim of the paper is also to review the shortcomings of the computational code and to discuss where it could be improved in the near future. Once the physical-numerical correlation was demonstrated to be satisfactory, the aim has been to upscale the simulation model to approximate natural phenomena and determine the influence of individual flow parameters on the properties of both the flow and subsequent deposits. The results presented in this article are a preliminary outcome of the comparative study and also address some of the issues concerning the calibration of the numerical model to laboratory experiments. The first results also shed light on a number of controversial issues regarding sediment transport in high-density turbidity currents and flow dynamics of turbidity currents in sinuous channels. The content of the paper is divided into an introduction to MassFLOW-3D™ followed by three sections each describing a separate case study. The first two cases focus on velocity and deposition in experimental high-density turbidity currents by comparing physical experiments with numerical simulations. In the last section, the MassFLOW-3D™ numerical code is applied to a natural-scale sinuous channel to explore flow behaviour around channel bends.

NUMERICAL APPROACH

CFD is a tool for numerical solution of the physical equations describing fluid flow and sediment transport. The method has been widely applied in the engineering branches of fluid mechanics but has thus far been little used in sedimentological research and reservoir studies in a fully 3D numerical scheme. The first CFD application to sediment gravity flows was pioneered by Harbaugh (1970) and, subsequently, by Tetzlaff & Harbaugh (1989). After that, a number of 2D and depth averaged 3D software packages have been developed in the last three decades, to calibrate numerical models to physical models (Groenenberg *et al.*, 2009), natural events (Dan *et al.*, 2007; Pirmez & Imran, 2003; Salles *et al.*, 2008), geomorphic features (Cartigny, 2012; Kostic, 2011) or outcrop observations (Groenenberg *et al.*, 2010). Only recently, fully 3D codes capable of simulating sediment transportation, deposition and erosion have been developed and applied (Aas *et al.*, 2010a; Aas *et al.*, 2010b). The deterministic process modelling software, MassFLOW-3D™, has been developed and used successfully to construct a three-dimensional model for the simulation of turbidity currents. All principal hydraulic properties of the flow (velocity, density, sediment concentration, apparent viscosity, turbulence intensity and bottom shear stress) and its responses to topography can be monitored continuously in three dimensions over the whole duration of the turbidity current. The present model has been calibrated and verified against laboratory flows and can readily be up-scaled to natural conditions (Aas *et al.*, 2010a; Aas *et al.*, 2010b).

The fluid motion of a sediment gravity flow in MassFLOW-3D™ is simulated by solving the fully 3D transient Reynolds-averaged Navier-Stokes equations (RANS) by a finite-volume-finite-differences method in a fixed Eulerian rectangular grid. The suspension of mono or poly-disperse, non-cohesive sediment is treated as a continuous phase and its variable spatial volumetric concentration is calculated. The effects of flow turbulence are simulated using the renormalisation-group (RNG) model (Appendix I), in which the equation constants are derived explicitly. This model is thought to be superior to the commonly used k– model in that it describes more accurately low intensity turbulence in liquid flows. The drift velocity (velocity needed to compute the transport of sediments due to drift relative to the fluid phase) of suspended particles is approximated by a non-linear multi-phase model, allowing the deposition of larger and faster drifting sediment to be simulated more accurately. Particle interactions in suspension (hindered settling, particle collisions and interlocking of grains) are approximated by the Richardson-Zaki correlation (Richardson & Zaki, 1954). Erosion, settling and lifting of sediment particles are calculated in terms of tensors superimposed on the sediment advection with the turbulent fluid. Entrainment (re-suspension, erosion) of sediment from the bed into the flow is calculated with the Mastbergen & Van den Berg model (Mastbergen & Van Den Berg, 2003), in which the critical Shields parameter is predicted using the Shield-Rouse equation (Guo, 2002). The erosion model also accounts for particle armouring by using the Egiazaroff formula (Kleinhans, 2002). The rolling and saltating motion of larger sediment near the packed bed interface (bed-load transport) is predicted according to the Meyer-Peter & Müller (1948) sediment transport equation. The governing equations describing these separate modules can be found in the MassFLOW-3D™ report (Basani & Hansen, 2009) and in Appendix I of the present paper.

FLOW DYNAMICS AND DEPOSITS OF TURBIDITY CURRENTS: SELECTED CASE STUDIES

Case study I: Flow structure of high-density turbidity currents

In this section, a comparison is made between the flow structures in our numerical modelling and physical experiments on high-density turbidity currents in a flume. The comparison is focused on the behaviour of the head of a high-density turbidity current (Case Study I-a, Table 1A) and its steady state velocity profile in the body of the turbidity current (Case Study I-b and c, Table 1B).

Laboratory set-up

The experiments were conducted in the flume facilities of the Department of Earth Sciences at Utrecht University. The set-up consisted of a straight, 4 m long inclined channel connected to a drainage basin. The width and depth of the channel was fixed to 0.07 m and 0.5 m, respectively (Fig. 1). To prevent major water entrainment at the inlet, an additional inlet piece was mounted (Fig. 1). A more thorough description of the experimental set-up can be found in Cartigny (2012). In a first run (Case Study I-a), a mixture of fine sand (d_{50} =160 µm) and tap water, prepared in a 1.1 m^3 mixing tank, was pumped through an electromagnetic discharge meter (Krohne Optiflux 2300) and released into the flume. The inlet gate through which the flow was released was 0.07 m high and

Table 1. (A) Case study I-a. Summary of the performed CFD simulations of Run 15 in the laboratory experimental study. (B) Case Study I-b-c. Summary of the performed CFD simulations of Run 36 in the laboratory experimental study; one grain size. The first five rows refer to a mono disperse mixture d50 = 50 µm; the last four rows to a bi-disperse mixture d50 = 50 µm and d50 = 144 µm).

A					
	Bedload coefficient	Entrainment coefficient	Turbulence Mixing Length (m)	Angle of Repose (deg)	Roughness (mm)
Run15a	8	0.018	0.0025	25	0.165
Run15b	8	0.018	0.0025	25	0
Run15c	0	0.018	0.0025	25	0.165
Run15d	4	0.018	0.0015	25	0.165
Run15e	12	0.018	0.0015	25	0.165
Run15f	8	0.01	0.0025	25	0.165
Run15g	8	0.025	0.0015	25	0.165
Run15h	8	0.018	0.0015	25	0.165
Run15i	8	0.018	dynamically computed	25	0.165
Run15j	8	0.018	0.0025	12	0.165

B					
	Bedload coefficient	Entrainment coefficient	Turbulence mixing length (m)	Angle of repose (deg)	Roughness (mm)
36a	8	0.018	0.002	25	0.144
36b	5	0.018	0.0025	25	0.144
36c	1	0.018	0.003	25	0.144
36d	0.2	0.018	0.0025	25	0.144
36e	0.01	0.018	0.0025	25	0.144
36 2 grains a	8	0.018	0.001	25	0.144
36 2 grains b	8	0.018	0.0015	25	0.144
36 2 grains c	8	0.018	0.002	25	0.144
36 2 grains d	8	0.018	0.001	25	0.25

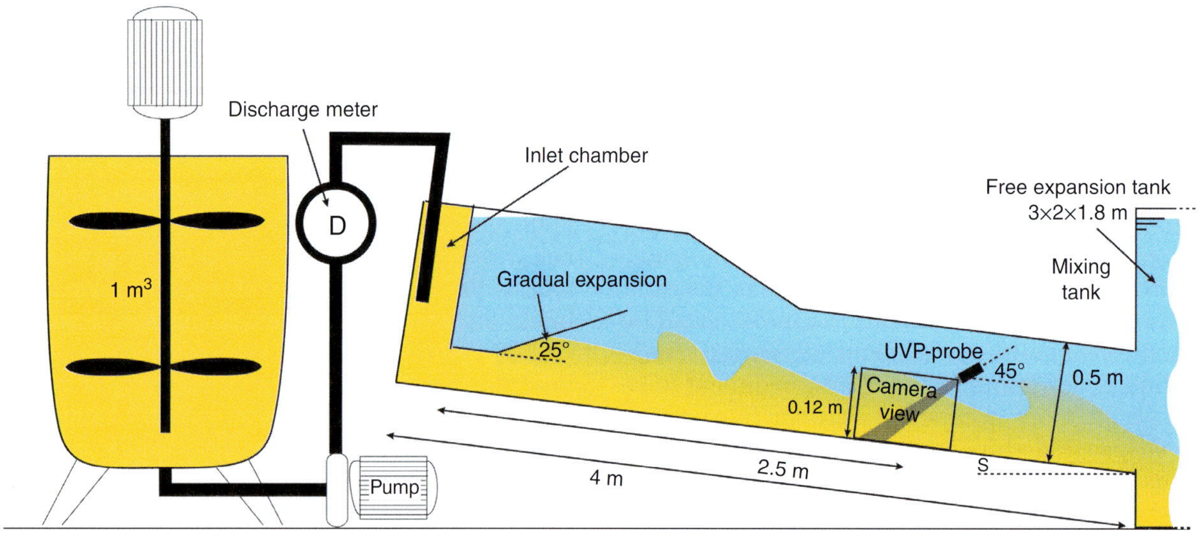

Fig. 1. Sketch for the experimental set-up used for the high concentration turbidity currents experiments.

the bed of was roughened by fixing 160 µm sand to the flume floor. In a second run (Case Study I-b), a mixture of coarse quartz silt (d_{50} = 50 µm) and water was utilised on a flume roughened with 144 µm sand. In a third run (Case Study I-c), another grain size (d_{50} = 144 µm) was added to the previous one (d_{50} = 50 µm), not altering the total sediment volume concentration, in a percentage of 50% fine sand and 50% coarse silt. Velocity data were collected by an Ultrasonic Velocity Profiler (MET-FLOW UVP-DOU MX) at 2.5 m from the inlet. The probe of the UVP was mounted 0.1 m above the bed, looking down at an angle of 45° (Case Ia) or 60° (Cases Ib-c) against the incoming flow (Fig. 1). At the same location a high-speed camera (Basler piA640-210 g m) captured the flow through the glass side-wall of the tank.

To produce a high-density turbidity current with a well-defined internal density stratification, the volumetric sediment concentration was set to 21% in Case Ia and 13% in Cases Ib-c. The inclination of the laboratory flume and the flow discharge were then adjusted to achieve the equilibrium slope profile of the turbidity current (i.e. slope at which the deposition rate balances erosion rate and no net-aggradation takes place. These conditions were achieved by setting the slope of the flume to 9.1° for Case Ia and 11.3° for Cases Ib-c. The volume discharge of all flows was 9.3 m^3/hr.

Numerical simulation set-up

The flume geometry was modelled and included in the simulation domain, which was set up using a fixed Eulerian rectangular computational mesh. The computational mesh consisted of 443,900 active cells. The resolution of the mesh was uniform in all three directions throughout the whole computational domain. The size of each cell was set to 0.0065 m × 0.012 m × 0.0036 m (XYZ).

The calibration of the numerical output with the laboratory experiment consisted of matching the velocity structure and spatial characteristics of the turbidity current. The four most influential parameters in the simulation model were evaluated in order to assess their effect on flow behaviour: the pick-up (entrainment or erosion of particles) coefficient, the bedload coefficient, the turbulent mixing length and the angle of repose. See Table 1A for the values of the parameters used in Case Study I-a and Table 1B for the Case Studies I-b and c.

Results and remarks

Two different regions of the turbidity current for the performed cases were analysed in detail: the head of the current for the first run and the body for the second and third runs.

Case Study I-a: Head of the flow – mono dispersed mixture

The head of the current, being the first part of the current to reach the control section (placed 2.5 m after the inlet, matching the location of the UVP probes in the physical model) can be recognised easily and analysed. Fig. 2 shows the velocity and visual data collected at the measurement location around the arrival of the head. The velocity probe location was fixed and the horizontal axis of Fig. 2 represents time as the turbidity currents passed by. The vertical axis represents the distance above the bed. In Fig. 3, some profiles of the head of the turbidity current, obtained from numerical simulations with varying settings of the key parameters, are displayed. All of the snapshots of Fig. 3 were taken 4.5 seconds after the release of the sediments from the mixing tank, which is the period of time that the flow travelled 2.3 m from the inlet to the control section. The primary shape of the sediment cloud is largely similar for all the simulations. However, the velocity profiles for the various cases are slightly different. The principal difference is the maximum velocity (red region in the snapshots) which was located at different heights within the turbidity current head, depending on the simulation. It must be kept in mind that, due to the highly turbulent and non-steady environment around the flow front, the maximum velocity region might shift abruptly from one region to another within the head of the flow; this might explain observed differences between simulations. Even so, the velocity maximum in the top half of the flow in Run d (Fig. 3B) seems unrealistic. Fig. 3C and E show a low velocity region above the flume floor, which might be attributed to deposition of a thin sediment layer by the head of the current under the respective settings. This deposition was not observed in the experiment. Notwithstanding these differences, for all the cases reported, the velocity of the cloud is in the range of 0.4 m s^{-1} to 0.7 m s^{-1} and the experimental front velocity was approximately 0.5 m s^{-1}.

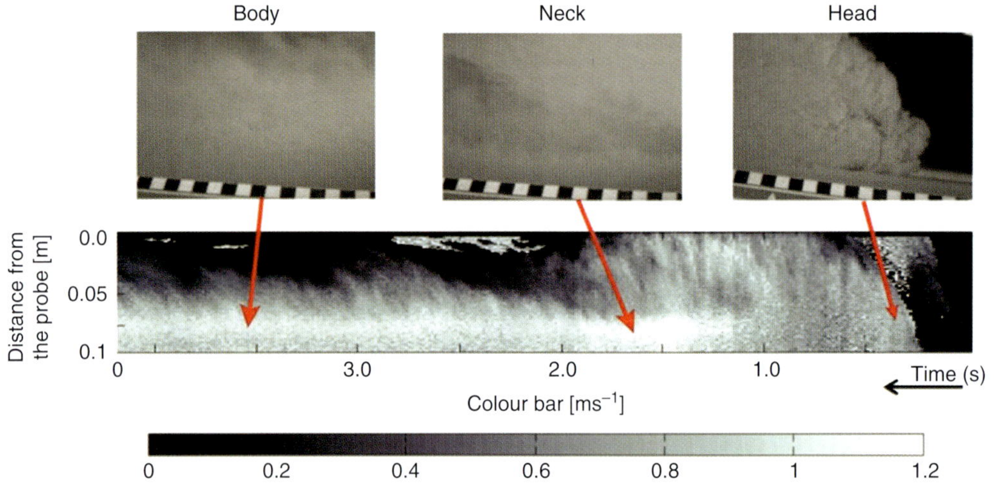

Fig. 2. Visual and velocity data of the passage of the head of the experimental turbidity current. (Top row) Three snapshots of the current taken with a high speed camera. The top row consists of three camera images of the high-density turbidity current. Black and white squares in the bottom of the image are 1 × 1 cm and the flow direction is from left to right. (Below) The velocity data collected by the UVP.

Figs 4A and B show a comparison of the velocities in the head of the high-density turbidity current in both the numerical and experimental set-up. Fig. 4C shows the sediment concentration for one of the performed simulations (Run 15 f in Table 1A). During the experiments, the UVP probe measured the velocities in the direction of the probe (45° with respect to the flume floor) at 128 bins spread over the full depth of the flow. One such cycle over 128 bins lasted for 14 ms. In order to compare directly the experimental velocity values with the numerical ones, the latter ones had to be calculated at the same angle as the UVP probes. The simulated velocity magnitudes in the head of the flow agree with the experimental results rather well (Fig. 4). Both datasets indicate a rather uniform velocity profile in the nose of the current. Furthermore, both datasets show that the maximum velocities occur in the back of the head (neck) where the body of the turbidity currents enters the head. Both simulations show that the incoming flows are decelerated in the head and start to detach from the bed. The shape and the velocity values for the numerical simulations match the observations from the laboratory. The highly-turbulent environment is reproduced well and the numerical parameters in the software only slightly change the solution of the performed runs (as shown in Fig. 3). This confirms the capability of the software to model, in a realistic way, a highly turbulent and transient environment at the front and head of a turbidity current.

Case Study I-b: Body of the flow – mono dispersed mixture

The body of the turbidity current follows the head of the flow (Meiburg & Kneller, 2010). While the head passes the control section in a limited time period, the body continues flowing for an extended time (approximately 1 minute in the experiments) and it is influenced by the presence of suspended and deposited sediments already in the system due to the passage of the head of the flow itself. Moreover, the body of the turbidity current experiences fluctuations in both concentration and velocity due to the instabilities on the interface between the current and the ambient fluid (Cartigny, 2012).

Time averaged velocity profiles were constructed to smoothen these instabilities and facilitate a comparison between the numerical and physical datasets.

In the physical model, a steady state velocity profile of the body of the high-density turbidity current was calculated by averaging the velocity profiles that were measured during the passing of the body under the UVP (Fig. 5). Fig. 6A shows the velocity profiles of the body of the turbidity currents for all the runs reported in Table 1B, measured 2.6 m from the inlet and averaged over a 53 s period, as well as the UVP profile from the physical experiment. In all the numerical simulations, the averaged velocity profiles have a peak of about 1.1 m s^{-1} approximately 2 cm above the flume floor. Above

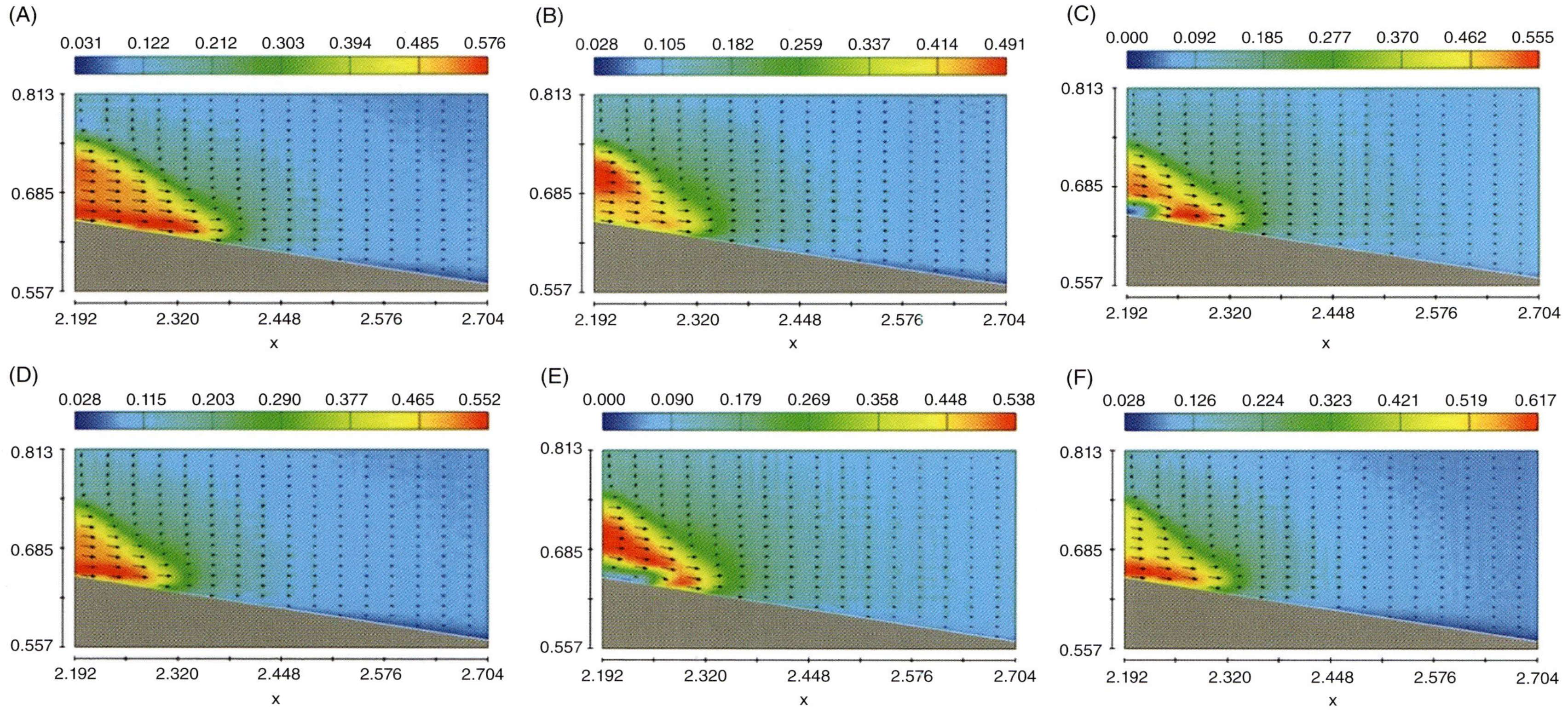

Fig. 3. 2D Shape of the head of the turbidity current and velocity values ($m\,s^{-1}$) measured after 4.5 s from the release from the mixing tank for (A) Run a of Table 1A. (B) Run d. (C) Run e. (D) Run f. (E) Run g. (F) Run h.

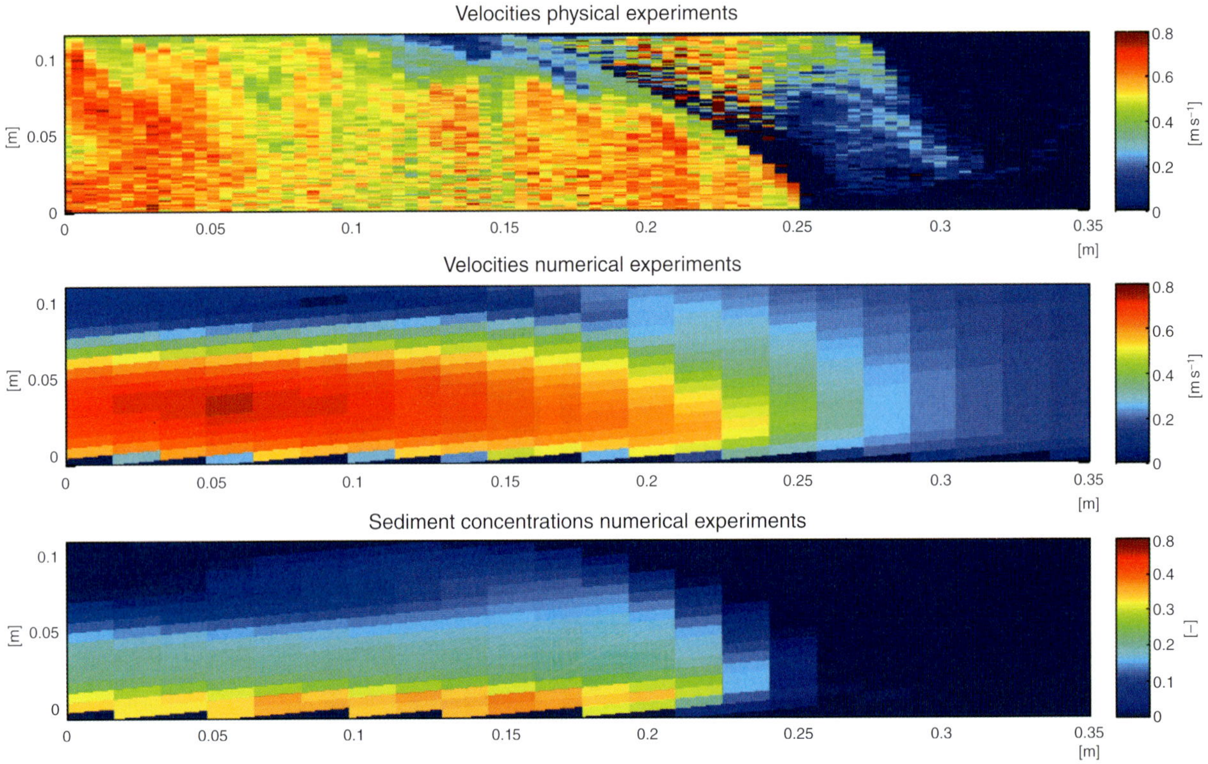

Fig. 4. Comparing numerical (Fig. 4B) and physical (Fig. 4A) experiments of the head of the turbidity current. Flow direction is from left to right. Magnitudes of the numerical 2D velocities are indicated by colours (see colour bar on the right). Horizontal time axis is replaced to indicate distance by multiplying the time values by an average head propagation velocity (0.3 m s^{-1}). Fig. 4C shows the sediment concentration at the head of the turbidity current, displaying how its shape matches well to the shape from the physical observation shown in Fig. 4A.

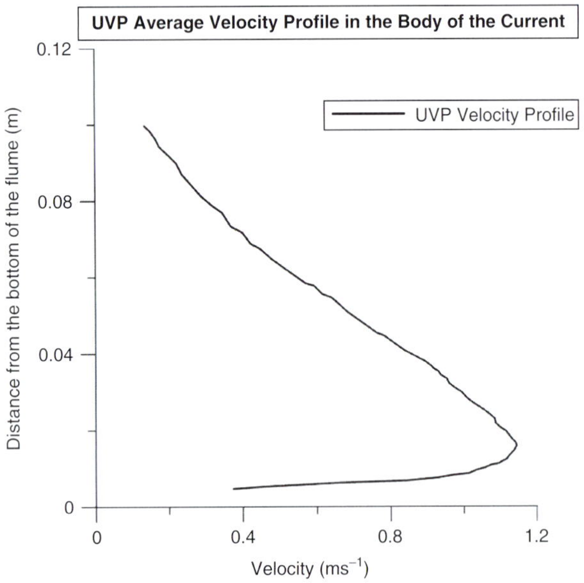

Fig. 5. The measured velocity profile of physical Case Ib.

that height, the velocity falls below 0.15 m s^{-1} at about 10 cm above the flume floor. For all numerical simulations the peak velocity is slightly lower than that observed in the physical experiments (approximately 3 cm s^{-1}, or 3%, slower). Furthermore, the highest velocity values in the physical experiments are located at a lower level than the numerical simulations (approximately 3 mm, approximately equal to the ΔZ of the computational mesh). The error between simulated cases and physical experiments is lower than 7% at all elevations. It is important to note that this misfit may be ascribed to either remaining uncertainties regarding parameters of the model or to experimental uncertainties. These experimental uncertainties could be due to a fluctuating flow discharge in the piping system, small imperfections in the measurement of the inlet size, unequal roughness of the sand glued to the flume floor, minor deviation of the grain size difference from the mean grain size (only a single grain size is used in the numerical simulations and not a distribution around d_{50}) and inaccuracies in

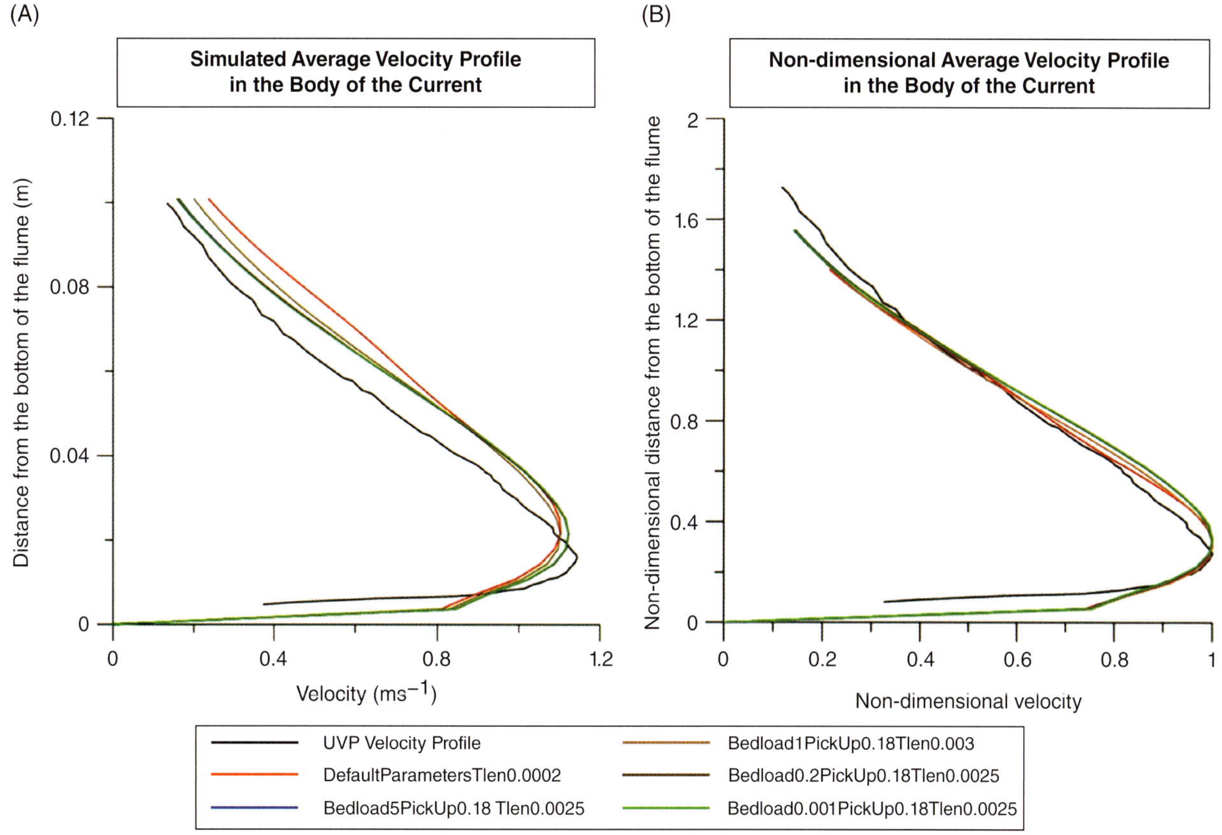

Fig. 6. (A) Velocity profiles for the body of a one-grain-size turbidity current, averaged on a 53 second period for the different numerical runs of Table 1B. (B) Non-dimensional velocity profiles for the body of a one-grain-size turbidity current, averaged on a 53 second period for the different numerical runs of Table 1B.

the UVP probe inclination. To illustrate that the remaining misfit is a problem of precision of the dimensionalisation of the boundary conditions and not due to the inability of the model to simulate a turbidity current, Fig. 6B shows a comparison between the UVP velocity profile and the simulation using the default parameter settings, where the velocity and spatial axes of both profiles have been non-dimensionalised with the maximum velocity and the $y_{0.5}$ (the height at which the velocity at the top of the flow decreases to 50% of the maximum velocity; see Kneller & Buckee, 2000). The observed fit in the shape of the velocity profile is near-exact.

Even without non-dimensionalisation, all the simulations, while changing some of the numerical parameters, show a similar shape and comparable velocity values. For all the cases listed above, changing the numerical parameters does affect the results; nevertheless, all the simulated velocity profiles match rather well to the observed values. This is a proof of the reliability and robustness of the software which, given the proper boundary conditions, is capable of reproducing the flow behaviour of turbidity currents with initial volume concentrations of 13% in a situation as described here.

Case Study I-c: Body of the flow – bi dispersed mixture

A crucial step for the numerical investigation is the analysis of a multi-species case, where a grain size mixture is present in the flow. This was performed by adding to the previously described runs another species with a grain size of 144 μm. Once again, the velocity profiles for the body of the current, averaged over a period of 53 seconds, are reported at the location of the UVP probes, 2.6 m from the inlet. The boundary conditions for this experiment are the same as the ones described above, as well as the total sediment concentration, with the exception that now there are 2 grain species in the mixture. The concentration is equal to Case Ib, reported above. The velocity profiles are shown in Fig. 7.

tank inlet and then gradually decreased towards the top of the water tank containing the expansion table. The smallest and largest cell had the dimensions of 0.025 m × 0.0075 m × 0.005 m and 0.1 m × 0.03 m × 0.118 m (XYZ format), respectively.

The simulation model requires input parameters generally unknown from the initial and final stages of laboratory experiments (e.g. turbulent mixing length, turbulent diffusion coefficient, sediment entrainment coefficient and bedload coefficient). These parameters were tuned empirically by comparing the depositional structures and spatial characteristics of the experimental and modelled turbidity current. The effect of some of these parameters on the sediment distribution within the flow and its deposits is discussed below.

Results and remarks

To compare the numerical model with the physical experiment, we focused on the expansion area directly downstream of the channel mouth. This area experienced profound deposition and is therefore well suited for testing the depositional capabilities of the simulation model.

The geometry of the resulting deposit was found to be highly sensitive to parameters included in the numerical models for sediment transport and fluid turbulence. In the sediment transport model, the pick-up (erosion) coefficient and the bedload coefficient proved to have a significant influence on the final run-out distance of the sediment load. The pick-up coefficient included in the entrainment model has a default value of 0.018 for a turbidity current travelling on a flat surface (Mastbergen & Van Den Berg, 2003). Increasing this coefficient yields a higher erosion rate, which is commonly translated into higher flow competence and longer run-out distance. The default value for this coefficient was found to yield the most satisfying results. The bedload coefficient, which determines the efficiency and competence of bedload transport, is assumed to have a default value of 8.0 (Meyer-Peter & Müller, 1948). As with the standard value for the pick-up coefficient, this default value was satisfactory.

Another parameter which has an impact on the deposit geometry is the turbulent mixing length. Turbulent mixing length (TLEN) is a physical quantity describing a characteristic distance a fluid parcel travels in a turbulent eddy before dispersing into the surrounding fluid (i.e. the length of travel

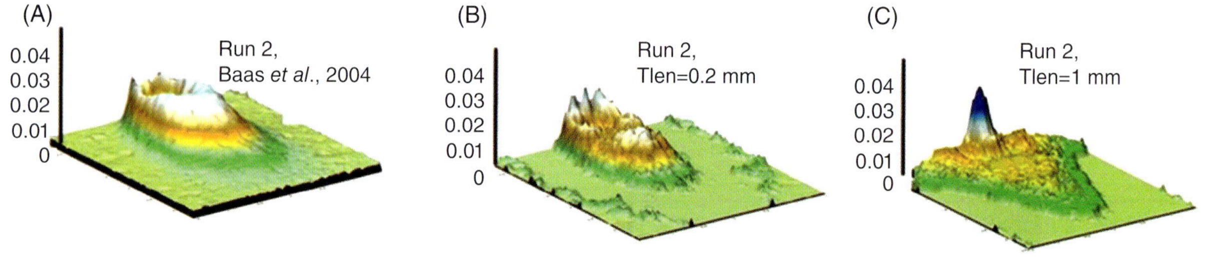

Fig. 9. Comparison between three-dimensional lobe shapes (dimensions expressed in m). (A) The three dimensional shape of the lobe for Run 2, obtained from Baas *et al.* (2004). (B) The three dimensional shape of the lobe for Run 2, obtained from MassFLOW-3DTM simulations with the turbulent mixing scale set to 0.15 mm. (C) The three dimensional shape of the lobe for Run 2, obtained from MassFLOW-3DTM simulations with the turbulent mixing scale set to 1 mm.

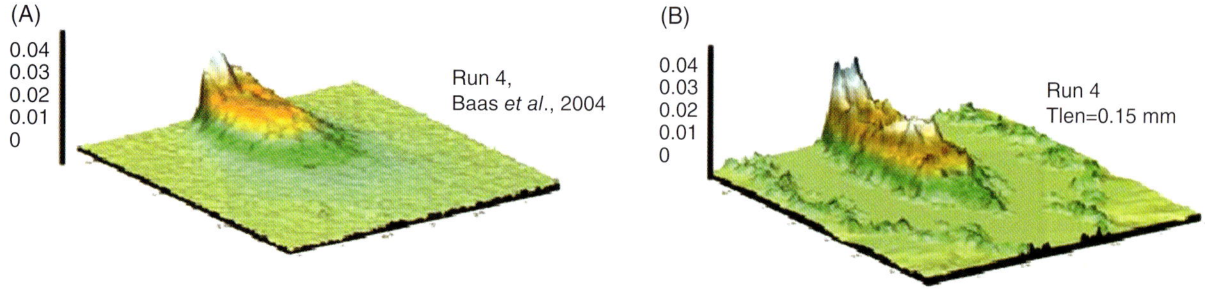

Fig. 10. Comparison between three-dimensional lobe shapes (dimensions expressed in m). A) The three dimensional shape of the lobe for Run 4, obtained from Baas *et al.* (2004). B) The three dimensional shape of the lobe for Run 2, obtained from MassFLOW-3DTM simulations with the turbulent mixing scale set to 0.15 mm.

of a turbulent eddy). This parameter is used by the turbulence model to limit the turbulent dissipation so that the turbulent viscosity does not become excessively large. If the mixing length is set to a high value, the turbulent viscosity (and consequently deposition) will be over-predicted.

Fig. 9 demonstrates the effect of TLEN on the geometry of the deposit. A TLEN of 1 mm renders the flow unable to sustain the sediment in suspension and results in premature deposition in the feeder channel (Fig. 9C). The amount of sediment reaching the expansion table is thus significantly reduced and the resulting lobe shape is inconsistent with the one obtained from the physical experiment. An appropriate mixing length for the particular turbidity current was found to be between 0.15 mm and 0.2 mm. This turbulent mixing length, along with the default erosion and bedload coefficient, provides the most genuine flow structure and gives a realistic distribution of sediment on the expansion table (Fig. 9C). Fig. 10 displays some of the results obtained for Run 4 in a similar way to the ones obtained for Run 2. Longitudinal and transverse sections of the resulting deposit for Run 2 show a good match with the measurements reported from laboratory experiments (Fig. 11). The numerical results match very well to the physical data, both on the longitudinal and the transversal cross sections, slightly underestimating the thickness of the deposits in the distal part of the lobe but getting the shapes and general architecture of the deposit right. It is possible to appreciate how the software is capable of reproducing both the depositional areas and the erosional zones, such as the one at the channel mouth in the laboratory experiments.

This case study suggests that the numerical model is capable of producing realistic depositional results. The parameters discussed above need to be adjusted to particular conditions and often vary from our theoretical assumptions from literature. The geometry of the simulated deposits in the present case do not seem to be affected by the discrepancies in the sediment transport mechanism.

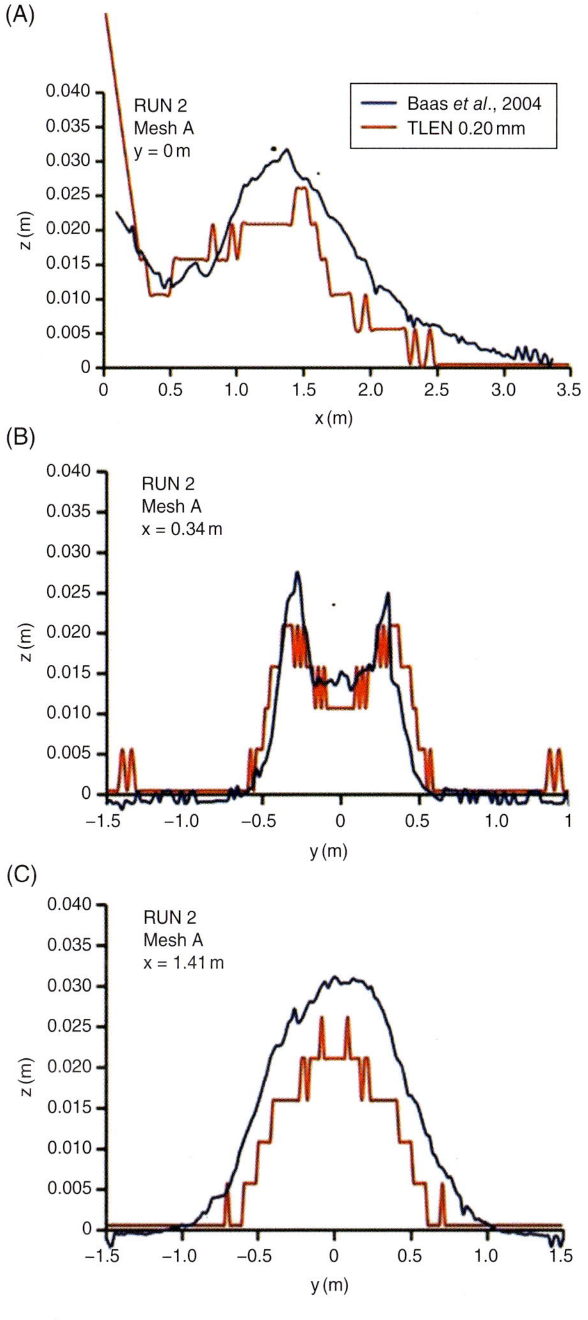

Fig. 11. Comparison between two-dimensional cross-sections from Baas *et al.* (2004) (blue line) and MassFLOW-3DTM results (red line). A) longitudinal sections through the centre line of the lobe. B) cross section shape of the lobe at 0.34 m from the channel mouth. C) cross section shape of the lobe at 1.41 m from the channel mouth. Note that the z axis is exaggerated.

Case study III: Flow structure of turbidity currents in sinuous channels

Deep-water sinuous channels show great variation in their transverse and plan-view geometry, infill sand net/gross, turbidite facies range and depositional architecture (Abreu *et al.*, 2003; Hubbard *et al.*, 2009; Kneller, 2003; Wynn *et al.*, 2007). However, the dynamics of their formative flows and causes of variation are poorly understood, as the results of laboratory and numerical experiments are contradictory and inconclusive (Corney *et al.*, 2006; Imran *et al.*, 2007; Peakall *et al.*, 2007). Further work is thus needed in order to establish a set of well-accepted ground rules for sedimentation in sinuous submarine channels. The final case study reports the preliminary results of a study of the flow structure of turbidity currents in a natural-scale sinuous channel in an attempt to contribute to the understanding of these heterogeneous systems.

Numerical simulation set-up

Turbidity currents were released into a non-erodible pre-defined channel consisting of an initial 1 km-long straight segment and subsequent three channel bends out of which two were 90° and one 180° (Fig. 12A). The study focused on the 2nd, a 180° bend, which had a radius of 600 m and a bend wavelength of 1800 m. The relatively long straight segment preceding the bends was introduced to stabilise the vertical velocity and density structure of the flow, so that the incoming flow at the domain boundary had a uniform vertical structure. Channel width and depth was set to 150 m and 10 m, respectively. Channel banks were constructed at the angle of repose (30°) and the surface was roughened with medium sand (ø = 0.5 mm). The geometry and dimensions of the flow inlet were identical to the cross-sectional geometry of the channel. Channel slope was set to 1°. Simulation domain, containing the channel surface, consisted of a Cartesian orthogonal mesh with a total number of 784,259 active cells. The mesh extended 265 m above the channel floor and had a pressure-specified top boundary to account for hydrostatic pressure at a water depth of 1 km. The resolution of the mesh was uniform up to 20 m above the channel floor and then gradually decreased towards the top. The smallest and largest cell had the dimensions of 14.8 m × 14.8 m × 1.5 m and 14.7 m × 14.7 m × 11 m (XYZ format), respectively.

Eight simulations runs were performed in total. In each run the turbidity current was released from the inlet for a duration of 1 hr and consisted of two separate grain sizes; medium sand (ø = 0.5 mm) and coarse sand (ø = 1.0 mm) with a density of 2648 kg m^{-3}. The medium/coarse sand ratio was kept at 4:1 during all the runs, independent of

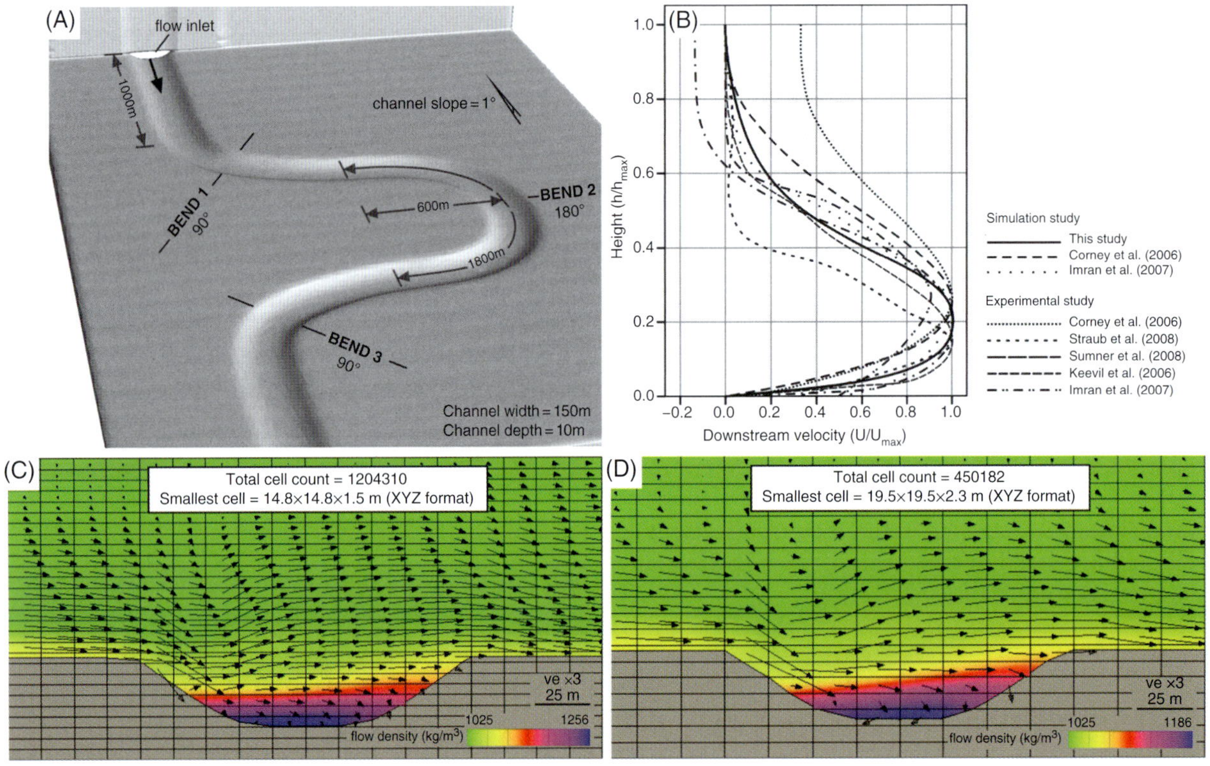

Fig. 12. (A) Dimensions and terminology of the pre-defined non-erodible channel surface used for simulations. (B) Plot showing the comparison of previously published downstream-velocity profiles with profile simulated in the present study. (C, D) Domain profiles demonstrating the independence of the simulated output from the resolution of the simulation grid. In D, the cell count is less than a half of the grid shown in C; however, there is virtually no difference in the calculated velocity and density (concentration) magnitude.

Table 3. Velocity and concentration values for the simulated cases of flow through a sinuous channel.

Run	Velocity at inlet (m/s)	Concentration at inlet (%)	Max. velocity at Bend 2 (m s^{-1})	Max. concentration at Bend 2 (%)
1	20	10	2.3	5
2	10	10	5	11
3	5	10	5	11
4	10	5	1.5	3.5
5	5	5	2.5	3.5
6	2	10	4	7
7	2	3	2.3	3.5
8	1	1	1.5	1.5

the overall sediment concentration. The initial flow velocity and sediment concentration were varied systematically for each run (Table 3) to determine their effect on the flow behaviour in channel bends.

The simulations in this particular case study have not been tuned to any specific physical experiments; however, the vertical velocity profile of the turbidity current (Run 2) measured before entering the first bend has been compared to previously published experimental and modelled data (Keevil *et al.* (2006), Straub *et al.* (2008) and Sumner *et al.* (2008)) indicating realistic flow properties (Fig. 12B). It should also be mentioned that the simulated results are independent of the grid resolution. Reducing the cell number of the computational grid by half, and the resolution by 35%, the model gives almost identical velocity and concentration values (Fig. 12C and D).

Results and remarks

In each of the simulated runs the turbidity current experienced significant super elevation at bend apexes accompanied by flow overspill (Fig. 13A and B; 14C and D). In flows with initial velocities of 5 m s^{-1} and higher, this effect was so prominent at the first bend that the velocities directly downstream dropped by 50% to 90% (Table 3). Super elevation and flow overspill decreased with decreasing initial velocity and increasing concentration, making the slower, higher-concentration flows more confined to the channel and more efficient in maintaining a constant velocity during the downstream transport. The high velocity and high concentration core of the flow travelled down channel attached to the outer bank in all three bends during all of the eight runs (see example of Run 2 and Run 7 in Fig. 13C to F). As a result, flow separation zones developed in the downstream part of the inner bends, where the turbulence intensified and velocities dropped to near zero values. The extent and position of separation zones varied between the runs. Flows with high velocities and concentrations developed extensive separation zones, located further downstream of bends than flows with low velocities and concentrations (Fig. 13C and D; 14A and B). Additionally, overbank flow re-entering the channel in the downstream part of Bend 2 and the upstream part of Bend 3 seem to have further accentuated the turbulence in these areas. The velocity and concentration changes also had an effect on the helicoidal circulation at channel bends. Flows with velocities of 5 m s^{-1} and higher displayed a weak basal-inward or basal-outward circulation at Bend 2 (Fig. 14C), whilst currents with velocities below 5 m s^{-1} all displayed basal-inward circulation cells (Fig. 14D). The ratio of maximum velocity versus flow height for the currents with a basal-inward helicity ranges from 0.19 to 0.41. The basal flow in helicoidal cells is generally oriented 2° away from the channel axis towards the inner bank and there can be a difference of up to 54° between the basal and the upper flow direction.

The behaviour of turbidity currents at the higher end of the velocity spectrum (velocity ≥ 5 m s^{-1}) can be compared to the laboratory experiments of Peakall *et al.* (2007), Amos *et al.* (2010) and Straub *et al.* (2011). The flows in these experiments showed significant super elevation at channel bends, profound overspill and an outward-directed secondary circulation cell. The high-velocity core of these currents travels attached to the inner bank in the upstream part of the bend and becomes detached from it at or directly downstream of the bend apex, with excessive

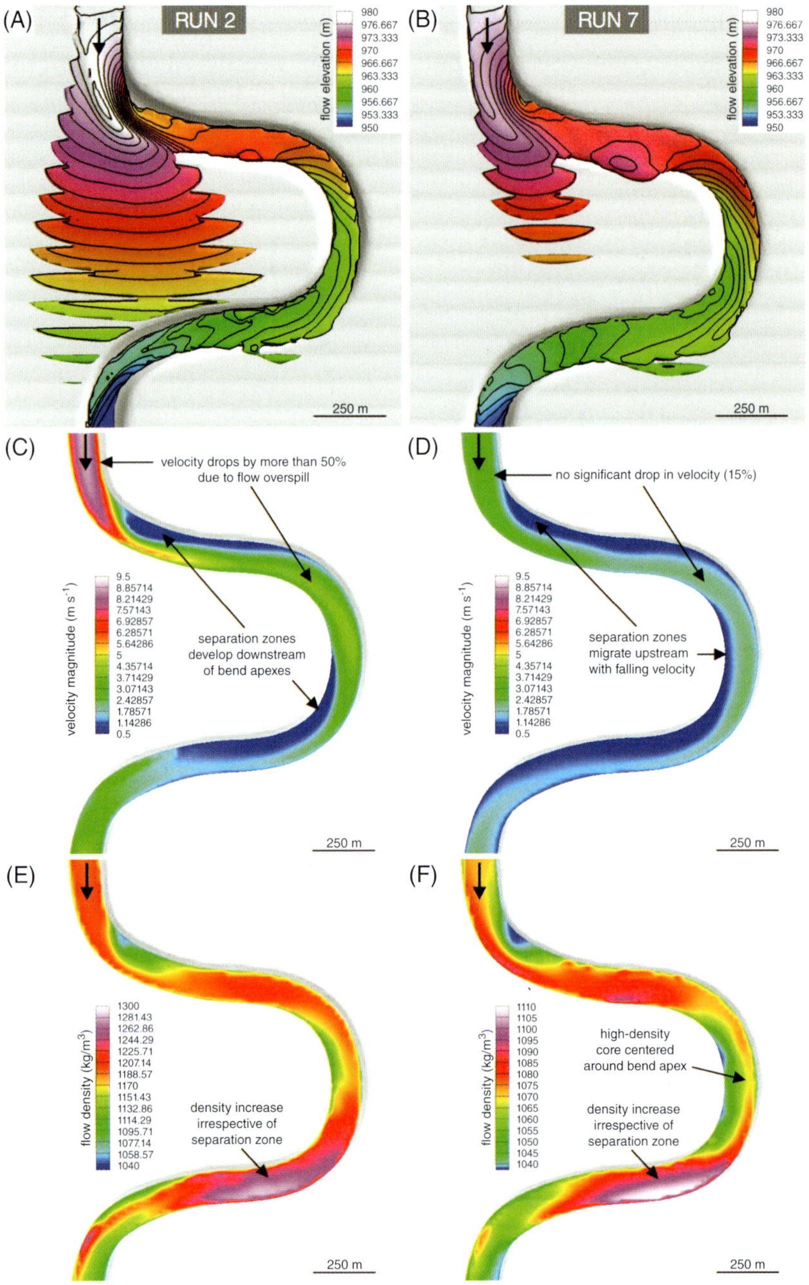

Fig. 13. Comparison of flow height, velocity and concentration levels between Run 2 and Run 7. (A) In Run 2, the flow experiences immense overspill at Bend 1 leading to significant reduction of flow height in Bend 2. (B) In Run 7, the overspill and its effect on flow height is reduced. Note the sup elevation of the flow at channel bends in both A and B. Flow surface visualised at 20% vol. concentration. (C) Slice 4 m above the channel floor showing velocity magnitude in Run 2. Overspill at Bend 1 reduces velocity of the remaining channelised flow by more than 50%. The high-velocity core travels attached to the outer banks developing low-velocity separation zones downstream of bend apexes. The velocity field in Bend 3 is greatly influenced by the re-entry of the overbank flow into the channel. (D) Slice 4 m above the channel floor showing velocity magnitude in Run 7. No significant reduction in velocity occurs downstream of Bend 1. Separation zones develop further upstream than in Run 2. The velocity field in Bend 3 does not seem to be affected by the overbank flow. (E, F) Density (concentration) slices at 4 m above the channel floor. The velocity separation zones do not seem to have an effect on the density levels, suggesting separation zones with high concentrations are prone to increased deposition. The high-density core in Run 7 appears to be more centred around the bend apex than that in Run 2.

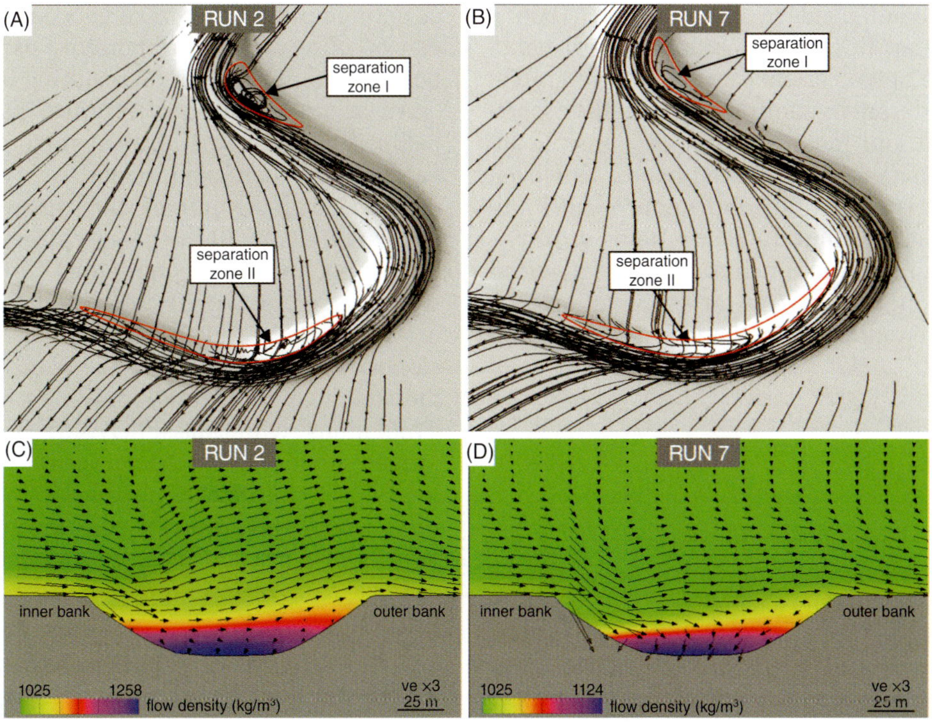

Fig. 14. (A, B) Stream traces illustrating the vector field flow in Run 2 and Run 7. The separation zones are recognised as areas of increased turbulence and chaotic flow. A) In Run 2, separation zones develop downstream of bend apexes and are further intensified by the overriding overbank flow. B) In Run 7, separation zones develop further upstream than in Run 2 and are less intense, possibly due to the less powerful overbank flow. (C) Density profile with 3D velocity vectors at the apex of Bend 2 (Run 2). The profile shows there is a weak inwardly-oriented secondary flow cell at the base of the channel, hinting at a cell reversal at higher flow velocities. The upper part of the channelised flow is characterised by an outwardly motion and flow overspill at the outer bend. The high-density core does not seem to experience super elevation and is confined to the centre of the channel. (D) Density profile with 3D velocity vectors at the apex of Bend 2 (Run 7). The secondary circulation cell is better developed than in C, showing a basal-inward orientation. Overspill is less pronounced and as in C, the high-density core is confined to the centre of the channel.

suspension fall-out deposition occurring in the downstream part of inner bend (Fig. 13C; 14A). These flows are therefore expected to be responsible for the downstream translation of submarine channels where the upstream part of the point bar is dominated by erosion and the downstream part by deposition (also see Amos *et al.*, 2010). If the current is in disequilibrium with the channel shape, such as the high-velocity currents in our simulations, significant overspill will occur at channel bends leading to substantial loss of velocity and concentration of the remaining channelised flow. These flows are thus inefficient in maintaining constant velocity and transporting the bulk sediment load to the deep basin. In contrast, flows at the opposite end of the velocity spectrum (velocity $\leq 5\,m\,s^{-1}$) show greater confinement to the channel, less extensive separation zones and an inward-directed circulation pattern at channel bends. The change in maximum velocity and concentration from bend to bend is minimal, hinting at a great efficiency of the transport system and a prevailing bypass, which is a general characteristic of sinuous channels migrating in the lateral domain (Kneller, 2003). The high-concentration core of these flows appears to be more centred at the apex of the bend (Figs 13D and 14B); indicating that deposition is more likely to take place in the central part of the inner bend, as in the case of bend expansion. This deposition is more probably dominated by traction as it occurs outside of the separation zone where the basal velocities are greatest and directed up the inner bank. The basal-inward circulation occurs even at very low maximum velocity/flow height ratios, as has been previously noted by Islam *et al.*

(2008) but challenged by Wynn *et al.* (2007). We suggest that flows with a basal-inward helicity, which do not experience excessive overspill at channel bends, are in equilibrium with the associated channel and most probably responsible for the lateral expansion of meander bends. However, further work is needed to confirm our results and interpretations. Non-erodible substrates in physical and numerical experiments should be replaced by erodible ones to account for sediment erosion and associated changes in channel geometry and position.

CONCLUSIONS

The present work shows the importance of the synergy between deterministic process-based modelling and laboratory experiments in order to achieve a better understanding of the intrinsic structure of turbidity currents.

Much of the physics of this very complex environment is still unknown and therefore laboratory experiments need to be carried on in order to have physical evidence of the processes involved in these currents. At the present stage, velocity measurements of the flows can be obtained quite easily, giving an accurate description of the current structure, whereas concentration measurements for such dense flows in the laboratory are still problematic. The interaction between the packed bed and the flow is still a crucial point to be investigated. Further testing and analysis of laboratory data needs to be carried out in order to achieve a better and more detailed understanding of the physics of these multi-phase currents, with particular attention to the high density regions and traction carpets resulting at the flume bottom.

The software MassFLOW-3D™ is capable of simulating a variety of flow situations involving sediment transport, erosion and deposition, taking into account the influence of the topography of the area. The software can reproduce accurately the behaviour of the turbidity currents when flowing along a submerged channel and can recreate the shape and size of lobes resulting from a turbidity current flowing from a channel into an expansion table, as well as the velocity and concentration distribution in the flow.

As observed from the numerical results, it is possible to appreciate how the resulting depositional structures match in a very good way that which can be observed from laboratory experiments in both size and shape of the resulting lobes.

An accurate tuning of physical parameters, as bedload coefficient, turbulent mixing length and entrainment coefficient, needs to be carried on in order to match the numerical results with the physical experiments. At the present stage, the software is still not capable of modelling fluid regions in which the grain-to-grain interaction (granular flow model) plays a fundamental role, such as traction carpets or regions close to the packed bed where the sediment concentration is very high. Further simulations with MassFLOW-3D™ will be performed in order to reproduce accurately the behaviour of low concentration turbidity currents and to further implement into the code the mathematical/numerical approach most able to deal with high concentration regions and traction carpets.

ACKNOWLEDGEMENTS

The authors would like to thank our partners in the MassFLOW-3D project, Det norske oljeselskap, Statoil, Noreco (Norwegian Energy Company), Wintershall Norge and Lotos E&P Norge for support. We thank Flow Science Inc. for co-operation in the project. We would also like to thank Nicole Clerx, Menno Hofstra and Roel Dirkx for their precious help in performing some of the laboratory experiments and numerical analysis. We further thank Evgeniy Tantserev for his efforts to write the governing equations of the mathematical model for the Appendix and Romain Rouzairol for his precious support.

APPENDIX I

Multi-phase flow modelling of transport, deposition and erosion of sediments

Suspended flow

The following system of equations describes multiphase flow behaviour for sediment gravity flows. Generalised equations about suspension flow modelling may be found in Ishii (1975) and Ungarish (1993) and the suspension flow model implemented in MassFLOW-3D can be further read about in FLOW-3D (2009) and Basani & Hansen (2009). This system consists of seven groups of equations

(1a, b)-6 and seven groups of unknowns $\bar{\mathbf{u}}, P, c_{s,i}, \mathbf{u}_{r,i}, \mathbf{u}_{drift,i}, K_i, \mathbf{u}_{lift,i}$ (i = 1,N; N- total number of sediment species):

$$\nabla \cdot \bar{\mathbf{u}} = 0, \quad \frac{\partial \bar{\mathbf{u}}}{\partial t} + \bar{\mathbf{u}} \cdot \nabla \bar{\mathbf{u}} = -\frac{1}{\bar{\rho}} \nabla P + \mathbf{g} + \nabla\left(\left(\nu + \nu_T\right) \cdot \nabla \bar{\mathbf{u}}\right) (1\text{a}, \text{b}) \tag{1}$$

$$\left(\frac{\partial c_{s,i}}{\partial t}\right)_x + \bar{\mathbf{u}} \cdot \nabla c_{s,i} = D\nabla^2 c_{s,i} - \mathbf{u}_{lift,i} \cdot \nabla c_{s,i} - \mathbf{u}_{drift,i} \cdot \nabla c_{s,i} \tag{2}$$

$$\mathbf{u}_{r,i} = \frac{\mathbf{g}}{K_i}\left(\rho_{s,i} - \bar{\rho}\right) f_{s,i} \tag{3}$$

$$\mathbf{u}_{drift,i} = \left(1 - f_{s,i}\right)\mathbf{u}_{r,i} - \sum_{j=1}^{N(-i)} f_{s,j}\mathbf{u}_{r,i} \tag{4}$$

$$K_i = \frac{3}{4}\frac{f_{s,i}}{d_{s,i}}\left(C_{D,i}\left\|\mathbf{u}_{r,i}\right\| + 24\frac{\mu_f}{\rho_f d_{s,i}}\right) \tag{5}$$

$$\mathbf{u}_{lift,i} = \alpha_i \mathbf{n}_s d_*^{0.3}\left(0_i - 0''_{cr,i}\right)^{1.5} \sqrt{\frac{\|\mathbf{g}\| d_{s,i}\left(\rho_{s,i} - \rho_f\right)}{\rho_f}} \tag{6}$$

1. a, b) $\bar{\mathbf{u}} \underline{\underline{def}} \left(1 - \sum_{j=1}^{N} f_{s,j}\right)\mathbf{u}_f + \sum_{j=1}^{N} f_{s,j}\mathbf{u}_{s,i}$ - the mean velocity of the fluid/sediment mixture (bulk flow), P –pressure.
2. $c_{s,i}$ – the concentration of the suspended sediment, in units of mass per unit volume.
3. $\mathbf{u}_{r,i} \underline{\underline{def}}\, \mathbf{u}_{s,i} - \mathbf{u}_f$ – the relative velocity.
4. $\mathbf{u}_{drift,i} \underline{\underline{def}}\, \mathbf{u}_{s,i} - \bar{\mathbf{u}}$ – the drift velocity to compute the transport of sediment due to drift.
5. K_i – drag function.
6. $\mathbf{u}_{lift,i}$ –the entrainment lift velocity (volumetric flux) of sediment.

Now we describe how this system of equations was derived and what it describes.

Equation (1a) is a mass balance equation for incompressible fluid-sediment mixture.

Equation (1b) is an Unsteady Reynolds-averaged Navier-Stokes equation for fluid-sediment mixture (bulk flow), where ν – molecular viscosity and ν_T – turbulent viscosity related to Reynolds stresses and $\nu_T \gg \nu$. Turbulent viscosity is calculated by the RNG turbulence model (Lyn, 2008). The RNG $k-\varepsilon$ model has several advantages over the standard $k-\varepsilon$ model. It is more accurate for rapidly strained flows and swirling flows and for lower Reynolds numbers (Re), the RNG model behaves better than the standard $k-\varepsilon$ which is only valid for high Reynolds number flows.

Short description of RNG model implemented in MassFLOW-3D™

The kinematic turbulent viscosity is computed from

$$\nu_T = 0.085\frac{k_T^2}{\varepsilon_T} \tag{7}$$

The two-equations renormalisation group (RNG) k-ε turbulence model is based on the turbulence viscosity hypothesis and solve two transport equations for the turbulent kinetic energy, k_T and the turbulent dissipation, ε_T.

The turbulent kinetic energy is defined as the energy of the turbulent velocity fluctuations:

$$k_T = \frac{1}{2}\left(\overline{u'^2} + \overline{v'^2} + \overline{w'^2}\right) \tag{8}$$

Where u', v' *and* w' are the component of the velocity fluctuations.

The transport equation for the turbulent kinetic energy k_T is

$$\frac{\partial k_T}{\partial t} + \left\{u\frac{\partial k_T}{\partial x} + v\frac{\partial k_T}{\partial y} + w\frac{\partial k_T}{\partial z}\right\} = P_T + G_T + D_K - \varepsilon_T \tag{9}$$

The transport equation of the turbulence dissipation ε_T is as follows:

$$\frac{\partial \varepsilon_T}{\partial t} + \left\{u\frac{\partial \varepsilon_T}{\partial x} + v\frac{\partial \varepsilon_T}{\partial y} + w\frac{\partial \varepsilon_T}{\partial z}\right\} = \frac{C_{\varepsilon 1} \cdot \varepsilon_T}{k_T}\left(P_T + C_{\varepsilon 3} \cdot G_T\right) + D_\varepsilon - C_{\varepsilon 2}\frac{\varepsilon_T^{\ 2}}{k_T} \tag{10}$$

The rate of turbulent energy dissipation, ε_T, in one-equation model is related to the turbulent kinetic energy k_T:

$$\varepsilon_T = 0.085\frac{\sqrt{3/2}k_T^{\ 3/2}}{TLEN} \tag{11}$$

The RNG model uses equations similar to the equations for the k-ε model. However, equation constants that are found empirically in the standard k-ε model are derived explicitly in the RNG model, like $C_{\varepsilon 2}$ is computed from the turbulent kinetic energy (k_T) and turbulent production (P_T) terms.

In order to prevent unphysical small dissipation rates, the minimum dissipation is limited by a maximum length scale, represented by TLEN in MassFLOW-3D™.

$$\varepsilon_{T,min} = 0.085 \frac{\sqrt{3/2} k_T^{3/2}}{TLEN} \tag{12}$$

TLEN and the initiation of the turbulent kinetic energy k_T are the two main turbulent parameters that are defined by the user during the setting up of a model. Equation (2) is a mass balance equation for each sediment species *i*, which describes the motion of the suspended sediment in the system by the advection-diffusion equation, with the addition of the effect of drifting and lifting of the sediment.

Equation (3) is obtained by subtracting the momentum balance for fluid-sediment mixture (1) from momentum balances for each sediment species *i*, with the use of relative velocity definition and assumptions: a) that the motion of sediment is nearly steady at the scale of the computational time and that the advection term is very small due to small gradients in the drift velocity, b) that the ratio of pressure gradient to mixture density is typically proportional to the acceleration of gravity, g.

Equation (4) is derived from the definitions for $\bar{\mathbf{u}}, \mathbf{u}_{r,i}$ *and* $\mathbf{u}_{drift,i}$.

Equation (5) is obtained by combining form drag and Stokes drag, (Clift *et al.*, 1978).

Equation (6) is modelled to calculate the entrainment lift velocity (volumetric flux) of sediment and the entrainment coefficient α_i, based on Mastbergen & Van Den Berg (2003).

Bed-load transport and packing of sediments

Bed-load transport describes the movement of large particles along the surface of the bed without being entrained into the bulk fluid flow. The Meyer-Peter & Muller formula predicts the volumetric flow rate of sediment per unit width over the surface of the packed bed.

The mass flux of sediment as computed across computational cell boundaries in MassFLOW-3D™ is a multiplication of the velocity of the sediment in each computational cell, bed load thickness, the volume fraction of the bed-load layer and density of the sediment species,

$$Q_{b,i} = \boldsymbol{u}_{bedload,i} \delta_i f_{b,i} \rho_{s,i} \tag{13}$$

Hindered settling is not important for the low concentration of the sediments.

Additional notations

ρ_f – fluid density.
$\rho_{s,i}$ – density of the sediment species.
$\bar{\rho}$ – mean density.
$f_{s,i}$ – volume fraction of sediment species *i*.
$d_{s,i}$ – diameter for sediment species *i*.
$C_{D,i}$ – drag coefficient for sediment species *i*.
P – pressure.

$$P_T = C_{SP}\left(\frac{\mu}{\rho V_f}\right)\left[\begin{array}{c} 2\left(\frac{\partial u}{\partial x}\right)^2 + 2\left(\frac{\partial v}{\partial y}\right)^2 + 2\left(\frac{\partial w}{\partial z}\right)^2 \\ +\left(\frac{\partial v}{\partial x}+\frac{\partial u}{\partial y}\right)^2 + \left(\frac{\partial u}{\partial z}+\frac{\partial w}{\partial x}\right)^2 + \left(\frac{\partial v}{\partial z}+\frac{\partial w}{\partial y}\right)^2 \end{array}\right]$$

– turbulent production term.

C_{SP} – turbulent parameter, whose default value is 1.0.

$$G_T = -C_\rho\left(\frac{\mu}{\rho^3}\right)\left(\frac{\partial \rho}{\partial x}\frac{\partial p}{\partial x} + \frac{\partial \rho}{\partial y}\frac{\partial p}{\partial y} + \frac{\partial \rho}{\partial z}\frac{\partial p}{\partial z}\right)$$ – the buoyancy production term.

C_ρ – has a default value of 0.0 unless the problem is thermally buoyant, in which case it takes a value of 2.5.

$$D_K = \frac{\partial}{\partial x}\left(\frac{\nu_T}{\sigma_k}\frac{\partial k_T}{\partial x}\right) + \frac{\partial}{\partial y}\left(\frac{\nu_T}{\sigma_k}\frac{\partial k_T}{\partial y}\right) + \frac{\partial}{\partial z}\left(\frac{\nu_T}{\sigma_k}\frac{\partial k_T}{\partial z}\right)$$

– diffusion term for the turbulent kinetic energy, σ_k as a value of 0.72 for the RNG model.

$C_{\varepsilon 1}, C_{\varepsilon 2}$ *and* $C_{\varepsilon 3}$ – non-dimensional parameters. The default value for $C_{\varepsilon 1}$ is 1.42. $C_{\varepsilon 2}$ is computed based on the value of k_T, ε_T and the shear rate. $C_{\varepsilon 3}$ has a value of 0.2.

$$D_\varepsilon = \frac{\partial}{\partial x}\left(\frac{\nu_T}{\sigma_\varepsilon}\frac{\partial \varepsilon_T}{\partial x}\right) + \frac{\partial}{\partial y}\left(\frac{\nu_T}{\sigma_\varepsilon}\frac{\partial \varepsilon_T}{\partial y}\right) + \frac{\partial}{\partial z}\left(\frac{\nu_T}{\sigma_\varepsilon}\frac{\partial \varepsilon_T}{\partial z}\right)$$

– diffusion term for the dissipation where σ_ε is equal to 0.72 for the RNG model.

$\mathbf{n}_s$ – outward pointing normal vector to the packed bed interface.

d_* – dimensionless mean particle diameter.
$\|\mathbf{g}\|$ – magnitude of the gravitational vector.
μ_f – fluid viscosity.

$\theta_i = \dfrac{\tau}{\|\mathbf{g}\| d_{s,i}\left(\rho_{s,i} - \rho_f\right)}$ – local Shields parameter at the bed interface (it is computed based on the local shear stress, τ).

$f_{b,i} = 0.18 \dfrac{f_{packed}}{d_*}\left(\dfrac{\theta_i}{\theta''_{cr,i}} - 1\right)$ – the volume fraction of the bed-load layer is predicted by van Rijn (1984).

$\dfrac{\delta_i}{d_{50}} = 0.3 d_*^{0.7}\left(\dfrac{\theta_i}{\theta''_{cr,i}} - 1\right)^{0.5}$ – bed load thickness, d_{50} – local mean particle size in the computational cell.

$\mathbf{u}_{bedload,i} = u_{bedload,i}\dfrac{\bar{u}}{\|\bar{u}\|}$ – the direction of the motion is determined from the motion of the liquid adjacent to the packed bed interface.

$u_{bedload,i} = \dfrac{q_{b,i}}{\delta_i f_{b,i}}$ – velocity of the sediment in each computational cell.

$q_{b,i} = \beta_i(\theta_i - \theta''_{cr,i})^{1.5} \cdot \left[\|\mathbf{g}\|\left(\dfrac{\rho_{s,i} - \rho_f}{\rho_f}\right) d_{s,i}^3\right]^{\frac{1}{2}}$ – volumetric bed-load transport rate per unit width.

β_i – bedload coefficient is typically equal to 8.0, (Van Rijn, 1984).

$$\theta''_{cr,i} = \theta'_{cr,i}\frac{\cos\psi \sin\beta + \sqrt{\cos^2\beta \tan^2\phi_i + \sin^2\psi \sin^2\beta}}{\tan\phi_i}$$

– the critical Shields parameter modified for sloping surfaces to include the angle of repose, (Soulsby 1997).

$\theta'_{cr,i} = \theta_{cr,i}\dfrac{1.666667}{\left(\log_{10}\left(19\dfrac{d_{s,i}}{d_{50}}\right)\right)^2}$ – the critical Shields parameter modified by the effect of armouring (Egiazaroff, 1965).

β – computed angle of the packed interface normal relative to the gravitational vector $\mathbf{g}$.

ϕ_i, ψ – user-defined angle of repose for sediment species i and angle between the flow and the upslope direction, respectively.

REFERENCES

Aas, T., Howell, J., Janocko, M. and **Jackson, C.A.L.** (2010a) Control of Aptian palaeobathymetry on turbidite distribution in the Buchan Graben, Outer Moray Firth, Central North Sea. *Mar. Petrol. Geol.*, **27**, 412–434.

Aas, T.E., Howell, J.A., Janocko, M. and **Midtkandal, I.** (2010b) Re-created Early Oligocene seabed bathymetry and process-based simulations of the Peïra Cava turbidite system. *J. Geol. Soc.*, **167**, 857–875.

Abreu, V., Sullivan, M., Pirmez, C. and **Mohrig, D.** (2003) Lateral accretion packages (LAPs): an important reservoir element in deep water sinuous channels. *Mar. Petrol. Geol.*, **20**, 631–648.

Amos, K.J., Peakall, J., Bradbury, P.W., Roberts, M., Keevil, G. and **Gupta, S.** (2010) The influence of bend amplitude and planform morphology on flow and sedimentation in submarine channels. *Mar. Petrol. Geol.*, **27**, 1431–1447.

Baas, J.H., Van Kesteren, W. and Postma, G. (2004) Deposits of depletive high-density turbidity currents: a flume analogue of bed geometry, structure and texture. *Sedimentology*, **51**, 1053–1088.

Basani, R. and Hansen, E.W.M. (2009) MassFLOW-3D Process Modelling; Three-dimensional numerical forward modelling of submarine massflow processes and sedimentary successions. Project report 2004–2008. CFD-R09-1-09.

Cartigny, M.J.B. (2012) Morphodynamics of supercritical high-density turbidity currents. *Utrecht Studies in Earth Sciences*, **10** (doctoral thesis), 153 pp.

Cartigny, M.J.B., Postma, G., van den Berg, J.H. and **Mastbergen, D.R.** (2011) A comparative study of sediment waves and cyclic steps based on geometries, internal structures and numerical modelling. *Mar. Geol.*, **280**, 40–56.

Clift, R., Grace, J.R. and **Weber, M.E.** (1978) *Bubbles, Drops and Particles*. Academic Press, New York.

Corney, R.K.T., Peakall, J., Parsons, D.R., Elliott, L., Amos, K.J., Best, J.L., Keevil, G.M. and **Ingham, D.B.** (2006) The orientation of helical flow in curved channels. *Sedimentology*, **53**, 249–257.

Dan, G., Sultan, N. and **Savoye, B.** (2007) The 1979 Nice harbour catastrophe revisited: Trigger mechanism inferred from geotechnical measurements and numerical modelling. *Mar. Geol.*, **245**, 40–64.

Duller, R.A., Mountney, N.P. and **Russell, A.J.** (2010) Particle Fabric and Sedimentation of Structureless Sand, Southern Iceland. *J. Sed. Res.*, **80**, 562.

Dykstra, M. and **Kneller, B.** (2009) Lateral accretion in a deep-marine channel complex: implications for channellized flow processes in turbidity currents. *Sedimentology*, **56**, 1411–1432.

Egiazaroff, I.V. (1965) Calculation of nonuniform sediment concentrations. *J. Hydr. Div.*, **91**, 225–247.

FLOW-3D (2009) *FLOW-3D User Manual Version 9.4.*

Groenenberg, R.M., Hodgson, D.M., Prelat, A., Luthi, S.M. and **Flint, S.S.** (2010) Flow-deposit interaction in submarine lobes: insights from outcrop observations and realizations of a process-based numerical model. *J. Sed. Res.*, **80**, 252.

Groenenberg, R.M., Sloff, K. and **Weltje, G.J.** (2009) A high-resolution 2-DH numerical scheme for process-based modelling of 3-D turbidite fan stratigraphy. *Comput. Geosci*, **35**, 1686–1700.

Guo, J. (2002) Hunter Rouse and Shields Diagram, In: *Proceedings of the 13th IAHR-APD congress, Singapore*, pp. 1096–1098.

Hadler-Jacobsen, F., Gardner, M.H. and **Borer, J.M.** (2007) Seismic stratigraphic and geomorphic analysis of deep-marine deposition along the West African continental margin, In: *Seismic Geomorphology: Applications to Hydrocarbon Exploration and Production* (Eds R.J. Davies, H.W. Posamentier, L.J. Wood and J.A. Cartwright), *Geol. Soc. London Spec. Publ.*, **277**, pp. 1–14.

Harbaugh, J.W. (1970) *Computer Simulation in Geology*, Stanford University California.

Hastings, D. (1986) Cretaceous stratigraphy and reservoir potential, mid Norway continental shelf. In: Spencer, A.M., Holter, E., Campbell, C.J., Hanslien, S.H., Nelson, P.H.H., Nysæther, E., and Ormaasen, E.G. (Eds.), *Proceedings of the Conference: Habitat of Hydrocarbons on the Norwegian Continental Shelf. Norwegian Petroleum Society/Graham & Trotman*, London, pp. 287–298.

Haughton, P., Davis, C., McCaffrey, W. and **Barker, S.** (2009) Hybrid sediment gravity flow deposits – classification, origin and significance. *Mar. Petrol. Geol.*, **26**, 1900–1918.

Heimsund, S. (2007) *Numerical Simulation of Turbidity Currents: A New Perspective for Small- and Large-Scale Sedimentological Experiments.* Unpublished Masters thesis, University of Bergen.

Huang, H., Imran, J. and **Pirmez, C.** (2005) Numerical model of turbidity currents with a deforming bottom boundary. *J. Hydraul. Eng.*, **131**, 283.

Hubbard, S.M., de Ruig, M.J. and **Graham, S.A.** (2009) Confined channel-levee complex development in an elongate depo-center: deep-water Tertiary strata of the Austrian Molasse basin. *Mar. Petrol. Geol.*, **26**, 85–112.

Imran, J., Islam, M.A., Huang, H., Kassem, A., Dickerson, J., Pirmez, C. and **Parker, G.** (2007) Helical flow couplets in submarine gravity underflows. *Geology*, **35**, 659.

Ishii, M. (1975) *Thermo-fluid dynamic theory of two-phase flow.* Eyrolles, Paris.

Islam, M.A., Imran, J., Pirmez, C. and **Cantelli, A.** (2008) Flow splitting modifies the helical motion in submarine channels. *Geophys. Res. Lett.*, **35**, L22603.

Keevil, G.M., Peakall, J., Best, J.L. and **Amos, K.J.** (2006) Flow structure in sinuous submarine channels: velocity and turbulence structure of an experimental submarine channel. *Mar. Geol.*, **229**, 241–257.

Kleinhans, M.G. (2002) *Sort Out Sand & Gravel: Sediment Transport and Deposition in Sand-Gravel Bed Rivers.* Unpublished Ph.D. Thesis, University of Utrecht.

Kneller, B. (2003) The influence of flow parameters on turbidite slope channel architecture. *Mar. Petrol. Geol.*, **20**, 901–910.

Kneller, B. and Buckee, C. (2000) The structure and fluid mechanics of turbidity currents: a review of some recent studies and their geological implications. *Sedimentology*, **47**(s1), 62–94.

Kostic, S. (2011) Modelling of submarine cyclic steps: Controls on their formation, migration and architecture. *Geosphere*, **7**, 294.

Lyn, D.A. (2008) Turbulence models for sediment transport engineering In: *Sedimentation Engineering: Processes, Measurements, Modelling and Practice.* (Ed. M.H. Garcia), **110**. American Society of Civil Engineers.

Mastbergen, D.R. and **Van Den Berg, J.H.** (2003) Breaching in fine sands and the generation of sustained turbidity currents in submarine canyons. *Sedimentology*, **50**, 625–637.

Meiburg, E. and **Kneller, B.** (2010) Turbidity currents and their deposits. *Annu. Rev. Fluid Mech*, **42**, 135–156.

Meyer-Peter, E. and **Müller, R.** (1948) Formulas for bed-load transport. In: *Proceedings of the 2nd Meeting of the International Association for Hydraulic Structures Research*, pp. 39–64.

Mulder, T. and **Alexander, J.** (2001) The physical character of subaqueous sedimentary density flows and their deposits. *Sedimentology*, **48**, 269–299.

Norwegian Petroleum Directorate (2009) The Resource Report 2009. http://www.npd.no/en/Publications/Resource-Reports/2009/

Peakall, J., Amos, K.J., Keevil, G.M., William Bradbury, P. and **Gupta, S.** (2007) Flow processes and sedimentation in submarine channel bends. *Mar. Petrol. Geol.*, **24**, 470–486.

Pirmez, C. and **Imran, J.** (2003) Reconstruction of turbidity currents in Amazon Channel. *Mar. Petrol. Geol.*, **20**, 823–849.

Postma, G., Nemec, W. and **Kleinspehn, K.L.** (1988) Large floating clasts in turbidites: a mechanism for their emplacement. *Sed. Geol.*, **58**, 47–61.

Richardson, J. and **Zaki, W.** (1954) Fluidization and Sedimentation-Part I. *Trans. Inst. Chem. Eng*, **32**, 38–58.

Salles, T., Lopez, S., Eschard, R., Lerat, O., Mulder, T. and **Cacas, M.** (2008) Turbidity current modelling on geological time scales. *Mar. Geol.*, **248**, 127–150.

Schmeeckle, M.W. and **Nelson, J.M.** (2003) Direct numerical simulation of bedload transport using a local, dynamic boundary condition. *Sedimentology*, **50**, 279–301.

Schwab, A., Tremblay, S. and **Hurst, A.** (2007) Seismic expression of turbidity-current and bottom-current processes on the Northern Mauritanian continental slope. *Geol. Soc. London Spec. Publ.*, **277**, 237.

Soulsby, R. (1997) *Dynamics of marine sands: a manual for practical applications.* Thomas Telford.

Straub, K.M., Mohrig, D., Buttles, J., McElroy, B. and **Pirmez, C.** (2011) Quantifying the influence of channel sinuosity on the depositional mechanics of channelized turbidity currents: A laboratory study. *Mar. Petrol. Geol.*, **28**, 744–760.

Straub, K.M., Mohrig, D., McElroy, B., Buttles, J. (2008) Interactions between turbidity currents and topography in aggrading sinuous submarine channels: a laboratory study. *GSA Bulletin* **120**, 368–385.

Sumner, E.J., Lawrence, A.A., Talling, P.J. (2008) Deposit structure and processes of sand deposition from decelerating sediment suspensions. *J. Sed. Res.*, **78**, 529–547.

Tetzlaff, D.M. and **Harbaugh, J.W.** (1989) *Simulating clastic sedimentation; Computer Methods in the Geosciences.* Van Nostrand Reinhold, New York, NY, USA.

Ungarish, M. (1993) *Hydrodynamics of Suspensions: Fundamentals of Centrifugal and Gravity Separation.* New York: Springer-Verlag.

Van Rijn, L.C. (1984) Sediment transport, part I: bed load transport. *J. Hydraul. Eng.*, **110**, 1431–1456.

Wynn, R.B., Cronin, B.T. and **Peakall, J.** (2007) Sinuous deep-water channels: Genesis, geometry and architecture. *Mar. Petrol. Geol.*, **24**, 341–387.

Int. Assoc. Sedimentol. Spec. Publ. (2014) **46**, 609–646

The Ormen Lange turbidite systems: sedimentary architectures and sequence structure of sandy slope fans in a sediment-starved basin

RODMAR RAVNÅS*†, ANDREW COOK*‡, KRISTOFFER ENGENES*, HARRY GERMS*§, MARTIN GRECULA*, JOSTEIN HAGA*, CRAIG HARVEY*‖ and JAMES A. MACEACHERN¶

**AS Norske Shell, PO Box 40, 4098 Tananger, Norway*
†Department of Petroleum Engineering, University of Stavanger, Stavanger, Norway
‡Current address: Shell International Exploration and Production, NL 2596 HR, den Haag, Netherlands
§Current address: Shell International Exploration and Production, TX 77025-1299 Houston, Texas, United States
‖Current address: Qatar Shell Service Company W.L.L,
¶Department of Earth Sciences, Simon Fraser University, Burnaby, BC, Canada, V5A 1S6

ABSTRACT

The reservoirs of the giant Ormen Lange field, located off the Møre coast, consist of deep-marine turbidites within the Upper Maastrichtian Springar Formation and the Danian Egga sandstone unit. The two sandstone units are interpreted in terms of inner to middle and outer parts of two, stacked lower slope sandy fan systems, respectively. The two fan systems were sourced from an uplifted hinterland; the tectonic rejuvenation of this is related to rift-margin uplift in response to phased late Campanian to Danian rifting within the western Møre Basin. Rift-margin uplift was accompanied by basinward tilting and draped folding over deep-seated basement highs along the Møre Margin, which produced a stepped slope margin with local slope minibasins and terraces. Each phase of repeated hinterland uplift, basin-margin tilting and sandy fan development represents a time span of a few million years. The Springar and Egga turbidite systems evolved through fill-spill processes on the slope terraces or minibasin(s). Within an individual minibasin or slope terrace, an early phase of 'infilling' or smoothing of inherited topography appears to have been followed by a more organised backfilling or upslope onlapping combined with forestepping or outbuilding of the sandy fan across the individual terraces or minibasins. Compensationally stacked channel storeys and lobe storeys form the fundamental architectural elements and probably reflect major flooding events in the hinterland or drainage basin. Multistorey channel belts represent further outbuilding of the local terrace or minibasin-filling fan, with sediment bypass and delivery to further downslope minibasins and terraces. The uppermost parts of the fan systems appear less organised and contain disorganised, slumped and gravitationally re-sedimented units alternating with multistorey channel-fills. These less-organised facies associations formed in response to an early stage or initial tilting of the basin margin slope. Final backstepping and return to predominantly hemipelagic to pelagic slope deposition is attributed to renewed upslope or hinterland structuring, which resulted in capture of sand in proximal subbasins.

Keywords: Slope topography, turbidite systems, megasequences, turbidite architectures, controlling factors.

INTRODUCTION

Slope turbidite systems possess many similarities to base-of-slope and inner basin-floor turbidite systems, especially in situations where the turbidite systems infill intra-slope subbasins or slope terraces ('steps') that are significantly larger than that of the turbidite systems themselves. However, slope turbidite systems also possess their own unique characteristics and variability that contrasts

From Depositional Systems to Sedimentary Successions on the Norwegian Continental Margin, First Edition.
Edited by A. W. Martinius, R. Ravnås, J. A. Howell, R. J. Steel, and J. P. Wonham.

with base-of-slope and inner basin-floor systems. Such unique characteristics can be attributed to controls imposed by: 1) the slope gradient; 2) the slope profile and topography evolution; 3) the proximity to and the type of delivery system(s) (e.g. Reading & Richards, 1994); and 4) the common association with erosional bypass and/or connection to exit channel features (e.g. Mayall & Stewart, 2001; McHargue, 2001; Booth *et al.*, 2002; Sinclair & Tomasso, 2002; Prather, 2003; Smith, 2004). In consort, this commonly results in a complex reservoir architecture and stratigraphy of slope turbidite infill.

Sediment delivery to the deep-water basins may be via deltas, shelf and upper slope gravitational sediment failure, or storm and shelfal current re-suspension of marine sediments (e.g. Piper & Normark, 2009 and references therein). Deltaic delivery systems may range from large and muddy to small and sandy deltas (Reading & Richards, 1994); the latter are more common in tectonically active basins and along immature passive margins, where shallow marine to shelfal sand is fed directly to upper slope canyons. Hyperpycnal flows generated from rivers-in-flood have recently been recognised as an important mechanism to generate prolonged or sustained sediment gravity flows (Mulder *et al.*, 2003; Bhattacharya & MacEachern, 2009; Piper & Normark, 2009). Upslope gravitational collapse and mass wasting is another important flow-initiation process, producing ensuing surge-like gravity flows (Piper &

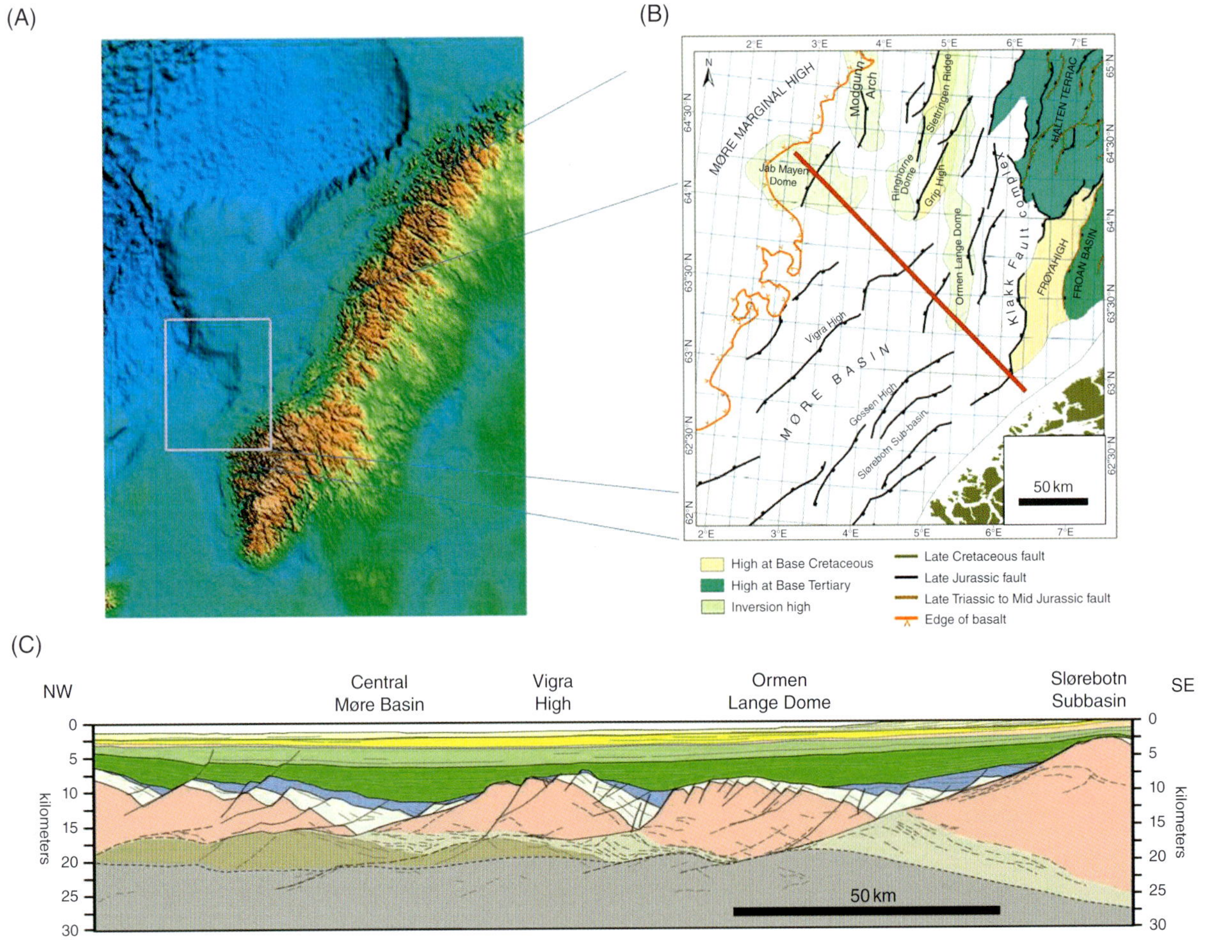

Fig. 1. (A) Location of Møre Basin offshore Mid-Norway. (B) Structural elements map of the Møre Basin with location of the Ormen Lange subbasin. (C) Interpreted section across the eastern Møre Basin (modified from Osmundsen & Ebbing, 2008). Basement is represented by various shades of purple, brown, grey and pink. Palaeozoic to Jurassic sediments are shown in pale and blue colours. Cretaceous is marked by clear (lower to middle) and pale green (upper), whereas the Palaeogene and Neogene are shown in orange-yellow and grey, respectively. The orange marks the Palaeocene. Line location is shown on the structural elements map (B) in red.

Normark, 2009). The resultant gravity flows, whether short-lived and surge-like or sustained but with spatially and temporarily variable flow patterns (e.g. Kneller, 1995; Kneller & Branney, 1995), may travel the steeper slope portions via focused routes such as canyons or deep channels, through a network of shallower channels, or as near-complete, unconfined flows (e.g. Reading & Richards, 1994). The slope topography, which may vary from unconfined through stepped to ponded (e.g. Meckel *et al.*, 2000; Booth *et al.*, 2002; Prather, 2003), normally changes through time as a result of the complex interplay between basin margin sedimentation and consequent shelf and slope progradation and slope topography creation due to structuring of the basins margins. Slope steps or subbasins may be connected through erosional knick-point channels (Pirmez *et al.*, 2000; Kneller, 2003; Heinö & Davies, 2007), via tortuous corridors (Smith, 2004), or some combination of the two. Basin margins of deep-water active basins, such as foreland-basins, rift-basins and immature passive margins, display most, if not all of these complex slope types and morphologies, which commonly leads to the trapping of coarse sediment within slope terraces and subbasins, normally with a complex fill-spill, bypass and backfill depositional pattern.

The reservoirs of the giant Ormen Lange Field (Fig. 1; Møller *et al.*, 2004), are contained within deep-marine, lower slope turbidite systems within the Upper Maastrichtian Springar Formation and the Danian Egga sandstone unit of the Tang Formation (Fig. 2). These reservoirs have previously been interpreted as basin-floor turbidites sourced by an inferred deltaic system (Gjelberg *et al.*, 2001, 2005; Sømme *et al.*, 2009) and as canyon-fed slope turbidites, with the canyons tapping into a shelfal sand source (Smith & Møller, 2003). These previous accounts interpreted the Ormen Lange reservoirs as lower slope or inner basin-floor sandy fan systems (following Piper & Normark, 2001), developed as laterally extensive channelised lobe to sheet systems with a relatively simple forestepping to backstepping infill pattern (Gjelberg *et al.*, 2001, 2005) coupled with some compensational stacking of individual turbidite complexes (Smith & Møller, 2003).

The aim of this contribution is firstly to present an update of the Ormen Lange depositional model, based on new information from additional production wells and new field and analogue studies. The focus has been on placing the Ormen Lange reservoir development into a basin-scale context, with emphasis on spatial and temporal changes in slope turbidite architectures as a response to slope topography and structuring. Secondly, a hierarchical subdivision of the Ormen Lange reservoirs is proposed, based on integration of sedimentological, ichnological and biostratigraphical studies. A novel approach applying detailed sedimentology with ichnofacies and microfacies studies has allowed for characterisation and distinction of surfaces embedding reservoir or depositional elements at various scales, analogous to studies of turbidite systems in outcrops and at the modern seafloor or in the shallow subsurface. Thirdly, an attempt is made to further decipher the potential controls on the reservoir development at various scales, as inferred from detailed studies of the Ormen Lange reservoirs combined with an integrated tectonostratigraphic evaluation of the greater Ormen Lange area.

Geological Setting

The Ormen Lange Field is located in the south-eastern part of the Møre Basin (Fig. 1; Blystad *et al.*, 1995); that is, at the transition from the Møre Basin proper to the Møre Margin to the south-east (Grunnaleite & Gabrielsen, 1995). The Møre Basin formed in response to Triassic, Middle to Late Jurassic and Late Cretaceous to Early Palaeogene rifting (Brekke, 2000; Færseth & Lien, 2002), with most of its present structural configuration reflecting intense Late Jurassic rifting and subsidence. Late Cretaceous and Early Palaeogene rifting was focused along the western margin of the Møre Basin, where tilting and compartmentalisation of fault-blocks occurred from the Coniacian until continental break-up commenced in the earliest Eocene (Fig. 2). In the eastern part of the Møre Basin, Late Cretaceous to Early Palaeogene rifting resulted only in minor reactivation of the deeper Jurassic structures (Grunnaleite & Gabrielsen, 1995; Færseth & Lien, 2002; Osmundsen & Ebbing, 2008) and associated drape-folding of the sedimentary cover over these deep-seated structures. The stepwise fault-block rotation in western Møre Basin and drape folding in eastern Møre Basin appear to have formed in response to a series of rift phases that can be recognised across the Mid-Norway deepwater area.

The Møre Margin, which formed the eastern part of the Late Jurassic rift basin (Grunnaleite & Gabrielsen, 1995; Jongepier *et al.*, 1996; Osmundsen

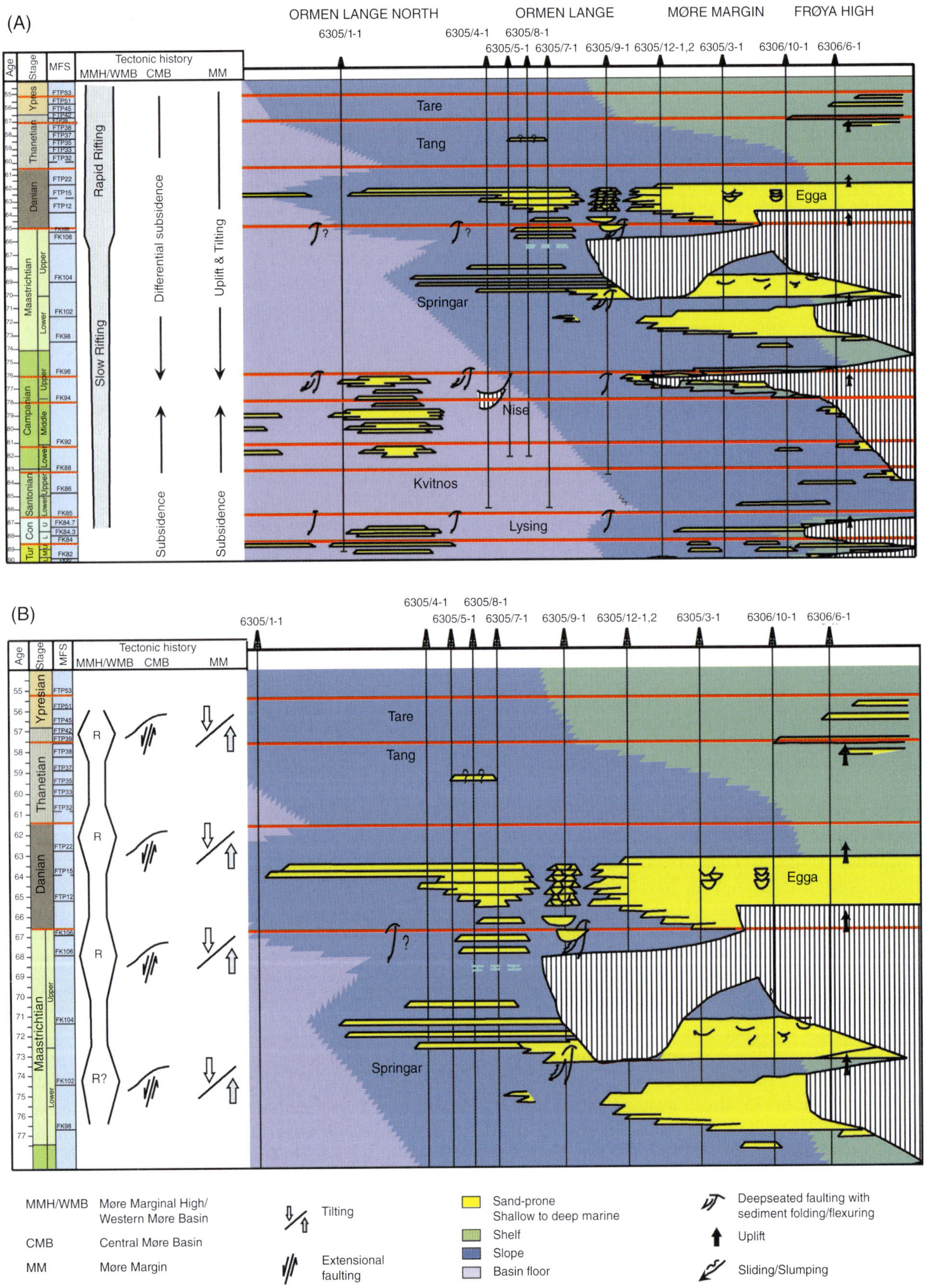

Fig. 2. (A) Upper Cretaceous and Lower Palaeogene tectono-stratigraphy of the Møre Basin. (B) Detailed Maastrichtian to Palaeocene tectono-stratigraphy of the Møre Margin and eastern Møre Basin.

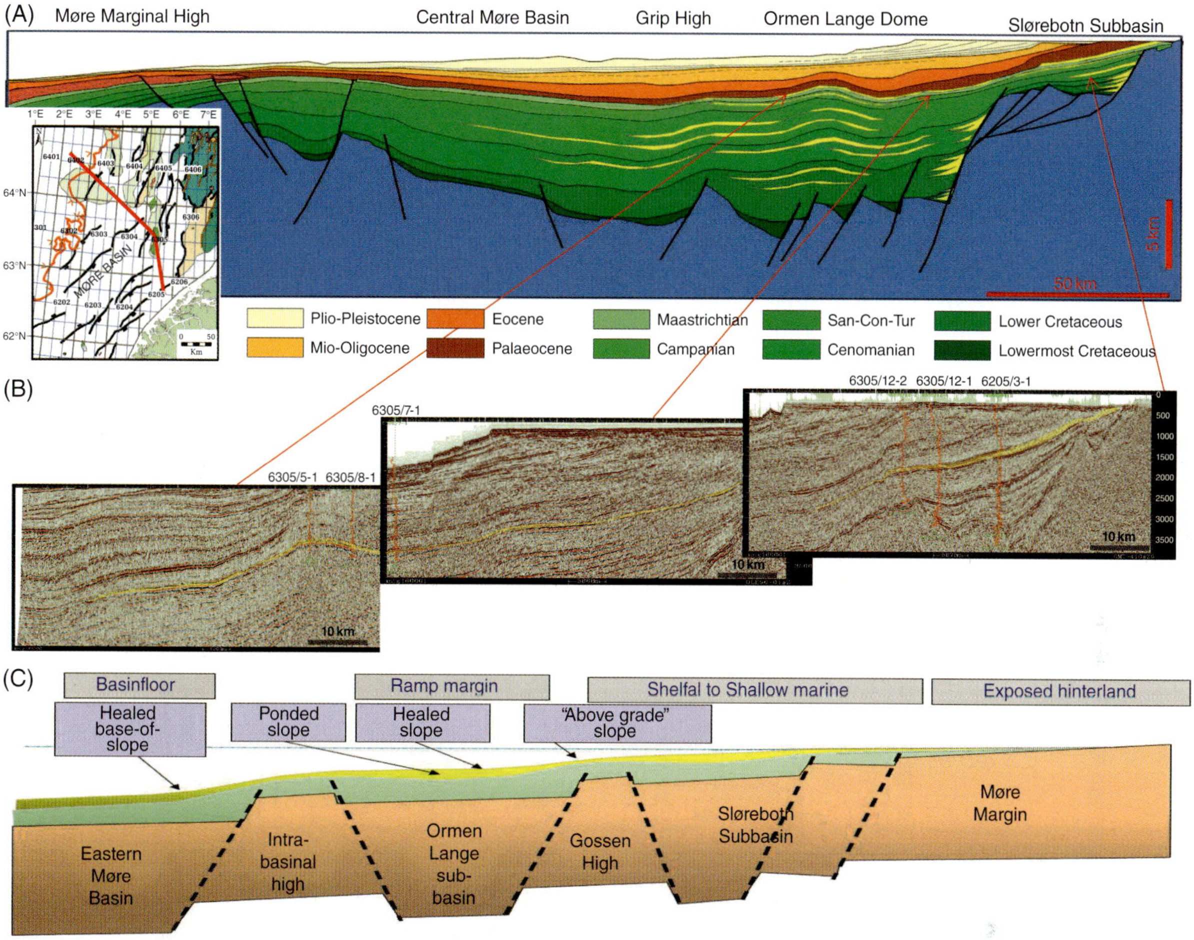

Fig. 3. Partly conceptualised geo-section (A) and seismic lines (B) across the Ormen Lange area. The geo-section runs from the Møre Margin (Slørebotn Subbasin) to the south-east and into the western Møre Basin to the north-west. Only the Cretaceous, Palaeogene and Neogene stratigraphy is shown. Note variations in thickness and seismic character and facies of the Upper Maastrichtian and Danian Ormen Lange reservoir sections from upslope minibasins to the east to the downslope Ormen Lange subbasin to the west. For location of structural elements shown in A, reference is made to Fig. 2B and (C) Inferred basin margin topography during the Maastrichtian and Danian.

& Ebbing, 2008), acted as the south-eastern slope-margin to the Møre Basin during most of the Late Cretaceous and Early Palaeogene (Brekke, 2000; Færseth & Lien, 2002; Martinsen *et al.*, 2005). During the Late Cretaceous to Early Palaeogene rift episode, the Møre Margin and eastern Møre Basin was progressively rotated basinward (i.e. tilted westward) in a stepwise manner during the successive rift phases. The stepwise tilting of the Møre Margin is attributed to temporal variations in subsidence rates of the Møre Basin, as well as uplift rates of the Norwegian hinterlands. This was a response to the successive rift phases that occurred along the western Møre Basin. Basinal subsidence rates increased from the Late Campanian onwards, as the Norwegian-Greenland Sea rift system gradually progressed from slow to rapid rifting rates as the rift evolved towards break-up. Uplift of the Norwegian hinterlands was probably already initiated in the Turonian or Coniacian (cf. Martinsen *et al.*, 1999) and probably was accentuated from the Maastrichtian onwards in response to the transition into the rapidly extending rift interval with increased intensity and degree of structuring.

Reactivation of or compaction over underlying Jurassic subbasins and highs is argued to have progressively changed the slope profile along the Møre Margin in the Late Cretaceous. During the Turonian and Coniacian, the Møre Margin formed the south-eastern, gently sloping, broad ramp margin to the Møre Basin. In the Santonian to Campanian, it changed to an increasingly steeper margin with a more pronounced shelf-slope profile. Increased rates of basinward tilting and structuring during the Maastrichtian to Ypresian resulted in yet another change into a stepped to mildly ponded slope profile. As argued below, stepped or terraced slope topography is favoured above a distinctly ponded slope profile (Fig. 3). Moreover, a lower slope environment is supported as the most probable setting for the Ormen Lange turbidite system (Smith & Møller, 2003), in contrast to Gjelberg *et al.* (2001, 2005), Martinsen *et al.* (2005) and Sømme *et al.* (2009), who favour an inner basin-floor setting.

Maastrichtian and Palaeocene stratigraphy

The Upper Cretaceous to Lower Palaeogene basin-fill of the Møre Basin is overall mud-prone, with the exception of some sandier intervals such as the Turonian to Coniacian Lange and Lysing Formations (Fugelli & Olsen, 2005; Martinsen *et al.*, 2005) and the Upper Maastrichtian to Danian sandstones of the Springar Formation and Egga sandstone unit (Gjelberg *et al.*, 2001, 2005; Smith & Møller, 2003; Sømme *et al.*, 2009), respectively (Fig. 2). The stratigraphic nomenclature applied here is the formal Mid-Norway lithostratigraphy of Dalland *et al.* (1988), in accordance with Smith & Møller (2003) but in contrast to Gjelberg *et al.* (2001, 2005), who favoured the northern North Sea lithostratigraphic scheme.

The Maastrichtian to Palaeocene succession along the Møre Margin has been re-evaluated biostratigraphically and a fully integrated and consistent biozonation of all wells in the eastern Møre Basin has been produced (Fig. 2B). An important outcome of this exercise was the re-dating of the

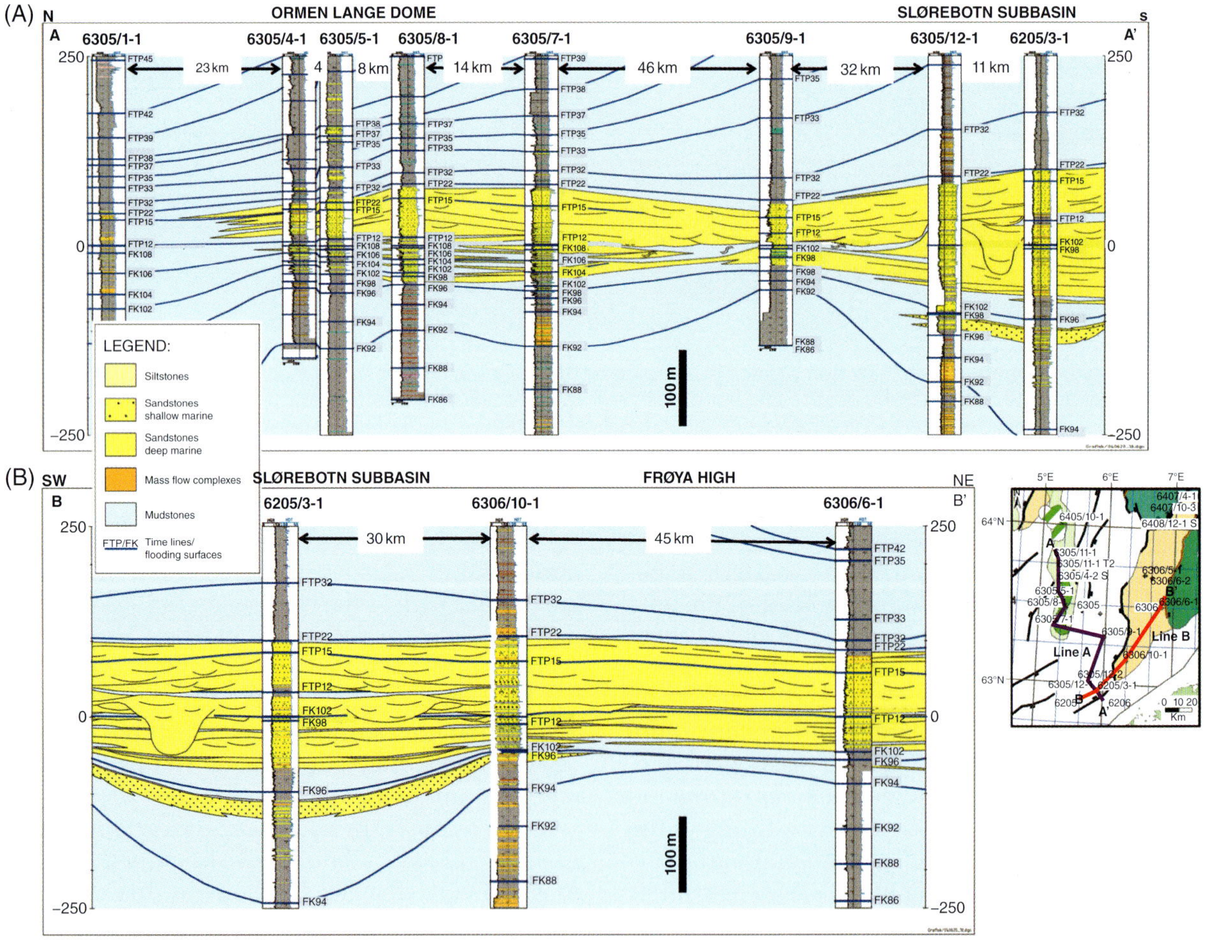

Fig. 4. Well correlations of the Maastrichtian and Palaeocene succession in the eastern Møre Basin. (A) Dip profile (line A on index map) from the upslope Slørebotn Subbasin to the Ormen Lange subbasin forming a downslope terrace. (B) Strike profile across the upslope Slørebotn Subbasin (line B on the index map). The correlations are flattened on Base Palaeocene. Stratigraphic positions of flooding surfaces (FKs and FTPs) are shown in Fig. 2.

lower part of the sandy succession in the Slørebotn Subbasin and on the Gossen High to the south-east of the Ormen Lange area. This succession has now been assigned to the Maastrichtian (as opposed to Danian by Gjelberg *et al.*, 2001, 2005) and should formally be included into the Springar Formation, similar to the Upper Maastrichtian sandstones in the Ormen Lange area. This re-dating has important implications for the tectonostratigraphic evolution of the area, as well as for reservoir correlations and development (Fig. 4 and see further discussion below).

The Upper Maastrichtian to Danian has been divided into two turbidite units or systems (sensu Mutti & Normark, 1987): 1) the Upper Maastrichtian Springar sandstones, including the lowermost Danian Tang Formation sandstones (i.e. the so-called Våle heteroliths); and 2) the Danian Egga sandstone unit (Fig. 4). These two turbidite systems represent a time-span of some 3 to 4 Myr each (Fig. 2B). Both turbidite systems are draped by pelagic claystones that can be correlated across the eastern Møre Basin and onto the Møre Margin and which represent prolonged intervals of sandstone deprivation and basinal sediment starvation. The two turbidite systems are, based on basin reconstructions, seismic stratigraphy as well as biostratigraphy and ichnofacies studies, interpreted as lower slope sandy fan systems (following Piper & Normark, 2001). Jointly they form an overall basinward stepping pattern, herein attributed to the linked basin margin structuring and slope steepening (see below).

Upper Maastrichtian & Danian sandy fan reservoir architectures

An extensive dataset, including field-wide 3D seismic data, 25 exploration and production wells and a total of 390 metres of core, is available from the Ormen Lange Field itself. This is supplemented by a relatively dense, regional 2D seismic grid, which allows direct correlation to other 3D covered areas and wells (some with additional cores) in the greater Ormen Lange (eastern Møre Basin) area.

Lithofacies

The core dataset of the Ormen Lange Field (Figs 5 and 6) has permitted recognition of a series of process-related 'lithofacies' of the muddier (Table 1) and sandier (Table 2) depositional units. The lithofacies identified in core have been grouped to form distinctive classes or units, based on log characteristics (especially from image logs) that have been used to classify and differentiate the uncored sections of the reservoir units. This approach has proven especially relevant for the interpretation of the production wells, which are generally uncored. Image logs have also proven valuable in identifying amalgamation and erosive surfaces. In addition, they have helped delineate the gross orientations of scoured or channelised surfaces and thereby the general orientation of the overall channelised or scoured architectural elements.

The bulk of the sandy facies of the Ormen Lange cored intervals have a massive appearance (Table 2; Gjelberg *et al.*, 2001, 2005; Smith & Møller, 2003), representing the Bouma (1962) T_a division or one or more of the Lowe (1982) S divisions. However, faint stratification defined by vague mineral alignments and alternating intervals of varying grain size populations are present even in these otherwise apparently massive facies, as also suggested by image logs. Accordingly, most of the sandy facies represent alternations of plane bed traction and migrating ripples and 'megaripples'. In consort, this supports deposition from sustained flows (e.g. Baas *et al.*, 2004; Kneller & Branney, 1995) rather than *en masse* dumping, for all of the coarser and sandier lithofacies population (Table 2). Sharp, erosive bases testify to flow erosion, which together with abundant candidate amalgamation surfaces recognised on image logs, support the notion that the thicker sandier units are formed by stacks of amalgamated beds. Bypass indicators are also provided by granule lag lithofacies, present both as channel-base/scour-base lags as well as scoured and rippled bed tops (e.g. see Kneller & McCaffrey, 2003). Subordinate sandy facies include varieties of shale-clast conglomerates (Table 2).

The finer-grained lithofacies population consists of different, commonly incomplete Bouma T_{abcde} turbidite beds (i.e. where only one or some of the Bouma (1962) divisions constitute the bed). These beds represent waning and depletive flow conditions, interpreted to constitute the peripheral to outer part of turbidite beds, bed elements and lobe storeys (see below for a further discussion of terminology). Rare muddy sandstones or debrites of hybrid flow origin (Baas *et al.*, 2009; see also Haughton *et al.*, 2009 and references therein) form localised bed or bedset cappings.

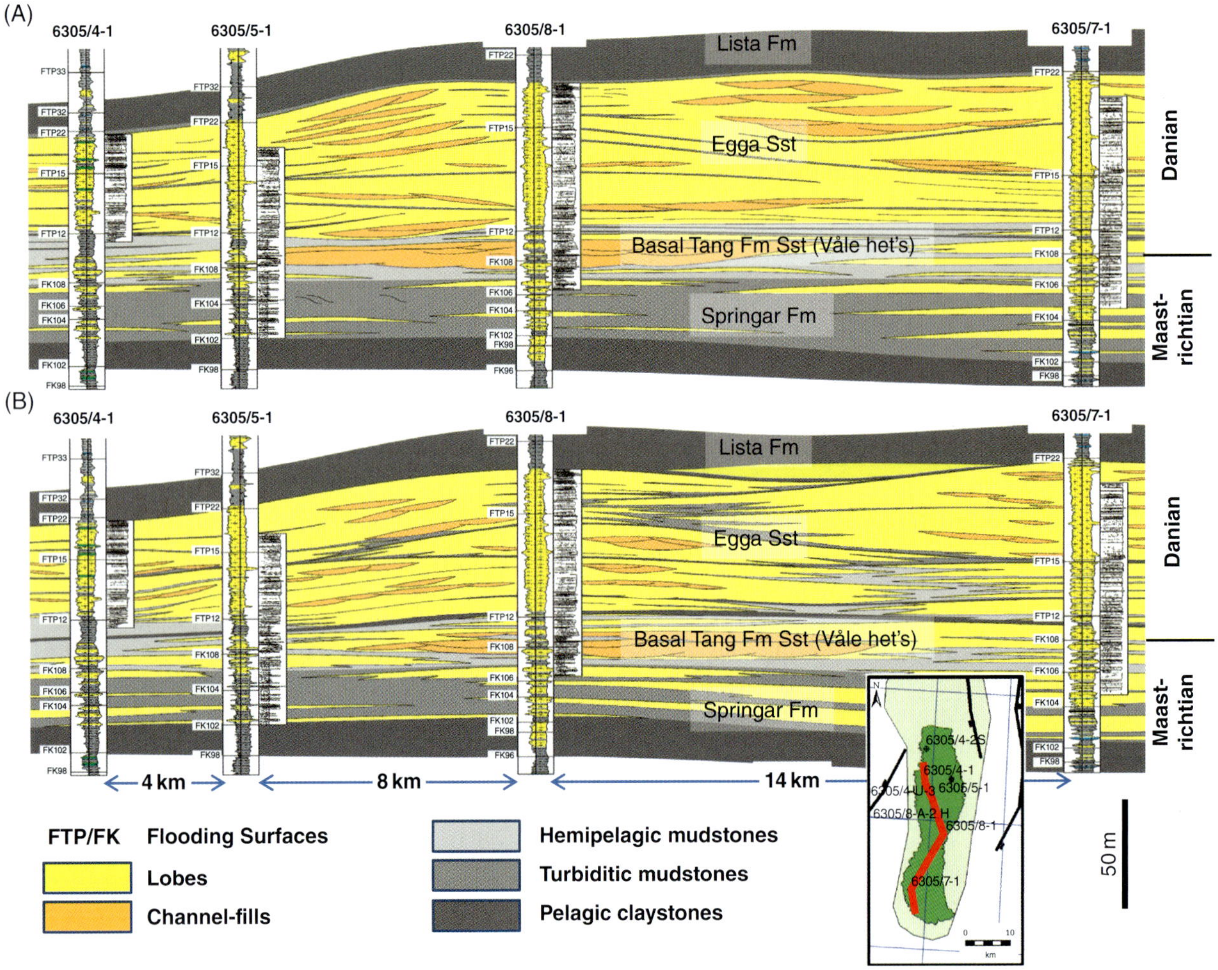

Fig. 5. Well correlation scenarios of the Maastrichtian and Palaeocene succession across the Ormen Lange Field. (A) Illustrates a scenario with laterally migrating lobe storeys and storey sets, whereas (B) illustrates a scenario with laterally shifting lobe storeys and storey sets. The correlations are flattened on Base Danian. Stratigraphic positions of flooding surfaces (FKs and FTPs) are shown in Fig. 2.

The muddier units have been characterised based on their combined lithological and textural features, including presence (or absence) of sedimentary structures, their style and intensity of bioturbation and their contained microfossil types and abundances (i.e. microfacies). Four different types of mudstones were identified; Pelagic, hemipelagic, contouritic and turbiditic. The identification of these various types proved fundamental in recognising architectural elements (see below) and provided further insight into the background depositional setting for the Ormen Lange fan systems. A differentiation and recognition of the origin and depositional processes inferred for this various mudstones were only achieved via the combined lithofacies, biofacies (ichnofacies) and microfacies approach.

Facies associations

The lithofacies have been grouped into a series of facies associations (Table 3) that represents transitional sub-environments or a facies tract continuum across an overall sandy turbidite fan (Fig. 7; see Piper & Normark, 2001). These facies associations include channel-belt sandstones, channel-to-lobe transition zone sandstones, central lobe sandstones, peripheral lobe strata and lobe fringe heteroliths. Two additional facies associations have been identified, consisting either of the anomalously thick and deformed lithofacies and/or the more disorganised lithofacies types. These occur at certain stratigraphic intervals and potentially represent repeated changes in the overall fan evolution, as discussed below.

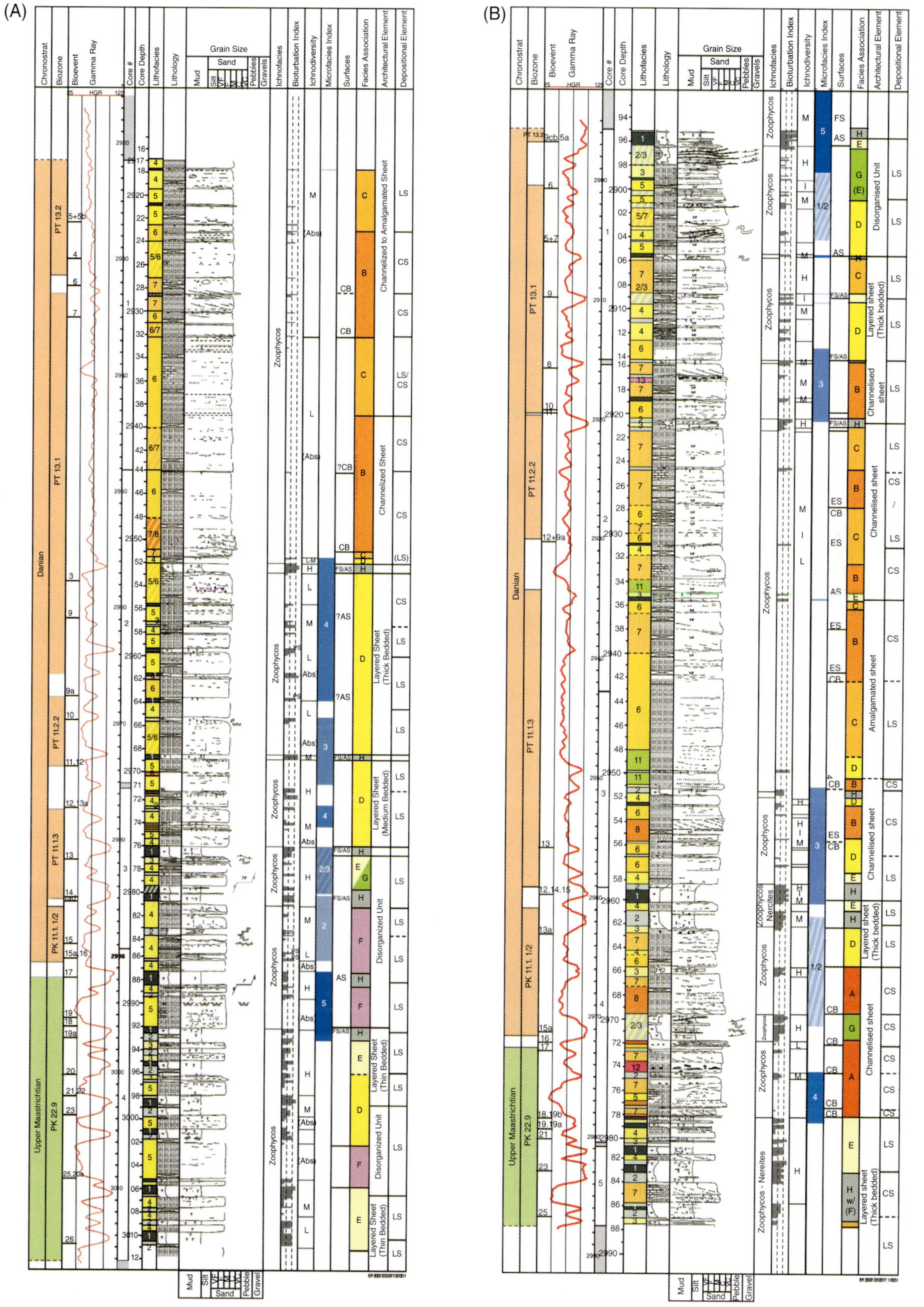

Fig. 6. Core descriptions of the Springar and Egga sandstones and their equivalents in the southern part (well 6305/7-1; A), the central part (well 6305/8-1; B) and the northern part (well 6305/1-1; C) of the Ormen Lange subbasin. Lithofacies coding refers

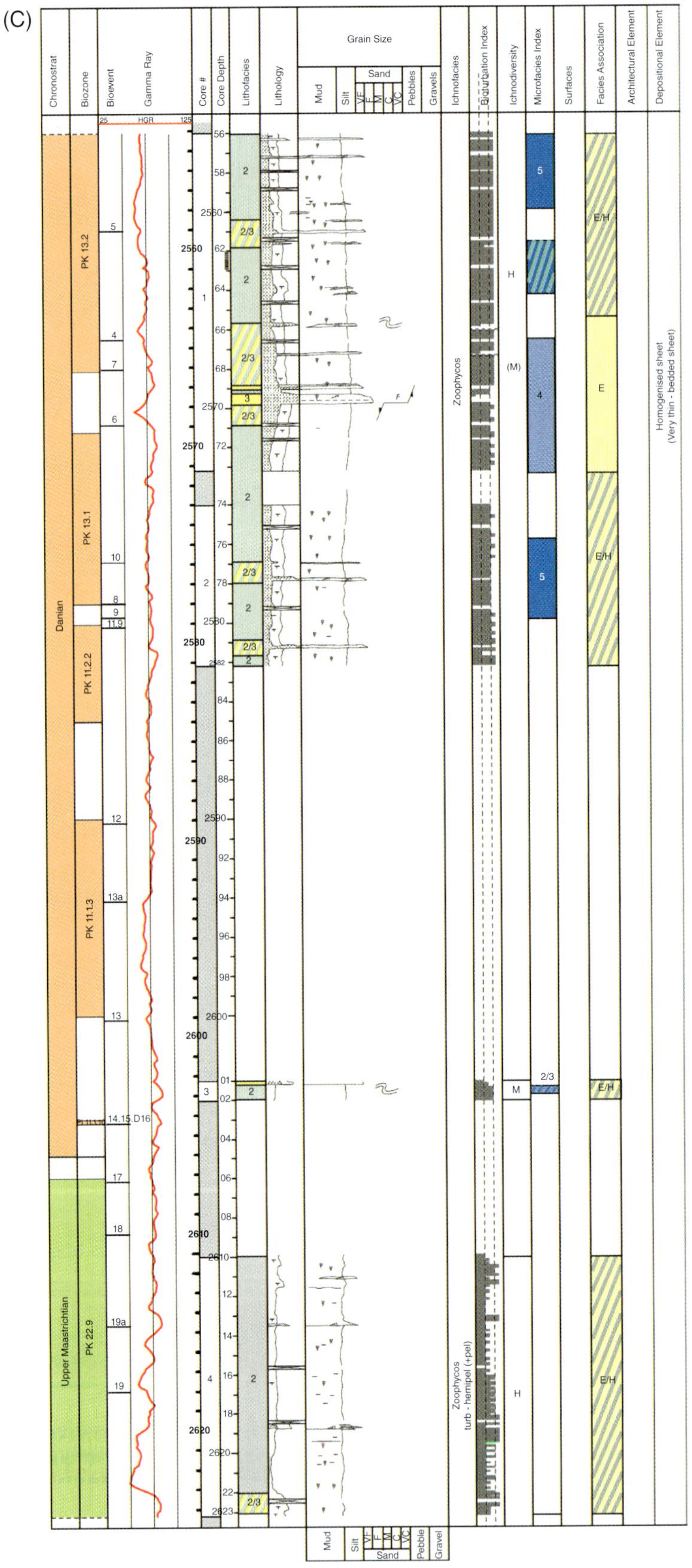

Fig. 6. (*Continued*) to Tables 1 and 2, whereas Facies Association coding refers to Table 3 Listed Bioevents are Ormen Lange specific field-wide events used to guide correlation and reservoir zonation across the Ormen Lange Field. Ichnodiversity is relative and are marked as low (L), medium (M) and high (H) reflecting less than five, five to ten, or more than ten different traces/species, respectively. (Abs) denotes intervals with no traces. Microfacies range from highly stressed, current or turbulence dominated environments represented by Microfacies type 1 to fully relaxed, quiet bottom conditions represented by Microfacies type 5. Microfacies types 2 through 4 represent transitional microfacies with gradual diminishing near sea-bed turbulence and current regimes. Identified surfaces are flooding surfaces (FS), abandonment surfaces (AS), erosional/amalgamation surfaces (es) and channel/channel complex bases (CB). Inferred depositional elements are channel storeys (CS) and lobe storeys (LS).

Table 1. Muddy lithofacies; their characteristics and inferred depositional processes. The mudstones are given genetic names to ease the cross-reference and readability of the text. Numbers in coloured boxes in left column refer to lithofacies coding in Fig. 6. The width of individual core slabs is 10 centimetres.

Shale Type	Lithology	Characteristics BI & Diversity	Microfacies Index	Core Appearance	Depositional Architecture
1 **Pelagic**	Dark to dark grey, rarely dark greenish grey claystones	High BI Medium to Low Diversity	High		Extensive blankets
Hemi-pelagic	Greyish and greenish grey claystones and mudstones	Medium to High BI Medium to High Diversity	Medium to High		Fairly extensive blankets
2 **Contouritic**	Green to grey mudstones and siltstones, sometimes ripple cross-laminated	Low to medium BI Medium to High Diversity	Low to Medium		Lobate blankets to patchy (erosional remnants)
Turbiditic	Grey and dark grey mudstones and siltstones, laminated	Low BI Low Diversity	Low		Lobate blankets

The channel-belt association consists of erosively based, commonly amalgamated bedsets (Table 3; Figs 8A and 9) of stratified and cross-stratified sandstones, ripple cross-laminated sandstone lags, granule sandstones and shale-clast conglomerates, as well as interbedded turbiditic and rare hemipelagic mudstones (see Tables 1 and 2 for interpretations of lithofacies). Locally, also, deformed and slumped strata are present. The bedsets normally form fining-upwards, or more rarely coarsening-to-fining upwards motifs. These are, based on the component facies, the mixture of coarse and heterolithic lithologies, as well as the vertical facies successions, interpreted to represent submarine channel-fills (Clark & Pickering, 1996). Such channel-fill motifs are common from sandy submarine fan systems in a variety of settings (Gardner & Borer, 2000; Gardner *et al.*, 2003; Grecula *et al.*, 2003a; Milli *et al.*, 2007; O'Byrne *et al.*, 2007; Sweet *et al.*, 2007), with individual motifs representing single-storey channel-fills. Basal boundary surface was probably concave upwards and erosional, as inferred from borehole dip-meter data. The common amalgamated character of these compound motifs suggests that they represent channel complexes (or storey sets; e.g. Eschard *et al.*, 2003; Grecula *et al.*, 2003a; Euzen *et al.*, 2007) at the outlet of slope tributary systems (e.g. Johnson *et al.*, 2001; Decker *et al.*, 2004; Gervais *et al.*, 2004, 2006; Hodgson *et al.*, 2007; Deptuck *et al.*, 2008). Channel sinuosity (as inferred from seismic data) is low (less than 1.2; e.g. see Figs 10 and 11).

The channel-to-lobe transition/braided fan association comprises amalgamated and scoured bedsets (Table 3; Figs 8B and 9) of stratified and cross-stratified sandstones, as well as rare shale-clast conglomerates. The bedsets are embedded in hemipelagic and/or contouritic mudstones, range from a few to more than nine metres-thick and define simple thinning-upward and fining-upward motifs or more complex thickening-upwards and coarsening-upwards to thinning-upwards and fining-upwards motifs. Interbedded claystones are relatively uncommon. Erosive surfaces was again concave upwards, although dipmeter data suggests a more gentler relief compared to the boundary surfaces of the proper channel storey fills (see

Table 2. Sandy lithofacies; their characteristics and inferred depositional processes. Numbers in coloured boxes in left column refer to lithofacies coding in Fig. 6.

Facies type	Characteristics	Log motif	Depositional process	Depositional geometry
3 **Fine-grained, normally-graded sandstones**	Sharp-based, normally graded, laminated to ripple cross-laminated, fine to very fine, thin-bedded sandstones and siltstones. Gradational tops – often bioturbated. Few cm to dm-thick lamina and beds.	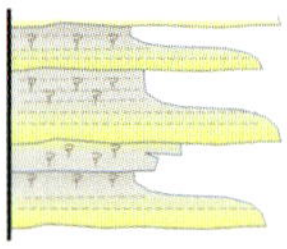	Fall-out from suspension from turbulent flow, followed by upper to lower flow regime, unidirectional tractional flows.	Tabular beds tapering out downflow.
4 **Normally-graded, massive to stratified sandstones**	Sharp erosively based, normally graded, laminated to ripple cross-laminated, medium to fine, thin to medium-bedded sandstones. Sharp or thin, gradational tops. 10+ cm to dm-thick beds.	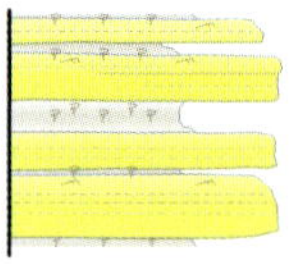	Fall-out from suspension from erosive turbulent flow, followed by tractional, waning depletive flow.	Tabular beds thinning downflow.
5 **Massive to dewatered sandstones**	Sharp, locally erosively-based medium to thick-bedded, massive, occasionally dewatered, medium to coarse, granule sandstones. Basal shear zones and internal small-scale normal faulting locally present. Dish structures and sandstones pipes/ pillars common in some beds. Dm to few m-thick beds to amalgamated units several metres thick.		Rapid fall-out from high-density sediment gravity flows, followed by post-depositional dewatering, fluidisation and remobilisation.	Tabular beds to elongate tongues; central to peripheral part of lobate beds.
6 **Ungraded, stratified sandstones**	Sharp, erosively-based, medium-bedded to thick-bedded, inversely graded, ungraded, and normally graded, vaguely stratified, medium to fine grained sandstones alternating with massive intervals. Plane-parallel to wavy stratifications and low-angle truncations. Beds are amalgamated or have sharp tops. Metres-thick amalgamated intervals.		Fall-out from sustained high-density sediment gravity flows as aggrading traction carpets (space stratification) and shear lamina.	Inner to central part of lobes and as scour fills.
7 **Cross-stratified plough & fill sandstones**	Erosively-based, alternating cross-stratified and massive, coarse to fine grained, sometimes pebbly sandstones. Cross-stratification normally defined by crude mineral alignments. Basal and internal scours are claystone draped.		Repeated scouring and ploughing followed by fall-out from sustained high-density sediment gravity flows with or without tractional movement as upper flow regime migrating megaripples.	Scour fills of channel-to-lobe transition area and central lobes.
8 **Cross-stratified sandstones**	Sharp, erosively based, inversely to normally graded, medium to thick-bedded, vaguely to clearly planar and trough cross-stratified, coarse to fine grained, sometimes pebbly sandstones. Beds/sets/cosets are amalgamated or have sharp tops. Dm-thick beds and bedsets, cosets up to few metres thick.	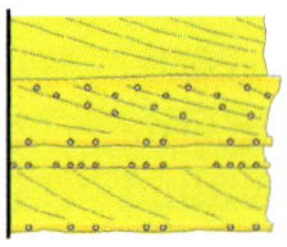	Fall-out from sustained high-density sediment gravity flows followed by tractional movement as upper flow regime migrating megaripples.	Outer distributary channel, channel-to-lobe transition megaripples to inner to central lobe scour fills.

Table 2. (*Continued*)

Facies type	Characteristics	Log motif	Depositional process	Depositional geometry
9 **Ripple cross-laminated sandstones**	Thin, erosively based, normally to ungraded, medium to fine grained, and occasionally coarse to granule sandstones that form thin caps to underlying sandstones.	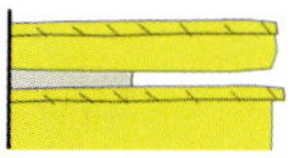	Tractional movement of ripple trains as bypass lags	Bed cappings in inner and central part of lobes; bypass indicator.
10 **Disorganised granule sandstones**	Sharp, erosively-based and sharp-topped, thin-bedded to medium-bedded, massive, disorganised granule sandstones. Sometimes ripple cross-laminated.	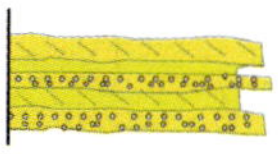	Rapid dumping and winnowing from high-density sediment gravity flows as bypass lags.	Channel- and scour-base lags in inner fan distributary channel and channel-to-lobe transition zone; bypass indicator.
11 **Disorganised mud-chip sandstones**	Sharp, erosively-based, thin-bedded to medium-bedded, massive, disorganised matrix-rich sandstones with abundant floating mud granules and mud chips. Bed/set tops are sharp or amalgamated.	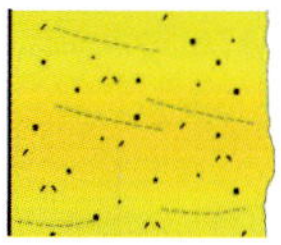	Rapid dumping from highly erosive high-density sediment gravity flows.	Channel-mouth, channel-to-lobe transition and inner lobe irregular units.
12 **Clay clast conglomerates – bed base clusters**	Clast-supported to matrix-supported conglomerates with cm-sized angular claystone rip-up clasts floating in a sandstone matrix composed of poorly sorted, ungraded, coarse to medium grained sandstone, often with abundant clay chips.		Rapid dumping from highly erosive high-density sediment gravity flows.	Channel-base lags and channel-margin collapse breccias.
13 **Clay clast conglomerates – clustered floating clasts**	Matrix-supported and clast-supported conglomerates with cm-sized subangular to rounded claystone rip-up clasts floating in upper part of beds or as bed cappings. Matrix composed of medium sorted, ungraded, coarse to medium grained, clay matrix-rich sandstone.		Rapid dumping from high-density sediment gravity flows; clasts kept aloft by dispersive pressure and high grain density.	Localised bed cappings and intra-bed pockets.
14 Mud-rich silty sandstones	'Dirty', commonly structureless, pebbly sandstones with abundant mudstone matrix. Pebbles are dominantly fragmented, angular to subrounded claystones and mudstones. Wispy lamination indicative of post-depositional dewatering and slumping locally present.	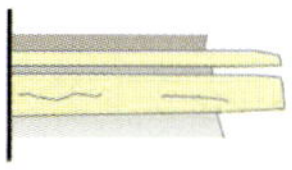	Frictional freezing of debris-flows deposits, or collapse of muddy flows to form so-called 'slurry-flows', in some cases followed by early post-depositional gravitational deformation as a 'flow-slide'.	Elongate tongues and fringes of lobate beds.

Table 3. Facies Associations; their characteristics, sedimentary architectures, geometries and inferred depositional sub-environment. Letters in coloured boxes in left column refer to facies association coding in Fig. 6. The length of an individual core slab is 1 metre.

Facies Association	Characteristics	Log motif	Sedimentary Architecture	Depositional sub-environment
A **Channel fills or channel-belts**	Erosive-based, amalgamated metres-thick bedsets. Fining upwards or rare coarsening-to-fining upwards motifs. Component facies are stratified and cross-stratified sandstones, ripple cross-laminated sandstones, granule sandstones, shale-clast conglomerates. Interbedded mudstone packages, sometimes comprised of slumped strata.	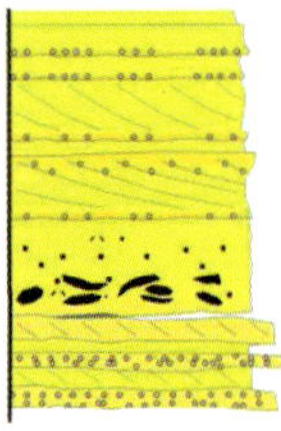	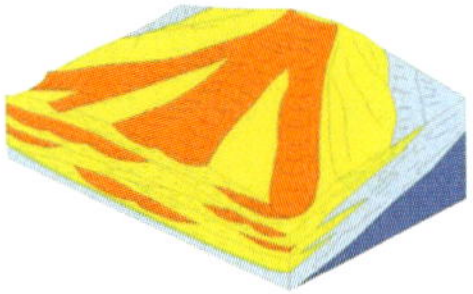	Inner fan distributary channels and channel-belts.
B **The Channel-to-Lobe transition/ Braided Fan association**	Layered to amalgamated or scoured bedsets, few to nearly nine metres-thick. Simple thinning and fining upward motifs, or more complex thickening and coarsening to thinning and fining upwards motifs. Component facies include stratified and cross-stratified sandstones as well as rare shale clast conglomerates embedded in hemipelagic and/or contouritic mudstones. Interbedded claystones range from relatively common to few.	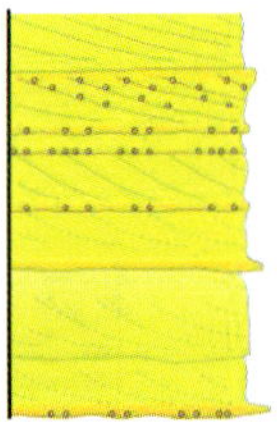	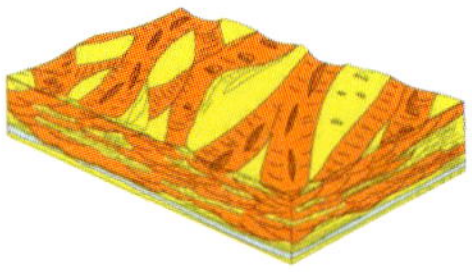	Inner-fan to mid-fan distributary channels and channel-belts as well as mid-fan channel-fills and multilateral channel storey sets. Channel to lobe transitional areas.
C **The Central Lobe association**	Layered to amalgamated bedsets of several metres-thick. Near symmetrical to asymmetrical coarsening and thickening to fining and thinning upwards motifs as well as irregular bedset motifs. Component facies are medium to thick-bedded stratified and cross-stratified, as well as some 'typical' turbidite Tabc and massive sandstone facies. Stratified and massive sand-stones are commonly amalgamated whereas hemipelagic mudstones separate basal Tabc beds and stratified sandstones.		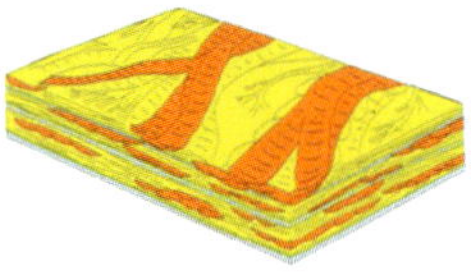	Channelised/ scoured to un-channelised lobes in mid-fan and outer-fan settings. Represents the central part of individual lobes (lobe storeys).

Table 3. (*Continued*)

Facies Association	Characteristics	Log motif	Sedimentary Architecture	Depositional sub-environment
D **The Peripheral to Marginal lobe association**	Layered, commonly coarsening and thickening to fining and thinning upwards motifs or bedsets up to more than five metres-thick. Component facies comprise thin to thick-bedded Tabcd and Tbcd turbidite beds as well as medium to thick-bedded stratified sandstones. Rare and thin matrix rich debris-flow or slurry flow deposits occur infrequent. The sandstones are separated by thin, eroded turbiditic and/or hemipelagic mudstones, or by rare amalgamation surfaces.		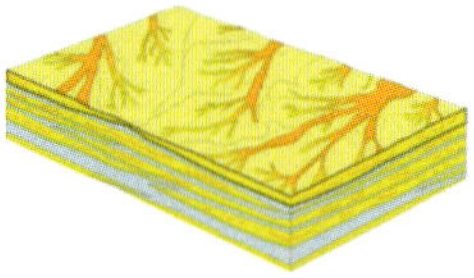	Peripheral part of individual lobes (lobe storeys) in mid-fan and outer-fan settings.
E **The Lobe Fringe association**	Layered, commonly coarsening and thickening to fining and thinning upwards motifs or bedsets up a few metres-thick. Component facies consist of thin to medium bedded Tbcd and Tcd turbidite beds. The sandstones are separated by thin to thick, turbiditic and hemipelagic mudstones.	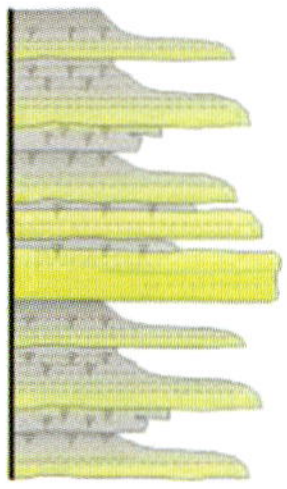	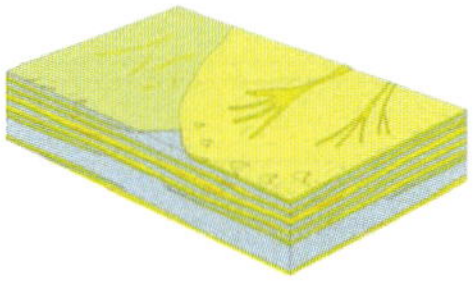	Outermost margin of individual lobes (lobe storeys) in outer-fan as well as mid-fan settings.
F **The Megabed association**	Variably thick layered packages of thick to very thick, massive and stratified sandstones, deformed, slumped sandstones, and some rare turbidites. The sandstone facies alternates with turbiditic, hemipelagic as well as pelagice mudstones and claystones, sometimes with sandstone injections or sandstones dikes cross-cutting mudstone laminations.	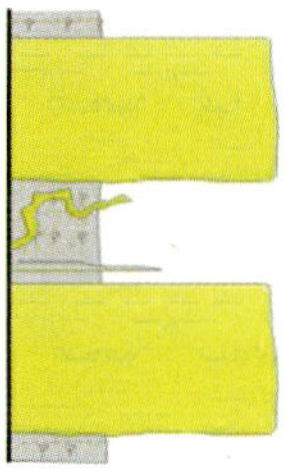	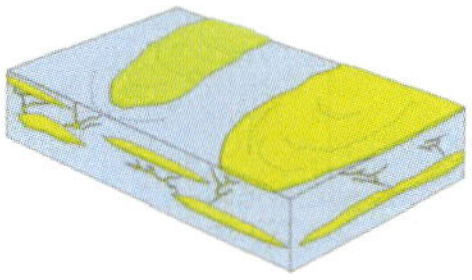	Debris flow tongues and other mass transport deposits in inner-fan to mid-fan settings.
G **The Slump-Bed association**	Metres-thick packages of deformed, slumped and thrust-faulted Bouma-type turbidites alternating with coarse to pebbly massive sandstones.			Slump-folded channel-mouth lobes or alternatively overbank or inter-channel crevasse splays.
H **The Back-ground Slope-to-Basin Mudstone associations**	Interbedded hemipelagic and turbiditic mudstones that alternates at centimetre to decimetre scale intervals. Millimetre to centimetre-thick silty to sandy (very fine) distal turbidite stringers and lamina are frequent in the turbidite mudstones. Other intervals are dominated by alternations of pelagic claystones and hemipelagic mudstones.	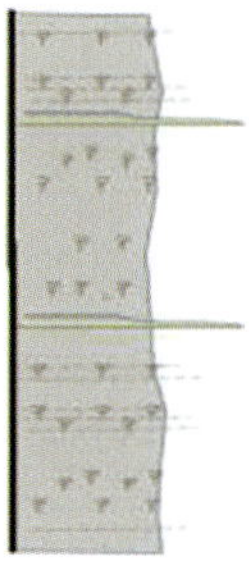	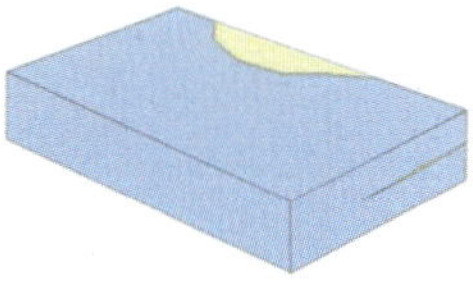	Slope to inner basinfloor mudstone and claystone sheets.

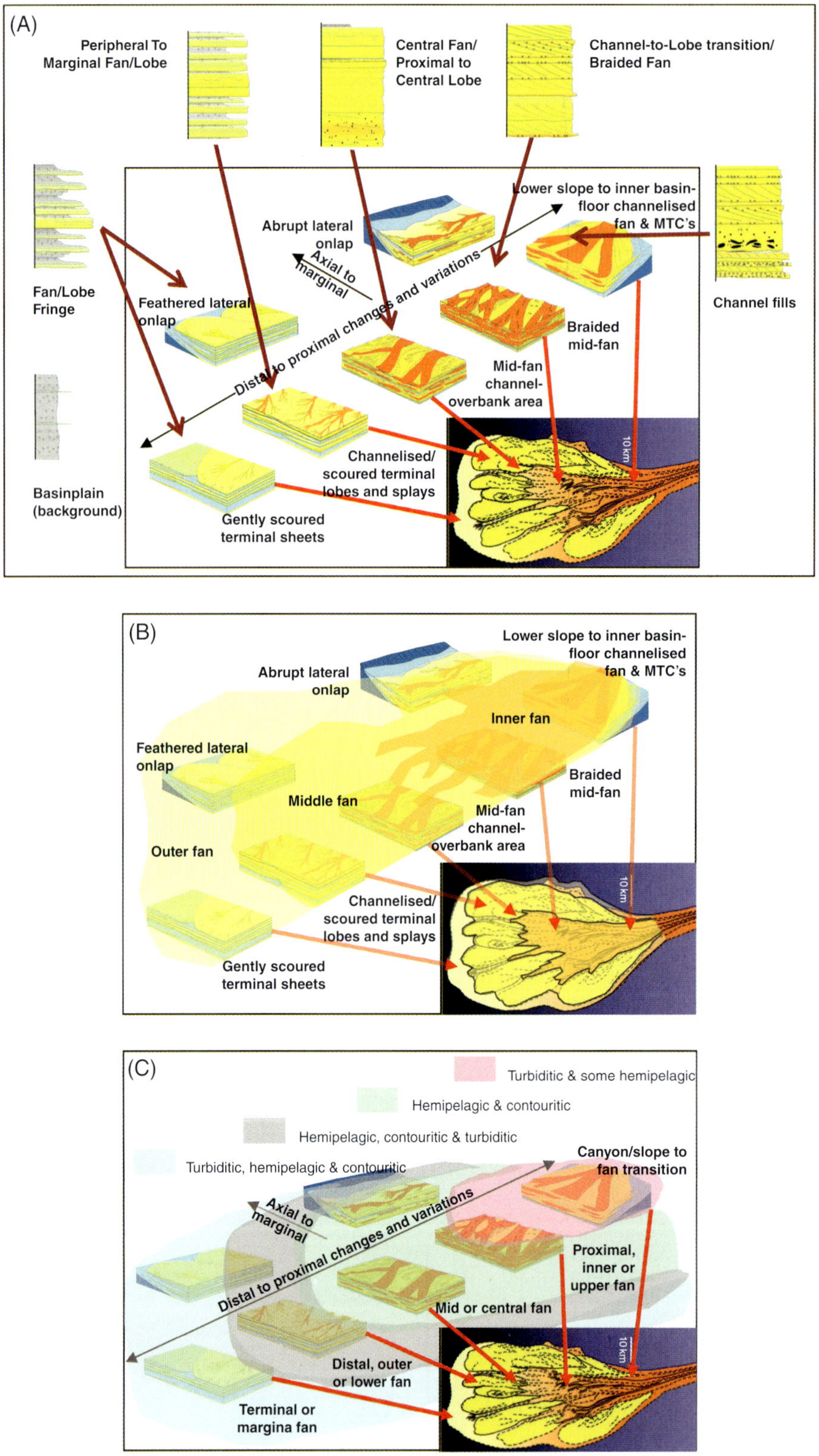

Fig. 7. (A) Idealised downfan partitioning of the sandy fan sub-environments and facies associations (See Table 3), with idealised facies motifs from the various sub-environments/facies associations. (B) Inferred fan sub-environments based on Ormen Lange cores and analogue datasets; and (C) mudstone partitioning inferred for a sandy slope fan based on the Ormen Lange core dataset.

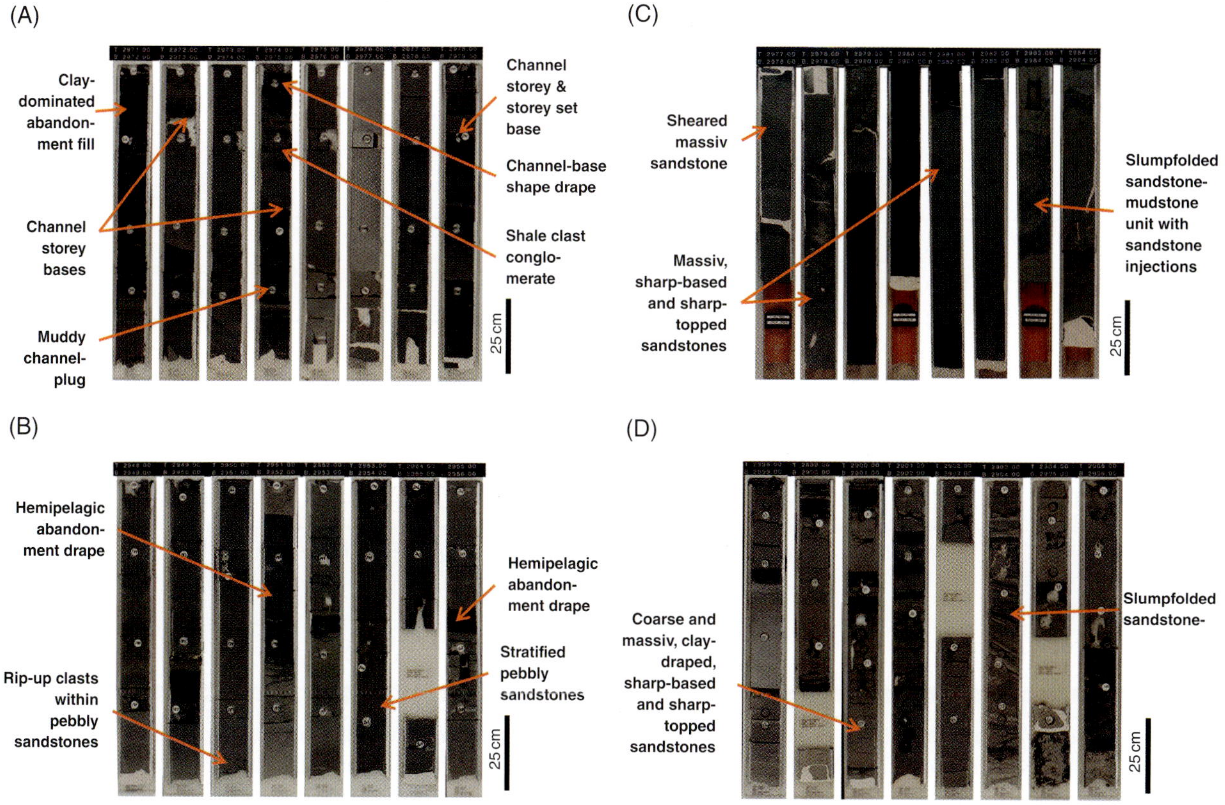

Fig. 8. Core photos illustrating typical features of the coarser and deformed facies associations. (A) Channel-belt association, with multistorey channel-fill unit, basal Danian or basal Tang Formation sandstones (i.e. the Våle heteroliths), 6305/8-1. (B) Layered braided fan or channel-to-lobe transition association comprised by coarse to pebbly sandstones, basal Danian Egga sandstone unit, well 6305/8-1. (C) Megabed association with sharp-based and sharp-topped, thick to very thick-bedded massive, medium-grained sandstones, some with sheared basal parts and interbedded slump-deformed sandstone-mudstone couplets with sandstone injectites, uppermost Springar Formation sandstones, well 6305/7-1. (D) Deformed central to peripheral lobe / crevasse splay facies alternating with clay-draped sharp-based and sharp-topped, thick, coarse-grained sandstones, the latter representing channel-to-lobe transition facies, Danian upper Egga sandstone unit, well 6305/8-1. Length of an individual core slab is 1 metre, whereas width is 10 centimetres.

above). The component facies, together with abundant scoured basal bed boundaries, point to deposition from repeated, shallow channelised or scouring flows that migrated relatively freely across the sediment (fan) surface. The channels were filled by and alternated laterally with down-current migrating bedforms or bars (Figs 8B and 10; Table 3). Such features are common in a 'braided' channelised lobe or the channel-to-lobe transition area (e.g. Gardner *et al.*, 1996; Johnson *et al.*, 2001; Lien *et al.*, 2003; Decker *et al.*, 2004; Gervais *et al.*, 2006; Milli *et al.*, 2007; Sweet *et al.*, 2007; Deptuck *et al.*, 2008: Jegou *et al.*, 2008), which is favoured as the overall setting for this association. The sandstone strata represent both channel (basal lags and barforms) as well as interchannel (overbank wedges and splays) deposits. Bedset motifs probably reflect variable scenarios of channel-filling and avulsion or lateral migration (e.g. Grecula *et al.*, 2003a; O'Byrne *et al.*, 2007).

The central lobe association consists of layered to amalgamated packages of medium-bedded to thick-bedded, stratified and cross-stratified turbiditic sandstones, as well as some Bouma-type T_a (thick, massive sandstones) and T_{ab} turbidites. Stratified and massive sandstones are commonly amalgamated, whereas thin hemipelagic mudstones separate T_a/T_{ab} beds and stratified sandstones. The facies are arranged in coarsening-upwards and thickening-upwards to fining-upwards and thinning-upwards as well as more irregular vertical stacking motifs, typically several metres to nearly ten metres-thick (Table 3; Fig. 9). This association represents a mixture of deposits from alternating channelised or

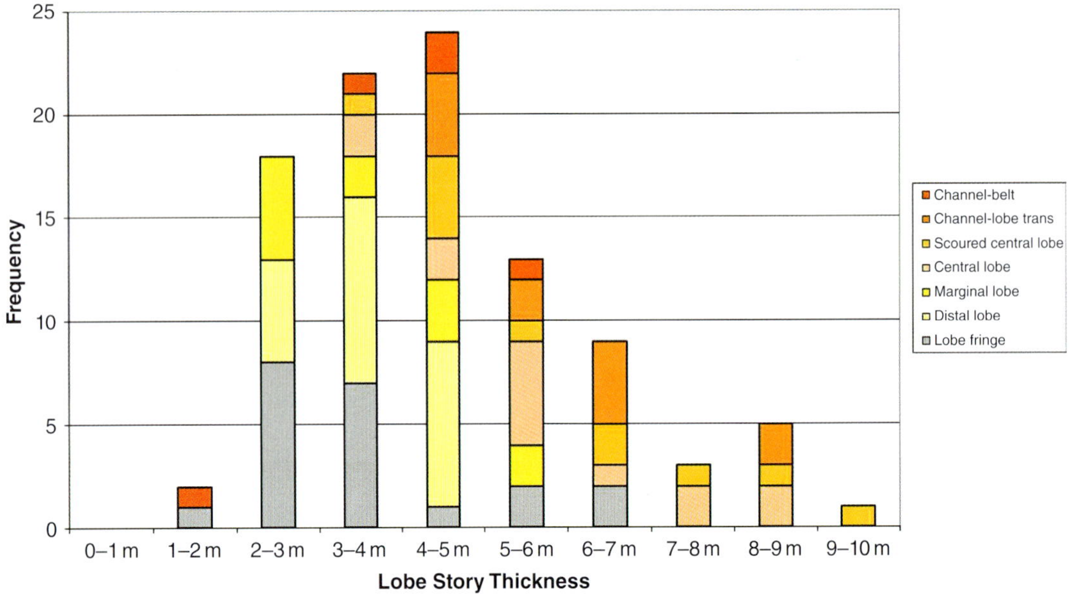

Fig. 9. Lobe storey thickness data from the Ormen Lange exploration wells core dataset.

scouring and waning-depletive flows (Kneller & Branney, 1995; Kneller and McCaffrey, 2003). The relative coarse nature of the sandstones and the high degree of bed amalgamation is typical of the central or middle lobe deposits (e.g. Johnson *et al.*, 2001; Decker *et al.*, 2004; Gervais *et al.*, 2006; Deptuck *et al.*, 2008; Jegou *et al.*, 2008; Prélat *et al.*, 2009; Mulder *et al.*, 2010; Marini *et al.*, 2011). Mounded geometry is suggested by vertical variations in bedding surfaces as indicated by dipmeter data.

The peripheral to marginal lobe association consists of layered packages up to 6 m thick (Table 3; Fig. 9), of thin-bedded to thick-bedded Bouma-type T_{abcd} and T_{bcd} beds, as well as medium-bedded to thick-bedded stratified sandstones. Rare, thin, matrix-rich debris-flow or slurry-flow deposits occur locally. The sandstones are separated by thin, erosionally truncated turbiditic and/or hemipelagic mudstones, or by rare amalgamation surfaces. The relatively finer grained and layered character, together with its component facies, suggest a more distal setting compared to the central lobe association (e.g. Mutti, 1992; Gardner *et al.*, 2003; Milli *et al.*, 2007; Marini *et al.*, 2011). The characteristics of this association is typical for deposits of middle-to-outer or marginal-distal parts of turbidite lobes (e.g. Johnson *et al.*, 2001; Decker *et al.*, 2004; Gervais *et al.*, 2006; Deptuck *et al.*, 2008; Jegou *et al.*, 2008; Prélat *et al.*, 2009; Mulder *et al.*, 2010; Marini *et al.*, 2011), further supporting such a setting for this association. Dipmeter data support an overall marginal lobe-setting with upper

Fig. 10. (A) Seismic line across the Ormen Lange field. Note thickened reservoir units (between Top Danian and Base Egga Sandstone and Base Egga Sandstone and Base Springar Sandstone) in the southern and northern part of the Ormen Lange subbasin. Note also changes in seismic facies between the two depositional thicks or depocentres; from a more layered and base-lapping high-reflective seismic facies across the southern to a more transparent seismic facies across the northern Ormen Lange depocentre. (B) Internal architectures in the southern part of the Ormen Lange subbasin: Proximal parts of the depocentre are dominated by localised thickened units with internal shingles, probably representing channel storeys or storey sets, whereas the intermediate parts are dominated by broad mounded features and low-relief shingles representing depositional lobes, in areas partly channelised. Note width of individual features of more than 6 km and thicknesses up to 50 ms TWT. Seismic thicknesses suggest that these features represent compensationally-stacked channel and lobe storey sets or complexes. (C) Illustrates a series of narrow as well as broader channel features (marked by arrows) which are incised into the middle and upper part of the Egga sandstone unit. The low-sinuosity channels and channel storeys/complexes, located across the southern depocentre or the southern part of the Ormen Lange Field, have a gross ESE–WNW orientation. Seismic sections are flattened on the Top Egga Sandstone reflector. Candidate channel features present across the northern depocentre have a more SE–NW orientation (not shown in the figure).

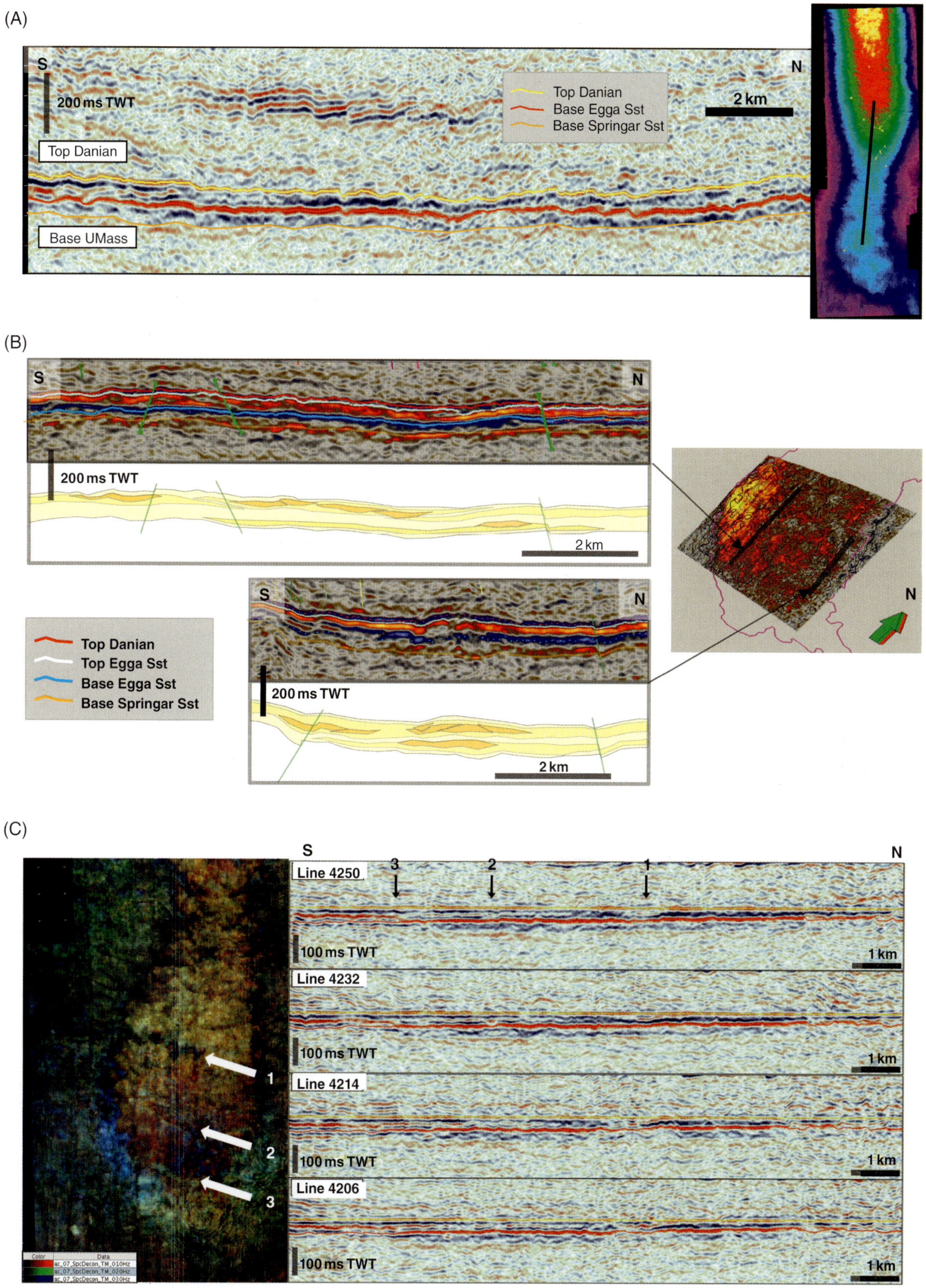
(A)
S
N
200 ms TWT
Top Danian
Base Egga Sst
Base Springar Sst
2 km
Top Danian
Base UMass
(B)
S
N
200 ms TWT
2 km
Top Danian
Top Egga Sst
Base Egga Sst
Base Springar Sst
(C)
Line 4250
Line 4232
Line 4214
Line 4206
100 ms TWT
1 km
1
2
3

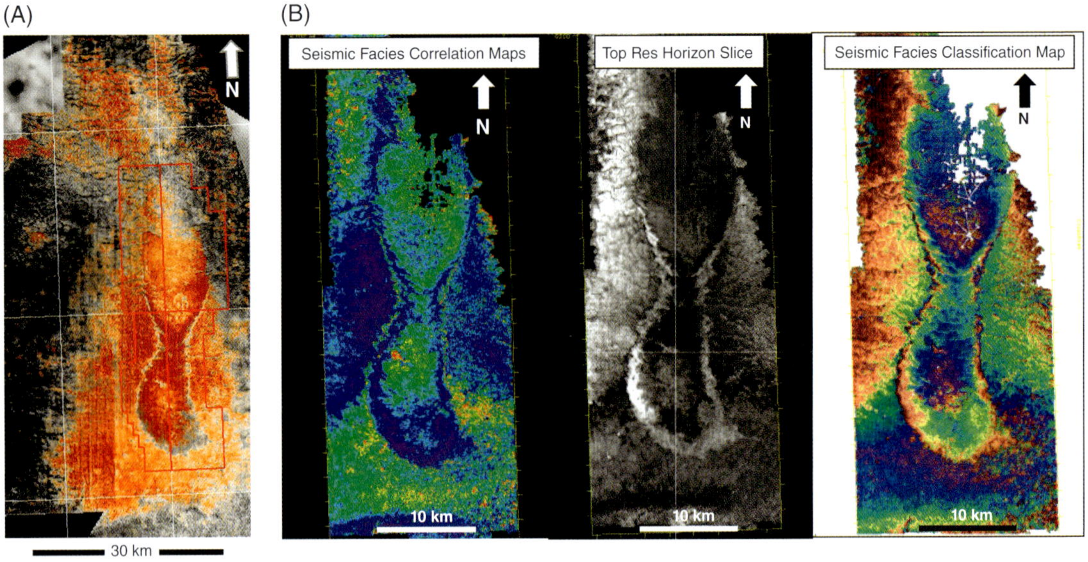

Fig. 11. (A) Seismic amplitude map illustrating the reservoir fairway of the Egga sandstone unit in the Ormen Lange subbasin. Note presence of ribbon-like features across depocentres in southern and northern part of the field (see also Fig. 11B) as well as extending towards the west into the adjacent downslope subbasin. These features probably represent low-sinuosity channel fills and fairways. Irregular to rounded margins of outer fairway is attributed to distal pinch- or shale-out of individual lobe storeys and complexes. The map represents partially transparent seismic amplitudes over a window covering the upper parts of the Danian reservoir play. The rate of transparency is gradually changing from less transparent on the positive amplitudes (red), to more transparent on the negative amplitudes (black). This methodology utilises the Shell proprietary 'Seismic Sculptor' software and is especially useful for visualisation and enhancements of laterally varying depositional patterns and geometries, such as seen for the Danian turbidite fairway (Egga Mbr) on Ormen Lange. (B) Seismic facies correlation map, top reservoir horizon slice semblance map and seismic facies classification maps. The seismic facies classification, correlation and horizon slices suggest lateral thickness and facies variations across the Ormen Lange field, with thicker and sandier facies representing two depocentres in the northern and southern parts of the field, respectively. Note however the strong imprint of fluid contact inhibiting direct correlation of seismic facies and amplitudes within to outside the Ormen Lange field area.

bounding surface, normally suggesting a broader and lower relief, convex upwards geometry of individual bedset motifs.

The Lobe Fringe association consists of isolated, thin-bedded to medium-bedded, incomplete (base-absent or top-absent) Bouma-type T_{abcd} turbidite beds, embedded in turbiditic and hemipelagic mudstones (Table 3). A gradational transition occurs from the near absence of turbidite beds (translational into the background turbiditic mudstone association) to well-layered packages of predominantly medium-bedded T_{abcd} and T_{bcd} turbidites alternating with background mudstone facies. The background facies of this association consists of nearly equal amounts of turbiditic and hemipelagic mudstones, as well as commonly abundant contouritic mudstones. The component facies and thinness of the interbedded sandstones in consort with the high proportion of background mudstones, is typical for deposit for the outermost parts of turbidite lobes (e.g. Twichell *et al.*, 1995; Johnson *et al.*, 2001; Remacha & Fernandez, 2003; Janocko, 2008; Prélat *et al.*, 2009; Marini *et al.*, 2011). This association is accordingly interpreted to represent marginal and distal lobe fringe deposits.

The Megabed association consists of layered packages of massive sandstones, deformed (sheared) and slumped sandstones, some stratified sandstones and rare Bouma-type T_{abc} turbidites (Table 3; Fig. 8C). Inter-well correlations of the massive sandstone beds are poor, suggesting limited lateral extent to the individual beds. The sandstone facies alternates with turbiditic, hemipelagic as well as pelagic mudstones and claystones, locally with sandstone injections or sandstone dikes cross-cutting mudstone laminations. This association represents rare incursions of event-type beds of predominantly debris flows origin (Lowe, 1979, 1982; Shanmugam, 2000) into the background depositional environment. The laterally confined nature of individual beds, potentially suggest a more tongue-like geometry. The co-occurrence with deformed strata and injectites suggest both basin-floor instability as well as post-depositional remobilisation. The presence of sandy injectites reflects post-depositional de-watering, suggesting initial high water content of the sandy parent bed, in turn reflecting rapid *en masse* dumping or frictional freezing of sandier flows (Lowe, 1982).

The slump-bed association consists of metres-thick packages of deformed, slumped and thrust-faulted Bouma-type (base-absent or top-absent) T_{abcd} turbidites (Table 3; Fig. 8D). The typically coarse-grained character of the slumped Ta and Tb divisions suggest proximity to channelised areas, similar to what was inferred for the central to peripheral lobe associations (see above). In turn, this favours an origin as channel-mouth lobes (e.g. Deptuck *et al.*, 2008) or alternatively, as overbank or inter-channel splay facies (e.g. Gardner *et al.*, 2003).

The background sediment associations consist of deep-marine mudstones and claystones (Table 1). The character and texture of these 'background' mudstones vary stratigraphically, from thoroughly bioturbated (BI 5 to 6) green-greyish mudstones to pin-striped laminated, virtually unbioturbated, dark claystones. Different combinations of lithofacies, ichnofacies and microfacies allowed a subdivision of these fine-grained lithologies into pelagic, hemipelagic, turbiditic and contouritic mudstones and claystones (Table 1).

The green-greyish variety consists predominantly of contouritic and hemipelagic mudstones as well as some pelagic mudstones and claystones, deposited in a well-oxygenated, lower slope setting. The contouritic variety is identified based on its lithology, which is similar but slightly siltier/sandier than the hemipelagic equivalent and microfacies type which reflects a stressed sea floor environment. The microfacies observed in the contouritic mudstones is similar to that observed in the turbiditic equivalents but very different to that of the hemipelagic and pelagic facies. The ghost sedimentary structures, suggesting an origin as up to 10 cm-high, slowly migrating (as can be inferred from bioturbation along foresets) ripples, is again different to what is observed in otherwise similar-looking turbidite beds (representing Bouma T_{cd} subdivisions) or deposits of hybrid-flow origin.

The pinstriped variety, by contrast, represents a mixture of distal turbiditic mudstones (Stow & Bowen, 1980; Janocko, 2008) with siltstone and sandstone interlamina alternating with bands of hemipelagic and pelagic mudstones and claystones. The pinstripe-bedded mudstones are interpreted to have been deposited in physico-chemically stressed environments, related to some combination of reduced oxygenation and increased sedimentation rates. Notably, the changes in the character of the background mudstones typically parallel stratigraphic variations in the sandy facies associations or fan types. The darker, pinstriped variety comprises the background sediment to the more disorganised or mass-flow-influenced fan facies associations or subtypes. Where both the sandy fan and background facies associations change in consort, an allogenic mechanism or change in the basin type or its configuration is favoured as the controlling factor(s).

Thicker intervals (or beds) of massive, dark pelagic and subordinate hemipelagic claystones form the caps of the sandier packages or bedsets. These pelagic claystones represent sand-starved conditions and accumulated during prolonged intervals when sandy fan deposition ceased, either due to sand trapping in upslope subbasins or to complete sand shut-off to the basinal areas.

Springar sandstones; reservoir architectures and stratigraphic development

The bulk of the Springar sandstone interval constitutes stratified units of interbedded turbiditic sandstones of the peripheral and lobe fringe facies associations, as well as rare central lobe associations with greenish-grey mudstones of mixed turbiditic and hemipelagic origin. Pelagic mudstones are common only at certain intervals, within cappings of individual bedsets. The upper part of the Springar sandstone interval is dominated by less well-organised sandstones of the megabed, central lobe and channel-belt facies associations.

The lower part of the Springar sandstone interval in the Ormen Lange area can be organised into packages up to 10 m-thick, although they typically do not exceed 6 metres in thickness (Fig. 9). These units commonly form bedsets composed of tabular and relatively extensive beds that define coarsening-to-fining and thickening-to-thinning upwards motifs (Figs 5 and 6). The basal coarsening-upwards and thickening-upwards trends are commonly

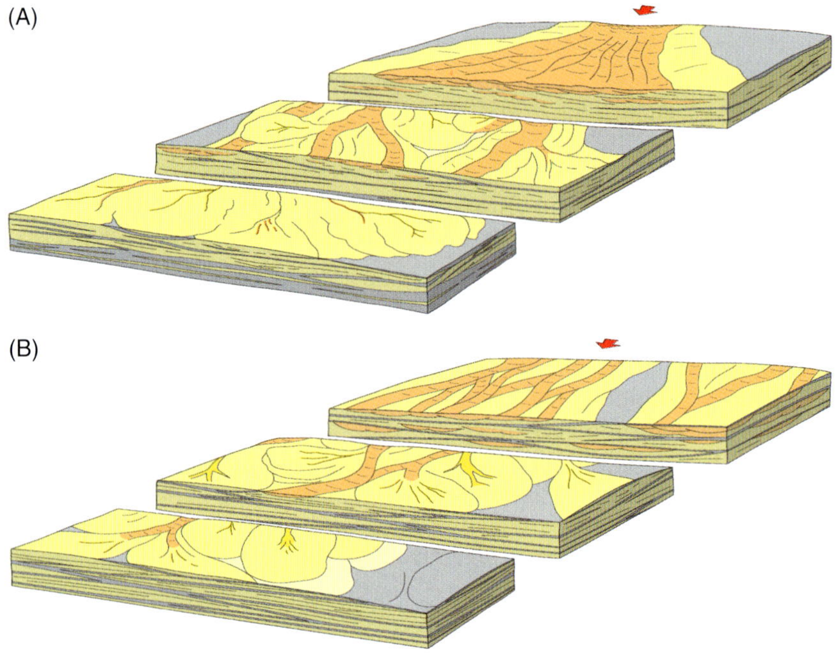

Fig. 12. Conceptual reservoir architectures for the organised (A) and less organised, or channel-mouth lobe (B) types of sandy fans.

accompanied by a change from a more distal to a more proximal facies association. In contrast, the upper fining-upward and thinning-upward trend is defined by the reverse change in facies associations. The motifs are underlain as well as capped by thicker mudstone intervals, commonly of mixed turbiditic-hemipelagic and pelagic origin. The motifs are interpreted to represent stacks of turbidite beds that jointly form the peripheral to the outer parts of a series of relatively wide and low-relief turbidite lobes; hence, the term 'lobe storey' is used for these metre-scale bedsets. These lobe storeys probably developed in the distal or outer (i.e. downfan) areas of the sandy fan environments (Fig. 12A; cf. Piper & Normark, 2001). Notably, the thicker lobe storeys also contain the more proximal facies association, that is, they comprise both lobe fringe and peripheral facies tracts. The layered development of the Springar sandstones in the Ormen Lange area form the downslope equivalents to thick and sandy packages in upslope subbasins, such as the Sløreboth Subbasin (Fig. 4). These upslope successions are uncored, but from well logs appear similar to the Egga sandstones.

The upper part of the Springar sandstones, including the so-called Våle heteroliths (Gjelberg *et al.*, 2001, 2005; Smith & Møller, 2003), consist of a variably developed package with marked changes in sediment architecture across the Ormen Lange subbasin. Abundant, thick-bedded sandstones of possible debris-flow origin stack in an apparent disorganised, but compensational manner in the southern part of the subbasin. In the central to northern part of the Ormen Lange subbasin, thick channel storey packages are present, commonly with interbedded slump units derived from upper slope settings (Fig. 12B). These channel storeys translate down-fan (or downslope) into relatively coarse-grained central and peripheral lobe deposits, probably representing isolated channel-mouth lobes. The channel-fills and lobe deposits are organised in gross fining-upwards and thinning-upwards and coarsening-to-fining-upwards or thickening-to-thinning-upwards motifs, representing single-storey channel-fills and lobe storey successions, respectively. It is tentatively postulated that a single-storey channel-fill translates down-fan into a single lobe storey motif, linking erosional-depositional events on the proximal fan to lobe formation and growth further down the depositional slope. The disorganised stack of megabeds, in contrast, appears to have formed by a series of gravitational re-sedimentation events. Their contained palynomorphs suggest derivation by reworking of upslope Upper Maastrichtian sands, probably via gravitational collapse of unconsolidated turbidite deposits. The Upper Maastrichtian in the Sløreboth Subbasin is locally top truncated, with deep, seismically visible gulley-fills incised into the

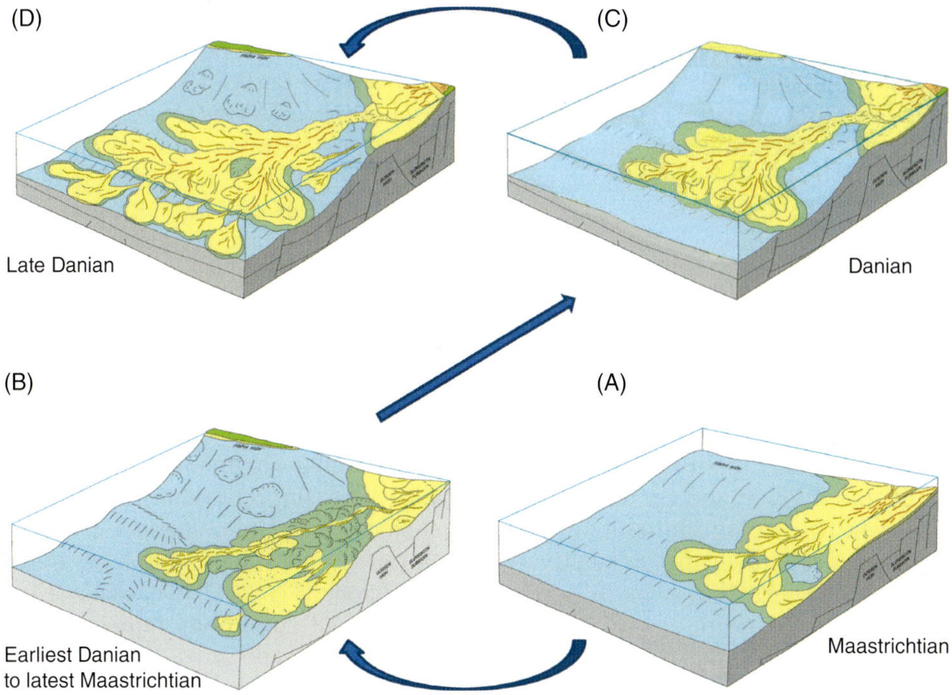

Fig. 13. Conceptual depositional settings for (A) the Upper Maastrichtian Springar sandstones, (B) the uppermost Maastrichtian to lowermost Danian Springar and basal Tang Formation sandstones (or Våle heteroliths), (C) the Danian Egga sandstones; and (D) the uppermost part of the Danian Egga sandstones. The bulk of the Springar and Egga sandstones form a series of organised sandy fan lobe complexes developed on several slope minibasins and terraces with an overall outbuilding of the fan systems from the Maastrichtian to the Danian. The uppermost part of the two fan systems formed during phases of basin margin structuring and tilting and represent a more disorganised style of sandy slope deposition.

background sandy fan units, whereas the same unit is represented by an erosional surface with overlying lag deposits in the intermediate slope area.

Stratigraphic development: The lower part of the Springar sandstones shows an early stage of fan arrival and outbuilding in the Ormen Lange area, probably via fill-spill from the upslope mini-basins such as the Slørebotn Subbasin (Fig. 13A). Thicknesses of turbidite beds suggest that only the larger flows were able to overspill from upslope mini-basins. The interbedded and thinner distal turbidite beds probably represent smaller-scale event beds, in turn suggesting upslope capture of the main body of the flow. The resultant geometry of the lobe storeys is therefore probably a broad, low-relief mounded to pseudo-sheet-like features (e.g. see Remacha & Fernandez, 2003). Deposition occurred as a series of compensational lobe-building events at the base-of-slope to inner basin floor, separated by intervals of predominantly hemipelagic deposition. The latter is related to lobe avulsion or upslope capture of active lobe deposition. Gross sedimentation rate was relatively low, suggesting that prolonged time intervals are represented by the hemipelagic-pelagic lobe storey caps. This well-organised style of terminal fan deposition was relatively abruptly replaced by a more disorganised style of deposition in the Ormen Lange subbasin, represented by the upper part of the Springar and the basal Tang Formation sandstones ('Våle heteroliths').

This subsequent (i.e. upper Springar and basal Tang) depositional stage involved upslope mass wasting with the downslope formation of 'clastic collapse wedges' constructed of slump or mass transport units and sandy debris flow deposits, herein jointly referred to as 'Mass Transport Deposits' or 'MTDs'. Concomitantly, there was a marked basinward shift in the channel facies and lobe facies tracts, with inner or proximal fan channel-belts placed upon outer fan lobes without or with only thin development of mid-fan channelised lobe deposits (Fig. 13B). Notably, the channel-belt seems to have been forced away laterally from the lower slope mass-transport deposits, feeding more localised, coarser-grained and potentially more elongate (Deptuck *et al.*, 2008) channel-mouth lobes or splays.

Shut-off of coarse sandy supply was relatively abrupt and replaced by a prolonged period of pelagic claystone deposition. These claystones form an extensive unit which separates the Springar and basal Tang Formation sandstones (i.e. the 'Våle heteroliths') from the overlying Egga sandstone unit.

Egga sandstone unit; reservoir architectures and stratigraphic development

In the Ormen Lange subbasin, the Egga sandstone unit predominantly consists of central to inner fan facies tracts. It forms an overall forestepping unit with a thin, backstepping uppermost part. The unit can be split into a lower and an upper part, potentially representing two individual lobe complexes, respectively, separated by a relatively thick pelagic mudstone unit that extends across the entire Ormen Lange subbasin.

The lower part of the Egga sandstone unit forms a high net-to-gross sandstone package, defined by a series of normally 3 to 7 (maximum 10) metres-thick bedsets (Fig. 9), again interpreted to represent individual lobe storeys (Fig. 12). The bedsets or lobe storeys are commonly separated by variably thick, mixed hemipelagic-pelagic mudstones. Each sandstone motif commonly shows a coarsening-to-fining and thickening-to-thinning upwards trend, typically paralleled by an associated upwards change from distal through successively more proximal facies, subsequently overlain by distal or marginal lobe facies tracts. The lobe storey motifs are commonly arranged so that they define an overall forestepping trend, suggesting an overall progradational or outbuilding of the sandy fan (Fig. 5). Such a simple stratigraphic organisation of facies tracts is best developed in wells to the south-west and north-west in the Ormen Lange subbasin (i.e. those located in more downfan positions). However, there are marked local departures from this generalised forestepping stacking pattern. Wells within the central-eastern part of the Ormen Lange subbasin have a basal, overall backstepping trend. This is defined by a vertical transition from relatively thick basal lobe storey(s), commonly consisting of coarse-grained channel-fill and channel-to-lobe transition facies types, overlain by central or mid-fan lobe storeys of alternating central to peripheral lobe deposits. Only subordinate channel-to-lobe transition facies types are associated. The basal coarser-grained lobe storeys are thicker where the underlying, disorganised, upper part of the Springar sandstones is thin and vice versa. Other production wells, located in more central parts of the subbasin, have a more irregular vertical facies trend, probably with vertical alternations between the various mid-fan facies associations. This is interpreted to reflect back-and-forth shifting of mid-fan depositional sub-environments (i.e. the channel-to-lobe transition, central lobe and peripheral lobe) through time (Fig. 13C). Wells off-axis of the main sand-fairway (e.g. well 6305/1-1 to the north; Figs 4 and 6) shows a relatively thick monotonous mudstone section, developed in predominantly contouritic mudstone facies.

The upper part of the Egga sandstone unit comprises another channelised package in the Ormen Lange subbasin, representing predominantly channel-belt, sandy inter-channel (sometimes slump-deformed) and channel-to-lobe transition facies associations (Fig. 13D). The change from the sediment-starved conditions represented by the field-wide pelagic drape across the lower part of the Egga sandstone unit to the channelised upper part of the unit is fairly sharp, suggesting a rapid return of inner fan environments and in places a fairly marked basinward shift in facies tracts. Seismic and core data suggest relatively shallow channels with sandy overbanks and inter-channel areas (Figs 6 and 10). The channelised unit is accordingly interpreted to represent a braided inner fan area (cf. Decker *et al.*, 2004), where channels could migrate relatively freely. However, the position of the channel-belts, as defined by the vertical stacking of channel-fill facies and seismic data, remained relatively fixed through time, their position probably controlled downcurrent from the outlet of fixed slope or ramp margin conduits (Fig. 11). Seismic data suggest that these channel storeys fed frontal lobe storeys or lobe storey sets/lobe complexes in the adjacent downflow subbasin, probably from channels incised via knickpoint erosion (see Pirmez *et al.*, 2000) into the fronting slope ramps. Within the Ormen Lange subbasin, the inter-channel areas are dominated by relatively coarse-grained sediments that probably represent overbank or crevasse channels and splays. Off-axis or distal sub-environments (well 6305/1-1; Fig. 6) are dominated by continuous mudstone deposition of contouritic origin, with localised interbedded, thin-bedded turbidites. Penecontemporaneous soft-sediment deformation formed in response to an unstable substrate, probably as a combination of higher local sedimentation rates coupled with basin-floor tilting. Locally, there is an increase in the amount of soft-sediment deformation towards the top of the unit. Final shut-off of fan deposition was again relatively rapid and left the Ormen Lange subbasin deprived of coarse-grained siliciclastic sediments until renewed progradation in the early Selandian of the basin marginal slope and deposition of the so-called 'Lista Fan' (Gjelberg *et al.*, 2005). This rapid return to a sediment-starved basin probably led to clay-dominated and mud-dominated fills of the channels incised into the uppermost part of the Egga sandstone unit (Figs 10 and 11).

Upslope, the Egga sandstone unit is represented by channel-fill and potential inter-channel facies associations, capped by channel-mouth lobes; the latter equivalent to the spill-cap of Gardner *et al.* (2000, 2003). In further upslope mini-basins (e.g. the Slørebotn Subbasin), the Egga sandstone unit is represented by a high net-to-gross sandstone unit, again with seismically identified channel-fills defined by both erosional and constructional architectures. By contrast, these are shallower and wider than are those incised into the underlying Springar sandstone. Candidate lobe storeys form the uppermost part of the Egga sandstone unit in such settings. Notably, the Egga sandstone unit terminates against an arcuate fault-plane along parts of the outboard edge of the Slørebotn Subbasin, suggesting local removal of the unit by late-stage gravitational collapse.

Stratigraphic development: The Egga fan represents the renewed arrival and infilling of a sandy fan in upslope mini-basins or slope terraces, followed by bypass and overspill into the neighbouring, downslope subbasin, such as the Ormen Lange subbasin (Fig. 13C). In the Danian, the Ormen Lange subbasin was positioned on the lower part of the slope. Initial-stage fan development was characterised by deposition of a series of isolated, channel-mouth, coarse-grained lobe storeys that infilled and overtopped the depositional topography produced during the late stage of Springar fan development. This initial or 'fan-arrival stage' was probably coupled with early aggradational infill (?backfill) of the lower stretch of the slope conduit(s) or channel fairways. As the inherited topography became depositionally smoothed, the subsequent phase of fan development was one of progradation (see also Gjelberg *et al.*, 2005), paralleled by the continued backfilling and onlap onto the slope of the proximal, channelised part of the Egga fan. As a result, an expansion of the Egga sandy fan probably occurred, with upslope partitioning of the coarser inner fan facies tracts, the development of a fairly extensive channel-dominated and lobe-dominated central or mid-fan area and an outbuilding, lobe-dominated outer fan area. In turn, this produced an apparent backstepping stacking pattern in the inner, proximal parts of the Ormen Lange subbasin, whereas a forestepping stacking pattern developed in the outer, more distal part of the subbasin. The central-fan or mid-fan area is dominated by alternating channel-fills and lobe storeys, which probably reflects compensational stacking of lobes (Fig. 10), as well as short-lived, intermittent outbuilding and retreat of the distributary channel fairway.

Intermittent shut-off of sandy supply left the Egga fan dormant for some time in the Ormen Lange subbasin, before a new phase of fan progradation (represented by the upper part of the Egga sandstone unit) occurred. During this stage, the channel-dominated area was rapidly shifted basinward and accompanied by a southward shift of the main sandy channel fairway. A sandy, broad 'braid-plain'-type channel-belt developed across the southern part of the Ormen Lange subbasin. However, the northern part of the subbasin was dominated by individual or narrower channel-belts flanked by sandy inter-channel or channel overbank areas; the latter dominated by crevasse sub-channels and splays (Figs 12 and 13D). The southern and northern channel-fairways fed laterally separate, frontal lobe complexes in the adjacent, downslope subbasin or terrace to the west and northwest, respectively. However, both channel-fairways probably originated from a common upslope feeder system or slope conduit to the south-east. The basinward shift in facies tracts recorded by the upper part of the Egga sandstone unit was accompanied by increased sediment instability in the uppermost part of the unit, related to tilting of the slope and basin floor (see below). Eventually, the basinward tilting of the slope area resulted in gravitational collapse of the Egga sandstones along the outer edge of upslope mini-basins, leading to ensuing downslope re-sedimentation of relatively coarse-grained sandy material.

Thus, a change from the laterally extensive mid-fan area present across the greater part of the Ormen Lange subbasin (represented by the lower part of the Egga sandstone unit) to an inner, potentially braided fan environment and finally to isolated inner fan channel-fairways in the upper part of the Egga sandstone unit, is envisaged. The lower part of the interval represents a sandy fan environment with suprafan lobes and intra-lobe channels that shifted laterally or avulsed across the fan surface (e.g. see Decker *et al.*, 2004), periodically spilling downslope into the next subbasin. The upper part, in contrast, represents two (or more) relatively stationary channel-fairways, fixed within sandy overbanks and which fed smaller and more locally developed frontal splays or lobe complexes in the adjacent downslope subbasin, probably via knickpoint channels incised into the frontal ramp that separated the two subbasins.

Springar and Egga fans – similarities and contrasts

The Upper Maastrichtian to Danian sandy fan systems, represented by the Springar sandstones and the Egga sandstone unit, respectively, can be split into a lower layered or 'well-organised' sandy fan (Piper & Normark, 2001), characterised by compensational stacking of lobe storeys and channel storeys and an upper less-organised to channelised part defined by a marked basinward shift in facies tracts (Fig. 13). In an individual subbasin or slope terrace, the lower part of the two fan systems represents different fan environments, with the older Springar sandstones consisting of facies associations that correspond to more distal environments than those of the younger Egga sandstone unit. Thus, the stratigraphic variability in sedimentary architectures from the Egga sandstone unit to the underlying Springar sandstones can probably be viewed as downfan or proximal through central to distal changes in architectural elements. Such variability is typical for the various depositional elements and thereby subenviroments that comprise a sandy submarine fan. All or only some of the facies associations may be present within an individual mini-basin or slope terrace.

In addition, both fan types evolved stratigraphically through an early stage of gradual progradation and infill, followed by a second stage of marked basinward shifts of facies tracts. Shut-off of sand deposition in the down-slope subbasin was relatively rapid for both of the systems. A common evolution of the two fan systems is envisaged (Fig. 13), which was probably a response to repeated slope and basin structural adjustments (discussed below).

Fan types; distribution and partitioning of depositional elements

The Ormen Lange fan types can be characterised as variants on the sandy fan (Piper & Normark, 2001) theme. Two sandy fan sub-types are recognised: one dominated by channel-fed lobe storeys and complexes and one with an increasing proportion of mass transport deposits (MTDs). There is sufficient stratigraphic variability in both the Springar and Egga fan successions to suggest that each fan system evolved from a sandy fan dominated by channel-fed lobe complexes to a one increasingly influenced by MTDs during the final stages of fan development, i.e. prior to shut-off of sandy supply. The transition between the two stages was gradual and was preceded by a marked basinward shift in facies tracts that can be recognised across the width of the fan.

Sandy fan – hierarchically ordered lobe complexes

The lower part of the Springar and the Egga fans represents the outer to middle and the middle to inner parts of a channelised sandy fan with suprafan lobes, respectively (Fig. 12A; see also Figs 5, 6 and 7). Based on the stacking and partitioning of facies associations and thus inferred depositional elements, an organised downfan arrangement of depositional elements can be discerned, from an incised slope feeder system through a braided channelised distributary area, into a lobe-dominated area. This is terminated downfan by broad and isolated, mounded to sheet-like lobe features or beds/bedsets (Fig. 14). Gjelberg *et al.* (2001, 2005) and Smith & Møller (2003) in their studies on the Egga fan system argued for a similar gross depositional environment, though without recognising facies tracts that reflect specific architectural elements or sub-environments. Modern canyon-fed, sandy fans with similar depositional elements include the Makassar fan in the Makassar Strait, off Borneo (Decker *et al.*, 2004), the Golo fan in the Tyrrhenian Sea (Gervais *et al.*, 2004, 2006; Deptuck *et al.*, 2008), the Orinoco fan in the Caribbean Sea (Belderson *et al.*, 1984; Ercilla *et al.*, 1998) and the Pochnoi fan in the Bering Sea (e.g. Kenyon & Millington, 1995). Outcrop analogues with similar gross structure and architectural elements include the carboniferous Ross Formation of western Ireland (Chapin *et al.*, 1994; Elliot, 2000; Lien *et al.*, 2003), the Permian Brushy Canyon of western Texas (Beaubouef *et al.*, 2000; Carr & Gardner, 2000; Gardner & Borer, 2000; Gardner *et al.*, 2003), the Permian Laingsburg and Tanqua-Karoo systems of south-western South Africa (Wickens & Bouma, 2000; Johnson *et al.*, 2001; Grecula *et al.*, 2003b; Sixsmith *et al.*, 2004; Hodgson *et al.*, 2006), the Maastrichtian Pab system of western Pakistan (Eschard, 2003, 2004)

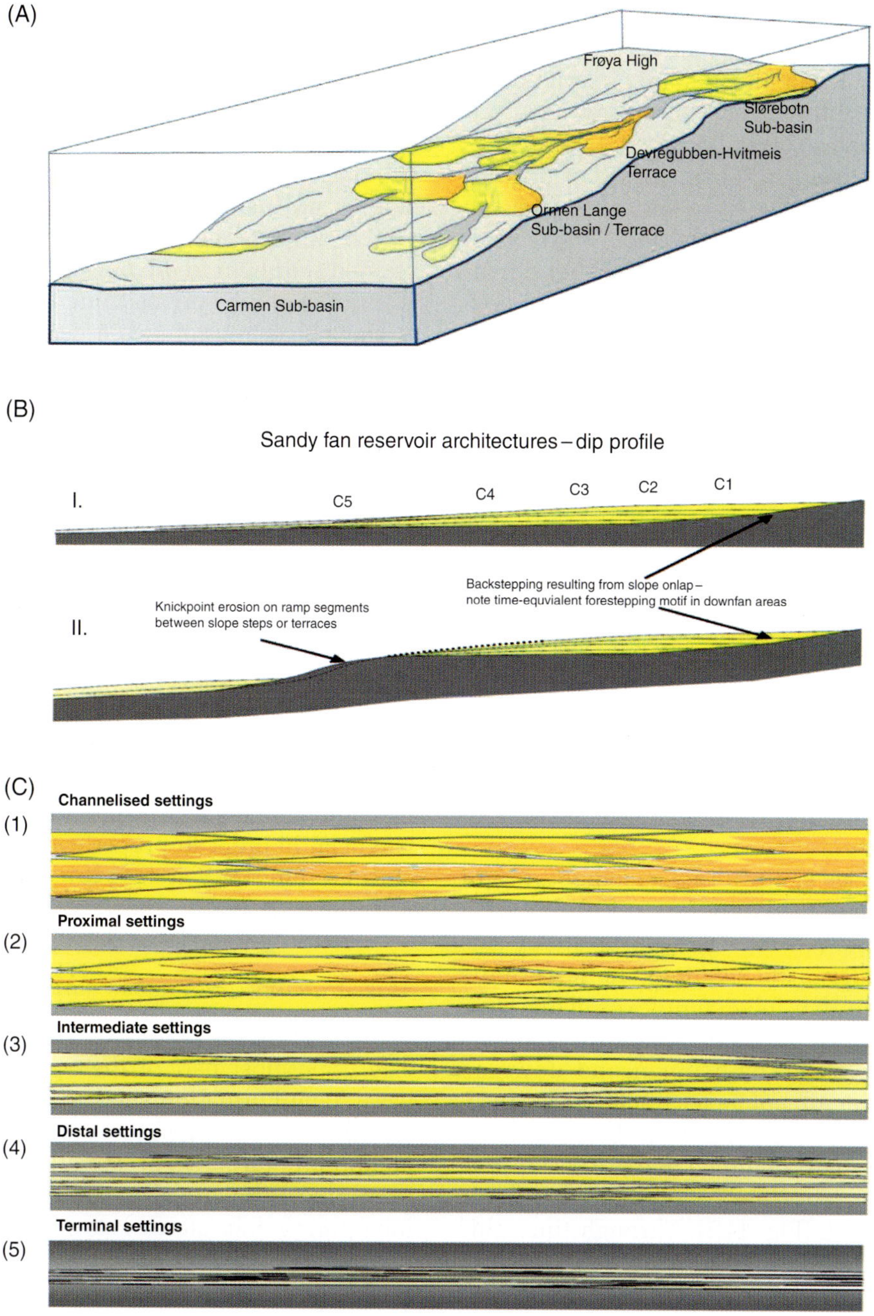

Fig. 14. (A) Examples of slope sandy turbidite fan types; I: ponded minibasin fill – the Slørebotn turbidite systems, II: at grade channelised turbidite fan – the inclined intermediate Hvitmeis-Dovregubben terrace, III: slope terrace turbidite fan – the Ormen Lange slope terrace and IV: lower slope isolate lobe complexes – the Carmen subbasin turbidite fills. (B) Examples of conceptual gross turbidite architectures along a dip profile across the Ormen Lange slope terrace turbidite system; I: without distal outlet knick point channel and II: with distal outlet knick point channel. (C) Conceptual sedimentary architectures of a series of strike profiles across an Ormen Lange type sandy slope turbidite system without outlet channel. Location of the profiles are indicated in dip profile in B-I.

and the Messinian Laga 1 system of central Italy (Milli *et al.*, 2007; Marini *et al.*, 2011).

Along the Møre Margin, the upslope feeder systems on the low-gradient ramp settings between the slope terraces and mini-basins most probably consisted of braided tributary systems, which were dominated by by-pass and lag deposition prior to deposition of spill-caps during retreat of the two fan systems. For the Ormen Lange fan systems, the inner fan was probably dominated by a braided, channelised or scoured distributary fairway that produced multilateral channel storeys as well as stacked multilateral channel belts, the latter representing channel storey sets and channel complexes, both bordered laterally by sandy overbank wedges. The central-fan or mid-fan area was dominated by a series of laterally shifting and avulsing channel-fed lobes, with intervening channels that fed lobes further downfan and in the outer fan area. The outer fan area was dominated by channel-fed terminal splays, either in the form of relatively wide and low-relief lobes or as fairly extensive sheets. The widths of the Ormen Lange slope mini-basins or terraces were probably smaller than the lengths that such sandy fans normally attain (Piper & Normark, 2001). Thus, on an individual slope terrace or mini-basin, there is normally a predominance of only some of the sandy fan sub-environments (or facies associations), with the remaining facies associations or facies tracts not developed or only forming volumetrically subordinate parts of the succession. The lower slope area appears to have been continuously affected and reworked by along-slope currents (i.e. contour currents), which probably prevented or inhibited the fall-out of suspended material.

An early phase of 'infilling' or smoothing of inherited topography is proposed, probably by a series of channel-fed lobe storeys or storey-sets which was similar to but smaller than those of the overlying central fan lobe complexes. This early stage of development gradually evolved into a more organised stage, with upfan and downfan migration or widening of facies tracts. As a consequence, this led to simultaneous backfilling or upslope onlapping combined with forestepping or outbuilding across the individual terraces or infilled mini-basin (Fig. 14B). Through time, this is interpreted to have produced a more elongate gross geometry, characterised by a wedge-shaped downflow thinning and feathering termination style of the various channel-and-lobe storey sets or complexes.

The Ormen Lange lobe-fill and channel-fill storeys are between 2 and 10 m thick (Fig. 9), with thinner lobe storeys normally occurring in more distal sub-environments. Lobe storey thickness data, combined with analogue (in-house proprietary outcrop and numerical modelling) data suggest widths on the order of 3 to 7 km and lengths of 5 to 10+ km, whereas outcrop data suggest widths and lengths of more than 10 km (Marini *et al.*, 2011). The lobe storeys stack in a shingled to compensational manner (Fig. 14C), probably reflecting upslope migration or avulsion of active distributary channel(s), respectively, as well as downfan translation and retreat of the braided distributary area. The lobe storeys and channel storeys are capped by hemipelagic mudstones, which reflect temporarily shut-off of coarse-grained sediment to the fan area or avulsion of the active site of lobe deposition.

The lobe storeys of the Ormen Lange area can be arranged in 5 to 15 m thick compound units, underlain and overlain by thicker, mixed hemipelagic and pelagic mudstones. These thicker packages consist of two or more lobe storeys or channel storeys and are referred to as lobe storey or channel storey sets or as channel complexes or lobe complexes (Fig. 14C), representing a higher order hierarchical unit. The pelagic character of the enclosing mudstone succession indicates a prolonged history of low 'local' deposition rates preceding and succeeding the formation of the storey sets or complexes. They are accordingly interpreted to represent larger-scale avulsions, or longer-term cessation of sand supply. Consequently, careful facies characterisation and discrimination of the background mudstones in core is crucial in order to properly construct the hierarchical sedimentary architecture for such sandy fan reservoirs.

Mass-transport influenced sandy lobe complexes

The uppermost part of the Springar fan, i.e. the basal Tang Formation sandstones or the Våle heteroliths and the uppermost part of the Egga fan, are composed of a series of laterally separate and distinct, locally vertically stacked channel-mouth lobe storeys, lobe storey-sets or, more rarely, candidate lobe complexes (from seismic) that alternate with megabeds, slumped units and slump blocks (i.e. the mass transport deposits or MTDs).

The channel-mouth isolated lobes (Fig. 12B) normally form channel-fed, erosively based, individual lobes or lobe elements (single or compound beds) of limited lateral extent. These are commonly thicker and composed of coarser-grained sediment than a typical lobe of the 'organised sandy fan'. Apart from that, they contain similar lithofacies and facies tracts. The two lobe types also show common partitioning of the lithofacies and facies tracts, as well as gross thickness distribution of bedsets or lobe storeys. Both systems likewise display compensational stacking of the lobe storeys. The smaller dimensions and coarser sediment of the inferred isolate channel-mouth lobes are consistent with deposition from smaller volume, sand-laden flows, potentially with short run-out distances from feeder channels (or slope conduits), which lead to narrow and elongate, possibly more tongue-like bodies (e.g. see Baas *et al.*, 2004; and cf. Mulder & Alexander, 2001; Al Ja'aidi *et al.*, 2004). The component facies suggest deposition from initial sand-laden, hyperconcentrated surge-type flows that later changed into more sustained flows (e.g. see Deptuck *et al.*, 2008).

Slump-blocks consist of laminated and bioturbated upper slope mudstones that are slump-folded and sheared. Slump blocks also show that they were thrust faulted and normally faulted whilst still in an unconsolidated state, most probably during downslope transport. They normally occur interbedded with channel-fill storeys and represent intermittent upslope gravitational mass-wasting events.

Megabeds comprise thick, single storey, massive or dewatered sandstones. The lithofacies characteristics suggest deposition by frictional freezing and points to deposition from highly concentrated, possibly unchannelised flows. Bed geometries are probably in the form of elongate, narrow and thick tongue-like bodies, although the potential for an unchannelised origin could suggest a broader, lobe-like geometry. Nevertheless, the poor correlation potential of packages containing 'megabeds' argue against laterally extensive sheet geometries and therefore against deposition from some exceptionally large flows.

Deformed units consist of sheared and thrust-faulted, normally faulted, slump-folded and disrupted beds or units, of which the parent architectures appear to have been originally medium-bedded to thin-bedded, layered facies tracts. Deformation typically occurred while the sediment was still unconsolidated, most probably shortly after deposition and prior to deposition of the overlying, undeformed strata. Deformation was probably gravitationally triggered, although only short distances of post-depositional remobilisation are inferred, in contrast to the interbedded slump blocks of upper slope origin. Thus, the deformed units were subject to penecontemporaneous *in situ* soft-sediment deformation, probably on an unstable slope.

The inner fan region of the MTD-influenced fan was dominated by channel-belts and inter-channel areas with unstable overbank units (deformed crevasse splays and sub-channels units), as well as mass-transport deposition from slides, slumps and 'megabed' flows. This passed downslope into a mid-fan area dominated by lobes and terminal parts of megabed flows and finally into an outer fan area where only the occasional large flows were able to reach and deposit their sediment load. Unstable basin-floor conditions were present along the entire fan profile. The background environment was dominated by pinstripe-bedded mudstones, suggesting a change to increasingly stressed environments, induced by a combination of reduced oxygenation and increased sedimentation rates.

Sediment feeder and delivery system(s)

The sandy facies, despite having a predominance of fine-grained to medium-grained sand, consist of varying and different grain sizes and textures (e.g. from fine to pebbly material; from angular to subrounded; and from poorly to well sorted) and include abundant detrital glauconite and chamosite (see also Gjelberg *et al.*, 2001, 2005; Smith & Møller, 2003). Grain size and bed thickness variations indicate that the mechanisms responsible for delivering sediments to the fronting basin varied in magnitude and frequency. The textural characteristics suggest mixing of first cycle and reworked sediment. Organic material, including plant remains and coal/coaly claystone clasts are abundant in the interbedded claystones and also occur as minute clasts in some sandier facies. The presence of plant fragments in both turbiditic mudstone and sandstone facies favour direct supply from a paralic or non-marine hinterland, with limited littoral and shelf storage. Jointly this suggests deltaic delivery systems, most probably from rivers in flood (see discussion below; see also Martinsen *et al.*, 2002, 2005; Sømme *et al.*, 2009), with some reworking and re-sedimentation of former deltaic, shallow-marine and shelfal sandstones. This is also supported by the character of the basal mudstones and sandstones of both lobe storeys and some single beds. Such units commonly show a subtle to pronounced coarsening-upward trend in the underlying turbiditic mudstones, which in turn are erosively truncated by the scoured basal surface of the overlying sandstones that commonly have a thin basal coarsening-upwards zone themselves. This is indicative of deposition from flows with increasing flow strength and velocity, typically argued to represent hyperpycnal flows derived directly from rivers in flood (Mulder *et al.*, 2003; Bhattacharya & MacEachern, 2009). The sustained character of the flows argued to have deposited most of the sandier facies also favours direct supply from a fluvial source (Mulder *et al.*, 2003; Piper & Normark, 2009).

The characteristics of the mass-transport deposits (the MTDs), particularly their downslope transport mechanisms (flow types) and contained reworking signal, favour initiation as upslope sediment failures (e.g. see Piper & Normark, 2009). Upslope sediment failure is the most probable origin for both the sandy ('megabeds') and the muddy (slump units) mass-transport deposits, attributed to failure of submarine fans on upslope terraces or mini-basins and to failure of the muddy slope substrate itself, respectively. The co-occurrence of channel-mouth lobes with mass-transport deposits, also suggest that these were sourced, at least in part, by re-working of older (and coarser) upslope lobe deposits or by flushing of the slope conduits.

Late Maastrichtian to Danian evolution of the Ormen Lange area

The Ormen Lange area saw repeated outbuilding and retreat of two fan systems during the Late Maastrichtian and the Danian (Fig. 12); the Springar and the Egga fan, respectively. Fan initiation and evolution was in response to preceding phases of hinterland uplift (see also Martinsen *et al.*, 1999, 2005; Sømme *et al.*, 2009), an uplift probably already initiated in the Late Campanian or even earlier. The two phases of fan progradation both experienced a late stage basinward shift of facies tracts, herein attributed to an early stage of gentle structuring and tilting of the basin-marginal slope area. These mild structural adjustments of the slope represented the initiation of two renewed phases of linked hinterland uplift and basinal subsidence, dated to the late Maastrichtian and late Danian, respectively. The resultant basin-margin structuring involved gross basinward tilting and thereby steepening of the basin's ramp-slope margin. As a result, a terraced or serrated slope profile formed with slope steps and subtle mini-basins where slope fans were deposited.

Hinterland and basin-margin structuring thus exerted also a first-order control on the fan development. Fan initiation and progradation occurred as a response to intermittent sediment supply from the upslope deltaic delivery system(s), which repeatedly were able to advance to the shelf edge and deliver their sediment load to the fronting slope and deeper-water basins. Repeated advances of the basin-marginal deltaic system(s) were, in turn, related to basin margin uplift as well as creation or rejuvenation of hinterland topography. Initial infilling of slope terraces and mini-basins smoothed any pre-existing basin floor topography and promoted the outbuilding of a broad and channelised braid-plain. Downslope shift of fan deposition and progradation occurred via fill-spill processes in upslope mini-basins. Repeated phases of slope structuring and tilting initially forced the fan further basinward, but also led to a change to smaller isolated and more localised channel-lobe systems, coupled with increased mass-wasting and gravitational re-sedimentation along upslope areas. Continued slope structuring eventually captured the sandy supply from the basin-marginal delivery systems in more upslope or basin-marginal (shelfal) subbasins, depriving the Møre Margin slope and deepwater basin further to the west of coarse clastic supply.

Controls on sedimentary architectures

Large-scale fan evolution

At the largest scale of resolution, the repeated development of slope fans along the Møre Margin reflects the ability of the deltaic delivery systems to advance to the shelf edge and deliver their loads to the fronting slope and basinal areas. This was probably promoted by the inferred narrow shelf that existed during Maastrichtian to Danian time (Osmundsen & Ebbing, 2008; Sømme *et al.*, 2009). Repeated advances and retreats of the basin-marginal deltas were, in turn, controlled by repeated basin-margin or rift-shoulder uplift and tectonic rejuvenation, tied to phases of active rifting within the western parts of the Møre Basin (Fig. 2). The development of the fan system along the Møre ramp margin was the product of sufficient sand supply to infill and overspill the upslope mini-basins and terraces and allow progradation of the sandy fan system down the gently tilted and mildly structured margin. Changes in fan style occurred in response to renewed structural readjustment to the basin and basinward tilting of the Møre ramp-margin. With time, the increased rate of structural adjustment and creation of basin margin topography resulted firstly in the capture of sand supply in upslope basins and secondly to full retreat of the deltaic delivery systems to the basin-margin, leaving the remainder of the basin in a sediment-starved state.

Bed, compound beds and lobe storeys

At the other end of the spectrum of stratigraphic resolution, individual beds and compound beds ('bed elements') probably reflect single major floods in the alluvial and fluvio-deltaic hinterland, at least during the progradational stages of fan development. The origin of lobe storeys is more speculative. However, the amalgamated character of the central lobe packages, coupled with their down-lobe and across-lobe storey feathering into layered packages of sandstone beds of inferred hyperpycnitic affinity and interbedded turbiditic mudstones, suggests either 1) prolonged but rather irregular delivery, or 2) repeated but frequent sediment delivery from hyperpycnal flows. An individual lobe storey is therefore interpreted to either represent a single major flood event of fluctuating intensity, or a series of frequent floods in the drainage basin. In that respect, it is interesting to note that the inferred dimensions of (and thus volume of) sediment within a single lobe storey is of the same order or slightly larger than that of deltaic mouth bars related to single flooding events (e.g. Fielding *et al.*, 2005). If a single flood were able to deliver all of its sediment load to the fronting deepwater areas, larger architectural elements would form, as additional sediment would be incorporated into the flow due to: 1) flow erosion into former delta-front sediments (e.g. Deptuck *et al.*, 2008); and 2) via conduit flushing *en route* to the receiving downslope basin (Piper & Normark, 2009).

The similar thickness of channel storeys and lobe storeys, combined with the facies characteristics of the channel storeys points to channel-fills originating from a combination of erosion, tractional deposition and lag aggradation during lobe construction and a later 'backfilling' stage, potentially reflecting waning flow conditions. The latter was probably promoted by the 'backstopping' effect of the depositional lobe topography at the channel 'mouth'. It is accordingly argued that architectural elements from single beds to lobe storeys (Fig. 14), in principle, represent climatic signals, either seasonal or periodic outsized flooding events. If, on the other hand, a channel storey to lobe storey reflects deposition during a series of recurring drainage basin flood events, a prolonged interval of time is required for the formation of a storey. In such a case, the alternating predominance of channel storeys to lobe storeys and hemipelagic/pelagic drapes may correspond to longer-term changes in flood frequencies in the drainage basin and thereby point to climatic cyclicity. In real world examples, lobe storeys probably form from single, exceptionally catastrophic, floods as well as via a series of floods of variable magnitude and frequency. These variations have the potential of leading to variable stacking patterns (e.g. vertical, compensational, or shingled) of the component beds.

Lobe storey set and lobe complexes

Avulsion of the site of active lobe deposition probably resulted from upfan lateral channel shifting or avulsion, in turn probably induced by (near-) complete channel-plugging from preceding flows or flow events. As a corollary, lateral migration may be favoured in instances where the bed or lobe storey produced during the preceding depositional event(s) were of insufficient height to promote channel backfilling. Lateral migration and short distance avulsion would have led to lateral shingling or compensational stacking of channel storeys and lobe storeys, respectively. The resultant stack of lobe storeys is referred to herein as 'lobe storey sets' or 'lobe complexes'. Borehole dipmeter data suggest that these lobe storey sets or lobe complexes were constructional features with significant depositional relief and topography to form an undulatory seafloor. Such a depositional topography would have had the potential to promote backfilling of the submarine fan distributary system and, with time, avulsion. Higher order avulsion, leading to longer distances lateral shift in the site of active fan deposition, resulted in the formation of a new channel storey to lobe storey set or complex. In either case, the stacking of bed, bedsets and lobe storeys to form channel storey and lobe storey sets and complexes are interpreted to represent autogenic processes on the fan surface and should be regarded as various types of 'self-organising' processes common to and intrinsic to any type of channel-fed depositional sub-environment (e.g. Piper & Normark, 2001).

Lobe storeys should ideally feather out laterally and distally through turbidite to hemipelagic mudstones. With increasing distance from the active site of lobe deposition, the mudstones should display an increasingly pelagic character. Pelagic or mixed hemipelagic-pelagic mudstones thus commonly mark a more significant stratigraphic event than do the hemipelagic lobe storey caps and probably reflect higher order, longer distance avulsions or longer-term cessation of sandy sediment supply. As such, the enclosing mudstones have the potential to be used as indicators of potential avulsion distance and thereby hierarchal order.

Intervening mudstones

The lack of, or thin, character of the interbedded mudstones is a conspicuous feature of the Springar (in upslope mini-basins) and Egga sandstone successions (see also Smith & Møller, 2003). Furthermore, there appears to be a characteristic partitioning of the various mudstone types across the fan surface (Fig. 7C). The character and distribution of the interbedded mudstones in the two sandstone units are here attributed to the inferred slope setting for the development of both the Springar and Egga fans. This setting promoted: 1) bypass of, and downslope deposition from the turbulent portions of the flows; 2) the continuous sweeping of the fan area by slope-parallel currents, which would have retarded and possibly at times prevented suspension fall-out over the inner to central parts of the fan area; and 3) some erosion by subsequent flows, evident by rip-up clasts and dispersed fragmented claystone chips. Mudstone fragments are normally more common in the basal beds of a lobe storey, suggesting that sandy depositional events within a lobe storey potentially occurred too frequently for suspension fall-out or mudstone partings to form, especially during the later stages of lobe storey formation. The broader implication of this is that mudstone deposition across the inner to central fan area mainly occurred during inter-flood intervals, forming candidate extensive drapes that envelope the more locally developed sandy reservoir elements (Figs 5 and 14). The thin character of the enclosing mudstones is attributed to the overall sediment-starved conditions for the eastern Møre Basin, especially during the Danian.

DISCUSSION

The sedimentological model and depositional history proposed above share some similarities and differences with those of earlier studies (e.g. Gjelberg *et al.*, 2001, 2005; Smith & Møller, 2003; Møller *et al.*, 2004; Sømme *et al.*, 2009) on the Ormen Lange fan systems. The most significant contrasts with previous studies are the subdivision of the Upper Maastrichtian to Danian section into two entirely separate fan systems, both of which developed across nearly the full length of the Late Cretaceous and Danian Møre ramp margin. Furthermore, the recognition of different depositional styles or fan types, and component depositional elements from the early through the late progradational stage to the final backstepping stage, has also not been recognised in previous studies. The repeated development of fan systems with a common evolutionary history for the Springar and Egga fans, linked to an integrated basinal structuring and sediment supply history, also appears to constitute a common theme in other tectonically active basins.

Vertical changes from an overall lobe-dominated outer to middle fan to a channelised inner fan environment is to be expected for any outbuilding submarine fan system. Such a simple fan progradation story may apply for the bulk of the Egga or Danian fan system. However, the change to isolated channels and channel-mouth lobes with interchannel areas dominated by scoured, unstable basin-floor during the final stages of fan system development (where soft-sediment deformation cannot be linked to levee or overbank failure), calls for allogenic controls, herein attributed to slope structuring and tilting. Slope tilting, causing changes in depositional style is also suggested by: 1) the relatively sharp transitions from outer to inner (Springar) and middle to inner (Egga) fan facies tracts; 2) the lack of extensive lobe complexes fronting the late-stage, fan channel complexes; 3) the facies characteristics of the late-stage MTD-influenced fan (e.g. the co-occurrence of frontal splays and intervening mass transport deposits); and 4) the change in background facies that accompanied the transition into the MTD-influenced fan stage.

The impact that the linked hinterland-basin structuring has on the development of basin-marginal deltaic delivery and, consequently on the fronting slope to basin-floor turbidite systems has been noted in nearly all types of tectonically active settings. The analogue Permian Laingsburg turbidite complex was also deposited on a terraced to mildly irregular slope subject to repeated deformation events (Grecula *et al.*, 2003b), although lying within a compressive, foreland setting. Interestingly, the Laingsburg turbidite complex also contains a series of stacked fan systems separated by soft-sediment deformed units reflecting repeated intervals of basin margin deformation and slope structuring. Basin-margin structuring also has been argued as a fundamental controlling factor on the repeated advances and retreats of sandy fans on the immature passive margin off western Africa (Hadler-Jacobsen *et al.* 2007) and for rift margin fans in the northern North Sea and northern Norwegian Sea (Ravnås *et al.* 2000).

The recognition of high-order architectural elements that represent specific depositional elements was previously speculative at best, but can now be better defined using the characteristics and significance of the various mudstone and claystone types. Importantly, this permits 1) reservoir architectures to better identified and more reliably delineated in the subsurface and 2) for architectural elements ranging from the bed scale through lobe storeys and lobe complexes to turbidite systems, to be organised hierarchically, similar to the approach applied to analogue outcrop turbidite systems (e.g. Mutti & Normark, 1987; Prélat *et al.*, 2009).

The inferred deltaic supply poses challenges for the postulated duration of the Ormen Lange fan systems. A ready source of sediment delivered to the shelf edge favours more rapid fan accumulation, yet it is well established that the fans span the Maastrichtian to Danian time. A plausible explanation is that significant intervals of time are recorded in the mudstone interval separating the two fan units, the drapes separating lobe storey set/-complexes and those separating the various lobe storeys within the fans. In turn, this implies that individual channel storeys to lobe storeys represent distinct depositional events, each producing a fundamental stratigraphic element, which in turn forms the basic building blocks for the depositional evolution of the sandy fan and its resultant sedimentary architectures.

The low sediment supply to the Møre Basin, especially in the Danian, is at odds with that of the Vøring Basin to the north (e.g. Kjennerud & Vergara, 2005). In contrast to the Vøring Basin, the Møre Basin was bordered by relatively low-relief hinterland of limited lateral extent (cf. Sømme *et al.*, 2009), formed partly by uplift of the former northern North Sea shelfal chalk sea (e.g. Brekke & Olaussen, 2007). The relatively low sediment supply to the Møre Basin during the Maastrichtian and Danian is thus attributed to the limited sediment yield potential from this newly formed and tectonically rejuvenated rift-shoulder hinterland. As such, the Maastrichtian and Danian fan systems sourced from the southern Norway uplands, such as the Springar and Egga fans along the Møre Margin and the Danian to Selandian Siri fan system in the Norwegian-Danish Basin (Hamberg *et al.*, 2005; Svendsen *et al.*, 2010), appear to represent a special case of fan development; sediment starved sandy fans with delivery of sand from a relatively mud-poor delivery system into an otherwise sediment starved basin.

The prevalence of slope-parallel currents along the Møre Margin resulted in alternating domains of inhibited mud deposition, such as across the Egga fan to the south and the development of sandy 'mud pits' in areas where the slope currents were sufficiently weak to permit net sediment accumulation (such as the northern part of the Ormen Lange subbasin).

CONCLUSIONS

1. The reservoirs of the giant Ormen Lange field, located off the Møre coast, are deep-marine turbidites within the Upper Maastrichtian Springar Formation and the Danian Egga sandstone unit. The Springar Formation represents the outer parts of a sandy fan system whereas the Egga sandstone unit is interpreted in terms of middle to inner parts of a similar sandy fan system.
2. The two fan systems were deposited on terraces or subtle minibasins on the lower slope. The two fan systems were sourced from an uplifted and tectonically rejuvenated hinterland, formed in response to Late Cretaceous to Early Palaeogene extension in the Norwegian-Greenland Sea rift.
3. The Springar and Egga turbidite systems evolved through a 'fill-spill' mechanism, with slope terraces and minibasins connected via a variety of channelised corridors.
4. Within an individual mini-basin or slope terrace, an early phase of 'infilling' or smoothing of inherited topography appears to have been followed by a more organised backfilling or upslope onlapping combined with forestepping or outbuilding of the sandy fan across the individual terraces or infilled mini-basin. Progradation of the fan resulted in basinward translation of the inner fan

channelised area and overspilling and sediment delivery to further downslope mini-basins and terraces.

5 Channel storeys and lobe storeys form the fundamental architectural element in the two turbidite fan systems and probably reflect major flooding events in the hinterland or drainage basin. Compensational stacking of the channel storeys to lobe storeys formed a series of channel complexes and lobe complexes related to autogenic, self-organising processes operating on the fan surface.

6 The uppermost part of the two fan systems represents a mass-transport influenced fan depositional style, formed as a response to an early phase or initial tilting of the basin margin slope. Cessation in fan growth was a response to subsequent structuring along the basin margin and its hinterland and sand capture in further upslope and proximal subbasins.

7 The limited siliciclastic supply is a characteristic feature to the Ormen Lange fan systems and has resulted in the overall sandy but thin nature of the two fan systems.

ACKNOWLEDGEMENTS

This study reflects the work of a team of Shell Exploration and Development geologists and geophysicists working over a period of several years on the Ormen Lange and neighbouring assets. The authors thank colleagues and partners in joint ventures for fruitful discussions over this period. Special thank goes to Cato Berge, Ciaran O'Byrne, Pete Mears, Carlos Pirmez, Ru Smith and Chris Townsend. Partners in the Ormen Lange Field (Statoil, ExxonMobil, DONG and Petoro) are acknowledged for permission to publish results from the Ormen Lange field model update completed in 2008. We would also like to thank I. Kane, D. Pyles and volume editor R.J. Steel for reviewing and suggesting improvements to this paper.

REFERENCES

Al Ja'aidi, O.S., McCaffrey, W.D. and **Kneller, B.C.** (2004) Factors influencing the deposit geometry of experimental turbidity currents: implications for sand-body architecture in confined basins. In: *Confined Turbidite Systems* (Eds S.A. Lomas and P. Joseph), *Geol. Soc. London. Spec. Publ.*, **222**, 45–58.

Baas, J., Kesteren, W.V. and **Postma, G.** (2004) Deposits of depletive quasi-steady high-density turbidity currents: a flume analogue of bed geometry, structure and texture. *Sedimentology*, **51**, 1053–1088.

Baas, J.H., Best, J.L., Peakall, J. and **Wang, M.** (2009) A phase diagram for turbulent, transitional and laminar clay suspension flows. *J. Sed. Res.*, **79**, 162–183.

Beauboeuf, R.T., Rossen, C., Sullivan, M.D., Mohrig, D.C. and **Jennette, D.C.** (2000) Deep-Water Sandstones, Brushy Canyon Formation, West Texas. AAPG Hedberg Field Research Conference, *AAPG Studies in Geology*, 1.2–3.9., 40 pp.

Belderson, R.H., Kenyon, N.H., Stride, A.H. and **Pelton, C.H.** (1984) A 'braided' distributary system on the Orinoco deep-sea fan. *Mar. Geol.*, **56**, 195–206.

Bhattacharya, J.P. and **MacEachern, J.A.** (2009) Hyperpycnal rivers and prodeltaic shelves in the Cretaceous seaway of North America. *J. Sed. Res.*, **79**, 184–209.

Blystad, P., Brekke, H., Færseth, R.B., Larsen, B.T., Skogseid, J. and **Tørudbakken, B.** (1995) Structural Elements of the Norwegian Continental Shelf – Part II: the Norwegian Sea Region. *Norw. Petrol. Direct. Bull.*, **8**, 44 pp.

Booth, J.R., Prather, B.E. and **Steffens, G.S.** (2002). Depositional models for ponded and healed-slope accommodation on above-grade slopes: implications for reservoir characterization. *AAPG Bull.*, **86**, A20–A21

Bouma, A.H. (1962) *Sedimentology of Some Flysch Deposits: a Graphic Approach to Facies Interpretation.* Elsevier, Amsterdam, 168 pp.

Brekke, H. (2000) The tectonic evolution of the Norwegian Sea continental margin with emphasis on the Vøring and Møre basins. In: *Dynamics of the Norwegian Margin* (Eds A. Nøttvedt *et al.*), *Geol. Soc. London Spec. Publ.*, **136**, 327–378.

Brekke, H. and **Olaussen, S.** (2007) High seas and low horizons. In: *The Making of a Land – Geology of Norway* (Eds I.B. Ramberg, I. Bryhni, A. Nøttvedt and K. Rangnes), 418–442.

Carr, M. and **Gardner, M.H.** (2000) Portrait of a basin-floor fan for sandy deepwater systems, Permian Lower Brushy Canyon Formation, West Texas. In: *Fine-Grained Turbidite Systems* (Eds A.H. Bouma and C.G. Stone), *AAPG Mem.*, **72**/*SEPM Spec. Publ.*, **68**, 215–232.

Chapin, M.A., Davies, P., Gibson, J.L. and **Pettingill, H.S.** (1994) Internal reservoir architecture of turbidite sheet sandstones in laterally extensive outcrops, Ross Formation, western Ireland. In: *Submarine Fans and Turbidite Systems: Sequence Stratigraphy, Reservoir Architecture and Production Characteristics – Gulf of Mexico and International* (Eds P. Weimer, A.H. Bouma and B.F. Perkins), Gulf Coast Section SEPM 15th Annual Research Conference, 53–68.

Clark, J.D. and **Pickering, K.T.** (1996) *Submarine Channels – Processes and Architecture.* Vallis Press, London, pp. 231.

Dalland, A., Worsley, D. and **Ofstad, K.** (1988) A lithostratigraphic scheme for the Mesozoic and Cenozoic succession offshore mid- and northern Norway. *Norw. Petrol. Direct. Bull.*, **4**, pp. 65.

Decker, J., Teas, P.A., Schneider, R.D., Saller, A.H. and **Orange, D.L.** (2004) Modern deep-sea sedimentation in the Makassar Strait: Insights from high-resolution multibeam bathymetry and backscatter, sub-bottom profiles and USBL-navigated cores. Extended Abstract, *IPA-AAPG Deepwater and Frontier Symposium, 2004.*

Deptuck, M.E., Piper, D.J.W., Savoye, B. and **Gervais, A.** (2008) Dimensions and architecture of late Pleistocene submarine lobes off the northern margin of Eastern Corsica. *Sedimentology*, **55**, 869–898.

Elliott, T. (2000) Depositional architecture of a sandrich, channelized turbidite system: The upper Carboniferous Ross Sandstone Formation, western Ireland. In: *Deep-water Reservoirs of the World* (Eds P. Weimer, R.M. Slatt, J. Coleman, N.C. Rossen, H. Nelson, A.H. Bouma, M.J. Styzen and D.T. Lawrence), Gulf Coast Section SEPM, 342–373.

Ercilla, G., Alonsoa, B., Barazaa, J., Casasa, D., Chioccib, F.L., Estradaa, F., Farrána, M., Gonthierc, E., Pérez-Belzuza, F., Pirmezd, C., Reedere, M., Torresf, J. and **Urgelesf, R.** (1998) New high-resolution acoustic data from the 'braided system' of the Orinoco deep-sea fan. *Mar. Geol.*, **146**, 243–250.

Eschard, R., Albouy, E., Deschamps, Euzen, T. and **Ayub, A.** (2003) Downstream evolution of turbidite channel complexes in the Pab Range outcrops (Maastrichtian, Pakistan). *Mar. Petrol. Geol.*, **20**, 691–710.

Eschard, R., Albouy, E., Gaumet, F. and **Ayub, A.** (2004) Comparing depositional architecture of basin floor fans and slope fans in the Pab sandstone, Maastrichtian, Pakistan. In: *Confined Turbidite Systems* (Eds S.A. Lomas and P. Joseph), *Geol. Soc. London Spec. Publ.*, **222**, 159–185.

Euzen, T., Eschard, R., Albouy, E. and **Deschamps, R.** (2007) Reservoir architecture of a turbidite channel complex in the Pab Formation, Pakistan. In: *Atlas of Deep-Water Outcrops* (Eds T.H. Nilsen, R.D. Shew, G.S. Steffens and J.R.J. Studlick), *AAPG Studies in Geology*, **56**, CD-ROM, 20 p.

Færseth, R.B. and **Lien, T.** (2002) Cretaceous evolution in the Norwegian Sea - a period characterized by tectonic quiescence. *Marine and Petroleum Geology*, **19**, 1005–1027.

Fielding, C.R., Trueman, J.D. and **Alexander, J.** (2005) Sharp-based, flood-dominated mouth bars from the Burdekin delta of Northeastern Australia: Extending the spectrum of mouth bar facies, geometry and stacking patterns. *J. Sed. Res.*, **75**, 55–66.

Fugelli, E.M.G and **Olsen, T.R.** (2005) Screening for deep-marine reservoirs in frontier basins: Part 1—Examples from offshore mid-Norway. *AAPG Bull.*, **89**, 853–882.

Gardner, J.V., Bohannon, R.G., Field, M.E. and **Masson, D.G.** (1996) The morphology, processes and evolution of Monterey Fan: a revisit. In: *Geology of the United States' Seafloor: The view from GLORIA* (Eds J.V. Gardner, M.E. Field, D.C. Twichell), Cambridge Univ. Press, New York, NY, pp. 193–220.

Gardner, M.H. and **Borer, J.M.** (2000) Submarine channel architecture along a slope to basin profile, Brushy Canyon Formation, West Texas. In: *Fine-Grained Turbidite Systems* (Eds A.H. Bouma and C.G. Stone), *AAPG Mem.*, **72**/*SEPM Spec. Publ.*, **68**, 195–214.

Gardner, M.H., Borer, J.M., Melick, J.J., Mavilla, N., Dechesne, M. and **Wagerle, R.N.** (2003) Stratigraphic process-response model for submarine channels and related features from studies of Permian Brushy Canyon outcrops, West Texas. *Mar. Petrol. Geol.*, **20**, 757–787.

Gervais, A, Savoye, B., Piper, D.J.W., Mulder, T., Cremer, M. and **Pichevin, L.** (2004) Present morphology and depositional architecture of a sandy confined submarine system: the Golo turbidite system, eastern margin of Corsica. In: *Confined Turbidite Systems* (Eds S.A. Lomas and P. Joseph), *Geol. Soc. London Spec. Publ.*, **222**, 59–89.

Gervais, A., Mulder, T. and **Savoye, B.** (2006) Sandy modern turbidite lobes: a new insight from high resolution seismic data. *Mar. Petrol. Geol.*, **23**, 485–502.

Gjelberg, J.G., Enoksen, T., Kjærnes, P., Mangerud, G., Martinsen, O.J, Roe, E. and **Vågnes, E.** (2001) The Maastrichtian and Danian depositional setting, along the eastern margin of the Møre Basin (mid-Norwegian Shelf): Implications for reservoir development of the Ormen Lange field. In: *Sedimentary Environments Offshore Norway—Paleozoic to Recent* (Eds O.J. Martinsen and T. Dreyer), *Norwegian Petroleum Society Special Publication*, **10**, 421–440.

Gjelberg, J.G., Martinsen, O.J., Charnock, M., Møller, N. and **Antonsen, P.** (2005) The reservoir development of the late Maastrichtian–early Palaeocene Ormen Lange gas field, Møre Basin, mid-Norwegian Shelf. In: *Petroleum geology: North-west Europe and Global Perspectives: Proceedings of the 6th Petroleum Geology Conference* (Eds A.G. Doré and B.A. Vining), Geological Society (London), 1165–1184.

Grecula, M., Flint, S.S., Wickens, H.D.V. and **Johnson, S.D.** (2003a) Upward thickening patterns and lateral continuity of Permian sand-rich turbidite channel fills, Laingsburg Karoo, South Africa. *Sedimentology*, **50**, 831–853.

Grecula, M., Flint, S.S., Potts, G., Wickens, H.D.V. and **Johnson, S.D.** (2003b) Partial ponding of turbidite systems in a basin with subtle growth-fold topography: Laingsburg-Karoo, South Africa. *J. Sed. Res.*, **73**, 603–620.

Grunnaleite, I. and **Gabrielsen, R.H.** (1995) The structure of the Møre Basin. *Tectonophysics*, **252**, 221–251.

Hadler-Jacobsen, F., Gardner, M.H. and **Borer, J.M.** (2007) Seismic stratigraphic and geomorphic analysis of deep-marine deposition along the West African continental margin. In: *Seismic Geomorphology: Applications to Hydrocarbon Exploration and Production* (Eds R.J. Davies, H.W. Posamentier, L.J. Wood and J.A. Cartwright), *Geol. Soc. London Spec. Publ.*, **277**, 47–84.

Hamberg, L., Dam, G., Wilhelmson, C. and **Ottesen, T.G.** (2005) Palaeocene deep-marine sandstone plays in the Siri Canyon, offshore Denmark-southern Norway. In: *Petroleum Geology: North-West Europe and Global Perspectives – Proceedings of the 6th Petroleum Geology Conference* (Eds A.G. Doré and B.A. Vining), Geological Society, London, 1185–1198.

Haughton, P., Davis, C., McCaffrey, W. and **Barker, S.** (2009) Hybrid sediment gravity flow deposits – Classification, origin and significance. *Mar. Petrol. Geol.*, **26**, 1900–1918.

Heinö, P. and **Davies, R.J.** (2007) Knickpoint migration in submarine channels in response to fold growth, western Niger Delta. *Mar. Petrol. Geol.*, **24**, 434–449.

Hodgson, D.M., Drinkwater, N.J., Flint, S.S., Luthi, S.M., Hodgetts, D., Johannessen, E.P., Wickens, H. DeV., Keogh, K.J., Kavanagh, J.P. and **Howell, J.A.** (2007) Statigraphy and Evolution of Fan 4, Skoorsteenberg Formation, South Africa. In: *Atlas of Deep-Water Outcrops* (Eds T.H. Nilsen, R.D. Shew, G.S. Steffens and J.R.J. Studlick), *AAPG Studies in Geology*, **56**, CD-ROM, 14 p.

Hodgson, D.M., Flint, S.S., Hodgetts, D., Drinkwater, N.J., Johannessen, E.P. and **Luthi, S.M.** (2006) Stratigraphic evolution of fine-grained submarine fan systems, Tanqua depocentre, Karoo Basin, South Africa. *J. Sed. Res.*, **76**, 20–40.

Janocko, M. (2008) *Pinchout Geometry of Sheet-like Sandstone Beds: a New Statistical Approach to the Problem of Lateral Bed Thinning Based on Outcrop Measurements.* M.Sc. thesis, University of Tromsø, 110 pp.

Jegou, I., Savoye, B., Primez, C. and **Droz, L.** (2008) Channel–mouth lobe complex of the recent Amazon Fan: the missing piece. *Mar. Geol.*, **252**, 62–77.

Johnson, S.D., Flint, S.S, Hinds, D. and **deV. Wickens, H.** (2001) Anatomy of basin floor to slope turbidite systems, Tanqua Karoo, South Africa: Sedimentology, sequence stratigraphy and implications for subsurface prediction. *Sedimentology*, **48**, 987–1023.

Jongepier, K., Rui, J.C. and **Grue, K.** (1996) Triassic to Early Cretaceous stratigraphy and structural development of the northeastern Møre Basin, off Mid-Norway. *Nor. Geol. Tidsskr.*, **76**, 199–214.

Kenyon, N.H. and **Millington, J.** (1995) Contrasting deep-sea depositional system in the Bering Sea. In: *Atlas of Deep-water Environments: Architectural Styles in Turbidite Systems* (Eds K.T. Pickering, R.N. Hiscott, N.H. Kenyon, F. Ricci Lucchi and R.D.A. Smith), Chapman & Hall, London, 196–203.

Kjennerud, T. and **Vergara, L.** (2005) Cretaceous to Palaeogene 3D bathymetry and sedimentation in the Vøring Basin, Norwegian Sea. In: *Petroleum Geology: North-West Europe and Global Perspectives – Proceedings of the 6th Petroleum Geology Conference* (Eds A.G. Dorè and B.A. Vining), Geological Society, London, 815–832.

Kneller, B.C. (1995) Beyond the turbidite paradigm; physical models for deposition of turbidites and their implications for reservoir prediction. In: *Characterization of Deep Marine Clastic Systems* (Eds A. Hartley and J. Prosser), *Geol. Soc. London Spec. Publ.*, **94**, 31–49.

Kneller, B.C. (2003) The influence of flow parameters on turbidite slope channel architecture. *Mar. Petrol. Geol.*, **20**, 901–910.

Kneller, B.C. and **Branney, M.J.** (1995) Sustained high-density turbidity currents and the deposition of thick massive sands. *Sedimentology*, **42**, 607–616.

Kneller, B.C. and **McCaffrey, W.D.** (2003) The interpretation of vertical sequences in turbidite beds: the influence of longitudinal flow structure. *J. Sed. Res.*, **73**, 706–713.

Lien, T., Walker, R.G. and **Martinsen, O.J.** (2003) Turbidites in the Upper Carboniferous Ross Formation, western Ireland: Reconstruction of a channel and spillover system. *Sedimentology*, **50**, 113–148.

Lowe, D.R. (1979) Sediment gravity flows: their classification and some problems of application to natural flows and deposits. In: *Geology of Continental Slopes* (Eds L.J. Doyle and O.H. Pilkey Jr), *SEPM Spec. Publ.*, **27**, 75–82.

Lowe, D.R. (1982) Sediment gravity flows II. Depositional models with special reference to the deposits of high-density turbidity currents. *J. Sed. Petrol.*, **52**, 279–297.

Martinsen, O.J., Bøen, F., Charnock, M.A., Mangerud, G. and **Nøttvedt, A.** (1999) Cenozoic development of the Norwegian margin 60-640N: sequences and sedimentary response to variable basin physiography and tectonic setting. In: *Petroleum Geology of Northwest Europe: Proceedings of the 5th Conference* (Eds A.J. Fleet and S.A.R. Boldy), *Geol. Soc. London*, 293–304.

Martinsen, O.J., Fonneland, H.C., Lien, T. and **Gjelberg, J.G.** (2002) On latest Cretaceous–Palaeocene drainage and deep-water sedimentary systems on-and offshore Norway. In: *Onshore-Offshore Relationships on North Atlantic Margins* (Ed. T. Thorsnes), *Norwegian Geological Society Abstracts and Proceedings*, **2**, 123–124.

Martinsen, O.J., Lien, T. and **Jackson, C.** (2005) Cretaceous and Palaeogene turbidite systems in the North Sea and Norwegian Sea basins: source, staging area and basin physiography controls on reservoir development. In: *Petroleum Geology: North-West Europe and Global Perspectives – Proceedings of the 6th Petroleum Geology Conference* (Eds A.G. Dorè and B.A. Vining), *Geol. Soc. London*, 1147–1164.

Marini, M., Milli, S. and **Moscatelli, M.** (2011) Facies and architecture of the Lower Messinian turbidite lobe complexes from the Laga Basin (central Appennines, Italy). *Journal of Mediterranean Earth Sciences*, **3**, 45–72.

Mayall, M. and **Stewart, I.** (2001) The architecture of turbidite slope channels. In: *Petroleum Geology of Deepwater Depositional Systems, Advances in Under-Standing 3D Architecture* (Eds S.I. Fraser, A. J. Fraser, H.D. Johnson and A.M. Evans), The Geological Society Conference March 20–22, 2001 Proceedings, The Geological Society, London.

McHargue, T. (2001) Recurring stacking pattern of reservoir elements in erosional slope valleys. In: *Petroleum Geology of Deepwater Depositional Systems, Advances in Understanding 3D Architecture* (Eds S.I. Fraser, A.J. Fraser, H.D. Johnson and A.M. Evans), The Geological Society Conference March 20–22, 2001 Proceedings. Burlington House, Piccadilly, London: The Geological Society.

Meckel, L.D., III, **Ibrahim, A.B.** and **Pelechaty, S.M.** (2000) Turbidite deposition in a muddy bypass system on the upper slope, offshore Brunei. *AAPG International Conference Abstracts CD-ROM*, 6p.

Milli, S., Moscatelli, M., Stanzione, O. and **Falcini, F.** (2007) Sedimentology and physical stratigraphy of Messinian turbidite deposits of the Laga basin (central Apeninnes, Italy). *Boll. Soc. Geol. Ital.*, **126**, 255–281.

Møller, N.K., Gjelberg, J.G., Martinsen, O.J., Charnock, M.A., Færseth, R.B., Sperrevik, S. and **Cartwright, J.A.** (2004) A geological model for the Ormen Lange hydrocarbon reservoir. *Norwegian Journal of Geology*, **84**, 169–190.

Mulder, T. and **Alexander, J.** (2001) The physical character of subaqueous sedimentary density currents and their deposits. *Sedimentology*, **48**, 269–299.

Mulder, T., Syvitski, J.P.M., Migeon, S., Faugeres, J.C. and **Savoye, B.** (2003) Marine hyperpycnal flows: Initiation, behavior and related deposits. A review. *Mar. Petrol. Geol.*, **20**, 861–882.

Mulder, T., Callec, Y., Parize, O., Joseph, P., Schneider, J.-L., Robin, C., Dujoncquoy, E., Salles, T., Allard, J., Bonnel, C., Ducassou, E., Etienne, S., Ferger, B., Gaudin, M., Hanquiez, V., Linares, F., Marchès, E., Toucanne, S. and **Zaragosi, S.** (2010) High resolution analysis of submarine lobes deposits: Seismic-scale outcrops of the Lauzanier area (SE Alps, France). *Sed. Geol.*, **229**, 160–191.

Mutti, E. (1992) *Turbidite Sandstones*. Agip-Istituto di Geologia Università di Parma.

Mutti, E. and **Normark, W.R.** (1987) Comparing examples of modern and ancient turbidite systems: problems and concepts. In: *Marine Clastic Sedimentology* (Eds J.K. Legget and G.G. Zuffa), Graham and Trotman, London, 1–38.

O'Byrne, C.J., Barton, M.D., Steffens, G.S., Pirmez, C. and **Buergisser, H.** (2007) Architecture of a laterally migrating channel complex — Isaac Formation, Windermere Supergroup, British Columbia, Canada. In: *Atlas of Deep-Water Outcrops* (Eds T.H. Nilsen, R.D. Shew, G.S. Steffens and J.R.J. Studlick), *AAPG Studies in Geology*, **56**, CD-ROM, 11 p.

Osmundsen, P.T. and **Ebbing, J.** (2008) Styles of extension offshore mid-Norway and implications for mechanisms of crustal thinning at passive margins. *Tectonics*, **27**, doi:10.1029/2007TC002242.

Piper, D.J.W. and **Normark, W.R.** (2009) Processses that initiate turbidity currents and their influence on turbidites: a marine geology perspective. *J. Sed. Res.*, **79**, 347–362.

Piper, D.J.W. and **Normark, W.R.** (2001) Sandy fans – from Amazon to Hueneme and beyond. *AAPG Bull.*, **85**, 1407–1438.

Pirmez, C., Beaubouef, R.T., Friedmann, S.J. and **Mohrig, D.** (2000) Equilibrium profile and base-level in submarine channels: examples from late Pleistocene systems and implications for the architecture of deepwater reservoirs. In: *Deep-Water Reservoirs of the World* (Eds P. Weimer, R.M. Slatt, J. Coleman, N.C. Rossen, H. Nelson, A.H. Bouma, M.J. Styzen), GCSSEPM Foundation 20th Annual Research Conference, pp. 782–805.

Prather, B.E. (2003) Controls on reservoir distribution, architecture and stratigraphic trapping in slope settings. *Mar. Petrol. Geol.*, **20**, 529–545.

Prélat, A., Hodgson, D.M. and **Flint, S.S.** (2009) Evolution, architecture and hierarchy of distributary deep-water deposits: a high-resolution outcrop investigation from the Permian Karoo Basin, South Africa. *Sedimentology*, **56**, 2132–2154.

Ravnås, R., Nøttvedt, A., Steel, R.J. and **Windelstad, J.** (2000) Syn-rift sedimentary architectures in the northern North Sea. In: *Dynamics of the Norwegian Margin* (Eds A. Nøttvedt *et al.*), *Geol. Soc. London Spec. Publ.*, **167**, 133–177.

Reading, H.G. and **Richards, M.** (1994) Turbidite systems in deep-water basin margins classified by grain size and feeder system. *AAPG Bull.*, **78**, 792–822.

Remacha, E. and **Fernández, L.P.** (2003) High-resolution correlation patterns in the turbidite systems of the Hecho Group (South-central Pyrenees, Spain). *Mar. Petrol. Geol.*, **20**, 711–726.

Shanmugam, G. (2000) 50 years of the turbidite paradigm (1950s–1990s): deep-water processes and facies models. *Mar. Petrol. Geol.*, **17**, 285–342.

Sinclair, H.D. and **Tomasso, M.** (2002) Depositional evolution of confined turbidite basins. *J. Sed. Res.*, **72**, 451–456.

Sixsmith, P.J., Flint, S., Wickens, H.D.V. and **Johnson, S.D.** (2004) Anatomy and stratigraphic development of a basin-floor turbidite system in the Laingsburg Formation, main Karoo basin, South Africa. *J. Sed. Res.*, **74**, 239–254.

Smith, R.D.A. (2004). Silled subbasins to connected tortuous corridors: sediment distribution systems on topographically complex subaqueous slopes. In: *Turbidite sedimentation in confined systems* (Ed. S. Lomas), *Geol. Soc. London Spec. Publ.*, **222**, 23–43.

Smith, R.D.A. and **Møller, N.** (2003) Sedimentology and reservoir modelling of the Ormen Lange field, mid Norway. *Mar. Petrol. Geol.*, **20**, 601–613.

Sømme, T.O., Helland-Hansen, W., Martinsen, O.J. and **Thurmond, J.B.** (2009) Predicting morphological relationships and sediment partitioning in source-to-sink systems. *Basin Res.*, **21**, 361–387.

Stow, D.A.V and **Bowen, A.J.** (1980) A physical model for the transport and sorting of finegrained sediments by turbidity currents. *Sedimentology*, **27**, 31–46.

Svendsen, J.B., Hansen, J.P.V. and **Friis, H.** (2010) An unconventional Source-to-Sink Setting – The Siri Fairway System, Norwegian-Danish North Sea. In: *From Depositional Systems to Sedimentary Successions on the Norwegian Continental Shelf*, NPF Abstract Volume, **67**.

Sweet, M.L., Campion, K.M. and **Beaubouef, R.T.** (2007) Deep-water-slope to Basin-floor-fan deposits of the Eocene Tyee Formation, Oregon. In: *Atlas of Deep-Water Outcrops* (Eds T.H. Nilsen, R.D. Shew, G.S. Steffens and J.R.J. Studlick), *AAPG Studies in Geology*, **56**, CD-ROM, 8 p.

Twichell, D.C., Schwab, W.C. and **Kenyon, N.H.** (1995) Geometry of sandy deposits at the distal edge of the Mississippi Fan, Gulf of Mexico. In: *Atlas of Deep-water Environments: Architectural Styles in Turbidite Systems* (Eds K.T. Pickering, R.N. Hiscott, N.H. Kenyon, F. Ricci Lucchi and R.D.A. Smith), Chapman & Hall, London, 282–286.

Wickens, H.D.V. and **Bouma, A.H.** (2000) The Tanqua fan complex, Karoo Basin, South Africa – outcrop analog for fine-grained, deep-water deposits. In: *Fine-grained turbidite systems* (Eds A.H. Bouma & C.G. Stone), *AAPG Mem.*, **72**/*SEPM Spec. Publ.*, **68**, 153–165.

Int. Assoc. Sedimentol. Spec. Publ. (2014) **46**, 647–676

Unravelling the nature of deep-marine sandstones through the linkage of seismic geomorphologies to sedimentary facies; the Hermod Fan, Norwegian North Sea

BJØRN KÅRE LOTSBERG BRYN*† AND MARK ANDREW ACKERS†*

* *Norwegian Energy Company, Nykirkebakken 2, 4013 Stavanger, Norway*
† *Present address: Centrica Energi, Rosenberggata 99, 4003 Stavanger, Norway*
(E-mail: bjornkaare.bryn@centrica.com)

ABSTRACT

The Late Palaeocene to Early Eocene deep-marine Hermod Sandstone Member of the Sele Formation in the Norwegian North Sea has been studied in a mid to outer submarine fan setting by using a combination of seismic reflection data and wells. A complex geological setting provided by a network of intricate sand fairways, in places affected by substantial post-depositional sediment remobilisation, makes the Hermod Member a challenging hydrocarbon reservoir to develop in the North Sea. The application of geophysical techniques resulted in convincing images of subtle details that are rarely documented in deeply buried submarine fan systems. These details include levee-confined channels, avulsions and braided frontal splays. Four main feeder channels enter the study area from west-north-west and interact differently with basin structure controlled by extensional faults. Structurally controlled channels make sharp turns, have wide avulsion angles, align with substratum faults and tend to build complex frontal splay zones. Non-confined channels cross substratum faults, build small frontal splays and display common avulsions close to the channel end, probably due to frequent channel shifting in response to slope advantage to the side of their frontal splays. Post-depositional sediment remobilisation structures have also been evaluated and candidate pockmarks were found to be closely associated with steep-sided mounds. It is proposed that blocking of pockmark craters resulted in the trapping of gas in the Hermod sands and the triggering of sand remobilisation. Tens of metres of clean sandstone, with no apparent systematic vertical grain-size grading and a sharp contact to underlying and overlying shale, occurs both in remobilised and apparently in situ sediments. It is further suggested that sand and mud became segregated during transport with clean sand concentrated in channels and frontal splays which later became parent beds for sand injection to mounds. The result is a mostly homogeneous sandstone type with a multifaceted 3D geometry.

Keywords: North Sea, seismic geomorphologies, submarine fans, turbidites, injectites, sediment remobilisation.

INTRODUCTION

Gravity-driven sand influxes during the Palaeocene and Eocene have deposited the reservoir bodies of several oil and gas fields in the North Sea (Ahamadi *et al.*, 2003 and references therein). The first Norwegian oil discovery was made in 1967 in this play type in what is today the Balder Field. Whereas some of the earliest discoveries came on stream in the 1970s (e.g. the Frigg Field), it took decades of further exploration, appraisal and characterisation before it could be concluded that some of the smaller fields were commercial. For instance, the Balder Field did not start production until 1999 and the British Gannet F structure, where oil was discovered in 1969, began production in 1997. The long delay before some of the earliest discoveries came on stream was not only a function of oil price and development cost but was also related to the time it took to gather

From Depositional Systems to Sedimentary Successions on the Norwegian Continental Margin, First Edition.
Edited by A. W. Martinius, R. Ravnås, J. A. Howell, R. J. Steel, and J. P. Wonham.

sufficient information to mature a subsurface geological model that captured the complex geology of these reservoirs (e.g. Bergslien, 2002).

The science of turbidite play analysis has evolved simultaneously with the development of the North Sea into a major petroleum province. The rapid gathering of extensive amounts of data resulted in a number of observations that could only be partly understood in the framework of existing models and several fields were re-interpreted as the science evolved. The first submarine fan model that related the vertical facies successions observed in outcrops to fan positions (Jacka *et al.*, 1968) was proposed one year after the first oil discovery and the wide implementation of such models arrived even later (Mutti & Ricci Lucchi, 1972). The recognition of high-density turbidites (Lowe, 1982) and linked debrites (Haughton *et al.*, 2003) and the awareness of sand-injectites as an important constituent in the petroleum system (see Hurst & Cartwright, 2007 and references therein) are other examples of advances in turbidite play analysis that have occurred contemporaneously with the production from and exploration for, turbidite reservoirs in the North Sea.

Seismic acquisition and processing techniques have also evolved rapidly over the last decades and the prolific parts of the North Sea Basin are gradually becoming blanketed with 3D seismic data. It is now widely documented that many reservoirs have a complex facies association that has resulted not only from a variety of primary depositional processes but also from different types of sediment remobilisation such as slumping, sliding or large-scale sand injection processes. However, the relative importance of these various processes may, in many cases, still be immeasurable due to the lack of core material and a seismic dataset that delineates the reservoir geometries clearly. This study targets an area in the Northern North Sea where both depositional elements and post-depositional sediment remobilisation can be inferred from seismic geomorphologies in a mid to outer submarine fan setting. The studied stratigraphic unit is the Late Palaeocene to Early Eocene Hermod Member of the Sele Formation (*sensu* Brunstad *et al.*, 2009). The Hermod Member is typically characterised in wells by thick successions of mainly homogeneous sandstones with no apparent traction structures or systematic grain size grading and a sharp contact to both overlying and underlying shale. This facies type is volumetrically the most important constituent in the area, but the origin of such entirely massive deep-marine sandstone beds is still a subject of intense research. The relevant working hypotheses include deposition under high-density turbidity currents (Lowe, 1982), debris flows (Shanmugam, 1995), or full scale remobilisation of the entire bed leading to obliteration of any primary vertical grain-size grading and depositional structures (Hiscott, 1979). A turbidite interpretation of the typical massive and non-graded sandstone tops with a sharp contact to overlying shale is cryptic because turbidity currents are expected to pass through a last waning phase when a graded and stratified top should form (Baas, 2004). Emplacement by sandy debris flows is an alternative interpretation, but laboratory experiments indicate that run out on gentle slopes (c. <5 degrees) requires presence of lubricating clay minerals in the flow (Marr *et al.*, 2001). The formation of massive non-graded sandstone beds by post-depositional sediment remobilisation is an interpretation that is particularly relevant for the Tertiary of the North Sea (Jonk *et al.*, 2005), but the fact that the occurrence of an abrupt transition to shale above is more the rule than an exception is still puzzling. This is a characteristic of the upper boundary of the Hermod Member, not only in the present study area, but throughout the Norwegian North Sea, as described in the standard definitions of the Hermod Member (Hardt *et al.*, 1989; Brunstad *et al.*, 2009 and Norwegian Petroleum Directorate Fact Pages at www.npd.no).

The combination of high-quality seismic images from a formation buried to a depth of approximately 2 kilometres where cores are (mostly) lithified represents a unique opportunity to compare and contrast the products of various geological processes in a submarine fan. This has led previous researchers to use the present study area as a case example to illustrate some key characteristics of this depositional environment. Blikeng & Fugelli (2000) identified a channel distributary pattern with channels reaching the outermost parts of the fan and pointed out the analogy to the modern Mississippi Fan and the Permian Brushy Canyon Formation, Delaware Basin, USA. Hadler-Jacobsen *et al.* (2005) presented data from the same study area in a broader discussion of how sediment delivery, gross shelf-to-basin relief, slope gradient and basin topography control deep-marine sedimentation. In this context, the outer Hermod Fan was classified as a high shelf-to-basin-relief system with a distinct slope set up by the western Viking Graben bounding fault some

120 kilometres to the west. Examples of the studied Hermod submarine fan have also been included in a more recent Norwegian offshore stratigraphic lexicon (Brunstad *et al.*, 2009).

The approach taken here has been to apply a combination of geophysical techniques to 3D seismic data to achieve enhanced imaging of geomorphologies over a larger area than presented in previous publications (i.e. Blikeng & Fugelli, 2000; Hadler-Jacobsen *et al.*, 2005; Brunstad *et al.*, 2009). The geophysical work flow has resulted in the visualisation of the Hermod Member sandstone with a level of detail that is rarely documented for ancient subsurface fans. The paper focuses on the analysis of depositional elements (*sensu* Mutti & Normark, 1991) and it also documents some pronounced sediment remobilisation structures that involve Palaeocene and Eocene stratigraphy above the Jurassic Frøy Field. The integration of this detailed analysis of the depositional and sand remobilisation structures with borehole data enabled an improved understanding of the distribution of this excellent hydrocarbon reservoir.

GEOLOGICAL SETTING

The Viking Graben (Fig. 1) was initiated as a rift arm in the Late Jurassic (Zanella & Coward, 2003). The rifting failed to develop to the point of break up and rapid thermal subsidence resulted in marine transgression over most of the Viking Graben by the latest Jurassic.

Continued subsidence resulted in the development of a broad deep-marine basin, bounded by major faults to the west that became the locus for deposition of thick successions of coarse clastics during episodic lowstands of relative sea-level in the Palaeocene and Early Eocene. The most extensive submarine fans are located along the western side of the basin and the inferred denudation of large volumes of sediments in the west may be related to pulses of regional uplift driven by mantle plume activity in the North Atlantic Igneous Province (White & Lovell, 1997). The mantle activity culminated in Atlantic break up (c. 56 to 53 Ma) and frequent ash eruptions produced a series of tuff beds which today represent an important stratigraphic marker bed (Balder Formation, Fig. 2). A period of rapid subsidence followed after the break up (Nadin & Kusznir, 1995) and the volumes of sediments supplied to deep sea fans decreased significantly (cf. Hermod, Frigg and Grid sandstone members in Fig. 2).

The depocentres of Tertiary basins (see e.g. Mudge & Copestake, 1992a, 1992b; Milton & Dyce, 1995; Mudge & Bujak, 1996) are roughly positioned above the axis of the Viking Graben and local elongation of sediment gravity flow deposits along the structural trends of this buried rift system suggests long-lived control on basin bathymetry (Fig. 1). Pronounced relief along the western bounding fault complex of the Viking Graben appears to have controlled the shelf edge position and formed a high relief shelf to basin profile that existed throughout the Palaeocene (Hadler-Jacobsen *et al.*, 2005). The supply of significant amounts of sandstone to the deep-marine basin is likely to have occurred primarily during periods when deltas prograded to the structurally controlled shelf edge (Dixon & Pearce, 1995). A well-defined canyon in the Beryl Embayment is interpreted to have formed during maximum regression of the Dornoch Delta as an incised valley that was subsequently linked into a submarine canyon that allowed for effective sediment transfer from shelfal/deltaic sources into the deep basin (Erik Imsland Wathne, pers.comm.). This canyon is inferred to be the main feeder of sediment to the studied Hermod Member submarine fan complex (Fig. 2).

Sandstone intrusions are widely distributed at Palaeocene and Eocene level in the northern North Sea (Huuse *et al.*, 2007). The intrusions connect sandstone reservoirs that are otherwise separated by shale and are therefore of key importance for aquifer support as well as hydrocarbon migration and trapping. A large sandstone injection complex which is thought to originate from a parent bed in the Hermod Formation forms the main reservoir in the Volund Field (de Boer *et al.*, 2007; Szarawarska *et al.*, 2010) and injectites are also important in the Greater Balder Area (Bergslien, 2002). A recent paper presented a thorough analysis of injectites and Palaeocene mounds in this area and proposed that the mounds are secondary, post-depositional structures caused by fluid influx (Wild & Breidis, 2010). The mounds in the Greater Balder Area are of similar age and bear many similarities to the mounds that exist over the Frøy Field that are presented herein. However, the Hermod Fan which is the main subject of the present study does not extend to the Volund or Balder Fields further to the south (Fig. 2; see also palaeogeographic maps by Ahmadi *et al.*, 2003 and Brunstad *et al.*, 2009).

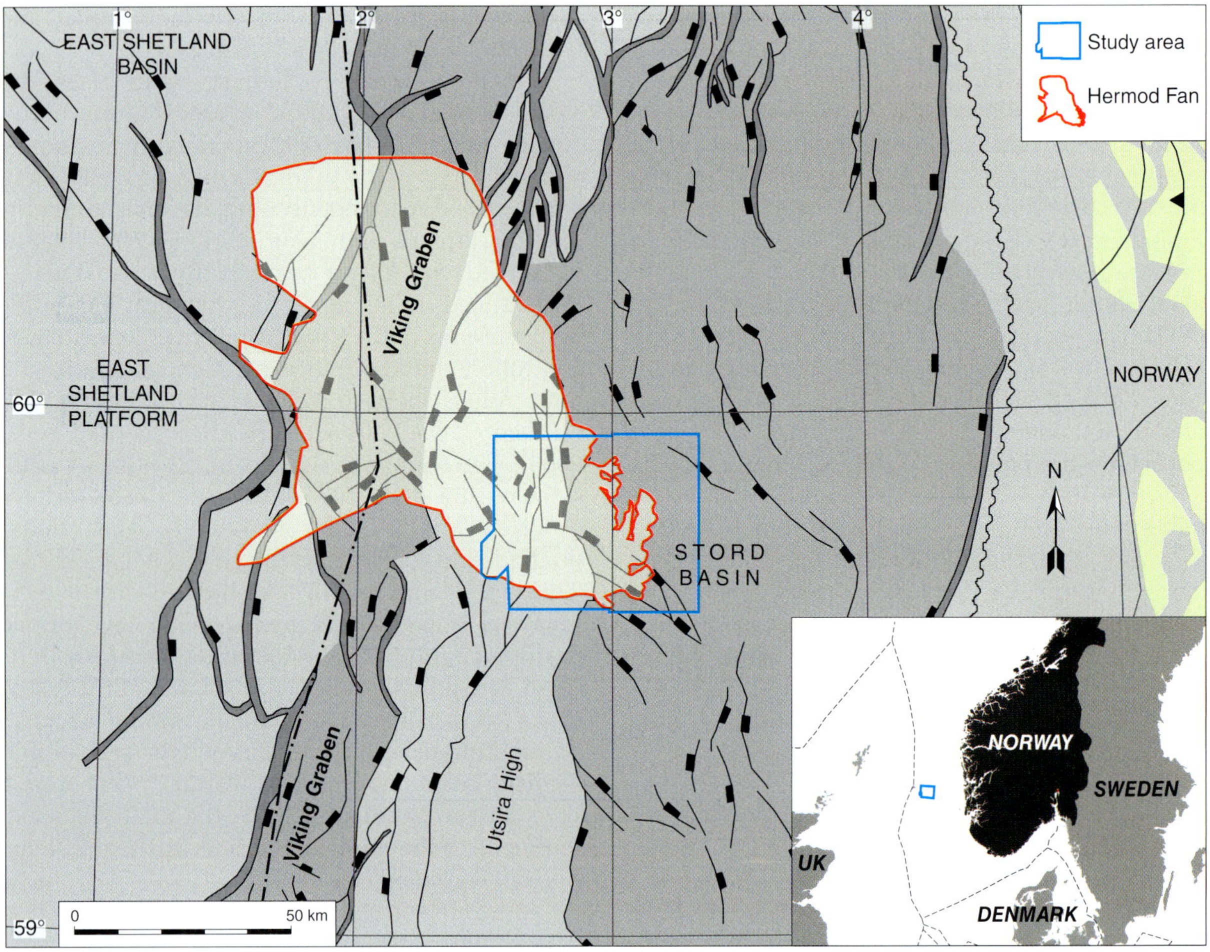

Fig. 1. Location map of the study area with main structural elements and gross outline of the studied Hermod Member submarine fan complex. The studied 'tongue-like' mid-fan to outer-fan follows the grain of a family of faults that concur with the northern rim of the Utsira High and a step-over zone in the segmented Viking Graben.

DATA AND METHODOLOGY

A standard subsurface work-flow, involving detailed core description tied to well logs, well correlations, well-seismic ties and seismic interpretation, was performed in the mid to outer fan area. The only Sele Formation core from this area is from well 25/6-3 and covers 45 metres of sandstone and shale. The seismic interpretation was performed on two 3D datasets; UHN98R07 and EN0101, which collectively cover a total area of 1734 square kilometres. These datasets were optimised for seismic stratigraphic analysis through the application of a combination of geophysical techniques including random and coherent noise cancellation, spectral enhancement, spectral decomposition (Partyka *et al.*, 1999) and Red-Green-Blue (RGB) volume blending in SVI Pro© (hereafter referred to as the RGB blend in text and figures). The noise-cancelled seismic data facilitated precise auto tracking of key seismic reflectors, resulting in detailed isochron maps that clearly mimic depositional elements and post-depositional sediment remobilisation structures.

The combination of the isochron maps and the RGB blend provided an efficient tool for the assessment of the internal geometries of the Hermod Member (Fig. 3). A combination of various SVI Pro© edge attributes was also applied to the seismic dataset as a quantitative approach for the detection of steep-sided mounds and pockmarks.

The optimal seismic imaging was achieved in the mid to outer fan area. The seismic representation of the more proximal parts of the fan tends to be rather poor, especially in places where the Frigg Sandstone Member overlies the Hermod

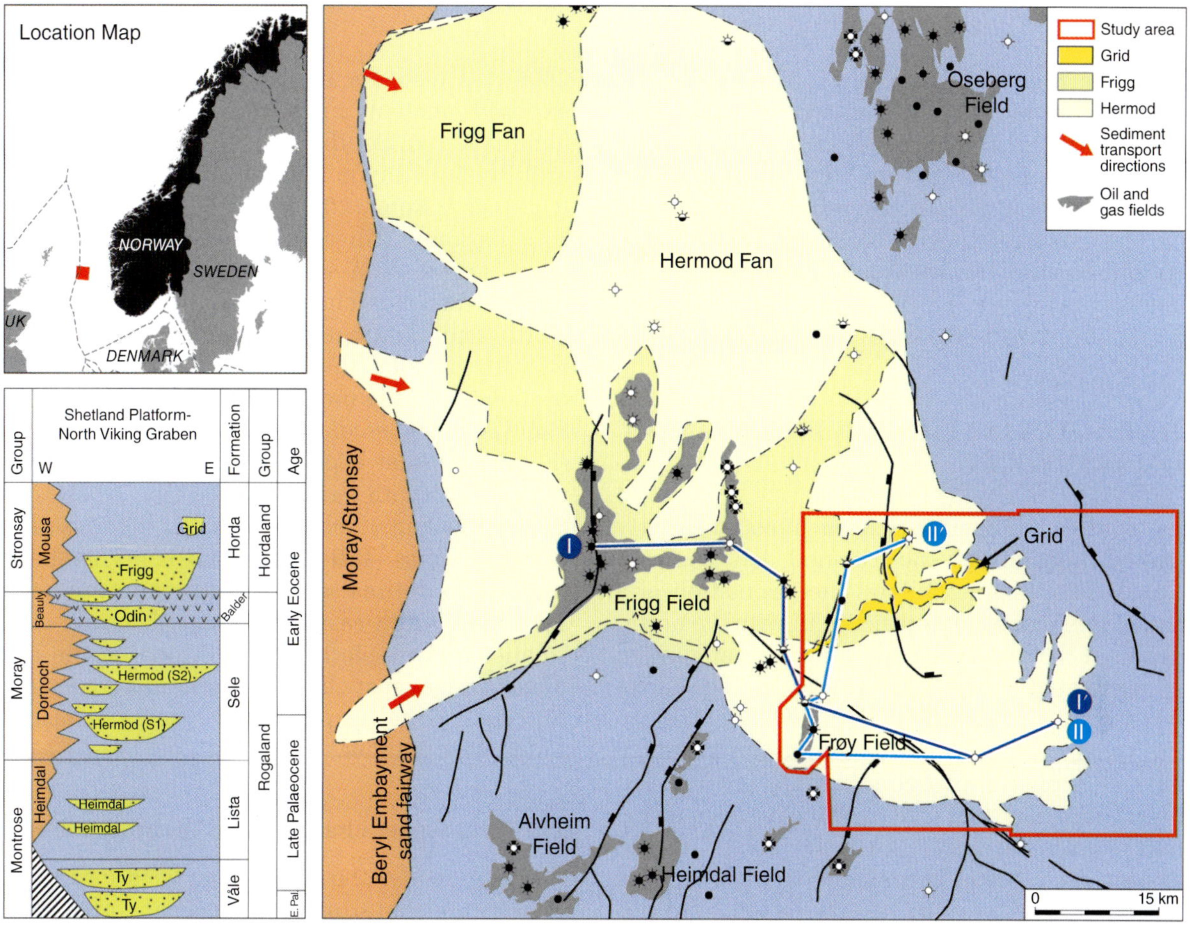

Fig. 2. Lithostratigraphic columns and the plan-view distribution of a Hermod Fan, a Frigg Fan and a younger channel complex assigned to the Grid Member. Line I-I' corresponds to well section of Fig. 15, line II-II' corresponds to well section of Fig. 4.

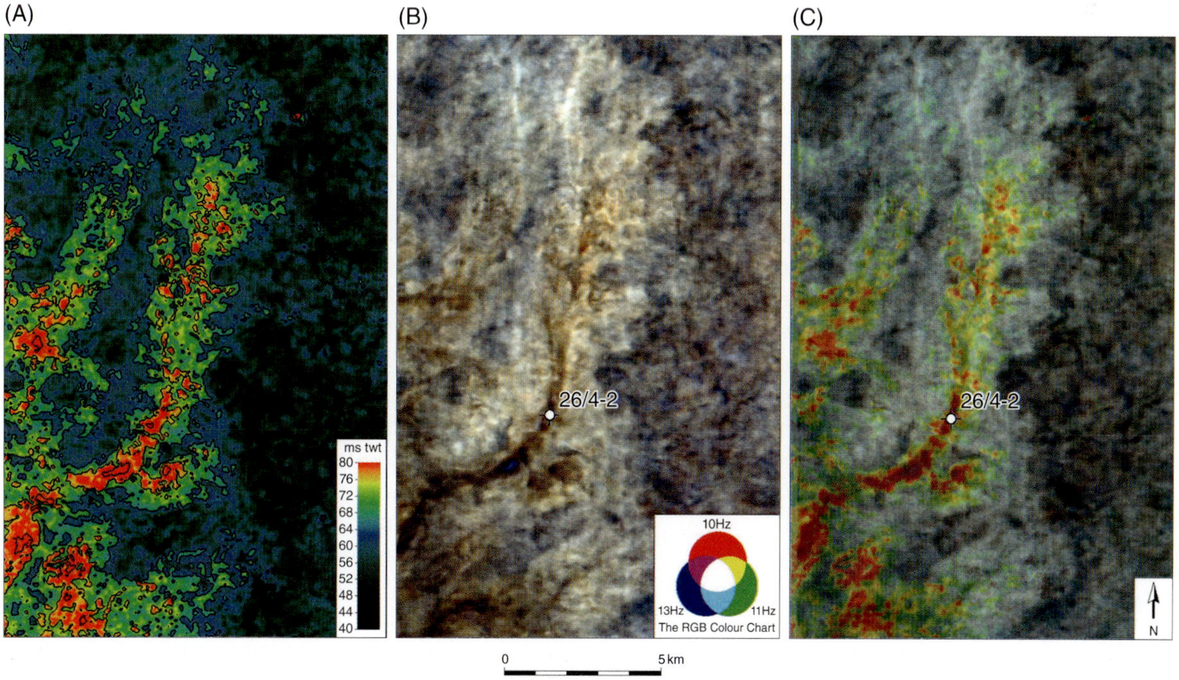

Fig. 3. (A) Isochron map between reflectors near the top and base Sele Formation. (B) Time slice at 34 ms below the (near) top Sele seismic pick in RGB blend of 10 Hz, 11 Hz and 13 Hz frequency components. (C) A blend of images A and B. Note close relationship between mapped thickness and imaged geometries in the RGB blend. TWT = two way transit.

Sandstone Member (see Fig. 2). The palaeogeographic maps are therefore based on the detailed mapping and tie of eight wells in the mid to outer fan area, supplemented by the correlation of twenty one more wells in the proximal part of the fan. The correlations were based on released well logs and biostratigraphic charts, calibrated to core wherever available (i.e. for wells 25/1-7, 25/6-3, 30/7-2, 30/8-2 and 30/10-7).

CHARACTERISATION OF THE HERMOD SUMARINE FAN

The studied submarine fan is a south-easterly directed 'tongue' of the Hermod fan complex (Fig. 2). The 'tongue' follows the orientation of a distinct fault zone that is concurrent with the northern rim of the Utsira High and a step-over zone in the segmented Viking Graben (Fig. 1). The large-scale stacking pattern of the Hermod fan relates to a third order cycle (*sensu* Mitchum & Van Wagoner, 1991) and phases of fan initiation, growth and retreat (*sensu* Gardner *et al.*, 2003) have been recognised within the larger 3rd order cycle (Hadler-Jacobsen *et al.*, 2005). The orientation of the studied Hermod 'tongue' suggests that the regional dip and overall direction of fan growth was from the north-west and towards the south-east (Fig. 1).

Hadler-Jacobsen *et al.* (2005) suggested that the older Hermod sandstones denoted S1 were deposited in the fan initiation and growth phases and that the younger S2 sandstones were deposited in the fan retreat phase. The S1-S2 subdivision (*sensu* Brunstad *et al.*, 2009) has also been adopted herein, but the diagnostic *Apectodinium* species acme has not been recognised in all wells and the correlations are therefore tentative. However, a vertical stacking pattern indicative of fan initiation, growth and retreat can be inferred for the two northernmost wells in the study area, whereas the well location furthest to the south-west appears to have received sand deposition only during the growth phase (Fig. 4). The stacking patterns in the wells agree with the assumption that the fan prograded from north-west towards south-east where deposition only occurred during maximum basinward progradation.

General fan outline

The Hermod fan consists of a complex stack of channels and splays (Fig. 5), with channels reaching the outermost part of the fan (Blikeng & Fugelli, 2000), as also documented for some very well exposed ancient submarine fans such as the Tanqua Karoo, South Africa (Johnson *et al.*, 2001) and Brushy Canyon Formation, West Texas, USA (Gardner & Borer, 2000). The Hermod fan has an irregular outer fringe formed by several branched front splays. The furthest basinward reaching channel makes a characteristic turn towards the North and terminates in a feather-like pattern which resembles the well-studied distal reaches of 'Lobe 8' in the modern Mississippi Submarine Fan (Twichell *et al.*, 1995) and the RGB blend (Fig. 3) hints at a possible frond-like fan fringe which is, with analogy to the Mississippi Submarine Fan, likely to consist of debrites rather than turbidites (Talling *et al.*, 2010). The characteristic fan outline seen in map view today, with an overall south-eastward oriented tongue and a smaller north-easterly oriented lobe on the most distal fringe, may be a pattern that repeatedly occurred as the fan prograded across the basin in a step-wise manner.

The main feeder channels and their branches

The term feeder channel is here used for the part of a channel that stretches from a canyon to a frontal splay. The transition from a feeder channel to a frontal splay coincides with the zone where a singular channel splits into multiple distributary channels with a marked reduction in channel width and depth (cf. Posamentier & Kolla, 2003). This is also the transition zone from confined to unconfined flow.

Four main feeder channels enter the study area from the west (numbered, without reference to chronology, C1 to C4 in Fig. 5), around 60 kilometres distant from the mouth of canyons along the eastern rim of the West Shetland Platform. Three of them split up into a northern and southern branch (labelled C for Channel, N for North and S for South, e.g. C1 in Fig. 5 splits into C1N and C1S) which again split into narrower channels in a downstream direction (Fig. 6). The channels are less than 0.5 km wide but can be traced for up to tens of kilometres. All identified channels display very low sinuosity ratios (<1.1) but portions of some channels may form distinct turns with sharp bends (up to 90°). Channel splitting that is likely to have resulted in the rapid abandonment of a channel and the initiation of a new channel with approximately similar width (i.e. avulsion) has been inferred at some locations. The angles at the 'Y-junctions'

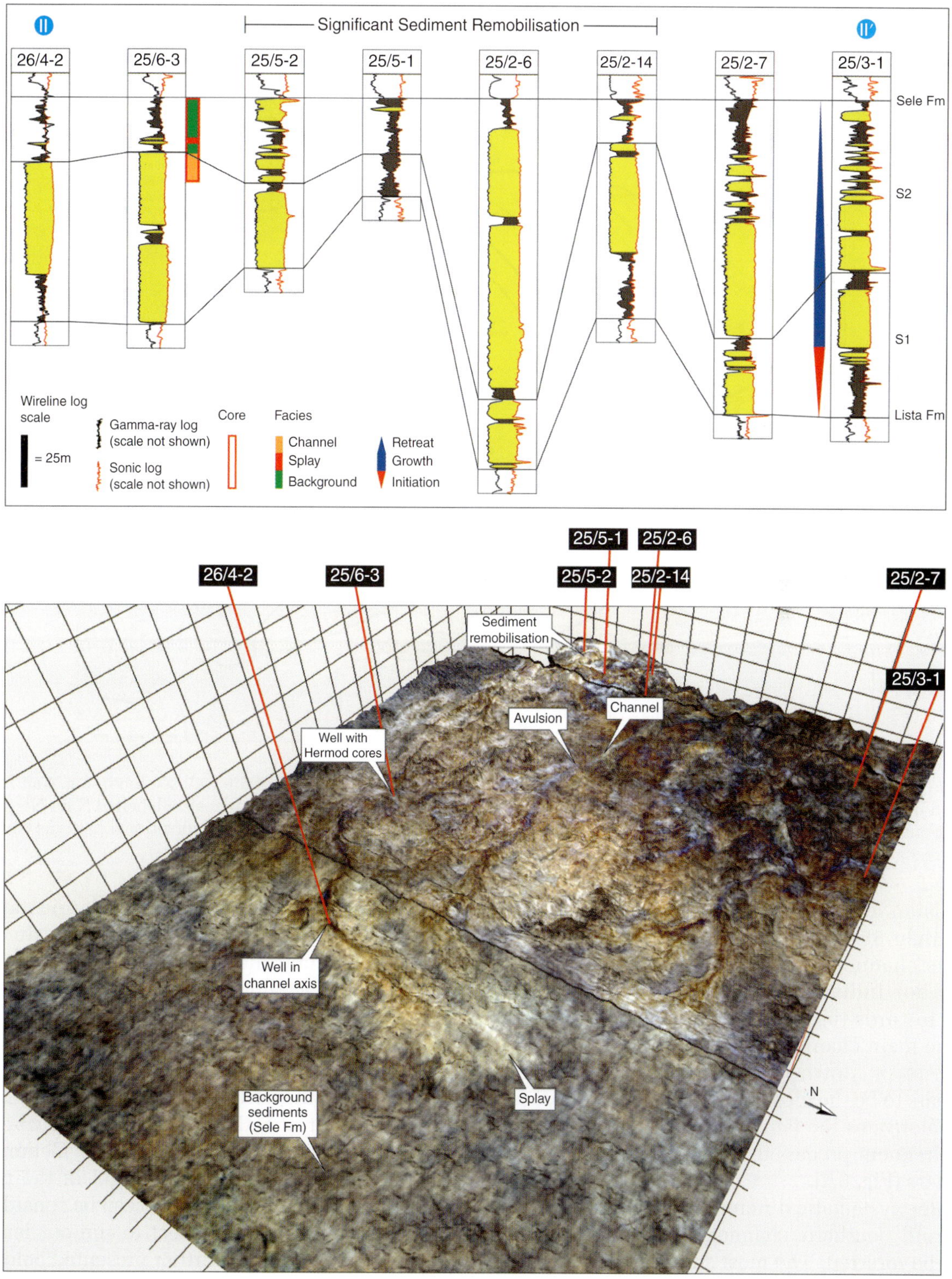

Fig. 4. Above: Hermod sandstones represented in yellow between sonic and gamma ray logs for the same well as displayed in the 3D view above. Well spacing ranges from well 22 km (from well 25/2-14 to well 25/3-1) to 2 km (the distance between the Hermod Member type well 25/2-6 and the nearest offset well 25/2-14). Note that some of the wells are drilled in an area where post-depositional sediment injection and evacuation has been interpreted and the vertical successions observed in these wells have not been used for analysis of sandstone stacking patterns. Only the two wells in the north-western corner of the study area (i.e. 25/2-7 and 25/3-1) and the two wells in the south-eastern corner (i.e. 25/6-3 and 26/4-2) are believed to represent (mostly) in situ sediments. Below: 3D illustration of the present day time-structure relief of the Top Sele Formation seismic reflector with colours extracted from the Hermod Member as it appears on a blended spectral decomposition volume (10/11/13Hz). The primary depositional architecture of the Hermod Member is dominated by channels with terminal splays and, locally, the occurrence of irregularly shaped mounds interpreted as sediment remobilisation structures. Note also the close link between structural relief and depositional architectures; the sand-rich channels and splays of the Hermod Member form positive reliefs that clearly stand out from the rather flat background shales of the Sele Formation (presented in a more monotonous blue colour).

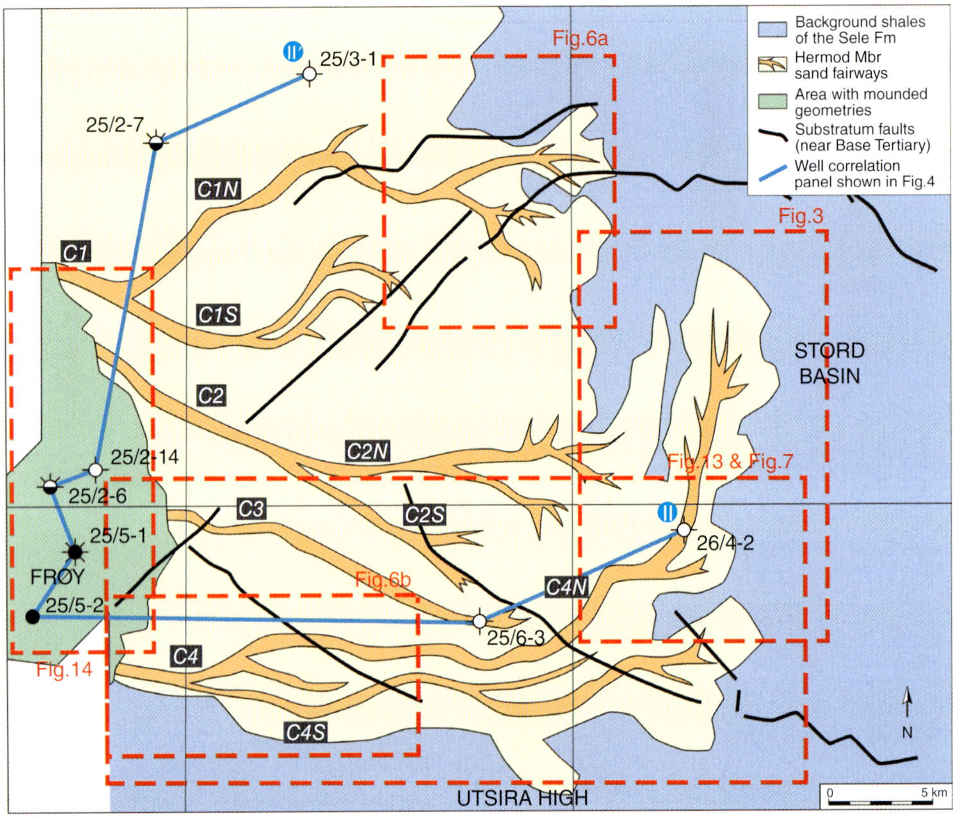

Fig. 5. Gross seismic geomorphologies of the outer Hermod fan (Fig. 1) with location of the well section in Fig. 4 and maps shown in other figures. The four main feeder channels are labelled C1 to C4 and their branches are labelled N for North and S for South. The green polygon indicates an area over the Jurassic Frøy Field where primary depositional patterns have not been recognised due to severe post-depositional sediment remobilisation.

(avulsion angles) are variable but in most cases less than 55°. Channel orientations of the main feeder channels are mainly towards east or south-east, but individual channels are locally deviated towards the north-west.

The main channels branch into narrower channels in a down-stream direction. Branching appears to be the most common, recurring process that elongates the fan and avulsion seems to be a less frequent process that initiates new branching clusters (Fig. 6A).

Some systematic differences between the northern and southern channels and their branches may be inferred. The most striking dissimilarities are firstly that the avulsion angles are lowest in the southernmost channel and highest in the northernmost channel and secondly that the northernmost channels make sharp turns near substratum faults and have sections that follow the two dominant fault orientations mapped in the underlying strata (i.e. north-west to south-east and south-west to north-east).

The link between channel and fault orientations suggests that the flows encountered a basin with subtle steps controlled by substratum faults. The avulsions had a tendency to direct flows in a more northerly direction at such steps, particularly in the northern part of the study area. It appears that the structural control on sand fairway orientations is strongest in the north-east (C1N). An exception is the distal part of C4N which is also structurally guided and turns towards the north in the outboard Stord Basin. The C1 channel in the north and the C4 channel in the south can be considered as end members with respect to run-out lengths and growth pattern, as further presented below.

Structural control on channel and splay evolution

C1S makes a 90° turn towards the left (in a down-stream direction) and splits into two almost parallel branches close to a substratum fault. Both branches run parallel to the fault (i.e. north-eastward) for

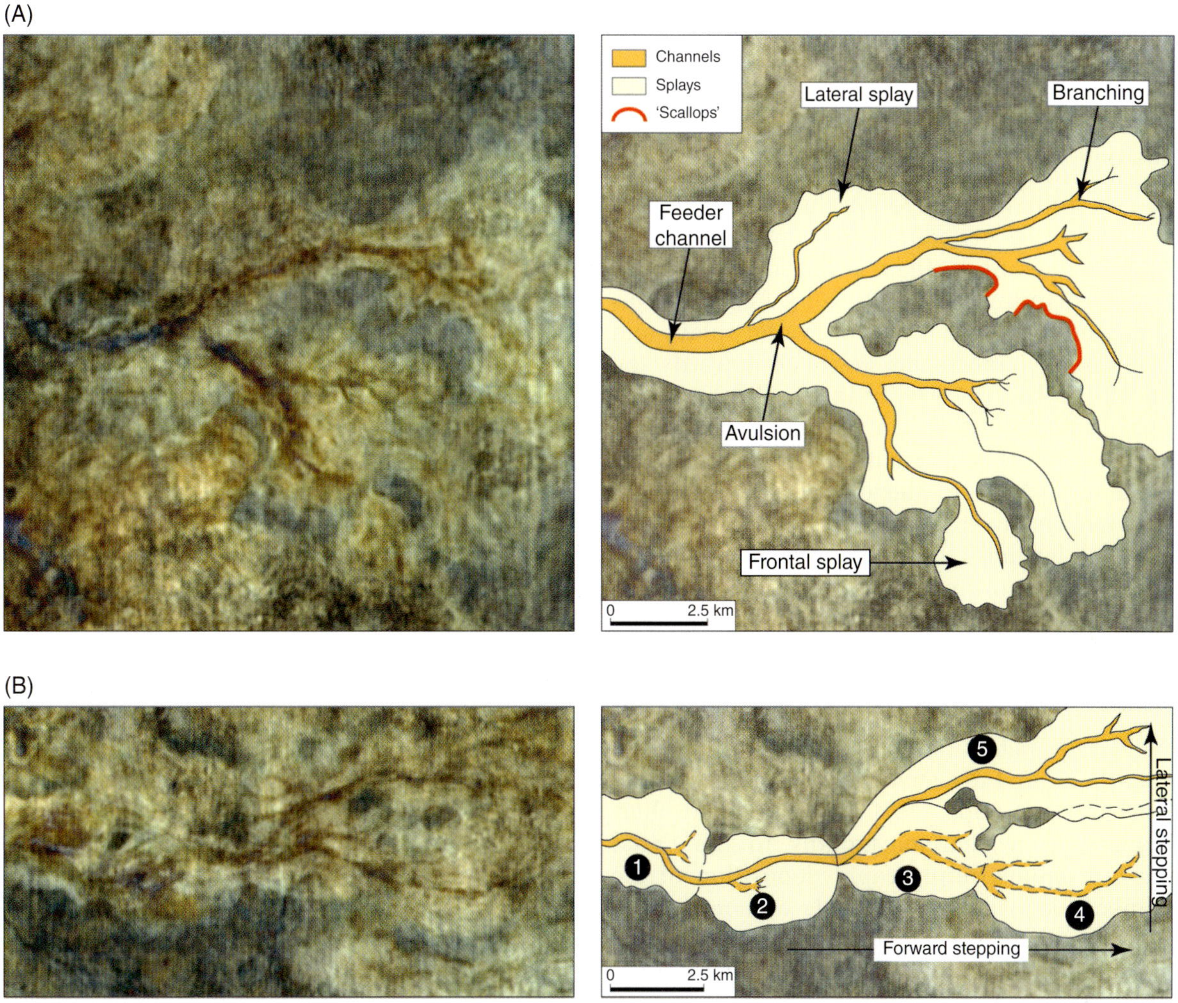

Fig. 6. Un-interpreted (left) and interpreted (right) RGB-blend time slices (10 Hz, 11 Hz and 13Hz). (A) Key stratigraphic elements in a channel-splay complex and (B) a series of basinward stepping transient fans. A dashed line annotated on channels indicates segments that have become abandoned due to avulsion. Repeated sediment blocking in the frontal splay region may have set the scene for a series of avulsions followed by new elongation of channels along a different path. The inferred youngest avulsion point occurred further up-stream and resulted in a lateral fan switch.

some 4 to 6 km. Both channels then turn 90° towards the right (i.e. back to the original orientation) and terminate more or less at the fault location. C1N follows a similar pattern on encountering a substratum fault, but the channel segment with a north-eastward orientation has a length of more than 10 km (Fig. 5). The frontal splays are characterised by a braided zone where the channels branch into narrower distributary channels downstream to a point where they approach the scale of seismic resolution and fringe out in a radial pattern (Fig. 6A). The length from the onset of channel branching to the point where the terminal splays thin to a scale below seismic resolution is typically less than 3 km.

The characteristic patterns described for the branches of C1S can be related to temporary deviations from the overall south-eastward fan progradation, due to subtle north-eastward oriented sea floor depressions. The fault structuration of sea floor morphology was probably negligible by the time the youngest and most distal branches re-gained their south-eastward orientation and spilled towards the south-east. A similar interpretation of a gradual south-eastward progradation of the Hermod Fan, interrupted by temporary fault-influenced deviations towards the northeast, can also apply to the C1N channel. The smaller scale channel branching within the frontal splay zones probably resulted from aggradation of sediments followed by stream bifurcation around evolving channel-front deposits.

Forward and lateral stepping channel-splay transition zones

C4N crosses faults mapped in the subsurface and displays a series of channels and fan shaped frontal splays that appear to be arranged in a forward to lateral stepping pattern (Fig. 6B). Avulsion sites are located in the axial part of frontal splays and the avulsed channels are deviated towards the side of the splays. The channel that has been inferred to be youngest can be traced with confidence for more than 35 km from the avulsion point shown in Fig. 6B to the outboard Stord Basin where it turns towards the north and terminates in a complex feather-like channelised splay.

C4N shows no deviation when crossing faults suggesting no direct structural control on channel orientation. The channel probably elongated step-wise towards the east in a series of discrete flow events or a train of pulses in a sustained flow. It appears that the frontal splays aggraded to form a relief that directed the elongating channel to the side (see steps 1 to 4 in Fig. 6B). The short distances from channel mouths to the distal fringes of splays imply rapid deposition of the sediment load once the currents became unconfined. The small size of splays suggests that the channel mouth was unstable and made frequent shifts controlled by rapidly changing depositional relief rather than long lived basin structure. The channel evolved during maximum fan growth and it represents the sand fairway that ultimately achieved the longest run-out length in the Hermod Fan.

Slope readjustment on the passage to the outboard Stord Basin

C4S is defined in the RGB blend by the same blue colour as the 'background sediments' of the Sele Formation (see Figs 4 and 6) and on isochron maps by thinning of the Sele Formation (Fig. 7). It can be inferred from some seismic sections that the thinning is associated with truncation of seismic reflectors under the Near Top Sele Formation seismic event. C4S is estimated to be around 40 m deep and 1 km wide at a central location and the depth appears to decrease from a proximal to distal position where the channel feeds frontal splay deposits in the Stord Basin (Fig. 7). The northern channel wall is ornamented in some places by scallop-like scars.

C4S is thought to be a fine-filled channel cut. The incision was probably initiated as flows encountered a deeper-floored base level in the outboard Stord Basin. The slope disequilibrium across the step to the outboard basin was then healed by erosion upstream of the step and deposition downstream of it (Pirmez *et al.*, 2000 and references therein).

The erosive part of the channel may have cut into previously deposited turbidites and the scallop-like structures could potentially be interpreted as collapse scars formed on the channel margin or overbank element, preferentially located on the northern flank where the channel may have eroded the levees of the adjacent C4N. Another interpretation could be megaflute erosion scars (*sensu* Elliot, 2000). However, the interpretation of the more delicate geometries is difficult because the area appears to have been affected by post-depositional sediment remobilisation after channel cutting (see subchapter on 'Sand remobilisation complexes' in this paper). A primary erosional origin is supported by the observation of truncated reflectors under the Near Top Sele Formation seismic event (see Fig. 7B).

Levee-confined feeder channels

Feeder channels are convincingly expressed by a discrete channel-shaped thickening in isochron maps that corresponds to a high-acoustic impedance event (Fig. 3). They are very well expressed in the RGB blend and further accentuated by draping the blend onto the Near Top Sele time-structure map or the Sele Formation isochron map (Figs 4 and 7). The channel-shaped sand bodies (dark brown in Fig. 7C) are narrower than the sands they overly (light brown to whitish areas in Fig. 7C).

The lateral thickness variations in the Sele Formation associated with a feeder channel can be best quantified in the area around well 26/4-2, where the top of a vividly imaged channel corresponds to a 47 m thick sandstone interval. In seismic cross sections, the top reflector is rather flat and the base reflector is concave upwards (Fig. 8). Seismic mapping indicates that the thickness of the 'background sediments' seen on either side of the drilled channel are similar to the total shale thickness of the Sele Formation at the well location (Fig.8). The thickest part of the Sele Formation contains the channel and also a deeper more

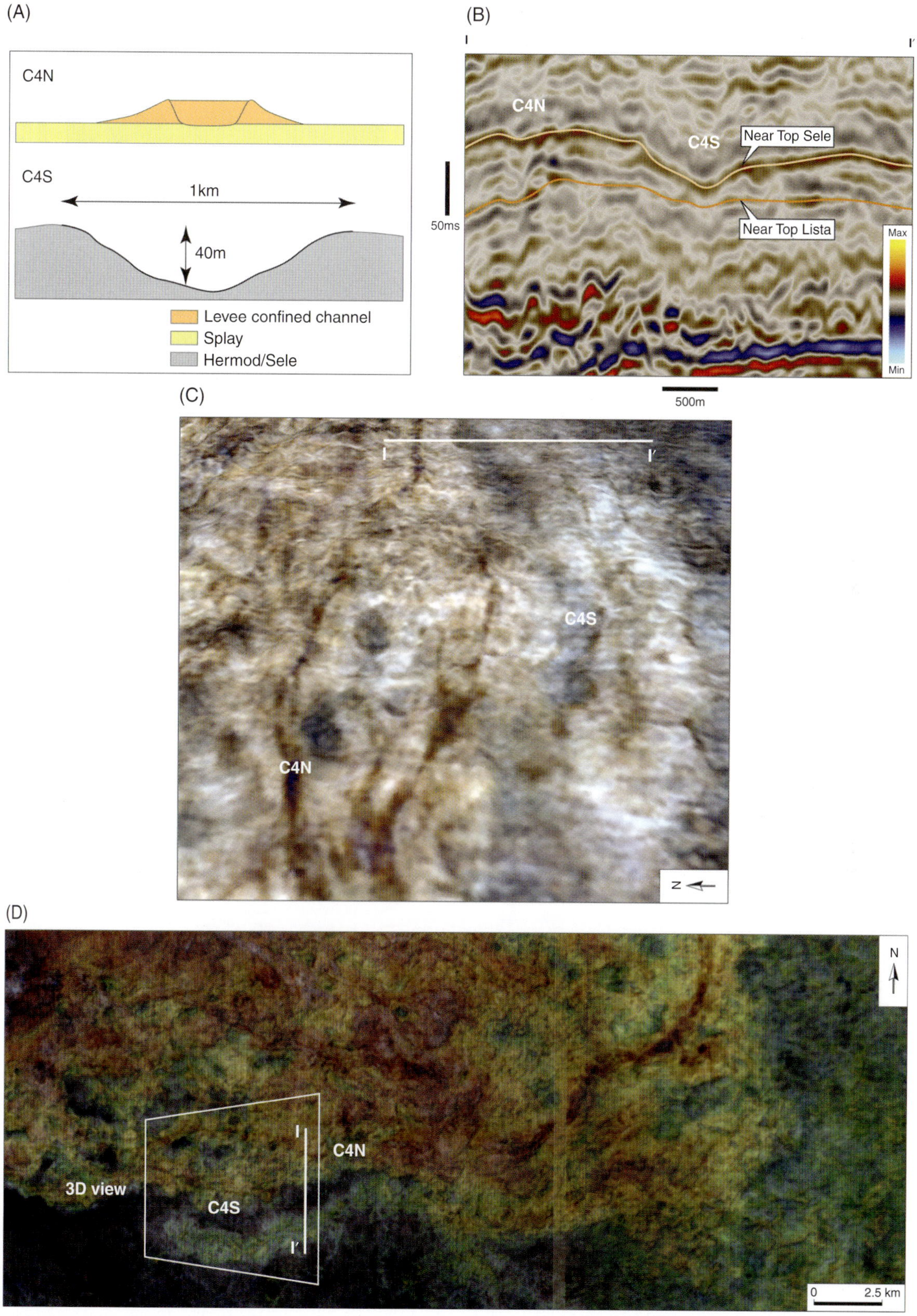

Fig. 7. (A) Sketches of channel interpretations for C4N and C4S; (B) seismic section across the two interpreted channel types (C4N and C4S); (C) oblique view on 3D image of RGB blend draped to isochron map plotted as z-value (i.e. 'isochron-relief map' with thick Sele Formation represented by a positive relief). Channel C4S is inferred to be youngest and positioned close to the leveed C4N to the north; (D) isochron map with warm colours representing thick areas. The C4S channel is thin (blue) but grades to a frontal splay towards the east (yellowish). Note that the area has also been affected by post-depositional sediment remobilisation (compare to Fig. 13).

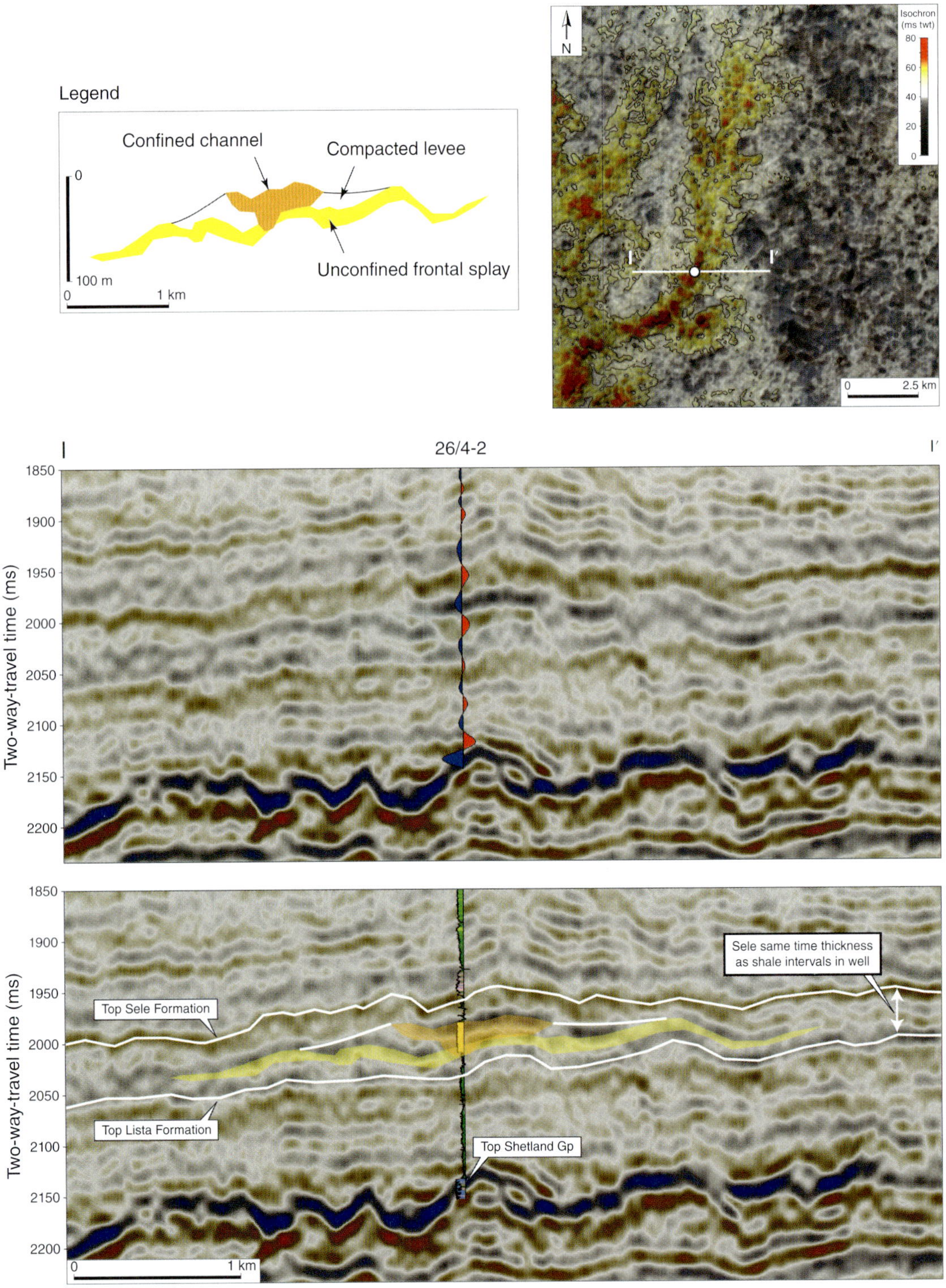

Fig. 8. Transect through channel drilled by well 26/4-2. Shows unmarked and interpreted seismic cross sections with an explanation of the interpretations and an isochron map at the top. The synthetic seismogram on the un-marked section is based on a 20 Hz Ricker wavelet. The section is believed to cross a frontal splay overlain by a leveed channel. The discrete narrow channel (Fig. 3) has a concave upwards base reflector but has not eroded significantly into the Sele Formation shales that appear to have constant thickness inside and outside the mapped sand fairway. The levees may be dominantly fine-grained and do not necessarily set up an acoustic impedance contrast to the background sediments and the splay could be thin and contain linked debrites (see also text and illustration in Fig. 7). Note also that the frontal splays are commonly reported in the literature as high amplitude reflector packages (HARP) and the levees are commonly referred to as 'gull wings' due to their shape in shallow high-frequency datasets. A similar pattern is not seen nor expected to be seen at 2 km depth because: (1) the dominant frequency is lower; (2) the signal-to-noise ratio is lower; (3) the acoustic impedance contrast between sand and shale is typically lower after differential compaction of sand and shale and (4) the geometry has changed due to differential compaction of sand and shale. The inferred frontal splay is therefore not a HARP but a thin and rather dim reflector and the levees are not gull wings but are collapsed and difficult to relate to a particular seismic top and base event. The frontal splay top reflector is only expected to be resolved on the sides of the channel.

sheet-like seismic reflector that can be seen near the base of the Sele Formation on both sides of the channel. Petrophysical logs (Fig. 4) demonstrate that the channel succession in well 26/4-2 is part of one singular interval of mainly homogeneous sandstone with a sharp contact to underlying and overlying shale.

The feeder channel drilled by well 26/4-2 was probably a conduit for many gravity driven flows and evolved over time in a process that involved several avulsions that ultimately routed gravity flows to the outboard Stord Basin. The response of gravity flows that enter a deeper-floored base level either by avulsion or elongation to an outboard basin is very well documented from modern fans. It has been found that avulsions normally result in the sequential development of sheets overlain by leveed channels (e.g. Twichell, *et al.*, 1996; Pirmez *et al.*, 2000; Kolla, 2007).

By analogy, the channel drilled by well 26/4-2, and other discrete, narrow feeder channels with similar characteristics, are candidate leveed feeder channels. The elevated position of the channel body, above a candidate frontal splay reflector, in combination with almost constant Sele Formation shale thickness, suggest that the deposition of the Hermod Member was not preceded by significant erosion into the Sele Formation background sediments. If the channel was fully erosionally confined, the encasing Sele Formation would have been thinner. The isochron map in combination with seismic cross-sections (Fig. 8) can be interpreted in terms of a tripartite architectural element association where the channel is flanked by levees and underlain by unconfined splay deposits. The interpretation of the frontal splay is supported by a sheet-like seismic reflector below the channel reflector. The presence of a levee is assumed from the overall geometry of the Sele Formation isochron in combination with the elevated position of the channel above the frontal splay and the common occurrence of the sheet-leveed channel succession in modern analogues. The original geometry of the leveed feeder channel is expected to be modified by differential compaction, with the more mud-rich levee exhibiting the largest effect.

SEDIMENTARY FACIES

Well 25/6-3 is the only well with Sele Formation cored in the seismic study area. Two 30 m thick Hermod Member sandstones have been drilled by this well. The core covers the uppermost part of the youngest sandstone and the overlying shales of the Sele Formation (Fig. 9). A convincingly imaged c. 500 m wide sinuous channel (channel C3) corresponds to a decrease in acoustic impedance towards the base of the oldest Hermod Member sandstone in well 25/6-3 (Fig. 9). The youngest sandstone in this well has also been interpreted as a channel because: (1) the channel-shaped relief seen on the top Sele Formation seismic map exceeds the thickness of the oldest sandstone (Fig. 10). The channel relief must therefore result from the combined effect of both sandstones in well 25/6-3; (2) the base reflector of the youngest sandstone has a concave upward shape (Fig. 10) and; (3) there is close similarity in log response between the cored interval of well 25/6-3 and the logged interval of well 26/4-2 where the sandstone is calibrated to an unambiguous channel on seismic reflection data.

The following sections describe the sedimentary facies observed in well 25/6-3 (Fig. 11).

Background sediments of the Sele Formation

The background sediments of the Hermod Member are by definition the Sele Formation (Brunstad *et al.*, 2009). The Sele Formation is widely distributed over the North Sea and consists of mudstones and siltstones with minor interbedded sandstones (Deegan & Scull, 1977). The upper part of the Sele Formation has been cored in well 25/6-3 where it overlies a thick succession of Hermod Member sandstones. The core displays dark grey to black mudstone without any visible bioturbation, interbedded with greyish siltstone laminae (<1 cm thick), interpreted as pelagic and hemi-pelagic deposits periodically interrupted by deposition under dilute but turbid silt clouds that may represent distal or lateral run-outs from the currents that deposited the Hermod sands. The dark shales are diagnostic for a period of anoxia during deposition of the Sele and Balder Formations (Brunstad *et al.*, 2009). Thin sandstone units that clearly cross-cut the shale are present at some locations (Fig. 9) and these have been interpreted as injectites that originate from the underlying Hermod Member sandstone.

Sandstone overlain by shale-clast conglomerates

Description: This facies consists of a fine-grained to medium-grained sandstone in m-scale units with varying proportions of shale clasts throughout the uppermost 2 m. The base of this 3 m thick

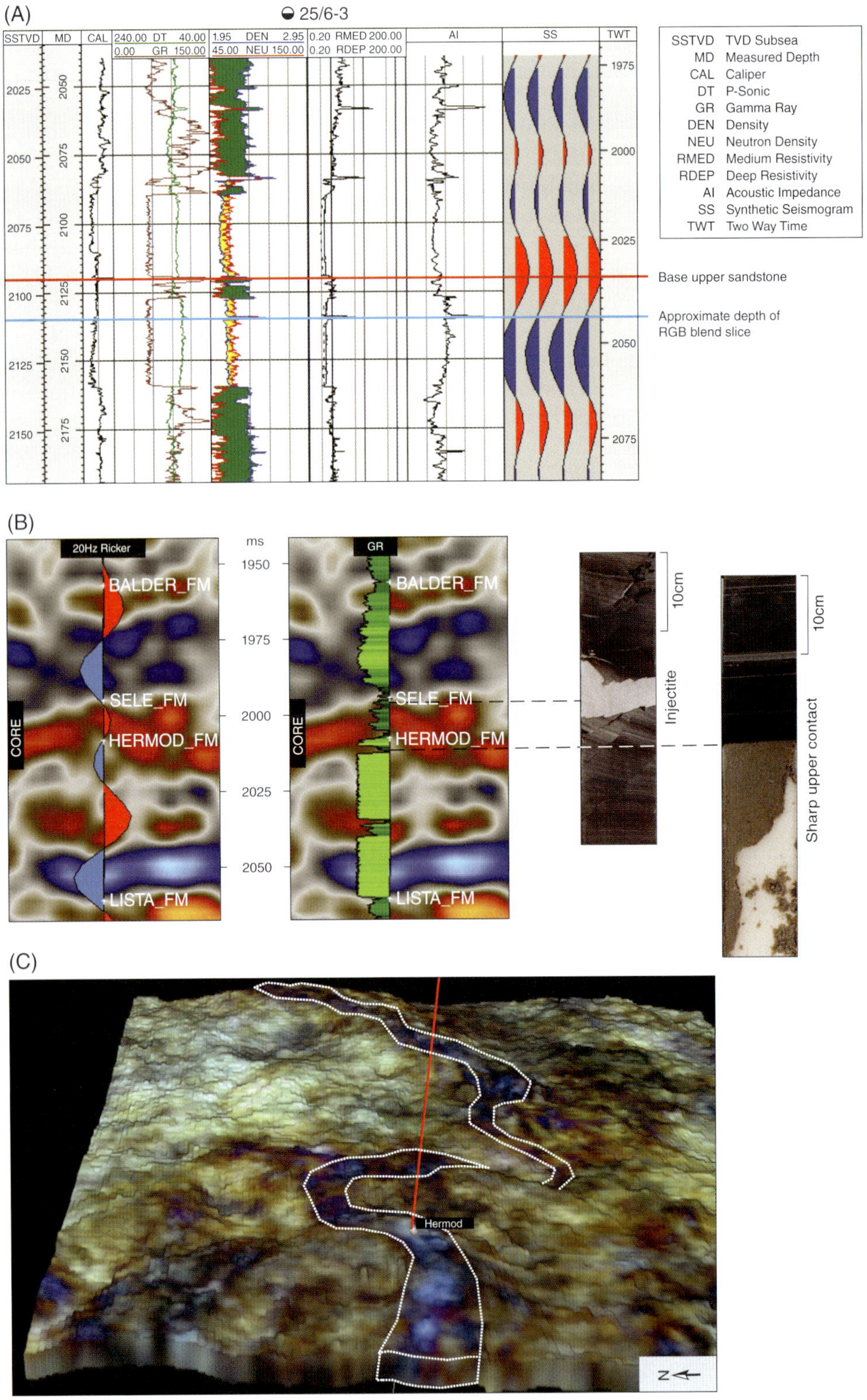

Fig. 9. (A) Petrophysical logs with calculated acoustic impedance and synthetic seismogram for well 25/6-3. The base of the youngest 30 m thick sandstone corresponds to an increase in acoustic impedance which sets up a red peak in the synthetic seismogram; (B) two identical seismic well sections through well 25/6-3, one with synthetic seismogram and one with gamma ray curve. The position of the core is shown on both sections. The core photos to the right document (left) a 3 cm to 5 cm thick injectite near the top of the Sele Formation and (right) the sharp transition from massive sandstone to dark grey claystone on top of the youngest of two 30 m thick sandstone intervals in well 25/6-3; (C) a sinuous channel (channel C3 previously discussed, Fig. 5) displayed by draping a time-slice through the RGB blend at the approximate level of the top of the oldest 30 m thick sandstone in well 25/6-3, on the near Top Hermod Formation seismic interpretation. Sub Sea True Vertical Depth (SSTVD) and MD are in metres. DT = Delta - T (interval transit time).

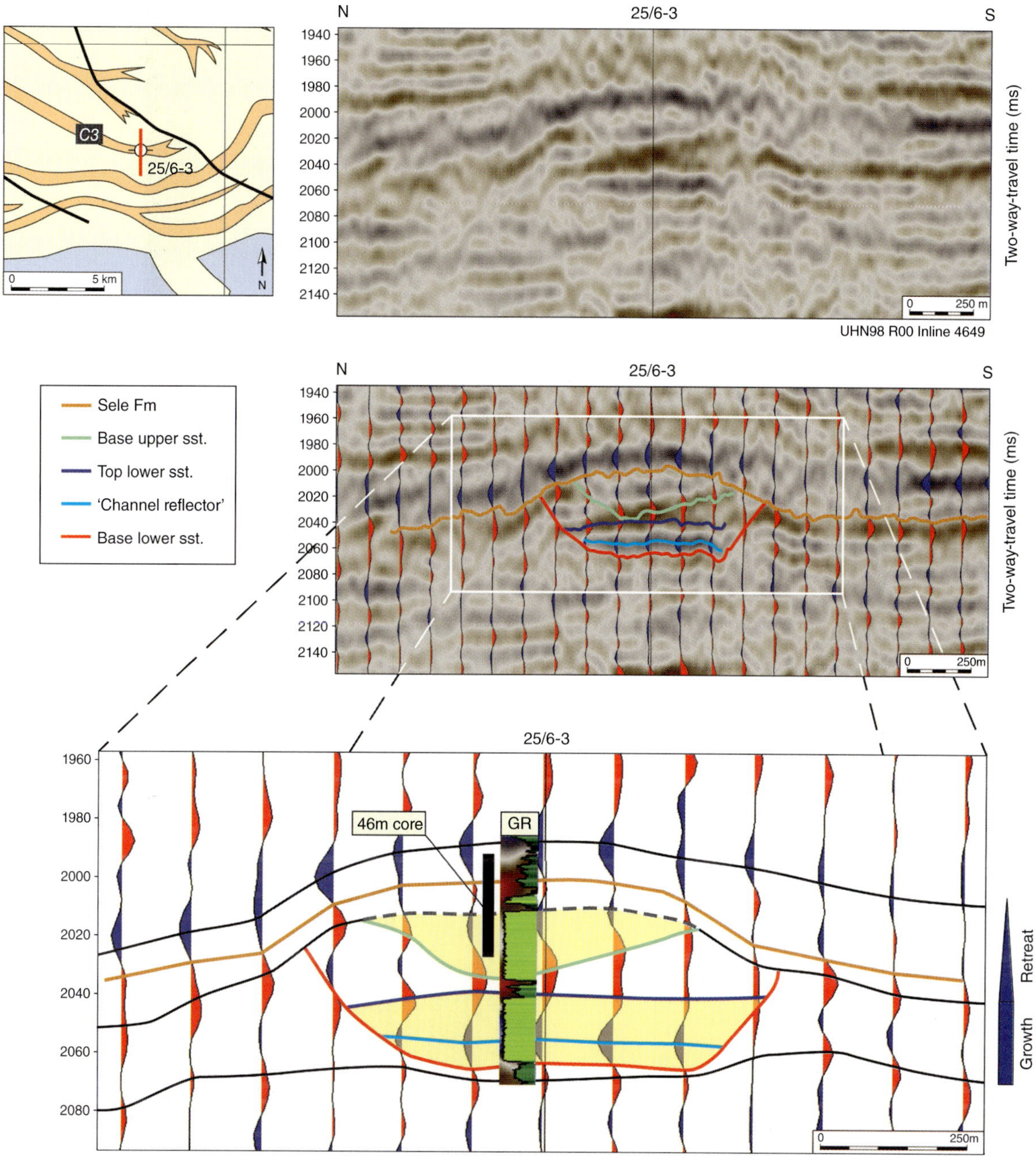

Fig. 10. Above: Interpreted and un-interpreted seismic section through well 25/6-3. The interpretations are based on the well tie shown in Fig. 9. The reflector labelled 'channel reflector' corresponds to a sinuous channel shape in the RGB blend (see Fig. 9C). Note that the seismic reflectors interpreted in cross-section form an 'eye shape' across the Hermod Member in well 25/6-3. The interpreted difference in thickness between this 'eye' and the surrounding area is 30 ms to 40 ms two way time (twt). Each of the two 30 m thick sandstones encountered by the well represents < 25 ms twt, suggesting that the 'eye shape' is set up by the combined effect of both sandstones. Below: Seismic wiggle display focused on the Hermod Member. The gamma ray log (GR) and the cored interval in well 25/6-3 are indicated. The 'eye shape' is interpreted as a channel complex with two 30 m thick sandstones. The base of the youngest 30 m thick sandstone ties (Fig. 9B) to a concave upward reflector. The channel complex is interpreted to have formed in the fan growth phase and the overlying more tabular unit is interpreted as deposits related to the fan retreat phase.

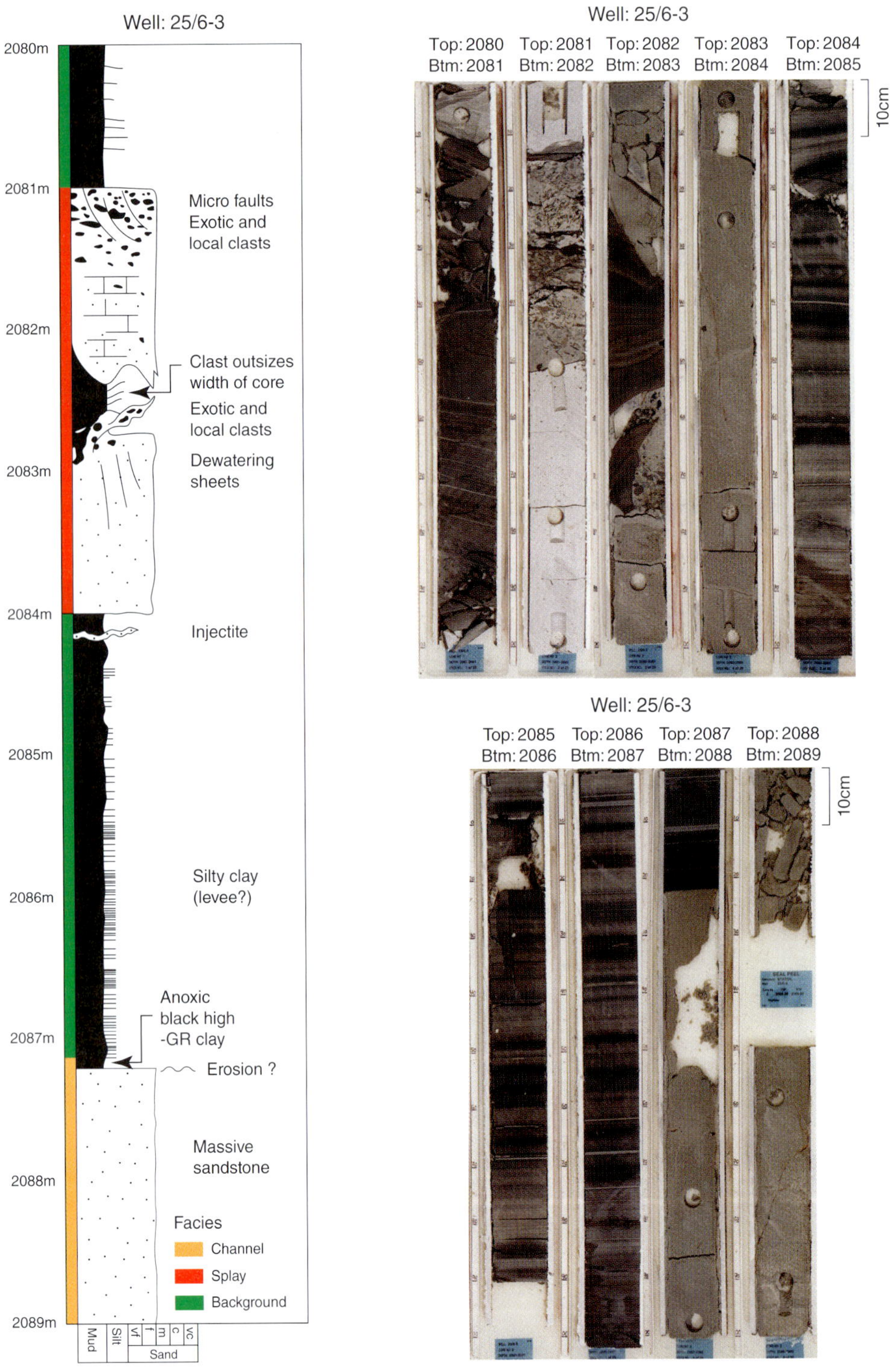

Fig. 11. A sedimentological core description of the cores shown in the photographs to the right. The core is from well 25/6-3 and is interpreted to be a vertical representation of the three main depositional elements observed in map view (Fig. 5). Criteria for distinguishing lateral (levee) and distal (splay fringe) run-outs within the gross 'Background' facies association have not been established, but the interval with abundant silt laminae represents the most probable levee candidate. It is also speculated that the anoxic dark shale capping the massive sandstone facies represents a channel abandonment phase. The dark shale corresponds to a high gamma-ray reading and a similar peak is also found at top of the channel drilled by well 26/4-2 (Fig. 4). The channel association may therefore be interpreted in a broader context as aggradation (sand/channel fill), upstream avulsion (black shale/abandonment), formation of a new channel lateral to the first channel (silty shale/levee) or, alternatively, the levees could be inner levees in a broader channel container (see Fig. 9). The sandstone towards the top of the core correlates to the S2 fan retreat phase (see Fig. 4) and has been interpreted as a basal turbidite with overlying debrites. The core width limits the confidence in the interpretation of big clasts. The two deformed shales around 2082.5 metres have here been interpreted as two similar looking clasts of different size. Note also the injectite and de-watering sheets that suggest significant post-depositional modification of the turbidite/debrite beds.

succession is a massive sandstone (1 m thick) with no apparent traction structures but over-steepened lamination evident towards the top. The sandstone passes gradually upwards to a zone with very poorly size-sorted shale clasts. This zone is overlain by a well-cemented interval with an upward increase in shale clast size and abundance. The top of the sandstone bed is capped by a laminated unit which is a few cm-thick. The largest shale clasts are of the same type as the in situ dark shales seen elsewhere in the core and are considered to be typical for the Sele Formation (as described by Brunstad *et al.*, 2009). Most of these clasts are 1 to 3 cm long, but some clasts outsize the circa 10 cm-wide core. Folded and micro-faulted silty lamination can be observed in the larger clasts. A second shale clast family consists of an association of mostly green and brown, small (<1 cm) angular clasts.

Interpretation and discussion: The basal sandy part of the cored interval in well 25/6-3 is interpreted as a turbidite bed. The presence of internal de-watering sheets and a thin injectite in the shale underneath suggest that sand may have been injected into the turbidite after deposition. The overlying shale-clast conglomerates are interpreted as debris-flow deposits. The small clasts of various colours are clearly distinct from the diagnostic dark shales of the Sele Formation which were deposited during a period of anoxia in the deeper parts of the Northern North Sea (see 'Background Sediments' above; and Brunstad *et al.*, 2009). The coloured clasts must therefore have been eroded from sediments in a shallower oxic part of the basin or from an older formation that was deposited prior to the period of anoxia.

Both scenarios imply long transportation and the fact that the clasts are still angular suggests that they were ripped up by deep erosion into a well lithified formation. It appears to be most likely, therefore, that these clasts have been transported at least from the deeply incised feeder canyon more than 60 km away. The larger clasts were eroded from the Sele Formation background sediments and became deformed in the sediment transport process. These clasts have not necessarily travelled very far in the flow. The debris-flow interpretation concerns the mechanism for sediment transportation immediately prior to primary deposition and does not preclude later sediment remobilisation. Soft-sediment deformation after deposition seems probable based upon the observation of injectites below (see Fig. 11) and some of the local shale clasts may thus have been plucked from the encasing shale during injection of sand (Johansson & Stow, 1995). The increase in clast size and abundance towards the top of the uppermost cemented interval is interpreted as an indication of reworking in a slightly turbulent flow where the clasts were free to move and become sorted. The topmost laminated layer was deposited from a low density turbidity current.

This facies is interpreted as a fan-fringe deposit overlying the transition from S1 to S2, which marks the turnaround from fan growth to fan retreat (Figs 4 and 10). The channel-splay transition zone is believed to be landward stacking in the fan retreat phase. Well 25/6-3 is drilled close to the terminus of a S1 seismic-scale channel (Fig. 5) and the overlying S2 sub-seismic scale sandstone unit could therefore represent a more distal facies type, deposited in the fan retreat phase. A systematic co-occurrence between turbidites and overlying debrites has been documented from the outermost splays of many ancient and modern submarine fans. The basal sandstone may be the deposit from a turbidity current which was a forerunner to the debris flow that deposited the shale-clast conglomerate on top of it (Haughton *et al.*, 2003). An alternative interpretation is that the big clasts and sandstones represent singular debrite intervals (Talling *et al.*, 2010).

Massive sandstone

Description: This facies consists of tens of metres-thick sandstones with 'blocky' petrophysical log responses. In well 25/6-3, the uppermost (c. 20 m) of seismically defined channel sandstone (Fig. 10) has been cored (Fig. 11). The cored interval consists of massive sandstone (i.e. sandstone with no visible traction structures). The grain size is mostly fine to medium (occasionally coarse to very coarse) and the grains are angular to sub-angular or sub-rounded and moderately sorted. No systematic vertical grain-size trends have been observed but some coarser grains appear to be concentrated as lags. Sedimentary structures are generally not visible; however, faint dish structures are present in some places. The upper boundary of this mostly massive sandstone is sharp and parallel to the layering in the overlying dark shales (Fig. 9). The shale that immediately overlies the sandstone has a dark colour and less associated silt laminae than the background sediments seen elsewhere in the core. The dark shale has a high gamma ray (GR) reading and a similar GR peak can also be seen immediately above the un-cored but convincingly imaged channel sandstone in well 26/4-2.

Interpretation and discussion: The process interpretation of massive sandstones of the type described from well 25/6-3 is difficult. Several alternative interpretations have been proposed over the years for massive deep-marine sandstones of this type (see overviews by Stow *et al.*, 1996; Nemec, 1997; Baas, 2004). The possible interpretations include deposition under a high-density turbidity current (Lowe, 1982); post depositional homogenisation caused by fluidisation (Hiscott, 1979; Duranti *et al.*, 2002; Duranti & Hurst, 2004; Jonk *et al.*, 2005) and 'freezing' of a sandy debris flow (Shanmugam, 1996; Shanmugam *et al.*, 1995; Marr *et al.*, 2001; cf. Hiscott *et al.*, 1997). The massive sandstone facies is interpreted here as the deposits of high-density turbidity currents. It is suggested that deposition occurred in a channel, primarily based on seismic geomorphology (Figs 9 and 10). The dark grey high GR clay directly above the sandstone is believed to result from a channel abandonment phase characterised by slow deposition in an anoxic environment. A similar GR response occurs on top of the channel drilled by well 26/4-2.

The lack of tractional structures within the sandstone could be explained as the product of rapid fall-out from suspension with insufficient time for development of an organised traction carpet (Walker, 1978). The general lack of vertical grain-size grading may be attributed to deposition under a sustained, steady flow, whereas the subtle variations in grain size observed in some places may relate to waxing and waning flow either during one flow event or by amalgamation of deposits from successive currents with slightly different flow power (Kneller & Branney, 1995; Kneller & McCaffrey, 2003). Cores with amalgamated massive sand collected from the active Congo submarine channel have recently been interpreted in terms of reworking of the sediment deposited in the channel floor by erosion and traction processes at the bases of turbidity currents (Babonneau *et al.*, 2010). The core from the Hermod Member may also consist of a series of amalgamated deposits from turbidity currents. This interpretation is supported by the network of channels and complex splays that can only be understood as the products of a series of flows travelling through feeder channels. The flows were most probably not identical in terms of grain size, density and thickness so that variations in these parameters could have resulted in readjustments of the slope equilibrium profile due to deposition and erosion (Kneller, 2003). Adjustments in the thalweg profile can also have been triggered by avulsions or flows encountering a structurally controlled sea floor relief (Pirmez *et al.*, 2000). It is therefore likely that repeated episodes of channel floor aggradation and erosion occurred before sand deposition ceased in such channels. Each turbidite bed may have been deposited with a mud cap, but this has been eroded by subsequent currents. The last turbidity current that travelled through the drilled feeder channels could theoretically have by-passed and eroded the top of previous turbidites before it deposited fines from suspension (see Baas, 2004, p. 308, example 3). These fines are eventually incorporated in the dark grey shale represented by a high GR reading directly on top of the sandy turbidite beds. The upward transition from dark grey shale to shale with silt laminae may represent the later overspill from adjacent flows.

The presence of a sharp upper contact with overlying shale may relate to a narrow grain-size range available during deposition, but the absence of traction structures towards the top is enigmatic. This enigma was first discussed by Walker (1965) who suggested that freezing of a traction carpet could cause by-pass of overriding fines. However, as thoroughly addressed by Baas (2004), the interpretation of the sharp upper transition to shale (Fig. 9) as the product of a turbidity current is problematic because turbidity currents inevitably pass through a final waning phase when bedforms should be produced. If beds with a massive top are deposited from turbidity currents, it seems more likely that the sharp upper contact resulted from secondary processes such as fluidisation or erosion (Baas, 2004).

An alternative interpretation is that the cored interval in well 25/6-3 was fluidised. This interpretation is supported by the recognition of injectites in the overlying shale (Fig. 11). Post depositional homogenisation could either occur as 'in situ homogenisation' or 'injection of the entire sand' (Hiscott, 1979). Abrupt upper contacts to shale, lack of primary depositional structure, common injectites in the shale above and steep-sided mounded geometries are some of the typical characteristics of injected beds (e.g. Jonk *et al.*, 2005, Hurst & Cartwright, 2007 and references therein, Huuse *et al.*, 2010 and references therein). However, full scale injection of the entire tens of metres-thick bed cored in well 25/6-3 is unlikely because of the lack of a mappable mounded geometry and the presence of undeformed high GR shale on top of the sand. In situ homogenisation is therefore a more likely explanation for the lack of primary depositional structures in core from well 25/6-3.

The lack of a unique interpretation for the massive sandstone facies that forms excellent reservoirs and is very common throughout the Palaeocene of the North Sea is challenging. It appears from seismic-scale observations that this sandstone type can have many different 3D geometries and modes of formation.

SAND REMOBILISATION COMPLEXES

Seismic geomorphologies indicative of intense post-depositional sediment remobilisation have been identified over the Frøy Field and the entire Rogaland Group appears to have been affected. In the Balder Formation, they occur as intrusive dykes and sills, collectively referred to as injectites. In the Sele Formation, the sediment remobilisation structures have sharp edges that mark an abrupt change in the following types of maps (Figs 10 to 12): (1) the elevation of the Near Top Sele time-structure map, (2) the isochron of the Sele Formation and (3) the thickness of the Hermod Member (confirmed at well locations). Discrete geometries that delineate low areas on map types 1 to 3 are referred to as depressions with an adjective that describes the plan-view shape (circular, oval, irregular). Discrete geometries that delineate elevated areas on map types 1 to 3 are called steep-sided mounds. The underlying Ty Member sandstones of the Våle Formation display complex seismic reflector configurations with common inclined reflectors that at various locations extend up into the Lista Formation. The disturbances in the Ty Member tend to occur under mounds in the Hermod Member, as further addressed in the interpretations below.

Depressions

Well 25/5-1 is drilled in a small sub-circular depression (c. 200 m wide and 40 m deep) which is connected to a larger (>1 km wide) and more irregularly shaped 'basin' and flanked by a steep-sided mound (Figs 10 and 12). This well is the only one in the study area with almost no sand in the Sele Formation (Fig. 4). A seismic cross-section through the well (Fig. 12) illustrates clearly a fault in the near top Sele reflector. The fault separates the circular depression from the adjacent mound. A flat Intra-Balder seismic reflector is juxtaposed to the top of the Sele Formation across the fault. The flat reflector in the Balder Formation ties to a circa 15 m thick sand with a blocky log response in well 25/5-1. Dipmeter data indicate that the sand is conformable with the tuff beds lying above and below. Current correlations suggest that this is the only well in the study area with a substantial amount of sand in the Balder Formation. The steep-sided depression drilled by well 25/5-1 displays four key characteristics (Fig. 10) that appear to be typical for similar structures in the area: (1) circular to oval map view shape; (2) contracted Sele Formation; (3) steep-sided contact to an elevated shoulder and (4) expanded Balder Formation with an extra high acoustic impedance reflector. The reflector(s) in the Balder Formation can be flat, inclined or complex (cf. Figs 10 and 11).

The steep-sided contact to an elevated shoulder (or mound) is in most places a fault. Some of these faults terminate above the Balder Formation, in the lower part of the Horda Formation.

The smallest mapped depressions are sub-circular in map view, with a diameter of ~ 200 m (Fig. 13). Some are isolated; some occur in clusters separated by delicate ridges, others form a trench with an adjacent ridge (Figs 10 and 11). Larger depressions display a more irregular shape and have an area of more than 2 km^2 (Fig. 14).

Steep-sided mounds

Steep-sided mounds have sub-circular to irregular shapes and are bounded by a trench or irregular depression on at least one side. The biggest mound has an area of more than 5 km^2 and a crest that is elevated more than 100 m above the deepest parts of the adjacent depressions. The type well for the Hermod Member (i.e. well 25/2-6) has been drilled into a mound with an irregular and complex geometry. Well 25/2-6 encountered 140 m of clean sandstones with minor interbedded shale (Fig. 4) and this makes it the well on the studied submarine fan with the highest net-to-gross sandstone ratio in the Sele Formation (out of 29 evaluated wells; Fig. 15). Trenches adjacent to elongated mounds can at some locations be seen to represent

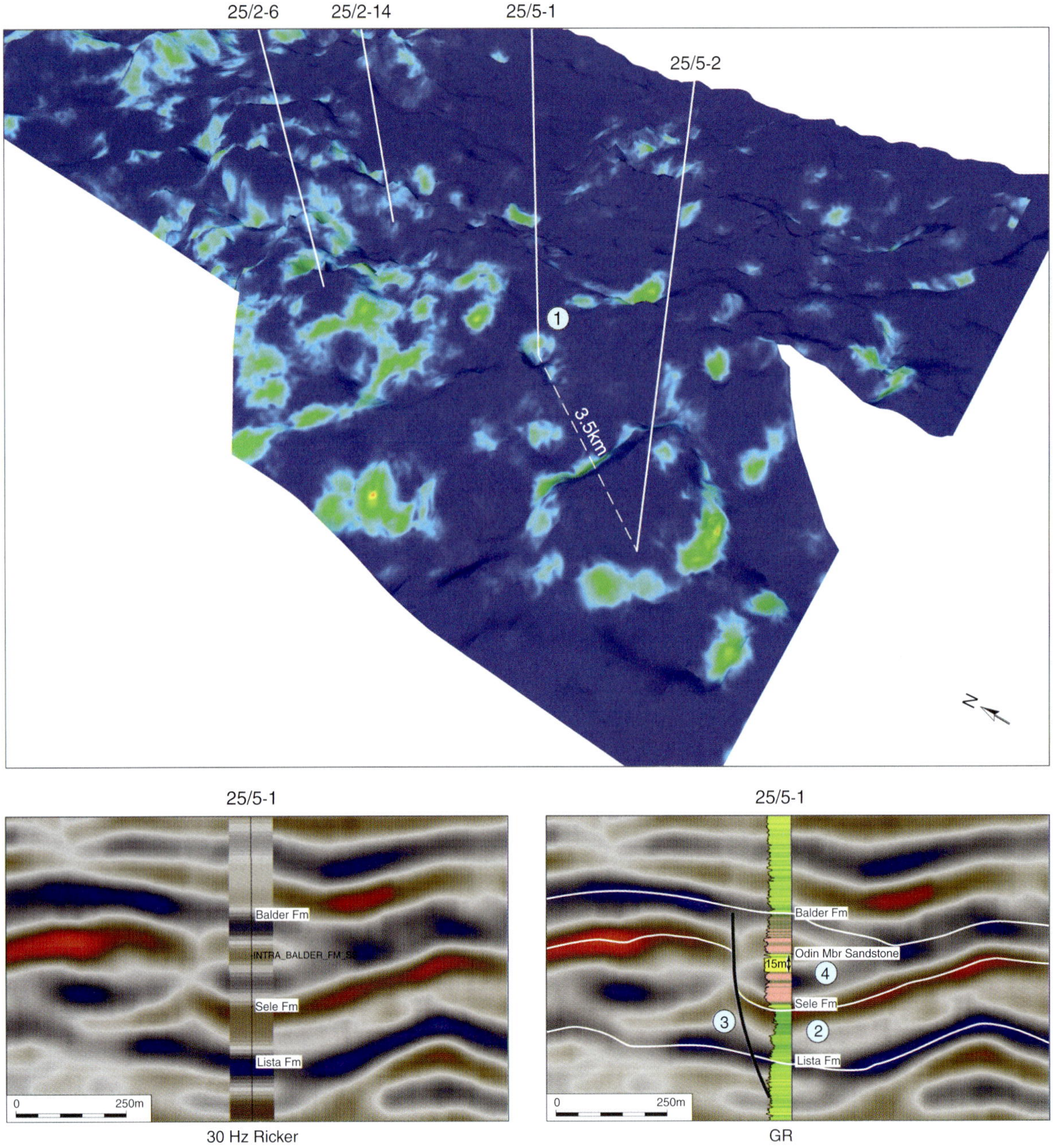

Fig. 12. Above: oblique 3D view of edge attribute map draped onto an auto-tracked Top Sele Formation time-structure map. The edge attribute is a combination of different quantitative measures of the dip and continuity of seismic reflectors. The green colours are therefore an objective illustration of edges. Well 25/5-1 is located in a small circular depression. Below: the seismic lines with well tie and lithology-filled GR curves and the edge map illustrate four characteristic features of the well 25/5-1 depression: (1) circular to oval map view shape; (2) contracted Sele Formation with no Hermod Member sandstone; (3) steep-sided contact to elevated shoulder and; (4) expanded Balder Formation with an extra high acoustic impedance reflector representative of Odin Member sandstone.

a nucleation of smaller, circular, depressions (Fig. 13). Seismic edge attributes suggest that small circular depressions can also define the steep sides of large sand-injection mounds (Fig. 12). The nucleation of circular depressions detected by the use of edge attributes can easily be mistaken for scallops in an isochron map (compare structure drilled by well 25/5-2 in Figs 12 and 14).

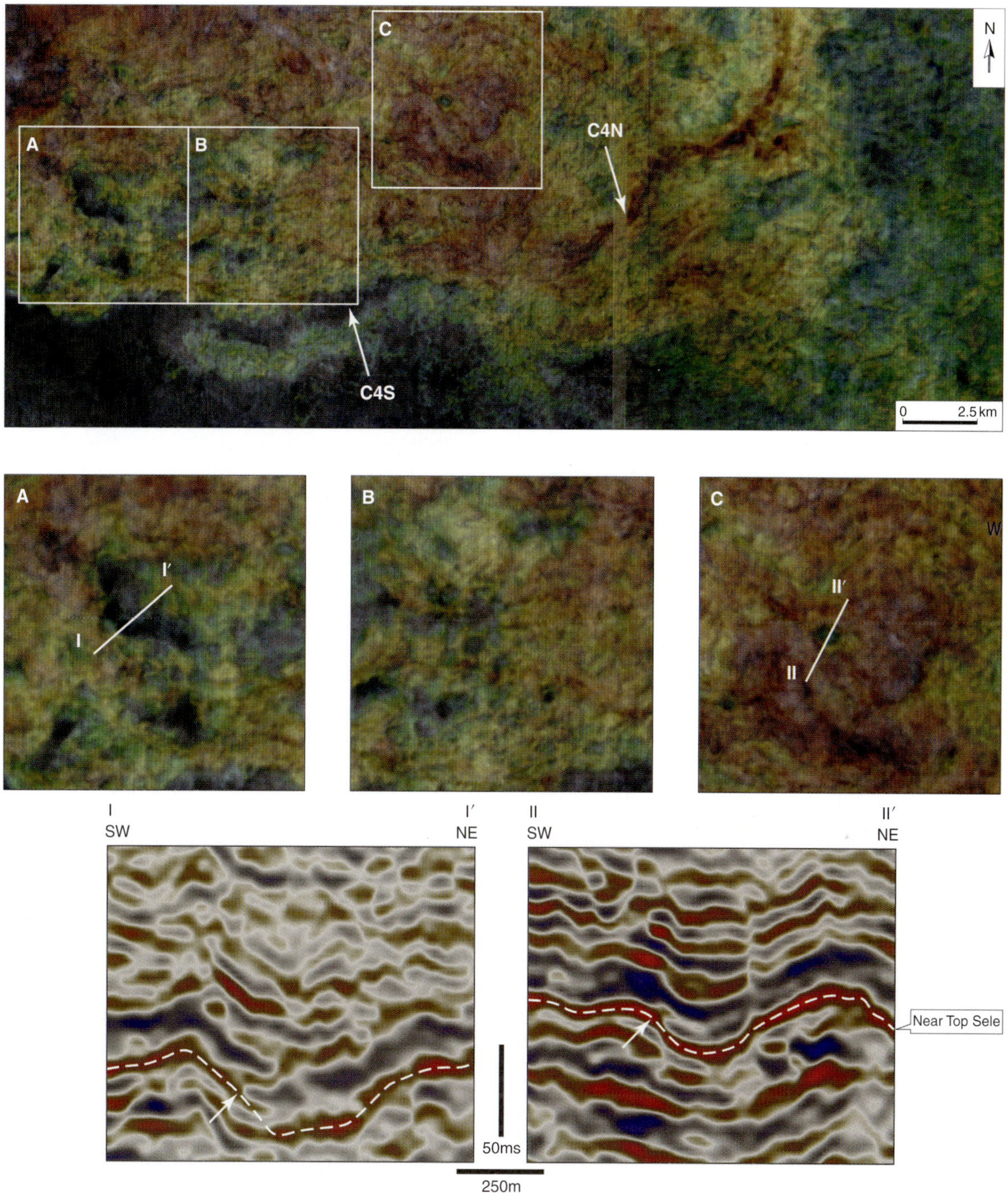

Fig. 13. The habitat of some small circular depressions. Above: RGB-blend time slice draped with a semi-transparent Sele Formation isochron map. Warm colours represent thick areas, cold colours represent thin areas. Channels C4N and C4S are indicated for reference. Small circular depressions may align and form a trench adjacent to a ridge (A). The seismic cross section illustrates the adjacent ridge, or elevated shoulder in 2D section, with arrow pointing at a break in the near top Sele reflector where the injectite in Balder is thought to evolve from. Small circular depressions may also be distributed more randomly and separated by more narrow ridges (B). Example (C) is a single isolated occurrence of a depression with a very similar size and the same four characteristics described for the well 25/5-1 depression (Fig. 12). Below: seismic sections at the top of the Sele Formation showing injectite occurrence at the edge of two depressions. The arrows point to breaks in the near top Sele reflector, with possible injectites originating from the breaks.

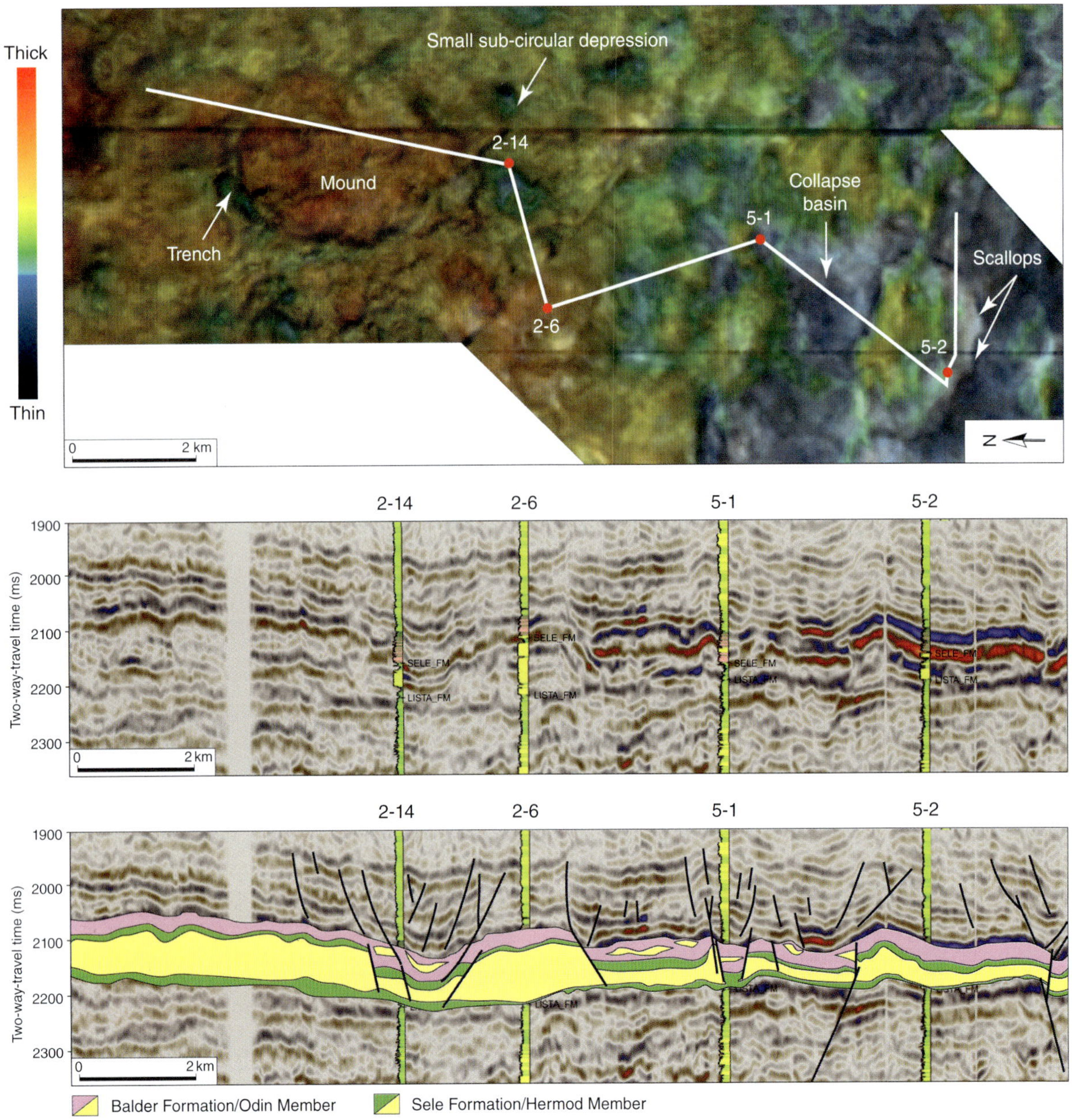

Fig. 14. Large-scale sand remobilisation in the Frøy area, evident in the Sele Formation isochron map and seismic section. The area can be divided into sand evacuation and sand injection zones. Faults that originate in the Hermod sand and terminate in the Horda Formation clays can be observed locally at the boundaries between these two zones. These faults are thought to have formed by roof collapse during sand evacuation. Hermod sand in well 25/5-1 appears to have been evacuated to a mound and thereafter fed back into the overlying Balder Formation across a roof collapse fault that juxtaposed Hermod to Balder.

INTERPRETATION

The lack of Hermod Member in a small circular depression drilled by well 25/5-1, within the broader Hermod sand fairway (Fig. 2), suggests that the depressions represent areas of post-depositional sand evacuation. Their close association to ridges or steep-sided mounds hint at a generic relationship and the most obvious interpretation is one of lateral transportation of Hermod sand from the depressions to the mounds which can have anomalously high net-to-gross sand content in the Sele Formation (i.e. well 25/6-1; Fig. 15). The large size of some discrete depressions or 'collapse basins'

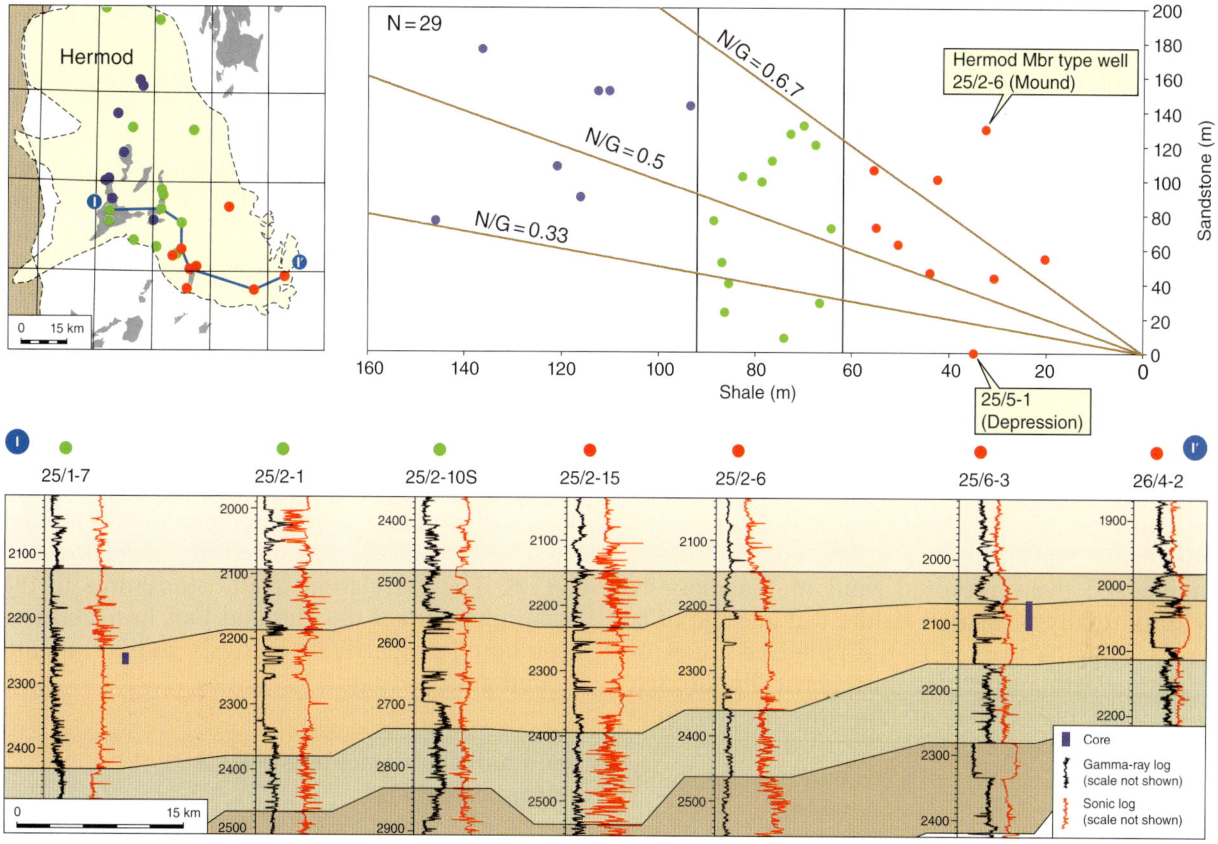

Fig. 15. Above: map distribution of the Hermod Member (Fig. 1) with wells colour coded according to where they plot in the thickness parameter plots to the right. The plot demonstrates the lack of a statistically significant relationship between shale thickness (i.e. claystone, siltstone and minor amounts of tuffs and limestone) and sandstone thickness. The wells plotted in blue tend to be positioned in the more proximal or axial parts of the Hermod Fan and contain a thick Sele Formation interval with low net-to-gross sandstone content compared to the wells plotted in red, which are all positioned in the more distal part of the Hermod Fan. The inferred sand evacuation from well 25/5-1 and injection to well 25/2-6 (see Fig. 13) are clearly anomalous compared to the other wells on the mapped Hermod Fan. Below: selected wells from a proximal to distal position relative to sediment sourced from the West Shetland Platform area. Main sandstone pulses indicated by the petrophysical logs in the Våle Formation (Ty Member Sandstone), Sele Formation (Sele Member Sandstone), Balder Formation (Odin Member Sandstone) and Horda Formation (Frigg Member Sandstone). The petrophysical logs do not distinguish stratified sandstone-shale successions from sandstones with large shale clasts and the serrated log responses seen in some wells may be caused by normal graded sandstones with mud cappings (as observed in core from well 25/1-7) or shale-clast conglomerates (as observed in core from well 25/6-3).

and the large cumulative size of smaller circular depressions (Fig. 12) further suggests that lateral movement of sand was a very important contributor to the growth of the mounds. The possible lateral transportation of sand from depressions to mounds has also been suggested for the Cecilie Field, Denmark, where cores and diagnostic zircon layers have been used to support this interpretation (Hamberg *et al.*, 2007).

The injectites in the Balder Formation are thought to have emanated from the mound shoulder, across faults that juxtaposed the mounded sand to the Balder tuff section. This interpretation is based on the observed common origin of flat, inclined or complex reflectors from a break on the Near Top Sele Formation at mound shoulders. Injection into the Balder Formation from Intra Sele Formation mounds is also described and recognised from the Balder-Ringhorne area, some 50 km south of Frøy (Wild & Breidis, 2010).

The timing of mound growth is restricted to a window defined by the deposition of Hermod sand (as this is believed to be a parent bed) and the upward termination of faults on the steep side of mounds (as these have contributed to the mound shape). It is unknown if the faults were also active during early differential compaction, after sand was injected into the mound and the time window for mound growth by sand injection could therefore potentially be narrower. However, injections in the Balder tuff documents a phase of post-depositional sand mobilisation no older than c. 54 Ma.

It appears from the above and the common co-existence of depressions, mounds and injectites, that their origins are connected. With reference to previous work on very similar mounds (Hamberg *et al.*, 2007) we propose that the mounds formed primarily by lateral movement of Hermod sand (with possible additional contribution from Ty sand). The faults on the steep-sided mounds were probably initiated by the evacuation of sand and further accentuated by mound growth and later post-depositional compaction. The faults juxtaposed over-pressured sand in the mounds with normally pressured Balder tuff and the tuffs in the hanging wall were the preferred zones of injection (rather than breaching vertically through the older claystones at the top of the Sele Formation). Wild & Breidis (2010) proposed fluid influx from the underlying structure as a probable trigger mechanism for sand remobilisation and formation of mounds in the Balder-Ringhorne area. The influx of fluids could supply the necessary overpressure to drive large-scale sand remobilisation Duranti (2007) and the presence of a hydrocarbon focusing Jurassic structure under the Frøy mounds could explain why the deformation is most intense in this area. It is speculated that the fluid phase in the present case is gas and both biogenic and thermogenic gas are possible as migration to the area from the adjacent source kitchen had already begun in the Cretaceous.

The disturbances in the older Ty Member under the Frøy mounds could be related to fluid influx along the same routes that fed fluid to the Sele and Balder Formations. Upward injection (Hurst *et al.*, 2003) of sand from the Ty Member to the Hermod Member is also possible, as suggested by the common occurrence of disturbances in the Ty Member under mounds in the Hermod Member and the mounds may thus have formed by the combined effect of vertical and lateral sand movement.

The circular depressions were probably initiated as pockmarks, generated by a similar sequence of events as forwarded by Hovland & Judd (1988). The pockmarks probably started to form when the Hermod reservoirs became blanketed by Sele Formation shale, but continued shale deposition and Balder ash bed deposition eventually blocked the craters. Subsequent gas seepage along the routes established to the pockmarks could then have provided the mechanism for the overpressure to drive the sand remobilisation that is interpreted to have taken place after the Balder tuffs were deposited. The mounds where then formed by sand withdrawal from the flank areas and re-deposition towards the mound centres (i.e. as proposed by Brooke *et al.*, 1995).

THE HERMOD MEMBER BEYOND THE SEISMIC STUDY AREA

Detailed seismic imaging of sand fairways has not been achieved for the proximal part of the Hermod Fan (i.e. outside the seismic study area shown in Fig. 2). The correlation of the study area to the remaining part of the Hermod Fan has therefore been based primarily on analysis and correlation of well data.

Sand and shale thicknesses from petrophysical logs

The number and thickness of sandstone and shale layers vary significantly from well to well (Figs 4 and 13) and no strong correlation between sand and clay content has been found at the scale of the entire fan. The lack of correlation between total sand and total shale content in wells in the Sele Formation suggests a lack of 'fan-scale causation' between these two variables. The mechanisms that focus sand to channels, steep-sided mounds and splays are different from the mechanisms that spread mud out over the entire basin. The thickness of mudstone intervals varies gradually and tends to decrease eastwards in the direction of diminishing accommodation space in the asymmetrical Viking Graben.

The highly variable sand content within the overall eastward thinning 3rd order fan cycle agrees with the complex sand framework geometry documented by the RGB blend (e.g. Fig. 4). The shale-sand thickness statistics underscore the existence of discrete and localised sand fairways and further document that the Hermod Fan consists of multiple channel-splay pairs of limited lateral extent that coalesce into a basinwide fan characterised by rather abrupt lateral changes in both number and thickness of sandstone beds (cf. Fig. 4).

A general tendency for increase in net-to-gross sand content from the proximal to distal part of the fan is observed. This is most probably related to the transition from a proximal by-pass dominated setting consisting mostly of channels to a more distal deposition dominated setting consisting mostly of splays (e.g. Piper & Normark, 2001). However, in well 25/2-6, there is an abnormal thickness of sandstone that is interpreted to result from sand injection. The sand in this case is thought to have been evacuated from the area drilled by well 25/5-1 (Fig. 13).

Why is the Hermod Member dominated by clean sandstone units overlain abruptly by mudstone?

The dominant lithology in all wells in the mid-fan to outer-fan is clean sandstone characterised by a blocky gamma-ray log pattern with smooth and almost constant low readings and abrupt changes to higher readings in the shale above and below (Fig. 4). The blocky log response defines sandstone units that are several tens of metres thick and this petrophysical log response represents 40% to 80% of the total Sele Formation thickness at the well locations. The thick, blocky, sandstone units are typically separated by a mudstone dominated interval that is < 10 metres thick. 1 to 3 blocky sandstone beds (>10 m thick) occur in each well, mainly in the older S1 part of the Sele Formation (Fig. 4). The high sandstone proportion and abrupt contact with overlying mudstone is not a peculiarity of the wells included in the present study but a common feature of the Hermod Member throughout the North Sea (Brunstad *et al.*, 2009).

By analogy with modern fans, it is thought that sand and mud can become segregated during transportation with the sand concentrated in channels and frontal splays (Piper & Normark, 2001) and it is probable that sufficient mud was advected by the time currents reached the outer fan that clean sand was deposited (e.g. Peakall *et al.*, 2000, p. 83).

Lopez (2001) argued that mixed load input to the Amazon submarine channels becomes segregated as fines build levees and coarser material becomes confined increasingly to the channel axis, ultimately resulting in clean sand in the channel and fine-grained, mostly muddy and silty levees. The argument was re-iterated by Droz *et al.* (2003) in relation to the Zaire (Congo) submarine fan, with leveed channels described as "powerful sand factories, able to carry, separate and clean sand from mixed load". It is worth noting that the vertical sequence of facies seen in core from well 25/6-3 is strikingly similar to the lateral facies variation documented in the active Congo turbidite channel some 200 km basinwards from the canyon (Babonneau *et al.*, 2010, compare Fig. 11 herein to their Figs 7 and 8); The massive, structureless, medium-grained to fine-grained sand is found in the axis of the Congo channel and the mud-silt successions are found in the levees.

Our interpretation of leveed channels in the Hermod Formation (Fig. 8) could imply that a similar segregation process has produced the homogeneous sand that typifies the Hermod Member. The very abrupt vertical change in lithology from massive sandstone to mudstone observed in the wells may relate to the rapid abandonment of discrete channels and their frontal splays giving way to deposition of deep-marine shales. This interpretation can account for the abundance of clean sandstone not only in channels but also in remobilised sediments. The clean sand seen in injectites and steep-sided mounds is most probably inherited directly from the parent bed and no significant additional sandstone homogenisation during remobilisation appears to have occurred. For example, the channel drilled by well 26/4-2 consists of equally clean sandstone as the steep-sided mound drilled by well 25/2-6.

The vertical facies succession in core from well 25/6-3 may be interpreted as a channel-levee association, where the massive sand represents the channel axis, the overlying dark shale represents upstream avulsion and abandonment and the silt-mud layers represent the lateral levee spill from an adjacent channel (Fig. 11). The very dissimilar thickness and number of sandstone and shale beds seen throughout the fan (Figs 4 and 13) provide evidence of avulsion and fan-switching generating sequences which form an apparently random overprint on the long term allocyclic patterns of initiation, growth and retreat.

IMPLICATIONS

Studies of 'seismic scale' outcrops and detailed images from modern systems have narrowed the gap in scale of observation between ancient and modern submarine fans. The complexities of these systems have now been documented in numerous papers but the limitations imposed by seismic resolution makes it difficult to compare and contrast them to more deeply buried submarine fans. The present paper was motivated by the excellent seismic imaging of subtle geometries that are normally not recognised in deeply buried submarine fans and it is therefore an example of how enhanced seismic imaging can strengthen the link between subsurface systems and their analogues.

Modern submarine fan channels generally have well developed levees and levees become increasingly recognised as an important constituent of ancient outcrop successions (e.g. Walker, 1985; Fergusson *et al.*, 1989; Di Celma *et al.*, 2011), but they are rarely observable in deeply buried subsurface fans. The interpretation of leveed channels, avulsions and fan-switching sequences herein suggest a strong autocyclic overprint of the long term allocyclic patterns of fan initiation, growth and retreat. Moreover, the suggested close association between sand and channels, as opposed to sand sheets that spread uniformly over large areas, have important implications for the lateral continuity of potential hydrocarbon reservoirs.

The possibility of sand being focused primarily in autocyclically-controlled levee-confined channels and frontal splays of limited size is an issue that should be accounted for in the estimation of petroleum migration, trapping and reservoir performance. Likewise, the erosion at the base of channels and the possible amalgamation of discrete sand fairways may become important for estimating the reservoir connectivity between discrete sand fairways. Post-depositional sand remobilisation can also be important for reservoir prediction as injectites may, in many cases, be the only connectors between sand fairways. An understanding of the relationships between possible fluid migration routes and connected elements in a set of deformation structures could be important for understanding the behaviour of hydrocarbons in these reservoirs. The recognition of sand-evacuation structures is also important as such depressions can indicate patches of non-reservoir in an otherwise excellent sandstone reservoir.

The basic building blocks of the sand framework are thought to be channels with connected splays, injectites and steep-sided mounds. Levees are expected to have high clay-silt content and may be indistinguishable from the distal run-outs from low-density turbidity currents in the outer parts of splays. The avulsion process is thought to have produced a complex stack of channels and splays, draped by inter-channel facies. Indications of incisions related to headward erosion from the nick-points set up by a structured palaeo sea floor have also been discussed (channel C4S previously discussed and Fig. 7). It is considered that such incisions could have eroded the levees of adjacent channels and later have evolved from erosional conduits to depositional channels. This process has probably promoted lateral communication between channel-splay pairs.

The observation of a vertical succession of tens of metres of clean sandstone in both steep-sided mounds and a leveed channel highlights the uncertainty associated with predicting sandstone geometry away from well control. The thick-bedded sandstones with excellent reservoir quality display a multifaceted 3D shape which can only be satisfactorily comprehended by the use of conditioned 3D seismic data. The findings of this study reflect the utility of tailored geophysical workflows and integrated core/well datasets for the optimal visualisation of sandstone geometries within the target interval.

CONCLUSIONS

The linkage of seismic geomorphologies to sedimentary facies has enabled the following conclusions to be drawn from this study:

1. The application of a combination of geophysical techniques applied to 3D seismic datasets have resulted in the recognition of some morphologies indicative of processes that are well known from modern submarine fans and outcrop studies, but rarely observable in deeply buried ancient fans. These morphologies include leveed channels, avulsions, braided frontal splays and pockmarks.
2. The regional dip and overall direction of fan growth was towards the south-east, largely controlled by the structural grain of an extensional transfer zone in the segmented Viking Graben. The fan evolved in a stepwise manner, interrupted by temporary channel deviations towards the north-east, following the grain of the Viking Graben. Deviations from the regional progradation trend suggest transient confinement by subtle, structurally controlled, bathymetric features.
3. Evidence for frequent avulsions suggests that only one part of the outer fan was active at a given time. Sand deposition occurred in active

parts of the fan, whilst background sedimentation dominated by claystone with silt occurred in inactive areas. The most distal south-eastern part of the fan was only active in the (maximum) growth phase of the fan, whereas the most proximal north-western part of the study area appears to have been active in the initiation, growth and retreat phases.

4 Sharp channel bends, wide avulsion angles and large, complex, frontal splays are associated with structurally confined sand fairways. Aggradation in the north-east-oriented structural confinements culminated in avulsions that directed channels back to the more regional south-east dip direction. The flow directions into and out of confinements were thus around 90° different from the flow directions within confinements. The turns were accomplished by sharp bends or wide avulsion angles. Unconfined channels cross substratum faults and tend to make frequent low-angle avulsions that originate close to the channel-splay-transition zone and are a response to slope advantage on the sides of frontal splays.

5 Flows entered a deeper-floored base level in the outboard Stord Basin and started to heal the break in slope by headward erosion and basinward splay deposition. The slope disequilibrium was further healed by aggradation of a levee-confined channel over the frontal splay.

6 Avulsions and fan switching appear to have resulted in a complex stack of channel-frontal splay pairs, commonly draped by background sediments deposited over abandoned fan branches. Amalgamation and erosion at the bases of channels is likely to have promoted communication between different channel-splay pairs. Injectites are documented from core and inferred from abnormal sand thickness in mounds and are expected to connect sand bodies otherwise separated by shale.

7 An area with large structures formed by post-depositional sediment remobilisation has been mapped in the Frøy Field area. The deformation of the Sele Formation is expressed as mounds and depressions separated by faults. The Hermod type well (i.e. 25/2-6) is situated on a mound, whereas well 25/5-1 is a 'type well' for the depressions.

8 The severe post-depositional sediment remobilisation over the Frøy Field occurred in the Early Eocene, probably due to gas seepage to blocked pockmark craters.

9 Clean sandstone occurs in channels, splays, injectites and steep-sided mounds. The abundance of clean and mostly homogeneous sandstone is attributed to the original sediment calibre that was delivered to the study area by feeder channels. The clean sand was remobilised post-depositionally and it is hypothesised that the deformation was caused by gas escaping from deeper strata to the sea floor.

ACKNOWLEDGEMENTS

We would like to acknowledge Per Øyvind Seljebotn and Snorre Heimsund who contributed to the early phases of this study. Ellen Lindland is acknowledged for technical advice on, and production of, all illustrations in this paper. We would also like to thank Erik Imsland Wathne for sharing with us documentation of some of his unpublished work from the Greater Frigg Area and for enlightening discussions over some relevant cores. The authors would also like to express our gratitude to Det Norske Oljeselskap, Bayerngas Produksjon Norge and Faroe Petroleum Norge for giving their permission to publish seismic data from Norwegian Production Licence 414 and for many fruitful discussions during our years of work in the area. Gratitude is also expressed to Norwegian Energy Company for giving us the opportunity to work on the Hermod Fan. The reviewers Trevor Elliot, Massimo Dall'Asta and Jonathan Wonham are acknowledged for constructive and useful feedback which improved the manuscript.

REFERENCES

Ahmadi, Z., Sawyers, M., Kenyon-Roberts, S., Stanworth, B., Kugler, K., Kristensen, J.B. and **Fugelli, E.** (2003) Palaeocene stratigraphy. In: *The Millenium Atlas: Petroleum Geology of the Central and Northern North Sea* (Eds D. Evans, C. Graham, A. Armour and P. Brathurst) pp. 235–259. Geological Society, London.

Baas, J.H. (2004) Conditions for formation of massive turbiditic sandstones by primary depositional processes. *Sed. Geol.*, **166**, 293–310.

Babonneau, N., Savoye, B., Cremer, M. and **Bez, M.** (2010) Sedimentary architecture in meanders of a submarine channel: detailed study of the present Congo turbidite channel (Zaiango Project). *J. Sed. Res.*, **80**, 852–866.

Bergslien, D. (2002) Balder and Jotun – two sides of the same coin? A comparison of two Tertiary oil fields in the Norwegian North Sea. *Petrol. Geosci.*, **8**, 349–363.

Blikeng, B. and **Fugelli, E.** (2000) Application of results from outcrops of the deep water Brushy Canyon

Formation, Delaware Basin, as analogues for the deep water exploration targets on the Norwegian Shelf. In: *GCSSEPM Foundation 20th Annual Research Conference , Deep-Water Reservoirs of the World* (Eds P. Weimer, R.M. Slatt, J. Coleman, N.C. Rosen, H. Nelson, A.H. Bouma, M.J. Styzen and D.T. Lawrence) *GCSSEPM Spec. Publ.*, **15**, 61–81.

Brooke, C.N, Trimble, T.J. and **Mackay, T.A.** (1995) Mounded shallow gas sands from the Quaternary of the North Sea; analogues for the formation of sand mounds in deep water Tertiary sediments? In: *Characterization of Deep Marine Clastic Systems* (Eds A.J. Hartley and D.J. Prosser) *Geol. Soc. Spec. Publ.*, **94**, 95–101.

Brunstad, H., Gradstein, F.M., Vergara, L., Lie, J.E. and **Hammer, Ø.** (2009) *A Revision of the Rogaland Group, Norwegian North Sea.* Norwegian Stratigraphic Lexicon.

de Boer, W., Rawlinson, P.B. and **Hurst A.** (2007) Successful exploration of a sand injectite complex: Hamsun prospect, Norway Block 24/9. In: *Sand Injectites: Implications for Hydrocarbon Exploration and Production* (Eds **A. Hurst** and **J. Cartwright**), AAPG Mem. **87**, 65–68.

Deegan, C.E. and **Scull, B.J.** (1977) A standard lithostratigraphic nomenclature for the Central and Northern North Sea. Inst. Geol. Sciences, Rep. 77/25, 36 pp.

Di Celma, C.N., Brunt, R.L., Hodgson, D.M., Flint, S.S. and **Kavanagh, J.P.** (2011) Spatial and Temporal Evolution of a Permian Submarine Slope Channel–Levee System, Karoo Basin, South Africa. *J. Sed. Res.*, **81**, 579–599.

Dixon, R.J. and **Pearce, J.** (1995) Tertiary sequence stratigraphy and play fairway definition, Bruce-Beryl Embayment, Quadrant 9, UKCS. In: *Sequence Stratigraphy on the Northwest European Margin* (Eds R.J. Steel, V.L. Felt, E.P. Johannesen, and C. Mathieu), *Norw. Petrol. Soc. Spec. Publ.*, **5**, 443–469. Elsevier, Amsterdam.

Droz, L., Marsset, T., Ondreas, H., Lopez, M., Savoye, B. and **Spy-Anderson, F.-L.** (2003) Architecture of an active mud-rich turbidite system: The Zaire Fan (Congo–Angola margin southeast Atlantic): Results from ZaıAngo 1 and 2 cruises. *AAPG Bull.*, **87**, 1145–1168.

Duranti, D. (2007) Large-scale sand injection in the Palaeogene of the North Sea: Modeling of energy and flow velocities. In: *Sand Injectites: Implications for Hydrocarbon Exploration and Production* (Eds A. Hurst and J. Cartwright), AAPG Mem. **87**, 129–139.

Duranti, D. and **Hurst, A.** (2004) Fluidisation and injection in the deep-water sandstones of the Eocene Alba Formation (UK North Sea). *Sedimentology*, **51**, 503–529.

Duranti, D., Hurst, A., Bell, C., Groves, S. and **Hanson, R.** (2002) Injected and remobilised Eocene sandstones from the Alba Field, UKCS. core and wireline log characteristics. *Petrol. Geosci.*, **8**, 99–107.

Elliot, T. (2000) Megaflute erosion surfaces and the initiation of turbidite channels. *Geology*, **28**, 119–122.

Fergusson, C.-L., Cas, R.A.F. and **Stewart, I.R.** (1989) Ordovician turbidites of the Hotham Group, eastern Victoria; sedimentation in deepmarine channel-levee complexes. *Aust. J. Earth Sci.*, **36**, 1–12.

Gardner, M.H. and **Borer, J.M.** (2000) Submarine channel architecture along a slope to basin profile, Brushy Canyon Formation, West Texas In: *Fine-grained turbidite systems* (Eds A.H. Bouma and C.G. Stone), AAPG Mem., **72**, 195–213.

Gardner, M.H., Borer, J.A., Melick, J.J., Mavilla, N., Dechesne, M. and **Wagerle, R.N.** (2003) Stratigraphic process–response model for submarine channels and related features from studies of Permian Brushy Canyon outcrops, West Texas. *Mar. Petrol. Geol.*, **20**, 757–787.

Hadler-Jacobsen, F., Johannessen, E.P., Ashton, N., Henriksen, S., Johnson, S.D. and **Kristensen, J.** (2005) Submarine fan morphology and lithology distribution: a predictable function of sediment delivery, gross shelf-to-basin relief, slope gradient and basin topography. In: *Petroleum Geology: North-West Europe and Global Perspectives* (Eds A.G. Doré and B.A. Vining), *Petrol. Geol. Conf. Series*, **6**, 1121–1145.

Hamberg, L., Jepsen, A.-M., Ter Borch, N., Dam, G., Engkilde, M.K. and **Svendsen, J.B.** (2007) Mounded structures of injected sandstones in deepmarine Palaeocene reservoirs, Cecilie Field, Denmark. In: *Sand Injectites: Implications for Hydrocarbon Exploration and Production* (Eds A. Hurst and J. Cartwright), AAPG Mem. **87**, 69–80.

Hardt, T., Holtar, E., Isaksen, D., Kyllingstad, G., Lervik, K.S., Lycke, A.S. and **Tonstad, K.** (1989) Revised Tertiary lithostratigraphic nomenclature for the Norwegian North Sea. In: *A Revised Cretaceous and Tertiary Lithostratigraphic Nomenclature for the Norwegian North Sea* (Eds D. Isaksen and K. Tonstad), *Norw. Petrol. Direct. Bull.*, **5**, 45–46.

Haughton, P.D.W., Barker, S.P. and **McCaffrey, W.D.** (2003) Linked debrites in sand-rich turbidite systems – origin and significance. *Sedimentology*, **50**, 459–482.

Hiscott, R.N. (1979) Clastic sills and dykes associated with deep-water sandstones, Tourelle formation, Ordovician Quebec. *J. Sed. Petrol.*, **49**, 1–10.

Hiscott, R.N., Pickering, K.T., Bouma, A.H., Hand, B.M., Kneller, B.C., Postma. G. and **Soh, W.** (1997) *Basin-floor fans in the north sea: sequence stratigraphic models vs. sedimentary facies: discussion. AAPG Bull.*, **81**, 662–665.

Hovland, M. and **Judd, A.G.** (1988) *Seabed Pockmarks and Seepages. Impact on Geology, Biology and Marine Environment.* Graham & Trotman, London.

Hurst, A. and **Cartwright, J.** (2007) Relevance of sand injectites to hydrocarbon exploration and production, In: *Sand Injectites: Implications for Hydrocarbon Exploration and Production* (Eds A. Hurst and J. Cartwright), AAPG Mem. **87**, 1–19.

Hurst, A., Cartwright, J. and **Duranti, D.** (2003) Fluidization structures produced by upward injection of sand through a sealing lithology. In: *Subsurface Sediment Mobilisation* (Eds P. Van Rensbergen, A.J. Maltman and C.K. Morley). *Geol. Soc. Spec. Publ.*, **216**, 123–137.

Huuse, M., Cartwright, J., Hurst, A. and **Steinsland, N.** (2007) Seismic characterization of large-scale sandstone intrusions. In: *Sand Injectites: Implications for Hydrocarbon Exploration and Production* (Eds. A. Hurst and J. Cartwright), *AAPG Mem.* **87**, 21–35.

Huuse, M., Jackson, C.A.-L., Rensbergen, P.v., Davies, R.J., Flemings, P.B. and **Dixon, R.J.** (2010) Subsurface sediment remobilisation and fluid flow in sedimentary basins: an overview. *Basin Res.*, **22**, 533–547.

Jacka, A.D., Beck, R.H., Germain, L.C.S. and **Harrison, S.C.** (1968) Permian deepsea fans of the Delaware Mountain Group (Guadalupian), Delaware Basin. In: *Permian Deepsea Fans of the Delaware Mountain Group (Ed. B.A. Silver), SEPM Permian Basin Spec. Publ.*, **68-11**, 49–90.

Johansson, M. and **Stow, D.A.V.** (1995) A classification scheme for shale clasts in deep water sandstones. *Geol. Soc. Spec. Publ.*, **94**, 221–241.

Johnson, S.D., Flint, S., Hinds, D. and **Wickens, H. De V.** (2001) Anatomy, geometry and sequence stratigraphy of basin floor to toe of slope basin turbidite systems, Tanqua Karoo, South Africa. *Sedimentology*, **48**, 987–1023.

Jonk, R., Hurst, A., Duranti, D., Parnell, J., Mazzini, A. and **Fallick, A.E.** (2005) Origin and timing of sand injection, petroleum migration and diagenesis in Tertiary reservoirs, south Viking Graben, *North Sea. AAPG Bull.*, **89**, 329–357.

Kneller, B. (2003) The influence of flow parameters on turbidite slope channel architecture. *Mar. Petrol. Geol.*, **20**, 901–910.

Kneller, B. and **Branney, M.J.** (1995) Sustained high-density turbidity currents and the deposition of thick massive sands. *Sedimentology*, **42**, 607–616.

Kneller, B. and **McCaffrey, W.** (2003) The interpretation of vertical sequences in turbidite beds: the influence of longitudinal flow structure. *J. Sed. Res.*, **73**, 706–713.

Kolla, V. (2007) A review of sinuous channel avulsion patterns in some major deep-sea fans and factors controlling them. *Mar. Petrol. Geol.*, **24**, 450–469.

Lopez, M. (2001) Architecture and depositional pattern of the Quaternary deep-sea fan of the Amazon: *Mar. Petrol. Geol.*, **18**, 479–486.

Lowe, D.R. (1982) Sediment gravity flows: II: Depositional models with special reference to the deposits of high-density turbidity currents: *J. Sed. Petrol.*, **52**, 279–297.

Marr, J.G, Harff, P.A., Shanmugam, G. and **Parker, G.** (2001) Experiments on subaqueous sandy gravity flows: The role of clay and water content in flow dynamics and depositional structures. *Geol. Soc. Am. Bull.*, **113**, 1377–1386.

Milton, N. and **Dyce, M.** (1995) Systems tract geometries associated with Early Eocene lowstands, imaged on a 3D seismic dataset from the Bruce area, UK North Sea. In: *Sequence Stratigraphy on the Northwest European Margin* (Eds R. J. Steel, V. L. Felt, E. P. Johannesen and C. Mathieu), *Norw. Petrol. Soc. Spec. Publ.*, **5**, 429–442.

Mitchum, R.M. and **Van Wagoner, J.C.** (1991) High-frequency sequences and their stacking patterns: Sequence-stratigraphic evidence of high-frequency eustatic cycles. *Sed. Geol.*, **70**, 131–160.

Mudge, D.C. and **Bujak, J.P.** (1996) An integrated stratigraphy for the Palaeocene and Eocene of the North Sea. In: *Correlation of the Early palaeogene in Northwest Europe* (Eds R.W.O`B. Knox, R.M. Corfield and R.E. Dunay), *Geol. Soc. Spec. Publ.*, **101**, 91–113.

Mudge, D.C. and **Copestake, P.** (1992a) Lower Palaeogene stratigraphy of the northern North Sea. *Mar. Petrol. Geol.*, **9**, 287–301.

Mudge, D.C. and **Copestake, P.** (1992b) Revised Lower Palaeogene lithostratigraphy for the Outer Moray Firth, North Sea. *Mar. Petrol. Geol.*, **9**, 53–69.

Mutti, E. and **Ricci Lucchi, F.** (1972) Le torbiditi dell'Appennino settentrionale: Introduzione all' analisi di facies. *Memori della Societa Geologica Italiana*, **11**, 161–199.

Mutti, E. and **W.R. Normark** (1991) An integrated approach to the study of turbidite Systems. In: *Seismic facies and sedimentary processes of submarine fans and turbidite systems* (Eds P. Weimer and M. H. Link), pp. 75–106, Springer-Verlag, New York.

Nadin, P.A. and **Kusznir, N.J.** (1995) Palaeocene uplift and Eocene subsidence in the northern North Sea Basin from 2D forward and reverse stratigraphic modeling. *J. Geol. Soc.*, **152**, 833–848.

Nemec, W. (1997) Deep-water massive sandstones. *Abstract, Norw. Geol. Soc. Stavanger Chapter Meeting*, Stavanger, 3 pp.

Partyka, G., Gridley, J. and **Lopez, J.** (1999) Interpretational applications of spectral decomposition in reservoir characterization, *The Leading Edge*, **18**, 353–360.

Peakall, J., McCaffrey, B.C., Kneller, B.C., Stelting, T.R. and **Schweller, W.J.** (2000) A process model for the evolution of submarine fan channels: implications for sedimentary architecture, In: *Fine-grained turbidite systems* (Eds A.H. Bouma and C.G. Stone), *AAPG Mem.*, *72/ SEPM Spec. Publ.*, **68**, 73–88.

Piper, D.J.W. and **Normark, W.R.** (2001) Sandy fans—from Amazon to Hueneme and beyond. *AAPG Bull.*, **85**, 1407–1438.

Pirmez, C., Beauboeuf, R.T., Friedmann, S.J. and **Mohrig, D.C.** (2000) Equilibrium profile and baselevel in submarine channels: examples from Late Pleistocene systems and implications for the architecture of deepwater reservoirs. In: *GCSSEPM Foundation 20th Annual Research Conference , Deep-Water Reservoirs of the World* (Eds P. Weimer, M. Slatt, J. Coleman, N.C. Rosen, H. Nelson, A.H. Bouma, M.J. Styzen and D.T. Lawrence) *GCSSEPM Spec. Publ.*, **15**, 782–805.

Posamentier, H.W. and **Kolla, V.** (2003) Seismic geomorphology and stratigraphy of depositional elements in deep-water settings. *J. Sed. Res.*, **73**, 367–388.

Shanmugam, G. (1996) High-density turbidity currents: are they sandy debris flows? *J. Sed. Res.*, **66**, 2–10.

Shanmugam, G., Bloch, R.B., Mitchell, S.M., Beamish, G.W.J., Hodgkinson, R.J., Damuth, J.E., Straume, T., Syvertsen, S.E. and **Shields, K.E.** (1995) Basin-floor fans in the North Sea: sequence stratigraphic models vs. sedimentary facies. *AAPG Bull.*, **79**, 477–512.

Stow, D.A.V., Reading, H.G. and **Collinson, J.D.** (1996) Deep Seas. In: *Sedimentary environments: processes, facies and stratigraphy* (Ed. H.G. Reading), pp 395–450. Blackwell Science, Oxford.

Szarawarska, E.P., Huuse, M., Hurst, A., de Boer, W., Lu, L., Molyneux, S. and **Rawlinson, P.** (2010) Three-dimensional seismic characterisation of large-scale sandstone intrusions in the lower Palaeogene of the North Sea: completely injected vs. in situ remobilised sandbodies. *Basin Res.*, **22**, 517–532.

Talling, P.J, Wynn, R.B., Schmmidt, D.N., Rixon, R., Sumner, E. and **Amy, L.** (2010) How did thin submarine debris flows carry boulder-sized intraclasts for remarkable distances across low gradients to the far reaches of the Mississippi fan? *J. Sed. Res.*, **80**, 829–851.

Twichell, D.C., **Schwab, W.C.** and **Kenyon, N.H.** (1995) Geometry of sandy deposits at the distal edge of the Mississippi Fan, Gulf of Mexico. In: *Atlas of Deep-Water Environments* (Eds K.T. Pickering, R.N. Hiscott, N.H. Kenyon, F. Ricci Lucci and R.D.A. Smith), pp 282–286. Chapman and Hall, London.

Twichell, D.C., **Schwab, W.C.**, **Kenyon, N.H.** and **Lee, H.J.** (1996) Breaching the Levee of a Channel on the Mississippi Fan. In: *Geology of the United States' Seafloor. The View from GLORIA* (Eds J.V. Gardner, M.E. Field and D.C. Twichell), pp 85–96. Cambridge University Press, New York.

Walker, R.G. (1965) The origin and significance of the internal sedimentary structures of turbidites. *Proc. Yorks. Geol. Soc.*, **35**, 1–32.

Walker, R.G. (1978) Deep-water sandstone facies and ancient submarine fans: models for exploration for stratigraphic traps. *AAPG Bull.*, **62**, 932–966.

Walker, R.G. (1985) Mudstones and thin-bedded turbidites associated with the Upper Cretaceous Wheeler Gorge conglomerates, California; a possible channel-levee complex. *J. Sed. Res.*, **55**, 279–290.

White, N. and **Lovell, B.** (1997) Measuring the pulse of a plume with the sedimentary record. *Nature*, **387**, 888–891.

Wild, J. and **Breidis, N.** (2010) Structural and stratigraphic relationships of the Palaeocene mounds of the Utsira High, *Basin Res.*, **22**, 533–547.

Yu, B., **Cantelli, A.**, **Marr, J.**, **Pirmez, C.**, **O'Byrne, C.** and **Parker, G.** (2006) Experiments on self-channelized subaqueous fans emplaced by turbidity currents and dilute mudflows. *J. Sed. Res.*, **76**, 889–902.

Zanella, E. and **Coward, M.P.** (2003) Structural framework. In: *The Millenium Atlas: Petroleum Geology of the Central and Northern North Sea* (Eds D. Evans, C. Graham, A. Armour and P. Brathurst) pp. 45–59. Geological Society, London.

Index

Note: page numbers in *italics* refer to figures, those in **bold** refer to tables

From Depositional Systems to Sedimentary Successions on the Norwegian Continental Margin, First Edition.
Edited by A. W. Martinius, R. Ravnås, J. A. Howell, R. J. Steel, and J. P. Wonham.